FORCE	*SI unit:* newton	*Abbreviation:* N
Other units	*Abbreviations*	*To convert to newtons multiply by*
pound	lb	4.448 N/lb
ton	—	8896 N/ton
dyne	—	10^{-5} N/dyne*

POWER	*SI unit:* watt	*Abbreviation:* W
Other units	*Abbreviations*	*To convert to watts multiply by*
horsepower	hp	746 W/hp
ton of refrigeration	—	3516 W/ton

WORK	*SI unit:* joule	*Abbreviation:* J
Other units	*Abbreviations*	*To convert to joules multiply by*
foot-pound	ft · lb	1.356 J/ft · lb
calorie	cal	4.184 J/cal
kilowatt-hour	kWh	3.6×10^6 J/kWh*
electron volt	eV	1.602×10^{-19} J/eV
erg	—	10^{-7} J/erg*
British thermal unit	Btu	1054 J/Btu

PRESSURE	*SI unit:* pascal	*Abbreviation:* Pa
Other units	*Abbreviations*	*To convert to pascals multiply by*
pound per square inch	lb/in.^{2}	6894 Pa/(lb/in.^{2})
standard atmosphere	atm	1.013×10^5 Pa/atm
torr (or millimeter of mercury)	(mm Hg)	133.3 Pa/torr
bar	—	10^5 Pa/bar*

OTHER USEFUL CONVERSIONS

A mass of 1 kg has a weight of about 2.21 lb.
The rest energy of 1 u is about 931 MeV.
A relative density of 1 is exactly 1000 kg/m^3.

COLLEGE PHYSICS

THIRD EDITION

JOHN J. O'DWYER

STATE UNIVERSITY OF NEW YORK
COLLEGE OF OSWEGO

BROOKS/COLE PUBLISHING COMPANY
PACIFIC GROVE, CALIFORNIA

To my father and mother, my first two teachers

Brooks/Cole Publishing Company A Division of Wadsworth, Inc.

Printed in the United States of America
10 9 8 7 6 5 4 3 2 1

Library of Congress Cataloging-in-Publication Data
O'Dwyer, John J. (John Joseph). 1925-
College physics / John J. O'Dwyer. —3rd ed.
p. cm.
Bibliography: p.
Includes index.
ISBN 0-534-11850-X
1. Physics. I. Title.
QC21.2.033 1990
530–dc20 89-30435
CIP

Sponsoring Editor: *Harvey Pantzis*
Project Development Editor: *Mary Arbogast*
Manufacturing: *Bill Bokermann, Vena Dyer*
Editorial Assistant: *Gerry Del Ré, Jennifer Kehr*
Production Editor: *Cece Munson, The Cooper Company*
Production Coordinator: *Joan Marsh*
Manuscript Editor: *Carol Dondrea*
Permissions Editor: *Carline Haga*
Cover and Interior Design: *Lisa Thompson*
Cover Neon Art: *Brain Coleman*
Art Coordinator: *Lisa Torri*
Interior Illustration: *Art by Ayxa*
Photo Editor: *Sue C. Howard*
Photo Researcher: *Stuart Kenter*
Typesetting: *The Clarinda Company*
Cover Printing: *The Lehigh Press*
Printing and Binding: *Arcata Graphics/Hawkins*

Chapter opening photographs, courtesy of the following: p. 1, 289, © Yoav/Phototake; p. 21, © Bob Daemmrich; p. 55, 145, NASA; p. 84 © Tom Branch/Science Source; p. 117, © Eric Meola; p. 168, © Robert A. Isaacs; p. 197, 378, © Tony Freeman/PhotoEdit; p. 221, © Doug Plummer; p. 254, © Richard Pasley/Stock, Boston; p. 320, Martin Dohrn/Science Photo Library; p. 353, Arthur Singer/Phototake; p. 413, © Fredrik B. Bodin/Stock, Boston; p. 442, © Will/Deni McIntyre; p. 467, © Keith Kent/Peter Arnold, Inc.; p. 490, 733, © Chuck O'Rear/WestLight; p. 528, 754, FourByFive; p. 559, Frank Siteman/Stock, Boston; p. 589, Pacific Gas & Electric; p. 620, © Fundamental Photographs; p. 654, © Mark McKenna; p. 678, © Photo Researchers/OMIKRON; p. 708, © Paramount/FourByFive; p. 789, UC, Berkeley, Radiation Laboratory; p. 819, © Science Source/Photo Researchers.

PREFACE

This text covers a two-semester or three-quarter college physics course for freshmen who intend to major in science. The treatment is non-calculus. Students who begin a serious study of physics in this way may be less distracted by the difficulties associated with calculus and bend more of their energies to understanding physics. High-school algebra provides the required mathematics, and the trigonometry needed is introduced in the text. Although high-school physics is a substantial asset, it is not a requirement.

As with the previous editions, the arrangement of the subject matter in this new edition of *College Physics* is traditional: kinematics, dynamics, fluids, vibrational motion and waves, heat and thermodynamics, electricity and magnetism, optics, and finally modern physics. The major changes in the third edition include the complete rewriting of the chapters on the second law of thermodynamics, electrostatics, magnetic fields and forces, and alternating current circuits. The second law of thermodynamics is now approached in a more traditional manner, and a section on entropy has been added. Electrostatics now occupies two chapters (one on point charges and the other on parallel plates) and features a more complete discussion of equipotential surfaces and lines of force. The introduction to magnetism has been altered and now begins with the magnetic force law and the Biot-Savart law. Alternating current circuits are now treated directly from the impedance concept; this limits the discussion to series circuits but completely avoids the more difficult ideas of the phasor diagram approach. The concepts of elementary particle physics have been expanded to cover several sections of the chapter on conservation laws. The review of trigonometry has been removed from the introductory chapter; the material now appears at the point where the physics first requires it.

Other relatively minor changes and additions have been dictated by experience with the previous editions. Numerous sections (almost always the final section of a chapter) can be omitted without any loss of continuity. The same is true of Chapter 15 (heat engines and the second law of thermodynamics), Chapter 21 (alternating current circuits), and Chapter 29 (conservation laws in nuclear and particle processes).

The presentation of the material emphasizes the main ideas simply and briefly, concentrating on the physics rather than on the mathematics. Basic laws and major definitions are prominently displayed in boxes throughout each chapter, together with explanations of any new units. At the conclusion of each chapter, the major ideas are restated in words as Key Concepts. The key concepts are followed by a set of Qualitative Questions and an extensive series of Problems at three levels of difficulty, thereby enabling students to test their mastery of the chapter contents. Throughout the problem sets there are many "matched pairs"—odd- and even-numbered problems (with and without supplied answers) that are essentially the same but call for the calculation of different problem variables.

The text uses the SI system of physical units. Not only is the SI the official system of scientific units, but it also frees the instructor from the vexing and confusing dual-units system in dynamics. In some instances non-SI units are used; this usually occurs only where dimensional analysis shows clearly that any convenient units system will suffice. For example, British units are used in some kinematics problems, calories are used in some heat problems, and electron volts are used in some atomic- and nuclear-energy problems. In any doubtful case the student should convert all data to SI by using the extensive conversion tables provided inside the front cover of this book.

My twofold hope for the students is that the material is presented both clearly and correctly. If this goal is achieved, students should have a good grasp of basic principles and harbor no misleading ideas.

My sincere thanks are due to colleagues in physics teaching who made numerous helpful suggestions and criticisms during the review process. They include William T. Achor, Western Maryland College; Don Chodrow, James Madison University; Lawrence B. Coleman, University of California, Davis; Miles J. Dresser, University of Pittsburgh; David J. Ernst, Texas A&M University; Ruth Howes, Ball State University; Stanford Kern, Colorado State University; James Kettler, Ohio University-Belmont; Marvin Morris, San Jose State University; James G. Potter, Florida Institute of Technology; James Purcell, Georgia State University; and Marllin Simon, Auburn University.

My most special thanks are due to Dr. R. A. Brown of SUNY Oswego for very many ideas to improve the clarity and accuracy of the material.

J. J. O'Dwyer

BRIEF CONTENTS

CONTENTS

APPENDICES

1

INTRODUCTION

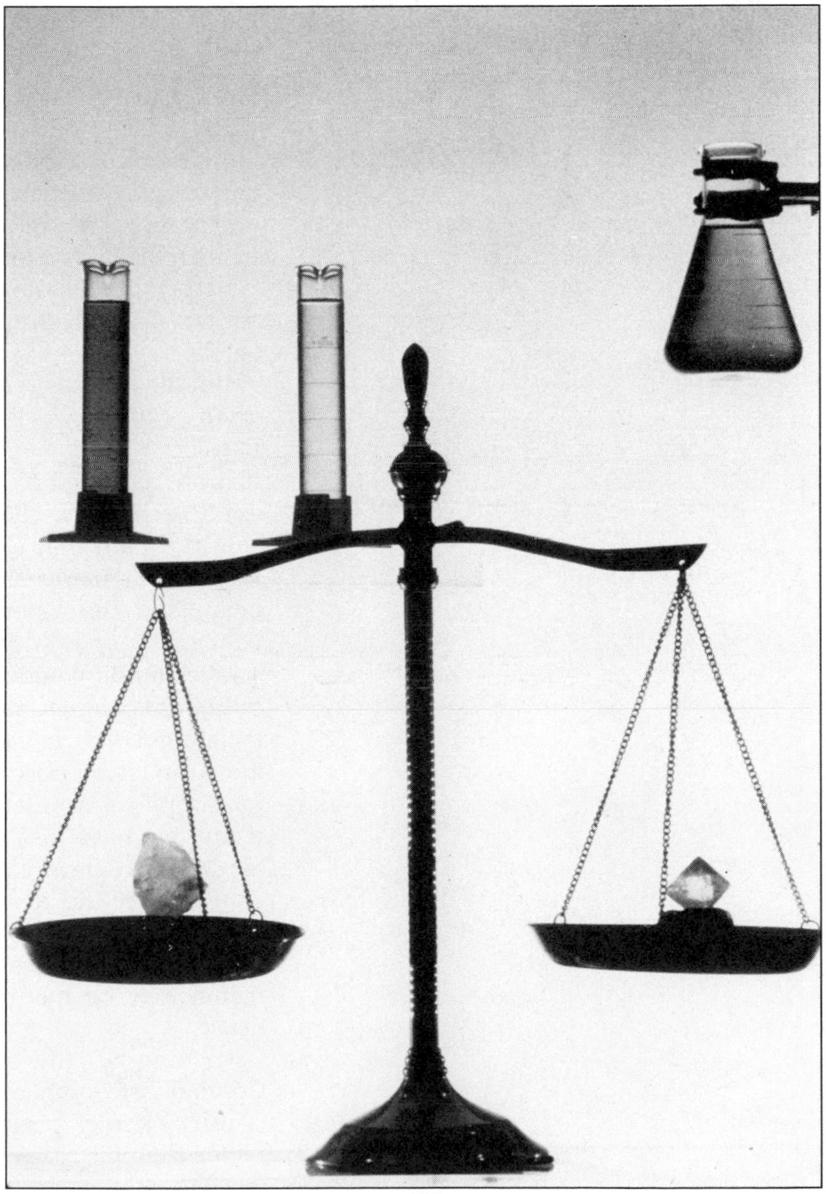

Defining an area of human knowledge at the beginning of a book on that subject is always difficult. It seems more logical to place the definition at the end of the book, where it would be fully appreciated. However, just as no one should go duck hunting without some idea of what a duck is, no one should begin a study of physics without some idea of what it is all about. At this point, then, we give a short description of the nature of physics and distinguish it from other sciences. The secondary goal of this introduction is to discuss some ideas that are necessary to begin the main part of the book.

1.1 The Nature of Physics

Physics is a science. This means that it seeks to understand the world of reality in terms of basic general principles. The search for basic principles is as old as civilization. The principles that we know today have been worked out in a long process involving observation, intuition, experimentation, debate, and reformulation. In all cases the ultimate court of appeal for the validity of a basic principle is the evidence—the observational and experimental data—which the principle either makes or fails to make intelligible.

But all sciences seek basic principles. What distinguishes physics from the rest? The distinction lies in how reality is conceived. Physicists build mental models *suitable for mathematical calculation* that are assumed to represent reality more or less faithfully. An example helps to make this clear. The earth moves around the sun in a path that is exceedingly complex—it is roughly oval in shape with many small disturbances due to the effects of the moon and the other planets. This statement about the earth's path in space is highly unsuitable for mathematical calculation: The term *oval* could apply to any squashed circle, and *small disturbances* is vaguer still. But this highly complex path is approximately circular. If we say that the earth (and the other planets) moves in circular orbits around the sun, we have a model suitable for simple mathematical calculations. In postulating this simple model we have lost something, since reality is much more complex. But we have gained a basis for calculations whose accuracy depends on the degree of correspondence between our model and the real world. We can work out consequences of the model and test them against further observations and experiments. The model may be modified or abandoned as required by this evidence.

Historically, the circular-orbit model was abandoned by the German astronomer Johannes Kelper in the early seventeenth century. Kepler proposed that the planetary orbits were elliptic rather than circular. The ellipse is a special type of oval that has an exact mathematical description. The change in the model was necessary because Kepler could not reconcile astronomical observations with the circular-orbit model, while the elliptic-orbit model agreed well with observational data.

But a more significant question lies beneath the surface. Could a single basic principle operate to cause the elliptic orbits of all the planets? Fifty years after Kepler announced his elliptic-orbit model, the English mathematician and physicist Isaac Newton proposed the principle of universal gravitation and his famous laws of motion. The gravitation principle gives specific mathematical form to a force of attraction between all material objects. Moreover, elliptic orbits are a mathematical consequence of the principle of universal gravitation and the laws of motion.

Newton's work went far beyond explaining planetary orbits. On a smaller scale, it provided the correct basis for calculating the motion of falling objects near the earth's surface. On a cosmic scale, it enabled astrophysicists to understand the initial steps in the formation of stars.

The principles of physics are not static truths; they must always be open to change as new evidence becomes available. Indeed, the history of physics shows just such a revision of ideas, with constant progress toward deeper and more universal understanding. The Greek philosopher Aristotle taught that the heavenly bodies move in circular orbits around the earth and that other objects (such as stones and snowflakes) fall toward the center of the earth. There is an element of truth in both of these "principles." However, 2000 years of observation, experimentation, and scholarship produced a point of view that completely superseded them. Newton's principle of universal gravitation, combined with his laws of motion, applies to the study of the motions of heavenly bodies, stones, and snowflakes. We no longer see any difference in principle between the orbit of a planet around the sun and the trajectory of a stone thrown by a child.

Our primary task in the study of physics is to understand its basic principles. Without doing so, we cannot begin to understand the many complex phenomena to be explained by science.

1.2 Fundamental Quantities and Standard Units

The suitable and correct definition of concepts is of prime importance in the study of physics. A definition is a statement that explains the unfamiliar in terms of the familiar. Since this process cannot continue indefinitely, we must come eventually to certain undefined basics. Physics provides a prescription for measuring these **fundamental quantities.**

The choice of fundamental physical quantities has a long history. Length and time have always been chosen as fundamental quantities because their nature is intuitively apparent. In some systems mass is the third fundamental quantity, while in others force plays that role. Electric charge, magnetic pole, and electric current have all been fundamental quantities in the physics of electromagnetism. Additional fundamental quantities have been introduced as needed to provide a complete framework for scientific measurements.

The selection of fundamental quantities is tied up with the selection of a system of units to be used as standards for measuring those quantities. Using length as a fundamental quantity is of no value unless a standard is available to science and industry which ensures that different measurements are strictly comparable. The choice of such standards has also varied greatly. The reasons for selecting different standards have been somewhat conflicting. On the one hand, it is desirable that the standard be accurately fixed; on the other hand, it is useful to have a convenient and easily reproducible standard.

The system of units to be used in this book is called the International System of Units (abbreviated SI for *Système International*). It is a modern version of the metric system that was recommended in 1960 at an international meeting known as the General Conference on Weights and Measures. In 1964 the U.S. National Bureau of Standards adopted SI units, and its use in scientific work continues to grow.

The architects and builders of this Mayan pyramid at Chichen Itza on the Yucatan peninsula must have had a standard unit of length that was readily accessible to all who were connected with the project. (Mexican National Tourist Council)

The fundamental quantities in the SI are:

1. Time
2. Length
3. Mass
4. Electric current
5. Temperature
6. Amount of a substance
7. Luminous intensity

There are also two supplemental geometrical quantities—plane angle and solid angle.

The first half of this book requires only the fundamental quantities of time, length, and mass. Let us look at their specifications and defer discussion of the remaining quantities until they are required.

Time

The assignment of a time standard requires some sort of regularly recurring natural event. Since ancient times the day was recognized as a convenient universal standard. In Roman times the daylight period was divided into a number of hours, but this did not constitute any sort of scientific standard since the hour was longer on a summer day than on a winter day. Not until the advent of mechanical timepieces in the fourteenth century were relatively accurate time standards possible. These clocks all indicated time units that are subdivisions of one day. A solar day is the time interval between successive passages of the sun across a meridian of longitude; the mean solar day is the yearly average of this time interval. The second was defined so that there were $(24 \times 60 \times 60)$ seconds in a mean solar day. Increasingly accurate scientific measurements have shown that the earth's period of rotation about its axis is slowly increasing, mainly as a result of the frictional forces associated with the tides. The rate of increase is about 0.001 second per century. In 1956 the General Conference on Weights and Measures defined the second as a certain fraction of the year 1900. In 1967 the second was redefined as 9 192 631 770 periods of vibration of a certain line in the spectrum of the cesium-133 atom. This standard is available to any scientific and commercial community having a modern scientific laboratory, and there is no known variation in this period.

Length

Standards of length have existed since ancient times. The cubit, based on the length of a man's forearm, was a basic unit of com-

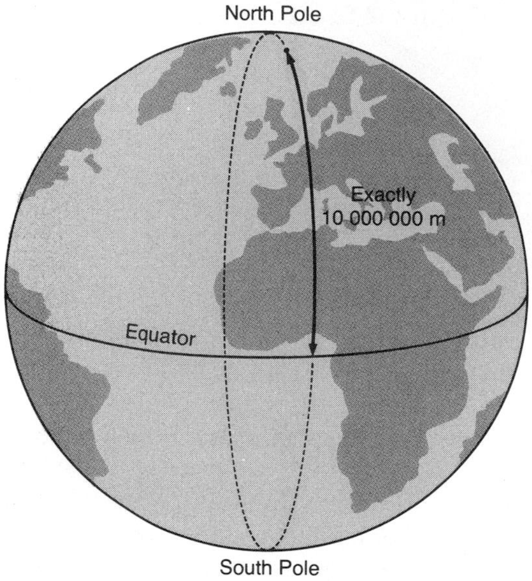

FIGURE 1.1 Illustration of the original definition of the meter. The meridian of longitude from the North Pole to the Equator and passing through Paris was assigned a length of 10 000 000 m.

merce in the Egyptian, Hebrew, and Roman civilizations. In medieval England the yard was the distance between the nose and the outstretched hand of King Henry I (1068–1135). With the growth of science and commerce more precise standards were required, and the English yard was fixed as the length of a certain bronze rod. In the closing years of the eighteenth century the French Academy of Sciences undertook the task of developing an internationally acceptable system of units. Their work produced the metric system, in which all multiples and submultiples of length and mass units are related to the unit by factors of powers of 10. To set a standard that was both fixed and universally accessible, they defined the meter with reference to places on the earth's surface. Imagine the nearly spherical earth to be intersected by a plane that passes through both the North and South Poles. The line of intersection is called a meridian of longitude. Figure 1.1 shows the meridian of longitude that passes through Paris. The French Academy assigned a length of exactly 10 000 000 meters to the portion of the meridian stretching from the North Pole to the Equator. This basis was difficult, if not impossible, to duplicate with any consistency. As a practical measure the idea was abandoned, and the meter was taken as the distance between two scratch marks on a certain platinum-iridium alloy bar kept at Sèvres, France, which is near Paris. Duplicates of this primary standard were made and kept in standards laboratories throughout the world for use in calibrating measuring instruments. In 1960 the General Conference on Weights and Measures defined the standard meter as 1 650 763.73 wavelengths in vacuum of a certain orange line in the spectrum of the krypton-86 atom. More recently, in 1983, the same body set the standard meter as the length of the path traveled by light in a vacuum during a time interval of 1/299 792 458 of a second.

Mass

The framers of the original metric system chose the gram as the unit of mass, which they defined as the mass of one cubic centimeter of water at the temperature of maximum density. To provide a convenient reference, a platinum-iridium cylinder was fabricated as a permanent standard. Since the gram is a rather small mass, this metal cylinder was made to be 1000 grams, or one kilogram. Later a small error was detected, so the metal cylinder was retained as the standard, and the correspondence with the mass of a certain volume of water was dropped. The standard kilogram was adopted in 1901 by the General Conference on Weights and Measures and has not been changed since that time. The primary standard is kept at Sèvres, and secondary standards are kept in various laboratories around the world.

Standard abbreviations are in use for all units. For the fundamental quantities introduced, the abbreviations are:

$$1 \text{ meter} = 1 \text{ m}$$

$$1 \text{ second} = 1 \text{ s}$$

$$1 \text{ kilogram} = 1 \text{ kg}$$

The standard units for the fundamental quantities are chosen to be of a convenient size for many human-related problems. An adult human being is about 2 m tall and has a mass between about 50 kg and 80 kg. In this case the standard units for length and mass lead to readily imaginable numbers. The time unit of 1 s is also easy to grasp. But physics is a universal science that applies to all known material objects, ranging from giant galaxies composed of billions of stars to the submicroscopic constituent particles of matter. Periods of time range from the age of the universe to the duration of one vibration of a high-frequency electromagnetic wave. The standard units are more or less in the center of the scale of things. Astronomical magnitudes are incredibly bigger, and atomic and nuclear magnitudes similarly smaller. Typical values of time intervals, lengths, and masses are shown in Table 1.1.

It is convenient to assign standard unit prefixes to designate multiples of units. Standard practice is to use prefixes that denote multiples of three powers of 10, both positive and negative, from the basic unit. Table 1.2 shows some of the prefixes commonly used in elementary physics, which cover the range from 1 billion times the base unit to one-billionth part of it. The prefix *centi-* is included since the centimeter is frequently used.

Centuries of growth of our complex civilization have produced an abundance of units for special purposes. Gem merchants, mariners, and physicians (to name but a few) all use standard units that are not related to SI units by factors of a power of 10. Despite international agreement on scientific units, many of these other units will undoubtedly persist for reasons of convenience. The mathematical treatment of topics in this book is such that all formulas are correct if the quantities are expressed in SI units—they may or may not be correct if other units are used. In the easier problems either the units cause no difficulty, or the data are quoted entirely in SI units. Mixed units are used on a few of the more difficult problems. Although it is true that experience in the solution of problems may permit one to avoid units conversion, it is always correct to proceed as follows:

1. Convert the basic data to SI units.

2. Solve the problem.

3. Convert the SI units answer to other units if required.

For this purpose an extensive listing of conversion factors is given inside the front cover.

TABLE 1.1 Some typical values (rounded off to the nearest power of 10) for times, lengths, and masses

	Some typical times
10^{17} s	Age of the earth
10^{13} s	Age of the human race
10^{9} s	Average human lifetime
10^{5} s	One day
10^{-3} s	Vibration period of a musical note
10^{-6} s	Vibration period of AM radio wave
10^{-15} s	Vibration period of visible light
10^{-23} s	Lifetime of unstable subatomic particle
	Some typical lengths
10^{25} m	Radius of observable universe
10^{16} m	Distance light travels in one year
10^{7} m	Diameter of the earth
10^{2} m	Length of a football field
10^{-6} m	Wavelength of visible light
10^{-10} m	Atomic diameter
10^{-15} m	Nuclear diameter
	Some typical masses
10^{30} kg	The sun
10^{25} kg	The earth
10^{8} kg	A supertanker
10^{3} kg	An automobile
10^{-6} kg	A grain of sand
10^{-15} kg	A DNA molecule
10^{-30} kg	An electron

TABLE 1.2 Some standard prefixes in common use

Prefix	*Abbreviation*	*Multiplying factor*
giga-	G	10^{9}
mega-	M	10^{6}
kilo-	k	10^{3}
centi-	c	10^{-2}
milli-	m	10^{-3}
micro-	μ	10^{-6}
nano-	n	10^{-9}

A glance at this front matter shows that most physical quantities can be expressed in terms of a staggering variety of units. The SI is only one of many metric systems, and there are also many British units. For our purposes, we will define all other units in terms of their SI equivalents. This practice is in accordance with most of the modern conventions. For example, the inch is now defined as 2.54 cm and not as some fraction of the standard yard. In this way we avoid having two different standards for the same fundamental quantity.

To begin with, let us be clear that we are really multiplying a quantity by unity (and, therefore, not changing it) when we make a units conversion. For example, Table 1.3 shows 0.9144 m/yd as the factor to use when converting yards to meters. The reason is that 1 yd = 0.9144 m and the quantity 0.9144 m/1 yd is therefore unity. To illustrate, let us calculate the number of meters in 100 yd:

$$\begin{aligned} 100 \text{ yd} &= 100 \cancel{\text{yd}} \times \frac{0.9144 \text{ m}}{1 \cancel{\text{yd}}} \\ &= 91.44 \text{ m} \end{aligned}$$

TABLE 1.3 Conversion factors for length units

Length
SI unit: meter
Abbreviation: m

Other units	*Abbreviations*	*To convert to meters multiply by*
inch	in.	0.0254 m/in.
foot	ft	0.3048 m/ft
yard	yd	0.9144 m/yd
mile	mi	1609 m/mi
nautical mile	—	1852 m/nautical mile
angstrom	Å	10^{-10} m/Å

The process involves arithmetical multiplication of the numbers and algebraic cancellation of the unit symbols.

How do we convert a quantity from SI units to non-SI units? We simply divide by the appropriate conversion factor. For example, let us calculate the number of yards in 100 m:

$$\begin{aligned} 100\ \text{m} &= 100\ \text{m} \div 0.9144\ \text{m/1 yd} \\ &= 100\ \cancel{\text{m}} \times \frac{1\ \text{yd}}{0.9144\ \cancel{\text{m}}} \\ &= 109.4\ \text{yd} \end{aligned}$$

Once again, we manipulate the numbers by the rules of arithmetic and the unit symbols by the rules of algebra.

The same approach works for other conversions. How do we convert 60 mi/h into meters per second?

$$\begin{aligned} \frac{60\ \text{mi}}{1\ \text{h}} &= \frac{60\ \cancel{\text{mi}} \times \dfrac{1609\ \text{m}}{1\ \cancel{\text{mi}}}}{1\ \cancel{\text{h}} \times \dfrac{3600\ \text{s}}{1\ \cancel{\text{h}}}} \\ &= 26.82\ \text{m/s} \end{aligned}$$

Once again, note the algebraic cancellation of units. In fact, this result is achieved more rapidly using the velocity conversion factor given inside the front cover:

$$\begin{aligned} 60\ \text{mi/h} &= 60\ \cancel{(\text{mi/h})} \times \frac{0.4470\ (\text{m/s})}{1\ \cancel{(\text{mi/h})}} \\ &= 26.82\ \text{m/s} \end{aligned}$$

Any complicated unit can be likewise converted by successive multiplications (or divisions) by unity.

1.3 Derived Quantities

We have so far considered only the fundamental quantities and their associated units. One of the chief tasks of physics is to define new quantities as required. Quantities that we define in terms of fundamental quantities are called **derived quantities.** Every derived quantity has a unit that follows from its definition. For example, we obtain an area when we multiply two lengths together. Area is a derived quantity whose SI units are meters multiplied by meters, or square meters. Similarly, volume is the derived quantity that follows from multiplying an area by a length. Its SI units are cubic meters.

When a unit to be converted is squared (or raised to a higher power), we require two (or more) applications of the conversion factor. For example, conversion of 10 ft^2 into square meters proceeds as follows:

$$\begin{aligned} 10\ \text{ft}^2 &= 10\ \cancel{\text{ft}^2} \times \frac{0.3048\ \text{m}}{1\ \cancel{\text{ft}}} \times \frac{0.3048\ \text{m}}{1\ \cancel{\text{ft}}} \\ &= 0.929\ \text{m}^2 \end{aligned}$$

Note once again that each time we multiply by the conversion factor we are really multiplying by unity. In the case above, two operations of the conversion factor are required so that the feet cancel in the right-hand side of the expression.

The definition of **density** offers us a slightly more complex example of a derived function. Density is one of the properties characteristic of a substance.* The density of a particular piece of some substance is its mass divided by the volume it occupies. Because of this, the SI unit of density is the kilogram per cubic meter. Let us summarize in display the information about density. The symbol for density is the Greek letter rho (ρ). Appendix B lists the commonly used Greek and Roman letters, together with the quantities that they symbolize. The same letter frequently stands for different things when used as a symbol in formulas or as an abbreviation for a unit. This is unfortunate but does not usually cause too much confusion. For example, m stands for mass when used as an algebraic symbol. As a units designation it is an abbreviation for meter, and as a prefix on a units designation it is an abbreviation for *milli-*.

*As we see later, density changes with changes of pressure and temperature. The change is very small for solids and liquids but quite substantial for gases.

TABLE 1.4 Densities of common substances at 20°C and 1-atm pressure

	Substance	*Density* (kg/m^3)
Solids	Aluminum	2.70×10^3
	Brass	8.44×10^3
	Copper	8.96×10^3
	Gold	19.3×10^3
	Iron or steel	7.86×10^3
	Lead	11.3×10^3
	Silver	10.49×10^3
	Oak	$0.64–0.77 \times 10^3$
	Pine	$0.43–0.65 \times 10^3$
Liquids	Methanol	0.810×10^3
	Ethanol	0.791×10^3
	Glycerin	1.26×10^3
	Gasoline	$0.66–0.69 \times 10^3$
	Mercury	13.6×10^3
	Water	1.00×10^3
	Seawater	1.025×10^3
Gases	Air	1.205
	Carbon dioxide	1.842
	Helium	0.1664
	Hydrogen	0.08375

■ Definition of Density

$$\text{Density} = \frac{\text{mass}}{\text{volume}}$$

$$\rho = \frac{m}{V} \tag{1.1}$$

SI unit of density: kilogram per cubic meter
Abbreviation: kg/m^3

Table 1.4 lists the densities of various substances at room temperature and pressure. The assignment of the kilogram mass unit places the density of water at 10^3 kg/m^3. A typical metal is approximately 10 times denser than water. The typical liquid has a density roughly similar to water, and air (at a temperature of 20°C and a pressure of 1 atm) is about 1000 times less dense. Comparing the densities of substances with the density of water is useful and convenient. To do this precisely, we define the **relative density** of a substance to be its density divided by the density of water. Relative density is also widely called specific gravity. Our definition of relative density is a quotient of two quantities. Since each quantity has the same units, the units cancel and produce a pure number.

■ Definition of Relative Density (Specific Gravity)

$$\text{Relative density} = \frac{\text{density}}{\text{density of water}}$$

$$\rho_R = \frac{\rho}{\rho_W} \tag{1.2}$$

Relative density is a pure number.

EXAMPLE 1.1

A cube of iron whose side is 15.25 cm has a mass of 27.8 kg. What are its density and its relative density?

SOLUTION To calculate the density from Eq. (1.1), we need to know both the mass and the volume of the cube. The mass is given in SI units, and the volume is readily found:

$$\begin{aligned}\text{Volume of cube} &= (15.25\ \text{cm} \times 10^{-2}\ \text{m/cm})^3 \\ &= 3.547 \times 10^{-3}\ \text{m}^3\end{aligned}$$

We can now calculate the density directly from Eq. (1.1):

$$\rho = \frac{m}{V}$$

$$= \frac{27.8 \text{ kg}}{3.547 \times 10^{-3} \text{ m}^3}$$

$$= 7.84 \times 10^3 \text{ kg/m}^3$$

Since the density of water is 10^3 kg/m^3, the relative density of iron, obtained by using Eq. (1.2), is:

$$\rho_R = \frac{\rho}{\rho_W}$$

$$= \frac{7.84 \times 10^3 \text{ kg/m}^3}{10^3 \text{ kg/m}^3}$$

$$= 7.84$$

Again, note that the relative density has no units because they cancel out in the calculation. ■

We have seen several examples of derived quantities, namely area, volume, density, and relative density; there will be many more in the study of physics. As we have also seen, each newly defined quantity brings with it a new unit. The new unit depends on the unit system being used; in SI the unit of area is square meters, while in other systems it could be square feet, square centimeters, or even acres. We can indicate the species of a new unit by noting that each physical quantity possesses a **dimension** that follows either from its fundamental nature or from its definition. We use capital letters inside square brackets [T], [L], and [M] to denote the dimensions of the fundamental quantities time, length, and mass, respectively. Since area is a length multiplied by a length, its dimension is $[L]^2$, and this remains true for any unit that one may choose to express area. Similarly, the dimension of volume is $[L]^3$, and that of density is $[M]\,[L]^{-3}$. As already mentioned relative density has no units and, therefore, no dimensions; it is a pure number.

The dimension of a physical quantity always indicates its nature, even if the expression is in exotic or mixed units. For example, a measurement may be quoted in milligrams per cubic inch. The milligram is a unit of mass, and the inch is a unit of length; it follows that the dimension of this quantity is $[M]\,[L]^{-3}$, and it must be a density. The utility of dimensional analysis is not limited to identifying the nature of quantities expressed in unfamiliar units. We shall see in later chapters that it can be used to check equations, or even to derive them within certain limits.

1.4 Rounding Off Numbers

Although the quantities 0.123 m and 0.123 00 m appear to be the same, from the point of view of physics they are quite different. The first might represent a measurement obtained with a meter stick; the second, a measurement obtained with a much more precise instrument, since the zeros are intended to be meaningful. In other words, we are taking the number of digits quoted as being an indication of the accuracy of the measurement. We frequently work problems with electronic calculators, which display answers containing up to ten digits. It is important to know how to handle this apparent abundance of knowledge.

To find the number of *significant digits* in a number, begin by locating the first nonzero digit. This digit and all those following (zero or otherwise) are significant digits. For example, the number 1.001 210 contains seven significant digits, but 0.001 210 contains only four.

The process of dropping significant digits is called *rounding off.* The necessity of rounding off arises after arithmetic operations on data. If we measure the lengths of the sides of a rectangle with a meter stick and find them to be 1.27 m and 2.13 m, the use of an electronic calculator to find the area of the rectangle gives:

$$\begin{aligned}\text{Area of rectangle} &= 1.27\ \text{m} \times 2.13\ \text{m} \\ &= 2.7051\ \text{m}^2\end{aligned}$$

The original measurements have three-digit accuracy, and it is unreasonable to suppose that the calculator can create accuracy that is not in the data. We would quote the answer as follows:

$$\text{Area of rectangle} = 2.71\ \text{m}^2$$

This is the closest three-digit number to the number appearing on the calculator. The general rule to follow when multiplying or dividing numbers is to keep the same number of significant digits in the answer as in the least precise factor.

The addition or subtraction of two numbers requires a slightly different treatment. Consider the addition of two lengths:

$$\begin{aligned} & 2.2348\ \text{m} \\ & +\underline{2.1\ \text{m}} \\ \text{Total length} = {} & 4.3348\ \text{m}\end{aligned}$$

The answer should be quoted as 4.3 m since the 2.1-m length cannot be written 2.1000 m, and the last three digits are therefore not significant. At first sight we appear to have followed the same

SPECIAL TOPIC 1 SCALING AND SKELETAL STRUCTURE

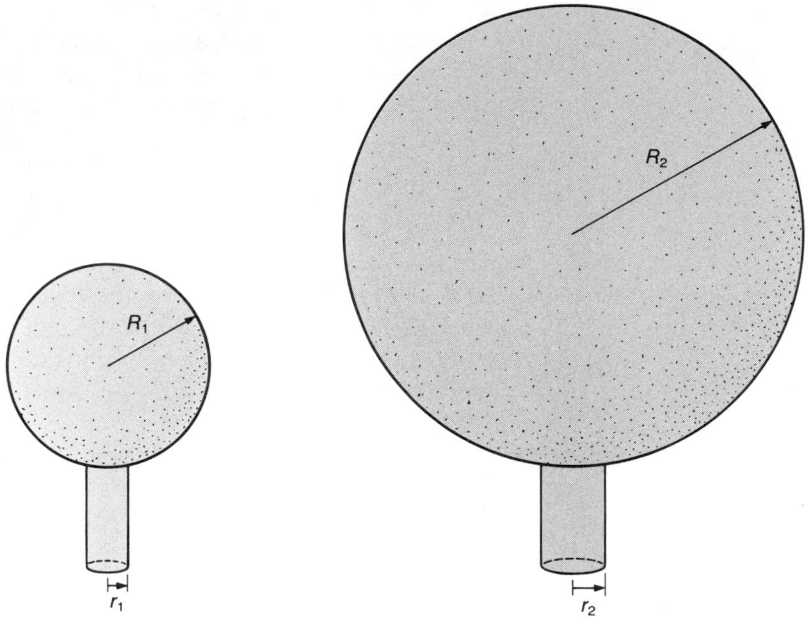

Two spheres of the same density but of different radii R_1 and R_2 supported by cylindrical struts of cross-sectional radii r_1 and r_2

Interspersed throughout this book will be a number of essays on scaling that consider its applications to animal structure and function. As the following discussion will show, the experimental results are accurate for the model under consideration; however, uncertainties arise when we attempt to apply the model and its assumptions to real situations.

Galileo was the first person to consider the problem of structures of the same shape and density (but of different size) supported by a strut or struts. Since the structures have the same shape, we can characterize their geometrical relationship to each other by a single dimensionless scale factor L. To take a very simple example, suppose the structures are spheres of the same density. We can then choose the dimensionless scale factor to be the ratio of their radii; the diagram shows two spheres in which the larger radius R_2 is three times as great as the smaller radius R_1. For this particular example, we have:

$$L = \frac{R_2}{R_1} = 3$$

rule as for multiplication and division, but this is accidental. Consider another example of the addition of two lengths:

$$\begin{array}{rr} & 2.2348\text{ m} \\ & +\underline{0.021\ \ \text{m}} \\ \text{Total length} = & 2.2558\text{ m} \end{array}$$

SPECIAL TOPIC 1 (continued)

The volume of a sphere is given by $(4\pi/3) \times (\text{radius})^3$; the ratio of the volumes will therefore be:

$$\frac{V_2}{V_1} = \frac{\left(\frac{4\pi}{3}\right) \times R_2^3}{\left(\frac{4\pi}{3}\right) \times R_1^3} = \left(\frac{R_2}{R_1}\right)^3 = L^3$$

In discussions on scaling we express this relationship by saying that the volume scales as L^3. For our specific example, the volume of the larger sphere is $3^3 = 27$ times the volume of the smaller one.

Since Eq. (1.1) shows that mass is directly proportional to volume if density remains constant, we conclude that the mass of the spheres also scales as L^3. In a subsequent chapter we will show that the weight of an object is proportional to its mass, and the weight of the spheres also scales as L^3.

Now let us consider the supporting strut; we make the assumption that its ability to support the weight of the sphere is proportional to its cross-sectional area. The spheres of our example are supported by cylindrical struts of circular cross section with radii r_1 and r_2, as shown in the diagram. Since the area of a circle is given by $\pi \times (\text{radius})^2$, our assumption requires that:

$$\frac{\pi r_2^2}{\pi r_1^2} = L^3 \qquad \text{that is,} \qquad \frac{r_2}{r_1} = L^{3/2}$$

The strut radius scales as $L^{3/2}$; for our example in which $L = 3$, we have:

$$\frac{r_2}{r_1} = 3^{3/2} \simeq 5.2$$

Note that the complete structure composed of sphere plus strut does not have a characteristic scale factor. In fact, we have just proved that this cannot be so—the radius of the strut increases at a faster rate than the radius of the sphere.

We can attempt to apply this result to animals of different species having about the same density and roughly similar shapes. An elephant and a small donkey have more or less the same shape, but they are supported by four legs rather than one strut. However, this just means that each leg supports one quarter of the weight; thus our scaling conclusions would stand. Suppose the body sizes of the two animals have a scale factor $L = 3$. This cannot be a ratio of the radii of spheres, but it could be the ratio of the animals' body lengths. Application of our scaling argument shows that the ratio of the leg-bone radii should be $L^{3/2} = 3^{3/2} \simeq 5.2$. Something of this nature is certainly observed, since the radius of the leg bone in larger animals is disproportionately larger than relative body size would indicate.

The scaling argument is accurate *provided we accept all the assumptions of the model*. In a future essay we shall see that the assumptions are indeed open to serious objection when the model is applied to animals. This objection will alter details of the scaling argument, but the general result remains true—that leg-bone radius must increase at a proportionately faster rate than animal body size.

The answer should be quoted as 2.256 m even though the 0.021-m length contains only two significant digits. It is the position of the significant digits relative to the decimal point that makes the difference. The general rule to keep in mind when adding or subtracting numbers is that undetermined digits in one number cannot be made significant by the addition or subtraction of a second number.

What procedure should we follow in problem solving? As a general rule we keep four significant digits while working through the problem and quote three in the answer. Adopting this routine ensures 1% overall accuracy in our calculations.

Problems refer to simulated situations, not to real physical measurements. For this reason we usually omit significant digits in quoting problem data. A genuine measurement of a mass to four significant figures would always appear as 3.000 kg. A mass quoted as 3 kg in a problem should be taken to mean 3.000 kg in spite of the lack of significant digits.

EXAMPLE 1.2

Calculate the volume of a circular cylinder 0.5 m in diameter and 1.35 m in height.

SOLUTION Instead of making a direct calculation, suppose we proceed in two steps. We first calculate the area of the circular cross section of the cylinder and then multiply by its height:

$$\begin{aligned}\text{Area of circular cross section} &= \pi \times \left(\frac{\text{diameter}}{2}\right)^2 \\ &= \pi \times \left(\frac{0.5\text{ m}}{2}\right)^2 \\ &= 0.1963\text{ m}^2\end{aligned}$$

(We keep four significant digits even though the data contain only one.)

$$\begin{aligned}\text{Volume of cylinder} &= \text{area of circular cross section} \times \text{height} \\ &= 0.1963\text{ m}^2 \times 1.35\text{ m} \\ &= 0.265\text{ m}^3\end{aligned}$$

(We keep only three significant digits in the final answer.) ■

Our discussion implies that highly accurate measurements are expressed by using a large number of significant digits. The definitions of the SI length and time units contain nine and ten digits, indicating the level of precision that standards laboratories attain in spectroscopic measurements.

1.5 Orders of Magnitude and Estimates

As already explained, we plan to use four significant digits in problem solving and retain three in the final answer. However, there can be circumstances in which we require only an approximate answer. Perhaps we require a rough check on an answer obtained by a more detailed approach, or a detailed approach may simply not be worth the time and energy.

A simple procedure to obtain an approximate result is to round off all numbers to one significant digit, and to retain only one in the final answer. This is called an order of magnitude estimate; the result is usually accurate to within a multiplicative factor of 2 or 3, and sometimes very much better. Let us illustrate with an example.

EXAMPLE 1.3

How many gallons do you estimate that a cylindrical container 36 cm in diameter and 68 cm high will hold?

SOLUTION The volume of a cylinder is the area of the base (πr^2) multiplied by the height (h):

$$\begin{aligned}\text{Volume} &= \pi \times r^2 \times h \\ &= \pi \times (18\text{ cm})^2 \times 68\text{ cm} \\ &\approx 3 \times (20\text{ cm})^2 \times 70\text{ cm} \\ &\approx 80{,}000\text{ cm}^3 \\ &= 80\ \ell\end{aligned}$$

A liter is approximately equal to a quart, and there are 4 qt in a gallon; it follows that the container will hold approximately 20 gal. A more exact calculation gives 18.3 gal for the capacity of the container, but you should not always expect 10% accuracy for such approximate calculations. Compensating errors combined in a fortuitous way to produce this fairly good result. ■

The problem that we have just worked depends on taking approximate values for accurately known quantities. Other types of estimate depend on educated guesswork, as we see in the following example.

EXAMPLE 1.4

How many spherical glass marbles each 1.7 cm in diameter would fit in the container of Example 1.3?

SOLUTION Only part of the volume of the container is occupied by the marbles, and the rest is air space. It seems to be a good guess that each marble will occupy about the same total space as a cube of side equal to the marble diameter. The volume of such a cube is given by:

$$(1.7\text{ cm})^3 \approx (2\text{ cm})^3 = 8\text{ cm}^3$$

Our estimate from Example 1.3 gives a volume of about 80,000 cm^3 for the container. We therefore estimate that the container will hold about 10,000 marbles. Note that we have no obvious way to check the accuracy of this answer since everything hinges on our guess of the total volume occupied per marble. ■

Study Review of a Chapter

Each chapter explains a number of concepts that should be mastered before proceeding. We list the most important ones as ''Key Concepts'' at the end of the chapter. Some of the key concepts are definitions of physical quantities, some are statements of physical laws, and others simply refer to important techniques or ideas. The list makes these distinctions. In most cases a review of the concepts should bring to mind the relevant formulas and units. If it does not, restudy the topic in question together with the worked examples. Nothing makes physics easier than a clear grasp of basic concepts.

A selection of conceptual questions offers an opportunity to think about the key concepts and to apply them qualitatively to various situations. Some of the questions have clear-cut answers, while others raise more difficult issues for discussion or debate.

The problems constitute the major test of comprehension of the material, and they are grouped in increasing order of difficulty. The easier problems at the beginning usually require only one key concept for solution. The chapter section containing the explanation of this concept follows in square brackets after the problem. The next group of problems are labeled as standard-level problems. They frequently require a combination of key concepts, or else present some difficulty of formulation that does not occur in the easier ones. A reference to a chapter section also accompanies a few of the standard-level problems. Finally, most of the problem sections contain a few advanced-level problems that offer a fairly stiff test of understanding of the material.

KEY CONCEPTS

The **fundamental quantities** introduced up to this point are time, length, and mass. Their basic units in the SI system are the second, the meter, and the kilogram, respectively.

Derived quantities are defined in terms of fundamental quantities. We determine their units from the units of the fundamental quantities concerned.

The **density** of a substance is a derived quantity. We define density as the mass of a piece of the substance divided by the volume it occupies. The SI unit of density is kilogram per cubic meter. See Eq. (1.1).

The **relative density** (or specific gravity) of a substance is the density of the substance divided by the density of water. See Eq. (1.2).

We assign **dimensions** [T], [L], and [M] to the fundamental quantities time, length, and mass, respectively. The dimensions of a derived quantity follow from its definition in terms of its component fundamental quantities.

QUESTIONS FOR THOUGHT

1. The various units systems are somewhat like different languages in which physicists express results. Is anything lost in the "translation" from one units system to another? What are the possible advantages in using different units systems for various purposes?

2. Units standards have evolved and changed over many centuries. For which of the fundamental quantities time, length, and mass was the establishment of an accurate and practical unit most difficult?

3. The laws of physics govern a vast range of phenomena in our universe, and special units are in use to describe very large and very small quantities. Make a list of special units of time, length, and mass used by people who work with either the very large or the very small.

4. Scientists believe that the period of rotation of the earth about its axis is slowly increasing. If the period of rotation were the basis for the unit of time, how could they know this?

5. A skeptic claims that the period of vibration of the spectral line from cesium-133 (as used in the SI designation of the second) is slowly changing. The implication is that the basis for the unit assignment is a poor choice. Make a defense in principle against this attack.

6. The length of the month is based on the motion of the moon. The *sidereal month* is the time taken by the moon to return to a given position relative to the distant stars. The *lunar month* is the time the moon takes to return to a given position relative to the sun. Which of these two intervals is longer, and by what approximate fraction?

7. How many significant digits would you retain in calculating the average gas mileage of your automobile on the basis of a single tankful of gasoline? Repeat the question if the average were calculated on the basis of 100 tankfuls of gasoline.

8. Give the dimensions of density and relative density. Do either of these quantities change in value with a change in the system of units?

9. You are told that the earth is about 400 times farther away from the sun than from the moon. Can you use this information to estimate the ratio of the sun's diameter to the moon's diameter?

PROBLEMS

A. Single-Substitution Problems

1. The distance by air from New York to London is 3500 mi. How far is this in kilometers? [1.2]

2. The distance between the sodium and chlorine ion centers in common salt is 2.81 Å. How many inches is this? [1.2]

3. How high is Mt. Everest (29 030 ft) in meters? [1.2]

4. The 1500 m race at the Olympics is often called the metric mile. What is the race distance in miles? [1.2]

5. In a timber-framed house a "2 × 4" is a piece of timber that is approximately 1⅝" × 3⅝" in cross section. What are its cross-sectional dimensions in millimeters? [1.2]

6. The United States has an area of 3 676 000 mi^2. What is its area in hectares? (A hectare is a metric unit of area that is equal to 100 m × 100 m.) [1.2]

7. What is the area of a football field (160 ft × 300 ft) in square meters? [1.2]

8. A commercial building offers 600 ft^2 of office space for rent. How many square meters are available? [1.2]

9. How many liters of gasoline are required to fill a 15-gal tank? [1.2]

10. By what percentage does a gallon differ from 4 liters? [1.2]

11. How many quarts are there in 1 m^3? [1.2]

12. A cubic block of metal 8 cm on a side has a mass of 4.32 kg. Calculate its density. [1.3]

13. Given that the density of copper is 8.96×10^3 kg/m^3, calculate the volume of a 60-g mass. [1.3]

14. The density of methanol is 810 kg/m^3. What is the mass of 1 ℓ of this liquid? [1.3]

15. The density of seawater is 1025 kg/m^3. Calculate the volume in liters of 10 kg of seawater. [1.3]

B. Standard-Level Problems

(Use the data of Table 1.4 as necessary.)

16. If gasoline sells for \$1 per gallon, what is the price per liter? [1.2]

17. The walls and ceiling of a room 14 ft × 12 ft × 8 ft are to be painted; the wet film of paint is 0.1 mm thick. How many gallons are required? [1.2]

18. A family consumes 180 gal of water per day and wishes to store a 10-week supply in a cylindrical cistern 8 ft high. What should be the diameter of the cistern in meters?

19. If the standard kilogram were a cube of brass, what would be the length of its edge in inches? [1.3]

20. What is the mass of water in a child's wading pool 6 ft × 6 ft × 8 in.? [1.3]

21. What is the mass of air (at 20°C and a pressure of 1 atm) in a room 10 m × 8 m × 3.5 m? [1.3]

22. A plank of wood 6 ft × 6 in. × ¾ in. has a mass of 3.4 kg. What is its relative density?

23. A concrete paving stone 3 ft × 3 ft × 2 in. has a mass of 100 kg. What is its relative density?

24. A 1-qt bottle has a mass of 640 g when empty. What is its mass when filled with glycerin?

25. A metal cube 4 in. on edge has a mass of 9.39 kg. What metal is it?

26. An empty bottle has a mass of 460 g; when filled with methanol, its mass is 975 g. What would be its mass when filled with mercury?

27. The great pyramid of Cheops has a base area of about 13 acres and originally stood about 480 ft high. The stone from which it was made has a relative density of about 2.1. What was the total mass of stone in the pyramid? (Volume of pyramid = ⅓ × area of base × height.)

28. The mass of the earth is 5.98×10^{24} kg, and its radius is 6.37×10^{6} m. What is the average relative density of the earth?

29. How many spherical raindrops each 1.2 mm in diameter are required for 1 kg of water? (Volume of sphere = $\frac{4}{3} \times \pi \times \text{radius}^3$.)

30. A student's desk made of white pine has a mass of 42 kg. What would be the mass of an otherwise identical desk made of white oak? (Relative density of white pine = 0.43, and relative density of white oak = 0.77.)

31. Make order of magnitude estimates of the following, and check your result by more exact calculation:

- **a.** The average density of the earth (Use data on the inside back cover.)
- **b.** The time taken to pump out, using a pump that can remove 120 gal/min, a basement of floor area 875 m^2 that is flooded to a depth of 90 cm. [1.5]

32. Make order of magnitude estimates of the following, explaining the guesses made in the process:

- **a.** The mass of water in the world's oceans
- **b.** The mass of air that a person inhales during a night's sleep. [1.5]

2

ONE-DIMENSIONAL KINEMATICS

The word *kinematics* comes from a Greek word meaning "motion"—the same Greek root gives us "cinema" ("pictures in motion"). In physics we restrict the meaning of kinematics to the study of the path of objects in motion without considering the forces that cause the motion. A figure skater's path across the ice is very complex, and the flight path of a seagull is even more complex. We need to begin with a simple model that permits us to arrive at the fundamental concepts. We do this by first limiting our study to motion in a straight line. In this case we can easily identify the basic concepts: displacement, velocity, and acceleration. The one-dimensional analysis provides us with a sound basis for expanding our treatment in subsequent chapters.

2.1 The Displacement of a Point Particle

It is not difficult for us to come up with examples of motion that are approximately in a straight line. A sprinter in a short dash, a cyclist or an automobile traveling along a straight road, and a stone falling from a height are all examples of this type of motion. However, even these motions have more complexities than we want to consider, since some parts of these bodies may not move in a straight line. For example, the legs of the sprinter and of the cyclist and the wheels of the cycle and of the automobile all exhibit a complex pattern of motion. Thus, we simplify still further and limit our model of motion to a *point particle moving backward or forward along a straight line.* With this model we cannot study the leg motion of the athlete or the wheel motion of the vehicle, but we do gain a basis for accurate understanding of the overall motion of the object in question.

In the first place we need a precise term for the change in position of the point particle. This term, **displacement,** simply means the difference between the original or starting position and the final or end position. The displacement might or might not be the same as the distance traveled. Let us express these ideas

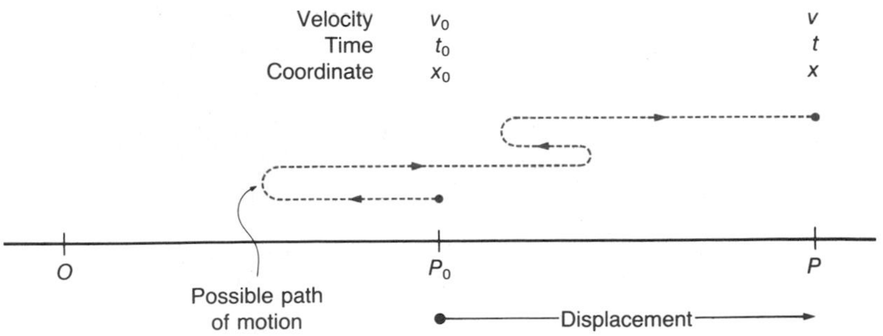

FIGURE 2.1 Diagram to illustrate the notation of one-dimensional kinematics.

mathematically. Consider a point particle moving on the straight line shown in Fig. 2.1. Suppose that the particle moves from the initial point P_0 to the final point P by a backward and forward motion, as shown on the diagram. At any instant we determine the particle position by its distance from the fixed reference point O; distances measured to the right of O we count positive, and those to the left negative. The point O is called the *origin of coordinates,* and the distance from O to the particle is called the *particle coordinate.* Let the coordinate of the initial position P_0 be x_0, and the coordinate of the final position P be x. We define the particle displacement for the motion from P_0 to P as follows:

$$\begin{aligned}\text{Displacement} &= OP - OP_0 \\ &= x - x_0 \end{aligned} \tag{2.1}$$

The displacement depends only on the final coordinate x and the initial coordinate x_0. It does not depend in any way on the back and forth motions of the particle as it travels from P_0 to P. Displacement is certainly not the total distance traveled between the beginning and the end points; it is not even the distance between the beginning and the end points measured along a straight line. This latter distance is only the *magnitude of the displacement.* A sign is also necessary to specify the *direction of the displacement.* We count displacements to the right as positive and those to the left as negative. With this convention, our definition of displacement as $(x - x_0)$ correctly includes both magnitude and direction.

A quantity that needs both magnitude and direction for its specification is called a **vector.** One-dimensional displacement is the first and simplest example of a large class of physical vector quantities. We recall that time, length, mass, and density do not have an associated direction. Such quantities, specified by magnitude alone, are called **scalars.** We will see that the rules for mathematical manipulation of vectors and scalars differ. In defining new quantities we must therefore carefully note whether they are vector or scalar.

Our definition of displacement plays a major role in future discussions. Let us finish the discussion by restating it briefly in display:

> *Displacement* is a vector quantity whose magnitude is the distance between the beginning and end points of the motion and whose direction is from the beginning point to the end point.

2.2 Definition of Average Velocity

One of the most obvious features of motion is how fast it is. Two measures of this quality are frequently used—namely, velocity and speed. Let us begin with velocity. Its definition follows directly from our concept of displacement, and it is of basic importance in future discussions. Returning to Fig. 2.1, let t_0 be the time when the point particle is at the initial point P_0, and t the time when it is at the final point P. The elapsed time for the motion is $(t - t_0)$. We define the **average velocity** for the motion by dividing the displacement by the elapsed time.

Definition of Average Velocity

$$\text{Average velocity} = \frac{\text{displacement}}{\text{elapsed time}}$$

$$\langle v \rangle = \frac{x - x_0}{t - t_0} \tag{2.2}$$

SI unit of velocity: meter per second
Abbreviation: m/s

In this definition, we have divided the vector displacement by the scalar time interval. The resulting average velocity takes its magnitude and units from the division of these quantities, but it takes its direction from the displacement alone. The average velocity is therefore a vector in the direction of the displacement. We use the brackets ⟨ ⟩ around the symbol for velocity to denote average.

The definition of speed is different from the definition of velocity. We define average speed in terms of total distance traveled rather than displacement:

$$\text{Average speed} = \frac{\text{total distance traveled}}{\text{elapsed time}} \tag{2.3}$$

Speed has the same units as velocity. The total distance traveled is simply the path length of the motion of the point particle. No direction is associated with the concept of path length, and the average speed is therefore a scalar. The total distance traveled may or may not be equal to the magnitude of the displacement. In Fig. 2.1, the total distance traveled is much larger than the magnitude of the displacement, and it follows that the average speed is greater than the magnitude of the average velocity. The only shared feature is the dimension, which is $[L]\,[T]^{-1}$ in both cases.

The omission of a boxed display or an algebraic representation for Eq. (2.3) implies that average speed is not as important a concept as average velocity. Let us illustrate the difference with a specific example.

EXAMPLE 2.1

A cyclist travels 10 km to the east along a straight road from his home in a time of 50 min. He then turns around and travels 3 km back toward his home in a time of 20 min. What are the average velocity and the average speed for the entire trip?

SOLUTION Let us first prepare the data of the problem in a form suitable for direct substitution in the equations defining average velocity and speed:

$$\text{Total distance traveled} = 10^4\ \text{m} + 3 \times 10^3\ \text{m}$$
$$= 1.3 \times 10^4\ \text{m}$$
$$\text{Displacement} = 10^4\ \text{m (east)} - 3 \times 10^3\ \text{m (west)}$$
$$= 7 \times 10^3\ \text{m (east)}$$

(East and west are just as convenient to describe the direction of a displacement as positive and negative signs—the essential point is the assignment of a direction.)

$$\text{Elapsed time} = 70\ \text{min} \times 60\ \text{s/min}$$
$$= 4.2 \times 10^3\ \text{s}$$

To calculate the average velocity we use Eq. (2.2):

$$\begin{aligned}\langle v \rangle &= \frac{\text{displacement}}{\text{elapsed time}} \\ &= \frac{7 \times 10^3 \text{ m (east)}}{4.2 \times 10^3 \text{ s}} \\ &= 1.67 \text{ m/s (east)}\end{aligned}$$

The average speed is given by Eq. (2.3):

$$\begin{aligned}\text{Average speed} &= \frac{\text{distance traveled}}{\text{elapsed time}} \\ &= \frac{1.3 \times 10^4 \text{ m}}{4.2 \times 10^3 \text{ s}} \\ &= 3.10 \text{ m/s}\end{aligned}$$

We can present many of the results of one-dimensional kinematics on a graph of displacement versus time. Figure 2.2 illustrates this for the data of Example 2.1. The point A on the diagram represents the cyclist's home, the point B his farthest point of travel to the east, and the point C his final position. Distances measured along the abscissa of the diagram represent times, and distances measured along the ordinate represent displacements. We define the slope of a line on such a diagram as the rise of the ordinate divided by the run of the abscissa. With this definition we have:

$$\text{Slope of line } AC = \frac{7 \times 10^3 \text{ m}}{4.2 \times 10^3 \text{ s}} = 1.67 \text{ m/s}$$

The slope of this line therefore represents the overall average velocity during the trip. Note that we do not identify the slope of

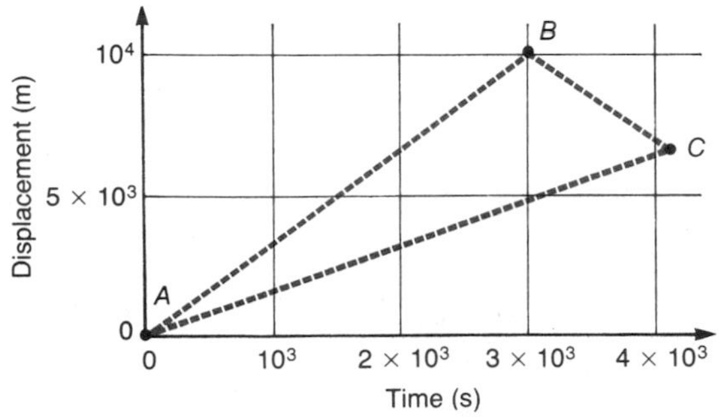

FIGURE 2.2 Graphical presentation of the data of Example 2.1.

the line with the tangent of the angle that the line makes with the abscissa because the units do not cancel, and the slope of this line represents a velocity. Changing the units on the axes would not change the slope of the line, but it would change the angle that the line makes with the abscissa.

Another advantage of the geometric approach to this problem is the additional information that is displayed on the diagram. Considering the lines joining A to B and B to C, we have:

$$\text{Slope of line } AB = \frac{10^4 \text{ m}}{3 \times 10^3 \text{ s}} = 3.3 \text{ m/s}$$

$$\text{Slope of line } BC = -\frac{3 \times 10^3 \text{ m}}{1.2 \times 10^3 \text{ s}} = -2.5 \text{ m/s}$$

Correctly interpreted, these figures give the average velocities for the outward and inward legs of the cyclist's journey. The positive direction of displacement on the diagram is to the east; negative displacement is therefore to the west. The positive slope of the line AB signifies an average velocity of 3.3 m/s (east) for the outward leg, and the negative slope of the line BC means an average velocity of 2.5 m/s (west) for the inward leg. The straight lines on Fig. 2.2 do not represent continuous data of displacement versus time—the only data points are A, B, and C. The sole significance of the lines is that their slopes represent average velocities. ■

2.3 Motion at Constant Velocity

A long-distance runner on a straight road, a jet airliner in level flight, or a train crossing the plains often moves with approximately constant velocity. Over any given time interval the magnitude of the displacement is equal to the distance traveled, and the magnitude of the velocity is therefore equal to the speed. In addition, the displacement divided by the elapsed time gives the same result for every pair of points along the path of motion. Since the velocity is the same at all points, it is unnecessary to use the average symbol. For the special case of constant velocity, Eq. (2.2) gives:

$$v = \frac{x - x_0}{t - t_0} \tag{2.4}$$

It is customary to modify Eq. (2.4) by agreeing to set the beginning coordinate point and the initial time instant at zero. This means that we select a coordinate origin so that $x_0 = 0$ m and start the clock at $t_0 = 0$ s. With this convention, the kinematic equation for constant velocity motion assumes a simpler form:

■ Kinematic Equation of Motion at Constant Velocity

$$\text{Displacement} = \text{velocity} \times \text{elapsed time}$$

$$x = vt \tag{2.5}$$

EXAMPLE 2.2

An automobile travels 2500 m north along a straight road at constant velocity. The elapsed time is 2 min. Calculate the velocity in kilometers per hour.

SOLUTION In solving kinematics problems it is a good practice to set out the data and the required quantities as follows:

$$x = 2500 \text{ m (north)}$$

$$t = 2 \text{ min} \times 60 \text{ s/min}$$

$$= 120 \text{ s}$$

$$v = ?$$

This Kansas grain train is hardly a point particle, but point-particle kinematics nevertheless gives a very useful model for calculating details of its motion. The reason is that the train moves as a complete unit, allowing us to fix our attention on one small part (say the engineer's window) whose displacement, velocity, and acceleration are the same as those of the whole train. (Courtesy of Missouri Pacific.)

Since the motion is at constant velocity, we use Eq. (2.5):

$$x = vt$$

$$\therefore v = \frac{x}{t}$$

$$= 2500 \text{ m}/120 \text{ s}$$

$$= 20.8 \text{ m/s (north)}$$

Now convert back to the required units:

$$v = 20.8 \text{ m/s} \div 0.2778 \text{ (m/s)/(km/h)}$$

$$= 74.9 \text{ km/h (north)}$$

We can also display the data of Example 2.2 on a graph as shown in Fig. 2.3. Point *A* of the diagram represents the starting point, and point *B* represents the end point of the motion. But note that the velocity has its constant value at all points of the motion, including the end points. This means that the automobile already traveling at 74.9 km/h crosses the starting line as the clock is started, and when it crosses the finishing line the clock is stopped. Since the velocity is constant, every point on the line *AB* represents a data point for the displacement of the automobile at some particular time. For example, we may read from the graph that the displacement after 50 s is approximately 1000 m [an exact calculation using Eq. (2.5) gives 1040 m]. In this respect the diagram differs from that of Fig. 2.2, in which only the end points are data points. ■

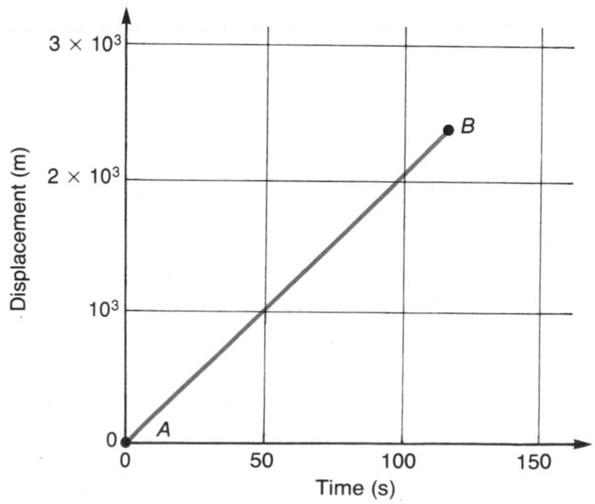

FIGURE 2.3 Graphical presentation of the data of Example 2.2.

We can summarize the meaning of displacement-versus-time graphs as follows:

> If motion is represented by a set of data points on a graph of displacement versus time, then the slope of a straight line joining any two of the points represents the average velocity between them. In addition, if the velocity is constant between any two points, then every point on the straight line joining them is a data point for the motion.

Converting some kinematics problems of everyday life to SI units makes a simple procedure very cumbersome. In deciding whether units present any difficulty, we simply follow their algebraic cancellation. Let us illustrate with an example.

EXAMPLE 2.3

A jet liner passes over St. Louis at 625 mi/h, heading straight toward Kansas City, which is 235 mi away. How much time elapses before the aircraft passes over Kansas City if it maintains a constant velocity?

SOLUTION Set up the data for the problem in the units in which it is quoted:

$$v = 625 \text{ mi/h}$$
$$x = 235 \text{ mi}$$
$$t = ?$$

Eq. (2.5) describes motion at constant velocity:

$$x = vt$$
$$235 \text{ mi} = 625 \text{ mi/h} \times t$$
$$\therefore t = 0.376 \text{ h}$$

We would normally quote this time interval in minutes:

$$t = 0.376 \text{ h} \times 60 \text{ min/h}$$
$$= 22.6 \text{ min}$$

■

2.4 Definition of Average Acceleration

Our everyday experience makes us aware of many instances in which velocity does not remain constant. A sprinter coming out of the starting block increases velocity from zero to the maximum in a short time. A driver reduces the velocity of an automobile to zero when approaching a stop sign. We say that the sprinter accelerates from the block and that the automobile decelerates as it approaches the stop sign. We need the concept of *instantaneous velocity* to define acceleration. Let us begin with an example of calculating average velocities.

EXAMPLE 2.4

An athlete begins a sprint from a starting block. Accurate distance-measuring and timing equipment provides the data shown.

Time (s)	*Displacement* (m)
0	0
0.5	0.5
1.0	2.0
1.5	4.5
2.0	8.0
2.5	12.5
3.0	17.5
3.5	22.5
4.0	27.5

Calculate the average velocities over consecutive 1-s intervals and also over consecutive 0.5-s intervals.

SOLUTION We require many applications of Eq. (2.2); for example, to determine the average velocity in the time interval between 2 s and 2.5 s, we have:

$$\langle v \rangle = \frac{x - x_0}{t - t_0}$$

$$= \frac{12.5 \text{ m} - 8 \text{ m}}{2.5 \text{ s} - 2 \text{ s}} = 9 \text{ m/s}$$

Tabulation of the results of a series of calculations gives the information shown in the following table.

	Time interval (s)	*Average velocity* (m/s)
1-*sec intervals*	0–1	2
	1–2	6
	2–3	9.5
	3–4	10
0.5-*sec intervals*	0–0.5	1
	0.5–1.0	3
	1.0–1.5	5
	1.5–2.0	7
	2.0–2.5	9
	2.5–3.0	10
	3.0–3.5	10
	3.5–4.0	10

The calculations show that the sprinter's average velocity increases up to a steady value of 10 m/s after an initial period of acceleration. During the acceleration period the average velocity is different over different time intervals. For example, the average velocity is 6 m/s between $t = 1$ s and $t = 2$ s, but it is 5 m/s between $t = 1$ s and $t = 1.5$ s. This raises a problem if we wish to determine the sprinter's velocity at the instant when the clock reads 1 s—the average velocity over any subsequent or prior time interval is of no use since this average depends on the size of the time interval. The key to the situation is to take the average velocity over a very small subsequent time interval. Differential calculus shows that an average calculated in this way, which is called the *instantaneous velocity*, is unique. ■

Let us develop a further insight into instantaneous velocity by presenting the data of Example 2.4 graphically. Figure 2.4 contains the first few data points on a displacement-versus-time graph at 0.5-s intervals. These points are joined by dashed lines (called secant lines), whose slopes are the average velocities over the time interval between the points; note, however, that the dashed lines do not provide data points for the entire motion. However, it is reasonable to suppose that a smooth curve joining the data points graphically indicates the displacement for every time point of the motion. We also show this smooth curve in Fig. 2.4. The two secant lines between 0.5 s and 1.0 s and also between 1.0 s and 1.5 s show different average velocities over these two time intervals. If we were to take smaller time intervals on each side of the 1.0-s point, the slopes of the secant lines would still be different; however, if we let the time interval in question approach zero, then each secant line becomes the tangent line to the curve at the 1.0-s point. The slope of this tangent line is the instantaneous velocity at the 1.0-s point. From measure-

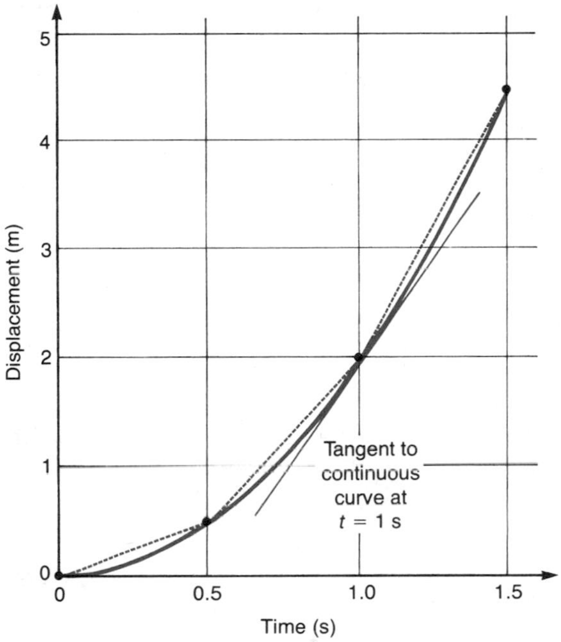

FIGURE 2.4 Graphical presentation of the first few data points of Example 2.4.

ments on the graph, we calculate that the slope of the tangent line is 4 m/s, which is the instantaneous velocity of the sprinter at this time.

> If motion is represented by a smooth curve on a displacement-versus-time graph so that every point on the curve is a data point for the motion, then the slope of the tangent to the curve at any point represents the instantaneous velocity at that point.

We have discussed in some detail a particular case of motion in which the velocity is not constant and have developed a graphical method for exhibiting the change in velocity. We can now define the precise measurement of a variation in a velocity that changes with time. Referring again to Fig. 2.1, let the instantaneous velocity at the initial point P_0 be v_0 at time t_0, and let the instantaneous velocity at point P be v at a later time t. With this notation we define the **average acceleration** as the change in velocity divided by the elapsed time. Let us set this definition in display.

■ Definition of Average Acceleration

$$\text{Average acceleration} = \frac{\text{change in velocity}}{\text{elapsed time}}$$

$$\langle a \rangle = \frac{v - v_0}{t - t_0} \qquad \textbf{(2.6)}$$

SI unit of acceleration: meter per second per second
Abbreviation: m/s^2

EXAMPLE 2.5

Calculate the average acceleration of the sprinter of Example 2.4 over the time interval from 0 s to 1 s.

SOLUTION To calculate the average acceleration, we need to know the instantaneous velocity at each end point of the interval. At 0 s the sprinter's instantaneous velocity is 0 m/s, and at 1 s it is 4 m/s (from the slope of the tangent in Fig. 2.4). Using Eq. (2.6) gives us the average acceleration:

$$\langle a \rangle = \frac{v - v_0}{t - t_0}$$

$$= \frac{4 \text{ m/s} - 0 \text{ m/s}}{1 \text{ s} - 0 \text{ s}}$$

$$= 4 \text{ m/s}^2$$

■

Note that time appears squared in the denominator of the units of acceleration. This can give rise to mixed units if we do not make appropriate conversions. An example illustrates this point.

EXAMPLE 2.6

An automobile accelerates from rest to 90 km/h in 10 s. Calculate average acceleration.

SOLUTION The use of Eq. (2.6) without any conversion of units gives:

$$\langle a \rangle = \frac{v - v_0}{t - t_0}$$

$$= \frac{90 \text{ km/h} - 0 \text{ km/h}}{10 \text{ s} - 0 \text{ s}}$$

$$= 9 \text{ km/h} \cdot \text{s}$$

Algebraic manipulation of the units designations has resulted in an acceleration unit of kilometers per hour per second. The dimensions of the unit are [LT^{-2}], which is the correct dimension for acceleration. Moreover, the answer is correct, but its form is unsuitable for further calculations. Rather than having conversions for any strange unit that may arise, it is easier to convert the initial data to SI units, which we can always do with the help of the conversion tables. However, in this particular case, we proceed by giving a simple equivalency—the speed of 90 km/h is exactly 25 m/s. Equation (2.6) then gives:

$$\begin{aligned}\langle a\rangle &= \frac{v - v_0}{t - t_0}\\ &= \frac{25 \text{ m/s} - 0 \text{ m/s}}{10 \text{ s} - 0 \text{ s}}\\ &= 2.5 \text{ m/s}^2\end{aligned}$$

■

Since acceleration is a vector velocity difference divided by a scalar time difference, it is a vector having the direction of the velocity difference. A one-dimensional acceleration that is in the opposite direction to the velocity is called a *deceleration*.

EXAMPLE 2.7

A bus driver applies the brakes and slows the bus from 90 km/h (25 m/s) to 54 km/h (15 m/s) over a time interval of 2 s. Calculate the average acceleration.

SOLUTION In previous examples the vector nature of acceleration caused no problem, but in this case we must be careful with the directions of the vectors. Let the direction of the velocity of the bus be positive. Then Eq. (2.6) gives for the average acceleration:

$$\begin{aligned}\langle a\rangle &= \frac{v - v_0}{t - t_0}\\ &= \frac{15 \text{ m/s} - 25 \text{ m/s}}{2 \text{ s} - 0 \text{ s}}\\ &= -5 \text{ m/s}^2\end{aligned}$$

The direction of the average acceleration is opposite to the direction of the instantaneous velocities. Thus, the bus is decelerating.

In a previous example we interpreted the slope of a straight line on a displacement-versus-time graph as a velocity. Figure 2.5 shows the data of Example 2.7 on a velocity-versus-time graph. The point A represents the velocity at the initial instant of time, and the point B represents the velocity 2 s later. The slope of the

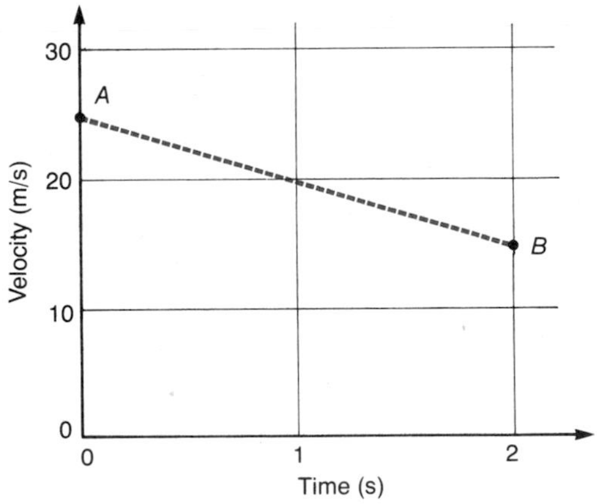

FIGURE 2.5 Graphical presentation of the data of Example 2.7.

line AB is an ordinate rise of -10 m/s divided by an abscissa run of 2 s. The result is -5 m/s^2, which is the average acceleration between A and B. In this case the points A and B are the only data points on the graph. If other data points are made available, they may or may not lie on the straight line. ■

2.5 Motion at Constant Acceleration

A particle moves with constant acceleration if the velocity change divided by the elapsed time yields the same figure for all points of the path of the motion. When acceleration is constant, we can remove the average symbol $\langle\ \rangle$ in Eq. (2.6) and rewrite it as:

$$a = \frac{v - v_0}{t - t_0} \tag{2.7}$$

As mentioned previously, we can always set the initial time instant t_0 as zero, but we must retain the velocity v_0, since it may not be zero at the initial time instant. Cross-multiplying Eq. (2.7) with $t_0 = 0$ gives:

$$v = v_0 + at \tag{2.8}$$

This is the basic equation for motion at constant acceleration, and as the argument above shows, it is really the definition of constant acceleration.

Many of the motions encountered in everyday life involve an acceleration that is approximately constant. Massive objects in free fall and the acceleration and braking of powered vehicles are instances of this behavior. Let us consider some examples.

EXAMPLE 2.8

A rapid transit car accelerates at a constant rate from 36 km/h (10 m/s) to 63 km/h (17.5 m/s) in 5 s. What is the acceleration?

SOLUTION Let us first set the data down:

$$v_0 = 10 \text{ m/s}$$

$$v = 17.5 \text{ m/s}$$

$$t = 5 \text{ s}$$

$$a = ?$$

Use Eq. (2.8) to calculate the acceleration:

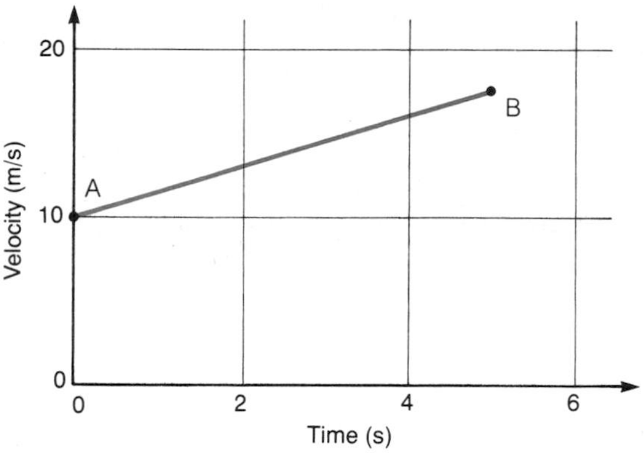

FIGURE 2.6 Graphical presentation of the data of Example 2.8.

$$v = v_0 + at$$

$$\therefore a = \frac{v - v_0}{t}$$

$$= \frac{17.5\ \text{m/s} - 10\ \text{m/s}}{5\ \text{s}}$$

$$= 1.50\ \text{m/s}^2$$

The data for this example are shown in Fig. 2.6 on a velocity-versus-time graph. The point A represents the initial velocity of 10 m/s, and the point B represents the final velocity of 17.5 m/s. The slope of the line joining these points represents the average acceleration. Since the acceleration is constant, all points on the line are data points for the motion. For example, we can use the graph to read off a velocity of about 13 m/s at the 2-s time point; an exact calculation using Eq. (2.8) gives just this answer. ■

Up to this point we have mainly used the velocity-versus-time graph to recognize that the slope of a line joining two points represents an acceleration. We can also use a curve on a velocity-versus-time graph to calculate the displacement, provided that all points on the curve are data points of the motion. To see how this works, let us consider an example.

EXAMPLE 2.9

A rapid transit car accelerates uniformly to a velocity of 90 km/h (25 m/s) in 15 s and then continues with constant velocity. How far does it travel in 1 min?

SOLUTION Plot the data on a velocity-versus-time graph as shown. The graph has two portions:

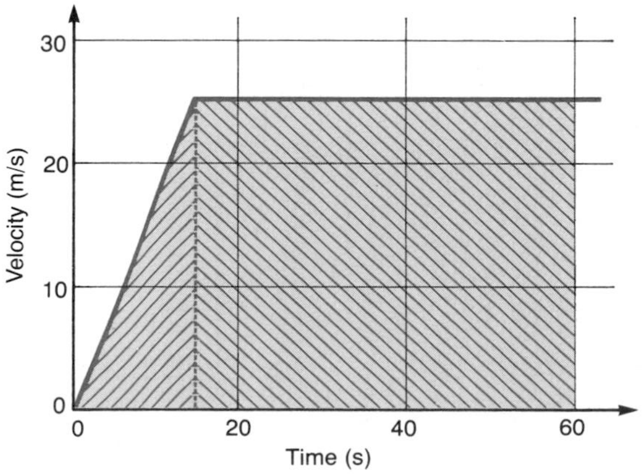

1. A straight line with positive slope corresponding to the initial period of constant acceleration
2. A straight line parallel to the abscissa corresponding to the period of constant velocity

First consider the displacement during the period of constant velocity, which lasts from 15 s to 60 s. The displacement during this period is given by Eq. (2.5):

$$\begin{aligned}\text{Displacement} &= \text{velocity} \times \text{elapsed time}\\ &= 25\ \text{m/s} \times 45\ \text{s}\\ &= 1125\ \text{m}\end{aligned}$$

Note that this is simply the area of the rectangle under the constant-velocity portion of the curve. We now state without formal proof that the area under any velocity-versus-time curve for some interval is the displacement that takes place during that interval. A rigorous proof of this statement requires calculus, but we can make a reasonable justification for the case in hand. The initial velocity is zero, and the velocity increases uniformly up to 25 m/s in the first 15 s. This means that the average velocity in the time interval up to 15 s is ½ × 25 m/s = 12.5 m/s. Multiplying this average velocity by the elapsed time gives the displacement:

$$\begin{aligned}\text{Displacement} &= \text{average velocity} \times \text{elapsed time}\\ &= 12.5\ \text{m/s} \times 15\ \text{s}\\ &= 187.5\ \text{m}\end{aligned}$$

But the portion of the velocity-versus-time curve up to 15 s is a triangle whose area is given by:

$$\begin{aligned}\begin{matrix}\text{Area of triangular part}\\ \text{of velocity-vs.-time curve}\end{matrix} &= \tfrac{1}{2} \times \text{base} \times \text{height}\\ &= \tfrac{1}{2} \times 15\ \text{s} \times 25\ \text{m/s}\\ &= 187.5\ \text{m}\end{aligned}$$

For this example the displacement is equal to the area under the curve in question. Finally, we have:

$$\begin{aligned}\text{Total displacement} &= 187.5\ \text{m} + 1125\ \text{m}\\ &= 1312.5\ \text{m}\end{aligned}$$

This is the total shaded area under the graph. ■

We can summarize our interpretation of velocity-versus-time graphs as follows:

If motion is represented by a set of points on a velocity-versus-time graph, then the slope of a straight line joining any two of these points represents the average acceleration between them. In addition, if the acceleration is constant between any two points, then every point on the straight line joining them is a data point for the motion. The area under the graph is the displacement during the time interval.

Uniformly accelerated motion is a topic of great importance in physics. For this reason it is useful to have a fully developed algebraic description of it. In the graphical treatment just described we calculated the displacement during uniform acceleration to be equal to the average velocity multiplied by the elapsed time. Let us express this result algebraically:

$$\begin{aligned} \text{Displacement} &= \text{average velocity} \times \text{elapsed time} \\ x &= \langle v \rangle\, t \\ &= \tfrac{1}{2}(v + v_0)t \end{aligned} \quad \textbf{(2.9)}$$

We have identified the average velocity $\langle v \rangle$ as one-half of the sum of the initial and final velocities. This averaging procedure is correct provided that the acceleration is uniform over the time interval considered. We can derive two additional equations by manipulation of Eqs. (2.8) and (2.9). To obtain the first, we substitute for v in Eq. (2.9) from Eq. (2.8):

$$\begin{aligned} x &= \tfrac{1}{2}(v_0 + v)t \\ &= \tfrac{1}{2}[v_0 + (v_0 + at)]t \\ &= v_0 t + \tfrac{1}{2}at^2 \end{aligned} \quad \textbf{(2.10)}$$

To find the second, we write Eq. (2.9) together with a cross-multiplied version of Eq. (2.8):

$$x = \tfrac{1}{2}(v + v_0)t$$

$$a = \frac{v - v_0}{t}$$

Multiplying left side by left side and right side by right side we have:

$$ax = \tfrac{1}{2}(v^2 - v_0^2)$$

Rearranging finally gives:

$$v^2 = v_0^2 + 2ax \quad \textbf{(2.11)}$$

The general utility of the equations of uniformly accelerated motion [Eqs. (2.8) through (2.11)] is so great that they are gathered together below for convenient reference:

Kinematic Formulas Describing Motion with Uniform Acceleration

$$v = v_0 + at \qquad \text{(a)}$$

$$x = \tfrac{1}{2}(v + v_0)t \qquad \text{(b)}$$

$$x = v_0 t + \tfrac{1}{2}at^2 \qquad \text{(c)} \qquad \text{(2.12)}$$

$$v^2 = v_0^2 + 2ax \qquad \text{(d)}$$

The initial conditions are chosen to be $x = 0$ and $v = v_0$ when $t = 0$.

These four equations form a useful set of relations between the five quantities x, t, v_0, v, and a that describe uniformly accelerated motion. Only two of these equations are independent; by this we mean that given any two equations, we could derive the remaining two. Each equation contains four of the five kinematic quantities; it follows that we must know three of these five quantities in order for a problem to be solvable. Let us illustrate the problem-solving technique with some examples.

EXAMPLE 2.10

A cyclist accelerates uniformly from rest at a rate of 2.1 m/s^2 until she attains a velocity of 36 km/h (10 m/s). How far does she travel, and what is the elapsed time during the period of acceleration?

SOLUTION The methodical approach to kinematics problems is to list the data and the unknowns:

$$v_0 = 0 \text{ m/s}$$

$$v = 10 \text{ m/s}$$

$$a = 2.1 \text{ m/s}^2$$

$$x = ?$$

$$t = ?$$

We then search Eq. (2.12) for equations to calculate the unknowns from the data. Eq. (2.12d) provides a calculation of the displacement:

$$v^2 = v_0^2 + 2ax$$

$$\therefore x = \frac{v^2 - v_0^2}{2a}$$

$$= \frac{(10\ \text{m/s})^2 - (0\ \text{m/s})^2}{2 \times 2.1\ \text{m/s}^2}$$

$$= 23.8\ \text{m}$$

Knowing v_0, v, a, and x, we could use any of the first three equations from Eq. (2.12) to calculate t. Since Eq. (2.12c) leads to a quadratic equation for t, it is less convenient than the first two. We use Eq. (2.12a):

$$v = v_0 + at$$

$$\therefore t = \frac{v - v_0}{a}$$

$$= \frac{10\ \text{m/s} - 0\ \text{m/s}}{2.1\ \text{m/s}^2}$$

$$= 4.76\ \text{s}$$

■

EXAMPLE 2.11

An automobile traveling at 90 km/h (25 m/s) undergoes uniform deceleration when the brakes are applied, slowing to 45 km/h (12.5 m/s) after 2.4 s.

a. What is the deceleration?

b. How far does it travel during this period?

c. How much farther does it travel before stopping if the deceleration remains constant?

SOLUTION In this problem we must remember that Eq. (2.12) refers to an elapsed time interval t. If we consider two different time intervals, we must adjust the data accordingly. The first two questions refer to the same elapsed time period, and we can handle them together:

$$v_0 = 25\ \text{m/s}$$

$$v = 12.5\ \text{m/s}$$

$$t = 2.4\ \text{s}$$

$$a = ?$$

$$x = ?$$

Equation (2.12a) offers immediate help:

$$v = v_0 + at$$

$$\therefore a = \frac{v - v_0}{t}$$

$$= \frac{12.5 \text{ m/s} - 25 \text{ m/s}}{2.4 \text{ s}}$$

$$= -5.208 \text{ m/s}^2$$

The positive signs given to v_0 and v imply that we have chosen the positive direction for vectors in the direction of motion of the automobile. The negative sign indicates that the acceleration vector has a direction that is opposite to that of the velocity vectors. We use Eq. (2.12b) to calculate the displacement during the time interval:

$$x = \tfrac{1}{2}(v + v_0)t$$

$$= \tfrac{1}{2} \times (12.5 \text{ m/s} + 25 \text{ m/s}) \times 2.4 \text{ s}$$

$$= 45.0 \text{ m}$$

To answer the third part of the question, we must note that it concerns a subsequent time interval. The initial velocity for the second time interval is equal to the final velocity for the first time interval, and the acceleration remains the same. We can write the data for the second time interval as follows:

$$v_0 = 12.5 \text{ m/s}$$

$$v = 0$$

$$a = -5.208 \text{ m/s}^2$$

$$x = ?$$

Equation (2.12d) is appropriate:

$$v^2 = v_0^2 + 2ax$$

$$\therefore x = \frac{v^2 - v_0^2}{2a}$$

$$= \frac{0 - (12.5 \text{ m/s})^2}{2 \times (-5.208 \text{ m/s}^2)}$$

$$= 15.0 \text{ m}$$

The analysis shows that a uniformly decelerating automobile travels 3 times as far when slowing from 90 km/h to 45 km/h as when it does when slowing from 45 km/h to a stop. ■

More complex problems involving periods of uniformly accelerated motion together with periods of constant velocity can also be handled by splitting up the time intervals.

EXAMPLE 2.12

The real kinematics of a sprint is very complicated, but it can be represented approximately as an interval of uniform acceleration followed by an interval of constant velocity to the finish line. Consider again the sprinter of Example 2.4, who accelerates from rest with uniform acceleration of 4 m/s^2 to a top speed of 10 m/s. Calculate her time for the 100-m dash.

SOLUTION Let us set up the data separately for the two different types of motion, using a single prime to indicate the first, uniform-acceleration interval and a double prime for the second, constant-velocity interval.

First interval:

$$v_0' = 0$$
$$v' = 10 \text{ m/s}$$
$$a' = 4 \text{ m/s}^2$$
$$t' = ?$$
$$x' = ?$$

Second interval:

$$v'' = 10 \text{ m/s}$$
$$t'' = ?$$
$$x'' = ?$$

The displacements in the two intervals are related to each other by the total displacement of the sprint:

$$x' + x'' = 100 \text{ m}$$

We can calculate the unknowns in the first interval using Eq. (2.12). Equation (2.12a) gives the elapsed time:

$$v' = v_0' + a't'$$

$$\therefore t' = \frac{v' - v_0'}{a'}$$

$$= \frac{10 \text{ m/s} - 0 \text{ m/s}}{4 \text{ m/s}^2}$$

$$= 2.5 \text{ s}$$

Equation (2.12b) gives the displacement in this interval

$$\begin{aligned} x' &= \tfrac{1}{2}(v_0' + v')t' \\ &= \tfrac{1}{2} \times (0 \text{ m/s} + 10 \text{ m/s}) \times 2.5 \text{ s} \\ &= 12.5 \text{ m} \end{aligned}$$

We can now calculate the displacement in the constant-velocity interval:

$$\begin{aligned} x'' &= 100 \text{ m} - x' \\ &= 100 \text{ m} - 12.5 \text{ m} \\ &= 87.5 \text{ m} \end{aligned}$$

Equation (2.5) gives the time interval for the constant-velocity motion:

$$\begin{aligned} x'' &= v''t'' \\ \therefore t'' &= \frac{x''}{v''} \\ &= \frac{87.5 \text{ m}}{10 \text{ m/s}} \\ &= 8.75 \text{ s} \end{aligned}$$

The total time of the sprint is the sum of the two time intervals:

$$\begin{aligned} t &= t' + t'' \\ &= 2.5 \text{ s} + 8.75 \text{ s} \\ &= 11.25 \text{ s} \end{aligned}$$

■

The problem raises an interesting point in the application of models to real situations. At a recent Olympic Games the winning time for the women's 100-m dash was 11.08 s. The simplest ideal model would give the sprinter a uniform velocity of 100 m/11.08 s = 9.025 m/s for the whole distance. The winning time for the 400-m relay was 42.55 s; the same model would imply a uniform velocity of 400 m/42.55 s = 9.400 m/s for the whole distance. This would mean that every woman in the relay event was faster than the winner of the individual race over the same distance. No sports lover would accept this conclusion. The difficulty with the simple model is the neglect of the time-consuming period of acceleration at the start of the race. In the relay race only the first runner experiences this—the zone for handing over the baton allows subsequent runners to reach full speed before taking the baton.

Now let us predict the times for sprinters all of the same ability who accelerate uniformly at 4 m/s^2 to a top speed of 10 m/s. The solution of Example 2.12 shows an individual time of 11.25 s. Since each subsequent runner in the relay would run the 100 m in 10 s, the time for the relay would be 41.25 s. The agreement with the Olympic times is not perfect, but the model does explain why the time for the 400-m relay is less than 4 times the winning time for the 100-m individual.

2.6 The Acceleration Due to Gravity

Aristotle taught that heavier bodies fall faster than lighter ones. He came to this conclusion by way of a limited set of observations that probably included watching stones falling in water. The study of falling objects continued to attract the attention of scholars, but the experimental difficulties are formidable. When we drop stones in air, they fall so rapidly that measuring their velocity at any point is difficult, even with modern equipment. No precision instruments were available either to Aristotle or to the Renaissance scientists who worked on the subject.

A complete history of kinematics and the application of kinematics to falling bodies would lead far beyond the scope of this book, but several outstanding contributions deserve mention. In the fourteenth century, a group of scholars at Merton College, Oxford, gave definitions for velocity and acceleration that closely parallel the definitions presented in the preceding sections. This group is known in the history of science as the Mertonians. They did not give a theory of falling bodies that would gain modern acceptance, but their analysis provided the basis for the work of Galileo Galilei (1564–1642), professor of mathematics at the University of Florence, Italy. Galileo realized the large role that air resistance plays in many instances of the free-fall of objects. In his experimental work he sought to minimize this effect by slowing down the time of the fall. To this end, he studied the motions of a pendulum bob and of an object rolling down an inclined plane. In both cases the object in question is falling, but the string of the pendulum and the slope of the inclined plane slow the motion and thus make the experimental work easier even if the theoretical analysis is more complex. The picturesque story that Galileo studied the acceleration of gravity by dropping weights from the top of the Leaning Tower of Pisa is probably false, since he certainly knew that air resistance would not be negligible for such a large drop. However, careful measurements led Galileo to conclude that *under suitable conditions* all bodies accelerate toward the earth at the same rate, regardless of their mass. We call this acceleration the **acceleration due to gravity,** and an average value for its magnitude is as follows:

SPECIAL TOPIC 2 GALILEO

Galileo was born at Pisa in northern Italy, in 1564. His father hoped that his gifted son would help restore the family fortune by studying medicine. However, the son found and studied the works of Archimedes; and physical science found one of its most brilliant exponents.

His early work led to the discovery of the constancy of gravitational acceleration for all objects for which buoyancy and air resistance play negligible roles. Experiments on pendulums and inclined planes played a major role in his discoveries, but he found great difficulty in measuring small time intervals. He used his pulse and a small water clock as timing devices. It is ironic that the Dutch physicist Huygens later turned the problem around and used the constancy of the period of a pendulum of given length (one of Galileo's discoveries) to control an accurate clock.

Galileo's work on mechanics also treated the strength of materials and structures. He showed that a structure of a given material that continued to grow by equal proportions in all dimensions would become weaker, and gave the first theoretical basis for the famous square-cube law that we discussed in Chapter 1.

Some of his greatest work was in astronomy. Hearing of the invention of optical lenses, he developed his own telescope, which had a magnification of 32×. Turning this instrument toward the heavens, he found the mountains on the moon, sunspots, the phases of Venus, and the four moons of Jupiter. The latter are still called Galilean satellites, and the instrument he used to find them is called a Galilean telescope (see Chapter 23).

Perhaps more than any other person, Galileo cemented the Copernican revolution firmly in place. Vindictive churchmen persecuted him on this account, resulting in his well-known prosecution by the Inquisition. The final years of his life were less than happy; he died at Florence in 1642.

Average Value of the Measured Acceleration Due to Gravity

$$g \approx 9.81 \text{ m/s}^2 \qquad \textbf{(2.13)}$$

Why does one quote an approximate average value? Can't the quantity be measured precisely? And what are the suitable conditions? To answer some of these questions, imagine that you have arranged an experiment to demonstrate the truth of Galileo's claim. The stage props are a lead slug, a feather, and a hydrogen-filled balloon. You have precision equipment to measure the acceleration of these objects after they are released, and are very gratified to find a value close to 9.81 m/s^2 for the acceleration of the lead slug. However, the feather falls in such a leisurely manner that even an untrained observer perceives its acceleration to be less than that of the lead slug, and upon its release the balloon floats upward—demonstrating a negative g! Clearly, to save the show we must vacuum-seal a building, pump the air out of it, and watch the experiment through a window. As a result of these steps, we measure a downward acceleration of about 9.81 m/s^2

for all three objects. Note, however, that the natures of the problems with the feather and the balloon are different. The feather accelerates more slowly than the lead slug because the frictional resistance of the air hampers its downward progress, but the balloon floats upward because of the relatively large buoyancy of the displaced air. We examine the whole question of buoyancy in detail in a subsequent chapter; for the purpose in hand, the intuitive notion is sufficient. The suitable conditions required to observe constant acceleration in a falling body are therefore that the frictional resistance to motion and the buoyancy of the air be negligible.

If we take the demonstration on the road, we find different values for g in various cities, as indicated in Table 2.1. Washington and Denver are at approximately the same latitude, and the acceleration of gravity is smaller at the higher elevation. London and Washington are both almost at sea level, and the acceleration of gravity is larger at the higher latitude. The earth is slightly flattened at the poles so that Washington is farther from the earth's center than London; Denver, with its higher altitude, is even farther away. These facts are consistent with g being due to the gravitational attraction of the earth, which decreases as one gets farther from the earth's center. However, the variation in the values of g cannot be entirely explained in this way. There is an important effect due to the earth's rotation. We postpone discussion of this effect to a subsequent chapter.

In summary, we regard g as the acceleration due to gravity with the following qualifications:

1. The value quoted in Eq. (2.13) is an average figure.

2. The value includes effects due to the earth's rotation.

3. The direction is always vertically downward.

4. The value excludes effects due to air resistance and buoyancy.

We therefore treat freely falling bodies as undergoing constant acceleration provided that they are sufficiently dense for air buoyancy to be negligible. In addition, they must not fall so far that air resistance and variation in the force of the gravitational attraction become significant factors.

TABLE 2.1 Values of g in various cities

City	*Elevation above sea level* (m)	*Latitude*	g (m/s^2)
Washington, D.C.	8	38°54′N	9.8008
Denver	1640	39°43′N	9.7961
London	30	51°30′N	9.8228

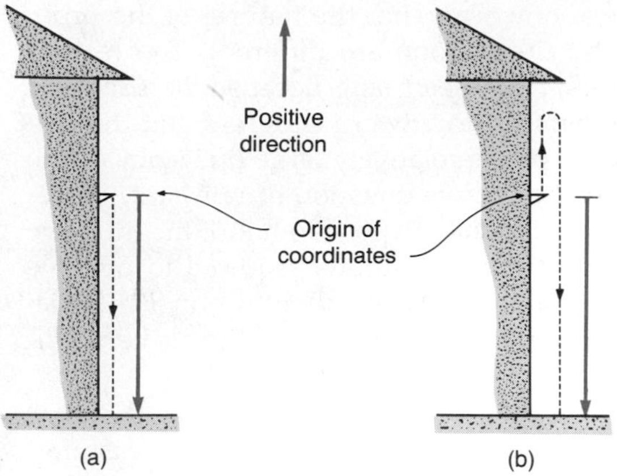

FIGURE 2.7 Illustration of choice of origin and sign conventions for Examples 2.13 and 2.14. The dotted line is the path of the motion, and the solid line is the displacement vector.

EXAMPLE 2.13

A stone falls from rest from a fourth-floor window that is 14 m above ground level. How long does it take to reach the ground? What is its velocity just before it strikes the ground?

SOLUTION In previous problems we assumed a suitable origin of coordinates and an agreed positive direction when they were not explicitly stated. In this problem (and some that follow) it is necessary to be explicit on these matters. The equations of uniformly accelerated motion, Eq. (2.12), include the understanding that $x = 0$ when $t = 0$. We must choose the origin of coordinates at the point from which the stone is dropped. The choice of the positive direction is arbitrary. The most usual convention sets the positive direction as upwards. We illustrate the choice of origin and the sign convention in Fig. 2.7(a). We can now write the problem data:

$$x = -14 \text{ m}$$
$$v_0 = 0 \text{ m/s}$$
$$a = -9.81 \text{ m/s}^2$$

(The displacement and acceleration vectors both point in the negative direction.)

$$t = ?$$
$$v = ?$$

Equation (2.12c) is appropriate for calculating the elapsed time:

$$x = v_0 t + \tfrac{1}{2}at^2$$
$$-14 \text{ m} = (0 \text{ m/s}) \times t + \tfrac{1}{2} \times (-9.81 \text{ m/s}^2) \times t^2$$
$$\therefore t^2 = 2.854 \text{ s}^2$$
$$t = \pm 1.689 \text{ s}$$

We select the positive sign, since the stone hits the ground later than $t = 0$. To find the final velocity, we use Eq. (2.12a):

$$v = v_0 + at$$
$$= 0 \text{ m/s} + (-9.81 \text{ m/s}^2) \times (1.689 \text{ s})$$
$$= -16.6 \text{ m/s}$$

The negative sign correctly indicates that the final velocity vector is in the downward direction. ■

EXAMPLE 2.14

A stone is thrown vertically upward with a velocity of 10 m/s from a fourth-floor window 14 m above ground level. How long does it take to reach the ground? What is its velocity just before striking the ground?

SOLUTION The problem is the same as the previous one except for the initial velocity. Let us make the same choice of origin and sign convention (Fig. 2.7b). Then the data become as follows:

$$\begin{aligned} x &= -14 \text{ m} \\ v_0 &= 10 \text{ m/s} \\ a &= -9.81 \text{ m/s}^2 \\ t &= ? \\ v &= ? \end{aligned}$$

Remember that x is the displacement and not the total distance traveled. For this problem the initial velocity vector points in the positive direction. Several procedures of calculation are possible. Let us choose the mathematically simplest one first by using Eq. (2.12d) to calculate the final velocity:

$$\begin{aligned} v^2 &= v_0^2 + 2ax \\ &= (10 \text{ m/s})^2 + 2 \times (-9.81 \text{ m/s}^2) \times (-14 \text{ m}) \\ &= 374.7 \text{ (m/s)}^2 \\ \therefore v &= \pm 19.36 \text{ m/s} \end{aligned}$$

We choose the negative sign since the final velocity vector is directed downward. To calculate the elapsed time we use Eq. (2.12a):

$$\begin{aligned} v &= v_0 + at \\ \therefore t &= \frac{v - v_0}{a} \\ &= \frac{-19.36 \text{ m/s} - 10 \text{ m/s}}{-9.81 \text{ m/s}^2} \\ &= 2.99 \text{ s} \end{aligned}$$

Another procedure is to calculate the elapsed time directly from the initial data by using Eq. (2.12c):

$$\begin{aligned} x &= v_0 t + \tfrac{1}{2}at^2 \\ -14 \text{ m} &= (10 \text{ m/s}) \times t + \tfrac{1}{2} \times (-9.81 \text{ m/s}^2) \times t^2 \end{aligned}$$

Rearrangement gives:

$$(-4.905\ \text{m/s}^2) \times t^2 + (10\ \text{m/s}) \times t + 14\ \text{m} = 0$$

This is a quadratic equation in t. Using the results of Appendix A we have:

$$t = \frac{(-10\ \text{m/s}) \pm [(10\ \text{m/s})^2 - 4 \times (-4.905\ \text{m/s}^2) \times (14\ \text{m})]^{1/2}}{2 \times (-4.905\ \text{m/s}^2)}$$

$$= \frac{-10\ \text{m/s} \pm 19.36\ \text{m/s}}{-9.81\ \text{m/s}^2}$$

$$= 2.99\ \text{s} \quad \text{or} \quad -0.954\ \text{s}$$

We choose the positive sign since the stone strikes the ground later than the initial time $t = 0$. To calculate the velocity we use either Eq. (2.12a) or Eq. (2.12d), leading to the same result as previously:

$$v = -19.36\ \text{m/s}$$

The existence of these alternate methods for solving the problem highlights the fact that the different parts of Eq. (2.12) are not independent of each other. They constitute a convenient set of formulas that may be used in different ways to solve problems.

In both the previous examples choices of sign arose, and we chose the correct sign on the basis of physical insight. Let us investigate the consequences of a "wrong" choice of sign. Consider the first method of solving Example 2.14, and pretend that we carelessly wrote $v = +19.36$ m/s for the final velocity. The use of Eq. (2.12a) to calculate the elapsed time now gives:

$$v = v_0 + at$$

$$\therefore t = \frac{v - v_0}{a}$$

$$= \frac{19.36\ \text{m/s} - 10\ \text{m/s}}{-9.81\ \text{m/s}^2}$$

$$= -0.954\ \text{s}$$ ■

This is precisely the discarded solution for the elapsed time found by the second method of solution to the problem. It is apparent that $v = +19.36$ m/s and $t = -0.954$ s give a correct mathematical solution, consistent with the initial data of the problem. The additional solution also has a physical meaning. A stone projected upward from ground level with a velocity of 19.36 m/s (positive sign for the upward velocity vector) attains a height of 14 m with an upward velocity of 10 m/s at a time 0.954 s later. The negative time corresponds to a prior motion of the stone that also fits the data. In a real physical situation we are interested in the subsequent motion of the stone and therefore select the solution that goes with positive elapsed time.

KEY CONCEPTS

The basic concept of kinematics is **displacement.** It is a **vector** whose magnitude is the distance between the beginning and end points and whose direction is from the beginning to the end point. See Eq. (2.1) and the display on p. 23.

Average velocity is displacement divided by elapsed time. It is a vector pointing in the direction of the displacement. See Eq. (2.2).

Average acceleration is change in velocity divided by elapsed time. It is a vector pointing in the direction of the velocity change. See Eq. (2.6).

Motion at constant acceleration is an important mathematical idea that many real systems closely reflect. The formulas describing uniformly accelerated motion are important for solving many types of problems. See Eq. (2.12).

The downward **acceleration due to gravity** is approximately the same for all objects if air resistance and air buoyancy are negligible. Although it changes slightly with a change in location on the earth's surface, it has an approximate average value of 9.81 m/s^2.

QUESTIONS FOR THOUGHT

1. Give an example of an object having zero velocity and nonzero acceleration.

2. Give an example of an object having zero acceleration and nonzero velocity.

3. An object moving along a straight line reverses the direction of its velocity. Does this require a reversal of the direction of its acceleration?

4. Is it possible for an object to move along a straight line indefinitely with its velocity and acceleration vectors always in opposite directions?

5. Is it correct to define average speed as the magnitude of the average velocity? Does this procedure differ from the definition that we have adopted for average speed? Give examples to clarify your answer.

6. What kinematic quantity does an automobile speedometer register?

7. We use a positive or negative sign to specify the direction of one-dimensional vectors. Think of other physical quantities that are sometimes positive and sometimes negative. Are all of these quantities one-dimensional vectors? If not, how does the choice of the signs for the two cases differ?

8. An innovative scientist suggests that the effect of air resistance should be included in quoted values for g (the acceleration due to gravity). He argues that the only effect would be a small reduction in the values. Refute this strange proposal.

9. Can you think of units of velocity or acceleration that might be introduced to make either or both of these quantities fundamental standards? Would such a proz cedure require us to drop any of the SI fundamental standards? If so, which ones?

PROBLEMS

A. Single-Substitution Problems

1. What is the time for a sprinter who does the 100-m race at an average velocity of 10.1 m/s? [2.2]

2. A bus travels in a straight line between two stops 620 m apart in 1 min 15 s. What is its average velocity? [2.2]

3. A stone falls from a wall and strikes the ground 1.68 s later; its average velocity during the fall is 8.24 m/s. How high is the wall? [2.2]

4. An automobile passes consecutive mileposts on a straight highway at times given as follows:

Begin	0 s
1 mile	67 s
2 miles	141 s
3 miles	203 s

Plot the data on a displacement-versus-time graph, and calculate the overall average velocity and the average velocity for each interval. [2.2]

5. A cyclist travels for 25 min in a straight line with an average velocity of 12.5 mi/h.

a. How far is he from the starting point at the end of this period?
b. Are the data adequate to calculate the total distance that the cyclist traveled? [2.2]

6. A hiker in the mountains observes that the echo of a sound returns from a cliff face in 6.2 s. If the velocity of sound is 340 m/s, how far away is the cliff? [2.3]

7. A motorist traveling along a highway at constant 55 mi/h glances aside at a map for 1.4 s. How many yards does he travel in this time? [2.3]

8. An Olympic competitor swims the 800-m freestyle in 8 min 48 s at approximately constant speed. What is the speed in kilometers per hour? [2.3]

9. A long-distance runner runs for 35 min in a straight line with constant velocity of 3 m/s.

a. How far is he from the starting point at the end of this period?
b. Are the data adequate to calculate the total distance traveled? [2.3]

10. A jet airliner moves from rest along a runway to its takeoff speed of 275 km/h in 13 s. What is its average acceleration during this period? [2.4]

11. An automobile overtaking slower traffic along a straight road increases its velocity from 28 mi/h (12.5 m/s) to 44.8 mi/h (20 m/s) in 1.7 s. What is its average acceleration during this period? [2.4]

12. Calculate the average acceleration of a flea that reaches 1 m/s within 1 ms after taking off from rest. [2.4]

13. A good sprinter is capable of an average acceleration of about 4 m/s^2 during the first part of a sprint. Compare this with the average acceleration of an automobile that takes 12 s to reach 60 mi/h. [2.4]

14. Calculate the uniform acceleration required for a bus to reach 14 m/s (a little over 30 mi/h) in 45 m starting from rest. [2.5]

15. If an athlete could accelerate at 4 m/s^2 throughout the whole of a 100-m dash, what would be the time for the event? [2.5]

16. An object moving with constant acceleration of 2.5 m/s^2 has a velocity of 15 m/s at a given instant. What is its velocity after it has traveled an additional 30 m? [2.5]

17. An object moving with constant acceleration of 1.8 m/s^2 has a velocity of 12.6 m/s at a given instant. What was its velocity 4.4 s earlier? [2.5]

18. What upward velocity should be imparted to a baseball so that it may be easily caught by a person at a window 9.2 m vertically above the point of projection? [2.6]

19. A heavy object falls from a point 24 m above the ground at a construction site. If a worker sees the object begin to fall, how much time does he have to avoid it? [2.6]

20. A stone thrown vertically upward returns to the point of projection 4.23 s later. To what height did the stone rise? [2.6]

21. A rock falls from a cliff top. What is the distance of fall in the first 4 s? [2.6]

B. Standard-Level Problems

22. A motorist sets out on a 150-km trip along a straight road and maintains 90 km/h for the first 50 km. She then reduces her speed to 60 km/h for the next 50 km, and then to 30 km/h for the remaining 50 km. What is the average speed for the trip? [2.3]

23. An express train is scheduled to maintain an average speed of 100 km/h between two stations that are 135 km apart. The train manages to average only 80 km/h over the first 60 km of the trip. What average speed is required over the remainder of the trip to maintain the schedule? [2.3]

24. A train 660 m long traveling at a constant speed of 80 km/h enters a tunnel. The elapsed time between entry of the leading locomotive and emergence of the end of the train is 1 min 12 s. How long is the tunnel?

25. A man fires a rifle at a target and hears the noise of the bullet striking 1.2 s after firing the rifle. If the velocity of the rifle bullet is 640 m/s, and the velocity of sound 340 m/s, how far away is the target?

26. The following data are the results of an acceleration test on a compact automobile:

Time (s)	*Velocity* (m/s)
0	0
1	3.82
2	7.15
3	10.02
4	10.97
5	13.58
6	15.96
7	18.35
8	20.62
9	21.54
10	23.42
11	25.21
12	26.87

a. What is the average acceleration over the whole 12-s period?
b. What is the greatest value of the average acceleration over a 1-s interval?
c. In what time intervals did gear shifting probably occur? [2.4]

27. A runner accelerates uniformly from rest at 3.8 m/s^2 until she has covered 12.8 m. Calculate:

a. the elapsed time
b. the final velocity [2.5]

28. An automobile accelerates uniformly from 8 m/s to 20 m/s over a distance of 85 m. Calculate:

a. the acceleration
b. the elapsed time [2.5]

29. A cyclist accelerates uniformly from an initial velocity of 2.8 m/s and travels 184 m in the ensuing 32 s. Calculate:

a. the final velocity
b. the acceleration

30. A motorcycle accelerates uniformly from rest to 20 m/s in 6.2 s.

a. What is its acceleration?
b. How far does it travel in this time?
c. How far does it travel in the next 2 s if the acceleration remains the same?

31. A rapid transit train accelerates at 2.2 m/s^2, cruises at 28 m/s, and decelerates at 3.2 m/s^2. How long does it take to travel from rest to rest between two stations 750 m apart?

32. Assume that a sprinter is capable of an acceleration of 4 m/s^2 over the first 14 m, and from then on runs at constant speed. What is his time for the 100-m dash?

33. An automobile starts from rest and accelerates uniformly at 1.85 m/s^2 until its velocity reaches 25 m/s. It travels at this constant velocity for 20 s and then stops, with uniform deceleration of 2.6 m/s^2. Calculate the total displacement.

34. An automobile starts from rest and accelerates uniformly at 2.1 m/s^2 for a period of 15 s, and then travels at constant velocity. How long does it take to travel 600 m?

35. A bus travels between two stops 1450 m apart. The driver accelerates at 1.6 m/s^2 over the first 80 m and then continues at constant velocity until 140 m before the stop.

a. What rate of uniform deceleration must the driver use from this point to stop correctly?
b. What is the total time between the stops?

36. An automobile is traveling on a wet road at 81 km/h when the driver notices a tree across the road 120 m ahead. The brakes can cause a maximum deceleration of 2.5 m/s^2. What is the longest reaction time that the driver can have and still avoid hitting the tree?

37. A stone is dropped from the top of a tall building. Ignoring air resistance, calculate the following:

a. the instantaneous velocity after it has fallen 20 m
b. the instantaneous velocity after it has fallen 40 m
c. the average velocity between the 20-m and 40-m marks [2.6]

38. A stone is thrown vertically downward from the top of a building 28 m above ground level with an initial velocity of 6 m/s. How long does it take to reach the ground? [2.6]

39. A stone is thrown vertically upward from ground level with an initial velocity of 26.8 m/s. At what subsequent times is it 25 m above ground level, and what are its velocities at these times?

40. A stone is thrown vertically upward with an initial velocity of 20 m/s. At what subsequent times is its speed 12 m/s, and where does this occur?

41. A group of students decide to measure the height of a building by projecting a stone vertically upward. The students find that the stone is level with the top of the building at 1.85 s and 2.73 s after being projected from ground level. How high is the building?

42. A heavy object dropped from a bridge passes close to a vertical flagpole just before striking the ground. The flagpole is 10.8 m high, and the object travels the distance from the top of the flagpole to the ground in 0.45 s. How high is the bridge?

43. A balloon that is ascending at a constant rate of 7.5 m/s releases a sandbag that strikes the ground 2.8 s later. How high was the balloon when the sandbag was released?

44. A ball is dropped from a height of 6 m above ground level, and rebounds to a height of 4.8 m. Calculate:

a. the velocity just before hitting the ground
b. the velocity just as it begins to rise on the rebound
c. the average velocity and the average acceleration between the instant of release and the top of the rebound

C. Advanced-Level Problems

45. A train standing between mile markers accelerates uniformly from rest and passes consecutive mile markers with speeds of 3.58 m/s and 8.94 m/s.

a. What is the acceleration of the train?
b. What is its speed at the next mile marker?
c. What is the elapsed time at the third mile marker?
d. What total distance has the train traveled to the third mile marker?

46. An automobile traveling at 45 km/h (12.5 m/s) has an acceleration capability of 2 m/s^2 and a braking capability of 6 m/s^2 from this speed. When it is 12 m from an intersection that is 10 m wide, the traffic lights turn amber and remain that way for 3 s before turning red. Can the automobile avoid being in the intersection when the lights go red?

47. Two police officers observe a motorist speeding by at 108 km/h (30 m/s). Assume that the police take 2 s to assess the situation. After this their patrol car accelerates at 2.5 m/s^2 for 15 s and then continues at uniform speed. How far from the observation point does the interception take place? Assume that the motorist maintains his constant speed.

48. A hiker drops a heavy stone over a cliff and hears it strike the cliff base 2.8 s after dropping it. If the velocity of sound is 335 m/s, how high is the cliff?

49. An automobile is 30 m behind a truck that is 20 m long, and both are traveling at 20 m/s. The automobile driver pulls out to overtake the truck, accelerates at 2.5 m/s^2 to 30 m/s, and maintains this speed until she is 40 m in front of the truck; at this point she pulls back into the right lane. If the length of the automobile is neglected:

a. How many seconds does the automobile take to complete this maneuver?
b. How far does the automobile travel in this time?

50. A man sprinting at 7.5 m/s is 32 m from a bus when it pulls away with an acceleration of 0.8 m/s^2.

a. Can he catch the bus?
b. What is the total distance of his sprint?
c. Why are there two answers to this problem?

51. The engineer of a fast express traveling at 36 m/s sees a slow freight train on the tracks 750 m ahead traveling at 18 m/s in the same direction as the express. What braking capability is required to avoid a collision if the freight train continues at a constant velocity?

3

TWO-DIMENSIONAL KINEMATICS

Up to this point we have considered only displacement, velocity, and acceleration vectors that point forward or backward along a straight line. However, real space is three dimensional, and vectors may point in any direction. The resulting mathematics can be complex. Nevertheless, important types of motion (notably projectile motion) exist that we can successfully analyze in two dimensions. The mathematics of these vector quantities is easier, and a separate study of two-dimensional kinematics is quite worthwhile.

3.1 Displacement as a Vector in Two Dimensions

For a swimmer constrained to swim in a lane or for a powered vehicle moving along a straight track, displacement can only be forward or backward relative to the lane or the track. The displacement, velocity, or acceleration vectors have either a positive or negative sign, as we saw in Chapter 2. However, for a runner on an open field or a boat on a lake, the displacement, velocity, and acceleration have directional properties that we cannot simply describe by a positive or negative number with respect to a single coordinate axis.

The first essential concept in one-dimensional kinematics is the displacement from an initial to a final position. In two dimensions the definition of displacement is exactly the same—it is the vector whose length is the straight-line distance from the initial point P_0 to the final point P, and whose direction is from P_0 toward P. We illustrate this in Fig. 3.1, where the curved line from P_0 to P indicates a possible path of motion, but the straight line (with arrowhead to indicate direction) is the displacement vector between the two points. What is different about two dimensions is that we now need two quantities to specify a vector. Let us make the simple choice of rectangular coordinate axes Ox and Oy, as shown in Fig. 3.1. The displacement from the initial point P_0 to the final point P results in coordinate changes $(x - x_0)$ and $(y - y_0)$ as the *rectangular components of the displacement* from P_0 to P. The vector displacement from P_0 to P is written **R** (boldface type to signify a vector quantity in a space of two or more dimensions), its rectangular components in the x and y directions are written R_x and R_y, and its magnitude is written R. The displacement **R** is completely equivalent to the displacement R_x in the direction of the x-axis followed by the displacement R_y in the direction of the y-axis.

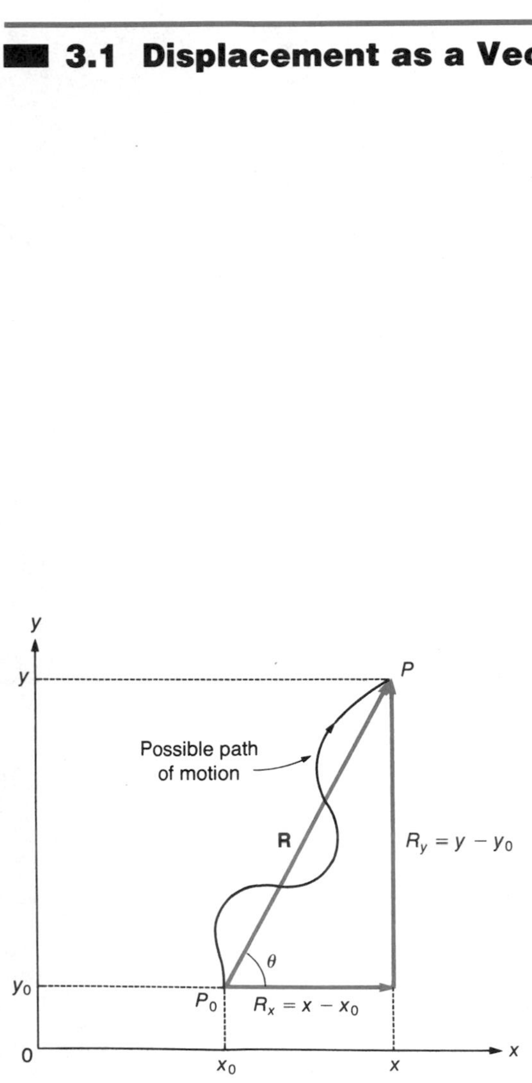

FIGURE 3.1 Diagram to illustrate the notation of two-dimensional kinematics.

The two components are not the only way to specify a vector displacement in two-dimensional space. Another possibility is to give the vector magnitude R (the length of the straight line from P_0 to P) together with the angle θ that the direction of the vector makes with the positive direction of the x-axis.

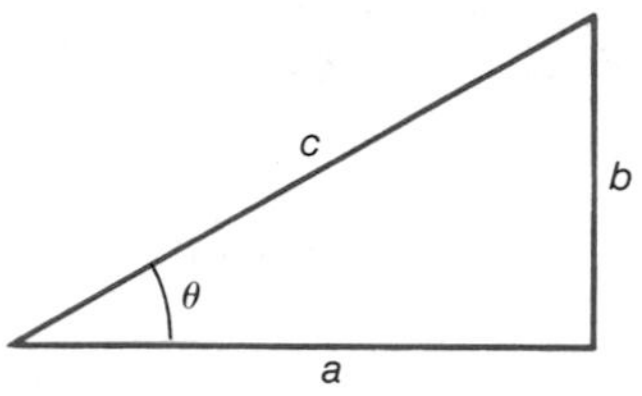

FIGURE 3.2 Diagram to illustrate the trigonometric relations for a right triangle.

To establish the important relationships between these two equivalent descriptions, we need to review the basic trigonometry of the right triangle. Referring to Fig. 3.2, the definitions of the three elementary trigonometric functions are as follows:

■ Definitions of Sine, Cosine, and Tangent

$$\sin\theta = \frac{\text{opposite side}}{\text{hypotenuse}} = \frac{b}{c}$$

$$\cos\theta = \frac{\text{adjacent side}}{\text{hypotenuse}} = \frac{a}{c} \qquad \textbf{(3.1)}$$

$$\tan\theta = \frac{\text{opposite side}}{\text{adjacent side}} = \frac{b}{a}$$

The trigonometric functions are all dimensionless numbers.

Since all three trigonometric functions are a length divided by a length, they are dimensionless numbers.

Transferring our attention to Fig. 3.1, we see that the results of Eq. (3.1) written in terms of the vector components R_x and R_y and the vector magnitude R become:

$$\sin\theta = \frac{R_y}{R}$$

$$\cos\theta = \frac{R_x}{R} \qquad \textbf{(3.1a)}$$

$$\tan\theta = \frac{R_y}{R_x}$$

This allows us to calculate the components of a vector from its magnitude and angle. Using the first and second parts of Eq. (3.1a), we have:

■ Vector Components Calculated from Magnitude and Angle

$$R_x = R\cos\theta$$
$$R_y = R\sin\theta \qquad \textbf{(3.2)}$$

Another well-known result for the right triangle is the Pythagorean theorem: referring to Fig. 3.2, we write this as:

Pythagorean Theorem for a Right Triangle

Square of the hypotenuse = sum of the squares of the other two sides

$$c^2 = a^2 + b^2 \qquad (3.3)$$

Once again transferring our attention to Fig. 3.1, the Pythagorean theorem for vector components and magnitude becomes:

$$R^2 = R_x^2 + R_y^2 \qquad (3.3a)$$

We can now calculate the vector magnitude and angle from its components using Eq. (3.3a) and the third part of Eq. (3.1a):

Vector Magnitude and Angle Calculated from Components

$$R = (R_x^2 + R_y^2)^{1/2}$$
$$\theta = \arc\tan\left(\frac{R_y}{R_x}\right) \qquad (3.4)$$

The notation "arc tan (R_y/R_x)" means the angle whose tangent is R_y/R_x. Let us illustrate these ideas with examples.

EXAMPLE 3.1

A man travels 3 km in a direction 32° E of N. What are the components of his displacement in north and east directions?

SOLUTION Since we have the magnitude and direction of the vector displacement, we need to find the components. Choose the east and north directions as the directions of the x- and y-axes, respectively. The angle θ in Eq. (3.2) is the angle between the direction of the vector and the x-axis. To find this angle, note that 32° E of N is the same direction as 58° N of E. Then Eq. (3.2) gives:

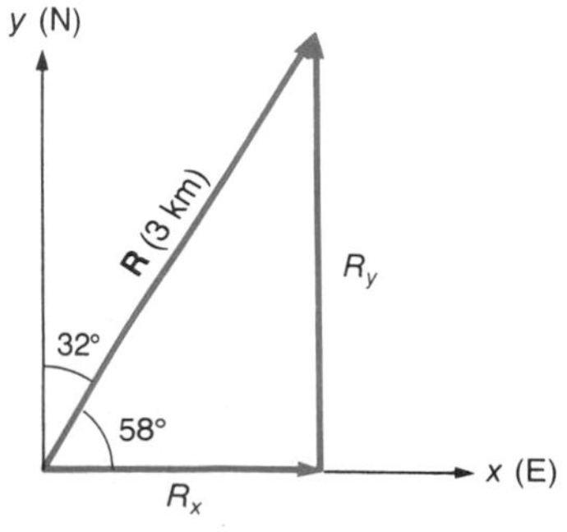

$$R_x = R\cos\theta$$
$$= 3 \times 10^3 \text{ m} \times \cos 58°$$
$$= 1590 \text{ m}$$
$$R_y = R\sin\theta$$
$$= 3 \times 10^3 \text{ m} \times \sin 58°$$
$$= 2540 \text{ m}$$

■

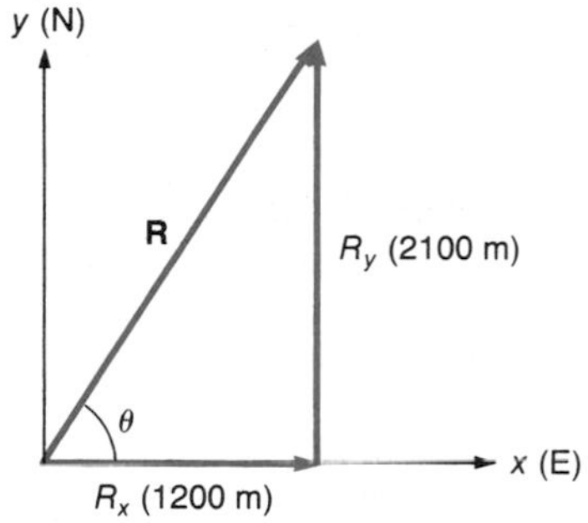

EXAMPLE 3.2

A hiker wishes to go to a point 1200 m east and 2100 m north of his present position. What is the distance in a straight line to the point in question, and in what direction should the hiker proceed?

SOLUTION Given the two components of a vector displacement, we need to find the magnitude and the direction. Again let us choose the east and north directions to be the x- and y-axes, respectively. Equation (3.4) gives the magnitude and direction:

$$\begin{aligned} R &= (R_x^2 + R_y^2)^{1/2} \\ &= [(1200\text{ m})^2 + (2100\text{ m})^2]^{1/2} \\ &= 2420\text{ m} \end{aligned}$$

$$\begin{aligned} \theta &= \arctan\left(\frac{R_y}{R_x}\right) \\ &= \arctan\left(\frac{2100\text{ m}}{1200\text{ m}}\right) \\ &= 60.3^\circ \end{aligned}$$

The distance is 2420 m, and the direction 60.3° N of E. ■

3.2 Addition and Subtraction of Vectors

We can easily add and subtract vectors in one dimension. A displacement of 3 km to the north followed by a displacement of 5 km to the south is equivalent to a displacement of 2 km to the south. No special mathematical method for solving this simple problem is necessary. However, when vector displacements can point in any direction on a plane, we need to consider the general technique for adding and subtracting them.

Consider the two displacements shown in Fig. 3.3. The first displacement, represented by the vector **T**, is from the origin O to the point P_1; the second displacement, represented by the vector **S**, is from the point P_1 to the point P_2. To add the two vector displacements, we need to find a single vector displacement that is entirely equivalent to the two individual displacements performed one after the other. For example, suppose that these displacements represent the path of a hiker on level ground. The net effect of proceeding from O to P_1 and then from P_1 to P_2 is the same as a single displacement from O to P_2. Let us represent the vector displacement from O to P_2 by **R**. The components of **R** are the sums of the components of **T** and **S**. In mathematical form we write this as follows:

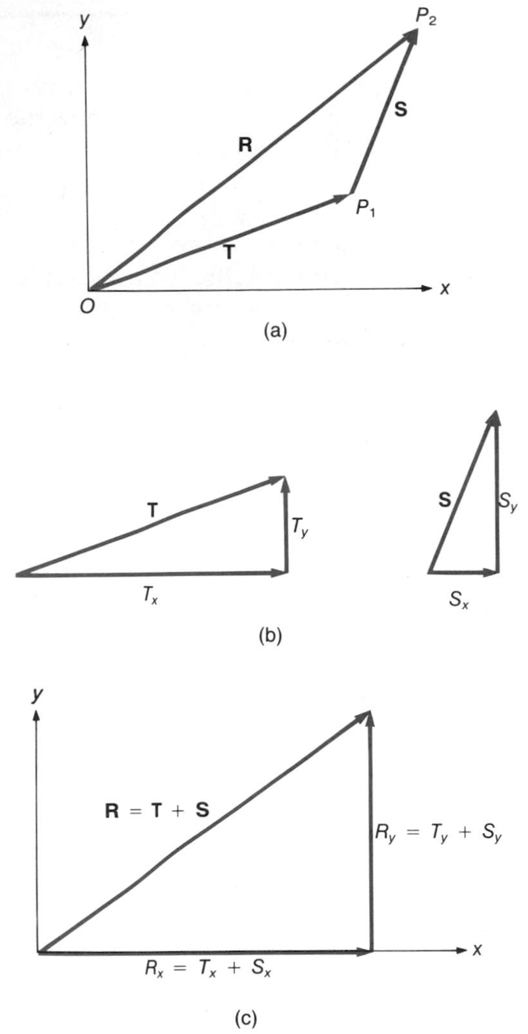

FIGURE 3.3 The addition of vector displacements: (a) The displacement **T** followed by the displacement **S** is equivalent to the displacement **R** = **T** + **S.** (b) The *x* and *y* components of the vectors **T** and **S.** (c) The sum of the *x* and *y* components of **T** and **S** gives the *x* and *y* components of **R.**

$$R_x = T_x + S_x$$
$$R_y = T_y + S_y$$

This pair of relations is often expressed succinctly as:

$$\mathbf{R} = \mathbf{T} + \mathbf{S}$$

This equation expresses the equivalence of the vector displacement **R** to the successive vector displacements **T** and **S.** We call the successive displacements the vector sum of **T** and **S.** The problem of vector addition is to calculate **R** from **T** and **S.** We can do this both graphically and analytically. Let us illustrate both methods with a specific example.

EXAMPLE 3.3

A canoeist travels 3 km across a lake in a direction 25° N of E and then 2 km in a direction 55° N of W. Calculate the displacement from the starting point.

GRAPHICAL SOLUTION Let us call the first displacement **T** and the second one **S.** In order to find the displacement **R** = **T** + **S** graphically, we simply plot the displacements on drawing paper, using a ruler and a protractor. Select a convenient scale, say 5 cm, to represent 1 km of displacement. The *x*-axis is the east direction and the *y*-axis the north. Begin by drawing a directed line 15 cm long at an angle of 25° with the positive direction of the *x*-axis. This directed line represents the vector **T.** From its head draw a directed line 10 cm long at an angle of 55° with the negative direction of the *x*-axis (this is the same as 125° with the positive direction of the *x*-axis). This directed line represents the vector **S.** The directed line labeled **R** then represents the displacement due to the displacement **T** followed by the displacement **S.** Using a ruler we find the length of **R** to be about 16.5 cm at an angle of about 61° with the positive direction of the *x*-axis. Since 5 cm rep-

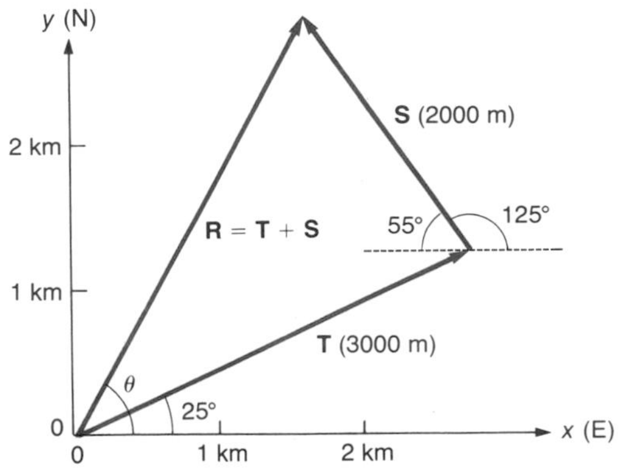

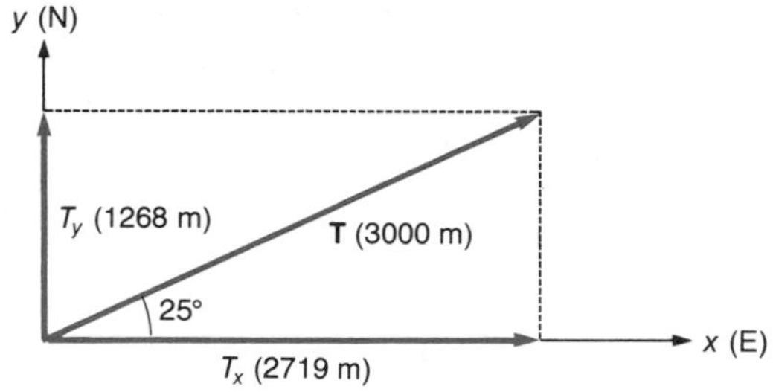

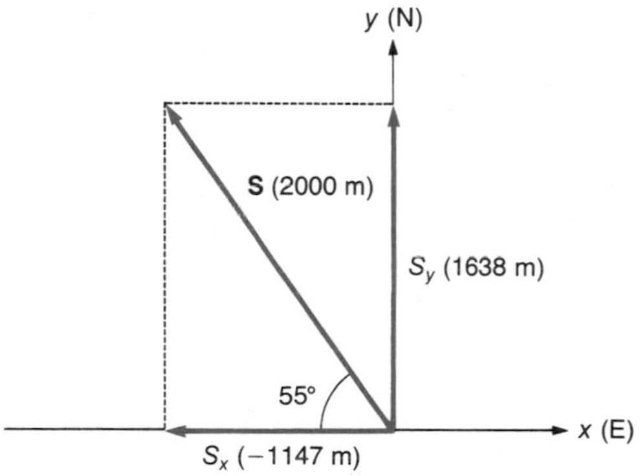

resents 1 km, the magnitude of the displacement **R** is about 16.5 cm ÷ 5 cm/km = 3.3 km. The vector angle is 61° N of E.

ANALYTICAL SOLUTION To calculate the components of **R**, we first calculate the components of **T** and **S**. With the help of the diagram and the use of Eq. (3.2), we find:

$$T_x = 3000 \text{ m} \times \cos 25^\circ$$
$$= 2719 \text{ m}$$
$$T_y = 3000 \text{ m} \times \sin 25^\circ$$
$$= 1268 \text{ m}$$
$$S_x = -2000 \text{ m} \times \cos 55^\circ$$
$$= -1147 \text{ m}$$
$$S_y = 2000 \text{ m} \times \sin 55^\circ$$
$$= 1638 \text{ m}$$

The components T_x, T_y, and S_y are all positive, but S_x points in the negative direction. Combining the components of **T** and **S** gives us the components of **R**:

$$R_x = T_x + S_x$$
$$= 2719 \text{ m} + (-1147 \text{ m})$$
$$= 1572 \text{ m}$$
$$R_y = T_y + S_y$$
$$= 1268 \text{ m} + 1638 \text{ m}$$
$$= 2906 \text{ m}$$

To calculate the magnitude and the angle of **R**, we use Eq. (3.4):

$$R = (R_x^2 + R_y^2)^{1/2}$$
$$= [(1572 \text{ m})^2 + (2906 \text{ m})^2]^{1/2}$$
$$= 3304 \text{ m}$$
$$\theta = \text{arc tan}\left(\frac{R_y}{R_x}\right)$$
$$= \text{arc tan}\left(\frac{2906 \text{ m}}{1572 \text{ m}}\right)$$
$$= 61.6^\circ$$

■

The technique for adding vectors in two dimensions is central to many problems in physics, and we use both the graphical and the analytical methods as convenient.

Subtracting vectors is a special case of addition. In order to subtract a vector **S** from another vector **T**, we simply add the vector −**S** and the vector **T**:

$$\mathbf{T} - \mathbf{S} = \mathbf{T} + (-\mathbf{S})$$

To find the negative of a vector graphically, we draw a vector of the same magnitude but pointing in the opposite direction. Analytically we simply change the signs of the rectangular components.

EXAMPLE 3.4

Two displacement vectors **S** and **T** have magnitudes and directions as follows:

	Magnitude	*Angle with positive x-axis*
S	5 m	30°
T	8 m	75°

Subtract the vector **S** from the vector **T**.

SOLUTION Let us choose the analytical method. We begin by using Eq. (3.2) to calculate the components of both **S** and **T**:

$$S_x = 5 \text{ m} \times \cos 30°$$
$$= 4.330 \text{ m}$$
$$S_y = 5 \text{ m} \times \sin 30°$$
$$= 2.500 \text{ m}$$
$$T_x = 8 \text{ m} \times \cos 75°$$
$$= 2.071 \text{ m}$$
$$T_y = 8 \text{ m} \times \sin 75°$$
$$= 7.727 \text{ m}$$

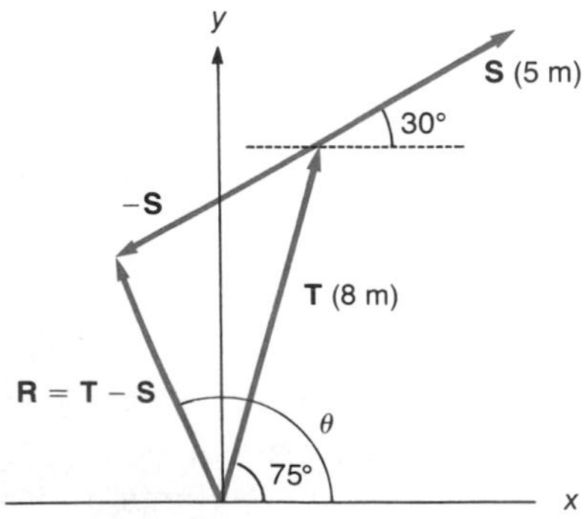

To subtract **S** from **T** we form the vector **R**:

$$\mathbf{R} = \mathbf{T} + (-\mathbf{S})$$

The components of **R** are:

$$R_x = 2.071 \text{ m} + (-4.330 \text{ m})$$
$$= -2.259 \text{ m}$$

$$R_y = 7.727 \text{ m} + (-2.500 \text{ m})$$
$$= 5.227 \text{ m}$$

The magnitude and direction of **R** are given by Eq. (3.4):

$$R = (R_x^2 + R_y^2)^{1/2}$$
$$= [(-2.259 \text{ m})^2 + (5.227 \text{ m})^2]^{1/2}$$
$$= 5.694 \text{ m}$$
$$\theta = \text{arc tan}\left(\frac{R_y}{R_x}\right)$$
$$= \text{arc tan}\left(\frac{5.227 \text{ m}}{-2.259 \text{ m}}\right)$$
$$= 113.4°*$$

■

We stress the addition and subtraction of displacement vectors because they are easy to visualize. In the foregoing examples we can imagine that each displacement vector represents a walk across a level field. With regard to adding or subtracting velocity and acceleration vectors, the procedure is exactly the same. We add and subtract all vectors, whatever they represent, in just the same way as we did displacement vectors.

3.3 Velocity and Acceleration Vectors

To define average velocity and acceleration in two dimensions, we closely follow the one-dimensional definitions. Recall our definition of average velocity in one dimension as the displacement divided by the elapsed time. To extend this result to two dimensions, consider the notation of Fig. 3.4. Let $\mathbf{R}_0$ be the position vector of a point particle at the initial time t_0, and **R** the position vector at time t. The displacement during the elapsed time interval $(t - t_0)$ is $(\mathbf{R} - \mathbf{R}_0)$. The definition of average velocity in two

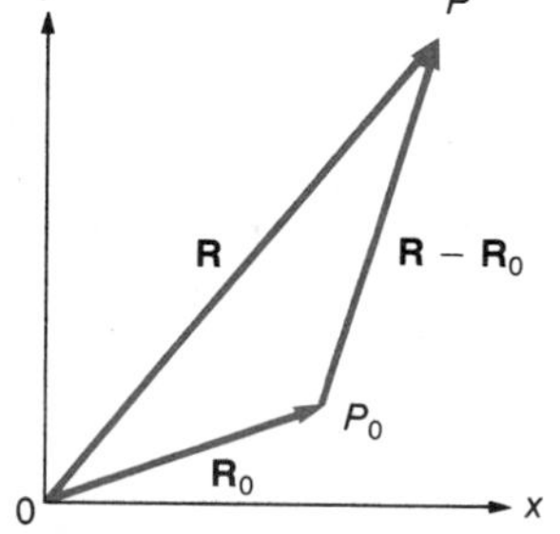

FIGURE 3.4 Diagram to illustrate the definition of average velocity.

*There is a small difficulty at this point concerning the use of calculators for inverse trigonometric functions. Most calculators will give −66.6° for the operation we have carried out rather than the quoted answer of 113.4°. The reason is that the calculator does not distinguish between arc tan (5.227 m/−2.259 m) and arc tan (−5.227 m/2.259 m); it simply calculates arc tan (−2.314) and arbitrarily gives a fourth quadrant answer. However, there is also a second quadrant result differing from the fourth quadrant result by 180°, and for our problem the second quadrant result is correct. In order to decide between the alternatives, it is essential to know whether R_x or R_y is negative. If R_x is negative and R_y is positive, the angle is in the second quadrant; if R_x is positive and R_y is negative, the angle is in the fourth quadrant. A similar difficulty occurs for the inverse tangent of a positive number. If both vector components are positive, the angle is in the first quadrant; if they are both negative, the angle is in the third quadrant.

dimensions is given in Eq. (3.5). The average velocity is the vector $(\mathbf{R} - \mathbf{R}_0)$ divided by the scalar $(t - t_0)$. This operation changes the units and the magnitude but not the direction. The vector $\langle \mathbf{v} \rangle$ therefore has the direction of $(\mathbf{R} - \mathbf{R}_0)$ and points along the direction of the particle displacement.

■ Definition of Average Velocity

$$\text{Average velocity} = \frac{\text{displacement}}{\text{elapsed time}}$$

$$\langle \mathbf{v} \rangle = \frac{\mathbf{R} - \mathbf{R}_0}{t - t_0} \qquad \textbf{(3.5)}$$

This definition is convenient to use if we know the magnitude and direction of the displacement, but an alternate form is preferable if we know the displacement components. To show this, we rewrite the vector equation in terms of its x and y components:

■ Definition of Average-Velocity Components

$$\text{Average-velocity component} = \frac{\text{displacement component}}{\text{elapsed time}}$$

$$\langle v_x \rangle = \frac{x - x_0}{t - t_0}$$
$$\langle v_y \rangle = \frac{y - y_0}{t - t_0} \qquad \textbf{(3.5a)}$$

We usually indicate the components of a vector by subscripts; for example, the components of **R** are usually written R_x and R_y. However, since **R** represents the vector displacement in this instance, its components are simply x and y.

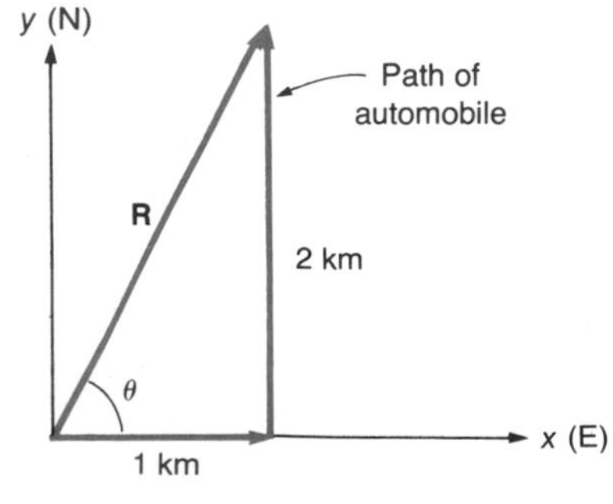

EXAMPLE 3.5

An automobile heads east for 1 km and then turns, traveling north for 2 km. The speed is constant at 45 km/h (12.5 m/s) throughout the whole trip, which takes 4 min. Calculate the average velocity over this period.

SOLUTION Select the starting point as the origin of coordinates, with the x direction east and the y direction north. The displacement components are:

$$x - x_0 = 1000 \text{ m}$$

$$y - y_0 = 2000 \text{ m}$$

The elapsed time is:

$$t - t_0 = 4 \text{ min} \times 60 \text{ s/min} = 240 \text{ s}$$

Use of Eq. (3.5a) immediately gives the average-velocity components:

$$\langle v_x \rangle = \frac{x - x_0}{t - t_0} = \frac{1000 \text{ m}}{240 \text{ s}}$$

$$= 4.17 \text{ m/s}$$

$$\langle v_y \rangle = \frac{y - y_0}{t - t_0} = \frac{2000 \text{ m}}{240 \text{ s}}$$

$$= 8.33 \text{ m/s}$$

We can find the magnitude and direction of the average velocity using Eq. (3.4):

$$\langle v \rangle = (\langle v_x \rangle^2 + \langle v_y \rangle^2)^{1/2}$$

$$= [(4.17 \text{ m/s})^2 + (8.33 \text{ m/s})^2]^{1/2}$$

$$= 9.32 \text{ m/s}$$

$$\theta = \text{arc tan} \left(\frac{\langle v_y \rangle}{\langle v_x \rangle} \right)$$

$$= \text{arc tan} \left(\frac{8.33 \text{ m/s}}{4.17 \text{ m/s}} \right)$$

$$= 63.4^\circ$$

The average velocity is 9.32 m/s in a direction 63.4° N of E. We can find the same result by first calculating the magnitude and direction of the displacement using Eq. (3.4):

$$|\mathbf{R} - \mathbf{R}_0| = [(x - x_0)^2 + (y - y_0)^2]^{1/2}$$

$$= [(1000 \text{ m})^2 + (2000 \text{ m})^2]^{1/2}$$

$$= 2236 \text{ m}$$

$$\theta = \text{arc tan} \left(\frac{y - y_0}{x - x_0} \right)$$

$$= \text{arc tan} \left(\frac{2000 \text{ m}}{1000 \text{ m}} \right)$$

$$= 63.4^\circ$$

The average velocity is given by Eq. (3.5):

$$\begin{aligned}\langle \mathbf{v} \rangle &= \frac{\text{displacement}}{\text{elapsed time}} \\ &= \frac{2236 \text{ m in direction } 63.4° \text{ N of E}}{240 \text{ s}} \\ &= 9.32 \text{ m/s in direction } 63.4° \text{ N of E}\end{aligned}$$ ■

We define the **average acceleration** analogously to the definition for one dimension [Eq. (2.6)]. Let $\mathbf{v}_0$ be the instantaneous velocity at the time t_0, and $\mathbf{v}$ the instantaneous velocity at a later time t. Recall that we subtract velocity vectors in exactly the same way as we subtract displacement vectors. To calculate the average acceleration we divide the vector $(\mathbf{v} - \mathbf{v}_0)$ by the scalar $(t - t_0)$, thus changing the vector magnitude but not its direction. The vector $\langle \mathbf{a} \rangle$ points in the same direction as $(\mathbf{v} - \mathbf{v}_0)$. In general, the velocity difference does not have the same direction as the displacement.

■ Definition of Average Acceleration

$$\text{Average acceleration} = \frac{\text{change in velocity}}{\text{elapsed time}}$$

$$\langle \mathbf{a} \rangle = \frac{\mathbf{v} - \mathbf{v}_0}{t - t_0} \qquad \textbf{(3.6)}$$

To write the corresponding component equations, we simply take x and y components of each side of Eq. (3.6) and achieve the result given in Eq. (3.6a). The quantities v_{0x} and v_{0y} are the x and y components of the initial velocity $\mathbf{v}_0$.

■ Definition of Average-Acceleration Components

$$\text{Average acceleration component} = \frac{\text{change in velocity component}}{\text{elapsed time}}$$

$$\langle a_x \rangle = \frac{v_x - v_{0x}}{t - t_0}$$

$$\langle a_y \rangle = \frac{v_y - v_{0y}}{t - t_0} \qquad \textbf{(3.6a)}$$

EXAMPLE 3.6

Calculate the average acceleration of the automobile in Example 3.5.

SOLUTION Let us work this problem analytically, using the component form for the average-acceleration equations. The initial velocity is 12.5 m/s east. Therefore, we have:

$$v_{0x} = 12.5 \text{ m/s} \quad \text{and} \quad v_{0y} = 0$$

The final velocity is 12.5 m/s north:

$$v_x = 0 \quad \text{and} \quad v_y = 12.5 \text{ m/s}$$

The elapsed time $(t - t_0) = 240$ s. Using Eq. (3.6a) we find for the average-acceleration components:

$$\begin{aligned} \langle a_x \rangle &= \frac{v_x - v_{0x}}{t - t_0} \\ &= \frac{0 \text{ m/s} - 12.5 \text{ m/s}}{240 \text{ s}} \\ &= -0.0521 \text{ m/s}^2 \\ \langle a_y \rangle &= \frac{v_y - v_{0y}}{t - l_0} \\ &= \frac{12.5 \text{ m/s} - 0 \text{ m/s}}{240 \text{ s}} \\ &= 0.0521 \text{ m/s}^2 \end{aligned}$$

An alternate method of solution employs the graphical construction shown in the diagram. Take a scale of, say, 1 cm to represent 2 m/s on the velocity vector diagram. Graphical construction gives $\mathbf{v} - \mathbf{v}_0 = 17.7$ m/s in a northwest direction. Then, using Eq. (3.6), we calculate the average acceleration:

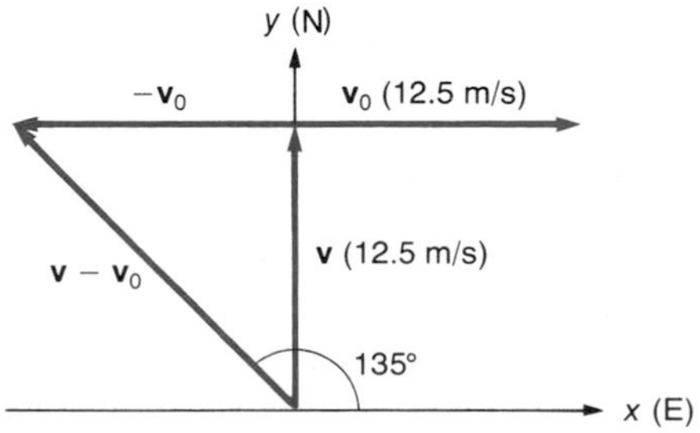

$$\begin{aligned} \langle \mathbf{a} \rangle &= \frac{\mathbf{v} - \mathbf{v}_0}{t - t_0} \\ &= \frac{17.7 \text{ m/s in NW direction}}{240 \text{ s}} \\ &= 0.0737 \text{ m/s}^2 \text{ in NW direction} \end{aligned}$$

We can readily verify that this vector has the same components as those obtained by using the analytical method. It may seem surprising that an automobile traveling at a constant 45 km/h undergoes acceleration. However, the essential feature of acceleration is change in velocity, which can occur even at a constant speed if the velocity changes direction. ■

3.4 Projectile Motion

The experiences of everyday life present us with an interesting case of motion in two dimensions. The flight of stones, arrows, and cannonballs has fascinated mankind since the dawn of history. Aristotle considered the philosophical cause of the motion of a javelin after it left the hand of the thrower. Medieval military engineers had all manner of fanciful formulas for predicting cannonball trajectories. However, it remained for Galileo, early in the seventeenth century, to take the first step in scientifically analyzing projectile motion. Galileo realized that the horizontal and vertical components of the motion are separate. The horizontal component of velocity of a projectile remains unchanged during the projectile's flight, since no acceleration acts in a horizontal direction. However, the acceleration due to gravity acts in a vertical direction, causing the vertical component of motion to be uniformly accelerated.

Let us apply Galileo's concept to a special case. Figure 3.5 illustrates the calculation of points on the flight path of a particle that is thrown horizontally from a point above the ground. Given that the initial velocity is 20 m/s in a horizontal direction, we calculate the particle's position on the curved flight path at 1-s intervals. The horizontal components of displacement are the same as those for a particle moving horizontally with a constant velocity of 20 m/s. The vertical components of displacement are the same as those for a particle dropped with zero initial velocity. The two component motions are independent of each other except that the displacement components must correspond to the elapsed flight time. In our example, the horizontal component of displacement after 4 s at a constant velocity of 20 m/s is equal to 80 m. Initially the vertical velocity component is zero. The vertical component of displacement after 4 s at constant acceleration of 9.81 m/s^2 downward is equal to 78.5 m downward from the point of projection. (Use the equations of uniformly accelerated motion to verify this.) The two displacement components serve to identify the displacement vector, and hence the position of the particle, after 4 s. In the same way we can calculate the particle's position for other time intervals. We show the positions on the diagram for the first six 1-s intervals. The curved line that joins these positions gives a continuous measure of the displacement of the particle from its point of projection.

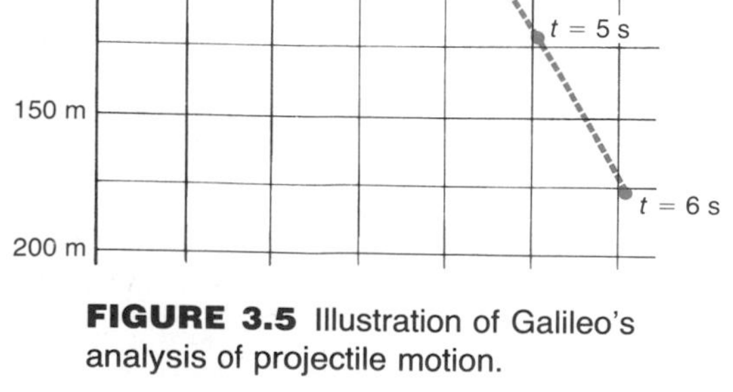

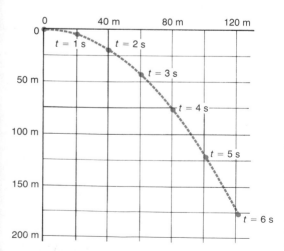

FIGURE 3.5 Illustration of Galileo's analysis of projectile motion.

We already have techniques for dealing with motion at both constant velocity and constant acceleration. It follows that we can

Aristotle's comments on the flight of a projectile were directed specifically at the javelin. It required the passage of some twenty centuries—culminating with the insight of Galileo—before this phenomenon was well understood.

write the formulas for projectile motion from our work in one-dimensional kinematics as long as we have a clear convention for separating the two components. Let x denote the horizontal component of displacement, and y the vertical component. We use the subscripts x and y on the symbols for velocity and acceleration to indicate horizontal and vertical components, respectively. With these conventions, we can write the kinematic equations for projectile motion. The horizontal component is Eq. (2.5) and the vertical component Eq. (2.12)—both modified to indicate component direction. Horizontally, the initial velocity component v_{0x} is the same as the velocity component v_x at subsequent time t. We could use either v_{0x} or v_x in Eq. (3.7a). The symbol a_x does not appear because the horizontal acceleration component is zero. Vertically, the initial velocity component v_{0y} is different from the velocity component v_y at subsequent time t. The acceleration component a_y is always $-9.81\ \text{m/s}^2$ if we choose the positive direction of the y-axis to represent upward directed vectors.

■ Equations of Projectile Motion

Horizontal component: Motion at constant velocity

$$x = v_x t \qquad \text{(a)}$$

Vertical component: Motion at constant acceleration

$$\left.\begin{aligned} v_y &= v_{0y} + a_y t && \text{(b)} \\ y &= \tfrac{1}{2}(v_y + v_{0y})t && \text{(c)} \\ y &= v_{0y}t + \tfrac{1}{2}a_y t^2 && \text{(d)} \\ v_y^2 &= v_{0y}^2 + 2a_y y && \text{(e)} \end{aligned}\right\} \quad (3.7)$$

Before considering further examples of projectile motion, we should look briefly at the limitations. Strictly speaking, Galileo's concept applies to a flat, airless, nonrotating earth. The two extreme cases shown in Fig. 3.6 illustrate this point. In Fig. 3.6(a) an athlete throws the shotput. The surface of the earth can be taken as flat, the acceleration of gravity is everywhere constant and perpendicular to this flat surface, the effect of the earth's rotation is negligible, air resistance to the shot can be neglected, and the wind does not cause any appreciable deviation. However, in Fig. 3.6(b) a battleship fires a shell 20 mi, and none of the above approximations are applicable. Intermediate cases result in more or less of a breakdown of these assumptions. For example, the effect of air resistance and wind drift is not negligible for the flight of a baseball or a football over a long distance. Calculations that ignore these effects give order-of-magnitude results rather than highly accurate answers. Let us work some examples using Galileo's method.

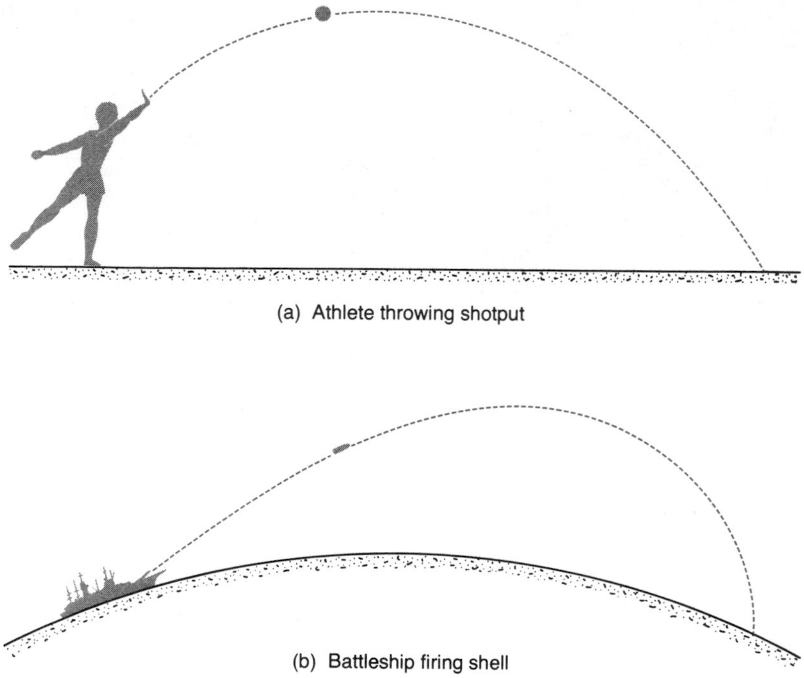

FIGURE 3.6 Extreme cases of projectile motion.

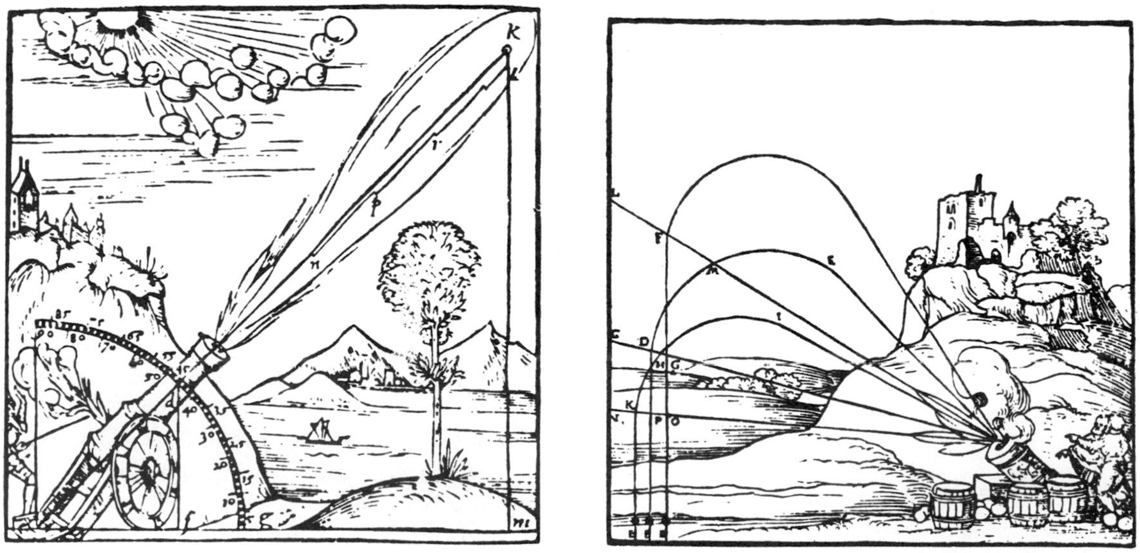

These prints illustrate the theories of early military engineers on cannonball trajectories. The print on the left shows a fanciful medieval theory in which the ball rises to its apex in a straight line and then falls vertically. The print on the right illustrates a theory due to Niccolo Tartaglia (1500–1557) of Brescia and shows a much better idea of the shape of the trajectories. (Historical Pictures Service, Inc.)

SPECIAL TOPIC 3 THE LONG JUMP

Consider a projectile whose initial velocity is v_0 at an angle θ above a flat plane surface. Its initial velocity components are $v_0 \cos\theta$ horizontally and $v_0 \sin\theta$ vertically. It follows that the range of the particle on a horizontal surface is given by:

$$\text{Range} = v_0 \cos\theta \times t$$

where t is the time of flight.

We can calculate t by considering the vertical component of the motion. The initial vertical velocity component is $v_0 \sin\theta$ and the final vertical velocity component is $-v_0 \sin\theta$. With the acceleration set at $-g$, Eq. (2.12a) gives:

$$-v_0 \sin\theta = v_0 \sin\theta - gt$$

$$\therefore t = \frac{2v_0}{g}\sin\theta$$

We then have

$$\text{Range} = v_0 \cos\theta \times \frac{2v_0}{g}\sin\theta$$

$$= \frac{v_0^2}{g}\sin 2\theta$$

(using the identity $\sin 2\theta = 2\sin\theta\cos\theta$).

This formula is of limited use, but the answer to part (a) of Example 3.7 could be found using it. It can also be used to find the angle of projection for maximum range on a level surface.

For a given magnitude of the initial velocity, the maximum range occurs for the maximum value of $\sin 2\theta$; this maximum occurs when $2\theta = \pi/2$; that is, $\theta = \pi/4$. It follows that we have:

$$\text{Maximum range} = \frac{v_0^2}{g}$$

corresponding to a 45° takeoff angle.

Consider now an athlete competing in the long jump. Let us make the very simple assumption that the proper technique is to approach the takeoff point running at full speed, and then to use one's "jumping ability" to change the velocity vector from horizontal to 45° above the horizontal. The record long jumps are about 8.2 m, and using this result in our equation for maximum range gives $v_0 \simeq 9$ m/s. This is not quite equal to an athlete's maximum running speed (which is closer to 11 m/s), but the agreement is sufficiently good to indicate that our simple model is on the right track.

Finally, we should notice that the "maximum range for a 45° projection angle" applies only when the takeoff and landing are at the same level. In the shotput event the shot is released from the athlete's hand some 2 m above the surface of the field. In this case the analysis is more difficult [see the articles listed in the references, Appendix D, especially S. K. Bose, "Maximizing the Range of the Shot Put without Calculus," Am. J. Phys. *51* (1983):458] and the maximum range is obtained for an angle of projection that is somewhat smaller than 45°.

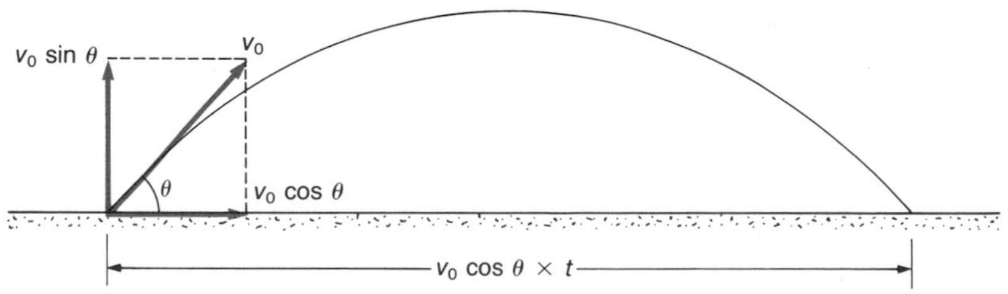

EXAMPLE 3.7

A golfer hits a ball on a flat fairway, giving it an initial velocity of 30 m/s at an angle of 40° with the horizontal.

a. What horizontal distance does the ball travel before striking the ground?

b. What is the maximum height of the ball above the fairway?

c. What is the velocity of the ball just before it strikes the ground?

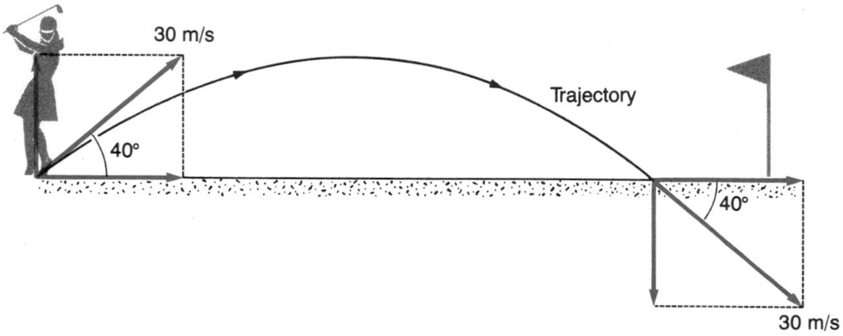

SOLUTION Before answering any of these questions, we must set the initial data in component form using Eq. (3.2):

$$v_{0x} = v_x = 30 \text{ m/s} \times \cos 40^\circ$$
$$= 22.98 \text{ m/s}$$
$$v_{0y} = 30 \text{ m/s} \times \sin 40^\circ$$
$$= 19.28 \text{ m/s}$$

Both initial velocity components are positive since they point in the positive direction of the axes. We can now set up the data for the vertical and horizontal motion components.

a.

Horizontal component	Vertical component
$v_{0x} = 22.98$ m/s	$v_{0y} = 19.28$ m/s
$x = ?$	$a_y = -9.81$ m/s^2
$t = ?$	$y = 0$ m
	$v_y = ?$
	$t = ?$

Note that the ball returns to the flat surface from which it was hit; thus, the vertical component of its displacement is zero. The time of flight is common to both the horizontal and vertical components. We cannot make any progress with the horizontal motion component, since two of the three quantities are unknown. We must therefore calculate the time of flight from the vertical motion component. Equation (3.7d) is suitable for this purpose:

$$y = v_{0y}t + \tfrac{1}{2}a_y t^2$$

$$0\ \text{m} = (19.28\ \text{m/s}) \times t + \tfrac{1}{2} \times (-9.81\ \text{m/s}^2) \times t^2$$

This is a simple form of a quadratic equation, which is discussed in Appendix A. Its solutions are:

$$t = 0\ \text{s} \quad \text{or} \quad 3.931\ \text{s}$$

Discarding the first solution as irrelevant to our problem (we already know that $y = 0$ at $t = 0$), we can now find the range from Eq. (3.7a):

$$\begin{aligned} x &= v_x t \\ &= 22.98\ \text{m/s} \times 3.931\ \text{s} \\ &= 90.3\ \text{m} \end{aligned}$$

b. Calculation of the maximum height of the ball concerns the vertical component of motion for part of the motion. At maximum height the instantaneous vertical velocity component is zero, and we can rewrite the problem data as follows:

$$\begin{aligned} v_{0y} &= 19.28\ \text{m/s} \\ a_y &= -9.81\ \text{m/s}^2 \\ y &= ? \\ v_y &= 0\ \text{m/s} \\ t &= ? \end{aligned}$$

To calculate the vertical displacement component we use Eq. (3.7e):

$$\begin{aligned} v_y^2 &= v_{0y}^2 + 2a_y y \\ \therefore y &= \frac{v_y^2 - v_{0y}^2}{2a_y} \\ &= \frac{(0\ \text{m/s})^2 - (19.28\ \text{m/s})^2}{2 \times (-9.81\ \text{m/s}^2)} \\ &= 18.9\ \text{m} \end{aligned}$$

c. We could calculate the velocity of the ball just before it strikes the ground by using the equations. But in this particular case, we can write it down after thinking about the problem. The horizontal velocity component remains unchanged throughout the motion; thus, it is the same just before striking the ground as it was at the beginning. The vertical velocity component just before striking the ground is equal in magnitude to the initial vertical velocity but points downward rather than upward. This follows since a ball thrown vertically upward returns to the ground at the same speed with which it was thrown. Since the velocity components just before striking the ground are of the same magnitude as the initial velocity components, the projectile strikes the ground with a velocity of 30 m/s. Since the vertical component is reversed in direction, the velocity points at an angle of 40° below the horizontal.

Many problems concerning projectiles do not refer to a horizontal plane surface. The principle of solution is exactly the same, but the details may be a little more complex. ■

EXAMPLE 3.8

A stone is thrown with a velocity of 40 m/s at an angle of 60° to the horizontal from the top of the vertical cliff 50 m high. How far out from the base of the cliff does the stone strike the water?

SOLUTION We first decompose the initial velocity into horizontal and vertical components. Using Eq. (3.2) we have:

$$v_{0x} = v_x = 40\text{ m/s} \times \cos 60° = 20\text{ m/s}$$

$$v_{0y} = 40\text{ m/s} \times \sin 60° = 34.64\text{ m/s}$$

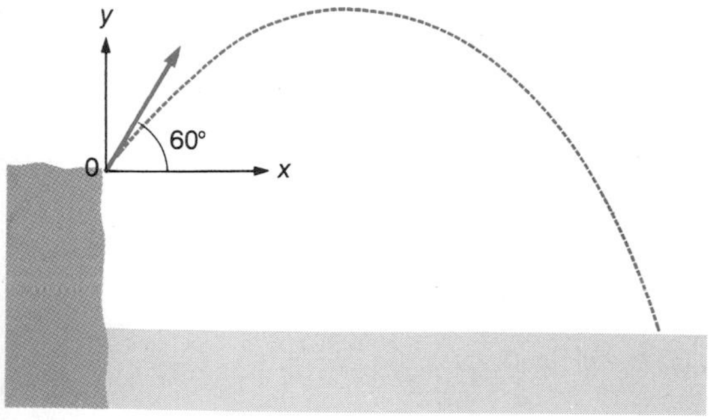

As usual, the positive direction on the x-axis is to the right, and the positive direction on the y-axis is upward. The data for the horizontal and vertical motion components become:

Vertical component	Horizontal component
$y = -50$ m	$x = ?$
$v_{0y} = 34.64$ m/s	$v_x = 20$ m/s
$a_y = -9.81$ m/s^2	$t = ?$
$v_y = ?$	
$t = ?$	

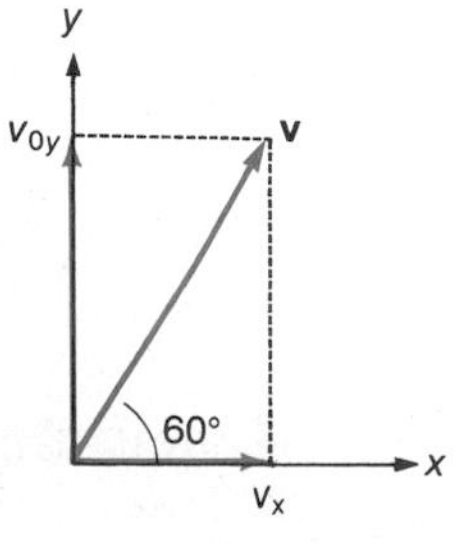

Initial velocity vector diagram

To calculate the time of flight from the vertical component data we use Eq. (3.7d):

$$y = v_{0y}t + \tfrac{1}{2}a_y t^2$$

$$-50\text{ m} = (34.64\text{ m/s}) \times t + \tfrac{1}{2} \times (-9.81\text{ m/s}^2) \times t^2$$

This is a quadratic equation for t, which simplifies to:

$$(-4.905\text{ m/s}^2)t^2 + (34.64\text{ m/s})t + 50\text{ m} = 0$$

From the results of Appendix A:

$$t = \frac{-34.64\text{ m/s} \pm [(34.64\text{ m/s})^2 - 4 \times (-4.905\text{ m/s}^2) \times 50\text{ m}]^{1/2}}{2 \times (-4.905\text{ m/s}^2)}$$

$$= \frac{-34.64\text{ m/s} \pm 46.69\text{ m/s}}{-9.81\text{ m/s}}$$

$$= 8.29\text{ s} \quad \text{or} \quad -1.23\text{ s}$$

We choose the positive time since the stone strikes the water after projection. Finally, to calculate the horizontal displacement component we use Eq. (3.7a):

$$x = v_x t$$

$$= 20\text{ m/s} \times 8.29\text{ s}$$

$$= 166\text{ m}$$

■

3.5 Relative Velocity

All velocities are measured relative to some coordinate system, and this is frequently taken to be a coordinate system fixed to the earth. However, some problems concerning *relative velocity* require us to take specific account of various coordinate reference systems. To begin with, let us consider a simple situation in one dimension. A flatbed truck (labeled A in Fig. 3.7) is traveling to the right with velocity $\mathbf{v}_{AG}$; the notation reads "the velocity of A relative to G (the ground)." A child on the bed of a truck drives a toy car B with a velocity $\mathbf{v}_{BA}$ relative to the truck A. We now find the velocity of the child's car relative to the ground as the sum of its velocity relative to the truck and the velocity of the truck relative to the ground. In symbols we write:

$$\mathbf{v}_{BG} = \mathbf{v}_{BA} + \mathbf{v}_{AG} \qquad \textbf{(3.8)}$$

Note the importance of the order of the subscripts; the velocity signified is that of the first mentioned object relative to the second one. Changing the order of the subscripts changes the sign of the relative velocity; thus we have:

$$\mathbf{v}_{BA} = -\mathbf{v}_{AB} \qquad \textbf{(3.9)}$$

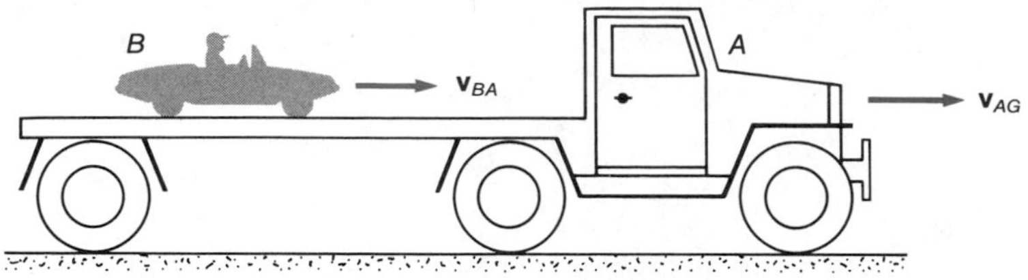

FIGURE 3.7 A flatbed truck A moves with velocity $\mathbf{v}_{AG}$ relative to the ground, and a child's car B moves with velocity $\mathbf{v}_{BA}$ relative to the truck.

Referring to our example, we see that the velocity of the truck relative to the child's car is of equal magnitude but opposite direction to the velocity of the child's car relative to the truck.

EXAMPLE 3.9

A moving sidewalk at an airport moves at 1.2 m/s. A child runs onto one end and runs at 3.0 m/s relative to the moving sidewalk and in the opposite direction. What is the child's velocity relative to the ground?

SOLUTION Let subscripts C, S, and G stand for the child, the sidewalk, and the ground, respectively. Also let the direction of motion of the sidewalk relative to the ground be the positive direction for velocity vectors. Our data are then $\mathbf{v}_{SG} = 1.2$ m/s and $\mathbf{v}_{CS} = -3.0$ m/s. Equation (3.8) becomes:

$$\begin{aligned}\mathbf{v}_{CG} &= \mathbf{v}_{CS} + \mathbf{v}_{SG}\\ &= -3.0\text{ m/s} + 1.2\text{ m/s}\\ &= 1.8\text{ m/s}\end{aligned}$$

The relative velocity in question is therefore 1.8 m/s in the opposite direction to which the sidewalk is moving. ■

The example just worked hardly requires the elaborate procedure that we have invoked to solve it, but problems in two (or even three) dimensions can be very much more confusing unless we have a well-disciplined method of approach. Once again the results expressed by Eqs. (3.8) and (3.9) carry over, with the velocity vectors now being understood to be in two dimensions.

EXAMPLE 3.10

An airliner A traveling due north at 250 m/s relative to the ground sights another airliner B at the same altitude and traveling due west at 250 m/s relative to the ground. What is the velocity of airliner B relative to airliner A?

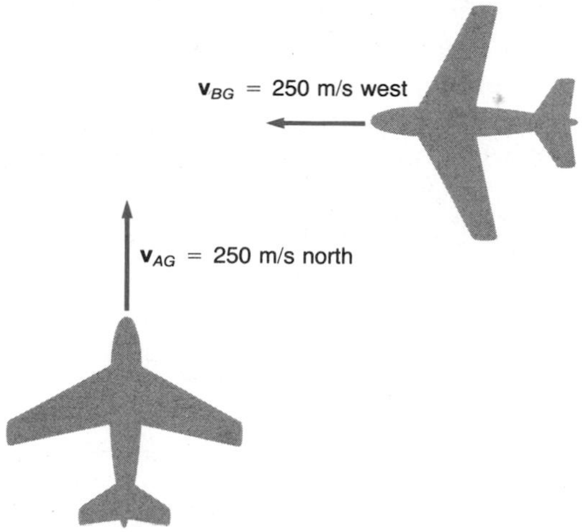

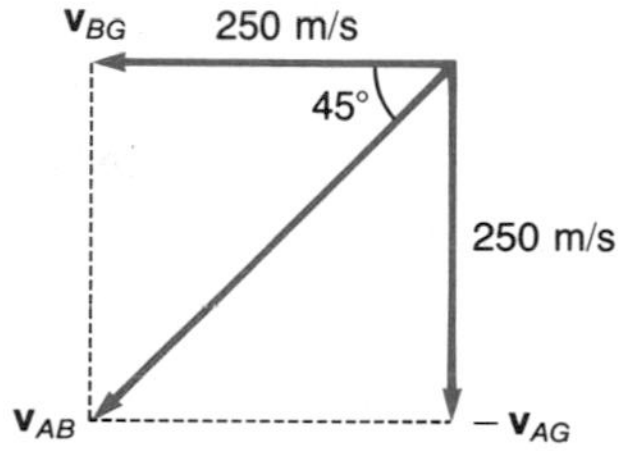

SOLUTION The data for the problem are:

$$\mathbf{v}_{AG} = 250 \text{ m/s (north)}$$
$$\mathbf{v}_{BG} = 250 \text{ m/s (west)}$$

Eq. (3.8) gives:

$$\mathbf{v}_{BG} = \mathbf{v}_{BA} + \mathbf{v}_{AG}$$

Transposing, this becomes:

$$\mathbf{v}_{BA} = \mathbf{v}_{BG} - \mathbf{v}_{AG}$$
$$= \mathbf{v}_{BG} + (-\mathbf{v}_{AG})$$

But $-\mathbf{v}_{AG}$ = 250 m/s (south) since the negative sign requires us to reverse the direction of the vector. The magnitude of the required velocity is found (as shown on the diagram) to be:

$$v_{BA} = 2 \times 250 \text{ m/s} \times \cos 45°$$
$$= 354 \text{ m/s}$$

and the direction is southwest. ■

Relative velocity problems also arise when an object is moving in a medium that is itself moving relative to some reference frame; examples of this type of situation are a boat or a swimmer in a flowing stream or an aircraft in a wind.

EXAMPLE 3.11

A swimmer in a river is heading due north relative to the water with a velocity of 1.2 m/s. The river is flowing northeast with a velocity of 0.6 m/s. Calculate the magnitude and direction of the velocity of the swimmer relative to the river bank.

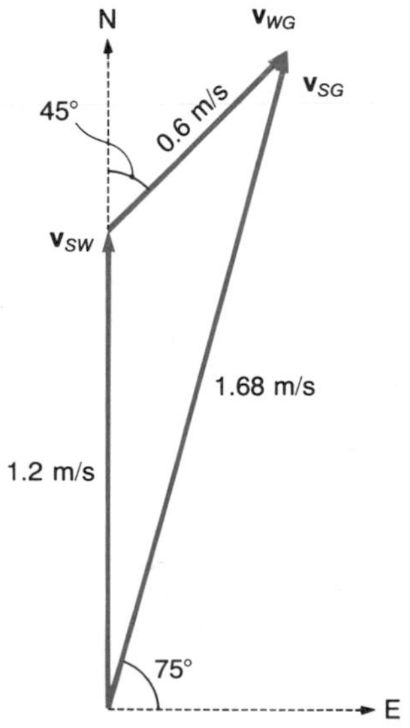

SOLUTION Using subscripts *S*, *W*, and *G* for the swimmer, the water, and the river bank, respectively, we have:

$$\mathbf{v}_{SW} = 1.2 \text{ m/s (north)}$$
$$\mathbf{v}_{WG} = 0.6 \text{ m/s (northeast)}$$

Eq. (3.8) becomes:

$$\mathbf{v}_{SG} = \mathbf{v}_{SW} + \mathbf{v}_{WG}$$

The addition of these vectors can be done either analytically or graphically. Let us use a graphical method and select 1 cm to represent 0.1 m/s on the velocity vector diagram. A directed line 12 cm long to the north then represents $\mathbf{v}_{SW}$, and a directed line 6 cm long to the northeast represents $\mathbf{v}_{WG}$. The swimmer's velocity relative to the bank is then $\mathbf{v}_{SG}$, which is measured from the diagram to be a line 16.8 cm long at 75° with the easterly direction. Since 1 cm represents 0.1 m/s, the swimmer's velocity relative to the bank is 1.68 m/s at 75° N of E. ■

KEY CONCEPTS

A vector quantity always requires **magnitude** and **direction** to be specified. A vector in two-dimensional space has two very useful descriptions:

1. The vector **magnitude** together with the **angular direction** relative to some fixed direction in the plane
2. The magnitudes and directions of the **rectangular components** in the plane

To calculate the components from the magnitude and direction, use Eq. (3.2). To calculate the magnitude and direction from the components, use Eq. (3.4).

Displacement is the most basic example of a vector in two dimensions. It is represented by a directed straight line from the beginning point to the end point of the motion. The magnitude of the displacement is the length of the line, and the direction is the angle that the line makes with a fixed direction.

Average velocity is displacement divided by elapsed time; see Eqs. (3.5) and (3.5a). **Average acceleration** is change in velocity divided by elapsed time; see Eqs. (3.6) and (3.6a).

The kinematics of a projectile are evaluated by decomposing the motion into horizontal and vertical components. The horizontal motion component has a constant velocity, and the vertical motion component experiences a constant downward acceleration due to gravity. See Eq. (3.7).

QUESTIONS FOR THOUGHT

1. Can the magnitude of a vector be less than the magnitude of one of its rectangular components?

2. What is the condition on two vectors for their sum and their difference to have equal magnitudes?

3. What is the condition on two vectors for their sum and their difference to point in the same direction?

4. Two vectors of equal magnitude add to produce a resultant that is equal in magnitude to each of them. What is the relative magnitude of the difference of the two original vectors?

5. Does an object that undergoes a constant acceleration necessarily move in a straight line?

6. Is it possible for a particle to have constant velocity with nonzero acceleration?

7. What are the relative directions of the velocity and acceleration vectors at the top point in the path of a projectile?

8. At what point on the path of a projectile does the speed have a minimum value? At what points does it have a maximum value?

9. You are asked to describe the changes in magnitude and direction of the displacement, velocity, and acceleration vectors of a projectile throughout its flight. Which of these three descriptions is so simple that you give it immediately?

10. Should a broad jumper favor a flat trajectory or one that rises higher? Give reasons to support both opinions and discuss the various factors involved.

PROBLEMS

A. Single-Substitution Problems

1. A woman walks east along a street for 800 m and then turns north; she walks north until the magnitude of her displacement from her starting point is 1400 m.

a. How far north does she walk?
b. What is her bearing from her original position? [3.1]

2. A man climbs a staircase in a building to a point 16 m vertically above his starting point; he then walks 28 m along a straight hallway. What is the magnitude of his displacement from his starting point? [3.1]

3. A ship is 8 mi east of a straight coastline that stretches in a north–south direction, and it observes that a lighthouse on the coast has a bearing of 47° W of N. The ship continues steaming, and at a later time the lighthouse is due west and 14 mi away. What is the displacement of the ship between these two sightings? [3.1]

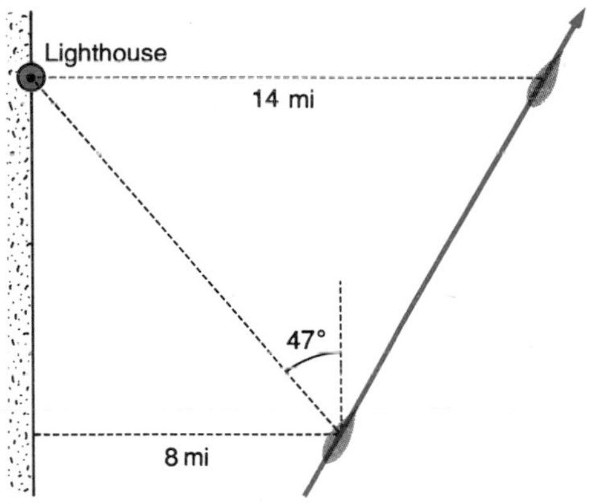

4. The components of two displacement vectors are given by:

$$R_{1x} = 5\text{ m},\ R_{1y} = 3\text{ m}$$

$$R_{2x} = 4\text{ m},\ R_{2y} = 9\text{ m}$$

Calculate the magnitude and direction of the vector $\mathbf{R}_1 + \mathbf{R}_2$. [3.2]

5. Calculate the magnitude and direction of the vector $\mathbf{R}_1 - \mathbf{R}_2$ for the components given in problem 4. [3.2]

6. Two vector displacements have magnitudes and angles (measured with respect to the positive direction of the x-axis) as follows:

$$R_1 = 8\text{ m},\ \theta_1 = 45°$$

$$R_2 = 10\text{ m},\ \theta_2 = 150°$$

Calculate the components of the vector $\mathbf{R}_1 + \mathbf{R}_2$. [3.2]

7. Calculate the components of the vector $\mathbf{R}_1 - \mathbf{R}_2$ for the magnitudes and directions given in problem 6. [3.2]

8. A point particle is displaced from the position

$$\mathbf{R}_1\ (R_1 = 6\text{ m},\ \theta_1 = 30°)$$

to the position

$$\mathbf{R}_2\ (R_2 = 12\text{ m},\ \theta_2 = 60°)$$

in an elapsed time interval of 2 s. Calculate the magnitude and direction of the average velocity. [3.3]

9. Calculate the magnitude and direction of the average velocity of a point particle that is displaced from position

$$\mathbf{R}_1\ (R_{1x} = 6\text{ m},\ R_{1y} = -3\text{ m})$$

to the position

$$\mathbf{R}_2\ (R_{2x} = -8\text{ m},\ R_{2y} = 4\text{ m})$$

in an elapsed time interval of 4 s. [3.3]

10. A sailboat making 3 m/s 30° E of N tacks and changes to a new course 30° W of N at the same speed. The whole operation takes 45 s. Calculate the magnitude and direction of the average acceleration. [3.3]

11. A zig-zagging automobile changes its velocity from 15 m/s in a northwesterly direction to 10 m/s in an easterly direction over a time interval of 8 s. Calculate the magnitude and direction of the average acceleration. [3.3]

12. A ball bearing rolls off a level surface 4 m above ground level at a speed of 2.5 m/s. What are the horizontal and vertical components of its velocity just before it strikes the ground? [3.4]

13. A stone thrown horizontally from a building 16 m high reaches the level ground 16 m from the base of the building. Calculate the initial velocity of the stone. [3.4]

14. A stone is thrown with a initial velocity of 15 m/s at an angle of 30° above a flat horizontal surface. To what maximum height does the stone rise? [3.4]

15. A stone thrown at an angle of 45° with the horizontal from a flat surface reaches a maximum height of 20 m above the surface. Calculate the magnitude of the initial velocity of the stone. [3.4]

B. Standard-Level Problems

16. A backpacker on an expedition needs to reach a point 5 km northwest of her position. In order to do so, she must walk 2 km northeast (because of rough terrain) before she can take a direct path to her destination.

a. What is the new bearing of her destination?
b. How far does she still have to walk? [3.2]

17. A girl cycles 1.5 km to the east and 0.5 km to the north; she then dismounts and walks 800 m in a direction 30° S of W. What are her distance and bearing from her original position? [3.2]

18. A sailboat travels at 3 m/s on a course 30° N of W for 3 min; it then tacks and maintains a course 30° W of N at the same speed for 4 min. Finally, it turns and runs due SW before the wind for 8 min at a speed of 4 m/s. What are its distance and bearing from the initial position? [3.2]

19. A running back runs straight ahead until he reaches the line of scrimmage. He then veers 40° to the left and runs for 5 yd; at this point he turns 70° to the right and is tackled after running 5 yd further. How many yards does he gain? [3.2]

20. At a racetrack the horses race ⅛ mi due west from the starting gate to the finish post and then 1 mi around the oval track to the finish post again. The winning time is 1 min 50 s. Calculate the average speed and the average velocity in miles per hour. [3.3]

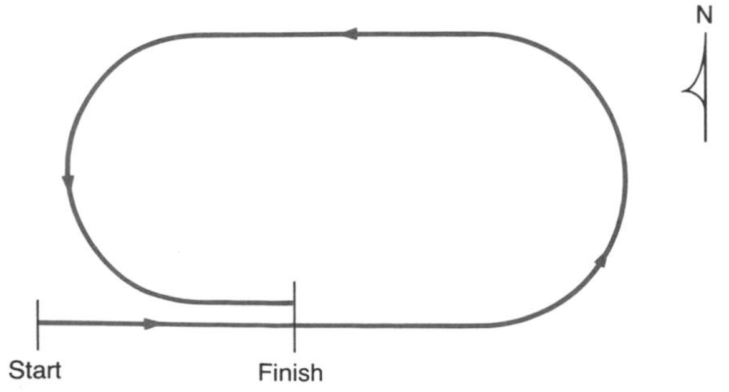

21. A 400-m running track consists of two straight 55.5-m sections and two circular end sections having a 46-m radius. A runner competing in the mile race passes the point P_0 at a constant speed of 6 m/s, which she maintains for the next lap (running counterclockwise). Calculate her average velocity and average acceleration over the interval from point P_0 to point P. [3.3]

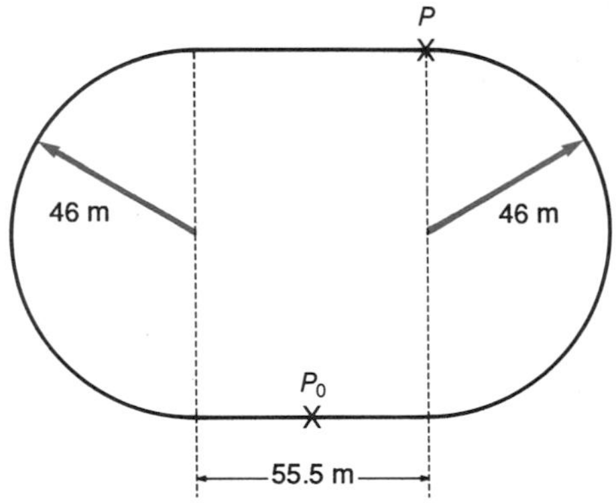

22. An automobile starts from rest and backs up 120 m in a southerly direction along a driveway to a street that runs east–west. The automobile turns and proceeds east on the street, reaching a speed of 50 km/h after traveling 200 m along the street. The time taken for the whole maneuver is 1½ min. What are the average velocity and the average acceleration during this interval?

23. A stone is thrown with a velocity of 20 m/s at an angle of 30° above the horizontal. Its velocity 1.5 s later is 17.95 m/s, and its path makes an angle 15.23° below the horizontal. Calculate the average acceleration over this period.

24. A stone is thrown from a bridge at an angle of 30° below the horizontal with a velocity of 25 m/s. If the stone hits the water 2.5 s later, how high is the bridge above the water level? [3.4]

25. Two tall buildings of equal height stand with vertical walls 40 m apart. A stone is thrown from the top of one at an angle 25° below the horizontal and directly at the other building with a velocity of 18 m/s. How far down from the top does it strike the opposite wall?

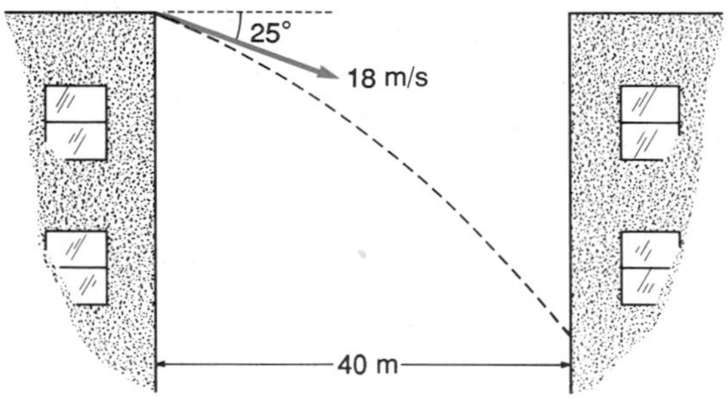

Figure for problem 25

26. A punter kicks a football that rises to a maximum height of 24 m and is caught 55 m downfield. What are the vertical and horizontal components of velocity when it leaves his boot? [3.4]

27. A ball thrown at an angle of 48° above the horizontal passes through an open window 11.4 m above the point of projection. If the ball is moving horizontally as it passes through the window, what is the velocity of projection?

28. A golfer wishes to chip a ball on level ground so that it will land 50 m in front of her. What should be the magnitude of the initial velocity of the ball if its direction is inclined at 50° with the horizontal?

29. A motorcycle stunt rider can reach her takeoff point at the top of a 10° slope with a speed of 108 km/h (30 m/s). How many buses, each 3.2 m wide, should she attempt to jump over if she lands on a point that is at the same height as her takeoff point? (Neglect air resistance—but she had better not!)

30. An athlete throws a shotput with a velocity of 14 m/s at an angle of 30° with the horizontal. If the point of release is 2 m above the ground, what is the distance of the throw?

31. A gun is fired from 3 m above the level ground at an angle of 5° above the horizontal. The velocity of the bullet is 300 m/s.

a. How long does the bullet take to hit the ground?
b. What is the horizontal range of the bullet? (Ignore air resistance.) [3.4]

32. A stone is thrown from the top of a vertical cliff at an angle of 30° above the horizontal. The cliff top is 26 m above water level, and the stone hits the water 56 m from the base of the cliff. Calculate the following:

a. The initial velocity of the stone
b. The total time of flight
c. The height of the stone above water level at the top of its trajectory

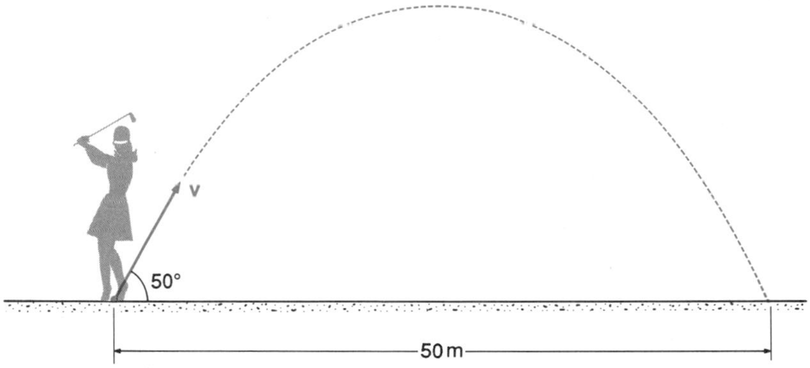

Figure for problem 28

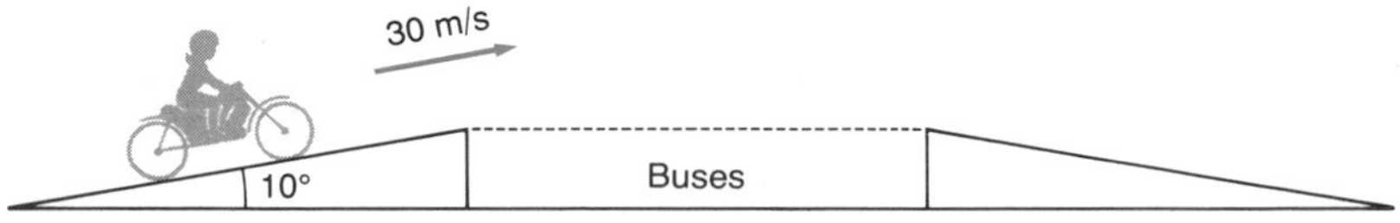

Figure for problem 29

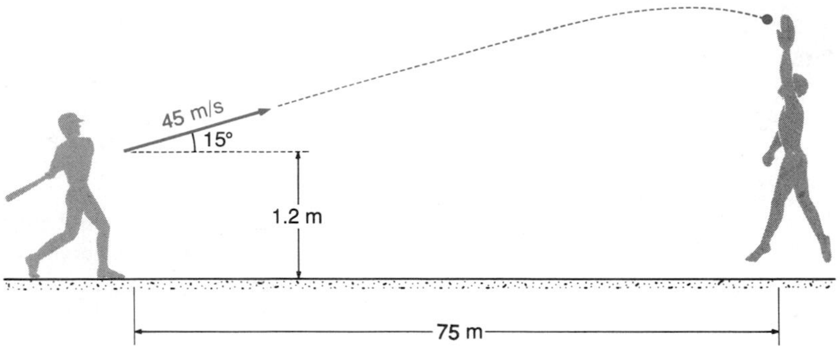

Figure for problem 33

33. A baseball is struck 1.2 m above ground level with a velocity of 45 m/s at a angle of 15° above the horizontal and straight toward a baseball fielder standing 75 m from home plate. By leaping he can catch a ball 3 m above ground level. Can he catch this ball without moving backward? (In working this problem ignore air resistance, but bear in mind that ignoring it is probably not a good idea.)

34. A particle moving on a flat horizontal surface has an initial velocity of 10 m/s in a northeasterly direction and a constant acceleration of 2 m/s^2 in a southerly direction. Calculate the magnitude and direction of its velocity 3 s later.

35. A jetliner's captain wishes to proceed due west. The cruising speed is 550 mi/h in still air, and the meteorology report indicates a 50-mi/h wind from the southeast.

a. In what direction should the course be set?
b. What is the rate of westerly progress?
c. How long does it take to reach a destination 1000 mi from the starting point?
d. How long will it take to make the round trip if the wind velocity and direction remain constant? [3.5]

36. A motorboat is capable of making 2.5 m/s in still water. The current in a river is flowing due south, and the helmsman finds he must steer the boat 40° N of W in order to make progress in a direction due west.

a. What is the speed of the current?
b. If the river is 600 m wide, how long does the boat take to cross over? [3.5]

37. A long straight section of a river is flowing uniformly at 0.8 m/s, and a swimmer on the bank notices a child's ball floating downstream directly opposite his position on the bank and 60 m out from the bank. The swimmer can make 1.2 m/s in still water.

a. How long does it take the swimmer to reach the ball by the most direct path?
b. When the swimmer reaches the ball, what is his displacement from his starting point?
c. Relative to the river bank, what is his velocity during the swim?

C. Advanced-Level Problems

38. The S.S. Seaspray heading due west at a speed of 8 knots sights the S.S. Seawind heading northwest at 10 knots. At the time of sighting, the ships are 7 nautical miles apart, and Seawind is observed to be 60° S of W from Seaspray.

a. What are the magnitude and direction of the velocity of Seawind relative to Seaspray?
b. At what time after the sighting does Seawind cross the course path of Seaspray?
c. How far ahead or astern of Seaspray is Seawind when it crosses Seaspray's course path?

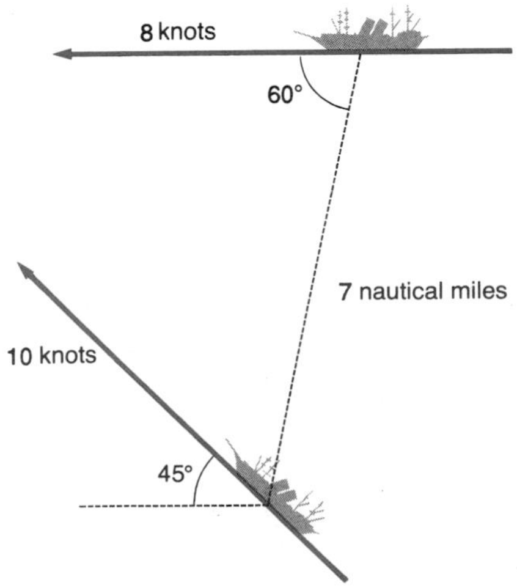

39. A golf ball is chipped with an initial velocity of 20 m/s along a level fairway.

a. What angle should the initial velocity make with the horizontal for the maximum height to be equal to the horizontal distance on the fly?

b. What is this horizontal distance?

40. A ball is thrown toward a vertical wall from a point 2 m above the ground and 3 m from the wall. The initial velocity of the ball is 20 m/s at an angle 30° above the horizontal. The collision with the wall is perfectly elastic; that is, the vertical velocity component remains unchanged and the horizontal velocity component is reversed. How far behind the thrower does the ball hit the ground?

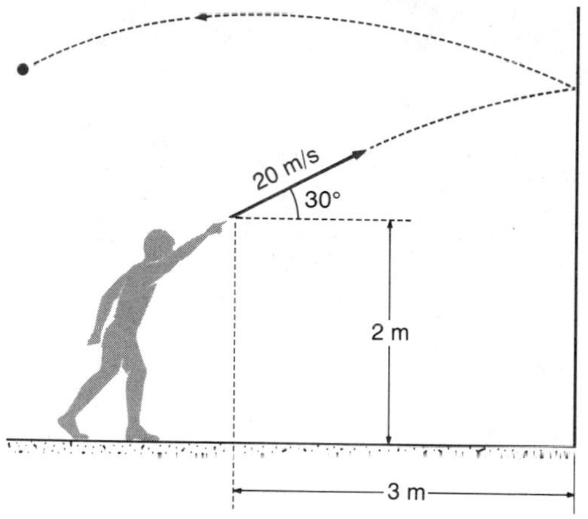

41. A skier takes off a ski jump with a velocity of 25 m/s at an angle of 30° above the horizontal. The takeoff point is 2 m above the slope, which falls away in front of her 15° below the horizontal. Treating the skier as a point particle and ignoring air resistance, at what distance down the slope does she land?

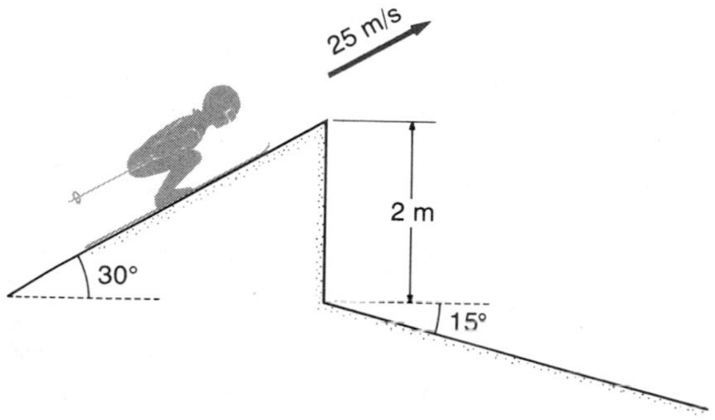

4

NEWTONIAN DYNAMICS

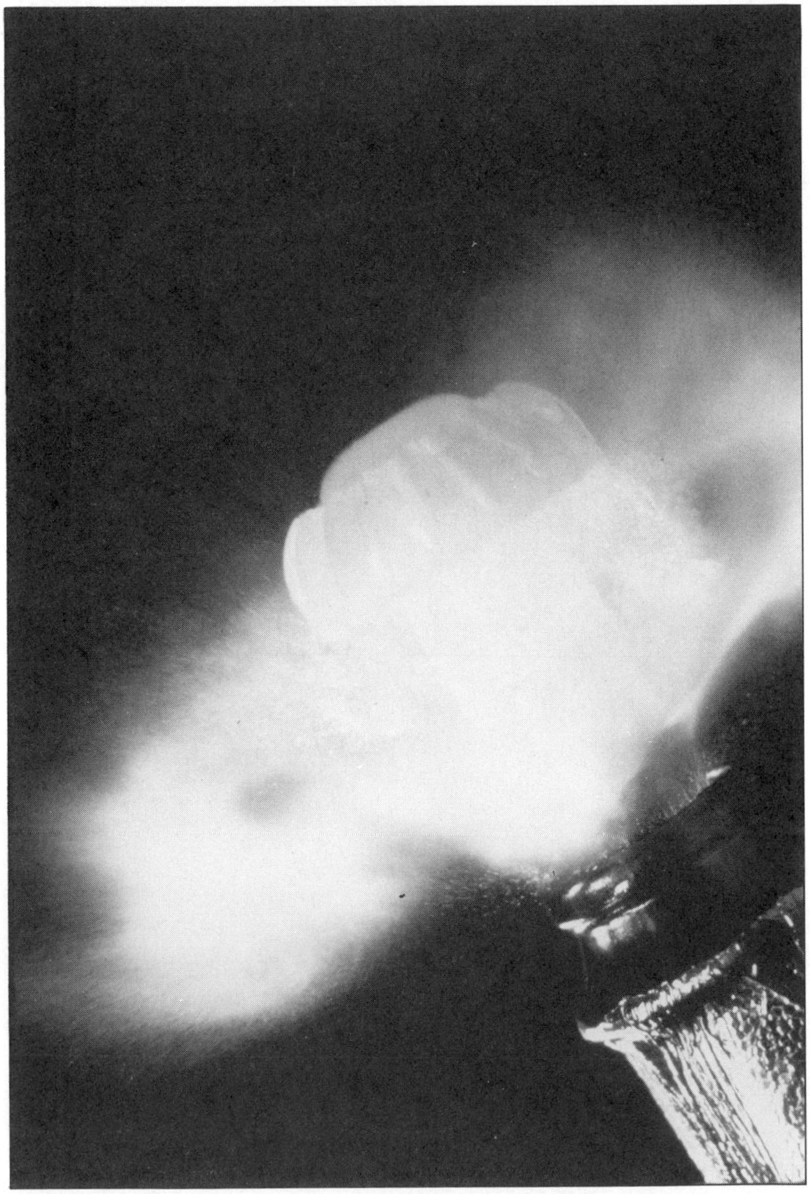

In kinematics we examined the motion of point particles without any reference to the cause of the motion. Beginning with the concepts of displacement and elapsed time, we formulated precise definitions of velocity and acceleration. However, experience teaches us that some external stimulus is needed to cause acceleration. A bicycle accelerates along the street only if we pedal it, and an automobile accelerates because of the propulsive effort of its engine. The study of the causes of motion is called "dynamics," from the Greek word for strength. We can summarize the whole point of Newton's dynamical laws of motion in a single phrase: "Force is the cause of acceleration."

4.1 The Law of Inertia

The dynamical laws of motion that bear his name were proposed by Isaac Newton (1642–1727). Newton was born close to the year in which Galileo died, but there is no doubt that his work owed much to his famous predecessor. Indeed, Newton himself attributed his extraordinary insight into dynamics to the groundwork laid by others: "If I have seen further than other men, it is because I stood on the shoulders of giants." Whatever may be the truth of Newton's humble remark, the system of dynamics that he proposed completely eclipsed all previous notions on the subject. Three centuries later it remains an underpinning of physics.

Anyone who has pedaled a bicycle or run a race is tempted to say that force is the cause of motion, but we must be wary of jumping to false conclusions. We already have precise kinematic concepts for displacement, velocity, and acceleration. To have a precise dynamics, we must be specific about the role of force relative to the kinematic quantities. Indeed, Newton began his dynamics with a law (sometimes called the **law of inertia**) that begins to clarify this point; it states a type of motion that requires no force:

Newton's First Law of Motion

> Every body continues in its state of rest or of uniform velocity in a straight line unless compelled to change that state by an external force.

It is intuitive that a state of rest continues in the absence of a force, but it is not nearly so obvious that a state of uniform velocity in a straight line persists without force. Many people think that a force is required to keep the movement going. Since in a sense they are right, the law of inertia warrants a closer look.

SPECIAL TOPIC 4 NEWTON

Isaac Newton was born at Woolsthorpe in Lincolnshire, England in 1642. He was raised by his grandparents, and sent to Cambridge University by his uncle. His great discoveries in gravitation were made during a long stay on the Lincolnshire farm that was necessitated by an outbreak of plague at Cambridge.

Before he was 30 years old, Newton was appointed professor of mathematics at Cambridge. He was coinventor of calculus with the German mathematician Leibnitz, each working independently; he invented the reflecting telescope (thereby doing away with chromatic aberration), and made many other discoveries in the field of optics. However, Newton's greatest claim to fame must rest with his three laws of motion that were formulated in his book *Mathematical Principles of Natural Philosophy*. Prior to his time, physics was called natural philosophy, and was a branch of that discipline. Newton probably did more than any other person in the history of science to establish the relationship between physics and mathematics.

In his later years he wrote a great deal of speculative theology, and eventually abandoned his scientific work to become master of the mint. Newton died in 1727, and was buried in Westminster Abbey, London, England.

Imagine a skater gliding on flat, smooth ice or a cyclist freewheeling along a flat road—both go effortlessly for a long way, but eventually both stop. Newton claims that the decrease of the original velocity to zero proves the existence of a deceleration and that a force is required to explain this deceleration. This force is the small frictional force between the skates and the ice or the air resistance and the road and bearing friction on the bicycle. The first law asserts that the uniform velocity would continue indefinitely if the friction were reduced to zero.

We see now that the first law refers to an ideal situation that can never quite occur in practice. The intuitive idea is partly correct—we cannot reduce friction to zero, so a truly force-free motion is not a practical possibility. Does this imply that Newton's law is only approximately true? By no means! It is an exact statement about a limiting situation that never quite occurs:

> Force is the cause of acceleration; uniform velocity requires no force.

The major significance of the law of inertia was to cast aside centuries of misapprehension and to set things straight so that calculations could begin. Aristotle speculated at some length on the cause of a javelin's motion after it left the hand of the thrower, and the thinking of later centuries became set in this mold, that is, seeking for the cause of motion. But the term *motion* is rather general, and we have carefully distinguished among displacement, velocity, and acceleration. With the help of the first law, we know that the acceleration of the javelin during its flight is due to external forces. Indeed, we have already studied this very topic in Galileo's analysis of projectile motion. The horizontal velocity component remains unchanged throughout the flight of a projectile (if air resistance is negligible) because there is no force component in a horizontal direction. On the other hand, the vertical velocity component changes because of the acceleration due to gravity, which implies a vertical force component acting on the projectile. The law of inertia therefore belongs at least partly, if not wholly, to Galileo.

4.2 The Definition of Force

Newton set the stage for his dynamics by naming force as the cause of acceleration. But what really is force? We can easily give examples, such as the downward pull of the earth's gravity on a stone, the push of the wind on a sail, or the resistance on the hull of a ship moving through water. A detailed explanation of any of these forces is immensely complex. Newton avoided this difficulty by precisely defining force in terms of the acceleration that it causes in the mass on which it acts. Newton's definition equates force to the product of the fundamental quantity mass and the derived quantity acceleration.

The motion of this catamaran is determined by Newton's laws as a consequence of the forces of the wind, the water, and gravity all acting on it. The catamaran problem is exceedingly complex, but we begin to understand Newton's laws by studying relatively simple cases. (Courtesy of Coast Catamaran Corporation)

Newton's Second Law of Motion

Definition of force

$$\text{Total net force on an object} = \text{mass of the object} \times \text{acceleration caused by the net force}$$

$$\mathbf{F} = m\mathbf{a} \qquad \textbf{(4.1)}$$

SI unit of force: kilogram-meter per second per second
New name: newton
Abbreviation: N

The net force is equal to the scalar mass multiplied by the vector acceleration. It is therefore a vector having the same direction as the acceleration vector. Moreover, the net force is the sum total of all the forces acting on the mass under consideration. Eventually we need methods for identifying and calculating all these forces, which are all vectors, and the net force is their sum.

Newton's second law assigns a crucial dynamical role to the fundamental quantity mass; in fact, the second law quantifies the

SPECIAL TOPIC 5 Scaling and Vertical Jumping

In Chapter 1 we used the concept of scaling to discuss the way in which the skeletal structures of animals of roughly similar shape change with size. We can employ the same line of reasoning to compare the vertical jumping ability of animals.

Consider a standing vertical jump in which the animal crouches and then leaps vertically. Let v be the takeoff velocity and h the height of the jump. After takeoff the animal experiences a downward acceleration g, and the use of Eq. (2.12d) readily gives $h = v^2/2g$. Since g does not change, the ability of different animals in standing vertical jumps should scale with the square of their takeoff velocity.

Let L be the characteristic dimensionless scale factor introduced in Chapter 1. We need to examine the act of crouching and achieving the takeoff velocity in terms of this scale factor. Let us assume that the strength of the animal to exert an upward force on itself and produce the required acceleration is proportional to the cross-sectional area of its muscle tissue. This cross-sectional area scales as L^2, and consequently so does the upward force that the animal can exert. However, the mass of the animal (proportional to volume for animals of roughly the same density) scales as L^3. Eq. (4.2) shows that acceleration is a force divided by mass, and it follows that acceleration scales as L^{-1}. This means, of course, that the larger animals produce smaller upward accelerations from the crouching position. Finally, we note that the displacement from the crouch to the takeoff position would be expected to scale as L. Combining this in Eq. (2.12d) with an acceleration that scales as L^{-1}, we find that the takeoff velocity is unaffected by scaling. It follows that the height attained in standing vertical jumps is also unaffected by scaling.

We have stipulated that animals be of roughly similar shape and size for scaling to apply, but in this instance we must reflect that the hind legs are of paramount importance in the crouching leap. Thus, we are not too surprised to learn that a man can make a standing jump of about 1 meter, whereas a giant kangaroo (with its very powerful hind legs) can jump about 2½ times as high. What is much more significant is that members of the small species of kangaroo can almost equal the leaping ability of their giant brethren, even though the dimensionless scale factor between the two species is about 5.

Our scaling argument again has the flaws that we discussed in Chapter 1. However, in spite of these flaws, it helps to explain why a small animal (such as a squirrel) and a large animal (such as a horse) can both make vertical leaps of approximately a meter.

first law. If there is no net force there is no acceleration, but a given net force produces an acceleration whose magnitude depends on the mass of the object. We must be careful in our interpretation of everyday experiences with respect to the second law. It is tempting to give instances in which a larger force is required to push a more massive object than to push a less massive one. Consider, for example, two large crates of merchandise standing on a level floor, one twice as massive as the other. Experience shows that a larger force is required to push the more massive crate across the floor. At this time, we must carefully reflect on what Newton's second law really implies. If two objects of different masses have the same *acceleration*, a larger *net force* is required for the more massive one. The horizontal forces acting on the packing crate are the push applied to move it and the opposing frictional force between the crate and the floor. Thus, identifying the net force on the crate is not as simple a matter as it might seem to be. For the moment, let us look at examples in which the actual nature of the net force is unspecified.

EXAMPLE 4.1

A certain net force acting on a 5-kg mass produces an acceleration of 2 m/s^2. Calculate the force.

SOLUTION The problem is one dimensional. Equation (4.1) relates the magnitudes of the force and the acceleration:

$$\begin{aligned} F &= ma \\ &= 5\text{ kg} \times 2\text{ m/s}^2 \\ &= 10\text{ kg m/s}^2 = 10\text{ N in the same direction as the acceleration} \end{aligned}$$

■

Newton's first law equates the absence of acceleration with zero net force, and indeed this situation is the simplest instance of Eq. (4.1). If $F = 0$, then $a = 0$ for any value of m. Another simple situation arises if the net force acting on a mass is constant. In this case the acceleration is also constant; Eq. (2.12) describes the motion kinematically. The equations of uniformly accelerated motion are therefore the kinematic equations that describe the motion of a mass subjected to a constant net force.

EXAMPLE 4.2

A cyclist and her machine have a combined mass of 80 kg. Calculate the constant net propulsive force required to produce a velocity of 12 m/s in 4 s when the machine starts from rest.

SOLUTION To calculate the force we need to know the acceleration. It is necessary to solve the kinematic problem first. (1) Set up the data for the motion:

$$\begin{aligned} v_0 &= 0 \\ v &= 12\text{ m/s} \\ t &= 4\text{ s} \\ a &= ? \end{aligned}$$

We calculate the acceleration using Eq. (2.12a):

$$v = v_0 + at$$

$$\therefore a = \frac{v - v_0}{t}$$

$$= \frac{12\ \text{m/s} - 0\ \text{m/s}}{4\ \text{s}}$$

$$= 3\ \text{m/s}^2$$

(2) We can now solve the dynamics with the help of Eq. (4.1):

$$F = ma$$

$$= 80\ \text{kg} \times 3\ \text{m/s}^2$$

$$= 240\ \text{N in the same direction as the acceleration}$$

The 240-N force is not the force that the cyclist exerts on the pedals. Rather, it is the net propulsive force on the cyclist and her machine, which arises from frictional interaction between the rear wheel and the road surface. ■

Strictly speaking, the form given for Newton's second law applies only when the accelerating object is of constant mass. Modification is needed for a body, such as a rocket, which sheds part of its mass as it accelerates. Newton himself gave the correct general definition. However, since we will treat only bodies of constant mass, the simplified form Eq. (4.1) will suffice.

4.3 The Equality of Action and Reaction

The real-life example of pushing a packing crate across a floor illustrates an important point about forces—the forces on the crate arise because of interaction of the crate with other bodies. The force that propels it across the floor comes from the interaction between the crate and the person pushing it, and the frictional force arises from the interaction between the crate and the floor. The situation illustrated by these examples is quite general. Any force that an object experiences always results from the action of some other object on it. Newton's third law deals with forces between objects.

Newton's Third Law of Motion

> If one object exerts a force on a second object, then the second object exerts a force on the first of equal magnitude but oppositely directed.

These equal magnitude forces between two objects are often called action-reaction pairs. One of the forces can be thought of as the action of one object on the other, and the other force as the reaction of the second object on the first. With this understanding we can write the third law as follows:

Newton's Third Law of Motion (Alternate Formulation)

> The two forces of an action-reaction pair are of equal magnitude and oppositely directed.

Note that the third law claims that **forces** occur only in pairs of equal magnitude and opposite direction—a single unopposed force is impossible in the world of reality. One may wonder how anything ever moves if all forces occur in equal and opposite pairs; the answer is that *the forces of action and reaction always act on different bodies.*

Consider Newton's own example, illustrated in Fig. 4.1. A horse pulls a stone forward by means of a horizontal rope at-

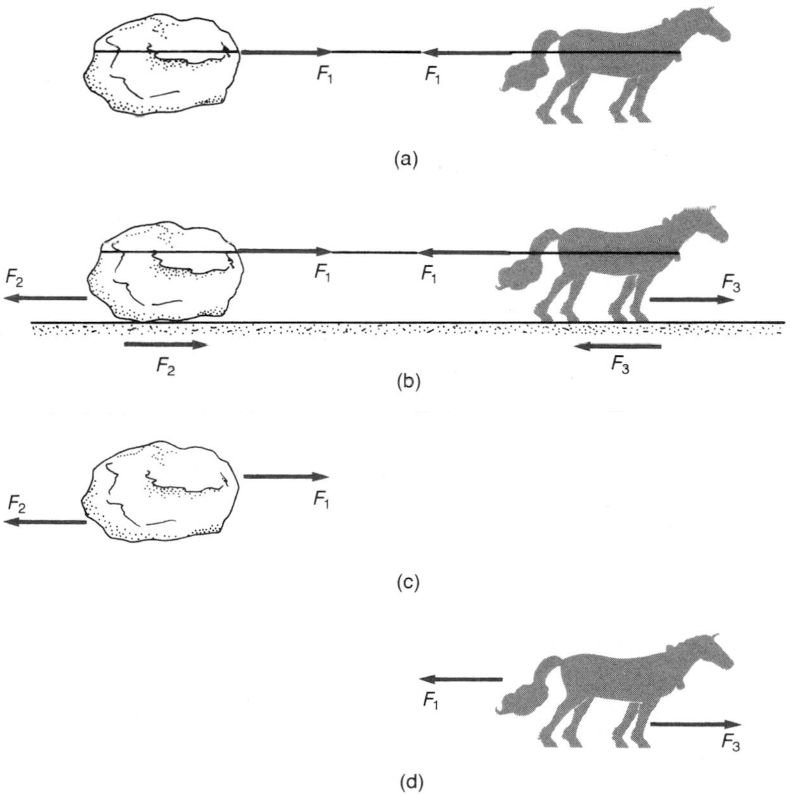

FIGURE 4.1 Diagram to illustrate the action-reaction law: (a) the horse and the stone; (b) the horse, the stone, and the earth; (c) the stone; and (d) the horse.

tached to the stone and the horse. The third law tells us that the horse pulls forward on the stone with a force equal to that which the stone exerts backward on the horse—this is an action-reaction pair of magnitude, say, F_1, as shown in Fig. 4.1(a). Then how does the horse ever move the stone? Experience may tempt you to say that the stone moves when the horse pulls a little harder, but this does not remove the difficulty, since the stone then pulls a little harder backwards. In fact, the horse never could move the stone if both were drifting in space, as Fig. 4.1(a) implies. In any real situation both the horse and the stone are resting on the earth, and we must consider all the action-reaction pairs between the various objects, including the frictional forces between the stone and the earth (of magnitude F_2) and between the horse and the earth (of magnitude F_3). We show all of these action-reaction pairs in Fig. 4.1(b). At this point we recall that the second law relates the acceleration of an object to the *net force* acting on it. We must now sketch each object separately and draw the forces acting on it. Figure 4.1(c) shows the forces acting on the stone, and Fig. 4.1(d) shows the forces acting on the horse. These diagrams are sometimes called *free-body force diagrams,* since they illustrate an isolated body subject only to the parts of action-reaction pairs that act on it.

We can now analyze the situation in terms of forces acting on each body individually. The net force on the horse is $(F_3 - F_1)$ in the forward direction. By pushing with its legs, the horse increases the magnitude of F_3 until $(F_3 - F_1)$ is positive. The horse than accelerates forward. At the same time, the net forward force on the stone, $(F_1 - F_2)$, also becomes positive. The stone accelerates forward at the same rate as the horse. We thus have an explanation for motion in terms of action-reaction force pairs. The action force is equal to the reaction force at all times. However, since both forces act on different objects, the net force on any one object may not be zero.

EXAMPLE 4.3

Two masses of 5 kg and 2 kg are in contact with each other on a smooth, frictionless surface. A force of 7 N acts on the 5-kg mass. Calculate the magnitude of the action-reaction force pair at the interface between the two masses.

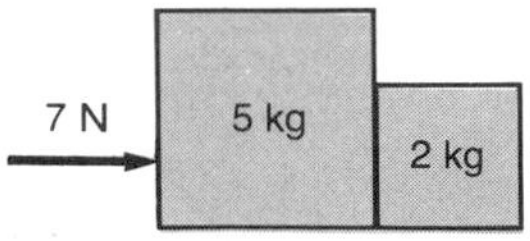

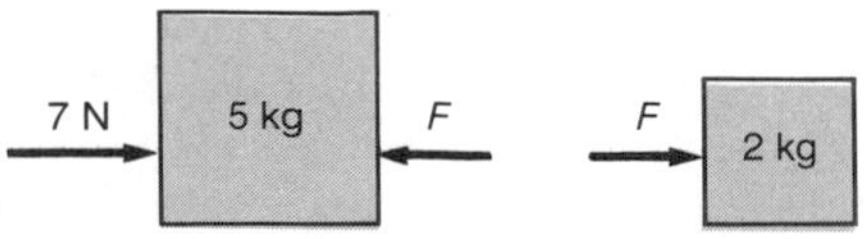

SOLUTION Draw the free-body diagram for each mass separately. Let F be the magnitude of the action-reaction force pair at the interface, and a the acceleration of the system to the right. Then Newton's second law, Eq. (4.1), applied to each block separately yields:

$$\text{Net force} = \text{mass} \times \text{acceleration}$$

$$7\text{ N} - F = 5\text{ kg} \times a \quad \text{for the 5-kg mass}$$

$$F = 2\text{ kg} \times a \quad \text{for the 2-kg mass}$$

Add these equations:

$$7\text{ N} = 7\text{ kg} \times a$$
$$\therefore a = 1\text{ N/kg}$$
$$= 1\text{ m/s}^2$$

Substitute this value of a in either of the original equations to find:

$$F = 2\text{ N}$$

for the magnitude of the action-reaction pair of forces. Note that the net force on the 5-kg mass is:

$$7\text{ N} - F = 7\text{ N} - 2\text{ N} = 5\text{ N}$$

This net force produces an acceleration of 1 m/s^2 in the 5-kg mass. For the 2-kg mass the net force is $F = 2$ N, and this produces the same acceleration of 1 m/s^2 in the 2-kg mass. ■

4.4 A Definition of Weight

Weight is a very useful concept in everyday life, even though it is not a fundamental idea in physics. However, it plays an enormous role in human affairs. We are all well aware of our own weight and the weight of objects that we carry. The weight acts to pull the objects downward toward the earth. We can use this simple notion as a starting point from which to develop a definition of weight.

In our study of the kinematics of falling objects, we discussed Galileo's discovery that all masses undergo the same downward acceleration due to gravity. However, Newton identified force as the cause of acceleration. From the second law we can say that the downward force is the mass of the object multiplied by the downward acceleration of gravity. We call this force the **weight.**

Weight is the downward force due to gravity.

Weight of an object = mass × acceleration due to gravity

$$w = mg \qquad \textbf{(4.2)}$$

SI unit of weight: newton (same as the unit of force)

The distinction between mass and weight is most important. The mass of any object is a fundamental property of that object,

The system of dynamics established by Newton enables engineers to predict the performance of this jetliner. The forces acting on it are its weight, the thrust supplied by the engines, the lift due to the wings, and the drag caused by air friction. The net force component in any direction is equal to the mass of the aircraft multiplied by the acceleration in that direction. In level flight at constant velocity the net force on the jetliner is zero. In this condition the thrust is equal and opposite the drag, and the lift is equal and opposite to the weight. (Courtesy of Trans World Airlines)

and it is directly related to the force required to cause a given acceleration. The mass remains the same whether an object is situated on the surface of the earth, on the surface of the moon, or even in interstellar space. On the other hand, the weight of an object is the force that the object experiences in some particular environment; therefore, the weight changes if the environment changes. The downward acceleration due to gravity on the surface of the moon is approximately one-sixth of that on the surface of the earth. It follows that the weight of an object on the surface of the moon is about one-sixth of its weight on earth.

However, there are also variations in weight as we move around the surface of the earth. In Table 2.1 we saw that the acceleration due to gravity varies by about 0.3% among three representative cities on the earth's surface. The weight variation corresponds exactly to the variation in the acceleration due to gravity. Since the change is small, the earth's surface is an approximately constant environment with regard to weight. Be-

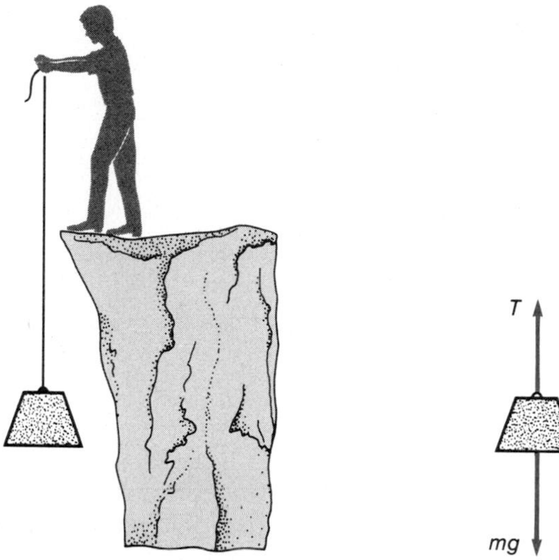

FIGURE 4.2 Free-body diagram analyzing vertical motion of an object partly supported by a rope.

cause of this approximate constancy, it is tempting to think of weight as an intrinsic property of an object. Since most of us spend our lives close to the earth's surface, everyday experience tends to support this illusion. However, as soon as our applications of physics leave the earth's surface, it becomes clear that weight is not an intrinsic property.

Physics is often applied to everyday life situations that occur near the earth's surface, and the weight in these situations is simply related to the mass by Eq. (4.2). However, an object does not fall with the acceleration g when forces other than the weight act on it. A typical situation is the raising (or lowering) of an object by means of a rope, where the tension in the rope provides an additional force on the object. We recall from our previous discussion of a horse pulling a stone that rope tension is not a single force but rather an action-reaction pair. Consider the example of a man raising (or lowering) an object of mass m by means of a rope (Fig. 4.2). The forces acting on the object are its weight mg downward and force T upward due to the tension in the rope. The net upward force on the object is $(T - mg)$. If $T = mg$, the net force is zero; Newton's first law tells us that the object is either at rest or moving with constant velocity either up or down. If T is greater than mg, the object is accelerating upward; if T is less than mg, it is accelerating downward. Let us consider a specific example.

EXAMPLE 4.4

A woman of mass 50 kg is lowered to the ground from an upper story window with a rope in which the tension is kept constant at 400 N. Calculate the acceleration of the woman during the descent.

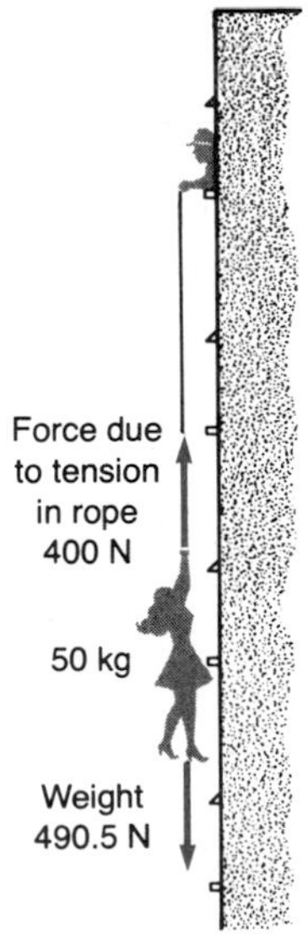

SOLUTION The free-body diagram shows the 50-kg woman subject to the downward force of her weight and the upward force due to the rope tension. We calculate the weight from Eq. (4.2):

$$\begin{aligned} w &= mg \\ &= 50\ \text{kg} \times 9.81\ \text{m/s}^2 \\ &= 490.5\ \text{N} \end{aligned}$$

Taking the upward direction as positive, the net force on the mass is given by:

$$\begin{aligned} F &= T - mg \\ &= 400\ \text{N} - 490.5\ \text{N} \\ &= -90.5\ \text{N} \end{aligned}$$

To calculate the acceleration caused by this net force, we use Newton's second law, Eq. (4.1):

$$F = ma$$

$$\therefore a = \frac{F}{m}$$

$$= \frac{-90.5\ \text{N}}{50\ \text{kg}}$$

$$= -1.81\ \text{m/s}^2$$

The negative sign reflects the downward acceleration. ■

We can use a similar technique to solve more difficult problems. In every case the key to solving dynamics problems lies in drawing a correct free-body force diagram for each body under consideration.

> The free-body force diagram must include all forces acting **on** the body, but not forces exerted **by** the body on other objects.

EXAMPLE 4.5

Two masses of 4 kg and 6 kg are connected by a light cord that passes over a light, frictionless pulley attached to a fixed support. Calculate the downward acceleration of the 6-kg mass and the tension in the cord.

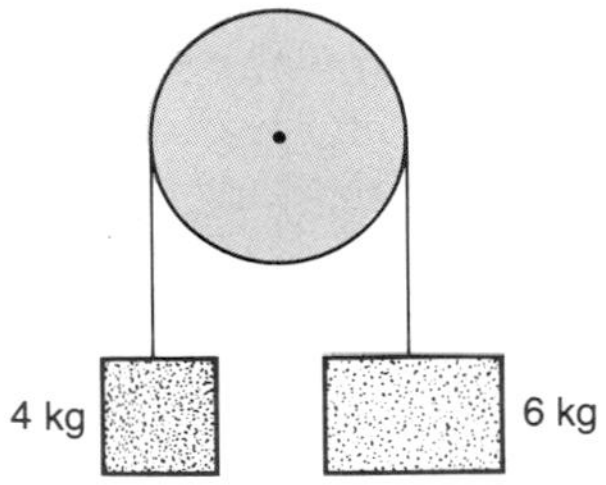

SOLUTION Since the 6-kg mass is accelerating downward, the tension in the supporting cord is not equal to its weight. The cord passes over a light, frictionless pulley, which does not change the magnitude of the tension. The same tension is therefore applied by the cord to the 4-kg mass. Let T be the tension in the cord. The free-body diagrams include an upward force T acting on each mass. The only other forces on the masses are their weights, and we use Eq. (4.2) to calculate them:

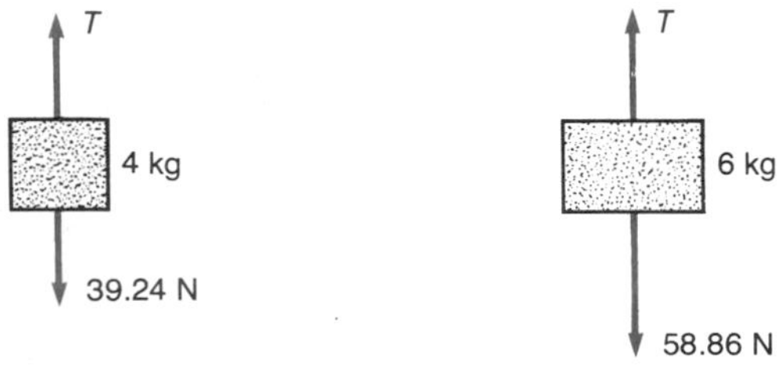

$$w = mg$$

$$\text{Weight of 4-kg mass} = 4\ \text{kg} \times 9.81\ \text{m/s}^2$$

$$= 39.24\ \text{N}$$

$$\text{Weight of 6-kg mass} = 6\ \text{kg} \times 9.81\ \text{m/s}^2$$

$$= 58.86\ \text{N}$$

Let a be the upward acceleration of the 4-kg mass; the downward acceleration of the 6-kg mass is $-a$ since the masses are connected by the cord. Let us adopt our usual convention that the positive direction is upward. Newton's second law [Eq. (4.1)] applied to each mass in turn gives us the following equations:

$$F = ma$$

$$T - 58.86\text{ N} = 6\text{ kg} \times (-a) \quad \text{for the 6-kg mass}$$

$$T - 39.24\text{ N} = 4\text{ kg} \times a \quad \text{for the 4-kg mass}$$

Subtracting the two equations gives:

$$19.62\text{ N} = 10\text{ kg} \times a$$

$$\therefore a = 1.96\text{ m/s}^2$$

Substituting this value in either of the original equations, we have:

$$T = 47.1\text{ N}$$

■

The apparatus described in Example 4.5 is called an Atwood machine after George Atwood (1746–1807), who used it to study falling bodies. The device reduces the value of the downward acceleration, making it easier to measure the time of fall.

Much of the confusion concerning weight arises from the fact that commercial transactions usually refer to mass in countries using metric units and to weight in countries using British units—butter is purchased by the kilogram in France but by the pound in the United States. For this reason it is useful to know the weight in pounds of 1 kg at an average point on the earth's surface. From Eq. (4.2):

$$\begin{aligned}\text{Weight of 1 kg} &= mg \\ &= 1\text{ kg} \times 9.81\text{ m/s}^2 \\ &= 9.81\text{ N} \\ &= 9.81\text{ N} \div 4.448\text{ N/lb} \\ &= 2.21\text{ lb}\end{aligned}$$

Although kilograms cannot be converted into pounds, since one quantity is a mass and the other a force, we can say that a mass of 1 kg has an average weight of 2.21 lb at the earth's surface.

Another term in frequent commercial use is *metric ton*, which refers to a mass of 1000 kg, in spite of the standard terminology of *megagram*. From the quasi-conversion described above, we see that a mass of 1 metric ton has a weight of about 2210 lb. The standard U.S. ton is a weight of 2000 lb; it is also in frequent commercial and engineering use.

Kinematics problems having some data in British units are not too difficult to handle because of the algebraic cancellation of units. However, in dynamics we need a whole new set of British units, with the 1-lb force as the standard. The SI system continues naturally into our study of electricity and magnetism, while the British system does not. For this reason we will abandon British units when working dynamics problems and use only SI units.

EXAMPLE 4.6

A 100-metric ton jet airliner reaches its takeoff speed of 280 km/h after traveling 1000 m of runway. What is the average thrust in pounds of each of its four jet engines?

SOLUTION We first need to determine the acceleration of the airliner by solving the kinematics problem:

$$\begin{aligned} v_0 &= 0 \text{ m/s} \\ v &= 280 \text{ km/h} \times 0.2778 \text{ (m/s)/(km/h)} \\ &= 77.78 \text{ m/s} \\ x &= 1000 \text{ m} \\ a &= ? \end{aligned}$$

Equation (2.12d) is suitable for calculating the acceleration:

$$\begin{aligned} v^2 &= v_0^2 + 2ax \\ \therefore a &= \frac{v^2 - v_0^2}{2x} \\ &= \frac{(77.78 \text{ m/s})^2 - (0 \text{ m/s})^2}{2 \times 1000 \text{ m}} \\ &= 3.025 \text{ m/s}^2 \end{aligned}$$

To calculate the force we use Eq. (4.1):

$$\begin{aligned} F &= ma \\ &= 10^5 \text{ kg} \times 3.025 \text{ m/s}^2 \\ &= 3.025 \times 10^5 \text{ N} \\ \text{Thrust from each engine} &= \tfrac{1}{4} \times 3.025 \times 10^5 \text{ N} \\ &= 7.562 \times 10^4 \text{ N} \\ &= 7.562 \times 10^4 \text{ N} \div 4.448 \text{ N/lb} \\ &= 1.70 \times 10^4 \text{ lb} \end{aligned}$$

■

4.5 Motion with Resisting Forces—Friction

When a parachutist jumps, she accelerates downward as long as her weight exceeds the upward force due to the parachute. When these two forces become equal in magnitude, the acceleration is zero, and she continues toward the ground with constant velocity.

When a body undergoes motion in the real world, it generally moves through some medium, and it frequently either slides or rolls across the surface of another medium. For example, a ship moves through the water, an airliner moves through the air, a stone slides across the ground, and an automobile rolls along the road. In all of these cases, the motion of the object gives rise to a force that opposes the motion, called a **frictional force,** which we have already mentioned in discussing the law of inertia. Not the ship, the airliner, the stone, or the automobile continues moving with constant velocity in a straight line unless a force is applied to overcome the frictional force. There is no general way of dealing with frictional force. It must be measured in each and every situation or described by some approximate relation suitable to the circumstances.

An object falling in the earth's atmosphere experiences a frictional retarding force due to air resistance, which in some circumstances may significantly affect the motion. Recall that we excluded air resistance and air buoyancy in the determination of g, and the same is true in the definition of weight. In practice air buoyancy causes an error of less than 1 part in 1000 in the weight of liquids and solids, and for many purposes this amount may be too small to worry about. For other objects (such as a helium-filled balloon) the air buoyancy force is greater than the total weight of the object, and so the object rises. For the moment we will treat only examples in which the air buoyancy force is negligible. However, air resistance may be a large factor in determining the downward acceleration of falling objects.

EXAMPLE 4.7

A skydiver of mass 60 kg finds herself falling with an acceleration of 1.8 m/s^2 downward at a given instant. Calculate the frictional force due to air resistance at this time.

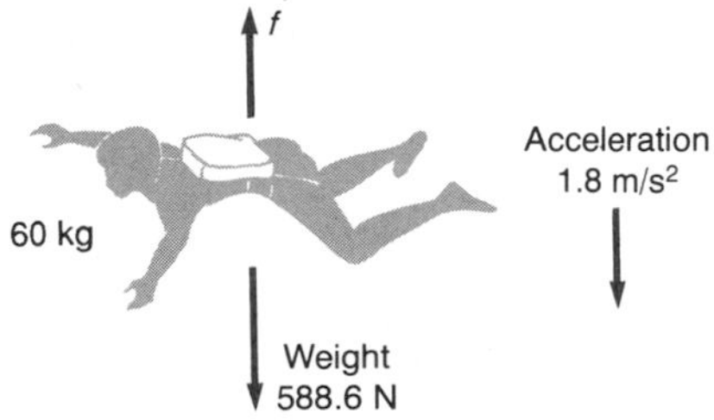

SOLUTION The free-body diagram shows the weight of the skydiver downward and the unknown frictional force upward. The weight is given by Eq. (4.2):

$$\begin{aligned} w &= mg \\ &= 60 \text{ kg} \times 9.81 \text{ m/s}^2 \\ &= 588.6 \text{ N} \end{aligned}$$

The net force is therefore (f − 588.6 N) in the upward direction. With our usual convention that positive vectors point upward, Newton's second law, Eq. (4.1), gives:

$$F = ma$$
$$f - 588.6\ \text{N} = 60\ \text{kg} \times (-1.8\ \text{m/s}^2)$$
$$\therefore f = 480\ \text{N}$$

■

In Example 4.7 we calculated the magnitude of the frictional force from a knowledge of the downward acceleration of the falling skydiver. There is no other simple way to estimate frictional forces when bodies traverse fluid media, such as air and water. However, if two solid bodies are in contact with each other at reasonably flat surfaces, we can make an approximate calculation of the frictional force resisting sliding motion of one surface across the other. The frictional force is the product of two factors: The first is a constant characteristic of the surfaces in contact, and the second is the magnitude of the normal component of one of the forces of the action-reaction force pair between the surfaces. The constant is called the **coefficient of friction,** and the magnitude of the normal component of one of the forces of the action-reaction force pair is often referred to briefly as the **normal reaction.** (The word *normal* used in this way means perpendicular to the flat surfaces.) The frictional force always opposes the sliding motion (or the tendency toward a sliding motion) and is parallel to the plane surfaces in contact.

In most cases the coefficient of friction is greater before an object begins to slide than after the motion has begun. We therefore need two coefficients of friction, one for the static case and the other for the kinetic. Let us summarize this discussion in display.

■ Definition of the Coefficient of Friction

Magnitude of maximum frictional force = coefficient of friction × magnitude of normal reaction

Static: $f_s \leq \mu_s R_n$ **(a)**

Kinetic: $f_k = \mu_k R_n$ **(b)**

(4.3)

μ_s and μ_k are pure numbers.

The force of friction between two flat surfaces is zero if no force is applied that tends to cause sliding. If we apply a gradually increasing force, then the force of friction also increases to oppose it up to the maximum value given by Eq. (4.3a). Once the sliding motion commences, the frictional force changes to the value given by Eq. (4.3b).

Note also that Eq. (4.3) is not a vector equation, but a relation between the magnitudes of two vectors. The frictional force vector is parallel to the plane of contact of the two surfaces, and the normal reaction is perpendicular to it.

TABLE 4.1 Typical values for coefficients of friction between various surfaces

	Coefficients of friction	
Surfaces	*Static*	*Kinetic*
Steel on ice	0.1	0.05
Metal on metal (lubricated)	0.1	0.05
Steel on Teflon	0.05	0.05
Wood on wood	0.5	0.3
Metal on metal (dry)	0.6	0.4
Glass on glass	0.9	0.4
Rubber tire on pavement		
Smooth tire, wet pavement	0.5	0.4
Grooved tire, wet pavement	0.8	0.7
Smooth or grooved tire, dry pavement	0.9	0.8

It is not possible to quote *accurate* values of μ_s and μ_k for given pairs of surfaces. Varying degrees of surface smoothness and surface lubrication due to small amounts of contaminating substances (such as oil or water) may greatly affect the value of the coefficient of friction. In addition, the value of the kinetic coefficient of friction may depend on the speed of the sliding motion. With these reservations in mind, Table 4.1 gives some typical values for coefficients of friction. Each entry has only one significant figure, reflecting the degree of uncertainty of these values. This uncertainty is very great for the values quoted for tires on pavement, since so many variable quantities affect the result.

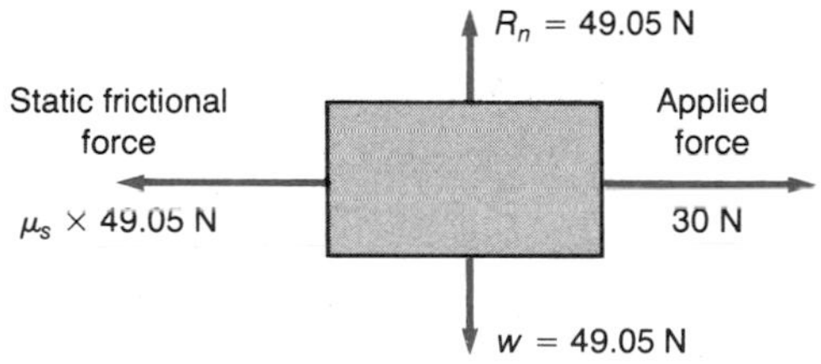

1. Block on the point of moving

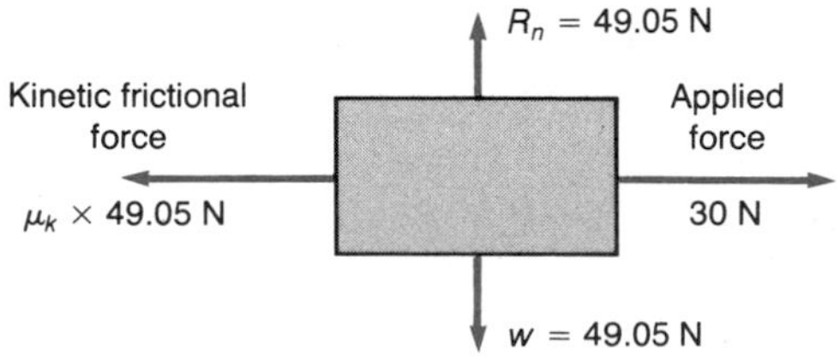

2. Block accelerating at 2.5 m/s²

EXAMPLE 4.8

A block of wood of mass 5 kg rests on a flat table, and a horizontally applied force of 30 N is just sufficient to make it slide. What is the coefficient of static friction between the block and the table? Application of the 30-N force is continued after the block starts to slide, and it accelerates at 2.5 m/s^2. What is the coefficient of kinetic friction between the block and the table?

SOLUTION The free-body diagram for the block includes forces in both the horizontal and vertical directions. Since the block always remains in contact with the table, the vertical acceleration and the net vertical force are both zero. The vertical forces on the block are the weight downward and force of the table upward. We calculate the weight from Eq. (4.2):

$$\begin{aligned} w &= mg \\ &= 5\text{ kg} \times 9.81\text{ m/s}^2 \\ &= 49.05\text{ N downward} \end{aligned}$$

It follows that the force of the table on the block is 49.05 N upward. This force, one of the action-reaction pairs of contact forces, is perpendicular to the surfaces in contact. The normal reaction is therefore given by:

$$R_n = 49.05 \text{ N}$$

In the horizontal direction there are the two different cases of static and kinetic friction.

1. If a force of 30 N is just sufficient to make the block begin to slide, the maximum static frictional force is 30 N, and Eq. (4.3a) is appropriate with the equality sign:

$$f_s = \mu_s R_n$$

$$\therefore \mu_s = \frac{f_s}{R_n}$$

$$= \frac{30 \text{ N}}{49.05 \text{ N}}$$

$$= 0.61$$

2. After the block begins to slide it accelerates, and the applied force of 30 N must exceed the force of kinetic friction. Using Eq. (4.3b) we have:

$$f_k = \mu_k R_n$$

$$= \mu_k \times 49.05 \text{ N}$$

The net horizontal force acting on the block is (30 N $-$ $\mu_k \times 49.05$ N), and Newton's second law, Eq. (4.1), gives:

$$F = ma$$

$$30 \text{ N} - \mu_k \times 49.05 \text{ N} = 5 \text{ kg} \times 2.5 \text{ m/s}^2$$

$$= 12.5 \text{ N}$$

$$\therefore \mu_k = \frac{30 \text{ N} - 12.5 \text{ N}}{49.05 \text{ N}}$$

$$= 0.36$$

(For the reasons mentioned previously, three significant figures are almost never justified for coefficients of friction. Thus, do not follow our usual practice in this problem; rather, quote only two significant figures in the answer.) ■

We have stressed that the magnitude of the frictional force is equal to the coefficient of friction multiplied by the normal component of the action-reaction force pair. In our previous example the magnitude of the normal reaction was equal to the weight of the object, but this is not always the case, as the following example shows.

EXAMPLE 4.9

A block of wood of mass 3 kg rests on a flat steel surface, and a force F is applied upward at an angle of 45° with the horizontal. The coefficients of static and kinetic friction between the wood and the steel are 0.45 and 0.25, respectively.

a. What is the least value of F that just causes the block to move?

b. What is the acceleration of the block if $F = 20$ N?

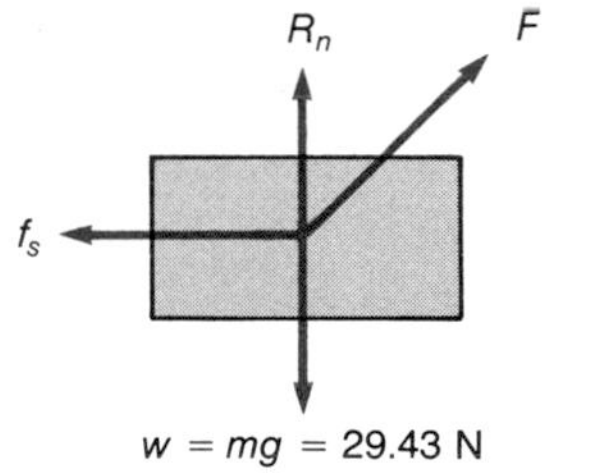

(a)

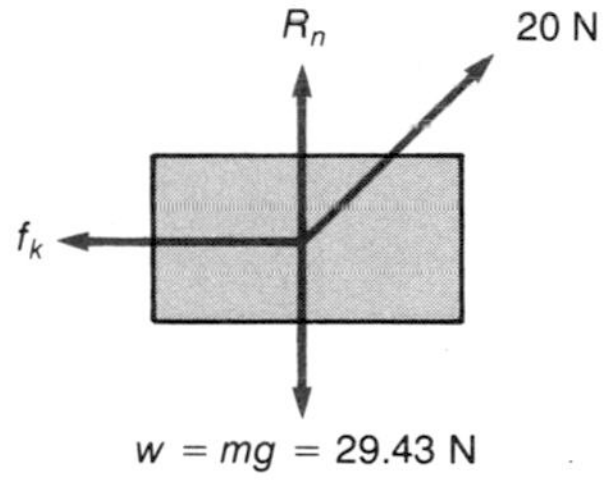

(b)

SOLUTION Each part of the problem requires a different free-body diagram. The common feature is the weight of the block, given by Eq. (4.2):

$$\begin{aligned} w &= mg \\ &= 3 \text{ kg} \times 9.81 \text{ m/s}^2 \\ &= 29.43 \text{ N} \end{aligned}$$

a. The free-body diagram shows the downward weight mg, the upward force R_n from the table, the static frictional force f_s, and the unknown force F inclined at 45° with the horizontal. We begin by decomposing F into horizontal and vertical components:

$$F_x = F \times \cos 45° = 0.7071\ F$$

$$F_y = F \times \sin 45° = 0.7071\ F$$

Since there is no vertical component of acceleration, the net vertical force is zero:

$$R_n + F_y - mg = 0$$

$$\therefore R_n = 29.43 \text{ N} - 0.7071 \times F$$

The maximum static frictional force is given by Eq. (4.3a):

$$\begin{aligned} f_s &= \mu_s R_n \\ &= 0.45 \times (29.43 \text{ N} - 0.7071 \times F) \end{aligned}$$

When the block is about to move, the horizontal component of F must equal the maximum static frictional force:

$$0.7071 \times F = 0.45 \times (29.43 - 0.7071 \times F)$$

$$\therefore F = 12.9 \text{ N}$$

b. If $F = 20$ N, its horizontal and vertical components are:

$$F_x = 20\text{ N} \times \cos 45° = 14.14\text{ N}$$

$$F_y = 20\text{ N} \times \sin 45° = 14.14\text{ N}$$

Since there is no vertical component of acceleration, the net vertical force is again zero:

$$R_n + F_y - mg = 0$$

$$\therefore R_n = 29.43\text{ N} - 14.14\text{ N}$$

$$= 15.29\text{ N}$$

The force of kinetic friction is given by Eq. (4.3b):

$$f_k = \mu_k R_n$$

$$= 0.25 \times 15.29\text{ N}$$

$$= 3.82\text{ N}$$

To calculate the horizontal acceleration of the block, we use the horizontal component of Eq. (4.1):

$$\text{Net horizontal force} = \text{mass} \times \begin{array}{c}\text{horizontal acceleration}\\ \text{component}\end{array}$$

$$F_x - f = ma_x$$

$$14.14\text{ N} - 3.82\text{ N} = 3\text{ kg} \times a_x$$

$$\therefore a_x = 3.44\text{ m/s}^2$$

■

The frictional force between automobile tires and pavement is of great practical importance. If an automobile accelerates or brakes gradually, the wheels roll without slipping. In this case the frictional force between the tires and the road is static friction. Too great an accelerating effort may produce wheel spin, and too great a braking effort may cause locked wheels. In both of these cases the tires slide on the pavement and the frictional force is kinetic friction. Since the coefficient of static friction is greater than the coefficient of kinetic friction, acceleration and braking are most effective if the tires do not slide on the road surface.

EXAMPLE 4.10

Calculate the minimum stopping distance for a 2000-kg automobile from 90 km/h (25 m/s) on a horizontal road surface

a. with grooved tires on a dry pavement and no slipping

b. with smooth tires on a wet pavement and locked wheels

(Use the values for the coefficients of friction given in Table 4.1.)

SOLUTION The only vertical forces acting on the car are the weight downward and the reaction force of the road upward. Since there is no vertical acceleration, these two forces are equal in magnitude and oppositely directed. For this problem, the magnitude of the normal reaction is equal to the magnitude of the weight.

a. Since the wheels do not slide, the magnitude of the frictional force is the force of static friction given by Eq. (4.3a):

$$\begin{aligned} f_s &= \mu_s R_n \\ &= \mu_s mg \\ &= 0.9 \times 2000 \text{ kg} \times 9.81 \text{ m/s}^2 \\ &= 1.766 \times 10^4 \text{ N} \end{aligned}$$

The frictional force is the only force in a horizontal direction; Newton's second law, Eq. (4.1), gives the deceleration:

$$\begin{aligned} F &= ma \\ \therefore a &= \frac{F}{m} \\ &= \frac{-1.766 \times 10^4 \text{ N}}{2000 \text{ kg}} \\ &= -8.84 \text{ m/s}^2 \end{aligned}$$

We can now solve the kinematics problem:

$$\begin{aligned} v_0 &= 25 \text{ m/s} \\ v &= 0 \\ a &= -8.84 \text{ m/s}^2 \\ x &= ? \end{aligned}$$

The appropriate equation of kinematics is Eq. (2.12d):

$$\begin{aligned} v^2 &= v_0^2 + 2ax \\ \therefore x &= \frac{v^2 - v_0^2}{2a} \\ &= \frac{(0 \text{ m/s})^2 - (25 \text{ m/s})^2}{2 \times (-8.84 \text{ m/s}^2)} \\ &= 35.4 \text{ m} \end{aligned}$$

b. The calculation is identical as for part (a) except that the coefficient of friction used is $\mu_k = 0.4$ instead of $\mu_s = 0.9$. The final result for the stopping distance is $x = 79.5$ m.

The stopping distances do not depend on the mass of the automobile, which cancels out in the equations. The calculations do not include driver reaction time but refer only to stopping distances after the brakes are applied. Moreover, the distances calculated are optimistic estimates since loss of driver control, bumps on the road surface, and other effects all extend stopping distance. ■

4.6 Motion on an Inclined Plane

Up to this point our examples of sliding friction have referred to motion on a horizontal plane, where both the velocity and acceleration are zero in the vertical direction. This implies a net zero vertical force component, and calculating the magnitude of the normal reaction is quite straightforward. However, we need to be more careful when we examine the sliding motion of a body on an inclined plane surface because the vector nature of the forces plays a major role in our analysis.

Consider a block that is dragged up an inclined plane by a rope parallel to the surface of the plane (see Fig. 4.3). To draw the free-body diagram for the block, we need to identify the forces acting on it. These forces are:

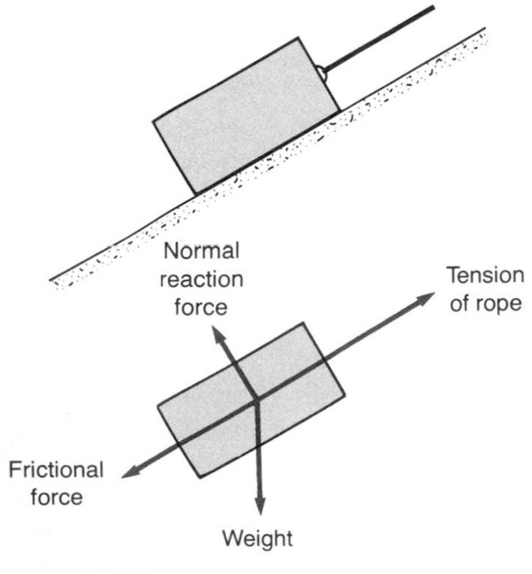

FIGURE 4.3 Illustration of the forces acting on a block being dragged up an inclined plane.

1. Its weight vertically downward
2. The force of the plane on the block. We usually quote this in terms of two components:
   a. The frictional force of the plane on the block acting in a direction opposite to the velocity of the block
   b. The component of the force of the plane on the block that is perpendicular to the surface of the inclined plane
3. The force due to the tension in the rope

Our standard method for dealing with vectors is to decompose them into components in two mutually perpendicular directions, usually the horizontal and the vertical. However, the special feature of a block sliding on an inclined plane is that the block maintains contact with the plane. There is no acceleration in a direction perpendicular to the surface of the inclined plane, and any acceleration is parallel to this surface. The components of Newton's second law in directions perpendicular and parallel to the inclined plane assume relatively simple forms:

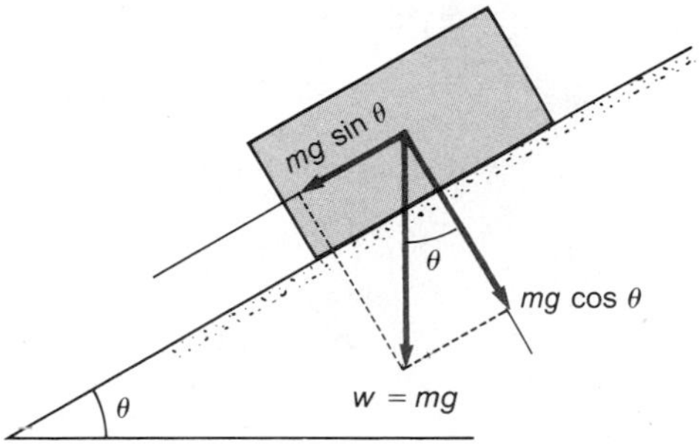

FIGURE 4.4 Diagram showing the decomposition of the weight into forces parallel and perpendicular to an inclined plane. (Other forces are omitted from this diagram.)

Net force component perpendicular to inclined plane	= 0		
			(4.4)
Net force component parallel to inclined plane	= mass ×	acceleration parallel to plane	

If we choose any other pair of mutually perpendicular directions for calculating the force components, we get two equations of accelerated motion because the only direction of zero acceleration is perpendicular to the plane. Our choice is therefore entirely one of convenience. Examining Fig. 4.3 we see that the weight of the block is the only force having components in both of the chosen directions. It is convenient to decompose the weight before attempting to solve detailed problems. We illustrate this decomposition in Fig. 4.4. Note that this diagram is simply Fig. 3.2 redrawn to suit our problem; the coordinate axes are no longer horizontal and vertical.

■ Block on a Plane Inclined at Angle θ to the Horizontal

Component of weight parallel to plane = $mg \sin \theta$

(4.5)

Component of weight perpendicular to plane = $mg \cos \theta$

EXAMPLE 4.11

A block of mass 10 kg is hauled up a plane inclined at 30° to the horizontal by a rope parallel to the plane. The coefficient of kinetic friction between the block and the plane is 0.25, and the tension in the rope is 100 N. What is the acceleration of the block?

SOLUTION The free-body diagram for the block includes all the forces shown in Fig. 4.3. We need to decompose the weight into components parallel and perpendicular to the plane. To find the weight of the block we use Eq. (4.2):

$$\begin{aligned} w &= mg \\ &= 10 \text{ kg} \times 9.81 \text{ m/s}^2 \\ &= 98.1 \text{ N} \end{aligned}$$

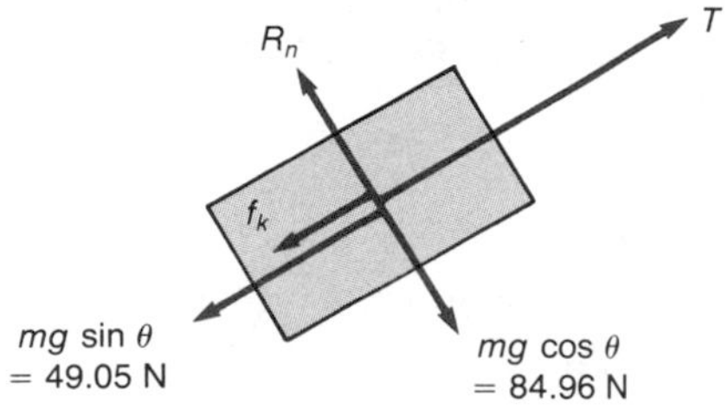

The components of the weight follow from Eq. (4.5):

$$\begin{aligned} \text{Component of weight parallel to plane} &= mg \sin \theta \\ &= 98.1 \text{ N} \times \sin 30^\circ \\ &= 49.05 \text{ N} \end{aligned}$$

$$\begin{aligned}\text{Component of weight perpendicular to plane} &= mg \cos\theta \\ &= 98.1\ \text{N} \times \cos 30^\circ \\ &= 84.96\ \text{N}\end{aligned}$$

The force of kinetic friction is given by Eq. (4.3b):

$$\begin{aligned}f_k &= \mu_k R_n \\ &= 0.25 \times R_n\end{aligned}$$

The other forces on the block are the normal force R_n and the force due to the tension in the rope. Let a be the acceleration of the block up the plane. We can now write the two component equations [Eq. (4.4)] as follows:

$$\begin{aligned}\text{Net force component perpendicular to plane} &= 0 \\ R_n - 84.96\ \text{N} &= 0\end{aligned}$$

$$\begin{aligned}\begin{matrix}\text{Net force component} \\ \text{parallel to plane}\end{matrix} &= \text{mass} \times \begin{matrix}\text{acceleration} \\ \text{parallel to plane}\end{matrix} \\ 100\ \text{N} - 49.05\ \text{N} - 0.25 \times R_n &= 10\ \text{kg} \times a\end{aligned}$$

(positive direction up the plane). The first of these equations gives $R_n = 84.96$ N. Substitution of this value in the second equation gives $a = 2.97\ \text{m/s}^2$. ■

EXAMPLE 4.12

A block of mass 3 kg rests on a flat horizontal surface. The coefficients of static and kinetic friction between the block and the surface are 0.6 and 0.4, respectively. The surface is slowly tilted until the block begins to slide. Calculate:

a. the angle with the horizontal when the block starts to slide

b. the acceleration of the block after it begins to move

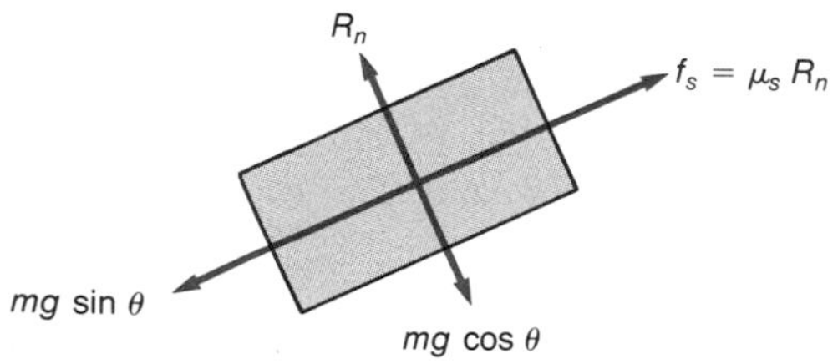

(a) Block on the point of sliding down the plane—no acceleration in any direction

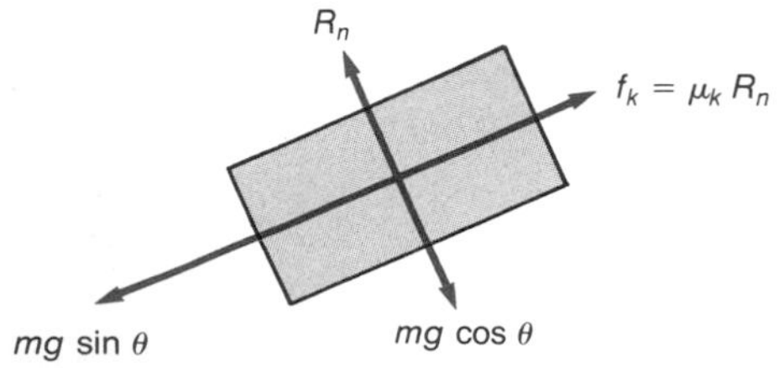

(b) Block sliding—acceleration down the plane

SOLUTION

a. Let θ be the angle between the surface and the horizontal when the block is about to slide. The free-body diagram for the block includes the two components of weight, the normal reaction force, and the frictional force. We use Eq. (4.2) to calculate the weight of the block:

$$\begin{aligned}w &= mg \\ &= 3\ \text{kg} \times 9.81\ \text{m/s}^2 \\ &= 29.43\ \text{N}\end{aligned}$$

The components of weight parallel and perpendicular to the plane are given by Eq. (4.5):

$$\text{Component of weight parallel to plane} = mg \sin\theta$$
$$= 29.43\ \text{N} \times \sin\theta$$
$$\text{Component of weight perpendicular to plane} = mg \cos\theta$$
$$= 29.43\ \text{N} \times \cos\theta$$

When the block is just about to move, the frictional force is given by Eq. (4.3a):

$$f_s = \mu_s R_n$$
$$= 0.6 \times R_n$$

With the acceleration zero in both directions, the two components of Eq. (4.4) become:

$$\text{Net force component perpendicular to plane} = 0$$
$$R_n - 29.43\ \text{N} \times \cos\theta = 0$$
$$\text{Net force component parallel to plane} = 0$$
$$29.43\ \text{N} \times \sin\theta - 0.6 \times R_n = 0$$

From the first equation we have:

$$R_n = 29.43\ \text{N} \times \cos\theta$$

Substituting in the second equation gives:

$$29.43\ \text{N} \times \sin\theta - 0.6 \times 29.43\ \text{N} \times \cos\theta = 0$$

$$\therefore \frac{\sin\theta}{\cos\theta} = \tan\theta = \frac{0.6 \times 29.43\ \text{N}}{29.43\ \text{N}}$$
$$= 0.6$$
$$\therefore \theta = 31°$$

b. After the block begins to move the frictional force is given by Eq. (4.3b):

$$f_k = \mu_k R_n$$
$$= 0.4 \times R_n$$

Let a be the acceleration of the block down the plane. The two components of Eq. (4.4) become:

$$\text{Net force component perpendicular to plane} = 0$$

$$R_n - 29.43 \text{ N} \times \cos 31° = 0$$

$$\begin{array}{c}\text{Net force component} \\ \text{parallel to plane}\end{array} = \text{mass} \times \text{acceleration parallel to plane}$$

$$29.43 \text{ N} \times \sin 31° - 0.4 \times R_n = 3 \text{ kg} \times a$$

(positive direction down the plane). The first equation gives R_n = 25.24 N. Substituting this value in the second equation gives:

$$29.43 \text{ N} \times \sin 31° - 0.4 \times 25.24 \text{ N} = 3 \text{ kg} \times a$$

$$\therefore a = 1.68 \text{ m/s}^2$$ ■

The example just worked contains an interesting result of general applicability. A block on an inclined plane is just on the point of sliding when the coefficient of static friction is equal to the tangent of the angle that the plane makes with the horizontal. To measure the coefficient of static friction between two substances, place a flat block of one substance on a plane surface of the other, and then tip the plane until the block just moves. The tangent of the angle of slope of the plane is the coefficient of static friction.

KEY CONCEPTS

The **law of inertia** assigns zero net force to a state of rest or a state of motion at constant velocity.

A mass that experiences **nonzero net force** undergoes an **acceleration** in the direction of the force. The net force is equal to the mass multiplied by the acceleration. See Eq. (4.1).

The **forces** of nature always occur in **equal and opposite pairs** between two objects. One part of the action-reaction pair acts on each of the two objects.

Weight is the force due to gravity acting on a mass. The weight is equal to the mass multiplied by the acceleration due to gravity. See Eq. (4.2).

Friction always opposes the motion of macroscopic bodies. For the case of flat surfaces in contact, the magnitude of the **frictional force** is equal to the coefficient of friction multiplied by the magnitude of the normal reaction. See Eq. (4.3).

When a body *slides on an inclined plane*

a. the net force component perpendicular to the plane is zero

b. the net force component parallel to the plane equals the mass multiplied by the acceleration. See Eqs. (4.4) and (4.5)

QUESTIONS FOR THOUGHT

1. "An object that is in a state of rest is not being acted on by any forces." Discuss this statement and correct it.

2. "An object that is instantaneously at rest is experiencing zero net force at that instant." Discuss this statement.

3. Suggest an experiment that would directly verify Newton's first law.

4. Explain the function of seat belts and head supports in preventing injury in automobile accidents.

5. The net force acting on a body is zero. What does this tell you about the magnitude and direction of its velocity and acceleration vectors?

6. "Force is the cause of acceleration," but in some instances a force acts on an object and no motion occurs. Modify the statement so that it more accurately conveys the idea that it attempts to express.

7. A heavy weight is suspended by a light string, and a second light string hangs below the weight. If you pull steadily on the lower string the upper string breaks, but if you pull with a sudden jerk the lower string breaks. Explain this strange result.

8. A heavy crate rests on a flat concrete floor, and a worker pushing horizontally does not succeed in moving it. Since the crate does not move, the frictional force resisting motion and the force applied by the worker are equal in magnitude and oppositely directed. Do these two forces constitute a Newtonian action-reaction force pair?

9. What factors influence the value of the coefficient of friction between two plane surfaces?

PROBLEMS

A. Single-Substitution Problems

1. A force of 7 N acts on a body of mass 2.3 kg. What is the magnitude of the acceleration? [4.2]

2. What force is required to accelerate an automobile of mass 1600 kg at 2.5 m/s^2? [4.2]

3. A sprinter of mass 70 kg accelerates at 4 m/s^2. What is the net accelerating force? [4.2]

4. Two blocks of masses 3 kg and 7 kg are in contact with each other on a frictionless horizontal surface. Calculate the magnitude of the action-reaction force at the interface between the blocks if a horizontal force of 18 N acts on the 3-kg block. [4.3]

5. Repeat problem 4 for the situation in which the 18-N force acts on the 7-kg block. [4.3]

6. Two blocks of masses 2 kg and 8 kg are in contact with each other on a frictionless horizontal surface. The magnitude of the action-reaction force at the interface between the blocks is 12 N. Calculate the magnitude of the external force on the 2-kg block that would cause this situation. [4.3]

7. Repeat problem 6 if the external force acts on the 8-kg block. [4.3]

8. A child has a mass of 18 kg. Calculate her weight (a) in newtons, (b) in pounds. [4.4]

9. A book of mass 3.5 kg rests on a flat table surface. Calculate the upward force of the table on the book. [4.4]

10. A man of mass 75 kg hauls himself off the floor up a vertical rope with an acceleration of 0.68 m/s^2. Calculate the tension in the rope while he maintains this acceleration. [4.4]

11. A loaded elevator of total mass 1650 kg is accelerating downward at 1.5 m/s^2. Calculate the tension in the supporting cable. [4.4]

12. A cyclist and machine (combined mass 75 kg) coast on a flat horizontal road with a deceleration of 0.72 m/s^2. Calculate the frictional force resisting motion. [4.5]

13. A large block of mass 42 kg rests on a horizontal surface. The coefficient of static friction between the block and the surface is 0.41. Calculate the minimum horizontal force necessary to move the block. [4.5]

14. A block of mass 12 kg rests on a horizontal surface; the coefficient of static friction between the block and the surface is 0.32.

a. How large is the force of friction on the block?
b. How large is the force of friction on the block if a horizontal force of 20 N is applied to it?
c. What is the minimum horizontal force needed to set the block in motion? [4.5]

15. A crate of mass 62 kg is being dragged across a level floor at constant velocity by rope in which the tension is 240 N. Calculate the coefficient of kinetic friction between the crate and the floor. [4.5]

16. An object of mass 3.2 kg is sliding down a frictionless inclined plane, and the accelerating force is 12 N parallel to the plane. Calculate the angle that the plane makes with the horizontal. [4.6]

17. A frictionless inclined plane makes an angle of 30° with the horizontal. An object of mass 8 kg is moving on the plane, and at a given instant its velocity is 5 m/s parallel to the plane in an upward direction. Calculate the magnitude and direction of the net force that the object experiences at this instant. [4.6]

18. A roller coaster car of mass 140 kg descends a slope inclined at 23° to the horizontal. Calculate the force accelerating the car down the slope, assuming that friction is negligible. [4.6]

B. Standard-Level Problems

19. What average braking force is required to stop an automobile of mass 1400 kg from a speed of 90 km/h (25 m/s) in 3.8 s? [4.2]

20. An automobile of mass 2000 kg is moving at 54 km/h (15 m/s). What average net forward force is required to accelerate the automobile to 90 km/h (25 m/s) over a distance of 85 m? [4.2]

21. A 1500-kg automobile is traveling at 90 km/h (25 m/s). What average braking force is required to stop it in 65 m? [4.2]

22. An automobile traveling at 15 m/s is involved in a collision and comes to rest while traveling a total distance of 2.2 m during the collision. What average force is applied by the seat belts to an occupant of 80-kg mass during the impact?

23. Referring to Fig. 4.1, let the mass of the horse be 450 kg; the mass of the stone, 80 kg; and the frictional force resisting the motion of the stone (F_2), 250 N. The rope by which the horse pulls the stone is horizontal, and the horse pulls so as to cause an acceleration of 1.2 m/s^2.

a. What is the tension in the rope?
b. What is the frictional force between the horse and the ground? [4.3]

24. Two blocks, one of mass 6 kg and the other of mass 2 kg, are placed in contact on a horizontal table, and a constant horizontal force is applied to the 6-kg mass. There is a constant frictional force of 18 N between the 6-kg mass and the table, but no frictional force between the 2-kg mass and the table. The blocks accelerate at 3 m/s^2.

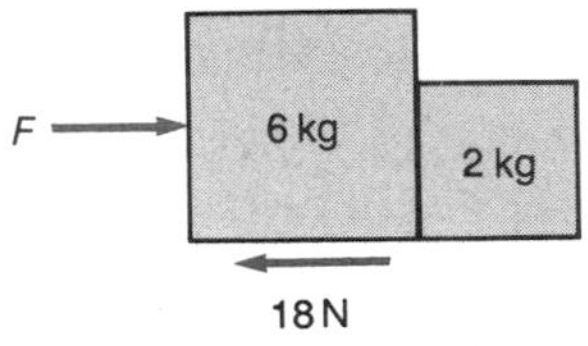

a. What is the propulsive force applied to the 6-kg mass?
b. What is the force at the interface between the two blocks? [4.3]

25. Two blocks, one of mass 4 kg and the other of mass 5 kg, are placed in contact on a horizontal table. A certain force applied horizontally to the 4-kg block causes an action-reaction force pair of 32 N at the interface between the blocks. There is a constant frictional force of 15 N between each block and the table.

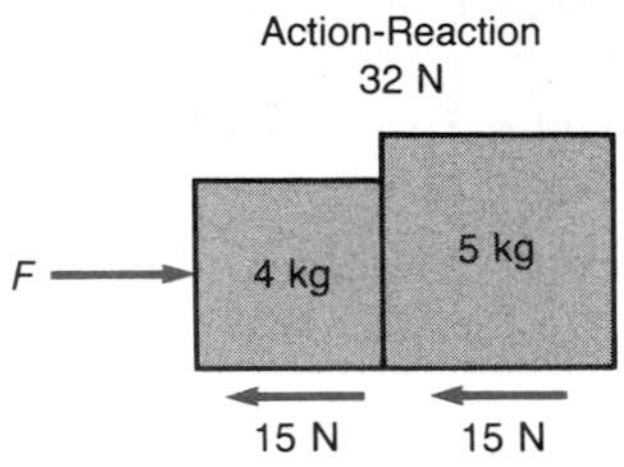

a. Calculate the magnitude of the applied force.
b. What is the acceleration of the blocks?

26. A 1600-kg automobile is capable of accelerating on the level from 45 km/h to 90 km/h in 5.5 s. A 420-kg trailer is now hitched to the automobile, and the trailer causes an average frictional drag of 460 N. Assume that the net average propulsion force remains unchanged.

a. What is the acceleration time from 60 km/h to 90 km/h with the trailer attached?
b. What is the force at the coupling during this acceleration?

27. A locomotive of mass 48 000 kg pulls six freight cars each having a mass of 8000 kg and each experiencing a frictional resistance of 3200 N. The locomotive accelerates the train at 0.85 m/s^2. Four more cars (each with the same mass and frictional resistance) are now coupled to the train. Assume that the net average propulsive force remains unchanged.

a. What is the time for acceleration from rest to 20 m/s?
b. What is the force at the coupling between the locomotive and the freight cars.

28. A vertical rope is attached to a boulder of mass 40 kg. What constant tension in the rope causes the boulder to have an upward velocity of 3.8 m/s at an instant 4 s after leaving the ground? [4.4]

29. A bag of cement weighs 400 N. What is the shortest time in which it can be hauled up to a ledge 9 m above the ground with a rope of breaking strength 600 N? [4.4]

30. A climber of mass 70 kg descends 18 m by sliding down a rope having a 550-N breaking strength. If the slide is carried out so skillfully that the rope is just about to break all the time, with what speed does the climber hit the ground?

31. A parachutist of mass 70 kg is descending at a steady speed of 6 m/s. What upward force does the parachute provide? The parachute is accidentally torn, and the downward speed increases from 6 m/s to 14 m/s during a 3-s interval. What average upward force does it provide over this period?

32. An 80-kg parachutist falls for 3.5 s and then opens her parachute. The canopy takes 1.8 s to open fully, and it reduces her downward velocity to 8 m/s. What is the average upward force exerted by the harness on the parachutist during the period in which the parachute opens? (Ignore air resistance forces during the first 3.5 s of fall.)

33. A 25-metric ton Navy jet whose engine develops an average thrust of 80 000 N is launched from a carrier by means of a steam-powered catapult. The jet takes 1.9 s to travel the 80-m track to takeoff.

a. What is the takeoff speed relative to the carrier deck?
b. What is the average thrust exerted by the catapult during launching?
(Assume that the acceleration is uniform.)

34. A tennis ball weighing 0.56 N acquires a speed of 175 km/h in a hard-hit service. If the time of contact between the ball and the racket is 10 ms, what average force is exerted by the racket on the ball?

35. A mass of 1.2 kg is suspended from a light cord, and a 3.4-kg mass is suspended under it by another cord. Calculate the tension in each cord when the two masses are lowered with an acceleration of 3.3 m/s^2 downward.

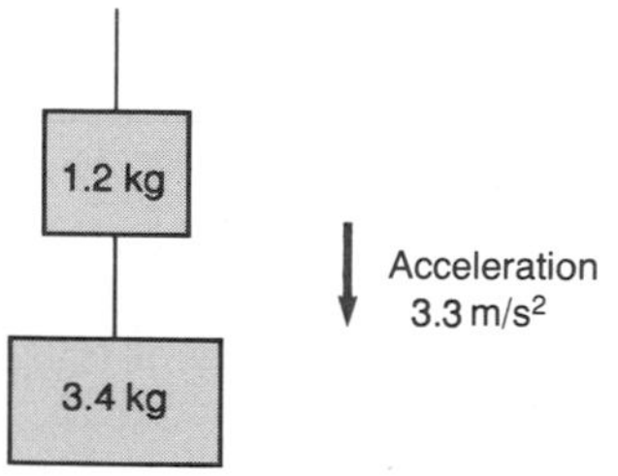

36. Two masses, each of 5 kg, are connected by a light cord passing over a frictionless pulley. One mass is free to slide on a frictionless horizontal surface, and the other falls vertically. Calculate the acceleration of each mass and the tension in the cord.

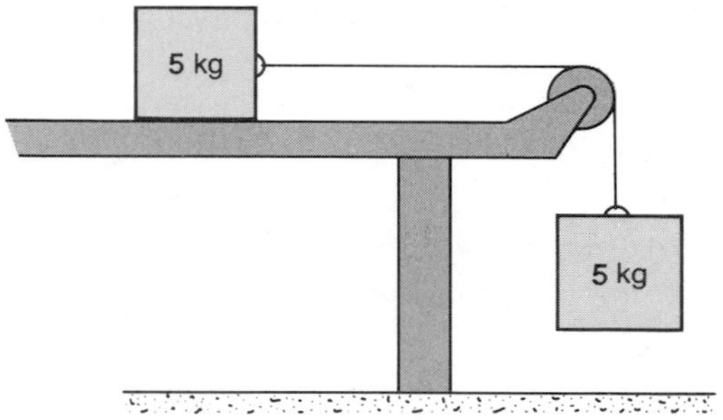

37. A 7-kg mass is attached to a cord that passes over a light frictionless pulley to another 7-kg mass on the other side. A 5-kg mass is attached by a cord to the bottom of one of the 7-kg masses. The whole system is free to move.

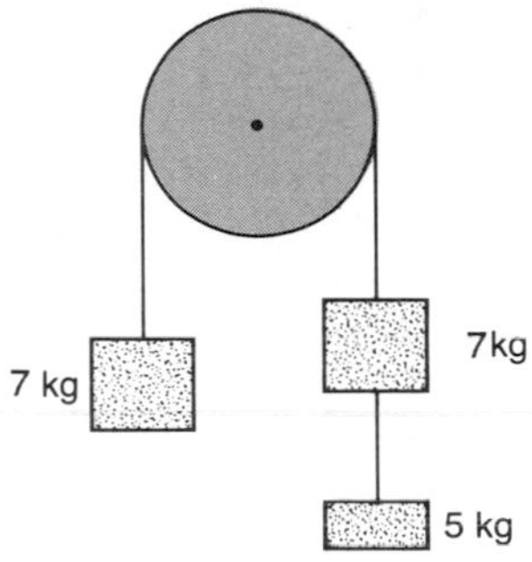

a. What is the upward acceleration of the single 7-kg mass?
b. What is the tension in each cord?

38. An 8-kg mass is attached to a cord that passes over a light frictionless pulley to a 6-kg mass on the other side. Another mass is attached by a cord to the bottom of the 6-kg mass. Calculate this mass if its attachment causes a downward acceleration of 2.1 m/s^2 on its side of the pulley. Also calculate the tension in each cord.

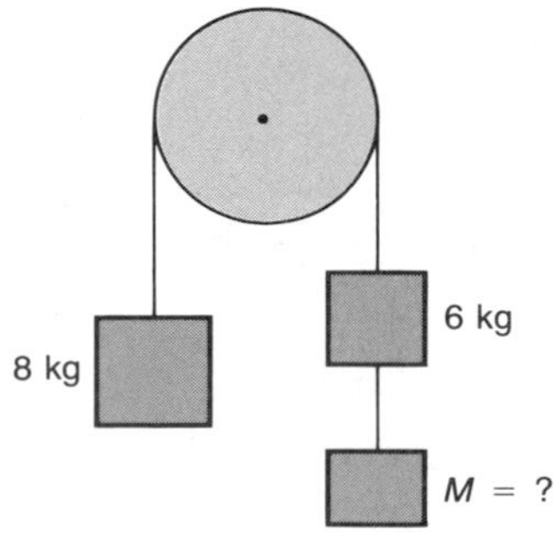

39. An aluminum ball of mass 400 g sinks in an oil bath with an acceleration of 4.8 m/s^2. What is the magnitude of the combined buoyancy and frictional force resisting its fall? [4.5]

40. A mass of 7 kg falling in deep water experiences an upward force (combined buoyancy and frictional resistance) of 11.5 N. What is the value of its downward acceleration? [4.5]

41. A block of mass 6 kg rests on a wooden floor. The coefficient of kinetic friction between the block and the floor is 0.28.

a. What horizontal force is necessary to accelerate the block across the floor at 1.6 m/s^2?
b. What is the largest value for the coefficient of static friction that would allow the same force to start the block sliding from rest?

42. A loaded sled of mass 60 kg is pushed onto the flat frozen surface of a lake with an initial velocity of 5.2 m/s.

a. How far does the sled travel before stopping if the coefficient of kinetic friction between the ice and the runners is 0.05?
b. If the coefficient of static friction is 0.10, what force is necessary to start the sled moving again?

43. A block of mass 2.5 kg is pushed across a rough horizontal surface (coefficient of kinetic friction 0.33) with an acceleration of 2 m/s^2. The applied force necessary to produce this acceleration is directed downward at an angle of 45° to the horizontal. Calculate the magnitude of the applied force.

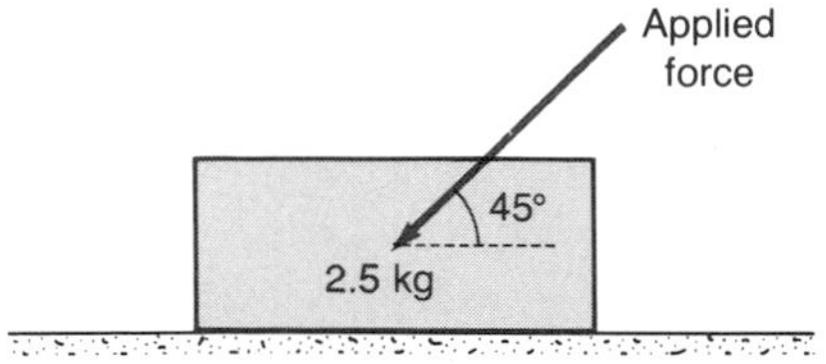

44. A block of mass 1.8 kg is pushed across a rough horizontal floor (coefficient of kinetic friction 0.29) by a force of 24 N directed at an angle of 30° with the horizontal. Calculate the acceleration of the block.

45. Two bags of cement, each weighing 400 N, are stacked on a horizontal floor. The coefficient of static friction between the two bags is 0.40, and that between the lower bag and the floor 0.30.

a. What horizontal force is required to slide the lower bag out if the top bag is restrained from moving by a horizontal force in the opposite direction?

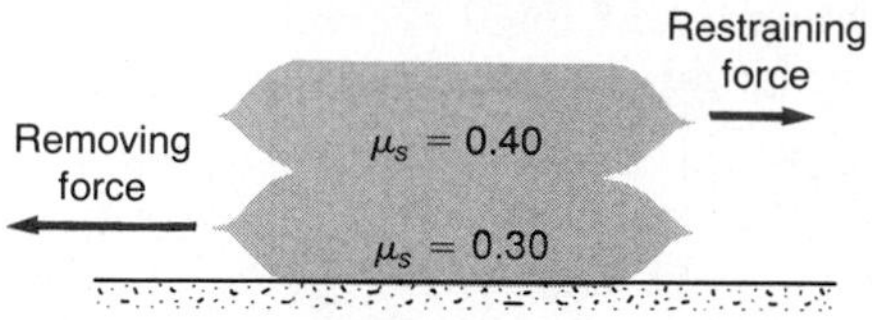

b. What is the magnitude of the restraining horizontal force on the top bag?

46. A cyclist and her machine together have a mass of 85 kg.

a. What average force between the tires and the road is necessary to stop from 27 km/h (7.5 m/s) in 3.5 s?
b. Can this stopping time be achieved if the coefficient of static friction between the tires and an oil-slicked road surface is 0.23?
c. What stopping time can be achieved on a good road surface with a coefficient of static friction 0.90?

47. A block on a horizontal table surface (coefficient of static friction 0.38) is connected by a light cord passing over a frictionless pulley to a mass of 2.5 kg that hangs vertically; under these conditions the block is just on the point of moving.

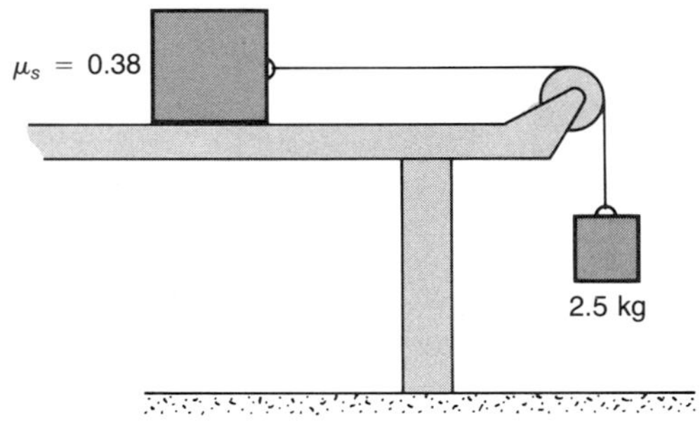

a. Calculate the mass of the block.
b. Just after the block moves, its acceleration is 1.2 m/s^2; calculate the coefficient of kinetic friction between the block and the table.

48. A 4-kg block rests on a horizontal table surface and is connected by a light cord passing over a frictionless pulley to a weight that hangs vertically. The coefficients of static and kinetic friction between the block and the tabletop are 0.35 and 0.21, respectively.

a. What weight is required on the end of the cord to just move the 4-kg block?

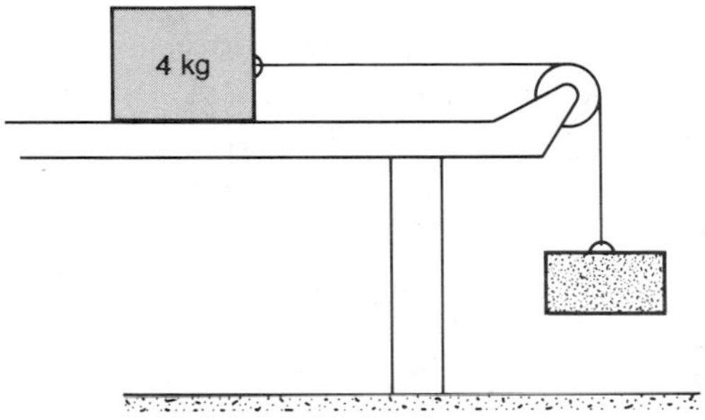

b. What is the acceleration of the block after it begins to move?

c. What is the tension in the cord after the block begins to move?

49. An 8-kg block rests on a 50° inclined plane and is restrained from slipping by a cord parallel to the plane. The coefficients of static and kinetic friction are 0.40 and 0.20, respectively.

a. What is the tension in the cord?

b. If the cord is cut, what is the acceleration of the block down the plane? [4.6]

50. A 35-kg child is playing on a 8° snow slope with a 12-kg toboggan. The coefficients of static and kinetic friction are 0.21 and 0.11, respectively.

a. What force must the child exert on the toboggan to pull it up the slope at constant speed? (Assume the pulling cord is parallel to the slope.)

b. If the child sits on the toboggan on the slope, what force is necessary to start it moving?

c. Once it begins to move, what is its acceleration down the slope? [4.6]

51. A 1200-kg automobile with smooth tires is descending a 6° hill in the rain at 90 km/h (25 m/s).

a. What is the stopping distance if the driver locks the wheels in an attempt to stop quickly?

b. If the automobile is ascending the hill at 90 km/h, what is the stopping distance if the wheels do not lock?

(Use the values of 0.45 and 0.35 for the coefficients of static and kinetic friction, respectively, between smooth tires and wet pavement.)

52. A packing crate of mass 320 kg is lowered down an inclined ramp 4 m long from the back of a truck that is 1.2 m high. The coefficient of kinetic friction between the crate and the ramp is 0.21.

a. How large a force acting parallel to the ramp is necessary to restrain the crate so it slides down the ramp at a constant velocity?

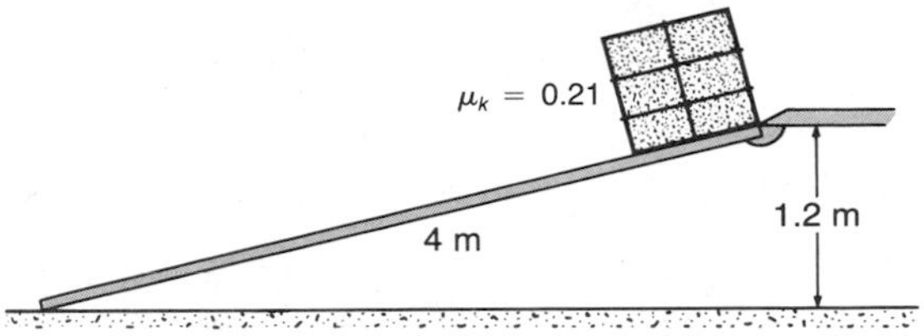

b. If the crate starts from rest at the top of the ramp and is allowed to slide without restraint, how fast is it moving when it reaches the bottom?

53. A wooden block of mass 1 kg rests on a horizontal wooden plank. If one end of the plank is raised, the block begins to slide when the angle with the horizontal is 34°.

a. What is the coefficient of static friction between the block and the plank?

b. After sliding begins, the acceleration of the block down the plane is 0.95 m/s^2; what is the coefficient of kinetic friction between the block and the plane?

c. Is it necessary to know the mass of the block to answer this problem?

54. A block of mass 3 kg is given an initial velocity of 12 m/s up a long inclined plane that makes an angle of 25° with the horizontal. The coefficient of kinetic friction between the block and the plane is 0.25.

a. Calculate the frictional force acting on the block.

b. How long does the block move up the plane before momentarily coming to rest?

c. How long does it take to slide from this position back to its starting point?

55. An automobile of mass 1200 kg starts from rest and accelerates uniformly to the top of a 125-m long incline in 16 s. The incline is at 7° to the horizontal, and the average frictional force resisting the motion is 450 N.

a. What is the speed of the automobile at the top of the incline?

b. What is the total propulsive force exerted on the driving wheels by the road?

56. A 75-kg skier starts from rest at the top of a 24° slope that is 100 m long. The coefficient of kinetic friction between his skis and the snow is 0.05. If he does not push, what is his speed at the bottom of the slope? (In this problem ignore air resistance, even though it plays an increasingly important role as the speed of the skier increases.)

57. An automobile towing a 600-kg trailer at a constant 90 km/h (25 m/s) on a level road exerts a force of

310 N in the tow coupling to maintain this constant speed. The automobile and trailer come to a hill inclined at 4° to the horizontal.

a. What is the force in the coupling if they maintain 90 km/h up the hill?
b. If the automobile and trailer decelerate at 0.4 m/s^2 on the hill, what is the force in the tow coupling?

C. Advanced-Level Problems

58. Two masses, one of 10 kg and the other of 5 kg, are suspended by a light cord passing over a frictionless pulley. Each mass is held 5 m above a horizontal floor and then released.

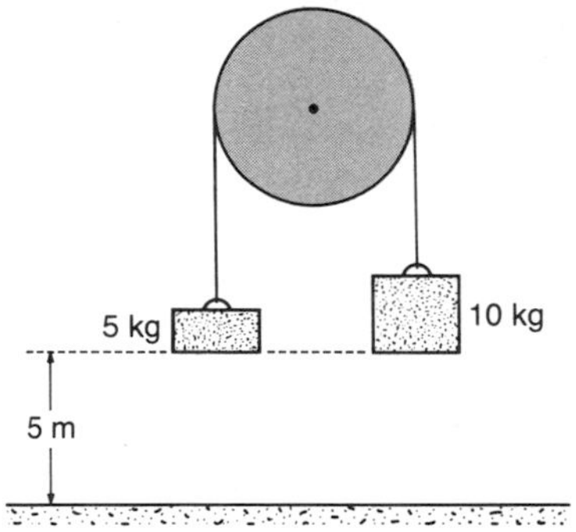

a. How long does the 10-kg mass take to hit the floor?
b. What maximum height above the floor does the 5-kg mass reach?

59. A balloon containing three people has an overall mass of 500 kg and is descending slowly with constant velocity. How much ballast should be thrown overboard to cause an upward acceleration of 0.2 m/s^2?

60. Two 1-kg blocks rest on opposite sides of the double inclined plane shown in the diagram. They are connected by a light cord passing over a frictionless pulley. What is the largest value of the coefficient of static friction (assumed to be the same for each block) at which slipping can just occur?

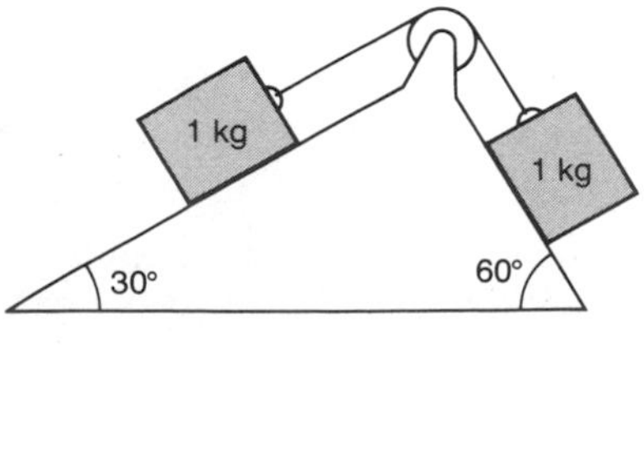

61. A block of mass 10 kg rests on a plane inclined at 30° to the horizontal. A light cord passing over a frictionless pulley attaches the block to a 7-kg mass, as shown in the diagram.

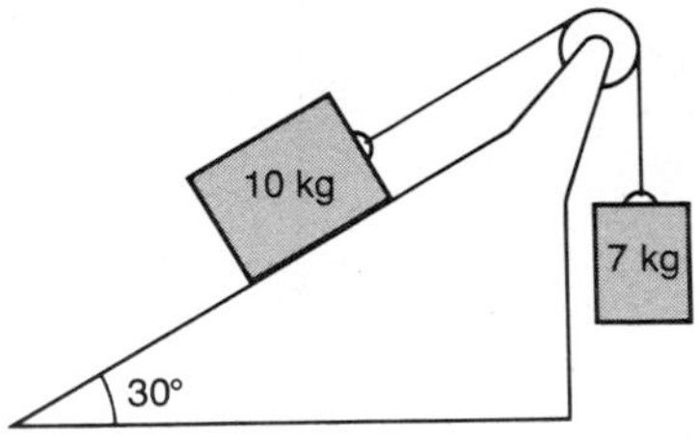

a. What is the tension in the cord if the coefficient of static friction between the block and the plane is 0.3?
b. What is the tension in the cord if the coefficient of kinetic friction is 0.15 and the block is sliding?
c. With the coefficient of kinetic friction 0.15, how long does the 7-kg block take to fall from rest through a distance of 1.2 m?

62. Two blocks, one of mass 3 kg and one of mass 5 kg, are in contact with each other and are held stationary on an inclined plane that makes an angle of 30° with the horizontal. The force holding them stationary is parallel to the plane and is just sufficient to prevent them from moving down the plane. The coefficient of static friction between the 3-kg block and the plane is 0.5, and that between the 5-kg block and the plane is 0.2.

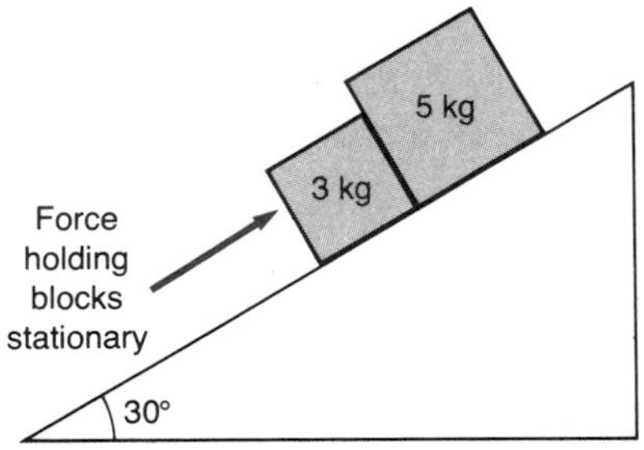

a. What is the magnitude of the force holding the blocks stationary?
b. What is the action-reaction force at the interface between the blocks?

63. A box of mass 90 kg rests on a plane inclined at 30° to the horizontal, and a force of 125 N applied parallel to the plane is required to prevent it from slipping.

a. What is the coefficient of static friction between the box and the plane?
b. How large a force applied horizontally is necessary to prevent the box from slipping?

5

STATICS

The name implies that statics is the study of bodies at rest under the action of forces. A massive weight hanging from several ropes, for example, is at rest under the action of the forces that are due to the tension in the ropes and its own weight, and the net total force on the weight is zero. However, a launch moving across a lake at constant velocity also experiences a zero net force, since its acceleration is zero. The study of bodies that experience zero net force, therefore, includes bodies at rest and bodies moving with constant velocity. Thus, we define statics as the study of bodies undergoing zero acceleration under the action of forces. When this condition holds, we say that the body is in equilibrium under the action of the forces. The large variety of practical problems is a chief motivation for a separate discussion of statics.

5.1 Statics of Concurrent Coplanar Forces

When three or more lines intersect at a common point, we say that the lines are concurrent. Three or more vectors representing displacements, forces, or any other physical vector quantity are concurrent if their lines of action meet at a common point. Forces that act on a point particle must be concurrent forces, since they must intersect at the position of the particle. The point particle model is one of the big simplifications used in previous chapters. We will continue to use it for the moment, but later in the chapter we will extend our considerations to objects of finite dimensions.

Consider a heavy object supported by cords as shown in Fig. 5.1. The forces due to the tensions in the three cords form concurrent forces that intersect at the knot. But recollect that the tension in a cord does not explicitly refer to a single force. In the lower cord the tension pulls up on the heavy object and down on the knot, while in the two upper cords the tension pulls down on the eyebolts and up on the knot (in directions along each cord). In dynamics problems we have seen the importance of drawing a free-body force diagram that includes the forces acting *on* a body, but not the forces exerted *by* it on other bodies. The statics of the problem in hand reduces to a study of the concurrent forces acting *on* the knot. The knot (and consequently the whole system) is in equilibrium if these forces produce zero acceleration, but the problem illustrated in Fig. 5.1 has another simplifying feature—the cords all lie in one plane. Thus, we can draw the force diagram accurately on a flat sheet of paper, since no force components come out of the paper or are directed into it. The forces are both **concurrent** and **coplanar.**

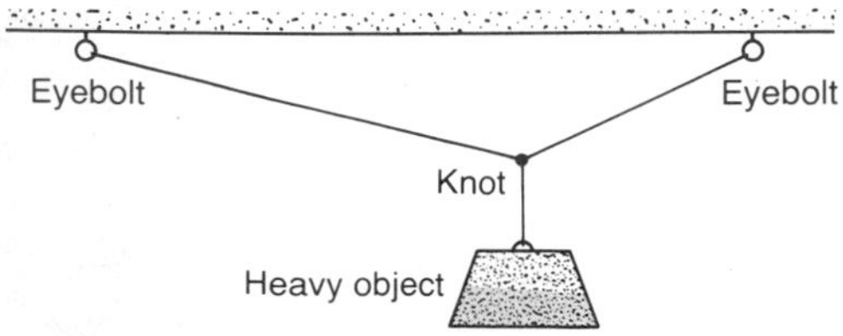

FIGURE 5.1 The forces on the knot due to the tensions in the ropes present an example of concurrent coplanar forces.

If the acceleration vector of the knot is zero, the acceleration component in any direction must be zero. It follows that the net force component in any direction must also be zero. This condition is met if the net force component in two arbitrary perpendicular directions is zero. We often choose horizontal and vertical directions (labeled by x and y coordinates), but we can make any convenient choice. Let us set the equilibrium conditions in display. The sign "Σ" means summation, and the first condition states that the sum of the x components of all the forces acting on the knot is zero. The second condition makes the same statement for y force components.

■ Conditions of Equilibrium of Concurrent Coplanar Forces

Net x component of force is zero: $\Sigma F_x = 0$ **(a)**
Net y component of force is zero: $\Sigma F_y = 0$ **(b)** **(5.1)**

EXAMPLE 5.1

A weight of 40 N hangs at the end of a cord knotted to two other cords making angles of 30° and 60° with the horizontal. The other ends of these two cords are tied to fixed supports. Calculate the tension in all three cords.

SOLUTION The basic features of this problem are explained above. We label the cords 1, 2, and 3, as shown in the diagram. Let T_1, T_2, and T_3 be the corresponding tensions in the three cords. Since cord 1 supports the 40-N weight, its tension is 40 N. The free-body force diagram for the knot shows the forces acting on the knot due to the tensions in the cords. We can calculate the horizontal and vertical components of these forces using Eq. (3.3). The three components are:

General situation diagram

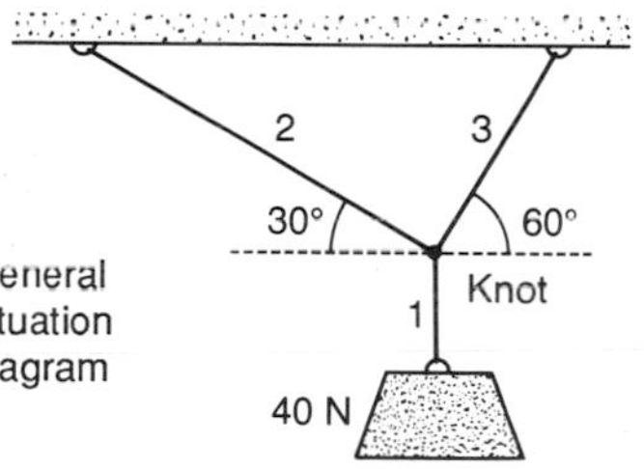

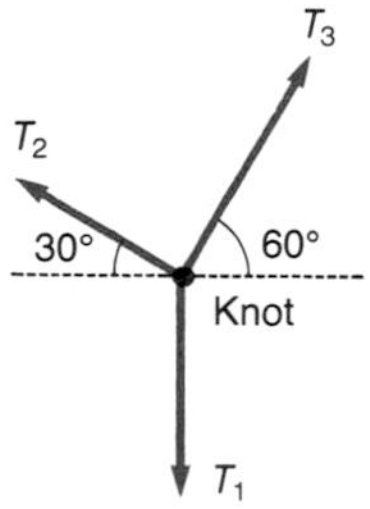

Free-body force diagram

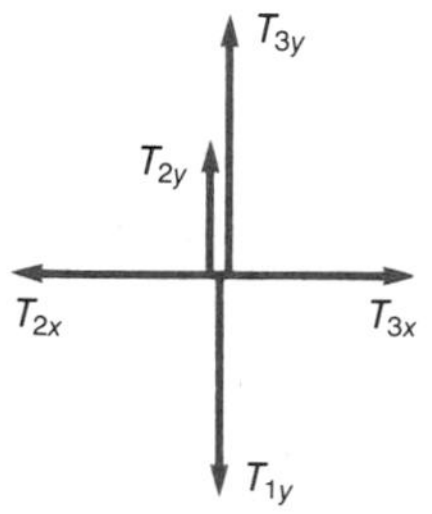

Horizontal and vertical components of forces

$$T_{1x} = 0$$

$$T_{2x} = -T_2 \times \cos 30° = -0.866T_2$$

$$T_{3x} = T_3 \times \cos 60° = 0.5T_3$$

The three vertical components are:

$$T_{1y} = -40 \text{ N}$$

$$T_{2y} = T_2 \times \sin 30° = 0.5T_2$$

$$T_{3y} = T_3 \times \sin 60° = 0.866T_3$$

The second part of the diagram shows the free-body force diagram for the knot with the three forces T_1, T_2, and T_3. In the third part of the diagram these forces are replaced by their components. We can now use the equilibrium conditions Eq. (5.1):

a. Net x component of force $= 0$

$$\Sigma F_x = T_{1x} + T_{2x} + T_{3x} = 0$$

$$0 - 0.866T_2 + 0.5T_3 = 0$$

b. Net y component of force $= 0$

$$\Sigma F_y = T_{1y} + T_{2y} + T_{3y} = 0$$

$$-40\text{ N} + 0.5T_2 + 0.866T_3 = 0$$

The two equations suffice to determine the two unknown quantities T_2 and T_3. From the first equation we have:

$$T_3 = 1.732T_2$$

Substituting this value in the second gives:

$$-40\text{ N} + 0.5T_2 + 0.866 \times 1.732T_2 = 0$$

Solving for T_2 and T_3 yields:

$$T_2 = 20\text{ N}$$

and $$T_3 = 34.6\text{ N}$$ ■

This bridge offers an example of an object in static equilibrium under the action of applied forces. The members must support their own weight together with that of the roadway and the traffic that it carries. The sometimes considerable force of the wind must also be included in designing the structure.

The preceding example is a good illustration of the procedure used to solve statics problems, which consists of the following series of steps:

1. Draw a general diagram of the problem situation.
2. Decide on the object that is in equilibrium, and draw a free-body force diagram of the forces acting *on* that object. Do not include the forces exerted *by* that object on other things.
3. Calculate the components of these forces in suitably chosen directions by using Eq. (3.3).
4. Use Eq. (5.1) to set the algebraic sum of these components equal to zero.
5. Solve the resulting equations for the unknown quantities.

EXAMPLE 5.2

A child weighing 250 N sits on a swing attached to an overhead support by a chain of negligible weight. The child is drawn aside

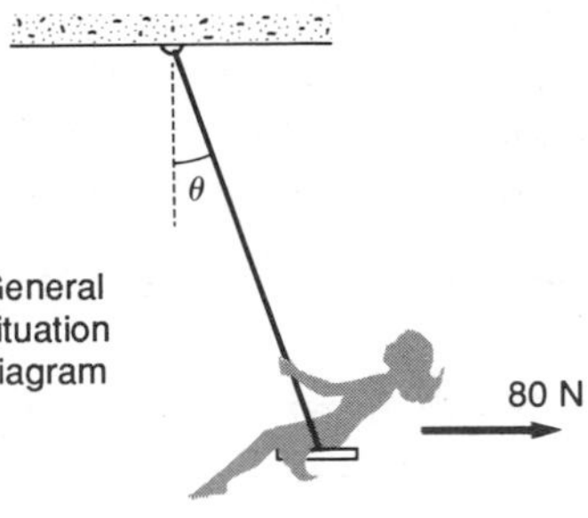

General situation diagram

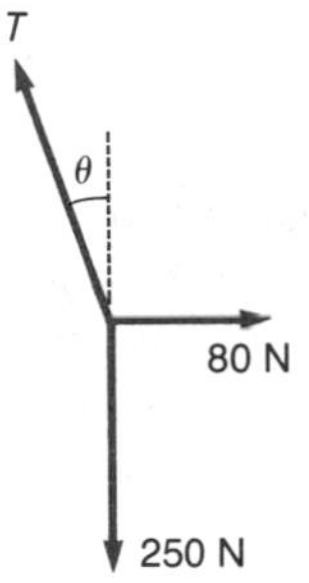

Free-body force diagram

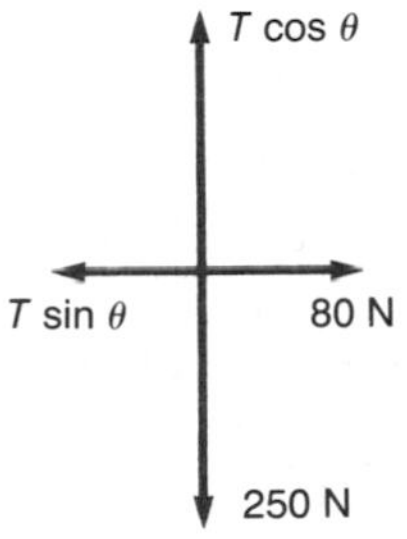

Horizontal and vertical components of forces

and held in equilibrium by a horizontal force of 80 N. Calculate the angle that the supporting chain makes with the vertical and the tension in the chain.

SOLUTION A basic assumption in this problem is that we can treat the child as a point particle. We can then draw the free-body force diagram as shown. Let T be the tension in the chain, and θ the angle that it makes with the vertical. The 250-N weight and the 80-N applied force act vertically and horizontally. The horizontal and vertical components of the force due to the chain tension are given by:

$$T_x = -T \sin \theta$$

$$T_y = T \cos \theta$$

Using Eq. (5.1) we have:

a. Net x component of force $= 0$

$$\Sigma F_x = 80 \text{ N} - T \sin \theta = 0$$

b. Net y component of force $= 0$

$$\Sigma F_y = T \cos \theta - 250 \text{ N} = 0$$

The two equations determine both T and θ. Equations containing unknown quantities in trigonometric functions can be very difficult to solve; writing them in various ways is often helpful to see if any simplifying idea suggests itself. We can rewrite this particular pair as follows:

$$T \sin \theta = 80 \text{ N}$$

$$T \cos \theta = 250 \text{ N}$$

The simple expedient of dividing one into the other clears one of the unknowns, and gives:

$$\tan \theta = \frac{80 \text{ N}}{250 \text{ N}} = 0.3200$$

$$\therefore \theta = \text{arc tan } 0.3200$$

$$= 17.7°$$

We can then calculate the tension from either of the original equations:

$$T \sin 17.7° = 80 \text{ N}$$

$$\therefore T = 262 \text{ N}$$

■

The essential feature in solving problems concerning the statics of a point particle is balancing the force components. Since all quantities in the equation have the same units, conversion of units is not required, provided we maintain consistency.

EXAMPLE 5.3

A tightrope walker weighing 160 lb moves to the center of the wire, which deflects at an angle of 6° with the horizontal on each side. Calculate the tension in the wire.

General situation diagram

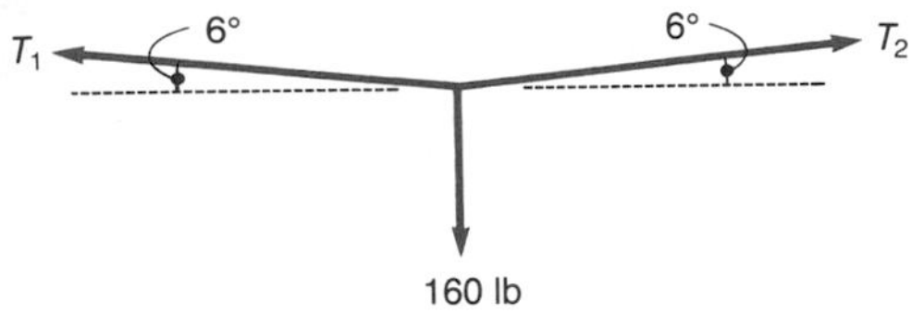

Free-body force diagram

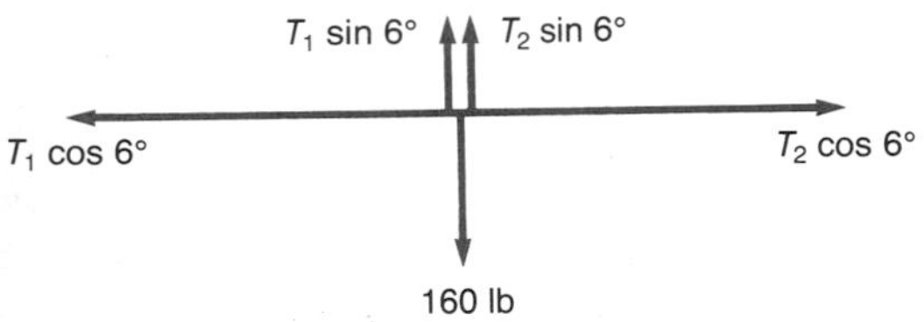

Horizontal and vertical components of forces

SOLUTION The point on the wire at which the performer places his feet is in equilibrium under the action of his weight and the forces due to the tension in the wire. Since the tensions in the sections of wire on each side of the performer may be different, let us call them T_1 and T_2. We now draw the free-body force diagram for the piece of wire directly under the performer's feet and decompose the forces into horizontal and vertical components. Substitution in Eq. (5.1) gives:

a. Net x component of force = 0

$$\Sigma F_x = T_2 \cos 6° - T_1 \cos 6° = 0$$

b. Net y component of force = 0

$$\Sigma F_y = T_1 \sin 6° + T_2 \sin 6° - 160 \text{ lb} = 0$$

The first equation shows that $T_1 = T_2$; let the value of each be T. The second equation then gives:

$$2T \sin 6° = 160 \text{ lb}$$

$$\therefore T = 765 \text{ lb}$$ ■

5.2 Torque and General Conditions of Equilibrium

Let us now consider the situation in which forces act on an object of considerable dimensions. In this case we must drop the point particle model and investigate the equilibrium of forces that may not be concurrent. Consider the example of two children on a seesaw. Suppose that the weights of the children are w_1 and w_2 and that the weight of the beam is negligible. The free-body force diagram in Fig. 5.2 shows the beam in a horizontal position. There are no horizontal forces, and the balancing of vertical force components shows that the force at the balance point is $(w_1 + w_2)$ vertically upward.

However, balancing horizontal and vertical force components is not sufficient to achieve equilibrium. Experience teaches that

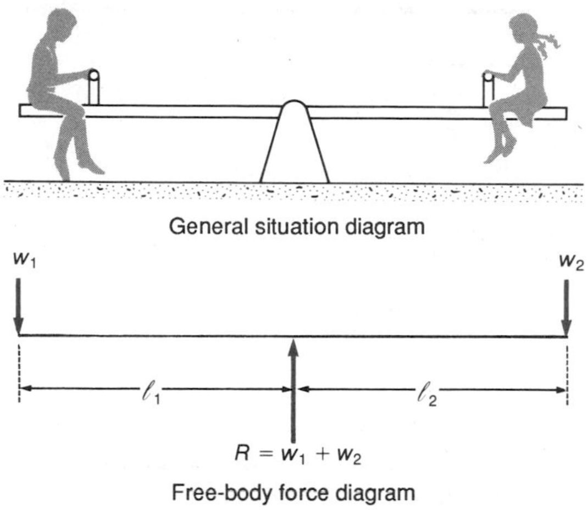

FIGURE 5.2 The force diagram for a seesaw.

the end of the seesaw that holds the heavier child drops if the balance point is at the center of the beam. To achieve a balance the support must be closer to the heavier child. To find the exact position for the support, we need to account for the twisting effect, called the **torque,** that the weight of each child causes about an axis through the hinge position. Let us proceed toward a general definition for torque by investigating the relatively simple seesaw problem in which the beam is horizontal and the forces that act on it are vertical. When a force acts on an object (for example, the weight of the child on the seesaw beam), the line of action of the force is a line drawn parallel to the force vector and passing through the point at which the force acts on the object. Thus, the lines of action of the weights of the two children are the vertical lines that pass through the points at which they sit on the beam. The torque of a force about an axis is the product of the force and the perpendicular distance from the axis to the line of action of the force. We call this distance the lever arm of the force about the axis in question. Let us select the axis for calculation of torques as the straight line through the balance point and perpendicular to the plane of the diagram. The lever arms of the weights about the axis are the distances ℓ_1 and ℓ_2 from the children to the balance point measured horizontally along the beam. The weight of each child produces a torque about the balance point; the torque due to the right-hand child tends to tilt the beam clockwise, and that due to the left-hand child tends to tilt it counterclockwise. By convention we assign a positive sign to counterclockwise torque and a negative sign to clockwise torque. With this convention, the beam balances if the net torque is zero. The weight of the first child exerts a counterclockwise torque and the weight of the second child a clockwise torque. The condition for zero net torque is:

$$w_1\ell_1 - w_2\ell_2 = 0$$

Later we will see that torque is the cause of angular acceleration, and the condition of zero torque is really a requirement for zero angular acceleration of the seesaw. Similar considerations apply to any extended object under the action of nonconcurrent forces.

We now proceed to set down a general definition for torque and to rewrite the conditions for static equilibrium. Figure 5.3 illustrates the calculation of torque when the force points in an arbitrary direction. The force vector **F** lies in the plane of the paper, and the axis about which we calculate the torque is perpendicular to the plane of the paper. We use the Greek letter tau (τ) as the symbol for torque, and $\ell_\perp$ as the symbol for lever arm. The subscript reinforces the idea of the lever arm as the perpendicular distance between the line of action of the force and the axis considered.

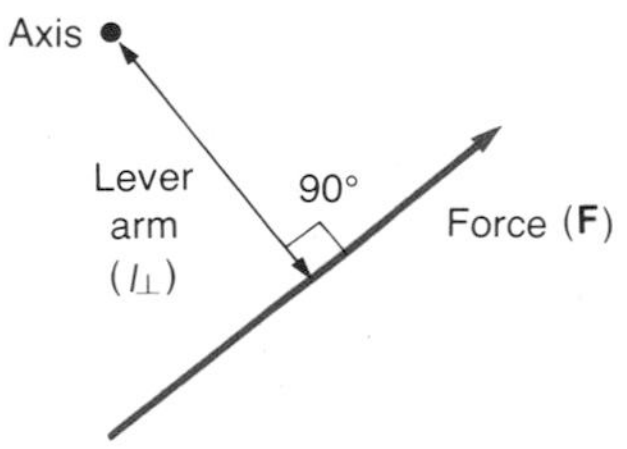

FIGURE 5.3 Diagram to illustrate the concept of lever arm of a force about a point.

■ Definition of Torque

$$\text{Torque} = \text{force} \times \text{lever arm}$$

$$\tau = F\ell_{\perp} \qquad (5.2)$$

SI unit of torque: newton-meter
Abbreviation: N · m

Let us rewrite the equations for static equilibrium so as to give mathematical form to this complete set of conditions.

■ Conditions of Equilibrium of Coplanar Forces

Net x component of force is zero: $\Sigma F_x = 0$ **(a)**

Net y component of force is zero: $\Sigma F_y = 0$ **(b)** **(5.3)**

Net torque is zero: $\Sigma \tau = 0$ **(c)**

EXAMPLE 5.4

Two children of weight 200 N and 250 N sit at opposite ends of a seesaw that is 4 m long. At what point should it be supported to achieve balance? (Neglect the weight of the seesaw beam.)

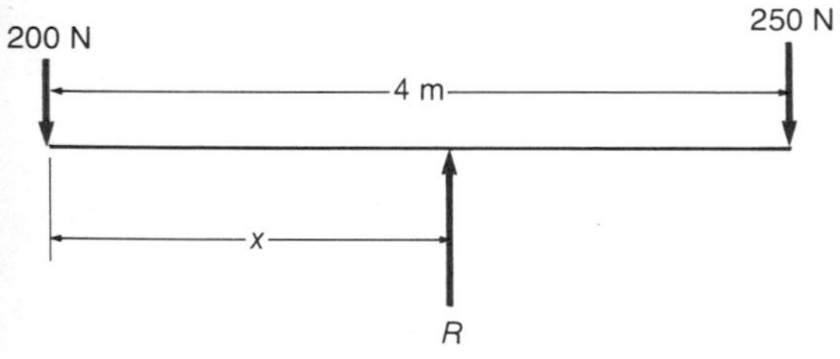

SOLUTION Let x be the distance from the child of weight 200 N to the support point, and R the force at the support. We apply the conditions of static equilibrium of Eq. (5.3):

a. Net x force component $= 0$

This equation is satisfied in a trivial way since there are no horizontal force components.

b. Net y force component $= 0$

$$\Sigma F_y = -200\ \text{N} - 250\ \text{N} + R = 0$$

$$\therefore R = 450\ \text{N}$$

c. Net torque $= 0$

Calculate the torques about an axis through the support point:

$$\Sigma \tau = 200\ \text{N} \times x - 250\ \text{N} \times (4\ \text{m} - x)$$

$$= 0$$

$$\therefore x = 2.22\ \text{m}$$

This gives the required position of the balance point. Our general requirement for zero net torque makes no stipulation about the position of the axis. An axis through the balance point is a natural choice, but we obtain exactly the same answer by requiring zero net torque about any axis. For example, let us calculate the torques about an axis through the point at which the 200-N force acts. The first two equilibrium conditions are exactly the same. The third condition gives:

c. Net torque $= 0$

$$\Sigma \tau = R \times x - 250 \text{ N} \times 4 \text{ m} = 0$$

but

$$R = 450 \text{ N}$$

$$\therefore x = \frac{1000 \text{ N} \cdot \text{m}}{450 \text{ N}}$$

$$= 2.22 \text{ m}$$

This is the same as the previous answer. In fact, we obtain the same result no matter what axis we use as a reference for the lever arm of the torques. In applying the equilibrium conditions Eq. (5.3), choosing an axis for calculating torques is entirely a matter of convenience. ■

EXAMPLE 5.5

A beam that is 5 m long is held horizontal by a smoothly hinged support at one end and by a cord at the other end; the cord makes an angle of 30° with the horizontal. A weight of 400 N is attached to the beam 3 m from the hinge. Calculate the tension in the cord and the horizontal and vertical components of the force at the hinge.

SOLUTION We begin by drawing the free-body force diagram. Let T be the tension in the cord, and let H_x and H_y be the horizontal and vertical components of the force at the hinge. (We arbitrarily selected the directions for the hinge force components since the algebra will supply a negative sign if the wrong direction is chosen.) The free-body force diagram is shown in (b), with the tension force inclined at 30° with the horizontal; the same diagram is repeated in (c) with the tension force given by way of its horizontal and vertical components. Proceeding with the series of steps of Eq. (5.3), we have

a. Net x force component $= 0$

$$\Sigma F_x = 0.866T + H_x = 0$$

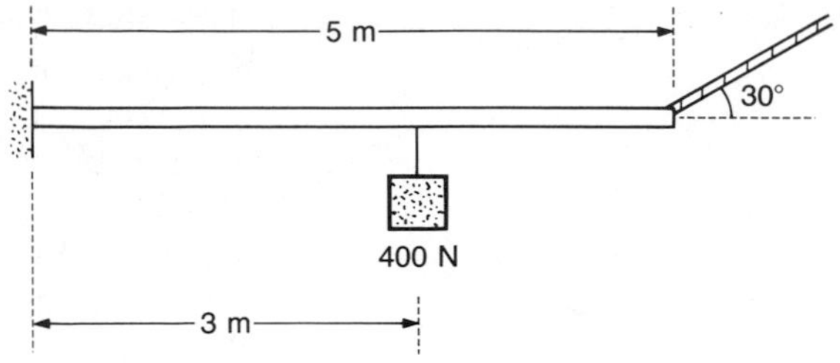

(a) General situation diagram

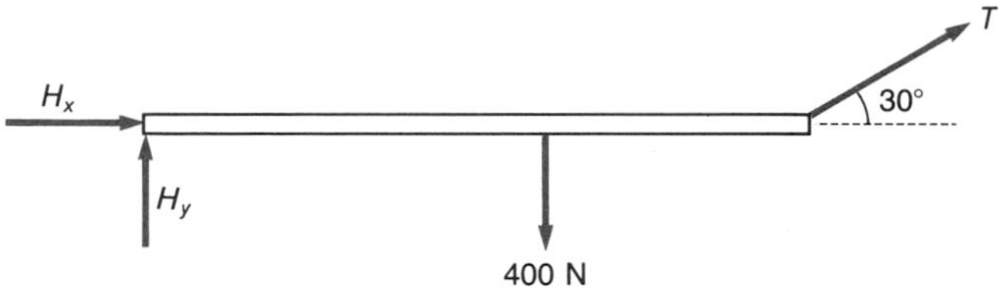

(b) Free-body force diagram

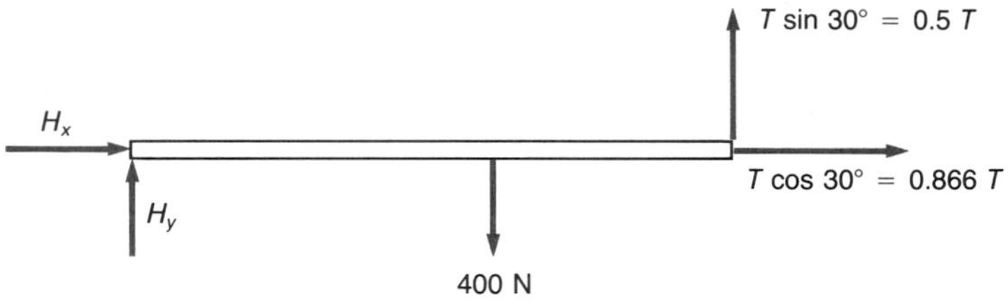

(c) Horizontal and vertical components of forces

b. Net y force component = 0

$$\Sigma F_y = H_y + 0.5T - 400 \text{ N} = 0$$

c. Net torque = 0

To apply this third condition, we need to select an axis for calculating torques. The hinge is convenient since the lever arms of both H_x and H_y about it are zero. In addition, inspection of the two equivalent free-body force diagrams shows an advantage in using (c) since the horizontal component of the tension also has zero lever arm about the hinge. The torque equation then becomes:

$$\Sigma \tau = 0.5T \times 5 \text{ m} - 400 \text{ N} \times 3 \text{ m} = 0$$

This equation at once gives us:

$$T = 480 \text{ N}$$

Substituting this value in the first and second equations gives:

$$H_x = -416\ \text{N}$$

and

$$H_y = 160\ \text{N}$$

Our solution shows that H_x is drawn on the diagram in the incorrect direction. However, it is not really appropriate to go back and make changes. We leave things as they are, with the direction shown for H_x on the diagram and the negative answer together providing correct information. ■

5.3 Center of Gravity

We could not treat the examples of the previous section by using a point particle model, but we introduced another fiction—beams of negligible weight. In any real situation the weight of the beam plays a role if the support point is not at the center. Complicated mathematics results if we include the weight of every small section of the beam in the free-body force diagram. A much more readily solvable problem emerges if we can represent the weight of the beam by a single force. The point through which the single force of weight acts regardless of the orientation of an object is called its **center of gravity** (abbreviated "cg"). In all cases of sufficiently high symmetry, the center of gravity is at the geometric center. A uniform beam, a uniform flat circular plate, and a uniform cube all have their centers of gravity at their geometric centers.

EXAMPLE 5.6

A decorative piece consists of a 20-cm-long uniform brass rod weighing 4 N and two ornaments, one weighing 6 N and one weighing 10 N, fixed to the ends. At what point should a supporting chain be attached so that the ornament balances?

SOLUTION Since the rod is uniform, its weight acts downward at the center of gravity, which is at its center. The general equilibrium conditions of Eq. (5.3) give:

a. Net x component of force $= 0$

There are no horizontal force components.

b. Net y component of force $= 0$

$$\Sigma F_y = R - 10\ \text{N} - 4\ \text{N} - 6\ \text{N} = 0$$

$$\therefore R = 20\ \text{N}$$

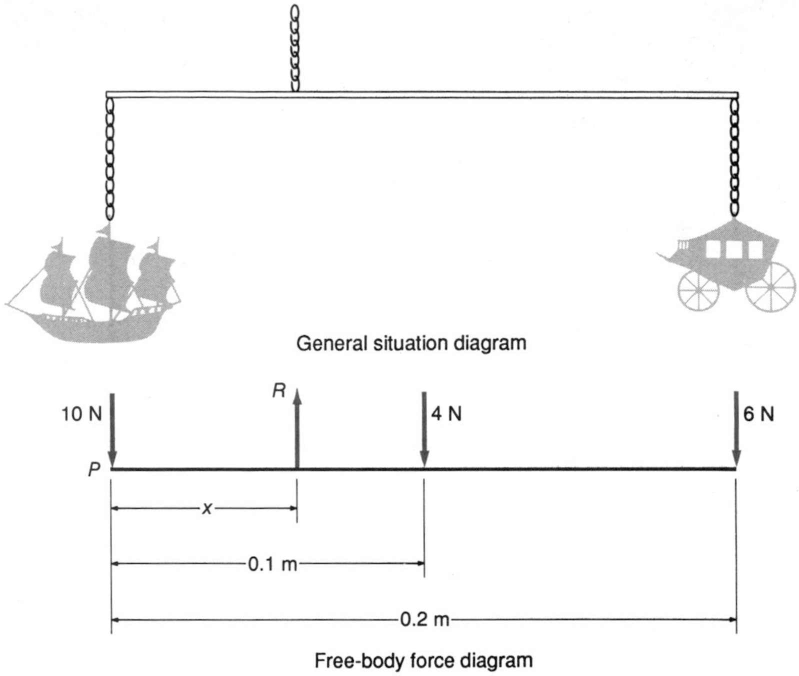

Let x be the distance of the supporting chain from the 10-N ornament. Let us calculate torques about an axis through the point P. The third equilibrium condition gives:

c. Net torque = 0

$$\Sigma \tau = R \times x - 4\text{ N} \times 0.1\text{ m} - 6\text{ N} \times 0.2\text{ m} = 0$$

but

$$R = 20\text{ N}$$

$$\therefore x = \frac{1.6\text{ N} \cdot \text{m}}{20\text{ N}}$$

$$= 0.08\text{ m}$$

We should place the support 8 cm from the heavier ornament. ■

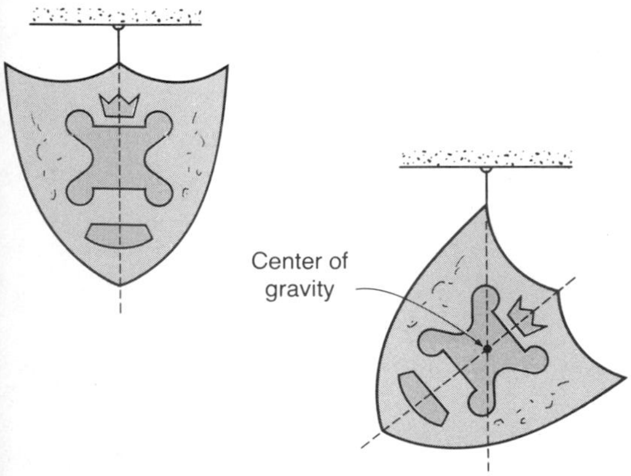

FIGURE 5.4 Find the position of the center of gravity of an object.

We can locate the center of gravity of a highly symmetrical object by inspection, as in the above example. However, for more complicated objects the center of gravity may be harder to find. If an object is flat we can find the position of the center of gravity experimentally. To do this we need only suspend it by a thread from several positions. Consider, for example, a wooden shield bearing a family coat of arms that is suspended by a thread from the top center, as shown in the first part of Fig. 5.4. The dotted line down the center of the shield indicates a continuation of the line of the supporting thread. The two forces acting on the shield are its weight downward, and the upward force due to the tension in the thread. Since the shield is in equilibrium, these two forces are of equal magnitude and act in opposite directions along the same line. It follows that the center of gravity of the shield lies on the new continuation of the line of the thread. We now hang the shield from another point (second part of the diagram); the center of gravity again lies on the dotted line of continuation of the supporting thread. The point of intersection of the two dotted lines is the center of gravity of the shield. If we suspend it from any other point, the continuation of the line of the thread passes through this point.

We can calculate the position of the center of gravity of a compound object if it is composed of symmetrical parts and has an axis of symmetry. Since the weight acts through the center of gravity, it is only necessary to find the point on the symmetry axis at which a support will balance the object. Let us illustrate by examples.

EXAMPLE 5.7

The pendulum of a grandfather clock is composed of a uniform mahogany rod 82 cm long that weighs 4 N and a brass disk 18 cm in diameter that weighs 10 N. Find the position of the pendulum's center of gravity.

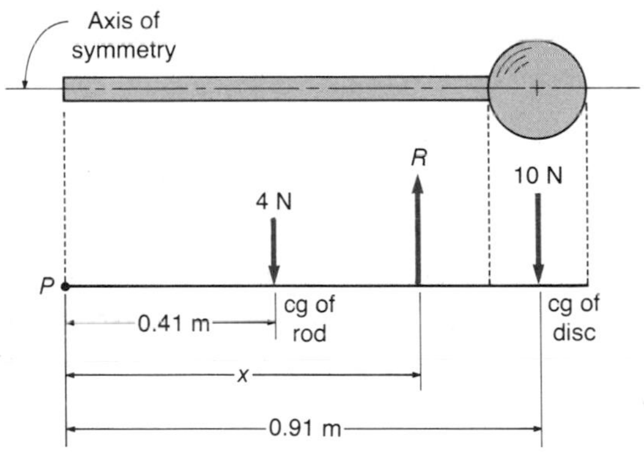

SOLUTION The center of gravity of the pendulum lies on its axis at the balance point. Imagine that we support the pendulum at a distance x from the mahogany rod end. The first condition of Eq. (5.3) is trivial since there are no horizontal force components. The second condition gives the support force as $R = 14$ N. To apply the third condition, we calculate torques about an axis through the end point of the pendulum (point P on the diagram):

$$\text{Net torque} = 0$$

$$\Sigma \tau = 14 \text{ N} \times x - 4 \text{ N} \times 0.41 \text{ m} - 10 \text{ N} \times 0.91 \text{ m} = 0$$

$$\therefore x = \frac{10.74 \text{ N} \cdot \text{m}}{14 \text{ N}}$$

$$= 0.767 \text{ m}$$

The center of gravity of the pendulum is 76.7 cm from the top end.

We can use the same method to make estimates even if composite parts are not highly symmetrical. ■

EXAMPLE 5.8

Assume that a man 1.80 m tall and weighing 830 N (187 lb) is composed approximately of the following uniform parts:

Body part	*Weight*	*Approximate shape and size*
Head	50 N	25-cm-diameter sphere
Trunk	440 N	65-cm-long cylinder
Arm (each)	45 N	58-cm-long cylinder
Thigh (each)	75 N	35-cm-long cylinder
Lower leg (each)	50 N	55-cm-long cylinder

Estimate the height of the center of gravity above ground level when the man is standing erect with his arms by his side.

SOLUTION Draw the model of the man, and mark the center of gravity of each of the uniform parts. The center of gravity of the whole man is his point of balance if he were turned on his side while remaining stiffly in the position shown. The free-body force diagram includes the weights of all the parts acting through the centers of gravity. The first two conditions of Eq. (5.3) simply give the supporting force $R = 830$ N. In applying the third condition, we calculate the torques of the forces about an axis through the position of the man's feet.

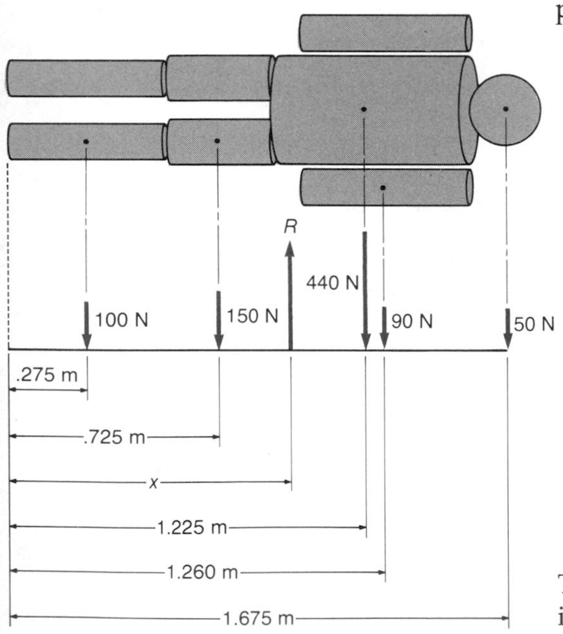

$$\text{Net torque} = 0$$

$$\begin{aligned} \Sigma\,\tau = {} & 830\text{ N} \times x \\ & - 100\text{ N} \times 0.275\text{ m} \\ & - 150\text{ N} \times 0.725\text{ m} \\ & - 440\text{ N} \times 1.225\text{ m} \\ & - 90\text{ N} \times 1.26\text{ m} \\ & - 50\text{ N} \times 1.675\text{ m} = 0 \end{aligned}$$

$$\therefore x = \frac{872.4\text{ N}\cdot\text{m}}{830\text{ N}}$$

$$= 1.05\text{ m}$$

The approximate nature of the model hardly justifies three significant digits in the answer. A man 1.8 m tall having the approximate proportions of the model has his center of gravity about 1 m above his feet for the position shown. Moreover, the calculation gives only the height of the center of gravity above the ground. A man is symmetrical side to side, so that the center of gravity is centrally positioned. However, since a man is not symmetrical back to front, calculating the exact position of the center of gravity in this direction requires a much more detailed model. ■

In all the examples above, the center of gravity is within the body. However, the center of gravity often lies outside the body, as in the example below.

EXAMPLE 5.9

A uniform wire 30 cm long and weighing 1.8 N is bent into three sides of a square. Find the position of the center of gravity.

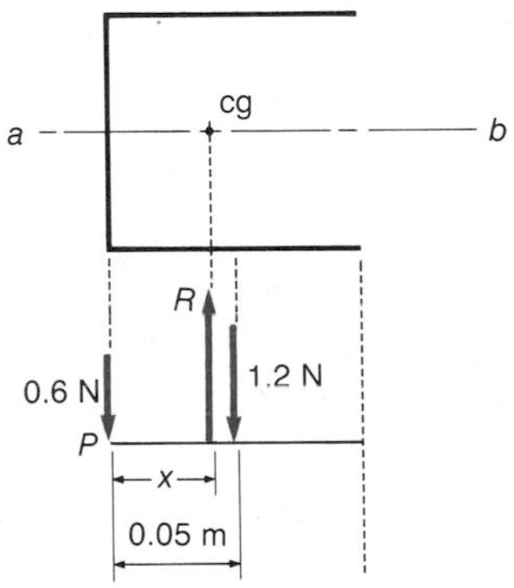

SOLUTION Each side of the object is 0.1 m long and weighs 0.6 N. By symmetry the center of gravity is on the dotted line marked *ab* on the diagram. To calculate the exact position, we find the balance point. The free-body force diagram includes the weights of all three sides of the object. The first two conditions of Eq. (5.3) give $R = 1.8$ N. If we calculate torques about an axis through the point P on the diagram, the third equilibrium condition gives:

$$\text{Net torque} = 0$$

$$\Sigma \tau = R \times x - 1.2 \text{ N} \times 0.05 \text{ m} = 0$$

but

$$R = 1.8 \text{ N}$$

$$\therefore x = \frac{0.06 \text{ N} \cdot \text{m}}{1.8 \text{ N}}$$

$$= 0.0333 \text{ m}$$

The center of gravity of the object is on the dotted line 3.33 cm from the wire. ■

5.4 Statics of Rigid Bodies

In our discussion of the seesaw problem, we assumed that the beam does not bend. This is another fictional model, since all bodies deform at least a small amount under the action of applied forces. We define a *rigid body* as one that does not deform. In some instances this fictional model is very good. In ordinary use, many strongly constructed objects of wood, metal, or plastic deform very little under the action of forces. Treating such objects as rigid is usually an excellent approximation when analyzing their behavior under the action of applied forces.

Many examples of rigid bodies are more complex than the uniform beams that we have considered so far. In general, the forces acting on a rigid body may have arbitrary directions. Provided we require these forces to be coplanar, the general conditions of static equilibrium, Eq. (5.3), continue to hold good. Many examples of rigid body statics are more difficult to solve than the problem of a rigid horizontal beam with all forces acting vertically.

EXAMPLE 5.10

A uniform, rectangular steel door 3 m high and 1 m wide weighs 500 N. Its hinges are 25 cm down from the top and 25 cm up from the bottom. The design is such that the lower hinge supports the entire weight of the door, while the upper hinge supplies only a horizontal force. Calculate the force that each hinge exerts on the door.

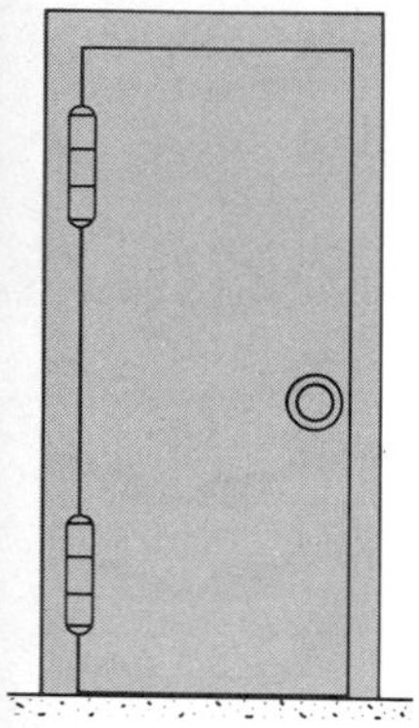

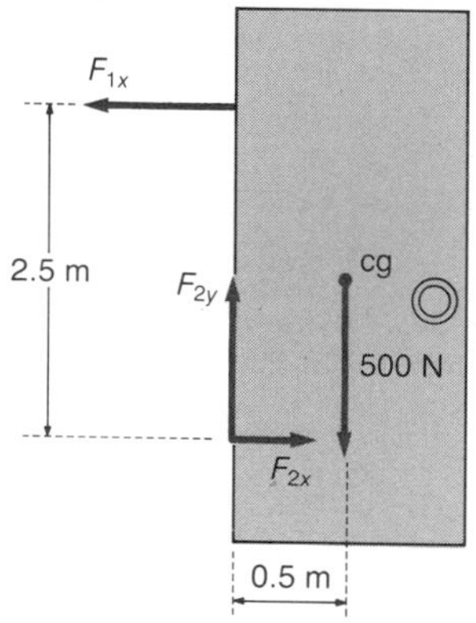

SOLUTION We begin by drawing the free-body force diagram for the door. The forces are the weight of 500 N vertically downward through the center of gravity, an unknown force $\mathbf{F}_1$ at the upper hinge, and another unknown force $\mathbf{F}_2$ at the lower hinge. Since the problem stipulates that $\mathbf{F}_1$ is horizontal, its vertical component F_{1y} is zero, and only F_{1x} appears on the force diagram. On the other hand, $\mathbf{F}_2$ may point in any direction, and it has both a horizontal component F_{2x} and a vertical component F_{2y}. We need all three equilibrium conditions of Eq. (5.3) to solve this problem:

a. Net x component of force $= 0$

$$\Sigma F_x = F_{2x} - F_{1x} = 0$$

b. Net y component of force $= 0$

$$\Sigma F_y = F_{2y} - 500\text{ N} = 0$$

$$\therefore F_{2y} = 500\text{ N}$$

Recall that we can calculate torques with reference to any convenient axis. If we use an axis through the lower hinge position, neither F_{2x} nor F_{2y} appears in the torque equation, since their lever arms about the lower hinge are zero.

c. Net torque $= 0$

$$\Sigma \tau = F_{1x} \times 2.5\text{ m} - 500\text{ N} \times 0.5\text{ m} = 0$$

$$\therefore F_{1x} = \frac{250\text{ N}\cdot\text{m}}{2.5\text{ m}}$$

$$= 100\text{ N}$$

From the equation for horizontal force components:

$$F_{2x} = 100\text{ N}$$

All the components of force at the hinges are now determined. ■

EXAMPLE 5.11

A 5-m-long ladder weighing 200 N has its center of gravity in the center. The top of the ladder rests against a *smooth* vertical wall, and the lower end on a *rough* floor; the coefficient of static friction between the floor and the ladder is 0.25. Calculate the largest angle that the ladder can make with the vertical before slipping.

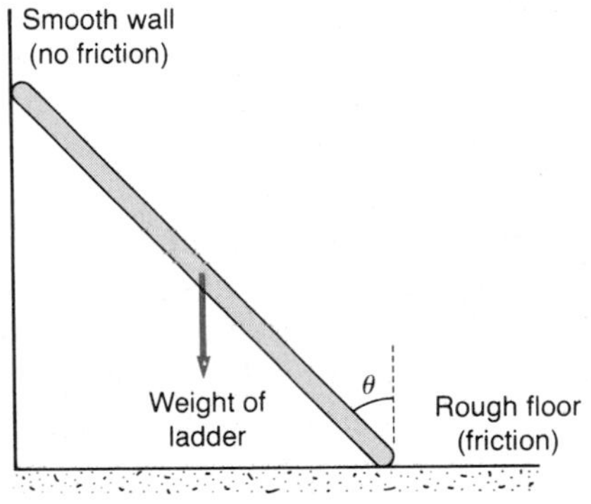

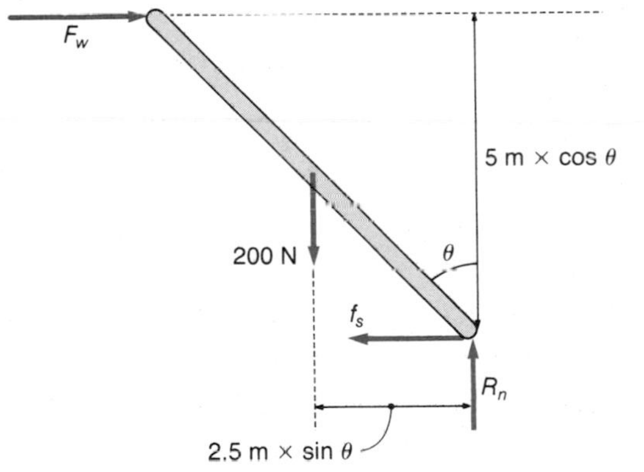

SOLUTION Let θ be the angle that the ladder makes with the vertical when it is on the point of slipping. Let F_W be the force of the wall on the ladder; this is a horizontal force since there is no friction. We must include both the horizontal and vertical components of the force of the floor on the ladder, which are the static frictional force f_s and the normal force R_n, respectively. When the ladder is on the point of slipping, we can use Eq. (4.3) to give $f_s = \mu_s R_n$. The only other force on the ladder is the weight of 200 N acting downward through the center of gravity. We show all of these forces on the free-body force diagram. We can now apply the conditions of static equilibrium Eq. (5.3):

a. Net x component of force $= 0$

$$\Sigma F_x = F_W - f_s = 0$$

b. Net y component of force $= 0$

$$\Sigma F_y = R_n - 200 \text{ N} = 0$$

From these equations we have:

$$R_n = 200 \text{ N}$$

$$F_W = f_s = \mu_s R_n$$

$$= 0.25 \times 200 \text{ N} = 50 \text{ N}$$

Let us calculate torques about an axis through the bottom of the ladder. The lever arm of the force F_W is 5 m × cos θ, and that of the 200-N weight is 2.5 m × sin θ. The third equilibrium condition gives:

c. Net torque $= 0$

$$\Sigma \tau = 200 \text{ N} \times 2.5 \text{ m} \times \sin\theta - 50 \text{ N} \times 5 \text{ m} \times \cos\theta = 0$$

$$\therefore \tan\theta = 0.5$$

$$\theta = 26.6°$$

■

We have already seen that units conversion is unnecessary when setting the net force component in some direction equal to zero—we need only keep the same units for every term in the equation. Similarly, we need not convert units in the condition for zero net torque if each term is expressed in the same units. Let us illustrate by an example.

EXAMPLE 5.12

A uniform steel beam weighing 100 lb is hinged to a vertical wall and supports a 400-lb weight at its outer end. The beam is held in position by a horizontal guy wire 5 ft long attached to a point on the wall 4 ft vertically above the hinge. Calculate the tension in the guy wire, and the magnitude of the force at the hinge.

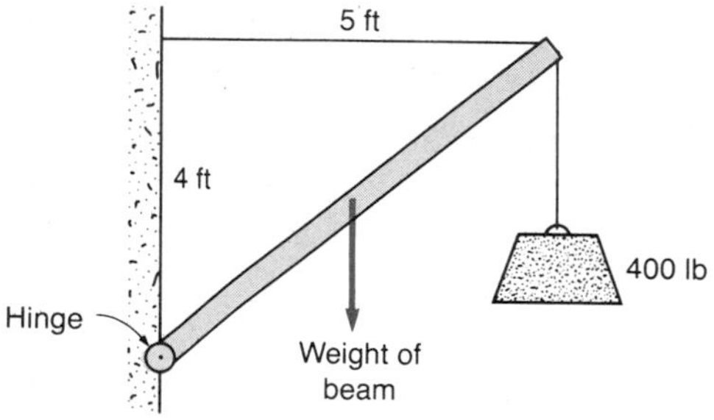

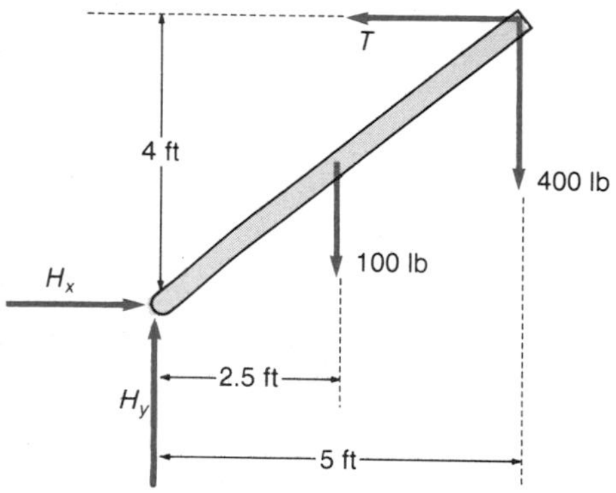

SOLUTION Let T be the tension in the guy wire, and H_x and H_y the horizontal and vertical components of the force at the hinge. The other forces on the free-body force diagram for the beam are its weight of 100 lb and the 400-lb weight at its outer end. We can now apply the conditions of static equilibrium, Eq. (5.3), to the beam:

a. Net x component of force = 0

$$\Sigma F_x = H_x - T = 0$$

b. Net y component of force = 0

$$\Sigma F_y = H_y - 100 \text{ lb} - 400 \text{ lb} = 0$$

$$\therefore H_y = 500 \text{ lb}$$

To apply the third condition, calculate the torques about an axis through the hinge. The lever arms of the forces H_x and H_y are zero about this axis, and the lever arms of the other forces are marked on the diagram.

c. Net torque = 0

$$\Sigma \tau = T \times 4 \text{ ft} - 100 \text{ lb} \times 2.5 \text{ ft} - 400 \text{ lb} \times 5 \text{ ft} = 0$$

$$\therefore T = 562 \text{ lb}$$

The first equation then gives:

$$H_x = 562 \text{ lb}$$

Note that we do not require units conversion, since every term in the equation is expressed in the same units. To calculate the magnitude of the force at the hinge we use Eq. (3.4):

$$\begin{aligned} H &= (H_x^2 + H_y^2)^{1/2} \\ &= [(562 \text{ lb})^2 + (500 \text{ lb})^2]^{1/2} \\ &= 752 \text{ lb} \end{aligned}$$

■

5.5 Muscle Forces in the Human Body

The human body produces its motion from a complicated arrangement of bones and muscles. The skeletal bones are linked to each other at joints that are more or less flexible, depending on the location. The muscles are usually attached to two different bones by means of tendons, which produce the forces that cause the relative motion of the bones.

Muscles are composed of a great number of long thin fibers that contract briefly under the stimulation of electrical impulses from the nerves. Each brief contraction is called a twitch, and a continued series of electrical impulses from the nerves causes a corresponding series of twitches in the muscle fibers. The whole process, with many electrical impulses acting on different fibers at different times, results in a smooth, overall contraction of the muscle. If the rate of electrical impulses from the nerves increases, then the force of contraction of the muscle increases. There is, however, a maximum muscle tension, which corresponds to approximately 35 N/cm^2 of muscle cross section. Increasing the rate of electrical impulses beyond that required for maximum muscle tension has no further effect.

The contracting muscle causes two pairs of action-reaction forces where the tendons at each end join the different bones. Many of the muscle-bone combinations in the human body are exceedingly complex, but the arm presents relatively simple examples of the principles involved.

We illustrate this by investigating the way in which forces in the muscles of the upper arm enable one to exert upward or downward forces with the hand. To simplify matters, let us assume that the upper arm is vertical and the forearm is horizontal. Figure 5.5(a) shows a weight held in the hand; the humerus in the upper arm is vertical and jointed at the elbow to the radius and the ulna in the horizontal lower arm. Ignoring all the forces in the wrist and fingers, we draw the simple model shown in Fig. 5.5(b). The forearm and the hand are represented by a rigid rod whose weight acts downward through its center of gravity; the weight in the hand produces a downward force on the end of the rod. The weight of the arm acts downward through its center of gravity, and static equilibrium is achieved by tension in the biceps together with the downward force of the humerus at the elbow joint. The rigid rod that represents the forearm is not a uniform rod since the arm is not uniform; its center of gravity is closer to the elbow joint than to the fingers.

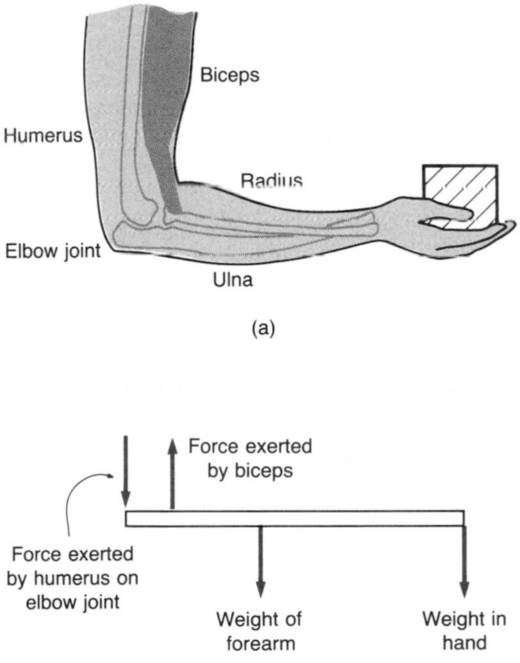

FIGURE 5.5 Supporting a weight in the hand with the forearm horizontal: (a) the muscles and bones involved; (b) simplified force diagram, with the whole forearm represented by a rigid rod.

EXAMPLE 5.13

A person holds a 6-kg mass in the manner indicated in Fig. 5.5. The distance between the center of gravity of the mass and the elbow joint is 36 cm. The arm has a weight of 15 N, and its center of gravity is 14 cm from the elbow joint; the biceps's tendon is

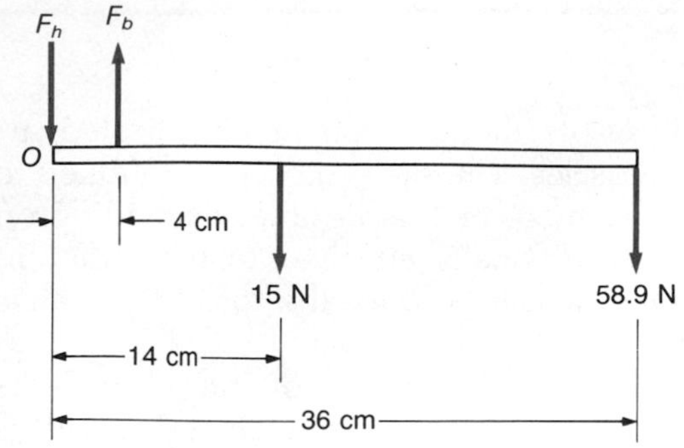

attached 4 cm from the elbow joint. Calculate the forces exerted by the biceps and the humerus on the forearm.

SOLUTION The weight of the 6-kg mass is found from Eq. (4.2):

$$\begin{aligned} w &= mg \\ &= 6 \text{ kg} \times 9.81 \text{ m/s}^2 \\ &= 58.9 \text{ N} \end{aligned}$$

Our diagram for the problem is essentially the same as that shown in Fig. 5.5(b); we have set F_b as the force due to the biceps and F_h as the force due to the humerus. There are no horizontal forces acting on the forearm, and the conditions for static equilibrium, Eq. (5.3), reduce to two. For vertical force components, we have:

$$F_b - F_h - 15 \text{ N} - 58.9 \text{ N} = 0$$

The elbow joint (point O on the diagram) is a convenient pivot point about which to calculate torques. We have:

$$F_b \times 0.04 \text{ m} - 15 \text{ N} \times 0.14 \text{ m} - 58.9 \text{ N} \times 0.36 \text{ m} = 0$$

From the second equation, we have:

$$F_b = 583 \text{ N}$$

Substitution in the first equation then gives:

$$F_h = 509 \text{ N}$$

Note that tension in the biceps and the force exerted by the humerus are both very much greater than the weight supported by the hand. Note also that we cannot use this model to make any calculations about muscle or bone forces in the forearm, wrist, and hand because we have represented all three by a single rigid rod. ■

Figure 5.6 shows the somewhat more complicated example of the whole arm held in a horizontal position by a combination of the deltoid muscle and the shoulder joint. The simplified force diagram shows the whole arm represented by a nonuniform rigid rod, with the weight acting downward through the center of gravity. The deltoid is attached to the humerus, and the tension in it is represented by a single force at some angle (usually in the range 15° to 18°) with the arm. The force exerted by the shoulder joint has, in general, both horizontal and vertical components.

Similar analyses can be carried out for other parts of the human body; increasing numbers of muscles and bones involved in an operation make the procedure more complicated, but no new principles are required.

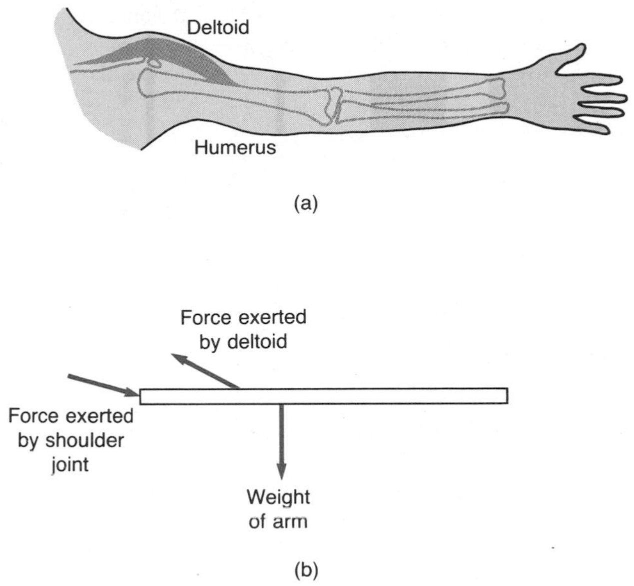

FIGURE 5.6 Holding the arm horizontal: (a) the muscles and bones involved; (b) simplified force diagram, with the whole arm represented by a rigid rod.

KEY CONCEPTS

Static equilibrium of a point mass requires zero acceleration in every direction. For **coplanar forces,** the net force component in any two mutually perpendicular directions must be zero. See Eq. (5.1).

The **torque** of a force about an axis is the product of the force and its lever arm about the axis. The torque is either **clockwise** or **counterclockwise** about the axis. See Eq. (5.2).

Static equilibrium of a rigid body requires zero acceleration in every direction and zero angular acceleration about any axis. For coplanar forces, the two conditions for a zero net force component must be supplemented by the condition of zero net torque. See Eq. (5.3).

The **center of gravity** of an extended body is the point through which the force of weight acts, regardless of the orientation of the body. For a highly symmetrical object, the center of gravity is the geometric center. Locating a balance point helps in finding the center of gravity of a less symmetrical object.

QUESTIONS FOR THOUGHT

1. "Statics is a special case of dynamics." Explain this statement.
2. Can a body be in motion and at the same time be in equilibrium? Can a body be at rest and at the same time not be in equilibrium?
3. A group of cheerleaders forms a human pyramid, with three standing on the bottom row, two standing on their shoulders, and one standing in the top row. Estimate the force exerted on the floor by each of the three cheerleaders in the bottom row.
4. The ends of a heavy steel cable are attached to the tops of two tall, adjacent buildings. Explain why it is not possible to pull the cable so taut that it is exactly horizontal at all points.

5. A water-skier is being towed across a lake at constant velocity. Is she in equilibrium under the action of all the forces acting on her? Draw a free-body force diagram, and discuss the effect of the skier's stance on maintaining her position.

6. A strong young man is asked to stand against a vertical wall with his feet together and his left foot hard against the wall. He is embarrassed to find that he cannot lift his right foot off the ground without falling over. Is this really a test of his strength?

7. Is your own center of gravity at a fixed position within your body, or are there simple things that you can do to change its position relative to the rest of your body?

PROBLEMS

A. Single-Substitution Problems

1. A child weighing 250 N sits on a swing attached to an overhead support. The swing is drawn aside by a horizontal force until the supporting rope makes an angle of 27° with the vertical.

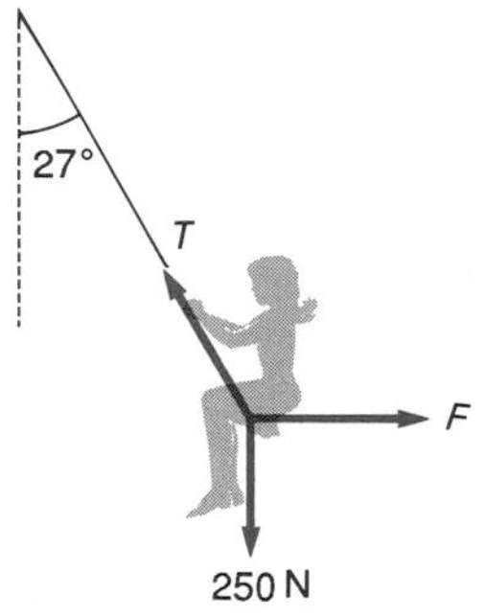

a. What force is required to pull the swing?
b. What is the tension in the rope? [5.1]

2. A traffic light is supported by a cable that makes an angle of 12° with the horizontal on each side of the light. The maximum safe tension in the cable is 620 N. What is the maximum safe weight for the light? [5.1]

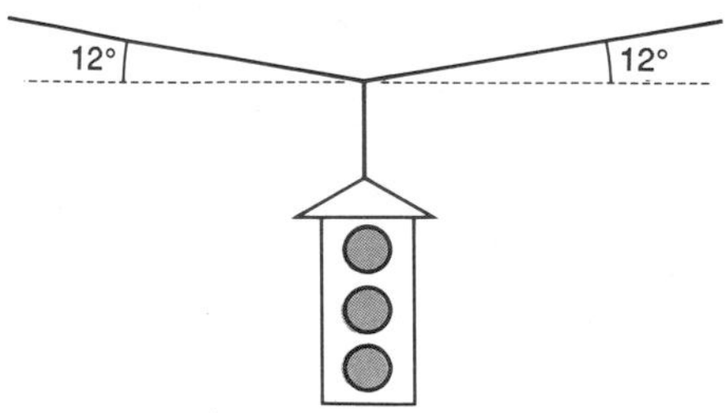

3. Calculate the weight of the suspended object, given that the tension in the horizontal cord shown in the diagram is 250 N. [5.1]

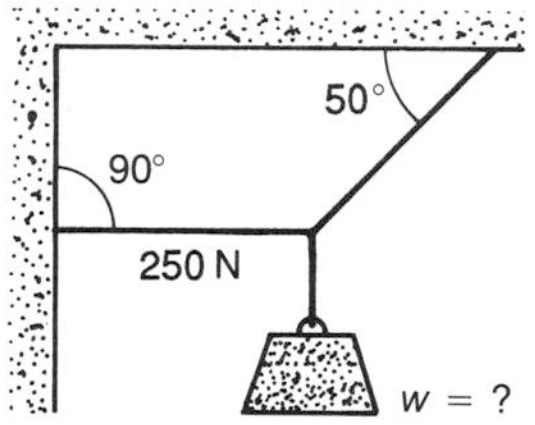

4. Calculate the tensions in the two cords that support the 460-N weight, as shown in the diagram. [5.1]

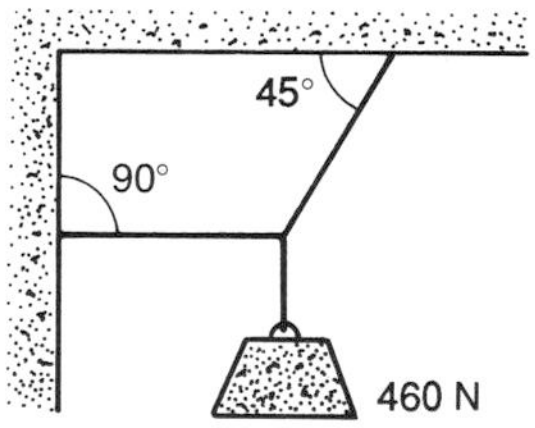

5. Can the three forces shown in the diagram be in equilibrium for any value of the angle θ? [5.1]

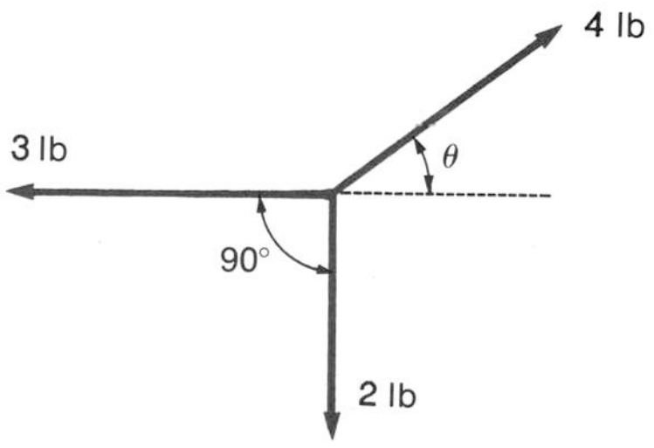

6. To pull an automobile from a ditch, a rope 20 m long is tied to the automobile and a post that is 18.6 m away. A force of 600 N is applied to the center of the rope, as shown in the diagram. Calculate the tension in the rope. [5.1]

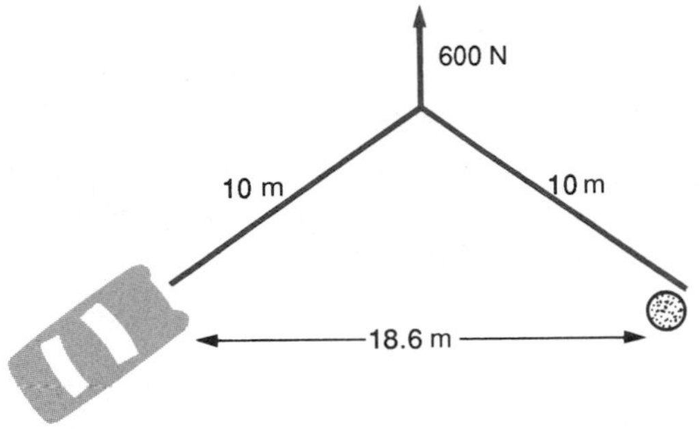

7. The steering wheel of an automobile is 50 cm in diameter, and the driver exerts a force of 7 N at the rim (and tangential to it) in order to turn a corner. What torque does she apply to the steering column? [5.2]

8. A mechanic uses a wrench 24 cm long to undo a nut. What force (applied perpendicular to the wrench) is required to exert a torque of 110 N · m on the nut? [5.2]

9. A rod of negligible weight is 26 cm long. With a mass of 3 kg attached to one end and an unknown mass at the other end, the rod balances 11 cm from the 3-kg mass. What is the unknown mass? [5.2]

10. Weights of 8 lb and 14 lb are attached to the ends of a rod that is 5 ft long and of negligible weight. At what point will the rod balance? [5.2]

11. A uniform rod of length 45 cm and mass 3 kg has a uniform spherical object of mass 4 kg and radius 10 cm attached to one end. Where is the center of gravity of the system? [5.3]

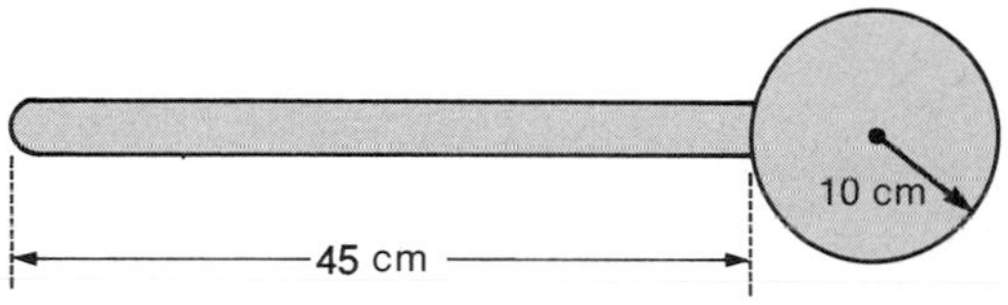

12. Two square steel plates of different thickness are welded together edge to edge, as shown in the diagram. One plate has a mass of 3 kg and dimensions 25 cm × 25 cm; the other plate has dimensions 40 cm × 40 cm. The center of gravity is right in the center of the welded joint. What is the mass of the second plate? [5.3]

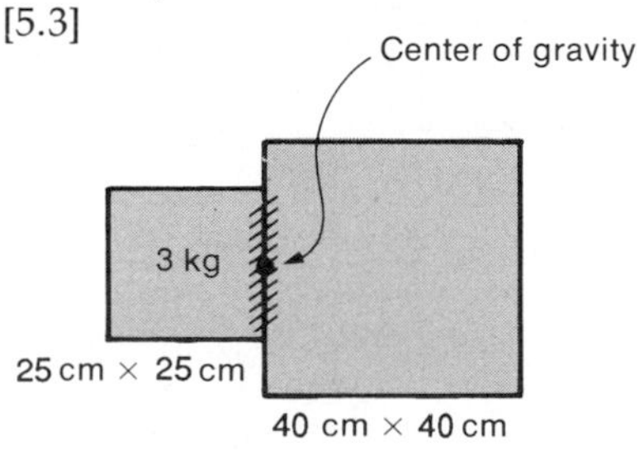

13. Two spheres of the same material having radii of 3 cm and 4 cm are fixed to the ends of a rod of negligible mass that is 16 cm long. Where is the center of gravity of the system? [5.3]

14. A square plate of edge length 40 cm and mass 8 kg has a 3-kg point mass attached to one corner. Where is the center of gravity of the system? [5.3]

B. Standard-Level Problems

15. Three bags of cement on a wooden platform (total weight 270 lb) are supported at rest by the hoisting device shown in the diagram. The weight of the pulley is negligible. Find the tension in the cord supporting the pulley and the angle that the cord makes with the vertical. [5.1]

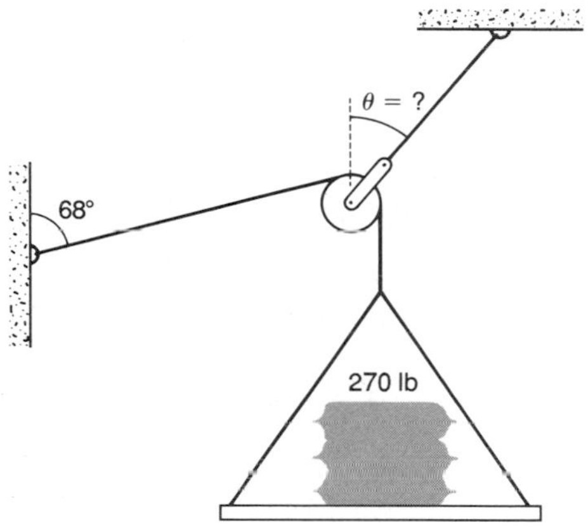

16. Calculate the tension in the cords shown in the diagram that support a weight of 14 lb. [5.1]

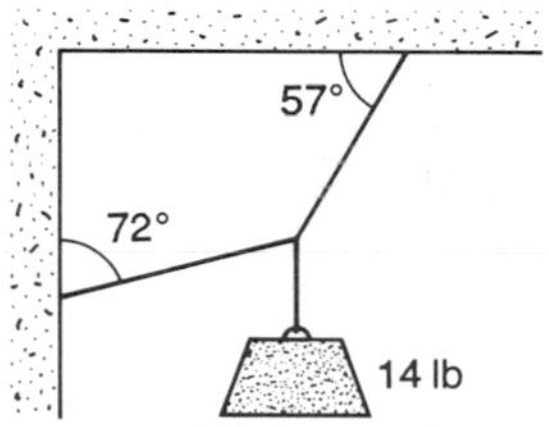

17. A tightrope walker whose mass is 65 kg stands at the midpoint of his wire, which is fixed to rigid supports 16 m apart. The wire sags a distance of 1.3 m below the support points.

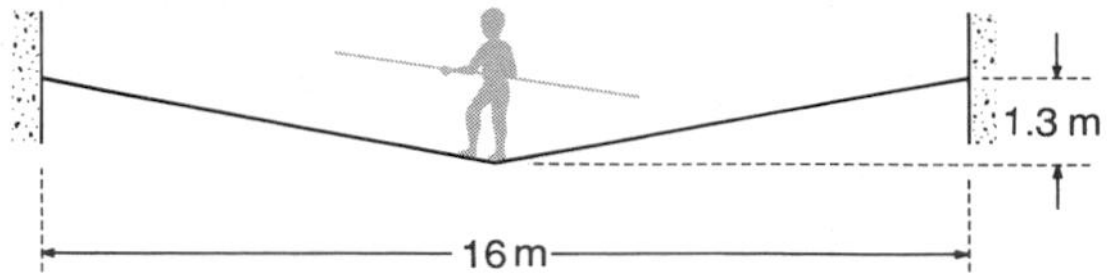

a. What is the length of the wire? (Assume that the wire does not stretch.)

b. What is the tension in the wire?

18. The diagram shows a hospital apparatus used for exerting traction force on a patient's leg. Calculate the magnitude of this force.

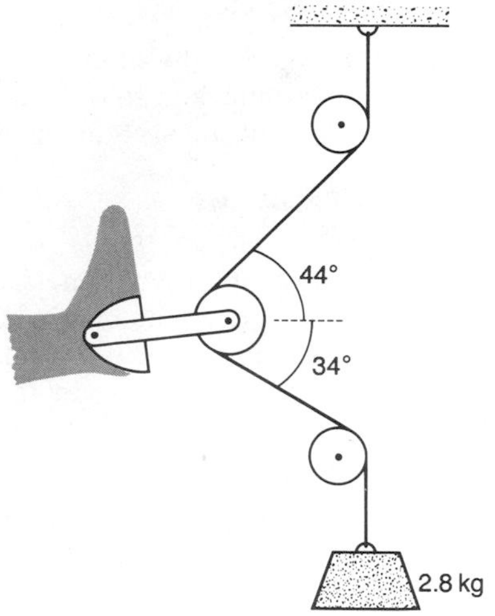

19. A steel ball bearing of mass 1.2 kg sits in a V groove, of which one wall is vertical and the other is inclined at 55° with the horizontal. There is no friction between the ball and either wall. Calculate the force of reaction between the ball and each of the walls.

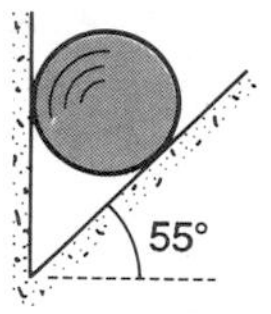

20. A toboggan is pulled across the snow by two horizontal ropes so that the net force is in the direction of travel. The 40-lb tension in one rope makes an angle of 25° with the direction of travel. The other rope makes an angle of 40° with this direction.

a. What is the tension in the second rope?
b. If the toboggan is moving at constant velocity, what is the frictional force resisting its motion?

21. A horizontal beam 6 m long has negligible weight and is supported by vertical ropes attached to each end.

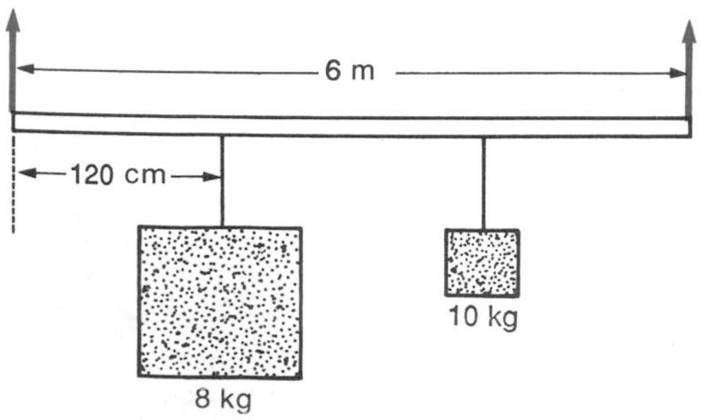

A mass of 8 kg is attached 120 cm from one end of the beam. Where should a mass of 10 kg be attached so that the tensions in the two ropes are equal? [5.2]

22. A 4.6-m-long beam of negligible weight is supported by a pair of vertical ropes 80 cm in from each end. A mass of 6 kg is hung from one end of the beam, and one of 4 kg from the other. What is the tension in the ropes? [5.2]

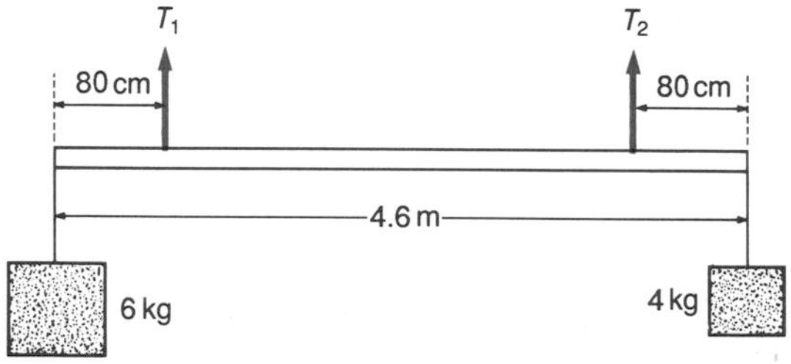

23. Two persons wish to carry a heavy uniform beam 6 m long and weighing 820 N; the beam has a mass of 60 kg placed 2 m from one end. One of them (who is to carry only one-third of the total weight) picks up the beam at the end farthest from the added mass. At what point should the other person hold the beam? [5.3]

24. A market gardener sets off on his rounds with 12 kg of vegetables in each basket at either end of a shoulder pole 3.2 m long with a mass of 2.8 kg. At his first stop he sells 3 kg from one basket. If he does not wish to equalize the loads in the baskets, by what distance must he shift the position of the pole on his shoulder to achieve a balance? [5.3]

25. A circus performer of mass 80 kg uses a 15-m beam weighing 1800 N to do the trick shown in the diagram. The beam is not held down in any way. If he wishes to walk to the very end of the beam, how far out over the edge can he place the end?

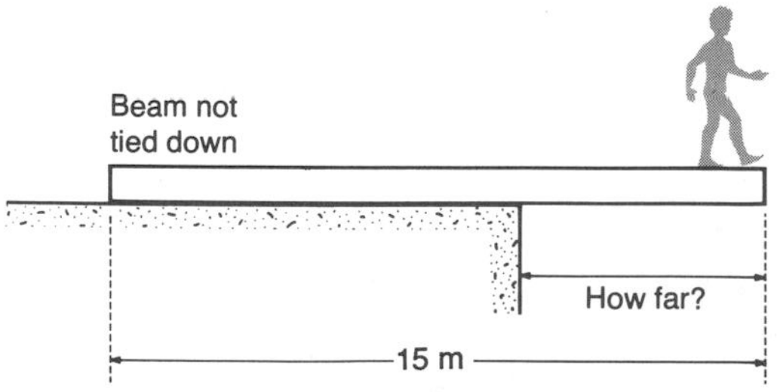

26. The dimensions of a mobile made from rods weighing 8 N per meter of length are shown in the diagram. If the first ornament weighs 3 N, what must the weight of the other two ornaments be in order to achieve a balance?

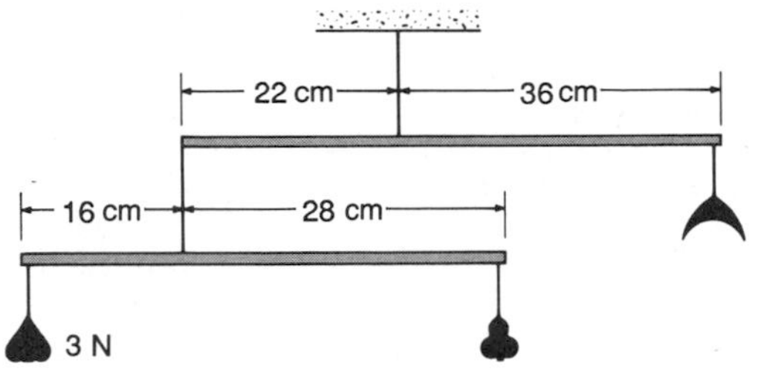

27. An 80-kg painter works on a plank 6.8 m long that has a mass of 35 kg. He wishes to place two supports so that the plank does not tip regardless of where he stands. What is the greatest distance in from the ends of the plank that the supports can be safely placed?

28. The bar of a barbell has a mass of 8 kg and is 1.6 m long. A weightlifter places a 25-kg mass 25 cm from one end and a 40-kg mass 25 cm from the other end. At what point should it be lifted so as to balance?

29. A uniform diving board 4.5 m long has a mass of 30 kg. It is fixed by two supports 60 cm and 120 cm from one end, which can supply supporting forces either upward or downward. What are the magnitude and direction of the forces at these support points if a 70-kg diver stands on the end of the board?

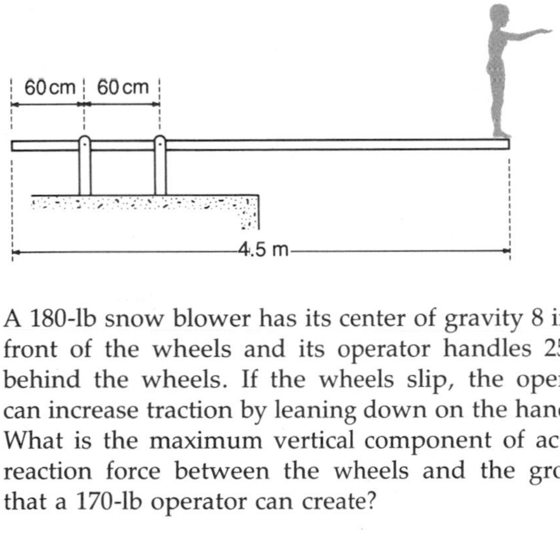

30. A 180-lb snow blower has its center of gravity 8 in. in front of the wheels and its operator handles 25 in. behind the wheels. If the wheels slip, the operator can increase traction by leaning down on the handles. What is the maximum vertical component of action-reaction force between the wheels and the ground that a 170-lb operator can create?

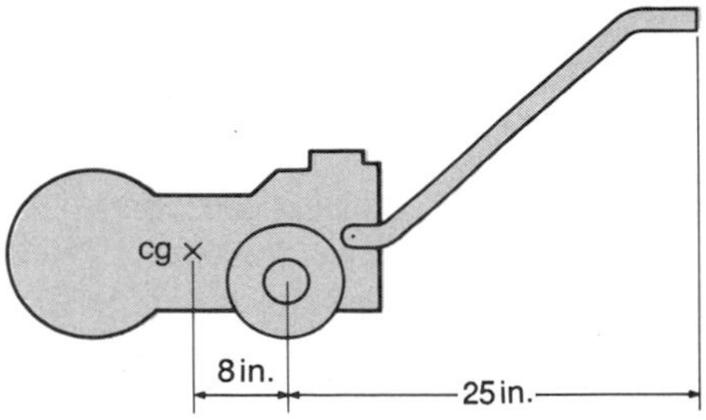

31. Two children construct a seesaw by gluing together two 6-ft planks, with an overlap of 18 in. Both planks are uniform, but one weighs 20 lb and the other 15 lb. The children, who weigh 40 lb and 50 lb, sit at the ends of the planks.

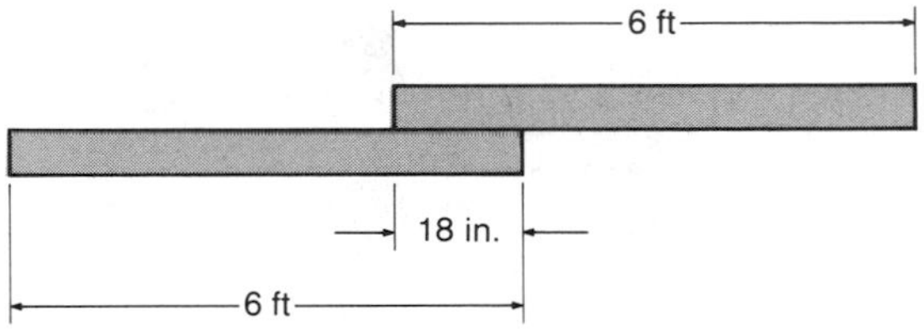

a. Where is the balance point for the seesaw that is closest to the center of the composite beam?
b. Is there another possible balance point?

32. A 1600-kg automobile has 63% of its weight on the front wheels when it is unloaded. The distance between the wheels is 3.6 m. What are the loads on the front and rear wheels after a 120-kg package is placed in the trunk with its center of gravity 80 cm behind the rear wheels?

33. A uniform circular disk of radius 35 cm has a mass of 8 kg. A point mass of 2 kg is attached to the outer edge of the disk. Where is the center of gravity of the combination?

34. A box consisting of four sides and a bottom but no top is in the form of a cube with 30-cm edges. The edges and the bottom are of uniform thickness and density. Where is the center of gravity of the box?

35. A piece of heavy wire of uniform cross section and density is bent into the shape shown in the diagram. Where is the center of gravity of the wire?

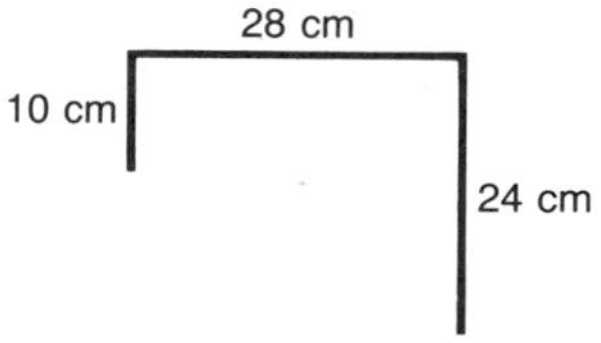

36. A crane boom makes an angle of 65° with the horizontal. A supporting steel cable attached at the top of the boom makes an angle of 45° with the horizontal. The 80-ft-long boom weighs 40 tons, and its center of gravity is at its midpoint. Find the horizontal and vertical components of the force exerted by the lower end of the boom on the main frame of the crane, and also find the tension in the supporting cable. [5.4]

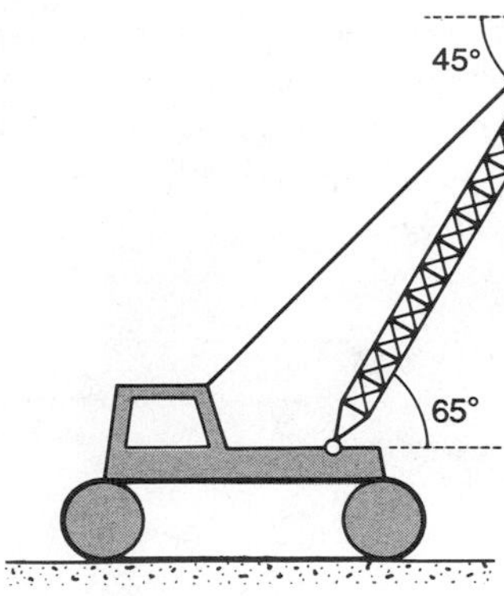

37. A heavy, uniform, steel trapdoor weighs 200 lb and measures 3 ft square. It covers an opening in a horizontal floor and is hinged along one edge.

a. What vertical force applied at the edge opposite the hinge is necessary to begin to open the door?
b. What vertical force applied at the same edge of the door is necessary to hold it open at a 45° angle? [5.4]

38. A trapeze artist of weight 600 N hangs by one hand from a point 40 cm from the end of a light horizontal rod 125 cm long. The rod is supported by two cords as shown in the diagram; the tension in the cord farthest from the performer's hand is 240 N. Calculate the tension in the other cord, and the angles that both cords make with the vertical.

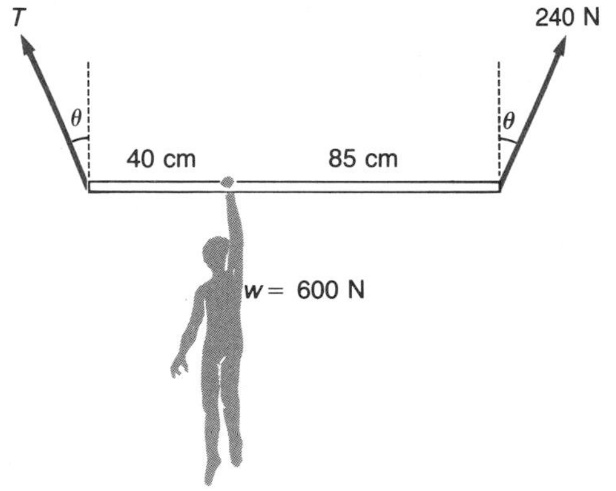

39. Two men are carrying a pool table downstairs and have it supported as shown, so that the diagonal of the table is horizontal. The upper support is a rope attached to the corner of the table that makes an angle of 20° with the vertical. The table itself has a mass of 110 kg and dimensions of 2.50 m by 1.40 m.

a. Calculate the tension in the rope.
b. Calculate the force that the lower man must apply to the corner to hold the table steady and clear of the stairs.

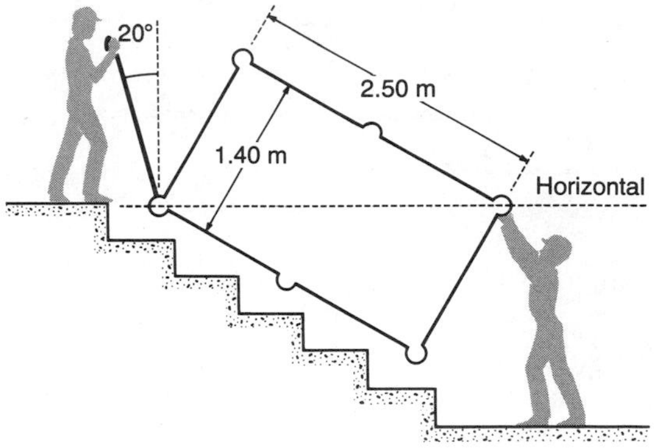

With the weights held steady, the weightlifter must supply an upward force equal to the weight of the lift. However, he must also ensure that the net torque on the bar (due to the weights and the forces supplied by his hands) is zero. See Problem 40 for a detailed example. (Courtesy of York Barbell Co.)

40. A novice weightlifter holds a 160-kg load slightly off center, as shown in the diagram. The line of his right forearm makes an angle of 15° to the vertical. Assume that the forces he exerts on the bar are directed along the line of his forearms. Calculate these forces and also the angle between his left forearm and the vertical. (The weight of the bar is assumed to be negligible.)

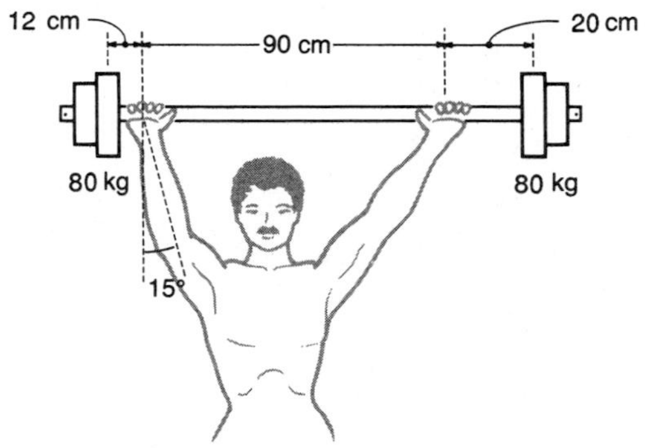

41. A 22-ft ladder weighing 48 lb rests at an angle of 30° with the vertical against a vertical wall. The top end of the ladder has small rollers so that the force exerted on it by the wall is horizontal. A 180-lb man stands on a rung that is 14 ft above the ground. What is the smallest value of the coefficient of static friction between the ladder and the ground that would prevent slipping?

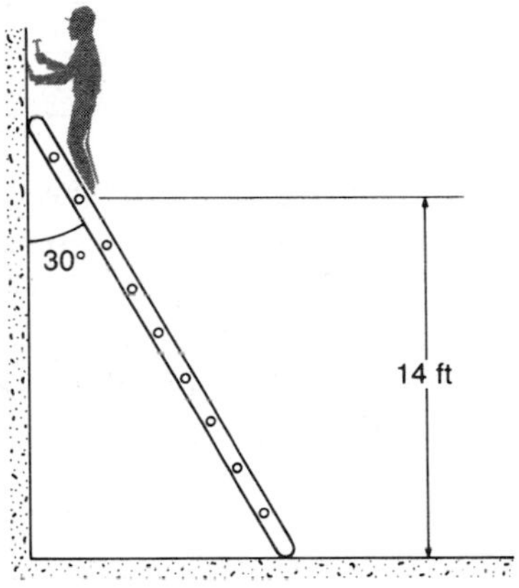

42. A person whose upper arm is vertical holds a 10-kg mass in the hand, the forearm being at 30° with the horizontal. The center of gravity of the weight is 40 cm from the elbow joint, and force due to the biceps is 5 cm from the elbow joint. The weight of the forearm is 18 N and its center of gravity is 15 cm from the elbow joint. Calculate the magnitudes of the forces exerted by the biceps and the elbow joint. [5.5]

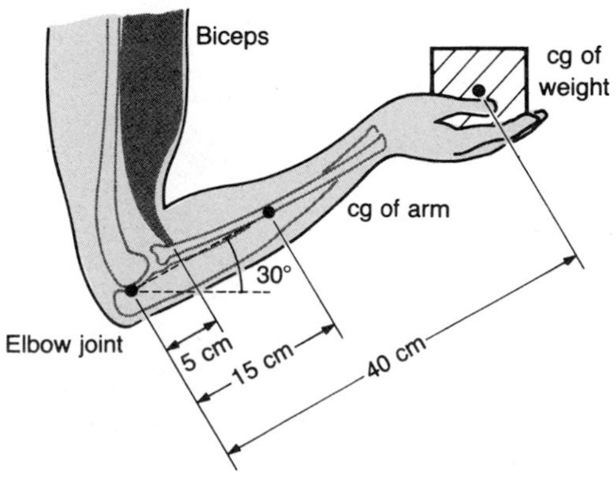

43. A person exerts a force of 200 N on the test arrangement shown in the diagram. Calculate the magnitude of the forces exerted by the biceps and the elbow joint. (Neglect the weight of the forearm.)

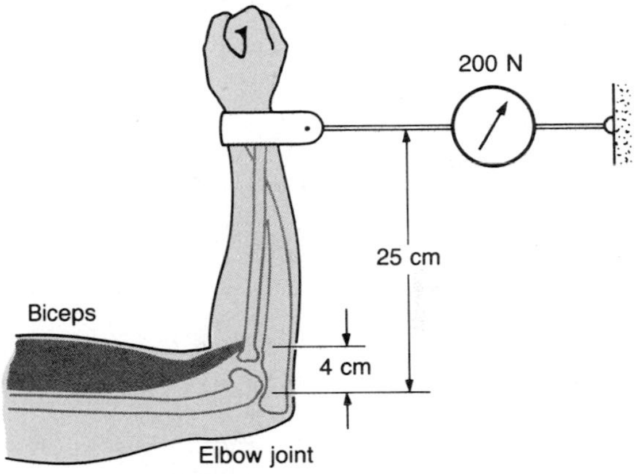

44. A person exerts a force of 100 N on the test apparatus shown in the diagram. The weight of the forearm is 16 N, and its center of gravity is 14 cm from the elbow joint. Calculate the magnitudes of the forces exerted by the triceps and the elbow joint.

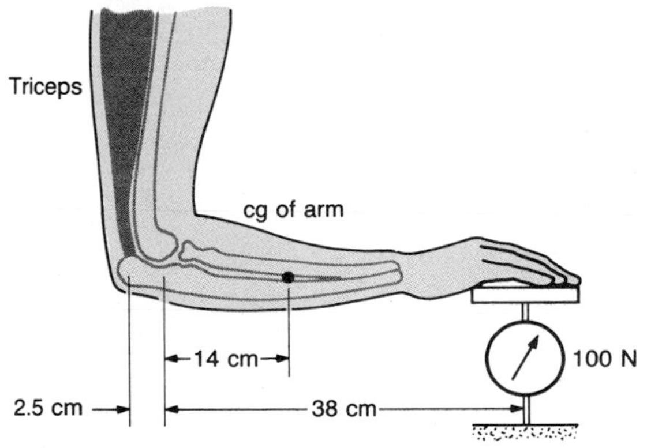

45. A person whose arm weighs 32 N holds it out horizontally (see Fig. 5.6); the center of gravity of the arm is 24 cm from the shoulder joint. The force exerted by the deltoid muscle acts in a direction 16° above the horizontal and at a point 12 cm from the shoulder joint. Calculate the magnitude of the force in the deltoid muscle, and the horizontal and vertical components of the force exerted on the humerus by the shoulder joint.

C. Advanced-Level Problems

46. A brass sign of mass 3 kg is supported as shown, by a uniform 1-m rod of mass 1 kg that is hinged at one end to a vertical wall. The center of gravity of the sign is 60 cm out from the wall. The sign is supported by a wire of breaking strength 50 N that is attached to the wall and the far end of the beam.

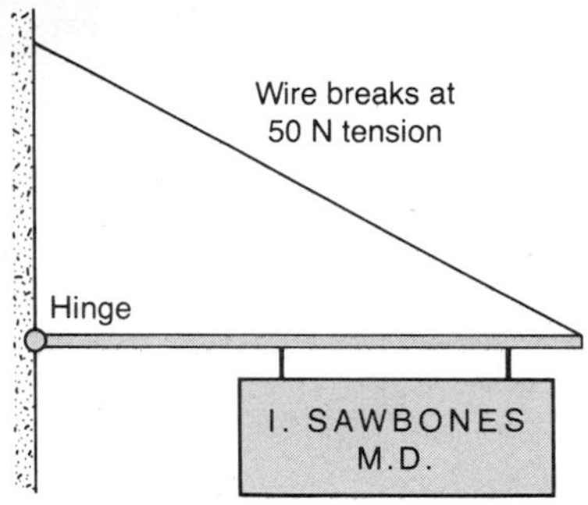

a. What is the minimum safe distance between the hinge and the point where the wire is attached to the wall?

b. What are the horizontal and vertical components of the force exerted by the hinge on the rod?

47. An 80-kg person climbs a 5-m-long ladder that has a mass of 12 kg. The foot of the ladder is 2 m from a wall that is perfectly smooth, so that there is no frictional force at the top of the ladder. The coefficient of static friction between the foot of the ladder and the ground is 0.35. How far up the ladder can the person climb before it slips?

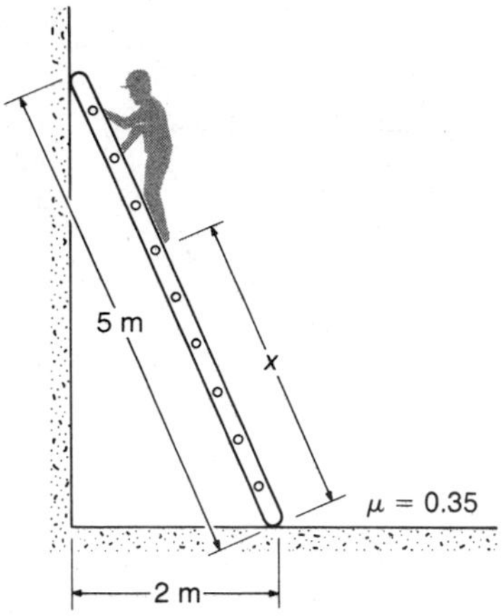

48. A 10-m-long beam weighing 100 N is subjected to the forces shown in the diagram.

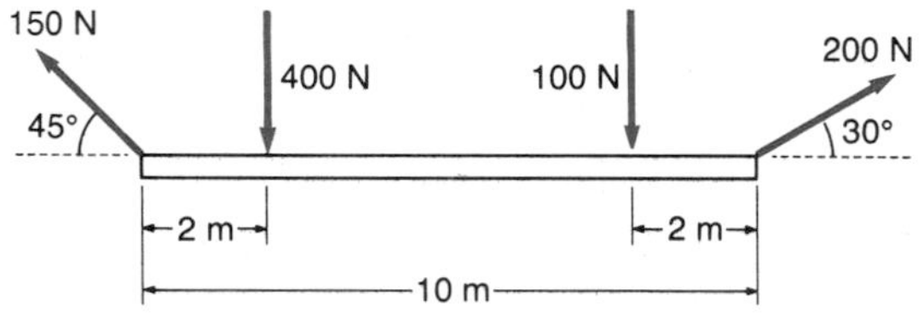

a. What are the magnitude and direction of the additional force required to establish equilibrium?

b. At what point must this force be applied?

49. An isosceles triangle made from a piece of uniform wire has two 20-cm sides and a third 10-cm side. Locate the center of gravity.

6

CENTRIPETAL FORCE AND UNIVERSAL GRAVITATION

Newton's three dynamical laws identify force as the cause of the acceleration of a mass. This definition enables us to calculate a force in terms of the effect that it produces—namely, the acceleration of the mass. However, Newton went much further than this for a special type of force and showed how to calculate the force of gravitation in terms of its causes.

Our definition of weight as the force that causes the acceleration due to gravity on a mass identifies weight only in terms of the acceleration that it produces. Newton showed that the force we know as weight in everyday life is a particular instance of a very general phenomenon. He identified the force of gravitation as the attraction between two masses, and calculated its magnitude in terms of the masses involved. He also applied the concept of gravitational attraction to all masses in the universe.

Newton's work was intimately connected with the kinematics and dynamics of an object moving in a circular path, and we begin by studying this situation.

6.1 Centripetal Acceleration

When a point mass moves at constant speed in a straight line, both the magnitude and direction of the velocity remain unchanged; the acceleration is zero because there is no velocity change. However, the situation is different if the path of motion is a curved line. Even if the speed is constant, the velocity vector changes direction. The change in velocity vector means that an acceleration must be associated with motion along a curved path. The simplest type of curved path is a circle, and a particle moving at constant speed in a circular path must experience acceleration. The approximately circular orbits of planets around the sun are important examples of this type of accelerated motion.

Before investigating the kinematics of circular motion, let us review the concept of the radian. To define the radian, consider the diagram of Fig. 6.1(a). An arc of a circle of length s separates two points P_0 and P on the circumference, and θ is the angle subtended by the arc P_0P at the center of the circle. The measure of θ in radians is equal to the length of the circular arc between P_0 and P, divided by the radius of the circle.

Definition of the Radian

$$\text{Angular measure in radians} = \frac{\text{length of circular arc}}{\text{radius of circle}}$$

$$\theta = \frac{s}{r} \tag{6.1}$$

The radian is a dimensionless number.

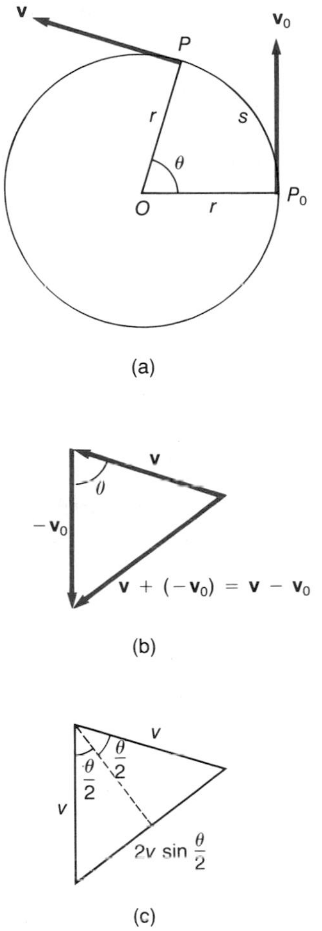

FIGURE 6.1 Diagram illustrating centripetal acceleration, (a) Particle moving at a constant speed in a circular path. (b) Vector composition of velocity difference. (c) Calculation of magnitude of velocity difference.

Since the radian measure of an angle is a length divided by a length, the units cancel and the radian is a dimensionless number. It is customary to write "radian" (abbreviated "rad") after a number in order to indicate clearly that it refers to an angle. In addition to radians, degrees and revolutions are other useful measures of angle. Useful conversions are:

$$1 \text{ rad} \simeq 57.29°$$

$$1 \text{ rev} = 360° = 2\pi \text{ rad}$$

To calculate the acceleration of a point particle traveling at constant speed in a circular path, refer again to Fig. 6.1(a). Let the particle be at point P_0 with velocity $\mathbf{v}_0$ at the initial time t_0; at a later time t it is at point P with velocity $\mathbf{v}$. Our requirement of constant speed means:

$$|\mathbf{v}_0| = |\mathbf{v}| = v$$

where v is the magnitude of the velocity vectors $\mathbf{v}_0$ and $\mathbf{v}$. To calculate the average acceleration over the time interval $(t - t_0)$, we first need to calculate the velocity change $(\mathbf{v} - \mathbf{v}_0)$. We do this with the help of the velocity vector diagram of Fig. 6.1(b). The vectors $\mathbf{v}_0$ and $\mathbf{v}$ are perpendicular to OP_0 and OP, respectively; it follows that the angle between $\mathbf{v}_0$ and $\mathbf{v}$ is the same as the angle between OP_0 and OP—that is, θ. The magnitude of $(\mathbf{v} - \mathbf{v}_0)$ is therefore equal to the side of an isosceles triangle opposite to the angle θ that encloses two equal sides each of length v. Referring to Fig. 6.1(c), we have:

$$|\mathbf{v} - \mathbf{v}_0| = 2v \sin\frac{\theta}{2} \qquad \textbf{(6.2)}$$

The direction of $\mathbf{v} - \mathbf{v}_0$ bisects the angle between OP_0 and OP and points toward the center of the circle. Since the particle is traveling at constant speed, we have [with the help of Eqs. (6.1) and (2.4)]:

$$\text{Arc length} = \text{speed} \times \text{elapsed time}$$

$$r\theta = v(t - t_0)$$

or

$$t - t_0 = \frac{r\theta}{v} \qquad \textbf{(6.3)}$$

Combining Eqs. (6.2) and (6.3) into the definition of average acceleration [Eq. (3.6)] gives, for the magnitude of the acceleration:

$$\langle a \rangle = \frac{|\mathbf{v} - \mathbf{v}_0|}{t - t_0}$$

$$= \frac{2v \sin \theta/2}{r\theta/v}$$

$$= \frac{2v^2 \sin \theta/2}{r\theta}$$

The average acceleration depends on the angle between OP_0 and OP, and its direction is toward the center of the circle bisecting this angle. We now turn to an approximation (proved in Appendix A5):

$$\sin\frac{\theta}{2} \approx \frac{\theta}{2} \qquad \text{if } \theta \ll 1$$

and provided that the angle is expressed in radians. This approximation is appropriate for the case in which the points P_0 and P are very close to each other. In this circumstance, we have:

$$\langle a \rangle = \frac{v^2}{r}$$

This expression does not depend on the angle θ between OP_0 and OP. The average acceleration between two points on the path that are very close together is the instantaneous acceleration. The instantaneous acceleration of a particle in uniform circular motion is called **centripetal acceleration.** Our display for Eq. (6.4) quotes the acceleration as if it were a scalar. However, we imply the direction in the word *centripetal,* which means pointing toward the center. For motion at uniform speed in a circle, the instantaneous-acceleration vector points radially inward at right angles to the path of motion.

The sight of a small trainload of people traveling upside down certainly requires an explanation. If the speed of the cars is great enough, then the weight cannot supply the force required for the centripetal acceleration; an additional force is needed—the rails push down on the cars, and the seats push down on the people. (Robert A. Isaacs 1983/Photo Researchers)

■ Acceleration in Constant-Speed Circular Motion

$$\text{Centripetal acceleration} = \frac{(\text{speed})^2}{\text{radius}}$$

$$a_c = \frac{v^2}{r} \qquad \textbf{(6.4)}$$

EXAMPLE 6.1

The moon revolves around the earth in an orbit that is approximately circular and of mean radius 3.84×10^8 m. The orbital period is 27.32 days. What is the centripetal acceleration of the moon?

SOLUTION Put the data in SI units:

$$27.32 \text{ days} = 27.32 \text{ days} \times 8.64 \times 10^4 \text{ s/day}$$
$$= 2.360 \times 10^6 \text{ s}$$

The speed of the moon in its orbit equals the circumference of the orbit divided by the orbital period:

$$v = \frac{2\pi \times \text{orbital radius}}{\text{orbital period}}$$

$$= \frac{2\pi \times 3.84 \times 10^8 \text{ m}}{2.360 \times 10^6 \text{ s}}$$

$$= 1022 \text{ m/s}$$

We use Eq. (6.4) to calculate the centripetal acceleration:

$$a_c = \frac{v^2}{r}$$

$$= \frac{(1022 \text{ m/s})^2}{3.84 \times 10^8 \text{ m}}$$

$$= 2.72 \times 10^{-3} \text{ m/s}^2$$ ■

In the latter part of the seventeenth century, Isaac Newton carried out this calculation of the centripetal acceleration of the moon. In subsequent sections we will see that his calculation played a leading role in introducing the theory of universal gravitation.

6.2 Dynamics of a Mass Moving in a Circular Path

We have just seen that an object moving at constant speed in a circular path experiences an acceleration directed toward the center of the circle—the centripetal acceleration. Newton's second law gives us a way to calculate the force that is required to cause the centripetal acceleration; in fact, it shows that motion in a circular path is impossible without the existence of such a force. Let us illustrate by means of an example.

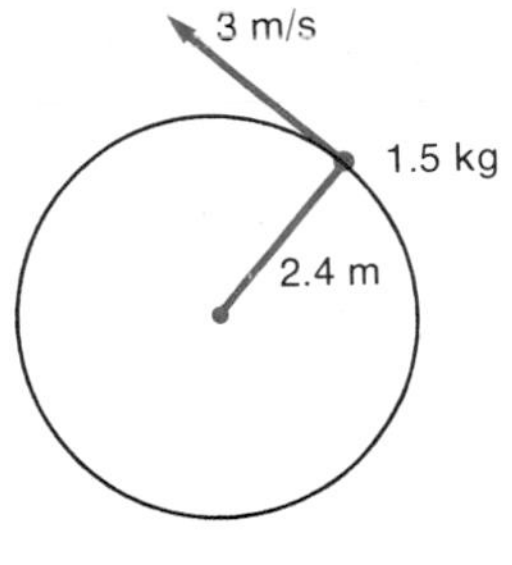

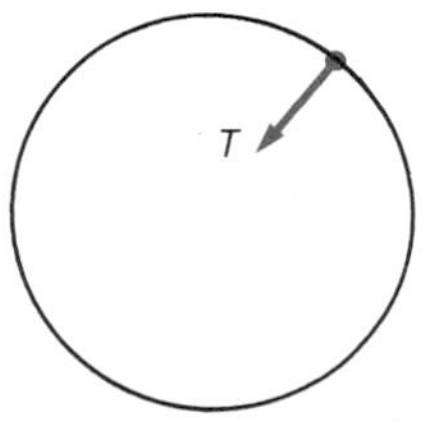

EXAMPLE 6.2

A mass of 1.5 kg moves in a circular path with a constant speed of 3 m/s on a horizontal frictionless surface. The mass is held to the circular path by a light cord 2.4 m long that has one end fixed and the other end attached to the mass. Calculate the tension in the cord.

SOLUTION The centripetal acceleration of the mass is given by Eq. (6.4):

$$a_c = \frac{v^2}{r}$$

$$= \frac{(3 \text{ m/s})^2}{2.4 \text{ m}}$$

$$= 3.75 \text{ m/s}^2$$

The direction of this acceleration is horizontal and always toward the center of the circle. The free-body force diagram shows only one force acting on the mass in a horizontal direction, and this force is the tension in the string. Newton's second law, Eq. (4.1), then gives:

$$F = ma$$

$$T = 1.5\ \text{kg} \times 3.75\ \text{m/s}^2$$

$$= 5.62\ \text{N}$$

■

We sometimes call the string tension a centripetal force since it appears as the force that supplies the centripetal acceleration. However, it is important to realize that centripetal force is not a new kind of force—it is simply any force (in this case a string tension) that causes centripetal acceleration.

Let us further illustrate the point with an example that puts the force of friction in the role of centripetal force.

EXAMPLE 6.3

An automobile of mass 1500 kg travels in a circular path of radius 20 m on a horizontal road surface at a speed of 45 km/h (12.5 m/s). What is the frictional force necessary to prevent slipping? What coefficient of friction between tires and road is required to supply this force?

SOLUTION The frictional force between the tires and the road must supply the centripetal acceleration. Newton's second law applied to this problem [with the use of Eq. (6.4)] states:

$$\text{Frictional force} = \text{mass} \times \text{centripetal acceleration}$$

$$f = \frac{mv^2}{r}$$

$$= 1500\ \text{kg} \times \frac{(12.5\ \text{m/s})^2}{20\ \text{m}}$$

$$= 1.172 \times 10^4\ \text{N}$$

To find the coefficient of friction, we need to know the normal force between the road and the automobile. The road surface is horizontal, and the normal force is equal to the weight given by Eq. (4.2):

$$R_n = w = mg$$

$$= 1500\ \text{kg} \times 9.81\ \text{m/s}^2$$

$$= 1.472 \times 10^4\ \text{N}$$

To calculate the required coefficient of friction, we use Eq. (4.3a):

$$f_s = \mu_s R_n$$

$$1.172 \times 10^4 \text{ N} = \mu_s \times 1.472 \times 10^4 \text{ N}$$

$$\therefore \mu_s = 0.80$$

■

It may appear surprising that the frictional force necessary to prevent skidding during a turn at relatively modest speed is such a large fraction of the weight of the automobile. Looked at from the other point of view, traffic turns of a fairly small radius that are typically encountered at urban intersections need to be made at a fairly low speed. High-speed turns require a large radius of curvature and banking of the track. Racetracks and well-engineered highways have these features.

The motion of an object in a vertical circle presents additional complications since the weight of the object also enters the problem. However, the difficulties are minimal if we consider either the highest or lowest points of the path.

EXAMPLE 6.4

A child whirls a stone of mass 0.5 kg in a vertical circle on the end of a string 40 cm long. At the lowest point of the circle the velocity of the stone is 3 m/s. Calculate the string tension at this point.

SOLUTION The centripetal acceleration that the stone experiences at the bottom of the circle is given by Eq. (6.4):

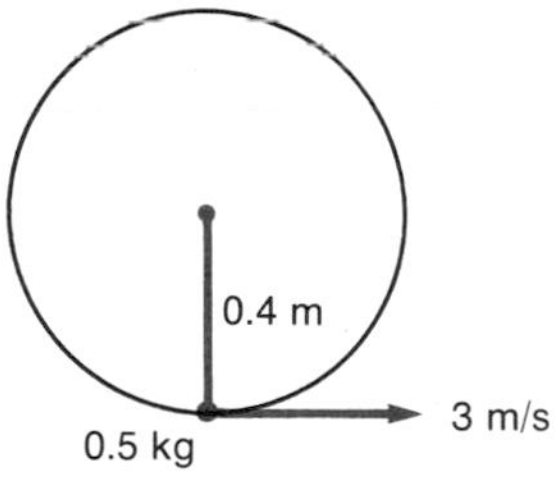

$$a_c = \frac{v^2}{r}$$

$$= \frac{(3 \text{ m/s})^2}{0.4 \text{ m}}$$

$$= 22.5 \text{ m/s}^2$$

The direction of this acceleration is upward—toward the center of the circle.

The free-body diagram shows the forces acting on the stone as its weight and the tension in the string. Using Eq. (4.2) we find:

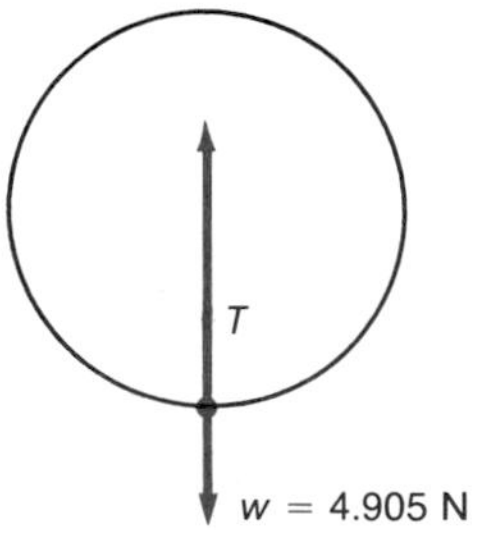

$$w = mg$$

$$= 0.5 \text{ kg} \times 9.81 \text{ m/s}^2$$

$$= 4.905 \text{ N}$$

Then Newton's second law [Eq. (4.1)] gives:

$$\text{Net force on stone} = \text{mass} \times \text{centripetal acceleration}$$
$$T - 4.905 \text{ N} = 0.5 \text{ kg} \times 22.5 \text{ m/s}^2$$
$$\therefore T = 16.2 \text{ N}$$

Note that in this case the centripetal force is the net force on the stone made up of the string tension and the weight combined vectorially. ■

6.3 Universal Gravitation and the Cavendish Experiment

According to a well-known story, Newton conceived the idea of universal gravitation while contemplating a falling apple. The story may well be false, but it is an excellent lead-in to the subject. Newton had a brilliant insight! The force that pulls the apple down to earth might be the same one supplying the centripetal acceleration that keeps the moon in its circular orbit.

We illustrate the data available to Newton in Fig. 6.2. The apple experiences an acceleration $g = 9.81 \text{ m/s}^2$ toward the center of the earth, and the moon experiences a centripetal acceleration (see Example 6.1) of $2.72 \times 10^{-3} \text{ m/s}^2$, also toward the center of the earth. If the same force is responsible for both accelerations, it is very much diminished at the position of the moon. The ratio of the two accelerations is:

$$\frac{\text{Centripetal acceleration of moon}}{\text{Acceleration of apple}} = \frac{2.72 \times 10^{-3} \text{ m/s}^2}{9.81 \text{ m/s}^2}$$
$$= 2.77 \times 10^{-4}$$

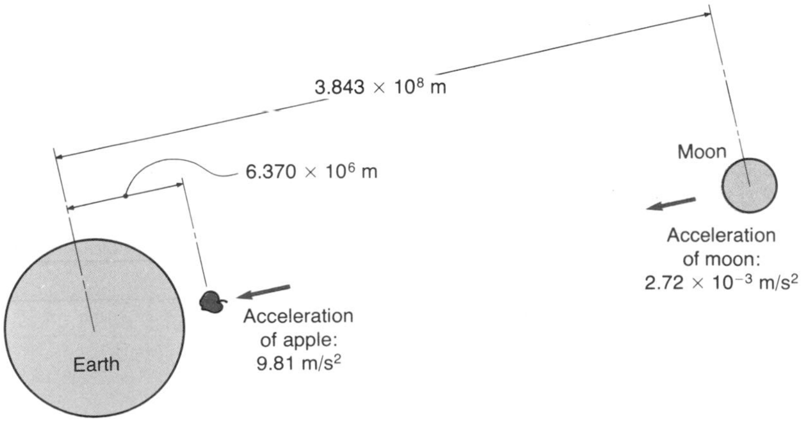

FIGURE 6.2 Newton's data on the moon and the apple.

SPECIAL TOPIC 6 CAVENDISH

Henry Cavendish was born in Nice, France, in 1731 during his parents' vacation. He was educated in England, and studied at Cambridge University but never actually graduated from that institution.

In addition to his famous experiment, known as the Cavendish experiment, for measuring the constant of gravitation, Cavendish worked on the properties of gases and on electricity. His work on the latter subject remained unpublished, and was rediscovered over several succeeding decades after his death.

Cavendish was the complete recluse—he probably came closer to the archetypal mad scientist than any other major figure of science. He spoke very little to men, and not at all to women. He died in London in 1810.

The Cavendish Laboratory at Cambridge (site of the discovery of the nuclear atom) was named in his honor.

We expect this decrease in acceleration to be related to the fact that the moon is much farther from the earth's center than the apple. The ratio of the two distances is:

$$\frac{\text{Radius of earth}}{\text{Radius of moon's orbit}} = \frac{6.370 \times 10^6 \text{ m}}{3.843 \times 10^8 \text{ m}} = 1.657 \times 10^{-2}$$

This is not the same factor as the ratio of the accelerations. However, the square of the distance ratio is very close to the acceleration ratio:

$$\left(\frac{\text{Radius of earth}}{\text{Radius of moon's orbit}}\right)^2 = 2.75 \times 10^{-4}$$

Newton referred to the force that the earth exerts on the moon and the apple as the force of gravity. For the reason explained, he recognized that the strength of the acceleration caused by the force of gravity decreases in proportion to the inverse square of the distance from the center of the earth. We can write this as an equation of proportionality:

$$\text{Acceleration due to force of gravity} \propto \frac{1}{(\text{distance from center of earth})^2}$$

Recall now that Newton's third law of dynamics requires every force to be a member of an action-reaction pair. The earth and the apple must exert equal and opposite forces on each other. Since the force on the apple is proportional to its mass, the force on the earth is also proportional to its mass. We can extend the proportionality equation by writing the mass product in the numerator:

$$\text{Force of gravitational attraction} \propto \frac{\text{mass of apple} \times \text{mass of earth}}{(\text{distance from center of earth})^2}$$

This relation refers only to the earth and the apple, but we could write a similar one for the earth and the moon. These special relations hardly deserve to be called universal, but Newton made a giant leap forward by postulating a force of gravitational attraction between every pair of masses in the universe. If the law is truly universal, the constant of proportionality must be the same for all pairs of masses. Let m_1 and m_2 be two point masses, and let r be the distance between them. We can write the law of universal gravitation in a simple algebraic form:

■ Force of Gravitational Attraction Between Two Point Masses m_1 and m_2 Separated by a Distance r

$$F = G\frac{m_1 m_2}{r^2} \tag{6.5}$$

G is the gravitation constant whose accepted modern value is

$$G = 6.672 \times 10^{-11}\ \mathrm{N \cdot m^2/kg^2}$$

Our statement of the law of universal gravitation applies to two point masses. In this case the meaning of the distance separating the two masses is quite clear. However, if either of the masses is of large extension, the distance between them may be difficult to determine. For example, it is not at all obvious that the distance between the earth and the apple should be put equal to the distance from the *center* of the earth to the apple. Newton delayed publication of his discovery for many years and invented integral calculus to cope with this difficulty. He proved that we can regard the whole mass of the earth as being concentrated at its center when using Eq. (6.5) to calculate the gravitational force between the earth and a small mass.

To use Eq. (6.5) to calculate gravitational forces, we must know the value of G. The most obvious way to determine it is to measure the attractive force between two known masses that are a known distance apart, which would supply experimental values for all quantities in Eq. (6.5) except G. However, this is a difficult experiment because the gravitational force between laboratory-sized objects is so small. Not until more than a century after Newton enunciated this law did Henry Cavendish (1731–1810) make the first direct determination of G. The equipment for the Cavendish experiment is shown in Fig. 6.3. A fiber (such as a thin bronze wire) supports a light rod having a small mass at each end. This arrangement is called a torsion balance. The rod twists through a readily observable angle under the action of a very small force. When other masses are far away, the rod oscillates back and forth, eventually coming to rest in some equilibrium po-

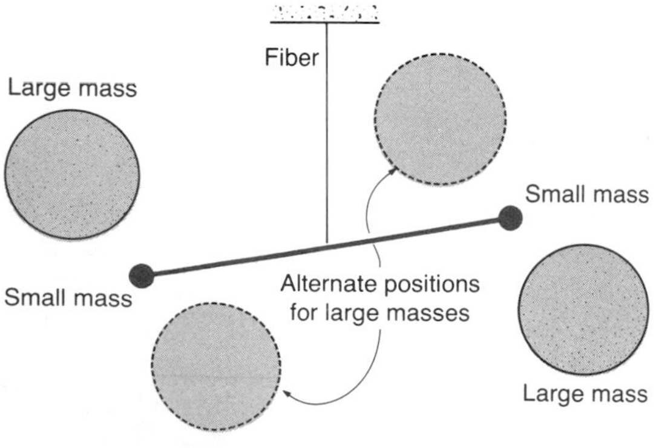

FIGURE 6.3 Cavendish experiment to measure G.

sition. Two large masses are now brought up close to the equipment, one on each side of the rod. Because of the forces of gravitational attraction between the large and small masses, the equilibrium position moves so that the rod is closer to the larger masses. Changing the large masses to the opposite sides of the rod moves the equilibrium position in the other direction. By measuring the equilibrium positions of the rod and by making independent measurements of the force necessary to deflect the rod, Cavendish found a value for G within 1% of the accepted modern value. His work pays great tribute to his skill and determination since the gravitational force that he measured is so small. Since the fiber was so sensitive that the rod's position was influenced by slight air currents, Cavendish enclosed the whole apparatus in a room and observed it through a telescope positioned in the wall. He determined average equilibrium positions from measurements made over a long period of time. The following example illustrates the small magnitude of the gravitational forces.

EXAMPLE 6.5

What is the force of gravitational attraction between two masses of 5 kg each whose centers are 0.15 m apart?

SOLUTION This is a straightforward calculation of gravitational attractive force from Eq. (6.5):

$$\begin{aligned} F &= G\frac{m_1 m_2}{r^2} \\ &= 6.672 \times 10^{-11}\ \mathrm{N \cdot m^2/kg^2} \times \frac{5\ \mathrm{kg} \times 5\ \mathrm{kg}}{(0.15\ \mathrm{m})^2} \\ &= 7.413 \times 10^{-8}\ \mathrm{N} \end{aligned}$$

This is an exceedingly small force. ■

We can now show that the force of universal gravitation is the main factor in the weight of objects near the earth's surface. It is probably easiest if we proceed by means of a specific example.

EXAMPLE 6.6

A man of mass 100 kg stands on a flat surface at the equator. Calculate:

a. the force of gravitational attraction between the man and the earth

b. the centripetal acceleration that the man experiences because of the earth's rotation about its axis

c. the magnitude of the action-reaction force between the man's feet and the flat surface

SOLUTION

a. This is a straightforward calculation of gravitational attraction between two masses using Eq. (6.5):

$$F_g = G\frac{m_1 m_2}{r^2}$$

$$= 6.672 \times 10^{-11}\ \text{N} \cdot \text{m}^2/\text{kg}^2 \times \frac{100\ \text{kg} \times 5.98 \times 10^{24}\ \text{kg}}{(6.37 \times 10^6\ \text{m})^2}$$

$$= 983.3\ \text{N}$$

Here, m_1 is the mass of the man, m_2 the mass of the earth, and r the radius of the earth (these data are taken from the inside back cover).

b. Because of the earth's rotation, the man moves around the circumference of the earth once in 24 h. His speed is therefore given by:

$$v = \frac{2\pi \times 6.37 \times 10^6\ \text{m}}{24\ \text{h} \times 3600\ \text{s/h}}$$

$$= 463.2\ \text{m/s}$$

The centripetal acceleration is given by Eq. (6.4):

$$a_c = \frac{v^2}{r}$$

$$= \frac{(463.2\ \text{m/s})^2}{6.37 \times 10^6\ \text{m}}$$

$$= 3.37 \times 10^{-2}\ \text{m/s}^2$$

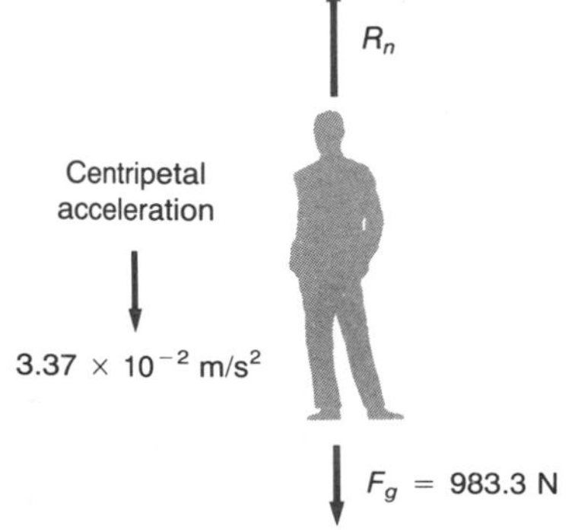

c. The free-body force diagram shows the forces $F_g = 983.3$ N and R_n acting on the man. The net force causes the centripetal acceleration. Using Eq. (4.2) we have:

$$F = ma_c$$

$$983.3\ \text{N} - R_n = 200\ \text{kg} \times 3.37 \times 10^{-2}\ \text{m/s}^2$$

$$\therefore R_n = 979.9\ \text{N}$$

If the man were standing on a scale platform, the scale would register as his weight the force of 979.9 N, which is about 0.3%

lower than the force of gravitation. Note that our calculations would be different for places other than the equator. At the North Pole, there is no centripetal acceleration and weight is equal to the force of gravitation. For other latitudes the centripetal acceleration is smaller than at the equator, and points in a different direction to the force of gravitation. However, in every case, our main conclusion is that the weight and the force of gravitation differ by a few tenths of 1% at most. ■

It is possible to define weight in several ways. Our choice has been that the weight of an object is the reading on a scale supporting the object and at rest in the frame of reference under consideration. This means that our definition $w = mg$ of Eq. (4.2) is exactly correct for objects at rest near the earth's surface. The main advantage of this definition is its operational nature; we can measure weight by a simple experiment. On the other hand, we could have defined weight to be the force of gravitation acting on the mass. This choice is often made, and it results in a different interpretation of Example 6.6 from the one we have given. If weight is defined as gravitational force, then there is no way to measure it directly on the earth's surface except for the special cases of the North and South Poles.

We can use the results of Example 6.6 to find an *approximate formula* giving the relation between g (the acceleration due to gravity) and G (the gravitation constant). Consider a mass m close to the earth's surface, and let m_E be the mass of the earth and r_E, its radius. If the weight is approximately equal to the force of gravitational attraction, then we have, from Eqs. (4.2) and (6.5):

$$mg \approx G\frac{mm_E}{r_E^2}$$

$$\therefore g \approx G\frac{m_E}{r_E^2} \tag{6.6}$$

Interestingly, Newton himself did not attempt a direct measurement of G, using instead Eq. (6.6) to estimate its value. Newton knew the acceleration due to gravity and the radius of the earth, but the earth's mass was not known in his time. Guessing that its average relative density was about 5, he then estimated its mass by multiplying this density by the volume of the spherical earth:

$$\begin{aligned} m_E &= \rho \times \tfrac{4}{3}\pi r_E^3 \\ &= 5 \times 10^3 \text{ kg/m}^3 \times \tfrac{4}{3} \times \pi \times (6.37 \times 10^6 \text{ m})^3 \\ &= 5.41 \times 10^{24} \text{ kg} \end{aligned}$$

Using this estimated value for the earth's mass in Eq. (6.6) gives:

$$G \approx \frac{g r_E^2}{m_E}$$

$$= \frac{9.81 \text{ m/s}^2 \times (6.37 \times 10^6 \text{ m})^2}{5.41 \times 10^{24} \text{ kg}}$$

$$= 7.4 \times 10^{-11} \text{ N} \cdot \text{m}^2/\text{kg}^2$$

Newton's guess was only about 10% off the mark!

6.4 Circular Satellite Orbits

We have already discussed how the moon's orbital motion led Newton to his law of gravitation. After using this particular example of orbital motion to develop a general law, Newton then turned the problem around and investigated all possible satellite orbits with a gravitational force. When a satellite orbits a spherical body much more massive than itself (e.g., a planet around the sun or a spacecraft around a planet), regarding the more massive body as a fixed attractive point center of gravitational force is a good approximation. Newton proved that the general closed-orbit solution to this problem is an ellipse having the attracting mass as one focal point. An ellipse is an oval-shaped figure with two focal points symmetrically placed. Just as rectangles can have various shapes, with a square being a special type of rectangle, ellipses can also have various shapes, with a circle being a special case. Two elliptic shapes are shown in Fig. 6.4; the more rounded ellipse has focal points that are closer together. For the special case of a circle, the two focal points coincide at one point—the center of the circle. The orbits of most planets and satellites are almost circular. We can use the circular-orbit model to give good approximations in some cases.

Consider the situation shown in Fig. 6.5, in which a satellite of mass m travels in a circular orbit of radius r around a large spherical mass M. Since the satellite is traveling in a circle, it is undergoing a centripetal acceleration caused by the gravitational force. The orbital speed of the satellite is the circumference of the orbit divided by the period of revolution T. We use this value of the orbital speed in Eq. (6.4) to calculate the centripetal acceleration:

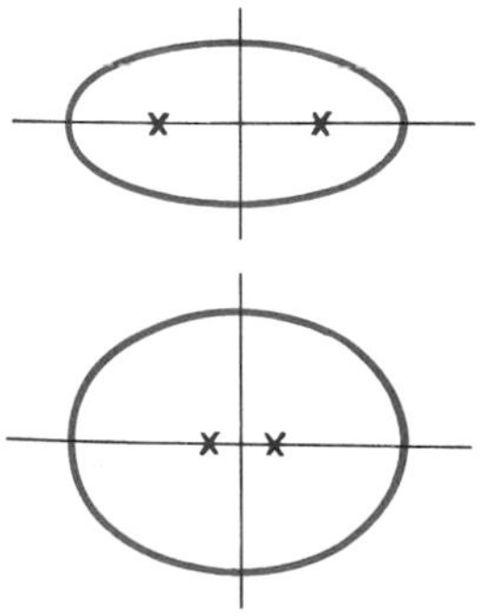

FIGURE 6.4 Two ellipses with their focal points marked.

$$a_c = \frac{v^2}{r} = \frac{(2\pi r/T)^2}{r} \tag{6.7}$$

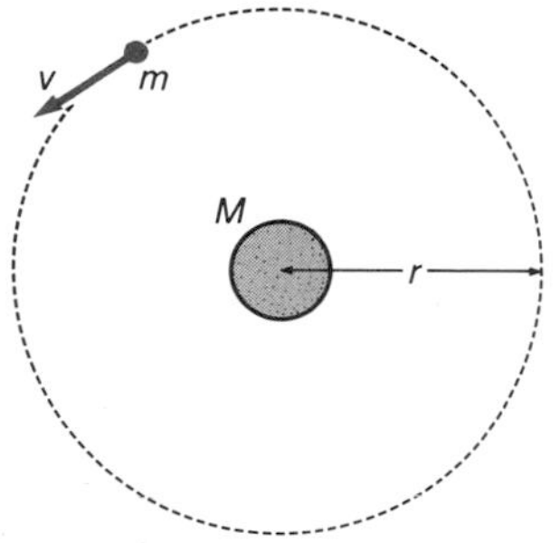

FIGURE 6.5 Diagram illustrating a circular satellite orbit.

The astronaut on a space walk and the spacecraft itself are both orbiting the earth with the same velocity. The condition for an object to be in circular orbit contains the gravitation constant, the mass of the large stationary mass, the orbital period, and the radius of the orbit. However, it does not contain the mass of the orbiting object. For this reason, the safety rope linking the astronaut to the spacecraft is slack, because they are both in the same orbit. (Courtesy of NASA)

Newton's second law, Eq. (4.1), applied to this problem states:

Gravitational force on satellite = mass of satellite × centripetal acceleration

$$G\frac{mM}{r^2} = m \times \frac{(2\pi r/T)^2}{r}$$

Rearranging the terms, we have:

$$GMT^2 = 4\pi^2 r^3 \tag{6.8}$$

This equation is quite famous in physics, and we can use it in different ways.

First consider its application to the motion of some planet in a circular orbit about the sun. With some rearrangement it becomes:

$$\frac{T^2_{\text{planet}}}{r^3_{\text{planet}}} = \frac{4\pi^2}{GM_{\text{sun}}} \tag{6.9}$$

The right-hand side of this equation does not depend on the planet chosen. This circumstance leads to a remarkable result:

> The square of a planet's year divided by the cube of its distance from the sun is the same for all planets.

This is the third law of planetary motion that was proposed by Johannes Kepler (1571–1630). Kepler was a German astronomer whose work preceded Newton's by some half a century. He arrived at his celebrated laws of planetary motion by analyzing the naked-eye observations of the Danish astronomer Tycho Brahe (1546–1601). A complete proof that Kepler's law follows mathematically from Newton's law of gravitation requires the consideration of elliptic orbits. However, we have at least verified the result for the special case of a circular orbit.

Second, Eq. (6.8) permits us to determine the mass of any celestial body that has a satellite of known period and orbital radius. To see this clearly, rearrange Eq. (6.8) to give:

$$M = \frac{4\pi^2 r^3}{GT^2} \tag{6.10}$$

Because the mass of the satellite does not appear in this equation, it is not needed in the calculation (but remember that it must be small compared with the mass M of the larger body). However, it is necessary to know the value of G. For this reason Cavendish's experimental determination of G made it possible to calculate the mass of the earth. The orbital radius and the period of the moon were both well known at that time.

EXAMPLE 6.7

Calculate the mass of the earth if the moon's mean orbital radius is 3.843×10^8 m and its orbital period is 27.3 days.

SOLUTION We first put the orbital period in SI units:

$$27.3 \text{ days} = 27.3 \text{ days} \times 86\,400 \text{ s/day}$$
$$= 2.359 \times 10^6 \text{ s}$$

Then Eq. (6.10) gives:

$$M = \frac{4\pi^2 r^3}{GT^2} = \frac{4 \times \pi^2 \times (3.843 \times 10^8 \text{ m})^3}{6.672 \times 10^{-11} \dfrac{\text{N}\cdot\text{m}^2}{\text{kg}^2} \times (2.359 \times 10^6 \text{ s})^2}$$
$$= 6.0 \times 10^{24} \text{ kg}$$

We give only two significant figures in this answer not because of a lack of precision in the data, but because of a breakdown in the assumptions under which we derived Eq. (6.10). Since the earth is only about 80 times more massive than the moon, it is not entirely accurate to consider the earth as a fixed attractive center about which the moon revolves. ■

Our discussion of satellites has somewhat of an astronomical flavor up to this point. The general principles governing the motion of earth satellites are exactly the same, but we now identify the large central mass as the earth. Thus, if a satellite of mass m orbits the earth (of mass M_E) in a circular orbit of radius r with speed v, we have:

$$\text{Gravitational force on satellite} = \text{mass of satellite} \times \text{centripetal acceleration}$$

$$G\frac{mM_E}{r^2} = \frac{mv^2}{r}$$

Rearranging gives:

$$v^2 = \frac{GM_E}{r} \qquad \textbf{(6.11)}$$

showing that the required circular orbital speed for an earth satellite depends only on the radius.

EXAMPLE 6.8

Calculate the speed of a satellite in circular orbit 2000 km above the earth's surface.

SOLUTION The radius of the circular orbit is equal to the earth's radius plus the altitude of the satellite giving:

$$r = 6.37 \times 10^6 \text{ m} + 2000 \text{ km}$$
$$= 8.37 \times 10^6 \text{ m}$$

Use of Eq. (6.11) then gives:

$$v^2 = \frac{GM_E}{r}$$
$$= 6.672 \times 10^{-11} \frac{\text{N} \cdot \text{m}^2}{\text{kg}} \times \frac{5.98 \times 10^{24} \text{ kg}}{8.37 \times 10^6 \text{ m}}$$
$$\therefore v = 6.90 \times 10^3 \text{ m/s}$$

■

It is interesting to reflect that astronauts endure two extreme situations during the course of a trip to earth orbit and back. Soon after blastoff, their rocket is providing an upward acceleration of almost $10g$. The net force on the astronaut's body is the vector sum of the contact force provided by the support couch (upward) and the gravitational pull of the earth (downward). It follows that the force supplied by the support couch is about 11 times the normal weight of the astronaut. If the astronaut were erect during this period of acceleration, the tissues holding the internal organs could be strained to the point of severe damage. The skeletal structure would not be damaged, since the bones and joints can safely endure the stresses involved. Moreover, blood would be forced out of the head and into the legs; the lack of blood in the brain could cause temporary loss of consciousness, and the excess in the legs could cause the rupture of small blood vessels. For these reasons, astronauts rest in the prone position during takeoff to minimize the physiological effects of the large acceleration.

During orbit the situation is entirely opposite. We have seen in Eq. (6.11) and Example 6.8 that the condition for orbital velocity does not contain the mass of the orbiting object. This means that the astronauts are in orbit just as much as the ship in which they are traveling. The gravitational force acting on them is exactly what is required to cause the centripetal acceleration appropriate to their orbit. The astronauts are weightless. Since the human body is accustomed to living with its weight, the condition of weightlessness could cause physiological or psychological problems. It is certainly clear, however, that humans can overcome these difficulties, as has been shown by astronauts who have taken part in orbital flights of several months.

6.5 Motion of a Vehicle on a Banked Track

We have already worked an example concerning the motion of an automobile in a circular path on a horizontal surface.

The dynamics of a vehicle traveling in a circular path around a banked track is a little bit more complicated. Experience teaches us that such a vehicle will tend to slide upward on the bank if it is traveling too fast; in this case, therefore, the frictional force will be parallel to the road surface and inward. Conversely, if the vehicle is traveling too slowly, it will tend to slide down the bank, and the frictional force will be outward. In between, there is a vehicle speed that is correct for the angle of bank in the sense that no frictional force is called into play to maintain the position of the vehicle. The corresponding free-body force diagrams are shown in Fig. 6.6.

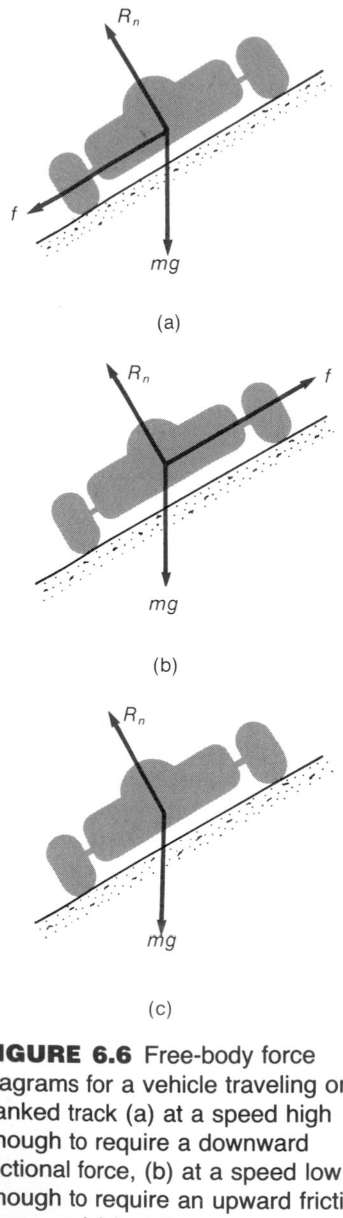

FIGURE 6.6 Free-body force diagrams for a vehicle traveling on a banked track (a) at a speed high enough to require a downward frictional force, (b) at a speed low enough to require an upward frictional force, and (c) at the correct speed for the angle of bank.

Let us begin by trying to find the correct speed (that is the one that requires no frictional force) for a vehicle on a banked track. In Fig. 6.7, we show the free-body force diagram with forces decomposed into vertical and horizontal components. The reason for this choice (rather than parallel to the plane and perpendicular to the plane) is that the vertical acceleration is zero, and the horizontal acceleration is just the centripetal acceleration v^2/r. The net vertical force is zero, and we have:

$$R_n \cos\theta - mg = 0$$

The net horizontal force is equal to the mass multiplied by the centripetal acceleration:

$$R_n \sin\theta = mv^2/r$$

Eliminating R_n from this pair, we find:

$$\tan\theta = \frac{v^2}{rg} \tag{6.12}$$

This condition is independent of vehicle mass. A track correctly banked for a given vehicle at a given speed is correctly banked for all vehicles at that speed.

EXAMPLE 6.9

A highway curve has a radius of 750 m. At what angle should it be banked so that traffic traveling at 90 km/h (25 m/s) experiences no sideways frictional force?

SOLUTION The condition in question is given by Eq. (6.12):

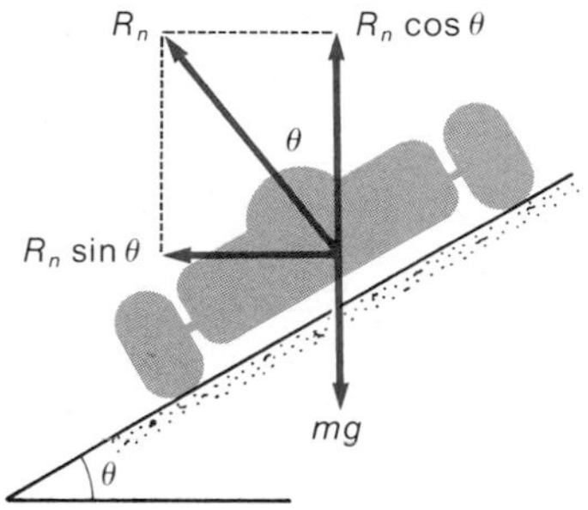

FIGURE 6.7 Horizontal and vertical components of forces acting on a vehicle on a banked track (case of no friction).

$$\tan\theta = \frac{v^2}{rg}$$

$$= \frac{(25 \text{ m/s})^2}{750 \text{ m} \times 9.81 \text{ m/s}^2}$$

$$= 0.08495$$

$$\therefore \theta = 4.85°$$ ■

If the condition of Eq. (6.12) is not met, then the frictional force cannot be zero. Provided that the speed does not deviate too greatly from the speed appropriate to the angle of bank, the vehicle will not slide given reasonable friction coefficients. To get an idea of the magnitudes involved, let us consider an example.

These race cars rounding a curve must experience a force to produce the required centripetal acceleration. The banking of the track is designed to allow the horizontal component of the normal reaction to provide at least part of the required force. (Courtesy of Indianapolis Motor Speedway)

EXAMPLE 6.10

A 2000-kg automobile travels on the curved track of Example 6.9 (750-m radius and 4.85° angle of bank). The coefficient of friction between the tires and the road is 0.5 (corresponding to smooth tires on a wet pavement). What is the maximum speed with which the automobile can negotiate the curve without sliding?

SOLUTION If the automobile is on the point of sliding, then the frictional force must have its maximum value of $\mu_s R_n$. This force together with the weight and the normal reaction are shown on

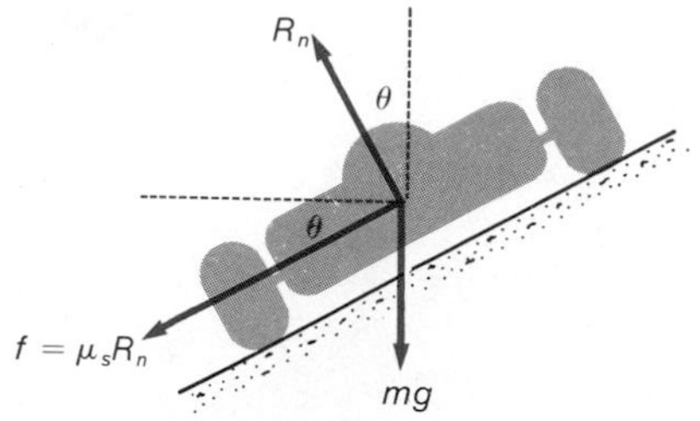

the diagram. Once again the vertical force components must add up to zero, since there is no acceleration in that direction:

$$R_n \cos\theta - mg - \mu_s R_n \sin\theta = 0$$

$$\therefore R_n = \frac{mg}{\cos\theta - \mu_s \sin\theta}$$

$$= \frac{2000\ \text{kg} \times 9.81\ \text{m/s}^2}{\cos 4.85^\circ - 0.5 \times \sin 4.85^\circ}$$

$$= 20\,560\ \text{N}$$

The horizontal force components must produce the centripetal acceleration:

$$R_n \sin\theta + \mu_s R_n \cos\theta = mv^2/r$$

$$\therefore v = \left[\frac{r}{m}(R_n \sin\theta + \mu_s R_n \cos\theta)\right]^{1/2}$$

$$= \left[\frac{750\ \text{m}}{2000\ \text{kg}}(20\,560\ \text{N} \times \sin 4.85^\circ + 0.5 \times 20\,560\ \text{N} \times \cos 4.85^\circ)\right]^{1/2}$$

$$= 67.0\ \text{m/s}$$ ■

The force diagrams in these examples show all the forces acting at a common point of the vehicle. However, the weight acts through the center of gravity, and the reaction forces on the tires are located at the four wheels. The conclusions we have reached are adequate as far as they go, but the whole approach would require upgrading for a more detailed analysis.

KEY CONCEPTS

A point particle traveling with uniform speed in a circular path experiences an acceleration directed radially inward toward the center of the circle. The acceleration is called *centripetal acceleration*. See Eq. (6.4).

Any mass moving at constant speed in a circular path requires a force (e.g., tension in a string or friction) to provide centripetal acceleration; the force is referred to as centripetal when it fills this role.

A force of mutual *gravitational attraction* exists between every pair of masses in the universe. The force is directly proportional to the product of the masses and inversely proportional to the square of their separation. See Eq. (6.5).

A small mass in closed orbit about a large spherical mass under the force of gravitational attraction describes an ellipse, with the center of the large mass at one focus. For the special case of a circular orbit, the center of the large mass is at the center of the circle. The gravitational force provides the centripetal acceleration.

QUESTIONS FOR THOUGHT

1. Assuming that the planets move in circular orbits, find the relationship between the orbital speed and orbital radius.

2. The lower the speed of an orbiting satellite, the larger the orbital radius. Because slight frictional forces experienced by a satellite in the upper atmosphere can reduce its speed, we might therefore expect the satellite to move to a higher orbit. However, the satellite moves to a lower orbit as its speed drops. Explain this apparent contradiction.

3. The determination of the gravitation constant by Cavendish made it possible to calculate the mass of the earth. Can we determine the mass of a planet (say Jupiter) from a knowledge of the gravitation constant, the radius of Jupiter's orbit, and the length of its year? What mass, if any, is it possible to determine from this data? What data do we need to determine the mass of Jupiter?

4. A research worker operates a centrifuge containing particles of different densities and of different masses suspended in a liquid. Does the centrifuge separate the particles of higher and lower mass or those of higher and lower density?

5. A cyclist leans over when making a turn. Explain the reason for this, using a free-body force diagram.

6. Does the gravitational force of the earth act on an astronaut working in an orbiting satellite? Is the astronaut weightless? Explain your answer.

PROBLEMS

(Use the data on the inside back cover as required.)

A. Single-Substitution Problems

1. At what speed must an object travel in a circle of radius 3 m to experience a centripetal acceleration of 9.81 m/s^2? [6.1]

2. An ice skater skates in a circular path of radius 6.5 m with a speed of 5 m/s. What centripetal acceleration does she experience? [6.1]

3. An automobile travels around a circular track of radius 160 m at a speed of 90 km/h (25 m/s). What is the centripetal acceleration? [6.1]

4. What is the smallest radius in which a cyclist should move at a speed of 10 m/s to avoid a centripetal acceleration of more than $0.25g$? [6.1]

5. A cord 2.8 m long has a breaking strength of 600 N. One end of the cord is fixed, and a 2-kg mass attached to the free end moves in a horizontal circular path on a frictionless level surface. What is the maximum speed if the cord is not to break? [6.2]

6. A rope 4 m long is fixed at one end, and a mass of 3 kg fixed to the other end moves in a horizontal circular path on a frictionless level surface with constant speed, completing one revolution per second. Calculate the tension in the rope. [6.2]

7. The turntable of a record player rotates so that the speed of a point on the rim (12 cm from the axis) is 40 cm/s. Calculate the frictional force required to keep a mass of 200 g at this position. [6.2]

8. Calculate the frictional force required in order that a runner of mass 80 kg can move with a speed of 8 m/s around a circular track of radius 18 m. [6.2]

9. Calculate the force of gravitational attraction between two lead spheres each of mass 100 kg and 10 m apart. [6.3]

10. If it were possible to have a gravitational attraction of 1 N between two equal masses 1 m apart, what would be the size of each mass? [6.3]

11. Calculate the force of gravitational attraction between (a) the sun and the earth, (b) the earth and the moon. [6.3]

12. What is the gravitational attraction between an electron and a proton that are separated by 0.529×10^{-10} m? (This corresponds to the first Bohr orbit in the hydrogen atom; the electrical attraction between the particles at this separation is 8.23×10^{-8} N.) [6.3]

13. Calculate the orbital period for a satellite that is one earth radius above the surface of the earth. [6.4]

14. Calculate the orbital speed for a satellite in orbit around the moon and just above its surface. [6.4]

15. The planet Mars has a moon that orbits with a period of about 7 h 40 min. The mean radius of its orbit (assumed to be circular) is 9.4×10^6 m. Calculate the mass of Mars. [6.4]

16. Is 10 000 mi/h a sufficient speed for an earth satellite just above the atmosphere? [6.4]

B. Standard-Level Problems

17. The earth's orbit around the sun is almost circular, with a mean radius of 1.496×10^{11} m. What centripetal acceleration does the earth experience? [6.1]

18. The hand of a stopwatch is 2.4 cm long (center to tip), and it makes one revolution of the dial every 2 s. Calculate the speed and the acceleration of a point on the tip of the hand. [6.1]

19. A dive-bomber pilot pulls out of a 140-m/s dive, causing his plane to travel in a path that is approximately the lower half of a vertical circle. What is the smallest radius of this circle if the centripetal acceleration is not to exceed 4 times the acceleration of gravity?

20. In order to test the physiological effects of a large acceleration, a person is strapped 4.4 m from the axis of a machine that rotates at 40 rev/min. How many times the acceleration of gravity is the centripetal acceleration?

21. The seats of a merry-go-round are 6.75 m from the axis. Calculate the maximum number of revolutions per minute if the centripetal acceleration on a rider is not to exceed $0.08g$.

22. A rough turntable rotates at 16.5 rev/min and frictional forces are able to supply a centripetal acceleration of 3.5 m/s^2 for an object placed on the surface. What is the maximum distance from the axis of rotation that an object can be placed without slipping?

23. A 2-kg block rests on a rough horizontal turntable at a distance of 60 cm from the axis of rotation. The block begins to slide when the turntable is rotating at 30 rev/min. Calculate the coefficient of friction between the block and the turntable. [6.2]

24. A block of mass 5 kg rests on a rough horizontal turntable at a point 125 cm from the axis of rotation. The coefficient of static friction between the block and the turntable is 0.4. How fast can the turntable rotate (in revolutions per minute) before the block begins to slide? Do you need to know the mass of the block to answer problems 23 and 24? [6.2]

25. A mass of 4 kg attached to a light rigid rod of length 85 cm is rotated at constant speed in a vertical circle. At what number of revolutions per minute is there zero force in the rod at the top of the circle?

26. A 70-kg man stands on a spring scale that is calibrated in newtons and placed on the floor of a van that is traveling at 90 km/h along a country road. What is the reading on the scale when the van crosses a sharp dip in the road whose vertical radius of curvature is 48 m?

27. A Ferris wheel has freely swiveling seats. The wheel has a radius of 8.6 m and rotates at constant speed in a vertical circle through one full revolution in 6.5 s. What is the action-reaction force between a 60-kg rider and the seat

a. at the top of the circle?
b. at the bottom of the circle?

28. How many meters above the surface of the earth is the force of gravity on a mass reduced by 25% from its value at the earth's surface? [6.3]

29. Calculate the acceleration due to gravity on the surface of the planet Jupiter. The mass of Jupiter is 1.90×10^{27} kg, and its radius is 6.99×10^4 km. (Neglect any effect due to rotation of the planet.) [6.3]

30. An astronaut of mass 70 kg weighs 687 N on the surface of the earth.

a. What is his mass on the surface of the moon?
b. What does he weigh on the surface of the moon? Use the following approximate values:

$$\text{Mass of moon} \approx \frac{1}{81}\ \text{mass of earth}$$

$$\text{Radius of moon} \approx \frac{3}{11}\ \text{radius of earth [6.3]}$$

31. Find the point between the earth and the moon at which the gravitational pulls of these two bodies on a spaceship exactly cancel one another.

32. A 2-kg mass has a weight of 8.5 N on the surface of a certain hypothetical spherical-shaped planet. The radius of the planet is 5×10^6 m. Use these data to calculate the average relative density of the planet. (Use the value for the universal gravitation constant from the inside back cover, and ignore any effect due to the rotation of the planet.)

33. Calculate the orbital period of a satellite in a circular orbit just above the planet Jupiter. (Assume that the radius of the orbit is approximately equal to the radius of Jupiter.) [6.4]

34. The period of revolution of an asteroid in circular orbit about the sun is 7.2 yr. What is its distance from the sun? Express your answer in earth orbit radii. [6.4]

35. A satellite in orbit just outside the earth's atmosphere has a period of about 1.4 h. At what orbital radius does a satellite have a period of 16 h? Express your answer in earth radii.

36. A communications satellite is to be placed in synchronous orbit above the equator; that is, its orbital period is to be such that it maintains its position at a given point above the earth's surface. Find the height of the satellite above the earth's surface. Express your answer in earth radii.

37. Apollo 8 orbited the moon in a circular orbit having a 120-min period at a distance of 112 km above the moon's surface. The radius of the moon is 1.74×10^3 km. Calculate the mass of the moon from these data and your knowledge of the gravitation constant.

38. Use the value of the gravitation constant 6.67×10^{-11} $N \cdot m^2/kg^2$, the length of the year 3.16×10^7 s, and the mean radius of the earth's orbit around the sun 1.50×10^{11} m to calculate the mass of the sun.

39. The banking of curves on a highway is not to exceed 5° and is designed for 90 km/h (25 m/s) traffic. What is the smallest permissible radius of curvature? [6.5]

40. A circular indoor cycling track has a mean radius of 45 m. The track is banked steeply at the top, and the angle of bank decreases by zero at the inner edge. What is the banking angle of that part of the track that is designed for vehicles going 72 km/h (20 m/s)? [6.5]

41. A light aircraft traveling at 85 m/s makes a turn of 1-mi radius. What is the correct angle of bank?

C. Advanced-Level Problems

42. A rapid transit system has a curve of radius 300 m that is correctly banked for 72 km/h (20 m/s). Each loaded car has a mass of 15 metric tons.

a. What is the angle of the bank?
b. If traffic conditions force a stop on the curve, what is the force due to each car on the inner rail tending to push it down the bank?
c. At what speed would the train have to negotiate the curve to produce the same force on the upper rail tending to push it up the bank?

43. A highway curve of radius 500 m is banked correctly for a speed of 90 km/h (25 m/s).

a. What is the angle of bank?
b. Friction between the tires and a dry road can provide a maximum force of 0.8 times the normal reaction between an automobile and the road. What is the highest speed at which an automobile can negotiate this curve without slipping?
c. In the rain the maximum frictional force is reduced to 0.55 times the normal reaction. What is the highest speed at which an automobile can negotiate an unbanked curve of the same radius in the rain?

44. A 2-kg mass is tied to one end of a string 50 cm long, and the other end is tied to a fixed support. The mass is traveling at constant speed in a horizontal circle of radius 30 cm whose center is vertically below the point of support. Calculate (a) the speed of the mass in its circular path, (b) the tension in the string.

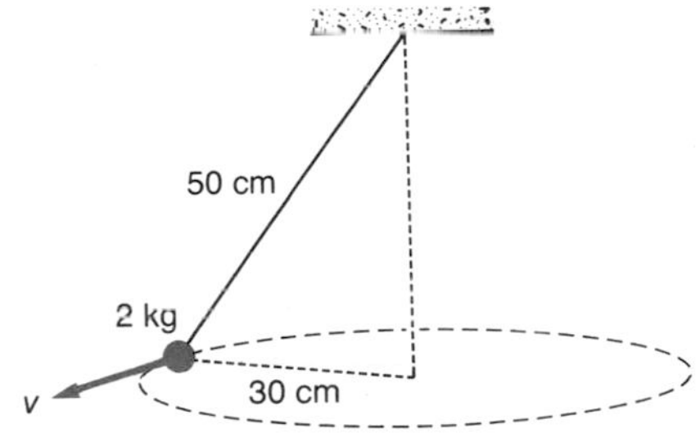

45. A model aircraft of mass 1.6 kg flies with a speed of 16 m/s in a horizontal circle of radius 25 m at the end of a control cable 30 m long. Calculate the tension in the cable. (Assume that the mass of the cable is negligible.)

7

WORK AND ENERGY

As we have seen in previous chapters, Newton's laws of motion provide us with a basis for solving dynamics problems. However, dynamics problems sometimes become so complex that solving them from Newton's laws is very difficult. In such cases, we can often make progress by using conservation laws. A conservation law identifies a physical quality of an isolated system that remains unchanged during a process in which other physical qualities of the system do change. The principle of energy conservation, for example, says that the energy of an isolated system remains the same while forces, velocities, and other physical features change.

The energy conservation principle has its roots in Newtonian dynamics. Indeed, in simple cases it produces the same results obtained by using Newton's laws. In advanced applications, however, it provides a more general viewpoint, incorporating ideas beyond those originating in dynamics.

We therefore begin our study of energy by using concepts from Newtonian dynamics. As we show, defining the work done by a force is a fruitful starting place that leads us to the definitions of various types of mechanical energy, and to an elementary form of the energy conservation principle.

7.1 The Definition of Work

In everyday usage the term **work** means the total amount of effort expended. A person does work when lifting an object from the floor to a shelf, a horse does work when pulling a plow, and a child does work when dragging a toboggan. One essential feature is the same in all these cases: A force displaces an object through a certain distance. However, do not be misled into defining work as the magnitude of the force multiplied by the magnitude of the displacement. This rather plausible idea misses another essential point. Consider the example of a child dragging a toboggan across a level snowfield [see Fig. 7.1(a)] causing a displacement $\mathbf{x}$ using a force $\mathbf{F}$ exerted by means of a tow rope that is inclined at an angle θ with the horizontal. In proceeding with the definition of work, we must recognize that it is only the force component parallel to the displacement that contributes to the work done by the force in the towing rope. We show the appropriate components in Fig. 7.1(b); the displacement vector $\mathbf{x}$ is horizontal and the component of $\mathbf{F}$ parallel to it is $F \cos \theta$. The work done is the product of x with $F \cos \theta$; the component of $\mathbf{F}$ that is perpendicular to the displacement ($F \sin \theta$) makes absolutely no contribution to the work.

With these preliminaries we set up our formal definition for the work done by a constant force:

Definition of Work

Work done by a constant force = component of force parallel to the displacement × displacement

$$W = Fx \cos \theta \quad \textbf{(7.1)}$$

SI unit of work: newton · meter
New name: joule
Abbreviation: J

The unit of work is named in honor of James Joule (1818–1889), an English physicist who did pioneering work in energy.

At first sight, the unit of work seems to be the same as the unit of torque since they are both newtons multiplied by meters. But remember that torque is force multiplied by lever arm, and a torque of 5 N · m implies that the distance is perpendicular to the force. On the other hand, work of 5 J implies that the distance is parallel to the force. For this reason, we always quote work in joules and never in newton-meters, which we reserve exclusively for describing torque.

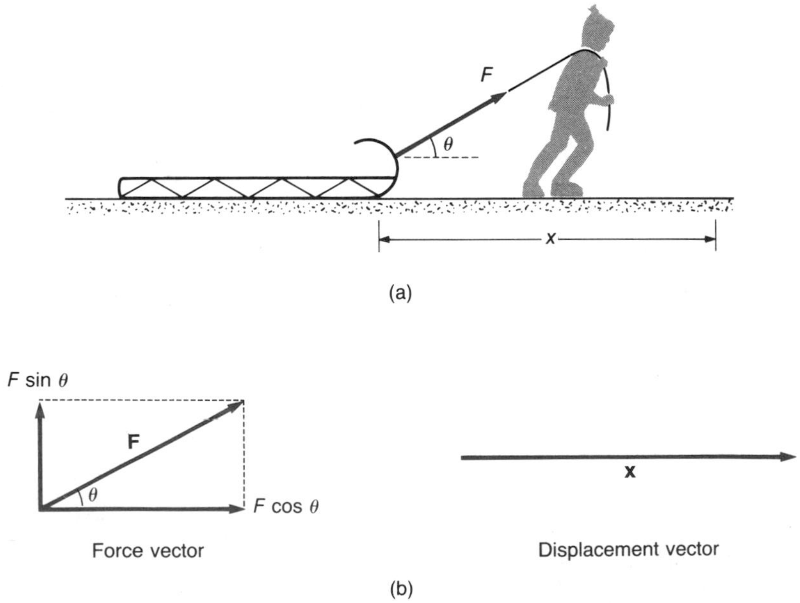

FIGURE 7.1 Diagram to illustrate the definition of work: (a) A child pulls a sled through a horizontal displacement x using a force F that is inclined at an angle θ with the horizontal, and (b) the components of **F** are $F \cos \theta$ parallel to the displacement vector and $F \sin \theta$ perpendicular to it.

We note also that our definition of work is, in fact, the product of the magnitudes of the force and displacement vectors multiplied by the cosine of the angle between them. A special notation is available to denote this product of two vectors; it is called the *scalar product* (or the *dot product*) and is written:

$$W = \mathbf{F} \cdot \mathbf{x} \tag{7.1a}$$

To calculate the scalar product we proceed exactly as indicated by Eq. (7.1).

Work is a scalar quantity even though its definition involves a special type of product of vectors. Physically, this means that we associate no direction with work even though force and displacement both have magnitude and direction.

EXAMPLE 7.1

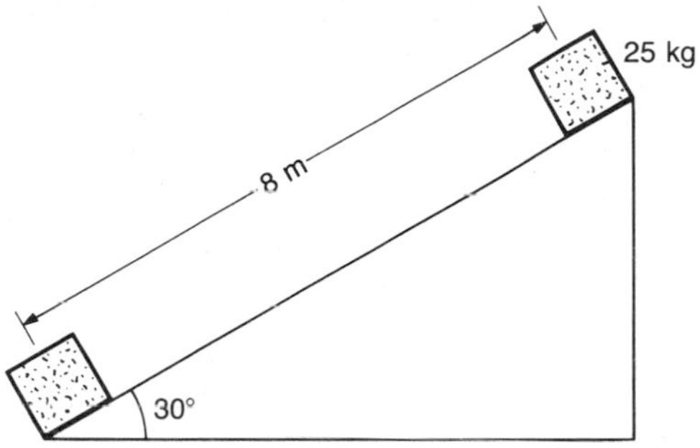

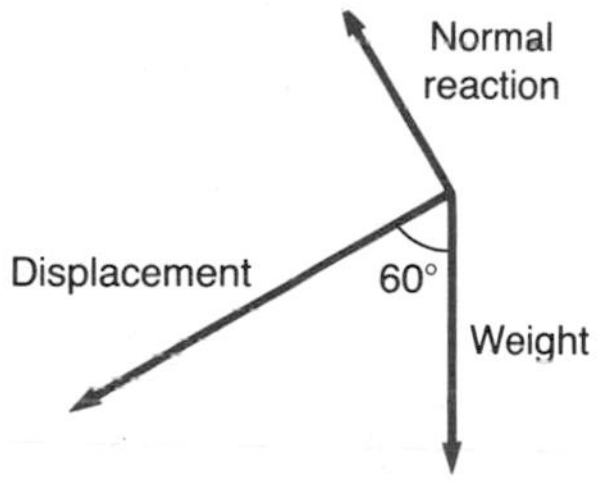

A block of mass 25 kg slides down a frictionless plane inclined at 30° with the horizontal through a distance of 8 m (measured parallel to the slope of the plane). How much work is done by the force of the weight during this slide?

SOLUTION If the angle of the plane with the horizontal is 30°, then the angle between the weight vector and the displacement vector is 60°. Equation (7.1) gives:

$$\begin{aligned} W &= Fx \cos \theta \\ &= 25 \text{ kg} \times 9.81 \text{ m/s} \times 8 \text{ m} \times \cos 60° \\ &= 981 \text{ J} \end{aligned}$$

Note that there is another force acting on the block besides the weight—namely, the normal reaction of the inclined plane. However, the normal reaction is perpendicular to the displacement and does no work during the slide. ■

7.2 Kinetic Energy and Gravitational Potential Energy

The basic notion of *energy* is *potential for doing work.* A tankful of gasoline, a heavy truck moving at speed, a charged automobile battery, and a pile driver lifted and ready to be dropped all possess energy. Why the energy exists and how it can perform work are different for each of these cases. To begin our study of energy, we will focus on the types of energy that can be explained in terms of dynamics.

The energy associated with a mass in motion is called kinetic energy, and the energy associated with a mass lifted to a height

This cross-country skier must work hard to gain gravitational potential energy on the uphill gradients; the payback on the downhill gradients is only partial, because some of the energy is dissipated. (© Tim Davis/Photo Researchers)

in the earth's gravitational field is called gravitational potential energy.

Let us consider kinetic energy first. In order to bring a freight car up to speed, a force must be applied for some distance, and this force does work on the freight car. In order to stop the car, a force must be exerted on it in the opposite direction, and the car does work on whatever applies the stopping force. For example, suppose that a locomotive pushes the freight car until it attains some speed. The locomotive performs work on the freight car, which now possesses that work as kinetic energy, and it can use this energy to do work itself. It we place wooden barricades across the track, the moving freight car could crash through them for some distance and do work in breaking them up—destructive work admittedly, but work nevertheless.

Intuitively, we feel that the kinetic energy of the car depends on both its mass and its speed. A heavy freight car moving at the normal cruising speed of a train possesses far more energy than a BB slug fired from an air rifle, even though the latter has a much higher speed. The exact relation is easily found. Consider a body of mass m acted on by a constant force F that produces a displacement x parallel to the direction of the force. During this displacement the velocity increases from an initial value v_0 to a final value v. From Eq. (7.1) the work done is:

$$
\begin{aligned}
W &= Fx \\
&= max \text{ [using Newton's second law Eq. (4.1)]} \\
&= m\frac{v^2 - v_0^2}{2} \text{ [using the kinematic relation Eq. (2.12d)]} \\
&= \tfrac{1}{2}mv^2 - \tfrac{1}{2}mv_0^2
\end{aligned}
$$

Before discussing this result further, we define the quantity called **kinetic energy** (abbreviated "KE"), which is equal to one-half the mass of an object multiplied by the square of its velocity.*

■ **Definition of Kinetic Energy**

$$\text{Kinetic energy} = \tfrac{1}{2} \times \text{mass} \times (\text{velocity})^2$$

$$\text{KE} = \tfrac{1}{2}mv^2 \qquad \textbf{(7.2)}$$

SI unit of kinetic energy: joule (same as the unit of work)

*Strictly speaking, this definition refers to *translational* kinetic energy; we shall generalize the concept to include *rotational* kinetic energy in a subsequent chapter.

SPECIAL TOPIC 7 SCALING AND THE RUNNING SPEED OF ANIMALS

Running is a very complicated activity that is difficult to explain in terms of Newtonian mechanics. When an animal is running at full speed in a straight line, there is no acceleration, and overly simplistic thinking might lead to the conclusion that no force is required to maintain speed. Not only would this conclusion ignore our own experience, but it also neglects the fact that a running animal must continually bring its legs up to its own speed (after the foot leaves the ground) and then slow them down to a stop (when the foot is placed on the ground). In fact, the major effort in running comes from the continual acceleration and deceleration of the legs.

Let us again consider animals of similar shape and size with scale factor L. At each step the work done by the leg muscles to bring the leg up to the running speed should equal the kinetic energy of the leg at that speed. The muscle force is proportional to the cross-sectional area of the muscle and scales as L^2. Assuming that the muscular-contraction distance scales as L, the work done by the muscle scales as L^3. The kinetic energy of the leg at running speed therefore also scales as L^3. But the mass of the leg is proportional to its volume and therefore scales as L^3. This accounts for the whole of the scaling of the kinetic energy; the velocity does not scale.

Although we should apply these ideas only to animals of similar shape, it is a remarkable fact that the running speed of a wide range of animals varies only between about 10 m/s and 20 m/s.

Generally speaking, the faster animals differ from the slower ones by having long thin legs (of relatively small mass) and thigh muscles that are very close to the body. This means that the continual speeding up and stopping is nearly all in the legs—and hardly at all in the more massive muscles that move them.

We can now state in words the formula just derived: The work done on an object by a constant force that produces a displacement in a parallel direction is equal to the difference between the final and initial values of the kinetic energy.

Since an increase in kinetic energy is equal to the work done, the unit of kinetic energy must be the same as the unit of work. Alternately, we can show that the formulas for work and kinetic energy lead to quantities having the same dimensions. Work has the dimensions of a force by a length or $[M]\,[L]\,[T]^{-2} \times [L] = [M]\,[L]^2\,[T]^{-2}$. Kinetic energy has the dimensions of a mass times the square of a velocity or $[M] \times \{[L]\,[T]^{-1}\}^2 = [M]\,[L]^2\,[T]^{-2}$. The two are the same.

We now have an explicit formula that shows exactly how mass and velocity determine the kinetic energy. Let us illustrate by means of an example.

EXAMPLE 7.2

Compare the kinetic energies of a 20 000-kg freight car moving at 90 km/h (25 m/s) and a 0.4-g BB traveling at 200 m/s.

SOLUTION In each case only direct application of the kinetic energy formula, Eq. (7.2), is required. For the freight car:

$$\begin{aligned} KE &= \tfrac{1}{2}mv^2 \\ &= \tfrac{1}{2} \times 20\,000 \text{ kg} \times (25 \text{ m/s})^2 \\ &= 6.25 \times 10^6 \text{ J} \end{aligned}$$

For the BB:

$$\begin{aligned} KE &= \tfrac{1}{2}mv^2 \\ &= \tfrac{1}{2} \times (4 \times 10^{-4} \text{ kg}) \times (200 \text{ m/s})^2 \\ &= 8 \text{ J} \end{aligned}$$

The result certainly verifies our intuition as to which object possesses more kinetic energy. ■

We derived the kinetic energy formula for a mass that experiences a constant force. If the idea of kinetic energy were limited in this way, it would be of very little use. Actually, Eq. (7.2) is the correct expression for the energy of motion regardless of the type of force that causes it. The proof of the **work-Energy theorem** for all cases is beyond the scope of this book, but we cannot overlook its immense practical importance.

■ Work-Energy Theorem

> The work done on an object by the net force is equal to the change in kinetic energy of the object.

EXAMPLE 7.3

A baseball has a mass of 0.15 kg, and a pitcher can throw a fast ball at about 180 km/h (50 m/s). How much work does the pitcher do on the ball?

SOLUTION The force the pitcher uses to impart velocity to the ball varies both in magnitude and direction. In fact, we can scarcely imagine how to begin detailed calculations of the force on the ball and its consequent acceleration. But the work-energy theorem gives us an approach that requires no detailed calculations involving forces:

$$\begin{aligned} \text{Work done on baseball} &= \text{change in kinetic energy} \\ &= \tfrac{1}{2}mv^2 \\ &= \tfrac{1}{2} \times (0.15 \text{ kg}) \times (50 \text{ m/s})^2 \\ &= 187 \text{ J} \end{aligned}$$

■

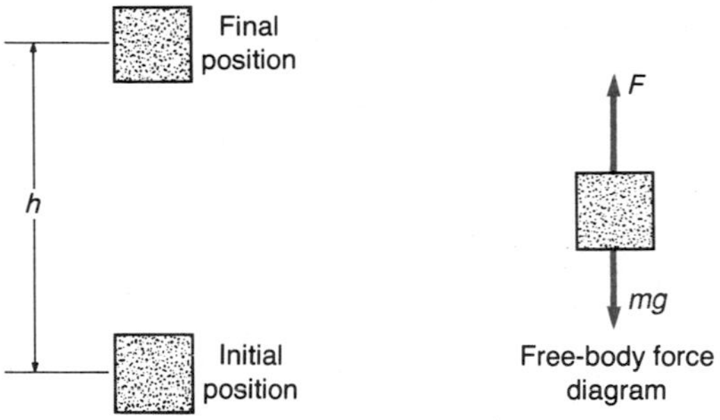

FIGURE 7.2 Diagram to illustrate the calculation of energy change for a mass forced upward by force.

The work-energy theorem tells us that the work done on an object by the net force is equal to its kinetic energy increase. But most of the movements of everyday life take place close to the surface of the earth, and the weight is one of the forces that contributes to the work done if there is a vertical component of displacement. Consider an object of mass m whose center of gravity is thrust vertically upward to a height h above its original level by a vertical force F (Fig. 7.2). The work-energy theorem tells us that the work done on the mass is equal to the change in kinetic energy. From the free-body force diagram, we see that the net upward force on the mass is $(F - mg)$. Since the vertical component of displacement is h, Eq. (7.1) permits calculations of the work done on the mass:

$$\text{Work done} = \text{force} \times \begin{array}{c}\text{component of displacement in a}\\ \text{direction parallel to the force}\end{array}$$

$$= (F - mg)h$$

By the work-energy theorem this is the same as the change in the kinetic energy:

$$Fh - mgh = \text{change in kinetic energy}$$

We can avoid continual explicit calculation of the work done by the weight by introducing a new energy form called **gravitational potential energy** (abbreviated "GPE"). The change in gravitational potential energy is equal to the weight multiplied by the vertical component of displacement of the center of gravity. It is important to note that h is the vertical component of displacement.

■ Definition of Gravitational Potential Energy Change

$$\begin{array}{c}\text{Gravitational}\\ \text{potential}\\ \text{energy change}\end{array} = \text{mass} \times \begin{array}{c}\text{acceleration}\\ \text{of gravity}\end{array} \times \begin{array}{c}\text{vertical component}\\ \text{of displacement of}\\ \text{center of gravity}\end{array}$$

$$\text{GPE} = mgh \qquad \textbf{(7.3)}$$

SI unit of gravitational potential energy: joule (same as the unit of work)

Horizontal motion during the lifting of a mass does not change the gravitational potential energy. Since the force of the weight acts vertically, the horizontal component of displacement does not contribute to the work done by the weight. With our definition of gravitational potential energy, the result of the work-energy theorem becomes:

$$Fh = \begin{array}{c}\text{change in}\\ \text{kinetic energy}\end{array} + \begin{array}{c}\text{change in gravitation}\\ \text{potential energy}\end{array}$$

Since Fh represents the work done by the applied forces (excluding the weight), we can modify the work-energy theorem as follows:

■ Modified Work-Energy Theorem

> The work done on an object by the net force (excluding the weight) is equal to the change in kinetic energy plus gravitational potential energy of the object.

Our proof of the modified work-energy theorem involved only a constant, vertical force, but once again the theorem is true for any type of force.

A special case occurs for an object that is lifted very slowly from rest, so that there is negligible change in the kinetic energy. When this happens all of the work done (by forces other than the weight) goes into increasing the gravitational potential energy. The object now possesses the capacity to do work because of its change in position relative to the earth. *Potential energy* is a general term that applies to energy possessed by virtue of position. Gravitational potential energy is the special case of energy due to position in the earth's gravitational field. In using Eq. (7.3), we always assume that the vertical displacement is not large enough to cause appreciable change in the acceleration due to gravity. Otherwise, a more elaborate analysis would be necessary.

EXAMPLE 7.4

A pile driver of mass 5000 kg is raised slowly to a height of 12 m above its original position. How much work is done by the raising mechanism?

SOLUTION The modified work-energy theorem allows this calculation directly:

Work done by forces excluding the weight (that is, by the raising mechanism) = change in KE + change in GPE

If the pile driver is raised slowly, the kinetic energy is negligible at all times; the use of Eq. (7.3) gives:

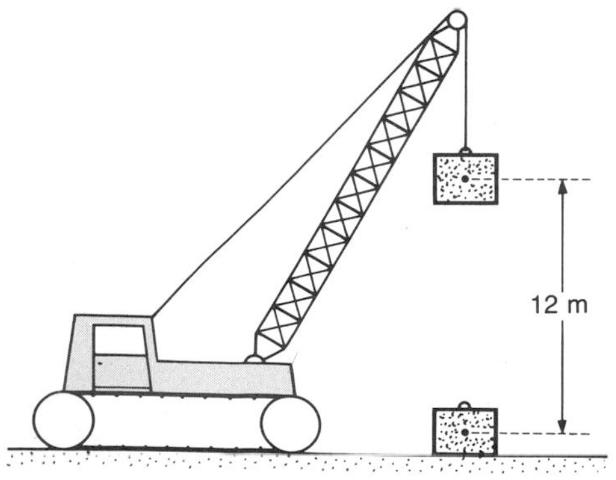

$$\begin{aligned}\text{Work done} &= \text{change in GPE}\\ &= mgh\\ &= 5000\ \text{kg} \times 9.81\ \text{m/s}^2 \times 12\ \text{m}\\ &= 5.89 \times 10^5\ \text{J}\end{aligned}$$

■

The work-energy theorem refers to *changes* in kinetic and gravitational potential energies, and not to absolute values of the energies. The reason is that *the zero of energy is a matter of convenient choice*—it is not possible to calculate absolute energies. Does a person driving an automobile at highway speed possess kinetic energy? You may answer, "Relative to the automobile, no, but relative to the road, yes." Does a stone on a rooftop possess gravitational potential energy? The answer is similar: "Relative to the rooftop, no, but relative to the ground, yes." We must always interpret the formulas of Eqs. (7.2) and (7.3) with this fact in mind.

We have already solved quite complex problems by using the principles of kinematics and dynamics, but often such problems are much easier from an energy viewpoint.

EXAMPLE 7.5

An automobile of mass 1500 kg approaches a hill inclined at 6° to the horizontal at a speed of 90 km/h (25 m/s). As it climbs a distance of 200 m along the incline, the net force on the automobile (propulsive force less frictional resisting force) is 250 N. Calculate the speed at the end of the climb.

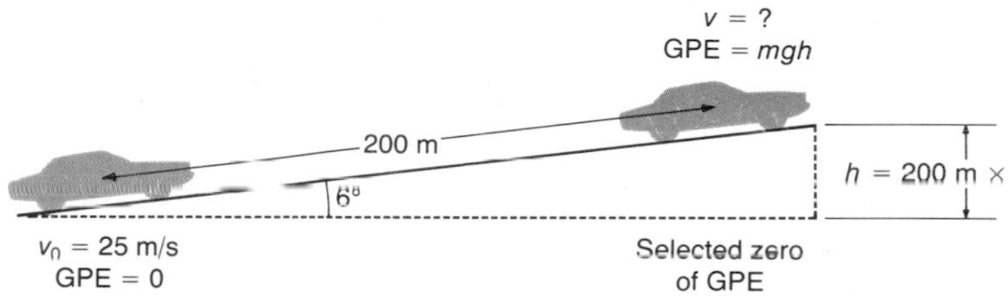

SOLUTION At the bottom of the hill the automobile has a kinetic energy corresponding to its speed of 25 m/s. We can set the gravitational potential energy at zero at this point. At the top of the hill the kinetic energy is unknown, and the gravitational potential energy is determined by the vertical height of the hill. The forces acting on the automobile (the weight excluded) are the propulsive and frictional forces, which act parallel to the incline. (There is also the normal reaction force, which is perpendicular to the incline and, thus, does no work.) We use Eq. (7.1) to calculate the work done by the propulsive and frictional forces:

$$\begin{aligned} W &= Fx \\ &= 250 \text{ N} \times 200 \text{ m} \\ &= 5 \times 10^4 \text{ J} \end{aligned}$$

The modified work-energy theorem gives us:

Work done by applied forces (excluding weight) = final (KE + GPE) − initial (KE + GPE)

Using Eqs. (7.2) and (7.3) for the kinetic and gravitational potential energies we have:

$$5 \times 10^4 \text{ J} = (\tfrac{1}{2}mv^2 + mgh) - (\tfrac{1}{2}mv_0^2 + 0)$$
$$= (\tfrac{1}{2} \times 1500 \text{ kg} \times v^2 + 1500 \text{ kg} \times 9.81 \text{ m/s}^2 \times 200 \text{ m} \times \sin 6°) - [\tfrac{1}{2} \times 1500 \text{ kg} \times (25 \text{ m/s})^2]$$
$$= (750 \text{ kg} \times v^2 + 3.076 \times 10^5 \text{ J}) - (4.687 \times 10^5 \text{ J})$$
$$\therefore v = 16.8 \text{ m/s}$$

■

7.3 Principle of Energy Conservation

The principle of energy conservation is one of the most general and far-reaching laws in physics. The history of the development of the principle parallels that of the evolution of the concept of energy. Isaac Newton, the originator of our system of dynamics, had not a single word to say on the subject of energy. The Industrial Revolution in the nineteenth century provided a practical motivation for studying the energy problem. Many scientists contributed, but James Joule (1818–1889) played a leading role in clarifying various energy forms and in identifying the flow of heat as a transfer of energy. In the early part of the twentieth century, the introduction of the theory of relativity again imperiled the energy conservation principle. It was saved by Albert Einstein (1879–1955), who proposed the famous mass-energy relationship. A precise definition of energy in general is probably impossible, but this does not detract from the validity of utility or the principle.

Energy Conservation Principle

> The energy of an isolated system is conserved.

The principle claims that the total energy of a system that remains *isolated* cannot be changed by any process taking place within it. The only way to change the total energy of any system is by interference from the outside; we shall return to this point in a subsequent chapter.

Let us begin with the simplest application of the energy conservation principle, involving only changes between kinetic and gravitational energy forms. In this and subsequent chapters we progressively enlarge the scope of application as the need arises. A restricted form of the energy conservation principle follows from the modified work-energy theorem. Recall that this theorem equates the work done by applied forces (the weight excluded) to the change in kinetic energy plus gravitational potential energy. This allows us to state a restricted energy conservation law that is usually called the **conservation of mechanical energy.** If applied forces (the weight excluded) do not work on an object, then its total kinetic energy plus gravitational potential energy remains constant. In adopting this terminology we count both kinetic energy and gravitational potential energy as mechanical energy forms.

EXAMPLE 7.6

The pile driver of Example 7.4, with a mass of 5000 kg, is dropped from a height of 12 m above its original position. Calculate the kinetic energy just as the pile driver reaches the original position; assume that no frictional force opposes the motion.

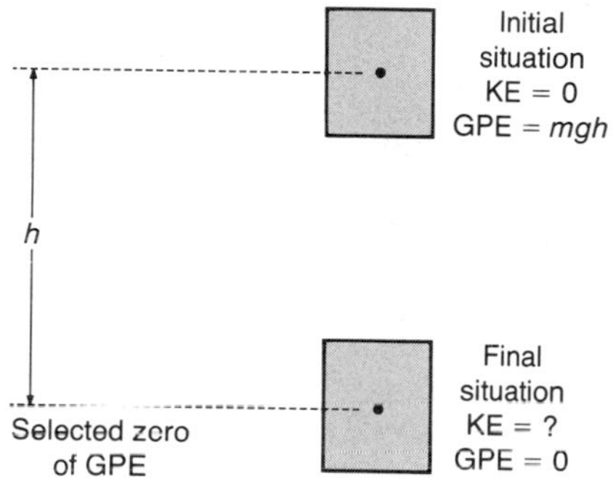

SOLUTION If no frictional force opposes the motion, the total mechanical energy is conserved. We select the zero of gravitational potential energy as shown in the diagram.

$$\text{Final (KE + GPE)} = \text{initial (KE + GPE)}$$

$$\text{Final (KE + 0)} = 0 + mgh$$

$$\text{Final (KE + 0)} = 0 + 5000\ \text{kg} \times 9.81\ \text{m/s}^2 \times 12\ \text{m}$$

$$\therefore \text{Final KE} = 5.89 \times 10^5\ \text{J}$$

■

The function of the pile driver is to convert the gravitational potential energy of a mass into kinetic energy to drive the pile. We could obtain the result of Example 7.6 by using other methods. We could first calculate the velocity that the pile driver acquires in free fall under the acceleration due to gravity and then substitute this velocity in Eq. (7.2) to calculate the kinetic energy. The result would be the same as we found by using the energy conservation principle. Since we derived the energy formulas from the kinematic and dynamic results of previous chapters, it is not surprising that energy conservation is assured in this simple case. You may even wonder what advantage the energy method has. The answer is that the *kinetic and gravitational potential energies depend only on the initial and final states* of the system in question. We can use the energy method when the corresponding kinematics problem is too complex to solve and also when other forms of energy enter the conservation equation. Let us examine some modifications of the pile driver problem to illustrate these points.

EXAMPLE 7.7

A pile driver of mass 5000 kg is dropped in curved guides so that the vertical component of its displacement is 12 m. The guides

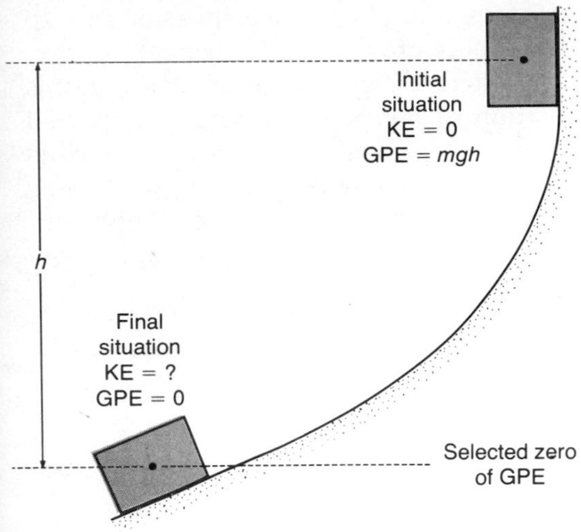

are well lubricated so that the frictional force opposing the motion is negligible. Calculate the kinetic energy of the pile driver at the bottom of the guides.

SOLUTION The weight is not the only force acting on the pile driver, which also experiences a force from the rails as it falls. Since friction is negligible, the force from the rails has no component parallel to the rails. Its only component is perpendicular to the rails at all points. This force component does no work, since the displacement is parallel to the rails at all points. The energy conservation principle still applies, and the answer to the problem is the same as the answer to Example 7.6. Note that we cannot solve the kinematics problem in this instance. We don't even know the shape of the curved guide rail, and we don't need it. The conservation of mechanical energy gives the required information concerning the initial and final states. ■

We can apply energy conservation with advantage even when applied forces (other than the weight) do work on a falling object. The work may transfer energy to other forms that are only vaguely specified. It could include heat generated by frictional forces, energy transferred to light or sound, or any other conceivable process. We simply lump all of these forms together as *energy dissipated.* With this understanding, one can write the energy conservation principle in yet another manner. The initial and final values of mechanical energy (kinetic plus gravitational potential) differ by the amount of energy dissipated. Note that the energy is not dissipated in the sense of being destroyed—that is forbidden by the general energy conservation principle. It simply goes into another form that is not available to contribute to the total mechanical energy. We can use another modification of the pile driver problem to illustrate this point.

EXAMPLE 7.8

The pile driver of Example 7.7 is dropped in curved guides that are inadequately lubricated. Consequently, frictional forces oppose the motion, and energy is dissipated, principally as heat in the guide rails. After the pile driver falls through a vertical distance of 12 m, its velocity is 12.8 m/s. Calculate the amount of energy dissipated during the fall.

SOLUTION The modified mechanical energy conservation law allows us to write:

$$\text{Initial (KE + GPE)} = \text{final (KE + GPE)} + \text{energy dissipated}$$

Using Eqs. (7.2) and (7.3) for the kinetic and gravitational potential energies, we have:

$$0 + mgh = \tfrac{1}{2}mv^2 + 0 + E_{diss}$$

$$5000 \text{ kg} \times 9.81 \text{ m/s}^2 \times 12 \text{ m} = \tfrac{1}{2} \times 5000 \text{ kg} \times (12.8 \text{ m/s})^2 + E_{diss}$$

$$5.886 \times 10^5 \text{ J} = 4.096 \times 10^5 \text{ J} + E_{diss}$$

$$\therefore E_{diss} = 1.79 \times 10^5 \text{ J}$$

■

The principle of energy conservation also offers us a way to make estimates in problems having very complex details.

EXAMPLE 7.9

A pole-vaulter whose mass is 75 kg reaches a velocity of 9 m/s at the end of his run. Estimate the height of his vault, assuming that his use of the pole converts all his kinetic energy into potential energy. (The actual situation is much more complex, since the vaulter can use his arms to lift himself on the pole.)

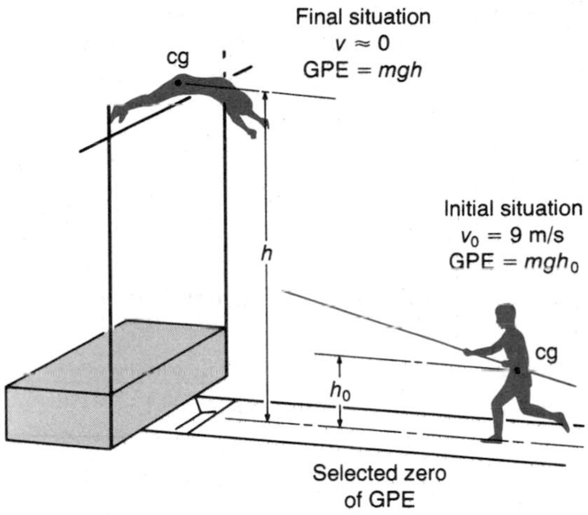

SOLUTION Let us take the initial situation as the one that occurs just before the vaulter leaves the ground. His velocity is 9 m/s, and we assume that his center of gravity is about 1 m above ground level. We select ground level as the convenient zero of gravitational potential energy. Take the final situation as his clearing the bar. We assume that his velocity at this point is negligible and that his center of gravity is approximately at the same height as the bar. (This is a bit restrictive; an agile vaulter might succeed in having his center of gravity pass under the bar.) The mechanical energy conservation equation then reads:

$$\text{Final (KE + GPE)} = \text{initial (KE + GPE)}$$

$$0 + mgh = \tfrac{1}{2}mv_0^2 + mgh_0$$

$$0 + 75 \text{ kg} \times 9.81 \text{ m/s}^2 \times h = \tfrac{1}{2} \times 75 \text{ kg} \times (9\text{m/s})^2 + 75 \text{ kg} \times 9.81 \text{ m/s}^2 \times 1 \text{ m}$$

$$\therefore h = 5.13 \text{ m}$$

The approximation involved in the assumption does not justify three significant figures in the answer. We should simply say that the energy method estimates a height of about 5 m for the vault. This estimate is about 15% below record levels, so the assumptions in the calculation are reasonable. ■

7.4 Power

In ordinary usage the terms *forceful, powerful,* and *energetic* are used almost interchangeably. In physics we have already given specific meanings to the terms *force* and *energy,* and we now pro-

SPECIAL TOPIC 8 SCALING AND POWER IN LONG-DISTANCE RUNNING

We have already mentioned the difficulty (or even the near impossibility) of setting up a theory of running based directly on Newton's laws. The discussion on scaling gives some insights, but an alternative approach based on energy considerations was given by Keller in "A Theory of Competitive Running" (*Phys. Today 26* (1973), p. 43). A detailed model was proposed using simple energy parameters that were estimated from a study of the track records of champion athletes.

The mathematical model is a bit complicated for sprints (which are defined to be races shorter than about 300 m), but the energy relations for the distance events can be set forth quite simply. Before a run, an amount of oxygen is available in the muscles that the body will eventually convert to available energy. In fact, it is easiest to think of the amount of oxygen as corresponding to amounts of energy available on reaction. Keller's figure for this quantity is 2403 J per kilogram body weight of the runner. As the run progresses, the oxygen is replenished by the respiratory and circulatory systems at a rate corresponding to a power of 41.51 W per kilogram of body weight. These are the assumed data for the energy and power available.

We now consider the power required to run at constant speed. In order to obtain good fits to track-record data, Keller had to assume that the power required per unit body mass was proportional to the square of the speed; that is:

$$P = Dv^2$$

where the constant D was found to be 1.121 $(\text{W/kg})/(\text{m}^2/\text{s}^2)$.

Let us use the figures quoted to do a sample calculation for a distance event. We will assume that the runner spends a couple of seconds accelerating to some constant speed and then runs at that speed thereafter. We ignore the small period of acceleration and seek to find the distance for which, say, 7 m/s is the optimum running speed. At this speed the power required is given by:

$$\begin{aligned} P &= Dv^2 \\ &= 1.121\ \frac{\text{W/kg}}{\text{m}^2/\text{s}^2} \times (7\ \text{m/s})^2 \\ &= 54.93\ \text{W/kg} \end{aligned}$$

Note that this is more than the power that the body can supply on an ongoing basis. The amount of the excess is:

$$54.93\ \text{W/kg} - 41.51\ \text{W/kg} = 13.42\ \text{W/kg}$$

This excess must be made up from the original stored energy of 2403 J/kg. If we make the extreme (and probably unjustified) assumption that the runner can go until the stored energy is completely exhausted, the elapsed time would be given by:

$$\frac{2403\ \text{J/kg}}{13.42\ \text{W/kg}} = 179.1\ \text{s}$$

Finally, we see that the distance run would be given by:

$$(7\ \text{m/s}) \times (179.1\ \text{s}) = 1254\ \text{m}$$

This is not one of the standard distance races. However, we find a time of about 136 s for 1000 m and 213 s for 1500 m, so our estimate is very reasonable.

We could have turned the whole calculation around and asked for the optimum speed for a 1500-m race. This gives a quadratic equation whose solution is $v = 6.80$ m/s, corresponding to a race time of 220 s (you should try it!).

Finally, we should note that these results did not depend on the mass of the runner—a conclusion that is very hard to accept. Moreover, there are some records for which the theory would predict that the runner spent more energy than he or she had available. Keller was well aware of these difficulties; nevertheless, he has provided an interesting insight into the physics of distance running.

ceed to do the same with *power*. The scientific meaning of **power** is the rate of doing work or the rate of transferring energy. A powerful engine is one that can perform a lot of work in a short time.

■ Definition of Average Power

$$\text{Average power} = \frac{\text{work done (or energy transferred)}}{\text{elapsed time}}$$

$$\langle P \rangle = \frac{W}{t} \tag{7.4}$$

SI unit of power: joule per second
New name: watt
Abbreviation: W

The name *watt* is given to the SI power unit in honor of James Watt (1736–1819), a Scot who was responsible for many improvements in the design of steam engines. The power capabilities of electrical equipment are usually quoted in watts, but for other purposes the horsepower is a common unit. One horsepower (abbreviation "hp") is approximately equal to 746 W. As the name implies, this unit of power had its origin in the working capability of horses. The first commercial steam engines in England were used to pump water out of mines, a job that was previously done by horses. To enable his customers to buy the size of steam engine they needed, James Watt took experimental data on the power capability of horses and branded his engines accordingly. However, he deliberately made an optimistic estimate of the power output of an average horse so that customers would not complain that a 10-hp steam engine was weaker than 10 horses.

EXAMPLE 7.10

An automobile of mass 1500 kg traveling at 25 m/s on a level road comes to an uphill gradient of 1 in 20. What additional horsepower is required to maintain speed on the hill?

SOLUTION Since the speed of the automobile is unchanged, the only additional work required from the motor is to supply the gravitational potential energy. At a speed of 25 m/s the vertical component of displacement in 1 s on a grade of 1 in 20 is given by 25 m/20 = 1.25 m. Gravitational potential energy gained in 1 s is therefore given by:

$$\begin{aligned} \text{GPE} &= mgh \\ &= 1500\ \text{kg} \times 9.81\ \text{m/s}^2 \times 1.25\ \text{m} \\ &= 1.84 \times 10^4\ \text{J} \end{aligned}$$

The energy required in joules per second is the same as the power in watts. Therefore:

$$\text{Extra power} = 1.84 \times 10^4 \text{ W}$$
$$= \frac{1.84 \times 10^4 \text{ W}}{746 \text{ W/hp}}$$
$$= 24.6 \text{ hp}$$

■

We have defined the average power over a given time interval in terms of the work done during the interval. If a constant force does work, we can relate the power to the force as follows: From Eq. (7.4) we have:

$$\langle P \rangle = \frac{W}{t}$$
$$= \frac{Fx_{\parallel}}{t} \text{ [using Eq. (7.1)]} \qquad \textbf{(7.5)}$$
$$= F \langle v_{\parallel} \rangle$$

The last step follows since the component of average velocity parallel to the applied force is given by $\langle v_{\parallel} \rangle = x_{\parallel}/t$. If we take the average on both sides of Eq. (7.5) over a short time interval, we can replace the average values by instantaneous values.

This equation shows that the power required to supply a constant force to an object increases as the velocity increases. Expressed in an alternate way, constant power cannot supply a constant force to an accelerating object.

■ Expression for Power in Terms of Force and Velocity

Instantaneous power = applied force × component of instantaneous velocity parallel to the force

$$P = Fv_{\parallel} \qquad \textbf{(7.6)}$$

EXAMPLE 7.11

With what constant velocity can a 2-hp motor raise a mass of 150 kg?

SOLUTION We begin by putting the data in SI units:

$$\text{Power of motor} = 2 \text{ hp} \times 746 \text{ W/hp}$$
$$= 1492 \text{ W}$$

To raise a mass at constant velocity requires an upward force equal to the weight of the mass. (Remember that the net force on the mass is zero if the mass is not acclerating.) We calculate the weight from Eq. (4.2):

$$\begin{aligned} w &= mg \\ &= 150 \text{ kg} \times 9.81 \text{ m/s}^2 \\ &= 1472 \text{ N} \end{aligned}$$

To find the power requirement in terms of force and velocity we use Eq. (7.6):

$$\begin{aligned} P &= Fv_{\parallel} \\ \therefore v_{\parallel} &= P/F \\ &= \frac{1492 \text{ W}}{1472 \text{ N}} \\ &= 1.01 \text{ m/s} \end{aligned}$$

■

EXAMPLE 7.12

A child is being towed in a toy car at a constant velocity of 2 m/s, and the steady tension in the tow cord is 35 N.

a. What is the power requirement for towing the car?

b. If the child and the car have a total mass of 25 kg, how much extra power is needed to begin accelerating at 0.2 m/s^2 from this speed?

SOLUTION

a. The first part of the problem is a simple application of Eq. (7.6):

$$\begin{aligned} P &= Fv_{\parallel} \\ &= 35 \text{ N} \times 2 \text{ m/s} \\ &= 70 \text{ W} \\ &= 70 \text{ W} \div 746 \text{ W/hp} \\ &= 0.0938 \text{ hp} \end{aligned}$$

(a) At constant 2 m/s.

(b) Accelerating from 2 m/s.

b. So long as the car moves at constant speed the tension in the cord and the frictional retarding force on the car are equal in magnitude and oppositely directed. When the car begins to accelerate the frictional drag force is unchanged (since the velocity of the car is still 2 m/s at the initial instant of acceleration), but the tension in the towing cord must be larger to provide the acceleration. In fact, Newton's second law [Eq. (4.1)] gives:

$$F = ma$$
$$T - 35\text{ N} = 25\text{ kg} \times 0.2\text{ m/s}^2$$
$$\therefore T = 40\text{ N}$$

Application of Eq. (7.6) to find the new power requirement gives:

$$\begin{aligned} P &= Fv_{\parallel} \\ &= 40\text{ N} \times 2\text{ m/s} \\ &= 80\text{ W} \\ &= 80\text{ W} \div 746\text{ W/hp} \\ &= 0.1072\text{ hp} \end{aligned}$$

The extra power requirement is therefore:

$$(0.1072\text{ hp} - 0.0938\text{ hp}) = 0.0134\text{ hp}$$

This is correct only when the acceleration begins; the power required will quickly increase since the increasing velocity of the car will result in increasing frictional drag, and hence a still greater tension in the towrope. ■

7.5 Simple Machines

In everyday usage the terms *machine* and *engine* are almost synonymous. However, science makes a distinction. We use the term **machine** to indicate a device for changing the form of mechanical work. Examples of machines in this sense are a crowbar used to lift a stone and a pulley arrangement used to lift a heavy load. The term *engine* (or preferably *heat engine*) is reserved for a device that transforms the flow of heat energy into mechanical work. We defer a discussion of heat engines until a later chapter.

Every machine requires a certain amount of input work; in return it performs output work. For example, the person who pushes on the free end of a crowbar does work by moving the crowbar through a certain distance. The work done by the person is the *input work,* and the force exerted is the *input force.* The other end of the crowbar does work in lifting the heavy stone, which is the *output work,* and the force exerted by the crowbar on the stone is the *output force.* The relationships between input and output work and force are the most important features of machines. We can begin with two general statements:

The automobile jack is an example of a machine having a high mechanical advantage. In the scissors-type jack illustrated, the operator can lift the automobile by exerting a relatively small force on the crank handle. Energy conservation requires that the work done turning the handle exceed the work done lifting the automobile by the amount dissipated because of friction. Lubrication of the screw thread and the joints reduces the energy dissipated and improves the efficiency of the jack. (Sandra G. Crosby)

1. The output force may or may not exceed the input force—it is a question of the design of the machine.
2. The output work can never exceed the input work—to say otherwise would violate the energy conservation principle.

These observations lead us to the definitions of two important quantities that apply to all types of machines. The **mechanical advantage** indicates the performance of the machine as a force magnifier.

$$\text{Mechanical advantage} = \frac{\text{magnitude of output force}}{\text{magnitude of input force}}$$

$$\text{MA} = \frac{F_{\text{out}}}{F_{\text{in}}} \tag{7.7}$$

Mechanical advantage is a pure number.

A mechanical advantage greater than unity indicates that the output force exceeds the input force; indeed, this is the designed purpose of many machines. If the mechanical advantage is less than unity, the input force exceeds the output force; some machines are designed to operate in this way.

The **efficiency** measures the performance of the machine as a user (and waster) of energy. Efficiency is denoted by the Greek letter eta (η). Since every practical machine dissipates some energy, usually because of frictional forces within the machine, the output work is always less than the input work. Only an ideal machine, free of energy-dissipating processes, would be 100% efficient.

$$\text{Efficiency} = \frac{\text{output work}}{\text{input work}}$$

$$\eta = \frac{W_{\text{out}}}{W_{\text{in}}} \tag{7.8}$$

Efficiency is a pure number.

Most machines are basically either inclined planes, levers, or pulleys. In each case the concepts of mechanical advantage and efficiency describe their utility.

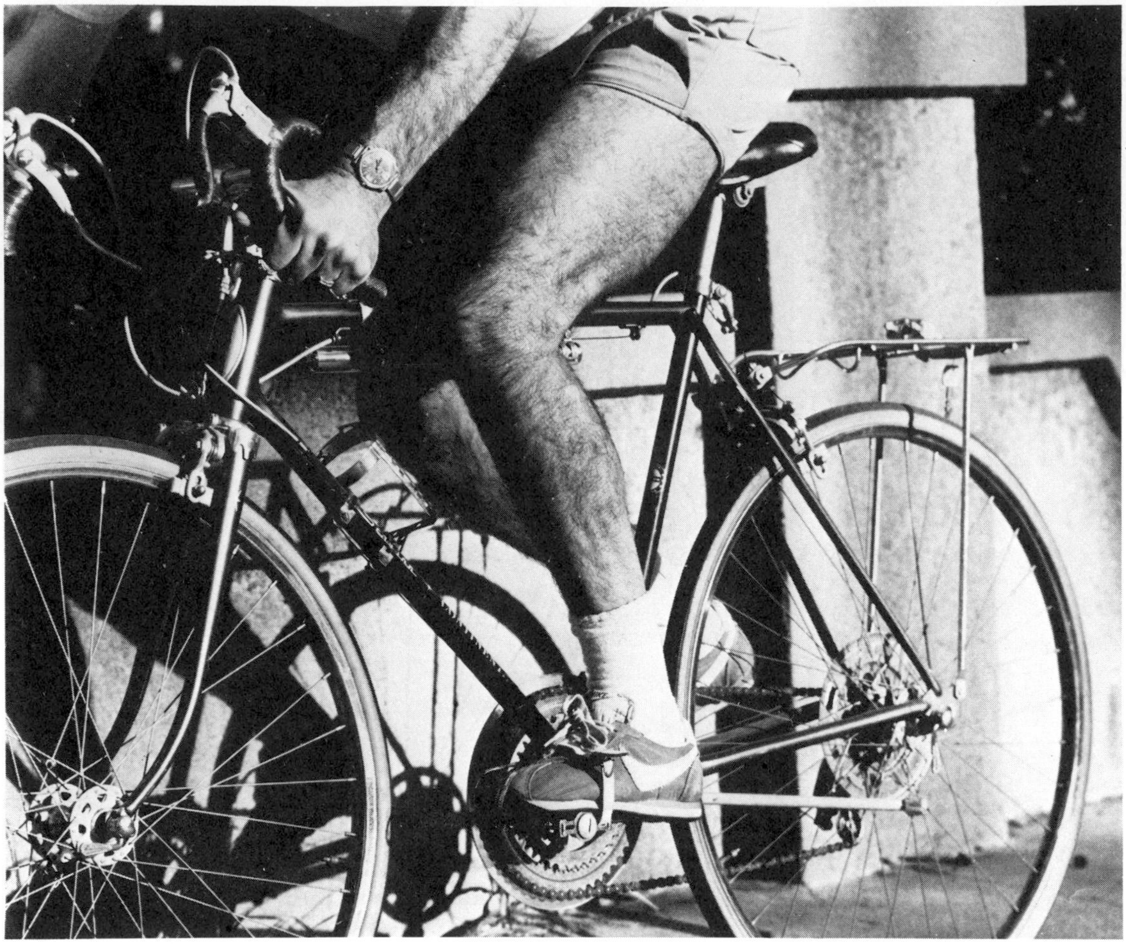

A bicycle is a complex machine, with the input force provided by the rider's foot on the pedal and the output force between the tire and the road. The efficiency is kept high by lubricating the chain and the bearings. The mechanical advantage can be altered by the rider using the gearshift. Low gear gives a large mechanical advantage for hill climbing, while higher gears give lower mechanical advantages for more comfortable cruising. (Sandra G. Crosby)

EXAMPLE 7.13

A weight of 400 N is pushed at a constant velocity up a frictionless plane inclined at 10° to the horizontal. Calculate the mechanical advantage and the efficiency. Repeat the calculation for a coefficient of kinetic friction 0.30 between the weight and the plane.

SOLUTION We use an inclined plane as a machine to lift a weight with a smaller force than would be needed to raise it ver-

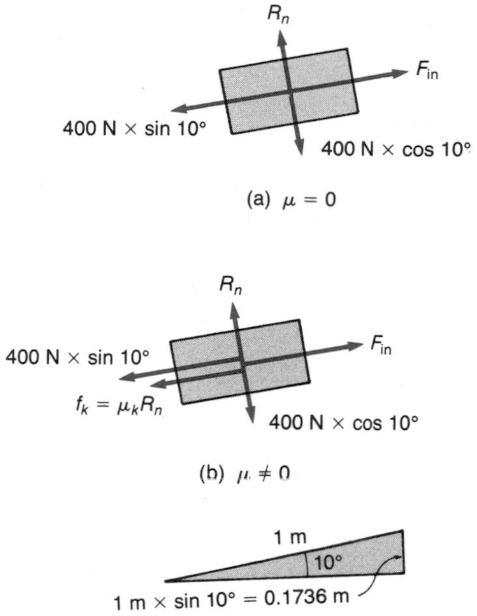

tically. The output force is the weight being lifted, and the input force is the force parallel to the plane required to push the weight. Since the weight moves up the plane with constant velocity, the equations of static equilibrium hold both parallel to the plane and perpendicular to it.

a. *The case* $\mu = 0$. The force required to push the block up the inclined plane is equal and opposite to the component of weight parallel to the plane. Using Eq. (4.5) we have:

$$F_{in} = mg \sin\theta$$
$$= 400 \text{ N} \times \sin 10°$$
$$= 69.46 \text{ N}$$

We can calculate the mechanical advantage from Eq. (7.7):

$$\text{MA} = \frac{F_{out}}{F_{in}}$$
$$= \frac{400 \text{ N}}{69.46 \text{ N}}$$
$$= 5.76$$

Since the plane is frictionless there is no energy dissipation, and the efficiency is 100%.

b. *The case* $\mu \neq 0$. The force pushing the block up the plane must overcome the component of weight parallel to the plane and the retarding frictional force. Using Eqs. (4.3) and (4.5) we have:

$$F_{in} = mg \sin\theta + \mu_k mg \cos\theta$$
$$= 400 \text{ N} \times \sin 10° + 0.3 \times 400 \text{ N} \times \cos 10°$$
$$= 187.6 \text{ N}$$

The mechanical advantage is again given by Eq. (7.7):

$$\text{MA} = \frac{F_{out}}{F_{in}}$$
$$= \frac{400 \text{ N}}{187.6 \text{ N}}$$
$$= 2.13$$

To find the efficiency, we must calculate both the input work and the output work. Suppose that we push the weight 1 m up the slope of the plane. The input work is given by Eq. (7.1):

$$W_{in} = Fx_{\parallel}$$
$$= 187.6 \text{ N} \times 1 \text{ m}$$
$$= 187.6 \text{ J}$$

The useful output work is the lifting of the weight of 400 N vertically through 1 m × sin 10° = 0.1736 m. The output work is equal to the increase in gravitational potential energy:

$$W_{out} = mgh$$
$$= 400 \text{ N} \times 0.1736 \text{ m}$$
$$= 69.46 \text{ J}$$

We calculate the efficiency from Eq. (7.8):

$$\eta = \frac{W_{out}}{W_{in}}$$
$$= \frac{69.46 \text{ J}}{187.6 \text{ J}}$$
$$= 37.0\%$$ ■

The example illustrates the general role played by friction in all machines. An ideal frictionless machine is 100% efficient and has the largest value of mechanical advantage that is possible for its particular design. The introduction of friction (without otherwise changing the design of the machine) lowers both the efficiency and the mechanical advantage.

EXAMPLE 7.14

Calculate the mechanical advantage of the nutcracker shown in the diagram. Assume that the nut is 2 cm from the pivot pin and that the force is applied to the handles 15 cm from the pivot pin.

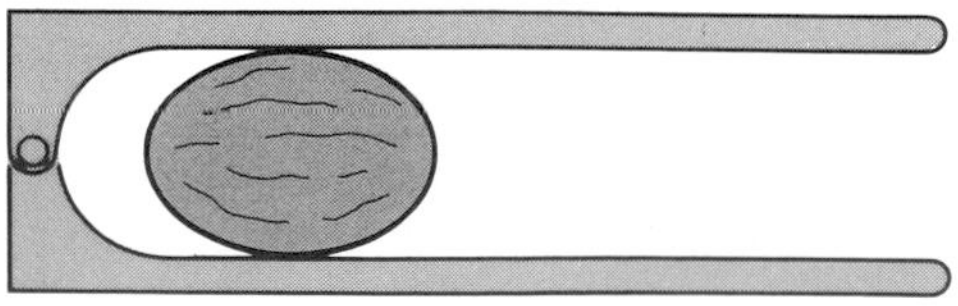

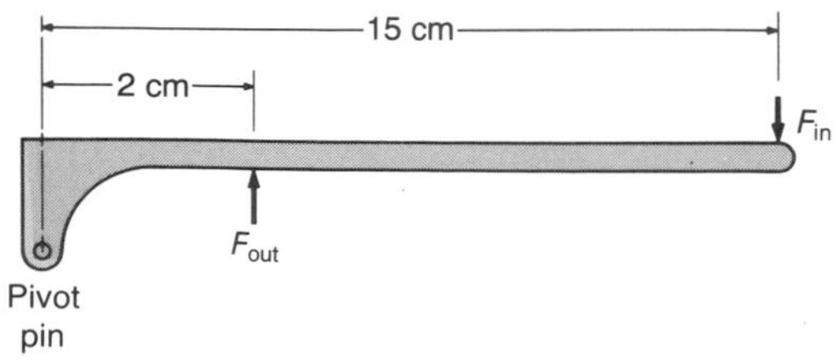

SOLUTION The nutcracker is an example of a lever. Every lever is a bar pivoted about some point with input and output forces at other points. We show the free-body force diagram for one arm of the nutcracker. The input force is applied to the handles, and the output force cracks the nut. The nutcracker arm is subject to the conditions of static equilibrium, Eq. (5.3). Since we do not need to know the force at the pivot pin, we need only apply the zero torque condition, Eq. (5.3c). Take torques about an axis through the pivot pin:

$$\Sigma \tau = F_{out} \times 2 \text{ cm} - F_{in} \times 15 \text{ cm} = 0$$

$$\therefore \text{MA} = \frac{F_{out}}{F_{in}} = \frac{15 \text{ cm}}{2 \text{ cm}} = 7.5$$

■

The efficiency of most lever systems is close to 100%, since little energy is usually dissipated at a simple pivot.

EXAMPLE 7.15

A load of 1000 N is lifted by the pulley system shown. A force of 550 N is required to lift the load. Find the mechanical advantage and the efficiency of the pulley system. If the load is lifted through a distance of 2 m, calculate the amount of energy dissipated in the pulley system.

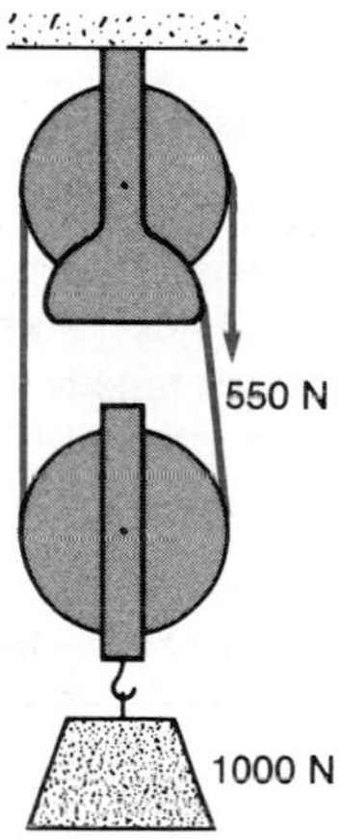

SOLUTION We calculate the mechanical advantage directly from Eq. (7.7):

$$\text{MA} = \frac{F_{out}}{F_{in}} = \frac{1000 \text{ N}}{550 \text{ N}} = 1.82$$

If the load rises 2 m, each side of the rope supporting the movable pulley is shorter by 2 m. This means that we must draw 4 m of rope over the fixed pulley, and the input force moves through 4 m. Equation (7.8) gives:

$$\eta = \frac{W_{out}}{W_{in}} = \frac{1000 \text{ N} \times 2 \text{ m}}{550 \text{ N} \times 4 \text{ m}} = 91\%$$

The energy dissipated during this lift is the difference between the input and output work:

$$\text{Energy dissipated} = 550 \text{ N} \times 4 \text{ m} - 1000 \text{ N} \times 2 \text{ m} = 200 \text{ J}$$

■

KEY CONCEPTS

Work is displacement multiplied by the force component in a parallel direction. See Eq. (7.1).

The **kinetic energy** of a body is equal to one-half its mass multiplied by the square of its velocity. See Eq. (7.2).

The **work-energy theorem** equates an increase in the kinetic energy of a body to the work done on it by the net force.

The change in **gravitational potential energy** of a body is equal to its weight multiplied by the vertical component of displacement of its center of gravity. See Eq. (7.3).

The **energy conservation principle** states that the energy of a closed system is conserved.

The **mechanical energy conservation law** requires that the kinetic energy plus the gravitational potential energy remain constant if applied forces (the weight excluded) do no work.

Power is the rate of doing work or supplying energy. See Eq. (7.5).

A **machine** is a device that produces a certain output force from an input force of different magnitude and direction. The **mechanical advantage** is the magnitude of the output force divided by the magnitude of the input force. See Eq. (7.7). The **efficiency** is the output work divided by the input work. See Eq. (7.8).

QUESTIONS FOR THOUGHT

1. The total energy of a massive body in motion remains constant. Does this mean that no forces act on it? Explain your answer.
2. An automobile moves along a straight road at constant velocity. The engine does work at a constant rate; energy results from the production of work. Where does this energy go to?
3. If air resistance is neglected, the time taken by a vertically thrown stone to reach its apex is the same as the time of its return to the point of projection. If air resistance is significant, which of these two times do you expect to be longer?
4. Is the energy crisis really an energy crisis, or is it a power crisis? Explain your answer.
5. An athlete sprinting at full speed is doing a substantial amount of work, but her kinetic energy is not increasing. Explain where the work goes. Can you suggest a way for the athlete to produce kinetic energy (by using her leg motion) that is more efficient than running?
6. Does the work done when you climb a steep hill depend on the path that you select or on the time taken? Does the average power expended depend on either or both of these factors?
7. Think of some examples of machines that have a mechanical advantage less than unity.
8. What changes occur in the mechanical advantage and the efficiency of a machine if it is operated in reverse? Illustrate your comments with a specific example.

PROBLEMS

A. Single-Substitution Problems

1. A loaded sled is pushed 100 m across a level snowfield by a constant horizontal force of 180 N. How much work is done on the sled? [7.1]
2. A heavy box is pushed across a level floor by a constant force of 200 N. How far can it be pushed with 1.2 MJ of work? [7.1]
3. A horse tows a barge in a canal by means of a rope that makes an angle of 15° with the direction of mo-

tion of the barge. The tension in the rope is 220 N. How much work is done in moving the barge 1000 m? [7.1]

4. What is the kinetic energy of a 1200-kg automobile traveling at 25 m/s? [7.2]

5. What is the kinetic energy of an 80-kg sprinter running at 10 m/s? [7.2]

6. What is the speed of a 9-g rifle bullet that has a kinetic energy of 1 kJ? [7.2]

7. Calculate the increase in gravitational potential energy of a 75-kg hiker who climbs an 800-m hill. [7.2]

8. An airliner of mass 100 metric tons takes off and increases its gravitational potential energy by 1 GJ. To what height has it climbed? [7.2]

9. What is the change in gravitational potential energy when a 4-kg brick falls from 28 m to the ground? [7.2]

10. A mass of 100 kg is hauled vertically upward from rest by a constant force of 1500 N through a distance of 20 m. The final kinetic energy of the mass is 6 kJ. How much energy was dissipated? [7.3]

11. A child of mass 20 kg descends a slide at an amusement park. The vertical component of her displacement is 4.2 m downward, and 440 J of energy are dissipated during the slide. Calculate her final velocity. [7.3]

12. A 35-kg crate is pushed 15 m across a rough horizontal floor by a constant force of 125 N. Its final speed is 4 m/s. How much energy was dissipated? [7.3]

13. A worker lifts 15-kg bags of apples from the ground to a bench 80 cm higher at the rate of 16 bags per minute. Calculate the power requirement for this operation. [7.4]

14. A pump drains a basement, raising the water through a vertical height of 3.2 m at the rate of 12 ℓ/min. Calculate the minimum power requirement for the motor. [7.4]

15. An electric motor drives a fan and supplies energy for 5 h at a rate of 0.32 hp. How many joules are supplied in this period? [7.4]

16. A small car encounters a frictional resisting force of 140 N when traveling at a constant speed of 25 m/s. Calculate the horsepower required. [7.4]

17. A block-and-tackle system has a mechanical advantage of 6. How large a mass can be raised vertically by a person exerting a force of 180 N? [7.5]

18. An input of 0.5 hp for 3 min results in the output of 50 000 J of work by a certain machine. Calculate the efficiency of the machine. [7.5]

19. A machine that is 80% efficient gives an output of 42 000 J of work during a 5-min time interval. Calculate the required input power. [7.5]

B. Standard-Level Problems

20. An athlete throws a shot of mass 7.2 kg with an initial speed of 14 m/s from the release point 2.1 m above ground level. Calculate the speed of the shot just before it hits the ground. [7.2]

21. A woman of 60-kg mass takes a running dive from a rigid diving platform 3 m above the water level. Her center of gravity is 1 m above the board, and her speed on leaving it is 2.5 m/s. What is her speed when she strikes the water? (Assume that her center of gravity is at water level when she hits the water.) [7.2]

22. A skier takes off from a jump with a speed of 30 m/s and lands at a point whose vertical displacement below the takeoff point is 20 m. Calculate the skier's speed on landing. (Ignore air resistance.) [7.2]

23. A roller coaster car crossing the top of a hill on its path has a speed of 8 m/s. In the bottom of the next valley its speed is 32 m/s. Ignoring friction, calculate the vertical height of the hill above the valley. [7.2]

24. A child of 20-kg mass sits on a swing with her center of gravity 4 m below the point of support. The swing is drawn aside until the supports make an angle of 30° to the vertical; then the swing is released. What is the child's velocity on passing through her lowest point? [7.2]

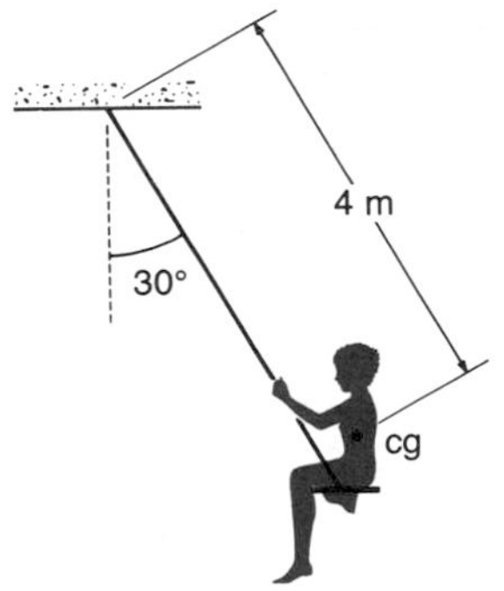

25. A punter kicks a football of mass 0.4 kg, which is caught 4 s later 65 yd downfield. During its flight, the ball attains a maximum height of 20 m. If frictional

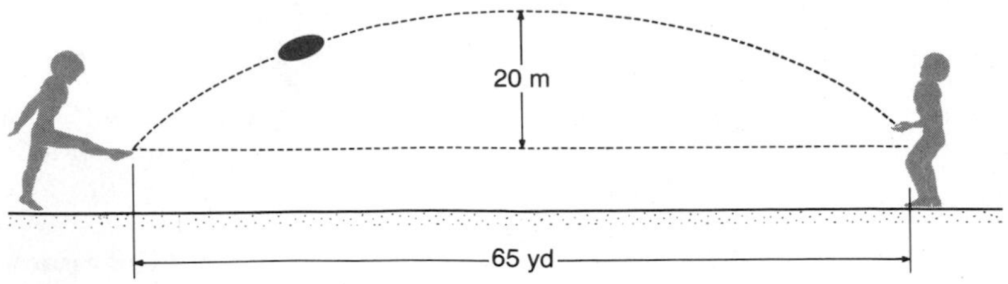

losses are neglected, how much energy does the punter impart to the football?

26. A discus thrower imparts 1200 J of energy to a discus of mass 2 kg. The distance of the throw is 65 m and the time of flight 2.60 s. To what height does the discus rise? (Neglect any effect due to air resistance.)

27. A tennis ball of mass 58 g is thrown vertically upward with a velocity of 18 m/s and returns to the ground with a velocity of 16 m/s.

a. How much energy is dissipated during its flight?
b. Assuming that half of the energy was dissipated during ascent and half during descent, to what height did the ball rise? [7.3]

28. A rubber ball of mass 40 g is thrown vertically to a height of 24 m.

a. If 1 J of energy was dissipated during the ascent, what was the initial velocity of the ball?
b. If the same amount of energy is dissipated during the descent, with what velocity does it strike the ground? [7.3]

29. A baseball weighing 1.5 N is thrown vertically upward with an initial velocity of 35 m/s and reaches a height of 55 m.

a. How much work does the thrower do in throwing the ball?
b. How much energy is dissipated on the way up?
c. If the ball loses the same amount of energy on the way down, what is its velocity when it returns to the projection point? [7.3]

30. A cyclist of mass 75 kg rides a 15-kg cycle. After coasting from rest down a 60-m-high hill the cycle speed is 80 km/h. How much energy is dissipated in the descent?

31. A 20-kg child slides down an amusement park slide 10 m long and inclined at 45° to the horizontal. The coefficient of kinetic friction between the child and the slide is 0.2. Calculate the speed of the child at the bottom and the energy dissipated during the slide.

32. A block of mass 2 kg starts at the top of a frictionless curved ramp with an initial speed of 8 m/s. The block slides down to a rough horizontal surface 5 m below its starting point. The coefficient of kinetic friction between the block and the horizontal surface is 0.32. How far does the block slide on the horizontal surface before coming to rest?

33. A 1.6-kg block slides from rest down a frictionless curved ramp to a horizontal surface 3 m below its starting point. The block slides across the horizontal surface 10 m before coming to rest. Calculate the coefficient of kinetic friction between the block and the horizontal surface.

34. A cross-country skier of mass 75 kg moves along level snow at 6 m/s and then comes to a downhill slope 300 m long, inclined at 5° to the horizontal. Coasting down the slope, the skier experiences an average frictional resisting force of 40 N. At what speed is the skier traveling at the bottom of the slope?

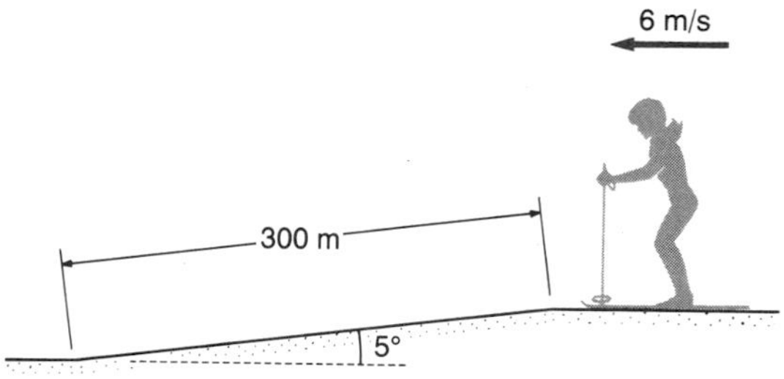

35. A 1000-kg compact automobile climbs a slope 300 m long, inclined at 4° to the horizontal. It exerts a propulsive force of 400 N and experiences an average frictional retarding force of 100 N. If its speed at the bottom of the slope is 25 m/s, what is its speed at the top?

36. Two football players, each of mass 100 kg and each traveling at 8 m/s, collide head-on and come to a full stop.

a. How much total energy do they absorb in the collision?
b. Assume that each player absorbs half the energy. From what height would a fall to the ground result in the same total energy being absorbed?

37. Two compact cars each of mass 1200 kg, each traveling in opposite directions, meet head-on and come to a complete stop.

a. If the total energy absorbed in the collision is 1 MJ, what was the speed of each car?
b. Assume that each car absorbs half the energy. From what height would a fall to the ground result in the same total energy absorption?

38. A 0.5-hp motor pumps water from a well 50 m deep. If 25% of the motor power is dissipated, how many liters of water does the pump deliver in 1 min? (Assume the kinetic energy given to the water is negligible in comparison with the potential energy.) [7.4]

39. A 1200-kg automobile can reach a speed of 30 m/s in 11 s after starting from rest. What is the average horsepower requirement of the motor if 20% of its power is dissipated? [7.4]

40. A cyclist of mass 80 kg riding a 15-kg cycle produces 1.2 hp for 15 s starting from rest. If 10% of the power is dissipated, what speed does he attain?

41. A locomotive of mass 200 metric tons pulls a freight train of 70 cars each having an average mass of 40 metric tons. What average horsepower is required to reach a speed of 45 km/h (12.5 m/s) within 4 min of starting from rest? (Neglect frictional losses.)

42. A 1000-kg automobile starts from rest and accelerates uniformly to 20 m/s up a gradient inclined at 4° with the horizontal in a time interval of 20 s. What average horsepower is required of the engine, if 15% is dissipated?

43. A 65-kg marathon runner traveling at a steady 4.5 m/s comes to a long uphill stretch inclined at 2° to the horizontal. How much extra power must he exert to maintain his speed?

44. A 100-metric ton jetliner climbs to 10 000 m and cruises at 1000 km/h within 20 min of beginning its takeoff run. What is the average power requirement from each of its three jet engines? (Neglect frictional losses.)

45. An elevator car of mass 300 kg can carry up to 12 persons averaging 75 kg each. The elevator is required to rise 25 floors (a total of 100 m) in 30 s. What horsepower must the motor have to drive it? (Assume that the change in kinetic energy is small relative to the gravitational potential energy change, and neglect frictional losses.).

46. An elevator meets all the specifications of the previous example but also has an 800-kg counterweight that falls as the elevator rises, and vice versa.

a. What horsepower is required to drive the loaded elevator up 25 floors under the conditions of the previous problem?
b. What horsepower is required to lower the unloaded elevator 25 floors in 30 s?

47. A ski-tow that can accommodate 50 skiers moves up a 20° slope at 2.5 m/s. If the skiers and their equipment average 80 kg each, how much power is required to run the operation? (Neglect the kinetic energy increase that each rider experiences when joining the tow.)

48. The floor of a basement is 60 m^2 and 3 m below ground level. The basement is flooded with water to a depth of 1 m.

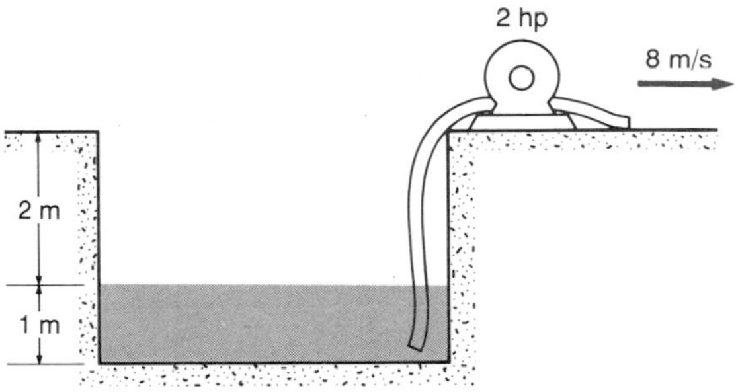

a. How much work must be done to raise the water to ground level?
b. If a 2-hp pump ejects the water at ground level with a speed of 8 m/s, how long does it take to pump the basement out? (Neglect frictional losses.)

49. Water flows from an upper level to a water turbine 30 m below at a rate of 25 m^3/min. The water commences with negligible velocity at the upper level and leaves the turbine with a velocity of 5 m/s. What is the power output of the turbine if energy dissipation is neglected?

50. A fish cruises at a speed of 30 cm/s and exerts an average power of 1/50 hp. What is the magnitude of the average force between the fish and the water?

51. A small boat is towed at a constant speed of 18 km/h by a towline in which the tension is 480 N. What is the required horsepower?

52. A 250-kg piano is pushed at a constant velocity up a ramp inclined at 12° to the horizontal by a force of 800 N.

a. Calculate the mechanical advantage and the efficiency of the system.
b. How much energy is dissipated in lifting the piano through a vertical distance of 4 m? [7.5]

53. A wheelbarrow is a lever in which the input force is exerted on the handles, the output force raises the load, and the pivot is the wheel axle. A load of 800 N requires a 300-N effort to raise it.

a. What is the mechanical advantage?
b. If the center of gravity of the load is 60 cm from the axle, how far are the handles from the axle? [7.5]

54. The pulley system shown is 85% efficient. What force is needed to lift a load of 500 kg? [7.5]

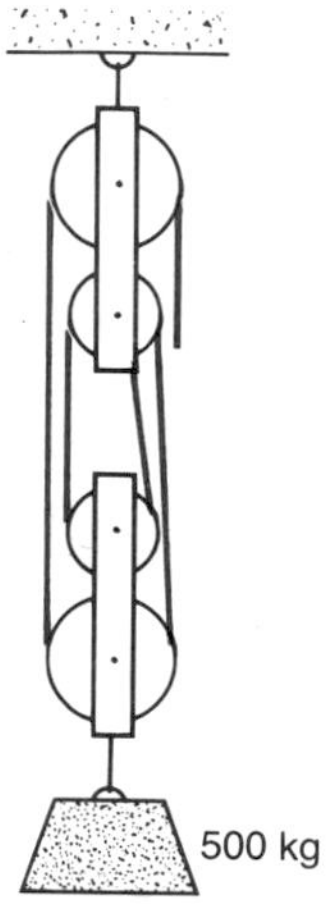

55. The coefficient of kinetic friction between a long plank and a 100-kg crate is 0.28. A worker who can exert a steady force of 480 N wishes to use the plank as an inclined plane to raise the crate at a constant velocity.

a. What is the maximum angle with the horizontal at which he can set the plank?
b. Calculate the mechanical advantage and the efficiency.

56. The coefficient of kinetic friction between a block and an inclined plane is 0.25. Regarding the inclined plane as a machine for raising the block at constant velocity, calculate the mechanical advantages when the plane is inclined at 10° and at 20° to the horizontal.

57. A 250-kg automobile engine is to be lifted by using a pulley system and a maximum available input force of 640 N. Design a suitable pulley system assuming 80% efficiency.

58. A dentist pumps her chair's foot pedal, which has a downward stroke of 20 cm, with a force of 40 N. If the device is 80% efficient, how many times must she push down on the pedal to lift an 80-kg patient at least 15 cm?

59. An automobile jack requires the operator to use a force of 200 N moved through 25 cm in order to raise the load 1 cm. If the jack is 80% efficient, what is the mechanical advantage?

C. Advanced-Level Problems

60. An automobile of mass 1500 kg requires 25 hp to drive it at a steady 90 km/h (25 m/s) along a level road.

a. What is the retarding frictional force at this speed?
b. How much horsepower is necessary to drive the automobile at 90 km/h down a hill inclined at 1° to the horizontal?
c. What is the inclination to the horizontal of a hill down which the automobile coasts at a steady 90 km/h?

61. A 1500-kg automobile is capable of accelerating from rest to 90 km/h in 10 s on a level road. If allowed to coast to a stop from 90 km/h (also on a level road), it travels 400 m before stopping. If the frictional force opposing the motion is constant throughout the tests, what is the average horsepower delivered by the engine during the acceleration period?

62. A mass of 5 kg is supported from a fixed point by a light cord 50 cm long. The mass is drawn aside until the cord makes an angle of 60° with the vertical and is then released. Calculate the tension in the cord when the 5-kg mass reaches its lowest position.

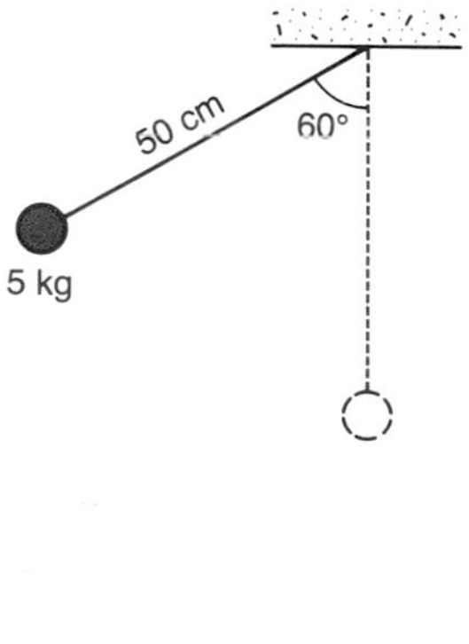

8

IMPULSE AND MOMENTUM

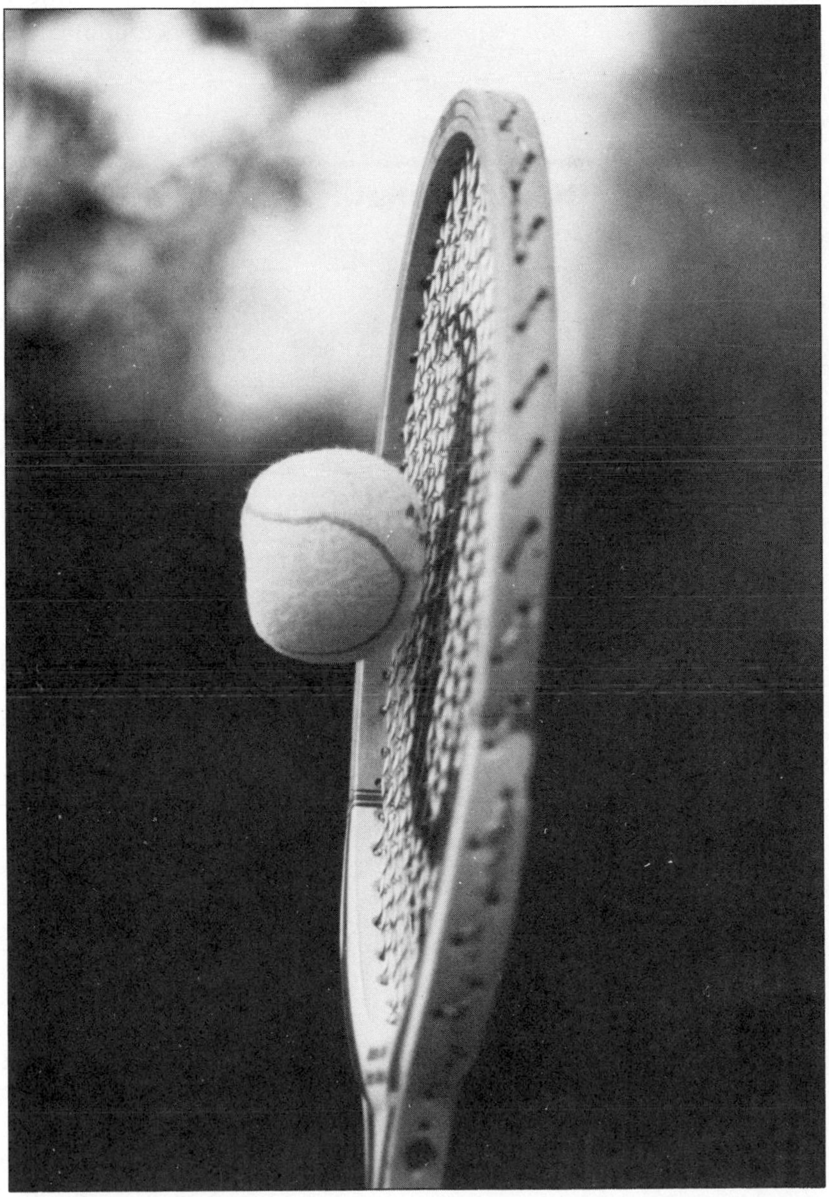

Isaac Newton introduced the phrase *quantity of motion* in his seventeenth-century formulation of the laws of dynamics. He defined the quantity of motion of a moving object as the mass multiplied by the velocity; today we call this quantity the momentum. Some of Newton's contemporaries used this same phrase, the "quantity of motion," to name the kinetic energy, and this ambiguity caused much confusion.

We have already seen that the energy conservation principle provides a basis for understanding many situations in dynamics. In this chapter we see that the linear momentum conservation principle provides a viewpoint that is particularly well suited to understanding collisions between two bodies. Our discussion of momentum begins with the definition of impulse, which naturally leads to the linear momentum conservation law.

8.1 Definitions of Impulse and Momentum

We began our discussion of energy with the definition of work. With the help of Newton's laws, the work-energy theorem quickly followed, assuming a central role in our discussion of energy. Once again we will begin with Newton's laws, but this time we will take a different approach.

Consider a constant mass m that experiences a net average force $\langle \mathbf{F} \rangle$ for a time interval t. The net force causes an average acceleration related to it by Newton's second law, Eq. (4.1), modified for average net force and average acceleration:

$$\langle \mathbf{F} \rangle = m\langle \mathbf{a} \rangle$$

But we defined average acceleration by Eq. (3.6):

$$\langle \mathbf{a} \rangle = \frac{\mathbf{v} - \mathbf{v}_0}{t}$$

Eliminating $\langle \mathbf{a} \rangle$ from this pair of equations gives:

$$\langle \mathbf{F} \rangle t = m\mathbf{v} - m\mathbf{v}_0 \qquad \textbf{(8.1)}$$

The content of this equation does not differ in any essential way from Newton's second law of motion. However, it leads to the definition of new concepts and ultimately to the introduction of a powerful conservation law.

Let us consider the two sides of Eq. (8.1) separately. The left-hand side is the average force multiplied by the time of action. This quantity is called the **impulse** of the force.

Definition of Impulse

$$\text{Impulse} = \text{average force} \times \text{time of action}$$
$$= \langle \mathbf{F} \rangle t \qquad \textbf{(8.2)}$$

SI unit of impulse: newton-second
Abbreviation: N·s

Impulse is a vector quantity having the same direction as the average applied force. There is no commonly accepted algebraic symbol referring specifically to impulse.

The right-hand side of Eq. (8.1) is the difference between the final and initial values of a quantity that we call the **linear momentum.** Linear momentum is a vector quantity whose direction is the same as the velocity direction. For one-dimensional problems, we sometimes write $p = mv$ instead of the explicit vector form of Eq. (8.3). Momentum and velocity are still vectors in one dimension, but we need only a positive or a negative sign to specify their direction. We include the term *linear* simply to distinguish linear momentum from angular momentum, which will be introduced in Chapter 9. In spite of the great importance of linear momentum as a physical quantity, the kilogram-meter per second unit has no new name.

Definition of Linear Momentum

$$\text{Linear momentum} = \text{mass} \times \text{velocity}$$
$$\mathbf{p} = m\mathbf{v} \qquad \textbf{(8.3)}$$

SI unit of momentum: kilogram-meter per second
Abbreviation: kg·m/s

With our definitions of impulse and momentum, we can restate the physical content of Eq. (8.1):

Impulse-Momentum Theorem

The change in linear momentum of a body is equal to the impulse of the average net force that causes the momentum change.

This statement is similar in form to the work-energy theorem, which states that the work done by an applied force is equal to

SPECIAL TOPIC 9 JET-PROPELLED ANIMALS

We are all familiar with aircraft that are propelled by jet engines. These engines take in air, heat it to a high temperature and pressure in the combustion chamber, and then expel it at high velocity. The thrust generated by the jet engine is the change in momentum per unit time that it causes in the air that flows through it.

Everyday experience shows that most animals use a form of propulsion that does not resemble this. In fact, most animals that move about a good deal are vertebrates or insects; they have rigid skeletons that react to muscle contractions to produce some form of coordinated movement. However, there is a remarkable exception—squids move by jet propulsion. The squid sucks water into a large mantle cavity through openings located around its head. Contraction of the mantle closes off these intake openings and squirts the water out in a jet through a funnel below the head; this produces a thrust that is similar to that produced with air by the jetliner. Moreover, this does not necessarily restrict the squid to slow local movement. One species migrates some 2000 km in several months, and can move in short bursts at speeds up to 3 m/s.

the change in kinetic energy. In our present discussion, the impulse of the applied force is equal to the change in linear momentum. The linear momentum and the kinetic energy are two quite distinct concepts, although both have some claim to be called the quantity of motion. Kinetic energy is a scalar quantity of motion, and it leads us to a mechanical energy conservation law. Eventually energy conservation leads far beyond the boundaries of dynamics. Linear momentum is a vector quantity of motion, and its conservation law is regarded as the very heart of dynamics. Before introducing momentum conservation, let us work some examples of the impulse-momentum theorem.

In real-life situations the impulse that changes the momentum may be due to a relatively weak force exerted over a long time interval. This happens when a weak frictional force slows an object. On the other hand, the momentum change may be a collision-type change, which happens very quickly. When a bat strikes a ball, a very large force operates for a short time interval. Let us consider some examples illustrating these points.

EXAMPLE 8.1

A woman of mass 50 kg is swimming with a velocity of 1.6 m/s. If she stops stroking and glides to a stop in the water, what is the impulse of the frictional force that stops her?

SOLUTION We calculate the initial momentum of the swimmer from Eq. (8.3):

$$\begin{aligned}\mathbf{p}_0 &= m\mathbf{v}_0 \\ &- 50\text{ kg} \times 1.6\text{ m/s} \\ &= 80\text{ kg}\cdot\text{m/s}\end{aligned}$$

The final momentum of the swimmer is zero, and we calculate the impulse of the force from Eq. (8.1):

$$\begin{aligned}\langle \mathbf{F}\rangle t &= m\mathbf{v} - m\mathbf{v}_0 \\ &= 0\ \text{kg}\cdot\text{m/s} - 80\ \text{kg}\cdot\text{m/s} \\ &= -80\ \text{kg}\cdot\text{m/s}\end{aligned}$$

The negative sign shows us that the impulse of the frictional force is oppositely directly to the original momentum.

Since impulse is the product of force and time, we can also write the answer as -80 N·s. The dimensions of impulse are $[MLT^{-2}] \times [T] = [MLT^{-1}]$; the dimensions of momentum are $[M] \times [LT^{-1}] = [MLT^{-1}]$. Since the two quantities have the same dimensions, their SI units are the same. Note that no details of a force are needed to calculate its impulse—we need only know the change of momentum that the force causes. ■

EXAMPLE 8.2

A baseball of mass 0.142 kg that moves with a velocity of 45 m/s is struck by a swinging bat so as to exactly reverse the direction of velocity without any change in magnitude. If the duration of the impact is 6 ms, what is the average force exerted on the ball?

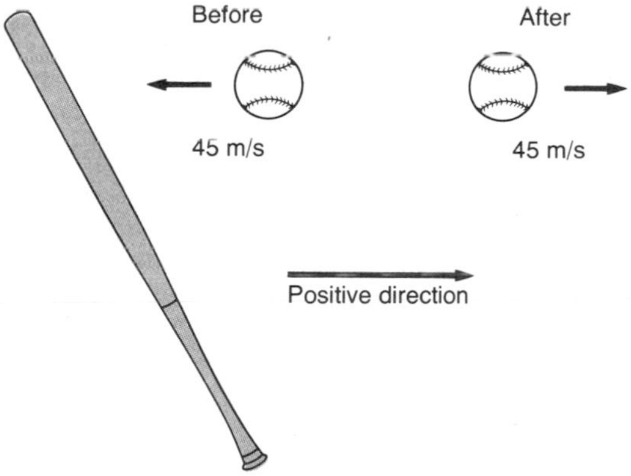

SOLUTION Select the positive direction indicated on the diagram:

$$\begin{aligned}\text{Change in linear momentum due to blow} &= m\mathbf{v} - m\mathbf{v}_0 \\ &= 0.142\ \text{kg} \times 45\ \text{m/s} - 0.142\ \text{kg} \times (-45\ \text{m/s}) \\ &= 12.78\ \text{kg}\cdot\text{m/s}\end{aligned}$$

Recall that linear momentum is a vector quantity. The magnitude of the linear momentum is the same before and after the blow from the bat, but its direction is reversed. This results in the momentum change that we have just calculated. To find the average force exerted on the ball, we use Eq. (8.1):

$$\begin{aligned}\langle \mathbf{F}\rangle t &= m\mathbf{v} - m\mathbf{v}_0 \\ \langle \mathbf{F}\rangle \times 6 \times 10^{-3}\ \text{s} &= 12.78\ \text{kg}\cdot\text{m/s} \\ \therefore \langle \mathbf{F}\rangle &= 2130\ \text{N}\end{aligned}$$

Note that we could find the same result using the methods of previous chapters. Equation (3.6) gives the acceleration of the ball during the blow:

$$\begin{aligned}\langle \mathbf{a} \rangle &= \frac{\mathbf{v} - \mathbf{v}_0}{t - t_0} \\ &= \frac{45 \text{ m/s} - (-45 \text{ m/s})}{6 \times 10^{-3} \text{ s}} \\ &= 1.5 \times 10^4 \text{ m/s}^2\end{aligned}$$

Then Eq. (4.1), modified for average values, gives:

$$\begin{aligned}\langle \mathbf{F} \rangle &= m\langle \mathbf{a} \rangle \\ &= 0.142 \text{ kg} \times 1.5 \times 10^4 \text{ m/s}^2 \\ &= 2130 \text{ N}\end{aligned}$$

■

You may wonder what help the impulse-momentum theorem really provides, since we could have worked either of the last two examples directly from kinematics and Newton's laws. As we have already mentioned, our major reason for defining impulse is to lead to linear momentum conservation.

Before proceeding in that direction, let us consider a type of problem involving *continuous momentum change,* whose solution follows readily from the impulse-momentum theorem.

EXAMPLE 8.3

Sand is pouring onto a level floor at the rate of 2 kg/s from a height of 4 m. Calculate the load on the floor that is additional to the weight of the sand.

SOLUTION In addition to supporting the weight of the sand, the floor must supply an upward force to change the sand's momentum to zero. We can select any time interval that is convenient, and calculate the momentum change during the interval. During a 1-s interval a mass of 2 kg reaches the floor, and we need to know its velocity just before striking the floor. This is easily achieved using energy conservation:

$$\begin{aligned}\text{Final KE} &= \text{initial GPE} \\ \tfrac{1}{2}mv^2 &= mgh \\ \tfrac{1}{2} \times 2 \text{ kg} \times v^2 &= 2 \text{ kg} \times 9.81 \text{ m/s}^2 \times 4 \text{ m} \\ \therefore v &= \pm 8.859 \text{ m/s}\end{aligned}$$

Let us select the upward direction as positive; since the sand is falling, we must take $v = -8.859$ m/s just before it strikes the floor. Just after the sand strikes the floor, its momentum is zero and the impulse-momentum theorem [Eq. (8.1)] gives:

$$\langle \mathbf{F} \rangle t = m\mathbf{v} - m\mathbf{v}_0$$

$$\langle \mathbf{F} \rangle \times 1 \text{ s} = 2 \text{ kg} \times 0 \text{ m/s} - 2 \text{ kg} \times (-8.859 \text{ m/s})$$

$$= 17.7 \text{ kg}\cdot\text{m/s}$$

$$\mathbf{F} = 17.7 \text{ N}$$

The positive sign indicates that the force exerted by the floor on the sand is upward. By Newton's third law, we see that the force exerted by the sand on the floor is of the same magnitude but directed downward. ■

8.2 Conservation of Linear Momentum

In Example 8.2 the momentum of a baseball is changed by a swinging bat. The change in momentum is due to the force exerted on the ball by the bat, and the bat is in the hands of a batter who is standing on the earth. This is a complex situation, and past experience shows that physics attains its insight from simple, idealized models. Imagine the baseball and the bat as isolated from all other influences—no batter and no gravity—and just colliding with each other. The ball undergoes a linear momentum change because of the average force exerted on it by the bat, and the bat undergoes a linear momentum change because of the average force exerted on it by the ball. But the force between the bat and the ball is an *action-reaction force pair* to which Newton's third law applies—they are equal and opposite forces. Their impulses are therefore equal and opposite, and so, too, are the changes in linear momentum that these impulses cause. The change in linear momentum of the ball is therefore balanced by an equal and opposite change in the linear momentum of the bat. Expressed somewhat differently, the total linear momentum of the bat-and-ball system remains unchanged. (The example chosen is complicated by possible angular momentum effects, but that is a separate question that does not affect linear momentum conservation.) The conclusion reached for the case of the bat and the ball can be generalized to describe any number of objects provided that no external net force acts on them. We call such a system an *isolated system*. We can now state the principle of conservation of linear momentum.

■ Linear Momentum Conservation Principle

The total linear momentum of an isolated system remains constant.

Since linear momentum is a vector, this is a vector conservation law. Both the *magnitude and direction* of the linear momentum remain constant. Expressed another way, all *components* of the linear momentum remain constant. In what follows, we will treat only problems involving two interacting bodies. It is helpful to write an equation expressing linear momentum conservation for the special case of only two bodies. Consider two bodies of mass m_1 and m_2 that collide with each other. Let the velocities before collision be $\mathbf{v}_1$ and $\mathbf{v}_2$ and those after the collision $\mathbf{v}_1'$ and $\mathbf{v}_2'$. The mathematical expression of linear momentum conservation reads as follows.

■ Linear Momentum Conservation Principle for Two Colliding Masses

$$m_1\mathbf{v}_1 + m_2\mathbf{v}_2 = m_1\mathbf{v}_1' + m_2\mathbf{v}_2' \tag{8.4}$$

The unprimed quantities $\mathbf{v}_1$ and $\mathbf{v}_2$ are velocities before the collision; the primed quantities $\mathbf{v}_1'$ and $\mathbf{v}_2'$ are velocities after the collision.

The boldface vector symbols in Eq. (8.4) emphasize that this is a vector conservation law. To visualize what this can mean, imagine two blocks of putty of equal masses traveling toward each other with equal speeds. When they collide, the whole mass comes to rest, and the total momentum is zero. But the total momentum was also zero before the collision, since it was the sum of two momentum vectors equal in magnitude but oppositely directed.

At first sight the principle of momentum conservation seems to apply to a highly fictional situation. Each physical system interacts with neighboring objects in a most complex way. However, in many cases we can isolate systems in which the parts exert very strong forces on each other compared with any extraneous forces. The baseball and the bat of Example 8.2 are a case in point. If the ball is struck at the correct place on the bat, the force on the batter's hands is negligible—you know if you have hit a baseball badly by the jarring feeling in your wrists! But what about the force of gravity—we can't escape that! Gravity exerts a force of $0.142\text{ kg} \times 9.81\text{ m/s}^2 = 1.39\text{ N}$ on the baseball, while the average force of the blow is 2130 N. The force of the blow is more than 1500 times greater than the force of gravity on the ball. But what about the bat? Its weight is about 75 times greater than the weight of the ball, so the force of the blow is about 20 times greater than the force of gravity on the bat.

In spite of the very complex interaction with the surroundings, regarding the baseball and the bat as an isolated system is a reasonably accurate model of the situation. This approximate model holds true only for the analysis of the collision. The subsequent motion of the baseball is very much influenced by gravity. Similar considerations hold for many other physical examples of collisions, at least for the duration of the collision.

EXAMPLE 8.4

An 80-g bullet is fired with a velocity of 280 m/s from a revolver whose mass is 3 kg. Calculate the recoil velocity of the revolver.

SOLUTION We assume that the explosion of the powder causes such a large action-reaction force pair between the bullet and the revolver that other forces are negligible. During the firing process the revolver and bullet are an isolated system. The problem is one dimensional, requiring only positive and negative signs to indicate vector directions. Let us choose the direction of the bullet's velocity as positive. The problem data are:

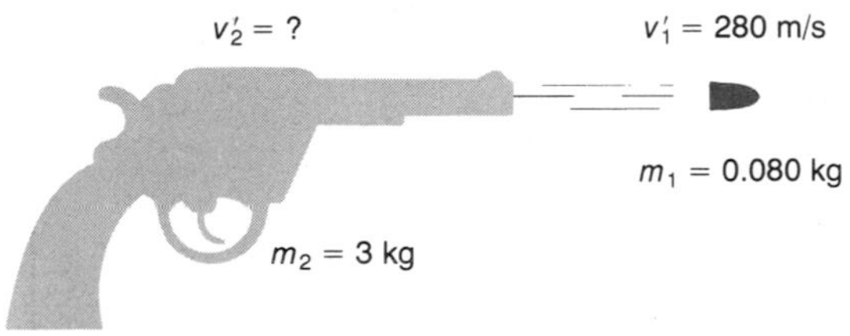

$$m_1 = 0.080 \text{ kg}$$
$$m_2 = 3 \text{ kg}$$
$$v_1 = 0 \qquad v_1' = 280 \text{ m/s}$$
$$v_2 = 0 \qquad v_2' = ?$$

The linear momentum conservation principle for two isolated bodies, Eq. (8.4), gives:

$$m_1 v_1 + m_2 v_2 = m_1 v_1' + m_2 v_2'$$
$$0 + 0 = 0.080 \text{ kg} \times 280 \text{ m/s} + 3 \text{ kg} \times v_2'$$
$$\therefore v_2' = -7.45 \text{ m/s}$$

This is a recoil velocity—it is in the opposite direction to the velocity of the bullet. ■

You may wonder about the status of energy conservation when two objects collide. Newtonian dynamics absolutely requires linear momentum conservation. There is no requirement on mechanical energy conservation, but we must have conservation of total energy. We can, in fact, use the linear momentum conservation law to calculate the amount of mechanical energy dissipated. Let us illustrate by an example.

EXAMPLE 8.5

A loaded freight car of mass 50 metric tons collides with a stationary empty car of mass 15 metric tons while moving at 18 km/h (5 m/s) on a level track. At the collision, the cars couple together. What is the subsequent velocity of the moving pair, and how

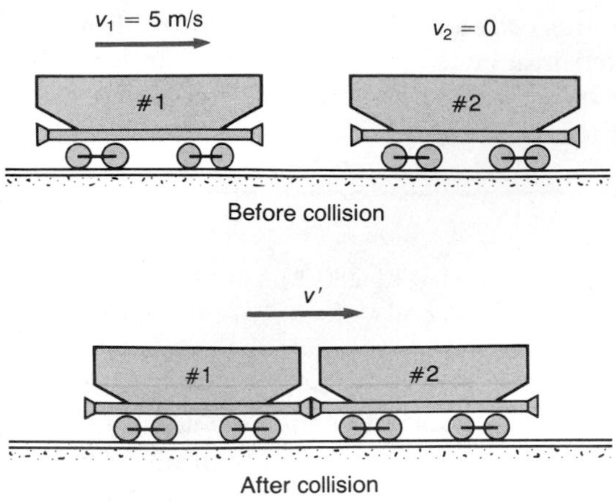

much energy is dissipated during the collision? (Ignore horizontal frictional forces.)

SOLUTION The force of gravity on the cars is not negligible during the collision, but it acts vertically. The upward force of the rails balances the force of gravity, and the vertical component of linear momentum is zero at all times. Since the only forces acting in the horizontal direction are the action-reaction force pair between the cars, it follows that the horizontal component of linear momentum is conserved. Let v' be the velocity of the cars after they couple together. Since the velocities are all horizontal, the problem is one dimensional. We select the positive direction as the original velocity direction for the loaded car. The problem data are as follows:

$$m_1 = 50 \text{ metric tons} = 5 \times 10^4 \text{ kg}$$

$$m_2 = 15 \text{ metric tons} = 1.5 \times 10^4 \text{ kg}$$

$$v_1 = 5 \text{ m/s}$$

$$v_2 = 0$$

$$v_1' = v_2' = v' = ?$$

We calculate v' by using the linear momentum conservation principle for two objects, Eq. (8.4):

$$m_1 v_1 + m_2 v_2 = m_1 v_1' + m_2 v_2'$$

$$5 \times 10^4 \text{ kg} \times 5 \text{ m/s} + 0 = 6.5 \times 10^4 \text{ kg} \times v'$$

$$\therefore v' = 3.846 \text{ m/s}$$

The cars are moving on a level track, and the gravitational potential energy does not change. We can write the energy conservation principle by using the kinetic energy definition of Eq. (7.2):

$$\text{KE before collision} = \text{KE after collision} + \text{energy dissipated}$$

$$\tfrac{1}{2}m_1 v_1^2 + \tfrac{1}{2}m_2 v_2^2 = \tfrac{1}{2}(m_1 + m_2)v'^2 + E_{\text{diss}}$$

$$\tfrac{1}{2} \times 5 \times 10^4 \text{ kg} \times (5 \text{ m/s})^2 + 0 = \tfrac{1}{2} \times 6.5 \times 10^4 \text{ kg} \times (3.846 \text{ m/s})^2 + E_{\text{diss}}$$

$$6.25 \times 10^5 \text{ J} = 4.808 \times 10^5 \text{ J} + E_{\text{diss}}$$

$$\therefore E_{\text{diss}} = 1.44 \times 10^5 \text{ J}$$

■

As we have already mentioned, Newton's laws require linear momentum conservation during collision but not kinetic energy conservation. The general law of total energy conservation requires that the kinetic energy change appear as another energy form, such as heat or sound.

So far we have considered linear momentum conservation in one dimension; that is, the motion is either backward or forward along a straight line. However, linear momentum is a vector quantity, and the concept applies equally well when we do not restrict the motion to one dimension.

EXAMPLE 8.6

An automobile of mass 2000 kg traveling at 15 m/s in a direction due south collides with a 1000-kg automobile traveling due east at 20 m/s. The automobiles lock together on impact, and the frictional force between the tires and the road is negligible compared with the force of the impact. With what velocity does the combined wreck begin to move immediately after the collision?

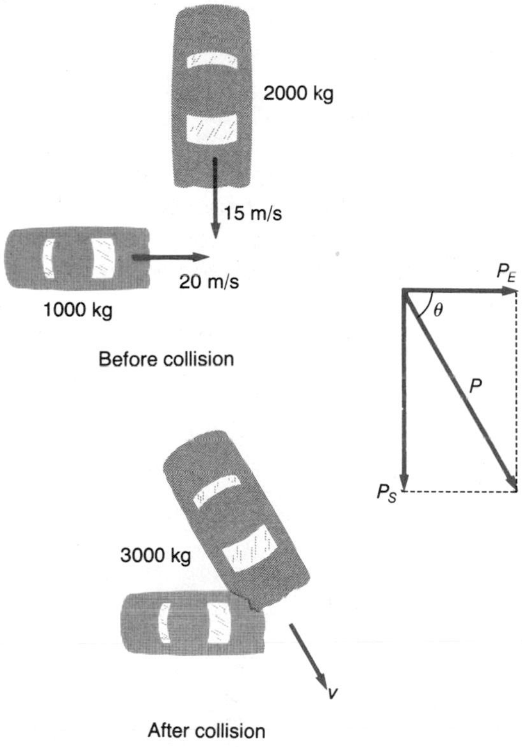

SOLUTION The frictional force between the tires and the road is negligible during the collision interval compared with the action-reaction forces of the collision, so it follows that linear momentum is conserved during the collision. The momentum components before collision are those of each car separately. Using subscripts *E* and *S* for east and south, respectively, we use Eq. (8.3) to calculate the momentum of each automobile immediately before the collision:

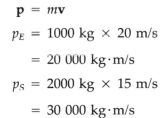

$$\begin{aligned} \mathbf{p} &= m\mathbf{v} \\ p_E &= 1000\ \text{kg} \times 20\ \text{m/s} \\ &= 20\ 000\ \text{kg}\cdot\text{m/s} \\ p_S &= 2000\ \text{kg} \times 15\ \text{m/s} \\ &= 30\ 000\ \text{kg}\cdot\text{m/s} \end{aligned}$$

The total momentum of the two automobiles together is the sum of these two components. We use Eq. (3.4) to find the magnitude and direction of a vector from its components:

$$\begin{aligned} p &= (p_E^2 + p_S^2)^{1/2} \\ &= [(20\ 000\ \text{kg}\cdot\text{m/s})^2 + (30\ 000\ \text{kg}\cdot\text{m/s})^2]^{1/2} \\ &= 36\ 060\ \text{kg}\cdot\text{m/s} \end{aligned}$$

The direction of the total momentum is:

$$\begin{aligned} \theta &= \text{arc tan}\ (p_S/p_E) \\ &= \text{arc tan}\ \frac{30\ 000\ \text{kg}\cdot\text{m/s}}{20\ 000\ \text{kg}\cdot\text{m/s}} \\ &= 56.3^\circ\ \text{S of E} \end{aligned}$$

Since momentum is conserved, the momentum of the combined wreck after the collision is equal to the momentum before the collision in both magnitude and direction. To calculate the velocity of the wreck, we use Eq. (8.3):

$$\mathbf{p} = m\mathbf{v}$$

$$36\,060\ \text{kg}\cdot\text{m/s} = 3000\ \text{kg} \times v$$

$$\therefore v = 12.0\ \text{m/s}$$

The final velocity has the same direction as the momentum—that is, 56.3° S of E. ■

8.3 Elastic and Inelastic Collisions

In Example 8.5 we saw that the collision of two freight cars can lead to a loss of kinetic energy to other energy forms. Collisions between macroscopic objects (as distinct from atomic or nuclear particles) always result in some loss of kinetic energy. Even a very elastic ball bouncing on a hard stone floor experiences a slow decrease in the height of successive rebounds, showing that each collision between the ball and the floor causes some energy dissipation.

We classify collisions between bodies into two extreme types, depending on the kinetic energy dissipation.

1. A collision is said to be **completely elastic** if no kinetic energy is dissipated. We can never quite achieve this ideal situation

The inelastic collision between the automobile and the barrier is not the one of primary concern in this test. There is a second inelastic collision—between the occupants and the interior of the vehicle—that is the major focus of the investigation. (Courtesy of U.S. Dept. of Transportation)

for collisions between macroscopic bodies, but it can occur in collisions between atomic and nuclear particles.

2. A collision is said to be **inelastic** if there is kinetic energy dissipation. A special case of inelastic collision between macroscopic objects occurs when the colliding bodies stick together after impact. Such a collision is called **completely inelastic.** A completely inelastic collision dissipates as much of the kinetic energy as the laws of dynamics permit.

The physics of collisions is frequently demonstrated by using air track equipment. Small carriages supported by jets of air glide almost without friction along a horizontal track. When two carriages collide, the horizontal component of total linear momentum is conserved since the only horizontal forces acting are the action-reaction force pair. Springs on the ends of the carriages give a reasonably close approximation to a completely elastic collision.

EXAMPLE 8.7

A carriage of mass 150 g moving with a velocity of 1.2 m/s collides with a 300-g carriage that is stationary on an air track. Calculate the subsequent velocity of each carriage if the collision is perfectly elastic.

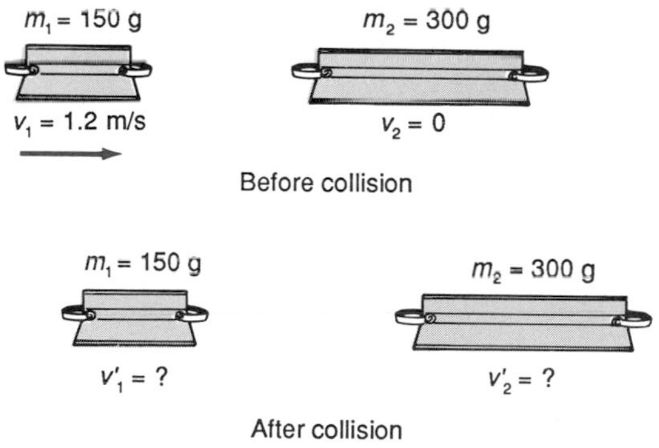

SOLUTION In every collision (whether elastic or inelastic) linear momentum is conserved. Using a subscript 1 for the first carriage and a 2 for the second one, we can write the data as follows:

$$m_1 = 150 \text{ g} = 0.15 \text{ kg} \qquad m_2 = 300 \text{ g} = 0.3 \text{ kg}$$

$$v_1 = 1.2 \text{ m/s} \qquad v_2 = 0$$

$$v'_1 = ? \qquad v'_2 = ?$$

Linear momentum conservation, Eq. (8.4), gives:

$$m_1 v_1 + m_2 v_2 = m_1 v'_1 + m_2 v'_2$$

$$0.15 \text{ kg} \times 1.2 \text{ m/s} + 0 = 0.15 \text{ kg} \times v'_1 + 0.3 \text{ kg} \times v'_2$$

We cannot solve this equation for either v'_1 or v'_2, but we can simplify it as follows:

$$v'_1 = 1.2 \text{ m/s} - 2.0 \times v'_2$$

Since the collision is completely elastic, there is no kinetic energy dissipation. Using the definition of kinetic energy, Eq. (7.2), we have:

Initial kinetic energy = final kinetic energy

$$\tfrac{1}{2}m_1v_1^2 + \tfrac{1}{2}m_2v_2^2 = \tfrac{1}{2}m_1v_1'^2 + \tfrac{1}{2}m_2v_2'^2$$

$$\tfrac{1}{2} \times 0.15\text{ kg} \times (1.2\text{ m/s})^2 + 0 = \tfrac{1}{2} \times 0.15\text{ kg} \times v_1'^2 + \tfrac{1}{2} \times 0.3\text{ kg} \times v_2'^2$$

Simplifying once again we have:

$$v_1'^2 = (1.2\text{ m/s})^2 - 2.0 \times v_2'^2$$

There are now two equations relating v_1' to v_2'—one from linear momentum conservation and the other from kinetic energy conservation. To solve, we square the first equation and subtract it from the second:

$$\begin{aligned} v_1'^2 &= 1.44\text{ (m/s)}^2 \qquad\qquad\qquad\qquad - 2.0 \times v_2'^2 \\ v_1'^2 &= 1.44\text{ (m/s)}^2 - 4.8\text{ m/s} \times v_2' + 4.0 \times v_2'^2 \\ \hline 0 &= 4.8\text{ m/s} \times v_2' - 6.0 \times v_2'^2 \end{aligned}$$

This is a simple type of quadratic equation (see Appendix A) whose solutions are:

$$v_2' = 0\text{ m/s} \quad \text{or} \quad 0.8\text{ m/s}$$

Substituting these values in the first relation between v_1' and v_2' gives the corresponding values of v_1':

$$v_1' = 1.2\text{ m/s} \quad \text{or} \quad -0.4\text{ m/s}$$

We must discard the first solution to the problem:

$$v_1' = 1.2\text{ m/s} \quad \text{and} \quad v_2' = 0\text{ m/s}$$

Although it is mathematically correct, it corresponds to the first carriage passing through the second one without touching it. The second solution is the physical solution to the problem:

$$v_1' = -0.4\text{ m/s} \quad \text{and} \quad v_2' = 0.8\text{ m/s}$$

Note that we do not need to know in advance the direction of the velocity vector for each carriage after the collision. We simply select both of these unknown quantities in the positive direction. The negative sign for the value of v_1' tells us that the first carriage rebounds in the negative direction. ■

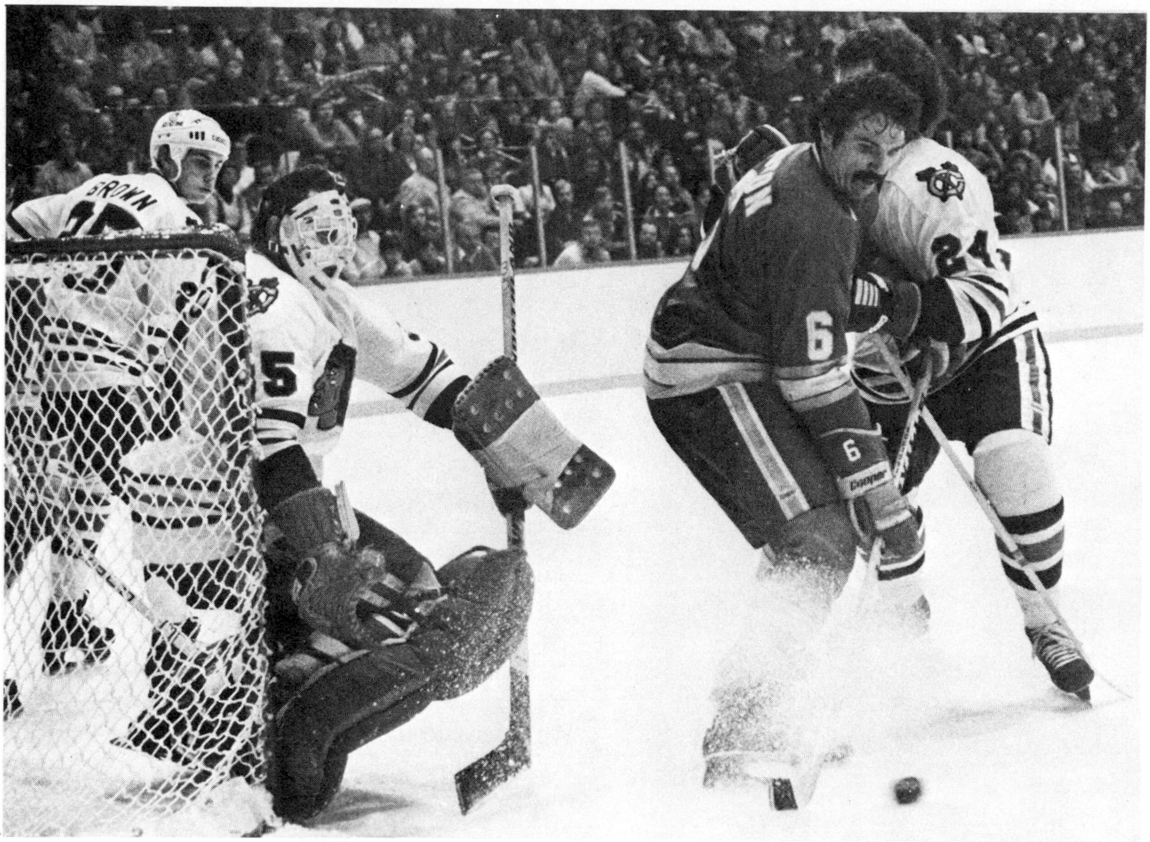

Number 6 and number 24 have just conserved momentum and total energy; however, they have dissipated kinetic energy into other energy forms. (Courtesy of Chicago Black Hawks)

EXAMPLE 8.8

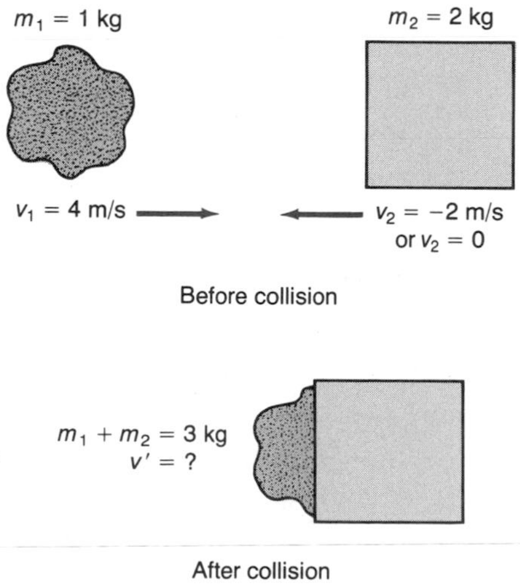

A piece of putty of mass 1 kg moving to the right at 4 m/s strikes a wooden block of mass 2 kg. Calculate the fraction of the original kinetic energy dissipated:

a. if the wooden block is initially moving to the left with a velocity of 2 m/s

b. if the wooden block is initially at rest

SOLUTION In each case linear momentum is conserved, and the final velocities of the putty and the block are the same. We treat the two cases separately.

a. $m_1 = 1$ kg $\qquad m_2 = 2$ kg

$v_1 = 4$ m/s $\qquad v_2 = -2$ m/s

$v_1' = v' = ?$ $\qquad v_2' = v' = ?$

Use of Eq. (8.4) gives:

$$m_1v_1 + m_2v_2 = m_1v_1' + m_2v_2'$$

$$1 \text{ kg} \times 4 \text{ m/s} + 2 \text{ kg} \times (-2 \text{ m/s}) = 3 \text{ kg} \times v'$$

$$\therefore v' = 0$$

The final velocity of the composite object is zero, and 100% of the initial kinetic energy is dissipated.

b. $m_1 = 1$ kg $\qquad m_2 = 2$ kg

$v_1 = 4$ m/s $\qquad v_2 = 0$ m/s

$v_1' = v' = ?$ $\qquad v_2' = v' = ?$

Again the conservation of linear momentum, Eq. (8.4), gives:

$$m_1v_1 + m_2v_2 = m_1v_1' + m_2v_2'$$

$$1 \text{ kg} \times 4 \text{ m/s} + 0 = 3 \text{ kg} \times v'$$

$$\therefore v' = 1.333 \text{ m/s}$$

We calculate the initial and final kinetic energies with the help of Eq. (7.2):

$$\begin{aligned} \text{Initial KE} &= \tfrac{1}{2}m_1v_1^2 + \tfrac{1}{2}m_2v_2^2 \\ &= \tfrac{1}{2} \times 1 \text{ kg} \times (4 \text{ m/s})^2 + 0 \\ &= 8 \text{ J} \\ \text{Final KE} &= \tfrac{1}{2}(m_1 + m_2)v'^2 \\ &= \tfrac{1}{2} \times 3 \text{ kg} \times (1.333 \text{ m/s})^2 \\ &= 2.667 \text{ J} \end{aligned}$$

The fraction of kinetic energy dissipated is given by:

$$\begin{aligned} \text{Fraction of KE dissipated} &= \frac{\text{initial KE} - \text{final KE}}{\text{initial KE}} \\ &= \frac{8 \text{ J} - 2.667 \text{ J}}{8 \text{ J}} \\ &= 66.7\% \end{aligned}$$

■

Both collisions in the preceding example are completely inelastic since the two objects stick together on colliding. However, all the kinetic energy can be dissipated only if the composite object is at rest after the collision. In this case the final momentum is zero, which requires the initial momentum to be zero. If the initial momentum is not zero, the final composite object cannot have zero velocity, and the kinetic energy cannot be completely dissipated.

The ballistic pendulum, which measures the speed of a bullet, puts completely inelastic collisions to practical use. A large block of wood is suspended by a light rod, and a bullet is fired into the block. The bullet lodges in the block, causing it to swing higher than its original position. We can calculate the velocity of the bullet by measuring the masses of the bullet and the block of wood and the vertical component of displacement caused by the inelastic collision.

EXAMPLE 8.9

A 20-g bullet strikes a ballistic pendulum of mass 4 kg and embeds in the block. Calculate the velocity of the bullet if the center of gravity of the block rises 12 cm vertically above its initial position.

SOLUTION Let us first reflect on the physics of the situation. A bullet possessing momentum and kinetic energy embeds in a stationary block. As a result of the inelastic collision, momentum is conserved, but kinetic energy is dissipated. After the collision the block swings upward. The forces acting on the block (the weight excluded) do no work, and the kinetic energy plus gravitational potential energy is conserved. Using the notation of the diagram we can write the data as follows:

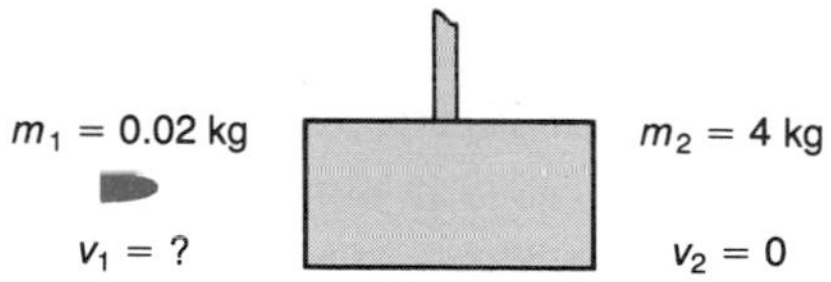

Before collision

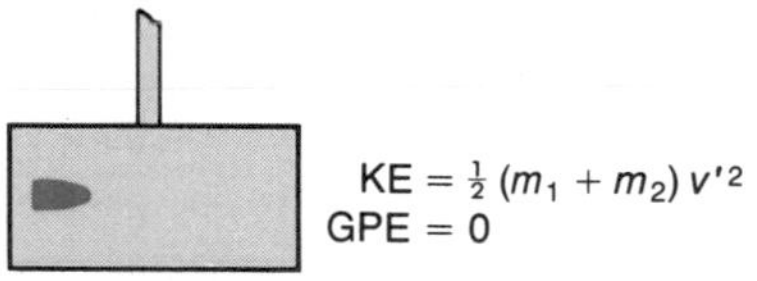

Immediately after collision

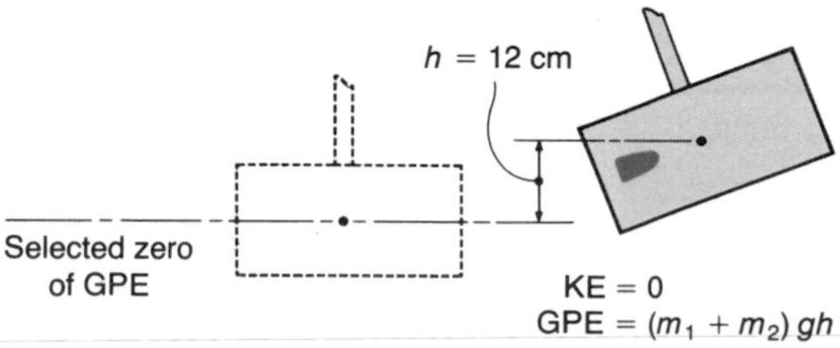

Block at maximum height

$$m_1 = 0.02 \text{ kg} \qquad m_2 = 4 \text{ kg}$$
$$v_1 = ? \qquad v_2 = 0$$
$$v_1' = v' = ? \qquad v_2' = v' = ?$$

The conservation of linear momentum, Eq. (8.4), gives:

$$m_1v_1 + m_2v_2 = m_1v_1' + m_2v_2'$$
$$0.02 \text{ kg} \times v_1 + 0 = 4.02 \text{ kg} \times v'$$
$$\therefore v_1 = 201 \times v'$$

After the collision, the mechanical energy conservation equation gives:

$$\text{Initial (KE + GPE)} = \text{final (KE + GPE)}$$
$$\tfrac{1}{2}(m_1 + m_2)v'^2 + 0 = 0 + (m_1 + m_2)gh$$
$$\tfrac{1}{2} \times 4.02 \text{ kg} \times v'^2 = 4.02 \text{ kg} \times 9.81 \text{ m/s}^2 \times 0.12 \text{ m}$$
$$\therefore v' = \pm 1.534 \text{ m/s}$$

The positive sign corresponds to the physical solution. Substituting in the relation derived from linear momentum conservation, we have:

$$v_1 = 201 \times v'$$
$$= 201 \times 1.534 \text{ m/s}$$
$$= 308 \text{ m/s}$$

■

8.4 Momentum of Objects of Varying Mass

Up to this point we have considered the application of linear momentum conservation to objects of fixed mass. However, sometimes objects are of varying mass. A rocket that is burning its fuel thrusts the products of combustion astern, losing mass as the fuel is expended. The application of linear momentum conservation to a rocket involves mathematics beyond the scope of this book, but we can examine a similar situation that illustrates the same physical principles. Consider an artillery shell that explodes at some instant during its flight. Note that the forces of the explosion are not external applied forces; instead, they give rise to action-reaction force pairs between the fragments. For this reason, linear momentum is conserved and the total linear momentum of all the fragments immediately after the explosion is the same as the linear momentum of the shell just before the explosion. Let us illustrate the consequences with a somewhat artificial example that only requires simple mathematics.

EXAMPLE 8.10

An artillery shell of mass 30 kg has a velocity of 250 m/s vertically upward. The shell explodes into two pieces; immediately after the explosion a fragment of mass 10 kg has a velocity of 120 m/s straight downward.

a. How high above the point of the explosion does the larger fragment rise?

b. If the shell had not exploded, how much higher would it have risen?

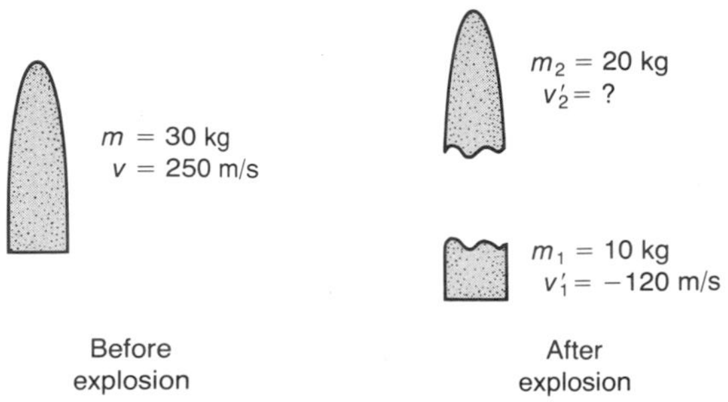

SOLUTION

a. The kinematics problem requires the application of linear momentum conservation to calculate the initial velocity of the larger fragment. Because the initial momentum of the shell has no horizontal component, the total horizontal component of momentum must be zero after the explosion. The horizontal momentum component of the smaller fragment is zero, so it follows that the same is true for the larger fragment. Only vertical momentum components enter our calculation. If we select the upward direction as positive and use the notation of the diagram, linear momentum conservation, Eq. (8.4), gives:

$$mv = m_1v_1' + m_2v_2'$$

$$30 \text{ kg} \times 250 \text{ m/s} = 10 \text{ kg} \times (-120 \text{ m/s}) + 20 \text{ kg} \times v_2'$$

$$\therefore v_2' = 435 \text{ m/s}$$

We have now calculated the upward velocity of the larger fragment immediately after the explosion. The figure is the initial velocity for our kinematics problem, which has the following data:

$$v_0 = 435 \text{ m/s}$$
$$a = -9.81 \text{ m/s}^2$$
$$v = 0 \text{ m/s}$$
$$x = ?$$

Equation (2.12d) is appropriate:

$$v^2 = v_0^2 + 2ax$$

$$(0 \text{ m/s})^2 = (435 \text{ m/s})^2 + 2 \times (-9.81 \text{ m/s}^2) \times x$$

$$\therefore x = 9640 \text{ m}$$

b. Had the shell not exploded, the data for the kinematics problem would be:

$$v_0 = 250 \text{ m/s}$$
$$a = -9.81 \text{ m/s}^2$$
$$v = 0 \text{ m/s}$$
$$x = ?$$

Equation (2.12d) again gives:

$$v^2 = v_0^2 + 2ax$$

$$(0 \text{ m/s})^2 = (250 \text{ m/s})^2 + 2 \times (-9.81 \text{ m/s}^2) \times x$$

$$\therefore x = 3190 \text{ m}$$

■

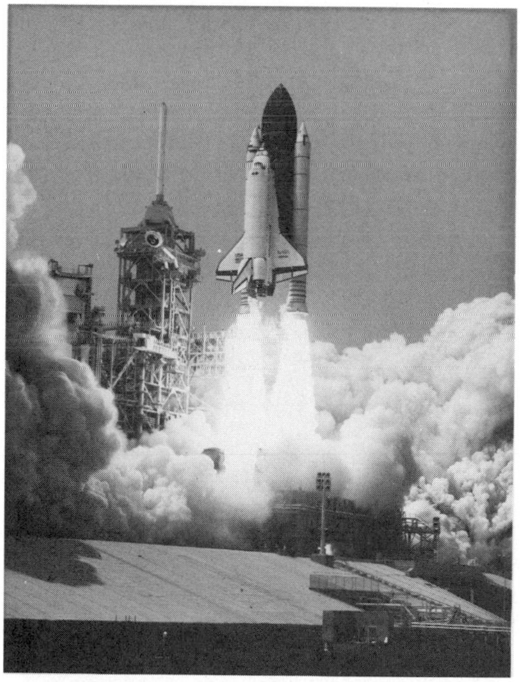

The rockets powering the space shuttle *Atlantis* show linear momentum conservation as the shuttle lifts off from the Kennedy Space Center. The total momentum of the space vehicle and the exhaust gases is zero both before and after liftoff. (Courtesy of NASA)

The explosion of the shell drives the upper fragment much higher than it would have gone otherwise. The burning rocket operates on exactly the same principle as the exploding shell except for two details. First, the rocket burns continuously rather than undergoing a single large explosion—in fact, we can say that the rocket engine is a device to produce a long, drawn-out explosion. Second, the fragments thrust downward from the rocket are the products of combustion of the explosion rather than clearly identifiable pieces of the rocket body. The products of combustion play the same role as the lower shell fragment of our example, thrusting up the body of the rocket just as the upper shell fragment is propelled.

As early as 1919 the American space pioneer Robert Goddard (1882–1945) proposed reaching the moon by rocket. Many critics in scientific circles ridiculed his proposal on the grounds that the rocket would cease to function once it left the earth's atmosphere, claiming that the propulsive force on the rocket depended on the exhaust gases pushing on the atmosphere. If this were true, the rocket could not be controlled in the vacuum of space. However, unlike his critics, Goddard had a clear understanding of Newton's law of equality of action-reaction force pairs. He carried out a series of small-scale experiments in vacuum that proved his point: The exhaust gases from the rocket need not push on the atmosphere—they do their job by pushing on the rocket!

KEY CONCEPTS

The **impulse** of a force is the average force multiplied by its time of action; it is a vector having the same direction as the average force. See Eq. (8.2).

Linear momentum is mass multiplied by velocity; it is a vector having the same direction as the velocity. See Eq. (8.3).

The action of a force on an object produces a linear momentum change equal to the impulse of the force. See Eq. (8.1).

The **linear momentum of an isolated system** (one that is subject to zero net external force) **is conserved.** This vector conservation law applies to all momentum components.

A collision in which kinetic energy is conserved is **completely elastic.** If kinetic energy is dissipated, the collision is **inelastic.** The special case of an inelastic collision in which objects stick together on colliding is called **completely inelastic.**

QUESTIONS FOR THOUGHT

1. Is it possible for a system of point masses to have momentum but no kinetic energy? Can the system have kinetic energy but no momentum? Explain your answers.

2. Two unequal masses have equal translational kinetic energies. Which one has the larger momentum?

3. Why does a shotgun kick harder if you hold it loosely than it does if you hold it tightly against your shoulder?

4. A bird is confined within a completely enclosed box. Is it possible to tell whether the bird is flying by weighing the box?

5. A canoeist in a calm lake has no paddle. Can the canoe be set in motion by standing up and walking from one end to the other?

6. Is it possible to propel a sailboat with a fan that is rigidly attached to the boat? If so, what requirements must be met?

7. Explain the application of the law of momentum conservation to a jet aircraft accelerating along a runway.

8. When a sailboat sails against the wind, its momentum has a component directly opposite to the direction of the wind. How do you reconcile this with the impulse-momentum theorem?

PROBLEMS

A. Single-Substitution Problems

1. Calculate the momentum of an 80-kg sprinter moving at 11 m/s. [8.1]

2. Calculate the momentum of a 1500-kg automobile moving at 25 m/s. [8.1]

3. What is the impulse of the accelerating force required to give a 500-μg flea a velocity of 1 m/s? [8.1]

4. A 0.41-kg football that is initially at rest acquires a velocity of 35 m/s when it is kicked. If the kicker's boot remains in contact with the ball for 0.012 s, what is the average force of the kick? [8.1]

5. A 25-g bullet acquires a muzzle velocity of 500 m/s as it travels through a rifle barrel. The time interval between firing the rifle and the emergence of the bullet is 2.4 ms. Calculate the average force of the explosion on the bullet. [8.1]

6. A baseball fielder stops a ball of mass 0.142 kg that is moving with a velocity of 35 m/s. By moving his fielding glove backward in a straight line, he takes the speed off the ball over an interval of 0.06 s. Calculate the average force exerted on his glove. [8.1]

7. A rifle of mass 6 kg fires a lead bullet of mass 32.5 g with a velocity of 800 m/s. If the rifle is held loosely, what is the recoil velocity? [8.2]

8. A loaded freight car of mass 25 000 kg moving at 3 m/s along a level track collides with an empty stationary car of mass 8000 kg, and the cars couple together on collision. Calculate the subsequent velocity of the cars. [8.2]

9. A loaded freight car of mass 20 000 kg moving at 2.8 m/s along a level track collides with an empty stationary car, and they couple together on collision. If the subsequent velocity of the pair is 1.8 m/s, what is the mass of the second car? [8.2]

10. A block of pine of mass 3 kg resting on a post is struck by a bullet of mass 20 g with a velocity of 600 m/s. The bullet emerges from the block with a velocity of 200 m/s. With what initial velocity does the block leave the post? [8.2]

11. A 25-g bullet strikes a 3-kg block of wood that is initially at rest, and embeds in the wood. If the subsequent velocity of the block is 3.8 m/s, calculate the initial velocity of the bullet. [8.2]

B. Standard-Level Problems

12. A fire hose delivers water at a speed of 15 m/s and a rate of 500 gal/min. Calculate the reaction force on the nozzle. [8.1]

13. A machine gun fires 30-g bullets with a muzzle velocity of 500 m/s at a rate of 180 per minute. Calculate the average recoil force. [8.1]

14. A 1200-kg automobile traveling at 30 m/s crashes into an unyielding obstacle. The front crumples and the car comes to rest in 0.04 s. Calculate:

a. the average force on the obstacle
b. the average force between a 75-kg passenger and the seat belt (Assume that the seat belt holds the passenger securely as the car comes to rest.)

15. A ventilating fan imparts a velocity of 6 m/s to 160 m^3 of air per minute. Calculate the reaction force exerted by the fan on its supports. (Take the air density as 1.2 kg/m^3.)

16. A picture window 2 m high by 3.5 m wide is subjected to a 90-km/h (25-m/s) gale that strikes it head-on. Assume that the window reduces the normal component of air velocity to zero. Calculate the force exerted by the gale on the window. (The density of the air is 1.2 kg/m^3.)

17. Sugar is poured from a height of 60 cm onto a scale pan at the rate of 200 g per second. By how much does the scale reading at any given time exceed the weight of the sugar on the pan?

18. A tennis ball of mass 57 g approaches a player with a velocity of 30 m/s. She returns the ball with a velocity of 30 m/s at a 160° angle to the direction of the incident velocity. The impact between the ball and racket lasts for 4 ms. Calculate the average force of the blow on the ball.

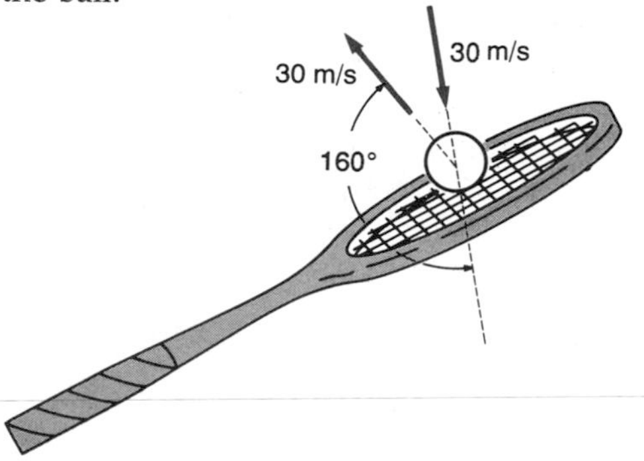

19. A canoe has a mass of 35 kg. Two occupants of masses 60 kg and 70 kg dive out simultaneously in opposite directions, one from each end of the canoe. The horizontal velocity component of each person is 3.5 m/s (relative to the water). Calculate the magnitude and direction of the initial velocity of the canoe after the divers leave. [8.2]

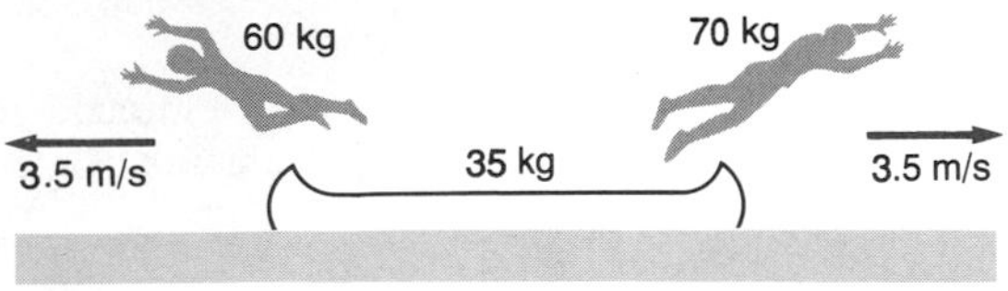

20. A 50-kg woman carrying two large stones, each of mass 10 kg, sits on a 12-kg sled that is at rest on frictionless ice. She throws the stones backwards off the sled with a velocity of 6 m/s horizontally relative to the sled. Calculate:

a. the velocity of the sled after she throws off the first stone.
b. the velocity of the sled after she throws off the second stone. [8.2]

21. A 100-kg running back moving due south at 8 m/s is hit by a 120-kg linebacker moving northwest at 7 m/s. Calculate the magnitude and direction of their velocity immediately after impact if the linebacker clasps the running back firmly. (Assume that forces between their feet and the turf are small compared with the force of the collision.) [8.2]

22. A cannon of mass 1.2 metric tons is mounted on wheels free to move on a horizontal surface. It fires a 10-kg ball with a muzzle velocity of 250 m/s at an angle of 15° above the horizontal. Calculate the recoil velocity of the cannon.

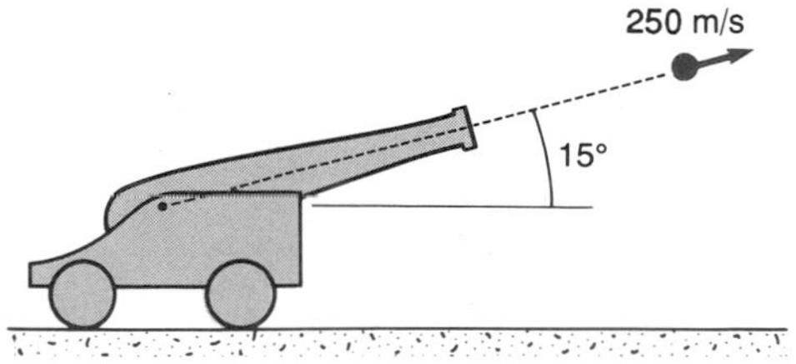

23. A 1200-kg automobile traveling north at 25 m/s collides with a 3600-kg truck traveling east at 20 m/s. Find the velocity just after the collision if the vehicles remain locked together.

24. A 1000-kg automobile traveling east at 25 m/s collides with a 4000-kg truck traveling south. The vehicles lock together, and it is determined from tire marks that the velocity of the wreck just after the collision was in the direction 11° E of S. Calculate the speed of the truck prior to the collision.

25. A stationary hand grenade blows into three parts. Two fragments have velocity directions at right angles to each other—a 400-g piece at 60 m/s and a 500-g piece at 64 m/s. The third fragment has a velocity of 80 m/s. Calculate the mass and the velocity direction of this fragment.

26. A loaded freight car of mass 40 000 kg is moving north along a level track at 4 m/s when it meets another car of mass 30 000 kg moving south on the same track at 3 m/s. The cars couple together on collision.

a. What are the magnitude and direction of the subsequent velocity?
b. How much energy is dissipated in the collision? [8.3]

27. A freight car of mass 30 000 kg moves east along a level track and meets another car of mass 25 000 kg moving west at 4 m/s. The cars couple together on collision, and the pair then moves east at 2.5 m/s.

a. What was the velocity of the first car before the collision?
b. How much energy was dissipated in the collision? [8.3]

28. A basketball of mass 0.23 kg is thrown horizontally against a rigid vertical wall with a velocity of 25 m/s. It rebounds with a velocity of 20 m/s. Calculate:

a. the impulse of the force of the wall on the basketball
b. the kinetic energy dissipated in the collision

29. A golf ball of mass 45 g is dropped from a height of 4 m above a concrete floor and rebounds to a maximum height of 3.1 m. Calculate:

a. the impulse exerted by the floor on the ball
b. the energy dissipated in the impact with the floor

30. A 3-kg steel ball moving at 8 m/s collides with a stationary steel ball of mass 2 kg. After the collision, which is along the line of the balls' centers, both balls move in the same direction. The velocity of the 3-kg ball is 4 m/s after the collision. Calculate:

a. the velocity of the 2-kg ball after the collision
b. the energy dissipated during the collision

31. A helium nucleus of mass 6.68×10^{-27} kg moving with a velocity of 2×10^5 m/s makes an elastic collision with a neutron of mass 1.67×10^{-27} kg that is at rest. After the collision, both particles move in the same direction. Calculate the velocity of each particle after the collision. [8.3]

32. A neutron of mass 1.67×10^{-27} kg moving with a velocity of 3×10^5 m/s makes an elastic collision with a helium nucleus of mass 6.68×10^{-27} kg that is moving along the same line and in the opposite direction. After the collision, the helium nucleus is at rest and the neutron is moving with a velocity of 5×10^5 m/s in the opposite direction to its original velocity. Calculate the velocity of the helium nucleus prior to the collision.

33. A freight car of mass 30 metric tons collides with another car at rest on a level track. The cars couple together on collision, which dissipates 30% of the kinetic energy. Calculate the mass of the stationary car.

34. A roller coaster of mass 1600 kg starts from rest, with its center of gravity 15 m vertically above a point on the track where an identical coaster is waiting at rest. After colliding, they stick together and continue up the next hill.

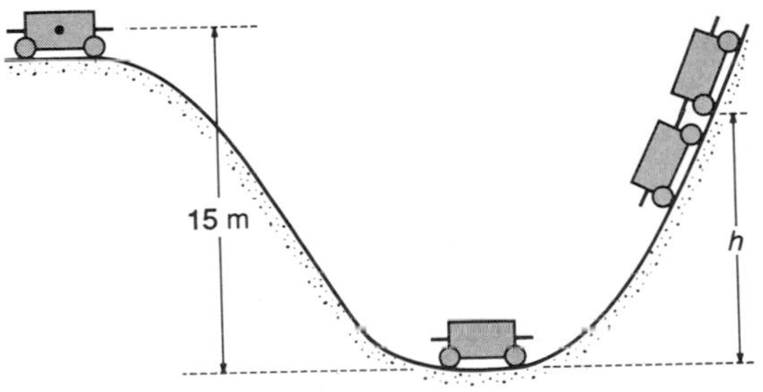

a. How high up the hill does the center of gravity of the combined pair of coasters rise? (Neglect frictional forces.)
b. How much energy is dissipated in the collision?

35. A 30-kg child running at 5 m/s jumps onto a stationary toboggan standing at rest on a level surface of ice. The toboggan has a mass of 12 kg. How far do the child and the toboggan slide if the coefficient of kinetic friction with the ice is 0.04?

36. A bullet of mass 25 g is fired horizontally into a 3-kg block of wood at rest on a rough horizontal surface. The bullet embeds in the block, which slides 4.5 m over the surface before coming to rest. The coefficient of kinetic friction between the block and the surface is 0.2. Calculate the initial velocity of the bullet.

37. A 4-kg block of wood at rest on a level floor is struck by a 25-g bullet traveling horizontally at 600 m/s. The bullet goes right through the block, which slides 0.8 m before coming to rest again. The coefficient of kinetic friction between the block and the floor is 0.25. Calculate the velocity of the bullet when it emerges.

38. A block of wood of mass 2 kg is placed over the muzzle of a rifle, and a bullet of mass 30 g is fired vertically upward. The bullet embeds in the wood, which rises to a maximum height of 3 m. Calculate the muzzle velocity of the bullet.

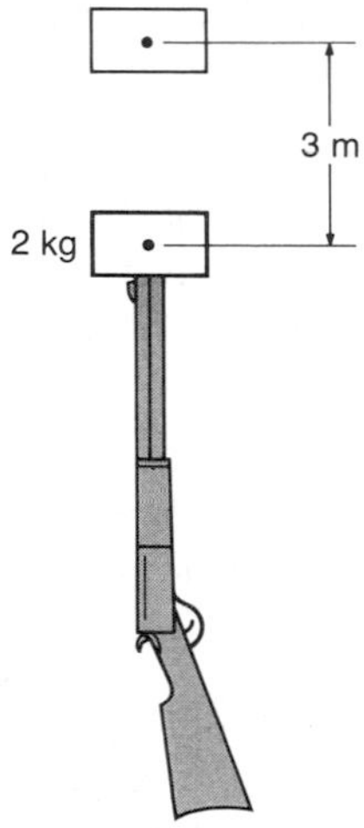

39. A 20-g rifle bullet is fired with a velocity of 600 m/s and embeds in a 5-kg ballistic pendulum. Calculate:

a. the vertical component of displacement of the pendulum
b. the kinetic energy dissipated

40. An 8-kg ballistic pendulum is used to measure the velocity of a 30-g pistol bullet. After firing, the bullet embeds in the block and the vertical component of displacement is 10 cm. Calculate the velocity of the bullet and the energy dissipated in the collision.

C. Advanced-Level Problems

41. A hopper empties crushed rock onto a horizontal conveyor belt at the rate of 5 metric tons per minute. How much power is needed to drive the belt at 3 m/s?

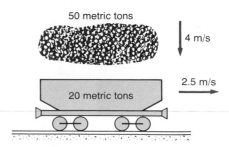

42. An empty freight car of mass 20 metric tons that is coasting at 2.5 m/s on a level track passes under a loading hopper that deposits 50 metric tons of coal into the car. Just before it hits the car, the coal's velocity is 4 m/s vertically downward. Calculate:

a. the velocity of the freight car after the loading is complete
b. the energy dissipated in the process

43. A wooden block of mass 1.2 kg rests on a plane inclined at 20° to the horizontal. A bullet of mass 20 g is fired into the block with a velocity of 500 m/s in a direction parallel to the slope of the plane and pointing downward. The bullet embeds in the block, which slides 17.5 m down the plane before stopping. Calculate the coefficient of friction between the block and the plane.

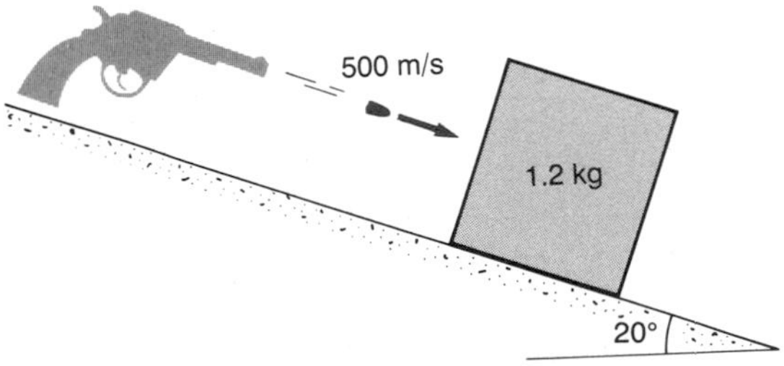

44. A 7-kg bowling ball moving with a velocity of 8 m/s strikes a single pin of mass 1.36 kg. The impact deflects the bowling ball 10° from its original path, and the initial velocity of the pin is at 45° to the path of the ball before collision. Calculate the velocities of the ball and the pin after the collision.

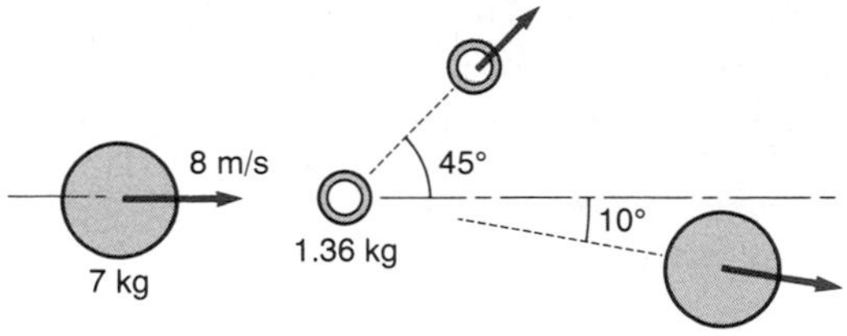

45. A billiard ball of mass 0.20 kg, which is sliding without rolling at 6 m/s, strikes another ball of equal mass that is initially at rest. After the collision the second ball moves with a velocity of 5 m/s at an angle of 30° with the original direction of motion of the cue ball.

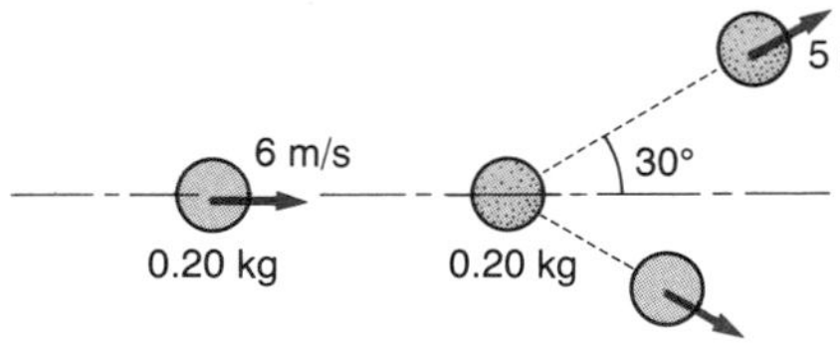

Calculate:

a. the magnitude and direction of the velocity of the cue ball after the collision
b. the percentage of the original kinetic energy lost in the collision

9

ROTATION OF RIGID BODIES

A Frisbee, a phonograph turntable, a spinning top, and a rolling wheel all show rotational motion, as distinct from the rectilinear, or straight-line, motion that we have studied so far. As always, we begin our study of rotational motion with the easiest case. A phonograph turntable rotates about a central spindle, which serves as the axis of rotation. This represents the simplest case of rotational motion, because the axis of rotation remains fixed. When a wheel rolls along a straight line, the axis of rotation advances with the wheel but does not change its direction. However, in our other two examples (the Frisbee and the spinning top), the direction of the rotation axis can change continuously. This situation introduces complexities that we want to avoid. In what follows we confine ourselves to a rigid body rotating about a fixed axis or a wheel rolling along a straight line. In both cases the direction of the axis of rotation remains unchanged.

Given a fixed direction for the rotation axis, the kinematics and dynamics of rotational motion closely resemble rectilinear kinematics and dynamics. Therefore, we can use much of our previous work by making suitable modifications.

9.1 Rotational Kinematics

The linear kinematics of a point particle describes the motion of the particle without considering the forces that cause the motion. In the same way we will begin our description of rotational motion with rotational kinematics, which ignores the causes of the motion. The important concepts in linear kinematics are displacement, velocity, and acceleration. In rotational kinematics the important concepts are angular displacement, angular velocity, and angular acceleration. Let us define the angular kinematic quantities by considering the simplest example—a disk-shaped wheel rotating about a fixed axis that goes through the wheel's center perpendicularly to its plane, as in Fig. 9.1. Imagine that at the initial time $t_0 = 0$, some fixed point on the rim of the wheel is at the point P_0. At a later time t, rotation of the wheel moves this point on the rim to position P, as shown in Fig. 9.1(a). The center of the wheel is the point O. The **angular displacement** of the wheel during this time interval is the angle between OP and OP_0, marked θ on the diagram. Although the SI unit of angular displacement is the radian (see Section 6.1), everyday angular measure is usually given in either degrees or revolutions. Since these units convert as follows:

$$1 \text{ rev} = 360° = 2\pi \text{ rad}$$

we can easily change all problem data to the SI radian measure.

It is important to note that angular displacement is not restricted to angles of 1 rev or less. The marked point in both Fig.

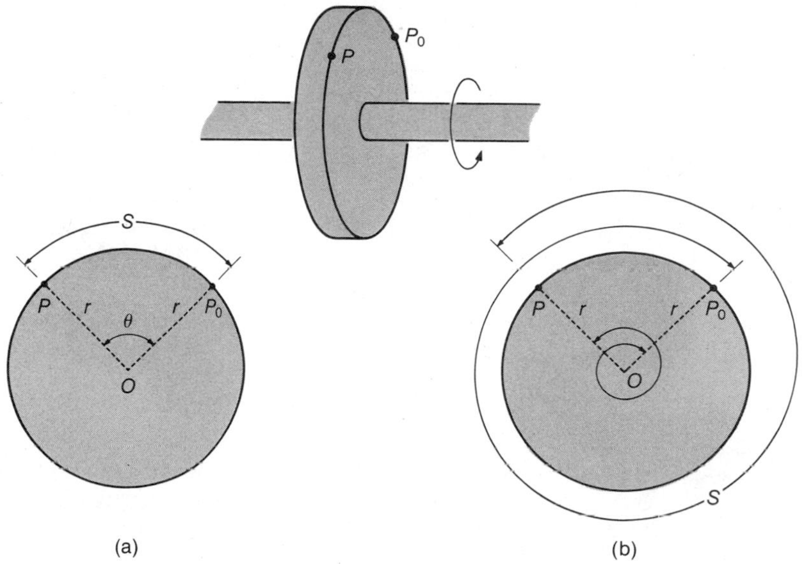

FIGURE 9.1 Diagram to illustrate the kinematics of a rotating wheel: (a) angular displacement less than 1 rev; (b) angular displacement in excess of 1 rev.

9.1(a) and Fig. 9.1(b) has the same initial and final positions. However, in Fig. 9.1(b) the wheel has turned through an additional revolution. For Fig. 9.1(a) we have:

$$\theta \approx \frac{1}{4}\text{ rev} = \frac{\pi}{2}\text{ rad}$$

and for Fig. 9.1(b) the value is:

$$\theta \approx 1\frac{1}{4}\text{ rev} = \frac{5\pi}{2}\text{ rad}$$

Proceeding in the same way as for linear kinematics, we define average **angular velocity** as the angular displacement divided by the elapsed time.

■ Definition of Average Angular Velocity

$$\text{Average angular velocity} = \frac{\text{angular displacement}}{\text{elapsed time}}$$

$$\langle\omega\rangle = \frac{\theta}{t} \qquad \textbf{(9.1)}$$

SI unit of angular velocity: radian per second
Abbreviation: rad/s

The symbol for angular velocity is the Greek letter omega (ω). If the average angular velocity is the same over all elapsed time intervals, the average symbol can be dropped from Eq. (9.1). In these circumstances the rotation is at a constant angular velocity, and cross-multiplication of Eq. (9.1) gives:

■ Equation Describing Rotational Motion at Constant Angular Velocity

$$\theta = \omega t \tag{9.2}$$

If the average angular velocity is different over different elapsed time intervals, we define the *instantaneous angular velocity* as an average taken over a small time interval that includes the point in question. Suppose now that the instantaneous angular velocity is ω_0 when the fixed point on the rim of the wheel is at position P_0 and ω when it is at position P. Once again we follow the approach of linear kinematics, defining average **angular acceleration** as the change in angular velocity divided by the elapsed time.

The symbol for angular acceleration is the Greek letter alpha (α). If the average angular acceleration is the same when calculated over all elapsed time intervals, the average symbol can be dropped from Eq. (9.3).

■ Definition of Average Angular Acceleration

$$\text{Average angular acceleration} = \frac{\text{change in angular velocity}}{\text{elapsed time}}$$

$$\langle\alpha\rangle = \frac{\omega - \omega_0}{t} \tag{9.3}$$

SI unit of angular acceleration: radian per second per second
Abbreviation: rad/s^2

All of our definitions of the angular kinematic quantities have exactly the same form as the corresponding rectilinear quantities. We can list the symbols for the rectilinear and rotational cases as follows:

■ Kinematic Analogues

	Linear	Angular
Displacement	x	θ
Velocity	v	ω
Acceleration	a	α

(9.4)

It also follows that the kinematic equations have the same algebraic form for both the rectilinear and the rotational cases. In particular, we can write down the equations for constant angular acceleration simply by replacing the linear kinematic quantities by rotational analogues. Using the kinematic analogues of (9.4) in Eq. (2.12), we find the following kinematic equations for constant angular acceleration:

■ Kinematic Equations for Rotational Motion at Constant Angular Acceleration

$$\begin{aligned} \omega &= \omega_0 + \alpha t && \text{(a)} \\ \theta &= \tfrac{1}{2}(\omega + \omega_0)t && \text{(b)} \\ \theta &= \omega_0 t + \tfrac{1}{2}\alpha t^2 && \text{(c)} \\ \omega^2 &= \omega_0^2 + 2\alpha\theta && \text{(d)} \end{aligned} \qquad (9.5)$$

The assumption for initial values in Eq. (9.5) is that $\theta = 0$ when $t = 0$, which corresponds to the assumption that $x = 0$ when $t = 0$, used in Eq. (2.12).

EXAMPLE 9.1

An automobile engine is idling at 500 rev/min. The accelerator is depressed, and after an elapsed time interval of 3 s, the engine speed is 4100 rev/min. Calculate the angular acceleration (assuming that it is constant) and the total angular displacement in revolutions.

SOLUTION We begin by putting the data in SI units:

$$\omega_0 = 500 \text{ rev/min} \times \frac{2\pi}{60} \text{ (rad/s)/(rev/min)}$$

$$= 52.35 \text{ rad/s}$$

$$\left(\text{Remember that 1 rev/min} = 2\pi \text{ rad/min} = \frac{2\pi}{60} \text{ rad/s}\right)$$

$$\omega = 4100 \text{ rev/min} \times \frac{2\pi}{60} \text{ (rad/s)/(rev/min)}$$

$$= 429.4 \text{ rad/s}$$

$$t = 3 \text{ s}$$

$$\alpha = ?$$

$$\theta = ?$$

To calculate the angular acceleration, we use Eq. (9.5a):

$$\omega = \omega_0 + \alpha t$$

$$\therefore \alpha = \frac{\omega - \omega_0}{t}$$

$$= \frac{429.4 \text{ rad/s} - 52.35 \text{ rad/s}}{3\text{s}}$$

$$= 125.7 \text{ rad/s}^2$$

The angular displacement is given by Eq. (9.5c):

$$\theta = \omega_0 t + \tfrac{1}{2}\alpha t^2$$

$$= 52.35 \text{ rad/s} \times 3 \text{ s} + \tfrac{1}{2} \times 125.7 \text{ rad/s}^2 \times (3 \text{ s})^2$$

$$= 722.7 \text{ rad}$$

$$= 722.7 \text{ rad} \div 2\pi \text{ rad/rev}$$

$$= 115 \text{ rev}$$ ■

Our discussion up to this point refers to an object rotating about a fixed axis. Let us apply these concepts to a wheel rolling without slipping along a straight line. If the wheel is fixed to a vehicle, the axis of rotation remains fixed relative to the vehicle but moves forward relative to the surface on which the wheel rolls. Consider the rolling wheel of radius r (Fig. 9.2) with a point P_0 on the wheel adjacent to a point P_0 on the ground at time $t_0 = 0$. At a later time t, another point P on the wheel is adjacent to another point P on the ground. At this time the point P_0 on the wheel has advanced around the circumference by an amount equal to the forward motion of the point of contact between the wheel and the ground. The forward displacement of the wheel is therefore equal to the tangential displacement of a point fixed on the rim. We express this algebraically with the help of Eq. (1.3):

$$x = S = r\theta \tag{9.6}$$

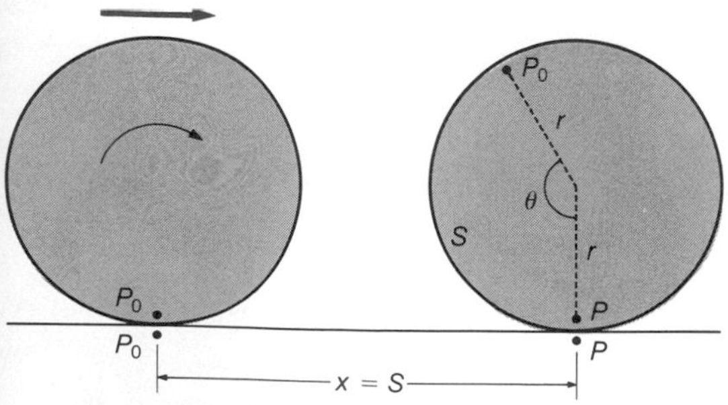

FIGURE 9.2 Diagram to illustrate kinematics of a rolling wheel.

In addition to this relation between the forward displacement of the wheel and its angular displacement about the rotation axis, we can find relations for the forward velocity and acceleration of the wheel. Using Eq. (2.2) to calculate the average forward velocity of the wheel over the time interval 0 to t, and Eqs. (9.1) and (9.6) to relate this to angular quantities, we have:

$$\langle v \rangle = \frac{x}{t} = \frac{r\theta}{t} = r\langle\omega\rangle \tag{9.7}$$

Since the velocity of the wheel is constant, we can drop the average symbols in Eq. (9.7) and write:

$$v = r\omega \tag{9.8}$$

This equation also expresses the relation between instantaneous forward velocity and instantaneous angular velocity of the wheel.

In a similar way, Eq. (2.6) gives for the average forward acceleration of the wheel:

$$\langle a \rangle = \frac{v - v_0}{t} = \frac{r(\omega - \omega_0)}{t} = r\langle \alpha \rangle \tag{9.9}$$

In deriving Eq. (9.9), we use Eqs. (9.3) and (9.8). For constant acceleration we can drop the average symbol:

$$a = r\alpha \tag{9.10}$$

The equations that relate linear kinematic quantities to corresponding angular ones apply to the rotation of any circular object and not just a rolling wheel. A rope unwinding without slipping from a circular cylinder is another example. The linear kinematic quantities for the motion of the rope are the same as the corresponding quantities measured around the circumference of the cylinder. The angular kinematic quantities for the cylinder therefore relate to the linear kinematic quantities in the same way as for the rolling wheel. Because of their importance, we write Eqs (9.6), (9.8), and (9.10) in display.

■ Relation Between Tangential and Angular Kinematic Quantities

$$\text{Displacement } x = r\theta \quad \text{(a)}$$
$$\text{Velocity } v = r\omega \quad \text{(b)} \tag{9.11}$$
$$\text{Acceleration } a = r\alpha \quad \text{(c)}$$

As we have already stressed, these relations apply only when the linear displacement x is equal to the circular arc that subtends the angular displacement θ. Furthermore, they illustrate the importance of using the radian as the measure of angle. The displacement relation Eq. (9.11a) is the same as Eq. (6.1) for the definition of angular measure in radians. Since the velocity and acceleration relations, Eqs. (9.11b) and (9.11c), are based directly on the displacement relation, we must use the radian per second as the angular velocity unit and the radian per second per second as the angular acceleration unit.

EXAMPLE 9.2

A wheel of radius 0.4 m starts from rest and accelerates without slipping with an angular acceleration of 1.5 rad/s^2 for 20 s. How far does the wheel travel in this time, and what is its final linear velocity?

SOLUTION Let us first treat the problem as one of rotational kinematics, and calculate the total angular displacement and the final angular velocity. The data are as follows:

$$\begin{aligned} \omega_0 &= 0 \\ \alpha &= 1.5 \text{ rad/s}^2 \\ t &= 20 \text{ s} \\ \omega &= ? \\ \theta &= ? \end{aligned}$$

Equation (9.5a) gives:

$$\begin{aligned} \omega &= \omega_0 + \alpha t \\ &= 0 + 1.5 \text{ rad/s}^2 \times 20 \text{ s} \\ &= 30 \text{ rad/s} \end{aligned}$$

We calculate the corresponding linear velocity from Eq. (9.11b):

$$\begin{aligned} v &= r\omega \\ &= 0.4 \text{ m} \times 30 \text{ rad/s} \\ &= 12 \text{ m/s} \end{aligned}$$

The total angular displacement is given by Eq. (9.5c):

$$\begin{aligned} \theta &= \omega_0 t + \tfrac{1}{2}\alpha t^2 \\ &= 0 + \tfrac{1}{2} \times 1.5 \text{ rad/s}^2 \times (20 \text{ s})^2 \\ &= 300 \text{ rad} \end{aligned}$$

The corresponding linear displacement is given by Eq. (9.11a):

$$\begin{aligned} x &= r\theta \\ &= 0.4 \text{ m} \times 300 \text{ rad} \\ &= 120 \text{ m} \end{aligned}$$

Note that we can work the whole problem as a linear kinematic one from the beginning. Equation (9.11c) gives the linear acceleration of the wheel:

$$\begin{aligned} a &= r\alpha \\ &= 0.4 \text{ m} \times 1.5 \text{ rad/s}^2 \\ &= 0.6 \text{ m/s}^2 \end{aligned}$$

We can now write the data for the problem as follows:

$$\begin{aligned} v_0 &= 0 \\ a &= 0.6 \text{ m/s}^2 \\ t &= 20 \text{ s} \\ v &= ? \\ x &= ? \end{aligned}$$

Equation (2.12a) gives the final velocity:

$$\begin{aligned} v &= v_0 + at \\ &= 0 + 0.6 \text{ m/s}^2 \times 20 \text{ s} \\ &= 12 \text{ m/s} \end{aligned}$$

Equation (2.12c) gives the linear displacement:

$$\begin{aligned} x &= v_0 t + \tfrac{1}{2}at^2 \\ &= 0 + \tfrac{1}{2} \times 0.6 \text{ m/s}^2 \times (20 \text{ s})^2 \\ &= 120 \text{ m} \end{aligned}$$

The approach to be used is entirely a matter of taste and convenience. The central idea is that Eq. (9.11) connects linear and rotational kinematics when a wheel rolls without slipping. ■

We have previously associated centripetal acceleration with a point moving in a circular path. Let us see how this association applies to a point on the rim of a rotating wheel. Consider a flywheel, as shown in Fig. 9.3, rotating about a shaft directed along its axis. The point P on the rim of the flywheel experiences two acceleration components. One is the tangential acceleration component that is related to the angular acceleration of the wheel by Eq. (9.11c):

$$a = r\alpha$$

In the following section we will see that this acceleration is a direct result of external torques acting on the wheel. But the point P is moving in a circular path, and it has a centripetal acceleration given by Eq. (6.3):

$$a_c = \frac{v^2}{r} = r\omega^2$$

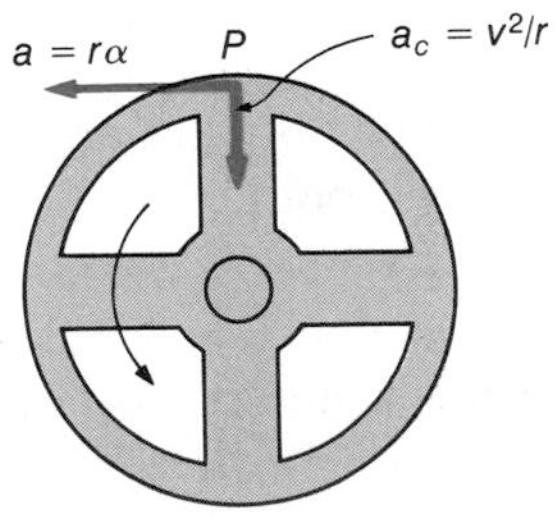

FIGURE 9.3 Tangential and centripetal components of acceleration for a point on the rim of a rotating wheel.

[We use the relationship of Eq. (9.11b) between the tangential velocity and angular velocity to write an equivalent formula for centripetal acceleration.] The cohesion of the material of the wheel supplies the force for this acceleration. As long as the wheel remains rigid, the centripetal acceleration has no externally visible effect. If the rotational velocity becomes sufficiently large, the wheel may break apart. Except for this contingency, we can ignore the centripetal acceleration component, since the tangential acceleration component is the only one relevant to the kinematics and dynamics of the situation.

9.2 Rotational Analogue of Newton's Second Law

After completing linear kinematics, we considered the cause of linear acceleration. Newton's second law gives a precise relation between force and the linear acceleration that it causes in a mass. In rotational dynamics, we seek a relation between the *torque* acting on a rotating object and the *angular acceleration* that it causes. To find this relation, we consider a rigid body as composed of a large number of point masses rotating in unison about the axis of rotation.

As a starting point, we will consider a single point mass m attached to one end of a light rigid rod of length r (Fig. 9.4). The other end of the rod is fixed to the point O but is free to rotate about it. Suppose that a force F acts on the mass in a direction that continually changes so as to remain perpendicular to the rod as the mass accelerates in a circular path. Newton's second law, Eq. (4.1), gives the relation between the force and the acceleration:

$$F = ma$$

Multiplying by r and rearranging, we have:

$$Fr = mr^2 \times \frac{a}{r} \tag{9.12}$$

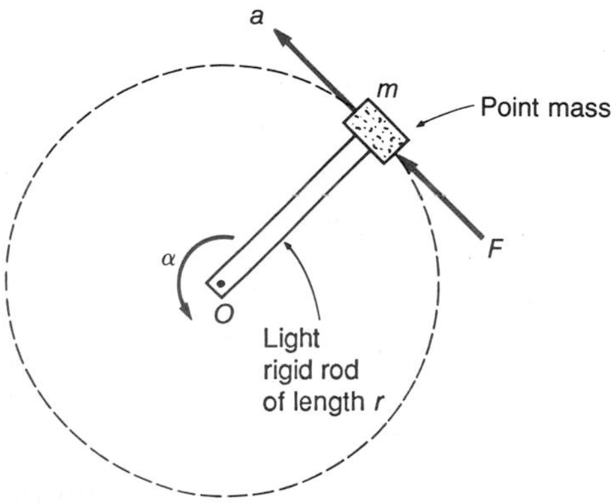

FIGURE 9.4 A point mass m attached to one end of a rigid rod of length r, which is rotating about a fixed point O to which it is attached.

We can write this equation in a different notation and with a different emphasis by noting the following points:

1. The quantity Fr is the **torque** τ acting on the point mass about the center of rotation O. The radial distance r is the lever arm of the force in the sense explained in Section 5.2 and Eq. (5.2).

2. We define the quantity mr^2 as the **moment of inertia** of the point mass about the center of rotation. (This idea is generalized below.)

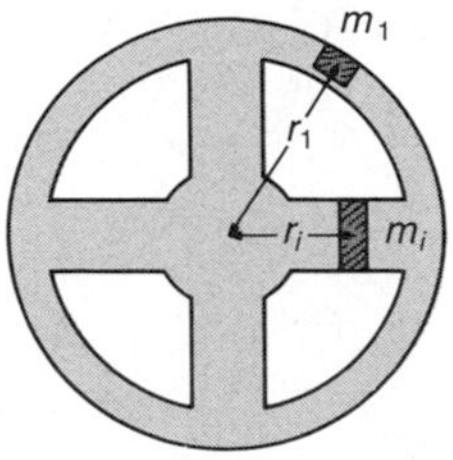

FIGURE 9.5 A rotating wheel regarded as an assembly of small masses.

3. Since a is the acceleration around the circumference, the quantity a/r is equal to the angular acceleration α about the center of rotation [from Eq. (9.11c)].

With these changes, we rewrite Eq. (9.12) as follows:

■ Rotational Analogue of Newton's Second Law

Torque = moment of inertia × angular acceleration

$$\tau = I\alpha \tag{9.13}$$

SI unit of torque: N · m [see Eq. (5.2)]
SI unit of moment of inertia: kg · m^2 [see Eq. (9.14)]
SI unit of angular acceleration: rad/s^2 [see Eq. (9.3)]

There would be little point to our elaborate presentation of Eq. (9.13) if the equation applied only to the movement of a point mass about a center of rotation. With some interpretation, we can apply it to the rotation of a rigid body about a fixed axis. To do this, consider a wheel rotating about a fixed axis (Fig. 9.5). We can regard the wheel as a collection of n small masses rigidly fixed with respect to each other by the strength of the material of which the wheel is made. The dynamics of each small mass individually is determined by Eq. (9.13). Since the masses are rigidly attached to each other, the angular acceleration of each one is the same. We have a total of n equations, each of the form of Eq. (9.13) and each containing the same angular acceleration:

$$\tau_1 = I_1\alpha$$
$$\tau_2 = I_2\alpha$$
$$\vdots$$
$$\tau_n = I_n\alpha$$

The torques $\tau_1, \tau_2, \ldots, \tau_n$ are the *external torques* acting on the point masses. We do not include torques that arise from the interaction of adjacent point masses for the reason explained below. Adding all these equations gives:

$$\sum_{i=1}^{n} \tau_i = \sum_{i=1}^{n} I_i\alpha$$

where I_i is the moment of inertia of the ith point mass, and τ_i is the net external torque acting on it. It is correct to omit the

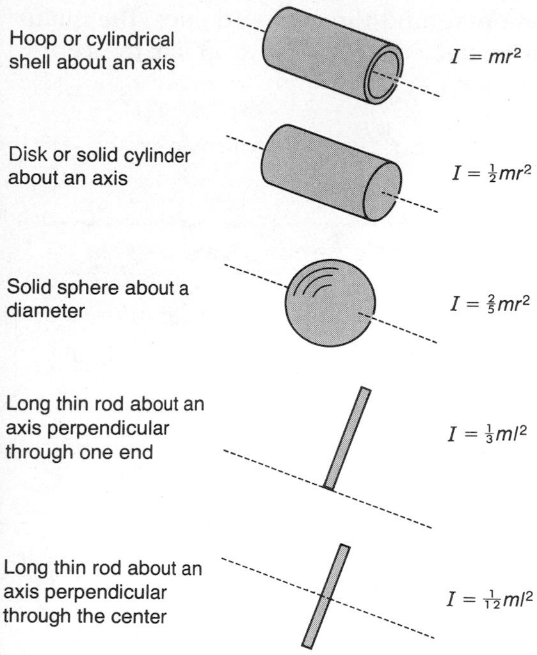

FIGURE 9.6 Moments of inertia of some simple objects of uniform composition. In each case the mass of the object is *m;* the radius of the hoop, disk, or sphere is *r;* and the length of the rod is *l*.

torques that arise from the interaction of the point masses with each other. Newton's third law permits us to cancel them in pairs in the summation of the left-hand side of the equation. We can state the content of the above equation in words:

Total net external torque	=	total moment of inertia	×	angular acceleration

This is of the same form as Eq. (9.13), where we now interpret τ as the net external torque acting on the wheel and I as the total moment of inertia. The total moment of inertia is the sum of the values of mr^2 calculated for each mass individually.

Definition of Moment of Inertia

Moment of inertia = sum of values mr^2 for point masses

$$I = \sum_{i=1}^{n} m_i r_i^2 \tag{9.14}$$

SI unit of moment of inertia: kilogram-meter^2
Abbreviation: $\text{kg} \cdot \text{m}^2$

The practical use of Eq. (9.14) to calculate moment of inertia is limited to cases in which an object can be regarded as a relatively small number of point masses. For most objects, calculating the moment of inertia requires integral calculus. The situation is not as complicated as it may appear, since many problems in rotational dynamics require only a few frequently used relationships, which are shown in Fig. 9.6.

The moment of inertia defined by Eq. (9.14) is the rotational analogue of mass. Remember that mass directly relates the net applied force to the acceleration that it causes via Newton's second law, Eq. (4.1). Similarly, the moment of inertia directly relates the net applied torque to the angular acceleration that it causes via Eq. (9.13). For this reason we describe Eq. (9.13) as the rotational analogue of Newton's second law. However, the rotational case has a complexity that does not appear in the rectilinear one. Consider, for example, a flat disk. We can give a definite value for its mass, but we cannot specify the moment of inertia unless we also give an axis. Two possible rotation axes are shown in Fig. 9.7. The moment of inertia about an axis through the center of the disk and perpendicular to its plane is given as $\frac{1}{2}mr^2$ in Fig. 9.6. The moment of inertia about a diametral axis is not given in Fig. 9.6, but its value is $\frac{1}{4}mr^2$. The moment of inertia is less in the second case, because more of the disk's mass is closer to the ro-

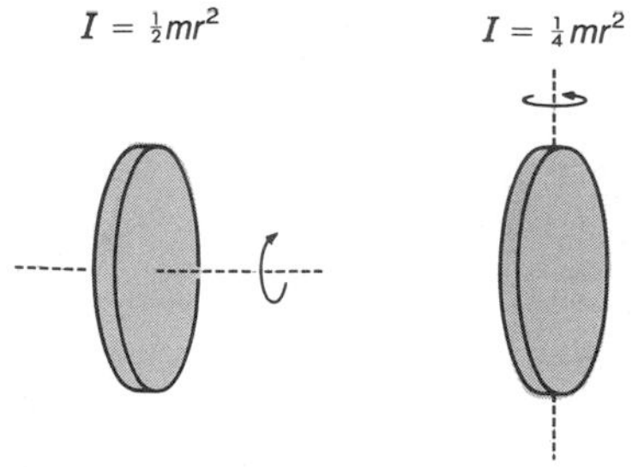

FIGURE 9.7 Diagram illustrating two of the many possible rotation axes for a flat disk.

tation axis than it is in the first case. The moment of inertia values given in Fig. 9.6 therefore refer not only to certain objects of simple shape but also to certain specified rotation axes. With the help of Fig. 9.6 and the definition of Eq. (9.14), we can easily calculate the moment of inertia of a variety of objects.

EXAMPLE 9.3

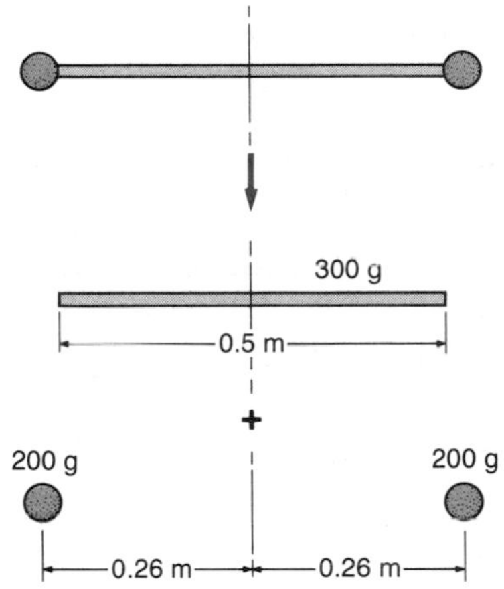

A baton consists of a 50-cm rod of mass 300 g with small metal spheres, each of mass 200 g, on either end. The centers of the metal spheres are 52 cm apart. Calculate the moment of inertia of the baton about an axis through the center of the rod and perpendicular to it.

SOLUTION We can consider the baton as a composite object, made up of a long, thin rod and a pair of point masses. We calculate the moment of inertia of the rod from the formula in Fig. 9.6:

$$\begin{aligned} I_{\text{rod}} &= \tfrac{1}{12}ml^2 \\ &= \tfrac{1}{12} \times 0.3 \text{ kg} \times (0.5 \text{ m})^2 \\ &= 6.25 \times 10^{-3} \text{ kg} \cdot \text{m}^2 \end{aligned}$$

Considering the metal spheres as point masses, we can calculate their moment of inertia from Eq. (9.14):

$$\begin{aligned} I_{\text{spheres}} &= \sum_{i=1}^{n} m_i r_i^2 \\ &= 0.2 \text{ kg} \times (0.26 \text{ m})^2 + 0.2 \text{ kg} \times (0.26 \text{ m})^2 \\ &= 2.704 \times 10^{-2} \text{ kg} \cdot \text{m}^2 \end{aligned}$$

The summation has only two terms since there are only two point masses. The total moment of inertia of the baton is given by:

$$\begin{aligned} I &= I_{\text{rod}} + I_{\text{spheres}} \\ &= 6.25 \times 10^{-3} \text{ kg} \cdot \text{m}^2 + 0.027 \text{ kg} \cdot \text{m}^2 \\ &= 0.0333 \text{ kg} \cdot \text{m}^2 \end{aligned}$$

■

We have already exploited the analogy between linear and angular kinematic quantities shown in Eq. (9.4). As a result, the equations of uniformly accelerated angular motion, Eq. (9.5), are identical in algebraic form to the equations of uniformly accelerated linear motion, Eq. (2.12). Newton's second law, Eq. (4.1), gives force as the cause of linear acceleration of a mass; the rotational analogue of this law, Eq. (9.13), gives torque as the cause of the angular acceleration of the moment of inertia. This permits us to display dynamic analogues. Further development of rotational dynamics supports this analogy.

Dynamic Analogues

Linear	Angular	
Force F	Torque τ	**(9.15)**
Mass m	Moment of inertia I	

EXAMPLE 9.4

A woman unwinds a rope from a uniform circular wooden cylinder by exerting a constant pull of 10 N on the rope. The cylinder is 25 cm in radius, has a mass of 40 kg, and is mounted horizontally in bearings of negligible friction. Initially the cylinder is at rest. Calculate the speed at which the rope is unwinding 10 s after the woman begins to pull.

SOLUTION We begin by calculating the moment of inertia of the cylinder using the formula in Fig. 9.6:

$$\begin{aligned} I &= \tfrac{1}{2}mr^2 \\ &= \tfrac{1}{2} \times 40 \text{ kg} \times (0.25 \text{ m})^2 \\ &= 1.250 \text{ kg} \cdot \text{m}^2 \end{aligned}$$

A rope unwinding from a circular cylinder provides a force that is always tangential to the cylindrical surface. The torque exerted by the rope is given by Eq. (5.2):

$$\begin{aligned} \tau &= Fl_{\perp} = Fr \\ &= 10 \text{ N} \times 0.25 \text{ m} \\ &= 2.5 \text{ N} \cdot \text{m} \end{aligned}$$

We can calculate the angular acceleration using Eq. (9.13):

$$\begin{aligned} \tau &= I\alpha \\ \therefore \alpha &= \frac{\tau}{I} \\ &= \frac{2.5 \text{ N} \cdot \text{m}}{1.250 \text{ kg} \cdot \text{m}^2} \\ &= 2.00 \text{ rad/s}^2 \end{aligned}$$

To calculate the final angular velocity of the cylinder, we use the kinematic equations. The data are as follows:

$$\omega_0 = 0$$
$$\alpha = 2.00 \text{ rad/s}^2$$
$$t = 10 \text{ s}$$
$$\omega = ?$$

Equation (9.5a) provides the result:

$$\begin{aligned}\omega &= \omega_0 + \alpha t \\ &= 0 + 2.00 \text{ rad/s}^2 \times 10 \text{ s} \\ &= 20 \text{ rad/s}\end{aligned}$$

The linear speed of the rope is the same as the tangential speed of a point on the surface of the cylinder, and Eq. (9.11b) gives:

$$\begin{aligned}v &= r\omega \\ &= 0.25 \text{ m} \times 20 \text{ rad/s} \\ &= 5.00 \text{ m/s}\end{aligned}$$

■

This example is conceptually identical to problems in linear dynamics. In the linear case, a given force accelerates a mass. After calculating the acceleration, we find the linear displacement from rest over a given time interval. In the rotational case, a given torque causes angular acceleration in a rotating object, which we use to calculate the angular displacement from rest over a given time interval.

The torque on a rotating wheel or cylinder is often caused by falling weights. This results in a more difficult problem than the one just discussed.

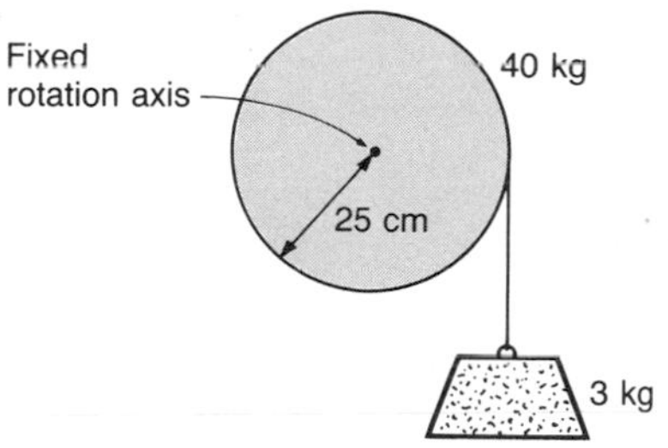

(a) General situation diagram

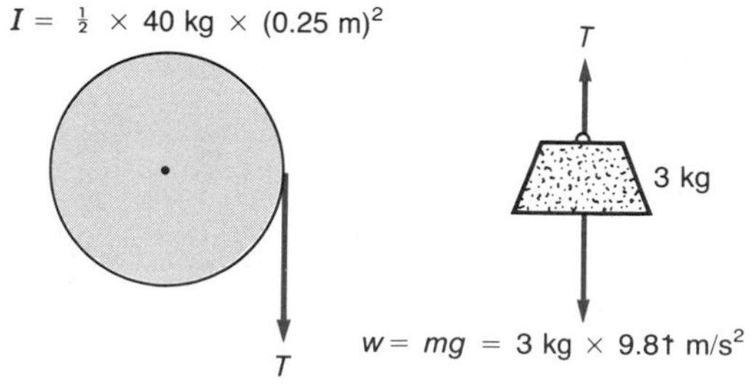

(b) Free-body force diagrams

EXAMPLE 9.5

Consider the cylinder of Example 9.4 undergoing angular acceleration under the influence of a 3-kg mass that is hanging from the end of a rope wound around the cylinder. Calculate the tension in the rope and the rate at which the rope unwinds 10 s after the mass falls from rest.

SOLUTION At first sight this may seem almost the same problem as the previous one. However, it is most important to remember that the tension in a rope is equal to the weight on the end of the rope only when that weight is not accelerating. We need to consider the dynamic equations for both the falling weight and the cylinder separately. Let T be the tension in the rope, and a the downward acceleration of the 3-kg mass. To find the weight of the mass, we use Eq. (4.2):

$$\begin{aligned} w &= mg \\ &= 3\ \text{kg} \times 9.81\ \text{m/s}^2 \\ &= 29.43\ \text{N} \end{aligned}$$

Using the free-body force diagram for the 3-kg mass, Newton's second law, Eq. (4.1), gives:

$$\begin{aligned} mg - T &= ma \\ 29.43\ \text{N} - T &= 3\ \text{kg} \times a \end{aligned}$$

Since this equation contains two unknown quantities, we can make no further progress by using it alone. The torque on the rotating cylinder is caused by the same rope tension that acts on the falling mass. Using Eq. (5.2), we find:

$$\begin{aligned} \tau &= Fl_{\perp} \\ &= T \times 0.25\ \text{m} \end{aligned}$$

The tangential acceleration of the rope is related to the angular acceleration of the cylinder by Eq. (9.11c):

$$\begin{aligned} a &= r\alpha \\ &= 0.25\ \text{m} \times \alpha \\ \therefore \alpha &= \frac{a}{0.25\ \text{m}} \end{aligned}$$

The moment of inertia of the cylinder is the same as that calculated in Example 9.4. We can now write the dynamic equation for the cylinder, Eq. (9.13):

$$\begin{aligned} \tau &= I\alpha \\ T \times 0.25\ \text{m} &= 1.250\ \text{kg} \cdot \text{m}^2 \times \frac{a}{0.25\ \text{m}} \\ \therefore T &= 20\ \text{kg} \times a \end{aligned}$$

The two equations permit us to determine the two unknowns. Substituting the second in the first gives:

$$\begin{aligned} 29.43\ \text{N} - 20\ \text{kg} \times a &= 3\ \text{kg} \times a \\ \therefore a &= 1.28\ \text{m/s}^2 \end{aligned}$$

We can now calculate the tension:

$$\begin{aligned} T &= 20 \text{ kg} \times a \\ &= 20 \text{ kg} \times 1.28 \text{ m/s}^2 \\ &= 25.6 \text{ N} \end{aligned}$$

As we expected, the tension is smaller than the weight of the downward accelerating mass. To calculate the velocity of the falling mass after a time interval, we use linear kinematics. The data are as follows:

$$\begin{aligned} v_0 &= 0 \\ t &= 10 \text{ s} \\ a &= 1.28 \text{ m/s}^2 \\ v &= ? \end{aligned}$$

Equation (2.12a) gives:

$$\begin{aligned} v &= v_0 + at \\ &= 0 + 1.28 \text{ m/s}^2 \times 10 \text{ s} \\ &= 12.8 \text{ m/s} \end{aligned}$$

■

9.3 Work and Rotational Kinetic Energy

The helicopter body would rotate in the opposite sense to the main rotor blades except for the opposing torque provided by the small propeller that is situated at the tail of the craft. (Courtesy of Bell Helicopter Textron)

The concepts of work and energy greatly simplify many problems in linear dynamics, and the same is true in rotational dynamics. We could develop each new concept of rotational dynamics from the beginning. However, it is simpler to write new rotational equations by using our lists of kinematic and dynamic analogues and then to check these equations for simple situations. We therefore adopt the following procedure:

1. Using the kinematic and dynamic analogues of (9.4) and (9.15), rewrite the linear result in rotational form.

2. Check the tentative equation thus obtained for the special case of a revolving point mass. A final step would be to demonstrate that the result obtained is true for a rigid body. We omit this, since it is similar to the argument of the preceding section for all cases considered.

A force does work when it causes linear displacement of a body in the direction of the force. Analogously, a torque does work when it causes angular displacement of a body about the axis of the torque. We begin by writing Eq. (7.1a) for the work

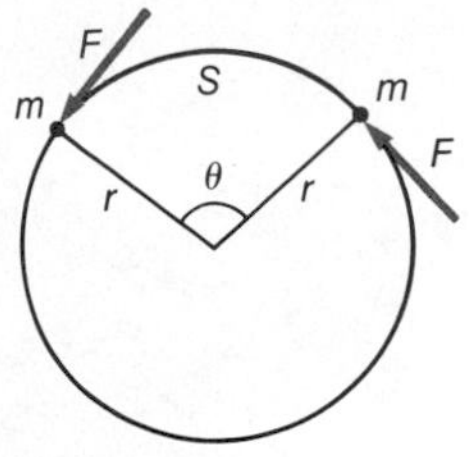

FIGURE 9.8 Illustration of the work done on a point mass by a constant tangential force F.

done by a force ($W = \mathbf{F} \cdot \mathbf{x}$) and changing it to the rotational case by using the analogue Eqs. (9.4) and (9.15). We replace F by its rotational analogue τ, and $\mathbf{x}$ by its rotational analogue θ. With these substitutions, we define the work done by a torque.

■ Definition of Work Done by a Torque

Work = torque × angular displacement about the axis of the torque

$$W = \tau\theta \tag{9.16}$$

To confirm our result, we again consider a point mass m attached to a fixed point by a light rod of length r (Fig. 9.8). Let the force F acting on the mass continually change direction so that it remains tangential to the circular path. Equation (7.1) gives the work done by the force F when the mass moves around a circular arc of length S:

$$W = FS$$
$$= Fr\theta \text{ [from Eq. (9.6)]}$$
$$= \tau\theta \text{ [from Eq. (5.2)]}$$

This early mill turns the kinetic energy of the flowing water into rotational kinetic energy of the main shaft, enabling the unit to do work on the grain between the millstones. (Historical Pictures Service, Inc.)

An extension of this argument would convince us that Eq. (9.16) is also true for a rigid body rotating about a fixed axis.

EXAMPLE 9.6

A flywheel whose moment of inertia is 3 kg · m^2 is mounted on frictionless bearings. Starting from rest, it is subjected to a torque of 6 N · m for a period of 20 s. Calculate the work done.

SOLUTION To calculate the work, we need to know the angular displacement of the flywheel. To obtain this, we must solve an angular kinematics problem. To calculate the angular acceleration, we use Eq. (9.13):

$$\tau = I\alpha$$

$$6\ \text{N} \cdot \text{m} = 3\ \text{kg} \cdot \text{m}^2 \times \alpha$$

$$\therefore \alpha = 2\ \text{rad/s}^2$$

The data for the kinematics problem are as follows:

$$\omega_0 = 0$$

$$\alpha = 2\ \text{rad/s}^2$$

$$t = 20\ \text{s}$$

$$\theta = ?$$

The angular displacement is provided by Eq. (9.5c):

$$\begin{aligned}\theta &= \omega_0 t + \tfrac{1}{2}\alpha t^2 \\ &= 0 + \tfrac{1}{2} \times 2\ \text{rad/s}^2 \times (20\ \text{s})^2 \\ &= 400\ \text{rad}\end{aligned}$$

To calculate the work done, we use Eq. (9.16):

$$\begin{aligned}W &= \tau\theta \\ &= 6\ \text{N} \cdot \text{m} \times 400\ \text{rad} \\ &= 2400\ \text{J}\end{aligned}$$

■

In linear kinematics, the definition of work led us quickly to the concept of kinetic energy, which is the energy that a mass possesses by reason of its velocity relative to surrounding objects. In a similar way we expect a rotating object to possess **rotational kinetic energy** because of its angular velocity. We begin with Eq. (7.2) for the linear kinetic energy (KE $= \frac{1}{2}mv^2$) and use the analogue displays (9.4) and (9.15). We replace m by its rotational analogue I, and v by its rotational analogue ω. This enables us to give a tentative definition of rotational kinetic energy.

■ Definition of Rotational Kinetic Energy

$$\text{Rotational kinetic energy} = \tfrac{1}{2} \times \text{moment of inertia} \times (\text{angular velocity})^2$$

$$\text{Rot KE} = \tfrac{1}{2}I\omega^2 \qquad \textbf{(9.17)}$$

We can check our formula for the case of a point mass m revolving in a circular path of radius r with tangential velocity v. The kinetic energy is given by Eq. (7.2):

$$\begin{aligned} \text{KE} &= \tfrac{1}{2}mv^2 \\ &= \tfrac{1}{2}m(r\omega)^2 \text{ [from Eq. (9.11b)]} \\ &= \tfrac{1}{2}I\omega^2 \text{ [from Eq. (9.14)]} \end{aligned}$$

Further investigation confirms Eq. (9.17) for describing a rigid body rotating about a fixed axis.

EXAMPLE 9.7

Solve Example 9.6 by using an energy method.

SOLUTION Calculating the final kinetic energy requires knowledge of the final angular velocity rather than the angular displacement. The angular acceleration of the flywheel is 2 rad/s^2, as calculated in Example 9.6. To calculate the final angular velocity, we set down the data:

$$\begin{aligned} \omega_0 &= 0 \\ \alpha &= 2 \text{ rad/s}^2 \\ t &= 20 \text{ s} \\ \omega &= ? \end{aligned}$$

Equation (9.5a) is appropriate:

$$\begin{aligned} \omega &= \omega_0 + \alpha t \\ &= 0 + 2 \text{ rad/s}^2 \times 20 \text{ s} \\ &= 40 \text{ rad/s} \end{aligned}$$

We calculate the rotational kinetic energy using Eq. (9.17):

$$\begin{aligned} \text{Rot KE} &= \tfrac{1}{2}I\omega^2 \\ &= \tfrac{1}{2} \times 3 \text{ kg} \cdot \text{m}^2 \times (40 \text{ rad/s})^2 \\ &= 2400 \text{ J} \end{aligned}$$

Since the bearings are frictionless, the final kinetic energy of the wheel is equal to the work done by the applied torque. ■

The formula of Eq. (9.17) gives the kinetic energy of a rigid body rotating about a fixed axis. In that case the rotational kinetic energy is the only kinetic energy under consideration. But a rolling wheel possesses kinetic energy due to its linear velocity in addition to its rotational kinetic energy. The correct way to calculate the total kinetic energy of a rolling wheel is not obvious, but we can state a rule that does not affront common sense.

$$\begin{matrix}\text{Total kinetic energy}\\ \text{of a rolling wheel}\end{matrix} = \begin{matrix}\text{kinetic energy}\\ \text{of linear motion}\end{matrix} + \begin{matrix}\text{kinetic energy of}\\ \text{rotation about the}\\ \text{center of the wheel}\end{matrix}$$

This result applies only for a wheel whose rotation axis is at its center of gravity.

EXAMPLE 9.8

A wheel of mass 5 kg and radius 40 cm rolls without slipping with an angular velocity of 10 rad/s. The moment of inertia of the wheel about its central axis is 0.65 kg · m². Calculate the fraction of the total kinetic energy in rotational form.

SOLUTION To calculate both the linear and rotational kinetic energies, we need to know both the linear and rotational velocities. The linear velocity is given by Eq. (9.11b):

$$\begin{aligned} v &= r\omega \\ &= 0.40 \text{ m} \times 10 \text{ rad/s} \\ &= 4 \text{ m/s} \end{aligned}$$

The linear kinetic energy is given by Eq. (7.2):

$$\begin{aligned} \text{KE} &= \tfrac{1}{2}mv^2 \\ &= \tfrac{1}{2} \times 5 \text{ kg} \times (4 \text{ m/s})^2 \\ &= 40 \text{ J} \end{aligned}$$

We can calculate the rotational kinetic energy from Eq. (9.17):

$$\begin{aligned} \text{Rot KE} &= \tfrac{1}{2}I\omega^2 \\ &= \tfrac{1}{2} \times 0.65 \text{ kg} \cdot \text{m}^2 \times (10 \text{ rad/s})^2 \\ &= 32.5 \text{ J} \end{aligned}$$

$$\begin{aligned} \begin{matrix}\text{Fraction of total kinetic energy}\\ \text{in rotational form}\end{matrix} &= \frac{\text{Rot KE}}{\text{Rot KE} + \text{KE}} \\ &= \frac{32.5 \text{ J}}{32.5 \text{ J} + 40 \text{ J}} \\ &= 44.8\% \end{aligned}$$

■

One of the major reasons for developing the concept of energy is to use it in the energy conservation law. If an object has rotational kinetic energy, we need to include this energy in any energy conservation equation concerning the object. For a rotating object the total energy is the sum of the linear kinetic energy, the rotational kinetic energy, and the gravitational potential energy. We must include the initial and final values of the sum of these three energy forms in the energy conservation equation.

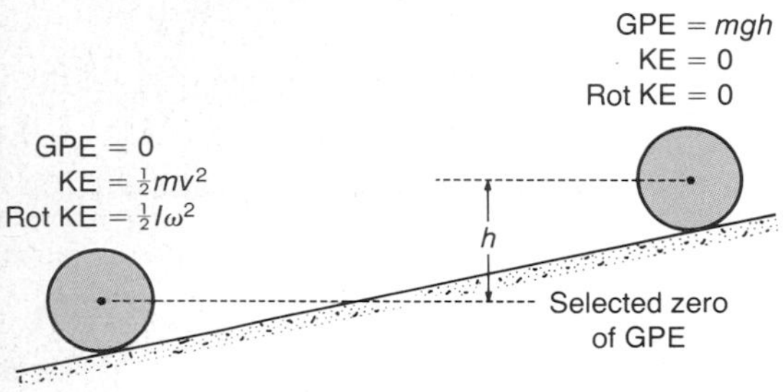

EXAMPLE 9.9

A uniform spherical bowling ball of mass 7.25 kg and radius 10.5 cm rolls without slipping on a floor that rises gradually. At a given position the velocity of the ball is 5 m/s. To what height above this position does the ball rise before coming momentarily to a stop? (Assume negligible energy dissipation.)

SOLUTION We can use the energy conservation principle to equate the initial and final total energies. To do this, we must know the moment of inertia of the ball and its angular velocity. From Fig. 9.6 we have:

$$\begin{aligned} I &= \tfrac{2}{5}mr^2 \\ &= \tfrac{2}{5} \times 7.25 \text{ kg} \times (0.105 \text{ m})^2 \\ &= 3.197 \times 10^{-2} \text{ kg} \cdot \text{m}^2 \end{aligned}$$

From Eq. (9.11b) we have (rolling *without* slipping):

$$\begin{aligned} v &= r\omega \\ \therefore \omega &= \frac{v}{r} \\ &= \frac{5 \text{ m/s}}{0.105 \text{ m}} \\ &= 47.62 \text{ rad/s} \end{aligned}$$

This completes the calculation of the preliminary data. We can now write the mechanical energy conservation equation, which must include the contribution from rotational kinetic energy:

$$\text{Initial (KE + GPE + Rot KE)} = \text{Final (KE + GPE + Rot KE)}$$

$$\tfrac{1}{2}mv^2 + 0 + \tfrac{1}{2}I\omega^2 = 0 + mgh + 0$$

$$\begin{aligned} &\tfrac{1}{2} \times 7.25 \text{ kg} \times (5 \text{ m/s})^2 + 0 \\ &+ \tfrac{1}{2} \times 3.197 \times 10^{-2} \text{ kg} \cdot \text{m}^2 = 0 + 7.25 \text{ kg} \times 9.81 \text{ m/s}^2 \times h + 0 \\ &\times (47.62 \text{ rad/s})^2 \end{aligned}$$

$$90.63 \text{ J} + 0 + 36.25 \text{ J} = 0 + 71.12 \text{ N} \times h + 0$$

$$\therefore h = 1.78 \text{ m}$$

■

We do not need to modify the basic concept of power defined in Eq. (7.4). The average power remains equal to the work done (or the energy transferred) divided by the elapsed time. However, Eq. (7.6), which relates instantaneous power to the applied force and the velocity ($P = Fv_{\parallel}$), must be replaced. Following our established procedure, we use the analogue displays (9.4) and (9.15) to make the change.

■ Relation Between Power, Torque, and Angular Velocity

Instantaneous power = torque × angular velocity about the axis of the torque

$$P = \tau\omega \tag{9.18}$$

Our new formula then equates the instantaneous power to the product of the torque and the angular velocity about the same axis. In applying this formula, we must remember that the same conditions apply as we noted in Eq. (7.6) for the linear case. Constant power cannot produce constant angular acceleration. The formula is most useful in cases in which a constant power is operating against a constant resisting torque.

EXAMPLE 9.10

An electric motor develops $\frac{1}{2}$ hp at 1500 rev/min. Calculate the torque delivered by the motor.

SOLUTION We begin by setting the data in SI units:

$$P = 0.5 \text{ hp} \times 746 \text{ W/hp}$$
$$= 373 \text{ W}$$

$$\omega = 1500 \text{ rev/min} \times \frac{2\pi}{60} \text{ (rad/s)/(rev/min)}$$
$$= 157.0 \text{ rad/s}$$

To calculate the torque, we use Eq. (9.18):

$$P = \tau\omega$$
$$\therefore \tau = \frac{P}{\omega}$$
$$= \frac{373 \text{ W}}{157 \text{ rad/s}}$$
$$= 2.38 \text{ N} \cdot \text{m}$$

■

9.4 Angular Momentum

Not surprisingly, there is a rotational quantity analogous to linear momentum, as well as a conservation principle to accompany it. We follow our standard procedure of using the analogue displays (9.4) and (9.15) on the appropriate linear equation. Linear momentum is mass multiplied by velocity [Eq. (8.3)]. It follows that we expect to define **angular momentum** as the moment of inertia multiplied by the angular velocity. The dimensions of I are $[ML^2]$, and those of ω are $[T^{-1}]$. This given $[ML^2T^{-1}]$ for the dimensions of L, which we can rewrite as $[ML^2T^{-2}] \times [T]$. Since these are the dimensions of energy and time, it follows that the joule-second is an appropriate unit for angular momentum.

Definition of Angular Momentum of a Rigid Body

Angular momentum = moment of inertia × angular velocity

$$L = I\omega \qquad \textbf{(9.19)}$$

SI unit of angular momentum: kilogram-meter² per second
Usual designation: joule-second
Abbreviation: J · s

Let us apply the definition of Eq. (9.19) to a point mass m moving with velocity v in a circular path of radius r. The definition gives:

$$\begin{aligned} L &= I\omega \\ &= mr^2\omega \text{ [from Eq. (9.14)]} \\ &= mvr \text{ [from Eq. (9.11b)]} \end{aligned} \qquad \textbf{(9.20)}$$

Equation (9.20) provides us with another insight into angular momentum. In fact, we could use it as an alternate definition. For a point particle, the angular momentum is equal to the product of the linear momentum and the lever arm about the rotation axis.

The law of conservation of angular momentum is also quite analogous to the linear momentum conservation law. Recall that the latter asserts the conservation of linear momentum for a system subject to zero external net force. For the angular case we expect conservation of angular momentum for a system that is subject to zero external net torque.

■ Angular Momentum Conservation Principle

> The angular momentum of a system of bodies is conserved provided that the external net torque on the system is zero.

Note that external forces may act on the system, but angular momentum is conserved provided those forces do not produce a net torque. The only torques allowed are those exerted by one part of the system on another part. An example clarifies this point.

EXAMPLE 9.11

A child of mass 40 kg runs at 6 m/s in a direction tangential to the edge of a playground merry-go-round that is at rest. The merry-go-round is in the form of a uniform disk of mass 160 kg and radius 3 m that is free to turn about a frictionless central axis. The child leaps onto the merry-go-round and clings to a fixed point at the edge. Calculate the subsequent angular velocity of the merry-go-round and the amount of energy dissipated in the "collision" between the child and the merry-go-round.

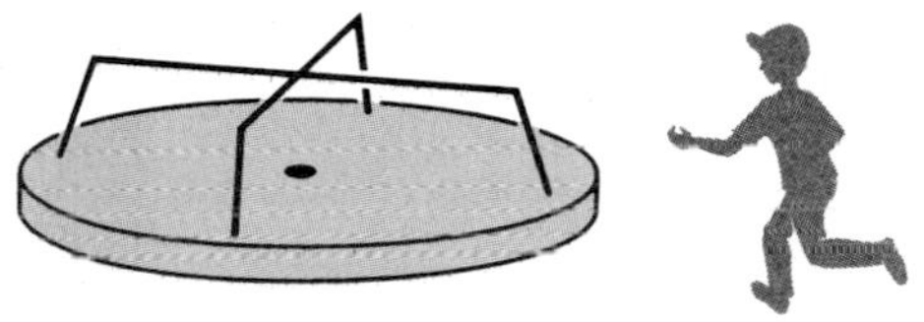

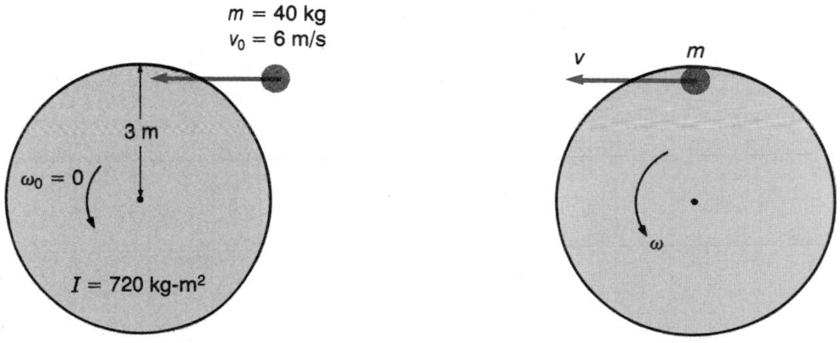

Before collision   After collision

SOLUTION At the time of impact, the child and the merry-go-round exert torques on each other. The only other sources of external force are gravity and the contact forces provided by the central bearing. Gravity exerts no torque, since its direction is parallel to the rotation axis and the bearing exerts no torque provided it is frictionless. The angular momentum of the child–merry-go-round system is therefore conserved. We can calculate the moment of inertia of the merry-go-round by using the appropriate formula from Fig. 9.6:

$$\begin{aligned} I &= \tfrac{1}{2}mr^2 \\ &= \tfrac{1}{2} \times 160 \text{ kg} \times (3 \text{ m})^2 \\ &= 720 \text{ kg} \cdot \text{m}^2 \end{aligned}$$

To calculate the angular momentum of the merry-go-round we use Eq. (9.19), which is suitable for rigid bodies of known moment of inertia. For the angular momentum of the child about the merry-go-round axis, we use Eq. (9.20), which treats the child as a point particle. We can now write the angular momentum conservation equation:

$$\begin{matrix}\text{Initial angular} \\ \text{momentum of child–} \\ \text{merry-go-round system}\end{matrix} = \begin{matrix}\text{final momentum of child–} \\ \text{merry-go-round system}\end{matrix}$$

$$mrv_0 + 0 = mrv + I\omega$$

$$40 \text{ kg} \times 3 \text{ m} \times 6 \text{ m/s} + 0 = 40 \text{ kg} \times 3 \text{ m} \times v + 720 \text{ kg} \cdot \text{m}^2 \times \omega$$

We can relate the tangential velocity of the child and the angular velocity of the merry-go-round after the collision by Eq. (9.11b):

$$\begin{aligned} v &= r\omega \\ &= 3 \text{ m} \times \omega \end{aligned}$$

Substituting this in the angular momentum conservation equation, we have:

$$720 \text{ J} \cdot \text{s} + 0 = 360 \text{ kg} \cdot \text{m}^2 \times \omega + 720 \text{ kg} \cdot \text{m}^2 \times \omega$$

$$\therefore \omega = 0.667 \text{ rad/s}$$

The final velocity of the child on the merry-go-round is given by:

$$\begin{aligned} v &= r\omega \\ &= 3 \text{ m} \times 0.667 \text{ rad/s} \\ &= 2 \text{ m/s} \end{aligned}$$

Since the whole process takes place without a change in height, no gravitational potential energy term enters the energy conservation equation. With the help of Eqs. (7.2) and (9.17), we write the total energy conservation as follows:

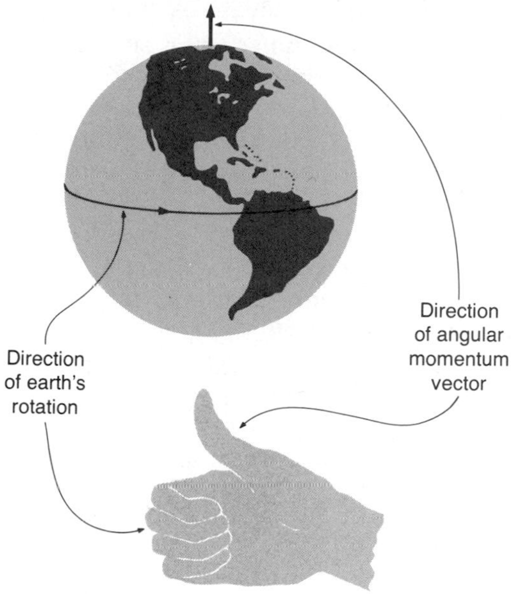

FIGURE 9.9 The determination of the direction of the angular momentum vector.

$$\begin{array}{c}\text{Initial kinetic energy}\\ \text{of child–merry-go-}\\ \text{round system}\end{array} = \begin{array}{c}\text{final kinetic energy}\\ \text{of child–merry-go-}\\ \text{round system}\end{array} + \text{energy dissipated}$$

$$\tfrac{1}{2}mv_0^2 + 0 = \tfrac{1}{2}mv^2 + \tfrac{1}{2}I\omega^2 + E_{\text{diss}}$$

$$\tfrac{1}{2} \times 40 \text{ kg} \times (6 \text{ m/s})^2 = \tfrac{1}{2} \times 40 \text{ kg} \times (2 \text{ m/s})^2 + \tfrac{1}{2} \times 720 \text{ kg} \cdot \text{m}^2 \times (0.667 \text{ rad/s})^2 + E_{\text{diss}}$$

$$720 \text{ J} = 80 \text{ J} + 160 \text{ J} + E_{\text{diss}}$$

$$\therefore E_{\text{diss}} = 480 \text{ J}$$ ■

In our treatment of linear kinematics and dynamics, we explicitly treated the vector nature of displacement, velocity, acceleration, force, and momentum. We have overlooked the vector nature of the corresponding rotational quantities because of our restriction to a fixed direction for the axis of rotation. In a sense, fixing the direction of the rotation axis is like restricting motion to one dimension in the linear case. One-dimensional kinematic and dynamic vectors point either in the positive or negative direction along a straight line; rotational kinematic and dynamic quantities about a fixed axis correspond to a clockwise or counterclockwise movement about the axis. This gives us the clue to assigning vector direction to the rotational quantities.

Let us discuss angular momentum, even though the same argument will apply to all the rotational quantities. The vector direction of angular momentum for an object rotating about a symmetry axis is along the axis of rotation. For example, the angular momentum vector of the earth points along the polar axis. To find the direction in which the vector points, we hold the right hand as shown in Fig. 9.9. If the fingers circle the thumb in the direction of the earth's rotation, the thumb points along the positive direction of the angular momentum vector. The angular momentum vector of the earth therefore points upward out of the North Pole.

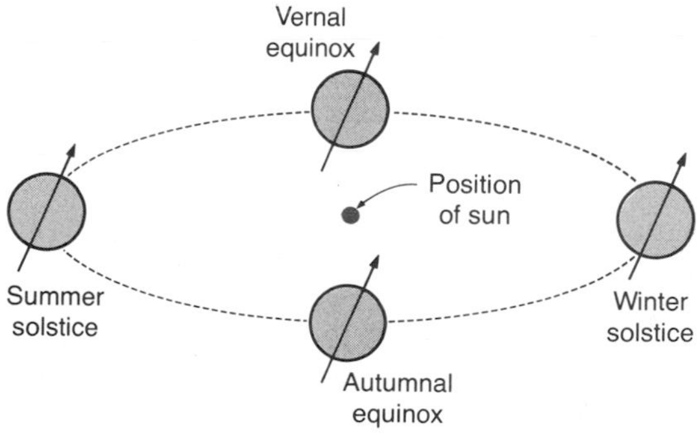

FIGURE 9.10 The fixed direction of the earth's angular momentum vector as it revolves about the sun gives rise to changing seasons (marked for the northern hemisphere).

The law of conservation of angular momentum requires that the *magnitude and direction* of the earth's angular momentum stay constant if the earth experiences zero net external torque. The regular recurrence of the seasons is evidence of the fixed direction of the earth's rotation axis. Figure 9.10 shows the earth in various positions in its orbit around the sun. At each position the angular momentum vector points toward the pole star. Note that this direction is not perpendicular to the plane of the orbit.

By now you are well aware that nothing is ever as simple as the ideal physical model would have us believe. The earth is not actually a rigid body sphere—in the first place we have all that water slopping around on the surface, and in the second place it is not quite spherical. These deviations from the rigid-body, spherical model permit the moon and the sun to exert small external torques on the earth. As a result, both the magnitude and direction of the earth's angular momentum change slowly. The slowing of the earth's rotation is increasing the length of the day by about 1 ms per century (see problem 52). The direction of the earth's polar axis is also shifting slowly in space, so that the present pole star will be about 55° away from the polar direction in 13 000 years. Change in direction of a rotation axis is called *precession,* and the slow shift in the direction of the earth's rotation axis is called the precession of the equinoxes. The existence of this effect was known to the ancient Babylonian astronomers, and its dynamic origin was explained by Isaac Newton. Many of the most beautiful examples of rotational dynamics involve precession, but their difficulty places them beyond the scope of this book.

KEY CONCEPTS

The basic concept of angular kinematics is **angular displacement.** See Fig. 9.1.

Average **angular velocity** is the angular displacement divided by the elapsed time. See Eq. (9.1).

Average **angular acceleration** is the change in angular velocity divided by the elapsed time. See Eq. (9.3).

The equations of rotational motion at constant angular acceleration have the same algebraic form as the equations of linear motion at constant acceleration with **rotational analogues** replacing the linear quantities. See Eqs. (9.4) and (9.15).

Tangential displacement, velocity, and acceleration are equal to the corresponding angular quantities multiplied by the radius of the circular arc. See Eq. (9.11).

Rotational dynamics about a fixed axis is described by equations analogous to those of linear dynamics. **Torque** is the rotational analogue of force, and **moment of inertia** is the rotational analogue of mass. See Eq. (9.13).

Work is torque multiplied by angular displacement. See Eq. (9.16).

Rotational kinetic energy and **angular momentum** are defined by relations analogous to the linear expressions. See Eq. (9.17) and Eq. (9.19).

The **angular momentum of a system is conserved** if the net external torque is zero. See the display on p. 245.

QUESTIONS FOR THOUGHT

1. Snow tires have a larger diameter than regular tires. How do snow tires affect the speedometer reading?

2. What are the directions and relative magnitudes of the instantaneous velocities for the following points on the rim of a rolling wheel:

  a. The point of contact between the wheel and the ground
  b. The point at the top of the wheel
  c. The two points (one on each side) that are halfway between the top of the rim and the point of contact.

3. Two solid spheres of the same mass are made from metals of different density. Which one has the larger moment of inertia about a diametral axis?

4. When a bowler releases a bowling ball, it usually skids for some distance and then begins to roll. Explain why these two phases of the ball's motion occur in this order.

5. Consider a massive rod balanced upright on a horizontal frictionless surface. If the balance is disturbed and the rod falls, what is the path of its center of gravity during the fall? How would the presence of friction between the foot of the rod and the horizontal surface modify your answer?

6. Helicopters frequently have a small propeller mounted on a transverse axis in addition to the large one that rotates about a vertical axis. Explain the reason for the second propeller.

7. An ice skater in a pirouette spins faster at the end of her movement than at the beginning. Explain how she accomplishes this.

8. A cat that is dropped upside down usually manages to land on its feet. Explain how the cat pulls off this apparent violation of angular momentum conservation.

PROBLEMS

A. Single-Substitution Problems

1. The wheels of an automobile are 55 cm in diameter. What is their angular velocity when the automobile is traveling at 90 km/h (25 m/s)? [9.1]

2. What is the tangential velocity of a point on the rim of a 12-in. phonograph record that is rotating at $33\frac{1}{3}$ rev/min? [9.1]

3. The blades of a small electric fan are 30 cm in diameter. Calculate the angular velocity in revolutions per minute if the peripheral speed of the tip of a blade is 50 m/s. [9.1]

4. Calculate the average angular acceleration of an electric motor that reaches a speed of 1400 rev/min in 3.8 s starting from rest. [9.1]

5. A motor accelerates from rest with an angular acceleration of 50 rad/s^2. What time interval is required for it to reach a speed of 800 rev/min? [9.1]

6. An automobile engine accelerates from rest to a speed of 2000 rev/min with constant angular acceleration in a time interval of 1.4 s. Through how many revolutions does it turn in this time interval? [9.1]

7. Two small metal balls of masses 400 g and 600 g are fixed to the opposite ends of a rod 120 cm long and of negligible mass. Calculate the moment of inertia about an axis perpendicular to the rod and passing through its center. [9.2]

8. Calculate the moment of inertia of a bowling ball about a diameter. The mass of the ball is 6.5 kg, and the radius 11.0 cm. [9.2]

9. Calculate the moment of inertia of a phonograph record (mass 0.15 kg and diameter 30.5 cm) about its axis. [9.2]

10. A flywheel 25 cm in radius mounted in frictionless bearings accelerates at 3.2 rad/s^2 when a constant force of 160 N is applied tangentially to the rim. Calculate the moment of inertia of the flywheel. [9.2]

11. A flywheel mounted in frictionless bearings has a moment of inertia of 2 kg·m^2. Calculate the angular acceleration when it is acted on by a torque of 60 N·m. [9.2]

12. How much work is done when a torque of 16 N·m turns a wheel through 20 rev? [9.3]

13. The rotating part of a small electric fan has a moment of inertia of 0.006 kg·m^2. At what angular velocity is the rotational kinetic energy equal to 0.25 J? [9.3]

14. Calculate the rotational kinetic energy of a child's top (moment of inertia 6×10^{-5} kg·m^2) that is spinning at 1400 rev/min. [9.3]

15. Calculate the required moment of inertia for a flywheel if it is to store 10^6 J of rotational kinetic energy when rotating at 1000 rev/min. [9.3]

16. Calculate the angular momentum of the following objects:

a. A cycle wheel ($I = 0.3$ kg·m^2) rotating at 250 rev/min
b. The earth ($I \approx 8 \times 10^{37}$ kg·m^2) about its axis [9.4]

B. Standard-Level Problems

17. An electric saw blade turning at 2200 rev/min comes to rest in 20 s with a uniform deceleration when the power is turned off. Calculate:

a. The angular deceleration
b. The number of revolutions it turns before stopping [9.1]

18. The power for an electric motor that is running at 500 rev/min is switched off. The motor slows down due to frictional losses in the bearings, and after 6 s its speed drops to 400 rev/min. If the angular deceleration remains constant, how many more revolutions does the motor make before stopping? [9.1]

19. An electric fan starts from rest and reaches its running speed of 300 rev/min with uniform angular acceleration after turning through 16 rev.

a. What is its angular acceleration?
b. How long does it take to reach its running speed? [9.1]

20. A stereo turntable starting from rest reaches its operating speed of $33\frac{1}{3}$ rev/min in 1 s after constant angular acceleration.

a. What is the angular acceleration?
b. Through what angle does the turntable turn in this 1-s interval? [9.1]

21. A flywheel accelerates with a constant angular acceleration of 4 rad/s^2. During a 4-s time interval it rotates through 35 rev. Calculate:

a. The angular velocity at the end of this 4-s time interval
b. The total number of revolutions turned from rest

22. A cabinetmaker's lathe is driven by a 1725-rev/min electric motor. A set of four pulleys on the motor shaft, which have diameters of 5 cm, 6.75 cm, 8.25 cm, and 10 cm, is opposite a set of pulleys on the drive shaft, which have diameters of 10 cm, 8.25 cm, 6.75 cm, and 5 cm, respectively. A belt is used to connect the opposing pulleys. A 3-in. diameter table leg is to be finished, and the recommended tool cutting speed is 1500 ft/min. Which set of pulleys should the craftsman select to be as close as possible to the recommended speed?

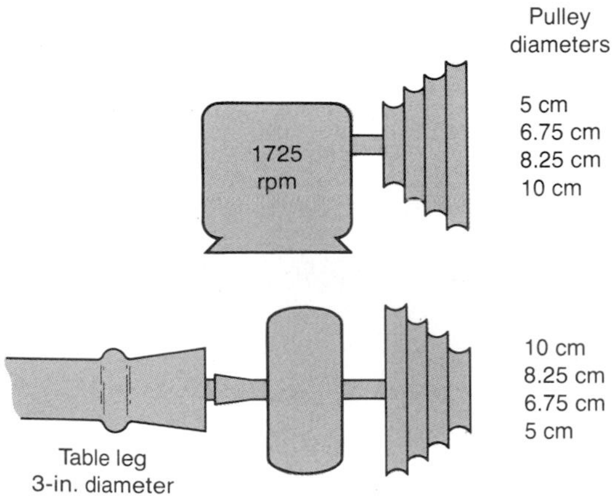

23. A wheel of diameter 0.8 m starts from rest and rolls without slipping along a level road with an acceleration of 1.8 m/s^2. Calculate:

a. The angular velocity of the wheel after 16 s
b. The number of revolutions the wheel turns in this time

24. A cyclist on a machine with 26-in. (66-cm) diameter wheels starts from rest and accelerates uniformly to a speed of 6 m/s in 10 s.

a. What is the angular acceleration of the wheels?
b. Through how many revolutions does each wheel turn?

25. The moment of inertia of a stereo turntable is 0.06 kg·m^2. What is the percentage increase in the moment of inertia when a record disk that is 1.3 mm thick and 30 cm in diameter is placed on the turntable? The relative density of the record is 1.45. [9.2]

26. Calculate the moment of inertia of a bicycle wheel regarded as being the sum of two contributions:

a. A hoop of mass 2.3 kg and radius 33 cm (the tires and the rim)
b. A disk of mass 0.4 kg and radius 33 cm (the spokes) [9.2]

27. A grindstone in the form of a uniform circular disk has a mass of 8 kg and a radius of 12 cm. It starts from rest with an angular acceleration of 2 rad/s^2. Calculate:

a. The moment of inertia of the grindstone
b. The torque required to produce the acceleration [9.2]

28. A stereo turntable has a moment of inertia of 8.5×10^{-3} kg·m^2. When the power is switched off, the turntable takes 40 s to come to rest from a speed of $33\frac{1}{3}$ rev/min. What is the average frictional torque in the bearings? [9.2]

29. A flywheel has a moment of inertia of 6 kg·m^2 and a radius of 30 cm. It is driven by a belt that supplies a constant force of 50 N tangentially to the rim of the wheel and accelerates the wheel uniformly from rest to 160 rev/min.

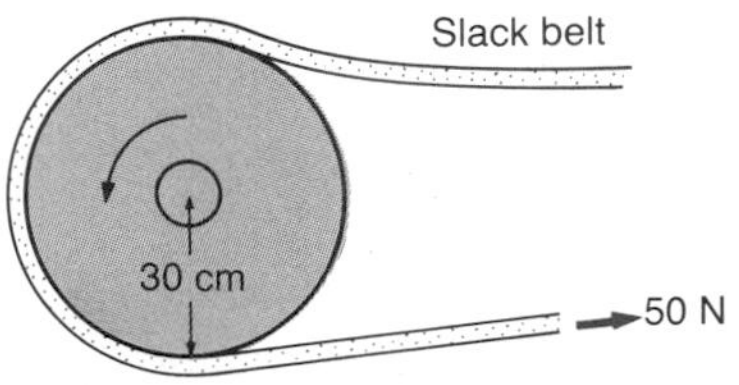

a. How long does this acceleration take?
b. What length of belt winds around the wheel during this time?

30. A wheel is mounted on an axle 10 cm in diameter that is free to rotate in frictionless bearings. A cord is wrapped around the axle and pulled with a constant force of 45 N. After the cord unwinds to a length of 4 m, the wheel is rotating at 160 rev/min. Find the moment of inertia of the wheel and axle.

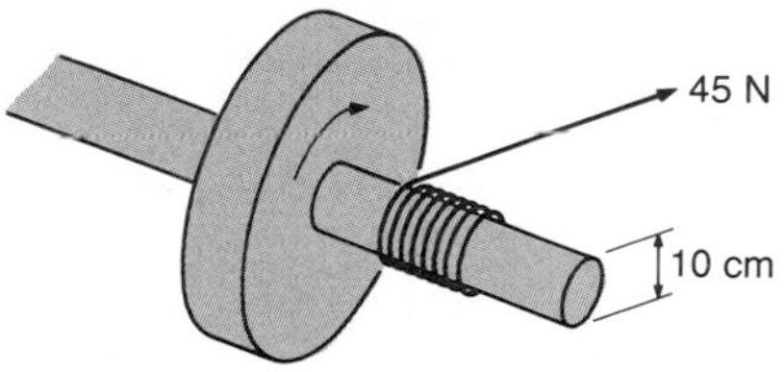

31. A flywheel is attached to a horizontal axle of radius 5 cm that is free to rotate in frictionless bearings. A mass of 10 kg hangs vertically on a cord wrapped around the axle. The moment of inertia of the wheel-axle unit is 0.4 kg·m^2. Calculate the angular acceleration of the wheel and the tension in the cord.

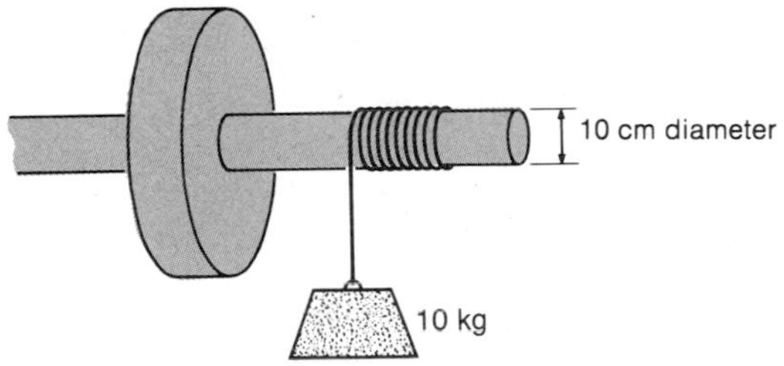

32. A bucket of water of mass 10 kg is attached to a rope that is wound on a uniform, cylindrical windlass 20 cm in radius and of mass 30 kg. The bucket is allowed to fall from rest to the water level 4 m lower in the well. Calculate:

a. The velocity of the bucket just before hitting the water

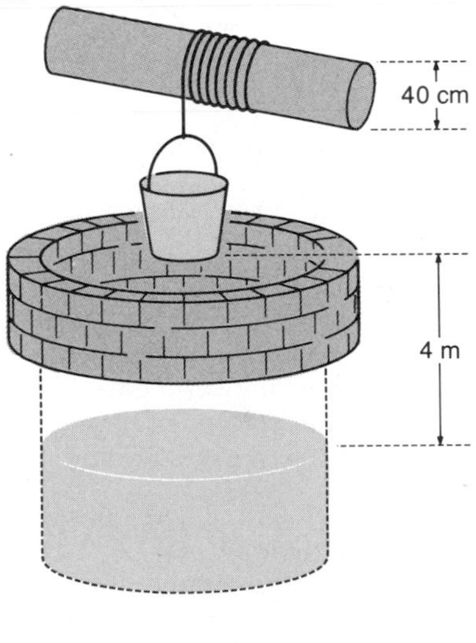

b. The angular acceleration of the drum as the bucket falls

c. The tension in the rope as the bucket falls (Ignore frictional torque in the windlass bearings.)

33. A flywheel turns in frictionless bearings on a horizontal axle of diameter 15 cm. An 8-kg mass is attached to a cord that is wrapped securely around the axle. The mass is released from rest and falls 2.4 m in 8 s. Calculate the angular acceleration and the moment of inertia of the flywheel and axle.

34. A flywheel turns in frictionless bearings on a horizontal axle of 10-cm diameter; the moment of inertia of the wheel and axle is 0.4 kg·m^2. A cord is wrapped securely around the axle and a mass attached to the cord falls freely. What value of the falling mass will cause an angular acceleration of 40 rad/s^2 in the flywheel?

35. The blade of a table saw is 25 cm in diameter and has a mass of 750 g. To calculate the moment of inertia, regard the saw as a uniform disk. How much kinetic energy is stored in the blade when it is rotating at 2100 rev/min? (The total rotational kinetic energy is larger because the blade is bolted to the rotating part of an electric motor.) [9.3]

36. A Frisbee has a complex shape, but it can be approximated by a hoop connected to a disk (see diagram). A certain Frisbee has a radius of 12 cm with 60 g of material in the rim and 40 g in the disk.

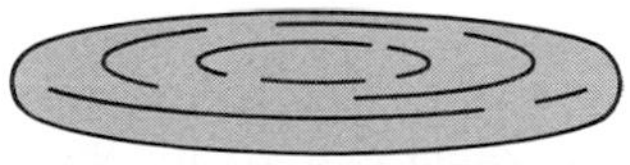

Approximate the moment of inertia of a Frisbee by adding the moments of inertia of two simple objects

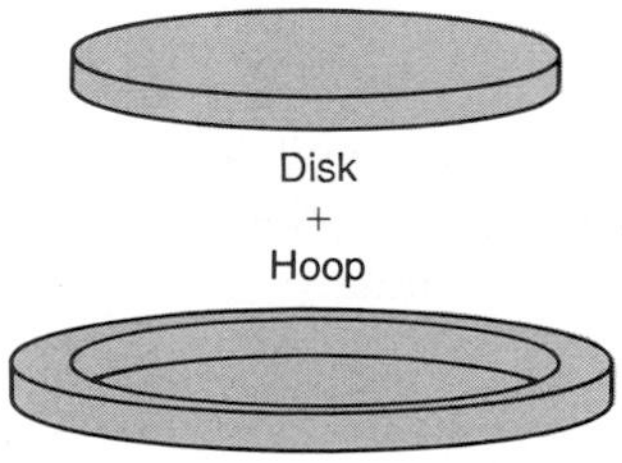

a. Calculate the moment of inertia about an axis through the center and perpendicular to the disk.

b. Calculate the rotational kinetic energy of the Frisbee when it is spinning at 400 rev/min. [9.3]

37. An automobile wheel has a mass of 18 kg, a radius of 35 cm, and a moment of inertia of 0.85 kg·m^2. How much kinetic energy is stored in the wheel when the automobile is traveling at 55 km/h? [9.3]

38. A disk of mass 8 kg and radius 25 cm rolls at constant velocity in a straight line. Calculate the velocity if the energy stored in the wheel is 750 J.

39. A flywheel of moment of inertia 100 $kg \cdot m^2$ is to be brought up to an angular velocity of 2000 rev/min in 5 min by an electric motor. What horsepower is required of the motor?

40. A stereo turntable of moment of inertia 0.02 $kg \cdot m^2$ accelerates uniformly to $33\frac{1}{3}$ rev/min in 1.2 s after being switched on. What power is required from the turntable motor?

41. It has been suggested that a small automobile might run on the energy stored in a rotating flywheel. Suppose that a half-ton flywheel could be made with a moment of inertia of 200 $kg \cdot m^2$ and brought up to a speed of 20 000 rev/min in an evacuated chamber. For how long a period could this flywheel deliver 10 hp before its speed fell to 5000 rev/min? (Ignore frictional losses.)

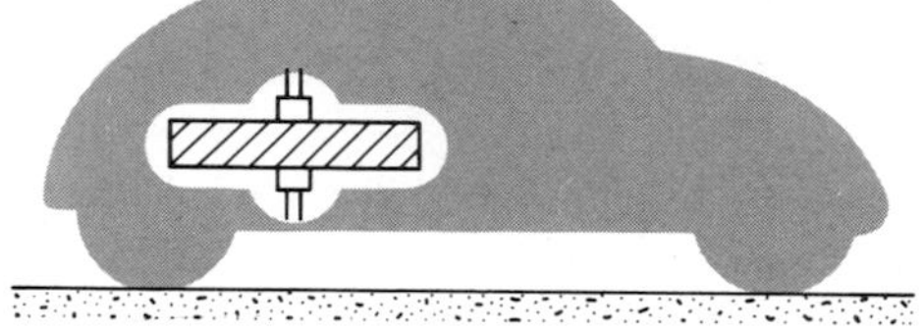

42. A press to punch holes in sheet metal operates in the following manner: A 1-hp electric motor drives a large flywheel to a speed of 800 rev/min. The energy stored in the wheel is used for the punching operation, which requires 6000 J of work over a 0.4-s interval. As a result of supplying this energy, the flywheel slows to 700 rev/min, and the driving motor speeds it up for the next operation. Calculate:

a. The average horsepower required for the punching operation
b. The moment of inertia of the flywheel
c. The number of punching operations per hour that this unit can handle

43. The ocean liner Conte de Savoia had three gyroscopic stabilizers. Each consisted of a giant flywheel of moment of inertia 4×10^5 $kg \cdot m^2$. The operating speed for the flywheels was 800 rev/min.

a. How long would it take a 100-hp motor to bring each flywheel up to speed starting from rest?
b. Rival shipping companies sought to frighten passengers by claiming that the energy stored in the ship's stabilizing gyroscopes was sufficient to lift the ship far out of the water if it could be used for

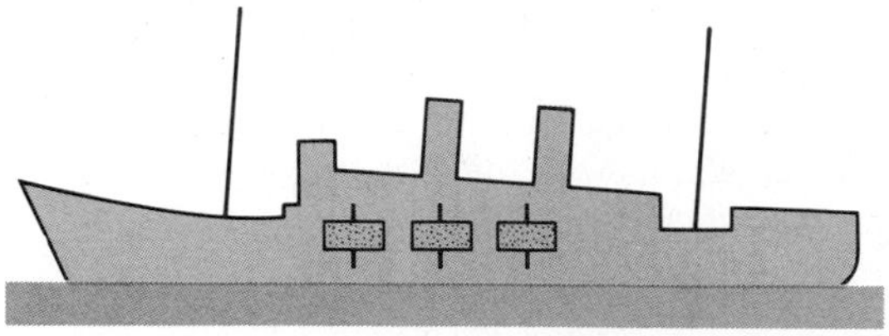

this purpose. The liner was about 48 000 metric tons. What distance of lift did the jealous rivals claim? (The company agreed with the calculations, pointing out that the energy stored in the boilers would be sufficient to lift the ship over a mile high.)

44. An electric motor running at 1800 rev/min is delivering 4 hp to a belt on its driving pulley. The moment of inertia of the rotating part is 0.5 $kg \cdot m^2$.

a. What torque is the motor delivering?
b. If the power fails, how long does it take the motor to stop? (Assume a constant retarding torque equal in value to the output torque of the motor when the power is on.)

45. A small grindstone in the form of a uniform circular disk is 12 cm in diameter. It is being driven at 1600 rev/min when a tool is pressed against the rim with a normal force of 18 N. The coefficient of kinetic friction between the tool and the stone is 0.85. Calculate:

a. The torque exerted by the tool on the grindstone
b. The extra power that the motor must supply to maintain the speed of 1600 rev/min when the tool is pressed against the wheel (Neglect frictional losses in the bearings.)

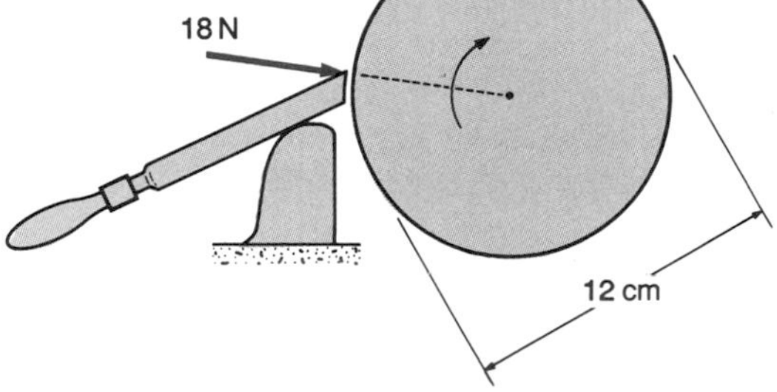

46. A park merry-go-round is in the form of a uniform circular disk of mass 160 kg and radius 2.5 m. It is rotating about a friction-free vertical axis at a speed of 1 rev in 3 s when a 30-kg child quickly sits at a point on the edge. Calculate the percentage decrease in the angular speed of the merry-go-round. [9.4]

47. Two flywheels, having moments of inertia of 4 $kg \cdot m^2$ and 3 $kg \cdot m^2$, are mounted on horizontal, frictionless shafts. The shafts are in line with each other and can be coupled together by a friction clutch. With the clutch disengaged, the first flywheel is brought up to a speed of 800 rev/min while the other remains at rest. The clutch is now engaged. Calculate:

a. The subsequent angular velocity of the wheels
b. The amount of energy dissipated in the clutch engagement process [9.4]

48. In an experiment designed to test conservation of angular momentum, a cannon of mass 600 kg is lashed firmly to the outer edge of a larger horizontal turntable in the form of a uniform circular disk 5 m in diameter. The disk has a mass of 2000 kg and is free to rotate about a central frictionless bearing. The cannon fires an 8-kg shot with a velocity of 240 m/s tangential to the rim of the disk. What value do you expect for the angular velocity of the turntable?

49. A uniform iron disk of radius 30 cm is rotating at 800 rev/min about a vertical axis in frictionless bearings. A permanent magnet of mass 1 kg is dropped onto the disk and sticks 12 cm from the rotation axis. The angular velocity reduces to 680 rev/min. Calculate the mass of the iron disk.

C. Advanced-Level Problems

50. A pole vaulter's pole is 5 m long and has a mass of 4.5 kg. The vaulter releases the pole in an almost vertical position, and it topples over, pivoting about the lower end. What is the velocity of the top end of the pole when it strikes the ground?

51. A cyclist and his machine together have a mass of 100 kg. Each wheel has a mass of 3 kg and a radius of 33 cm. In order to calculate the moment of inertia of the wheels, assume that all of the mass is concentrated in the rim.

a. What percentage of the total kinetic energy is stored as rotational kinetic energy of the wheels at a speed of 15 m/s?
b. Is it accurate to ignore the rotational kinetic energy of the wheels in problem 30 of Chapter 7?

52. Consider the earth as a uniform sphere of radius 6.38×10^6 m and average relative density 5.50.

a. Calculate the moment of inertia of the earth about an axis.

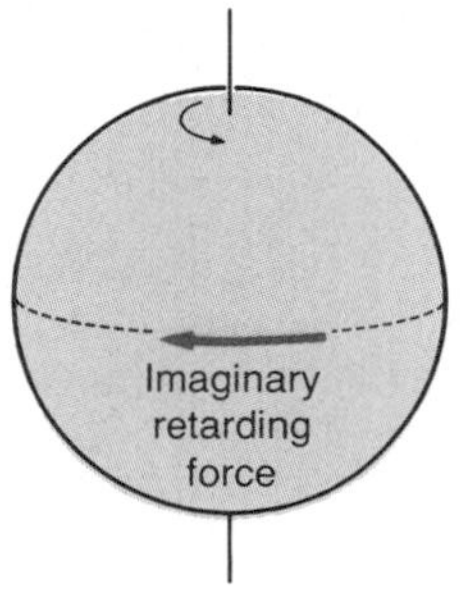

b. An increase in the length of the day by 1 ms per century can be shown to be an angular deceleration of 2.67×10^{-22} rad/s^2. What frictional torque would cause this change?
c. If the frictional torque in question could be supplied by a single force on the earth acting tangentially at the equator, how large would this force be? (The tides are believed to be responsible for a frictional retarding torque on the earth of approximately the magnitude calculated in this problem.)

53. A spherical iron cannonball of mass 18 kg (relative density 7.86) rolls without sliding down a 50-m slope inclined at 10° to the horizontal. Calculate:

a. The velocity of the ball at the bottom of the slope
b. The amount of work needed to stop the ball after it reaches the bottom of the slope

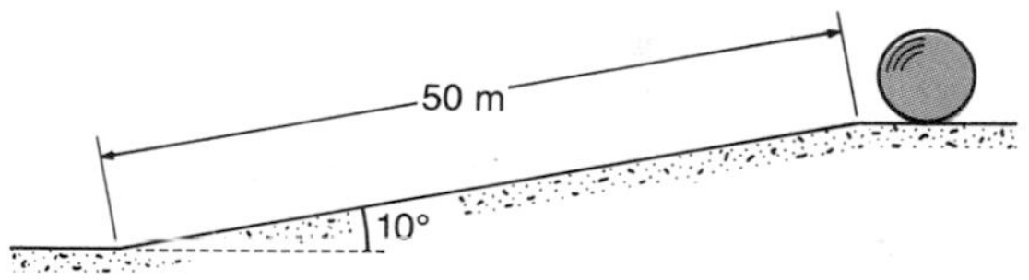

54. A colonial corn mill is driven by water power. The water wheel is 3 m in radius and receives water from the mill race at 5000 gal/min. Assume that the wheel collects the water at the top point of its motion and spills it out after turning 90°. Assume also that the kinetic energy of the water coming from the mill race does not contribute to the power of the mill and that spillage is negligible.

a. What is the horsepower of the mill under the conditions stated?
b. If the wheel turns 1 rev in every 12 s, what torque does it exert?

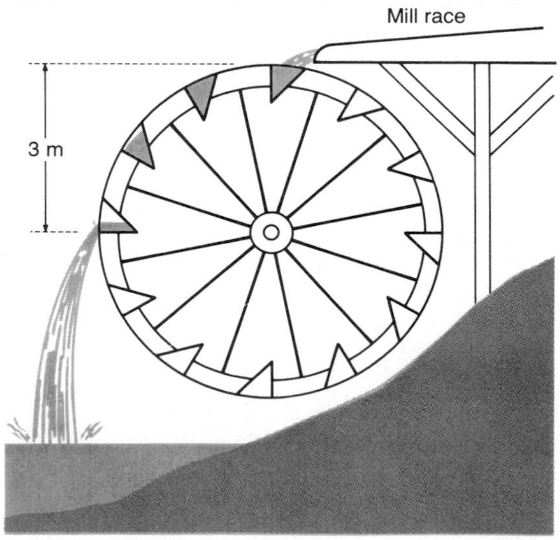

10

FLUIDS

In previous chapters we concentrated on two simple ideal models—namely, the point particle and the rigid body. However, a large class of substances, called fluids, exists that are not even remotely like a point particle or a rigid body. The water in a river, the blood in our bodies, and the air of the earth's atmosphere are well-known examples of fluids. The common characteristic of these substances is the ability to flow, and for this reason they are called fluids. Because of this ability, a fluid does not have any definite size or shape unless it is held in a container. Of the three substances mentioned, water and blood are liquids, and have a free horizontal surface in an open container. Air is a gas and must be contained in a closed vessel since a gas shows no free surface.

Fluids share at least one feature with point masses and rigid bodies—they possess mass. Thus, we can use Newton's laws to describe fluid motion. We begin with the study of fluids at rest, which is called hydrostatics, and its primary concept is hydrostatic pressure. We then develop a simplified model of a fluid in motion and derive a form of the energy equation that is particularly useful for describing many fluid flow phenomena.

10.1 Fluid Pressure

Pressure is a basic concept in the physics of fluids in somewhat the same way that force is in the dynamics of point particles and rigid bodies. The concepts of air pressure in an automobile tire, blood pressure, atmospheric pressure, and pressure at the bottom of a swimming pool are all familiar in everyday life. There are various reasons for the existence of the pressures mentioned. The air in an inflated automobile tire is at an elevated pressure because a pump forces the air into a confined space surrounded by strong walls. The pressure of the blood is primarily due to the continual pumping action of the heart. The pressure of the atmosphere is due to the weight of the air above, and the additional pressure at the bottom of a swimming pool is due to the weight of water.

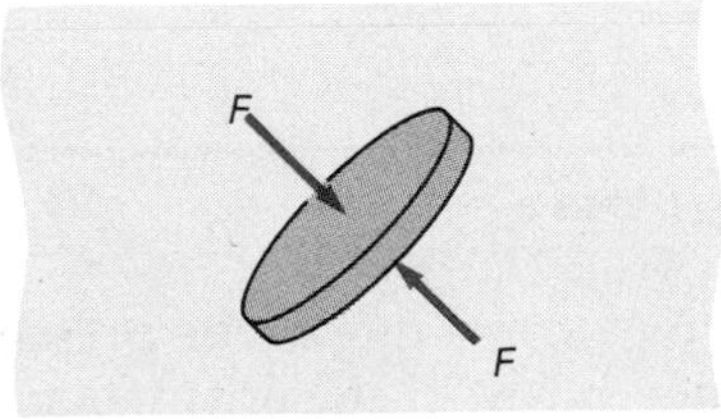

FIGURE 10.1 Determination of pressure from the force exerted on the face of a thin, hollow disk.

Whatever the reason for the existence of pressure, we need a definition that suitably covers all the particular cases. In order to develop some physical feel for the pressure, let us consider a hollow disk introduced into a fluid (Fig. 10.1). Experience tells us that the pressure of the fluid causes forces that tend to crush the disk faces together, as shown in the diagram. We define the **pressure** as the magnitude of the crushing force divided by the area of the disk face.

Definition of Pressure

$$\text{Pressure} = \frac{\text{magnitude of fluid force perpendicular to an imaginary area}}{\text{area}}$$

$$p = \frac{F_\perp}{A} \qquad (10.1)$$

SI unit of pressure: newton per square meter
New name: pascal
Abbreviation: Pa

Of course, it is not necessary to have a real disk in the fluid in order for the pressure to exist; the disk simply assists our imagination. The pascal is so-called in honor of the French mathematician and physicist Blaise Pascal (1623–1662). In the following section we will describe one of Pascal's more spectacular experiments on fluids. The pascal is an extremely small unit [the conversion tables show 6894 Pa/(lb/in.^{2})], and the kilopascal (kPa) is a frequently used, practical unit. Many other practical units for pressure exist, and we will explain their origins as they arise in the discussion.

Pressure is by definition a scalar quantity (it is the magnitude of the fluid force perpendicular to an area divided by the area). Let us imagine that we slowly change the orientation of the disk in Fig. 10.1 while keeping its center at the same place in the fluid. The force on the face of the disk is certainly a vector; this force points perpendicularly inward to the disk face, and its magnitude does not change as we alter the disk orientation. In fact, the disk is only a help to our imagination; if we do not imagine it to be there, then the physical condition of the fluid is described by the scalar pressure. We express this important notion as a basic law of fluid pressure.

First Basic Law of Fluid Pressure

The magnitude of the force on an area due to the pressure at any point in a fluid is independent of the orientation of the area.

This law is often expressed somewhat loosely as: "The pressure at any point in a fluid acts equally in all directions." Of course, pressure is a scalar quantity that does not have a direction, and we must interpret the loose statement as indicated. Experience in swimming underwater supports this law. Both eardrums feel the pressure, and the situation is not altered by twisting the head around. If a swimmer places his head sideways, both the upward and downward facing ears experience the effect of water pressure. In fact, the downward facing ear experiences a slightly greater pressure, since it is further below the free surface of the water.

Let us now calculate the pressure below the free surface of a liquid. We proceed by first calculating the pressure due to the liquid and then considering the effect of the atmospheric pressure (or other pressure) at the free surface. Consider the situation shown in Fig. 10.2, in which a cyclindrical flask of uniform cross-sectional area A contains a column of liquid of density ρ and height h. The volume of liquid in the flask is given by:

$$V = hA$$

We now use Eq. (1.1) to calculate the total mass of the liquid:

$$\begin{aligned} m &= \rho V \\ &= \rho h A \end{aligned}$$

The weight of the liquid is given by Eq. (4.2):

$$\begin{aligned} w &= mg \\ &= \rho h A g \end{aligned}$$

The base of the flask supports the weight of the liquid. We can use the definition of Eq. (10.1) to calculate the pressure on the base of the flask:

$$\begin{aligned} p &= \frac{F_{\perp}}{A} \\ &= \frac{\rho h A g}{A} \\ &= \rho h g \end{aligned}$$

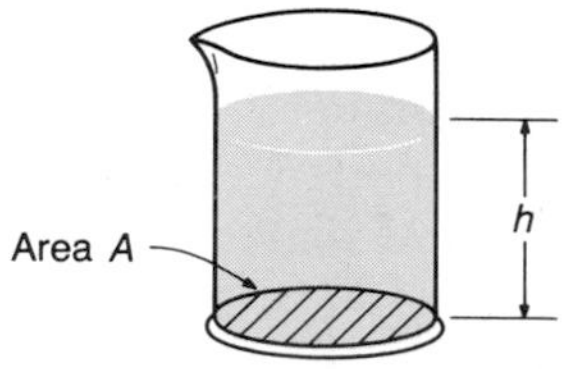

FIGURE 10.2 Diagram to illustrate the calculation of pressure due to a column of liquid.

This equation turns out to be true under far less restrictive conditions than those quoted in the derivation. It correctly calculates the pressure difference between any two points in a fluid of uniform density ρ and vertical height separation h. Because of the great utility of this result, we set it in display.

■ Formula for Pressure Difference Within a Uniformly Dense Fluid

$$\text{Pressure difference} = \text{fluid density} \times \text{vertical height separation} \times \text{acceleration of gravity}$$

$$p = \rho h g \qquad \textbf{(10.2)}$$

We should emphasize that the pressure calculated by Eq. (10.2) is the pressure due to the fluid column alone. If the flask of Fig. 10.2 is open to the atmosphere, the pressure at the free surface of the liquid is the pressure of the atmosphere. The total pressure at any point in the liquid is the pressure of the liquid column plus the atmospheric pressure. If p_a is the atmospheric pressure, the total pressure at a depth h below the free surface is given by:

$$p = p_a + \rho h g \qquad \textbf{(10.3)}$$

The pressure of Eq. (10.3) is called the **absolute pressure,** as distinguished from the pressure of Eq. (10.2), which is called the **gauge pressure.** Most pressure gauges read the difference between the absolute pressure and the pressure of the surrounding atmosphere, and this accounts for the nomenclature.

EXAMPLE 10.1

Calculate the gauge pressure at the bottom of a swimming pool 3 m deep.

SOLUTION The gauge pressure is given by Eq. (10.2):

$$\begin{aligned} p &= \rho h g \\ &= 10^3 \text{ kg/m}^3 \times 3 \text{ m} \times 9.81 \text{ m/s}^2 \\ &= 29.4 \text{ kPa} \end{aligned}$$

■

The deep sea submersible *Trieste II* has an immensely strong steel hull to withstand the pressures of deep dives. The pressure increase is about 1 atm for each 10 m of the dive. (U.S. Navy Photograph)

The human ear is very sensitive to rapid pressure changes. If the pressure in the middle ear does not have time to adjust, any exterior change of pressure results in a force on the eardrum that causes discomfort or even pain. A swimmer diving rapidly to the bottom of the pool of Example 10.1 experiences a pressure differential on the eardrum of the amount calculated.

Our example concerned the calculation of gauge pressure below the free surface of a liquid. But Eq. (10.2) gives the pressure difference between points at different heights in any fluid of uniform density. We can use it to find the pressure difference between two points in a liquid or between two points in a gas. A gas has no free surface, but it must have uniform density over the height range in question for Eq. (10.2) to be applicable. Let us illustrate by an example.

EXAMPLE 10.2

The density of air is given in Table 1.4 as 1.205 kg/m^3 under certain conditions of temperature and pressure. Calculate the air pressure difference between a point at ground level and a point 5 m higher.

SOLUTION Since the air density does not change substantially over a height of 5 m, we are justified in using Eq. (10.2) to calculate the pressure difference:

$$\begin{aligned} p &= \rho h g \\ &= 1.205\ \text{kg/m}^3 \times 5\ \text{m} \times 9.81\ \text{m/s}^2 \\ &= 59.1\ \text{Pa} \end{aligned}$$

■

Note that we cannot use this approach to calculate the air pressure difference over a large height interval, since Eq. (10.2) holds true only for a fluid of uniform density. The atmosphere is not uniformly dense, mainly because the lower air layers are compressed by the weight of the air above them.

Part of our task in this chapter is to develop an ideal fluid model. The examples of this section illustrate its first characteristic, namely, *uniform density*. Under most circumstances liquids are uniformly dense, but large bodies of gas may not be so. We will add other characteristics of the ideal fluid as the physics of the situation requires.

10.2 Pascal's Principle

The simple method for calculating fluid pressure explained in the previous section has a long history. Pascal first investigated the problem of calculating the pressure due to liquid in a container of nonuniform cross section. Pascal lived in a wine-producing area in central France. The people of the region were very familiar with large, strong wine-storage casks and were therefore most impressed with his experimental approach. He inserted a long glass tube vertically into the top of a wine cask filled with water (Fig. 10.3). The joint between the tube and the cask was securely sealed. He then added water to the glass tube, and the cask burst when the water level in the tube was about 12 m. This is an extraordinary outcome. A large cask contains several hundred liters of water, and it can be handled without any danger of the cask bursting. Yet adding several more liters to the glass tube destroyed the strong cask. We need an explanation.

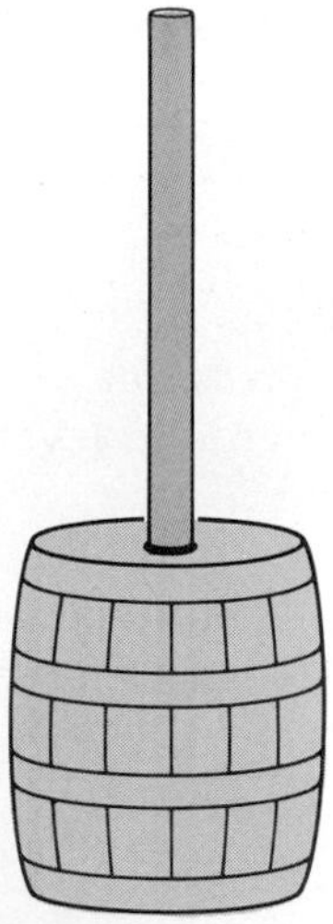

FIGURE 10.3 Pascal's experiment illustrating the pressure of a liquid column.

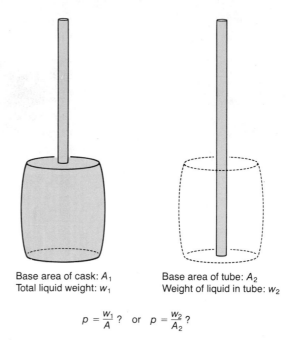

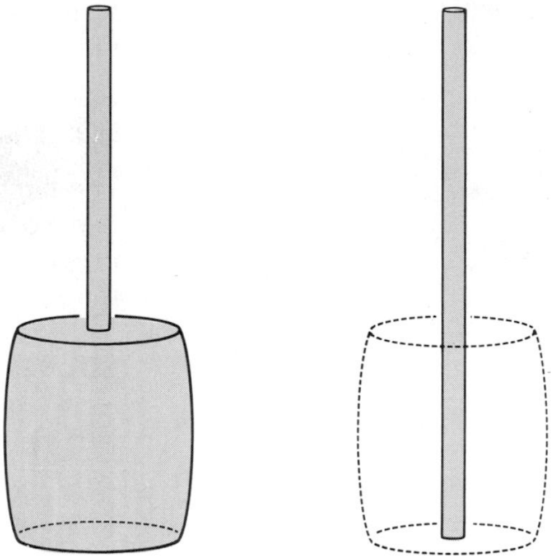

FIGURE 10.4 How should we calculate the pressure in Pascal's cask experiment?

Two apparently reasonable methods exist for calculating the water pressure near the base of Pascal's cask. In the first method, illustrated on the left of Fig. 10.4, we calculate the pressure by dividing the total weight of liquid in the cask and the glass tube by the area of the base of the cask. However, Eq. (10.2) implies that we should calculate the weight of a cylinder of liquid of the same cross section as the glass tube and divide by the cross-sectional area of the tube, which is illustrated on the right in Fig. 10.4. Since these two procedures lead to different answers, one of them must be wrong. The key to the situation is contained in the famous **Pascal's principle,** which Pascal proposed to explain his experimental results.

■ Second Basic Law of Fluid Pressure—Pascal's Principle

> A change of pressure exerted at any point in a body of fluid at rest is transmitted undiminished to every point in the fluid.

To see how this principle helps with the problem in hand, consider its application to a small-scale model of Pascal's cask.

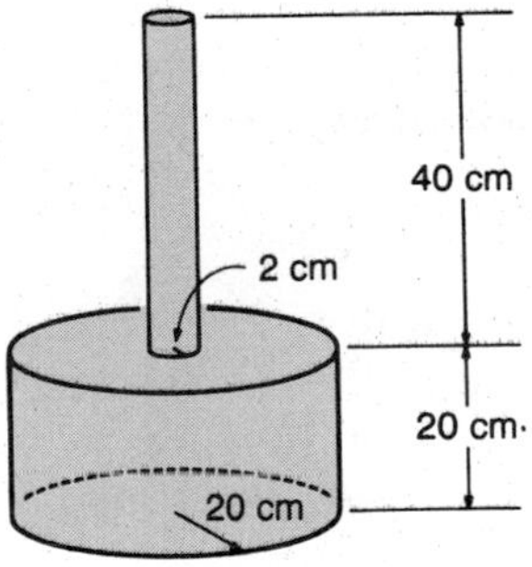

EXAMPLE 10.3

A nonuniform flask consists of a lower portion that is a circular cylinder 20 cm in radius and 20 cm high. The upper portion is also a circular cylinder, which is 2 cm in radius and 40 cm high. The flask is filled with water. Calculate:

a. The total weight of water in the flask

b. The gauge pressure of the water at the bottom of the flask

c. The force exerted on the base of the flask due to the gauge pressure

SOLUTION

a. We calculate the mass of water in the flask by multiplying its volume by the density of water:

$$\text{Volume of flask} = \begin{matrix}\text{volume of}\\ \text{lower cylinder}\end{matrix} + \begin{matrix}\text{volume of}\\ \text{upper cylinder}\end{matrix}$$

$$= \pi \times (0.20\text{ m})^2 \times 0.20\text{ m} + \pi \times (0.02\text{ m})^2 \times 0.40\text{ m}$$

$$= 2.564 \times 10^{-2}\text{ m}^3$$

$$\text{Mass of water} = \text{volume} \times \text{density}$$

$$= 2.564 \times 10^{-2}\text{ m}^3 \times 10^3\text{ kg/m}^3$$

$$= 25.64\text{ kg}$$

The weight of the water is given by Eq. (4.2):

$$w = mg$$

$$= 25.64\text{ kg} \times 9.81\text{ m/s}^2$$

$$= 251.5\text{ N}$$

b. To calculate the gauge pressure at the base of the uniform upper cylinder of water, we use Eq. (10.2):

$$p = \rho h g$$

$$= 10^3\text{ kg/m}^3 \times 0.40\text{ m} \times 9.81\text{ m/s}^2$$

$$= 3.92 \times 10^3\text{ Pa}$$

The pressure difference between the top and bottom of the uniform lower cylinder is also given by Eq. (10.2):

$$\begin{aligned} p &= \rho h g \\ &= 10^3\ \text{kg/m}^3 \times 0.20\ \text{m} \times 9.81\ \text{m/s}^2 \\ &= 1.96 \times 10^3\ \text{Pa} \end{aligned}$$

By Pascal's principle, the pressure due to the upper cylinder is transmitted undiminished throughout the water in the lower cylinder. The gauge pressure at the base of the lower cylinder is just the sum of the two pressures calculated above, namely, 5.88×10^3 Pa.

c. To calculate the force on the bottom of the flask due to this gauge pressure, we use Eq. (10.1):

$$\begin{aligned} p &= \frac{F_\perp}{A} \\ \therefore F_\perp &= pA \\ &= 5.88 \times 10^3\ \text{Pa} \times \pi \times (0.20\ \text{m})^2 \\ &= 739\ \text{N} \end{aligned}$$

The total force on the bottom of the flask due to the water is almost 3 times the weight of water in the flask. This can only mean that the presence of the water causes a downward force greater than its weight. The pressure due to the water in the neck of the flask acts at all points in the water in the lower part of the flask and in all directions. In particular, the water exerts an upward force on the upper surface of the large lower cylinder, which is balanced by an equal and opposite one downward on the water. As a result, the force on the base of the flask exceeds the weight of the water. ■

Returning to Fig. 10.4, we see that Pascal's law specifies the second method as the correct way to calculate the pressure in the flask. This means that the application of Eq. (10.2) is not restricted to flasks of uniform cross section; the equation gives the correct pressure for a containing vessel of any shape.

In the previous section we distinguished absolute pressure from gauge pressure by appealing to physical intuition. Pascal's law provides a more complete understanding. Consider, for example, the calculation of absolute pressure at the bottom of a swimming pool. We first calculate the pressure due to the water column that extends from the bottom of the pool up to the free surface of the water, which is the gauge pressure at the bottom of the pool. We then simply add the atmospheric pressure to calculate the absolute pressure at the bottom of the pool. It is correct to add the atmospheric pressure, since Pascal's law tells us that this pressure (which exists at the water surface) is transmitted undiminished to every point of the water in the pool.

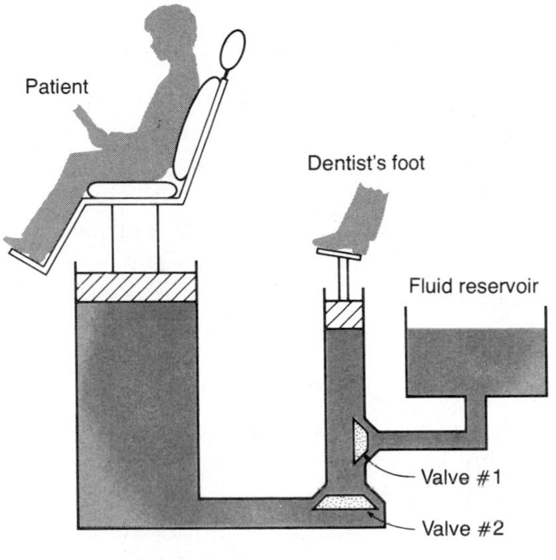

FIGURE 10.5 Operation of a hydraulic press.

We can also use Pascal's principle to explain the operation of the hydraulic press, which is used in automobile hoists, industrial stamping operations, and dentists' chairs. The essential elements of the hydraulic press used in a dentist's chair are shown in Fig. 10.5. Two uniform cylinders of different cross-sectional areas are joined at their bases and have close-fitting pistons in their upper portions. The whole system is filled with liquid that is replenished by a reservoir-and-valve system. When the dentist presses down on the small piston, the force from his foot causes a pressure in the liquid that closes valve #1 to the supply reservoir, opens valve #2, and raises the patient, who is seated above the larger piston. When the dentist removes his foot, the weight of the chair and patient causes a pressure in the liquid that closes valve #2. The small piston is drawn up the cylinder by a spring, and valve #1 opens to allow liquid from the supply reservoir to refill the small cylinder. The valves and the reservoir simply keep the system full of liquid. The essential physics of the operation follow from Pascal's principle. The pressure generated in the liquid by a relatively small force on the smaller piston is transmitted undiminished to the larger piston, where it causes a relatively large upward force.

EXAMPLE 10.4

An old-fashioned dentist's chair uses a hydraulic press in which the radius of the smaller piston is 3 cm. The dentist wishes to obtain a total upward force on the larger piston of 250 lb (1112 N) with a 10-lb (44.5-N) force on the smaller one. Neglecting frictional losses, what radius is required for the larger piston?

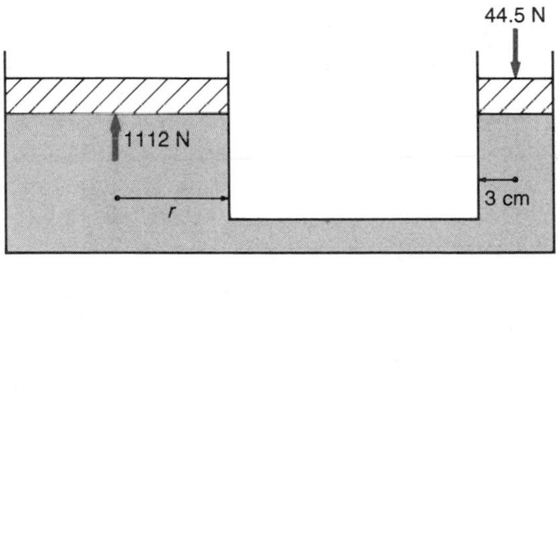

SOLUTION When the dentist pushes down on the smaller piston, he causes a pressure in the liquid given by Eq. (10.1):

$$p = \frac{F_\perp}{A}$$

$$= \frac{44.5\text{ N}}{\pi \times (0.03\text{ m})^2}$$

$$= 1.574 \times 10^4\text{ Pa}$$

By Pascal's principle this pressure is transmitted undiminished to the lower surface of the larger piston. Let r be the radius of the larger piston. The relation between the force on this piston and the pressure in the liquid is again given by Eq. (10.1):

$$p = \frac{F_\perp}{A}$$

$$1.574 \times 10^4\text{ Pa} = \frac{1112\text{ N}}{\pi r^2}$$

$$\therefore r = 0.150\text{ m}$$

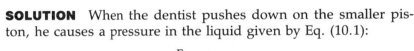

■

10.3 Pressure Gauges and Pressure Units

The measurement of pressure is important in many everyday situations as well as in scientific work, and a wide variety of instruments are available for this purpose. The simplest form of pressure gauge consists of a U-tube containing a liquid, as shown in Fig. 10.6. The top ends of the U-tube are connected to vessels #1 and #2, which contain fluids at different pressures. The liquid in the U-tube is frequently mercury, and the fluids in vessels #1 and #2 are usually gases. This type of pressure gauge is called a *manometer*, and it directly measures the pressure difference between its two vessels. Let ρ be the density of the liquid in the U-tube, and p_1 and p_2 the pressures in vessels #1 and #2, respectively. Let h_1 be the height of the liquid column in the arm of the U-tube connected to vessel #1, and h_2 the corresponding quantity on the other side. Let the absolute pressure in the liquid at the bottom of the U-tube be p_0. Then, using Eq. (10.2) for the liquid column in each arm of the U-tube in turn, we have:

$$p_0 = p_1 + \rho h_1 g$$

$$p_0 = p_2 + \rho h_2 g$$

Subtracting these equations gives:

$$\begin{aligned} p_2 - p_1 &= \rho(h_1 - h_2)g \\ &= \rho h g \end{aligned} \qquad \textbf{(10.4)}$$

The difference in height between the liquid levels is a direct measure of the pressure difference between the two vessels. If one arm of the manometer is open to the atmosphere, the height difference between the liquid levels is a direct measure of the gauge pressure in the vessel connected to the other arm. Mercury is often used in manometers because its large density leads to reasonably small height differences of the fluid levels; consequently, pressure difference is often measured by the height difference between mercury columns. A convenient unit of pressure for work involving manometers is based on the height difference expressed in millimeters of mercury. The pressure difference measured by a column of mercury 1 mm high is called 1 torr in honor of the Italian physicist Evangelista Torricelli (1608–1647). The conversion from the torr to the SI unit is given by 1 torr = 133.3 Pa.

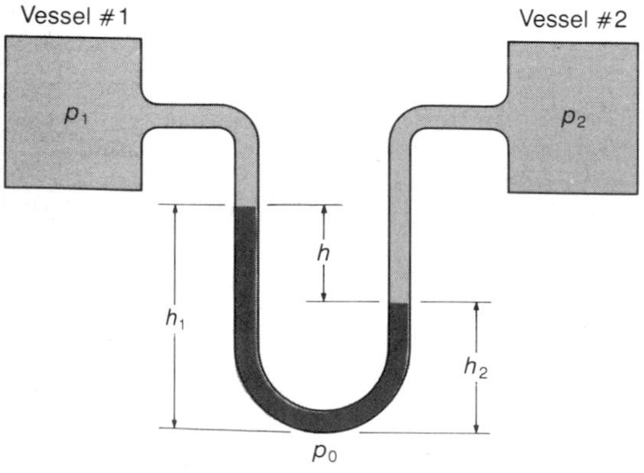

FIGURE 10.6 A manometer.

The torr is commonly used for measuring blood pressure, and the usage originates from the use of mercury in glass manometers to measure the pressure. The heart is a pump that circulates blood through the blood vessel system of the body. The pumping action of the heart results from successive contractions and relaxations of the heart muscles. As a result, the pressure of the blood varies rhythmically from a high value when the heart contracts (called

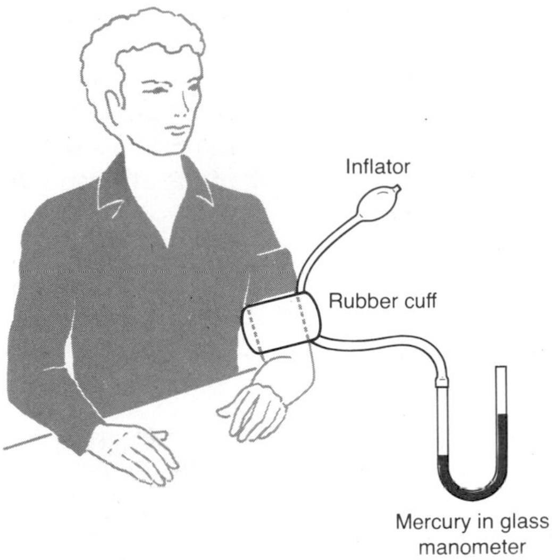

FIGURE 10.7 The measurement of blood pressure.

the systolic pressure) to a low value when the heart relaxes (called the diastolic pressure). The systolic and diastolic absolute pressures are both slightly in excess of atmospheric pressure. In order to take the blood pressure, the physician fastens an inflatable rubber cuff around the patient's upper arm (Fig. 10.7). A small inflator raises the air pressure in the rubber cuff until all blood supply to the lower arm ceases. Due to leakage the air pressure in the cuff slowly falls. By listening with a stethoscope, the physician can determine the point at which blood flow just recommences at maximum pressure. This corresponds to the systolic pressure, which the manometer records as gauge pressure directly in torr. As the pressure falls still further, a point is reached at which the blood flow continues even when the heart relaxes. At this point the manometer reads the diastolic pressure. Blood pressure varies with age and physical condition. In young adults a normal systolic pressure is about 120 torr, and a normal diastolic pressure is about 80 torr. To ensure that the measured pressure is the same as the pressure in the aorta, blood pressure is measured while the patient's elbow is at the same height as the heart.

Because of the importance of the earth's atmosphere for human existence, it is not surprising to find units based on atmospheric pressure. The pressure of the atmosphere continually changes at all points on the earth's surface, and it is an important piece of meteorological data. In spite of its variability, the pressure of the atmosphere is the basis for a unit of pressure. A *normal atmosphere* (abbreviation "atm") is defined as the pressure that supports a 760-mm column of mercury. The conversions of units are:

$$1 \text{ atm} = 760 \text{ torr (by definition)}$$
$$= 1.013 \times 10^5 \text{ Pa}$$

Noting that 1 atm equals approximately 100 kPa gives us a feel for the size of this unit.

Another common unit of pressure, used especially in meteorology, is the *bar*, which is defined as exactly 10^5 Pa. The bar is less than a normal atmosphere by a little more than 1%.

An instrument that measures atmospheric pressure directly is called a barometer. A simple form of barometer consists of a long glass tube closed at one end and filled with mercury. The tube is inverted so that its open end is below the free surface of mercury in a bowl (Fig. 10.8). The mercury falls in the tube, leaving a space between the top of the mercury column and the sealed end of the tube. This space is a vacuum, and the absolute pressure at the top of the mercury column is zero, while the pressure at the free surface of mercury in the bowl is atmospheric pressure. Thus, the height of the mercury column in millimeters gives the atmospheric pressure directly in torr.

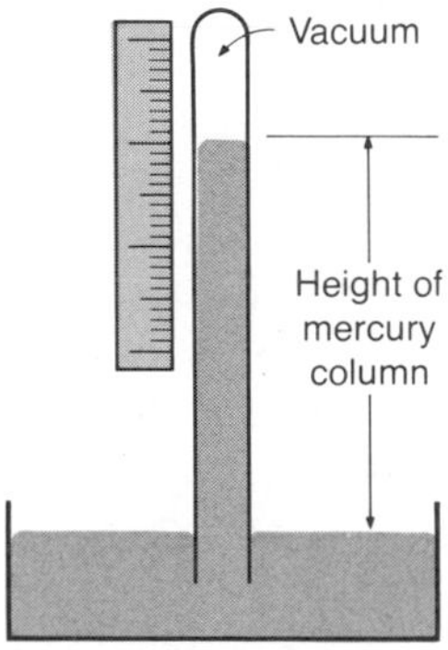

FIGURE 10.8 A mercury-in-glass barometer.

This type of barometer was invented by Torricelli, who succeeded Galileo as professor of mathematics at the University of Florence, Italy. In the intellectual climate of the early seventeenth century, the classical doctrine of nature's abhorrence of a vacuum was held in high regard. Torricelli's instrument was considered more remarkable for its production of a vacuum than for its ability to measure atmospheric pressure. However, both Torricelli and his contemporary Pascal realized the scientific importance of the invention and used it to show that the pressure of the atmosphere does not remain constant, but changes on a daily or even an hourly basis. Pascal also showed that atmospheric pressure decreases with increasing altitude.

EXAMPLE 10.5

A person blows into a rubber tube connected to one arm of a manometer containing glycerin of relative density 1.260. The other arm of the manometer is open to the atmosphere. If the difference in liquid level is 480 mm, what is the gauge pressure of the air in the person's lungs? Express the answer in normal atmospheres.

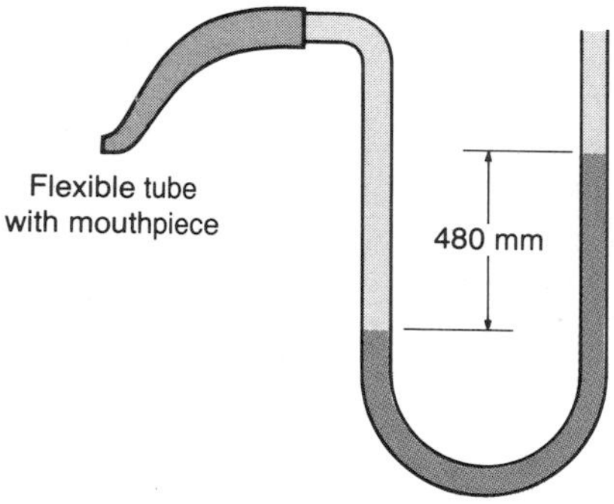

SOLUTION To calculate the pressure difference measured by the manometer, we use Eq. (10.4):

$$\begin{aligned} p_2 - p_1 &= \rho h g \\ &= 1.260 \times 10^3 \text{ kg/m}^3 \times 0.480 \text{ m} \times 9.81 \text{ m/s}^2 \\ &= 5933 \text{ Pa} \\ &= 5933 \text{ Pa} \div 1.013 \times 10^5 \text{ Pa/atm} \\ &= 0.0586 \text{ atm} \end{aligned}$$

Since one side of the manometer is open to the atmosphere, this value is the required gauge pressure. ■

The instruments described above have the great advantage of measuring pressure directly in terms of an easily observed height of a mercury column, but from a practical point of view they have the serious disadvantage of being fragile and bulky. Thus, instruments of more robust construction are needed. Pressure in a steam boiler or an air compressor is usually measured by a Bourdon gauge, shown in Fig. 10.9(a). A metal tube AB of oval cross section is bent into a nearly circular arc. The end B is closed and free to move, and the end A is open to the pressure vessel. The pressure tends to cause the bent tube to unwind to a straighter shape, and in doing so actuates a pointer through a gear-and-lever system.

The usual type of portable barometer is the aneroid barometer, shown in Fig. 10.9(b). A sealed, evacuated metal container with bellowslike sides contracts or expands with variations in the external pressure. This contraction or expansion of the container is

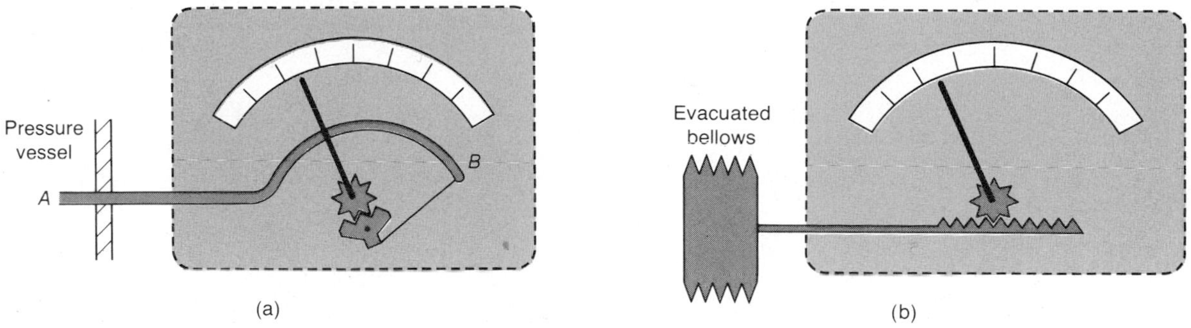

FIGURE 10.9 (a) A Bourdon pressure gauge; (b) an aneroid barometer.

transmitted to a pointer by a system of gears and levers. Neither of these instruments permits direct calculation of the pressure, and both must be calibrated with an instrument that does permit a direct pressure calculation. For instance, a Bourdon pressure gauge can be calibrated against a mercury-in-glass manometer, and an aneroid barometer can likewise be calibrated against a mercury-in-glass barometer. Many other highly specialized pressure gauges have been developed, especially for measuring extremely low pressures.

10.4 Archimedes' Principle

The force of buoyancy in a fluid is a very common phenomenon. A ship floating on the water and a helium-filled balloon floating in the air both experience an upward force that counteracts their weight. The buoyancy principle was discovered several centuries before the Christian era by the great Greek physicist and mathematician Archimedes (287–212 B.C.). Let us state **Archimedes' principle** and then demonstrate that it follows from laws previously stated.

Archimedes' Principle

> A body that is wholly or partly immersed in a fluid experiences a buoyant force equal to the weight of the fluid displaced.

As we have implied, the buoyancy principle is not distinct from other principles of physics. To show this, we consider a cube of side l submerged in a liquid of density ρ with two of its faces horizontal. The upper horizontal face is a depth h below the free

The famous engineer Isambard Brunel came in for a lot of criticism for his design of the *Great Eastern* that is shown leaving Sheerness to lay the Atlantic cable. Opponents claimed that the ship should be constructed of wood rather than iron because of the buoyancy of wood floating in water. However, Brunel's understanding of Archimedes' principle triumphed, and his great ship was the forerunner of the giant ocean liners. (Historical Pictures Service, Inc.)

surface of the liquid (Fig. 10.10). Each face of the cube experiences a force due to the pressure of the liquid. The forces on the four vertical faces cancel in pairs, since each is balanced by an equal and opposite force on the opposite cube face. However, the forces on the top and bottom faces of the cube do not balance, resulting in a net force due to the liquid pressure. The absolute pressures on the top and bottom faces of the cube are given by Eq. (10.3):

$$p_1 = p_a + \rho hg$$
$$p_2 = p_a + \rho(h + l)g$$

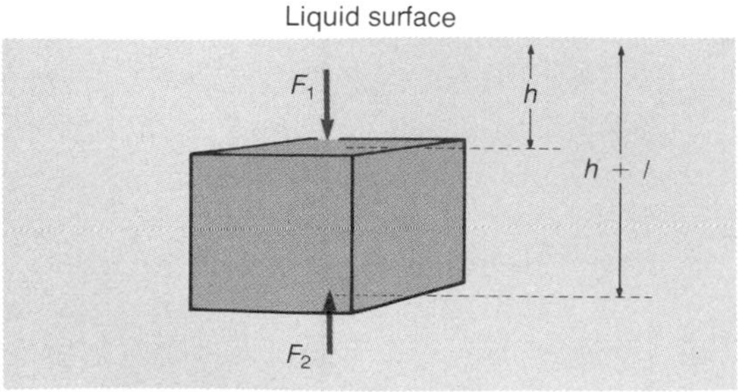

FIGURE 10.10 Calculation of the buoyancy force on a cube due to the pressure in the liquid.

The forces on the upper and lower faces are equal to the pressures multiplied by the area of the face. The upper face experiences a downward force given by:

$$F_1 = (p_a + \rho hg)l^2$$

SPECIAL TOPIC 10 ARCHIMEDES

Archimedes was born in Syracuse, Sicily, in about 287 B.C. He was the greatest scientist of the ancient world—one whose equal did not appear for another two millenia.

In addition to discovering the principle of buoyancy, he explained the operation of levers, produced a helical cylinder that served as a water pump, calculated a value of π to four significant digits, and invented a form of exponential notation for large numbers.

He is famous for many inventions and devices that helped hold the Romans at bay during their 3-year siege of his native city. Eventually, however, the Romans prevailed, and Archimedes was killed in the ensuing sack, which took place in about 212 B.C.

In a similar way, the upward force on the lower face is:

$$F_2 = [p_a + \rho(h + l)g]l^2$$

The difference between these two forces is the net buoyancy force exerted by the liquid:

$$\begin{aligned}\text{Net buoyancy force} &= F_2 - F_1 \\ &= [p_a + \rho(h + l)g]l^2 - (p_a + \rho hg)l^2 \\ &= \rho l^3 g \\ &= \text{weight of displaced liquid}\end{aligned}$$

We see that Archimedes' principle follows from the laws of Newtonian dynamics for the case of a cube immersed in a uniform liquid. This result is true in general for a body of arbitrary shape immersed in any fluid.

The story is told that Archimedes made his discovery while reflecting on the problem of determining whether the king's crown was made of pure gold. He was bathing at the time and presumably thinking of the buoyancy force that caused his body to float in the bath. In his excitement he rushed unclothed from the bath shouting "Eureka"—which means "I have discovered it." Archimedes knew the density of gold and could measure the weight of the crown, but he lacked a way to determine the volume of the irregularly shaped crown without destroying it. To show how he resolved this dilemma, consider the following example.

EXAMPLE 10.6

A modern-day Archimedes finds the weight of a gold crown to be 60.21 N. When suspended by a light thread and totally immersed in water, the tension in the thread is 56.53 N. What value

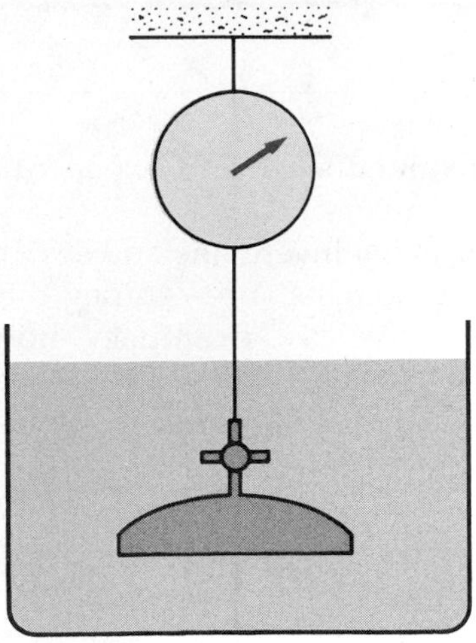

(a) General situation diagram

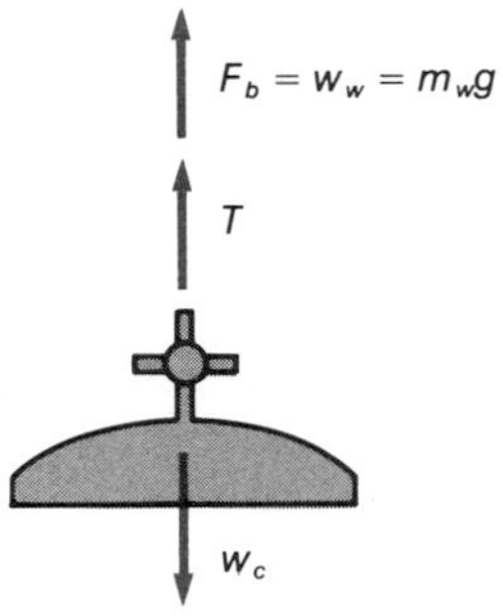

(b) Free-body force diagram

does he calculate for the average density of the material in the crown?

SOLUTION We can calculate the mass of the crown from its weight by using Eq. (4.2):

$$\begin{aligned} m_c &= \frac{w_c}{g} \\ &= \frac{60.21 \text{ N}}{9.81 \text{ m/s}^2} \\ &= 6.138 \text{ kg} \end{aligned}$$

(The subscript c refers to the crown.) From the free-body force diagram, we see that the tension in the thread plus the buoyant force must equal the weight of the crown:

$$\begin{aligned} T + F_b &= w_c \\ 56.53 \text{ N} + F_b &= 60.21 \text{ N} \\ \therefore F_b &= 3.68 \text{ N} \end{aligned}$$

This buoyant force is equal to the weight of the displaced water by Archimedes' principle. We again use Eq. (4.2) to find the mass of the displaced water from its known weight:

$$\begin{aligned} m_w &= \frac{w_w}{g} \\ &= \frac{3.68 \text{ N}}{9.81 \text{ m/s}^2} \\ &= 0.3751 \text{ kg} \end{aligned}$$

(The subscript w refers to the water displaced by the crown.) We calculate the volume of the displaced water by using Eq. (1.1) and the value of the density of water from Table 1.4:

$$\begin{aligned} V_w &= \frac{m_w}{\rho_w} \\ &= \frac{0.3751 \text{ kg}}{10 \text{ kg/m}^3} \\ &= 3.751 \times 10^{-4} \text{ m}^3 \end{aligned}$$

Since the crown is totally immersed, its volume is equal to the volume of the displaced water. Then Eq. (1.1) for the crown gives:

$$\rho_c = \frac{m_c}{V_c}$$

$$= \frac{6.138 \text{ kg}}{3.751 \times 10^{-4} \text{ m}^3}$$

$$= 1.636 \times 10^4 \text{ kg/m}^3$$

A comparison with the figures of Table 1.4 shows that this density is substantially less than the density of pure gold. ■

The story of Archimedes' exploit may be fanciful, but the circumstances lend a ring of truth to it. Archimedes lived in an age when accurate scientific instruments were unknown. The only data that he had to solve this problem were two accurate values of weight. As long as the human race has been in the business of buying and selling gems and precious metals, accurate weighing instruments have been available!

EXAMPLE 10.7

A small balloon and its load have a mass of 10 kg (with the balloon uninflated) and an enclosed volume of 20 m^3 when inflated. Calculate the upward acceleration of the unit when filled with helium. (Take the densities of air and helium at 1.2 kg/m^3 and 0.17 kg/m^3 under the conditions obtaining.)

This mode of air transportation relies entirely on the buoyancy force to keep it aloft. The conventional jetliner relies on its engine power and wing lift; engine failure does not cause an airship to fall to the earth. (Courtesy of The Goodyear Tire & Rubber Co.)

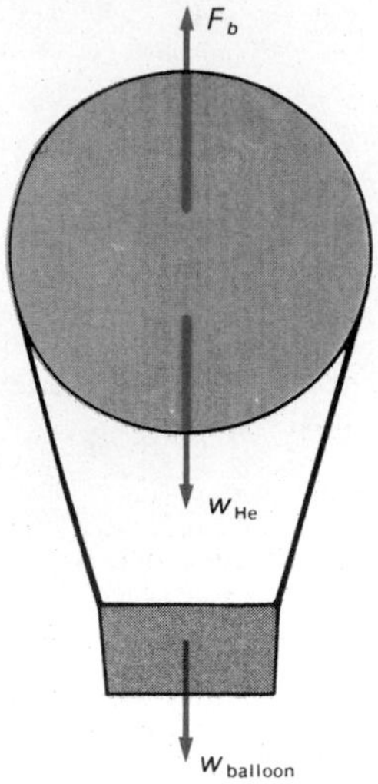

SOLUTION The forces acting on the balloon are its own weight plus the weight of the enclosed helium acting downward and the buoyancy force due to the displaced air acting upward. Since the balloon volume is 20 m^3, the mass of helium it contains is given by 0.17 kg/m^3 × 20 m^3 = 3.4 kg, and the mass of displaced air is given by 1.2 kg/m^3 × 20 m^3 = 24 kg. Using Archimedes' principle, and Eq. (4.2), we can calculate the forces:

$$\begin{aligned} w_{\text{balloon}} + w_{\text{He}} &= (m_{\text{balloon}} + m_{\text{He}})g \\ &= (10 \text{ kg} + 3.4 \text{ kg}) \times 9.81 \text{ m/s}^2 \\ &= 131.5 \text{ N} \\ F_b = w_{\text{air}} &= m_{\text{air}}\, g \\ &= 24 \text{ kg} \times 9.81 \text{ m/s}^2 \\ &= 235.4 \text{ N} \end{aligned}$$

The net upward force is found from the free-body force diagram:

$$\begin{aligned} F_b - (w_{\text{balloon}} + w_{\text{He}}) &= 235.4 \text{ N} - 131.5 \text{ N} \\ &= 103.9 \text{ N} \end{aligned}$$

Newton's second law [Eq. (4.1)] allows us to calculate the upward acceleration:

$$\begin{aligned} a &= \frac{F}{m} \\ &= \frac{103.9 \text{ N}}{13.4 \text{ kg}} \\ &= 7.75 \text{ m/s}^2 \end{aligned}$$

■

The previous examples showed how to use the free-body force diagram in conjunction with Archimedes' principle to solve buoyancy problems. Let us consider a slightly more difficult example.

EXAMPLE 10.8

A cubic block of pine of density 600 kg/m^3 and 20 cm on edge floats in water; the pine block supports a cube of copper 5 cm on edge by a thread attached to the center of its bottom face. Calculate the depth to which the pine sinks in the water.

SOLUTION We begin with free-body force diagrams showing the thread tension, the weight, and the buoyant force as it affects each block. (The subscripts p and c stand for pine and copper,

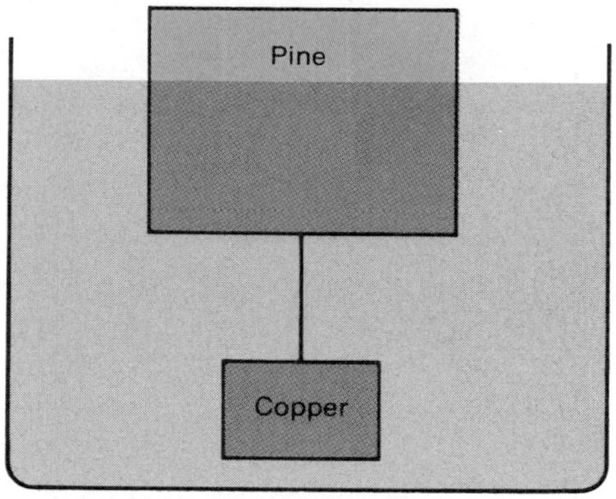

(a) General situation diagram

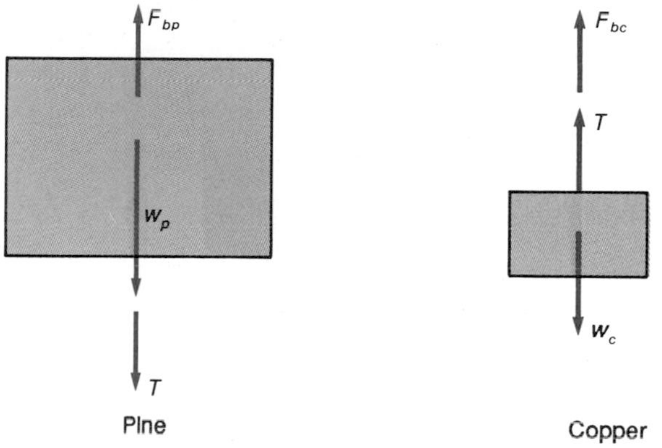

(b) Free-body force diagrams

respectively.) We begin with the copper block; its volume is $(0.05 \text{ m})^3 = 1.25 \times 10^{-4} \text{ m}^3$. From Table 1.4 the density of copper is $8.96 \times 10^3 \text{ kg/m}^3$, and Eq. (1.1) enables us to calculate the mass:

$$\begin{aligned} m_c &= \rho_c V_c \\ &= 8.96 \times 10^3 \text{ kg/m}^3 \times 1.25 \times 10^{-4} \text{ m}^3 \\ &= 1.12 \text{ kg} \end{aligned}$$

The weight of the copper is then given by Eq. (4.2):

$$\begin{aligned} w_c &= m_c g \\ &= 1.12 \text{ kg} \times 9.81 \text{ m/s}^2 \\ &= 10.99 \text{ N} \end{aligned}$$

Since the copper is totally submerged it displaces a volume of water equal to its own volume. By Archimedes' principle, the weight of this water is equal to the buoyant force on the copper:

$$F_{bc} = \rho_w V_c g$$
$$= 10^3 \text{ kg/m}^3 \times 1.25 \times 10^{-4} \text{ m}^3 \times 9.81 \text{ m/s}^2$$
$$= 1.226 \text{ N}$$

Static equilibrium of the copper block then yields:

$$T + F_{bc} = w_c$$
$$T + 1.226 \text{ N} = 10.99 \text{ N}$$
$$\therefore T = 9.764 \text{ N}$$

We now turn our attention to the pine block, and calculate its mass and weight, just as we did for the copper block. Its volume is $(0.20 \text{ m})^3 = 8 \times 10^{-3} \text{ m}^3$, and using Eq. (1.1) we have:

$$m_p = \rho_p V_p$$
$$= 600 \text{ kg/m}^3 \times 8 \times 10^{-3} \text{ m}^3$$
$$= 4.8 \text{ kg}$$

From Eq. (4.2), the weight of the pine is:

$$w_p = m_p g$$
$$= 4.8 \text{ kg} \times 9.81 \text{ m/s}^2$$
$$= 47.09 \text{ N}$$

We can now use the condition of static equilibrium on the pine block:

$$F_{bp} = T + w_p$$
$$= 9.764 \text{ N} + 47.09 \text{ N}$$
$$= 56.85 \text{ N}$$

By Archimedes' principle, this must be equal to the weight of water displaced by the pine block. Let V_p' be the volume of water displaced by the pine block; then we have:

$$F_{bp} = \rho_w V_p' g$$
$$56.85 \text{ N} = 10^3 \text{ kg/m}^3 \times V_p' \times 9.81 \text{ m/s}^2$$
$$\therefore V_p' = 5.795 \times 10^{-3} \text{ m}^3$$

Finally let d be the depth to which the pine sinks in the water. Then we have:

$$V_p' = (0.20 \text{ m})^2 \times d$$
$$5.795 \times 10^{-3} \text{ m}^3 = 4 \times 10^{-2} \text{ m}^2 \times d$$
$$\therefore d = 0.145 \text{ m} = 14.5 \text{ cm}$$

■

10.5 Bernoulli's Equation

Up to this point we have considered the physics of fluids at rest—the study of hydrostatics. But in many important instances fluids are in motion, and we call this study **hydrodynamics.** The wind, the circulation of the blood, and the flow of a river are all instances of fluid flow. The physical complexity of hydrodynamic problems is apparent from these examples, but we can find interesting and widely applicable results from a special form of the energy equation. In order to apply the energy equation to specific problems, we must restrict ourselves to a particular type of fluid flow and a special model of a fluid. The necessary restrictions are best illustrated by examples.

A fast-flowing shallow stream of water frequently exhibits eddies that change position even though the flow of water is constant, but a deep and slowly flowing stream shows a steady motion that is free of turbulence and eddies. The type of flow typified by the fast-flowing stream is called *turbulent flow,* which we exclude because of energy dissipation involved. We will concern ourselves with fluid flow of the latter type, which is called *streamline flow.* However, another energy-dissipating process is possible in a fluid, even in streamline flow, if different layers of fluid experience frictional forces when moving relative to each other. Fluid friction is called *viscosity,* and the corresponding force is a viscous force. To ensure that there is no energy dissipation, we will also exclude viscous flow. The ideal fluid also has the uniform density that we assumed in calculating hydrostatic pressure. An incompressible fluid is uniformly dense since it does not change volume when the pressure changes, and most liquids approximate this ideal. Let us sum up by stating all the requirements for the ideal fluid flow: The fluid is *nonviscous, incompressible,* and moving in *streamline motion.*

Before stating the energy theorem, we need a method for calculating the velocity changes that occur when a fluid flows in a pipe of variable cross section. Consider an ideal fluid in streamline flow in a pipe of variable cross section (Fig. 10.11). Let the cross-sectional areas be A_1 and A_2 at positions #1 and #2. We assume the velocity of the fluid is v_1 at all points on the cross section A_1 and v_2 at all points on the cross section A_2. Let the fluid on the cross section A_1 displace by x_1 during a time interval t, and the fluid on the cross section A_2 displace x_2 during the same time interval. The volume of fluid flow per unit time at position #1 is:

$$Q_1 = \frac{A_1 x_1}{t} = A_1 v_1$$

At position #2 the volume of fluid flow per unit time is:

$$Q_2 = \frac{A_2 x_2}{t} = A_2 v_2$$

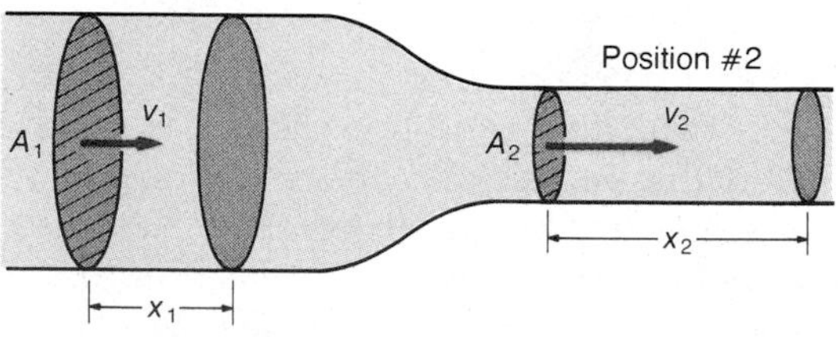

FIGURE 10.11 Flow of an ideal fluid in a pipe of variable cross section.

SPECIAL TOPIC 11 SCALING AND THE FLIGHT OF BIRDS

When birds fly, they rely on their wings to provide lift and on the motion of their wings to provide propulsion. The details are exceedingly complex. However, we can use scaling arguments to compare the gliding ability of birds of similar shape. The restriction to gliding means that we are considering the wings only insofar as they provide lift.

Bernoulli's equation shows that the pressure difference between the upper and lower wing surfaces is proportional to the difference of the squares of the air velocity over those surfaces. If we assume that the air velocity over each wing surface is proportional to the speed of the bird, then the difference in question is proportional to the square of the bird's speed. But the lift is the pressure difference multiplied by the area of the wings. Once again we use the characteristic scale factor L. Wing area scales with L^2. This means that lift scales with L^2v^2, where v is the bird's speed. In level flight the lift must equal the weight, and this latter scales with L^3. We then have

$$L^2v^2 \propto L^3$$

so that the bird's speed scales with $L^{1/2}$.

Thus, a large bird must have a higher flying speed than a small one. House sparrows are airborne after a step or two, but Canada geese require a takeoff run (either on land or in water). Researchers surmise that the large winged reptiles that existed during the dinosaur age would have had to throw themselves off clifftops to reach the high flying speed that their size required.

Since the fluid is incompressible, the flow is the same at all points in the pipe. Hence $Q_1 = Q_2$, and we can drop the subscripts on this symbol. The resulting equation for flow rate is called the **continuity equation.**

Continuity Equation for Streamline Flow of an Ideal Fluid

$$\text{Volume of fluid flow per unit time} = \text{cross-sectional area} \times \text{fluid velocity}$$

$$Q = A_1v_1 = A_2v_2 \qquad \textbf{(10.5)}$$

SI unit of flow: cubic meter per second
Abbreviation: m^3/s

Our derivation of the continuity equation is for ideal fluid flow, but it also applies under less restrictive conditions. Blood is a reasonably incompressible fluid that moves through the blood vessels in a streamline flow. However, blood is decidedly viscous, and this property causes a velocity that is high in the center of an artery but less toward the walls. In these circumstances we must interpret the velocity in the continuity equation as an average velocity.

EXAMPLE 10.9

In a normal adult at rest, the heart pumps blood at the rate of about 5 ℓ/min. Calculate the average flow velocity in an aorta of radius 9 mm.

SOLUTION We notice that the SI unit of flow rate is inconveniently large for use in blood circulation. Liters per minute is a more convenient unit, so we use the conversion tables to change this to SI units:

$$Q = 5\ \ell/\text{min} \times 1.667 \times 10^{-5}\ (\text{m}^3/\text{s})/(\ell/\text{min})$$
$$= 8.333 \times 10^{-5}\ \text{m}^3/\text{s}$$

The cross-sectional area of the aorta is given by:

$$A = \pi r^2$$
$$= \pi \times (9 \times 10^{-3}\ \text{m})^2$$
$$= 2.545 \times 10^{-4}\ \text{m}^2$$

We can now use Eq. (10.5) with the understanding that the velocity is the average velocity:

$$Q = A\langle v\rangle$$
$$\therefore \langle v\rangle = \frac{Q}{A}$$
$$= \frac{8.333 \times 10^{-5}\ \text{m}^3/\text{s}}{2.545 \times 10^{-4}\ \text{m}^2}$$
$$= 0.327\ \text{m/s}$$

■

The continuity equation is kinematic in nature, but is a necessary preliminary to developing the energy conservation equation for streamline flow in an ideal fluid. The energy equation is called **Bernoulli's equation** after Daniel Bernoulli (1700–1782). He was a member of a famous Swiss family who provided three generations of professors of mathematics and physical sciences at the University of Basel during the seventeenth and eighteenth centuries. To derive Bernoulli's equation, we consider the situation illustrated in Fig. 10.12. An ideal fluid flows through a pipe of nonuniform cross section. We select the two positions in the pipe labeled #1 and #2, and their heights above the gravitational potential energy reference level are h_1 and h_2, respectively. We imagine that the fluid flow is accompanied by the movement of two imaginary frictionless pistons at the two selected positions. During a given time interval, the first piston of area A_1 displaces x_1, and the second piston of area A_2 displaces x_2. The net effect

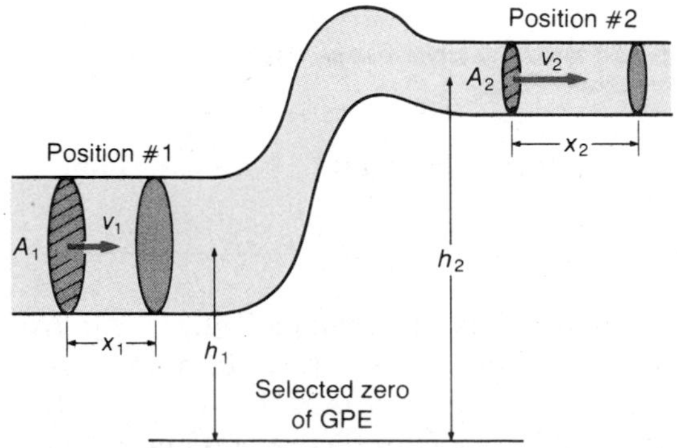

FIGURE 10.12 Illustration of the derivation of Bernoulli's equation.

of the displacement of the imaginary pistons is the transfer of a mass m of the fluid of density ρ from position #1 to position #2. The mass of displaced fluid is equal to the density multiplied by the volume:

$$m = \rho A_1 x_1 = \rho A_2 x_2$$

The force required to push the first piston is equal to the pressure at position #1 (p_1) multiplied by the area of the piston (A_1):

$$F_1 = p_1 A_1$$

Similarly, the force exerted on the second piston is given by:

$$F_2 = p_2 A_2$$

We can now state the work-energy relation:

$$\begin{matrix}\text{Work done } \textit{on} \\ \text{first piston}\end{matrix} - \begin{matrix}\text{work done } \textit{by} \\ \text{second piston}\end{matrix} = \text{final (KE + GPE)} - \text{initial (KE + GPE)}$$

Using the expressions for F_1, F_2, and m, we have:

$$p_1 A_1 x_1 - p_2 A_2 x_2 = \tfrac{1}{2}\rho A_2 x_2 v_2^2 + \rho A_2 x_2 g h_2 - \tfrac{1}{2}\rho A_1 x_1 v_1^2 - \rho A_1 x_1 g h_1$$

Cancelling common factors, transposing, and using the result $A_1 x_1 = A_2 x_2$ give:

$$p_1 + \tfrac{1}{2}\rho v_1^2 + \rho g h_1 = p_2 + \tfrac{1}{2}\rho v_2^2 + \rho g h_2$$

The functions on the right and left sides of the equation are the same. This function therefore has a constant value at all positions in the pipe.

Bernoulli's Equation for Streamline Flow of an Incompressible, Nonviscous Fluid

$$p + \tfrac{1}{2}\rho v^2 + \rho g h = \text{constant} \qquad \textbf{(10.6)}$$

Each term in Bernoulli's equation has dimensions $[ML^{-1}T^{-2}]$. We associate this dimension with either pressure (unit of newton per square meter) or energy per unit volume (unit of joule per cubic meter). We can regard Bernoulli's equation as expressing energy conservation per unit fluid volume. The three terms are the fluid pressure energy, the kinetic energy, and the gravitational potential energy—all per unit fluid volume.

EXAMPLE 10.10

A Venturi tube is a narrowed constriction or throat in an otherwise uniform pipeline. The approaches to the throat are carefully designed to preserve streamline flow. Water is flowing in a horizontal streamline flow at the rate of 12 ℓ/min in a circular pipe 2 cm in diameter that has a Venturi throat 1 cm in diameter.

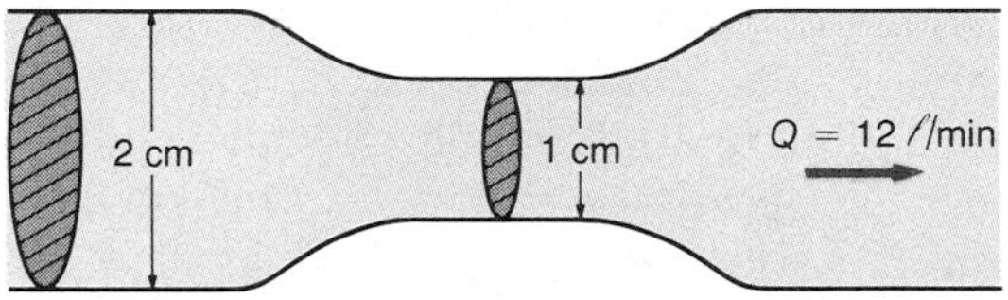

a. What is the water velocity in the pipe and in the Venturi throat?

b. If the water has a gauge pressure of 80 kPa in the pipe, what is the gauge pressure in the Venturi throat?

SOLUTION The problem is basically one of fluid dynamics. We must be careful to set the data in SI units:

$$Q = 12\ \ell/\text{min} = \frac{12 \times 10^{-3}\ \text{m}^3}{60\ \text{s}} = 2 \times 10^{-4}\ \text{m}^3/\text{s}$$

a. We can calculate the water velocities from the continuity equation, Eq. (10.5):

$$Q = A_1 v_1 = A_2 v_2$$

$$\therefore v_1 = \frac{2 \times 10^{-4}\ \text{m}^3/\text{s}}{\pi \times (10^{-2}\ \text{m})^2} = 0.6367\ \text{m/s}$$

and

$$v_2 = \frac{2 \times 10^{-4}\ \text{m}^3/\text{s}}{\pi \times (10^{-2}\ \text{m})^2} = 2.547\ \text{m/s}$$

b. To calculate the pressure in the Venturi throat, we first find the absolute pressure in the pipe by adding atmospheric pressure to the gauge pressure:

$$p_1 = 8 \times 10^4\ \text{Pa} + 1.013 \times 10^5\ \text{Pa} = 1.813 \times 10^5\ \text{Pa}$$

Since the pipe is horizontal, the Bernoulli equation has no gravitational potential energy term, and Eq. (10.6) gives:

$$p_1 + \tfrac{1}{2}\rho v_1^2 = p_2 + \tfrac{1}{2}\rho v_2^2$$

$$1.813 \times 10^5\ \text{Pa} + \tfrac{1}{2} \times 10^3\ \text{kg/m}^3 \times (0.6367\ \text{m/s})^2 = p_2 + \tfrac{1}{2} \times 10^3\ \text{kg/m}^3 \times (2.547\ \text{m/s})^2$$

$$\therefore p_2 = 1.783 \times 10^5\ \text{Pa}$$

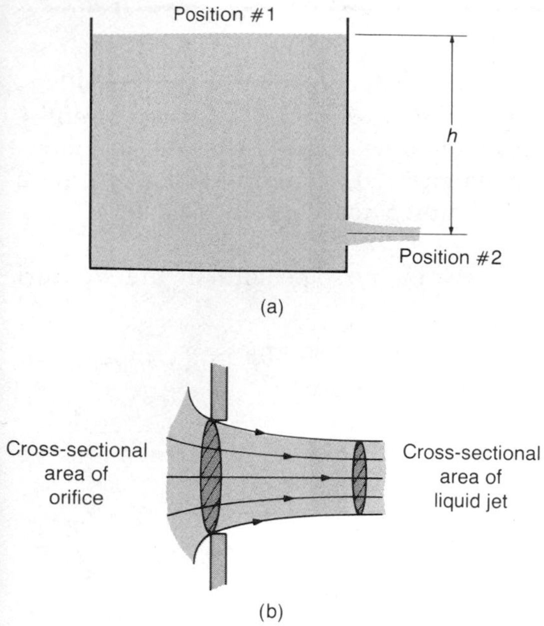

FIGURE 10.13 (a) Illustration of Torricelli's theorem; (b) contraction of the emerging jet of liquid.

We obtain the gauge pressure in the Venturi throat by subtracting atmospheric pressure:

$$\text{Gauge pressure in Venturi} = 1.783 \times 10^5 \text{ Pa} - 1.013 \times 10^5 \text{ Pa}$$
$$= 77.0 \text{ kPa}$$ ■

The Venturi tube is an example of the application of Bernoulli's equation to nonviscous streamline flow in a pipe. However, the physical presence of a pipe is not really necessary; all that is required is a nonviscous streamline flow. Bernoulli's equation finds application in many special problems of this type. One of the most famous concerns the flow of a liquid through a small hole situated below the free surface of a large reservoir (Fig. 10.13a). The two positions selected for evaluating the Bernoulli function of Eq. (10.6) are at the free surface of the liquid in the reservoir and in the liquid jet just after escape through the hole. Let us take the gravitational potential energy reference level at position #2, and let h be the height of the free surface above the reference level. Let v be the velocity of the liquid at position #2. If the area of the reservoir is large compared with the area of the hole, the liquid velocity is approximately zero at position #1. The pressure at both positions is equal to the atmospheric pressure p_a. We now calculate the Bernoulli function at each position:

$$p + \tfrac{1}{2}\rho v^2 + \rho g h = p_a + 0 + \rho g h \text{ at position \#1}$$
$$= p_a + \tfrac{1}{2}\rho v^2 + 0 \text{ at position \#2}$$

Combining these equations we have:

$$v = (2gh)^{1/2} \tag{10.7}$$

The result for the velocity of efflux of a liquid through a hole situated at depth h below the free surface is the same as that acquired by a body falling freely from a height h; this relationship is known as *Torricelli's theorem*. The velocity is independent of the density of the stored liquid. Calculation of the liquid flow from the hole is complicated by the fact that converging streamlines cause the liquid jet to contract to a smaller cross-sectional area than the area of the hole. We show a sketch of the converging streamlines in Fig. 10.13(b).

EXAMPLE 10.11

A large water tank has a hole 3 mm in diameter and 4 m below the free surface of the water. The escaping water jet is contracted so that its cross-sectional area is 65% of the area of the hole. What is the rate of loss of water in liters per minute?

SOLUTION We first calculate the efflux velocity using Torricelli's theorem, Eq. (10.7):

$$
\begin{aligned}
v &= (2gh)^{1/2} \\
&= (2 \times 9.81\ \text{m/s}^2 \times 4\ \text{m})^{1/2} \\
&= 8.859\ \text{m/s}
\end{aligned}
$$

The cross-sectional area of the jet is 65% of the area of the circular orifice:

$$
\begin{aligned}
\text{Cross-sectional area of water jet} &= 65\% \times \text{area of orifice} \\
&= 0.65 \times \pi \times \left(\frac{3 \times 10^{-3}\ \text{m}}{2}\right)^2 \\
&= 4.594 \times 10^{-6}\ \text{m}^2
\end{aligned}
$$

The rate of leakage of water is given by Eq. (10.5):

$$
\begin{aligned}
Q &= Av \\
&= 4.594 \times 10^{-6}\ \text{m}^2 \times 8.859\ \text{m/s} \\
&= 4.070 \times 10^{-5}\ \text{m}^3\text{/s} \\
&= 4.070 \times 10^{-5}\ \text{m}^3\text{/s} \div 1.667 \times 10^{-5}\ (\text{m}^3\text{/s})/(\ell/\text{min}) \\
&= 2.44\ \ell/\text{min}
\end{aligned}
$$

■

Another interesting application of Bernoulli's theorem is the Pitot tube, which is used to measure fluid velocities. A center tube (Fig. 10.14) connects to one arm of a manometer, and an enclosed outer tube connects to the other arm. The outer tube has a series of small openings around its periphery at position #2. We choose position #1 in the entry to the center tube. The tube is set up in a fluid streaming from left to right as shown. At position #1 the fluid velocity is zero since a dead area arises at the inlet to the center tube. At position #2 the fluid velocity is the full stream velocity. The gravitational potential energy term of the Bernoulli equation does not contribute, and the remaining terms give:

$$p_1 = p_2 + \tfrac{1}{2}\rho v^2$$

The manometer measures the pressure difference $(p_1 - p_2)$. This enables us to calculate the velocity directly from the manometer reading and the density of the streaming fluid:

$$v = \left[\frac{2(p_1 - p_2)}{\rho}\right]^{1/2} \qquad \textbf{(10.8)}$$

Large errors arise if the design of the Pitot tube causes eddies in the fluid. Since no fluid is entirely nonviscous, the effect of viscosity must also be considered in precision measurements.

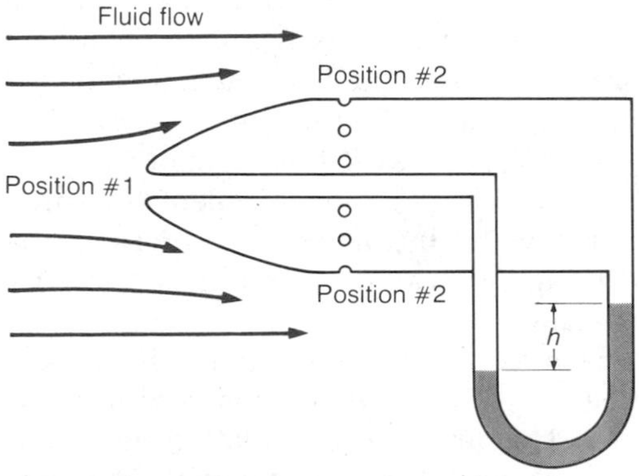

FIGURE 10.14 The Pitot tube.

EXAMPLE 10.12

The airspeed of a light aircraft is measured with a Pitot tube. The measured pressure difference is 2.24 kPa when the air density is 1.25 kg/m^3.

a. Calculate the airspeed.

b. What column height in a mercury manometer corresponds to the measured pressure difference?

SOLUTION

a. The airspeed calculation is a straightforward application of Eq. (10.8):

$$v = \left[\frac{2(p_1 - p_2)}{\rho}\right]^{1/2}$$

$$= \left(\frac{2 \times 2.24 \times 10^3 \text{ Pa}}{1.25 \text{ kg/m}^3}\right)^{1/2}$$

$$= 59.9 \text{ m/s}$$

b. The height of a mercury column in a manometer corresponds to the pressure measurement in torr. We only need a units conversion:

$$\text{Pressure difference} = 2.24 \times 10^3 \text{ Pa} \div 133.4 \text{ (Pa)/torr}$$

$$= 16.8 \text{ torr}$$

The pressure corresponds to a column of mercury 16.8 mm high. ■

Bernoulli's equation also finds application in calculating the lift of an aircraft wing. Figure 10.15 shows a typical pattern of streamlines around the cross section of the wing. The design is such that the air velocity above the wing is higher than the air velocity below it. Application of Bernoulli's equation shows a lower pressure above the wing than below it, which results in a net upward force on the wing called lift. If the angle of the wing relative to the airflow direction is increased, turbulent flow may occur in the region above the wing. In this case the pressure difference between the two wing surfaces may be reduced to such an extent that the aircraft stalls.

In this last example, we have applied Bernoulli's theorem to airflow problems, but experience shows that air is very far from being an incompressible fluid. However, Bernoulli's equation expresses conservation of energy, and its use is legitimate when no energy is dissipated even if the fluid is compressible.

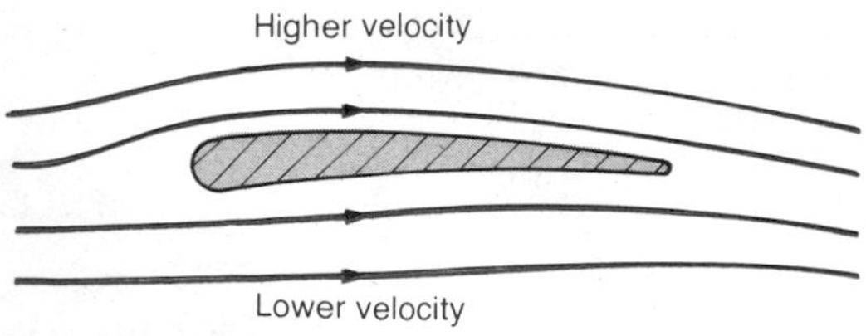

FIGURE 10.15 Streamlines around the cross section of an aircraft wing.

KEY CONCEPTS

Pressure is equal to the force normal to an area divided by the area. See Eq. (10.1).

Pascal's principle states that a pressure change in a body of fluid is transmitted undiminished to every point in the fluid.

Archimedes' principle equates the buoyancy force on an object to the weight of fluid that it displaces. See p. 267.

The **continuity equation** for streamline flow of an ideal fluid in a tube expresses the flow rate as the product of the cross-sectional area of the tube and the fluid velocity. See Eq. (10.5).

Bernoulli's equation expresses the energy conservation law for streamline flow of an ideal fluid. It includes terms for the pressure, kinetic, and gravitational potential energies per unit volume. See Eq. (10.6).

QUESTIONS FOR THOUGHT

1. A weather balloon rises to a definite height in the atmosphere and then remains at more or less the same height. Can you propose an object that would sink to a definite depth in the ocean and then remain there? If not, what is the difference between the two cases?
2. Metal cans containing liquid often leak when taken aloft in the cargo space of airliners. Do you think it would help to leave a little air inside the can and to seal the stopper firmly?
3. Give reasons why water-in-glass barometers are less popular than the mercury-in-glass models.
4. We live at the bottom of a large sea of air (the atmosphere). Make a rough estimate of the extent to which the buoyant force of the air affects your weight.
5. When going on an ocean trip, would you prefer a ship whose center of gravity was higher or lower than the center of gravity of the displaced water? Explain your preference.
6. Scientists claim that the melting of polar ice would substantially increase the main sea level in the world's oceans. Would the change occur because of melting of the Arctic or the Antarctic ice caps, or would they contribute more or less equally? Explain your answer.
7. The cross-sectional area of a stream of water falling from a faucet often becomes smaller as the water falls. How do you explain this phenomenon?
8. For what reasons might a baseball pitcher prefer to pitch with the wind behind him, and for what reasons might he prefer to pitch into it?
9. A carnival booth offers a prize for anyone who can blow a Ping-Pong ball vertically upward out of a funnel. Explain why all of the contestants fail.
10. The cross section of an aircraft wing is designed so that the airflow produces lift in level flight. If this is so, how can an aircraft fly upside down?
11. Two small boats moving parallel to each other in the same direction are sucked together. Explain this phenomenon.
12. Use Bernoulli's equation to explain why the flow of water from a faucet decreases when another faucet in the same dwelling is turned on.
13. The flat roof of a large building is ripped off in a gale, and the owner proposes to place heavy weights on the new roof to prevent the accident from recurring. Can you propose a better idea?

PROBLEMS

A. Single-Substitution Problems

1. At what depth in the ocean is the absolute pressure 1000 kPa? (Take the relative density of seawater at 1.025.) [10.1]
2. Calculate the pressure in an ocean trench 11 000 m deep. (Take the relative density of seawater at 1.025.) [10.1]

3. By what amount is the pressure at the top of a hill 400 m high lower than at the base? (Take the air to have uniform density of 1.2 kg/m^3 over the interval in question.) [10.1]

4. A circular plate of radius 10 cm on a piece of underwater equipment can withstand a maximum inward force of 5000 N. Calculate the depth at which the plate would fail. (Assume atmospheric pressure inside the equipment.) [10.2]

5. A water tower has a circular hole of diameter 3 cm at a position 15 m below the free surface of the water. Calculate the force required to hold a plug in position to stop the hole. [10.2]

6. A hydraulic hoist has a piston diameter of 36 cm and uses oil at a maximum gauge pressure of 1100 kPa. Calculate the largest mass that the hoist can raise. [10.2]

7. A hydraulic press is used to raise a 1500-kg automobile. The maximum available gauge pressure is 650 kPa. Calculate the smallest diameter for a piston that would raise the load. [10.2]

8. If water was used instead of mercury in the barometer of Fig. 10.8, how long would the glass tube need to be? [10.3]

9. A mercury manometer has one arm open to the atmosphere and the other connected to a closed vessel in which the pressure is 26 atm. What is the height difference of the mercury levels in the manometer arms? [10.3]

10. What relative density is required for the fluid in a manometer if a pressure differential of 80 Pa between the two arms causes a 1-cm difference in fluid levels? [10.3]

11. The average relative density of a certain object is 0.91. Calculate the relative proportions of a water–methanol mixture in which the object will remain immersed but suspended. (The relative density of methanol is 0.810.) [10.4]

12. How many cubic meters of seawater (relative density 1.025) does an 80 000-metric ton aircraft carrier displace? [10.4]

13. A block of oak (density 750 kg/m^3) has dimensions 40 cm $\times$ 30 cm $\times$ 10 cm. Calculate the buoyancy force on the block

a. if it floats freely in water
b. if it is held completely submerged in water [10.4]

14. A block of wood has dimensions 80 cm $\times$ 40 cm $\times$ 20 cm and floats in water, with 4×10^4 cm^3 of its volume submerged. Calculate the density of the block. [10.4]

15. An ideal fluid is flowing in streamline flow in a pipe of radius 5 cm. To what radius should the pipe change to effect a 70% reduction in the fluid velocity? [10.5]

16. An ideal fluid is flowing in streamline flow with a velocity of 4 m/s in a pipe of radius 8 cm. The pipe gradually reduces to 6 cm in radius. What is the velocity of flow in the reduced pipe? [10.5]

17. A large tank contains water with the top open to the atmosphere. Calculate the velocity of the water flowing from a leak 4.2 m below the free surface of the water. [10.5]

18. A tank of water has its top open to the atmosphere. Calculate the height of water required in the tank to obtain a velocity of 15 m/s from an open valve in the base. [10.5]

B. Standard-Level Problems

19. A conical-shaped flask has a base diameter of 18 cm and contains methanol whose free surface is 24 cm above the base. The top of the flask is open to the atmosphere. Calculate the force on the base of the flask due to the presence of the methanol. [10.2]

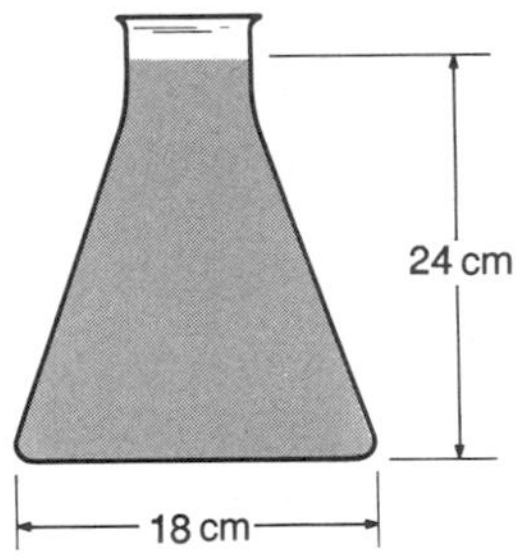

20. Consider the following data for Pascal's wine cask experiment. A long tube of internal radius 4 mm was fixed into the top of a wine cask full of water. The cask was 2 m high and the radius of the cask was 40 cm. Water was slowly added to the vertical tube, and the cask burst when the water level was 12 m above the cask top. See the figure on page 285. Calculate:

a. The weight of water in the tube
b. The pressures at the top and bottom of the cask
c. The total force on the top and bottom faces [10.2]

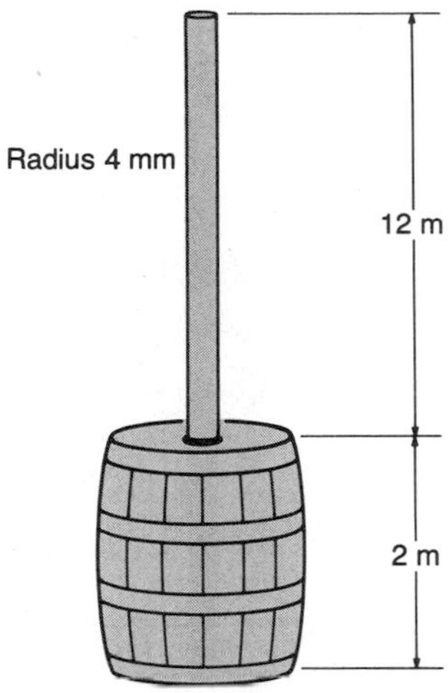

Figure for Problem 20

21. An automobile climbs a hill 250 m high, and the pressure in the middle ear of one of the passengers does not change during the climb. Calculate:

a. The pressure differential on the passenger's eardrum
b. The force on the eardrum if its area is 6.5×10^{-5} m^2 (Take the average density of the atmosphere over the height in question as 1.2 kg/m^3.)

22. A commercial airliner flies at an altitude of 12 000 m, with the cabin pressurized to the air pressure that exists at 1500 m. Estimate the force on a window 25 cm $\times$ 15 cm in area using the following assumptions:
i. The air pressure is negligible at 12 000 m.
ii. The air density is constant at 1.2 kg/m^3 from ground level up to 1500 m.

23. Air is contained in two vessels *A* and *B* that are connected to a horizontal tube *CD*. The vertical tube *DE* is open to the atmosphere at its upper end. Mercury is in the tubes, and the heights of the mercury surface in the various tubes are shown in the diagram. Calculate:

a. The gauge pressure in the tube *CD*
b. The air pressures in the vessels *A* and *B*

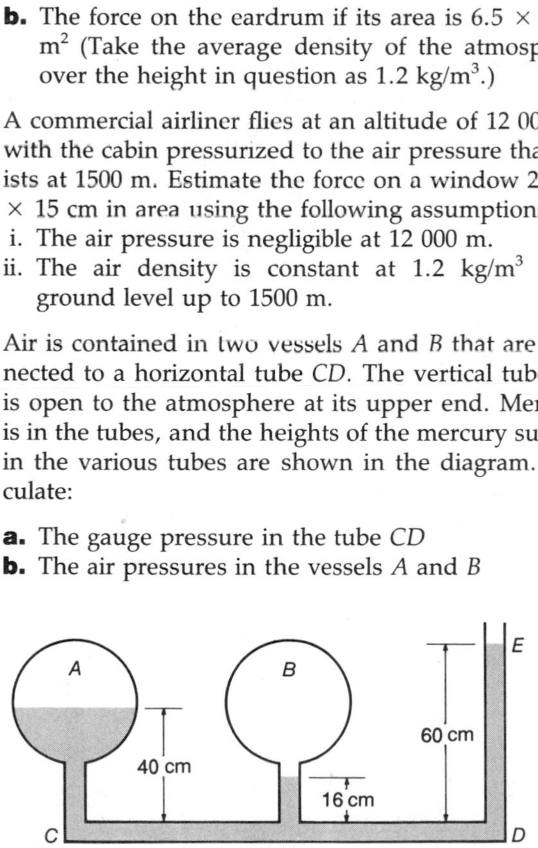

24. A hydraulic press consists of two vertical cylinders of circular cross-sectional areas 6 cm^2 and 120 cm^2 that are joined at the base as shown. The press contains oil of relative density 0.93, and the friction between the pistons and the cylinder walls is negligible. Initially the pistons are at the same level.

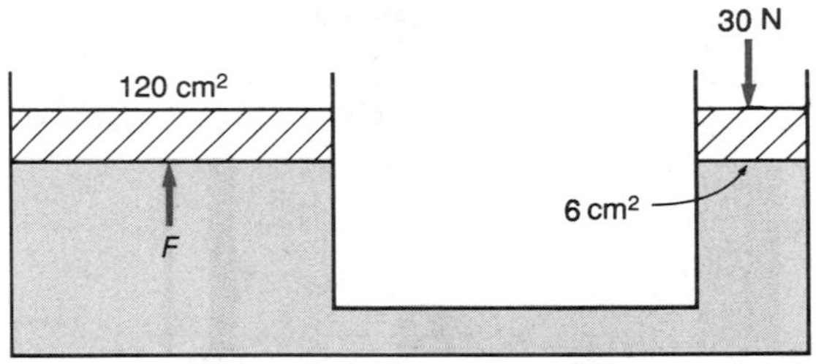

a. Calculate the upward force on the larger piston due to a downward force of 30 N on the smaller one.
The smaller piston is pushed down through a distance of 60 cm.
b. Through what distance does the larger piston rise?
c. What is now the upward force on the larger piston due to a downward force of 30 N on the smaller one?
(In working this problem assume that the masses of the two pistons are negligible.)

25. Water is poured into a U-tube until it stands 25 cm high in each arm. Oil is added slowly to one arm until the water in the other arm stands 30 cm high. The free surface of the oil then stands at 32.1 cm. Calculate the relative density of the oil. [10.3]

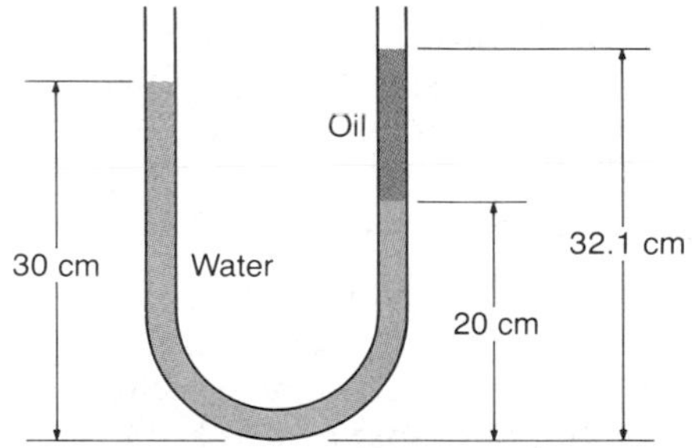

26. A U-tube contains mercury in the bottom, and columns of oil and water each 20 cm high in its two arms. The relative density of the oil is 0.86, and the arm containing oil is connected to a pressure vessel. The arm containing water is open to the atmosphere. What is the gauge pressure in the vessel if the mercury in the arm containing water is 2 cm higher than it is in the other arm? (Take the relative density of mercury as 13.6.) See the figure on page 286. [10.3]

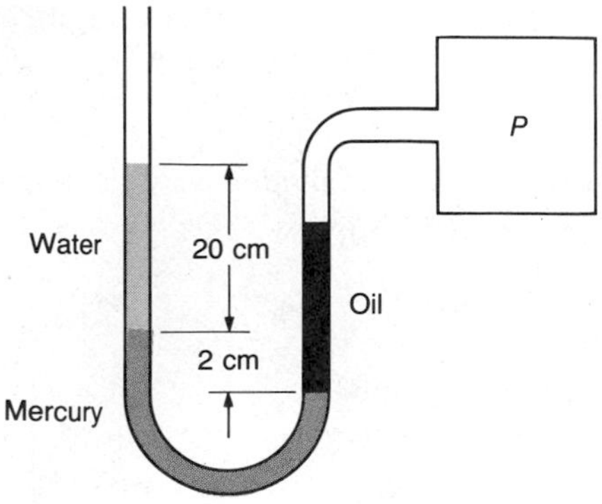

Figure for Problem 26

27. By sucking inward, a man can develop a gauge pressure of about -75 torr in his lungs. What is the maximum length straw he can use to drink water?

28. The normal atmosphere is defined to be exactly the pressure at the bottom of a column of mercury 760 mm high at a place where $g = 9.80665$ m/s^2, and at a temperature for which the density of mercury is 13 595.1 kg/m^3. Express the normal atmosphere in kilopascals to six significant digits.

29. What percentage of an iceberg is submerged? Take the relative densities of ice and seawater as 0.917 and 1.025, respectively. [10.4]

30. An empty canoe floats in fresh water, and its shape at the waterline is approximately 2.8 m × 70 cm rectangle. How much deeper does it sink if two persons, each of mass 80 kg, climb in and take up positions that keep the canoe on an even keel? [10.4]

31. A raft is made of oak logs, each 5 m × 35 cm × 35 cm, of relative density 0.77. How many logs are required for the raft to support ten people of average mass 80 kg in a freshwater lake to the point where 85% of the raft is submerged? [10.4]

32. A metal block is supported by a cord. The tension in the cord is 17.81 N when the metal is in air and 16.76 N when it is completely immersed in water. Identify the metal with the help of Table 1.4. (Ignore the small effect of air buoyancy.) [10.4]

33. A cord supports a metal block; the cord tension is 16.14 N with the metal in air. When the metal is completely immersed in water, the cord tension is 10.16 N, and it is 11.51 N when the metal is completely immersed in oil. Calculate the relative densities of the metal and the oil. (Neglect the small effect of air buoyancy.)

34. A block of white pine 3 m × 50 cm × 10 cm floats in fresh water.

a. What mass of brass must be placed on top of the pine so that its top surface is just level with the water?

b. What mass of brass must be suspended under the pine so that its top surface is just level with the water?

(The relative densities of white pine and brass are 0.43 and 8.44, respectively.)

35. The mass of a balloon and its associated load is 400 kg, and the volume of the inflatable enclosure is 700 m^3. Calculate:

a. The net upward force if the balloon is helium-filled

b. The additional upthrust gained by using hydrogen instead of helium

(Use the data of Table 1.4 for densities at atmospheric pressure and a temperature of 20°C.)

36. An object of mass 1200 kg and relative density 7.86 that is submerged in water is to be raised by using the buoyancy force of completely submerged air-filled balloons. Under the conditions of the work, the air density in the submerged balloons is 8.5 kg/m^3. What is the minimum volume that a balloon must have to begin raising the object?

37. A balloon is inflated with air and is held completely submerged in a tank of water by a cord that is tied to an aluminum block resting on the bottom of the tank. Under the conditions of submergence, the volume of the balloon is 0.014 m^3, the density of the enclosed air is 4.50 kg/m^3, and the mass of the aluminum block is 26 kg. Calculate the magnitude of the action-reaction force between the aluminum block and the bottom of the tank. (Assume that the bottom of the tank is rough so that there is water under the block and it experiences a buoyant force.)

38. A beaker of mass 0.8 kg contains 1.6 kg of water and sits on a scale calibrated in newtons. A block of aluminum of mass 3 kg is suspended from a spring balance (also calibrated in newtons) and is completely immersed in the water. The relative density of the aluminum is 2.70. What is the reading on the scale and the spring balance? See the figure at the top of page 287.

39. A forced-air ventilating system is required to change the air in a hall 28 m × 14 m × 5 m every 20 min through eight circular ducts. What duct radius is required if the air velocity is not to exceed 2 m/s? [10.5]

40. A horizontal glass tube 4 cm in radius narrows gradually to 2 cm in radius. Water flows in the tube at a rate of 10 ℓ/min. Calculate the flow velocity in each part of the tube. [10.5]

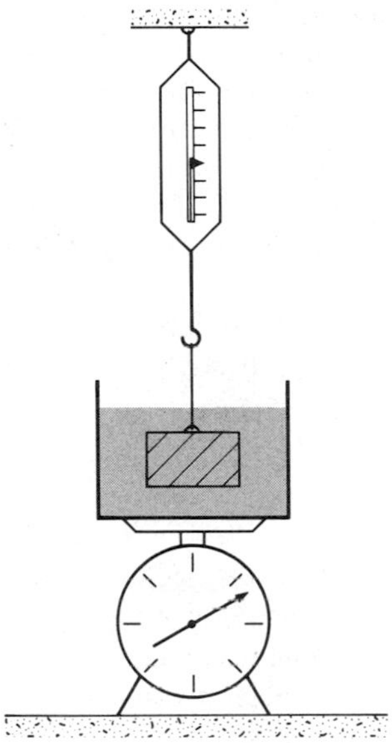

Figure for Problem 38

41. A pipe carries water in a streamline flow down a slope that is 2 m high. At the top of the slope the speed is 4 m/s, and at the bottom of the slope the cross-sectional area of the pipe has doubled. Calculate the pressure difference between the water at the top of the slope and the water in the expanded pipe at the bottom of the slope. [10.5]

42. Water flows in a horizontal pipe at a rate of 200 ℓ/min. At a point where the cross section is 40 cm^2, the gauge pressure is 200 kPa. What cross section is required in a Venturi throat to reduce the gauge pressure to 180 kPa? [10.5]

43. A liquid of relative density 0.95 is contained in a hypodermic syringe. The interior diameters of the needle and the body of the instrument are 0.2 mm and 2.1 cm, respectively. What force on the plunger is necessary to cause the liquid to eject horizontally from the needle into the atmosphere at a rate of 0.1 mℓ/s?

44. A large tank is partly full of water, and a gauge pressure of 4 atm is maintained in the upper part of the tank by compressed air. A small hole is drilled in the tank 7 m below the free surface of the water. Calculate the velocity of the water emerging from the hole.

45. A cylindrical tank standing on end contains water with its free surface 15 m above the floor level. The top is open to the atmosphere. A hole 1 mm in diameter is drilled horizontally 4 m below the surface of the water.

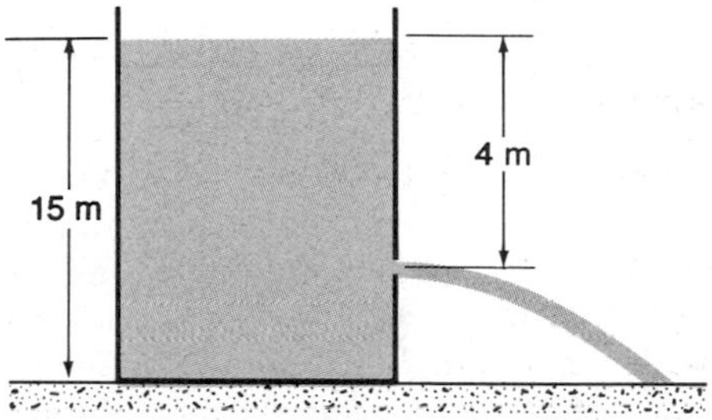

a. How far from the bottom of the tank does the jet of water strike the floor?
b. If the streamlines contract the area of the emerging water jet to 65% of the area of the hole, what is the flow in gallons per minute?

46. Water is flowing in a pipeline of nonuniform cross section. At a certain point the velocity is 3 m/s, the gauge pressure is 300 kPa, and the cross-sectional area of the pipe is 16 cm^2. At a second point in the line 12 m lower, the cross section is 10 cm^2. Calculate:

a. The water velocity at the second point
b. The gauge pressure at the second point
c. The flow rate in the pipe

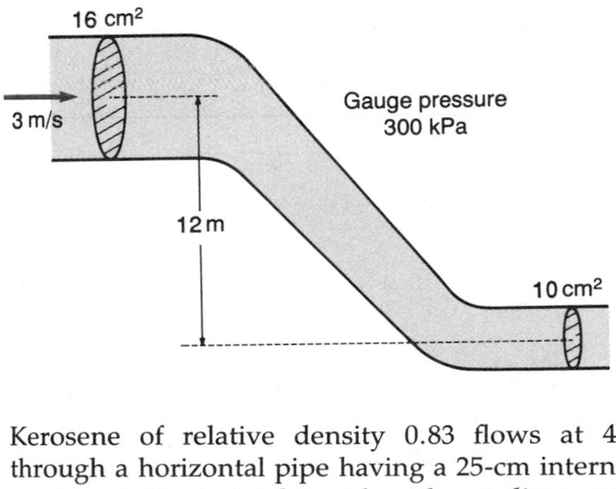

47. Kerosene of relative density 0.83 flows at 4 m/s through a horizontal pipe having a 25-cm internal diameter. A Venturi tube with a throat diameter of 8 cm is installed in the pipe. Calculate:

a. The flow of kerosene in liters per minute
b. The pressure difference between the pipe and the Venturi throat

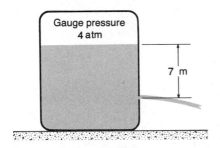

48. Water enters a building through a main supply pipe of large internal diameter and is delivered to a bathroom faucet 6 m higher on the second floor through a pipe of 1.8-cm internal diameter. Assume that the water in the main remains at 45 lb/in.^{2} (310 kPa) gauge pressure whether the water is flowing or not. Calculate:

a. The gauge pressure at the faucet if no water is being drawn
b. The gauge pressure at the faucet if water is being drawn at the rate of 30 gal/min (1.892 ℓ/s)

49. Water flows steadily at a rate of 3000 gal/min (189.2 ℓ/s) through a pipe of variable cross section. At the lower position the pipe diameter is 30 cm and the gauge pressure is 60 lb/in.^{2} (207 kPa). At the second position, which is 6 m higher, the pipe diameter is 40 cm. Calculate the gauge pressure at the second position.

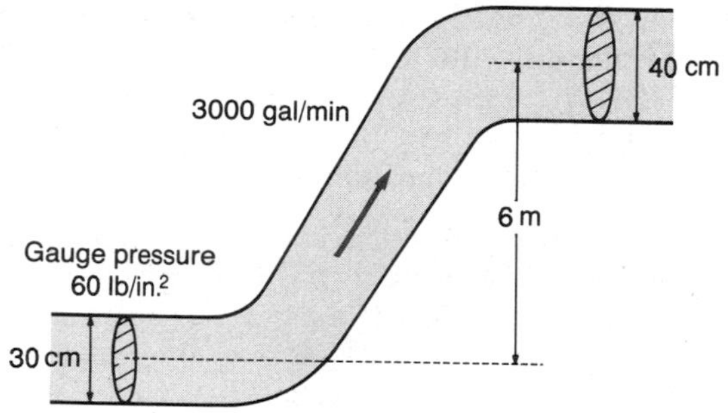

50. The velocity of flowing water is measured by a Pitot tube that records a pressure difference of 5 kPa. What is the water velocity?

51. A Pitot tube is mounted on an aircraft to measure the airspeed and records a pressure difference of 35 torr. What is the airspeed if the air density is 0.95 kg/m^3 when the measurement is taken?

C. Advanced-Level Problems

52. A precision measurement is made of the mass of water in a beaker by using a beam balance and brass masses:

$$\text{Mass of beaker alone} = 101.783 \text{ g}$$

$$\text{Mass of beaker full of water} = 454.691 \text{ g}$$

The balancing is done in air at 20°C, for which the density is 1.205 kg/m^3. The relative density of brass is 8.44. Calculate the mass of water in the beaker, taking explicit account of air buoyancy forces.

53. A barge of mass 25 000 kg is floating in a closed lock and carrying a 35 000-kg load of iron (relative density 7.86). The surface area of the lock is 500 m^2, and the waterline area of the barge is 300 m^2. If the load is dumped overboard, what is the change in the water level in the lock?

54. An aspirator pump produces a partial vacuum by utilizing the pressure drop that occurs when water is forced through a confined opening. Water flows at 3 m/s and with a gauge pressure of 250 kPa in the entrance pipe to such a pump. What reduction in diameter is necessary to achieve an absolute pressure of 10 kPa in the vacuum chamber?

11

ELASTICITY AND SIMPLE HARMONIC MOTION

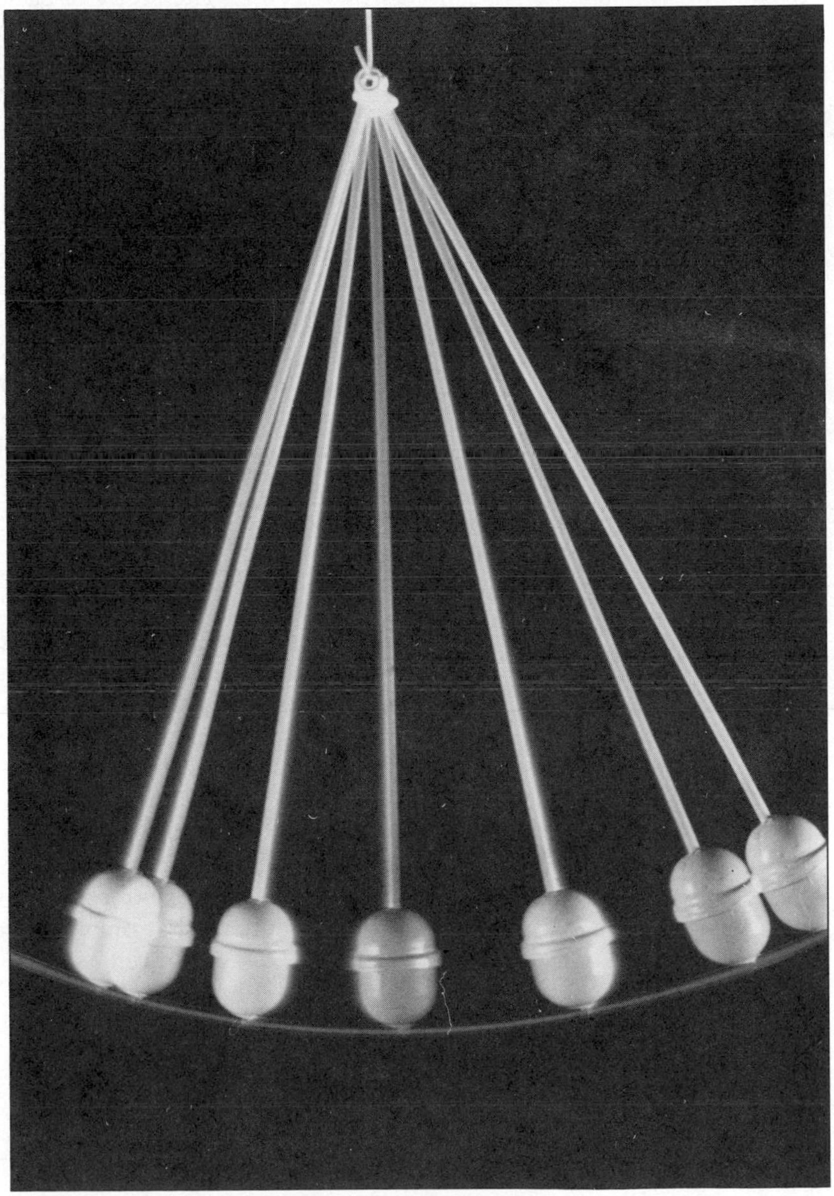

In our treatment of dynamics, we introduced the concept of a rigid body whose deformation is negligible when forces act on it. Because most solid objects undergo only small deformation even when quite large forces are applied, the rigid body idealization is excellent for many problems in the statics and dynamics of solid objects. However, even the strongest materials deform slightly when forces are applied. The branch of physics that studies the relationship between solid body deformations and the forces that cause them is called elasticity. The properties of a specific substance that enable us to calculate this relationship are called the elastic moduli of the substance. We shall see that different types of deformation involve different elastic moduli.

The study of elasticity leads naturally to the study of vibratory motion. Many of the vibrations that we encounter in everyday life are very complex. However, as happens so often in physics, there is an ideal simple type of vibration that we call simple harmonic motion. We conclude the chapter with a study of simple harmonic motion.

11.1 Hooke's Law and Elastic Moduli

When a force or set of forces is applied to some object, the object deforms in a way that is more or less complex. For example, a weight hung on a rubber cord stretches the cord and at the same time decreases the cord's cross-sectional area. Pressure applied uniformly to an air-filled rubber balloon (by taking it to the bottom of a swimming pool) shrinks its volume. A twisting force applied to each end of a long, thin steel bar bends the bar into a curved arc. In all of these cases the deformation is quite visible, and the object in question is not even close to being rigid under the forces acting on it, but deformation also occurs in cases where it is much less visible. A steel rod supporting a weight stretches slightly, and a cast iron block taken to the bottom of a swimming pool shrinks slightly under the increased pressure.

The basic physical quantities for describing deformation under the action of applied forces are stress and strain. We will see below that different physical deformations require different definitions of stress and strain, but we will start with general definitions that only need to be more specific for each special case.

In defining **stress** we relate the magnitude of the force applied to an object with the area of the object on which the force acts.

Definition of Stress

$$\text{Stress} = \frac{\text{applied force}}{\text{area}} \tag{11.1}$$

SI unit of stress: newton per square meter

The unit is the same as the pascal, which we introduced previously as the unit of hydrostatic pressure.

In the general definition we do not specify the direction of the applied force relative to the area over which it acts. For different types of stress, such as stretching stress or twisting stress, the force is directed differently toward the area. However, in each case the basic physical nature of stress is the force divided by the area. We also note that hydrostatic pressure is a special type of stress whose effect on solid objects is one of the special cases of elasticity that we will introduce below.

Strain is a measure of the deformation that stress causes. In defining strain, we compare the deformation with the original size of the object. Strain has no unit, since the same unit appears in both the numerator and the denominator of the defining relation. We must define *strain* more specifically for each of the different types of deformation.

Different substances behave differently under stress. A rubber cord stretches under stress and returns to its original size when the stress is removed (provided the stress is not too large), while bubble gum stretches under stress and more or less retains its new shape and size when the stress is removed. True, these examples are extreme cases, but they illustrate an important distinction. A substance that recovers its shape and size upon removal of stress is called *elastic*, but note that nothing is elastic without reservation, since all bodies break if subjected to enough stress. The point at which an object begins to be so heavily stressed that it does not recover is called the *elastic limit*. We will deal exclusively with elastic substances that are not stressed beyond their elastic limit.

Definition of Strain

$$\text{Strain} = \frac{\text{change in a geometric dimension}}{\text{original value of the dimension}} \tag{11.2}$$

Strain is a pure number.

In addition to being elastic over a wide range of stress, many materials possess another important property. Up to a certain value of stress, the ratio of the stress to the strain is a constant characteristic only of the material. This constant is called the **elastic modulus,** and the fact that such a region of stress exists is called **Hooke's law.**

Let us illustrate these ideas by sketching possible stress-strain diagrams for different materials. Figure 11.1 shows three cases. The arrows on the lines indicate the direction of changing stress. In Fig. 11.1(a) a stress is applied and then removed, but a permanent strain remains; the material is nonelastic. In Fig. 11.1(b) the material is elastic since the strain becomes zero after the stress is removed. Behavior typical of Hooke's law is shown in Fig. 11.1(c). Not only is the material elastic, but the stress is proportional to the strain. In what follows we concern ourselves exclusively with the third situation, which represents an ideal elastic solid. This is not to say that Hooke's law elastic materials are the rule. Many everyday materials are not in this category, but we will confine ourselves to those that are.

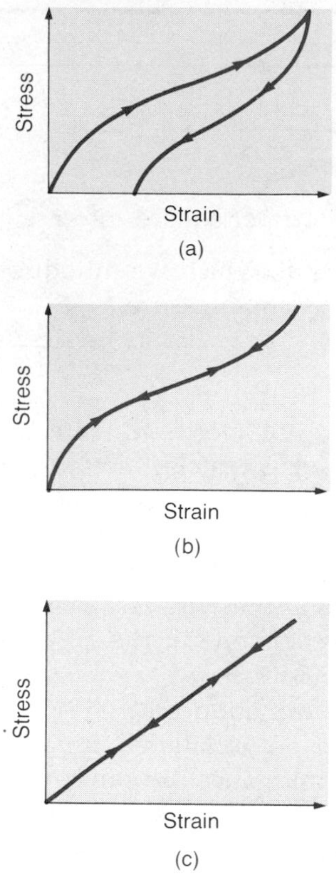

FIGURE 11.1 Stress-strain diagrams: (a) nonelastic behavior; (b) elastic behavior; (c) Hooke's law elastic behavior.

■ Hooke's Law of Elastic Behavior

$$\text{Elastic modulus} = \frac{\text{stress}}{\text{strain}} \qquad \textbf{(11.3)}$$

SI unit of elastic modulus: newton per square meter (same as the unit of stress)

Robert Hooke (1635–1703) was a contemporary of Isaac Newton. He had an outstanding professional career, being curator of experiments to the Royal Society in London and later professor of geometry at Oxford. However, a lot of his energy was dissipated in disputes about priority. He originally published his law of elastic behavior in code. Apparently he wished to claim the law but was not too keen on others knowing about it. His original experimental work involved the stretching of coiled springs and long iron wires. However, we now take his law in a general sense as referring to any type of stress-and-strain situation.

11.2 Young's Modulus, Bulk Modulus, and Shear Modulus

There are many ways to stress and strain a body, but three simple types have many practical applications. For each of these three types, we must make the general stress and strain definitions of Eqs. (11.1) and (11.2) more specific. The three elastic moduli that

SPECIAL TOPIC 12 THE ELASTIC PROPERTIES OF BIOLOGICAL MATERIALS

Biological materials are usually complex mixtures of substances, and their elastic properties do not follow the relatively simple laws that we have set forth for homogeneous, isotropic materials. For example, bone consists of living cells embedded in a structure made up largely of collagen fibers and hydroxyapatite crystals. Collagen is a protein that occurs in all connective tissues, and it is responsible for the strength of bone under tensile stress. Hydroxyapatite is an inorganic salt, and it is responsible for the strength of bone under compressive stress. (Note that a similar situation occurs in man-made mixtures such as reinforced concrete; in that case, the steel rods supply the strength under tensile stress, and the concrete supplies the strength under compressive stress.) It comes as no surprise, therefore, that the Young's modulus for bone under tensile stress is different from its value under compressive stress. In fact, the tensile modulus and the compressive modulus are about one-twelfth and one-twentieth of the values for steel.

For soft connecting tissues, the situation is quite different. In the space between the living cells, these tissues contain fibers composed of collagen, which give tensile strength, and other fibers of elastin, which permit the tissue to withstand large strains. Elastin is a protein and a member of a class of substances called elastomers. In the absence of strain, the molecules of an elastomer are coiled up, but as the substance is strained the molecules uncoil and permit large strain without breaking. Quite commonly, biological soft tissues can endure strains of 100%—a doubling of their original length. Moreover, there is in general no proportionality between stress and strain in these tissues. This means that concepts such as Young's modulus do not apply, and complete stress-strain data—and not simply an elastic constant—must be known.

arise from these three stress types are called the *principal elastic moduli,* and their importance lies in the fact that any stress (and its consequent strain) may be built up from them.

The first basic stress type is tension or compression of a uniform rod along its axis. Consider a rod whose unstressed length is l_0 and whose unstressed cross-sectional area is A. Figure 11.2 illustrates the effect of longitudinal tension or compression. In tension the rod extends to a greater length than it had originally, while in compression it contracts. However, the distortion is not only in the direction of the applied force. As one might expect, the cross-sectional area of an extended rod decreases, while that of a contracted rod increases. These are the qualitative features of the situation. To make our ideas precise, we will use the general definitions of the previous section.

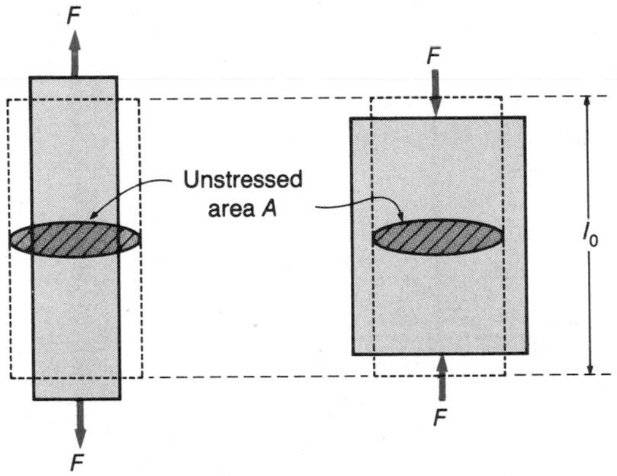

FIGURE 11.2 Longitudinal tension and compression of a uniform rod. The dotted lines show the original dimensions of the rod, and the solid lines show the strained dimensions.

Longitudinal stretching or compression requires definitions of longitudinal stress and strain. We refer to both cases as *tensile* to avoid continual distinction between stretching and compression. The definition of tensile stress follows from Eq. (11.1) as applied to the situation of Fig. 11.2. We include the subscript on the symbol for area as a reminder that it is normal to the direction of the tensile force.

Definition of Tensile Stress

$$\text{Tensile stress} = \frac{\text{tensile force}}{\text{cross-sectional area normal to the direction of the force}} = \frac{F}{A_{\perp}} \tag{11.4}$$

Note that it is always the unstrained cross-sectional area that is used in the definition of stress.

The definition of tensile strain follows from the general definition in a similar way. Let l_0 and l be the unstretched and the stretched lengths of the rod, respectively. Using the general definition of Eq. (11.2), we define tensile strain as follows:

Definition of Tensile Strain

$$\text{Tensile strain} = \frac{\text{change in length}}{\text{original length}} = \frac{l - l_0}{l_0} \tag{11.5}$$

For a Hooke's law elastic material, Eq. (11.3) now permits us to define the tensile elastic modulus, which is called **Young's modulus.**

Definition of Tensile Elastic Modulus

$$\text{Young's modulus} = \frac{\text{tensile stress}}{\text{tensile strain}}$$

$$Y = \frac{F/A_{\perp}}{(l - l_0)/l_0} \tag{11.6}$$

Table 11.1 lists the Young's modulus for several materials within their elastic limits. Of course, exact value for a given material depends on the material's previous history of heat treatment and mechanical working. The few significant figures quoted in the table reflect the uncertainty in the values.

Table 11.1 Representative values of elastic constants of different materials

Material	*Young's modulus* Pa (N/m^2)	*Bulk modulus* Pa (N/m^2)	*Shear modulus* Pa (N/m^2)	*Poisson's ratio*
Steel	2.0×10^{11}	1.6×10^{11}	7.7×10^{10}	0.29
Cast iron	1.1×10^{11}	8.3×10^{10}	4.7×10^{10}	0.25
Copper	1.1×10^{11}	1.2×10^{11}	4.0×10^{10}	0.35
Aluminum	6.9×10^{10}	6.9×10^{10}	2.6×10^{10}	0.33
Lead	1.6×10^{10}	7.7×10^{9}	5.6×10^{9}	0.43
Brass	9.8×10^{10}	8.4×10^{10}	4.1×10^{10}	0.34
Glass	5.5×10^{10}	5.2×10^{10}	2.2×10^{9}	0.25
Rubber	4.0×10^{6}	2.4×10^{9}	—	0.50
Water	—	2.2×10^{9}	—	—
Benzene	—	1.2×10^{9}	—	—
Acetone	—	8.0×10^{8}	—	—
Methanol	—	9.4×10^{8}	—	—

Returning to Fig. 11.2, let us consider the change in lateral dimensions that the strained rod undergoes. It has been found experimentally that the lateral strain bears a constant relation to the tensile strain. The constant is called **Poisson's ratio.**

■ Definition of Poisson's Ratio

$$\text{Poisson's ratio} = -\frac{\text{lateral strain}}{\text{tensile strain}} \qquad \textbf{(11.7)}$$

The negative sign in the definition reflects the fact that a longitudinal extension is associated with a lateral contraction, and vice versa. We do not state Eq. (11.7) in symbols, since it applies to rods of arbitrary cross section. For a rod of circular cross section, we can define the lateral strain as the change in radius divided by the original radius; for a rod of square cross section, we can use the change in side length divided by the original side length. In short, a suitable definition of lateral strain depends on the exact shape of the rod. Let us consider an example to illustrate the very small strains that may be associated with quite large stresses.

EXAMPLE 11.1

A steel tie rod 50 cm long and 1 cm in radius is subjected to a tensile force of 1 ton (8896 N). By what amount does the rod stretch, and what is the decrease in its radius? (The changes in dimension shown in the diagram are exaggerated.)

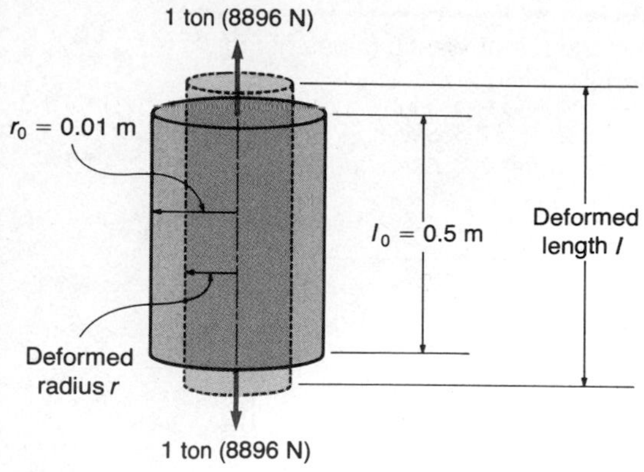

SOLUTION To calculate the stress in the rod, we need to know the cross-sectional area normal to the direction of the applied force:

$$\begin{aligned} A_\perp &= \pi r^2 \\ &= \pi \times (10^{-2}\ \text{m})^2 \\ &= 3.142 \times 10^{-4}\ \text{m}^2 \end{aligned}$$

The tensile stress is given by Eq. (11.4):

$$\begin{aligned} \text{Tensile stress} &= \frac{F}{A_\perp} \\ &= \frac{8896\ \text{N}}{3.142 \times 10^{-4}\text{m}^2} \\ &= 2.831 \times 10^7\ \text{Pa} \end{aligned}$$

The Young's modulus of steel is 2.0×10^{11} Pa from Table 11.1. We can now calculate the tensile strain from Eq. (11.6):

$$\begin{aligned} \text{Young's modulus} &= \frac{\text{tensile stress}}{\text{tensile strain}} \\ 2.0 \times 10^{11}\ \text{Pa} &= \frac{2.831 \times 10^7\ \text{Pa}}{\text{tensile strain}} \\ \therefore \text{Tensile strain} &= 1.416 \times 10^{-4} \end{aligned}$$

The extension is given by Eq. (11.5):

$$\begin{aligned} \text{Tensile strain} &= \frac{\text{change in length}}{\text{original length}} \\ 1.416 \times 10^{-4} &= \frac{\text{change in length}}{0.50\ \text{m}} \\ \therefore \text{Change in length} &= 7.08 \times 10^{-5}\ \text{m} \end{aligned}$$

Despite such a large applied force, the extension of the rod is less than 0.1 mm. We can calculate the decrease in the radius of the rod by using the value of Poisson's ratio for steel. Equation (11.8) gives:

$$\begin{aligned} \text{Poisson's ratio} &= -\frac{\text{lateral strain}}{\text{tensile strain}} \\ 0.29 &= -\frac{\text{lateral strain}}{1.416 \times 10^{-4}} \\ \therefore \text{Lateral strain} &= -4.105 \times 10^{-5} \end{aligned}$$

Since the rod is of circular cross section, the radius is a suitable lateral dimension. With the help of Eq. (11.2) we write:

$$\text{Lateral strain} = \frac{\text{change in radius}}{\text{original radius}}$$

$$-4.105 \times 10^{-5} = \frac{\text{change in radius}}{0.01 \text{ m}}$$

$$\therefore \text{Change in radius} = -4.105 \times 10^{-7} \text{ m}$$

The negative sign reflects the fact that the rod contracts laterally. The amount of contraction is less than 1 μm. ■

Note that our original calculation of stress used the unstrained area of cross section. A check shows that the difference between the unstrained and strained areas is about 1 part in 10 000, which is quite negligible. The convention in problems on elasticity is to use the unstrained dimensions to calculate the stress.

Although tensile stress and strain may seem to offer the simplest example of distortion under an applied force, lateral strain complicates the situation. In reality, a simpler situation results when the distortion is the uniform compression or dilation associated with a hydrostatic pressure stress. Figure 11.3 illustrates this situation for a spherical-shaped body. The difference between this and the tensile case is that we now have no longitudinal and lateral directions. The stress presses in (or pulls out) equally from every direction and is identical in nature with the hydrostatic pressure defined by Eq. (10.1). Under pressure, the volume of the body changes from its original value V_0 to a strained value V. We define volumetric strain with the help of Eq. (11.2):

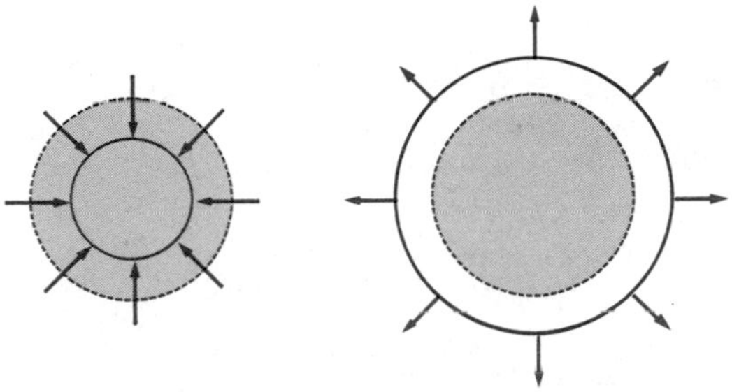

FIGURE 11.3 Uniform compression or dilation of a spherical body under a uniform pressure. The dotted lines show the original sphere, and the solid lines show the strained sphere.

■ **Definition of Volumetric Strain**

$$\text{Volumetric strain} = \frac{\text{change in volume}}{\text{original volume}} = \frac{V - V_0}{V_0} \tag{11.8}$$

The constant ratio of pressure to volumetric strain is called the **bulk modulus.** With the help of Eq. (11.3), our definition of bulk modulus follows:

■ **Definition of Bulk Modulus**

$$\text{Bulk modulus} = -\frac{\text{volumetric stress (pressure)}}{\text{volumetric strain}}$$

$$B = -\frac{p}{(V - V_0)/V_0} \tag{11.9}$$

The negative sign reflects the fact that an increase in pressure always causes a decrease in volume. Table 11.1 shows typical values of the bulk modulus for common materials. Liquids and gases also undergo volumetric strain when subjected to a uniform pressure, and Table 11.1 also includes typical values of the bulk modulus for common liquids. However, because the elastic properties of gases are strongly temperature dependent, we will treat the topic separately in a subsequent chapter.

EXAMPLE 11.2

A block of rubber, originally at atmospheric pressure, is placed in an evacuated enclosure. Calculate the volumetric strain.

SOLUTION The rubber is subject to a *pressure reduction* of 1 atm. We take this into account in writing the data:

$$p = -1 \text{ atm} \times 1.013 \times 10^5 \text{ Pa/atm}$$
$$= -1.013 \times 10^5 \text{ Pa}$$
$$B = 2.4 \times 10^9 \text{ Pa} \qquad \text{from Table 11.1}$$

We use Eq. (11.9) to calculate the volumetric strain:

$$\text{Bulk modulus} = -\frac{\text{pressure}}{\text{volumetric strain}}$$
$$2.4 \times 10^9 \text{ Pa} = -\frac{-1.013 \times 10^5 \text{ Pa}}{\text{volumetric strain}}$$
$$\therefore \text{Volumetric strain} = 4.2 \times 10^{-5}$$

The volume of the rubber increases by a few one-hundredths of 1%. ■

The third type of distortion is called *shearing*. We illustrate the nature of shear stress by referring to the cube of material shown in Fig. 11.4. A set of forces parallel to four faces of the cube is required to produce the shear strain shown in the diagram. The end face of the cube (which was originally square) is distorted into a parallelogram. If the original area of a face of the cube is *A*, the appropriate definition of shear stress follows from Eq. (11.1).

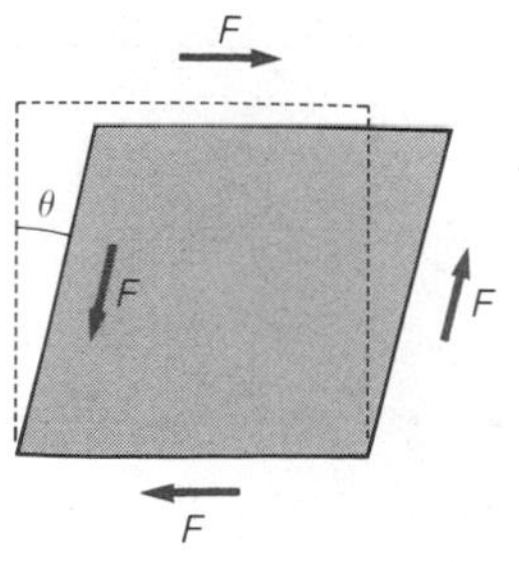

FIGURE 11.4 A cube of material (viewed edge on) twisted out of shape by a shear stress. The dotted lines show the original shape, and the solid lines show the strained shape.

■ Definition of Shear Stress

$$\text{Shear stress} = \frac{\text{shear force}}{\text{area parallel to the direction of the force}}$$
$$= \frac{F}{A_{\parallel}} \qquad \textbf{(11.10)}$$

The subscript on the symbol for area is a reminder that the shear force is parallel to the area in question. For shear strain, we require a dimensionless geometric measure of the distortion of the cube. This is given by the angle (marked θ on the diagram) by which two adjacent faces deviate from their original perpendicular condition. Equation (11.3) then serves to define the **shear modulus.**

■ Definition of Shear Modulus

$$\text{Shear modulus} = \frac{\text{shear stress}}{\text{shear strain}}$$

$$S = \frac{F/A_{\parallel}}{\theta} \qquad \textbf{(11.11)}$$

Typical values of the shear modulus are also shown in Table 11.1.

EXAMPLE 11.3

An aluminum cube, 10 cm on a side, is subject to shear stress. A force of 50 000 N acts on each of four faces, as shown in Fig. 11.4. Calculate the shear strain.

SOLUTION We use Eq. (11.10) to calculate the shear stress:

$$\text{Shear stress} = \frac{F}{A_{\parallel}}$$

$$= \frac{50\ 000\ \text{N}}{0.1\ \text{m} \times 0.1\ \text{m}}$$

$$= 5 \times 10^6\ \text{N/m}^2$$

From Table 11.1 the shear modulus of aluminum is 2.6×10^{10} N/m^2. From Eq. (11.11) we get:

$$\text{Shear modulus} = \frac{\text{shear stress}}{\text{shear strain}}$$

$$2.6 \times 10^{10}\ \text{N/m}^2 = \frac{5 \times 10^6\ \text{N/m}^2}{\text{shear strain}}$$

$$\therefore \text{Shear strain} = 1.92 \times 10^{-4}$$

We recall that the shear strain is the angle (measured in radians) by which the corners of the end face differ from a right angle. The forces on the cube faces are in excess of 5 tons, but the angle of distortion is only 1.92×10^{-4} rad $\approx 0.01°$. ■

Shear stress is important in problems concerning the twisting of rods and fibers. However, these problems involve advanced mathematics that is beyond the scope of this book.

We noted above that all substances fail to obey Hooke's law if the stress exceeds a certain point. In fact, everything breaks if subjected to a large enough force. When this occurs, the Hooke's law elastic moduli of Table 11.1 no longer apply. Can we assign any sort of approximate limit to the use of the elastic moduli? For most metals, elastic behavior is limited to a *strain not exceeding about 0.1%*. For other substances there is little that we can say. Some substances are very brittle and can tolerate even less strain. On the other hand, rubber may easily retain its elastic properties with strains exceeding 100%.

11.3 Simple Harmonic Motion

Experience provides us with many examples of vibratory motion. The bob of a pendulum clock moves from side to side, a mass suspended from a spring oscillates up and down, and a wooden block floating in water bobs up and down. The common feature in these examples of vibratory motion is a regular and recurrent reversal of the direction of motion. Following the well-proven technique of physics, we will introduce an idealized type of vibration and then apply the conclusions to real situations.

In previous chapters we considered various basic types of motion for a point mass, and in every case found that the type of motion goes hand in hand with a specific acceleration. We have studied motion at constant acceleration in some detail, and we have also studied uniform motion in a circular path, which requires a centripetal acceleration of constant magnitude. Similarly, we will define our idealized type of vibratory motion by identifying the type of acceleration that causes it.

We define the ideal vibratory motion, which is called *simple harmonic motion* (abbreviated "SHM"), as follows:

Definition of Simple Harmonic Motion

A body that moves in a straight line with an acceleration whose magnitude is proportional to its distance from a fixed point on the line and directed toward the fixed point is executing simple harmonic motion.

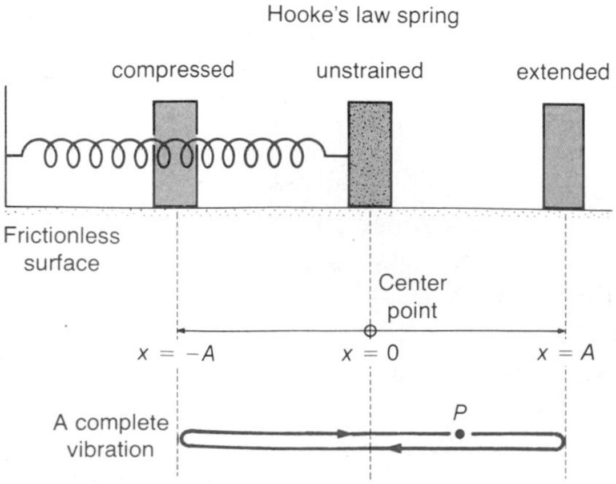

FIGURE 11.5 Illustration of the definition of simple harmonic motion.

Consider the situation illustrated in Fig. 11.5. A massive body undergoes a vibratory motion on a frictionless horizontal surface under the influence of a *Hooke's law spring of negligible mass.* We define a Hooke's law spring as one for which the spring force is proportional to the amount of extension (or compression) from the unstressed position. A Hooke's law elastic solid is certainly a Hooke's law spring within its elastic limit, but for a solid the strain is very small, and any back-and-forth motion is almost undetectable. When we wind the appropriate type of metal into a coil, the resulting spring may well obey the Hooke's law condition for very large displacements from its unstressed position. The constant of proportionality between the spring force and the displacement from the unstressed position is the **Hooke's law spring constant.**

■ Definition of the Spring Constant of a Hooke's Law Spring

Spring force = −spring constant × displacement from unstressed position

$$F = -kx \qquad \textbf{(11.12)}$$

SI unit of spring constant: newton per meter
Abbreviation: N/m

The negative sign expresses the fact that the force exerted by the spring is in the opposite direction to its displacement from the unstressed position.

We can now show that a vibrating body that is subject only to the force of a massless Hooke's law spring (or its equivalent) moves with SHM. To see this, we suppose that the object of Fig. 11.5 has a mass m. The net force is related to the acceleration (in a horizontal direction) by Newton's second law [Eq. (4.1)], $F = ma$; but Eq. (11.12) gives $F = -kx$ for the Hooke's law spring force. It follows by equating these expressions that the acceleration of the mass is given by:

$$a = -(k/m)x \qquad \textbf{(11.13)}$$

This result agrees with our general definition of SHM. The acceleration is proportional to the distance from the center point of the motion ($x = 0$) and the constant of proportionality is the spring constant divided by the mass (k/m). Moreover, the negative sign assures us that the acceleration is always directed toward the center point (if x is positive, the acceleration points in the negative direction and vice versa).

It is important to note that a real spring is not necessary for SHM, although it is sometimes associated with it. In the pendulum of a grandfather clock, the "spring" is the force of gravity; in the case of a wooden block bobbing in the water, the forces of buoyancy and gravity provide the spring. The essential element for SHM is proportionality between the force on the vibrating body and its displacement from the center point of the motion.

We will continue by defining several important quantities relating to SHM.

1. A **complete vibration** is one complete passage through every part of the motion. From point P of Fig. 11.5, the particle undergoes a complete vibration when it travels to one end, then back to the other end, and finally returns to the original point. The particle is then back in its original position *and* moving in its original direction. A special case of a complete vibration starts at one end, goes to the other end, and then returns to the starting point.
2. The **period** is the time taken for a complete vibration, which is independent of the position chosen for the starting point of the complete vibration.
3. The **frequency** is the number of complete vibrations in 1 s; it is the reciprocal of the period (provided the period is measured in seconds).

■ Definition of Frequency

$$\text{Frequency} = \frac{1}{\text{period}}$$

$$\nu = \frac{1}{T} \qquad \textbf{(11.14)}$$

SI unit of frequency: cycle per second
New name: hertz
Abbreviation: Hz

The symbol for frequency is the Greek letter nu (ν). The cycle per second is named hertz in honor of the German physicist Heinrich Hertz (1857–1894), who made the first experimental discovery of radio frequency electromagnetic waves.

4. The **amplitude** is absolute value of the maximum displacement from the center point. The total distance between the end points of the motion is twice the amplitude.

In this section we have defined the type of acceleration that causes SHM and introduced the definitions of some kinematic quantities associated with it. In the following sections we will investigate the dynamics of SHM and derive some useful equations relating to it.

11.4 The Period of Simple Harmonic Motion

To learn more about the dynamics of SHM, we will now find a basic relation that expresses the period in terms of the spring constant and the mass of the vibrating body. For this purpose, we begin by investigating the special relationship between the centripetal acceleration experienced by a point moving in a circular path and the acceleration of a point moving in SHM.

Consider a point traveling with constant angular velocity ω in a circle of radius A centered at the origin O (Fig. 11.6). Suppose that the point is at P' on the circumference of the circle at some instant of time. Let P be the foot of a perpendicular drawn from P' to the x-axis, and let the angle $\angle OP'P$ be θ. We wish to prove that the point P describes SHM back and forth across the diameter of the circle as point P' revolves uniformly around the circumference. The motion of P is vibratory since it moves between the two extremes $x = A$ and $x = -A$. The vibration is simple harmonic if we can show that the acceleration is proportional to the distance from the center and directed back toward it. The tangential velocity is related to the angular velocity by Eq. (9.11b):

$$v = r\omega = A\omega$$

since the radius of this particular circle is A. We can calculate the centripetal acceleration required to keep the point revolving in a circle by using Eq. (3.9):

$$\begin{aligned} a_c &= \frac{v^2}{r} = r\omega^2 \\ &= A\omega^2 \end{aligned}$$

The horizontal component of this acceleration is:

$$\begin{aligned} a &= -A\omega^2 \sin\theta \\ &= -\omega^2 x \end{aligned}$$

This acceleration is proportional to the distance from the center and directed toward it; it follows that the point P is performing SHM. Substituting in Eq. (4.1) we find:

$$\begin{aligned} F &= ma \\ -kx &= m(-\omega^2 x) \\ \therefore\ \omega &= \left(\frac{k}{m}\right)^{1/2} \end{aligned} \qquad \textbf{(11.15)}$$

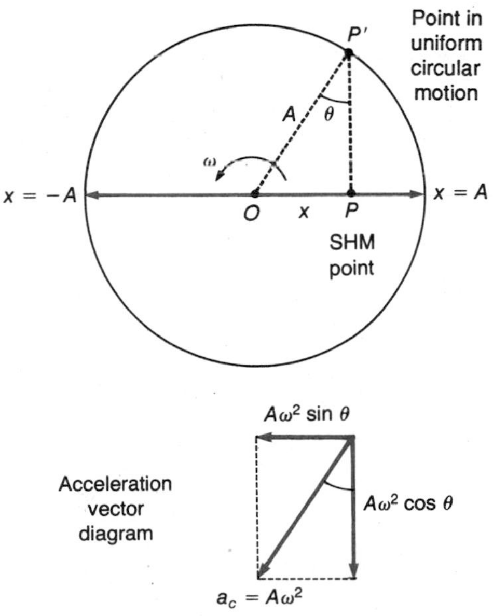

FIGURE 11.6 Relation between uniform motion in a circle and SHM.

We can look at SHM in another way. At all times the position of a point executing SHM is at the foot of a perpendicular from a phantom particle that revolves in a circle with angular velocity given by Eq. (11.15). From this point of view, the particle P is the primary motion, and the particle P' is an imaginary particle associated with it. The circle on which P' moves is called the **circle of reference** for the SHM. The period of the SHM is the same as the time required for one complete revolution of the phantom particle on the circle of reference. A displacement of 2π rad with angular velocity ω requires a time given by:

$$T = \frac{2\pi}{\omega} \qquad \textbf{(11.16)}$$

Combining Eqs. (11.15) and (11.16) yields the following for the period of the SHM:

■ Equation Relating the Period, the Mass, and the Spring Constant in SHM

$$\text{Period of SHM} = 2\pi\left(\frac{\text{mass}}{\text{spring constant}}\right)^{1/2}$$

$$T = 2\pi\left(\frac{m}{k}\right)^{1/2} \qquad \textbf{(11.17)}$$

Note that the period is independent of the amplitude of the motion, depending only on the mass and the spring constant.

EXAMPLE 11.4

A mass of 50 g vibrates in SHM with a frequency of 150 Hz. Calculate the spring constant associated with this motion.

SOLUTION The period, the mass, and the spring constant are related by Eq. (11.17). To set the data in convenient form we use Eq. (11.14) to find the period:

$$\nu = \frac{1}{T}$$

$$\therefore T = \frac{1}{\nu}$$

$$= \frac{1}{150\ \text{Hz}}$$

$$= 6.67 \times 10^{-3}\ \text{s}$$

We can now calculate the spring constant using Eq. (11.16):

$$T = 2\pi\left(\frac{m}{k}\right)^{1/2}$$

$$\therefore k = \frac{4\pi^2 m}{T^2}$$

$$= \frac{4\pi^2 \times 0.05 \text{ kg}}{(6.67 \times 10^{-3} \text{ s})^2}$$

$$= 4.44 \times 10^4 \text{ N/m}$$ ■

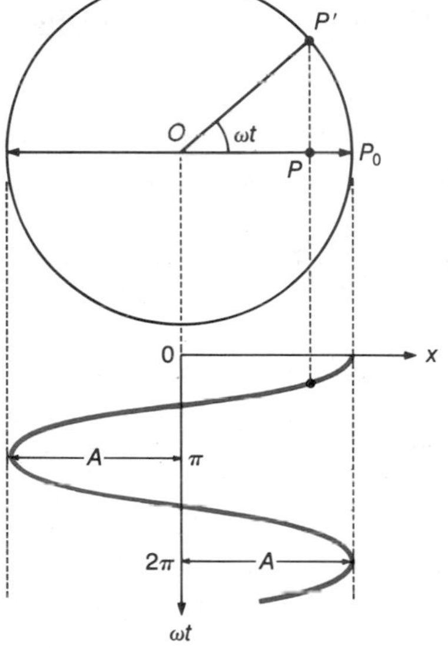

FIGURE 11.7 Illustration of the displacement in SHM as a cosine function of time.

The idea of the phantom particle moving with uniform speed on the circle of reference permits a simple graphical display of SHM displacement. Consider the diagram of Fig. 11.7, in which a point mass vibrates horizontally in SHM of amplitude A about the center point O. Correspondingly, a phantom particle moves with angular velocity ω around the circle of reference of radius A. At $t = 0$, let the vibrating point mass be at the end point of the motion P_0. At this instant the phantom particle is in the same position. At a later time t the phantom particle has moved with constant angular velocity ω to the position P' and its angular displacement from the initial position is ωt. At the same time the real point mass has moved to the position P. We can express the position of the real point mass as follows:

$$OP = OP' \cos \omega t$$

or

$$x = A \cos \omega t \tag{11.18}$$

In the lower part of the diagram we show the displacement of the real point mass from the center position expressed as a function of time. With $t = 0$ at the end point of the motion, the curve is a cosine function. If we had chosen $t = 0$ at the center point of the motion, we would have obtained a sine curve. There is no important physical distinction between the two; it is merely a matter of selecting the initial time instant.

EXAMPLE 11.5

A point mass on a Hooke's law spring is pulled aside 10 cm from the center position and then released. The resulting SHM has a period of 0.4 s. Calculate the displacement of the point mass when $t = 0.55$ s.

SOLUTION The situation corresponds exactly to Fig. 11.7. When the point mass is released at $t = 0$, it is at the end point of the

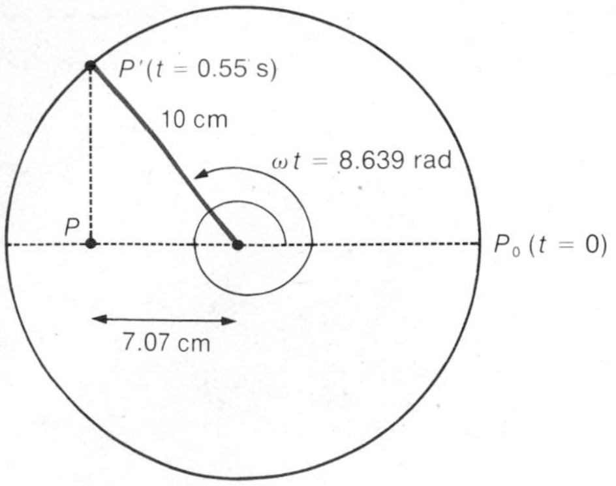

motion. We can therefore set $A = 10$ cm and calculate the angular velocity of the phantom particle using Eq. (11.16):

$$T = \frac{2\pi}{\omega}$$

$$\therefore \omega = \frac{2\pi}{T}$$

$$= \frac{2\pi}{0.4 \text{ s}}$$

$$= 15.71 \text{ rad/s}$$

We now have all the data required to calculate the displacement from Eq. (11.17):

$$x = A \cos \omega t$$

$$= 10 \text{ cm} \times \cos (15.71 \text{ rad/s} \times 0.55 \text{ s})$$

$$= 10 \text{ cm} \times \cos 8.639 \text{ rad}$$

$$= -7.07 \text{ cm}$$

The negative sign indicates that the particle is to the left of the center position at the time in question. The diagram illustrates the solution in terms of the motion of the phantom particle on the circle of reference. ■

Graphical representation of SHM is especially important in discussing wave motion, which we will do in some detail in the following chapter.

11.5 The Simple Harmonic Motion Energy Equation

Energy conservation is an important concept in all branches of dynamics, and SHM is no exception. A point mass vibrating horizontally as in Fig. 11.5 comes to rest momentarily at each end of its path of motion—its kinetic energy at these points is zero. At other points on the path, the velocity and the kinetic energy are not zero. The energy conservation principle requires that some other energy be associated with the motion since kinetic energy is not conserved. Including gravitational potential energy does not help, since the particle moves back and forth along a horizontal path with no change in this quantity. However, the spring is alternately compressed and expanded, and in these states of strain the spring can do work by returning to its unstrained length. The energy equation must therefore contain a term describing the **elastic potential energy** in the spring. We assume

SPECIAL TOPIC 13 THE EXTRAORDINARY JUMPING ABILITY OF THE FLEA

The common flea is able to jump to a height of more than 100 times its own size. Research with high-speed cameras indicates that the initial velocity of the flea is about 1 m/s, and that this velocity is attained during 1 ms prior to takeoff. The mass of the flea is about 500 μg, so the takeoff kinetic energy as given by Eq. (7.2) is 0.25 μJ. Since this energy is acquired during the takeoff interval of 1 ms, Eq. (7.4) gives the average power requirement as 0.25 mW.

It is also known that the leg muscles of the flea account for about one-fifth of its total mass, and that insect muscles can develop a maximum power of about 60 W/kg. This means that the leg muscles of the flea can contribute about 6 μW during takeoff; the required power of 0.25 mW is more than 40 times greater.

The secret lies in the ability of the flea to store energy as elastic potential energy in a biological material called resilin. A small pad of this material is located in the large hind leg of the flea. Prior to jumping, the resilin pad is slowly compressed until the required 0.25 μJ of energy is stored; at this point a small ratchetlike device catches in a latch and keeps the resilin pad in its compressed state without further exertion by the flea. In order to jump, another small muscle releases the catch and makes the stored elastic potential energy available.

Finally, we note that storing 0.25 μJ of energy using muscles that develop 6 μW of power requires about 40 ms; even though this is a very short time interval, it is much longer than the 1-ms takeoff period.

that the mass of the spring is negligible so that no kinetic energy is associated with the movement of the spring itself—the only energy associated with the spring is the elastic potential energy, which depends on the extension of the spring, and not on the velocity of the point mass. Consider the spring shown in Fig. 11.8, whose unstressed length is l_0. Application of a force F stretches the spring to a length $l_0 + x$. If it is a Hooke's law spring, the force required to hold it at this new length is given by Eq. (11.12):

$$F = kx$$

The force required to extend the spring is therefore not a constant, but begins at zero and increases up to the value quoted. The average force to extend the spring is the mean of 0 and kx. We therefore have for the average spring force:

$$\langle F \rangle = \tfrac{1}{2}kx$$

The work done on the spring is the average force multiplied by the extension of the spring:

$$\begin{aligned} W &= \langle F \rangle x \\ &= \tfrac{1}{2}kx^2 \end{aligned}$$

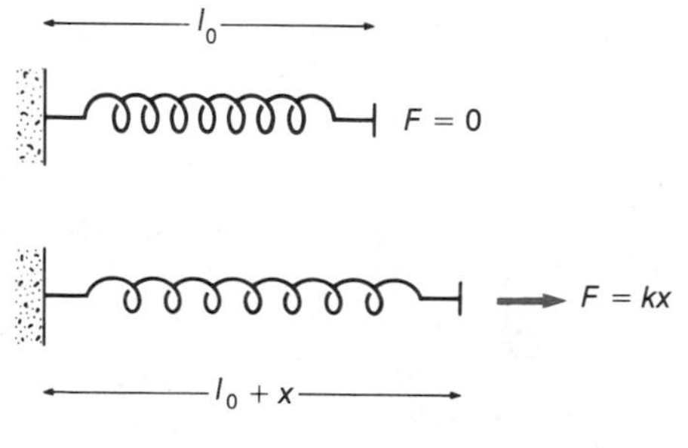

Average force = $\frac{1}{2}(0 + kx) = \frac{1}{2}kx$

FIGURE 11.8 Gradual extension of a Hooke's law spring under an applied force.

The work done in stretching the spring is equal to the energy stored in the form of elastic potential energy:

■ Formula for Elastic Potential Energy of a Hooke's Law Spring

$$\text{Elastic potential energy} = \tfrac{1}{2} \times \text{spring constant} \times (\text{spring extension})^2$$

$$\text{Elas PE} = \tfrac{1}{2}kx^2 \qquad \textbf{(11.19)}$$

Since the spring extension is squared in Eq. (11.19), the elastic potential energy is positive for both negative and positive values of x. Thus, energy is stored in the spring during both extension and compression.

During its vibration the spring continually interchanges kinetic with elastic potential energy. At the end points of the path the energy is all elastic potential stored in the spring, and at the mid-point the energy is all kinetic stored in the vibrating mass. At any other point the energy is partly kinetic and partly elastic potential. These considerations permit us to write a formula for the total energy of a point mass undergoing SHM. Consider a particle of mass m at some point (displacement x) between the end point and the center point of the motion. Let the velocity of the particle at this point be v. The kinetic energy of the mass is given by Eq. (7.2), and the elastic potential energy in the spring by Eq. (11.19). Combining these results, we have a formula for the total energy associated with the SHM.

■ Energy Equation for SHM

$$\text{Total energy} = \text{kinetic energy} + \text{elastic potential energy}$$

$$E_{\text{SHM}} = \tfrac{1}{2}mv^2 + \tfrac{1}{2}kx^2 \qquad \textbf{(11.20)}$$

Two special cases of Eq. (11.20) arise if the particle is at the end point of the motion ($v = 0$) or at the center point ($x = 0$). In the first case the energy is entirely elastic potential energy, and in the second case it is entirely kinetic. The energy equation applied to these two special points gives the following results:

At the end point of the path ($x = A; v = 0$)

$$E_{\text{SHM}} = \tfrac{1}{2}kA^2 \qquad \textbf{(11.20a)}$$

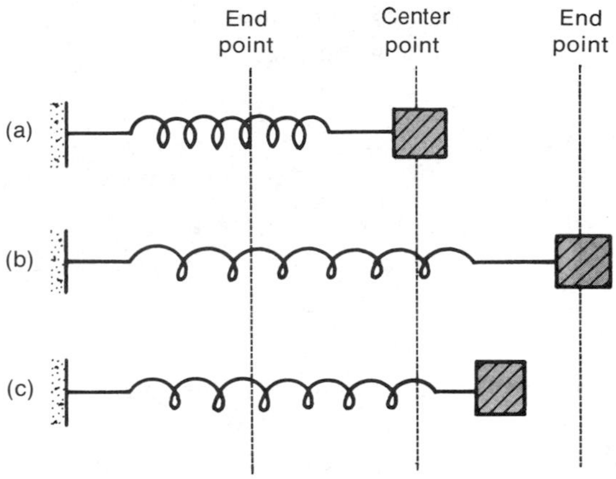

	Displacement	Velocity	Elas PE	KE
(a)	0	v_{max}	0	$\frac{1}{2} mv_{max}^2$
(b)	A	0	$\frac{1}{2} kA^2$	0
(c)	x	v	$\frac{1}{2} kx^2$	$\frac{1}{2} mv^2$

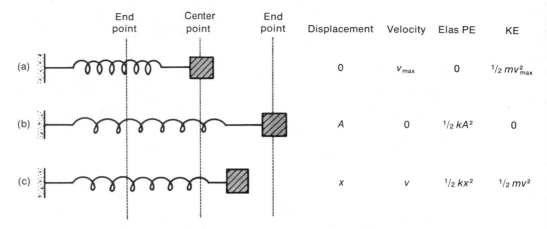

FIGURE 11.9 The division of the total energy of vibration between kinetic energy of the particle and elastic potential energy of the spring (a) at the center point, (b) at the end point, and (c) at some intermediate point of the motion.

At the center point of the path ($x = 0$; $v = v_{max}$)

$$E_{SHM} = \tfrac{1}{2}mv_{max}^2 \tag{11.20b}$$

We write the velocity at the center point of the path as v_{max} since it has its greatest value at that point. Moreover, the vibrating mass on the spring is an isolated system. Energy conservation requires that the energies at either end point, the center point, and any in-between point all be equal. Figure 11.9 illustrates the division of the total energy between kinetic energy of the mass and elastic potential energy of the spring at various points.

The period of SHM does not depend on the amplitude, but Eq. (11.20a) shows that the energy depends on the square of the amplitude. A point mass oscillating with SHM continues indefinitely with the same amplitude if no frictional force dissipates the energy.

EXAMPLE 11.6

A mass of 100 g vibrates horizontally in SHM with a frequency of 20 Hz and an amplitude of 15 cm. Calculate:

a. The total energy of the motion

b. The velocity of the mass when it is 10 cm from the center point

SOLUTION

a. We can use the energy equation to calculate the velocity at a given point if the other quantities are known. To find the spring constant for the motion, we proceed in two steps. Equation (11.14) gives the period:

$$T = \frac{1}{\nu}$$

$$= \frac{1}{20\ \text{Hz}}$$

$$= 0.05\ \text{s}$$

We can then calculate the spring constant from Eq. (11.17):

$$T = 2\pi\left(\frac{m}{k}\right)^{1/2}$$

$$\therefore k = \frac{4\pi^2\, m}{T^2}$$

$$= \frac{4\pi^2 \times 0.1\ \text{kg}}{(0.05\ \text{s})^2}$$

$$= 1580\ \text{N/m}$$

With this information the total energy of the vibration is given by Eq. (11.20a):

$$E_{\text{SHM}} = \tfrac{1}{2}kA^2$$

$$= \tfrac{1}{2} \times (1580\ \text{N/m}) \times (0.15\ \text{m})^2$$

$$= 17.78\ \text{J}$$

b. We can now calculate the velocity at the point $x = 0.1$ m by using Eq. (11.20):

$$E_{\text{SHM}} = \tfrac{1}{2}mv^2 + \tfrac{1}{2}kx^2$$

$$17.78\ \text{J} = \tfrac{1}{2} \times 0.1\ \text{kg} \times v^2 + \tfrac{1}{2} \times 1580\ \text{N/m} \times (0.1\ \text{m})^2$$

$$= 0.05\ \text{kg} \times v^2 + 7.90\ \text{J}$$

$$\therefore v = \pm\ 14.1\ \text{m/s}$$

The positive and negative signs correspond to the different possible directions of the velocity to the right and left of the center point. ■

When a mass vibrates vertically on a Hooke's law spring, the gravitational potential energy changes as the position of the mass changes. Because of this continual change in the gravitational potential energy, we need to determine the correct form for the energy equation. Consider a massless Hooke's law spring hanging

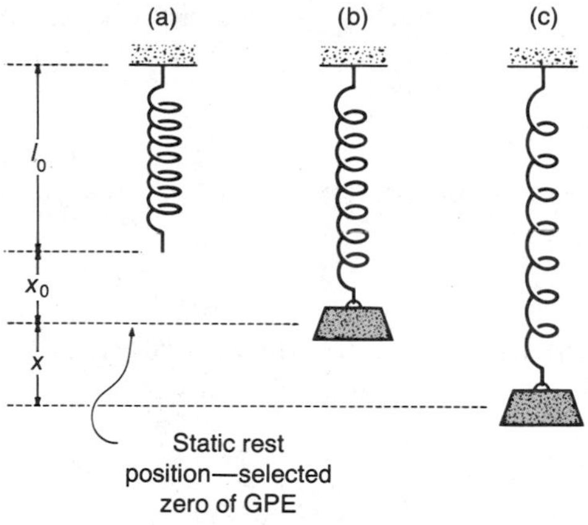

FIGURE 11.10 Mass vibrating on a vertical Hooke's law spring: (a) no mass attached; (b) static position with mass attached; (c) attached mass vibrating.

vertically (Fig. 11.10). With no mass attached the length of the spring is l_0; with a mass m attached to the lower end, the extension of the spring is x_0. The force extending the spring is the weight of the mass mg, and the restoring force is kx_0. The static rest position corresponds to the equality of these two forces:

$$mg = kx_0$$

Let us now set the mass vibrating and write down the energy when the extension is x below the static rest position. Three elements comprise the total energy. The kinetic energy is given by Eq. (7.2), KE $= \frac{1}{2}mv^2$. The elastic potential energy follows from Eq. (11.19), with the extension of the spring set equal to $(x + x_0)$; this gives Elas PE $= \frac{1}{2}k(x + x_0)^2$. To calculate the gravitational potential energy we must select a zero reference level. Let us take the static rest position of the mass as the reference zero for gravitational potential energy. In the extended position the mass is a distance x *below* the reference level. Using Eq. (7.3), we find GPE $= -mgx$. Adding all three terms, we obtain the total energy:

$$\begin{aligned}\text{Total energy} &= \text{KE} + \text{Elas PE} + \text{GPE} \\ &= \tfrac{1}{2}mv^2 + \tfrac{1}{2}k(x + x_0)^2 + (-mgx) \\ &= (\tfrac{1}{2}mv^2 + \tfrac{1}{2}kx^2) + (kxx_0 - mgx) + (\tfrac{1}{2}kx_0^2)\end{aligned}$$

Let us consider the three bracketed terms in order.

1. The first term is the same as the expression for the total energy of a horizontal spring, Eq. (11.20).
2. The second term is zero, since:

$$mg = kx_0$$

3. The third term is a constant for a given mass and a given spring. We can ignore this term in calculations involving *changes* in kinetic and elastic potential energies, since $\frac{1}{2}kx_0^2$ does not change during SHM.

We are left with Eq. (11.20) for the energy of a mass oscillating vertically on a Hooke's law spring. The gravitational potential energy does not appear explicitly, since it is canceled out by a term in the elastic potential energy. We ignore another constant term in the elastic potential energy, which does not contribute to energy changes.

EXAMPLE 11.7

A Hooke's law spring of negligible mass hanging vertically from a support is extended 10 cm by a mass of 800 g attached to its lower end. The mass is then pulled down an additional distance and released to carry out a vertical SHM. The velocity on passing through the static rest position (center point of SHM) is 2 m/s.

a. Calculate the energy of the motion.

b. What distance below the static rest position was the mass pulled down before release?

SOLUTION With the conventions explained above for selecting the zero of energy, we can use the same energy equations that apply to horizontal SHM. The total energy follows immediately from Eq. (11.20b):

$$\begin{aligned} E_{SHM} &= \tfrac{1}{2}mv_{max}^2 \\ &= \tfrac{1}{2} \times 0.8 \text{ kg} \times (2 \text{ m/s})^2 \\ &= 1.6 \text{ J} \end{aligned}$$

To make further progress we need to know the spring constant. The weight of a mass of 800 g stretches the spring 10 cm. Equating the weight to the restoring force at the static rest position, we have:

$$\begin{aligned} mg &= kx_0 \\ 0.8 \text{ kg} \times 9.81 \text{ m/s}^2 &= k \times 0.1 \text{ m} \\ \therefore k &= 78.48 \text{ N/m} \end{aligned}$$

We can now calculate the amplitude of the SHM using the alternate energy equation, Eq. (11.20a):

$$\begin{aligned} E_{SHM} &= \tfrac{1}{2}kA^2 \\ 1.6 \text{ J} &= \tfrac{1}{2} \times 78.48 \text{ N/m} \times A^2 \\ \therefore A &= 0.202 \text{ m} \end{aligned}$$

The amplitude is the same as the maximum distance below the static rest position. It follows that the mass was pulled down an additional 20.2 cm. ■

11.6 The Simple Pendulum

One of the best-known regular vibratory motions is the motion of a pendulum. The usual pendulum of a grandfather clock consists of a metal disk at the end of a wooden rod, but this construction makes it difficult to calculate the period of vibration. As usual, we postulate a simple idealized model that turns out to be a good approximation to many real cases. The model consists of a point mass at the end of an inextensible cord of negligible mass. We

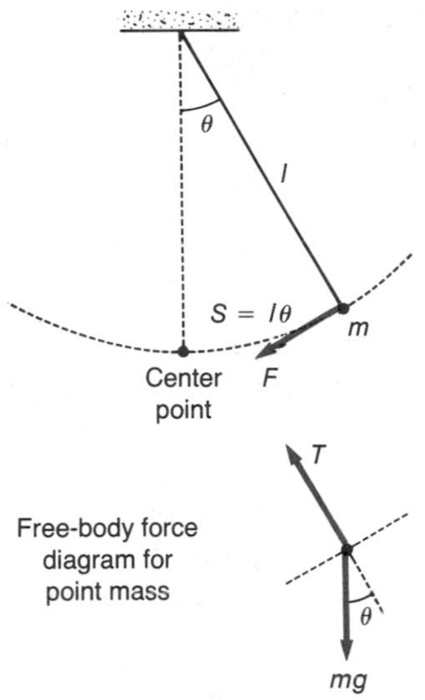

FIGURE 11.11 The simple pendulum.

call this model a **simple pendulum.** Consider a simple pendulum vibrating back and forth between two end points (Fig. 11.11). At some point in the path the cord makes an angle θ with the vertical, and the forces acting on the point mass are its weight and the tension in the cord.

We take the component of net force perpendicular to the cord as the restoring force, and we have:

$$\text{Restoring force on point mass} = -mg \sin \theta$$

$$F \approx -mg\,\theta$$

In the approximation we have used the result of Appendix A5, which requires θ to be much smaller than 1 rad. Multiplying and dividing by l, we can rewrite the expression for the restoring force:

$$F \approx -\frac{mg}{l}\, l\theta \tag{11.21}$$

The quantity $S = l\theta$ is the length of the arc separating the point mass from the center point. The pendulum does not move exactly in a straight line, but if the angle θ is very small, the path of motion is approximately straight. If we regard the arc length S as the displacement from the center point, comparison of Eqs. (11.12) and (11.21) gives us the spring constant of the motion:

$$\text{Spring constant} \approx \frac{\text{mass of pendulum bob} \times \text{acceleration due to gravity}}{\text{pendulum length}}$$

$$k \approx \frac{mg}{l} \tag{11.22}$$

Acceptance of these arguments requires some good will. The pendulum bob does not move in a straight line, the restoring force does not point toward the center, and the magnitude of the restoring force is not exactly proportional to the distance from the center. All of these objections become less serious, however, as the angle of displacement of the pendulum from the vertical becomes smaller. Granted the approximation, the equations of SHM all apply. In particular, we can calculate the period of the pendulum by substituting Eq. (11.22) into Eq. (11.17).

Period of a Simple Pendulum

$$\text{Period} = 2\pi\left(\frac{\text{pendulum length}}{\text{acceleration due to gravity}}\right)^{1/2}$$

$$T = 2\pi\left(\frac{l}{g}\right)^{1/2} \tag{11.23}$$

Note that the period of a simple pendulum does not depend on either the mass of the bob or the amplitude of the vibration (provided that the amplitude is small). In spite of the fact that Eq. (11.23) is approximate, we can use it to make very accurate measurements of g with simple everyday equipment.

EXAMPLE 11.8

A student wishing to measure the value of g in her home town ties a heavy lead sinker to the end of a long fishing line and makes a simple pendulum by hanging the sinker from a point on a high ceiling. She measures the distance from the support point to the center of the sinker as 723.4 cm and then times 300 complete vibrations by using the sweep second hand of her wristwatch. The time observed is 26 min 59 s. What value of g does the student find?

SOLUTION The period is the time of one complete vibration:

$$T = \frac{26 \text{ min } 59 \text{ s}}{300} = \frac{1619 \text{ s}}{300}$$

$$= 5.397 \text{ s}$$

Use Eq. (11.23) to calculate g:

$$T = 2\pi\left(\frac{l}{g}\right)^{1/2}$$

$$\therefore g = \frac{4\pi^2 l}{T^2}$$

$$= \frac{4\pi^2 \times 7.234 \text{ m}}{(5.397 \text{ s})^2}$$

$$= 9.805 \text{ m/s}^2$$

The student is probably well justified in keeping the fourth significant digit of this answer if the measurements are taken carefully and the pendulum is pulled aside no more than about 20 cm. ■

KEY CONCEPTS

Stress is the force divided by the area, and **strain** is the change in the dimension divided by the original dimension. See Eqs. (11.1) and (11.2).

The stress is proportional to the strain for a **Hooke's law** substance within the elastic limit. The constant of proportionality is called the **elastic modulus.** See Eq. (11.3).

The elastic moduli for tensile, volumetric, and shear strains are called **Young's modulus, bulk modulus,** and **shear modulus,** respectively. See Eqs. (11.6), (11.9), and (11.11).

The **Hooke's law spring constant** is the spring force divided by the displacement from the unstressed length. See Eq. (11.12).

The motion of a point mass subject to the action of a Hooke's law spring is called **simple harmonic motion.** See the display on p. 301.

With regard to simple harmonic motion:

1. The **period** is the elapsed time for a complete vibration.
2. The **frequency** is the number of complete vibrations in 1 s.
3. The **amplitude** is the magnitude of the maximum displacement from the center point.

Every simple harmonic motion is related to the uniform motion of a particle on the **circle of reference.** See discussion on p. 303.

The **period** of a simple harmonic motion is determined by the mass and the spring constant. See Eq. (11.17). The period is independent of the amplitude.

The **total energy** of a simple harmonic motion is partly **kinetic energy** and partly **elastic potential energy.** The proportion of each type in the total changes periodically with time. See Eqs. (11.20), (11.20a), and (11.20b).

The motion of the bob of a **simple pendulum** is approximately simple harmonic for small angular displacements. The pendulum period depends only on its length and the acceleration due to gravity. See Eq. (11.23).

QUESTIONS FOR THOUGHT

1. "The Young's modulus of a material is the stress required to produce unit strain." Explain this statement together with any reservations that you have about it.

2. Give examples of material for which the Young's modulus exceeds the breaking stress and also those for which it does not.

3. Prepare a list of vibratory motions that seem to be approximately simple harmonic. Also list some vibratory motions that are clearly not simple harmonic, and explain why they are not.

4. Can you predict the period of oscillation of an unknown mass on a spring of unknown force constant by simply measuring the extension of the spring that occurs when the mass is suspended from it?

5. Would it be possible to use the properties of a simple pendulum to designate the values of any of the fundamental units of mass, length, and time? If so, how many of these fundamental quantities could be designated in this way, and which ones?

6. Estimate the walking speed of a human being by using a model that treats the swing of the leg in one stride as being one-half of the periodic vibration of a simple pendulum. What advantage does your model give to people with long legs?

7. The amplitude of any real vibratory motion decreases with time because frictional forces dissipate the energy. This effect is called *damping* of the vibratory motion. Give examples of practical devices in which the damping of a vibratory motion is made as small as possible and also examples in which large damping is present.

PROBLEMS

A. Single-Substitution Problems

(Use the data of Table 11.1 for problems 3 through 6.)

1. Calculate the tensile stress in a copper wire 6 mm in radius that supports a mass of 200 kg. [11.2]

2. A piece of rubber originally 12 cm long is stretched to a new length of 18 cm. Calculate the tensile strain. [11.2]

3. An aluminum rod is 80 cm long and of circular cross section 1 cm in diameter. Calculate the extension of the rod under a tensile load of 4500 N. [11.2]

4. A copper tube 30 cm long has inner and outer radii of 2 cm and 3 cm, respectively. How large a compressive load is required to cause a reduction of 0.14 mm in the length of the rod? [11.2]

5. A cast iron cylinder is 7.5 cm in diameter. How many metric tons can it support if the compressive strain is not to exceed 0.50%? [11.2]

6. A steel rod of circular cross section supports a mass of 10 metric tons. What should be its radius if the compressive strain is not to exceed 0.06%? [11.2]

7. Acetone in a container is subjected to a pressure of 30 atm. What is the fractional change in volume? [11.2]

8. What pressure in atmospheres must be applied to a block of rubber to reduce its volume by 8%? [11.2]

9. The valve spring from an automobile compresses 0.5 mm under a load of 140 N. Calculate the spring constant. [11.3]

10. What force is required to produce an extension of 2 cm in a spring whose spring constant is 120 N/m? [11.3]

11. Calculate the period of a vibration whose frequency is 440 Hz. [11.3]

12. A mass undergoing SHM takes 0.16 s to travel from one end point to the other. The end points are 20 cm apart. Calculate the period, the frequency, and the amplitude of the motion. [11.3]

13. A Hooke's law spring has a spring constant of 3000 N/m. What mass will have an SHM vibration frequency of 60 Hz on this spring? [11.4]

14. A vertical Hooke's law spring extends 12 cm when a 2.5-kg mass is attached to it. What is the period of vibration of the mass? [11.4]

15. What spring constant is required of a Hooke's law spring for a 1-kg mass to have a vibrational frequency of 8 Hz? [11.4]

16. A vertical Hooke's law spring stretches 5 cm when a 4-kg mass is attached to it. How much elastic energy is stored in the spring? [11.5]

17. A force of 40 N extends a Hooke's law spring by 10 cm. What extension of the spring would be required to store 10 J of elastic potential energy? [11.5]

18. What is the length of a simple pendulum that has a period of 1 s at a place where $g = 9.808\ \text{m/s}^2$? [11.6]

19. A simple pendulum 6.1 m long has a period of 4.953 s. What is the local value of g? [11.6]

B. Standard-Level Problems

(Use the data of Table 11.1 for problems 20 through 37.)

20. A steel wire originally 3 m long and 1 mm in diameter supports a mass of 6 kg. Calculate:

a. The stress in the wire
b. The extension of the wire
c. The change in diameter of the wire [11.2]

21. Calculate the maximum load that a steel wire 1 mm in radius can support if the stress is not to exceed $6 \times 10^7\ \text{N/m}^2$. Also calculate the change in diameter of the wire under this maximum load.

22. A steel I-beam column 8 m long has the cross-sectional dimensions shown in the diagram. What load causes a 5-mm reduction in the height of the column?

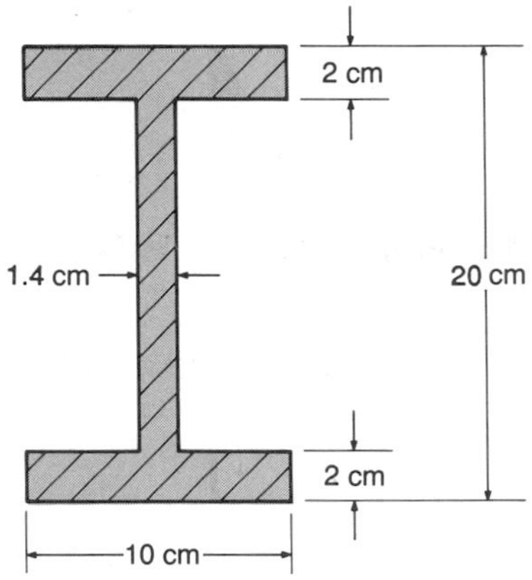

23. A steel pipe column of outside radius 4 cm and inside radius 3.2 cm supports a ceiling beam in a basement. What load does the pipe support if the linear strain is 3.4×10^{-4}?

24. A piece of rubber 40 cm long and of rectangular cross section 9 mm × 6 mm is used to support a 300-g mass. Calculate:

a. The extension of the rubber
b. The reduction in the dimensions of the cross section

25. The beam and small masses of a Cavendish apparatus (see Fig. 6.3) have a total mass of 80 g and are supported by a glass fiber of original diameter 0.1 mm. The original length of the glass fiber was 1.6 m. Calculate:

a. The extension of the fiber when supporting the beam and masses
b. The change in diameter of the fiber under load

26. An elevator car has a mass of 800 kg and carries ten people who average 85 kg. Its maximum upward acceleration is $2\ \text{m/s}^2$. What is the strain in the steel supporting cable of cross-sectional area $6\ \text{cm}^2$ when the car is rising at maximum acceleration?

27. Two rods, one of steel and the other of aluminum, are each 40 cm long and of square cross section 1.5 cm × 1.5 cm. The rods are placed side by side, and the combination subjected to a tensile force of 7000 N. Assuming that the two rods each experience the same tensile strain, calculate the stress in each rod.

28. A uniform concrete beam 16 m long and of mass 8 metric tons is supported at the ends by two vertical steel cables each of 2.5-cm radius. An additional load of 12 metric tons is situated 3 m from one of the cables. Calculate the strain in each cable.

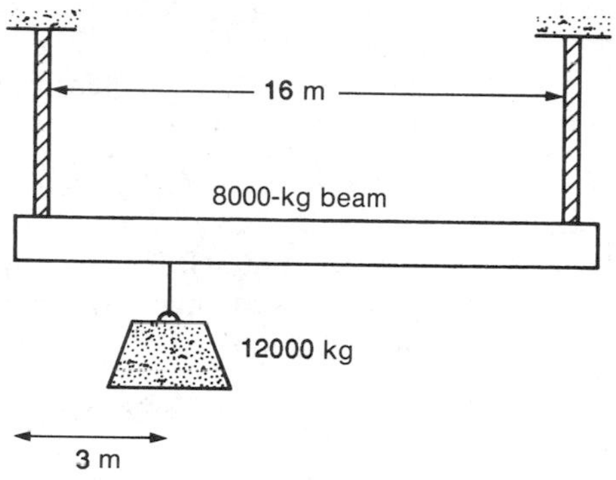

29. A brass cube originally of 4 cm edge is subject to a uniform pressure that decreases the length of each side by 1 part in 800. Calculate the pressure in atmospheres.

30. A piece of rubber in the shape of a sphere has a radius of 4 cm. Calculate the change in radius if it is immersed in water at the bottom of the wall of the Hoover dam (180 m deep).

31. The deepest known ocean trench is south of Guam, 9750 m deep. What is the percent increase in the water density at this depth due to the pressure? The bulk modulus of seawater is 2.1×10^9 N/m^2.

32. A 3-ℓ jug full of methanol has a cylindrical plug 1 cm in diameter. Assuming that the jug and the plug remain rigid, what force is required on the plug to reduce the volume of methanol to 2999 cm^3?

33. The density of methanol at room temperature and atmospheric pressure is 810.0 kg/m^3. A spherical steel container of internal radius 10 cm is filled with methanol under pressure, and careful measurements give the mass of methanol in the container as 3.435 kg. Calculate the pressure in the container assuming that its dimensions do not change.

34. A circular cylinder of inner radius 8 cm is closed at one end and has a tightly fitting piston at the other end. The cylinder contains 2000 cm^3 of oil, and a force of 7×10^4 N on the piston causes it to move inward 1 mm. Calculate the bulk modulus of the oil. (Assume that the cylinder remains rigid.)

35. A brass cube 20 cm on edge is subjected to shear forces of 4×10^5 N, as shown in Fig. 11.4. Calculate the value of the shear strain.

36. A steel cube 30 cm on edge is subjected to forces as shown in Fig. 11.4. Calculate the magnitude of each force that is required to cause a shear strain of 0.025°.

37. Calculate the shear modulus of a material if a cube 25 cm on edge undergoes a shear strain of 0.01° when subjected to forces as shown in Fig. 11.4, each of magnitude 8×10^4 N.

38. A Hooke's law spring of original length 36 cm is compressed to 32 cm in length by a force *F*; an additional force of 60 N compresses it to a length of 30.5 cm. Calculate the value of the force *F*. [11.3]

39. A Hooke's law spring hanging vertically supports a mass of 8 kg and is 80 cm long when supporting this mass. The addition of further 4 kg causes an additional 2.5-cm extension of the spring.

a. What is the spring constant?
b. What is the length of the spring in an unstressed condition? [11.3]

40. A mass of 2 kg is carrying out SHM with an amplitude of 8 cm and a frequency of 4 Hz. Calculate:

a. The velocity at the center point of the motion
b. The acceleration at an end point of the motion [11.4]

41. A certain mass is carrying out SHM on a Hooke's law spring with amplitude 5 cm. The acceleration of the mass at the end point of the motion is 18 m/s^2. Calculate the period of the SHM. [11.4]

42. A vertical spring whose upper end is fixed has a mass of 4 kg attached to the lower end; the period of vertical oscillation is 0.36 s. Given that the unstretched length of the spring is 20 cm, calculate its equilibrium length if the oscillatory motion is stopped. [11.4]

43. A vertical spring whose upper end is fixed is extended 8 cm when a mass of 2 kg is attached to its lower end.

a. What mass must be attached to the end of the spring so that the period of oscillation is 1.5 s?
b. If the amplitude of the oscillation is 6 cm, where is the mass 2 s after passing through its lowest position?

44. Two springs whose spring constants are 50 N/m and 80 N/m are attached to the opposite sides of a block of mass 2 kg resting on a frictionless surface; the other ends of the springs are held fixed. Calculate the frequency of horizontal oscillation of the block.

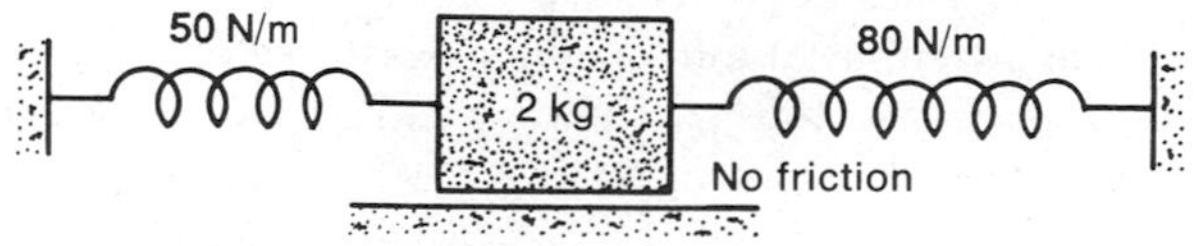

45. A 1500-kg automobile sinks 6 cm on its springs when four persons, each of 85 kg, enter it. Calculate:

a. The frequency of vertical oscillation of the unloaded automobile
b. The frequency of vertical oscillation of the automobile with the four persons on board

46. A mass of 80 g vibrates in SHM with a frequency of 30 Hz and an amplitude of 12 cm. What is the maximum velocity of the mass? [11.5]

47. A certain mass vibrates in SHM with an energy of 36 J, a period of 0.04 s, and an amplitude of 12 cm. What is the value of the mass? [11.5]

48. A mass of 250 g vibrates in SHM with a period of 1.6 s. What amplitude is required for the energy of the vibrational motion to be 0.30 J? [11.5]

49. At what position is the energy of a mass executing SHM stored equally as kinetic energy in the mass and elastic potential energy in the spring? Express your answer as a fraction of the amplitude. [11.5]

50. A certain mass executing SHM has 30% of its total energy kinetic at a point 3 cm from the center of the motion. What is the amplitude of the motion?

51. A suspended 4-kg mass stretches a Hooke's law spring by 10 cm. The mass is then set vibrating in SHM, and its velocity at the center point is 5 m/s. Calculate:

a. The period of the SHM
b. The amplitude of the SHM
c. The velocity at a point 15 cm above the center point of the motion

52. The pan of a spring balance has a mass of 1.5 kg, and the balance is calibrated to read zero with the pan empty. A 10-kg sack of potatoes causes a downward deflection of 16 cm when placed on the pan. After the balance settles, the pan is pushed downward and then released.

a. Calculate the vibration period of the sack of potatoes.
b. How far was the pan pushed down if the velocity of the sack is 0.8 m/s when the scale reading is 10 kg?

53. The up-and-down motion of an automobile piston in the cylinder is approximately SHM. In a certain engine the piston has a mass of 0.5 kg and the stroke (twice the amplitude) is 9 cm. The engine is turning at 3500 rev/min. Calculate:

a. The effective spring constant of the SHM
b. The acceleration of the piston at the top point of its motion

54. A Hooke's law spring of negligible mass, 320 N/m spring constant, and 60 cm in length is suspended vertically. A mass of 4 kg is attached to the lower end of the spring, held at rest momentarily with the spring unextended, and then released.

a. How far does the 4-kg mass fall before coming momentarily to rest again?
b. What is the period of the SHM?
c. What is the velocity of the mass at a point 15 cm below its original position?

55. A child swings on a swing whose cables are 3.8 m long. Treating the child and the swing as a simple pendulum, calculate the number of times per minute that she passes through the position in which the chains are vertical. [11.6]

56. A pendulum has a period of exactly 2 s at a place where $g = 9.8010\ \text{m/s}^2$. What change in length must be made to keep the period of 2 s if the pendulum is moved to a place where $g = 9.8210\ \text{m/s}^2$? [11.6]

57. A pendulum is adjusted to have a period of exactly 1 s in Washington, D.C., where $g = 9.8008\ \text{m/s}^2$. Calculate its period in Denver, where $g = 9.7961\ \text{m/s}^2$.

58. A pendulum has a period of 1 s on the surface of the earth. What is the period of this pendulum if it is taken to the surface of the moon? (The earth is about 81.4 times more massive than the moon, and its radius is about 3.67 times larger.)

C. Advanced-Level Problems

59. **a.** A cube of copper of edge 10 cm is subjected to a linear compression force of 5×10^4 N across a pair of opposite faces. Calculate the change in volume of the cube.
b. The same cube is subjected to a uniform pressure that exerts a total inward force of 5×10^4 N on each face. Calculate the change in volume of the cube.

60. A block of oak of relative density 0.71 and dimensions 30 cm × 30 cm × 10 cm floats on water with the 10-cm dimension vertical. The block is pushed into the water 1 cm deeper and then released. Calculate:

a. The effective spring constant of the SHM
b. The period of the motion
c. The maximum vertical velocity of the block

(Ignore frictional forces.)

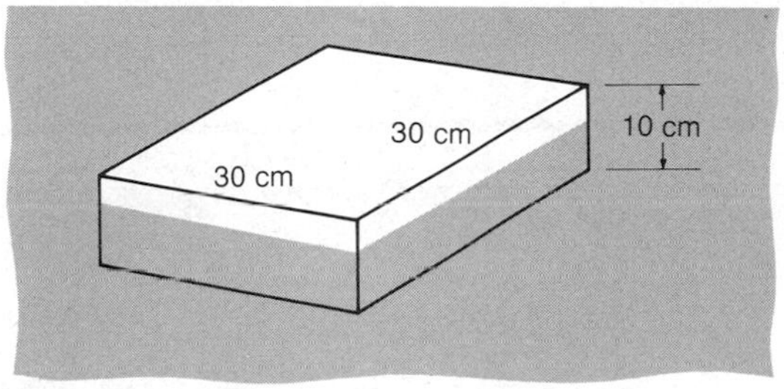

61. A 4-kg block rests on a frictionless surface and is attached to a vertical wall by a Hooke's law spring of spring constant 3000 N/m. The block is struck by a 25-g bullet traveling horizontally at 500 m/s that embeds in the block. Calculate:

a. The amplitude of the resulting SHM
b. The vibration frequency

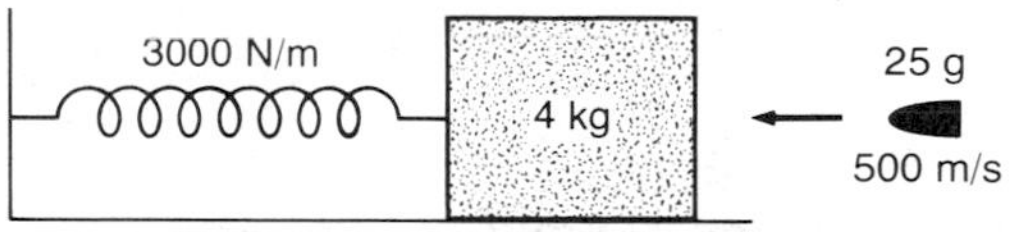

12

WAVE MOTION

The waves of the ocean, the boom of a supersonic aircraft, the vibrations of a guitar string, and the electromagnetic waves from a TV transmitter are all examples of wave motion. Although there are obvious differences, certain basic characteristics identify all of these examples as waves:

1. A disturbance that propagates—in most instances, in an identifiable physical medium
2. The capability of transmitting energy from place to place without transmission of the medium between the two places

Every example of wave motion has these features; differences among the examples cited arise from other reasons. Deep water waves are periodic disturbances that continually transmit energy across the surface of the water. They are called periodic waves. A sonic boom is a disturbance in the air that transmits a single shot of energy—this nonrepetitive type of wave is called a pulse. The vibration of a guitar string is a special type of wave called a standing wave. Although the disturbance on a guitar string does not travel along the string, it can be shown to be equivalent to two equal periodic waves traveling in opposite directions. Electromagnetic waves carry energy from the transmitting antenna to the receiving antenna. These waves do not require an identifiable physical medium, but are transmitted through space, which is a vacuum apart from their presence.

12.1 Simple Harmonic Waves

We will begin by studying a special type of continuous wave called a **simple harmonic wave.** To fix our ideas, consider a wave traveling along a very long guitar string. The wave on such a string is somewhat like a wave in the middle of the ocean—the boundaries of the medium are so far away that we can study features of the wave as if the boundaries did not exist.

We illustrate the special characteristics of a simple harmonic wave in Fig. 12.1. It is easy to imagine how we can determine the space variation of the wave shape—we just take a photograph of the string at some given time instant. For a simple harmonic wave, the result shown in Fig. 12.1(a) is a sinusoidal curve, which exhibits a series of crests, or places at which the string has a maximum displacement in one direction from its undisturbed position. There is also a series of troughs that correspond to the maximum displacement in the other direction. However, this is only a photograph of the instantaneous shape of the wave. As time goes by, the whole shape moves in the direction of the wave propagation, not by bodily movement of the string in the direction of propagation, but by a vibratory motion of each part of the string transverse to the propagation direction.

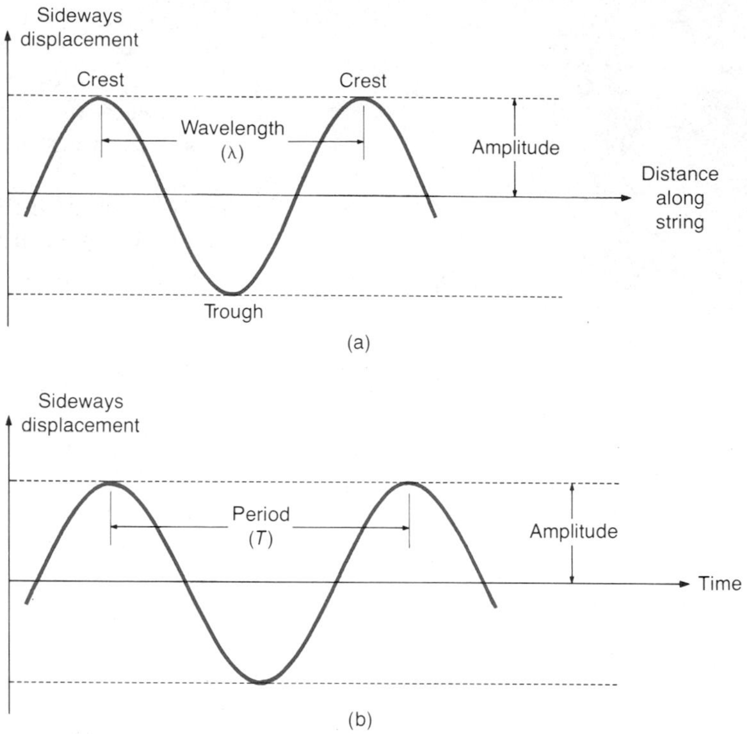

FIGURE 12.1 Illustration of a simple harmonic wave on a string: (a) disturbance on the string at some instant of time; (b) displacement of a given point as a function of time.

Consider the string as being composed of a large number of small parts. Then the wave is a large number of SHMs, each one progressively a little out of step with the adjacent ones. Suppose the wave shape of Fig. 12.1(a) is traveling to the right. The piece of string that is at a crest has reached the top end point of its SHM. The piece of string immediately to the right of a crest has not quite reached its top end point but will do so when the wave crest arrives. Proceeding in this way along the string, we see that each piece of the string is carrying out SHM of the same amplitude, but just a little out of step with its immediate neighbors. This effect continues until the piece of string at the next crest is again exactly in step with the previous crest, since it has also reached the top end point of its SHM. In Fig. 12.1(b) we show the simple harmonic displacement of a fixed piece of the string as a function of time.

Referring to Fig. 12.1, we now define the important quantities that characterize a simple harmonic wave.

1. The wave **amplitude** A is the maximum displacement of the string from the center, or undisplaced, position. It is the amplitude of the SHM carried out by each piece of the string.

2. The wave **period** T is the period of the SHM carried out by each piece of the string. It is the time interval between corresponding points on the displacement-versus-time graph of Fig. 12.1(b).

3. The wave **frequency** ν is the number of vibrations per second by each piece of the string. It is related to the period by:

$$\nu = \frac{1}{T} \tag{12.1}$$

This is exactly the same result as Eq. (11.14) for SHM.

4. The **wavelength** λ (Greek letter lambda) is the distance between corresponding points on the wave shape (see Fig. 12.1a). Although two crests are usually chosen, any corresponding points can be used.

5. The **wave velocity** v is the velocity of progression of the wave shape. The distance from crest to crest is λ, and the time of a complete vibration is T. The wave velocity is therefore the wavelength divided by the wave period.

■ Formula Relating Wave Velocity, Wavelength, and Period

$$\text{Wave velocity} = \frac{\text{wavelength}}{\text{wave period}}$$

$$v = \frac{\lambda}{T} \tag{12.2}$$

or

$$v = \lambda\nu$$

The alternate form of Eq. (12.2) follows from Eq. (12.1).

Although the reasoning given for Eq. (12.2) is based on a simple harmonic wave, the result remains true for any periodic wave regardless of its shape.

EXAMPLE 12.1

An AM radio station broadcasts electromagnetic waves at a frequency of 550 kHz. The velocity of electromagnetic waves in air is 3×10^8 m/s. Calculate the wavelength of the waves.

SOLUTION We can calculate the wavelength using Eq. (12.2):

$$v = \lambda\nu$$

$$3 \times 10^8 \text{ m/s} = \lambda \times 550 \times 10^3 \text{ Hz}$$

$$\therefore \lambda = 545 \text{ m}$$

■

12.2 Transverse and Longitudinal Waves

For a wave traveling along a stretched string, the displacement of each part of the string is sideways, and the propagation of the wave shape is along the string. We call such a wave **transverse** since the direction of SHM of the medium is transverse to the direction of propagation. A wave is called **longitudinal** if the SHM of the medium occurs in the same direction as the direction of propagation. Sound waves in air or compression waves in a metal bar are examples of longitudinal waves; in such cases we are dealing with the propagation of compressions and expansions. For example, if the longitudinal wave is a sound wave, then a compression is a region in which the air is momentarily compressed to a higher pressure and an expansion is a region that is momentarily expanded to a lower pressure. These regions of higher and lower pressure of a longitudinal wave propagate in a manner analogous to the propagation of the crests and troughs of a transverse wave. We can give a graphical illustration of a longitudinal wave, as shown in Fig. 12.2. The sinusoidal shape of Fig. 12.2(a) plots the density of the medium as a function of po-

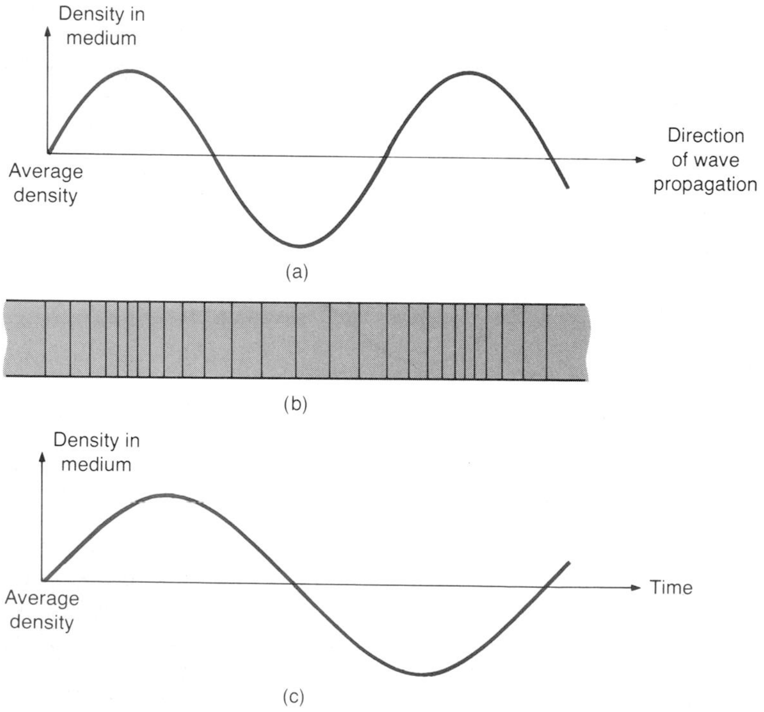

FIGURE 12.2 Illustration of a longitudinal simple harmonic wave: (a) density variation with position at a fixed time; (b) representation of density variation by shading lines—the higher density corresponds to the more closely spaced lines; (c) variation of density with time at a fixed position.

sition. The density variations are represented by shaded lines in Fig. 12.2(b); places where the vertical shading lines are close together are compressions, and places where they are widely separated are expansions. The density variation as a function of time at some fixed position is illustrated in Fig. 12.2(c). These diagrams are remarkably similar to those of Fig. 12.1, and our definitions of period, frequency, wavelength, and wave velocity all apply. The difference is simply that Fig. 12.1 represents the crests and troughs of a transverse wave, while Fig. 12.2 represents the compressions and expansions of a longitudinal wave.

Wave velocities for various types of waves can be calculated in terms of the physical properties of the medium and of outside forces. We can avoid the difficulties of rigorous derivations of wave velocity formulas by using dimensional arguments. To do this we construct formulas that arrange the relevant physical quantities in such a way as to make the units correct. Consider the case of transverse waves on a string of mass m and length l that is subject to a tension T. The physical variables, their dimensions, and their SI units are as follows:

Variable	*Dimension*	*SI Unit*
Tension (T)	$[MLT^{-2}]$	N
Mass (m)	[M]	kg
Length (l)	[L]	m

The problem is to combine these quantities to produce a velocity with dimensions $[LT^{-1}]$ or SI unit meters per second. Since the dimension [M] must disappear from the final result, we begin by dividing the tension by the mass. This leaves us with a quantity whose dimensions are $[LT^{-2}]$. To obtain a velocity we need only multiply by the length of the string (giving a quantity of dimensions $[L^2T^{-2}]$) and then take the square root. The dimensions of $(Tl/m)^{1/2}$ are $[LT^{-1}]$. We rewrite this simple formula in display, making a slight change in emphasis that identifies the importance of the mass per unit length of string.

■ Velocity of a Transverse Wave on a Stretched String

$$\text{Velocity} = \left(\frac{\text{string tension}}{\text{mass per unit length}}\right)^{1/2}$$

$$v = \left(\frac{T}{m/l}\right)^{1/2} \tag{12.3}$$

Our dimensional argument for finding the wave velocity could be in error by a multiplicative numerical constant. If we multiply the right-hand side of Eq. (12.3) by some pure number (such as π or $\sqrt{3}$), the dimensions remain $[LT^{-1}]$. We must rely on the results of more rigorous analysis to assure us that Eq. (12.3) is correct without multiplicative factors.

Compression waves on a metal bar cause density variations that give rise to local elastic strains that are longitudinal, and the appropriate elastic modulus is Young's modulus. The physical variables that should determine the wave velocity are the Young's modulus and the density of the bar.

The dimensions and SI units of these physical variables are:

Variable	*Dimension*	*SI Unit*
Young's modulus (Y)	$[ML^{-1}T^{-2}]$	N/m^2
Density (ρ)	$[ML^{-3}]$	kg/m^3

The dimensions of $(Y/\rho)^{1/2}$ are $[LT^{-1}]$ with SI unit of meters per second.

■ Velocity of a Longitudinal Wave on a Bar

$$\text{Velocity} = \left(\frac{\text{Young's modulus}}{\text{density}}\right)^{1/2}$$

$$v = \left(\frac{Y}{\rho}\right)^{1/2} \tag{12.4}$$

A more elaborate analysis shows that no multiplicative factor is required.

EXAMPLE 12.2

Calculate the wavelength of a longitudinal compression wave of frequency 800 Hz in a brass rod. (Use the Young's modulus from Table 11.1 and the density from Table 1.4.)

SOLUTION Looking up the required data for brass, we find:

$$Y = 9.8 \times 10^{10} \text{ Pa}$$

$$\rho = 8.44 \times 10^3 \text{ kg/m}^3$$

The velocity of the longitudinal compression wave is given by Eq. (12.4):

$$v = \left(\frac{Y}{\rho}\right)^{1/2}$$

$$= \left(\frac{9.8 \times 10^{10}\ \text{Pa}}{8.44 \times 10^{3}\ \text{kg/m}^{3}}\right)^{1/2}$$

$$= 3.407 \times 10^{3}\ \text{m/s}$$

To calculate the wavelength, we use Eq. (12.2):

$$v = \lambda\nu$$

$$3.407 \times 10^{3}\ \text{m/s} = \lambda \times 800\ \text{Hz}$$

$$\therefore \lambda = 4.26\ \text{m}$$

■

A longitudinal compression wave can also propagate in a fluid. The velocity formula is similar to Eq. (12.4) except that the bulk modulus replaces the Young's modulus.

■ Velocity of a Longitudinal Compressive Wave in a Fluid

$$\text{Velocity} = \left(\frac{\text{bulk modulus}}{\text{density}}\right)^{1/2}$$

$$v = \left(\frac{B}{\rho}\right)^{1/2} \qquad \textbf{(12.5)}$$

Applying this formula to liquids causes no problems, but applying it to gases raises a difficulty. The bulk modulus for a gas depends strongly on the temperature and also on whether the gas temperature is controlled as the volume changes. Not only are several types of bulk modulus possible, but they are strongly temperature-dependent. Sound waves in air are an important example of longitudinal compressive waves in a gas, but application of Eq. (12.5) requires the use of the correct bulk modulus. For the moment let us avoid this difficulty by simply quoting the experimental value for the velocity of sound in air at 15°C as approximately 340 m/s. Longitudinal compressive waves in media other than air are sometimes called sound waves. A swimmer under the water can hear sounds, but these "sounds" are detected because of longitudinal waves in the water.

Perhaps the best known of all waves are deep ocean waves. They are in fact very complex, having both longitudinal and transverse character. This means that an object floating on the surface has to-and-fro SHM in addition to the more obvious up-and-down SHM. Surface ripples are also possible in water. If we drop a small stone into a still pond, these ripples are readily seen. They differ from deep ocean waves and from compression waves in the body of the fluid. The possibility of three different, complex wave motions in a body of water is sufficient evidence of the difficulties that arise in some types of wave motion.

Shock waves from an earthquake provide us with another example of a complex wave system. A disturbance deep underground sends both longitudinal and transverse waves through the earth. Since these waves have different velocities, they arrive at distant observation points at different times. Although a detailed analysis is greatly complicated by variation of the earth's elastic properties from place to place, the measurement of the time difference between the arrival of the different wave types is used to help determine the center of the earthquake.

12.3 Standing Waves

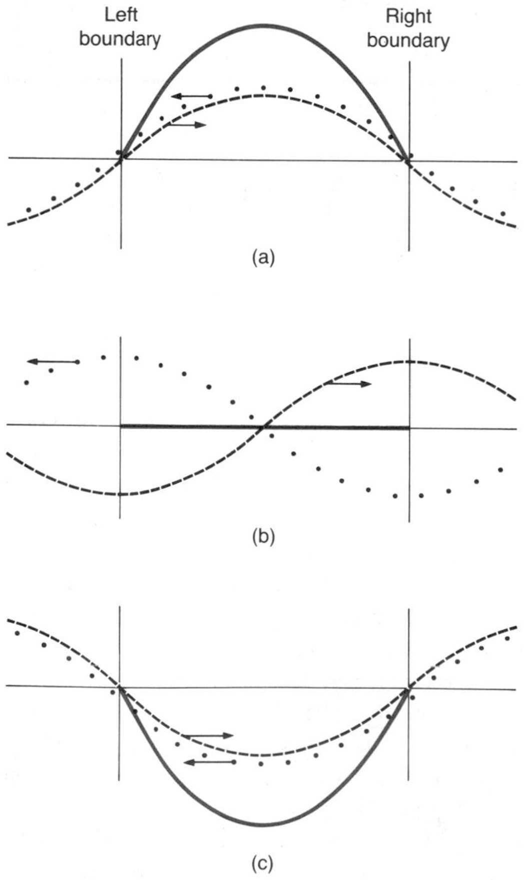

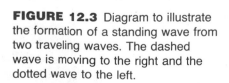

FIGURE 12.3 Diagram to illustrate the formation of a standing wave from two traveling waves. The dashed wave is moving to the right and the dotted wave to the left.

Up to this point we have restricted our discussion to waves in a medium whose boundaries are very far away. A boundary causes reflection of a wave. Imagine a long rope tied to a post. You pick up one end of the rope and give it a sharp jerk, causing a wave to travel toward the post. A short time later you feel the wave arrive back at your hand after reflecting off the post. The same situation arises in many other cases of wave motion. A sound wave reflected from a cliff causes an echo, and a radar wave reflected from an obstacle causes a signal in the receiver. The single reflection of a wave may be quite complex, but, paradoxically, multiple reflection often results in a simple situation.

Consider a guitar string fixed firmly at the bridge and the nut but otherwise free to vibrate. If we set up a vibration, the exact nature of the displacement depends on where we pluck the string. Moreover, the vibration shape of the plucked string is clearly visible, having the general form of a wave, but the shape does not travel along the string as it would for a *single* traveling wave. For this reason, we call the vibration of the guitar string a **transverse standing wave.**

Let consider a specific example to show that two equal traveling waves going in opposite directions add together to produce a standing wave. In Fig. 12.3 we show two traveling waves of equal amplitude, the dashes traveling to the right and the dots to the left. The vertical lines on the diagram (labeled right and left boundary) represent fixed points, and we wish to find out what sort of disturbance exists between these boundaries when the two traveling waves are added. Beginning with the situation of Fig. 12.3(a), in which all parts of the waves are in the same corresponding position (crest alongside crest, trough alongside trough, and so on), the sum produces the positive half-sinusoid shown as a solid colored line. Figure 12.3(b) shows the situation at a later time when each traveling wave has moved one-quarter of a wavelength; the sum is now the solid colored line of zero displace-

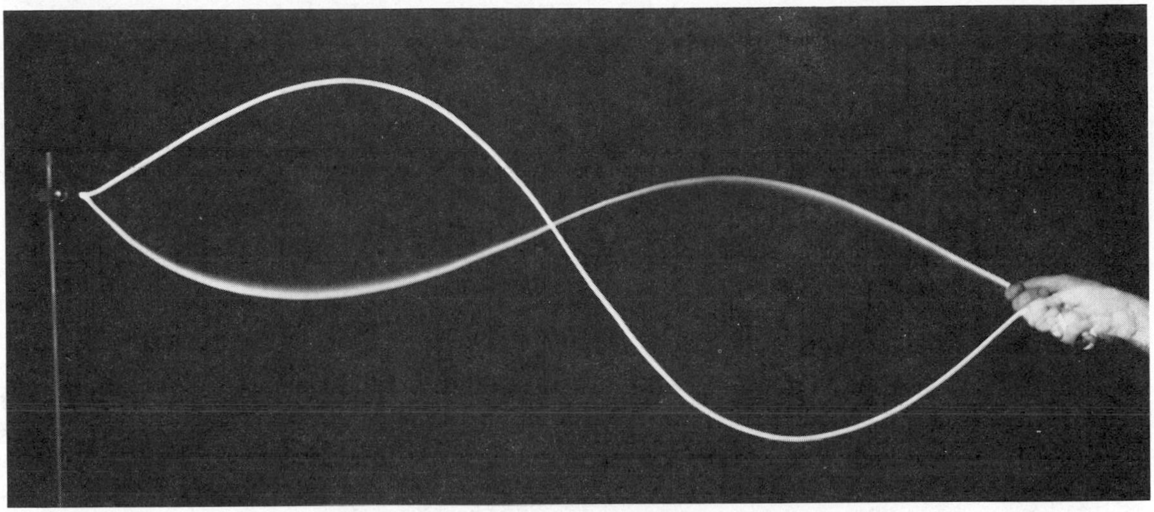

A standing wave usually includes a mixture of the various vibrational modes, but carefully controlled circumstances can result in one mode only. The hand movement is occurring at just the right frequency to excite the second harmonic of this string. (Fundamental Photographs, New York)

ment. After moving another quarter of a wavelength the traveling waves are now in the positions shown in Fig. 12.3(c), and their sum is the solid colored negative half-sinusoid. We therefore conclude that the two traveling waves produce a vibrating half-sinusoid pattern between the two boundaries. This pattern does not travel—more succinctly, it is a standing wave.

The construction of a standing wave from two traveling waves is important, because it allows us to associate a wave velocity with a standing wave, even though the wave shape is not moving. The wave velocity of a standing wave is the same as that of a traveling wave in the same medium. For a guitar string, we can calculate the wave velocity from Eq. (12.3); for compression waves along a rod, Eq. (12.4) is appropriate.

A few additional comments should be made about our construction of a standing wave. In the first place, in Fig. 12.3 we have shown only three relative positions of standing waves, and we simply assert that other relative positions lead to standing wave shapes intermediate to those illustrated. Secondly, the two boundaries are not really necessary since the two traveling waves add up to a standing wave at all points. We have introduced the boundaries as a help to the imagination. They could, for example, be the nut and the bridge of a guitar; the solid colored line would then represent a standing wave on a guitar string.

If we follow along with the idea that the nut and bridge of a guitar are boundaries, then the only firm requirement on the standing wave shape is that the ends do not move. Points of zero displacement for a standing wave pattern are called **nodes.** Wave

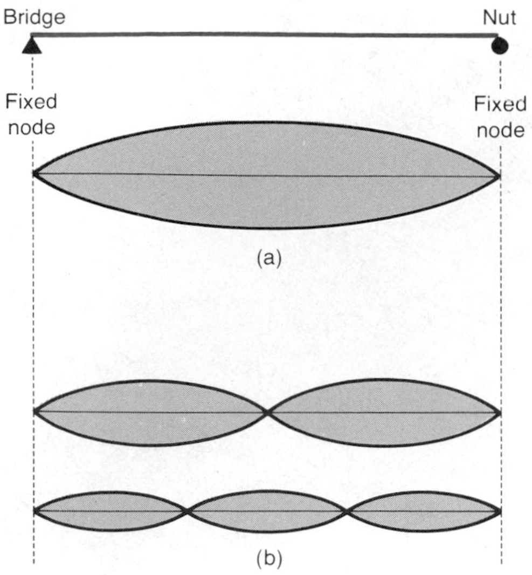

FIGURE 12.4 The modes of vibration of a guitar string: (a) the fundamental mode; (b) the second and third modes.

patterns that are simple harmonic in shape and that have a node at each end are called the **modes of vibration** of the guitar string. The modes are classified in order of decreasing wavelength; the mode having the longest wavelength is called the **fundamental mode.** The construction of Fig. 12.3 shows the fundamental mode with the string oscillating up and down between the two extreme positions. We refer to the upper and lower boundaries as the envelope of the vibration, and we see that for the fundamental mode the envelope is a one-half-wavelength pattern. This half-wavelength envelope pattern for the fundamental mode is illustrated in Fig. 12.4(a). For the fundamental mode the string length is exactly one-half wavelength.

This discussion of the properties of the fundamental mode gives us the clue for how to correctly go about constructing the standing wave patterns for higher modes; it is formulated in the following:

> The distance between adjacent nodes in a standing wave pattern is a one-half-wavelength pattern.

Since there must always be nodes at the ends of the string, the second mode pattern will have just one additional node at an intermediate point. It must, therefore, consist of two half-wavelength patterns, as shown in Fig. 12.4(b), with the additional node at the center of the string, and for this mode the string length is exactly one wavelength. The third mode will have yet another node in its pattern; it consists of three half-wavelength patterns with the additional nodes at the third points of the string. It follows that the string length for the third mode is exactly three-halves of a wavelength. Higher modes have an additional half-wavelength pattern for each increase in the mode number.

We can use this analysis to calculate the frequencies of the modes of a guitar string.

EXAMPLE 12.3

A nylon guitar string is under a tension of 160 N and has a mass density of 7 g/m. The distance between end supports is 90 cm. Calculate the frequencies of its first four modes of vibration.

SOLUTION We begin by identifying the envelope shapes that correspond to the various modes; this is done in the first two columns of the diagram. Next, we equate the length of the string to the number of half-wavelength patterns for the mode in question, as shown in the third column of the diagram. This permits calculation of the wavelength of each mode, as shown in the

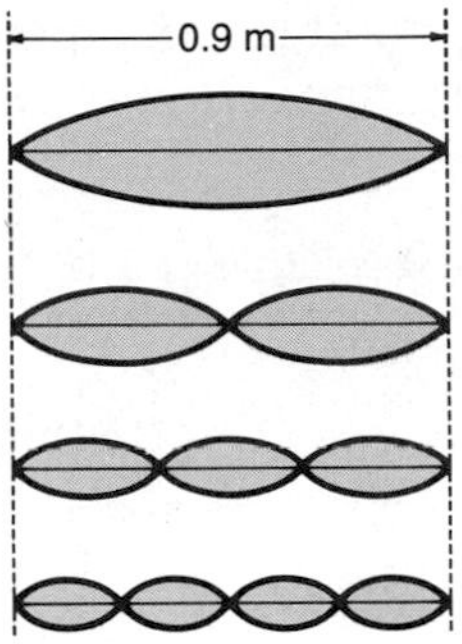

Mode number (number of half-wavelength patterns)	Relation between string length and wavelength	Wavelength λ	Frequency $\nu = v/\lambda = (151.2\ \text{m/s})/\lambda$
1	0.9 m = 1 (λ/2)	1.8 m	84.0 Hz
2	0.9 m = 2 (λ/2)	0.9 m	168 Hz
3	0.9 m = 3 (λ/2)	0.6 m	252 Hz
4	0.9 m = 4 (λ/2)	0.45 m	336 Hz

fourth column. To determine the frequency, we must first find the wave velocity with the help of Eq. (12.3):

$$
\begin{aligned}
v &= \left(\frac{T}{m/l}\right)^{1/2} \\
&= \left(\frac{160\ \text{N}}{7 \times 10^{-3}\ \text{kg/m}}\right)^{1/2} \\
&= 151.2\ \text{m/s}
\end{aligned}
$$

We can now calculate the frequency for each mode using Eq. (12.2). Take the fundamental mode first:

$$
\begin{aligned}
v &= \lambda\nu \\
\therefore \nu &= \frac{v}{\lambda} \\
&= \frac{151.2\ \text{m/s}}{1.80\ \text{m}} \\
&= 84.0\ \text{Hz}
\end{aligned}
$$

In a similar way, the frequencies of the second, third, and fourth modes are 168 Hz, 252 Hz, and 336 Hz, respectively, as shown in the fifth column of the diagram. ■

A mode whose frequency is an integral number of times the fundamental frequency is called a *harmonic*. In the example of the guitar string, the second mode is also the second harmonic, since its frequency is twice the frequency of the fundamental; the third mode is the third harmonic, and so on. We will see in a later example that this very simple relation between the modes and the harmonics does not always follow. In fact, sometimes none of the higher modes are harmonics of the fundamental mode.

One has only to pluck a guitar string closer to the bridge than to the nut to see that the shape of the standing vibration is not a simple harmonic mode. Then why are the simple harmonic

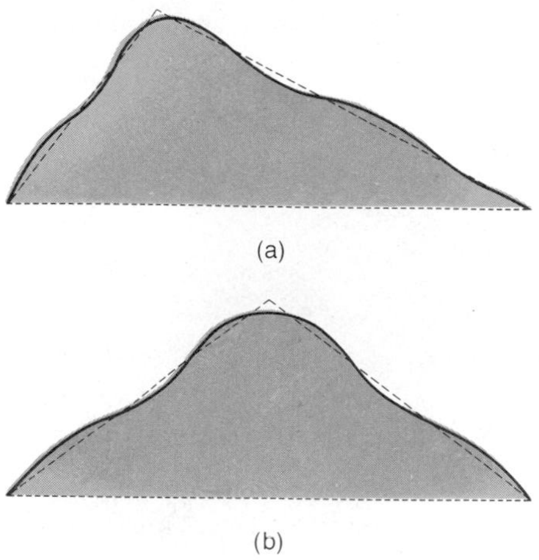

FIGURE 12.5 Possible standing wave shapes of a guitar string: (a) 100% fundamental mode, 35% second harmonic, 11% third harmonic, and 4% fifth harmonic; (b) 100% fundamental mode, 11% third harmonic, and 4% fifth harmonic. (The dotted lines represent the exact shape obtained by including harmonics of all orders.)

modes presented as such important concepts? The answer lies in an incredible theorem of mathematics known as Fourier's theorem, which is applicable to many more things than just the vibrations of a guitar string. For our purpose the following expression of the theorem is sufficient.

> A vibration of any wave shape can be accurately represented by a sum of simple harmonic wave shapes.

It is true that the Fourier sum may contain a very large number of wave shapes, but the application of the theorem to the guitar string problem entails very few terms. Figure 12.5(a) shows the wave shape calculated from a 100% fundamental mode, 35% second harmonic, 11% third harmonic, and 4% fifth harmonic. The wave is almost exactly the shape of vibration that occurs if we pluck the string at a point one-fourth the distance from the bridge to the nut. In Fig. 12.5(b), the shape illustrated contains 100% fundamental mode, 11% third harmonic, and 4% fifth harmonic, which corresponds to plucking the string halfway between the bridge and the nut. The two examples show the overwhelming importance of the fundamental mode, which contributes the major share of the vibration for any standing wave shape of the string. The different shapes arise from various admixtures of low-order harmonics. The addition of higher-order harmonics to the wave shapes of Fig. 12.5 brings the shapes even closer to the dotted line configuration, but the diagram makes it clear that the role of the higher-order harmonics is very small. In each case the dotted lines show the shape that results from adding all harmonics.

So far, we have examined in detail the transverse standing wave on a guitar string. It is an important example, illustrating the essential features of any standing wave with a node at each end. However, not all examples of standing waves have a node at each end. Consider a metal rod firmly clamped at one end and free at the other (Fig. 12.6). We can set up a longitudinal standing wave in this rod by striking a longitudinal blow to begin the vibration. (The plucking of the guitar string is a transverse blow that sets up transverse vibrations.)

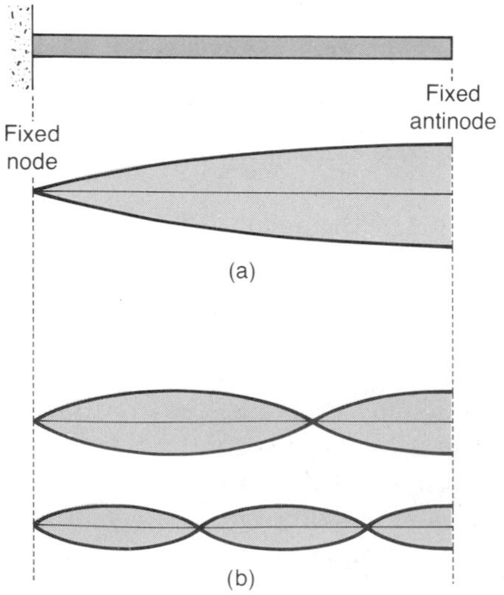

FIGURE 12.6 Longitudinal modes of vibration of a metal rod clamped at one end: (a) the fundamental mode; (b) the second and third modes.

We have already noted that a longitudinal wave is both a density wave and a displacement wave with nodes in different positions. In discussing the mode patterns we must be careful to relate them to the correct physical quantity. Since the fixed end of the rod cannot have any longitudinal motion, it must be a node of the displacement wave. On the other hand, the free end of the rod is a position of maximum longitudinal displacement for any simple harmonic mode. We call such a position an *antinode*.

Since the simple harmonic modes of a metal rod must have a displacement node at one end and an antinode at the other, their displacement mode patterns will differ from those of a guitar string. The fundamental mode pattern is shown in Fig. 12.6(a), and comparison with Fig. 12.1 shows that it is a one-quarter-wavelength pattern. We must, therefore, make a slight generalization of our previous display.

> The distance between a node and an adjacent antinode in a standing wave pattern is a one-quarter-wavelength pattern.

For the fundamental mode, the length of the rod is exactly one-quarter-wavelength.

The method of constructing standing wave patterns for higher modes now follows. There must be a node at one end of the rod and an antinode at the other. The second node will, therefore, have an additional node between the ends, and its pattern will be a one-half-wavelength pattern together with a one-quarter-wavelength pattern, as shown in Fig. 12.6(b). The additional nodal point occurs at the two-thirds point of the rod. For the second mode, the length of the rod is exactly three-quarters of a wavelength. In a similar manner, we must add yet another half-wavelength pattern to make up the standing wave pattern for the third mode. Inspection of Fig. 12.6(b) shows that the length of the rod is exactly five-quarters of a wavelength for the third mode.

EXAMPLE 12.4

A steel rod of length 50 cm is tightly clamped at one end and is struck a longitudinal blow on its free end. Calculate the frequencies of the first four modes of simple harmonic longitudinal vibration. (Use the Young's modulus from Table 11.1 and take 7.86×10^3 kg/m^3 as the density of steel.)

SOLUTION Once again, we begin by determining the envelope patterns and using this information to calculate the wavelength of each mode. The first three columns of the diagram show the en-

Mode shape

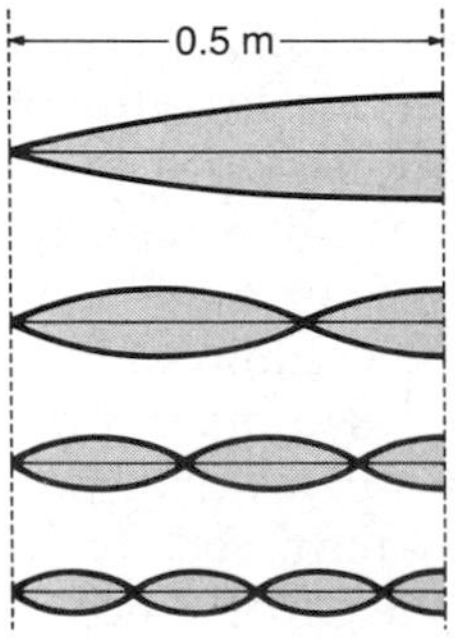

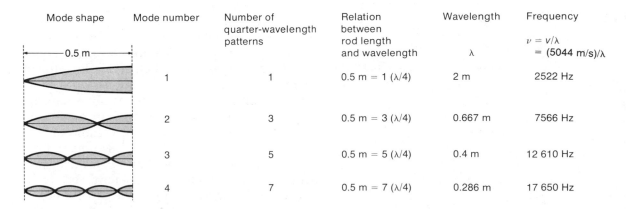

Mode number	Number of quarter-wavelength patterns	Relation between rod length and wavelength	Wavelength λ	Frequency $\nu = v/\lambda = (5044\ \text{m/s})/\lambda$
1	1	$0.5\ \text{m} = 1\ (\lambda/4)$	2 m	2522 Hz
2	3	$0.5\ \text{m} = 3\ (\lambda/4)$	0.667 m	7566 Hz
3	5	$0.5\ \text{m} = 5\ (\lambda/4)$	0.4 m	12 610 Hz
4	7	$0.5\ \text{m} = 7\ (\lambda/4)$	0.286 m	17 650 Hz

velope patterns, the mode number, and the number of quarter-wavelength patterns in the first four modes. The fourth column equates the length of the rod to the number of quarter-wavelength patterns, and the fifth column uses this information to calculate the mode wavelength. To find the mode frequencies, we need the longitudinal wave velocity, and this is found with the help of Eq. (12.4):

$$\begin{aligned} v &= \left(\frac{Y}{\rho}\right)^{1/2} \\ &= \left(\frac{2.0 \times 10^{11}\ \text{Pa}}{7.86 \times 10^{3}\ \text{kg/m}^{3}}\right)^{1/2} \\ &= 5044\ \text{m/s} \end{aligned}$$

To calculate the frequencies, we use Eq. (12.2). For the fundamental mode, we have:

$$\begin{aligned} v &= \lambda\nu \\ \therefore\ \nu &= \frac{v}{\lambda} \\ &= \frac{5044\ \text{m/s}}{2.00\ \text{m}} \\ &= 2522\ \text{Hz} \end{aligned}$$

Similarly the frequencies of the second, third, and fourth modes are 7566 Hz, 12 610 Hz, and 17 650 Hz, respectively, as shown in the sixth column of the diagram. ■

Another difference between the guitar string and the rod is now apparent. The vibration frequency of the second mode of the rod is 3 times the fundamental frequency. The second mode is therefore a third harmonic. Likewise, we find the third and fourth modes to be fifth and seventh harmonics, respectively. In contrast, a guitar string, which has a node at each end, has harmonics of every order. It is evident that the even-order harmonics of the longitudinal vibrations are suppressed in the rod. The rod illustrates the essential features of any standing wave pattern with a node at one end and an antinode at the other.

Still other possibilities for various types of envelope patterns arise in standing wave vibrations. If a stream of air is blown against an edge of the open end of a pipe, then longitudinal standing waves are set up in the column of air within the pipe. This describes an organ pipe; a closed organ pipe is one in which the other end is closed, and an open organ pipe has both ends open. The mode patterns are determined by conditions on the standing wave at each end of the pipe. Since the air cannot vi-

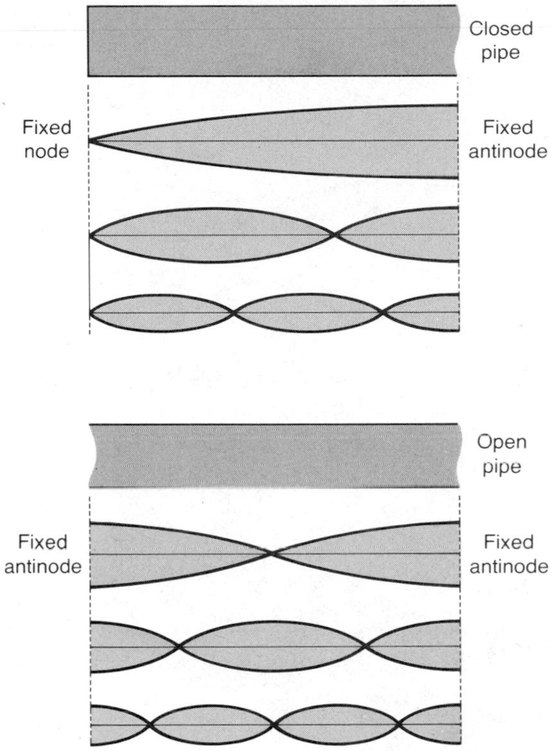

FIGURE 12.7 The fundamental and first two higher-order modes of longitudinal vibration of the air column in closed and open organ pipes.

brate longitudinally against a closed end, a closed pipe end must be a node for the standing wave of displacement. An open end is an antinode (at least approximately so) for the standing wave of displacement. Figure 12.7 shows the possible vibration modes of closed and open organ pipes. A closed pipe has identical mode patterns with those for the longitudinal vibration of a metal rod clamped at one end. An open organ pipe has both ends open, and its mode patterns require an antinode at each end of the pipe. This results in patterns that are different from any of the previous cases. Let us illustrate by an example.

EXAMPLE 12.5

Calculate the frequencies of the first four modes of an open organ pipe 66 cm long. Use 340 m/s for the velocity of the longitudinal standing waves (the velocity of sound in air).

SOLUTION The envelope shapes for the modes, the mode numbers, and the number of quarter-wavelength patterns are shown in the first three columns of the diagram. In the fourth column, the pipe length is set equal to the number of quarter-wavelength

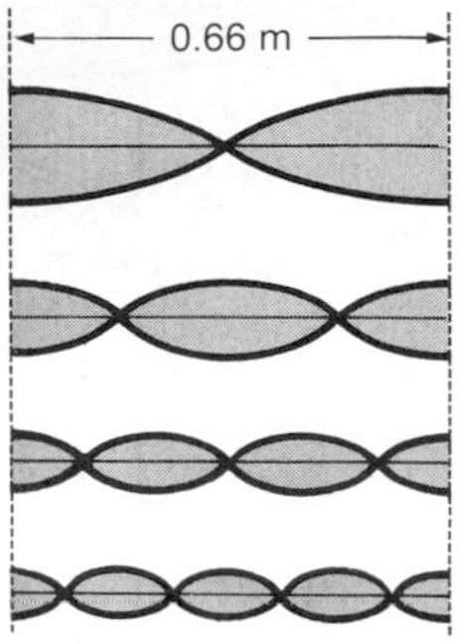

Mode shape	Mode number	Number of quarter-wavelength patterns	Relation between air column length and wavelength	Wavelength λ	Frequency $\nu = v/\lambda = (340\text{ m/s})/\lambda$
	1	2	$0.66\text{ m} = 2\left(\frac{\lambda}{4}\right)$	1.32 m	258 Hz
	2	4	$0.66\text{ m} = 4\left(\frac{\lambda}{4}\right)$	0.66 m	515 Hz
	3	6	$0.66\text{ m} = 6\left(\frac{\lambda}{4}\right)$	0.44 m	773 Hz
	4	8	$0.66\text{ m} = 8\left(\frac{\lambda}{4}\right)$	0.33 m	1030 Hz

patterns, and this equation is solved for the wavelength in the fifth column. Finally, we use Eq. (12.2) to determine the frequencies of the modes. For the fundamental mode:

$$v = \lambda\nu$$

$$\therefore \nu = \frac{v}{\lambda}$$

$$= \frac{340\text{ m/s}}{1.32\text{ m}}$$

$$= 258\text{ Hz}$$

Similarly, the frequencies of the second, third, and fourth modes are 515 Hz, 773 Hz, and 1030 Hz, respectively, as shown in the sixth column of the diagram. These modes are also second, third, and fourth harmonics of the fundamental. ■

The different lengths organ pipes are responsible for the notes of different fundamental frequency. The long pipes provide the low-frequency and the short pipes the high-frequency notes from the organ. (Courtesy of M. P. Moeller, Inc., Hagerstown, Maryland)

Although the mode patterns are different for the guitar string and the open organ pipe, the relation between mode number and harmonic number is the same for both cases. The mode patterns are the same for the rod clamped at one end and for the closed organ pipe; it follows that the relation between mode number and harmonic number is the same for both cases.

In all the examples we have considered, the modes are simple harmonic standing waves, and all of the higher-order modes are harmonics of the fundamental. This simple result does not hold true for more complex systems. The standing waves that are the modes of the circular membrane of a drum do not have sinusoidal shape, and the higher-order modes are not harmonics of the fundamental. However, the drum membrane does possess modes of vibration, which have exactly the same central role in the analysis of vibrations of the drum that the simple harmonic modes have in the systems studied in this section.

12.4 Resonance

Modes of vibration are important for yet another reason. Let us imagine that we have a guitar that is properly tuned. Each of the six strings has a fundamental vibration mode whose frequency is that of the corresponding musical note. What happens if we strike a tuning fork and hold it close to the guitar strings? (A tuning fork is a small metal instrument used by professional tuners. It emits a note of a particular frequency, which is usually noted on the handle.) In general, nothing happens unless a special circumstance holds: If the frequency of the tuning fork is equal to the frequency of the fundamental mode of one of the strings, that string begins to vibrate. The very small influence of the nearby tuning fork is sufficient to cause the string to vibrate *provided* that the tuning fork is vibrating at the fundamental mode frequency of the string. This phenomenon is an example of **resonance.** Resonance is not necessarily restricted to fundamental modes. The tuning fork would elicit a resonant response from the guitar string at the frequency of any of its vibrational modes, but the effect is much larger for the fundamental mode.

The guitar string and the tuning fork are special examples of a quite general situation. Whenever a periodic disturbance acts on an object, the object responds with large-amplitude vibrations if the frequency of the applied disturbance is equal to a vibrational mode frequency of the object. The effect is especially large if the frequency in question is the fundamental mode frequency of the object. In this connection, we note that a point mass has only one natural vibration frequency. A child sitting on a swing can be regarded as a simple pendulum whose fundamental (and only) natural vibration frequency is the frequency of the pendulum. An adult can cause vibrations of very large amplitude by exerting a small force exactly in synchronism with the vibrating swing.

The resonance response of a vibrating body to an external disturbance may be either *sharp* or *broad*. A sharp resonance is one in which the frequency of the external stimulus must be almost exactly equal to the natural vibration frequency of the excited object to set up any appreciable displacement. However, when the frequency is almost exactly correct, the resulting displacement is very large. For example, the sound of a musical note must be almost exactly equal to the fundamental vibrational mode of a wine glass to have any effect. When equality of frequency is achieved, the resulting displacement is very large and the glass breaks. On the other hand, a broad resonance is one for which a certain amount of displacement occurs for frequencies of the external stimulus that are equal or only reasonably close to the natural vibration frequency. In this case,. the displacement may not be outstandingly large for any frequency, and the resonance phenomenon is greatly suppressed. An automobile can be regarded as an object that is connected to the road surface by the springs in the suspension units. External stimuli provided by the bumps on the road cause only small displacement of the automobile over a wide range of frequency.

The basic physical reason for the distinction between a sharp and a broad resonance is the extent of energy dissipation by the vibrating object. If the energy dissipation per cycle of the vibratory motion is negligible compared with the energy of the vibration, then the resonance is sharp; in the opposite extreme, it is broad. There is very little energy dissipation per cycle in the vibration of the wine glass, and the resonance is sharp. However, the shock absorbers in the automobile provide large dissipation of the energy. This results in a broad resonance, which ensures relatively small displacement of the vibrating object for all frequencies of the external stimulus.

12.5 The Interference of Waves—Beats

When two or more waves travel in the same medium, they may meet and interfere. The wave displacement at the point of interference is the sum of the displacements due to each wave separately. In general, the overall displacement pattern of a medium due to interfering waves may be very complex, but several simple and important cases deserve our attention.

In Section 12.4 we studied standing waves in various systems. The standing wave pattern arises from the interference of two waves of equal amplitude and frequency but traveling in *opposite* directions. We do not press this point of view in the study of standing waves, since we are more concerned with finding the possible wave patterns that fit the required nodal and antinodal points.

If two waves of equal amplitude and frequency travel in the *same* direction, the resulting wave displacement depends on the phase relationship between the component waves. Figure 12.8 illustrates two special phase relationships. In Fig. 12.8(a) the component waves are exactly in step, and we say that these waves are **in phase.** The wave that results from their interference has twice the amplitude of either component separately and travels in the same direction with the same frequency. This situation is called **constructive interference.** On the other hand, waves that are exactly out of step are said to be **out of phase,** and their interference results in zero net displacement, as shown in Fig. 12.8(b). This situation is called **destructive interference.** The two phase relationships illustrated lead to resultant waves that have the largest and smallest possible amplitudes. Any phase relationship other than these two produces a wave whose amplitude is greater than zero but less than twice that of each component.

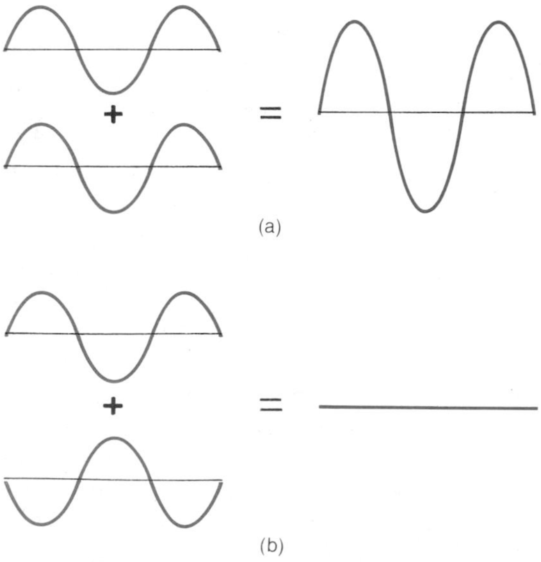

FIGURE 12.8 Interference between waves of equal amplitude and frequency traveling in the same direction: (a) constructive interference from waves in phase; (b) destructive interference from waves out of phase.

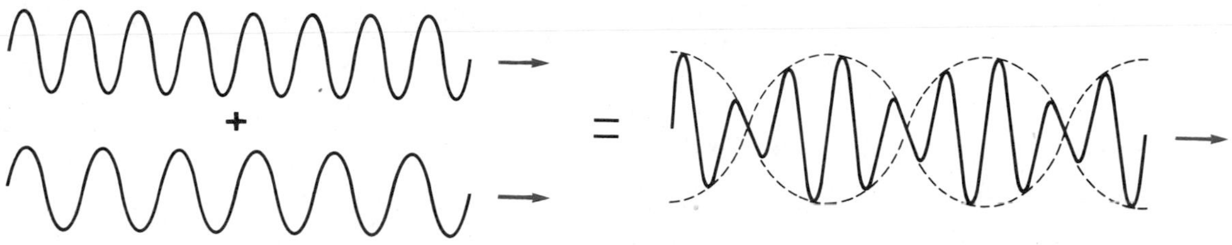

FIGURE 12.9 Beats arise from the addition of two waves of slightly different frequency traveling in the same direction.

A special case with a very distinctive result arises when two waves of equal amplitude but slightly different frequency interfere. Figure 12.9 illustrates the relatively complex interference wave pattern that results from the addition of two simple harmonic waves. We can regard the pattern as a simple harmonic wave of relatively slowly varying amplitude. The amplitude reaches a peak value of twice the amplitude of either component wave when the component waves are exactly in phase. The amplitude falls to zero when the component waves are exactly out of phase. The periodic variations in amplitude are called **beats,** and the number of amplitude maxima per second is called the **beat frequency** of the composite wave. The beat frequency is equal to the difference of the wave frequencies of the two component waves. The frequency of the composite wave is the average of the frequencies of the component waves.

■ Addition of Two Harmonic Waves of Frequencies ν_1 and ν_2 Results in a Composite Wave.

$$\text{Beat frequency} = \nu_1 - \nu_2$$

$$\text{Wave frequency} = \frac{\nu_1 + \nu_2}{2} \tag{12.6}$$

EXAMPLE 12.6

Two speakers are emitting simple harmonic sound waves of frequency 258 Hz and 256 Hz. Calculate the beat and wave frequencies.

SOLUTION The addition of the two sound waves produces beats whose characteristics are given by Eq. (12.6):

$$\begin{aligned} \text{Beat frequency} &= \nu_1 - \nu_2 \\ &= 258 \text{ Hz} - 256 \text{ Hz} \\ &= 2 \text{ Hz} \\ \text{Wave frequency} &= \frac{\nu_1 + \nu_2}{2} \\ &= \frac{258 \text{ Hz} + 256 \text{ Hz}}{2} \\ &= 257 \text{ Hz} \end{aligned}$$

■

The phenomenon of beats can be easily demonstrated by playing the same note on two identical musical instruments, one of which is slightly out of tune.

However, a human person listening to notes played on musical instruments does not necessarily "hear" exactly what we might predict by adding two mathematical expressions. For two musical notes that have harmonics, the ear actually hears the two separate notes, even when the beat frequency is small. In the absence of harmonics (that is, pure tones) the ear still hears two separate notes, provided that the beat frequency is above a certain value.

12.6 The Doppler Effect

The frequency of a fixed-frequency siren on an emergency vehicle seems to change as the vehicle passes a listener standing by the roadside, and the frequency of light from a distant galaxy seen by astronomers differs from the light's emission frequency. These are instances of a general frequency effect known as the **Doppler effect,** which depends on the relative motion of the emitter and the receiver of waves. The Austrian physicist Christian Doppler (1803–1853) first used the idea to explain color changes in double stars. His calculations were somewhat in error, but his basic idea was correct.

The details of the Doppler effect for light waves and sound waves are different because light waves travel through empty space, while sound waves require a medium (the air) for their propagation. Let us first consider the Doppler effect for sound waves in air. The velocity of sound depends only on the physical properties of the air through which it passes, and not on the velocity of the source that generates the sound.

Consider a fixed source emitting simple harmonic sound waves of frequency ν. The velocity of the sound waves in the air is v. A listener who receives the sound wave and who is moving either directly toward the source or directly away from it with velocity v_R measures the sound velocity as $(v + v_R)$ (Fig. 12.10a). (We adopt the convention of counting approach velocity as positive and recession velocity as negative.) Since motion of the receiver

SPECIAL TOPIC 14 DOPPLER

Christian Doppler was born at Salzburg in 1803 and became a professor of mathematics at the University of Prague.

His major contribution to science was the explanation of the phenomenon in which the observed pitch of a sound wave depends on the relative motion of the source and the receiver. Most interesting is the almost comic-opera manner in which his equations were tested experimentally. For several days a locomotive pulled flat cars along a track at various speeds. On the cars were trumpeters who played various musical notes, and standing by the track-bed were musicians who possessed absolute pitch (that is, they could unfailingly identify the note they heard). The musicians made records of the notes they heard, both as the train approached and as it receded; Doppler's equations proved to be well founded.

He died in Venice when only 50 years of age.

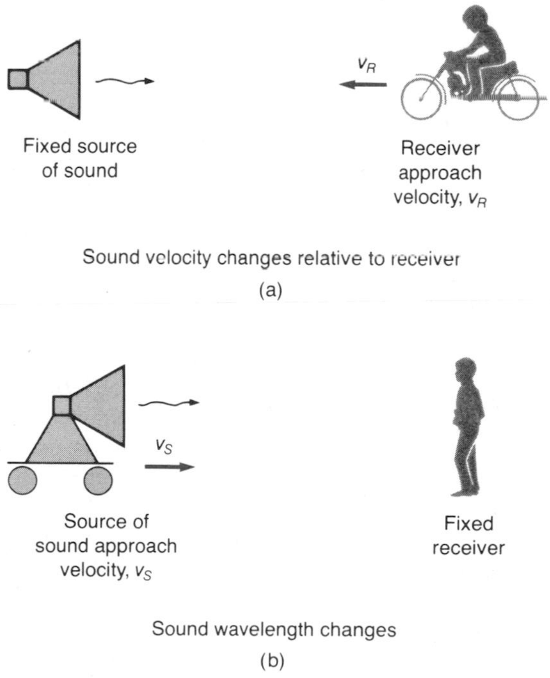

FIGURE 12.10 Illustration of the Doppler effect: (a) a receiver moving with velocity v_R; (b) a source of sound moving with velocity v_S.

does not change the wavelength of the sound, we can calculate the observed frequency to the receiver by using Eq. (12.2):

$$\nu' = \frac{v + v_R}{\lambda} = \nu\left(\frac{v + v_R}{v}\right) \tag{12.7}$$

The result shows that the observed frequency increases if the receiver is moving toward the source of sound and decreases if the receiver is moving away.

The Doppler effect is not the same if the source of sound is moving and the receiver is stationary. A moving source causes a change in the wavelength of the sound wave. Let us consider a simple example to illustrate this point. An insect swimming on the surface of a pond sends out ripples. Figure 12.11 illustrates the pattern of the ripples with circles representing each wave crest. Each circle is centered on the point occupied by the insect when it originated the wave crest that the circle represents. The wavelength of the ripples is shorter in the direction in which the insect is swimming and longer in the opposite direction. The example of the swimming insect is easy to visualize because we can see the ripples, but the change in wavelength of sound emitted by a moving source is no less real. Consider the situation of Fig. 12.10(b), in which a source of sound moves with velocity v_S and the receiver is fixed. As in the previous case, we count approach velocity as positive and recession velocity as negative. At some instant of time, the source emits a wave crest that travels forward in the air with a velocity v. After a time interval of one period ($T = 1/\nu$), the source emits the second crest. At this time the first crest has traveled a distant vT and the source has traveled a distance $v_S T$. The distance between the two crests is the difference

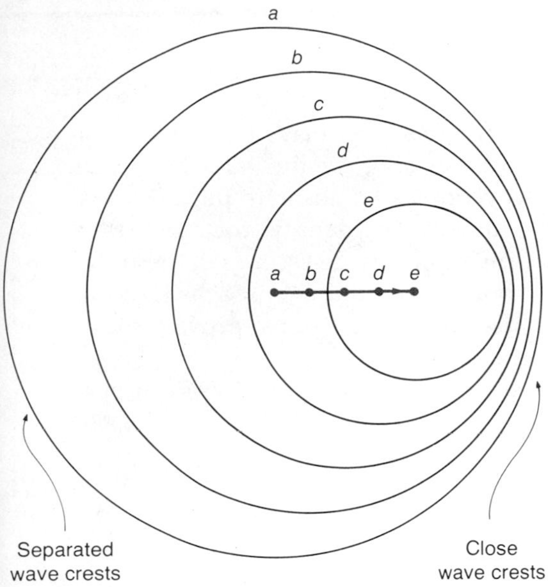

FIGURE 12.11 Diagram to illustrate the change in wavelength when a source moves in the medium in which the waves propagate. Each circular wave crest is centered at the position of the source when the crest was emitted.

$(vT - v_S T)$, and this distance is a wavelength. We can therefore write:

$$\lambda = vT - v_S T$$
$$= \frac{1}{\nu}(v - v_S)$$

But the observed frequency to the stationary receiver is:

$$\nu' = \frac{v}{\lambda}$$
$$= \nu\left(\frac{v}{v - v_S}\right) \qquad \textbf{(12.8)}$$

The observed frequency increases if the source of sound is moving toward the receiver and decreases if it is moving away. We can summarize the results for the Doppler effect by writing Eqs. (12.7) and (12.8) in the same display and intuitively extending the concept to the case in which the source and the receiver both move. Although the arguments presented use sound waves as the example, the results are true for any wave in a material medium.

■ Doppler Effect for Waves in a Material Medium

1. Receiver moving
Wave source stationary
Apparent change in wave velocity

$$\nu' = \nu\left(\frac{v + v_R}{v}\right) \qquad \text{(a)}$$

2. Receiver stationary
Wave source moving
Real change in wavelength

$$\nu' = \nu\left(\frac{v}{v - v_S}\right) \qquad \text{(b)} \qquad \textbf{(12.9)}$$

3. Receiver moving
Wave source moving
Apparent change in wave velocity
Real change in wavelength

$$\nu' = \nu\left(\frac{v + v_R}{v - v_S}\right) \qquad \text{(c)}$$

In every case, the signs of v_R and v_S are positive for approach velocity and negative for recession velocity.

EXAMPLE 12.7

An emergency vehicle uses a siren that emits a sound wave of frequency 800 Hz. What change in the frequency of the sound does a person standing at the roadside hear when the vehicle passes at 126 km/h (35 m/s)? (Use 340 m/s for the velocity of sound in air.)

SOLUTION To calculate the observed frequency of sound, we use Eq. (12.9b):

$$\nu' = \nu\left(\frac{v}{v - v_S}\right)$$

a. As the vehicle approaches the listener:

$$v_S = 35 \text{ m/s}$$

$$\therefore \nu' = 800 \text{ Hz} \frac{340 \text{ m/s}}{340 \text{ m/s} - 35 \text{ m/s}}$$

$$= 892 \text{ Hz}$$

b. As the vehicle recedes from the listener:

$$v_S = -35 \text{ m/s}$$

$$\therefore \nu' = 800 \text{ Hz} \frac{340 \text{ m/s}}{340 \text{ m/s} + 35 \text{ m/s}}$$

$$= 725 \text{ Hz}$$

The observed frequency changes from 892 Hz as the vehicle approaches to 725 Hz after it passes. This is a large change that is readily apparent even to an unmusical listener. ■

The Doppler effect discussed up to this point concerns waves in a material medium. Radio waves and light waves are both examples of electromagnetic waves. The electromagnetic wave vibration is not a vibration of a material medium, and the arguments used in deriving Eq. (12.9) do not apply to this case. The Doppler formula for electromagnetic waves requires the theory of special relativity and leads to the following result:

■ Doppler Effect for Electromagnetic Waves

$$\nu' = \nu\left(\frac{1 + v_{RS}/c}{1 - v_{RS}/c}\right)^{1/2} \quad (12.10)$$

v_{RS} is the relative approach velocity of the source and the receiver. c is the velocity of electromagnetic waves (3×10^8 m/s in free space).

There is a real difference of principle between the Doppler formulas for electromagnetic waves and for sound waves. In the case of sound waves, the source and the receiver are assigned velocities independently of each other that are relative to the medium in which the sound propagates. A listener who stands stationary in still air has zero velocity for the purposes of Eq. (12.9), since his velocity is zero relative to the medium in which the sound wave propagates. However, we cannot assign zero velocity to the same person receiving electromagnetic waves on a radio set because there is no material medium for propagation of the electromagnetic waves. The only physically meaningful velocity that enters the problem is the relative velocity of the source and the receiver.

EXAMPLE 12.8

A hydrogen discharge tube in the laboratory emits an ultraviolet radiation of frequency 2.467×10^{15} Hz. The same vibration from the hydrogen in the quasar 3C9 has a frequency of 8.197×10^{14} Hz, as observed in a laboratory on earth. Calculate the velocity of the quasar relative to the earth on the assumption that the frequency change is a Doppler shift.

SOLUTION We need only to use Eq. (12.10) to calculate the relative velocity of the source (quasar 3C9) and the receiver (the earthborne laboratory):

$$\nu' = \nu\left(\frac{1 + v_{RS}/c}{1 - v_{RS}/c}\right)^{1/2}$$

$$8.197 \times 10^{14}\ \text{Hz} = 2.467 \times 10^{15}\ \text{Hz} \times \left[\frac{1 + v_{RS}/(3 \times 10^8\ \text{m/s})}{1 - v_{RS}/(3 \times 10^8\ \text{m/s})}\right]^{1/2}$$

Squaring both sides and cross-multiplying lead to:

$$1 + \frac{1 + v_{RS}}{3 \times 10^8\ \text{m/s}} = 0.1104\left(1 - \frac{v_{RS}}{3 \times 10^8\ \text{m/s}}\right)$$

$$\therefore v_{RS} = -2.4 \times 10^8\ \text{m/s}$$

The negative sign means that the quasar is receding. However, it is not at all certain that the whole of this large frequency change is due to the Doppler effect, and the matter is still being actively investigated. ■

12.7 Sound Intensity and the Decibel

The level of a sound wave is a matter of some practical importance. The concept has both objective and subjective aspects. The objective level relates to the energy transferred by the longitudi-

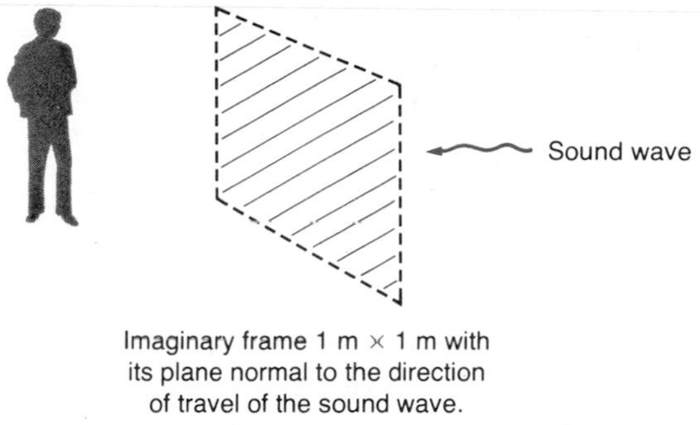
Sound wave
Imaginary frame 1 m × 1 m with its plane normal to the direction of travel of the sound wave.

FIGURE 12.12 Illustration of the SI unit of sound intensity as the watt per square meter.

nal wave in the air, which is a matter for exact scientific measurement. The subjective noise level is a measure of how a sound seems to a human listener after it is processed by the hearing system. The two are quite different because the human ear is not equally sensitive to sounds of the same energy having different frequencies. In order to distinguish, we refer to the objective noise level as **sound intensity** and the subjective noise level as **loudness.**

Let us begin with the idea of sound intensity. Consider a sound wave traversing an imaginary area of 1 m^2 whose plane is normal to the direction of propagation of the sound wave (Fig. 12.12). The SI unit of sound intensity is the quantity of sound wave energy that flows through this imaginary unit area in a unit time interval. The unit is therefore the joule per square meter per second, or, more briefly, the watt per square meter. The faintest sound of any frequency that a normal human can hear has an intensity of about 10^{-12} W/m^2, and the strongest sound that can be tolerated without pain has an intensity of about 1 W/m^2. Human hearing ability covers a range of 12 decades of intensity. The SI unit (or another unit differing only by a multiplicative factor) is inconvenient because human hearing involves such a large range. A mathematical function that measures the **sound intensity level** narrows the range and offers a useful practical unit, the **decibel** (abbreviation "dB").

Definition of the Decibel as a Unit of Sound Intensity Level

$$\beta = 10 \log_{10} (I/I_0) \qquad (12.11)$$

β = sound intensity level in dB

I = sound intensity in SI units

I_0 = threshold of hearing intensity of 10^{-12} W/m^2

The decibel is a pure number.

TABLE 12.1 Approximate intensities and intensity levels of typical sounds

Sound	*Intensity* (W/m^2)	*Intensity level* (dB)
Hearing threshold	10^{-12}	0
Rustling leaves	10^{-11}	10
Whisper	10^{-10}	20
Conversation	10^{-6}	60
Loud radio	10^{-4}	80
Jackhammer	10^{-2}	100
Discomfort	1	120

The definition of the decibel shows that it is essentially a comparative unit. It is a measure of how much a given sound intensity exceeds a reference level, usually selected to be the threshold of hearing. Table 12.1 shows the intensity levels in decibels of some typical sounds. The table shows that the effect of the logarithm in the definition (12.11) is to compress audible sounds of range 10^{-12} W/m^2 to 1 W/m^2 into the range of 0 to 120 dB. We do not list the conversion from SI units to decibels in the conversion tables, since it involves a functional relation rather than a simple multiplicative factor. For this reason the SI unit is called *intensity*, and the decibel is referred to as *intensity level.*

We have already discussed the wave produced by the interference of two sound waves that have a fixed phase relationship. If two sources of sound have no fixed phase relationship, they are said to be *incoherent*. Incoherent sources are not necessarily unpleasantly noisy—technically, two musical instruments are just as incoherent as two jackhammers. To calculate the intensity due to two incoherent sources, we simply add their intensities.

EXAMPLE 12.9

A jackhammer causes a sound intensity level of 95 dB at a point where a person is standing. A second jackhammer, adjacent to the first, then begins to operate. What is the sound intensity level at the position of the listener?

SOLUTION The physical reality is the sound intensity, which is doubled by the addition of the second hammer. To calculate the sound intensity due to one hammer, we use Eq. (12.11):

$$\beta = 10 \log_{10}\left(\frac{I}{I_0}\right)$$

$$95 \text{ dB} = 10 \log_{10}\left(\frac{I}{10^{-12} \text{ W/m}^2}\right)$$

$$\therefore I = 3.162 \times 10^{-3} \text{ W/m}^2$$

The sound intensity due to the two hammers is twice this amount; that is, 6.324×10^{-3} W/m^2. Use of Eq. (12.11) gives the decibel level for this intensity:

$$\beta = 10 \log_{10}\left(\frac{I}{I_0}\right)$$

$$= 10 \log_{10}\left(\frac{6.234 \times 10^{-3} \text{ W/m}^2}{10^{-12} \text{ W/m}^2}\right)$$

$$= 98.0 \text{ dB}$$

However, we can work this example much more quickly by taking advantage of the functional form of the intensity level unit.

ALTERNATE SOLUTION Let I_1 be the intensity of a single hammer. Its intensity level is given by Eq. (12.11) as:

$$\beta_1 = 10 \log_{10}\left(\frac{I_1}{I_0}\right)$$

The intensity level of two hammers is:

$$\beta_2 = 10 \log_{10}\left(2\frac{I_1}{I_0}\right)$$

Subtract these equations:

$$\beta_2 - \beta_1 = 10 \log_{10}\left(2\frac{I_1}{I_0}\right) - 10 \log_{10}\left(\frac{I_1}{I_0}\right)$$
$$= 10 \log_{10} 2$$
$$= 3.01 \text{ dB}$$

It follows that doubling the sound intensity adds about 3 dB to the intensity level regardless of the intensity of the original sound. The intensity level with two hammers is 98 dB, as previously calculated. ■

Let us consider another problem on sound intensity and sound intensity level.

EXAMPLE 12.10

A single person produces a sound intensity level of 60 dB by talking. How many people (all talking in the same way) are needed to produce a sound intensity level of 80 dB?

SOLUTION The sound intensity due to one person producing 60 dB can be calculated from Eq. (12.11):

$$\beta = 10 \log_{10}\left(\frac{I}{I_0}\right)$$
$$60 \text{ dB} = 10 \log_{10}\left(\frac{I}{10^{-12} \text{ W/m}^2}\right)$$
$$\therefore I = 10^{-6} \text{ W/m}^2$$

Similarly, we find that the sound intensity due to a crowd of people producing 80 dB is:

$$I = 10^{-4} \text{ W/m}^2$$

This is 100 times the sound intensity that is produced by one person, and we, therefore, require a crowd of 100 people.

Our rule that a 3-dB increase in intensity level corresponds to an approximate doubling of intensity only permits a rough estimate in this case. An increase of intensity level of 20 dB (from 60 dB to 80 dB) lies between 18 dB and 21 dB, and hence somewhere between 6 and 7 doublings of the intensity. This requires a crowd of people between $2^6 = 64$ and $2^7 = 128$; our exact answer of 100 people does indeed fall within this range. ■

The subjective aspect of sound concerns the apparent loudness to humans. Young persons with good hearing can hear simple

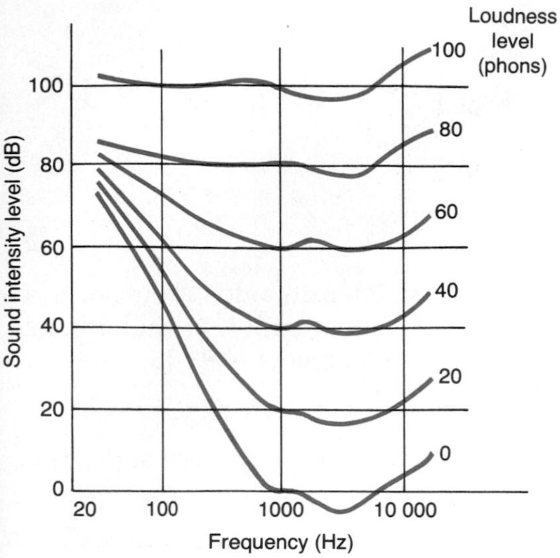

FIGURE 12.13 Subjective sound level as a function of frequency.

harmonic sound waves from a low frequency of about 20 Hz up to a high of about 20 kHz. However, sound waves of different frequency with the same intensity level do not have equal loudness. The intensity level in decibels of a 1000-Hz simple harmonic sound wave is taken as a reference level, and human subjects are asked to match its loudness with other simple harmonic sound waves of different frequency. The results are shown in Fig. 12.13 with lines of constant subjective loudness marked in phons. The subjective loudness level in phons is the sound intensity level in decibels at 1000 Hz. For example, the diagram shows that a 40-dB sound at 1000 Hz seems to be of the same loudness as a 62-dB sound at 100 Hz—both have a subjective loudness level of 40 phons. The diagram shows the ear to be most sensitive to sound in the range between 1000 Hz and 5000 Hz, and markedly less sensitive at the edges of the frequency range of audible sound.

KEY CONCEPTS

A **wave** is a disturbance that transmits energy without transmission of the material medium. A **simple harmonic wave** occurs when the disturbance possesses the following features:

1. The wave shape at a fixed instant of time is sinusoidal.
2. The displacement at a fixed position is simple harmonic. See Fig. 12.1.

The parameters characteristic of a simple harmonic wave are **amplitude, period, frequency, wavelength,** and **wave velocity.** See the definitions on p. 323.

A wave is **transverse** if the displacement associated with the disturbance is perpendicular to the direction of propagation of the wave. It is **longitudinal** if the displacement direction is parallel to the direction of propagation.

The **wave velocities** for waves in specific systems depend on the physical properties of the system in question. See Eqs. (12.3), (12.4), and (12.5).

A **standing wave** is the vibration of an extended object that results in a nonpropagating wave shape. The simplest standing waves are the **vibrational modes.** The modes are simple harmonic for the transverse vibrations of a stretched string, the longitudinal vibrations of an elastic rod, and the longitudinal vibrations of the air column in an organ pipe.

A **node** of a standing wave is a point of zero displacement, and an **antinode** is a point of maximum displacement. The simple harmonic modes are the sinusoidal waves that fit the node-antinode pattern imposed by the state of the object. See Examples 12.3 through 12.5.

Two propagating waves of nearly equal frequency produce **beats.** See Eq. (12.6).

When a wave source and a wave receiver are in relative motion, the frequencies observed by the source and the receiver differ. This frequency shift is called the **Doppler effect.** See Eq. (12.9) for waves in a material medium and Eq. (12.10) for electromagnetic waves.

Sound intensity is the energy transmitted per unit time through a unit area normal to the direction of propagation of the sound wave.

Sound intensity level is a measure of the sound intensity relative to the intensity at the threshold of hearing. The **decibel** is the measure of the sound intensity level. See Eq. (12.11).

QUESTIONS FOR THOUGHT

1. Would you expect a guitar or an organ to suffer a change in fundamental note frequency because of a rise in temperature? Would you expect the frequency of the note to be higher or lower in each case?

2. When a guitar string vibrates in its fundamental mode, it assumes a straight line configuration twice during each vibration period, which is exactly the same configuration that it would have if it were not vibrating at all. What has become of the energy of the vibrational mode when the string is exactly straight?

3. Why can you easily distinguish between an unseen person speaking in an adjacent room and a radio playing in the same room?

4. An old method for estimating the distance to a lightning stroke is to count the number of seconds between seeing the flash and hearing the thunder and then to divide the result by 5. In what units is this distance estimated?

5. What observational evidence can you cite to refute the suggestion that the speed of sound waves in air might be markedly different for waves of different wavelength?

6. A large office room has carpeted floors and drapes on the windows. Another similar room has bare windows and tiled floors. Would you expect one room to be noisier than the other when many people are working in it? Explain your answer.

7. A wind is blowing from a siren toward a person listening to it. Does the wind make any contribution to the Doppler frequency shift

a. if the siren is moving toward the listener?
b. if the listener is moving toward the siren?

In each case do you expect a lowering or a raising of the frequency because of the presence of the wind?

PROBLEMS

A. Single-Substitution Problems

(Use the following data for solution of the problems: speed of sound in air = 340 m/s at 15°C; elastic moduli from Table 11.1; densities from Table 1.4.)

1. Calculate the wavelength in air of a musical note of frequency 440 Hz. [12.1]

2. Deep ocean waves are observed to have a wavelength of 500 m and a period of 16 s. Calculate the wave velocity. [12.1]

3. An electromagnetic wave in free space has a velocity of 3×10^8 m/s and a wavelength of 1.8 mm. Calculate the frequency of the wave. [12.1]

4. Calculate the speed of transverse waves in a string of mass density 3 g/m that is experiencing a tension of 110 N. [12.2]

5. A guitar string has a mass density of 3.5 g/m. What tension is required in the string to produce a transverse wave velocity of 320 m/s? [12.2]

6. A string that is 85 cm long experiences a tension of 220 N; the velocity of transverse waves is 280 m/s. Calculate the mass of the string. [12.2]

7. Calculate the speed of a longitudinal compression wave in water. [12.2]

8. Compare the speeds of longitudinal compression waves along aluminum and lead rods. [12.2]

9. Calculate the wavelength of a 7000-Hz longitudinal compression wave in copper. [12.2]

10. Calculate the frequency of a longitudinal compression wave that has a wavelength of 16 cm in methanol. [12.2]

11. A guitar string 65 cm long has a fundamental vibration frequency of 440 Hz. Calculate the velocity of transverse waves in the string. [12.3]

12. Calculate the fundamental frequency of a closed organ pipe that is 19.3 cm long. [12.3]

13. What should be the length of an open organ pipe so that its fundamental mode has a frequency of 256 Hz? [12.3]

14. Three tuning forks have frequencies of 256 Hz, 259 Hz, and 263 Hz. Calculate the beat frequencies that can be heard by using any two of these instruments. [12.5]

15. An A string of a guitar (440 Hz) produces a beat frequency of 3 Hz when it is plucked at the same time as a similar string on another guitar. Calculate the possible fundamental frequencies of the string of the second instrument. [12.5]

16. Calculate the approach velocity of a siren toward a stationary observer that will produce an 8% increase in observed frequency in still air. [12.6]

17. Calculate the approach velocity of an observer toward a stationary siren that will produce an 8% increase in observed frequency in still air. [12.6]

18. A boy walking at 1.6 m/s toward a stationary listener whistles a 256-Hz note. Calculate the frequency of the sound heard by the listener. [12.6]

19. A passenger in an automobile is traveling at 20 m/s toward a stationary siren. The most intense component of the noise from the siren is 260 Hz. What frequency does the passenger hear? [12.6]

20. Calculate the intensity level for a sound whose intensity is 6×10^{-4} W/m^2. [12.7]

21. Calculate the intensity of a sound whose intensity level is 78 dB. [12.7]

22. Calculate the sound intensity in an automobile at highway speed if the intensity level is 64 dB. [12.7]

23. An architect requires that the sound intensity in an auditorium be kept below 2 μW/m^2. To what intensity level does this correspond? [12.7]

B. Standard-Level Problems

24. A guitar string 75 cm long has a fundamental vibration frequency of 268 Hz when the tension is 200 N. Calculate the mass of the string. [12.3]

25. Calculate the frequencies of the first three modes of a guitar string that is 65 cm long, and has a mass of 2.4 g and a string tension of 500 N. [12.3]

26. The fundamental vibration frequency in a violin string of length 32 cm and mass density 1.4 g/m is 440 Hz. Calculate the tension in the string. [12.3]

27. A wire fixed at both ends has a mass density of 4 g/m and is subject to a tension of 360 N. The frequency of one of its vibrational modes is 324 Hz, and the next higher mode has a frequency of 405 Hz. Calculate:

a. The frequency of the fundamental mode
b. The length of the wire

28. A steel piano wire 60 cm long has a mass of 2.4 g and is subject to a 650-N tension. Calculate:

a. The frequency of its fundamental mode of vibration
b. The wavelength of a sound wave generated by the wire vibrating in its fundamental mode

29. A brass wire 0.8 mm in diameter is held fixed at one end and the other end passes over a small pulley and has a 5-kg aluminum mass attached. The distance from the fixed end to the pulley is 80 cm.

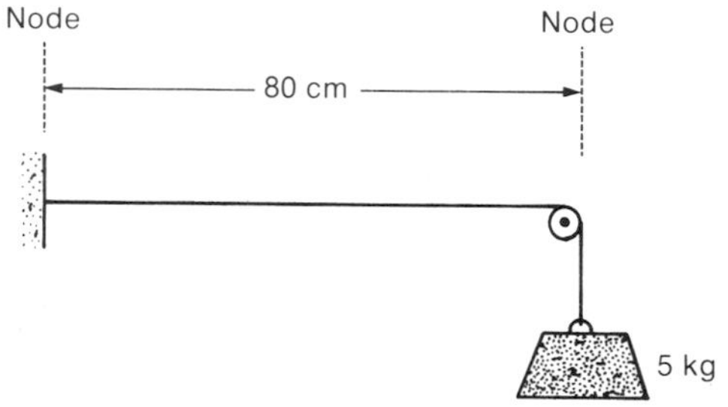

a. Calculate the frequency of the fundamental mode of transverse vibration of the wire. (The nodes are at the fixed end and at the pulley.)
b. The 5-kg mass is totally immersed in water; what is the new frequency of the fundamental mode?

30. A steel wire 0.12 mm in radius and 80 cm long is subject to a tensile strain of 3×10^{-4}. Calculate the frequency of the fundamental mode of transverse vibration.

31. A closed organ pipe has a fundamental vibration frequency of 64 Hz. Calculate:

a. The length of the pipe
b. The frequencies of the second and third modes of vibration

32. An open organ pipe is 58.4 cm long. Calculate the frequency of the fundamental, the second, and the third modes.

33. An organ pipe has three successive harmonics of frequencies 783, 1305, and 1827 Hz.

a. Is the pipe open or closed?
b. What is the frequency of its fundamental mode?
c. What is the length of the pipe?

34. A brass rod 140 cm long is clamped firmly at one end. Calculate:

a. The frequency of its fundamental mode of longitudinal compressional vibration

b. The frequencies of its second and third vibrational modes. What order harmonics of the fundamental are these modes?

35. How many beats per second are heard when two closed organ pipes, one 60 cm long and the other 61.20 cm long, are sounded? [12.5]

36. Two organ pipes simultaneously emit notes whose fundamental wavelengths are 176 cm and 180 cm. Calculate the number of beats per second. [12.5]

37. Two simple harmonic sound waves have wavelengths of 80 cm and 81 cm. Calculate:

a. The beat frequency
b. The beat frequency heard under water from two longitudinal compression waves having the same wavelengths

38. A 256-Hz tuning fork produces 4 beats per second in conjunction with the fundamental mode of a violin string. The string is 44 cm long and has a mass of 1.6 g. The beat frequency decreases when the tension of the violin string is increased slightly. Calculate:

a. The frequency of the fundamental mode of the string before the change in tension
b. The original tension in the string

39. Two identical guitar strings 15 cm long have a fundamental frequency of 110 Hz under a tension of 155 N. The tension of one string is kept fixed. What possible tensions in the other string would give rise to 2 beats per second when the strings are played together?

40. An automobile horn has a frequency of 400 Hz. Calculate:

a. The apparent frequency change when a listener drives past the stationary horn at 90 km/h
b. The frequency change if the horn drives past a stationary listener at 90 km/h [12.6]

41. A pedestrian standing by the roadside hears an automobile horn of apparent frequency 375 Hz. He knows that the horn frequency is 420 Hz. What is the speed of the automobile, and is it approaching or receding? [12.6]

42. A high-speed locomotive traveling at 125 km/h emits a 420-Hz note on its horn. Calculate:

a. The frequency change heard by a stationary listener as the locomotive passes
b. The frequency change heard by a listener in another train traveling at 90 km/h in the opposite direction on parallel tracks

43. An astronaut who delights in precise measurements has a favorite radio station that broadcasts on a wavelength of 3.17562 m. On his way to the moon the astronaut's spaceship is receding from the transmitting antenna at 20 000 mi/h. To what wavelength should he tune his radio receiver?

44. A distant galaxy is moving away from the earth at a speed that causes an apparent decrease of 40% in the frequency of light waves emitted from it. Calculate the relative speed of recession of the galaxy.

45. Deep ocean waves of wavelength 750 m and period 22 s are traveling from west to east.

a. What is the period of pitch of a cruise liner that is making 25 knots due west?
b. If the ship changes course to due east, what is the new period of pitch?

46. A piece of electronic audio equipment is advertised as having "a signal-to-noise ratio of 44 dB." Calculate the intensity ratio for the signal and the noise. [12.7]

47. If the intensity of a sound wave is increased sixfold, by what amount does the intensity level change? [12.7]

48. Sound from a busy industrial site is coming normally through an open window 1.2 m × 0.8 m at an 85-dB level. How much energy per hour comes through the window? [12.7]

49. An automobile horn produces an 82-dB sound intensity level at a given place. Calculate the intensity level of 12 such horns all sounding together.

50. By listening carefully an observer detects a 30% increase in the intensity of a certain sound. What is the corresponding intensity level increase?

51. Acoustic insulation in a certain room lowers the average sound intensity level by 25 dB. What percentage of the original sound intensity is present after installing the insulation?

52. Three sources of noise produce sound intensity levels of 58 dB, 69 dB, and 71 dB when acting separately. Calculate the sound intensity level of all three noise sources acting together.

C. Advanced-Level Problems

53. A stationary open organ pipe 66.4 cm long resonates in its fundamental mode with a 248-Hz sound source that is in motion. With what velocity must the source move for resonance to occur, and is it approaching or receding?

54. An underwater sonar system generates 50-kHz compression waves in seawater. After reflection from a target, the waves are detected on their return to the original sending point. Calculate:

a. The wavelength of the transmitted waves
b. The frequency of the returning waves after reflection from the hull of a submarine that is receding at 25 knots (Assume that the hull of the submarine receives the waves with an apparent frequency shift and then retransmits them, causing another frequency shift.)

55. A musician walks toward a large, smooth wall at 1.6 m/s, blowing a pipe with a fundamental frequency of 440 Hz. How many beats per second does she hear between the direct sound of the pipe and the sound reflected from the wall? (Consider the hint for solving the previous problem.)

13

TEMPERATURE AND THE IDEAL GAS LAW

In our study of dynamics, we developed methods to calculate energy. An object may have kinetic energy, gravitational potential energy, or elastic potential energy. (This list is not exhaustive; we are naming only the energy forms previously discussed.) To calculate the total energy of a collection of objects, we would simply add up the contributions made by each object. However, many physical systems contain an impossibly large number of separate objects. For instance, a jug of water contains an enormous number of water molecules, and we can hardly calculate the energy of the water by adding up the contributions from each molecule. What can we say about the energy of such a system and the energy exchanges that occur when it interacts with other systems? In this chapter we consider these questions. Moreover, we seek to do so using only **macroscopic** physical properties; by this we mean measurable properties of the bulk system that take no heed of its **microscopic** constituents. The part of physics that studies this question is called **thermodynamics.** We shall concentrate on the first and second laws of thermodynamics, and their implications for the behavior of physical systems. A necessary preliminary to which we now address ourselves is the development of the concept of temperature.

13.1 Temperature and Thermometers

Up to this point we have required only three fundamental quantities (time, length, and mass) in our study of mechanics and related topics. Moreover, all derived quantities such as energy, momentum, and pressure were defined in terms of the three fundamentals. We now consider a new class of phenomena called **thermal** phenomena that are governed by the laws of thermodynamics. Our understanding of them requires the introduction of a fourth fundamental quantity called **temperature.** We begin with the intuitive concept of temperature that is based on our sensation of hotness or coldness. The hotter something is, then the higher its temperature; the colder it is, then the lower its temperature.

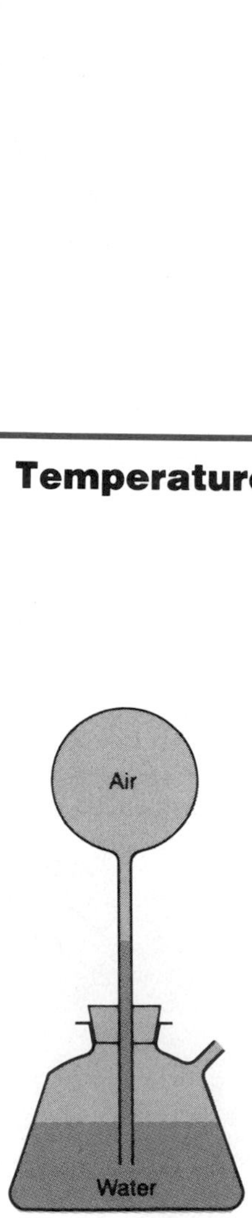

FIGURE 13.1 Galileo's temperature indicator. When the air in the flask becomes hotter, the water level in the tube drops; when the air becomes colder, the water level rises.

In order to make a scientific measurement of temperature, we must go beyond the realm of our subjective sensations. An instrument to do this is called a **thermometer.** Any system that has properties that change with temperature in a reproducible manner can be made to serve as a thermometer. Galileo constructed a crude device that indicated rising temperature from the expansion of heated air, as illustrated in Fig. 13.1. Air is contained in an inverted flask with a long thin neck that dips below the water level in a second flask that is open to the atmosphere. A temperature rise of the air in the inverted flask causes a drop of the water level in the thin neck, and conversely. There are many temperature-sensitive devices in common use. A well-known instance

SPECIAL TOPIC 15 FAHRENHEIT

Daniel Fahrenheit (1686–1736) was a Polish-born instrument maker who spent most of his working life in Holland. He knew that changing temperature caused a small change in the height of the mercury column in a barometer, and he reasoned that mercury in a fine capillary tube might be a good temperature-sensitive system. Fahrenheit constructed his scale by assigning 0°F to the temperature of an ice and salt mixture, and 96°F to the normal temperature of human blood. Not only are these fixed points poorly reproducible, but the values assigned to them were influenced by a misunderstanding of previous work. However, Fahrenheit was the inventor of the mercury-in-glass thermometer, which was the first reasonably accurate temperature-measuring instrument. He used it to show that many liquids had fixed characteristic boiling points under normal conditions, and also discovered that boiling points change with changes in pressure. Fahrenheit later modified his scale slightly to set the boiling point of water at 212°F.

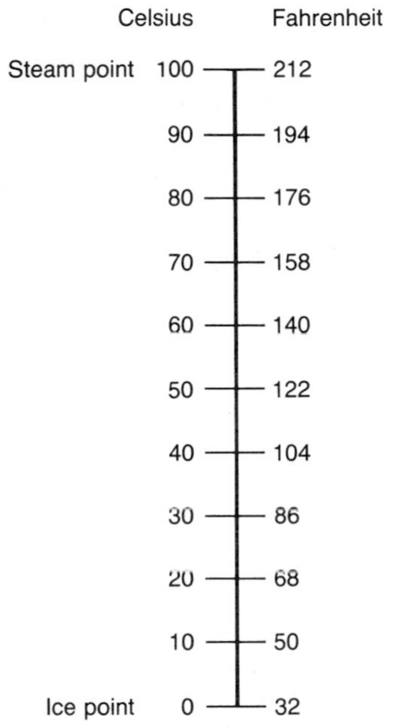

Some equivalencies on the Celsius and Fahrenheit temperature scales

Today we define the Fahrenheit scale by setting the ice point at 32°F and the steam point at 212°F. Since there are 180 Fahrenheit degrees between these fixed points, we can convert Fahrenheit temperatures to Celsius by using the relation

$$T_C = \frac{5(T_F - 32)}{9}$$

where T_F and T_C are the Fahrenheit and Celsius temperatures, respectively. The accompanying diagram shows some of the equivalent temperatures.

is the column of mercury or alcohol in a thin capillary tube attached to a small reservoir of the liquid; the level in the thin capillary rises or falls with temperature change. Yet another is the bimetallic strip that bends in one direction or the other as temperature changes.

To turn a temperature-sensitive device into a thermometer, we must select **fixed points,** assign numerical values to them, and suitably divide the interval between them. A fixed point is some natural event that is readily reproducible, and whose occurrence always corresponds to the same configuration of the temperature-sensitive device. Newton made up a large list of fixed points. Some of them, such as the boiling point of water, were in fact capable of fairly accurate reproduction, while others, such as the human body temperature or the temperature of coals in a small fire, are not sufficiently reproducible to be of much value.

For many years the ice point and the steam point were the fundamental fixed thermometric points. The ice point is the temperature at which ice and water coexist in equilibrium at atmospheric pressure; the steam point is the temperature at which water vapor and water coexist in equilibrium at atmospheric pressure. The designation of the ice point as 0 and the steam point as 100 determines a scale of temperature. The scale was known as centigrade, but for most practical purposes it is the same as the scale we now call Celsius. The temperature at the ice point is written 0°C, and at the steam point 100°C. This scale is named in honor of the Swedish astronomer Anders Celsius (1701–1744), who was the first to popularize its use. It is an interesting historical fact that Celsius designated the ice point as 100 and the steam point as 0—the reverse designations from the present Celsius scale! There could hardly be a more convincing example to illustrate the arbitrariness in our assignment of certain numbers as measures of temperature.

A Celsius thermometer suitable for measuring temperatures between the ice and steam points can be made from a temperature-sensitive device that is marked "0°C" at the ice point and "100°C" at the steam point, and that has the intervening interval divided into one hundred equal parts. We have, of course, no guarantee that thermometers made from different physical systems (for example, mercury in glass, or alcohol in glass) will give the same readings when used to measure the temperature of an object that is intermediate between the fixed points. However, a class of thermometers of outstanding accuracy does exist whose temperature readings do not depend on the measuring substance. These are the gas thermometers—modern, sophisticated models of Galileo's original temperature indicator. Gas thermometers are discussed fully in Section 13.2.

For most practical purposes, the problem of measuring temperatures that lie between fixed points reduces to finding thermometers that agree closely with gas thermometers. The mercury-in-glass and the platinum resistance thermometers are both satisfactory in this respect. A thermometer for extremely accurate work should be calibrated against a gas thermometer in a standards laboratory. The gas thermometers themselves are bulky, fragile devices that take a long time to come to thermal equilibrium and are therefore most unsuitable for practical everyday use.

The mercury-in-glass thermometer covers only a limited range of temperatures, since mercury freezes at −38.9°C and boils at 357°C. Because the freezing point of alcohol is lower than −100°C, an alcohol-in-glass thermometer is more useful for low temperatures. The platinum resistance thermometer covers an extraordinarily wide range of temperature and is very useful as a secondary standard.

TABLE 13.1 Some fixed points of the International Practical Temperature Scale

Fixed point	*Celsius temperature*
NBP of neon	−246.05
NBP of oxygen	−182.96
Triple point of water	0.01
Steam point	100.00
NMP of zinc	419.58
NMP of silver	961.93
NMP of gold	1064.43

Note: NBP = normal boiling point; NMP = normal melting point.

To measure temperatures outside the range 0°–100°C, we can mark the scales of the thermometers with additional equal intervals below the ice point and above the steam point. However, such an extrapolation is not satisfactory for temperatures far removed from the two fixed points. As a matter of practicality, other fixed points are necessary for making accurate thermometers. Table 13.1 lists some of the fixed points of the International Practical Temperature Scale, most of which are normal melting points (NMP) and normal boiling points (NBP) of various substances. The word *normal* means that the measurement is done at a pressure of exactly 1 atm. An accurate high-temperature thermometer is calibrated by using the melting points of zinc, silver, and gold. The low-temperature calibration points are the boiling points of liquefied gases. The study of thermometry is a living art, and the fixed points change as new research is performed and more careful measurements are made.

13.2 The Absolute Temperature Scale and the Ideal Gas Law

We now address the task of setting up the absolute temperature scale—one that is available as a standard to check the accuracy of any practical thermometer. It is necessary to have such a standard since, as we have already mentioned, there is nothing so far to ensure that different thermometers agree at temperatures between the fixed points. The standard is the constant-volume gas thermometer illustrated in Fig. 13.2. A flask of gas connects to a mercury manometer that is equipped with a reservoir so that the general mercury level in the manometer can be adjusted. The gas flask is immersed in the substance whose temperature is being measured, and the supply in the reservoir is adjusted until the left-hand mercury column aligns with the fixed volume point. The difference between the two mercury levels is a direct measure of the gauge pressure of the gas in the flask. The gas pressure varies with temperature and therefore gives a measure of the temperature of the gas.

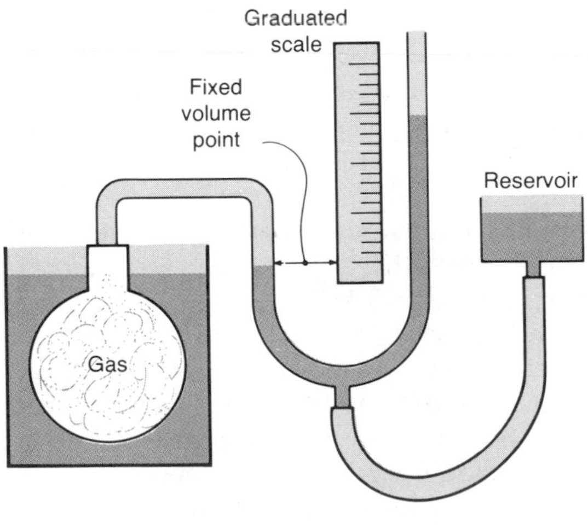

FIGURE 13.2 Diagram of a constant-volume gas thermometer.

A gas thermometer that is calibrated at the ice point and the steam point gives two pressure values at these points, as shown graphically in Fig. 13.3. A straight line drawn backward through the calibration points intersects the temperature axis at −273.16°C, and this remains true regardless of the gas used, provided that the gas density is kept low. This cannot imply that the gas pressure would drop to zero if the temperature were reduced to this point, since every known gas liquefies at a temperature higher than −273.16°C. However, this experimental fact did lead to the suggestion that −273.16°C was the lowest temperature possible, and many subsequent experiments confirm this. The temperature in question is called the absolute zero, and we choose it as one of the fixed points on the absolute temperature

SPECIAL TOPIC 16 KELVIN

Lord Kelvin was born William Thomson in Belfast in 1824. Appointed professor at Cambridge University at the age of 22, he was responsible for much of the pioneering work in the new area of thermodynamics. He proposed that −273°C be an absolute zero of temperature, and the degrees on the absolute scale have since been named for the title that was later conferred on him.

In the middle of the nineteenth century, the "Great Eastern" was laying the Atlantic cable. Thomson made many improvements in cables and instruments; without his work, it is doubtful that the cable could ever have functioned properly.

As an old man he was made Baron Kelvin of Largs—a title made up from the Kelvin River and the town of Largs through which it flows in Ayrshire, Scotland. He died in 1907 and was buried next to Newton in Westminster Abbey.

scale. For the other fixed point, we choose the triple point of water; this is the temperature at which all three phases of water (solid, liquid, and gas) coexist in equilibrium. The pressure at the triple point of water is 4.58 torr, and the temperature is 0.01°C. We shall discuss the triple point in greater detail in the following chapter. For our present purpose we simply need to know that it is unique; there is no other combination of temperature and pressure for which the three phases of water coexist in equilibrium.

Having chosen the fixed points, we must now define the scale. Before doing this, we note that the absolute temperature scale is identical with the Kelvin thermodynamic scale, which we discuss in a subsequent chapter, and we need not distinguish between temperatures on the two scales. The temperature designations on the scale are named kelvin (in honor of Lord Kelvin), the abbreviation is "K," and the degree symbol is not used. We then define the absolute temperature scale by:

$$T = \left(\frac{p}{p_{TP}}\right) 273.16\ \text{K} \qquad (13.1)$$

where p is the gas pressure at temperature T, and p_{TP} is the pressure at the triple point. Note that this sets the absolute zero at 0 K and the triple point of water at 273.16 K.

The Celsius scale is not defined in terms of the ice and steam points, but rather is defined by setting the triple point of water at 0.01°C, and the degree size the same as the kelvin degree. This means that the ice and steam points are only approximately 0°C and 100°C, but the approximation is so close that it need not concern us for practical purposes. For reference, we tabulate various equivalent temperatures on the Celsius and Kelvin scales.

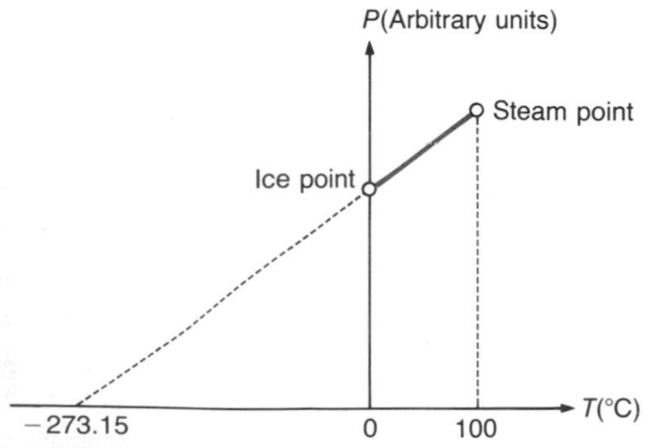

FIGURE 13.3 Location of the absolute zero of temperature using a constant-volume gas thermometer.

	Celsius temperature (°C)	*Kelvin temperature* (K)
Absolute zero	−273.15	0
Ice point	0	273.15
Triple point of water	0.01	273.16
Steam point	100	373.15

We express the relation between Celsius and Kelvin temperatures as follows:

$$T(\text{K}) = T(°\text{C}) + 273.15 \tag{13.2}$$

We return now to consider the requirements on the gas in the gas thermometer. The following conditions are verified experimentally:

1. The gas must be of low density. As a practical guide, a density less than that of air under normal conditions will suffice.
2. The gas must be at a considerably higher temperature than its liquefaction point.

Suitable gases for gas thermometers over a fairly wide range of conditions include hydrogen, helium, nitrogen, oxygen, argon, and neon. Historically, the discovery of relations between pressure, volume, and temperature of a gas occurred long before the advent of accurate gas thermometers. Robert Boyle (1627–1691), a contemporary of Newton and one of the founders of the Royal Society, investigated the elastic properties of air and discovered that the volume of air in a container varies inversely with the applied pressure. Today, we know this relation as Boyle's law, and we add the stipulation that the temperature must be kept constant. The scientific measurement of temperature was just beginning when Boyle carried out his experiments. Undoubtedly, he did his work under conditions in which the temperature change was very small.

Almost a century later, the French scientist J. A. C. Charles (1746–1823) proposed the gas law that bears his name. He discovered that the volume of a gas varies directly with the temperature if the pressure remains constant. Charles was a designer of hot air balloons, and his work on gases was strongly motivated by his interest in balloons. However, his work was not very exact. His compatriot, Joseph Gay-Lussac (1778–1850), really deserves the credit for the experimental work that put the law on a sound footing.

Boyle's law and Charles' law taken together imply that we can write the relation between pressure, volume, and temperature in the following form:

$$pV = \text{constant} \times T \tag{13.3}$$

If we hold T constant in this relation, then V is inversely proportional to p (Boyle's law). If we hold p constant, then V is directly proportional to T (Charles' law). The combination of the gas laws given by Eq. (13.3) does not hold true for every gas under all possible conditions of temperature and pressure. In fact, the gas must satisfy the conditions that we have given for the gas in a gas thermometer. This leads us once again to propose an ideal model of reality—this time a model of gas behavior—and use Eq. (13.3) as its basis. An **ideal gas** is, by definition, a gas that exactly obeys Eq. (13.3). The gases that are suitable for gas thermometers are close to ideal over a wide range of conditions. However, the equation is not accurate for any gas near its liquefaction point; since all gases liquefy at sufficiently low temperatures, no gas is ideal for all possible conditions.

We can directly relate the constant in Eq. (13.3) to the quantity of gas. Every gas is composed of a number of molecules in random motion that occasionally collide with each other and with the walls of the container. It is an experimental fact that the constant of Eq. (13.3) is proportional to the number of molecules in the gas. This fact allows us to write the *equation of state* of an ideal gas.

■ Equation of State Defining an Ideal Gas

$$pV = NkT \tag{13.4}$$

p is the absolute pressure.
V is the volume.
N is the number of molecules.
T is the Kelvin temperature.
k is the Boltzmann constant, 1.381×10^{-23} J/K.

An equation that relates pressure, volume, temperature, and quantity of gas is called an equation of state. The quantities p, V, and T are the state variables for the quantity of gas under consideration.

The Boltzmann constant is named for the Austrian physicist Ludwig Boltzmann (1844–1906), who is famous as a pioneer in statistical physics. The constant is one of the fundamental constants of physics, and it is intimately related to the universal gas constant, which we can use in an alternate statement of the ideal gas equation of state.

We can write the form of Eq. (13.4) that is preferred in chemistry from the following considerations. In dealing with macroscopic amounts of a substance, it is convenient to have a measure of quantity other than the total number of molecules. The mass is one such measure, but it is not proportional to the number of molecules.

The SI unit of quantity (which is proportional to the number of molecules) is the *mole* (abbreviation: mol), and it is defined to be the amount of any pure substance that contains Avogadro's number of molecules. Avogadro's number is the number of carbon atoms in 12 g of carbon-12.* The accepted modern value for Avogadro's number is $N_A = 6.022 \times 10^{23}$/mol. Amedeo Avogadro (1776–1856) was professor of mathematical physics at the University of Turin, Italy. Ironically, Avogadro had little idea of the number of molecules in 1 mol, but he had a firm understanding of why it was the same for all pure substances.

The mass of 1 mol of a pure substance is called the *molecular (or atomic) mass*. The molecular mass is frequently called the molecular weight in chemistry, but we will avoid this terminology to emphasize the distinction between mass and weight which is so necessary in dynamics. If we have n moles of a substance, then the number of molecules present is given by $N = nN_A$. This permits us to write Eq. (13.4) as:

$$\begin{aligned} pV &= nN_AkT \\ &= nRT \end{aligned} \qquad \textbf{(13.4a)}$$

where we have written:

$$\begin{aligned} R &= N_Ak \\ &= (6.022 \times 10^{23}\text{/mol}) \times (1.381 \times 10^{-23}\ \text{J/K}) \\ &= 8.314\ \text{J/K·mol} \end{aligned}$$

This quantity is called the *universal gas constant*. We can regard it as the Boltzmann constant of a mole, or alternately, we can regard the Boltzmann constant as the universal gas constant of one molecule. Both constants are expressions of the same physical reality.

Since the thermal expansion of gases is very large compared with that of either liquids or solids, it is convenient to set up standard conditions for comparing gas density.

The conditions of pressure and temperature:

$$p = 1\ \text{atm} = 1.013 \times 10^5\ \text{Pa} = 760\ \text{torr}$$

$$T = 0°\text{C} = 273.15\ \text{K}$$

are referred to as standard temperature and pressure (STP).

---

*The carbon-12 atom has a nucleus that contains 6 neutrons and 6 protons; it accounts for about 99% of all naturally occurring carbon.

EXAMPLE 13.1

Calculate the density of oxygen at STP. The molecular mass of oxygen is 32 g, and it closely obeys the ideal gas equation of state under these conditions.

SOLUTION Consider a quantity of 1 mol of oxygen that has a mass of 32 g and contains 6.022×10^{23} molecules. We can calculate the volume of the gas from either Eq. (13.4) or (13.4a):

$$pV = NkT$$

$$1.013 \times 10^5 \text{ Pa} \times V = \begin{array}{r} 6.022 \times 10^{23} \times 1.381 \times 10^{-23} \text{ J/K} \\ \times\ 273.15\text{K} \end{array}$$

$$\therefore V = 22.4 \times 10^{-3} \text{ m}^3$$

To calculate the density, we use Eq. (1.1):

$$\rho = \frac{m}{V} = \frac{32 \times 10^{-3} \text{ kg}}{22.4 \times 10^{-3} \text{ m}^3} = 1.43 \text{ kg/m}^3$$

The first part of the solution to this problem is of wider validity. The calculation of the volume is correct for 1 mol of any gas that is approximately an ideal gas at STP. The volume of 1 mol of any gas at STP is $22.4 \times 10^{-3} \text{ m}^3 = 22.4\ \ell$ provided it obeys the ideal gas equation of state. ■

We can write another useful equation for changes in pressure, temperature, and volume of a fixed quantity of gas. Consider a fixed quantity of gas having initial pressure p_1, temperature T_1, and volume V_1; and final values of p_2, T_2, and V_2, respectively. Since the quantity of gas is constant, then N of Eq. (13.4) or n of Eq. (13.4a) remains constant. This means that the quantity pV/T also remains constant as the state variables change from their initial to their final values.

■ Useful Formula Relating the State Variables of a Fixed Quantity of an Ideal Gas

$$\frac{p_1V_1}{T_1} = \frac{p_2V_2}{T_2} \qquad \textbf{(13.5)}$$

The subscripts $_1$ and $_2$ refer to initial and final conditions.
Units of T_1 and T_2: kelvin
p_1 and p_2: any units of *absolute* pressure
V_1 and V_2: any units of volume

Note that the temperature must be in kelvin, but that the absolute pressure and the volume can be in any units as long as they are the same on both sides of Eq. (13.5).

EXAMPLE 13.2

A flexible rubber balloon contains air at a gauge pressure of 150 torr and a temperature of 20°C. The balloon is immersed in boiling water at 100°C, and the volume increases by 15%. What is the new gauge pressure of the air in the balloon?

SOLUTION Since air is approximately an ideal gas under these conditions and the quantity remains constant, Eq. (13.5) is appropriate. Set the units in the required form.

a. The pressure must be absolute pressure, but any unit will suffice. Since 1 atm = 760 torr, we have:

$$p_1 = 150 \text{ torr} + 760 \text{ torr}$$
$$= 910 \text{ torr}$$
$$p_2 = ?$$

b. We can state the volume in any unit provided we retain the same unit for the initial and final states:

$$V_1 = 1 \text{ rubber balloon full}$$
$$V_2 = 1.15 \text{ rubber balloon full}$$

c. The temperature must be in kelvin. We convert with the help of Eq. (13.2):

$$T(\text{K}) = T(°\text{C}) + 273.15$$
$$T_1 = 20°\text{C} + 273.15 = 293.15 \text{ K}$$
$$T_2 = 100°\text{C} + 273.15 = 373.15 \text{ K}$$

Substitute in Eq. (13.5):

$$\frac{p_1 V_1}{T_1} = \frac{p_2 V_2}{T_2}$$

$$\frac{910 \text{ torr} \times 1 \text{ rubber balloon full}}{293.15 \text{ K}} = \frac{p_2 \times 1.15 \text{ rubber balloon full}}{373.15 \text{ K}}$$

$$\therefore p_2 = 1007 \text{ torr}$$

$$\text{New gauge pressure} = 1007 \text{ torr} - 760 \text{ torr}$$
$$= 247 \text{ torr}$$

Note again the cancellation of the units of pressure and volume from each side of Eq. (13.5). ■

EXAMPLE 13.3

The gas in the cylinder of a diesel engine is originally at a temperature of 47°C and a pressure of 1 atm. After compression the temperature is 700°C and the pressure is 48.5 atm. What is the ratio of the initial volume to the final volume?

SOLUTION The quantity of gas remains constant, and we can use Eq. (13.5) if the ideal gas equation of state is approximately true. Set the units in required form:

a. The pressures are absolute pressures in the same units:

$$p_1 = 1 \text{ atm}$$

$$p_2 = 48.5 \text{ atm}$$

b. We use Eq. (13.2) to find the Kelvin temperatures:

$$T(\text{K}) = T(^\circ\text{C}) + 273.15$$

$$T_1 = 47^\circ\text{C} + 273.15 = 320.15 \text{ K}$$

$$T_2 = 700^\circ\text{C} + 273.15 = 973.15 \text{ K}$$

Substitute in Eq. (13.5):

$$\frac{p_1 V_1}{T_1} = \frac{p_2 V_2}{T_2}$$

$$\frac{1 \text{ atm} \times V_1}{320.15 \text{ K}} = \frac{48.5 \text{ atm} \times V_2}{973.15 \text{ K}}$$

$$\therefore \frac{V_1}{V_2} = 15.9$$

This quantity is called the compression ratio of the diesel engine. ■

13.3 Thermal Expansion of Solids and Liquids

With very few exceptions solids and liquids expand with increasing temperature. As we have already seen, many practical thermometers make use of this fact. A steel rod becomes slightly longer if it is heated, and its cross-sectional area and volume increase as well. A flask full of alcohol spills over when heated because the volume of the alcohol increases. This example is complicated by the fact that the glass flask also expands on heating, but it expands much less than the alcohol does.

In a previous chapter we studied the small changes in length and volume that accompany the action of external forces. We described these changes as elastic strains and set down the general concept of strain in Eq. (11.2). The changes of length and volume due to temperature change are called **thermal strains.** For many substances the thermal strain is proportional to temperature change, and the constant of proportionality is called the **coefficient of thermal expansion.**

Since strain is a pure number, the quantities on the right-hand side of Eq. (13.6) are also pure numbers. The coefficients of thermal expansion therefore have the unit of a reciprocal temperature difference. We refer to the temperature difference as Celsius degrees and use the abbreviation "C°." This practice distinguishes a temperature difference from a temperature, which is measured in degrees Celsius with the abbreviation "°C."

■ Definition of Coefficient of Thermal Expansion

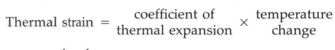

$$\text{Thermal strain} = \begin{matrix}\text{coefficient of}\\ \text{thermal expansion}\end{matrix} \times \begin{matrix}\text{temperature}\\ \text{change}\end{matrix}$$

$$\frac{l - l_0}{l_0} = \alpha(T - T_0) \qquad \textbf{(a)} \qquad \textbf{(13.6)}$$

$$\frac{V - V_0}{V_0} = \beta(T - T_0) \qquad \textbf{(b)}$$

SI unit for coefficient of thermal expansion: (Celsius degree)$^{-1}$
Abbreviation: (C°)$^{-1}$

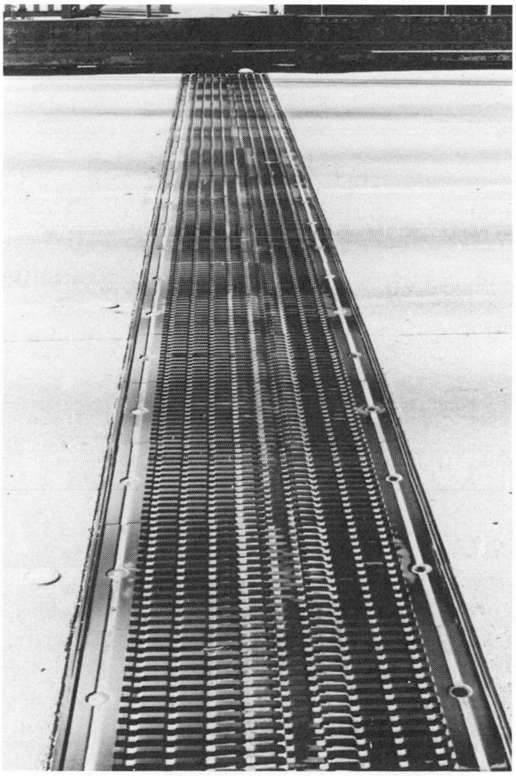

This articulated joint in the road surface on the Golden Gate Bridge permits thermal expansion to occur without producing a damaging distortion to the structure of the bridge. (Courtesy of Golden Gate Bridge, Highway and Transportation District)

It is convenient to distinguish between linear and volumetric thermal strains (we made a similar distinction for elastic strains). The coefficients α and β defined by Eq. (13.6) relate to linear and volumetric strain, respectively. Table 13.2, which lists the values of thermal expansion coefficients of various substances, follows the usual practice of quoting only the volumetric coefficient for liquids and both coefficients for solids. Under fairly general conditions, the coefficient of volumetric expansion can be shown to be 3 times the coefficient of linear expansion. It is also evident from Table 13.2 that liquids tend to have much higher coefficients of thermal expansion than solids.

The coefficients of thermal expansion are constant only over a very limited range of temperature. For higher temperatures the thermal expansion coefficients are slightly higher than the values listed; for lower temperatures they become very much lower, eventually falling to zero. For the purpose of working problems, we will take the values listed as approximately correct over a limited temperature range.

TABLE 13.2 Coefficients of thermal expansion of solids and liquids at 0°C

	Thermal expansion coefficient $[(C°)^{-1}]$	
Substance	*Linear*	*Volumetric*
Aluminum	23×10^{-6}	69×10^{-6}
Brass	19×10^{-6}	57×10^{-6}
Copper	16×10^{-6}	48×10^{-6}
Iron or steel	12×10^{-6}	36×10^{-6}
Lead	29×10^{-6}	87×10^{-6}
Silver	18×10^{-6}	54×10^{-6}
Soda glass	9×10^{-6}	27×10^{-6}
Pyrex glass	3.3×10^{-6}	10×10^{-6}
Fused quartz	0.5×10^{-6}	1.5×10^{-6}
Methanol	—	1130×10^{-6}
Ethanol	—	753×10^{-6}
Glycerin	—	480×10^{-6}
Mercury	—	182×10^{-6}
Benzene	—	1240×10^{-6}
Gasoline	—	560×10^{-6}

EXAMPLE 13.4

A steel rail is exactly 25 m long when the temperature is −5°C. By how much does it expand when the temperature rises to 30°C?

SOLUTION The problem is one of linear thermal expansion involving Eq. (13.6a):

$$\frac{l - l_0}{l_0} = \alpha(T - T_0)$$

$$\frac{l - 25 \text{ m}}{25 \text{ m}} = 12 \times 10^{-6}(\text{C°})^{-1} \times [30°\text{C} - (-5°\text{C})]$$

$$= 4.2 \times 10^{-4}$$

$$\therefore l - 25 \text{ m} = 0.0105 \text{ m}$$

The expansion of the rail is 1.05 cm. Even though the thermal strain is very small, the total expansion of a long rail is considerable, necessitating the gaps that exist between rail sections that are laid on ties. It is a relatively simple matter to calculate the thermal stress in a heated object that is not allowed to expand.

■

EXAMPLE 13.5

The 25-m-long steel rail of the previous example is laid without provision for expansion when the temperature is −5°C. The cross-sectional area of the rail is $6.5 \times 10^{-3}\ \text{m}^2$. Calculate the compressive force in the rail when the temperature rises to 30°C.

SOLUTION If the rail is not permitted to expand, the thermal strain must be equal in magnitude and opposite in sign to the elastic strain. In Example 13.4, we calculated a thermal strain of 4.2×10^{-4}. The elastic strain is therefore compressive and equal to this value. The Young's modulus for steel from Table 11.1 is 2.0×10^{11} Pa. Using these data in Eq. (11.6), we have:

$$Y = \frac{F_{\perp}/A}{(l - l_0)/l_0}$$

$$2.0 \times 10^{11}\ \text{Pa} = \frac{F_{\perp}/(6.5 \times 10^{-3}\ \text{m}^2)}{4.2 \times 10^{-4}}$$

$$\therefore F_{\perp} = 5.46 \times 10^5\ \text{N}$$

This force is more than 60 tons, but the situation would never come to this in practice. The rail would simply relieve itself by buckling. The very large value of the thermally induced force in a constrained rail underscores the need for expansion gaps. ■

It is relatively easy to visualize the thermal expansion of a solid rod, but what happens if one heats a solid that has a hole in it? For example, a steel washer is a circular disk with a circular hole cut out in the center. It is tempting to say that heating the washer causes the metal to expand outward at the outer edge and inward at the inner edge, thereby decreasing the diameter of the hole. But this is not correct, as the following consideration shows. Imagine that we begin with a circular disk of steel and mark the circle to represent the hole to be cut out. Now, before cutting out the hole, we heat the whole disk; thermal expansion increases the diameter of the whole disk and also the diameter of the marked circle. This means that our proposed hole now has a larger diameter, as indeed would also be the case if it had been cut out before heating. The same considerations hold for holes within volumes, such as the interior of bottles—the capacity of a bottle increases on heating exactly as if the vacancy were made of the material of the bottle.

EXAMPLE 13.6

A Pyrex glass flask is filled with exactly 1 ℓ of benzene at 10°C. How much of the liquid spills out if the flask and contents are heated to 30°C?

SOLUTION The problem concerns a difference between two volumetric expansions. The glass flask expands but the benzene expands even more and overflows. The original volume of the glass flask and benzene is:

$$V_0 = 1\ell \times 10^{-3}\ \text{m}^3/\ell = 10^{-3}\ \text{m}^3$$

Equation (13.6b) applied to the flask gives:

$$\frac{V - V_0}{V_0} = \beta(T - T_0)$$

$$\frac{V - 10^{-3}\ \text{m}^3}{10^{-3}\ \text{m}^3} = 10 \times 10^{-6}\ (\text{C}°)^{-1} \times (30°\text{C} - 10°\text{C})$$

$$\therefore (V - 10^{-3}\ \text{m}^3)_{\text{Pyrex}} = 2 \times 10^{-7}\ \text{m}^3$$

The same equation applies to the benzene:

$$\frac{V - V_0}{V_0} = \beta(T - T_0)$$

$$\frac{V - 10^{-3}\ \text{m}^3}{10^{-3}\ \text{m}^3} = 1240 \times 10^{-6}\ (\text{C}°)^{-1} \times (30°\text{C} - 10°\text{C})$$

$$\therefore (V - 10^{-3}\ \text{m}^3)_{\text{benzene}} = 2.48 \times 10^{-5}\ \text{m}^3$$

The amount of overflow is the difference between these quantities:

$$\begin{aligned}\text{Overflow} &= 2.48 \times 10^{-5}\ \text{m}^3 - 2 \times 10^{-7}\ \text{m}^3\\ &= 2.46 \times 10^{-5}\ \text{m}^3\\ &= 24.6\ \text{cm}^3\end{aligned}$$

■

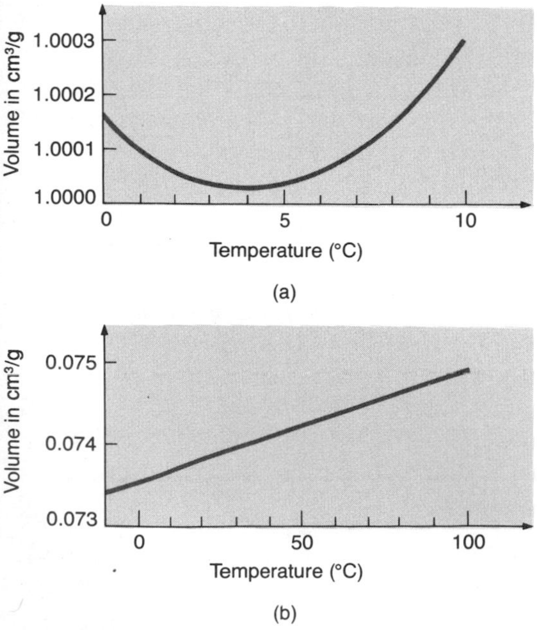

FIGURE 13.4 Volume of a gram of liquid as a function of temperature: (a) water; (b) mercury.

As mentioned before, most substances expand on heating. A few, such as germanium and silicon, show small regions of thermal contraction with increasing temperature at very low temperatures, but the outstanding example of a substance possessing a negative coefficient of thermal expansion is water. For the temperature interval 0°C to 4°C, water actually contracts as the temperature increases. We illustrate the magnitude of this effect in Fig. 13.4(a), which shows data for the volume of 1 g of water over the temperature range 0°C through 10°C. The maximum density for water occurs at about 4°C. The diagram also shows that a water-in-glass thermometer would be a poor choice. Between 0°C and 8°C two different temperatures correspond to each volume of water. For comparison, Fig. 13.4(b) shows the volume of 1 g of mercury over the temperature range −10°C through 100°C. Mercury is an excellent choice for a thermometer liquid.

The strange thermal expansion behavior of water is of immense importance for many living things. When the fall comes, a freshwater lake begins to get cooler. As the surface layer cools, it becomes denser and sinks, pushing up warmer water that is in turn cooled at the surface. The process continues until the whole lake reaches a temperature of about 4°C. Further cooling of the surface layer makes it less dense, and it therefore stays on the surface until winter temperatures turn it to ice. If the lake is sufficiently deep, the water at the lower depths never freezes, enabling fish and other marine life to survive the winter.

13.4 Kinetic Theory of Gases

The idea that material substances consist of large numbers of tiny particles has its roots in ancient Greek philosophy. The philosophical theory is called *atomism,* but the corresponding physical theory is usually known as *kinetic theory.* The name comes from our understanding of the properties of a substance in terms of the rapid, random motion of its constituent particles. The first successful application of kinetic theory to gases was given by Daniel Bernoulli in his famous treatise on hydrodynamics, published in 1738.

A gas consists of molecules (or atoms), which can be regarded as relatively small masses traveling in random directions. Occasionally the molecules collide with each other or with the walls of the container. In a collision with the wall the normal component of molecular momentum is reversed, and the wall must provide the impulse necessary to cause this momentum change. Since the wall turns all the molecules back into the body of the gas, the molecules exert a net outward force on the wall. The forces of the collision between a molecule and the wall are a Newtonian action-reaction pair. The wall is therefore responsible for an equal net inward force on the molecules, which, if divided by the area of the wall, gives the pressure of the gas.

The central problem of the kinetic theory of gases is calculating the pressure of a gas from the properties of its individual molecules. The simplest case is calculating the pressure of an ideal gas. From the point of view of kinetic theory, an ideal gas is one in which the energy of attraction between individual molecules is negligible compared with the molecular kinetic energy. Since attractive forces between molecules are responsible for liquefaction, this condition is equivalent to requiring that the gas be far from its liquefaction point.

We can now use our knowledge of Newtonian dynamics (in particular the impulse momentum theorem) to derive an equation for the pressure of an ideal gas from kinetic theory. Consider a single molecule of mass m and speed v in a cubic box whose side length is L. We assume that the molecule undergoes elastic collisions with the walls of the box. This means that the velocity components parallel to the walls of the box do not change, although the component perpendicular to the wall simply reverses in direc-

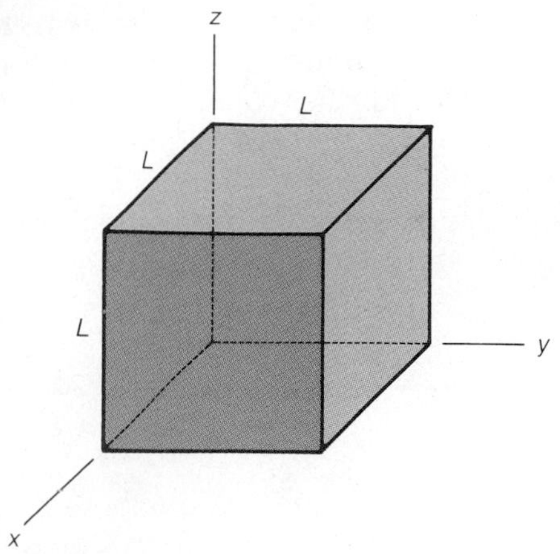

FIGURE 13.5 Diagram to illustrate calculation of the pressure of an ideal gas contained in a cube of side L.

tion. The molecular speed, therefore, remains constant. Let v_x be the magnitude of the x component of the molecular velocity; we see from Fig. 13.5 that the molecule has to travel a distance $2L$ in the x direction between successive collisions with the shaded face of the box. Hence, the number of collisions per second with the shaded face is $v_x/2L$. Since the x component of momentum is reversed during a collision, the momentum change in each collision is $2mv_x$. From the impulse-momentum theorem, the force on the face is the momentum transfer per second:

$$\begin{aligned}\text{Force on shaded face} &= (2mv_x) \times (v_x/2L)\\ &= mv_x^2/L\end{aligned}$$

Now let us suppose that the box contains N molecules which, according to the ideal gas assumption, collide with each other very infrequently. The force on the shaded face must include a contribution from each molecule, so that we have:

$$\text{Force on shaded face} = N\frac{m}{L}\langle v_x^2\rangle$$

where $\langle v_x^2\rangle$ is the mean value of the square of the x component of molecular velocity. But the force on the shaded face is the gas pressure multiplied by the area of the face; it follows that:

$$pL^2 = N\frac{m}{L}\langle v_x^2\rangle$$

or

$$pV = Nm\langle v_x^2\rangle$$

since $V = L^3$ is the volume of the box. Similar results apply to faces perpendicular to the y and z axes, giving:

$$pV = Nm\langle v_y^2\rangle$$

and

$$pV = Nm\langle v_z^2\rangle$$

Adding these equations gives:

$$\begin{aligned}3pV &= Nm\{\langle v_x^2\rangle + \langle v_y^2\rangle + \langle v_z^2\rangle\}\\ &= Nm\langle v^2\rangle\end{aligned}$$

Hence

$$pV = \tfrac{1}{3}Nm\langle v^2\rangle \qquad \textbf{(13.7)}$$

We now have two results for the product of the pressure and volume of an ideal gas. The first is the equation state, Eq. (13.4), and the second is the kinetic theory result, Eq. (13.7). Let us write them together for easy comparison:

$$pV = NkT$$

$$pV = \tfrac{1}{3}Nm\langle v^2\rangle$$

$$= \tfrac{2}{3}N\langle \tfrac{1}{2}mv^2\rangle$$

It is correct to put $\frac{1}{2}m$ inside the averaging sign since it is a constant that does not affect the averaging process. We see at once that the mean molecular kinetic energy is simply related to the gas temperature and the Boltzmann constant.

■ Relation Between Temperature and Molecular Properties for an Ideal Gas

$$\langle \tfrac{1}{2}mv^2\rangle = \tfrac{3}{2}kT \qquad \textbf{(13.8)}$$

Originally we introduced temperature as the quality of hotness. Equation (13.8) gives additional insight from the molecular viewpoint. The **mean translational kinetic energy** of an ideal gas molecule is proportional to the **absolute temperature.** It is important that the velocity in the kinetic energy calculation be *random velocity* and not velocity of *uniform flow*. The velocity of gas molecules streaming along in uniform flow does not contribute to their temperature.

The relationship between temperature and mean random energy can be extended to substances other than ideal gases. Temperature is a measure of mean random energy for all substances, whether they be gaseous, liquid, or solid. However, the actual expression takes the simple form of Eq. (13.8) only for an ideal gas.

Equation (13.8) can be used to calculate a molecular velocity that is not the mean velocity but the root-mean-square (rms) velocity. An example of such a calculation will make the meaning clear.

EXAMPLE 13.7

Calculate the rms velocity of oxygen molecules at STP.

SOLUTION The velocity calculation from Eq. (13.8) does not require the pressure, but STP conditions ensure that the oxygen is close to being an ideal gas. The molecular mass of oxygen is 32 g, and 1 mol contains Avogadro's number of molecules. The mass of a molecule of oxygen is therefore given by:

$$m = \frac{32 \times 10^{-3}\ \text{kg}}{6.022 \times 10^{23}}$$
$$= 5.314 \times 10^{-26}\ \text{kg}$$

The use of Eq. (13.8) gives:

$$\langle \tfrac{1}{2}mv^2 \rangle = \tfrac{3}{2}kT$$
$$\langle \tfrac{1}{2} \times 5.314 \times 10^{-26}\ \text{kg} \times v^2 \rangle = \tfrac{3}{2} \times 1.381 \times 10^{-23}\ \text{J/K} \times 273.15\ \text{K}$$
$$\therefore \langle v^2 \rangle = 2.130 \times 10^5\ (\text{m/s})^2$$

Taking the square root yields:

$$(\langle v^2 \rangle)^{1/2} = 461\ \text{m/s}$$ ■

We now see the reason for referring to this velocity as root-mean-square. To obtain it from individual molecular velocities, we first square each velocity, then take the mean of this set of results, and finally extract the square root. This procedure does not give the same result as averaging the magnitudes of the individual molecular velocities; detailed analysis for an ideal gas shows that the rms velocity is about 10% higher than the mean speed.

The kinetic theory of gases also provides us with a good understanding of Dalton's law of partial pressures. John Dalton (1766–1844) was a contemporary of Charles and Gay-Lussac, who did extensive work on the gas laws and is responsible for a law concerning mixtures of different gases.

Consider a mixture of two or more gases that obeys the ideal gas equation. The pressure is due to the collisions that the various atoms and molecules have with the containing walls. For an ideal gas the molecules have negligible interaction with each other, and each component gas behaves as if the other were not present. In particular, each constituent gas exerts the same pressure that it would exert if it alone occupied the container. Dalton's law is a formal statement of this effect.

■ Dalton's Law of Partial Pressure

The total pressure in an ideal gas mixture is the sum of the pressures that each component gas would exert if it alone occupied the whole volume.

The use of Dalton's law permits calculation of the pressure of mixtures of ideal gases.

EXAMPLE 13.8

A gas mixture contains 40 g of nitrogen (molecular mass 28 g) and 10 g of hydrogen (molecular mass 2 g). The volume of the mixture is 36.8 ℓ at 273 K. Calculate the pressure of the gas mixture.

SOLUTION We use the ideal gas equation in conjunction with Dalton's law to find the total pressure. Since the molecular mass of nitrogen is 28 g, we have:

$$\text{\# of moles of nitrogen} = \frac{40\text{ g}}{28\text{ g}} = 1.429$$

We find the partial pressure of the nitrogen from Eq. (13.4a):

$$\begin{aligned} pV &= nRT \\ p \times 36.8 \times 10^{-3}\text{ m}^3 &= 1.429\text{ mol} \times 8.314\text{ J/K}\cdot\text{mol} \times 273\text{ K} \\ \therefore p &= 8.811 \times 10^4\text{ Pa} \end{aligned}$$

The molecular mass of hydrogen is 2 g, so we have:

$$\text{\# of moles of hydrogen} = \frac{10\text{ g}}{2\text{ g}} = 5$$

We again use Eq. (13.4a) to calculate the partial pressure of the hydrogen:

$$\begin{aligned} pV &= nRT \\ p \times 36.8 \times 10^{-3}\text{ m}^3 &= 5\text{ mol} \times 8.314\text{ J/K}\cdot\text{mol} \times 273\text{ K} \\ \therefore p &= 3.084 \times 10^5\text{ Pa} \end{aligned}$$

From Dalton's law, the total pressure is equal to the sum of the partial pressures:

$$\begin{aligned} p &= 8.811 \times 10^4\text{ Pa} + 3.084 \times 10^5\text{ Pa} \\ &= 3.965 \times 10^5\text{ Pa} \end{aligned}$$

Note that the hydrogen contributes only 20% to the total mass of the gas mixture but about 78% to the total pressure. ■

KEY CONCEPTS

Temperature is the measure of hotness; an instrument for measuring temperature is called a **thermometer.**

Practical thermometers are calibrated with the help of the **fixed points** on the International Practical Temperature Scale. See Table 13.1.

The **Celsius** and **Kelvin** (or absolute) temperature scales are in common scientific usage. The fixed point is the triple point of water; the fixed points for practical use are the ice point and the steam point. See Eq. (13.2).

An **ideal gas** is defined by the equation of state, Eq. (13.4). Real gases approximate ideal gases if they have low density and are far removed from the liquefaction temperature.

Most substances expand as the temperature increases. The **coefficient of thermal expansion** is the measure of this effect. See Eq. (13.6).

The **mean translational kinetic energy** of an ideal gas molecule is directly proportional to the **absolute temperature.** See Eq. (13.8).

QUESTIONS FOR THOUGHT

1. What type of thermometer would you select to meet each of the following requirements separately:

a. Extreme accuracy
b. Portability
c. Ability to measure very high and very low temperatures
d. Inexpensiveness

2. You hold a mercury-in-glass thermometer with its bulb just above the flame of a Bunsen burner. The mercury at first falls slightly and then rises quickly. Explain this effect.

3. Explain the operation of a thermometer in which the temperature-sensing device consists of two different strips of metal riveted to each other.

4. Make a list of the difficulties that would arise from using a water-in-glass thermometer to measure air temperature outside your house.

5. Two copper cubes have the same external dimensions, but one is solid and the other is hollow. If they are both heated to the same temperature, will the changes in the outside dimensions of the two blocks differ? Does the volume of the cavity in the hollow block become larger or smaller when the block is heated?

6. The air in a room consists mainly of oxygen and nitrogen molecules. Which of the two has the higher average thermal velocity and by what approximate percentage?

7. "Substances that are gaseous at normal temperature and pressure usually have low relative molecular (or atomic) mass." Justify this rule and see if you can think of a notable exception.

PROBLEMS

A. Single-Substitution Problems

Use the coefficients of thermal expansion given in Table 13.2 and the elastic moduli from Table 11.1. Use the following values for atomic and molecular masses:

Oxygen	32 g
Nitrogen	28 g
Carbon dioxide	44 g
Air (average)	29 g
Helium	4 g
Hydrogen	2 g
Argon	40 g
Ozone	48 g

1. Calculate the pressure due to 3 mol of oxygen occupying a volume of 10 ℓ at a temperature of 27°C. [13.2]

2. A 12-ℓ container holds 1.5 mol of hydrogen at a pressure of 230 kPa. Calculate the temperature. [13.2]

3. Calculate the volume occupied by 2 mol of helium at a temperature of 27°C and a pressure of 180 kPa. [13.2]

4. A certain quantity of argon occupies 4.2 ℓ at 30°C and 280 kPa pressure. What volume does it occupy at 80°C and 460 kPa pressure? [13.2]

5. A certain quantity of air has a volume of 8.5 ℓ at 20°C and 1 atm pressure. What is its volume at 140°C and 2.5 atm pressure? [13.2]

6. A steel bridge member is exactly 12 m long at 28°C. By how much does it contract if the temperature falls to −10°C? [13.3]

7. Calculate the length of a brass rod that expands 1 mm after being transferred from a mixture of ice and water to a container of boiling water. [13.3]

8. The Eiffel Tower in Paris is a steel structure 300 m high. Calculate the temperature change necessary to cause a change of 10 cm in its height. [13.3]

9. Ten gallons of benzene is measured out at a temperature of 5°C. At what temperature is the volume larger by 1 quart? [13.3]

10. A container of mercury holds 50 mℓ at 40°C. By what amount does the volume of the mercury change when the temperature falls to 5°C? [13.3]

11. Calculate the molecular mass for gas whose molecules have an rms velocity of 500 m/s at 8°C. [13.4]

12. Calculate the rms velocity for argon atoms at 23°C. (Take the mass of the argon atom to be 6.68×10^{-26} kg.) [13.4]

13. Calculate the rms velocity for carbon dioxide molecules in air at 67°C. (Take the mass of a carbon dioxide molecule to be 7.35×10^{-26} kg.) [13.4]

14. At what temperature do hydrogen molecules have an rms velocity of 1000 m/s? (Take the mass of the hydrogen molecule to be 3.34×10^{-27} kg.) [13.4]

B. Standard-Level Problems

15. The density of air at 100°C is 2 kg/m^3. What is the pressure? [13.2]

16. Calculate the density of argon at a pressure of 8 atm and a temperature of 127°C. [13.2]

17. Calculate the mass of ozone that occupies 16 ℓ at a pressure of 300 torr and a temperature of 7°C. [13.2]

18. The gauge pressure of air in an automobile tire is 200 kPa (29 lb/in.^{2}) when the temperature is 20°C. After fast highway driving, the temperature rises to 80°C. Calculate the new gauge pressure of the air in the tire. (Neglect any change in the volume of the tire.) [13.2]

19. Helium is stored in 25-ℓ steel cylinders at a gauge pressure of 25 atm. How many cylinders are required to fill a balloon of volume 12.5 m^3 at atmospheric pressure? [13.2]

20. A 3-ℓ flask contains hydrogen at a gauge pressure of 3 atm and a temperature of 27°C. The flask is heated until the gauge pressure is 4 atm. Calculate:

a. The final temperature
b. The number of grams of hydrogen contained in the flask

21. A good vacuum has a pressure of about 10^{-10} torr. How many air molecules are present in 1 ℓ of this vacuum at 27°C?

22. How many molecules are present in 1 ℓ of nitrogen at 27°C and 760 torr?

23. At what pressure is there 1 μg of oxygen in 1 ℓ at 0°C?

24. A mercury-in-glass barometer is made from a tube with a circular bore 2.5 mm in radius. Because of faulty manufacture, the upper end of the tube (which should be evacuated) contains some nitrogen. On a day on which the atmospheric pressure and temperature are 760 torr and 27°C, the barometer reads 731 torr, with a space of 32.5 cm between the mercury and the top of the tube. How many grams of nitrogen are contained in the instrument?

25. A 25-ℓ flask contains 1 kg of oxygen at a temperature of 30°C. Overnight leakage reduces the pressure by 400 torr and the temperature by 3 C°. How many grams of oxygen leaked out?

26. A football of volume 1.12 ℓ is pumped up with air to a gauge pressure of 172 kPa (25 lb/in.^{2}) when the temperature is 18°C.

a. What mass of air does the ball contain?
b. If the temperature of the ball rises to 36°C, what mass of air must be released to bring the pressure back to 172 kPa? (Neglect changes in the volume of the ball.)

27. An air bubble of volume 20 cm^3 rises from the bottom of a freshwater lake 45 m deep where the temperature is 4°C. What is the volume of the bubble just before it reaches the lake surface where the temperature is 18°C. (Ignore effects due to water vapor in the bubble.)

28. A hot-air balloon has a volume of 700 m^3 and a total mass of 175 kg, including the passenger. Assume that the heater causes a uniform temperature inside the balloon and that inside and outside pressures are equal.

a. What inside temperature is required to lift the balloon on a day when the atmospheric temperature and pressure are 27°C and 760 torr?
b. At an altitude of 1500 m the atmospheric temperature and pressure are 5°C and 630 torr. What inside temperature is required to maintain this altitude?

29. Aluminum rivets for aircraft have a slightly larger diameter than the holes in which they fit. The rivets are cooled in dry ice at −78°C and then slipped through the holes to ensure a tight fit. A rivet hole is 0.375 in. in diameter at 20°C. How much larger than the diameter of the hole should the rivet diameter be at 20°C in order to ensure a snug fit when cooled? [13.3]

30. A wooden wagon wheel has a diameter of 1.25 m. A steel tire is made to fit snugly when the steel is heated to 550°C. By what amount is the inner diameter of the tire smaller than the diameter of the wheel at 20°C? [13.3]

31. A steel surveyor's tape, which is correctly marked at a temperature of 25°C, is used to measure the width of a building lot on a day when the temperature is −5°C. The reading on the tape is 16.41 m. What is the measuring error? [13.3]

32. A bar of aluminum 20 cm long is joined end to end with a bar of iron that is 40 cm long. What is the thermal expansion coefficient of the composite rod?

33. What size gap should be left between steel rail sections 80 ft long if they are laid at 15°C, and if the gap should just close when the rail temperature rises to 55°C?

34. The density of mercury is 13.558×10^3 kg/m^3 at 15°C. Calculate its density at 35°C.

35. A copper sphere has a radius of 8 cm at −10°C. Calculate the change in volume of the sphere when it is heated to 35°C.

36. A thermometer is constructed of mercury in fused quartz. The volume of the bulb is 0.25 cm^3, and the capillary is 0.15 mm in radius. Calculate the distance between the 0°C and 100°C marks on the thermometer. (Neglect the expansion of the quartz, and assume that the volume of the mercury at 0°C is equal to the volume of the bulb.)

37. A glass flask made of soda glass has a volume of exactly 2 ℓ at 25°C and is filled with mercury at this temperature and then cooled to 10°C. How many milliliters of mercury are required to top the flask up again?

38. A motorist fills his 16-gal steel tank with gasoline in the early morning when the temperature is 5°C. The automobile is left standing and the temperature increases to 28°C. How much gasoline spills out the overflow?

39. A copper cylinder, which has a diameter of exactly 12 cm at a temperature of 15°C, is 0.003 cm too large to go through a circular hole in an aluminum plate at the same temperature.

a. To what temperature must the aluminum plate be heated so that the copper cylinder will fit through the hole?
b. If both the aluminum plate and the copper cylinder are heated, what temperature is necessary so that the fit can occur?

40. A structural steel beam 8 m long is placed in position when the temperature is 15°C. The cross-sectional area of the beam is 100 cm^2, and the ends are rigidly fixed. Calculate the outward force exerted when the temperature rises to 40°C.

41. A steel wire 1 mm in diameter and 50 cm long is stressed by a tensile force of 60 N, and clamped in the stressed condition. What temperature change in the wire is necessary to reduce the stress to zero, and should the wire be heated or cooled?

42. An aluminum sphere of radius 4 cm is heated from an initial temperature of 20°C. After the heating, a pressure of 500 atm returns the radius of the sphere to its original value. What was the final temperature of the sphere?

43. A copper cube of side 15 cm is heated from 40°C to 85°C. Calculate the pressure necessary to prevent the cube from expanding.

44. Calculate the rms velocity of nitrogen molecules in air at 0°C. [13.4]

45. At what temperature do helium atoms have an rms velocity of 30 000 km/h? [13.4]

46. The rms speed of the molecules in a tank of nitrogen is 450 m/s at a pressure of 850 torr. Calculate the density of the nitrogen.

47. The density of oxygen in a tank is 2 kg/m^3 and the pressure is 1250 torr. Calculate the rms velocity of the molecules.

48. The rms velocity of argon atoms in a tank is 400 m/s, and the gas density is 1.5 kg/m^3. Calculate the pressure in atmospheres.

49. The temperature of the interior of the sun is believed to be about 10^8 K. Calculate the rms velocity of a hydrogen nucleus (mass 1.67×10^{-27} kg) at this temperature.

50. The temperature of interstellar space is 2.7 K. The principal occupants of the space are electromagnetic radiation and hydrogen atoms. It is believed that the density of hydrogen atoms near the edge of a galaxy is about 1 per 10^5 cm^3. Calculate:

a. The pressure of the hydrogen gas
b. The rms velocity of the hydrogen atoms

51. A 20-ℓ tank contains 9.75 g of carbon dioxide and 10 g of nitrogen at 350 K. Calculate the pressure in the tank.

52. A tank contains 10 kg of nitrogen at a pressure of 2 atm and a temperature of 300 K. Another tank at the same temperature contains 3.5 m^3 of hydrogen at a pressure of 1 atm. A valve connecting the two tanks is opened, and their contents mix without temperature change. Calculate the final pressure.

C. Advanced-Level Problems

53. An aluminum cube of side 10 cm at 10°C is clamped so that the distance between two opposite faces is held to exactly 10 cm. The cube is heated to 130°C.

a. What force must be applied by the clamp to hold the 10-cm dimension?
b. Calculate the change in length of the other edges of the cube.

54. The pendulum of a clock has a period of exactly 2 s at a temperature of 22°C. The pendulum is a long, thin brass rod with a heavy bob on the end and can be regarded as a simple pendulum. How many seconds per day does the clock gain or lose if the temperature rises to 35°C?

55. a. A large balloon of volume 5000 m^3 is filled with helium at atmospheric pressure and 20°C. The mass of the balloon, rigging, ballast, and cabin is 3000 kg. What useful payload can the balloon lift?
b. The balloon is fitted with a device to equalize the helium pressure and the outside air pressure. When it rises to an altitude of 3000 m, the air pressure and temperature are 525 torr and -5°C. What mass of ballast must be thrown overboard to maintain the same useful payload?

14

HEAT AS A FORM OF ENERGY

We have already anticipated the first law of thermodynamics to some extent in the discussion of the energy conservation principle. We accounted for any imbalance of mechanical energy in the conservation equation by introducing an extra term for energy dissipation. The first law of thermodynamics expresses the relationship more formally by explicitly introducing a heat flow term into the energy conservation law. This clarification of energy conservation, which was one of the great triumphs of nineteenth-century physics, arose almost by consensus from the work and thoughts of many leading scientists of the time. The first law sets the stage for a detailed investigation of the temperature changes that occur in bodies due to energy transfer between them.

14.1 The First Law of Thermodynamics

The concept of temperature amounts to putting the intuitive idea of hotness on a scientific basis. We begin our discussion of heat energy using an intuitive approach. Consider a red-hot iron nail and a large container of hot water. Everyone agrees that the nail is hotter than the water, which our study of temperature confirms, but there is also agreement that the water contains more heat energy than the nail. A large amount of water must be heated for a long time to become hot, while a small nail becomes red-hot very quickly. But suppose that we heat the water and the nail by rubbing them back and forth for a long time on a rough surface, without ever taking them near any fire. Do we then say that the water contains more work than the nail? Truly, there is something to be said for both the heat and the work statements, but the hot water heated by the fire is in the same condition as the water heated by rubbing on a rough surface. As a first step to cleaning up the terminology, we will not talk about either heat content or work content, but use the term **internal energy** content. The internal energy is the sum total of the energy of the molecular constituents of the body in question, and this energy includes translational, vibrational, and rotational kinetic energy of the molecules as well as any intermolecular potential energy. When a substance undergoes an increase in temperature, there is a corresponding increase in internal energy. The final state of higher internal energy is the same no matter how the temperature increase comes about. We reserve the term **heat** to mean the energy that flows from one body to another because of a temperature difference between them. We use the term **work** in the sense already explained in dynamics.

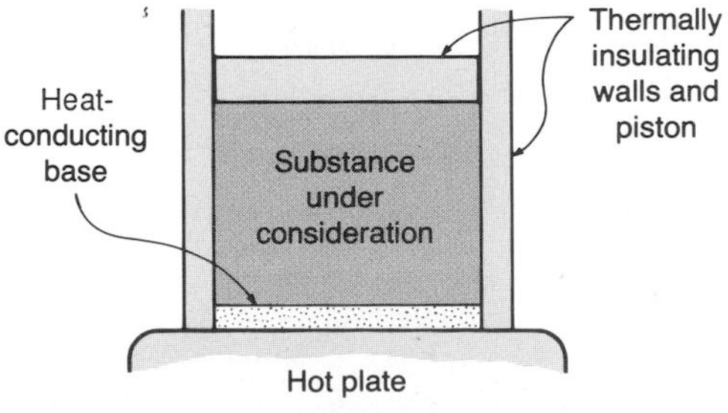

FIGURE 14.1 Diagram to illustrate the first law of thermodynamics.

The examples so far considered are from everyday experience. In order to set up the first law of thermodynamics, it is helpful to consider the more formal situation shown in Fig. 14.1. Consider a substance (which could be either gas, liquid, or solid) contained in a cylinder fitted with a frictionless piston. The piston and the cylinder walls are thermally insulating and permit no flow of heat. The base of the cylinder sits on a hot plate, which permits heat flow to the substance in the cylinder. This flow of heat tends to increase the store of internal energy. At the same time, the substance in the cylinder may expand, pushing up the piston and thereby doing work at the expense of its stored internal energy. The **energy conservation** principle that exhibits the energy balance for these effects is known as the **first law of thermodynamics.**

■ First Law of Thermodynamics

$$\text{Quantity of heat input to a substance} = \text{increase of internal energy} + \text{work done by the substance}$$

$$Q = (U - U_0) + W \qquad \textbf{(14.1)}$$

All terms of the equation are in joules, because the joule is the SI unit of heat. The signs require careful thought. With the formulation given, we count the following quantities as positive: heat *input*, internal energy *increase*, and work done *by* the substance. Correspondingly, we take the following quantities as negative: heat *output*, internal energy *decrease*, and work done *on* the substance. Unfortunately, there has been no general agreement on the assignment of sign to the work terms. As a result, some books have a different sign of W in Eq. (14.1) along with a different convention for the meaning of the term.

EXAMPLE 14.1

A quantity of gas in a cylinder of the type shown in Fig. 14.1 receives 1600 J of heat from the hot plate. At the same time 800 J of work are done on the gas by outside forces pressing down on the piston. Calculate the change in internal energy of the gas.

SOLUTION The problem is a simple application of the first law of thermodynamics.

$$\text{Heat input} = 1600\ \text{J}$$

$$\text{Work done by the gas} = -800\ \text{J}$$

(The sign is negative since the 800 J of work is done on the gas.) The first law, Eq. (14.1), states:

$$Q = (U - U_0) + W$$
$$1600 \text{ J} = (U - U_0) - 800 \text{ J}$$
$$\therefore U - U_0 = 2400 \text{ J}$$

Since the sign is positive, the internal energy increases. ■

The first law of thermodynamics and the example of its application appear almost trivial *after we understand them.* But the simple formulation of Eq. (14.1) represents a giant step forward in understanding heat. Throughout most of the eighteenth century a fluid theory of heat was in vogue. Early experiments on the transfer of heat from hot to cold bodies had shown that the heat lost by the hot body was equal to the heat gained by the cold one, and heat was identified with a rather elusive fluid named caloric; conservation of caloric appeared to be a basic law of heat exchange processes. The caloric theory was quite successful in some applications, but it received a death blow from experiments that developed from the famous observations on the boring of cannons by Count Rumford (1753–1814). Rumford was born Benjamin Thompson in Woburn, Massachusetts, but sailed to Europe because of his Royalist sympathies in the War of Independence. A scientifically and technically skilled man, he became director of the Bavarian Arsenal and observed the phenomenon of heat generation in the boring of cannons. As the metal from the cannon was cut into small chips by the boring tool, the cooling water became very hot. According to the prevailing theory, the small chips lost their ability to retain caloric and transferred it to the water. However, a blunt boring tool that was unable to cut chips from the bore of the cannon still made the cooling water very hot, and Thompson correctly surmised that the generation of heat came from the motion of the boring tool. Caloric was certainly not conserved—you could make all the caloric you needed by boring away with a blunt tool! Thompson received the title of count from the Elector of Bavaria for the scientific and technical services rendered.

14.2 The Specific Heat of Solids and Liquids

When an iron pot is placed on a fire, its temperature increases due to the flow of heat from the fire. If a mass of water equal to the mass of the iron is placed in the pot, the temperature rises much more slowly. In fact, it takes much more than twice as long to raise the temperature of the pot containing water than of the pot alone. For a given temperature increase, the iron pot requires less heat input than the same mass of water. Looked at ... m another viewpoint, we can say that a given amount of heat per unit mass causes a larger temperature increase in iron than in water. The **specific heat** is the property of a substance that relates its temperature change to the heat flow. We define specific heat as follows:

■ Definition of Specific Heat

$$\text{Heat flow} = \begin{matrix}\text{mass of}\\ \text{substance}\end{matrix} \times \begin{matrix}\text{specific}\\ \text{heat}\end{matrix} \times \begin{matrix}\text{temperature change}\\ \text{caused by the}\\ \text{heat flow}\end{matrix}$$

$$Q = mC(T - T_0) \qquad \textbf{(14.2)}$$

SI unit of specific heat: joule per kilogram per Celsius degree
Abbreviation: J/kg · C°

For a positive heat flow (heat input), the final temperature T is greater than the initial temperature T_0; for a negative flow (heat output), the final temperature is less than the initial temperature. We express the temperature difference $(T - T_0)$ in Celsius degrees. Numerically, the difference is the same whether the temperatures are expressed in degrees Celsius or in kelvin.

The first law of thermodynamics tells us that heat input to a substance goes partly into internal energy increase and partly into work done by expansion of the substance. If the substance in the cylinder of Fig. 14.1 is either a solid or a liquid, the volume change is small and the work done is therefore negligible; almost all of the heat input goes to increase the internal energy of the substance. Since the internal energy is directly related to the temperature, almost all the heat input to a solid or a liquid goes into causing a temperature increase. Table 14.1 lists representative values of specific heats of various solids and liquids. The figures are approximate room temperature values. The specific heat depends on temperature, decreasing strikingly and falling to zero as the temperature falls toward absolute zero. As the temperature increases above room temperature, the specific heats of most of the substances listed increase only slightly. The figures quoted are therefore reasonably accurate for higher temperatures, but grossly inaccurate for substantially lower temperatures.

TABLE 14.1 Specific heats of solids and liquids at room temperature

	Specific heat	
Substance	*J/kg·C°*	*kcal/kg·C° (or cal/g·C°)*
Aluminum	900	0.215
Brass	384	0.092
Copper	387	0.093
Iron or steel	450	0.108
Lead	128	0.031
Silver	235	0.056
Soda glass	840	0.20
Ice (0°C to −10°C)	2100	0.50
Water	4184	1.000
Methanol	2550	0.61
Ethanol	2490	0.59
Glycerin	2430	0.58
Mercury	139	0.033
Benzene	1750	0.42

EXAMPLE 14.2

An aluminum pot of mass 0.7 kg contains 1.8 kg of water at 20°C. The pot is placed on a hot plate that raises the temperature to 100°C. Calculate the heat input required for this temperature increase.

SOLUTION The heat input required to raise the temperature of liquids and solids is given by Eq. (14.2) used in conjunction with the specific heat values of Table 14.1. Since both the aluminum and the water undergo an increase in temperature, we write a separate term for each substance:

$$
\begin{aligned}
Q &= mC(T - T_0)_{\text{water}} + mC(T - T_0)_{\text{aluminum}} \\
&= \begin{array}{l} 1.8\ \text{kg} \times 4184\ \text{J/kg} \cdot \text{C}^\circ \times (100^\circ\text{C} - 20^\circ\text{C}) + 0.7\ \text{kg} \\ \qquad \times\ 900\ \text{J/kg} \cdot \text{C}^\circ \times (100^\circ\text{C} - 20^\circ\text{C}) \end{array} \\
&= 6.025 \times 10^5\ \text{J} + 5.040 \times 10^4\ \text{J} \\
&= 6.53 \times 10^5\ \text{J}
\end{aligned}
$$

■

In a practical situation, the total energy input to the hot plate may far exceed the calculated heat transfer to the aluminum and the water, because the hot plate itself and the surroundings require additional heat input to raise their temperatures.

The study of the physics of heat was well under way before the first law of thermodynamics was fully appreciated. Units for quantity of heat transfer had already been introduced based on the heat input required to raise the temperature of water. The small calorie (abbreviated "cal") was defined as the quantity of heat input required to raise the temperature of 1 g of water through 1 C°. The kilocalorie (abbreviated "kcal"), which is the quantity of heat required to raise the temperature of 1 kg of water through 1 C°, is commonly used by nutritionists in assigning energy values to various foods.* The definition of the calorie in terms of the heating of water has been dropped, and it is now defined simply by its SI conversion, 1 cal = 4.184 J. In British engineering practice, the British thermal unit (abbreviated "Btu") was defined as the quantity of heat required to raise the temperature of 1 lb of water through 1 F°. It, also, is now defined by its SI conversion, 1 Btu = 1054 J. In Table 14.1, we show specific heat values in caloric units as well as SI units.

EXAMPLE 14.3

A 70-kg person eats a 1200-kcal meal. Calculate the increase of body temperature if all of this energy were used to that end. Take the average specific heat of the human body to be 0.85 kcal/kg · C°.

SOLUTION If the energy of the food goes entirely to raising the body temperature, Eq. (14.2) is appropriate. If we express the en-

*Nutritionists frequently refer to kilocalories as "Calories." However, we will use the SI prefix rather than the initial capital letter.

ergy units on *both sides* of the equation in kilocalories, there is no need to convert to SI units:

$$Q = mC(T - T_0)$$

$$1200 \text{ kcal} = 70 \text{ kg} \times 0.85 \text{ kcal/kg} \cdot \text{C}° \times (T - T_0)$$

$$\therefore T - T_0 = 20.2 \text{ C}°$$

Since such a relatively immense temperature rise does not occur, most of the energy of the food must be diverted to other uses. ■

EXAMPLE 14.4

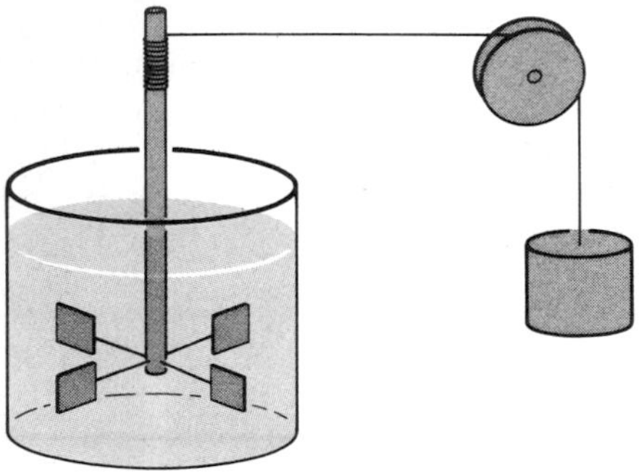

A copper container of mass 100 g contains 500 g of water. The water is stirred by steel paddles of mass 85 g that are operated by a falling mass of 15 kg acting on a cord-and-pulley system. Calculate the temperature rise of the water, the copper container, and the paddles if the mass falls 2.40 m. (Ignore losses to the surroundings, and assume that the falling mass moves so slowly that its kinetic energy gain is negligible.)

SOLUTION The amount of energy transferred to the water, the container, and the paddles is equal to the gravitational potential energy lost by the falling mass. We can calculate this energy transfer with the help of Eq. (7.3):

$$\begin{aligned} \text{GPE} &= mgh \\ &= 15 \text{ kg} \times 9.81 \text{ m/s}^2 \times 2.4 \text{ m} \\ &= 353.2 \text{ J} \end{aligned}$$

This amount of energy transfer to the water produces the same temperature rise as the same quantity of heat input. We can calculate the corresponding temperature rise from Eq. (14.2) (one term each for the water, the container, and the paddles):

$$\begin{aligned} \text{Energy input} &= mC(T - T_0)_{\text{water}} + mC(T - T_0)_{\text{container}} \\ &\quad + mC(T - T_0)_{\text{paddles}} \\ 353.2 \text{ J} &= 0.5 \text{ kg} \times 4184 \text{ J/kg} \cdot \text{C}° \times (T - T_0) \\ &\quad + 0.1 \text{ kg} \times 387 \text{ J/kg} \cdot \text{C}° \times (T - T_0) \\ &\quad + 0.085 \text{ kg} \times 450 \text{ J/kg} \cdot \text{C}° \times (T - T_0) \\ \therefore T - T_0 &= 0.163 \text{ C}° \end{aligned}$$

It is convenient to use the SI specific heat units in this application, since the energy input is in joules. ■

With the help of the first law of thermodynamics, we can calculate the temperature rise due to the operation of the apparatus described in Example 14.4. Historically, the situation was reversed. James Joule used equipment of this type to amass experimental evidence for the validity of the first law by measuring the temperature change caused by the falling weight. Joule was one of the leading figures in establishing the first law of thermodynamics. While he and his wife spent their honeymoon in the French Alps, he took measurements of the water temperature at the top and bottom of many of the waterfalls that abound in the region. Joule reasoned that the gravitational potential energy loss of the falling water should result in an increase in temperature at the foot of the falls. Unfortunately, the experiment is complicated by other factors (such as evaporation of the water as it falls), and he obtained no significant results from the work.

14.3 Change of Phase and Latent Heats

Water is one of the most common of everyday substances, occurring as a solid (ice), a liquid (water), and a gas (steam). Experience teaches that temperature is the main determinant of which form actually occurs, but we shall see that pressure is an equally important factor. Most pure substances can occur as gases, liquids, or solids, and these forms are called phases of the substance. Things such as wood, which are complex, structured mixtures of many pure substances, do not have clearly identifiable phases.

The basic information about phases of a substance is usually given on a phase diagram. Figure 14.2 shows a phase diagram for

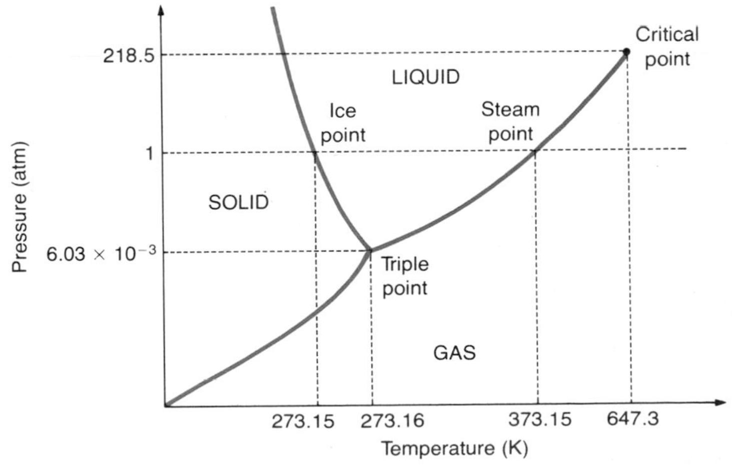

FIGURE 14.2 Phase diagram for water (not to scale).

water. The axes are pressure and temperature, and three intersecting lines divide the diagram into three regions, corresponding to the solid, liquid, and gas phases. Every point on the plane is a possible thermal equilibrium state of the phase for the region in question, and every point on the lines separating the phases is a possible thermal equilibrium state for a mixture of the two respective phases. The point of intersection of the three lines is called the **triple point**—which occurs only at the values of pressure and temperature for which all three phases can coexist in thermal equilibrium. The triple point of water is the fixed point of the Kelvin scale of temperature, occurring at 273.16 K and a pressure of 6.03×10^{-3} atm. (Note that Fig. 14.2 is not drawn to scale, so that the qualitative features of the diagram are clearer.) Every point on the line separating the solid and liquid phases is a melting point, and the diagram shows that the melting temperature of ice decreases as the pressure increases. The normal melting point (ice point at a pressure of 1 atm) occurs at 273.15 K. In a similar way, boiling points occur on the line separating the liquid and gas phases. The normal boiling point of water (steam point at a pressure of 1 atm) occurs at 373.15 K. The points on the line separating the solid and gas phases are called sublimation points. *Sublimation* is the direct change from solid to gas phase, without an intervening liquid phase. The diagram shows that sublimation points for water occur only if the temperature is lower than 273.16 K *and* the pressure is lower than 6.03×10^{-3} atm.

Melting, boiling, and sublimation are **first-order phase transitions.** In a first-order phase transition, the internal energy changes by a large amount without any corresponding temperature increase. Consider an experiment in which we take ice at a very low temperature and heat it at a pressure of 1 atm. After measuring the specific heat over successive small temperature intervals, we plot the results, as shown in Fig. 14.3. The specific heat of ice is very low near absolute zero and rises to about 2100 J/kg · C° (0.5 cal/g · C°) near the normal melting point. At the melting point the specific heat becomes infinite. An input of 3.35×10^5 J/kg (80 cal/g) is necessary just to melt the ice without causing any temperature increase. This energy is called the **latent heat of fusion.** Continuing the experiment shows that the specific heat of water is approximately constant at 4184 J/kg · C° (1 cal/g · C°) up to the steam point. At this point the specific heat again becomes infinite, and an energy input of 2.26×10^6 J/kg (539 cal/g) is needed to vaporize the water without any temperature increase. This energy is called the **latent heat of vaporization.** Most pure substances behave similarly when heated from a very low temperature at atmospheric pressure: The solid, liquid, and gas phases occur in that order with a latent heat at each phase transition. We define the latent heat at a first-order phase transition as the heat per unit mass that is required to effect the phase change.

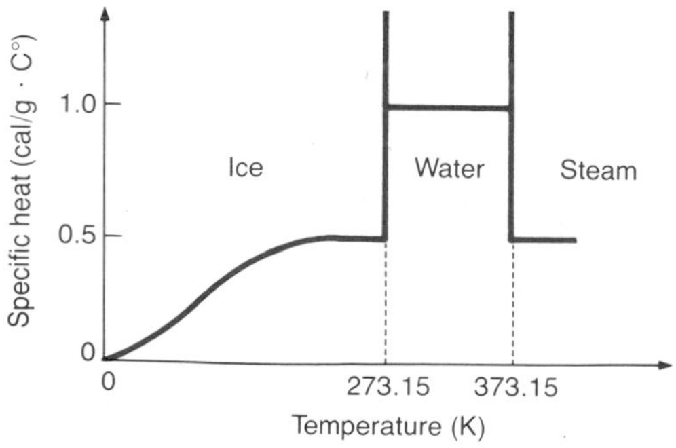

FIGURE 14.3 Specific heat of water at a pressure of 1 atm.

■ Definition of Latent Heat

Heat transfer in a phase change = mass of substance × latent heat

$$Q = mL \tag{14.3}$$

SI unit of latent heat: joule per kilogram
Abbreviation: J/kg

A subscript f for fusion usually designates the solid-liquid phase change, and a v for vaporization of the liquid-gas phase change. Note the use of the word *transfer* to describe the heat exchange. When ice melts or water boils, a heat input is necessary, but when steam condenses or water freezes, heat flows out of the substance. The latent heats of fusion and vaporization for some common substances are shown in Table 14.2; for convenience, we give the values in both J/kg and cal/g.

TABLE 14.2 Normal melting points and boiling points with associated latent heats for some common substances

Substance	*Normal melting point (K)*	*Latent heat of fusion*		*Normal boiling point (K)*	*Latent heat of vaporization*	
		J/kg	*kcal/kg (or cal/g)*		*J/kg*	*kcal/kg (or cal/g)*
Hydrogen	13.8	5.86×10^4	13.8	20.3	4.52×10^5	108
Oxygen	54	1.39×10^4	3.3	90	2.13×10^5	51
Nitrogen	63	2.55×10^4	6.1	77	2.00×10^5	48
Ethanol	156	10.1×10^4	26	351	8.57×10^5	205
Mercury	234	1.17×10^4	2.7	630	2.94×10^5	70
Water	273.15	33.5×10^4	80	373.15	22.6×10^5	539
Lead	600	2.45×10^4	5.9	2023	8.70×10^5	208
Aluminum	932	39.6×10^4	94.5	2740	105×10^5	2500

EXAMPLE 14.5

An aluminum pot of mass 0.7 kg contains 1.8 kg of ice at −10°C. If the pot is placed on a hot plate, how much heat input is required to convert the ice to steam at 100°C? What percentage of this heat input is required to convert the boiling water to steam?

SOLUTION The aluminum pot does not undergo a phase transition in the temperature range −10°C to 100°C. Calculation of the heat input requires one term of the type of Eq. (14.2). The ice undergoes two phase transitions, each of which requires heat input described by a term of the type of Eq. (14.3). It is also heated in two distinct phases, each requiring a term of the type of Eq. (14.2). Let L_f and L_v be the latent heats of fusion and vaporization, respectively. We write the total required heat input as follows:

$$\begin{aligned} Q &= mC(T - T_0)_{\text{aluminum}} + mC(T - T_0)_{\text{ice}} + mL_{f\ \text{ice}\rightarrow\text{water}} \\ &\quad + mC(T - T_0)_{\text{water}} + mL_{v\ \text{water}\rightarrow\text{steam}} \\ &= 0.7\ \text{kg} \times 900\ \text{J/kg} \cdot \text{C}° \times [100°\text{C} - (-10°\text{C})] \\ &\quad + 1.8\ \text{kg} \times 2100\ \text{J/kg} \cdot \text{C}° \times [0°\text{C} - (-10°\text{C})] \\ &\quad + 1.8\ \text{kg} \times 3.35 \times 10^5\ \text{J/kg} \\ &\quad + 1.8\ \text{kg} \times 4184\ \text{J/kg} \cdot \text{C}° \times (100°\text{C} - 0°\text{C}) \\ &\quad + 1.8\ \text{kg} \times 2.26 \times 10^6\ \text{J/kg} \\ &= 5.53 \times 10^6\ \text{J} \end{aligned}$$

In each term the final and initial temperatures are the extreme temperatures of a single phase. The percentage of the heat input that goes to vaporizing the boiling water is:

$$\frac{1.8\ \text{kg} \times 2.26 \times 10^6\ \text{J/kg}}{5.53 \times 10^6\ \text{J}} \times 100 = 73.5\%$$

■

The latent heat between two phases of a substance is different for different points on the line separating the phases. As the pressure and the corresponding boiling point increase, the latent heat of vaporization falls to zero, and the line separating the liquid and gas phases terminates at a point called the **critical point.** The critical point for water occurs at a temperature of 647.3 K and a pressure of 218.5 atm. If either the temperature or the pressure exceeds the respective value, there is no real distinction between water and steam.

The orderly progression from solid to liquid to gas, which is typical of most substances that are heated at atmospheric pressure, depends on the triple point being at a lower pressure than 1 atm. This is not the case for a relatively small number of common substances, of which the best-known example is carbon dioxide. The phase diagram for carbon dioxide is shown in Fig. 14.4. The triple point occurs at a temperature of 216.55 K and a pressure of 5.10 atm. A liquid phase of carbon dioxide is impossible at a lower pressure. If we heat solid carbon dioxide from a very low temperature at atmospheric pressure, it goes directly to the gas phase at a temperature of 194.6 K, and the latent heat of sublimation is about 5.8 × 10^5 J/kg (138 cal/g). The critical point for carbon dioxide occurs at a temperature of 304.2 K and a pressure of 72.9 atm.

Cooling of the human body by perspiration depends on the body supplying the heat to evaporate the water produced. An example illustrates the magnitude of the effect.

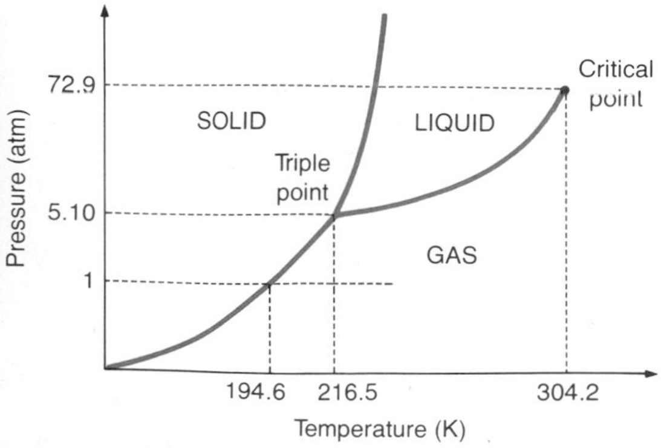

FIGURE 14.4 Phase diagram for carbon dioxide (not to scale).

EXAMPLE 14.6

In the absence of noticeable exertion or perspiration, evaporation of water from the skin and lungs of the human body amounts to about 25 g/h. Calculate the heat loss in kcal/h. Assume a body temperature of 37°C, at which temperature the latent heat of vaporization of water is 24.2×10^5 J/kg.

SOLUTION The result is given immediately using Eq. (14.3); the heat loss in 1 h is:

$$\begin{aligned} Q &= mL \\ &= 0.025 \text{ kg} \times 24.2 \times 10^5 \text{ J/kg} \\ &= 6.05 \times 10^4 \text{ J} \\ &= 6.05 \times 10^4 \text{ J} \div 4184 \text{ J/kcal} \\ &= 14.5 \text{ kcal} \end{aligned}$$

This amount represents the lowest level of heat loss by perspiration. Under the stated conditions of negligible exertion, it is minor compared with other mechanisms of heat loss that we discuss below. Under heavy exertion, the perspiration loss may be more than 50 times larger; this would give a heat loss of about 725 kcal/h. In these circumstances the heat loss by perspiration would be the major mechanism of body cooling. ■

We are now ready to look at the problem that gave birth to the caloric theory of heat. When liquids and solids at different temperatures are mixed, a common final temperature is attained. Provided that *no external work* is done on or by the mixed substances, and provided there is *no chemical reaction,* we can calculate the final temperature by using the data on specific heats and latent heats. The heat lost by the hot bodies is equal to the heat gained by the cold ones as they all come to their final common temperature.

EXAMPLE 14.7

A block of lead of mass 0.4 kg at a temperature of 95°C is dropped into 2 kg of water originally at 20°C that is contained in a copper pot of mass 0.15 kg. Calculate the final temperature if no heat is lost to the surroundings.

SOLUTION Let T be the final temperature of the mixture. Since none of the substances undergo a phase change within the temperature range, each term of the heat balance equation is of the form of Eq. (14.2). Using specific heat values from Table 14.1, we have:

$$\begin{matrix}\text{Heat lost} \\ \text{by lead}\end{matrix} = \begin{matrix}\text{heat gained} \\ \text{by water}\end{matrix} + \begin{matrix}\text{heat gained} \\ \text{by copper}\end{matrix}$$

0.4 kg × 128 J/kg · C° × (95°C − T) (lead at 95°C to lead at temperature T)

= 2 kg × 4184 J/kg · C° × (T − 20°C) (water at 20°C to water at temperature T) + 0.15 kg × 387 J/kg · C° × (T − 20°C) (copper at 20°C to copper at temperature T)

Solving the equation gives T = 24.1°C. ■

In this example we used SI units of specific heat. Using the calorie as a unit of energy is acceptable provided that we put every term of the heat balance equation in the same units. We can also apply the method of this example to cases involving phase transitions.

EXAMPLE 14.8

How much steam at 100°C must be mixed with 0.1 kg of ice at −10°C to produce a final product of water at 10°C? (Neglect any heat transfer to the container.)

SOLUTION Both substances undergo a phase transition, and a corresponding term must appear in the heat balance equation. Let m be the required mass of steam. Each heat transfer term is of the form of Eq. (14.2) or Eq. (14.3). Using the specific heats and latent heats from Tables 14.1 and 14.2, we have:

Heat lost by steam = heat gained by water

m × 539 cal/g (steam at 100°C to water at 100°C) + m × 1 cal/g · C° × (100°C − 10°C) (water at 100°C to water at 10°C) = 100 g × 0.50 cal/g · C° × {0°C − (−10°C)} (ice at −10°C to ice at 0°C) + 100 g × 80 cal/g (ice at 0°C to water at 0°C) + 100 g × 1 cal/g · C° × (10°C − 0°C) (water at 0°C to water at 10°C)

Solving the somewhat lengthy equation gives m = 15.1 g. ■

14.4 Specific Heats of Ideal Gases

Gases differ from liquids and solids in having a very large thermal expansion coefficient. Suppose that the substance that is heated in the cylinder of Fig. 14.1 is a gas. As we transfer heat to the gas, the gas expands, and the work done in the expansion process is not negligible. In fact, the amount of work done depends on how the pressure varies during the expansion; the specific heat likewise depends on the pressure variation. In principle, there are as many specific heats as there are ways of altering the gas pressure during the expansion.

If we hold the pressure constant, the volume increases as the gas is heated. The specific heat obtained in this way is designated C_p. If we hold the volume constant, heating the gas increases the pressure, and this specific heat is called C_v. The specific heats at constant pressure and constant volume are called the **principal specific heats** of the gas. Since the quantity of a gas is often specified in moles, it is convenient to modify the specific heat definition of Eq. (14.2) accordingly.

■ Definition of Principal Molar Specific Heats of a Gas

$$\text{Heat flow} = \begin{matrix}\text{number of}\\ \text{moles}\end{matrix} \times \begin{matrix}\text{specific}\\ \text{heat}\end{matrix} \times \begin{matrix}\text{temperature change}\\ \text{caused by heat flow}\end{matrix}$$

$$Q = nC_p(T - T_0) \quad \textbf{(a)} \qquad \textbf{(14.4)}$$

$$Q = nC_v(T - T_0) \quad \textbf{(b)}$$

SI unit of molar specific heat: joule per mole per Celsius degree

Abbreviation: J/mol · C°

For solids and liquids, *specific* refers to unit mass. For gases we refer to the molar quantity rather than the mass. The cases (a) and (b) correspond to constant pressure and constant volume, respectively. Table 14.3 gives the principal molar specific heats for a number of common gases at a temperature of 15°C and a pressure of 1 atm.

The specific heats of an ideal gas have simple and important implications. Consider n moles of an ideal gas heated at constant pressure p_0. Figure 14.5 illustrates the expansion of the gas from an initial volume and temperature V_0 and T_0 to final volume and temperature V and T. Let A be the cross-sectional area of the cylinder containing the gas, and d the displacement of the piston that occurs during the expansion. Since the pressure remains constant at p_0, the force of the gas on the piston is constant at p_0A. We can calculate the work done by the gas during the expansion by using Eq. (7.1):

$$\begin{aligned} X &= Fx \\ &= p_0Ad \\ &= p_0(V - V_0) \end{aligned}$$

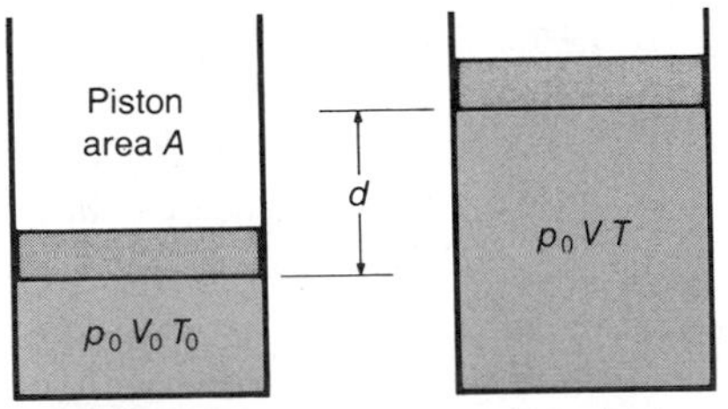

FIGURE 14.5 Expansion of a gas at constant pressure.

TABLE 14.3 Specific heat data for some common gases at 15°C and 1 atm pressure

Gas	Molecular mass (g)	Principal molar specific heats (J/mol · C°) C_p	C_v	Molecular specific heat difference (J/mol · C°) $C_p - C_v$	Specific heat ratio ($\gamma = C_p/C_v$)
Helium	4	20.8	12.5	8.3	1.667
Argon	40	20.8	12.5	8.3	1.667
Hydrogen	2	28.7	20.4	8.3	1.405
Oxygen	32	29.5	21.1	8.4	1.396
Nitrogen	28	28.9	20.6	8.3	1.401
Air (dry)	29	28.6	20.4	8.4	1.403
Carbon dioxide	44	36.8	28.3	8.5	1.300
Ethylene	28	42.9	34.3	8.6	1.250

The equation of state for an ideal gas, Eq. (14.4a), gives:

$$p_0V_0 = nRT_0$$

and

$$p_0V = nRT$$

Subtracting these two results, we have:

$$p(V - V_0) = nR(T - T_0)$$

The work done by the gas during its constant-pressure expansion is therefore given by:

$$W = nR(T - T_0) \tag{14.5}$$

We can now use the first law of thermodynamics to calculate the difference between the final and initial values of the internal energy of the gas. Rearrangement of Eq. (14.1) gives:

$$\begin{aligned} U - U_0 &= Q - W \\ &= nC_p(T - T_0) - nR(T - T_0) \end{aligned} \tag{14.6}$$

We have used Eq. (14.4a) for the heat input at constant pressure, and Eq. (14.5) for the work done by the expanding gas. For a gas heated at constant volume, the external work is zero and the energy difference is:

$$\begin{aligned} U - U_0 &= Q - W \\ &= nC_v(T - T_0) - 0 \end{aligned} \tag{14.7}$$

In this case we have used Eq. (14.4b) for the heat input to the gas at constant volume. However, the internal energy differences at constant pressure and constant volume are the same provided that the initial and final temperatures are the same, because the internal energy of an ideal gas depends only on the temperature and not on the pressure or volume. Equating the values of $U - U_0$ from Eqs. (14.6) and (14.7) gives a general relation between principal molar specific heats of an ideal gas.

Relation Between Principal Molar Specific Heats of an Ideal Gas

$$C_p - C_v = R \qquad (14.8)$$

R is the universal gas constant.

The fourth column of Table 14.3 shows the difference of the molar specific heats for the gases listed. The near equality of this figure to the universal gas constant ($R = 8.314$ J/mol · K) clearly confirms Eq. (14.8).

Examining the values for molar specific heats in Table 14.3 shows a remarkable fact. In spite of the big disparity in relative molecular mass, the molar specific heats of the monatomic gases helium and argon are the same. Continuing down the table, the diatomic gases are approximately the same. More advanced work on the kinetic theory of ideal gases explains the reason for this agreement. A convenient index of the effect is the ratio of the principal molar specific heats C_p/C_v, which is usually designated by the Greek letter gamma (γ). For monatomic ideal gases, kinetic theory predicts $\gamma = 5/3$. The fifth column of Table 14.3 shows excellent agreement with the kinetic theory prediction. For diatomic ideal gases, kinetic theory gives $\gamma = 7/5$ within a restricted temperature range. The diatomic gases listed have γ values remarkably close to $7/5$ at 15°C. For gases that have more than two atoms per molecule, kinetic theory predicts values that are closer to unity, the exact value depending on the number of atoms in the molecule for a limited range of temperature.

EXAMPLE 14.9

A cylinder containing 20 mol of oxygen at a temperature of 275 K is fitted with a frictionless piston that maintains a pressure of 1 atm on the gas. Calculate the heat energy supplied and the external work done when the gas is heated to 425 K.

SOLUTION We can calculate the heat absorbed at constant pressure by using Eq. (14.4a) and the molar specific heat value from Table 14.3:

$$\begin{aligned} Q &= nC_p(T - T_0) \\ &= 20 \text{ mol} \times 29.5 \text{ J/mol} \cdot \text{C}^\circ \times (425 \text{ K} - 275 \text{ K}) \\ &= 8.85 \times 10^4 \text{ J} \end{aligned}$$

(Remember that the difference between two temperatures in kelvins is equal to their difference in Celsius degrees.) We can calculate the change in internal energy of the gas by calculating the heat energy required to produce the same temperature increase at constant volume. To this end we use Eq. (14.4b):

$$\begin{aligned} U - U_0 &= nC_v(T - T_0) \\ &= 20 \text{ mol} \times 21.1 \text{ J/mol} \cdot \text{C}^\circ \times (425 \text{ K} - 275 \text{ K}) \\ &= 6.33 \times 10^4 \text{ J} \end{aligned}$$

We now calculate the work done by the gas from the first law of thermodynamics, Eq. (14.1):

$$\begin{aligned} Q &= (U - U_0) + W \\ 8.85 \times 10^4 \text{ J} &= 6.33 \times 10^4 \text{ J} + W \\ \therefore W &= 2.52 \times 10^4 \text{ J} \end{aligned}$$

■

In a previous chapter we postponed determining an expression for the velocity of sound in an ideal gas because the ratio of principal specific heats enters the problem. As previously, we find the correct functional form for the equation from dimensional considerations. The physical quantities that describe the gas are the pressure p, the density ρ, and the temperature T. The only way to make a velocity from these quantities is to set:

$$v = \text{constant} \times \left(\frac{p}{\rho}\right)^{1/2}$$

In previous applications of this type, the constant was frequently unity, but in this case a detailed analysis shows that it is the square root of the ratio of principal specific heats. Therefore, we have for the velocity of sound in an ideal gas:

$$v = \left(\frac{\gamma p}{\rho}\right)^{1/2} \qquad \textbf{(14.9)}$$

This equation appears to make the sound velocity dependent on pressure, but this is not really the case, as we can see by converting Eq. (14.9) to an alternate form. Let M be the mass in kilograms of 1 mol of the gas. If n moles occupy a volume V, the gas density is given by:

$$\rho = \frac{nM}{V}$$

The pressure is given by the ideal gas equation, Eq. (13.4a):

$$p = \frac{nRT}{V}$$

Substituting these results in Eq. (14.9) gives:

$$v = \left(\frac{\gamma RT}{M}\right)^{1/2} \tag{14.10}$$

(In this equation M stands for the mass of 1 mol measured in kilograms.)

This alternate expression for sound velocity shows that temperature is the really meaningful physical parameter. The quantities γ, R, and M are all constants for a given gas.

EXAMPLE 14.10

Calculate the velocity of sound in air at 15°C. (The average molecular mass of air is 29.)

SOLUTION We find the ratio of principal specific heats $\gamma_{air} = 1.403$ from Table 14.3. The temperature in kelvin is given by Eq. (13.2):

$$\begin{aligned} T(\text{K}) &= T(°\text{C}) + 273.15 \\ &= 15 + 273.15 \\ &= 288.15 \text{ K} \end{aligned}$$

The mass of 1 mol of air is 29 g. This means that M of Eq. (14.10) is given by:

$$M = 29 \times 10^{-3} \text{ kg/mol}$$

We can now calculate the sound velocity from Eq. (14.10):

$$\begin{aligned} v &= \left(\frac{\gamma RT}{M}\right)^{1/2} \\ &= \left(\frac{1.403 \times 8.314 \text{ J/mol} \cdot \text{K} \times 288.15 \text{ K}}{29 \times 10^{-3} \text{ kg/mol}}\right)^{1/2} \\ &= 340 \text{ m/s} \end{aligned}$$

■

14.5 Heat Transfer Processes

A sailplane soaring over east-central Washington state. Rising air currents, such as those producing the cumulus clouds in the background, enable modern sailplanes to remain aloft as long as daylight or the pilot's endurance hold out. When columns or rising air reach the condensation level, the water vapor condenses as visible cloud; for this reason, sailplane pilots watch for the first wisps of forming cloud and pursue them as thermal markers, or signposts marking the location of the otherwise invisible columns of rising air. (Courtesy of George Uveges, Soaring Society of America)

Flow of heat due to a temperature difference occurs by means of various mechanisms. When we put a kettle of water on a hot plate, heat flows from the hot plate through the metal base of the kettle into the cold water. This type of heat flow is called **thermal conduction,** or simply conduction. The temperature difference between the two sides of the kettle base causes the heat flow but not any flow of the material of the base.

When a woman standing in a cold room turns on baseboard heaters, she feels the flow of warm air from them. This flow occurs because of the temperature difference between the air adjacent to the heaters and the air in the remainder of the room. The reduction in density of the heated air results in a buoyancy force that causes the heated air to rise relative to the colder surrounding air. Thus, there is a flow of the air as well as a flow of heat. This mechanism of heat transfer is called **thermal convection.**

The heat reaching the earth from the sun cannot be transferred either by conduction or convection since the space between the earth and the sun has no material medium. The energy is carried by electromagnetic waves that do not require a material medium for propagation. This type of heat transfer is called **thermal radiation.** We can summarize the characteristics of these three basically different types of heat flow as follows:

Heat transfer process	*Material medium required*	*Flow of material medium occurs*
Conduction	Yes	No
Convection	Yes	Yes
Radiation	No	—

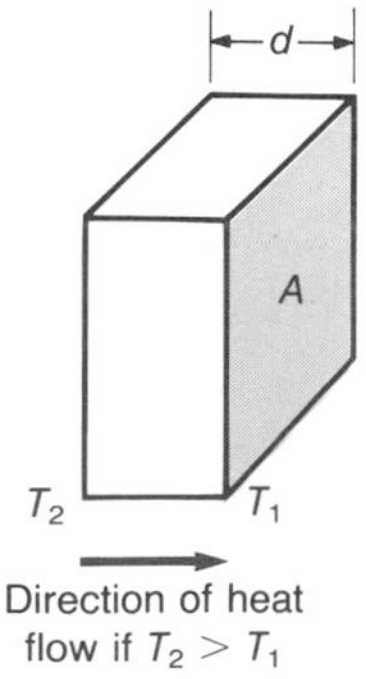

FIGURE 14.6 Diagram to illustrate the notation of the thermal conduction equation.

Let us begin with a theory of thermal conduction. Consider the conduction of heat through a slab of material in which two faces, each of area A, separated by a distance d, are maintained at temperatures T_1 and T_2, illustrated in Fig. 14.6. Experiments have shown that the rate of heat flow is proportional to the area of the faces and the temperature gradient, which is defined as the temperature difference between the two faces divided by their distance of separation. Our definition of **thermal conductivity** follows from the proportionality of the rate of heat flow to the cross-sectional area and the temperature gradient. The symbol for thermal conductivity is the Greek letter κ (kappa).

The thermal conductivities of some common materials are listed in Table 14.4. The values are for room temperature and are reasonably accurate at higher temperatures. Because the thermal conductivity of most substances decreases sharply at very low

TABLE 14.4 Typical room temperature values of thermal conductivity of common substances

	Substance	*Thermal conductivity (W/m · C°)*
Good conductors	Silver	415
	Copper	380
	Aluminum	205
	Brass	85
	Iron	60
	Mercury	8.5
Poor conductors	Ice	2.2
	Glass	1.1
	Brick masonry	0.60
	Plaster	0.60
	Hardwood	0.20
	Rubber	0.15
	Asbestos	0.085
	Wool felt	0.052
	Rock wool	0.050
	Styrofoam	0.050
	Corkboard	0.045

temperatures, figures for low-temperature applications should be obtained from detailed experimental results.

■ Definition of Thermal Conductivity

Heat flow = thermal conductivity × area of cross section × temperature gradient × time

$$Q = \kappa A \frac{T_2 - T_1}{d} t \qquad \textbf{(14.11)}$$

SI unit of thermal conductivity: watt per meter per Celsius degree

Abbreviation: W/m · C°

The substances mentioned in Table 14.4 are grouped as good and poor conductors, but there is no clear line of demarcation. Silver conducts heat about 10 000 times more effectively than corkboard, and the other substances listed have intermediate values. In a later chapter we will list the electrical conductivities of various substances. As a general rule, good thermal conductors are good electrical conductors, and poor thermal conductors are good electrical insulators.

EXAMPLE 14.11

An iron poker is 0.8 m long and 1 cm in diameter. One end is put in a fire where the temperature is 450°C, and the other end is at the room temperature of 25°C. Calculate the rate of heat flow along the poker.

SOLUTION We begin by preparing the data in SI units:

$$\kappa = 60 \text{ W/m} \cdot \text{C}° \qquad \text{(from Table 14.4)}$$

$$\begin{aligned} A &= \pi r^2 \\ &= \pi \times \left(\frac{0.01 \text{ m}}{2}\right)^2 \\ &= 7.854 \times 10^{-5} \text{ m}^2 \end{aligned}$$

$$\begin{aligned} \frac{T_2 - T_1}{d} &= \frac{450°\text{C} - 25°\text{C}}{0.8 \text{ m}} \\ &= 531.2 \text{ C}°/\text{m} \end{aligned}$$

$$t = 1 \text{ s}$$

The heat transfer is given by Eq. (14.11):

$$\begin{aligned} Q &= \kappa A \frac{T_2 - T_1}{d} t \\ &= 60 \text{ W/m} \cdot \text{C}° \times 7.854 \times 10^{-5} \text{ m}^2 \times 531.2 \text{ C}°/\text{m} \times 1 \text{ s} \\ &= 2.50 \text{ J} \end{aligned}$$

■

In engineering applications the British thermal unit is frequently used as a heat unit. An example illustrates the usage.

EXAMPLE 14.12

A solid brick wall is 9 in. thick, 9 ft high, and 25 ft long. The inside temperature is 25°C and the outside temperature is −5°C. Calculate the heat loss by conduction through the wall in Btu/h.

SOLUTION Set the data in SI units in order to calculate the heat conducted through the wall in 1 h:

$$\kappa = 0.60 \text{ W/m} \cdot \text{C}° \qquad \text{(from Table 14.4)}$$

$$\begin{aligned} A &= (9 \text{ ft} \times 0.3048 \text{ m/ft}) \times (25 \text{ ft} \times 0.3048 \text{ m/ft}) \\ &= 20.90 \text{ m}^2 \end{aligned}$$

$$\frac{T_2 - T_1}{d} = \frac{25°C - (-5°C)}{9 \text{ in.} \times 0.0254 \text{ m/in.}}$$

$$= 131.2 \text{ C°/m}$$

$$t = 1 \text{ h} \times 3600 \text{ s/h}$$

$$= 3600 \text{ s}$$

The heat conducted is given by Eq. (14.11):

$$Q = \kappa A \frac{T_2 - T_1}{d} t$$

$$= 0.60 \text{ W/m} \cdot \text{C°} \times 20.90 \text{ m}^2 \times 131.2 \text{ C°/m} \times 3600 \text{ s}$$

$$= 5.92 \times 10^6 \text{ J}$$

$$= 5.92 \times 10^6 \text{ J} \div 1054 \text{ J/Btu}$$

$$= 5610 \text{ Btu}$$

■

There are many important examples of thermal convection. The operation of a sailplane depends on upward air currents (thermal updrafts), which occur because of convection effects. These are often caused by certain features of the landscape, such as freshly plowed fields, which absorb heat from the sun more readily than their surroundings do. The air over these features then becomes relatively hot and rises, causing the updraft that allows a sailplane to gain altitude without any power source of its own. Another example concerns the placement of inlet and outlet air ducts in the rooms of heated and air-conditioned buildings. The ducts should be situated to avoid the formation of stagnant air pockets. Since heated air rises, heating inlet ducts should be close to the floor, and return ducts should be high on the walls. Since cool air falls, air conditioning inlets should be high on the walls, and return ducts near the floor. In practice, however, two ducting systems are rarely used because of the additional cost.

We have already mentioned the freezing of rivers and lakes from the surface downward in connection with the anomalous thermal expansion properties of water. This is also an example of convection, since the cooler surface water sinks if its temperature is above 4°C but remains on the surface otherwise.

There is no really satisfactory theory of thermal convection, but we can write a formula analogous to that for thermal conduction and use it to give approximate results. The heat transferred in a convection process is given by:

$$Q = KA(T_2 - T_1)t \tag{14.12}$$

where K is the convective heat transfer coefficient, A is the area of the surface, T_2 is the temperature of the surface, T_1 is the temperature of the convecting medium at a distant place, and t is the time of the heat transfer.

Note that this formula uses a temperature difference rather than the temperature gradient used in the thermal conduction case. However, more importantly, we cannot give a table of convective transfer coefficients K. They must be determined empirically for various situations, and they depend not only on the convective medium but on the type and shape of surface, the surface orientation, and the magnitude of the temperature difference. Bodily heat loss by convection is important for humans, as the following example shows.

EXAMPLE 14.13

The area of the skin of an adult human being is about 2 m^2. If the skin temperature of an unclothed person is 31°C when the surrounding air is 24°C, what is the rate of heat loss in kilocalories per hour. (Assume a convective heat transfer coefficient of 7.1 $W/m^2 \cdot C°$.)

SOLUTION Using Eq. (14.12) we have for the heat loss in one hour:

$$
\begin{aligned}
Q &= KA(T_2 - T_1)t \\
&= 7.1\ \text{W/m}^2 \cdot \text{C}° \times 2\ \text{m}^2 \times (31°\text{C} - 24°\text{C}) \times 3600\ \text{s} \\
&= 3.58 \times 10^5\ \text{J} \\
&= 3.58 \times 10^5\ \text{J} \div 4184\ \text{J/kcal} \\
&= 85.5\ \text{kcal}
\end{aligned}
$$

Since a resting human generates approximately 70 kcal/h by metabolism, we see that the situation envisaged in this problem would result in a net heat loss. Note also that even a small ongoing draft of air would greatly increase the heat loss by convection. ■

The final method of heat transfer is radiation. It is common experience that the heat radiated from a hot body increases with increasing temperature, but the heat radiated is also influenced by the physical condition of the radiating surface. The relevant surface condition, called *emissivity*, is determined by many factors. A matte black surface radiates heat far more effectively than a bright shiny surface. In addition, a black surface absorbs the radiant energy that falls on it, while a shiny surface reflects the radiation. The combination of the most effective radiator and the best possible heat absorber is called a *blackbody*, and its emissivity is unity; the opposite extreme is called a *perfect reflector*, and its emissivity is zero. Once again, these concepts are idealized models of reality, since every real body radiates and reflects to some extent. We account for this intermediate state by assigning an emissivity that lies between the extreme values of zero and unity. However, the blackbody is an important practical concept, since many bodies that are not even approximately matte black come very close to having unit emissivity. The sun is a leading case in point—it hardly has a matte black appearance, but its emissivity is close to unity. The exact form of the thermal radiation law was discovered by the Austrian physicist Joseph Stefan (1835–1893), who knew of experimental work in which the heat radiated by a hot wire increased by a factor of 11.7 as the temperature of the wire was raised from 525°C to 1200°C. These temperatures are 798.15 K and 1473.15 K, and Stefan noticed the remarkable fact that:

$$\left(\frac{1473.15\ \text{K}}{798.15\ \text{K}}\right)^4 = 11.6$$

Stefan's Law of Thermal Radiation

Energy radiated = emissivity × Stefan's constant × (absolute temperature)^4 × area of radiator × time

$$Q = e\sigma T^4 At \qquad \textbf{(14.13)}$$

Emissivity is a pure number.
The value of Stefan's constant is

$$\sigma = 5.670 \times 10^{-8}\ \text{W/m}^2 \cdot \text{K}^4$$

This led him to postulate that the amount of radiant energy emitted is proportional to the fourth power of the absolute temperature, a relationship known as **Stefan's law.**

Determining the value of Stefan's constant, which was readily obtained from laboratory experiments, set the stage for calculating the surface temperature of the sun.

EXAMPLE 14.14

By measuring the heat energy received on earth from the sun, it is inferred that the output of thermal radiation from the sun's surface is 6.14×10^7 W/m^2. Assuming that the sun is a blackbody, calculate its surface temperature.

SOLUTION Since the radiation is 6.14×10^7 W/m^2, we set:

$$Q = 6.14 \times 10^7\ \text{J},\ A = 1\ \text{m}^2,\ \text{and}\ t = 1\ \text{s}$$

in Stefan's law, Eq. (14.13):

$$Q = e\sigma T^4 At$$

$$6.14 \times 10^7 \text{ J} = 1 \times 5.670 \times 10^{-8} \text{ W/m}^2 \cdot \text{K}^4 \times T^4 \times 1 \text{ m}^2 \times 1 \text{ s}$$

$$\therefore T = 5740 \text{ K}$$ ■

We can also use Stefan's law to calculate the thermal energy absorbed by an object from blackbody surroundings in terms of their temperature. If the object in question is at the same temperature as the surroundings, the energy radiated to the surroundings exactly balances the energy absorbed from them. If the temperatures are not the same, there is a net flow of radiant energy between the object and the surroundings. In using Stefan's law to calculate energy absorption, we take the emissivity appropriate

These solar panels at the Department of Energy test facility in New Mexico can be rotated to keep the plane of the panel normal to the direction of the sun's rays. This assures the greatest amount of radiant power falling on each panel. (Courtesy of U.S. Department of Energy)

to the absorbing object and the temperature of the blackbody surroundings.

EXAMPLE 14.15

The area of the skin of an adult human being is about 2 m^2. If the skin temperature of an unclothed person is 31°C when the surroundings are 24°C, what is the rate of heat loss in kcal/h? (Assume an emissivity of 95%.)

SOLUTION We use Stefan's law, Eq. (14.13), to calculate the heat radiated from the person's body and also the heat absorbed from the surroundings. In both cases Eq. (14.13) is appropriate:

$$Q = e\sigma T^4 At$$

$$\text{Radiated heat} = 0.95 \times 5.670 \times 10^{-8}\ \text{W/m}^2 \cdot \text{K}^4 \times [(273.15 + 31)\ \text{K}]^4 \times 2\ \text{m}^2 \times 3600\ \text{s}$$

$$= 3.312 \times 10^6\ \text{J}$$

$$\text{Absorbed heat} = 0.95 \times 5.670 \times 10^{-8}\ \text{W/m}^2 \cdot \text{K}^4 \times [(273.15 + 24)\ \text{K}]^4 \times 2\ \text{m}^2 \times 3600\ \text{s}$$

$$= 3.023 \times 10^6\ \text{J}$$

$$\text{Net loss in 1 h} = 3.312 \times 10^6\ \text{J} - 3.023 \times 10^6\ \text{J}$$

$$= 2.89 \times 10^5\ \text{J}$$

$$= 2.89 \times 10^5\ \text{J} \div 4184\ \text{J/kcal}$$

$$= 69.1\ \text{kcal}$$

■

14.6 Temperature Regulation of the Human Body

The temperature in deep tissues of the human body—the so-called core temperature—remains constant at about 37°C to within about 0.5 C°, irrespective of external circumstances (except for the case of illness). The skin temperature, however, may rise or fall by several Celsius degrees depending on circumstances. It follows that the body must have methods of controlling its temperature by retaining (or losing) extra heat if the core temperature begins to fall (or rise).

The skin, subcutaneous tissues, and fatty tissues are the thermal insulation system for the body. Fat is an effective insulator whose thermal conductivity is only about one-third that of other tissues. The amount of insulation thus varies a great deal from person to person, but with no blood flowing in the surface tissues the average insulating properties approximate a suit of clothing.

Small blood vessels come through the insulating tissues and supply the continuous venous plexus located immediately be-

SPECIAL TOPIC 17 SCALING AND METABOLISM

In previous chapters we have used scaling arguments to discuss the way in which various functions of similarly shaped animals vary with size. To this end we used a single dimensionless scaling factor L (the ratio of corresponding body lengths) and found the way in which the function under consideration scaled with L.

We will now use this approach to predict the way in which animal metabolism should scale. Since metabolic rate is directly dependent on oxygen absorption, it should scale in the same way as the area of lung surface. Moreover, since excess heat is lost through the exterior surface of the body, an equilibrium body temperature should require that the metabolic rate scale in the same way as the exterior body surface area. In our simple model with a single scaling factor L, all body areas (that is, both the lung surface and the exterior surface) scale is L^2. We therefore expect that the metabolic rate should scale as L^2.

For our present discussion, it is convenient to introduce a dimensionless scale factor M, which is the ratio of animal mass, rather than the scale factor L, which we have used up to now. We have shown (for example, in our discussion on scaling in Chapter 1) that animal mass should scale with L^3; it follows that animal length should scale with $M^{1/3}$. Finally, both body surface area and metabolic rate should scale with the square of length; that is, $M^{2/3} \approx M^{0.67}$. Careful experimental work on mammals ranging in size from a mouse to an elephant shows that body surface area scales with $M^{0.63}$ and metabolic rate with $M^{0.75}$. Not only are these indices different from each other, but neither is equal to 0.67, which our simple scaling model would require.

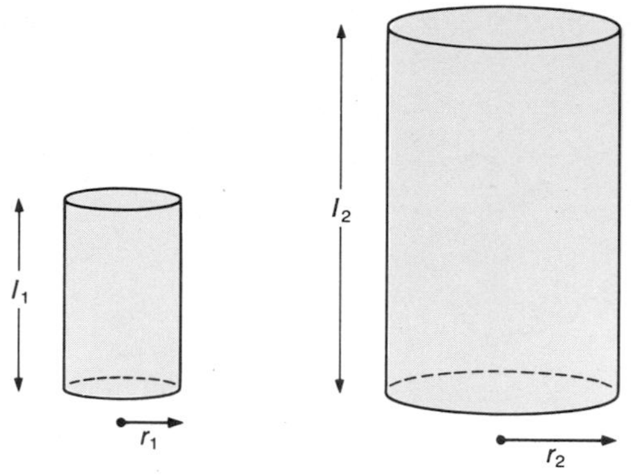

Cylinders of differing length and radius. Stability against buckling under their own weight requires $(r_2/r_1) = (l_2/l_1)^{3/2}$.

In order to address this discrepancy, McMahon [*Science* 179 (1973):1201] proposed a new structure with two characteristic lengths rather than one. The model is based on the fact that various parts of the body tend to be almost cylindrical in shape; this leads to two characteristic lengths: the cylinder length L and the cylinder radius R. The diagram shows two cylinders of lengths l_1 and l_2 and radii r_1 and r_2. In our previous, simple model, the ratio of the radii would be equal to the ratio of the lengths, and all of the preceding conclusions would still hold. However, instead of simply requiring that the cylinders have the same shape, McMahon proposed the further requirement that each cylinder be stable against buckling under its own weight. The analysis is beyond the scope of this book, but it is shown that this criterion requires:

$$\frac{r_2}{r_1} = \left(\frac{l_2}{l_1}\right)^{3/2}$$

neath the skin. A huge variation in blood supply to the venous plexus is possible, ranging from almost nothing up to about one-third of the total cardiac output. The flow rate is controlled by constriction or dilation of the blood vessels. A high flow rate associated with a vasodilated state corresponds to a large transfer of heat from the core to the skin, while a low flow rate associated

SPECIAL TOPIC 17 (continued)

If we take $l_2/l_1 = L$, the dimensionless scale factor, then the cylinder radii scale with $L^{3/2}$. The volume ratios are:

$$\frac{V_2}{V_1} = \frac{\pi r_2^2\, l_2}{\pi r_1^2\, l_1} = \left(\frac{l_2}{l_1}\right)^4 = L^4$$

It follows that volume and, consequently, mass scale with L^4. Turning the relation around and using the dimensionless mass scale factor, we conclude that cylinder length scales with $M^{1/4}$ and cylinder radius with $M^{3/8}$.

If we think of the body as made up of cylindrical parts, then most of the area is the curved surfaces of the cylinders, since the ends are usually joined to other cylinders. The area of the curved surface of a cylinder is $2\pi rl$; it follows that surface area should scale with the product of cylinder length and radius—that is, with $M^{5/8} = M^{0.625}$. This is very close indeed to the experimental index of 0.63.

In order to have the metabolic rate scale as a different power of the mass, it is necessary to abandon the idea that it scales with surface area. Instead, we assume that it scales with the power used to contract muscle, which [from Eq. (7.6)] is the muscular force multiplied by the velocity of contraction. Experimentally, all mammals exert about the same force per unit muscle cross-sectional area, and all have about the same muscle contraction velocity. This means that power scales with muscle cross-sectional area—that is, with the square of the cylinder radius. It follows that the power and (in view of our assumption) the metabolic rate should scale with $M^{3/4} = M^{0.75}$, in exact agreement with the experimental results. Since metabolic processes use oxygen absorbed in the lungs, the lung surface area should also scale as $M^{0.75}$. This result is well verified experimentally.

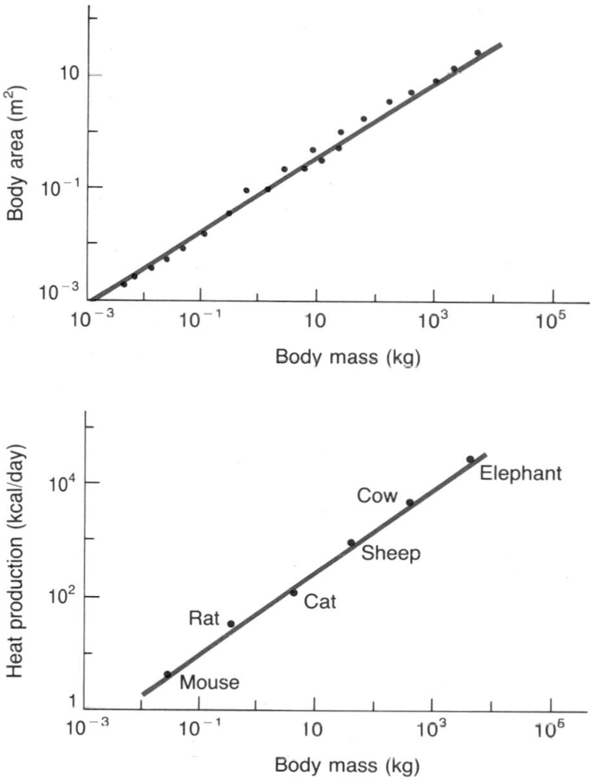

The body surface areas and the metabolic rates of mammals of various mass. The solid lines are the predictions of the scaling model based on buckling strength. [Diagrams from T. McMahon, "Size and Shape in Biology," *Science* 179 (1973):1203.]

McMahon's scaling model, regarding animals as being composed of cylindrical segments whose shape is dependent on a buckling criterion, gives better agreement with experiments than the simple scaling model in which all animals are assumed to be of similar shape. Clearly we must be cautious with scaling analysis, since reasonable assumptions (such as those of Chapter 1) can lead to erroneous conclusions.

with a vasoconstricted state greatly reduces the heat flow. The range of thermal conductance in the thermal insulation system varies by a factor of about eight between the fully vasoconstricted and the fully vasodilated states.

Heat production in the body results from the chemical metabolism of the dietary intake. Metabolism provides about 70 kcal/h when the body is resting; this rate is called the basal metabolism, and the figure quoted is a rough average since the rate varies quite a lot from person to person. Moreover, the metabolic rate becomes 5 to 10 times greater during light to medium exercise, and up to 20 times greater during heavy exercise.

The major means of heat loss from the surface of the body are radiation, convection, and evaporation. Radiation from an object whose temperature is about 300 K consists mainly of infrared radiation in the 10-μm wavelength range. As we have seen, there is also absorption of radiation from surrounding objects, and the difference between the heat radiated and absorbed is the net radiative heat transfer. Interestingly, most clothing is quite transparent to radiation in the 10-μm range.

Convection can occur only after the air next to the skin has been heated by conduction. Convection is obviously very much affected by air currents, and is greatly increased by even a light breeze. Examples 14.13 and 14.15 show that radiation and convection loss play roughly equal roles under the same conditions.

Water evaporates continually from the skin and the lungs at the rate of about 25 g per hour. The latent heat of vaporization for water at 37°C is about 24.2×10^5 J/kg, which is about 8% larger than the heat of vaporization at 100°C. This causes an ongoing (and uncontrolled) heat loss of about 14.5 kcal/h, as we calculated in Example 14.6. In addition, the body is able to increase the rate of evaporation by sweating. Sweat is not exactly water since it contains a variety of other substances, but its latent heat of vaporization is about the same as that of water. Note that if the temperature of the surroundings is greater than the skin temperature, the only mechanism by which the body can lose heat is by evaporation.

Despite the state of physical exertion and the temperature of the surroundings, the body seeks to balance the heat supply and the various heat transfer processes to maintain a core temperature of 37°C under all circumstances. The body's main thermostat is called the hypothalamus; it is located in the brain and responds very quickly to changes in temperature.

Other receptors monitor the core temperature in specific parts of the body—the spinal cord, the large veins, and the abdomen. Still others are distributed in the skin. It is believed that most of these secondary temperature receptors detect coldness rather than warmth, so that their major function is to prevent hypothermia.

Falling temperatures produce various reflex effects that either conserve or increase the supply of heat. First, vasoconstriction makes the thermal conductivity of the insulating layer as low as possible; second, shivering causes an increase in the rate of heat production; and third, sweating is inhibited to whatever extent possible.

Rising temperatures produce the opposite results. First, vasodilation produces the greatest possible thermal conductivity in the insulating layer, and, second, profuse sweating is initiated. As we saw in Example 14.6, sweating can account for a rate of heat loss up to about 10 times the basal metabolism rate.

In addition to the involuntary processes, there is a voluntary method for inhibiting heat loss that is the most effective method of all: wearing clothing. Clothing inhibits heat loss by creating a warm layer of air next to the skin, and since this air can convect away only very slowly, the heat loss by convection is greatly reduced. Of course, it may happen that a strong breeze penetrates the clothing and continually sweeps the warm air away. For this reason, windproof clothing is effective in reducing convection losses in windy conditions. In addition to loss by convection, we must also consider loss by radiation. As we have mentioned, most clothing does not stop the radiation loss, and for this reason modern arctic clothing is made with a thin gold film that reflects thermal radiation from the body. For arctic conditions the most effective clothing must be both windproof and radiation-proof.

Factors under the control of the individual could also play a major role in increasing the rate of heat loss. Removing clothing helps to increase both convective and radiative loss since it does away with the layer of warm air between the clothes and the body. Even more effective is any kind of breeze, since this directly increases convective losses. It also helps to make sweating more effective since it removes moisture-laden air close to the skin, and replaces it with drier air that more effectively evaporates the sweat. Note that this last comment is not relevant if the whole environment is excessively humid. Indeed, there are valleys near the shore of the Red Sea where human beings have never dwelt, since the continual oppressive humidity makes any sort of exertion virtually impossible.

KEY CONCEPTS

Heat flow is a form of energy transfer.

The **energy conservation** equation written to include heat flow is called the **first law of thermodynamics.** See Eq. (14.1).

Specific heat measures the quantity of heat required to cause a unit temperature rise in a unit mass of a substance. See Eq. (14.2).

The three phases of pure substances are solid, liquid, and gas. A transition between two phases is a **first-order phase transition.**

A first-order phase transition is associated with a heat transfer that is not accompanied by a temperature change. The heat transfer per unit mass is called the **latent heat.** See Eq. (14.3).

The **principal specific heats** of a gas are those for constant pressure and constant volume. See Eq. (14.4).

The velocity of sound in an ideal gas depends on the ratio of the principal specific heats. See Eqs. (14.9) and (14.10).

The physical processes of heat transfer are **conduction, convection,** and **radiation.**

Heat conduction through a medium is determined by its **thermal conductivity.** See Eq. (14.11).

Heat radiation by a body is proportional to the product of its emissivity and the fourth power of its absolute temperature. This is **Stefan's law** of radiation. See Eq. (14.13).

QUESTIONS FOR THOUGHT

1. The specific heat at constant pressure of an ideal gas is always greater than its specific heat at constant volume. Can you envisage a substance whose constant-volume specific heat exceeds the specific heat at constant pressure? What is necessary for this situation to occur?

2. Communities living near the Great Lakes experience a moderation in temperature compared with those located some distance inland. Explain this effect, and say whether you expect it to be more pronounced on the eastern or western shores of the lakes.

3. In a previous chapter we stated that a vacuum exists above the mercury column in a mercury-in-glass barometer. Is this statement exactly correct? If not, how would you modify it?

4. Can ice and steam coexist in equilibrium with each other under any conditions?

5. Can you offer a reason for the increase of boiling point with increasing pressure that is observed in all substances?

6. Explain why a pressure cooker can cook food at a temperature that is well above the normal boiling point of water.

7. The terms *evaporation*, *sublimation*, and *boiling* all refer to the passage of atoms (or molecules) into the gaseous phase. What distinguishes the three processes from each other?

8. People who live in hot, dry climates store drinking water in canvas bags. After some hours of storage, the water in the bag is considerably cooler than the surrounding air. Explain how this cooling effect occurs.

9. On a very cold day a tool with a metal handle feels much colder to the touch than one with a wooden handle. Explain why this is so.

10. Why do superior quality saucepans have stainless steel sides and copper bases?

11. A vacuum flask maintains the temperature of its contents by minimizing the rate of heat energy transfer by conduction, convection, and radiation. Explain how this is achieved.

12. Why do automobile manufacturers frequently insist on tinted glass for models equipped with air conditioning?

13. The top of Mt. Everest is closer to the sun during the daytime than the valley of the Ganges River. Why is it colder on the mountaintop than in the river valley?

PROBLEMS

A. Single-Substitution Problems

(Use the data in Table 14.1, 14.2, 14.3, and 14.4 for solving these problems.)

1. A system absorbs 11 500 J of heat and at the same time performs 8000 J of external work. Calculate the change in the internal energy of the system. [14.1]

2. A system suffers a 7000-J decrease in internal energy and at the same time gives off 2500 J of heat. Calculate the work done by the system. [14.1]

3. Calculate the number of kilocalories required to raise the temperature of 2 kg of lead from 30°C to 70°C. [14.2]

4. A block of metal of mass 500 g absorbs 1200 cal of heat, and its temperature increases by 42.8°C. Calculate the specific heat of the metal. [14.2]

5. How many joules are required to raise the temperature of 1 ℓ of water from 25°C to 100°C? [14.2]

6. A block of ice at −2°C loses 1 MJ of heat and its temperature falls to −9°C. What is the mass of the ice? [14.2]

7. A 400-g block of aluminum at 80°C is placed in 800 g of water at 15°C. Calculate the final temperature of the mixture. (Ignore the heat capacity of the container.) [14.3]

8. A piece of brass at 70°C is placed in 600 g of water at 10°C, and the final temperature of the mixture is 18°C. Calculate the mass of the brass. (Ignore the heat capacity of the container.) [14.3]

9. How much energy is required to change 20 mol of oxygen from a liquid at its boiling point to a gas at the same temperature? [14.3]

10. How much energy is required to melt a lead bullet of 30-g mass having initial temperature of 40°C? [14.3]

11. How much energy is required to increase the temperature of 3 kg of carbon dioxide from 5°C to 45°C if the pressure is maintained constant at 1 atm? [14.4]

12. How much heat must be extracted from 2 kg of nitrogen to cool it from 70°C to 20°C if the volume is kept constant? [14.4]

13. Calculate the speed of sound in helium gas at 40°C. [14.4]

14. Calculate the speed of sound in air at a temperature of 100°C. [14.4]

15. Calculate the rate of flow of heat along an iron poker; one end is held in a fire at 650°C, and the other end is kept at 25°C. The poker is 70 cm long and of square cross section 1.5 cm × 1.5 cm. [14.5]

16. A wool felt pad 30 cm × 30 cm and 1 mm thick separates two flat metal plates whose temperatures are 0°C and 80°C. Calculate the rate of heat flow through the pad. [14.5]

17. The bottom of an aluminum kettle is 4 mm thick and 25 cm in diameter. Calculate the rate of heat flow to water in the kettle at 60°C from a hot plate at 60.5°C. [14.5]

18. A cube of iron of side 10 cm is heated in a fire to 750°C (a bright cherry red). Assuming blackbody radiation, find the rate of energy radiation from the iron. [14.5]

19. At what temperature is the heat radiation from a blackbody surface equal to 1 W/m^2? [14.5]

20. The roof of a house has an area of 250 m^2 and its temperature rises to 57°C on a hot day. At what rate does the roof (assumed blackbody) radiate energy? [14.5]

B. Standard-Level Problems

21. How much energy is required to raise the temperature of 2 ℓ of glycerin from 20°C to 45°C? [14.2]

22. A mercury-in-glass thermometer contains 0.32 cm^3 of mercury, and the glass has a mass of 160 g. How much heat is absorbed by the whole instrument in a 10 C° rise in temperature? [14.2]

23. The Niagara Falls are about 60 m high. If all of the energy of the falls went to heating the water, what temperature difference would you expect to find between the top and bottom of the falls?

24. A lead bullet of mass 30 g is fired at 500 m/s through a piece of hardboard and emerges on the other side with a speed of 320 m/s. Calculate the temperature rise of the bullet if it stores half of the kinetic energy lost.

25. A steel wool pad is rubbed on a steel plate of mass 800 g in strokes 16 cm long at the rate of 3 strokes per second. The kinetic friction coefficient between the surfaces is 0.85, and the normal force between pad and plate is kept constant at 15 N. By how much does the temperature of the plate increase in 1 min? (Assume no heat loss to the surroundings, and that all the heat goes to increase the temperature of the plate.)

26. A 75-kg person has a basal metabolic rate of 70 kcal/h. What would be the temperature rise in one hour due to this generation of heat if it were all applied to that end? (Take the average specific heat as approximately equal to that of water.)

27. An electric heater with a 1200-W heating coil is advertised as capable of boiling 1 ℓ of water in 7 min. Assume that the initial water temperature is 15°C. Calculate the percentage of the heat input lost to the surroundings.

28. An electric heater is immersed in 450 g of a liquid contained in an insulated copper calorimeter of mass 180 g. Electrical energy is dissipated for 10 s at a constant rate of 3.5 kW. The temperature of the liquid and the calorimeter increases from 22.3°C to 51.8°C. Calculate the average specific heat of the liquid over this temperature range. (Neglect heat losses.)

29. A copper container of mass 800 g contains 0.7 ℓ of water at 15°C. A 300-g block of aluminum is put into the water, and the final temperature is 20°C. What was the initial temperature of the aluminum block? [14.3]

30. A piece of lead of mass 1200 g is heated to 80°C and dropped into 0.75 ℓ of water in a glass beaker of mass 200 g initially at 10°C. Calculate the final temperature of the various components assuming no heat loss to the surroundings. [14.3]

31. A glass beaker of mass 500 g contains 200 g of water at 20°C. When 350 g of another liquid at 70°C are added to the water, the final temperature is 40°C. Calculate the specific heat of the second liquid. [14.3]

32. A 250-g aluminum container holds 600 g of water at 50°C, and a 400-g copper container holds 500 g of methanol at 0°C. Calculate the final temperature:

a. if the water is poured into the methanol
b. if the methanol is poured into the water

33. A copper calorimeter of mass 600 g contains 150 g of ice and 400 g of water at 0°C. A 1-kg block of aluminum at 80°C is added to the calorimeter. Calculate the final temperature.

34. Heat transfer of 3.5×10^5 J is required to change a block of ice at −10°C to water at 10°C. What was the mass of the ice?

35. An insulated vessel of negligible heat capacity contains 300 g of ice at −10°C. How much steam at 100°C and atmospheric pressure must be introduced into the vessel so that the final content is water at 40°C? [14.3]

36. An ice cube of mass 60 g at −10°C is dropped into a glass of water at 15°C; the mass of the glass is 150 g, and the glass contains 450 g of water. Calculate the final temperature of the system assuming no heat loss to surroundings. (Assume that all of the ice melts.)

37. An aluminum kettle of mass 550 g contains 800 g of water at 25°C and is placed on a 1.2-kW hot plate.

a. How long does it take for the water to boil?
b. How much longer does it take for the kettle to boil dry?
c. How much later does the aluminum begin to melt?

(Assume no heat losses to the surroundings.)

38. A house heated by solar energy must have a heat storage capability for use on overcast days. In a certain design 5×10^6 Btu are stored in sealed containers that are heated from 25°C to 42°C on a sunny day.

a. If the heat is stored in water, what total container volume is required?
b. What volume of container is required to store the heat in Glauber's salt?

Data for Glauber's salt	
Density	1464 kg/m^3
Melting point	32°C
Specific heat of solid	1.93×10^3 J/kg · C°
Specific heat of liquid	2.85×10^3 J/kg · C°
Latent heat of fusion	2.43×10^5 J/kg

39. An unheated room contains 100 crates of apples, each of mass 40 kg at 10°C. A drop in the outside temperature causes a loss of 1000 Btu/h of heat from the room.

a. How long does it take for the temperature in the room to fall to −1°C? (Assume that the specific heat of the apples is 3800 J/kg · C°, and that the apples do not freeze.)
b. How long does it take for the same temperature drop to occur if three casks of water, each of 800 kg, are placed in the room with the apples?

40. A piece of copper of mass 300 g is cooled from 10°C to the boiling point of liquid nitrogen by immersing it in a large container of that liquid at its boiling point. What mass of nitrogen is vaporized? (Take the average specific heat of copper as 0.08 cal/g · C° over the temperature range involved.)

41. One mole of carbon dioxide is heated at constant volume from 0°C to 100°C. Calculate the heat added, the work done, and the increase in internal energy of the gas. [14.4]

42. One mole of nitrogen is heated from 27°C to 177°C in a constant pressure process. Calculate the heat added, the work done, and the increase in internal energy of the gas. [14.4]

43. A gas expands at constant pressure of 2×10^5 Pa until its volume is twice the initial value. During the expansion 8000 J of heat are added to the gas and its internal energy increases by 2000 J. Calculate the final volume.

44. A gas expands at constant pressure of 2×10^5 Pa from an initial volume of 10 ℓ to a final volume of 26 ℓ. During the expansion 9000 J of heat are added to the gas. Calculate the change in internal energy.

45. A cylinder contains 20 mol of argon (atomic mass 40 g) at 0°C and a pressure of 1 atm. The gas is heated at constant pressure and its volume increases by 250 ℓ.

a. What is the final temperature of the gas?
b. How much heat is supplied during the expansion?

46. A room measuring 5 m × 3 m × 2.8 m high contains dry air at a pressure of 1 atm and a temperature of 10°C.

a. How many British thermal units of heat energy are required to raise the air temperature to 25°C if the room is sealed at constant volume?
b. If the room could expand without any air leaking out in such a way as to maintain the pressure at 1 atm, how much heat energy is required? (Additional energy is required to raise the temperature of the walls and contents of the room, but ignore this fact in this problem.)

47. The temperature of the atmosphere is frequently about 15°C at sea level and falls by about 10°C for every 1500-m increase in altitude up to about 12 000 m. Calculate the speed of sound at 5000-m and 10 000-m altitudes.

48. Calculate the frequency of the fundamental mode of an open organ pipe 30 cm long that is filled with carbon dioxide at 27°C.

49. A windowpane is made of glass 120 cm × 180 cm × 3 mm thick. The outside temperature is −10°C, and the temperature inside the room is 25°C. How many British thermal units are lost through the glass in an hour? [14.5]

50. A lake is covered with ice 15 cm thick. By how many millimeters does the ice thickness increase in 1 h due to thermal conduction during a winter night when the air temperature is −10°C, and the water temperature immediately below the ice is 0°C? Assume that the ice remains approximately 15 cm thick for the time period involved; the relative density of ice is 0.917. [14.5]

51. An uninsulated copper pipe 3 m long and 5 cm in radius has walls that are 2.5 mm thick. The pipe carries water at 75°C through a cooling bath that maintains the outside of the pipe wall at 70°C. Calculate the rate of heat transfer from the water in the pipe to the cooling bath. [14.5]

52. The ceiling of a room 4 m × 5 m is covered with plasterboard 18 mm thick. The temperature is 25°C in the room and −5°C in the attic.

- **a.** How many British thermal units of heat per hour must be supplied to the room to make up for the loss through the ceiling?
- **b.** What percentage of this heat is saved by insulating the ceiling with 15 cm of rock wool? (In working this part of the problem, assume that both faces of the plaster are at 25°C. In fact, the plaster does make a contribution to the thermal insulation with the rock wool in place, but it is negligible.)

53. A picnic ice chest has dimensions 75 cm × 40 cm × 30 cm, and the walls are insulated with 2 cm of Styrofoam. The chest is filled with 6 kg of ice and cans of soda, all at 0°C. It is placed in a shady spot where the outside air temperature is 30°C. How long a time elapses before the temperature of the soda begins to rise?

54. An unclothed person with a skin area of 1.6 m^2 and a skin temperature of 33°C is losing heat by convection at a rate of 120 kcal/h. Calculate the air temperature. (Assume the convective heat transfer coefficient is 7.1 $W/m^2 \cdot C°$.)

55. How much heat (in kcal/h) will an unclothed person with a skin area of 1.6 m^2 and a skin temperature of 32°C lose by convection if the air temperature is 10°C. (Assume a convective heat transfer coefficient of 7.1 $W/m^2 \cdot C°$.)

56. A blackbody radiator has its temperature raised from 15°C to 75°C. By what factor is the radiated power changed?

57. An iron sphere of radius 5 cm is heated to 1200°C (white hot).

- **a.** What is the rate of energy loss by radiation?
- **b.** At what temperature is the rate of radiant loss one-half of the above value?
- **c.** Is the radiant energy absorbed from room temperature surroundings a significant fraction of the radiant energy loss referred to in (a)? (Assume an emissivity of 65%.)

58. A tungsten filament in a light bulb, which has a temperature of 2450°C, radiates energy at the rate of 60 W. Calculate the surface area of the filament if its emissivity is 0.44.

59. A hot water room radiator has a total surface area of 8500 cm^2 and is coated with aluminum paint of emissivity 0.20. The water in the radiator is at 75°C and the surroundings in the room are at 18°C.

- **a.** How many British thermal units of energy are transferred to the room by radiation in 1 min? (Assume blackbody surroundings.)
- **b.** How many British thermal units of energy are transferred to the room by convection in 1 min? (Assume a convective heat transfer constant of 8 $W/m^2 \cdot C°$.)

60. An unclothed person of body area 2 m^2 and skin temperature 37°C is in a sauna where the temperature is 45°C. Calculate the net rate of radiant heat gain by the person. (Assume an emissivity of 0.95.)

61. Calculate the net loss of heat (in kcal/h) by radiation from an unclothed person of body area 2 m^2 whose skin temperature is 31°C in surroundings at a temperature of 17°C. (Assume an emissivity of 0.95.)

C. Advanced-Level Problems

62. A metal plate is made from a sheet of aluminum and a sheet of brass that are in contact with each other. The temperature on the aluminum side of the plate is 100°C, and on the brass side 0°C.

- **a.** Calculate the temperature at the interface if the plates are of equal thickness.
- **b.** If the temperature of the interface is 50°C, calculate the ratio of the thicknesses of the two plates.

63. An insulated vessel of negligible heat capacity contains 500 g of ice at $-10°C$, and a controlled amount of steam at 100°C and atmospheric pressure can be added.

a. What is the final result of adding 50 g of steam?
b. What is the final result of adding an additional 50 g of steam?

64. A black iron kettle of mass 2 kg contains 2 kg of water at a temperature of 100°C. The external surface area of the kettle is 500 cm^2, and the temperature of the surroundings is 25°C. Calculate the drop in temperature of the kettle in 1 min due to radiant energy transfer between it and blackbody surroundings.

65. The output of radiant energy from the surface of the sun is 6.14×10^7 W/m^2. Use values of the sun's radius, the earth's radius, and the radius of the earth's orbit to estimate the temperature of the earth. (Assume that the earth is a blackbody radiator, radiating all of the energy it receives from the sun back into space in all directions.)

15

HEAT ENGINES

The first heat engines to be used extensively were steam engines. The early designs consumed extravagant amounts of fuel because thermodynamic principles governing efficiency were not understood. However, the work of the French engineer Sadi Carnot (1796–1832) at the beginning of the nineteenth century established a sound basis for understanding heat engine efficiency. Unlike the concept of temperature and the application of the energy conservation principle to heat flow, little about the efficiency of heat engines can be understood intuitively.

Like many great scientific advances, the work of Carnot had ramifications well beyond its original area of investigation. Far from being limited to the efficiency of steam engines, the principles of thermodynamics apply in every situation involving heat transfer.

15.1 Thermodynamic Efficiency of Heat Engines

Before looking at the factors that determine heat engine efficiency, let us consider the operation of a steam engine so that we can focus attention on features common to all heat engines. The first step is to raise steam by heating water in a boiler. To do this, we can use the chemical energy stored in a fuel, such as coal or oil. We next transfer the steam from the boiler to a mechanical device (such as a turbine or a piston-cylinder-crank arrangement) that performs mechanical work at the expense of the internal energy of the steam. After being removed from the machine, the steam is condensed by a cooling agent (such as the water in a nearby river) and pumped back into the boiler, and the whole process is repeated. We illustrate such a system in Fig. 15.1(a).

The essential parts of any **heat engine** are:

1. A hot reservoir and a cold reservoir. In the case of the steam engine, the hot reservoir is the boiler and the cold reservoir is the cooling water. The temperatures of the reservoirs are kept constant as the engine operates. We call the temperatures of the hot and cold reservoirs T_H and T_C, respectively.

2. A working substance that carries heat energy from the hot reservoir to the cold reservoir. In our example, steam is the working substance.

3. A machine that utilizes the changes in the working substance to do external work. The turbine or piston-cylinder-crank arrangement fills this role.

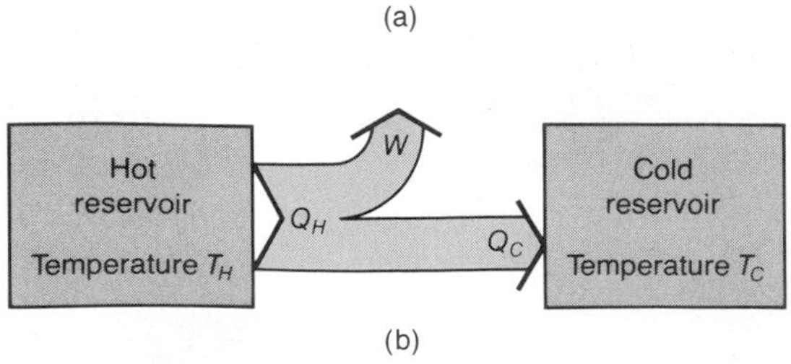

FIGURE 15.1 The operation of a steam engine: (a) schematic diagram; (b) essential thermodynamic features.

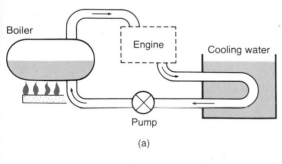

The features of the machine are a matter of mechanical design and do not affect the thermodynamic arguments. The essential thermodynamic point is the heat energy transfer from the hot to the cold reservoir. The working substance performs external work as it transfers heat energy. We show this concept schematically in Fig. 15.1(b). A quantity of heat energy Q_H flows from the hot reservoir, the machine performs an amount of work W, and heat energy Q_C flows into the cold reservoir. If we neglect losses, the conservation of energy relates these quantities as follows

$$Q_H = W + Q_C \tag{15.1}$$

This is essentially the first law of thermodynamics applied to the heat engine.

The heat energy Q_H comes from the combustion of the fuel, the work W is the useful output of the engine, and the heat energy Q_C is lost to the cold reservoir. For a given value of Q_H, we would like to make W as large as possible. In fact, nothing in the first law prevents us from putting $Q_H = W$ and $Q_C = 0$. In such a case all the heat energy from the fuel would go into useful work; but we can never achieve this desirable situation. As shown below, the second law of thermodynamics sets a strict limit on the fraction of Q_H that can be converted into useful work W. We define this fraction as the **thermodynamic efficiency** of the heat engine.

Definition of Thermodynamic Efficiency

$$\text{Thermodynamic efficiency} = \frac{\text{useful work done}}{\text{heat input from hot body}}$$

$$\eta_{\text{thermo}} = \frac{W}{Q_H} = \frac{Q_H - Q_C}{Q_H} \tag{15.2}$$

The higher the efficiency, the more effective is the engine in converting heat input into work.

EXAMPLE 15.1

A steam turbine producing 10 kW of mechanical power is operating at 32% thermodynamic efficiency. Calculate the heat input per hour to the engine.

SOLUTION A power of 10 kW is 10^4 J/s, so Eq. (15.2) allows us to calculate the heat input per second to the engine:

$$\eta_{\text{thermo}} = \frac{W}{Q}$$

$$Q = \frac{W}{\eta_{\text{thermo}}}$$

$$= \frac{10^4 \text{ J/s}}{0.32}$$

$$= 31\ 250 \text{ J/s}$$

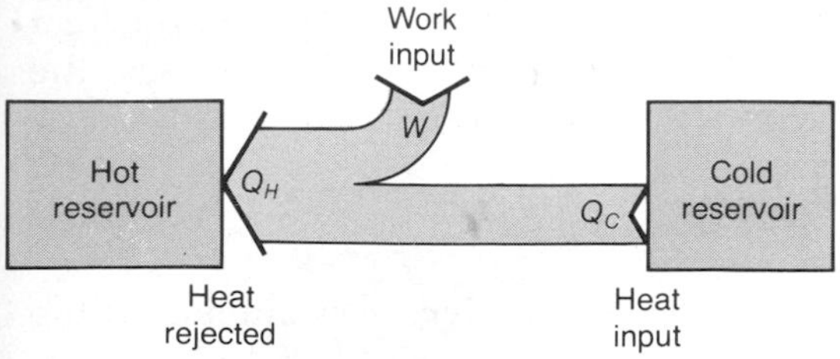

FIGURE 15.2 Essential thermodynamic features of a refrigerator.

Then the heat input per hour is given by 31 250 J/s × 3600 s/h = 1.12×10^8 J/h. ■

Let us now turn to an engine that is the very opposite of what we have just been considering. A refrigerator is a device that takes in mechanical work and transfers heat energy from a cold reservoir to a hot reservoir. In fact, a refrigerator is just a heat engine running in reverse. We have simply to reverse the directions indicated on Fig. 15.1(b) to produce a schematic diagram for a refrigerator. This is illustrated in Fig. 15.2; an amount of work W is input to the refrigerator, which transfers heat Q_C from the cold reservoir and rejects heat Q_H to the hot reservoir. Neglecting losses, the first law of thermodynamics gives:

$$W = Q_H - Q_C \tag{15.3}$$

For heat engines, the efficiency is a figure of merit. For refrigerators, we use the coefficient of performance, which is defined to be the heat extracted from the cold reservoir divided by the work input.

As long as we use heat engines, there is no escape from so-called "thermal pollution"— the only choice is where we put it. Many installations use the water in a lake or a river as the cold reservoir and reject heat to it. This cooling tower reminds us of another option, namely using the atmosphere as the cold reservoir. (Courtesy of U.S. Department of Energy)

■ Definition of Coefficient of Performance for a Refrigerator

$$\text{Coefficient of performance} = \frac{\text{heat extracted from cold reservoir}}{\text{work input}}$$

$$\text{COP} = \frac{Q_C}{W} \tag{15.4}$$

A strange unit has come into common use for refrigerators and air conditioners. A refrigerator has a 1-ton capacity if it can extract the same amount of heat energy from the cold reservoir that would be required to change 1 ton of water at 0°C into 1 ton of ice at 0°C in a 24-h period. Since this is a quantity of heat energy divided by a time interval, the ton of refrigeration is essentially a unit of power that is equivalent to 3516 W.

EXAMPLE 15.2

A certain 5-ton refrigeration unit operating continuously requires a 7.5-hp motor. Calculate the coefficient of performance.

SOLUTION We begin by putting the data into SI units:

$$\text{Rate of heat energy extraction from cold reservoir} = 5 \text{ tons} \times 3516 \text{ W/ton}$$

$$= 1.758 \times 10^4 \text{ W}$$

$$\text{Rate of work input} = 7.5 \text{ hp} \times 746 \text{ W/hp}$$
$$= 5.595 \times 10^3 \text{ W}$$

The use of Eq. (15.4) for a time period of 1 s gives

$$\text{COP} = \frac{Q_C}{W}$$
$$= \frac{1.758 \times 10^4 \text{ J}}{5.595 \times 10^3 \text{ J}}$$
$$= 3.14$$

■

At first sight refrigerators seem to possess a remarkable feature—the heat extracted from the cold reservoir exceeds the work input! The first law of thermodynamics may appear to have been violated, but the diagram of Fig. 15.2 shows that this is not so. The COP simply reflects the ratio of two energy components that, taken together, make up the heat energy rejected to the hot reservoir.

15.2 The Second Law of Thermodynamics

The second law of thermodynamics sets a limit on the performance of a heat engine that is not implied by the first law. As we have already said, there is nothing in the first law to prohibit a heat engine from extracting a certain amount of heat from a hot reservoir and converting all of it into work with no rejection of heat to the cold reservoir. The impossibility of having a heat engine perform in this manner forms the basis for one of the various equivalent statements of the second law of thermodynamics.

Second Law of Thermodynamics (First Form)

> It is impossible for a heat engine working in a cyclic process to absorb heat from a hot reservoir and convert it completely into mechanical work.

Referring to Eq. (15.2), the second law claims that a 100% efficient heat engine is impossible. Note that the law asserts that nature operates down a type of one-way street. If we set up an engine that produces a certain amount of mechanical work, then we can use all of this work to produce heat. This could be achieved, for example, by having the engine drive a disk whose motion was impeded by brake pads and the work output of the engine would appear as heat in the disk and the pads. While there is thus no law to prevent the total conversion of work into heat, the second law prohibits total conversion in the opposite direction. If we try to convert the heat generated by the disk and the brake pads back into mechanical work, then it is possible to recover only some of it.

We have seen that a refrigerator is a heat engine running in reverse, and it is therefore not surprising that we can give an alternate statement of the second law based on refrigerators. Once again, we begin by considering a situation that does not violate the first law—namely, a device that takes a certain amount of heat from a cold reservoir and transfers all of it to a hot reservoir without the input of any work. Our statement of the second law to prohibit this is as follows:

■ Second Law of Thermodynamics (Second Form)

It is impossible for any device to produce as its sole result the flow of heat from a cold to a hot reservoir.

Referring to Eq. (15.4), we see that the second law in this form claims that a refrigerator with an infinite coefficient of performance is impossible.

There are equivalent and more homely statements of this version of the second law. For example, we could say that the spontaneous direction of heat flow is from the hotter to the colder, and not the reverse. This statement is probably closest to the experience of everyday life, but it is not clear at first glance how it has anything to do with the efficiency of heat engines. However, it is easy to show that if one had a refrigerator of infinite coefficient of performance (that is, spontaneous heat flow from the cold reservoir to the hot one) and a heat engine whose efficiency was less than 100%, then the two operating together would constitute a 100% efficient heat engine. We illustrate this in Fig. 15.3; the heat engine takes heat Q_H from the hot reservoir, produces an amount of mechanical work W, and rejects heat Q_C to the cold reservoir. Alongside it, the refrigerator with the infinite coefficient of performance takes heat Q_C from the cold reservoir and transfers it all to the hot reservoir without any input of work. The net result of the two devices working together is the absorption of heat $(Q_H - Q_C)$ from the hot reservoir and its complete conversion into mechanical work W. The composite device is therefore a 100% efficient heat engine. Using similar arguments we could show that a 100% efficient heat engine operating alongside a refrigerator of finite coefficient of performance is equivalent to a refrigerator of infinite coefficient of performance. This means that any device that violates one form of the second law could be used to construct a device that violates the other form of that law. We

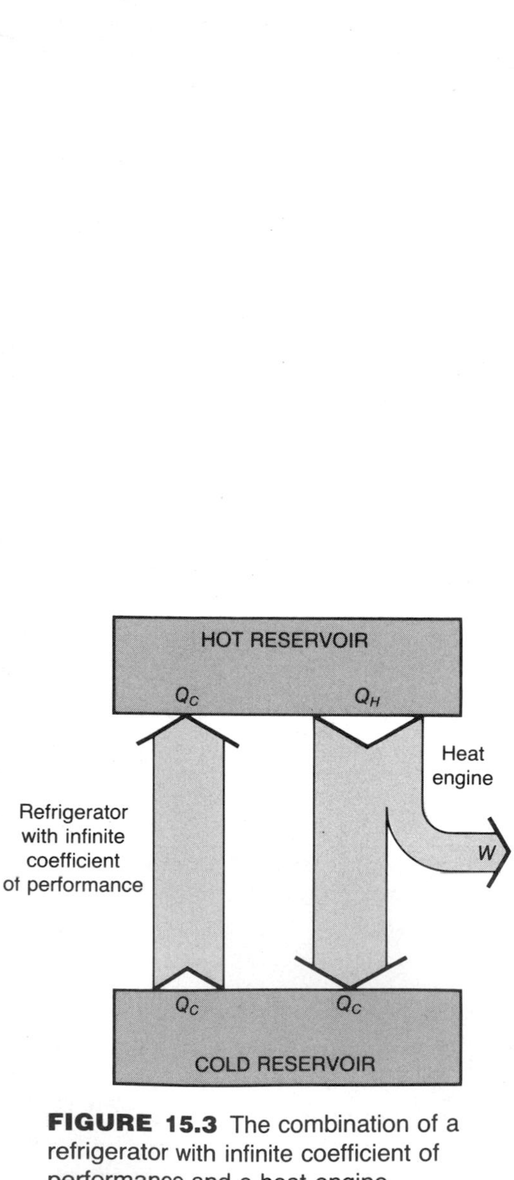

FIGURE 15.3 The combination of a refrigerator with infinite coefficient of performance and a heat engine amounts to a 100% efficient heat engine.

thus conclude that the two forms of the second law that we have so far considered are equivalent.

Before discussing details of heat engine efficiency, we must consider some of the important processes that a working substance may undergo.

15.3 Processes in Gaseous Working Substances

There is no stipulation about the nature of an engine's working substance—it may be a gas, a liquid, or even a solid. In a steam engine, it is a gas that is far from being ideal. In a gasoline or a diesel engine, the working substance is essentially air behaving much like an ideal gas. For practical reasons, the working substance in most types of heat engines is a gas. It is therefore appropriate to classify the important types of processes that a gas undergoes as it extracts heat from the hot reservoir, does work in a machine, and rejects heat to the cold reservoir.

Three basic processes correspond to constancy of the three state variables of a gas—pressure, volume, and temperature:

1. The **isobaric** process maintains constant pressure while volume and temperature change.
2. The **isochoric** process maintains constant volume while pressure and temperature change.
3. The **isothermal** process maintains constant temperature while volume and pressure change.

In addition to these processes defined from the constancy of a state variable, a fourth important process involves the absence of heat energy flow:

4. An **adiabatic** process has no heat energy flow either into or out of the substance. All three state variables change in an adiabatic process.

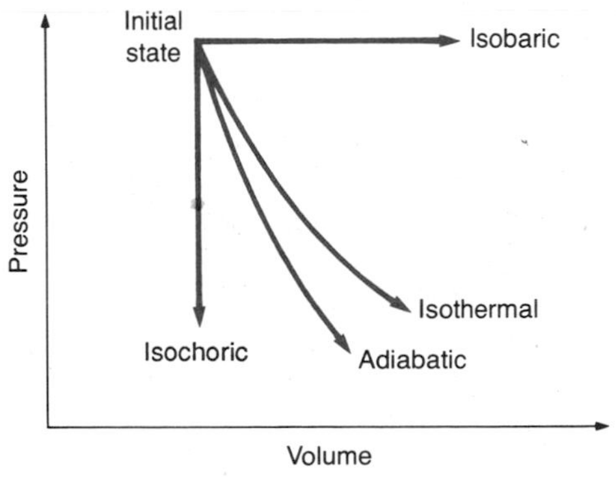

FIGURE 15.4 Important basic processes in a gaseous working substance displayed on a pressure-versus-volume diagram.

Figure 15.4 illustrates the four processes on a pressure-versus-volume diagram. Starting from some initial state, the isobaric and isochoric processes are represented by lines parallel to the volume axis and the pressure axis, respectively. The curve representing an isothermal expansion slopes downward to the right, since the pressure drops with increasing volume if the temperature remains constant. The curve of adiabatic expansion drops more sharply still, since there is no heat input and the temperature falls as the volume increases. As mentioned above, steam is not an ideal gas under the normal operating conditions of a steam engine, and so no simple algebraic description of the basic processes is possible. However, gasoline and diesel engines use air as a working substance under conditions in which the air approximates an ideal gas. With the help of the ideal gas law [Eq. (13.4a)]:

$$pV = nRT$$

we can write equations describing the changes in state variables for a given quantity of ideal gas.

Equations of Ideal Gas Processes

Isobaric process

$$p = \text{constant} \qquad \therefore \frac{T}{V} = \text{constant} \qquad \textbf{(a)}$$

Isochoric process

$$V = \text{constant} \qquad \therefore \frac{T}{p} = \text{constant} \qquad \textbf{(b)}$$

Isothermal process **(15.5)**

$$T = \text{constant} \qquad \therefore pV = \text{constant} \qquad \textbf{(c)}$$

Adiabatic process

$$pV^{\gamma} = \text{constant} \qquad \textbf{(d)}$$

γ is the ratio of the principal specific heats.

Equations for the first three processes follow directly from the ideal gas law. The derivation of Eq. (15.5d) is beyond the scope of this book.

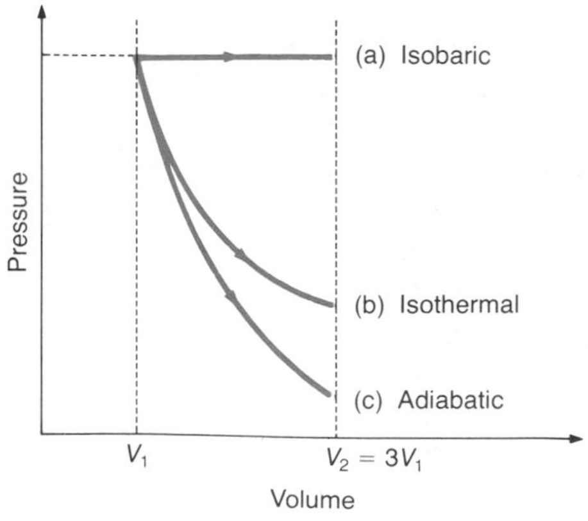

EXAMPLE 15.3

A quantity of oxygen at a pressure of 8 atm and a temperature of 350 K expands to 3 times the original volume. Calculate the final pressure and temperature for:

a. isobaric expansion

b. isothermal expansion

c. adiabatic expansion

SOLUTION Cancellation of units in the equations permits us to use any unit for volume and any absolute unit for pressure. The initial conditions are as follows:

$$p_1 = 8 \text{ atm}$$
$$T_1 = 350 \text{ K}$$

In every case the relation between the final and initial volumes is:

$$V_2 = 3V_1$$

We now consider the three processes separately.

a. For an isobaric process, Eq. (15.5a) gives:

$$\frac{T_1}{V_1} = \frac{T_2}{V_2}$$
$$\frac{350 \text{ K}}{V_1} = \frac{T_2}{3V_1}$$
$$\therefore T_2 = 1050 \text{ K}$$

b. For an isothermal process, Eq. (15.5c) gives:

$$p_1V_1 = p_2V_2$$
$$8 \text{ atm} \times V_1 = p_2 \times 3V_1$$
$$\therefore p_2 = 2.67 \text{ atm}$$

c. For an adiabatic process, Eq. (15.5d) and Table 14.3 (for the value of γ) give:

$$p_1V_1^\gamma = p_2V_2^\gamma$$
$$8 \text{ atm} \times V_1^{1.396} = p_2 \times (3V_1)^{1.396}$$
$$\therefore p_2 = 1.73 \text{ atm}$$

With the final pressure known, we can use Eq. (13.5) to calculate the final temperature:

$$\frac{p_1V_1}{T_1} = \frac{p_2V_2}{T_2}$$
$$\frac{8 \text{ atm} \times V_1}{350 \text{ K}} = \frac{1.73 \text{ atm} \times 3V_1}{T_2}$$
$$\therefore T_2 = 227 \text{ K}$$

Note that we do not need to know the quantity of oxygen present to work the problem. ■

Before we calculate the efficiency of an ideal gas cycle, we will derive a relation for the work done by a gas that expands at con-

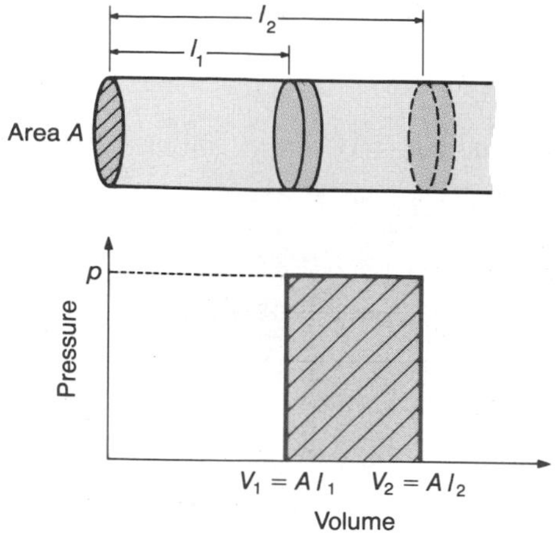

FIGURE 15.5 The work done by an expanding gas is equal to the area under the expansion curve on the indicator diagram.

stant pressure. Consider the situation illustrated in Fig. 15.5. An ideal gas expands at constant pressure p in a cylinder of cross-sectional area A. In so doing, it forces the piston from an initial volume $V_1 = Al_1$ to a final volume $V_2 = Al_2$, as illustrated on the accompanying pressure-versus-volume diagram. As the piston moves out, it does work. Since the gas is at constant pressure, the force on the piston is a constant equal to the gas pressure multiplied by the piston area:

$$F = pA$$

The work done is the force multiplied by the piston displacement:

$$\begin{aligned} W &= F(l_2 - l_1) \\ &= pA(l_2 - l_1) \\ &= p(V_2 - V_1) \end{aligned}$$

Thus, the work done *by* the gas is equal to the area under the line on the pressure-versus-volume diagram representing the expansion process. If the gas contracts at constant pressure, the work done *on* it is equal to the area under the line representing the contraction process. The diagram, called an indicator diagram, offers a very convenient way of representing a heat engine cycle. We can use the result just obtained together with the ideal gas laws to calculate the thermodynamic efficiency of an engine that operates in some specified manner. Let us illustrate the procedure with an example.

EXAMPLE 15.4

A working substance of 1 mol of nitrogen operates an engine that follows the cycle illustrated in the diagram. Calculate:

a. the work done per cycle

b. the heat input per cycle

c. the thermodynamic efficiency of the engine

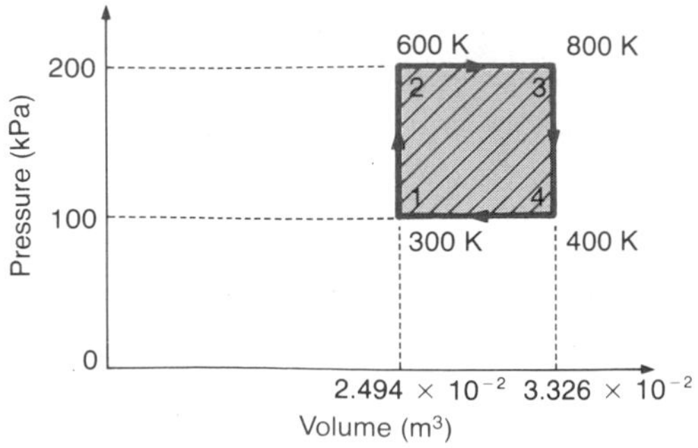

SOLUTION The cycle consists of two isobaric processes ($2 \rightarrow 3$ and $4 \rightarrow 1$) and two isochoric processes ($1 \rightarrow 2$ and $3 \rightarrow 4$). The temperatures (calculated from the ideal gas equation of state for 1 mol) are marked at the four corners of the indicator diagram.

a. The work done *by* the gas during the expansion process is the area under the line $2 \rightarrow 3$:

$$\begin{aligned} W_{2\rightarrow 3} &= 2 \times 10^5 \text{ Pa} \times (3.326 \times 10^{-2} \text{ m}^3 - 2.494 \times 10^{-2} \text{ m}^3) \\ &= 1664 \text{ J} \end{aligned}$$

Similarly, the work done *on* the gas during the contraction is the area under the $4 \rightarrow 1$ line:

$$W_{4\rightarrow 1} = 10^5 \text{ Pa} \times (3.326 \times 10^{-2} \text{ m}^3 - 2.494 \times 10^{-2} \text{ m}^3)$$
$$= 832 \text{ J}$$

It follows that the net work done *by* the gas per cycle is the difference of these two figures:

$$W = W_{2\rightarrow 3} - W_{4\rightarrow 1}$$
$$= 1664 \text{ J} - 832 \text{ J}$$
$$= 832 \text{ J}$$

b. Heat input occurs while the gas is being heated, which occurs during the processes $1 \rightarrow 2$ and $2 \rightarrow 3$. We use the specific heats at constant volume and constant pressure from Table 14.3 in Eq. (14.4):

$$Q_H = 1 \text{ mol} \times 20.4 \text{ J/mol} \cdot \text{C}° \times 300 \text{ C}° \text{ (for the process } 1 \rightarrow 2) + 1 \text{ mol} \times 28.9 \text{ J/mol} \cdot \text{C}° \times 200 \text{ C}° \text{ (for the process } 2 \rightarrow 3)$$
$$= 11\,900 \text{ J}$$

c. To calculate the thermodynamic efficiency we use Eq. (15.2):

$$\eta_{\text{thermo}} = \frac{W}{Q_H}$$
$$= \frac{832}{11\,900 \text{ J}}$$
$$= 7.0\%$$

■

This example brings up several interesting points. The work done per cycle [part (a)] is equal to the shaded area enclosed by the lines that represent the cycle on the indicator diagram. Our proof that this is correct applies only to the example considered, which involves only isobaric and isochoric processes. However, the result is of general validity.

> The net work done during a reversible closed cycle in a clockwise direction is equal to the area enclosed by the process curve on the indicator diagram.

A second major point of interest is the very low value of the thermodynamic efficiency of the cycle. What is wrong with the cycle of Example 15.4 that makes it so inefficient? To answer this question we must look at some of the implications of reversibility and yet another alternate statement of the second law of thermodynamics.

15.4 The Carnot Cycle

We have already seen that no heat engine can be 100% efficient, but we have not yet answered the question as to what the greatest possible efficiency is, or what the physical parameters are that determine this maximum efficiency. The French engineer Sadi Carnot (1796–1832) answered these questions in the early nineteenth century. Carnot was minister of war under Napoleon, but he devoted much of his time to science after the battle of Waterloo. He was only 27 years old when he published his famous work, "The Motive Power of Heat." His early death from cholera a few years later delayed the publication of much of his subsequent work.

Carnot studied the ideal model of a perfectly reversible engine, which we call the **Carnot engine.** All processes involving the working substance must be reversible, and all mechanical operations of the engine must also be reversible. Carnot showed that no engine working between fixed-temperature hot and cold reservoirs can be more efficient than a reversible engine. To see that this is so, we begin by noting that a Carnot engine is quite indifferent to being an engine or a refrigerator; since the cycle is reversible, the engine can be run backward as a refrigerator. Figure 15.6 illustrates schematically the heat flow and work output/input for both cases. Consider such a Carnot engine running as a refrigerator with a work input W, heat absorption from the cold reservoir Q_C, and heat rejection to the hot reservoir Q_H. Now suppose that we have an engine that is more efficient than the Carnot engine. For a work output W, it will require heat absorption Q_H' from the hot reservoir, and $Q_H' < Q_H$ since the engine is more efficient than the Carnot engine. The more efficient engine would reject heat Q_C' to the cold reservoir, and the first law shows us that $Q_C' < Q_C$. We now set the more efficient engine to drive the Carnot refrigerator, as shown in Fig. 15.7. The net result is a device that takes a positive amount of heat $(Q_C - Q_C')$ from the cold reservoir and delivers the positive amount of heat $(Q_H - Q_H')$ to the hot reservoir. This is in contradiction to the second form of the statement of the second law. It follows that no engine working between given temperature limits can be more efficient than a perfectly reversible engine.

Since heat flow from a higher to a lower temperature reservoir is irreversible (by the second law), the working substance of a Carnot engine must at all times be at the same temperature as a reservoir with which it exchanges heat. This means that any pro-

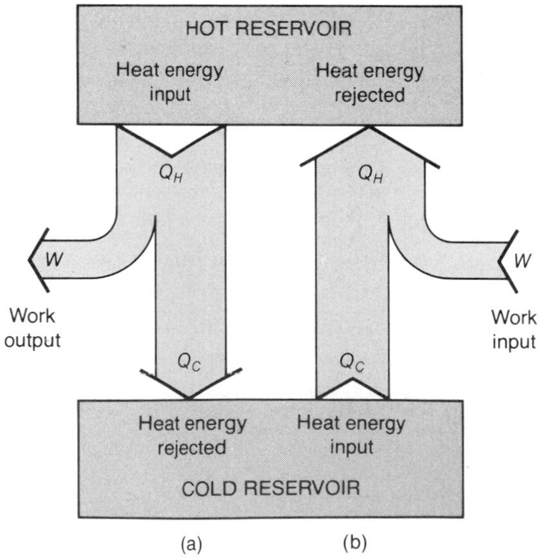

FIGURE 15.6 Illustration of a perfectly reversible heat engine: (a) working as an engine; (b) working as a refrigerator. In both cases the first law energy balance $Q_H = W + Q_C$ is the same.

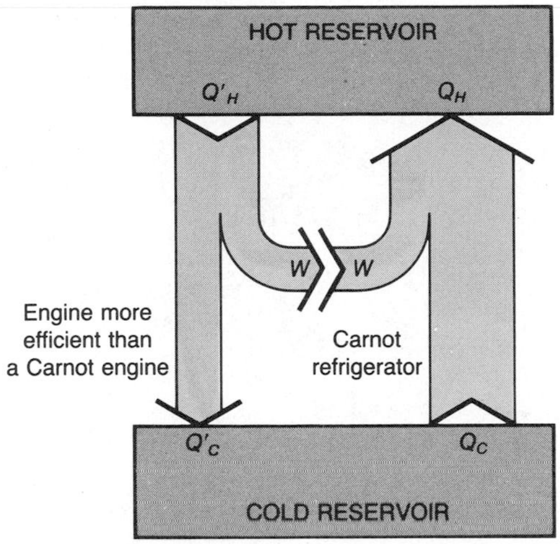

FIGURE 15.7 An engine more efficient than a reversible Carnot engine that is arranged to drive a reversible Carnot engine as a refrigerator would violate the second law.

cess involving heat flow must be isothermal. In processes that involve temperature change, there can be no heat flow if reversibility is to be maintained; such processes must therefore be adiabatic. The sequence of processes in the working substance of a Carnot engine is called the Carnot cycle, and we illustrate this cycle for an ideal gas in Fig. 15.8. At point 1 the working substance is at the temperature of the cold reservoir, and it undergoes adiabatic compression until its temperature reaches the hot reservoir temperature at point 2. Since the process is adiabatic, no heat flows to or from the working substance, and we can make this step reversible without violating the second law. Between points 2 and 3 the working substance takes in heat from the hot reservoir in an isothermal expansion. Since the working substance is at the same temperature as the hot reservoir, this step is also reversible. This is followed by reversible adiabatic expansion to point 4 at the cold reservoir temperature, and then reversible isothermal compression to point 1 to complete the cycle. Carnot used the ideal gas law to calculate the efficiency of this cycle using an ideal gas as working substance. His statement of the second law is as follows:

■ Second Law of Thermodynamics (Third Form)

The most efficient possible heat engine operating between fixed-temperature hot and cold reservoirs is a perfectly reversible engine. All such engines have the efficiency:

$$\eta_{\text{Carnot}} = \frac{T_H - T_C}{T_H} \qquad (15.6)$$

Temperatures are expressed in kelvins.

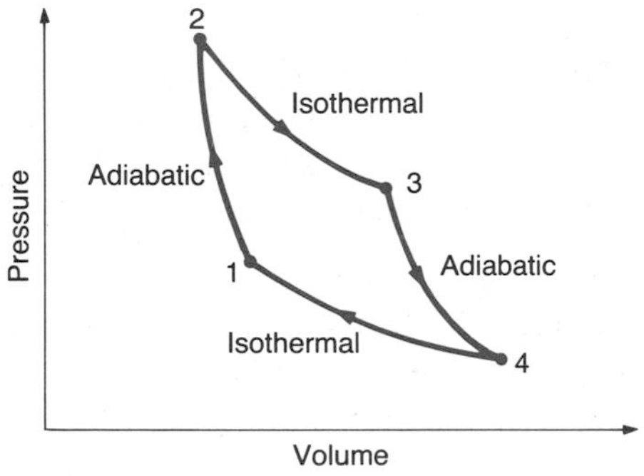

FIGURE 15.8 Illustration of a Carnot cycle on a pressure-volume diagram.

The maximum efficiency, which is called the **Carnot efficiency,** is a remarkably simple function of the temperatures of the hot and cold reservoirs. Moreover, the efficiency does not depend on the working substance; it applies to any Carnot engine operating between given temperature extremes, regardless of the nature of the working substance.

EXAMPLE 15.5

A small steam engine operates with a boiler temperature of 200°C and a condenser water temperature of 45°C. Calculate the Carnot efficiency for an engine operating between these temperature extremes.

SOLUTION We need only express the temperature extremes in kelvins and calculate the Carnot efficiency. Using Eq. (13.2) to convert to kelvins we have:

$$T(\mathrm{K}) = T(^\circ\mathrm{C}) + 273.15$$

$$T_H = 200 + 273.15 = 473.15\ \mathrm{K}$$

$$T_C = 45 + 273.15 = 318.15\ \mathrm{K}$$

The Carnot efficiency follows from Eq. (15.6):

$$\eta_{\text{Carnot}} = \frac{T_H - T_C}{T_H}$$

$$= \frac{473.15\ \mathrm{K} - 318.15\ \mathrm{K}}{473.15\ \mathrm{K}}$$

$$= 0.328$$

The result is interesting. No amount of improvement in design can ever raise the efficiency of this engine above 32.8% if the boiler temperature and the condenser temperature remain the same. ■

A Carnot engine is not in any sense practical, since the isothermal heat transfers between the reservoirs and the working substance would take an inordinate amount of time. In order for the heat energy to flow at a reasonable rate, the temperature of the working substance and the temperature of the reservoir with which it exchanges heat energy must differ appreciably. All practical heat engines suffer from irreversibility for this reason. In the case of the steam engine, the fire is hotter than the water in the boiler, and the cooling water is colder than the exhaust steam in the condenser. However, even though the Carnot engine is no more than an idealized concept, it is important because it indicates the upper limit of the efficiency of an engine working between given temperature extremes.

Like the Carnot heat engine, the Carnot refrigerator is an ideal that no actual device can ever match in performance. The coefficient of performance of the Carnot refrigerator is also a function only of the temperatures of the hot and cold reservoirs.

■ Coefficient of Performance of a Carnot Refrigerator

$$\mathrm{COP}_{\text{Carnot}} = \frac{T_C}{T_H - T_C} \qquad \textbf{(15.7)}$$

EXAMPLE 15.6

Calculate the COP for a Carnot refrigerator operating between 60°C and 5°C.

SOLUTION We use Eq. (13.2) to convert the temperatures to the Kelvin scale:

$$T(\text{K}) = T(^\circ\text{C}) + 273.15$$
$$T_H = 60 + 273.15$$
$$= 333.1 \text{ K}$$
$$T_C = 5 + 273.15$$
$$= 278.1 \text{ K}$$

The Carnot COP is given by Eq. (15.7):

$$\text{COP}_{\text{Carnot}} = \frac{T_C}{T_H - T_C}$$
$$= \frac{278.1 \text{ K}}{333.1 \text{ K} - 278.1 \text{ K}}$$
$$= 5.05$$

Any real refrigerator operating between the same temperature extremes must have a smaller COP. ■

15.5 The Absolute Zero of Temperature and the Absolute Kelvin Thermodynamic Temperature Scale

In calculating the efficiency of a Carnot engine, we have used the zero on the ideal gas temperature scale as the zero of temperature. But we can now assign a deeper meaning to the zero of temperature. Consider a Carnot engine operating between hot and cold reservoirs at temperatures T_H and T_C, respectively. The efficiency is given by Eq. (15.6):

$$\eta_{\text{Carnot}} = \frac{T_H - T_C}{T_H}$$

If $T_C = 0$, then the Carnot engine is 100% efficient for any value of T_H. Under such circumstances all of the heat energy taken from the hot reservoir is converted into mechanical work, and no heat energy is rejected to the cold reservoir. A cold reservoir cannot have a lower temperature, for the efficiency would then be greater than 100% and thus violate the first law of thermodynamics. Hence, the zero of the ideal gas scale is an absolute thermodynamic zero. This definition of absolute zero does not depend on the properties of any particular substance, since all Carnot engines have the same efficiency between given temperature limits regardless of their working substances.

> The absolute zero of temperature is the temperature of a reservoir to which a Carnot engine rejects no heat energy.

Even though we assign a finite number to the absolute zero of temperature (0 K, or -273.15°C), the temperature is a physical infinity in the sense that it can never be attained. The lowest temperature that has been obtained for liquid or solid substances is in the vicinity of 10^{-3} K. Temperatures of about 10^{-6} K have been attained for systems of nuclear spins, but this temperature is not communicable to a macroscopic substance.

In addition to proposing the satisfactory definition of absolute zero, Lord Kelvin introduced an absolute thermodynamic temperature scale. Consider a hot reservoir at some temperature above absolute zero that supplies heat energy to the first of a series of Carnot engines. The heat rejected by the first engine forms the supply for the second one, and so on down the chain. The engines are arranged so that they all do equal amounts of work, and the cold reservoir for the final engine is at absolute zero. Finally, the operating temperature range of all the Carnot engines is the same. All that we require to define the absolute thermodynamic temperature scale are the upper fixed point and the number of Carnot engines in the chain. The number of Carnot engines in the chain is a figurative way of assigning the size of the degree. We accomplish all this by setting the triple point of water at 273.16 K. Since the absolute thermodynamic scale is identical with the ideal gas scale of a previous chapter, for practical purposes we make no distinction between them.

15.6 Entropy

The history of the concept of entropy begins with the German physicist Rudolf Clasius (1822–1888). He discovered that the quantity whose value is equal to the ratio of the flow of heat energy to the absolute temperature was of great importance in thermodynamics; he named this quantity the **entropy.**

■ Definition of Entropy Change (for Isothermal Process)

$$\text{Change in entropy} = \frac{\text{heat flow}}{\text{absolute temperature}}$$

$$S - S_0 = \frac{Q}{T} \qquad \textbf{(15.8)}$$

SI unit of entropy: J/K

Note that our definition is restricted to an isothermal heat flow. We require calculus whenever the heat flow causes a temperature change, but, as we shall see below, excellent approximations can often be obtained with a simple averaging procedure.

EXAMPLE 15.7

Calculate the entropy change of 1 kg of water at 100°C that is changed into steam at the same temperature.

SOLUTION Since this is an isothermal process, we have a simple application of Eq. (15.8). To calculate the heat added to the water, we use Eq. (14.3) along with the latent heat of vaporization from Table 14.2:

$$\begin{aligned} Q &= mL \\ &= 1\ \text{kg} \times 22.6 \times 10^5\ \text{J/kg} \\ &= 22.6 \times 10^5\ \text{J} \end{aligned}$$

Then from Eq. (15.8) we have:

$$\begin{aligned} S - S_0 &= \frac{Q}{T} \\ &= \frac{22.6 \times 10^5\ \text{J}}{373\ \text{K}} \\ &= 6060\ \text{J/K} \end{aligned}$$

The water absorbs heat in this process, so we set Q positive, and the change to steam is accompanied by an increase in entropy. If we had considered a kilogram of steam condensing into a kilogram of water, then the entropy change would have been negative since heat is extracted in this case. ■

The example problem that we have just worked leads us to a much more important type of problem—namely, the overall change in entropy that occurs when two substances at different temperatures are mixed together. We face a difficulty here because Eq. (15.8) holds only for an isothermal process. However, we shall simply approximate the calculation by using the average of the initial and final temperatures for the substance whose entropy change is being calculated. This approximation is excellent whenever the temperature change is small compared with the absolute temperature at any point in the process.

EXAMPLE 15.8

Calculate the change in entropy that occurs when 1 kg of water at 20°C is mixed with 1 kg of water at 100°C in a perfectly insulated container to form 2 kg of water at 60°C. (By perfectly insulated we mean that the contents of the container are a completely isolated system.)

SOLUTION We must find the entropy change for each half of the system separately, and then add these results for the total change of entropy of the system. Let us begin with the hot water; from Eq. (14.2) and Table 14.1 we find:

$$\begin{aligned} Q_H &= mC(T - T_0) \\ &= 1\text{ kg} \times 4184\text{ J/kg} \cdot \text{C}^\circ \times (60^\circ\text{C} - 100^\circ\text{C}) \\ &= -1.674 \times 10^5\text{ J} \end{aligned}$$

As it cools from 100°C to 60°C, the average temperature of the hot water is 80°C = 353 K; the temperature difference of 40°C is certainly much smaller than the absolute temperature of the water so we use Eq. (15.8) with the average temperature:

$$\begin{aligned} (S - S_0)_H &= \frac{Q_H}{\langle T_H \rangle} \\ &= \frac{-1.674 \times 10^5\text{ J}}{353\text{ K}} \\ &= -474.2\text{ J/K} \end{aligned}$$

Likewise, we find for the cold water:

$$\begin{aligned} Q_C &= mC(T - T_0) \\ &= 1\text{ kg} \times 4184\text{ J/kg} \cdot \text{C}^\circ \times (60^\circ\text{C} - 20^\circ\text{C}) \\ &= 1.674 \times 10^5\text{ J} \end{aligned}$$

and

$$\begin{aligned} (S - S_0)_C &= \frac{Q_C}{\langle T_C \rangle} \\ &= \frac{1.674 \times 10^5\text{ J}}{313\text{ K}} \\ &= 534.8\text{ J/K} \end{aligned}$$

(Remember that the average temperature of the cold water is 40°C = 313 K.) Then the total entropy change for the isolated system is given by:

$$S - S_0 = (S - S_0)_H + (S - S_0)_C$$
$$= -474.2\ \text{J/K} + 534.8\ \text{J/K}$$
$$= 60.6\ \text{J/K}$$

■

Our example shows that there is a positive entropy change for the system even though it is isolated. This property of the entropy stands in marked contrast to the energy, which latter is conserved in an isolated system. So long as the two lots of water are kept separate from each other (for example, by a perfectly insulating wall), both the total energy and the total entropy do not change. When they are allowed to mix, the energy still does not change but there is an increase of entropy accompanying the irreversible mixing process. In fact, what we have seen is an example of yet another statement of the second law:

■ Second Law of Thermodynamics (Fourth Form)

The entropy of an isolated system can never decrease.

It is not surprising to find that the Carnot cycle occupies a unique position with regard to entropy change. From Eq. (15.2) the thermodynamic efficiency of any heat engine cycle is given by:

$$\eta_{\text{thermo}} = \frac{W}{Q_H} = \frac{Q_H - Q_C}{Q_H}$$

By Eq. (15.6) the Carnot cycle efficiency is given by:

$$\eta_{\text{Carnot}} = \frac{T_H - T_C}{T_H}$$

Setting the two efficiencies equal to each other, we readily find that for a Carnot cycle:

$$\frac{Q_H}{T_H} - \frac{Q_C}{T_C} = 0 \qquad \textbf{(15.9)}$$

The first term is the entropy increase of the working substance during the reversible isothermal expansion, and the second is the entropy decrease during the reversible isothermal compression (processes 2→3 and 4→1 on Fig. 15.8). During the reversible adiabatic compression and expansion (processes 1→2 and 3→4 on

Fig. 15.8) there is no entropy change since there is no heat flow in the adiabatic process. Equation (15.9) then tells us that the total entropy change of the working substance during a complete cycle of the Carnot engine is zero. Likewise, we can show that the total entropy change of the two reservoirs is zero; the hot reservoir loses and the cold one gains equal amounts of entropy. The total entropy change associated with a complete cycle of the Carnot engine is therefore zero, and this is really the same thing as saying that the Carnot engine is perfectly reversible. Any other engine cycle must be both irreversible and accompanied by an entropy increase.

Surely all of the statements of the second law of thermodynamics must mean that the law is difficult to understand or that its depth of meaning is very great; probably both of these are true. Moreover, we have not yet begun to exhaust our understanding of it. In addition to the thermodynamic concept of entropy that we have discussed up to now, there is also a statistical concept of entropy. In statistical thermodynamics the entropy is defined as a measure of the disorder of a system, and the second law is set down as a statement of the natural tendency of an isolated system to move toward the state of greater disorder.

■ Second Law of Thermodynamics (Fifth Form)

> The order of an isolated system can never increase.

Once again the insulated container with the hot and cold water offers a good example. Before mixing is permitted, a certain degree of order (hot water at one end and cold at the other) exists. After mixing occurs, there is simply uniformly warm water. It would not violate energy conservation for the water to rearrange itself spontaneously back to the original configuration, but it would violate the second law, which does not permit spontaneous increase in the order of an isolated system. Note that there is no claim of absolute impossibility here; we could insert a dividing wall into the middle of the warm water, connect a refrigerator to the system, and reseparate the water into hot and cold. But the introduction of the refrigerator has removed the property of isolation of the system. We can always decrease the entropy (or increase the order) in a system, but this will always be at the expense of neighboring systems.

15.7 Practical Heat Engines

The heat engines that have most affected our civilization are the steam engine and the internal combustion engine. The steam engine has a long history that began in the eighteenth century, when it was first used on a commercial scale to pump water from mines. Soon after it was used to power the machinery in the factories of the industrial revolution. In the early nineteenth century the first successful steam locomotive was built and operated by George Stephenson (1781–1848) in the north of England. About the same time Robert Fulton (1765–1815) ran the first successful steamship, on the Hudson River in New York. Nineteenth-century transportation is the story of the steam locomotive replacing the horse-drawn carriage and the steamship displacing the square-rigged sailing ship. These modes of transport have since largely given way to others, but the steam engine is still used as the heat engine in many large electricity-generating plants; in fact, the giant steam turbines that drive electrical generators are the largest heat engines ever built.

The first internal combustion engines, which appeared in the middle of the nineteenth century, used the same gas that was used in streetlights, and were widely applied for industrial purposes. In the latter part of the century, the invention of the carburetor permitted the internal combustion engine to run on liquid gasoline. This type of engine has been used to power the hundreds of millions of lawn mowers, small boats, and automobiles that have since been built. The large internal combustion engine used to power buses, trucks, and ships is usually a diesel engine, which is similar to the gasoline engine but runs on less refined fuel.

Most practical heat engines differ in important ways from the true heat engine of our previous discussion. We have considered a working substance that is placed in contact with heat reservoirs (from which it absorbs, or to which it rejects, energy) in a cyclic manner. The working substance returns to its initial state when the cycle is completed, and is therefore never expended. Large steam turbines usually function in this way, but the same is not true for all steam engines. Steam locomotives simply exhaust the working substance to atmosphere at the end of the expansion stroke; as a consequence, they are forced to continually replenish their water supply. The working substance is also exhausted to the atmosphere in the internal combustion engine. In addition, there is a second reason why this is not a true heat engine: In the internal combustion engine the working substance does not acquire heat energy from a hot reservoir, but from a highly exothermic chemical reaction—that is, from the ignition of the fuel vapor and the oxygen in the cylinder. The theory of heat engines is applied to such engines as an approximation.

As we have already mentioned, practical engines do not operate on a Carnot cycle. Engineers use specific cycles to approximate the behavior of the working substance in the various engine types. All of these cycles, operating between the same temperature extremes as the Carnot cycle, are less efficient than the Carnot cycle. In addition, other factors reduce the efficiency of practical engines, and we can place them into two general categories:

1. Mechanical losses are caused by friction in the bearings and other moving parts. Friction always acts in opposition to a motion. When a heat engine works as an engine, friction reduces the work output. When it works as a heat pump, friction increases the required work input. Although we can never eliminate friction, careful mechanical design and lubrication of the bearings reduce it to such extent that its effect is small.

2. Thermal losses are a much more important factor in reducing efficiency. One obvious source is heat that is lost from the system without ever being transferred to the working substance. The hot gases from the smokestack of a steam plant or the tail pipe of an automobile are examples of this type of loss.

A final figure that indicates the effectiveness of an engine relates the work done by the engine to the energy value of the fuel consumed. We can find the energy value of the fuel by measuring the total heating effect produced in some substance by complete combustion of the fuel. This measurement is also called the **energy of combustion,** and Table 15.1 lists some typical values for common fuels. The energy of combustion in all cases is of the order of tens of megajoules per kilogram of fuel. We can now define **overall efficiency** of a heat engine.

TABLE 15.1 Energies of combustion of various fuels

Fuel	*Energy of combustion (MJ/kg)*
Coal	27–33
Wood	8–15
Natural gas	54
Crude oil	45
Gasoline	48
Kerosene	46
Propane	50
Methanol	23
Ethanol	30

■ Definition of Overall Heat Engine Efficiency

$$\text{Overall efficiency} = \frac{\text{work done by heat engine}}{\text{energy of combustion of fuel consumed}}$$

$$\eta_{\text{overall}} = \frac{W}{E_{\text{combust}}} \qquad (15.10)$$

(The work done and the energy of combustion are measured over the same time interval.)

To get some feeling for the magnitudes involved, let us consider some actual examples.

EXAMPLE 15.9

A small steam engine burns 90 kg of coal per hour while delivering a constant output of 100 hp. The heat obtained from complete combustion of the coal is 28 MJ/kg. Calculate the overall efficiency of the engine.

SOLUTION We need to calculate the work done by the engine and the energy of combustion of the fuel consumed over some given time interval. Let us select an hour as a suitable period and set the data in SI units:

$$\text{Engine power} = 100 \text{ hp} \times 746 \text{ W/hp}$$
$$= 7.46 \times 10^4 \text{ W}$$
$$\text{Time} = 3600 \text{ s}$$

The work done in 1 h is found from Eq. (7.4):

$$\langle P \rangle = \frac{W}{t}$$
$$\therefore W = \langle P \rangle t$$
$$= 7.46 \times 10^4 \text{ W} \times 3600 \text{ s}$$
$$= 2.686 \times 10^8 \text{ J}$$

The energy of combustion of the fuel consumed in 1 h is:

$$\text{Energy of combustion} = 90 \text{ kg} \times 28 \times 10^6 \text{ J/kg}$$
$$= 2.52 \times 10^9 \text{ J}$$

We can now calculate the overall efficiency from Eq. (15.10):

$$\eta_{\text{overall}} = \frac{W}{E_{\text{combust}}}$$
$$= \frac{2.686 \times 10^8 \text{ J}}{2.52 \times 10^9 \text{ J}}$$
$$= 0.107$$

■

EXAMPLE 15.10

An automobile engine, which delivers an average of 25 hp while cruising at 90 km/h (25 m/s), travels 30 km per gallon of gasoline. Calculate the overall efficiency. (Take the density of gasoline as 2.55 kg/gal and the energy of combustion as 48 MJ/kg.)

SOLUTION At a speed of 90 km/h, the automobile travels 30 km and consumes 1 gal of gasoline in 20 min. It is therefore convenient to calculate the work output and the energy of combustion of the fuel consumed over a 20-min period:

$$\text{Engine power} = 25 \text{ hp} \times 746 \text{ W/hp}$$
$$= 1.865 \times 10^4 \text{ W}$$

We can calculate the work done by the engine in 20 min by using Eq. (7.4):

$$\langle P\rangle = \frac{W}{t}$$

$$\therefore W = \langle P\rangle t$$

$$= 1.865 \times 10^4 \text{ W} \times 20 \text{ min} \times 60 \text{ s/min}$$

$$= 2.238 \times 10^7 \text{ J}$$

The energy of combustion of 1 gal of gasoline is given by:

$$\text{Energy of combustion} = 48 \times 10^6 \text{ J/kg} \times 2.55 \text{ kg}$$

$$= 1.224 \times 10^8 \text{ J}$$

We can now calculate the overall efficiency by using Eq. (15.10):

$$\eta_{\text{overall}} = \frac{W}{E_{\text{combust}}}$$

$$= \frac{2.238 \times 10^7 \text{ J}}{1.224 \times 10^8 \text{ J}}$$

$$= 0.183$$ ■

Let us set down the main points of our discussion about heat engines:

1. The most basic characteristic of any heat engine is the range of temperature of its operation. Given the hottest and the coldest temperatures attained by the working substance, we can cal-

The drawing shows Newcomen's steam engine of the early eighteenth century. Its operation is explained on page 437 with reference to Fig. 15.9. The photograph below shows a modern-day successor to the Newcomen engine—a steam turbine that is about 100 times more efficient than its earlier counterpart. (Drawing: Culver Pictures, Inc.; photograph: courtesy of U.S. Department of Energy)

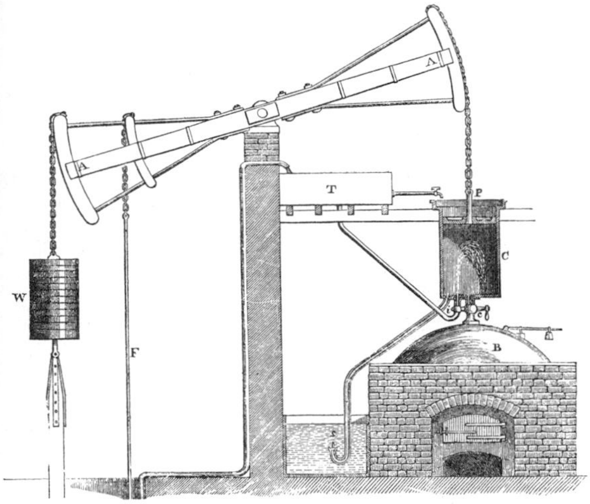

culate the Carnot efficiency, which is the maximum possible efficiency for the engine.

2. Because of the specific nature of the working substance and the design of the machine that does the external work, there may be substantial deviations from the Carnot cycle. For particular cases, ideal reversible cycles can be set up that approximate the processes occurring in the actual engine. The thermodynamic efficiencies for these cycles are lower than the Carnot cycle efficiency between the same temperature limits.

3. The actual engine falls short of attaining the thermodynamic efficiency of its ideal cycle because of irreversible processes such as frictional losses, heat losses, and heat flow between bodies at different temperatures. The overall efficiency is less than the thermodynamic efficiency of the actual engine cycle, which is in turn less than Carnot cycle efficiency.

If all of this seems to be gloomy, consider the workings of the first commercial steam engines in the early eighteenth century. These were beam-type engines that forced an up-and-down rocking motion in a beam connected to a water pump, as illustrated in Fig. 15.9. The operating procedure consisted of the following steps:

1. With valve B closed, the operator opened valve A, admitting steam from the boiler to the uninsulated cylinder (the hot cylinder walls lost much energy to the surroundings). This forced the piston to the top of the cylinder and raised the left-hand end of the beam.

2. The operator then closed valve A and squirted a jet of cold water into the hot cylinder. (It is hard to imagine a more irreversible process.) The steam condensed to water and the air pressure forced the piston down, thereby lowering the left-hand end of the beam.

3. The operator opened valve B to let the water drain out, and then the process started over again.

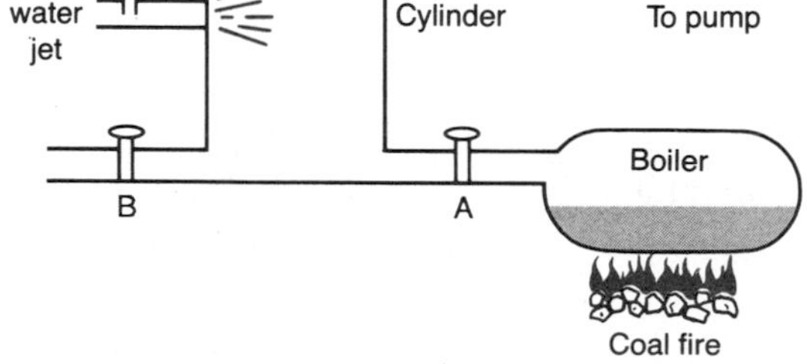

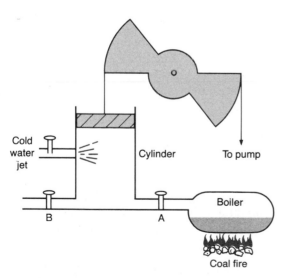

FIGURE 15.9 Diagram of an early beam-type steam engine used to pump water from coal mines.

The boiler for these engines produced steam at a pressure slightly above atmospheric and at a temperature of about 100°C. If the cold reservoir temperature equals that of the cooling water, the Carnot efficiency is approximately 15%. The overall efficiency was probably about 2%, and these engines burned an enormous amount of coal.

The figures quoted for the 100-hp steam engine in Example 15.9 yields a Carnot efficiency of about 30% and an overall efficiency of about 10%. These figures are typical of a small coal-fired reciprocating steam engine (with 200°C boiler temperature) of the early twentieth century. A modern oil-fired steam turbine generating station uses steam at about 350°C with a corresponding

Carnot efficiency of about 50%. A typical overall efficiency is about 35%.

The evolution of the steam engine is an excellent illustration of the historical quest for improved efficiency. The first step was to increase the corresponding Carnot efficiency. Since the temperature of the cold reservoir is essentially determined by the temperature of available cooling water, the Carnot efficiency could be increased only by raising the temperature of the hot reservoir. The efficiency improvements in the engine went hand in hand with increased steam temperature. As a result, the engine was constructed from materials that could operate satisfactorily at elevated temperatures. However, none of these engines operates on a Carnot cycle, and simply increasing the temperature of the hot reservoir does not guarantee improved performance. The design must also be improved so that the actual cycle is as close to being reversible as possible. For example, cold water is not pumped directly into the hot boiler, since this is a highly irreversible process. Instead, the water is preheated in several stages so that it is quite hot when it finally reaches the boiler.

KEY CONCEPTS

A **heat engine** is a device that takes heat energy from a hot reservoir, rejects heat energy to a cold reservoir, and performs external work.

The **thermodynamic efficiency** of a heat engine is equal to the work done by the engine divided by the heat input from the hot reservoir. See Eq. (15.2).

A **refrigerator** is a device that takes in mechanical work and transfers heat energy from a cold reservoir to a hot reservoir.

The **coefficient of performance** of a refrigerator is equal to the heat extracted from the cold reservoir divided by the work input. See Eq. (15.4).

A **Carnot engine** is an ideal, perfectly reversible heat engine operating between two temperature extremes. The Carnot engine efficiency is equal to the reservoir temperature difference divided by the hot reservoir temperature. See Eq. (15.6). The Carnot refrigerator COP is equal to the cold reservoir temperature divided by the reservoir temperature difference. See Eq. (15.7).

Thermodynamically, **entropy** change is defined as heat flow divided by the absolute temperature. See Eq. (15.8).

Statistically, **entropy** is a measure of disorder.

The **second law of thermodynamics** has many equivalent forms. It prohibits all of the following:

1. A cyclic heat engine that transforms all of the absorbed work into heat
2. A spontaneous flow of heat from cold to hot
3. A cyclic heat engine that is more efficient than a Carnot engine working between the same temperature extremes
4. A decrease in the entropy of an isolated system
5. An increase in the order of an isolated system

Most heat engines employ gaseous working substances. The basic processes are **isobaric, isochoric, isothermal,** and **adiabatic.** For an ideal gas, the equations describing the changes in state variables for these processes assume simple forms. See Eq. (15.5).

The **energy of combustion** of a fuel is the total amount of energy produced by complete combustion of that fuel. Experimental values for the energies of combustion of various fuels are given in Table 15.1.

The **overall efficiency** of a heat engine is the external work done divided by the energy of combustion of the fuel consumed. See Eq. (15.10).

QUESTIONS FOR THOUGHT

1. A gas undergoing adiabatic expansion performs external work. Identify the energy supply that decreases in the performance of this work. Is your answer the same if the gas expansion is isothermal?

2. Consider the isothermal expansion of an ideal gas. All of the heat energy supplied to the gas is converted into external work, and so the process is 100% efficient. Explain why this process does not violate the second law of thermodynamics.

3. Three identical cylinders with frictionless pistons contain equal quantities of an ideal gas at the same initial volume and temperature. The gases in the three cylinders undergo isobaric, isothermal, and adiabatic expansions to the same final volume. In which case is the external work done the greatest, and in which case is it the least? In which case is the final temperature the greatest, and in which case is it the least?

4. Why is the efficiency of practical heat engines usually so low? Focus your answer on those factors that would be difficult, if not impossible, to improve.

5. It has been proposed that we use the heat energy contained in the ocean to produce electricity. Under what circumstances could such a scheme work? List any difficulties that you can see in building a practical system based on this idea.

6. If an electric refrigerator operates in an enclosed room, does it warm or cool the room? Would your answer change if the refrigerator door was kept open?

PROBLEMS

(Use the tabulated data of Chapter 14 for specific heats and latent heats.)

A. Single-Substitution Problems

1. A heat engine performs 4×10^6 J of work, and rejects 8×10^6 J to the cold reservoir. Calculate its thermodynamic efficiency. [15.1]

2. A heat engine working with a thermodynamic efficiency of 35% rejects 3.2×10^8 J of heat to the cold reservoir. Calculate the work done by the engine. [15.1]

3. A refrigerator requires the input of 2×10^6 J of work, and rejects 8×10^6 J of heat to the hot reservoir. Calculate its COP. [15.1]

4. A refrigerator with a COP of 2.5 absorbs 5×10^6 J of heat from the cold reservoir. How much heat does it reject to the hot reservoir? [15.1]

5. Calculate the work done when 6 ℓ of an ideal gas expands isobarically to 24 ℓ at a pressure of 4 atm. [15.3]

6. A gas is compressed isobarically at a pressure of 5 atm from an initial volume of 12 ℓ. Calculate the final volume if the work done on the gas is 3200 J. [15.3]

7. An ideal Carnot engine operates between hot and cold reservoirs at 400°C and 30°C, respectively. Calculate its efficiency. [15.4]

8. An ideal Carnot engine operates at 60% efficiency with a cold reservoir temperature of 40°C. Calculate the hot reservoir temperature. [15.4]

9. An ideal Carnot refrigerator has a COP of 5 and a hot reservoir temperature of 40°C. Calculate the cold reservoir temperature. [15.4]

10. Calculate the COP for an ideal Carnot refrigerator if the hot and cold reservoir temperatures are 20°C and 0°C, respectively. [15.4]

11. Calculate the entropy change that occurs in 400 g of water at 0°C when it is frozen to ice at the same temperature. [15.6]

12. Calculate the entropy change that occurs in 400 g of lead at 327°C when it is melted at the same temperature. [15.6]

For problems 13–15 use the energies of combustion from Table 15.1.

13. Calculate the overall efficiency of a gasoline engine that does 5×10^8 J of work while using 45 kg of gasoline. [15.7]

14. A gasoline engine has 18% overall efficiency; calculate the work done in a time interval in which it uses 15 kg of gasoline. [15.7]

15. A small model steam engine burns ethanol with an overall efficiency of 6%. Calculate the work done during a time in which it uses 8.2 mg of methanol. [15.7]

B. Standard-Level Problems

16. An internal combustion engine using air as the working substance has a compression ratio of 8:1. At the beginning of the adiabatic compression stroke, the pressure is 1 atm and the temperature is 37°C. Calculate the pressure and temperature at the end of the adiabatic compression. [15.3]

17. A mass of 600 g of nitrogen at a pressure of 1 atm and a temperature of 27°C is heated isochorically until the pressure is 1.6 atm. It is then heated isobarically until the volume doubles. Calculate the total heat energy absorbed by the nitrogen. [15.3]

18. Two identical cylinders each contain 20 mol of oxygen at a pressure of 10 atm and a temperature of 450°C. The gas in both cylinders is then allowed to expand to atmospheric pressure, adiabatically in one case and isothermally in the other. What is the ratio of the final volumes?

19. A quantity of nitrogen at a temperature of 280 K undergoes a process that triples the value of the pressure. Calculate the final temperature, and the final volume ratio if the process is (a) isothermal, (b) adiabatic, (c) isochoric.

20. An ideal Carnot heat engine has a thermodynamic efficiency of 30% and rejects heat to a cold reservoir at 25°C.

a. What is the temperature of the hot reservoir?
b. By how much must the temperature of the cold reservoir be lowered to increase the thermodynamic efficiency to 50% if the temperature of the hot reservoir stays unchanged? [15.4]

21. A nuclear steam-generating plant operates at 30% overall efficiency and the cold reservoir temperature is 15°C. What is the lowest possible steam temperature required by the second law of thermodynamics? [15.4]

22. An air conditioner has a cooling capacity of 4000 Btu/h between temperature extremes of 2°C and 37°C. If it operates at 40% of the COP of an ideal Carnot refrigerator between the same temperature limits, how much horsepower is required to drive the unit?

23. Calculate the COP of an ideal Carnot refrigerator operating between 3°C and 35°C. If a real refrigerator operates at 35% of this COP and requires 3 hp to drive it, calculate the cooling capacity in tons of refrigeration.

24. The air conditioner in an automobile is rated at 3 tons when the outside air temperature is 35°C and the low temperature of the cycle is 3°C. The unit operates at 28% of the COP of an ideal Carnot refrigerator between the same temperature limits. How much horsepower must the engine supply to drive the air conditioner?

25. A refrigerator contains 800 g of water in a bowl of negligible heat capacity at 0°C while the temperature in the room outside is 25°C. The unit is powered by a ⅓-hp motor, and the COP is 21% of the value for an ideal Carnot refrigerator working between the same temperature limits. How long does it take the water to freeze?

26. The cylinder of a heat engine contains 1 mol of air at 300 K and a pressure of 1 atm. The air is taken reversibly around the cycle shown in the diagram. Calculate:

a. The work done by the air
b. The thermodynamic efficiency of the cycle
c. The thermodynamic efficiency of an ideal Carnot engine working between the same temperature extremes

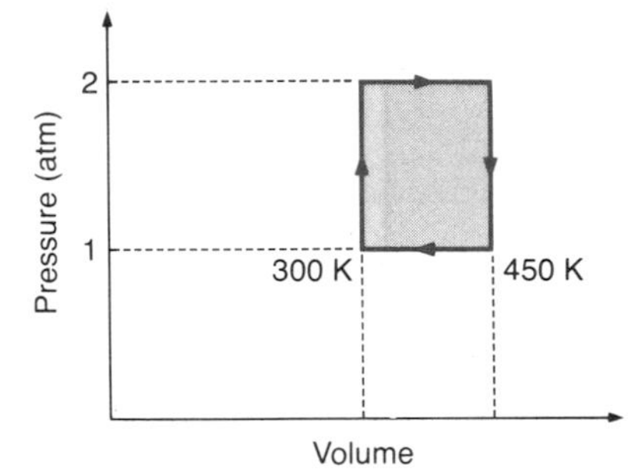

27. A heat engine takes 2 mol of nitrogen reversibly around the cycle shown in the diagram. Calculate:

a. The temperatures at the four corner points of the cycle
b. The thermodynamic efficiency of the cycle
c. The thermodynamic efficiency of an ideal Carnot engine working between the same temperature extremes

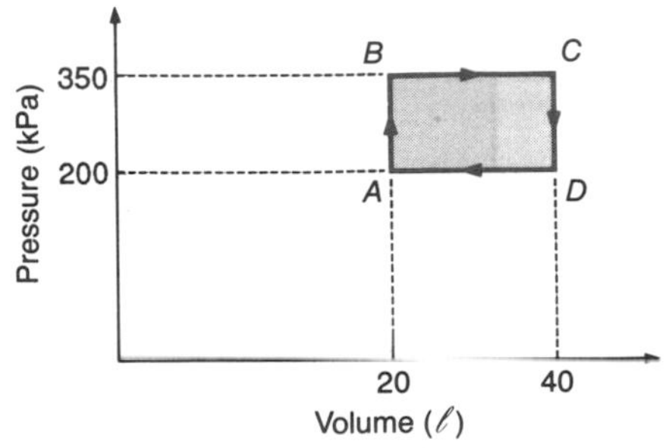

Use the approximation of Example 15.8 in the solution of problems 28–31.

28. Calculate the change in entropy when 800 g of water initially at 20°C is mixed with 500 g of water initially at 40°C in an insulated container. [15.6]

29. A copper container of mass 320 g initially at 20°C is filled with 280 g of water initially at 80°C in an insulated enclosure. Calculate the entropy change. [15.6]

30. How much ice initially at −15°C is required to freeze 1 kg of water initially at 15°C so that the final mixture is ice at 0°C? What is the entropy change if the process takes place in an insulated container?

31. How much steam at 100°C must be mixed with 10 kg of water initially at 50°C so that the final mixture is water at 80°C? What is the entropy change if the process takes place in an insulated container?

For problems 32–37 use the energies of combustion from Table 15.1.

32. A steam turbine whose overall efficiency is 38% produces 10 000 hp. Calculate the rate of supply of oil to the boiler in metric tons per day. [15.7]

33. An automobile gasoline engine delivers 15 hp with a gasoline consumption of 2.2 gal/h. Calculate the overall efficiency. [15.7]

34. A steam turbine uses steam at 430°C and condenser cooling water at 30°C. The unit burns oil at the rate of 85 metric tons per day, and operates at 50% of the ideal Carnot thermodynamic efficiency. Calculate the horsepower delivered.

35. An internal combustion engine operates with a maximum cylinder temperature of 900°C, and the surrounding air is 27°C. The engine delivers 25 hp while consuming gasoline at the rate of 3.2 gal/h. What fraction of the ideal Carnot thermodynamic efficiency is the overall efficiency of this engine?

36. An 800-MW steam plant uses 320 metric tons of coal per hour. (Assume the energy of combustion is 30 MJ/kg.)

a. Calculate the overall efficiency.
b. The condenser is cooled by river water that flows through the plant at 5×10^6 kg/min. Calculate the temperature rise of the water if 70% of the lost heat energy is rejected in the condenser.

37. A 750-MW steam plant uses steam at 360°C and cooling water at 40°C. The heat energy supplied from the fuel is 3.74×10^6 J and the heat energy loss is 2.33×10^6 J for each kilogram of steam used. Calculate:

a. The overall efficiency
b. The steam used in kilograms per second
c. The ideal Carnot thermodynamic efficiency for an engine operating between the same temperature extremes

16

ELECTROSTATICS

POINT CHARGES

The Greeks knew that amber, when rubbed with some materials, attracts small pieces of straw and other light objects, even without touching them. Little serious interest was taken in this phenomenon for many centuries. One of the first to do experimental work on the strange properties of amber was William Gilbert (1540–1603), court physician to Queen Elizabeth I of England. Gilbert identified substances other than amber that also possess the strange property of causing an attractive force at a distance. He called these substances electric (the Greek word for amber is *elektron*). Strange to say, a century passed before the French scientist Charles Dufay (1698–1739) observed a force of repulsion between electrified substances. Dufay suggested that there are two kinds of electric fluid: vitreous and resinous. Two objects possessing the same kind of fluid repel each other, and objects possessing different kinds attract each other. The rubbing motion, which electrifies bodies, somehow transfers some of the fluids. Benjamin Franklin (1706–1790) suggested that there is only one electric fluid, and a body having too much is positively charged, while one that has too little is negatively charged. Today we recognize two species of electric charge, but we retain Franklin's terminology.

During the early eighteenth century, many ingenious devices were constructed to produce continual high-speed rubbing of substances such as amber and glass; such devices could cause surprisingly large sparks in air. Apparatus to store the electricity (devices that we now call capacitors) were also constructed.

However, the French military engineer Charles Augustin Coulomb (1736–1806) laid the first scientific basis for the study of electricity by using a torsion balance to study the force between two stationary charges. Electrostatics is the study of charges at rest; we begin with the famous law that Coulomb discovered, and continue with a discussion involving only point charges.

16.1 Electric Charge

Simple experiments show that some bodies acquire a capacity to exert forces on each other after undergoing a rubbing process. A comb run through hair on a dry day will pick up small pieces of paper, and a rubber balloon can be lifted by a glass rod that has been rubbed with a silk cloth. These forces are plainly not gravitational, since they do not occur without the rubbing action. We attribute the force observed in these simple experiments to **electric charge;** the rubbing action effects the charge transfer from one body to the other.

The force of gravity between two masses is always attractive, but experiment shows that electrostatic force may be either attractive or repulsive. The traditional substances used to demonstrate this are ebony rubbed with fur and glass rubbed with silk. Sup-

SPECIAL TOPIC 18 COULOMB

Charles Augustin de Coulomb was born at Angoulême in 1736, and began his professional career as an engineer in the French army. After some years in the West Indies, he returned to France and settled in the small town of Blois in the Loire valley. Here he began his scientific work, and made his name with the invention of the torsion balance. This instrument consists of a horizontal rod suspended at its midpoint by a thin wire. It is capable of measuring a very small lateral force on the rod by observing the angular displacement produced. Coulomb used this delicate device to establish the inverse square law of force between electric charges, which is still called Coulomb's law.

Secluded in Blois, he escaped the terrors of the French revolution. Napoleon restored him to the posts that he had lost during the revolution, and he died in Paris in 1806.

pose we rub the ebony rod briskly with the fur and then suspend it by a fine thread. After being rubbed with the silk, the glass rod causes the end of the ebony rod to swing toward it, exhibiting a force of attraction. If we bring another charged ebony rod near the suspended rod, the suspended rod swings away, showing a force of repulsion. If a charged glass rod is suspended by the thread, a charged ebony rod shows an attractive force, while another charged glass rod shows a repulsive force. The act of rubbing also charges the fur and the silk. A charged silk cloth attracts a charged glass rod, and a charged fur repels it. The experiment demonstrates the existence of two species of charge and a qualitative feature of the law of force between charges.

> Charges of the same species repel; charges of different species attract.

We follow Franklin's description and call the two charge species positive and negative. The question of identity is entirely a matter of convention. Again we follow Franklin—the glass and fur are positively charged, and the ebony and the silk are negatively charged.

Although we defer a treatment of atomic structure until a later chapter, it is convenient to anticipate some of the results in order to understand the experiments just described. The basic constituents of matter are positively charged protons, uncharged neutrons, and negatively charged electrons. The charges on the proton and the electron are equal in magnitude and differ only in sign. Combinations of these particles form approximately 100 different atoms that have the following common features:

1. A densely packed core of radius somewhat larger than 10^{-15} m made up of protons and neutrons. We call this core the nucleus.

2. A surrounding cloud of electrons disposed so that the diameter of the complete atom is somewhat larger than 10^{-10} m.

3. For a neutral atom, the number of protons in the nucleus is equal to the number of surrounding electrons. The net charge on the atom is therefore zero since the positive and negative contributions exactly cancel.

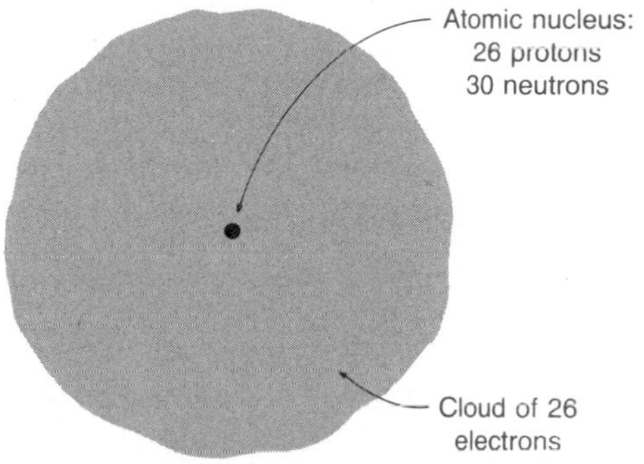

FIGURE 16.1 Diagram representing an atom of iron. The diagram is not to scale. The diameter of the nucleus is about 10^5 times smaller than the atomic diameter.

As an example of atomic composition, we show a schematic diagram of the normal iron atom in Fig. 16.1. The nucleus contains 26 protons and 30 neutrons, and the surrounding electron cloud contains 26 electrons. Atoms may combine chemically to form compounds. Solid, liquid, and gaseous bodies are composed of atoms, compounds, or mixtures. In every case a material body is composed of an array of nuclei and a cloud of electrons. If the body is electrically neutral, the negative charge of the electron cloud exactly balances the positive nuclear charge.

The rubbing action that produces an electric charge transfers some of the electrons from one substance to the other. In the example given, electrons are transferred from the glass to the silk cloth and from the fur to the ebony. The silk cloth and the ebony then have an electron surplus and are negatively charged. The glass and the fur have an electron deficit, which results in a net positive charge on each.

A basic electrical property of bodies is also explained in terms of structure. A glass rod rubbed with silk is readily charged, but a metal rod so rubbed shows no sign of a charge. The glass rod holds its charge even if held in the hand, but a metal rod rapidly conducts any charge that it acquires to the body of the person holding it.

Most metals are excellent conductors, and substances such as glass and polymers are good insulators. In a metallic body one or more electrons from every atom are free to move throughout the whole body, but in an insulator the electrons are strongly bound to a locality, and transport of electric charge is greatly impeded. In a subsequent chapter we will see that the division of different materials into these two categories is not absolute, since every substance conducts electric charge to some extent. Good insulators are just exceedingly poor conductors.

> Substances that readily conduct electric charge are called *conductors*, and those that inhibit charge flow are called *insulators*.

In addition to rubbing, we can also charge objects by *contact* with other charged bodies and by a process called *induction*. We can clearly demonstrate the differences between these processes with the help of a gold leaf electroscope, which consists of a thin metal stem having a pair of suspended gold leaves at its lower end and a metal ball at the other end. The stem enters a trans-

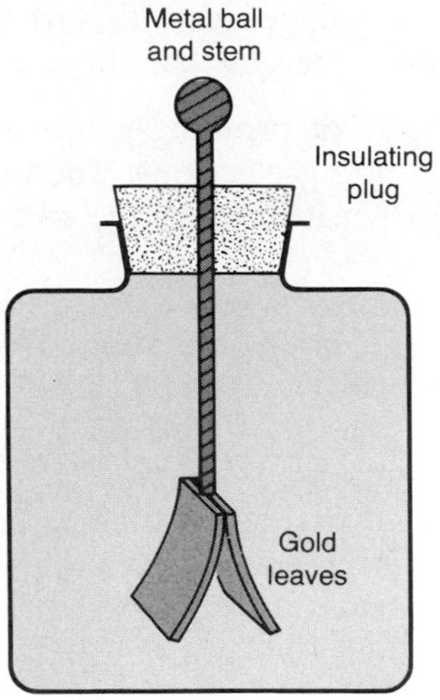

FIGURE 16.2 A gold leaf electroscope.

parent container through an insulating plug (Fig. 16.2), so that the leaves are visible and yet protected from air currents. With no charge on the ball-stem-leaves system, the leaves hang vertically side by side. If the electroscope is charged, the ball, stem, and leaves all share the charge; the electrostatic repulsion between the charges on the leaves forces them apart, in spite of the gravitational force that tends to pull them down.

To show the difference between contact charging and charging by induction, let us suppose that we have a glass rod positively charged by rubbing with silk. To charge the electroscope by contact, we simply touch the glass rod to the metal ball. Electrons from the metal flow into the surface of the glass rod in an attempt to neutralize its positive charge. The loss of electrons leaves the metal positively charged, and the leaves separate because of repulsion between their positive charges. We illustrate the process in Fig. 16.3(a).

To charge the electroscope by induction, we proceed with the following series of steps:

1. We hold the glass rod next to the metal ball but not touching it. Negative electrons in the metal are attracted into the ball, leaving an electron deficiency in the leaves. The positively charged leaves separate.

2. We ground the metal ball by touching it to a much larger body that has a copious supply of electrons to share. (Grounding can be done by touching the ball with the finger or connecting it by a wire to any large metal object.) The electrons in the ball are held there by the rod, but other electrons flow from the ground into the conductor, neutralizing the positive charge on the leaves. The uncharged leaves collapse.

3. We break the ground connection without moving the glass rod. There is no change in the charge distribution.

4. Finally, we remove the glass rod. The negative charge on the electroscope is now shared between the metal ball and the leaves, and the negatively charged leaves separate.

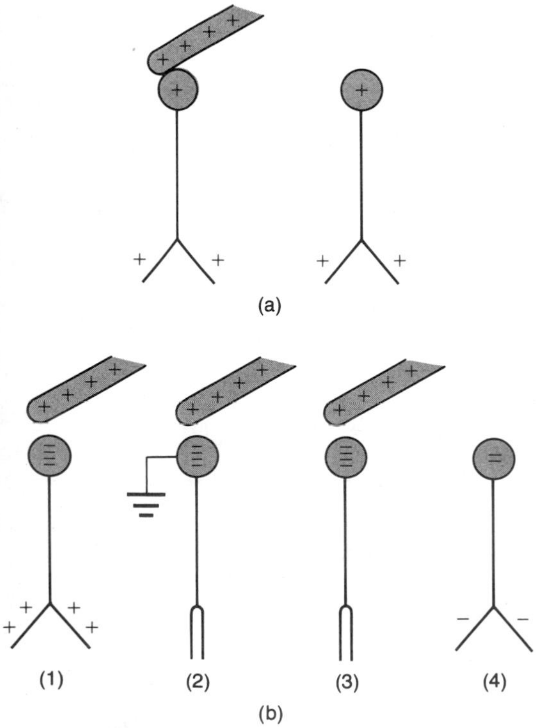

FIGURE 16.3 Charging a gold leaf electroscope (a) by contact and (b) by induction.

The sequence of steps in charging by induction is illustrated in Fig. 16.3(b). If we begin with a negatively charged ebony rod and carry out the steps of charging by induction, we finish with a positively charged electroscope.

We can use an electroscope to determine whether an object is charged and what the sign of the charge is. We begin by charging the electroscope with a charge of a known sign, say negatively, by using a glass rod. Another object brought into the vicinity of the metal electroscope ball causes little change in the position of the leaves if that object is uncharged. However, if the object is negatively charged, the leaves separate much more because the negative charge on the object repels negative charge in the metal ball, driving it down the stem into the leaves. The leaves now have a larger negative charge and therefore separate farther. A positively charged object held near the metal ball collapses the leaves, since it attracts the negative charge from them up the stem into the ball.

16.2 Coulomb's Law

Up to this point our discussion of forces due to electric charges has been qualitative. We have already seen that an attractive force exists between charges of opposite sign and a repulsive force between charges of the same sign. The details of the basic force law were discovered by Coulomb, one of the great figures of eighteenth-century science and engineering. He invented the torsion balance that Cavendish used in determining the universal gravitation constant, and Coulomb used the instrument in the discovery of the basic force law, known as **Coulomb's law.**

We began dynamics with the simple case of point masses. Coulomb made a similar simplification when investigating the electrostatic force law. He used two charged pith balls that were widely separated in comparison with the ball radii. This permitted him to regard the pith balls as *point charges*. Using his torsion balance, he discovered that the force of attraction (or repulsion) was proportional to the product of the two point charges q_1 and q_2 and inversely proportional to the square of their separation r.

Coulomb's Law of Force Between Electric Charges

$$\text{Electric force} \propto \frac{\text{product of the two charges}}{\text{square of the separation}}$$

$$F = \frac{1}{4\pi\epsilon_0}\frac{q_1 q_2}{r^2} \qquad \textbf{(16.1)}$$

SI unit of charge: coulomb
Abbreviation: C

Permittivity of free space $\epsilon_0 = 8.854 \times 10^{-12}\ \mathrm{C^2/N \cdot m^2}$

A useful result is:

$$\frac{1}{4\pi\epsilon_0} = 8.988 \times 10^9\ \mathrm{N \cdot m^2/C^2}$$

$$k \approx 9 \times 10^9\ \mathrm{N \cdot m^2/C^2}$$

The force is attractive for charges of opposite sign and repulsive for charges of the same sign. A positive result from the equation (obtained if q_1 and q_2 are either both positive or both negative) means that the force is repulsive; a negative result (corresponding to different signs for q_1 and q_2) means that the force is attractive.

There is a good deal of information in the display above. In the first place, we notice the close formal relationship with the law of universal gravitation [Eq. (6.3)]:

$$F = G\frac{m_1 m_2}{r^2}$$

In both cases the force is proportional to the product of the properties of the two point objects that cause it. These are the two point charges in the electrostatic case, and the two point masses in the gravitational case. Again, in both cases the force is proportional to the inverse square of the separation of the two point objects.

However, the treatment of the proportionality constants is slightly different. In the gravitational case we introduced the gravitation constant G, and we could proceed in a similar way in the electrostatic case by introducing an electrostatic force constant denoted by a single symbol. The clumsy expression $1/4\pi\epsilon_0$ is retained because it results in simpler forms for subsequent equations. The quantity ϵ_0 is the **permittivity of free space.** Its value leads to the convenient approximation:

$$\frac{1}{4\pi\epsilon_0} \approx 9 \times 10^9 \text{ N} \cdot \text{m}^2/\text{C}^2$$

So far, we have no definition of the SI charge unit. The basic SI electrical unit is a unit of electric current named the ampere, which is defined in a later chapter. For the time being we will define the coulomb as the quantity of charge transferred by a current of 1 ampere in 1 s. A coulomb is therefore the same as an ampere-second. The whole of electricity and magnetism is full of different possibilities for the unit names of newly defined quantities. We see from Eq. (16.1) that the SI unit for ϵ_0 is (coulomb)^2/newton-(meter)^2. This is an awkward combination that is rarely used. We shall obtain the frequently used name in the following chapter. All SI electrical and magnetic units are chosen so that mechanical quantities remain in their familiar units, and the force calculated from Coulomb's law is correctly expressed in newtons.

The coulomb is an immensely large unit of charge, as the following example shows.

EXAMPLE 16.1

Calculate the repulsive force between two charges, each 1 C, which are 0.5 m apart.

SOLUTION We have only to use Coulomb's law, Eq. (16.1):

$$F = \frac{1}{4\pi\epsilon_0}\frac{q_1 q_2}{r^2}$$

$$= 9 \times 10^9 \text{ N} \cdot \text{m}^2/\text{C}^2 \times \frac{1 \text{ C} \times 1 \text{ C}}{(0.5 \text{ m})^2}$$

$$= 3.6 \times 10^{10} \text{ N}$$

This is a very large force. In fact, there is no way to store a charge of 1 C on a small object whose dimensions are of the order of centimeters. Thus, the microcoulomb is a more realistic unit for describing the charge on small objects. ■

We have written Coulomb's law, Eq. (16.1), in a scalar form, although it is essentially vector in nature. The direction of the force vector is given by the attractive or repulsive character of the interaction. Let us consider an example in which the use of Coulomb's law requires vector treatment.

EXAMPLE 16.2

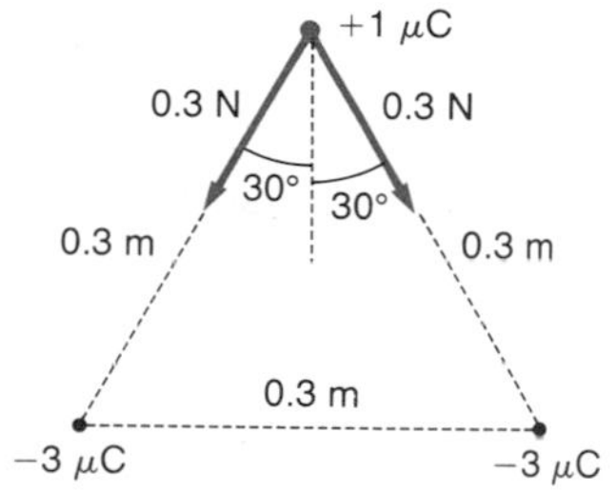

Three charges +1 μC, −3 μC, and −3 μC are placed at the corners of an equilateral triangle as shown in the diagram. Calculate the magnitude and direction of the force on the +1 μC charge.

SOLUTION The charges of −3 μC each exert a force on the +1 μC charge given by Eq. (16.1):

$$F = \frac{1}{4\pi\epsilon_0}\frac{q_1 q_2}{r^2}$$

$$= 9 \times 10^9 \text{ N} \cdot \text{m}^2/\text{C}^2 \times \frac{(-3 \times 10^{-6} \text{ C}) \times (1 \times 10^{-6} \text{ C})}{(0.3 \text{ m})^2}$$

$$= -0.3 \text{ N}$$

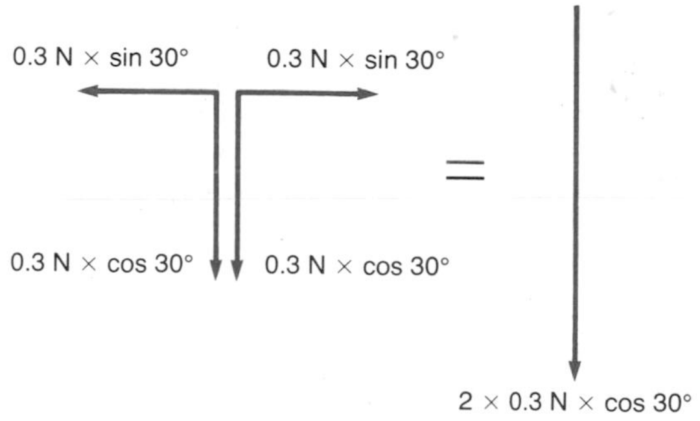

Since the sign is negative, the force is attractive. The total force on the +1 μC charge is the vector sum of the two 0.3-N forces of attraction. We show the components of these forces in the diagram. The resultant total force is given by:

$$F = 2 \times 0.3 \text{ N} \times \cos 30°$$

$$= 0.520 \text{ N directed straight downward}$$

Note that each charge experiences a force due to the presence of the other two. We could have calculated the force of repulsion between the two −3 μC charges, but this makes no contribution to the force on the +1 μC charge and does not help solve the problem. In working such a problem, we assume that the point charges are held fixed in position. ■

In our introductory discussion of electric charge, we explained the acquisition of positive or negative charge in terms of the transfer of electrons from one object to another. The basic electric charges are the charges on electrons and protons. In terms of the SI unit, these charges are as follows:

$$\text{Electron charge} = -1.602 \times 10^{-19}\ \text{C}$$

$$\text{Proton charge} = +1.602 \times 10^{-19}\ \text{C}$$

No smaller charges are known to exist in a free state, and the charge on any body is an integral number of these basic charges. The accepted theory of more massive fundamental particles postulates the existence of quarks. Quarks carry charges of one-third and two-thirds the size of the basic electronic charge. However, they seem to occur only in combinations that add up to an integral number of electronic charges, and they have not been detected in the free state.

EXAMPLE 16.3

A glass rod carries a charge of $+0.35\ \mu\text{C}$ after being rubbed with a silk cloth. How many electrons were removed from the rod?

SOLUTION The number of electrons removed is equal to the excess of protons on the rod. This number is given by:

$$n = \frac{\text{charge on rod}}{\text{magnitude of each proton charge}}$$

$$= \frac{0.35 \times 10^{-6}\ \text{C}}{1.602 \times 10^{-19}\ \text{C}}$$

$$= 2.18 \times 10^{12}$$

This is an immense number of electrons, but even so it is a very tiny fraction of the total number of electrons in a typical glass rod, which is in the vicinity of 10^{25}. ■

In a previous chapter we saw that the gravitational force between masses holds the components of the solar system in their orbits. Likewise, the force of electrical attraction between the nucleus and electrons holds atomic constituents together. In the atom, gravitational forces are utterly negligible, as the following example shows.

EXAMPLE 16.4

The hydrogen atom consists of a proton of mass 1.673×10^{-27} kg at an average separation of 5.29×10^{-11} m from an electron of mass 9.11×10^{-31} kg. Calculate the forces of electrical and gravitational attraction.

SOLUTION The force of electrical attraction is given by Eq. (16.1):

$$
\begin{aligned}
F &= \frac{1}{4\pi\epsilon_0}\frac{q_1 q_2}{r^2} \\
&= \frac{8.988 \times 10^9\ \mathrm{N \cdot m^2/C^2} \times (1.602 \times 10^{\;19}\ \mathrm{C}) \times (-1.602 \times 10^{\;19}\ \mathrm{C})}{(5.29 \times 10^{-11}\ \mathrm{m})^2} \\
&= -8.24 \times 10^{-8}\ \mathrm{N}
\end{aligned}
$$

The negative sign indicates an attractive force. The force of gravitational attraction is given by Eq. (6.3):

$$
\begin{aligned}
F &= G\frac{m_1 m_2}{r^2} \\
&= \frac{6.672 \times 10^{-11}\ \mathrm{N \cdot m^2/C^2} \times (1.673 \times 10^{-27}\ \mathrm{kg}) \times (9.11 \times 10^{-31}\ \mathrm{kg})}{(5.29 \times 10^{-11}\ \mathrm{m})^2} \\
&= 3.63 \times 10^{-47}\ \mathrm{N}
\end{aligned}
$$

The gravitational force is about 10^{40} times weaker than the electrical force! The gravitational force really is negligible on the atomic scene. ■

Because of the negligible size of the gravitational force in an atom, we must pause and wonder whether the attractive forces that are responsible for the planetary orbits could arise from relatively small electric charges on the sun and the planets. Suppose, for example, that the sun was positively charged. To explain the planetary orbits we would need a suitable negative charge on each planet, and the orbits would then be just the same as if the attractive force was a gravitational one. But is there a catch somewhere? In fact there is! When two planets pass in close proximity they cause slight disturbances in each other's orbits. The disturbances are observed to be caused by *attractive forces*, but if all the planets were negatively charged, the small disturbances of orbit at close proximity would be due to *repulsive forces*. This conflicts with our observations. The inverse square law of force that holds the solar system is attractive only and cannot be electrostatic.

16.3 The Electrostatic Field

Suppose that we bring a small test charge into the vicinity of a collection of electric charges all fixed in position. Wherever we place the test charge, it experiences an electrostatic force of a certain magnitude and direction. It is convenient to use the force

We normally think of air as an electrical insulator—several meters of air insulates wires on high-voltage transmission lines at hundreds of thousands of volts. However, this stroke of lightning gives vivid illustration that air can conduct if the electric field strength is large enough; moreover, the length of the discharge gives some idea of the enormous voltage difference between cloud and ground. (Culver Pictures, Inc.)

data on the test charge to indicate the electrostatic effect of the set of fixed charges. We say that the fixed charges are responsible for an **electrostatic field** in their vicinity.

Moreover, we can measure the electrostatic field quantitatively by defining the electrostatic field strength, which we do by dividing the force on the test charge by the size of the test charge.

Definition of Electrostatic Field Strength

$$\text{Electrostatic field strength} = \frac{\text{force on test charge}}{\text{test charge}}$$

$$\mathbf{E} = \frac{\mathbf{F}}{q'} \qquad \textbf{(16.2)}$$

SI unit of electrostatic field strength: newton per coulomb
Abbreviation: N/C

Although the defining equation for electrostatic field strength contains the size of the test charge in the denominator, the field strength itself is independent of the test charge. The reason is that the force on the test charge (in the numerator of the equation) is also proportional to the size of the test charge. To see how this works, let us consider a special case—the problem of finding the electrostatic field strength due to a single point charge q. We follow the procedure implied by Eq. (16.2) and place a test charge q' at a distance r from the point charge q. By Coulomb's law, Eq. (16.1), the magnitude of the force between the two charges is:

$$F = \frac{1}{4\pi\epsilon_0}\frac{qq'}{r^2}$$

Substitution in Eq. (16.2) cancels q' and leads to the following result:

Formula for the Electrostatic Field Strength at a Point Distant *r* from a Point Charge *q*

$$E = \frac{1}{4\pi\epsilon_0}\frac{q}{r} \qquad \textbf{(16.3)}$$

The field direction is radially outward from a positive charge, and radially inward toward a negative one. A positive result from the equation therefore indicates a radially outward field, and a negative result corresponds to a radially inward field.

Since $\mathbf{F} = q'\mathbf{E}$, the vectors **F** and **E** point in the same direction if q' is positive, but in opposite directions if q' is negative. But **F** reverses direction when the sign of the test charge changes; it follows that the direction of **E** does not change when the sign of the test charge changes. Moreover, the magnitude of the field strength is the force that a unit test charge would experience. It follows that the electrostatic field strength vector has a magnitude and direction that do not depend on either the size or the sign of the test charge.

We can use Eq. (16.3) to calculate the field due to several point charges.

EXAMPLE 16.5

Calculate the electrostatic field due to two point charges $+3\ \mu\text{C}$ and $-6\ \mu\text{C}$ that are 0.6 m apart:

a. at a point midway between them

b. at a point 0.5 m from each charge

SOLUTION The field at any point is the sum of the fields due to each charge alone. Let us refer to the $+3\ \mu\text{C}$ charge as q_1 and the $-6\ \mu\text{C}$ charge as q_2.

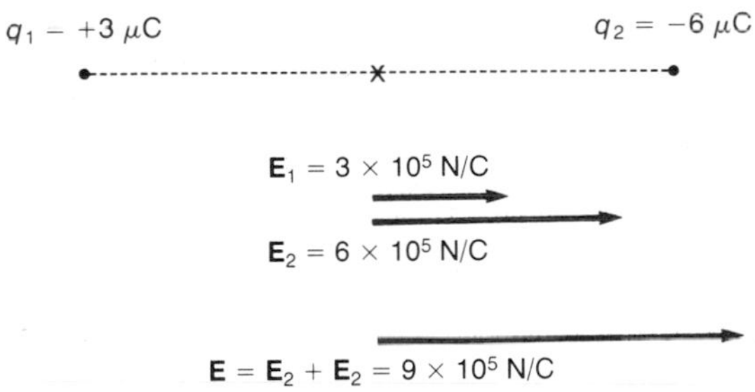

Case (a)

a. The point under consideration is 0.3 m from each charge, and Eq. (16.3) gives the contributions to the field:

$$E_1 = \frac{1}{4\pi\epsilon_0}\frac{q_1}{r^2}$$

$$= 9 \times 10^9\ \text{N} \cdot \text{m}^2/\text{C}^2 \times \frac{3 \times 10^{-6}\ \text{C}}{(0.3\ \text{m})^2}$$

$$= 3 \times 10^5\ \text{N/C}$$

The positive sign indicates a field directed radially outward from the charge q_1.

$$E_2 = \frac{1}{4\pi\epsilon_0}\frac{q_2}{r^2}$$

$$= 9 \times 10^9\ \text{N} \cdot \text{m}^2/\text{C}^2 \times \frac{(-6 \times 10^{-6}\ \text{C})}{(0.3\ \text{m})^2}$$

$$= -6 \times 10^5\ \text{N/C}$$

The negative sign indicates a field directed radially inward toward the charge q_2.

Since the two field contributions point in the same direction, the vector sum is the arithmetic sum:

$$E = E_1 + E_2$$
$$= 3 \times 10^5 \text{ N/C} + 6 \times 10^5 \text{ N/C}$$
$$= 9 \times 10^5 \text{ N/C pointing from the } +3\ \mu\text{C and toward the } -6\ \mu\text{C}$$

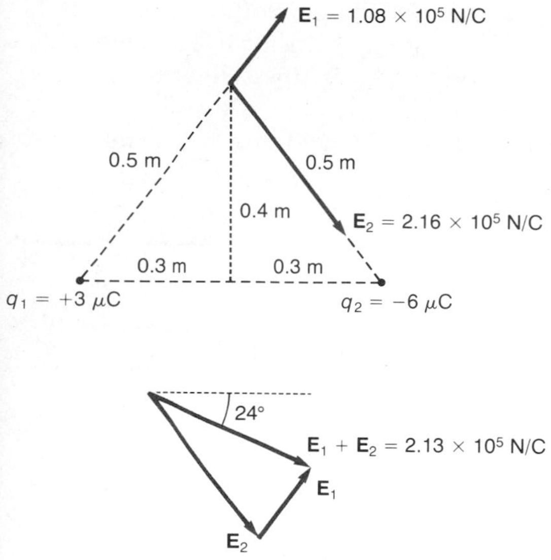

Case (b)

b. Once again, we calculate the field contributions from Eq. (16.3):

$$E_1 = \frac{1}{4\pi\epsilon_0}\frac{q_1}{r^2}$$
$$= 9 \times 10^9 \text{ N}\cdot\text{m}^2/\text{C}^2 \times \frac{3 \times 10^{-6}\text{ C}}{(0.5\text{ m})^2}$$
$$= 1.08 \times 10^5 \text{ N/C}$$

The positive sign indicates that the field is outward from q_1.

$$E_2 = \frac{1}{4\pi\epsilon_0}\frac{q_2}{r^2}$$
$$= 9 \times 10^9 \text{ N}\cdot\text{m}^2/\text{C}^2 \times \frac{(-6 \times 10^{-6}\text{ C})}{(0.5\text{ m})^2}$$
$$= -2.16 \times 10^5 \text{ N/C}$$

The negative sign indicates that the field is inward toward q_2. Because the two contributions in this case are not in the same direction, we must add them vectorially, as shown on the accompanying diagram. Carrying out the calculation (or making a graphical solution) shows that the total field is 2.13×10^5 N/C pointing in a direction making an angle of 24.0° with the line joining the two charges. ■

Calculation of the electrostatic field at a large number of points due to several point charges is a most tedious process. It is necessary to use Eq. (16.3) to calculate the magnitude of the field due to each of the point charges at every point for which it is required, and then to make a vector sum of the results.

16.4 Electrostatic Potential

Our experience shows that energy methods are frequently very useful for solving difficult problems in dynamics, and the same is true for electrostatics. Recall that the energy change in some system can be calculated as the work done on the system between its initial and final states. Using this idea we can find the energy change for a pair of point charges that shift their relative positions by calculating the work done to accomplish the shift. Consider two positive point charges $+q_1$ and $+q_2$ at an initial separation

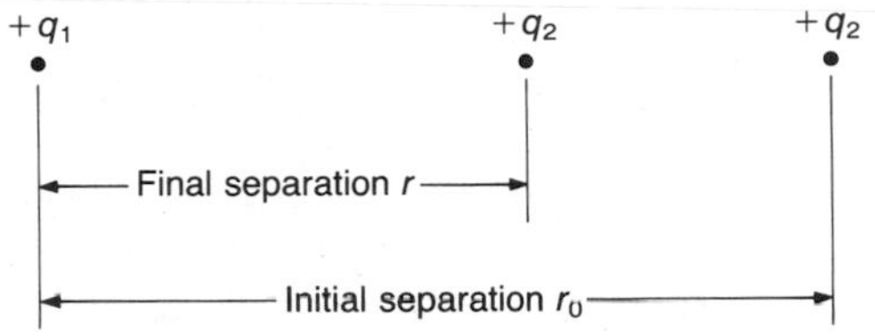

FIGURE 16.4 Illustration for calculation of the change in potential energy of a pair of point charges.

r_0, as shown in Fig. 16.4. The work done on the charge $+q_2$ in moving it to a final position that is a distance r from $+q_1$ is the average electric force multiplied by the displacement (there is no need to worry about components of either vector since they are parallel). The force on the charge q_2 at any position is given by Coulomb's law, Eq. (16.1):

$$F = \frac{1}{4\pi\epsilon_0} \frac{q_1 q_2}{r^2}$$

We would require calculus to calculate the correct average for this expression between the points r_0 and r, but the final result is the fairly simple expression:

$$\langle F \rangle = \frac{1}{4\pi\epsilon_0} \frac{q_1 q_2}{r_0 r}$$

The displacement of the charge is $(r_0 - r)$, and we can calculate the work done to move the charge $+q_2$ by using Eq. (7.1):

$$\begin{aligned} W &= Fx \\ &= \frac{1}{4\pi\epsilon_0} \frac{q_1 q_2}{r_0 r} (r_0 - r) \\ &= \frac{1}{4\pi\epsilon_0} \left(\frac{q_1 q_2}{r} - \frac{q_1 q_2}{r_0} \right) \end{aligned} \tag{16.4}$$

We identify this work as the change in the electrostatic potential energy of the charge pair between the initial and the final positions.

We are always at liberty to make a convenient choice for the zero potential energy. For the problem in hand, we choose r_0 so that the energy of the charges is zero in the initial position; this means that the appropriate choice is an infinitely large value of r_0. The final term then vanishes in Eq. (16.4), and the electrostatic potential energy of a pair of point charges is given by the following relatively simple formula:

■ Formula for the Electrostatic Potential Energy of a Pair of Point Charges q_1 and q_2 Separated by a Distance r

$$U = \frac{1}{4\pi\epsilon_0} \frac{q_1 q_2}{r} \tag{16.5}$$

The assigned energy zero is for an infinite charge separation.

Note that Eq. (16.5) yields a positive value for either a pair of positive or a pair of negative charges; this is correct since work must be expended to bring these charges toward one another against the repulsive force. For a positive and a negative charge the energy is negative; once again this is correct since the charges do work as they approach each other under the influence of the attractive force. A negative value for the potential energy simply means that work must be done on the system to bring it up to the assigned energy zero.

EXAMPLE 16.6

Three charges $+1\ \mu C$, $-3\ \mu C$, and $-3\ \mu C$ are initially separated by great distances. Calculate the work required to place them on the sides of an equilateral triangle of side 0.3 m.

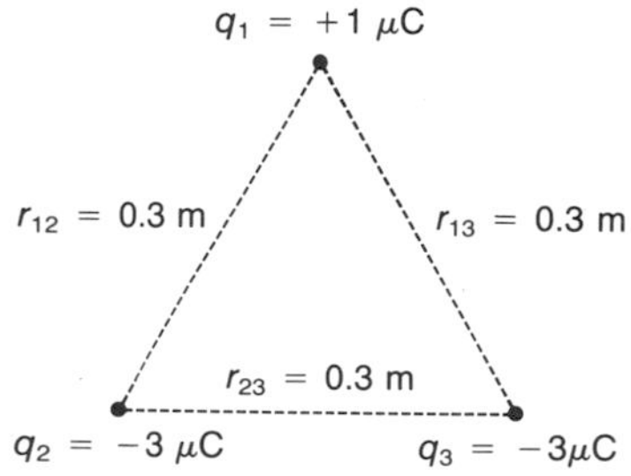

SOLUTION Direct calculation of the work done by the electrostatic forces would be a formidable problem of integral calculus, but we have seen how to avoid this by simply calculating the difference between the initial and final values of the electrostatic potential energy. The total initial potential energy is zero since all of the charges are initially at a large distance apart. In the final configuration, we must use Eq. (16.5) to calculate the energy of each pair of charges. The total final potential energy is then the scalar sum of these three terms since potential energy is a scalar. Let the three charges be q_1, q_2, and q_3, as indicated on the diagram, and let r_{12}, r_{13}, and r_{23} be final separations. Using Eq. (16.5) the final electrostatic energy is given by:

$$U = \frac{1}{4\pi\epsilon_0}\frac{q_1q_2}{r_{12}} + \frac{1}{4\pi\epsilon_0}\frac{q_1q_3}{r_{13}} + \frac{1}{4\pi\epsilon_0}\frac{q_2q_3}{r_{23}}$$

$$= 9 \times 10^9\ \text{N}\cdot\text{m}^2/\text{C}^2\left\{\frac{(+1 \times 10^{-6}\ \text{C}) \times (-3 \times 10^{-6}\ \text{C})}{0.3\ \text{m}}\right.$$

$$+ \frac{(+1 \times 10^{-6}\ \text{C}) \times (-3 \times 10^{-6}\ \text{C})}{0.3\ \text{m}}$$

$$\left. + \frac{(-3 \times 10^{-6}\ \text{C}) \times (-3 \times 10^{-6}\ \text{C})}{0.3\ \text{m}}\right\}$$

$$= -0.09\ \text{J} - 0.09\ \text{J} + 0.27\ \text{J}$$

$$= +0.09\ \text{J}$$

The final energy of the charges is higher than the initial energy by 0.09 J, so that an expenditure of 0.09 J of work is required to bring the charges to their final position. ■

EXAMPLE 16.7

Three charges of $+1\ \mu C$, $-3\ \mu C$, and $-3\ \mu C$ are at the corners of an equilateral triangle of side 0.3 m. Calculate the work done in moving the $+1\ \mu C$ charge from its initial position to a point on the line joining the other two charges and 0.1 m from one of them.

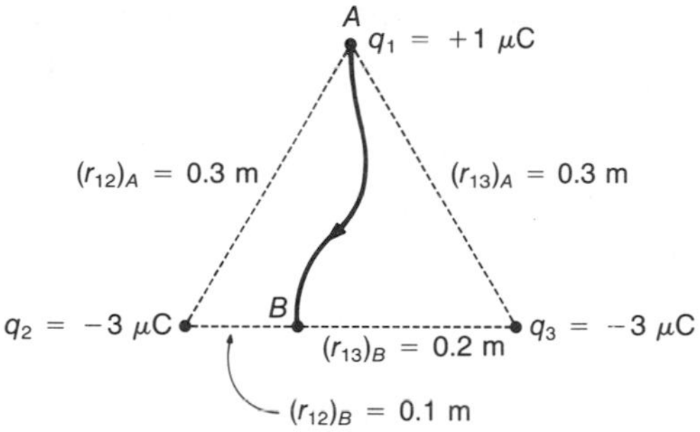

SOLUTION Let the initial and final configurations be denoted by the subscripts A and B, respectively. The energy U_A is the same as the final energy in the previous example—that is, $U_A = 0.09$ J. We therefore require only U_B, which is given by:

$$\begin{aligned} U_B &= \frac{1}{4\pi\epsilon_0}\frac{q_1q_2}{(r_{12})_B} + \frac{1}{4\pi\epsilon_0}\frac{q_1q_3}{(r_{13})_B} + \frac{1}{4\pi\epsilon_0}\frac{q_2q_3}{(r_{23})_B} \\ &= 9 \times 10^9\ \text{N} \cdot \text{m}^2/\text{C}^2 \left\{ \frac{(+1 \times 10^{-6}\ \text{C}) \times (-3 \times 10^{-6}\ \text{C})}{0.1\ \text{m}} \right. \\ &\quad + \frac{(+1 \times 10^{-6}\ \text{C}) \times (-3 \times 10^{-6}\ \text{C})}{0.2\ \text{m}} \\ &\quad \left. + \frac{(-3 \times 10^{-6}\ \text{C}) \times (-3 \times 10^{-6}\ \text{C})}{0.3\ \text{m}} \right\} \\ &= -0.27\ \text{J} - 0.135\ \text{J} + 0.27\ \text{J} \\ &= -0.135\ \text{J} \end{aligned}$$

The work required to move the charge q_1 is the energy difference:

$$\begin{aligned} U &= U_B - U_A \\ &= -0.135\ \text{J} - 0.09\ \text{J} \\ &= -0.225\ \text{J} \end{aligned}$$

The negative sign indicates that work is done by the charge q_1 in moving to its final position.

We could have used a more direct approach that does not depend on the result of Example 16.6. We begin by noting that the mutual potential energy of charges q_2 and q_3 does not change as the charge q_1 moves from its initial to its final position. We therefore need only to calculate the potential energy change of the charge q_1 in the field of the other two. Carrying out this calculation for the initial and final configurations would yield the same value for the energy difference that we calculated above. ■

Electrostatic potential energy changes depend only on the initial and final configurations—the path of motion between the two configurations is irrelevant. The situation is the same as that of a backpacker who climbs a mountain peak. The change of potential energy does not depend on whether the path leads directly to the summit or circles slowly upward. A force field is said to be conservative if the potential energy change does not depend on the path, and we see that both the gravitational field and the electrostatic field are examples of conservative fields.

In the previous section, we introduced the concept of electrostatic field to describe the influence of a set of electric charges. In essence, our definition of electrostatic field strength, Eq. (16.3), is the force per unit positive charge. In a similar way, we now define electrostatic potential difference as the energy change per unit positive charge. Consider a set of electric charges and a test charge q' that is moved from some initial point A to some final point B. We use the energy change resulting from this change of configuration to define the electrostatic potential difference between the two points.

■ Definition of Electrostatic Potential Difference

$$\text{Energy change of test charge} = \text{test charge} \times \text{electrostatic potential difference}$$

$$U_B - U_A = q'(V_B - V_A) \tag{16.6}$$

SI unit of electrostatic potential difference: joule per coulomb
New name: volt
Abbreviation: V

The unit of electrostatic potential difference is the joule per coulomb, which is named the volt in honor of the Italian physicist Alessandro Volta (1745–1827). The assignment of the volt permits us to rewrite the unit of electric field strength:

$$\begin{aligned}\text{Electric field strength} &= \text{newton/coulomb}\\ &= \text{newton} \cdot \text{meter/coulomb} \cdot \text{meter}\\ &= \text{joule/coulomb} \cdot \text{meter}\\ &= \text{volt/meter}\end{aligned}$$

The volt per meter (abbreviated V/m) is the preferred designation for electric field strength, and we use it henceforth.

In our discussion of two point charges, we set the energy zero at infinite separation. For a test charge q' placed at a distance r from a charge q, Eq. (16.5) gives for the energy:

$$U = \frac{1}{4\pi\epsilon_0}\frac{qq'}{r}$$

This is the energy difference between two configurations—one in which the charge separation is r and one in which it is infinite. Using Eq. (16.6), we see that potential difference is equal to energy difference divided by the test charge. We can therefore write a formula for the electrostatic potential due to a point charge.

Formula for Electrostatic Potential at a Point Distant *r* from a Point Charge *q*

$$V = \frac{1}{4\pi\epsilon_0}\frac{q}{r} \qquad (16.7)$$

The zero of potential is an infinite distance from the charge.

Because we have selected the energy zero at infinite separation between the charge q and the test charge q', the zero of potential occurs at an infinite distance from the charge q. In writing Eq. (16.7) we have dropped the symbol $V_B - V_A$ (which explicitly denotes a potential difference) and replaced it by the symbol V. However, we must not forget that electrostatic potential is always a difference, no matter how we choose to represent it by symbols.

Calculation of the potential due to a set of point charges is a simpler task than calculation of the field. This is because the potential is a scalar quantity, and we need only make a scalar sum of the potentials due to the different charges.

EXAMPLE 16.8

Two charges of $+1\ \mu C$ and $-3\ \mu C$ are separated by 30 cm. Calculate the electrostatic potential at a point that is 30 cm distant from each charge.

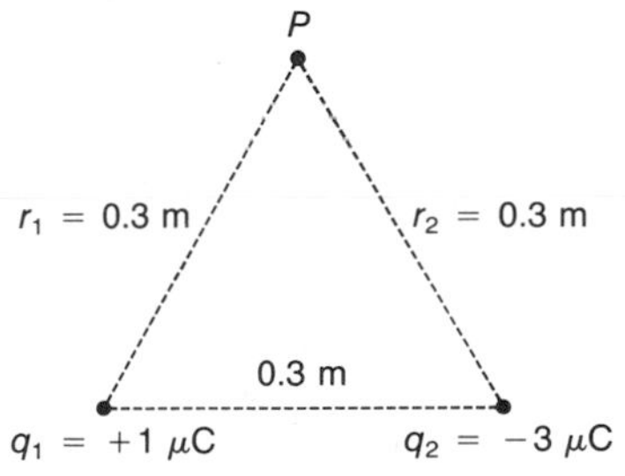

SOLUTION Let us call the two charges q_1 and q_2, as shown in the diagram, and let P be the point at which we require the potential. Scalar addition of two terms [each of the type of Eq. (16.7)] gives:

$$\begin{aligned} V &= \frac{1}{4\pi\epsilon_0}\frac{q_1}{r_1} + \frac{1}{4\pi\epsilon_0}\frac{q_2}{r_2} \\ &= 9 \times 10^9\ \mathrm{N \cdot m^2/C^2}\left\{\frac{(+1 \times 10^{-6}\ \mathrm{C})}{0.3\ \mathrm{m}} + \frac{(-3 \times 10^{-6}\ \mathrm{C})}{0.3\ \mathrm{m}}\right\} \\ &= 3 \times 10^4\ \mathrm{V} - 9 \times 10^4\ \mathrm{V} \\ &= -6 \times 10^4\ \mathrm{V} \end{aligned}$$

We must continually bear in mind that this is the potential difference between the point P and a point at an infinite distance. Thus, the potential is 6×10^4 V lower than the potential at a very distant point. ■

16.5 Lines of Force and Equipotential Surfaces

We now examine a technique that is frequently used to exhibit the character of an electrostatic field with the help of a diagram. Since the field is a vector, we could follow our past practice of drawing a directed line segment to represent it. However, if we wanted to draw a diagram that indicated the value of the field at a large number of points, the diagram could be very confusing. We avoid this by drawing a line-of-force diagram that allows us to make a quick estimate of the general features of the field. A **line of force** is a line drawn so that it points in the direction of the field at all places. Two general consequences follow from this definition:

1. The field strength vector is tangential to the line of force at each point on it.
2. Lines of force cannot intersect since the direction of the electric field vector must be unique.

For the simple case of isolated point charges, the lines of force are straight lines, as shown in Fig. 16.5. For a positive point charge, the lines all point radially outward, and for a negative point charge they point radially inward. This observation offers us a clue to another characteristic of lines of force—they must all originate on positive charge and terminate on negative charge. This means that the diagrams of Fig. 16.5 are to some extent incomplete; we cannot have isolated charges of a single sign. The lines originating on the positive point charge must terminate on negative charge. In drawing such a diagram, we understand that the lines radiating out from the positive charge terminate on negative charge that is located at a very large distance away. Likewise, the lines coming in to the negative charge are assumed to originate on very distant positive charge.

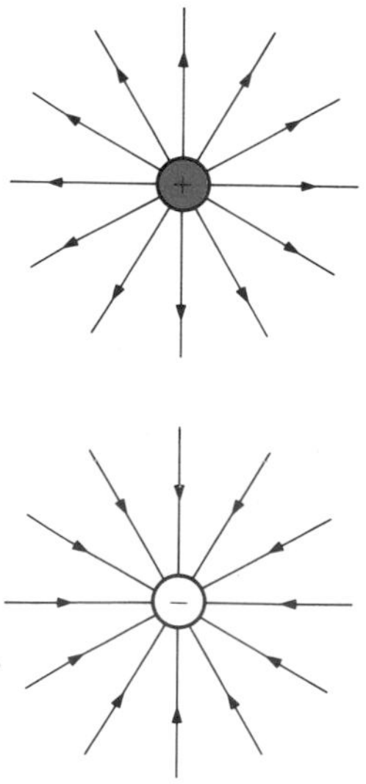

FIGURE 16.5 The lines of electric force for isolated positive and negative point charges.

The number of lines of force that appear on a diagram is a matter of some choice. There are an infinite number of straight lines passing through a point; this means that there are an infinite number of lines of force originating on the positive charge, or terminating on the negative one. We choose to draw only a selected number in order to keep the diagram clear, and simply imagine the others.

We can also use line-of-force diagrams to estimate relative field strengths. Equation (16.3) shows us that the field near a point charge (small values of r) is larger than at distant points (large values of r). Figure 16.5 shows the lines of force crowded close together near the point charge, and more spread out at distant points. We therefore conclude that the field is large where lines of force cluster together, and small where they are far apart. Moreover, Eq. (16.3) shows that the field strength near a point

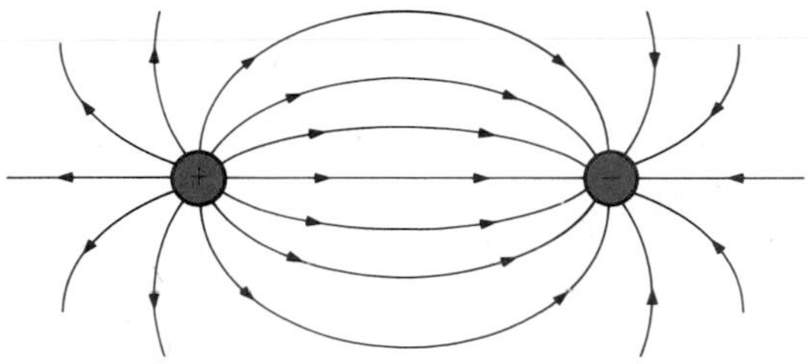

FIGURE 16.6 The lines of electric force for a pair of point charges of equal magnitude but opposite sign.

charge is proportional to the size of the charge. This means that we should make the lines of force originating (or terminating) on a point charge proportional to the charge size whenever we are drawing a line-of-force diagram for a system of two or more point charges. In this way we ensure that the lines of force cluster more strongly near the larger charge than near the smaller one, with a corresponding difference in electric field strength.

Let us illustrate with some examples. Figure 16.6 shows the line-of-force diagram for a pair of charges of equal magnitude but opposite sign. In view of our convention, the number of lines of force originating on the positive charge will be equal to the number terminating on the negative one. It follows that none of the lines of force on this diagram will terminate on charge at infinity. Lines originating on the positive charge, but not shown on the diagram as terminating, do all terminate on the negative charge. This situation does not always hold, as we see in Fig. 16.7; here we illustrate the line-of-force diagram for a pair of charges of opposite sign, but with the magnitude of the positive charge twice that of the negative one. Again following our convention, we must have twice as many lines of force originating on the positive charge as we have terminating on the negative one. This means that only one-half of the lines of force originating on the positive charge can terminate on the negative one; the other half must terminate on distant negative charge.

There is yet another way to display information about the electrostatic field graphically, and it is perhaps even more informative than the line-of-force diagram. This consists in the display of **equipotential surfaces;** these are defined as surfaces on which the value of the potential is constant. We explain this method by considering the equipotential surfaces of a single point charge. Equation (16.7) shows that the potential due to a single point charge is constant if r is constant; it follows that the equipotential surfaces for a point charge are spheres centered on the point. We illustrate this in Fig. 16.8 for a point charge of $\frac{1}{9}$ μC; the circles shown represent spheres around the point charge for five selected potentials of 1, 2, 3, 4, and 5 kV. There are an infinite number of equipotential surfaces (one for each of the infinite number of possible values of the potential), but we have elected to show just five of them at equal potential spacings. Comparison with the line-of-force diagram shows that the lines of force are everywhere normal to the equipotential surfaces. This is an important characteristic that is true for every electrostatic field.

It is also clear from the diagram that the equipotential surfaces are clustered together more closely where the field is strong, and are more spread out where it is weak. Given a set of equipotentials of known values of potential, we can easily make quantitative estimates of the electric field. In fact, our definitions of electric field and electric potential verify the following intuitive conclusion:

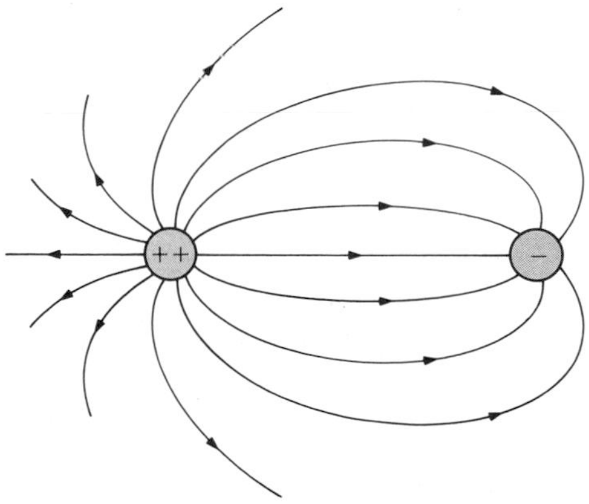

FIGURE 16.7 Lines of force for a pair of point charges with the positive charge twice the magnitude of the negative one.

Estimation of Average Field from Equipotential Surfaces

To estimate the electrostatic field at some point, calculate the difference in potential between the two adjacent equipotential surfaces and divide by the separation of the surfaces measured along a line of force. The result is the average electrostatic field along the chosen line of force, and it is directed from the higher to the lower potential.

Let us illustrate this procedure for point P on Fig. 16.8. The adjacent equipotential surfaces are 3 kV and 2 kV, and the separation between these surfaces (measured along the line of force) is 0.5 m − 0.333 m = 0.167 m. We conclude that the average field between these two equipotential surfaces on the line of force passing through P is given by:

$$\langle E \rangle = \frac{3\ \text{kV} - 2\ \text{kV}}{0.167\ \text{m}} = 6.0\ \text{kV/m}$$

pointing outward (from the higher toward the lower potential). Our example shows that accurate estimation of the field at any point requires us to have a great many equipotential surfaces drawn on the diagram; it would then be possible to find two equipotentials on opposite sides of any point and fairly close to it.

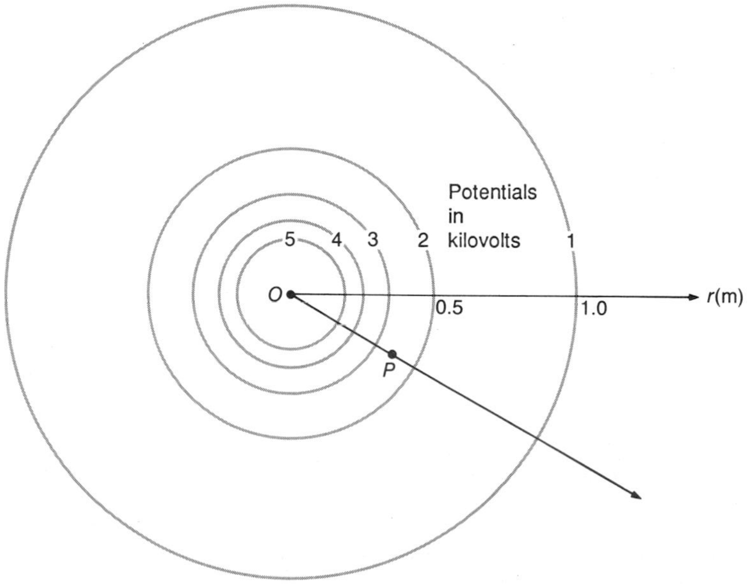

FIGURE 16.8 Some of the equipotential surfaces for a charge of +1/9 μC. The circles represent spheres centered on the point charge, and the values of the potentials are marked in kilovolts.

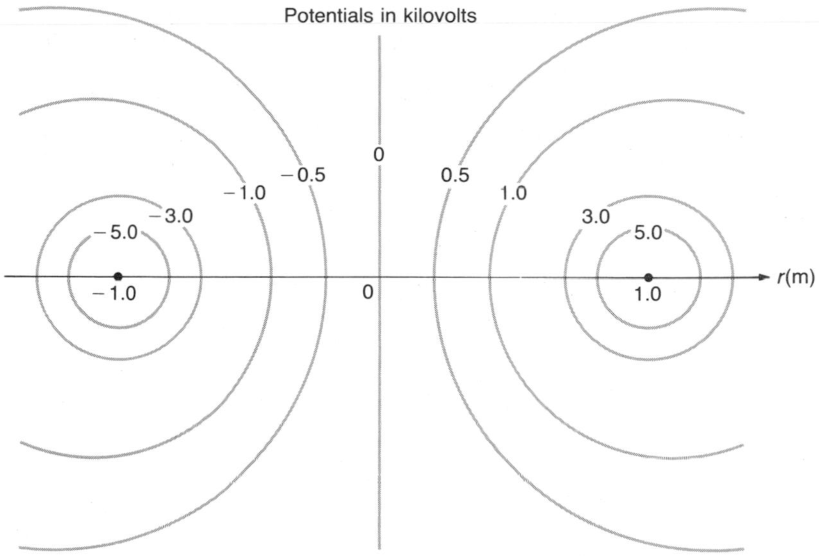

FIGURE 16.9 Some of the equipotential surfaces for charges $\pm \frac{1}{9}$ μC separated by 2.0 m. The surfaces are only approximately spherical.

This procedure is not in fact very helpful for a single point charge, since we could have used Eq. (16.3) to make an exact calculation of the field at any given point. However, in more complicated cases, the equipotential diagram is of great help in quickly estimating the magnitude and direction of the field at any point. Figure 16.9 shows some of the equipotentials for charges of $\pm \frac{1}{9}$ μC that are 2 m apart. The equipotentials are approximately spherical, and a procedure exactly the same as the one we have just used would permit the estimation of the field at any point that lies between two equipotential surfaces.

KEY CONCEPTS

Electric charge occurs in two species: positive and negative. The smallest known free negative charge is the charge on the electron, and the smallest known free positive charge is that on the proton. The electron and proton charges are of equal magnitude but of opposite sign.

Charges of like sign repel and charges of opposite sign attract. **Coulomb's law** governs the magnitude of the forces of attraction or repulsion between a pair of point charges. See Eq. (16.1).

The **electrostatic field** at a point is the force per unit positive charge on a small test charge at the point; see Eq. (16.2). A formula for the electrostatic field of a single point charge is given by Eq. (16.3). The electric field due to two or more point charges is the vector sum of the fields due to each charge separately.

Two or more point charges possess **electrostatic potential energy** when placed in proximity to one another. The energy of two point charges is given by Eq. (16.5). The energy of more than two point charges is the sum of the energies of all of the pairs.

The **electrostatic potential** between two points is the work per unit positive charge required to move a small test charge from one point to the other; see Eq. (16.6). A formula for the potential due to a single point charge is given by Eq. (16.7). The potential due to two or more point charges is the sum of the potentials due to each charge separately.

The general nature of electrostatic fields can be exhibited by line-of-force and/or equipotential-surface diagrams. A line of force is such that the field is everywhere tangential to it; an equipotential surface is one on which the potential has a constant value. Lines of force are perpendicular to equipotential surfaces at all points.

QUESTIONS FOR THOUGHT

1. Two objects in a laboratory are observed to attract each other. Does this prove they are both charged? Does it prove that at least one of them is charged? Explain your answer. What if the two objects are observed to repel each other?

2. A stationary electric charge experiences zero net force in a certain region of space. Does this prove conclusively that the electrostatic field is zero in that region?

3. Can a point charge increase its electrostatic potential energy by moving to a region of lower electrostatic potential?

4. A point charge of a certain mass is released from rest in a nonuniform electrostatic field. In general, does its trajectory follow a line of force? Is there a special case for which the answer is different?

5. Electrostatic lines of force run from positive to negative charge. Do they also run from regions of higher electrostatic potential to lower, or vice versa, or is there no connection? Explain your answer.

6. Is it possible for the electrostatic potential energy of two charges in proximity to be zero? Is this possible for three charges in proximity?

7. Is there any circumstance in which the electrostatic potential at some point could be positive in the field of a single negative point charge?

8. Consider two point charges of opposite sign. Do any points exist in their combined electrostatic field for which any of the following statements are true?

a. Potential is zero and field strength is not.
b. Field strength is zero and potential is not.
c. Both potential and field strength are zero.

Repeat the question for two point charges of the same sign.

PROBLEMS

A. Single-Substitution Problems

1. Two point charges $+4\ \mu C$ and $+10\ \mu C$ are separated by 12 cm. Calculate the force that each exerts on the other. [16.2]

2. A pair of point charges $+8\ \mu C$ and $-10\ \mu C$ are 15 cm apart. Calculate the force that each exerts on the other. [16.2]

3. A charge of $+10\ \mu C$ experiences a force of 20 N attraction from another charge of $-8\ \mu C$. Calculate the distance separating the two charges. [16.2]

4. A charge of $+10\ \mu C$ experiences a 20-N force of repulsion from another charge that is 25 cm away. What are the sign and magnitude of this charge? [16.2]

5. Calculate the magnitude and direction of the electric field strength at a point 25 cm from a point charge of $-15\ \mu C$. [16.3]

6. At what distance from a $-10\ \mu C$ charge is the magnitude of the electric field strength equal to 2000 N/C? [16.3]

7. The electric field strength at a point 40 cm from an isolated point charge is 1200 N/C directed away from the charge. Calculate the value of the point charge. [16.3]

8. Calculate the electrostatic potential energy due to a pair of point charges $+6\ \mu C$ and $-12\ \mu C$ that are 18 cm apart. [16.4]

9. Calculate the electrostatic potential energy due to a pair of point charges $-4\ \mu C$ and $-10\ \mu C$ that are 45 cm apart. [16.4]

10. What must be the separation of two point charges each of $+5\ \mu C$ if the electrostatic potential energy is 0.5 J? [16.4]

11. Two point charges that are separated by 50 cm have an electrostatic potential energy of 0.8 J. One of the charges is $-2.5\ \mu C$. What are the sign and magnitude of the other? [16.4]

12. What is the electrostatic potential energy associated with the helium nucleus if the two protons in the nucleus are assumed to be point charges separated by 1.8×10^{-15} m? [16.4]

13. Calculate the electrostatic potential at a point 50 cm distant from a point charge of $+8\ \mu C$. [16.4]

14. At what distance from a point charge of 10 μC is the electrostatic potential 10 kV? [16.4]

15. The electrostatic potential is -8 kV at a distance of 12 cm from a certain point charge. What are the sign and magnitude of the charge? [16.4]

B. Standard-Level Problems

16. Three point charges, $+3\ \mu C$, $-5\ \mu C$, and $+8\ \mu C$, are placed in that order along a straight line with 18 cm separating adjacent charges. Calculate the force on each charge. [16.2]

17. Two charges, one $+8\ \mu C$ and the other $-8\ \mu C$, are placed 40 cm apart. Another charge, placed midway on the line joining them, experiences a force of 12 N toward the $+8\ \mu C$ charge. Calculate the value of the third charge. [16.2]

18. Four charges, each of $+9\ \mu C$, are placed at the corners of a square of side 20 cm. Calculate the force on each charge.

19. Three equal positive charges placed at the corners of an equilateral triangle of side 15 cm each experience a force of magnitude 30 N. Calculate the value of the charges.

20. Two point charges each of $-10\ \mu C$ are placed at two corners of an equilateral triangle of 30 cm side length. Calculate the magnitude and direction of the electrostatic force on a $-8\ \mu C$ charge placed:

a. at the vacant corner of the triangle
b. at the centroid of the triangle (the point equidistant from all three vertices)

21. Three point charges each $+16\ \mu C$ are located at three corners of a square of side 20 cm. Calculate the magnitude and direction of the electrostatic force on a $-10\ \mu C$ charge placed:

a. at the vacant corner of the square
b. at the center of the square

22. a. Calculate the magnitude and direction of a vertical electric field that supports a proton.
b. Repeat the calculation for an electron. [16.3]

23. A spherical water drop in a fog is suspended motionless. The radius of the drop is 1.5 μm, and the electric field of the earth is 275 V/m downward. How many excess electrons does the water drop carry? [16.3]

24. A particle of mass 20 μg that carries a charge of $+6.4\ \mu C$ is released from rest in an electric field of 8 V/m. What is the acceleration of the particle due to the electric field?

25. A particle carrying a charge of $+12\ \mu C$ experiences an acceleration of 750 m/s^2 in an electric field of 24 V/m. Calculate the mass of the particle.

26. A charge of $+5\ \mu C$ is placed 40 cm from another charge, and the electrostatic field at the point midway on the line joining them is 5×10^5 N/C directed away from the $+5\ \mu C$ charge. Calculate the value of the second charge.

27. Three charges $+6\ \mu C$, $-6\ \mu C$, and $+12\ \mu C$ are placed in order along a straight line with 20 cm between adjacent charges. Calculate the electric field at the following positions:

a. midway between the $+6\ \mu C$ and $-6\ \mu C$ charges
b. on the extension of the straight line and 20 cm beyond the $+12\ \mu C$ charge

28. Three point charges, each of $-15\ \mu C$, are placed at the vertices of an equilateral triangle of side 30 cm. Calculate the electric field strength:

a. at the centroid of the triangle (the point equidistant from all three vertices)
b. at the midpoint of one of the sides

29. Four charges, each $-10\ \mu C$, are placed at the corners of a square of side 20 cm.

a. Calculate the electric field at the center of the square.
b. Calculate the electric field at the center point of one of the sides.

30. Calculate the electrostatic potential energy of four charges, each of $+3\ \mu C$, placed at the corners of a square of side length 27 cm. [16.4]

31. Three charges, each of $+5\ \mu C$, are placed on three of the corners of a square of side length 18 cm. What charge must be placed on the vacant corner so that the electrostatic potential energy is zero? [16.4]

32. What is the electrostatic potential energy associated with the nucleus of a beryllium atom if the four protons are assumed to be located at 2.3×10^{-15} m from each other (that is, at the corners of a regular tetrahedron)?

33. How much work is required to take one of the four charges of problem 30 from its corner position to the midpoint of the square?

34. Two point charges of $+5\ \mu C$ and $-10\ \mu C$ are located 10 cm apart. How much work is necessary to increase their separation by (a) a factor of 3, and (b) a factor of 10?

35. Three point charges, two of $-4\ \mu C$ and one of $+4\ \mu C$, are placed at the corners of an equilateral triangle of side length 20 cm. What is the electrostatic potential at the midpoint of each of the sides?

36. Four charges, $\pm 8\ \mu C$ and $\pm 4\ \mu C$, are placed at the corners of a square of side length 45 cm in the order shown. Calculate the electrostatic potential:

a. at the center of the square
b. at the midpoint of the side between $+8\ \mu C$ and $-4\ \mu C$ charges
c. at the midpoint of the side between $+4\ \mu C$ and $-4\ \mu C$ charges

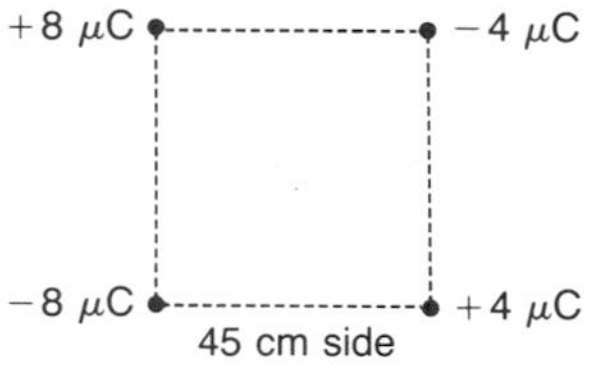

C. Advanced-Level Problems

37. A right-triangle has sides of length 30 cm, 40 cm, and 50 cm. Charges of $+20\ \mu C$, $-40\ \mu C$, and $+80\ \mu C$ are placed at the vertices of the triangle in the order shown. Calculate:

a. the magnitude and direction of the force on the $-40\ \mu C$ charge

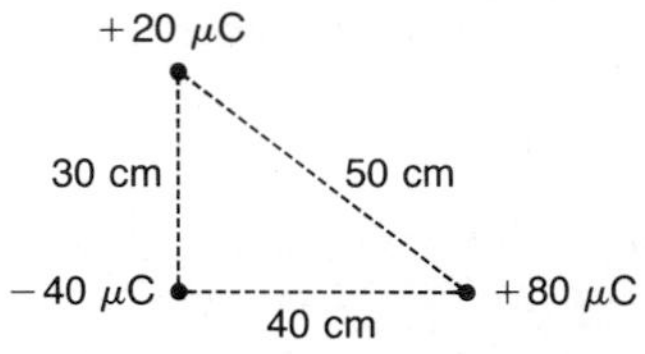

b. the magnitude of the electric field strength at the midpoint of the hypotenuse of the triangle

38. Suppose that the moon had an electric charge due to a quantity of 1 g of excess electrons, and the earth a positive charge of the same magnitude.

a. Regarding the earth and moon as point charges, calculate the force of electrostatic attraction.
b. Does this fictional electrostatic force exceed the actual gravitational force between the earth and the moon? (Use the data inside the back cover.)

39. Two small plastic balls, each of mass 6 g, are suspended from the same point by light threads each 30 cm long. When the balls are given equal positive charges, they take up an equilibrium position, with their centers 20 cm apart. What is the charge on each ball?

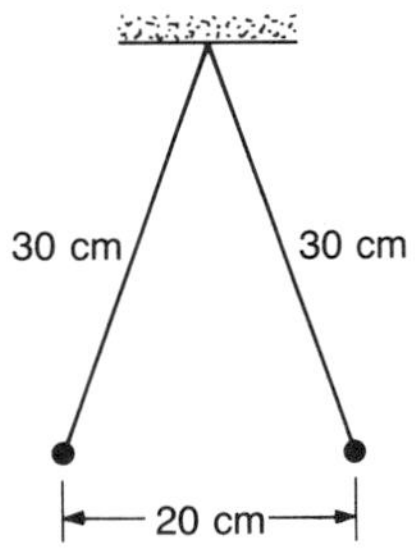

40. The simplest Bohr model for atomic hydrogen viewed the electron as a point charge revolving in a circular path of radius 5.29×10^{-11} m about the stationary proton. The necessary centripetal force is provided by the electrostatic force of attraction between the two charges. Calculate:

a. the electrostatic potential energy of the electron and the proton
b. the kinetic energy of the electron (Use the data inside the back cover.)

17

ELECTROSTATICS

PARALLEL PLATES

The major problem of electrostatics is the calculation of fields and potentials due to various configurations of charge. In the previous chapter, we discussed this problem for point charges, since it is the consideration of point charges (beginning with Coulomb's law) that forms the conceptual underpinning of the subject. In this chapter, we turn to the consideration of the fields of charged conductors—in particular to the problem of a pair of flat parallel metal plates carrying charges of opposite sign but equal magnitude. This problem has major practical application in every type of electrical circuit.

17.1 The Electrostatic Field of Charged Parallel Conducting Plates

We begin with some very general observations that apply to any conductor that carries electrostatic charge. Recall that we define a conductor as an object that contains plentiful mobile charges, and this property leads us to the first characteristic of a charged conductor in electrostatic equilibrium:

I. The field in the interior of the conductor is zero.

If this were not so, the nonzero field within the conductor would cause a flow of charge, and thereby violate our assumption of electrostatic equilibrium.

The second property also concerns the field:

II. The field at the surface of the conductor (and just outside) is normal to the surface.

The reason for this is that any field component parallel to the surface would cause a flow of charge along the surface and once again violate the electrostatic equilibrium assumption.

The third property relates to the location of the charge:

III. All of the charge is on the surface of the conductor.

Let us first be sure we understand just what this implies. The whole of the conductor contains mobile charges, and it would seem that all parts of the conductor must therefore be charged. However, this is not so—the mobile charges in the interior of the conductor are neutralized by other charges of opposite species; it is only on the surface that an excess of one species of charge exists. Consider for example a negatively charged block of copper. The mobile charges in copper are electrons. In the interior of the copper both the mobile and immobile electrons are exactly neutralized by the immobile positive charges of the copper nuclei. Only at the surface of the copper does an excess of electrons exist. When we say that the interior of the conductor is uncharged, we mean that the net charge is zero. If this were not so, then lines of force would originate (or terminate) on the interior charge, leading to the existence of a field within the conductor. We have already seen that this cannot be so, and it follows that there can be no net charge within the conductor.

Our final property concerns the equipotential surfaces:

IV. The conducting surface is an equipotential surface, and all points in the interior are at the same potential as the surface.

This follows since a potential difference between any two points of the conductor would imply an average electrostatic field between those points, and hence a flow of charge. This cannot happen in electrostatic equilibrium, and all points on the surface and within the interior of the conductor must be at the same potential.

These four properties are not unconnected properties of the field and equipotentials of a charged conductor, but rather connected consequences of Coulomb's law applied to this problem.

One of the simplest systems that we can imagine is a charged conducting sphere. By symmetry, the charge is uniformly distributed on the surface of the sphere, the equipotentials (external to the conductor) are concentric spheres, and the lines of force are radial straight lines passing through the center of the sphere. We illustrate the charge distribution and the lines of force in Fig. 17.1. When we compare this figure with Fig. 16.5, we see that the field outside the conductor is the same as that due to a point charge at the center, but inside the conductor it is not the same, because the field there is zero.

Next we consider two equal conducting spheres carrying charges of equal magnitude but opposite sign. It is tempting to say that the field outside the conductors is the same as that for the corresponding point charge problem illustrated in Fig. 16.6. But this is not true; a new factor that we have not so far considered intrudes in this problem. The positive and negative charges on the two conductors attract each other and cause a redistribution of surface charge so that most of it is on adjacent parts of the spheres. (Remember that there can be no redistribution of charge if the charge is all concentrated at a point.) Since the number of lines of force originating (or terminating) on positive (or negative) charge is proportional to the amount of charge, the lines of force will be strongly clustered between the two conductors. This implies a strong field in that region, and a much weaker field elsewhere. Figure 17.2 shows the line-of-force diagram, which bears only a rough qualitative correspondence to that for the point charge problem shown in Fig. 16.6. Note that all the properties for the electrostatic fields of conductors are still true; in particular both the conducting surfaces are equipotentials even though the charge is not distributed uniformly on either of them.

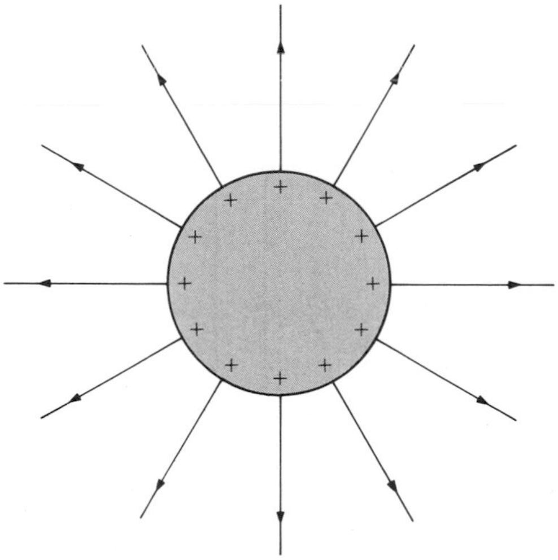

FIGURE 17.1 The electrostatic field of a charged spherical conductor.

We are, at last, ready to consider the problem of flat parallel conducting plates carrying charges of equal magnitude but oppo-

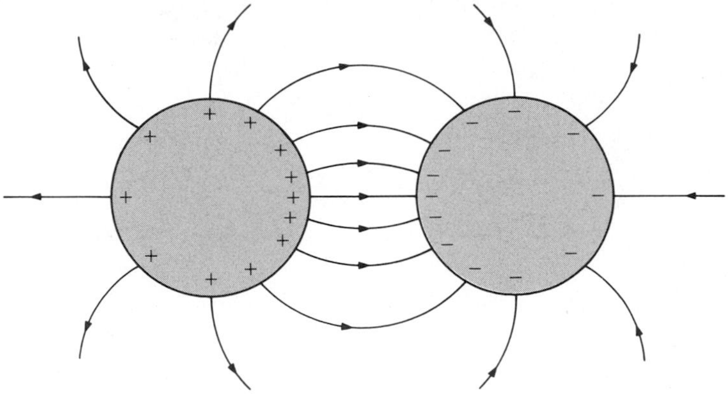

FIGURE 17.2 The electrostatic field of a pair of spherical conductors carrying charges of equal magnitude but opposite sign.

site sign. The line-of-force diagram for the problem of two spheres convinces us that the lines of force for the pair of flat plates should look something like the situation illustrated in Fig. 17.3(a). We expect an even more pronounced tendency for the positive and negative charges to attract each other to the adjacent surfaces. We now go one step farther, and approximate the actual line-of-force diagram by the idealized one shown in Fig. 17.3(b). The idealized lines of force are uniformly spaced and normal to the plates; the equipotential surfaces are flat planes parallel to the plates (which are themselves equipotential). Geometrically, this is the simplest line-of-force/equipotential-surface diagram—a set of parallel straight lines normal to a set of parallel flat planes. The approximation is valid so long as every lateral dimension of the conducting plates greatly exceeds the separation. More briefly, we refer to this situation as being closely spaced parallel plates.

Since all the lines of force in the ideal model are uniformly spaced between the plates, the electric field strength is constant in that region and zero elsewhere. Recall also that the number of lines of force originating (or terminating) on positive (or negative) charge is proportional to the charge strength. It follows that increasing the plate separation (without going so far as to disturb the validity of the idealized model) does not alter the number of lines of force provided that the charges on the plates remain constant. The density of lines of force between the plates (and hence the electric field strength) is proportional to the charge density on the plates, and does not depend on the plate separation. For plates of area A carrying charges $\pm q$, the magnitude of the charge density is q/A. Comparison with Eq. (16.3) shows that dividing the charge density by the permittivity of free space yields a quantity whose dimensions are that of electric field strength. A calculus-based derivation would show that this is in fact the field between the plates—no additional dimensionless constant is required.

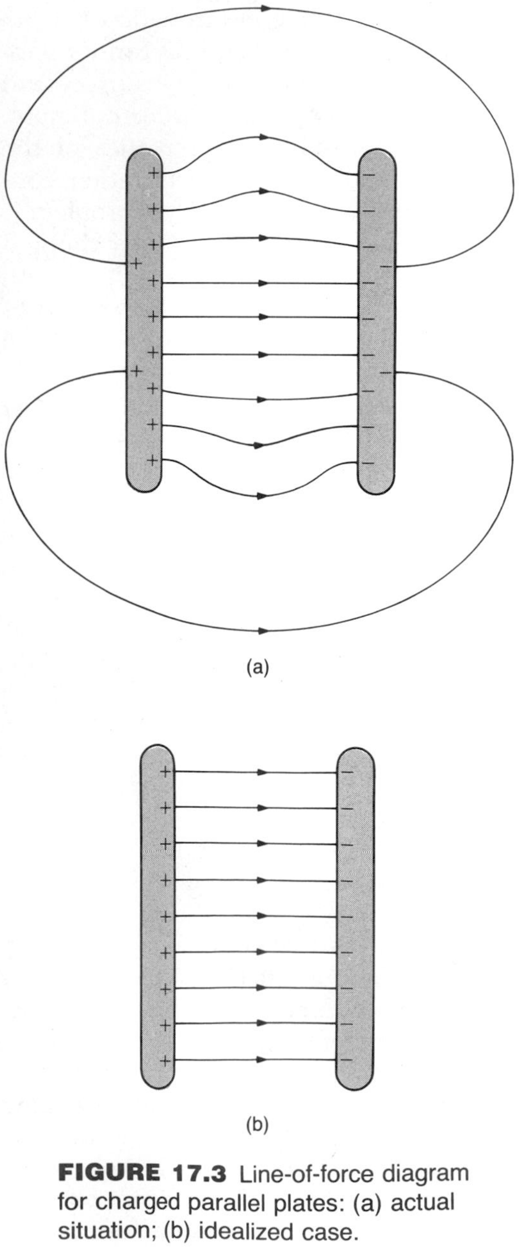

FIGURE 17.3 Line-of-force diagram for charged parallel plates: (a) actual situation; (b) idealized case.

■ Formula for the Electrostatic Field Strength Between Charged Parallel Plates

$$E = \frac{q}{\epsilon_0 A} \tag{17.1}$$

The field direction is from the positive to the negative plate.

EXAMPLE 17.1

A pair of rectangular copper plates 40 cm × 25 cm are separated by 3 mm. What charge should be placed on the plates to cause a field of 800 V/m between them?

SOLUTION Since the plates are very close compared with their linear dimensions, we are justified in using the result of the ideal model. Equation (17.1) gives:

$$E = \frac{q}{\epsilon_0 A}$$

$$\therefore q = E\epsilon_0 A$$

$$= 800 \text{ V/m} \times 8.854 \times 10^{-12} \text{ C}^2/\text{N} \cdot \text{m}^2 \times 0.40 \text{ m} \times 0.25 \text{ m}$$

$$= 7.08 \times 10^{-10} \text{ C}$$

This is the magnitude of the charge on each plate, positive on one and negative on the other. Note that there is nothing to prevent us from having charges of any sign and magnitude on parallel plates. If we were to place a positive charge on one plate and a negative charge of smaller magnitude on the other, charges of magnitude equal to the smaller would face each other across the plate separation; this must be so in order to produce the correct line-of-force diagram. The surplus positive charge would be on the rear surface of the positive plate, and lines of force originating on it would terminate on distant negative charge. ■

Another simple result for parallel plates follows from our previous identification of the average field strength between two equipotential surfaces as the difference in potential between them divided by their separation. Each parallel plate is an equipotential, and since the field is constant there is no need to refer to an average. This gives the following simple result for the field between parallel plates of separation d at potential difference V.

■ Formula for the Field Strength Between Parallel Plates at a Given Potential Difference

$$E = \frac{V}{d} \tag{17.2}$$

The field direction is from the higher to the lower potential.

This result shows us that the field strength is independent of plate area, and depends only on plate separation provided that the potential remains constant. This is not in conflict with our previous claim that the field strength is independent of plate separation, and depends only on the plate area provided the charge remains constant. To keep the charge constant, the plates must be isolated; to keep the potential constant, the plates must be connected to a source of constant potential such as a battery (see following chapter).

EXAMPLE 17.2

A pair of parallel conducting plates, each 40 cm × 50 cm, have a separation of 5 mm and carry charges of $\pm 0.5\ \mu C$. Calculate the potential difference between the plates.

SOLUTION We first use Eq. (17.1) to calculate the electric field between the plates:

$$\begin{aligned} E &= \frac{q}{\epsilon_0 A} \\ &= \frac{5 \times 10^{-7}\ \text{C}}{8.854 \times 10^{-12}\ \text{C}^2/\text{N} \cdot \text{m}^2 \times 0.4\ \text{m} \times 0.5\ \text{m}} \\ &= 2.824 \times 10^5\ \text{V/m} \end{aligned}$$

We can now use Eq. (17.2) to calculate the potential difference:

$$\begin{aligned} V &= Ed \\ &= 2.824 \times 10^5\ \text{V/m} \times 5 \times 10^{-3}\ \text{m} \\ &= 1410\ \text{V} \end{aligned}$$

The example makes clear the preferred usage of volts per meter for the electric field unit. ■

17.2 Capacitance

We have already mentioned the capacitor as a device for storing the charges generated by frictional electrostatic generators. This charge-storing ability has made it an essential element in power and electronic engineering.

A capacitor consists of a pair of adjacent conductors that carry charges of equal magnitude but opposite sign, and that are separated by insulating material. Since the charges on the conductors are equal and opposite, the total charge on the whole device is zero. However, when we refer to a capacitor as having a charge q, we mean that q is the magnitude of the charge on either conductor. With this understanding, we define the **capacitance** of a pair of conductors as the ratio of the charge to the accompanying potential difference.

Definition of Capacitance of a Pair of Conductors

$$\text{Capacitance} = \frac{\text{charge}}{\text{potential difference}}$$

$$C = \frac{q}{V} \tag{17.3}$$

SI unit of capacitance: coulomb per volt
New name: farad
Abbreviation: F

The unit of capacitance is named in honor of Michael Faraday (1791–1867). Faraday had no formal education and began work as a bookbinder's apprentice. His native talent was recognized by Sir Humphry Davy, who gave him a post as laboratory assistant at the Royal Institution in London. From this modest beginning, Faraday's discoveries, which we will investigate in a subsequent chapter, set the groundwork for modern-day electrical engineering.

The definition of capacitance applies to any pair of conductors. The simplest capacitor is a pair of closely spaced parallel conducting plates. We can discover some features common to all capacitors by investigating this simple case. Consider a pair of parallel plates, each of area A and separated by a distance d, carrying charges $+q$ and $-q$. The electrical field between the plates follows from Eq. (17.1):

$$E = \frac{q}{\epsilon_0 A}$$

We can now calculate the potential difference between the plates from Eq. (17.2):

$$V = Ed$$
$$= \frac{qd}{\epsilon_0 A}$$

This equation enables us to write the capacitance as the ratio q/V.

Formula for the Capacitance of Closely Spaced Parallel Plates

$$C = \frac{\epsilon_0 A}{d} \tag{17.4}$$

ϵ_0 = permittivity of free space
A = area of each plate
d = plate separation

Generally, the capacitance of any pair of conductors is equal to the permittivity constant multiplied by a geometrical factor whose dimension is [L]. For the case of parallel plates the geometrical factor is A/d.

The parallel plate capacitor formula also allows us to introduce a new units designation for ϵ_0, the constant in Coulomb's law. Substituting SI units for C, A, and d in Eq. (17.4) gives the unit of farad per meter for ϵ_0, which is the commonly used unit for this constant.

EXAMPLE 17.3

Calculate the capacitance of two parallel conducting plates, each 1 m^2 in area, separated by 1 mm.

SOLUTION The problem involves direct use of Eq. (17.4):

$$C = \frac{\epsilon_0 A}{d}$$
$$= \frac{8.854 \times 10^{-12}\ \text{F/m} \times 1\text{m}^2}{10^{-3}\ \text{m}}$$
$$= 8.85 \times 10^{-9}\ \text{F}$$

The farad is an exceedingly large unit of capacitance. Very common practical units are the microfarad (μF) and the nanofarad (nF). (See the metric prefix abbreviations in Table 1.2.) The capacitance just calculated might well be written $C = 8.85$ nF. ■

Since a capacitor has two charges in close proximity, our previous work on the energy associated with point charges leads us to the conclusion that energy storage goes hand in hand with charge storage in a capacitor. However, there are differences in the two situations. When calculating the energy of a pair of point charges, the charge size is given and the variable quantity is the separation of the point charges. Under these circumstances the natural choice for the energy zero is infinite charge separation. When calculating charge storage in a capacitor, the capacitor is given and the variable quantity is the amount of charge stored on it. The natural choice for the energy zero is the uncharged capacitor, and we calculate the energy of a charged capacitor by calculating the work that must be done to charge it. Begin with a pair of uncharged plates and gradually transfer small amounts of charge from one plate to the other until the plates carry charges of $+q$ and $-q$. When the first small quantity of charge is transferred, the potential difference between the two plates is close to zero. However, by the time the last small contribution to the charge crosses from one plate to the other, the potential difference is nearly up to the final value V. The average potential difference during the charge transfer is $V/2$, and the work done in transferring the charge q against this average potential difference is given by $qV/2$. This is also the value of the **electrostatic potential energy** stored in the capacitor; we can use the definition of capacitance [Eq. (17.3)] to write this result in alternate ways.

■ Formulas for the Energy Storage in a Charged Capacitor

$$U = \frac{qV}{2} = \frac{CV^2}{2} = \frac{q^2}{2C} \qquad \textbf{(17.5)}$$

EXAMPLE 17.4

Calculate the charge and the potential difference between the plates that would result in an energy storage of 1 J in a 0.2-μF capacitor.

SOLUTION The third form of Eq. (17.5) is suitable for finding the charge:

$$U = \frac{q^2}{2C}$$

$$\therefore q = \pm(2CU)^{1/2}$$

$$= \pm(2 \times 2 \times 10^{-7}\ \text{F} \times 1\ \text{J})^{1/2}$$

$$= \pm\ 0.632\ \text{mC}$$

We can find V from either of the remaining forms of Eq. (17.5); the first one is convenient:

$$U = \frac{qV}{2}$$

$$\therefore V = \frac{2U}{q}$$

$$= \frac{2 \times 1\ \text{J}}{0.632 \times 10^{-3}\ \text{C}}$$

$$= 3160\ \text{V}$$ ■

The charge and the potential difference calculated in Example 17.4 are both fairly large. In the next section, we shall discuss dimensions for a typical capacitor that would be appropriate for the situation described.

17.3 Dielectrics

Up to this point we have said nothing of the medium in which the electric fields exist. The assumption is that the parallel plates are separated by air. We should actually specify a vacuum in the region between the plates, but for practical purposes the difference between air and a vacuum is negligible in electrostatics. However, an insulating medium other than air or vacuum increases the capacitance of a pair of conductors significantly. Moreover, the factor by which the capacitance increases is determined only by the nature of the insulating material and not by the shape of the conductors or the amount of charge on them. An insulator in this role is referred to as a *dielectric,* and the property of the insulator that increases the capacitance is called the **dielectric constant.**

TABLE 17.1 Dielectric constants of some common solid and liquid dielectrics at room temperature

Substance	*Dielectric constant*
Solids	
Paper	2–4
Glass	5–9
Mica	4.6
Rubber	2–3
Polyethylene	2.3
Sodium chloride	5.6
Liquids	
Silicone oil	2.2–2.8
Water	81
Methanol	31
Ethanol	25

Definition of Dielectric Constant

$$\text{Dielectric constant} = \frac{\text{capacitance with dielectric in place}}{\text{vacuum capacitance}}$$

$$K = \frac{C_{\text{diel}}}{C_{\text{vac}}} \qquad \textbf{(17.6)}$$

The dielectric constant is a pure number.

We show the dielectric constants for some common insulating materials in Table 17.1. The capacitance of a parallel plate capacitor whose plates are separated by a dielectric of dielectric constant K is found with the help of Eqs. (17.4) and (17.6).

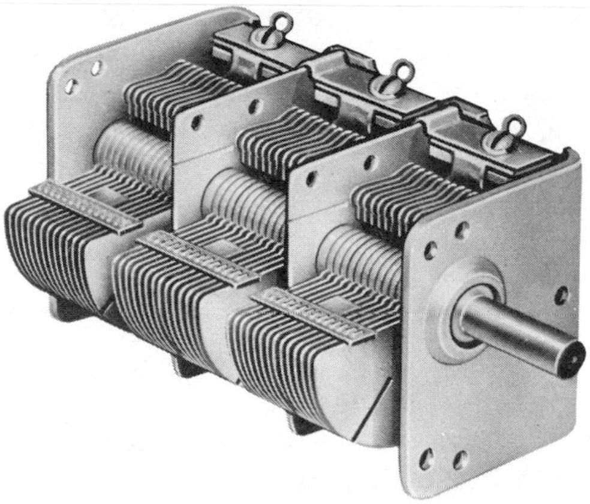

This variable capacitor has a number of flat plates on a shaft and other plates fixed to the body of the device. The plates on the shaft can be moved between the fixed plates by turning the shaft; the net effect is to alter the area of plate that directly faces a neighboring plate and thereby vary the capacitance. (Courtesy of Bell Industries, J. W. Miller Division)

Formula for the Capacitance of Parallel Plates with Dielectric Spacing

$$C = \frac{K\epsilon_0 A}{d} \quad (17.7)$$

K = dielectric constant of the medium between the plates
ϵ_0 = permittivity of free space
A = area of each plate
d = plate separation

We can gain at least a qualitative understanding of the way in which the insertion of a dielectric between capacitor plates increases the capacitance. All material bodies, both conductors and dielectrics, are composed of atoms or molecules that are in turn composed of electrons and positive nuclei. In a metallic conductor, some of the electrons are mobile and can produce a continuing flow of charge in the presence of an electric field. In a dielectric, none of the charges are mobile, but an electric field can displace them to some extent from the positions that they would occupy in its absence. This phenomenon is called polarization of the dielectric. The slight shift of electric charge that the field produces results in a charge on the surface of the dielectric, and this induced surface charge is called polarization charge.

Let us show in detail how this works. Figure 17.4(a) shows the line-of-force diagram for the familiar parallel plate capacitor, and Fig. 17.4(b) shows the same capacitor with dielectric inserted between the plates. The field induces a negative polarization charge adjacent to the positive plate and a positive polarization charge adjacent to the negative plate. Recalling that lines of force originate on positive and terminate on negative charge, we see that the introduction of the dielectric has altered the line-of-force pattern between the capacitor plates. Some of the lines of force from the positive plate continue through to the opposite negative plate, but some of them terminate on the negative polarization charge that is in very close proximity. (A similar conclusion holds for the situation near the negative plate.) This causes the lines of force to be less dense within the dielectric than they would be if it were not there. This in turn means that the electric field between the plates is lower in the presence of the dielectric, provided that the charge on the plates remains constant. Lowering the electric field means lowering the potential difference between the plates [see Eq. (17.2)], and this means increasing the capacitance [see Eq. (17.3)]. The polarization charge must always be less than the charge on the plates; in Fig. 17.4 the polarization charge is shown as one-half the charge on the plates. This would lead to an electric field that is one-half, and a capacitance that is twice, the value they would be in the absence of the dielectric. This situation

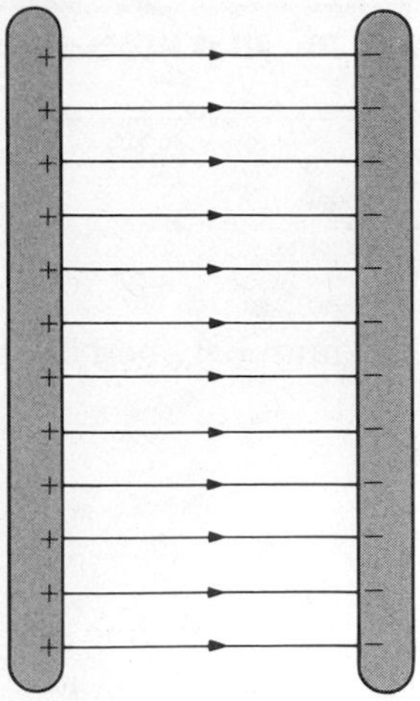

(a)

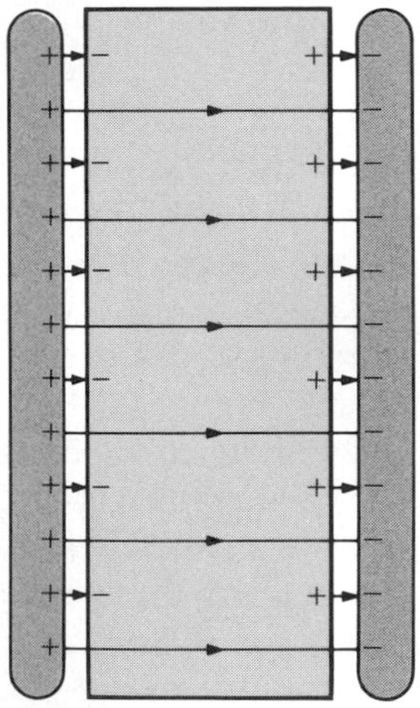

(b)

FIGURE 17.4 A parallel plate capacitor, showing lines of force for: (a) vacuum (or air) between the plates; (b) dielectric between the plates.

would correspond to a dielectric constant of exactly 2. For larger values of the dielectric constant, the induced charge would be a larger fraction of the charge on the plates.

In addition to increasing the capacitance, the dielectric often ensures the physical separation of the capacitor plates. Paper and thin polymer sheets are examples of dielectrics that have been used in the manufacture of capacitors.

EXAMPLE 17.5

A parallel plate capacitor is made from metal foil strips, each 4 cm wide, and separated by a polyethylene sheet 0.02 mm thick. What length of foil is required for a 0.02-μF capacitor?

SOLUTION We can calculate the area of each foil using Eq. (17.7):

$$C = \frac{K\epsilon_0 A}{d}$$

$$\therefore A = \frac{Cd}{K\epsilon_0}$$

$$= \frac{2 \times 10^{-7}\ \text{F} \times 2 \times 10^{-5}\ \text{m}}{2.3 \times 8.85 \times 10^{-12}\ \text{F/m}}$$

$$= 0.1965\ \text{m}^2$$

Since each foil is a rectangle of width 0.04 m, the required length is given by:

$$\ell = \frac{0.1965\ \text{m}^2}{0.04\ \text{m}}$$

$$= 4.91\ \text{m}$$

The capacitor of this example would not be left as a thin narrow strip almost 5 m long, but would be rolled up into a small cylinder about the size of a person's thumb. It would then be not strictly "parallel plate," but the very close proximity of the foils would ensure approximate correspondence with the simple ideal model. ■

All insulating materials break down if the electric field in them becomes too large. We are all familiar with the spark discharge through the air known as a lightning stroke. All gaseous, liquid, and solid insulation fails in a similar way at high field strength. When the discharge occurs, the material becomes highly conducting—at least momentarily. Table 17.2 lists the approximate values of electric field strength that break down the insulating properties of various substances.

TABLE 17.2 Approximate values of the dielectric breakdown strength of various substances at room temperature

Substance	*Dielectric breakdown strength (MV/m)*
Dry air	3
Transformer oil	60
Glass	30–250
Sodium chloride	150
Polyethylene	750
Mica	800

EXAMPLE 17.6

Could the 0.2-μF capacitor of Example 17.5 be used to store 1 J of energy?

SOLUTION Example 17.4 shows that 1 J of energy in a 0.2-μF capacitor requires a potential difference of 3160 V. Since the dielectric is 2×10^{-5} m thick, the field is given by Eq. (17.2):

$$E = \frac{V}{d} = \frac{3160 \text{ V}}{2 \times 10^{-5} \text{ m}} = 1.58 \times 10^{8} \text{ V/m}$$

Comparing this with Table 17.2, we see that this is about 21% of the ultimate breakdown strength. Bearing in mind that various external factors (such as moisture) can cause a very substantial reduction in the breakdown strength of a dielectric, we see that this field is too high for safety. The capacitor would have to be redesigned with thicker dielectric, or else it could have its energy storage rating set lower than 1 J. ■

17.4 Combinations of Capacitors

Capacitors are often connected to each other in various ways by conducting wires. Two simple, basic types of connection—series and parallel connections—are useful in analyzing more complex arrangements; they are illustrated in Fig. 17.5. For capacitors wired to each other in series or parallel, we wish to calculate the effective series capacitance C_s or the effective parallel capacitance C_P, for each of the units considered as a whole.

Consider first the series wiring of Fig. 17.5(a). A charge $+q$ applied to the left-hand plate of the first capacitor induces a charge $-q$ on the plate immediately opposite. But the whole conductor enclosed by dotted lines (two plates and a connecting wire) is isolated and has zero net charge. It follows that the left-hand plate of the second capacitor must carry a charge $+q$ and induce a charge $-q$ on the plate immediately opposite it. Continuing this argument shows that all capacitors in a series arrangement carry the same charge. Moreover, this charge is the same as the equivalent charge on the system as a whole, since the extreme right-hand plate also carries a charge $-q$. Turning to the potential difference, we see that each system enclosed by dotted lines is equipotential, since each is a single conducting object. It follows that the potential difference between the extreme left- and extreme right-hand plates is the sum of the potential differences

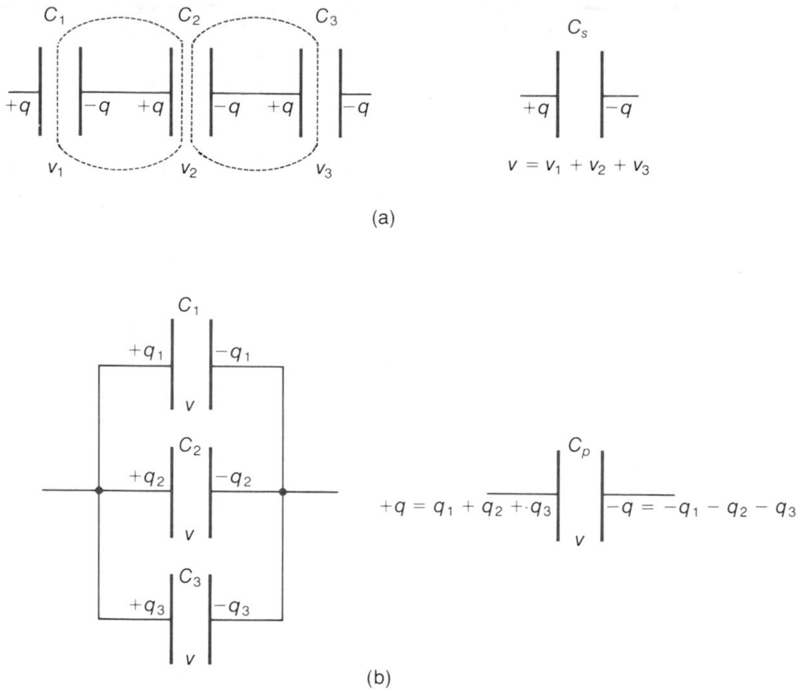

FIGURE 17.5 Series and parallel arrangements of capacitors: (a) for capacitors in series, the potential differences add; (b) for capacitors in parallel, the charges add.

across each of the capacitors. As far as charge and potential difference are concerned, we can replace the series arrangement by a single capacitor C_s carrying the same charge as each member of the series, and having a potential difference equal to the sum of the potential differences. Equation (17.3) gives for the equivalent series capacitance:

$$C_s = \frac{q}{V}$$

$$= \frac{q}{V_1 + V_2 + V_3}$$

$$\therefore \frac{1}{C_s} = \frac{V_1}{q} + \frac{V_2}{q} + \frac{V_3}{q}$$

$$= \frac{1}{C_1} + \frac{1}{C_2} + \frac{1}{C_3} \tag{17.8}$$

If the capacitors are wired in parallel, as shown in Fig. 17.5(b), then all of the left-hand plates with their connecting wires form an equipotential; so also do all of the right-hand plates. It follows that the potential difference across each capacitor is the same, and equal to the potential difference for the system as a whole. Turning to the charge, we see that the total charge on the left side of the system is the sum of the charges on the left-hand plates; the same is true for the right-hand plates. We can therefore replace the parallel arrangement by a single capacitor, C_p, having the same potential difference as each member of the combination, and having a charge equal to the sum of the individual charges. Equation (17.3) gives for the equivalent parallel capacitance:

$$\begin{aligned} C_p &= \frac{q}{V} \\ &= \frac{q_1 + q_2 + q_3}{V} \\ &= \frac{q_1}{V} + \frac{q_2}{V} + \frac{q_3}{V} \\ &= C_1 + C_2 + C_3 \end{aligned} \qquad \textbf{(17.9)}$$

The formulas for equivalent series and parallel capitance, Eqs. (17.8) and (17.9), are written for three capacitors, but they hold true for any number of units. The formulas are also useful for analyzing arrangements that are combinations of the two types of arrangements.

EXAMPLE 17.7

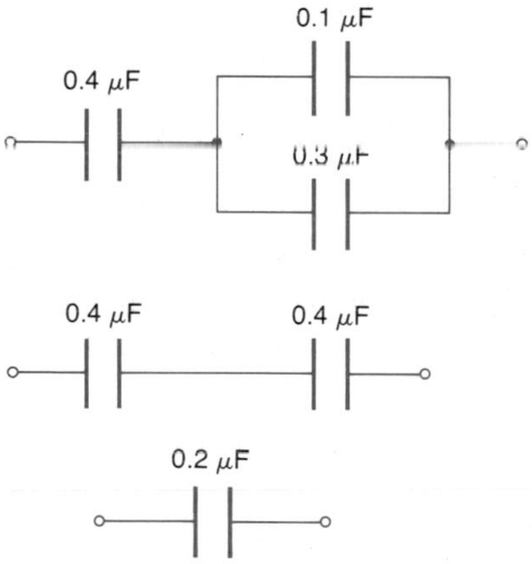

Three capacitors, 0.1 μF, 0.3 μF, and 0.4 μF, are wired as shown in the top part of the diagram. Calculate the equivalent capacitance of the system.

SOLUTION We can analyze arrangements of this type in steps by recognizing that the parts are either simple series or parallel combinations. None of the capacitors are in a simple series arrangement, but the 0.1-μF and 0.3-μF capacitors are in parallel. Their equivalent capacitance follows from Eq. (17.9):

$$\begin{aligned} C_p &= C_1 + C_2 \\ &= 0.1\ \mu\text{F} + 0.3\ \mu\text{F} \\ &= 0.4\ \mu\text{F} \end{aligned}$$

Combining this equivalent capacitance with the remaining capacitor gives the second part of the diagram, which shows two 0.4-μF capacitors in series. We use Eq. (17.8) to find the total equivalent capacitance:

$$\begin{aligned} \frac{1}{C_s} &= \frac{1}{C_1} + \frac{1}{C_2} \\ &= \frac{1}{0.4\ \mu\text{F}} + \frac{1}{0.4\ \mu\text{F}} \\ \therefore C_s &= 0.2\ \mu\text{F} \end{aligned}$$

This is the equivalent capacitance of the whole arrangement. ■

EXAMPLE 17.8

A potential difference of 100 V is applied to the system of capacitors of Example 17.7. Calculate the charge on the 0.1-μF capacitor.

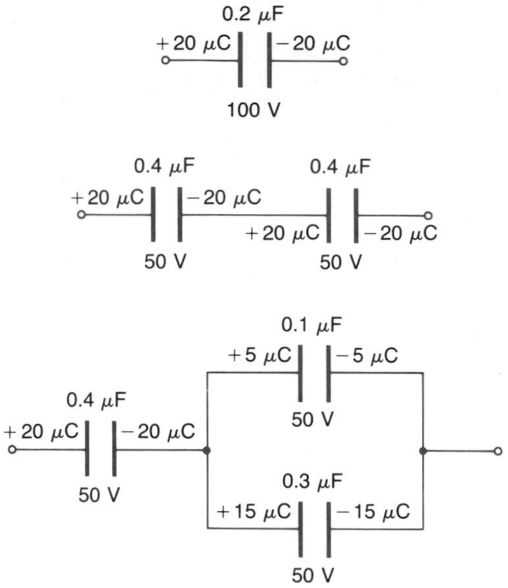

SOLUTION To solve this problem, we must work backward through the equivalent systems. Beginning with the final equivalent system, Eq. (17.3) gives:

$$\begin{aligned} q &= CV \\ &= 2 \times 10^{-7}\ \mathrm{F} \times 100\ \mathrm{V} \\ &= 20\ \mu\mathrm{C} \end{aligned}$$

This is the charge on the system as a whole, and it is also the charge on each two series capacitors of the intermediate equivalent system. The potential difference across each capacitor of the intermediate system is given by Eq. (17.3):

$$\begin{aligned} V &= \frac{q}{C} \\ &= \frac{20 \times 10^{-6}\ \mathrm{C}}{4 \times 10^{-7}\ \mathrm{F}} \\ &= 50\ \mathrm{V} \end{aligned}$$

This is also the potential across each of the capacitors that is in parallel in the original arrangement. Equation (17.3) applied to the 0.1 μC gives:

$$
\begin{aligned}
q &= CV \\
&= 10^{-7}\ \text{F} \times 50\ \text{V} \\
&= 5\ \mu\text{C}
\end{aligned}
$$

It is easy to verify that solution of one charge easily gives all of the other charges, as shown in the diagram. ■

17.5 Motion of a Point Charge in a Constant Electrostatic Field

A point charge q' in a constant electrostatic field E experiences a force $\mathbf{F} = q'\mathbf{E}$ by Eq. (16.2). If the point charge has mass m, then it experiences a constant acceleration $\mathbf{a} = q'\mathbf{E}/m$, given by Eq. (4.1). We can therefore use our previous work on the motion of a point mass subject to constant acceleration to solve the problem of a charged point mass in an electrostatic field. The two equations just mentioned supply all the necessary information, but an energy method can often lessen the work required to solve a problem. In this connection, we recall that Eq. (16.6):

$$U_B - U_A = q'(V_B - V_A)$$

gives the potential energy difference for the charge q' between points of potential V_B and V_A.

Before proceeding with illustrative problems, we mention two approximations that are often made:

1. A uniform field is provided by charges that are uniformly distributed on parallel plates. Introducing a point charge into the region between the plates disturbs the uniform charge distribution on the plates and thereby alters the field that the point charge experiences. We shall always assume that the point charge is so small that it causes a negligible disturbance to the charges on the plates.
2. A charged point mass experiences both the force of its weight and the electrostatic force. The weight is frequently neglected as negligible, but this is not always true and should be checked out for the conditions of the problem.

EXAMPLE 17.9

A pair of charged parallel conducting plates have a potential difference of 75 V between them. An electron that is close to the negative plate has an initial velocity of 3×10^6 m/s. Calculate its velocity just before it strikes the positive plate.

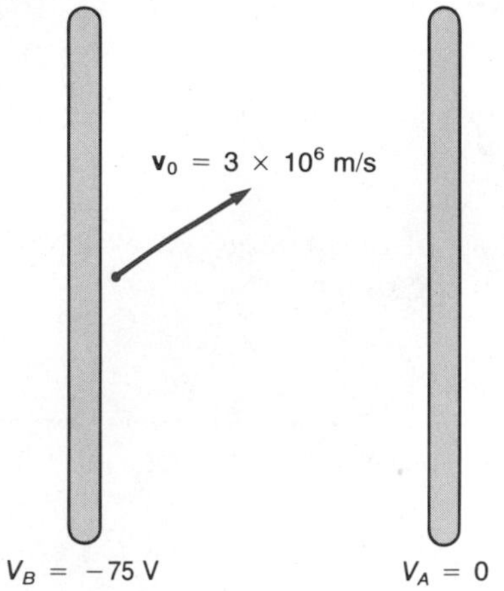

SOLUTION The electrostatic potential is highest near the positive plate and lowest near the negative plate. But since the electron carries a negative charge, its potential energy is highest near the negative plate and lowest near the positive plate. It is convenient to set the final potential energy (near the positive plate) at zero, and this means setting the potential of the positive plate at zero. The potential of the negative plate will therefore be −75 V, and the potential energy of the electron near it is given by:

$$\begin{aligned} U_B - U_A &= q'(V_B - V_A) \\ &= (-1.6 \times 10^{-19}\ \text{C}) \times (-75\ \text{V} - 0\ \text{V}) \\ &= 1.202 \times 10^{-17}\ \text{J} \end{aligned}$$

The energy equation now enables us to calculate the final velocity:

$$\text{Initial (KE + Elec.PE)} = \text{Final (KE + Elec.PE)}$$

$$\tfrac{1}{2}mv_0^2 + q'V = \tfrac{1}{2}mv^2 + 0$$

$$\tfrac{1}{2} \times 9.11 \times 10^{-31}\ \text{kg} \times (3 \times 10^6\ \text{m/s})^2 + 1.202 \times 10^{-17}\ \text{J}$$

$$= \tfrac{1}{2} \times 9.11 \times 10^{-31}\ \text{kg} \times v^2$$

$$\therefore v = 5.95 \times 10^6\ \text{m/s}$$

At first sight we might suspect an error in this calculation, since it might seem that the final velocity should depend on the direction of the initial velocity. However, we can work the example from force and acceleration considerations to show that the components of the final velocity depend on the direction of the initial velocity, but that its magnitude does not. ■

EXAMPLE 17.10

The electron of Example 17.9 has its initial velocity making an angle θ with the line perpendicular to the plates. The plate separation is 1 cm. Calculate the components of velocity parallel and perpendicular to the plates just before the electron strikes the positive plate.

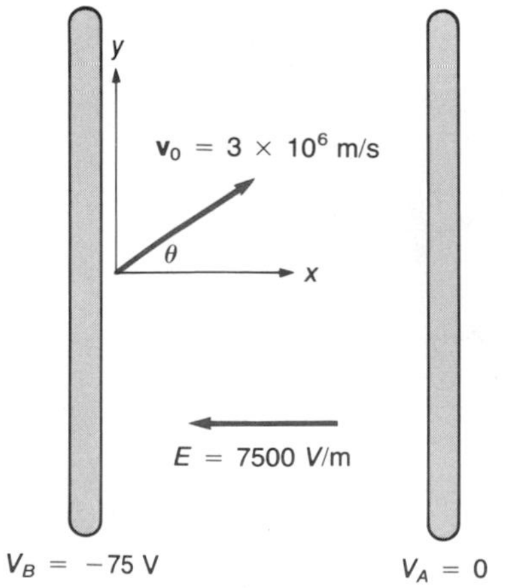

SOLUTION Let the x-direction be perpendicular to the plates, and the y-direction parallel to them. Then the initial velocity components are

$$v_{0x} = 3 \times 10^6\ \text{m/s} \times \cos\theta$$

$$v_{0y} = 3 \times 10^6\ \text{m/s} \times \sin\theta$$

Since the electric field is uniform between the plates and perpendicular to them, the electron experiences no force in the y-direction, and its final velocity component in this direction is 3×10^6 m/s $\times \sin\theta$.

To calculate the electric field in the x-direction, we use Eq. (17.2):

$$\begin{aligned} E_x &= \frac{V}{d} \\ &= \frac{-75 \text{ V}}{10^{-2} \text{ m}} \\ &= -7500 \text{ V/m} \end{aligned}$$

The force on the electron is given by Eq. (16.2) (modified for x components):

$$\begin{aligned} F_x &= q'E_x \\ &= (-1.602 \times 10^{-19} \text{ C}) \times (-7500 \text{ V/m}) \\ &= 1.202 \times 10^{-15} \text{ N} \end{aligned}$$

(At this point we recollect that the weight of the electron is given by $w = mg = 9.11 \times 10^{-31}$ kg $\times$ 9.81 m/s^2 = 8.93×10^{-30} N; a figure that is utterly negligible compared with the electrostatic force.)

The acceleration is given by Eq. (4.1) (modified for x components):

$$\begin{aligned} F_x &= ma_x \\ \therefore a_x &= \frac{F_x}{m} \\ &= \frac{1.202 \times 10^{-15} \text{ N}}{9.11 \times 10^{-31} \text{ kg}} \\ &= 1.319 \times 10^{15} \text{ m/s}^2 \end{aligned}$$

We can calculate the final velocity component v_x with the help of Eq. (2.12d) (modified for x components):

$$\begin{aligned} v_x^2 &= v_{0x}^2 + 2a_x x \\ &= (3 \times 10^6 \text{ m/s} \times \cos\theta)^2 + 2 \times 1.319 \times 10^{15} \text{ m/s}^2 \times 10^{-2} \text{ m} \\ \therefore v_x &= \{9 \times 10^{12} \times \cos^2\theta + 2.638 \times 10^{13}\}^{1/2} \text{ m/s} \end{aligned}$$

Recollecting that:

$$v_y = \{3 \times 10^6 \times \sin\theta\} \text{ m/s}$$

we see that both components depend on the angle θ. However, the magnitude of the final velocity is [using Eq. (3.4):

$$v = (v_x^2 + v_y^2)^{1/2}$$

$$= \{9 \times 10^{12} \times \cos^2 \theta + 2.638 \times 10^{13} + 9 \times 10^{12} \times \sin^2 \theta\}^{1/2} \text{ m/s}$$

Using the trigonometric identity $\sin^2 \theta + \cos^2 \theta = 1$, we have finally:

$$v = 5.95 \times 10^6 \text{ m/s}$$

This is the same answer that we derived much more easily in Example 17.9. The energy method works quickly and easily provided that one does not require the fine details. If fine details are required, the more tedious methods based on forces and accelerations will always provide them. ■

KEY CONCEPTS

A pair of **charged parallel plate conductors** has lines of force that are straight, uniformly spaced, and normal to the conductors; the equipotentials are uniformly spaced flat planes that are parallel to the conductors.

The **electrostatic field strength** between charged parallel plates can be calculated in terms of the **charges** on the plates [Eq. (17.1)], or in terms of the **potential difference** between them [Eq. (17.2)].

The **capacitance** of a pair of conductors that carry equal and opposite charges is equal to the magnitude of the charge divided by the potential difference that it causes; see Eq. (17.3). For parallel plates there is a simple formula for the capacitance [Eq. (17.4)].

The **electrostatic potential energy** of a charged capacitor can be written in terms of any two of the following quantities: charge, capacitance, and potential difference; see Eq. (17.5).

The **dielectric constant** of insulating material is the factor by which the capacitance increases when it is placed between the plates of a capacitor that originally had air insulation; see Eq. (17.6).

The **equivalent capacitance** of capacitors that are wired in **series** or in **parallel** can be calculated from the individual capacitances; see Eqs. (17.8) and (17.9).

QUESTIONS FOR THOUGHT

1. Give an example of a conducting object that has negative charge on some parts and positive charge on other parts. Can lines of force go from positive to negative charges on the surface of the same conductor?

2. Why must the electrostatic field inside a conductor always be zero? Does this mean that the electrostatic potential inside a conductor must also be zero?

3. Is it possible to define the capacitance of a single conductor? (*Hint:* Where do the lines of force from it terminate?)

4. A thin flat metal plate is placed between the plates of a parallel plate capacitor and parallel to them. Does this alter the capacitance between the two outer plates?

5. A parallel plate capacitor is connected to a source of constant potential (such as a battery), and a slab of dielectric is partly inserted between the plates. Is the dielectric drawn further into the region between the plates, is it rejected, or is there no force on it?

6. Two capacitors of different capacitance are wired in series and connected to a source of constant potential. In which of the two capacitors is the stored energy greatest? Does your answer change if the capacitors are wired in parallel?

7. What desirable feature of a capacitor do you associate with a high value of the dielectric constant? With a high value of the dielectric breakdown strength?

PROBLEMS

A. Single-Substitution Problems

1. A pair of closely spaced square metal plates of side 40 cm carry equal and opposite charges. If the electric field between the plates is 2500 V/m, calculate the magnitude of the charges. [17.1]

2. The electric field strength between a pair of closely spaced parallel plates is 800 V/m. Calculate the plate area if the charges are ± 6 μC. [17.1]

3. Two closely spaced parallel metal plates, each 20 cm in radius, carry charges of ± 3 μC. What is the electric field strength between them? [17.1]

4. A pair of conducting parallel plates are separated by a 1.5-mm air gap. Calculate the electric field strength between the plates if the potential difference between them is 2000 V. [17.1]

5. A pair of parallel conducting plates 1.2 cm apart have a uniform electric field of 8000 V/m between them. Calculate the potential difference between the plates. [17.1]

6. Calculate the potential difference required to cause a charge of 25 μC on a 0.2-μF capacitor. [17.2]

7. Calculate the charge required to cause a potential difference of 120 V on a 0.04-μF capacitor. [17.2]

8. Calculate the plate area required to form a 0.1-μF capacitor, the parallel plates being separated by an 0.8-mm air gap. [17.2]

9. Two parallel conducting plates, each 60 cm × 40 cm, are separated by a 1.25-mm air gap. Calculate the capacitance of the parallel plates. [17.2]

10. How much energy is stored in a 0.1-μF capacitor that has a potential difference of 1200 V between the plates? [17.2]

11. What capacitance is required to store 250 mJ of energy at 300 V potential difference? [17.2]

12. Calculate the charge on a 0.25-μF capacitor that stores 450 mJ of energy. [17.2]

13. A 0.01-μF capacitor consists of two parallel plates, each of area 0.4 m^2 and separated by 1.5 mm. Calculate the dielectric constant of the insulating material. [17.3]

14. A parallel plate capacitor uses polyethylene (dielectric constant 2.3) as the insulating material. Calculate the capacitance if the plate area is 0.75 m^2 and the separation is 0.25 mm. [17.3]

15. What plate area is required to form a capacitance of 5 nF with mica insulation 0.16 mm thick? (The dielectric constant of mica is 4.6.) [17.3]

16. A certain glass has a dielectric constant of 5.6. What thickness of this insulation is required to form a 0.12-μF capacitor with plates 0.04 m^2 in area? [17.3]

17. Calculate the equivalent capacitance for three 0.1-μF capacitors that are wired (a) in series, (b) in parallel. [17.4]

18. Three 0.5-μF capacitors are wired, as shown in the diagram. Calculate the equivalent capacitance. [17.4]

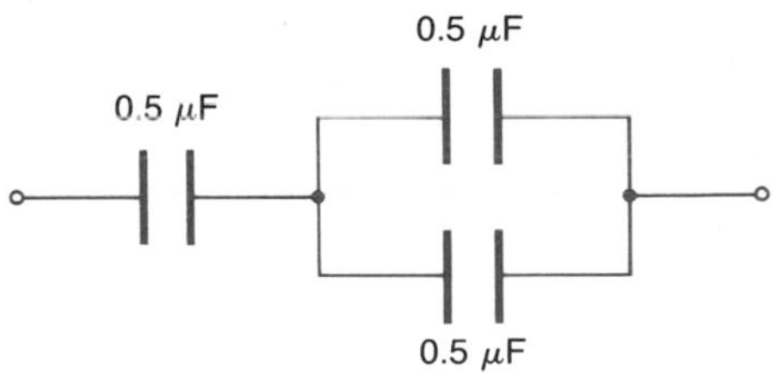

19. What potential difference is required to accelerate an electron from rest to a speed of 10^6 m/s? [17.5]

20. What potential difference is required to accelerate a proton from rest to a speed of 10^6 m/s? [17.5]

B. Standard-Level Problems

21. Two parallel metal plates, each of area 1500 cm^2, are separated by 1 mm. Calculate the charge on each plate if the potential difference between them is 640 V. [17.1]

22. Two square parallel conducting plates of side length 75 cm are separated by 1.6 mm.

a. What potential difference is caused by charges of $\pm 60\ \mu C$ on the plates?
b. What charge is required to cause a potential difference of 1000 V between the plates? [17.1]

23. Two circular metal plates are separated by 1.8 mm. Calculate the plate radius if charges of $\pm 5\ \mu C$ cause a potential difference of 800 V between them.

24. Two parallel circular metal plates each of radius 80 cm carry charges of $\pm 1.2\ \mu C$. What is the plate separation if the potential difference is 1.2 kV?

25. A parallel plate capacitor has plates 20 cm $\times$ 20 cm spaced 1.5 mm apart. Calculate the stored energy if the potential difference is 500 V. [17.2]

26. The plates of a parallel plate capacitor are 1.2 mm apart. Calculate the plate area if a potential difference of 120 V stores 5 μJ of energy. [17.2]

27. A photoflash unit consists of a xenon-filled glow tube and a capacitor that is slowly charged to a potential difference of 12 V. The glow tube requires 40 mJ of energy for the flash. What is the value of the capacitance?

28. The electrostatic potential energy stored in a thunderstorm can be estimated by considering the storm as a closely spaced parallel plate capacitor. A storm cloud covers an area 12 km $\times$ 12 km at a height of 0.8 km above the earth's surface. Taking the cloud to be one plate of a capacitor and the earth the other plate, estimate the stored energy if the average electric field between the cloud and the earth is 8×10^5 V/m.

29. Two aluminum strips 5 cm wide and a polyethylene strip of the same width and 0.1 mm thick are used to make a parallel plate capacitor; the capacitor is formed by sandwiching the polyethylene between the metal strips. What length of strip is required for a 0.2-μF capacitor? [17.3]

30. Two aluminum strips 4 cm wide and 2 m long are used to make a parallel plate capacitor with paper insulation 0.08 mm thick. Calculate the dielectric constant of the paper if the capacitance is 0.03 μF. [17.3]

31. A mica capacitor is made by sandwiching a sheet of mica 0.035 mm thick between square metal plates. What is the side length of the metal plates if the capacitor stores 6 mJ of energy at a potential difference of 1.2 kV?

32. A 0.5-μF capacitor has mica as the insulating material. The capacitor is connected to a 50-V supply voltage, and fully charged. Calculate:

a. the charge on the capacitor
b. the energy stored in the capacitor

The supply voltage is now disconnected (leaving the charge on the capacitor) and the mica insulator is withdrawn. Calculate:

c. the energy stored in the capacitor
d. the potential difference between the plates

33. A capacitor of 0.1 μF with air insulation is connected to a supply voltage of 120 V and fully charged. Calculate:

a. the charge on the capacitor
b. the energy stored in the capacitor

The supply voltage is now disconnected (leaving the charge on the plates) and a sheet of polyethylene insulation is inserted. Calculate:

c. the energy stored in the capacitor
d. the potential difference between the plates

34. Calculate the equivalent capacitance of the four capacitors 0.1 μF, 0.3 μF, 0.25 μF, and 0.6 μF wired as shown in the diagram. [17.4]

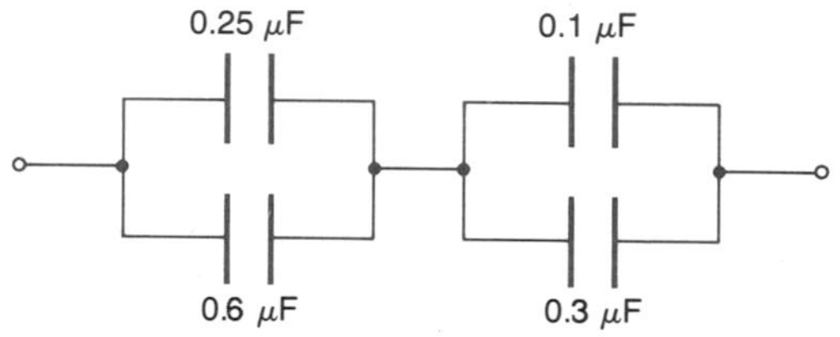

35. The equivalent capacitance of the four capacitors shown on the diagram is 0.15 μF. Calculate the capacitance of capacitor *C*. [17.4]

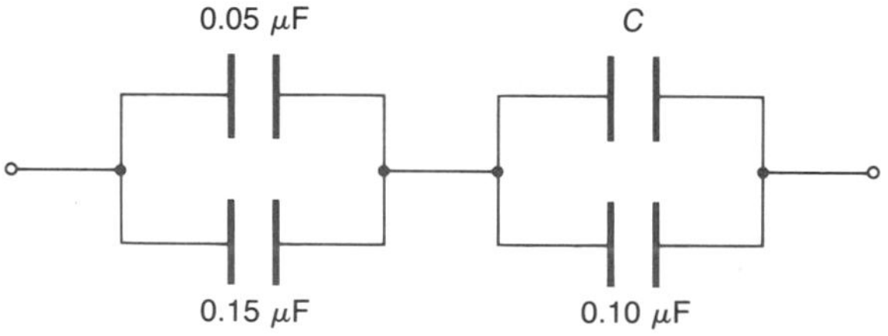

36. Three capacitors are connected as shown in the diagram. Calculate the value of the capacitance *C* if a potential difference of 20 V between points *A* and *B* causes a charge of $\pm 28\ \mu C$ on the system.

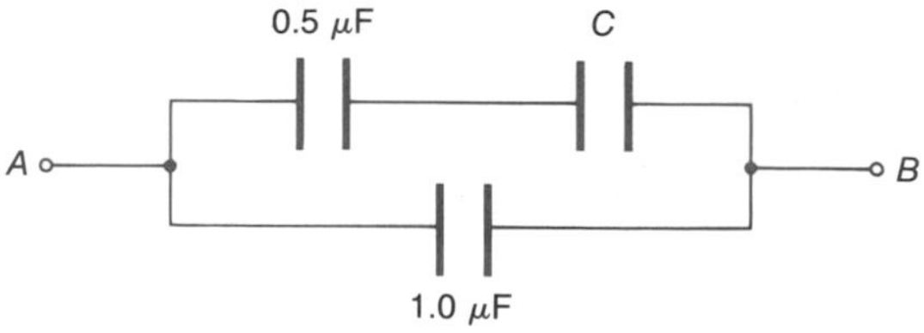

37. Three capacitors are connected as shown in the diagram.

a. Calculate the equivalent capacitance.
b. A potential difference of 30 V is maintained between points A and B. Calculate the charge and the potential difference across the 4-μF capacitor.

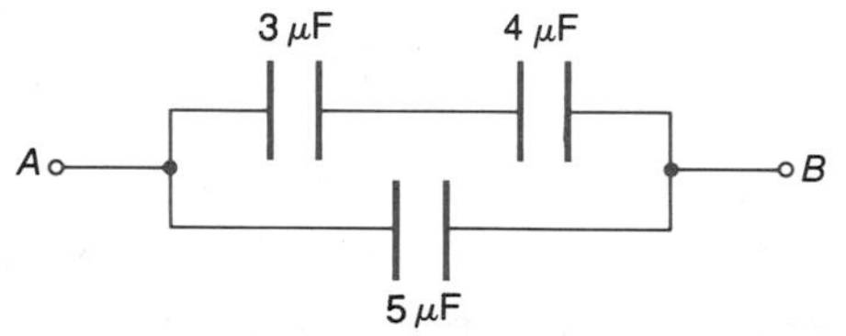

38. Two parallel metal electrodes in a vacuum carry equal and opposite charges, and each has an area of 4000 cm^2. The plates are 1.8 cm apart. An electron adjacent to the negative plate has an initial velocity of 2×10^6 m/s, and just before striking the positive plate its velocity is 6×10^6 m/s. Calculate the magnitude of the charges on the plates. [17.5]

39. Two parallel metal plates in vacuum each have an area of 3000 cm^2 and carry charges of ± 6 μC. The plates are 1.6 cm apart. A proton adjacent to the positive plate has an initial velocity of 1.25×10^6 m/s at an angle of 30° with the surface of the plate. Calculate:

a. the speed of the proton just before it strikes the negative plate
b. the component of the proton's velocity that is perpendicular to the negative plate just before it strikes it

C. Advanced-Level Problems

40. Two parallel metal plates of area 1.2 m^2 and separation 0.75 mm carry charges of ± 60 μC. A small plastic ball of mass 15 g hangs by a thread from the upper (positively charged) plate. How many excess electrons does the plastic ball carry if the tension in the thread is 0.05 N?

41. From a box of 10 capacitors, each 0.1 μF and rated at 500 V, construct circuits with the following characteristics:

a. a capacitance of 0.1 μF rated at 1000 V
b. a capacitance of 0.25 μF rated at 500 V

42. A parallel plate capacitor is designed with mica insulation. The capacitance is to be 0.25 μF rated at 1500 V. What should be the area of each plate if the field in the mica is not to exceed 1.8×10^7 V/m?

18

DIRECT CURRENT CIRCUITS

The flow of electric charge from one place to another is called electric current. This phenomenon may occur in gaseous, liquid, or solid materials. Air carries electric current during a stroke of lightning; the liquid in an electrolytic tank is the vehicle for electric current; and the metal wire carrying power to a house is also a carrier of electric current. An arrangement of objects in a continuous closed loop that carries electric current is called an electric circuit. Many products of modern technology, such as television sets, clinical X-ray machines, and microwave ovens, contain complex masses of electric circuits. In some parts of these circuits the electric current is in low-pressure gas, but in the great majority of cases it exists in solid circuit elements. In order to establish our understanding, we shall focus our attention on the very important case of electric current in solid materials.

18.1 Electric Current

Consider two objects charged so that an electrostatic potential difference exists between them. For example, we could charge two blocks of copper positively and negatively by induction, using ebony and glass rods, respectively. If we now connect the copper blocks to each other by means of a copper wire, electric charge flows in the wire for a short period of time and then ceases. We call the flow of electric charge *electric current*. However, practical applications usually require some form of continuing charge flow. Depending on the nature of the application, the continuing charge flow may be steady and persistent in one direction, or it may continually reverse direction. The first case is called a *direct current* (abbreviated ''dc'') and the second an *alternating current* (abbreviated ''ac''). We will postpone the study of alternating currents and concentrate on direct currents.

The first practical device capable of causing a direct current in a circuit was the Voltaic pile, invented by Alessandro Volta (1745–1827). According to Volta, his pile was composed of several dozen pieces of copper or silver, each in contact with a piece of tin or zinc, and immersed in salt water or lye. The pile did not produce the spectacular sparks associated with frictional electrostatic generators; instead, it produced electric current for a prolonged period. Volta was greatly honored by Napoleon, who correctly predicted that his invention would be of immense practical importance.

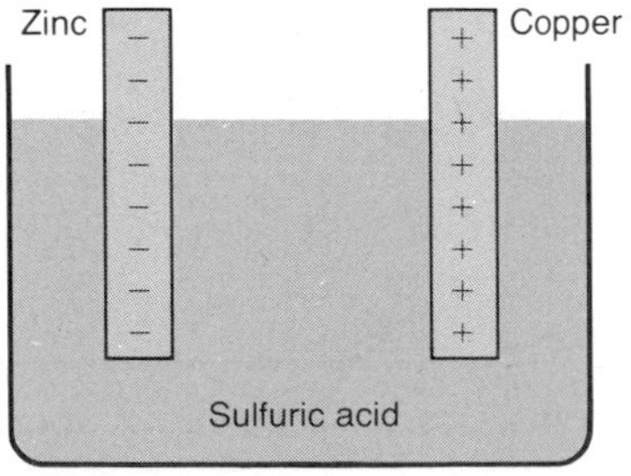

FIGURE 18.1 The elementary cell of a Voltaic pile.

Figure 18.1 shows a single element of a Voltaic pile. Zinc and copper electrodes rest in a solution of dilute sulfuric acid. As a result of chemical action, the copper electrode becomes positively charged, and the zinc electrode negatively charged. This establishes a potential difference between the two electrodes. If an external circuit (such as a piece of wire) is connected between the two electrodes, an electric charge flows because of the potential

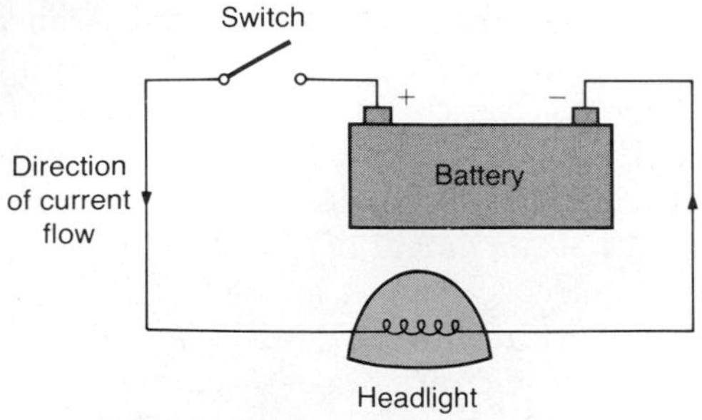

FIGURE 18.2 The supply of current to an automobile headlight from the battery is a simple example of a direct current circuit.

difference. In addition, the chemical action maintains the potential difference and causes persistent electric current in the wire.

In modern terminology a Voltaic pile is called a *battery,* and various chemical systems are used in batteries for different applications. A typical example is an automobile battery supplying direct current to the headlight filament, as shown in Fig. 18.2. The headlight terminals are connected to the battery terminals by means of metal wires, and a switch is inserted in one wire. When the switch is open, there is no current; when the switch is closed, the circuit is completed and there is current in the headlight filament.

With these preliminaries, we can now set up a quantitative relationship between the passage of charge and **electric current.** The SI current unit is named in honor of André Marie Ampère (1775–1836), a French physicist famous for his discoveries in electromagnetism. If the current is steady, we can drop the averaging symbol from Eq. (18.1).

■ Relationship Between Electric Current and Charge Flow

$$\text{Average electric current} = \frac{\text{charge flow}}{\text{elapsed time}}$$

$$\langle I \rangle = \frac{q}{t} \qquad \textbf{(18.1)}$$

SI unit of electric current: coulomb per second
New name: ampere
Abbreviation: A

EXAMPLE 18.1

A steady 5-A current exists in a copper wire. How many electrons pass a fixed point in the wire in 1 min?

SOLUTION We first calculate the quantity of charge passing a point in the wire in 1 min by using Eq. (18.1):

$$I = \frac{q}{t}$$

$$\therefore q = It$$

$$= 5\ \text{A} \times 60\ \text{s}$$

$$= 300\ \text{C}$$

Each electron carries a charge of -1.6×10^{-19} C. The number of electrons required to cause a charge flow of 300 C is given by:

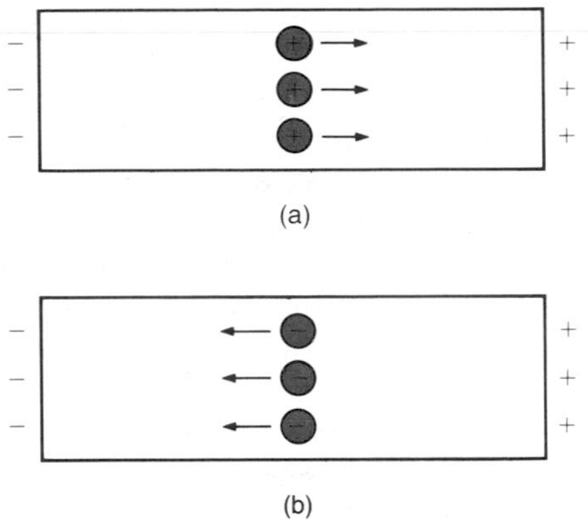

FIGURE 18.3 Transport of positive charge to the right in (a) produces the same effect as transport of negative charge to the left in (b). In both cases the positive direction of conventional current flow is from left to right.

$$n = \frac{300\ \text{C}}{1.6 \times 10^{-19}\ \text{C}}$$
$$= 1.87 \times 10^{21}\ \text{electrons}$$

We have disregarded the negative sign in calculating the number of electrons, because its importance lies only in relating the direction of current to the direction of motion of the charge carriers, as shown in Fig. 18.3. ■

Because electric current is defined as a quantity of charge divided by an elapsed time, it is clearly a scalar. Strictly speaking, it is therefore incorrect to speak of current in a certain direction, since a scalar quantity does not have a direction. However, the existence of the current means transport of electric charge. The charge carriers may carry either positive or negative charge. In p-type semiconductors the carriers are predominantly positive; in metals the electron charge carriers are negative, and in gas discharges both positive and negative carriers contribute to the charge flow. We define *conventional current* as a quantity having the magnitude of the current, and the direction of positive charge flow. Note that the direction of the conventional current does not depend on the sign of the charge carrying particles. We illustrate this point in Fig. 18.3. In part (a) of the diagram, positive charge flows to the right in a conductor, making this end of the conductor positive and the left end negative. In part (b) negative charge flows to the left, producing the same effect. Henceforth, we shall simply refer to the current direction with the understanding that we mean the direction of the conventional current.

Battery terminals are usually marked positive (red) and negative (black). The current direction *in the external circuit* is from the positive to the negative terminal. Since the charge flows in a complete circuit, the positive direction of current *within the battery* is from the negative to the positive terminal.

18.2 Ohm's Law

We have already discussed the current in a circuit element (such as an automobile headlight) caused by a battery. In the external circuit the current direction is from positive to negative, so that the equivalent positive charge flow is from higher to lower potential. But the charge flow is continuous, and within the battery the equivalent positive charge flow is from lower to higher potential. Any device that can cause an equivalent positive charge flow from lower to higher potential is said to be capable of producing an *electromotive force* (abbreviated "emf"). Every dc circuit must contain at least one source of emf, since it is impossible that equivalent positive charge should flow from higher to lower potential for all intervals of a circuital path in the absence of any source of emf.

Well-known sources of emf are batteries, electrical generators, thermocouples, and photo-cells. In each case, they cause an equivalent positive charge flow against a certain potential difference; the emf is, therefore, measured in volts. We use the script symbol $\mathscr{E}$ to mean emf—for example, $\mathscr{E} = 12$ V for most automobile batteries. Chemical action within the battery maintains a potential difference of 12 V across the external circuit under normal service conditions.

Consider now a circuit element that is part of a complete circuit supplied by a battery (or other source of emf). Let the potential difference across the circuit element be V, and let the current in it be I. The ratio of potential difference to current is called the *resistance* of the circuit element. The unit of resistance is named in honor of George Simon Ohm (1789–1854), a Bavarian who did pioneering work in the theory of dc circuits. The standard abbreviation for ohm is the capital Greek omega (Ω). It is essential to note that the voltage quoted in the definition of resistance is the *voltage difference* between the ends of the circuit element in question.

■ Definition of Electrical Resistance

$$\text{Resistance} = \frac{\text{potential difference across circuit element}}{\text{current in circuit element}}$$

$$R = \frac{V}{I} \tag{18.2}$$

SI unit of resistance: volt per ampere
New name: ohm
Abbreviation: Ω

EXAMPLE 18.2

The high-beam filament of an automobile headlight carries a current of 4.5 A, and the voltage difference between its terminals is 12 V. Calculate the resistance of the filament under these operating conditions.

SOLUTION The problem is a simple calculation involving Eq. (18.2):

$$R = \frac{V}{I} = \frac{12\ \text{V}}{4.5\ \text{A}} = 2.67\ \Omega$$

■

SPECIAL TOPIC 19 OHM

Georg Simon Ohm was born in Bavaria in 1787, and his father took care to see that his son received a good education in science. As a young man, Ohm had a high school teaching appointment, but his ambition was to teach in a university. He decided to do experimental work in the new field of current electricity, and as a relatively impoverished teacher, he had to make a good deal of his own equipment. His work established the basis for the definitions of electrical resistivity and resistance. For some reason, he aroused opposition and resentment in his own country to such an extent that he even had to resign his high school teaching position.

However, the importance of his work was slowly realized abroad. In 1841 the Royal Society of London awarded him the Copley Medal and made him a member. Eventually Bavarians recognized his fame, and he was appointed professor at the University of Munich, which allowed him to spend the closing years of his life in the position he had aspired to as a young man.

His name is used twice in assigning units; the electrical resistance unit is the ohm, and the unit of electrical conductance (which is the reciprocal of the resistance) is the mho.

Generally speaking, the resistance of a circuit element depends on the voltage applied between its terminals. For example, the high-beam filament of Example 18.2 would exhibit a different resistance if it was connected to a 6-V battery. In this instance, the change in resistance is due to the temperature change that the current causes, but the resistance in many types of circuit elements depends directly on voltage even for negligible temperature change.

However, a very large class of circuit elements exists in which the resistance does not depend on the applied voltage. Such elements are called **ohmic,** and they obey **Ohm's law.**

Ohm's Law

For a large class of circuit elements, the resistance does not depend on the voltage or the current over a wide range of operating conditions.

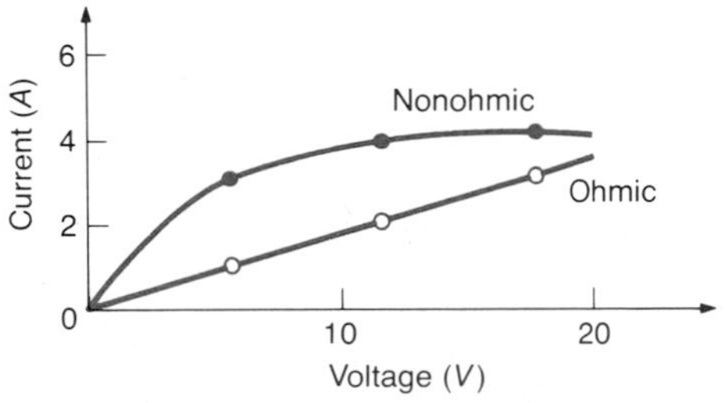

FIGURE 18.4 Typical data points for voltage and current of ohmic and nonohmic circuit elements.

Samples of typical current-voltage characteristics of an ohmic and a nonohmic circuit element are shown in Fig. 18.4. For the nonohmic circuit element the current and voltage measurements at each experimental point give different values for the resistance. On the other hand, every experimental point yields the same resistance for the ohmic circuit element. For this reason, we call an ohmic circuit element a **resistor,** and its circuit properties are determined by the value of its **resistance.** For the present we will concentrate on ohmic circuit elements.

EXAMPLE 18.3

Calculate the current when a 12-V battery is connected across a 4-Ω resistor.

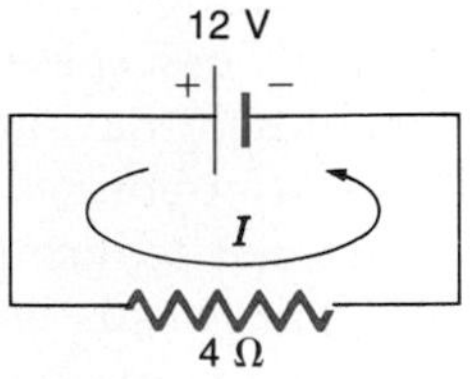

SOLUTION The diagram shows the conventional ways of representing a battery and a resistor. A pair of thick parallel lines of unequal length and thickness represent the battery; the longer and thinner of the two lines is the positive side. The zigzag line represents a resistor, and the straight lines are electrical connections of negligible resistance. The resistor is ohmic, and Eq. (18.2) gives:

$$R = \frac{V}{I}$$

$$\therefore I = \frac{V}{R}$$

$$= \frac{12\ \text{V}}{4\ \Omega}$$

$$= 3\ \text{A}$$

■

In this simple example the battery emf is equal to the voltage across the resistor because the resistor is connected straight to the battery terminals. If the resistor is not so connected, the voltage across it is equal to the value of the resistance multiplied by the current. This follows from the cross-multiplication of terms in Eq. (18.2) to give:

$$V = RI \qquad \textbf{(18.2a)}$$

Likewise, the current in a resistor is determined by the resistance and the voltage across it:

$$I = \frac{V}{R} \qquad \textbf{(18.2b)}$$

The physical content of Eqs. (18.2), (18.2a), and (18.2b) is the same, but the emphasis differs.

18.3 Electrical Resistivity

We can calculate the electrical resistance of circuit elements made from pure substances from the geometrical dimensions of the circuit element and a characteristic constant of the substance. Consider, for example, a wire of uniform cross-sectional area A and

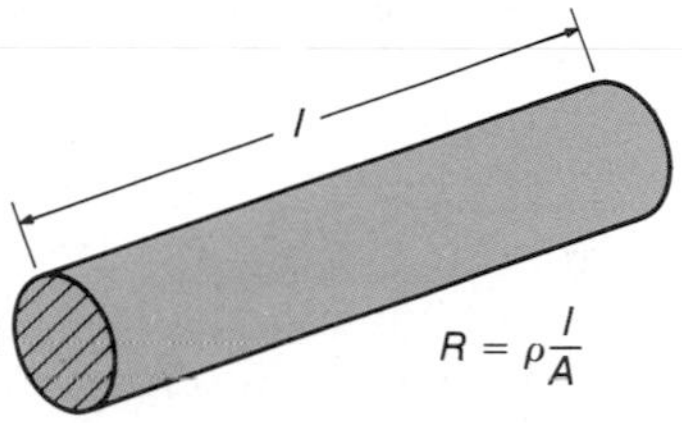

FIGURE 18.5 Illustration of the calculation of resistance from resistivity.

length l, as shown in Fig. 18.5. The resistance of the wire is directly proportional to its length and inversely proportional to its cross-sectional area. We incorporate these experimental facts into the definition of the **resistivity** of the wire material.

■ Definition of Electrical Resistivity

$$\text{Resistance} = \frac{\text{resistivity} \times \text{length}}{\text{cross-sectional area}}$$

$$R = \rho\frac{l}{A} \qquad \textbf{(18.3)}$$

SI unit of resistivity: ohm-meter
Abbreviation: $\Omega \cdot \text{m}$

The symbol for resistivity is the Greek letter rho (ρ), which is also used for density [see Eq. (1.1)], but this should not cause confusion. Table 18.1 shows the resistivities of several substances. The first group of substances are pure metals and alloys, all of which are good conductors with relatively low resistivities. Germanium and silicon are semiconductors, whose resistivity depends strongly on impurities; the values quoted indicate the order of magnitude only. The third group shows the enormous resistivities associated with high-class insulators. Finally, the resistivities of some common liquids illustrate the large effect of an impurity (such as salt) on the resistivity of water.

TABLE 18.1 Resistivities and temperature coefficients of resistivity for various materials

Substance	*Resistivity at 20°C* ($\Omega \cdot$ m)	*Temperature coefficient of resistivity* ($(C°)^{-1}$)
Conductors		
Silver	1.58×10^{-8}	3.8×10^{-3}
Copper	1.72×10^{-8}	3.8×10^{-3}
Gold	2.44×10^{-8}	3.4×10^{-3}
Aluminum	2.82×10^{-8}	3.9×10^{-3}
Tungsten	5.6×10^{-8}	4.5×10^{-3}
Nickel	6.6×10^{-8}	3.4×10^{-3}
Platinum	1.1×10^{-7}	3.9×10^{-3}
Constantan	4.9×10^{-7}	5×10^{-6}
Nichrome	1.1×10^{-6}	4×10^{-4}
Semiconductors		
Germanium	10	—
Silicon	3000	—
Insulators		
Glass	10^{10}–10^{13}	—
Polyethylene	10^{14}	—
Teflon	10^{16}	—
Fused quartz	10^{17}	—
Liquids		
Seawater	0.25	—
Ethanol	3300	—
Distilled water	5000	—

EXAMPLE 18.4

A 5-Ω resistor is made from nichrome wire 1 mm in diameter. What length of wire is required?

SOLUTION The resistivity of nichrome is $1.1 \times 10^{-6}\ \Omega \cdot \text{m}$ from Table 18.1. We can calculate the length of wire required by using Eq. (18.3):

$$R = \rho\frac{l}{A}$$

$$\therefore l = \frac{RA}{\rho}$$

$$= \frac{5\ \Omega \times \pi \times (0.5 \times 10^{-3}\ \text{m})^2}{1.1 \times 10^{-6}\ \Omega \cdot \text{m}}$$

$$= 3.57\ \text{m}$$

■

SPECIAL TOPIC 20 SUPERCONDUCTIVITY

The resistivity of metals decreases with decreasing temperature and falls to a very low value near absolute zero. In 1911, Kamerlingh Onnes, professor of experimental physics at the University of Leiden, discovered that the resistivity of some metals falls to an immeasurably small value below a certain critical temperature. In this condition of essentially zero resistivity, the metal is called a superconductor. The first superconductor discovered was mercury, which has a critical temperature of 4.15 K. Later, several dozen other elements were found to be superconducting, and all of them had critical temperatures below 10 K. The quantum mechanical theory of superconductivity depends on explaining the creation of superconducting electron pairs, and this theory was not put in a satisfactory form until nearly half a century after the first experimental work.

In electrical equipment, conductors of zero resistivity would offer many advantages. For example, there would be no energy dissipation in Joule heating; but, even more important than the saving of energy, there would be other advantages based on this absence of dissipation. For instance, electrical power transmission lines could be run at a lower voltage; this would mean that they would be smaller, less dangerous, and easier to install underground. Electric motors could be smaller and lighter, and computers could be faster—just to mention a few of the technical benefits.

To reach the temperature required for superconductivity in the first substances that exhibited this behavior, liquid helium had to be used. However, the relatively large expense of liquid helium refrigeration precluded nearly all applications except for laboratory magnets, which were used to create high magnetic fields for research purposes. In 1967, the critical temperature for a niobium–aluminum–germanium alloy was raised above 20 K. The refrigeration for this superconductor could be produced using liquid hydrogen, but from a commercial viewpoint this represents only a slight advantage.

In 1986, Georg Berdnorz and Alex Muller discovered a superconducting oxide of a rare earth–barium–copper alloy with a critical temperature above 30 K. In subsequent research, which won them the Nobel prize, they found critical temperatures as high as 90 K in oxides of similar alloys. At this point, the possibility of using liquid nitrogen refrigeration became feasible since the liquefaction point of nitrogen is 77 K. This was a great advance, because liquid nitrogen is much less expensive and easier to handle than either liquid helium or liquid hydrogen. The ideal, however—a room temperature superconductor that will remove the refrigeration problem altogether—is still being sought.

The problem of an easily attainable critical temperature is only one of those facing researchers in the field of superconductivity. Superconducting materials must also have mechanical properties that make them suitable for the application being considered. For example, a superconductor that is both brittle and mechanically weak will pose difficulties in applications such as motors or transmission lines, although it can be used in computers.

Limited commercial use of superconductors may be possible in the near future, but whether they will ever enjoy widespread use is still in doubt.

For the semiconducting and insulating substances listed in Table 18.1, the resistivity is strongly dependent on both the impurity content and temperature—an increase in either factor sharply decreases the resistivity. On the other hand, the resistivity of the metallic conductors is relatively insensitive to impurity content and increases slowly with rising temperature. The change in the resistivity of metals with increasing temperature is described approximately by the **temperature coefficient of resistivity** for temperatures near room temperature.

■ Definition of Temperature Coefficient of Resistivity

Fractional change in resistivity = temperature coefficient of resistivity × temperature change

$$\frac{\rho - \rho_0}{\rho_0} = \alpha(T - T_0) \tag{18.4}$$

Alternate form:

$$\frac{R - R_0}{R_0} = \alpha\,(T - T_0)$$

SI unit of temperature coefficient of resistivity: (Celsius degree)$^{-1}$
Abbreviation: (C°)$^{-1}$

The alternate form of Eq. (18.4) follows since resistance is proportional to resistivity for any given resistor. The formula does not hold at sufficiently low temperatures, where fractional resistivity change is not proportional to temperature change. At sufficiently high temperatures, the resistor material melts, and the formula is again invalid. Table 18.1 shows values for the temperature coefficients of resistivity of metals and alloys.

EXAMPLE 18.5

Two resistors, one made from copper wire and the other from wire of the alloy constantan, each have the same resistance at 20°C. Calculate the percentage changes in their resistance if the temperature is raised to 150°C.

SOLUTION The temperature coefficients of resistivity for copper and constantan are 3.8×10^{-3} (C°)$^{-1}$ and 5×10^{-6} (C°)$^{-1}$, respectively, from Table 18.1. The alternate form of Eq. (18.4) is suitable for our problem:

$$\begin{aligned}\frac{R - R_0}{R_0} &= \alpha(T - T_0)\\ &= 3.8 \times 10^{-3}\,(\text{C}°)^{-1} \times (150°\text{C} - 20°\text{C})\\ &= 49\% \text{ for copper}\\ &= 5 \times 10^{-6}\,(\text{C}°)^{-1} \times (150°\text{C} - 20°\text{C})\\ &= 0.065\% \text{ for constantan}\end{aligned}$$

The example shows a significant change in the resistance of the copper resistor and only a very small change for the constantan. This alloy was developed (and named) for use in situations requiring constant resistance as the temperature changes. ■

Temperature coefficients of resistivity are not shown for insulators or semiconductors because such figures would be very large, negative, and strongly dependent on impurity content. The resistivity figures shown are only approximate but do permit order-of-magnitude calculations of current.

EXAMPLE 18.6

A Teflon spacer 2 mm thick and 1 cm in radius insulates two metal plates from each other. Calculate the leakage current in the insulator if the plates are at a potential difference of 12 kV.

SOLUTION The resistivity of Teflon at 20°C is $10^{16}\ \Omega \cdot \text{m}$ from Table 18.1. The resistance of the spacer is given by Eq. (18.3):

$$R = \rho \frac{l}{A}$$

$$= 10^{16}\ \Omega \cdot \text{m} \times \frac{2 \times 10^{-3}\ \text{m}}{\pi \times (10^{-2}\ \text{m})^2}$$

$$= 6.37 \times 10^{16}\ \Omega$$

The current in the Teflon insulator is given by Eq. (18.2b):

$$I = \frac{V}{R}$$

$$= \frac{12 \times 10^3\ \text{V}}{6.37 \times 10^{16}\ \Omega}$$

$$= 1.88 \times 10^{-13}\ \text{A}$$

This extremely small current shows why insulators are often called nonconductors of electricity. ■

18.4 Resistors in Series and in Parallel

Many electric circuits are far more complex than the simple single-resistor circuit of Example 18.3. As a first step toward analyzing more difficult circuits, we will consider two basic arrangements.

Resistors connected end to end, as shown in Fig. 18.6, are said to be in *series*. The diagram shows three resistors in series with

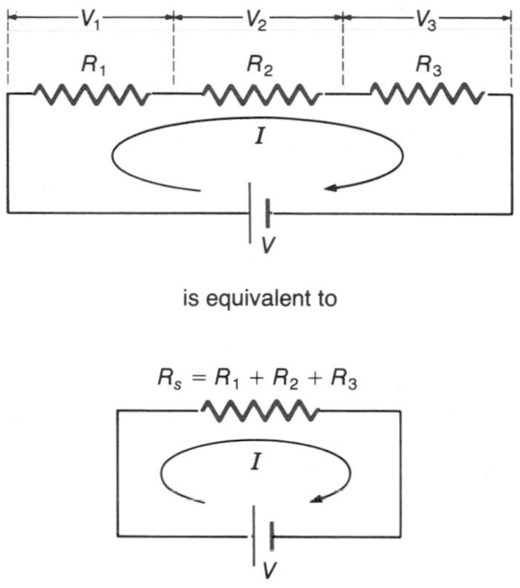

FIGURE 18.6 Separate resistors in series have the same resistance as a single equivalent resistor.

resistances R_1, R_2, and R_3. The current I causes potential differences across each of the resistors that must add up to the battery voltage V. This follows from our definition of potential difference as the energy change per unit charge. A charge passing through resistors wired in series gives rise to energy changes that must be added to give the total energy change. Let V_1, V_2, and V_3 be the potential differences across the three resistors. Then we have:

$$V = V_1 + V_2 + V_3$$

Dividing by the current gives:

$$\frac{V}{I} = \frac{V_1}{I} + \frac{V_2}{I} + \frac{V_3}{I}$$
$$= R_1 + R_2 + R_3$$

from Eq. (18.2). But V/I is simply the equivalent resistance of the whole circuit as seen looking out from the battery. We therefore write:

$$R_s = R_1 + R_2 + R_3 \tag{18.5}$$

This gives the equivalent resistance of three resistors wired in series. The general formula, which applies to any number of resistors in series, is simply the sum of the resistances of each resistor.

The other basic configuration is resistors in *parallel*. In this case the resistors are side by side and wired to a battery, as shown in Fig. 18.7. With this configuration, the same potential difference exists between the ends of each resistor, which for the circuit shown is the battery voltage V. However, the current from the battery splits into three parts through each of the three resistors. Let these currents be I_1, I_2, and I_3. The total current from the battery is given by:

$$I = I_1 + I_2 + I_3$$

Dividing each side of the equation by the battery voltage V gives:

$$\frac{I}{V} = \frac{I_1}{V} + \frac{I_2}{V} + \frac{I_3}{V}$$
$$= \frac{1}{R_1} + \frac{1}{R_2} + \frac{1}{R_3}$$

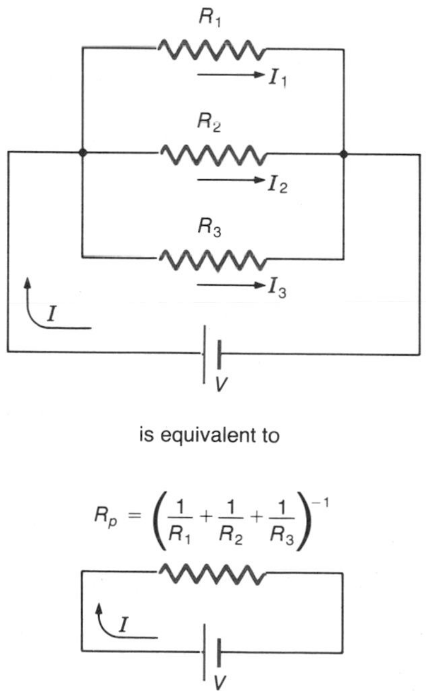

FIGURE 18.7 Separate resistors in parallel have the same resistance as a single equivalent resistor.

from Eq. (18.2). But I/V is the reciprocal of the resistance of the whole circuit as seen from the battery. Therefore, we can write:

$$\frac{1}{R_p} = \frac{1}{R_1} + \frac{1}{R_2} + \frac{1}{R_3} \tag{18.6}$$

This gives the equivalent resistance of three resistors wired in parallel. For each additional parallel resistor, an extra term appears in Eq. (18.6).

We can summarize our discussion of series and parallel resistors as follows:

Resistors in Series

The current in each resistor is the same.
The total voltage is the sum of the voltages across each resistor separately.

Resistors in Parallel

The voltage across each resistor is the same.
The total current is the sum of the currents through each resistor separately.

EXAMPLE 18.7

Calculate the equivalent resistance of the circuit illustrated.

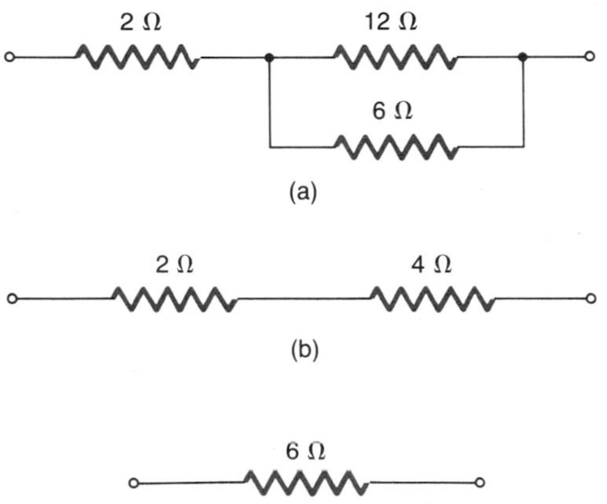

(a) (b) (c)

SOLUTION We can find the equivalent resistance of the combination by successive applications of the formulas for series and parallel resistors.

a. Inspection of the circuit shows that the 12-Ω and the 6-Ω resistors are in parallel, and using Eq. (18.6) gives:

$$\frac{1}{R_p} = \frac{1}{R_1} + \frac{1}{R_2}$$

$$= \frac{1}{12\ \Omega} + \frac{1}{6\ \Omega}$$

$$\therefore R_p = 4\ \Omega$$

This allows us to reduce the circuit to that shown in (b).

b. The second diagram shows the 2-Ω and 4-Ω resistors in series, and Eq. (18.5) gives:

$$R_s = R_1 + R_2$$

$$= 2\ \Omega + 4\ \Omega$$

$$= 6\ \Omega$$

The total equivalent resistance is 6 Ω, as illustrated in the final diagram. In deciding whether resistors are in series or parallel, remember that the general shape of the diagram does not necessarily help. Resistors are in series when there is the same current in each one, and they are in parallel when the same potential difference exists across each one. ■

EXAMPLE 18.8

Calculate the current and the potential difference for each of the resistors of Example 18.7, if a 12-V battery is connected across the combination.

SOLUTION We proceed by making our way backward through the equivalent circuit diagrams.

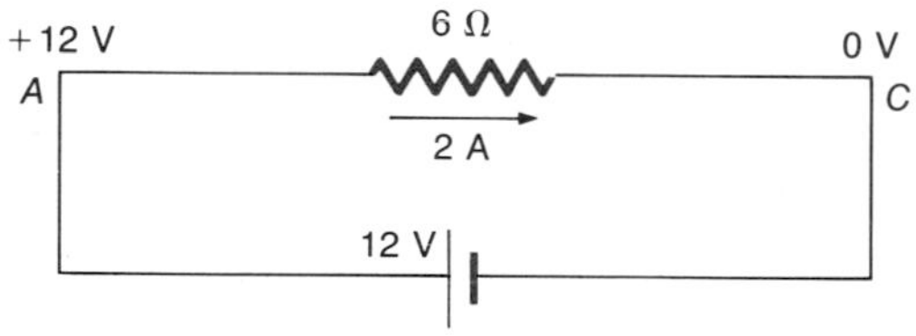

(a)

a. The battery maintains a 12-V potential difference between points A and C, with point A being at the higher potential. It is convenient to set $V_A = +12$ V and $V_C = 0$ V. The current in the final equivalent 6-Ω resistor is given by Eq. (18.2b):

$$I = \frac{V}{R} = \frac{12\text{ V}}{6\ \Omega} = 2\text{ A}$$

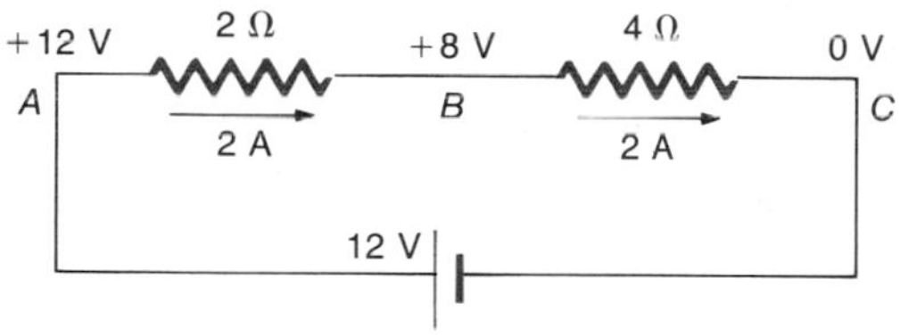

(b)

This resistor represents the whole complex between points A and C, and is not identified with any resistor of the original circuit.

b. In the next step the current in both the 2-Ω and 4-Ω resistors is 2 A, and the potential drop across the 4-Ω resistor is given by Eq. (18.2a):

$$V = RI = 4\ \Omega \times 2\text{ A} = 8\text{ V}$$

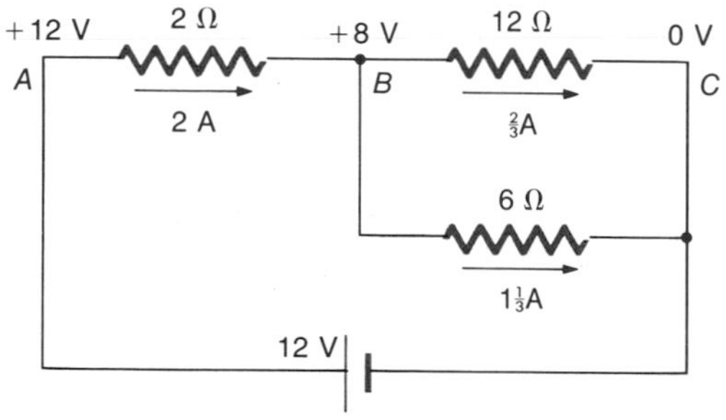

(c)

Since the point B is at the higher potential, $V = +8$ V. The 2-Ω resistor of this diagram is the 2-Ω resistor of the original circuit. The potential difference across it is 12 V − 8 V = 4 V and the current in it is 2 A. The 4-Ω resistor represents the complex between points B and C, but is not identified with either resistor of the original circuit.

c. In this step, we see that there is a potential difference of 8 V across both the 12-Ω and 6-Ω resistors. The current in the 6-Ω resistor is given by Eq. (18.2b):

$$I = \frac{V}{R} = \frac{8\text{ V}}{6\text{ A}} = 1\frac{1}{3}\text{ A}$$

Similarly, the current in the 12-Ω resistor is ⅔ A. This completes the calculation of the currents and voltages of the three resistors of the original circuit. ■

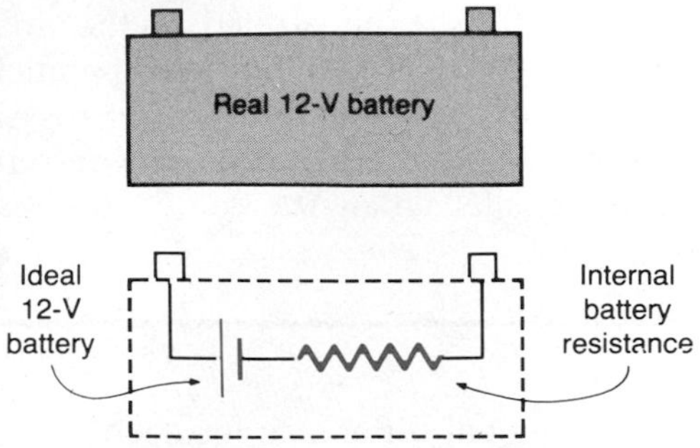

FIGURE 18.8 A real battery is represented by a model consisting of an ideal battery in series with a resistor.

After a little experience in measuring currents and voltages, it may seem surprising that Ohm's law is so prominent. In the early experiments in current electricity, metal strips were used as resistors, which had quite low values of resistance. Under these circumstances the battery voltage divided by the current is not constant, and the metal strips do not appear to have constant resistance. This rather confusing situation was clarified by Ohm, who reevaluated the role of the battery in the circuit. Up to this point we have considered only an **ideal battery,** which is one that is capable of maintaining a constant emf regardless of the current drawn from it. However, the internal parts of a battery carry the current that it supplies, and they create a certain amount of resistance. We therefore represent a real battery by an ideal battery in series with a resistor, as shown in Fig. 18.8. The voltage between the terminals is equal to the voltage of the ideal battery only when there is no current; otherwise the terminal voltage is less than the ideal battery voltage. For this reason the ideal battery voltage is frequently called the *open-circuit battery voltage,* or the *emf of the battery.* The potential difference across a resistor connected to the battery is not the open-circuit battery voltage but the actual terminal voltage. Let us illustrate with an example.

EXAMPLE 18.9

A battery has an open-circuit voltage of 12 V and an internal resistance of 0.6 Ω. Calculate the current

a. when a 1-Ω resistor is connected to the battery.

b. when a 2-Ω resistor is connected to the battery.

SOLUTION We represent the battery by an ideal 12-V battery in series with its internal resistance of 0.6 Ω. The external resistance is also in series with the internal resistance. With the help of Eq. (18.5) we see that the equivalent resistance is simply the sum of the battery resistance and the external resistance.

a. For the 1-Ω external resistance, Eq. (18.2b) gives:

$$I = \frac{V}{R}$$

$$= \frac{12\ \text{V}}{1\ \Omega + 0.6\ \Omega}$$

$$= 7.5\ \text{A}$$

b. For the 2-Ω external resistance, the same equation gives:

$$I = \frac{V}{R}$$

$$= \frac{12 \text{ V}}{2\ \Omega + 0.6\ \Omega}$$

$$= 4.62 \text{ A}$$ ■

On the surface, the results of Example 18.9 give little reason to accept Ohm's law, since a 12-V battery does not create twice the current in a 1-Ω resistor that it creates in a 2-Ω resistor. In his classic investigation, Ohm showed that the fallacy lay in supposing that a 12-V battery applies a 12-V potential difference to any resistor to which it is connected.

EXAMPLE 18.10

Calculate the voltages at the battery terminals for each of the two cases in Example 18.9.

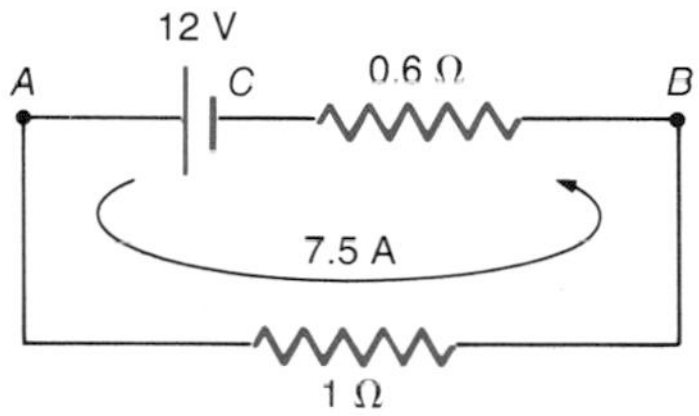

(a)

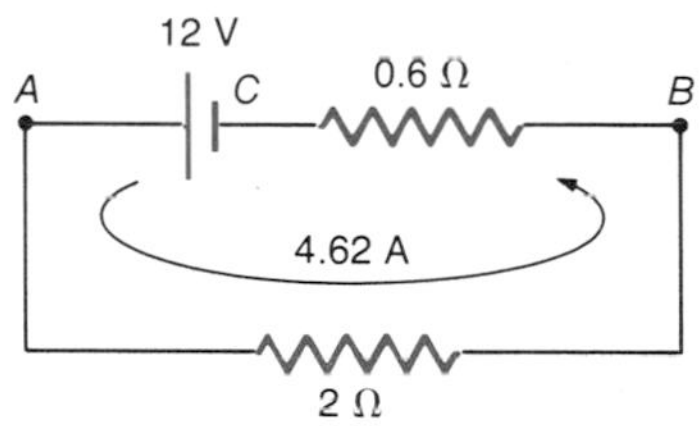

(b)

SOLUTION Let A and B be the positive and negative terminals of the real battery, and C the negative terminal of the ideal battery. Since the open-circuit voltage is 12 V, we can set $V_A = +12$ V and $V_C = 0$ V.

a. With a 1-Ω external resistor, the current is 7.5 A, and the potential difference across the 1-Ω resistor is given by Eq. (18.2a):

$$V = RI$$

$$= 1\ \Omega \times 7.5 \text{ A}$$

$$= 7.5 \text{ V}$$

Since the current direction is from A to B, we have $V_B = 12$ V − 7.5 V = 4.5 V. The potential difference between the battery terminals is:

$$V_A - V_B = 12 \text{ V} - 4.5 \text{ V} = 7.5 \text{ V}$$

b. With a 2-Ω external resistor, the current is 4.62 A, and Eq. (18.2a) gives the potential difference across the resistor:

$$V = RI$$

$$= 2\ \Omega \times 4.62 \text{ A}$$

$$= 9.24 \text{ V}$$

It follows that $V_B = 12$ V − 9.24 V = 2.76 V, and the potential difference between the battery terminals is:

$$V_A - V_B = 12 \text{ V} - 2.76 \text{ V} = 9.24 \text{ V}$$

The terminal voltage is less than the open-circuit voltage, and the difference increases with increasing current drain from the battery. ■

18.5 Power in dc Circuits

When a battery causes electric current in an external circuit, it transfers energy from the battery to the circuit. Many familiar examples demonstrate this energy transfer. For instance, a piece of resistance wire connected between the terminals of a battery becomes hot as the energy supplied by the battery increases the temperature of the wire. Alternatively, an electric motor may use the energy from a battery to perform mechanical work. Other examples may be more complex, but in each case the supply of energy from the battery goes hand in hand with the transfer of charge through the external circuit or device.

Consider an ideal battery (having negligible internal resistance) that maintains a potential difference V between its terminals. The battery is connected to some device, as shown in Fig. 18.9(a). Let q be the amount of charge flowing from the battery through the device during a time interval t. We now assert that the energy supplied to the device is equal to the product of the charge flow and the battery voltage:

$$U = qV \qquad \textbf{(18.7)}$$

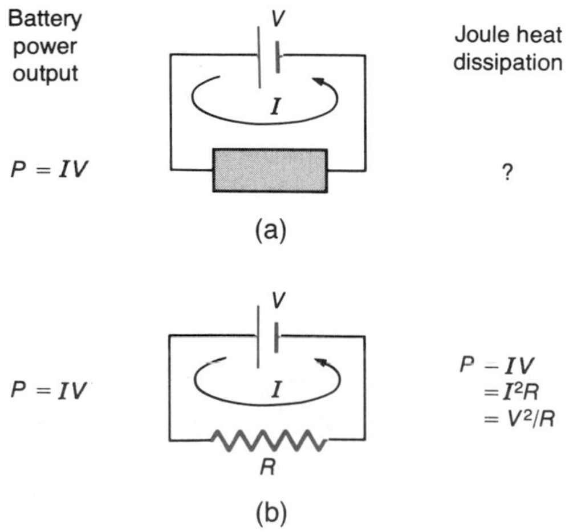

FIGURE 18.9 Power relationships when an ideal battery supplies a constant current to an external circuit. (a) The external circuit is a device of unknown characteristics. (b) The external circuit is a resistor.

Recollect that Eq. (16.6) defines electrostatic potential in a similar way:

$$U_B - U_A = q'(V_B - V_A)$$

In this case the energy supplied to the test charge q' is equal to the product of the test charge and the electrostatic potential difference $(V_B - V_A)$ through which it moves. The energy relation of Eq. (18.7) is the same except that the battery voltage replaces the electrostatic potential difference. Now divide both sides of Eq. (18.7) by the time t:

$$\frac{U}{t} = \frac{q}{t} V$$

Using Eq. (7.4) for average power and Eq. (18.1) for average current, we can rewrite this result in the following form:

Formula Expressing the Power Output of a Battery

Average power output = average current × battery voltage

$$\langle P \rangle = \langle I \rangle V \qquad \textbf{(18.8)}$$

The SI power unit ampere-volt is the same as the watt.

We can easily show that the ampere-volt is equal to the watt:

$$\begin{aligned} 1\ \text{A} \cdot \text{V} &= 1\ \text{C/s} \times 1\ \text{J/C} \\ &= 1\ \text{J/s} \\ &= 1\ \text{W} \end{aligned}$$

If the battery current is constant, the power output is also constant, and we can omit the averaging symbols in Eq. (18.8).

How battery power is utilized depends on the type of device being powered. In every instance, however, heat energy increases in at least some parts of the device. The dissipation of electrical energy as heat is frequently called **Joule heating** to honor the pioneering experiments of James Joule in this area. In general, only a fraction of the battery power output is dissipated as Joule heat.

Formulas for Joule Heat Power Generated in a Resistor

Joule heat power = current × potential difference across resistor

$$\begin{aligned} P &= IV && \textbf{(a)} \\ &= I^2R && \textbf{(b)} \qquad \textbf{(18.9)} \\ &= \frac{V^2}{R} && \textbf{(c)} \end{aligned}$$

The SI power units ampere-volt, (ampere)^2-ohm, and (volt)^2 per ohm are all identical to the watt.

However, a different situation arises when the external device is a resistor, as illustrated in Fig. 18.9(b), in which case the entire power output of the battery goes to Joule heating. The equality of the battery power output to the Joule heat gives special formulas that apply to current in a resistor. Equations (18.2a) and (18.2b) yield the second and third alternate formulas from the first formula.

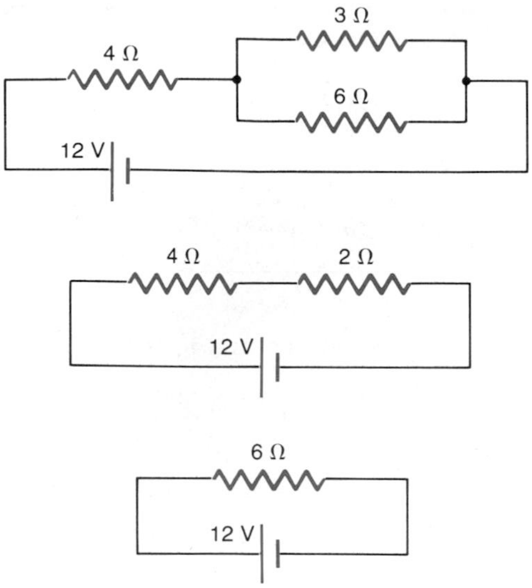

EXAMPLE 18.11

A 12-V ideal battery is connected across the resistor circuit shown in the top diagram. Calculate the power output from the battery and the Joule heating in each resistor.

SOLUTION Calculating the power output of the battery requires a knowledge of the battery current. To this end, we calculate the equivalent resistance of the circuit by using the formulas for series and parallel reduction. The 3-Ω and 6-Ω resistors are in parallel, and Eq. (18.6) gives:

$$\frac{1}{R_p} = \frac{1}{R_1} + \frac{1}{R_2}$$

$$= \frac{1}{3\ \Omega} + \frac{1}{6\ \Omega}$$

$$\therefore R_p = 2\ \Omega$$

The equivalent 2-Ω resistor is in series with the 4-Ω resistor, and Eq. (18.5) gives:

$$R_s = R_1 + R_2$$

$$= 4\ \Omega + 2\ \Omega$$

$$= 6\ \Omega$$

The battery current is given by Eq. (18.2b):

$$I = \frac{V}{R}$$

$$= \frac{12\ \text{V}}{6\ \Omega}$$

$$= 2\ \text{A}$$

The power output of the battery follows from Eq. (18.8):

$$P = IV$$

$$= 2\ \text{A} \times 12\ \text{V}$$

$$= 24\ \text{W}$$

The Joule heating in each resistor must be calculated individually. The 4-Ω resistor carries the whole 2-A current and Eq. (18.9b) gives the Joule heat dissipation:

$$P = I^2R$$

$$= (2\ \text{A})^2 \times 4\ \Omega$$

$$= 16\ \text{W}$$

The voltage drop across the 4-Ω resistor follows from Eq. (18.2a):

$$\begin{aligned} V &= RI \\ &= 4\ \Omega \times 2\ \text{A} \\ &= 8\ \text{V} \end{aligned}$$

Since the battery maintains 12 V across the whole circuit, the potential difference across each resistor in the parallel combination is 4 V. Equation (18.9c) gives the Joule heating in each resistor in turn:

$$\begin{aligned} P &= \frac{V^2}{R} \\ &= \frac{(4\ \text{V})^2}{3\ \Omega} \\ &= 5\tfrac{1}{3}\ \text{W in the 3-}\Omega\text{ resistor} \\ &= \frac{(4\ \text{V})^2}{6\ \Omega} \\ &= 2\tfrac{2}{3}\ \text{W in the 6-}\Omega\text{ resistor} \end{aligned}$$

The total electrical power dissipated in all resistors is:

$$16\ \text{W} + 5\tfrac{1}{3}\ \text{W} + 2\tfrac{2}{3}\ \text{W} = 24\ \text{W}$$

which is equal to the power output of the battery. ■

Let us turn now to the case of a nonideal battery so that we can represent the actual situation more realistically. The battery now has an internal resistance, and some of the power produced goes to increase the battery temperature by Joule heat production within the battery itself. We can illustrate this by means of an example.

EXAMPLE 18.12

A battery having an open-circuit voltage of 6 V and an internal resistance of 0.2 Ω is connected across a 3-Ω resistor. Calculate the power produced by the battery and the power dissipated in the 3-Ω resistor.

SOLUTION Since the external resistance and the battery resistance are in series, we can find the total equivalent resistance from Eq. (18.5):

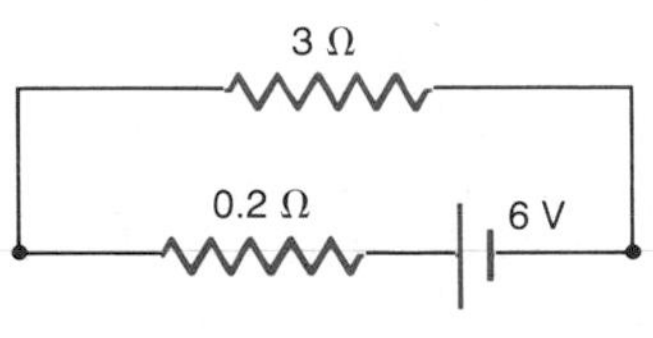

$$\begin{aligned} R_s &= R_1 + R_2 \\ &= 3\ \Omega + 0.2\ \Omega \\ &= 3.2\ \Omega \end{aligned}$$

The battery current follows from Eq. (18.2b):

$$I = \frac{V}{R}$$

$$= \frac{6\ \text{V}}{3.2\ \Omega}$$

$$= 1.875\ \text{A}$$

The total power produced by the battery is given by Eq. (18.8):

$$P = IV$$

$$= 1.875\ \text{A} \times 6\ \text{V}$$

$$= 11.25\ \text{W}$$

To calculate the power dissipation in the 3-Ω resistor, we use Eq. (18.9b):

$$P = I^2R$$

$$= (1.875\ \text{A})^2 \times 3\ \Omega$$

$$= 10.55\ \text{W}$$

Since the power production of the battery is greater than the power dissipated in the external resistor, energy conservation requires that the balance be dissipated in the internal resistance of the battery. We can verify this by using Eq. (18.9b):

$$P = I^2R$$

$$= (1.875\ \text{A})^2 \times 0.2\ \Omega$$

$$= 0.70\ \text{W}$$

The total power produced by the battery goes to heating the external resistor and to heating the battery itself. ■

18.6 Kirchhoff's Laws

Determining equivalent resistances by combining resistors that are in series or in parallel works only for certain types of circuits. We need a more powerful technique to handle complicated circuits that cannot be handled by the simpler analysis.

Let us begin by clarifying certain terms used in describing circuits. A *circuit junction* is a point at which three or more wires come together, such as the points B and D in Fig. 18.10. A *circuit loop* is any closed path that returns to the same point. The circuit

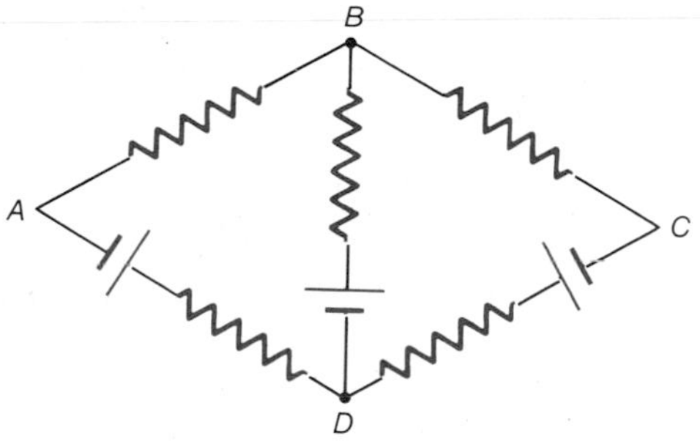

FIGURE 18.10 Illustration of the terms *circuit junction*, *circuit loop*, and *circuit branches*.

loops in Fig. 18.10 are *ABDA*, *BCDB*, and *ABCDA*. A *circuit branch* is the path between any two adjacent junction points. The branches in Fig. 18.10 are *DAB*, *BD*, and *BCD*.

The laws for analyzing circuits were set down in the middle of the nineteenth century by Gustav Kirchhoff (1824–1887), a German physicist who made notable contributions to many areas of physics.

Kirchhoff's Laws

1. The sum of the currents entering any junction equals the sum of the currents leaving the junction.
2. The sum of the battery emf's around any loop equals the sum of the potential differences across the resistors around the loop.

Kirchhoff's laws are not basic new laws of physics—they simply form an expression of conservation laws applied to electric circuits. In a steady state, charge cannot accumulate or diminish at any place, and the first law expresses conservation of electric charge by requiring that the charge flow into any junction be equal to the charge flow out. The second law is simply the requirement of energy conservation. To see this, we imagine that a small positive test charge is carried around a circuit loop. The work done is the product of the charge and the potential difference that it traverses; this must total zero for the whole loop since the test charge is returned to its original starting point. However, sources of emf do work on a test charge when it moves from lower to higher potential. On the other hand, the voltage drop across a resistor does work on the test charge when it moves from higher to lower potential. For this reason, we put the battery emf's and resistor voltages on opposite sides of the Kirchhoff second law. In practical application, Kirchhoff's laws provide more equations than necessary for solving problems. In addition, we must exercise considerable care to ensure that the signs are correct when interpreting the second law. For these reasons, it is a good idea to have a standard operating sequence when performing a circuit analysis from Kirchhoff's laws.

1. Select a symbol and a direction for the current in each circuit branch. It does not matter which direction we select for the current. If our selected direction is correct, the equations give a positive value for the current. If the actual current is opposite to our choice, the equations yield a negative value.
2. Write the Kirchhoff first-law equations for all the circuit junctions except one. (We can use the equation for the final junction as a check on the result of the calculation.)

3. Write the Kirchhoff second-law equations for as many loops as necessary to give enough equations to solve the problem. On the question of signs, we observe the following conventions:

a. Count a battery emf as positive when that battery alone would send a clockwise current around the loop in question.

b. Count the potential difference across a resistor as positive if the current in the resistor is clockwise with respect to the loop in question.

We should emphasize that our procedural steps are not the only way to use the Kirchhoff laws to solve a circuit. For complex circuits, engineers usually employ unknown loop currents in place of the unknown branch currents that we employ; in addition, it is possible to use a variety of sign conventions. However, we must choose some particular procedure. Let us illustrate with an example.

EXAMPLE 18.13

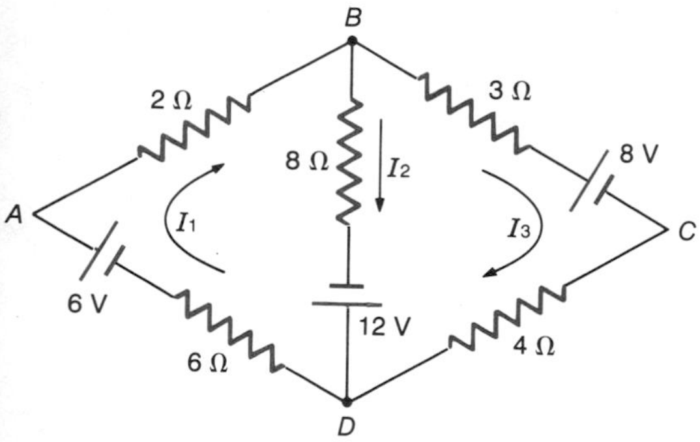

Calculate the current in each branch of the circuit shown in the diagram.

SOLUTION We follow the procedural steps outlined above.

1. Let I_1, I_2, and I_3 be the currents in the three branches of the circuit with directions selected arbitrarily as shown on the diagram.

2. The Kirchhoff first-law equation for the junction B is:

$$I_1 = I_2 + I_3$$

Since the circuit only has two junctions, we need write only one junction equation.

3. Since there are three unknown currents, we need a total of three equations. So, we need two loop equations in addition to the junction equation already written. Let us write the Kirchhoff second-law equations for the loops *ABDA* and *CDBC*. When we use Eq. (18.2a) to calculate the potential differences across the resistors (with proper attention to the signs), the Kirchhoff second-law equations give:

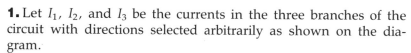

$$6\text{ V} + 12\text{ V} = 2\ \Omega \times I_1 + 8\ \Omega \times I_2 + 6\ \Omega \times I_1 \text{ for loop } ABDA$$

$$-8\text{ V} - 12\text{ V} = 3\ \Omega \times I_3 + 4\ \Omega \times I_3 - 8\ \Omega \times I_2 \text{ for loop } CDBC$$

Note that it is not possible to describe a branch current or a battery as positive or negative except with respect to a particular loop. The current I_2 is positive with respect to the loop *ABDA* but negative with respect to the loop *CDBC*. Similarly, the 12-V battery has a positive emf around one loop and a negative emf around the other loop. To solve the equations, we substitute for I_1 from the junction equation into the loop equations. After collecting terms we have:

$$18\ \text{V} = I_2 \times 16\ \Omega + I_3 \times 8\ \Omega$$

$$-20\ \text{V} = -I_2 \times 8\ \Omega + I_3 \times 7\ \Omega$$

Solving the pair of simultaneous linear equations gives:

$$I_2 = 1.625\ \text{A and } I_3 = -1\ \text{A}$$

The junction equation immediately yields:

$$I_1 = 0.625\ \text{A}$$

The currents I_1 and I_2 have the direction selected on the diagram, and the current I_3 is opposite to the direction indicated.

This example is typical of problems that we cannot solve from the equations for equivalent series and parallel resistances. ■

We can also calculate the power relations in a circuit containing batteries in opposition; let us illustrate with an example.

EXAMPLE 18.14

A battery having an open-circuit voltage of 12 V and an internal resistance of 0.4 Ω is connected (positive to positive) to a battery having an open-circuit voltage of 6 V and an internal resistance of 0.2 Ω. The connecting leads contain a 2-Ω resistor. Calculate the power produced by the 12-V battery and account for it in detail.

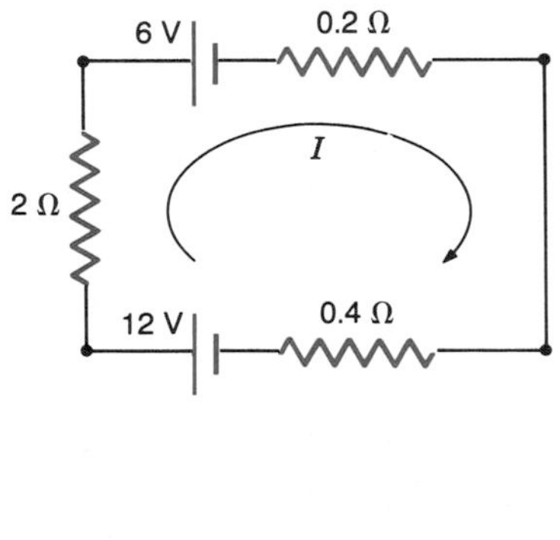

SOLUTION Let I be the current in the circuit in a clockwise direction. Applying Kirchhoff's second law to the loop gives:

$$12\ \text{V} - 6\ \text{V} = 2\ \Omega \times I + 0.2\ \Omega \times I + 0.4\ \Omega \times I$$

$$\therefore I = 2.308\ \text{A}$$

The power produced by the 12-V battery is given by Eq. (18.8):

$$\begin{aligned} P &= IV \\ &= 2.308\ \text{A} \times 12\ \text{V} \\ &= 27.69\ \text{W} \end{aligned}$$

The use of Eq. (18.9b) gives the power dissipation in the various resistors:

$$\begin{aligned} P &= I^2R \\ &= (2.308\ \text{A})^2 \times 2\ \Omega \\ &= 10.65\ \text{W in the 2-}\Omega\text{ resistor} \end{aligned}$$

Also

$$\begin{aligned} P &= (2.308\ \text{A})^2 \times 0.4\ \Omega \\ &= 2.13\ \text{W in the internal resistance of the 12-V battery} \end{aligned}$$

Also

$$\begin{aligned} P &= (2.308\ \text{A})^2 \times 0.2\ \Omega \\ &= 1.07\ \text{W in the internal resistance of the 6-V battery} \end{aligned}$$

The total power dissipated in all resistors (including the internal resistances of the batteries) is:

$$10.65\ \text{W} + 2.13\ \text{W} + 1.07\ \text{W} = 13.85\ \text{W}$$

This falls short of the 27.69 W produced by the 12-V battery, and the difference, 27.69 W − 13.85 W = 13.84 W, is the power supplied to the 6-V battery but not dissipated in its internal resistance. This power goes to charge the battery, and we can calculate the charging power input by using Eq. (18.8):

$$\begin{aligned} P &= IV \\ &= 2.308\ \text{A} \times (-6\ \text{V}) \\ &= -13.85\ \text{W} \end{aligned}$$

The negative sign reflects an electric power input rather than a power output. A battery supplies power when the direction of current is out from the positive plate, and the battery is discharging. If the current is into the positive plate, the battery is receiving power and is charging. ■

18.7 Circuit Measurements

The basic instrument for dc circuit measurements is called a **galvanometer.** A closely wound coil of wire situated in a magnetic field experiences a magnetic force when it carries current, and the resultant deflection of the coil is displayed by a pointer moving along a scale calibrated to measure electric current directly. The operation of the instrument is explained in a subsequent chapter. For the moment we will take the galvanometer's current-measuring capability for granted and show how we can use it to measure voltage and resistance as well. A typical inexpensive galvanometer has a coil resistance of about 200 Ω and shows a full-scale deflection of the pointer for a current of 1 mA. To aid in our understanding, let us investigate the circuits for which this instrument can be used for a range of current and voltage measurements.

If our galvanometer has a full-scale deflection when the current is 1 mA, we must modify it to measure larger currents. To do this we use a resistor in parallel with the meter. Such a resistor is called a **shunt,** and it causes the greater part of a current to bypass the galvanometer coil. Let us illustrate the usage of a shunt by working an example.

EXAMPLE 18.15

A galvanometer has a coil resistance of 200 Ω and shows a full-scale deflection for a current of 1 mA. Design a shunt that enables the instrument to measure 10 A at full-scale deflection.

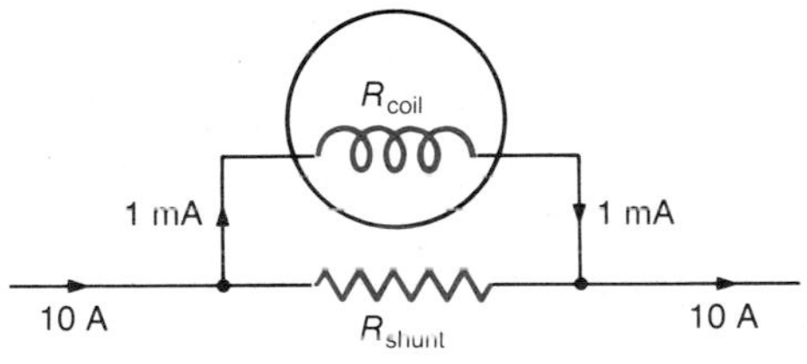

SOLUTION We connect the shunt resistor in parallel with the meter coil as shown in the diagram. The shunt is to be designed so that the meter gives a full-scale deflection when the current is 10 A in the leads entering the parallel combination. For this to occur we must have 1 mA in the meter coil and (10 A − 1 mA) = 9.999 A in the shunt. The potential difference across the meter coil when it carries 1 mA is given by Eq. (18.2a):

$$\begin{aligned} V &= R_{coil} \times I_{coil} \\ &= 200\ \Omega \times 10^{-3}\ \text{A} \\ &= 0.2\ \text{V} \end{aligned}$$

This same potential difference must cause a current of 9.999 A in the shunt. We can calculate the shunt resistance by using Eq. (18.2):

$$\begin{aligned} R_{shunt} &= \frac{V}{I_{shunt}} \\ &= \frac{0.2\ \text{V}}{9.999\ \text{A}} \\ &= 0.0200\ \Omega \end{aligned}$$

Note that it is not necessary to use different symbols for the coil and shunt potential differences, since they are both the same. Since the function of the shunt is to draw current away from the coil, no shunt arrangement can cause this galvanometer to deflect full scale for a current smaller than 1 mA. ■

Frequently a galvanometer is mounted in a box along with a selection of shunts that enable it to measure currents over a wide range. A galvanometer set up in this way is called an **ammeter,** since it is organized to measure current in amperes.

Our second basic circuit quantity is the potential difference between two points. When a galvanometer is arranged to measure potential difference, we call the instrument a **voltmeter.** A **series resistor** is required to set the galvanometer up for this purpose. Let us again illustrate with an example.

EXAMPLE 18.16

Design a series resistor that will enable the galvanometer of Example 18.14 to be used as a voltmeter measuring 100 V at full-scale deflection.

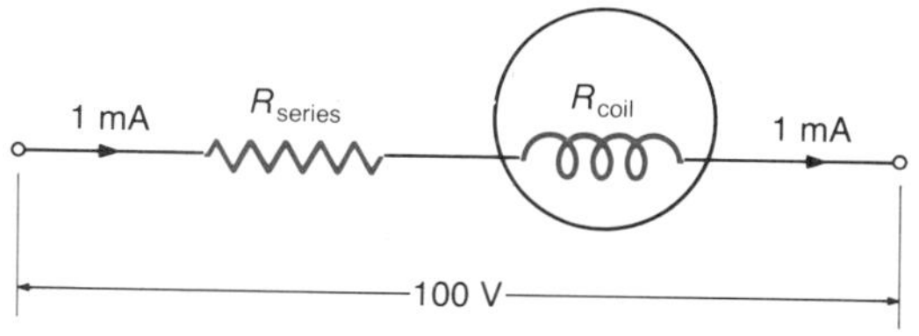

SOLUTION The series resistor must be such that the current is 1 mA in the galvanometer when the potential difference across the outside terminals is 100 V. Since the series resistor and the resistance of the galvanometer coil are in series, Eq. (18.2) gives:

$$R_{\text{series}} + R_{\text{coil}} = \frac{V}{I}$$

$$R_{\text{series}} + 200\ \Omega = \frac{100\ \text{V}}{10^{-3}\ \text{A}}$$

$$= 10^5\ \Omega$$

$$\therefore R_{\text{series}} = 99\ 800\ \Omega$$ ■

A typical voltmeter contains a galvanometer mounted with a selection of series resistors that permit a wide range of voltage measurements. The examples show that a typical ammeter is basically a low-resistance instrument, and a voltmeter a high-resistance instrument.

With instruments to measure both voltage and current, the measurement of resistance would seem to be a simple matter. We need only set up a circuit containing the unknown resistance and wire an ammeter and voltmeter into the circuit to measure the potential difference and the current in the resistor. Figure 18.11 shows that this simple goal cannot be attained. We can always

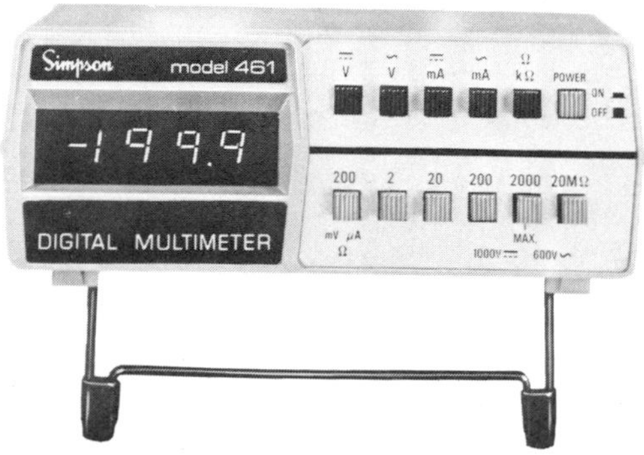

A multimeter is an instrument designed to measure voltage, current, or resistance. The heart of the upper model is a moving coil galvanometer; the rotary switch selects series resistors, shunts, or other circuits to enable the galvanometer to perform the various functions. The lower model is solid state with a digital display; the various functions are selected by pushbutton switches. (Courtesy of Simpson Electric)

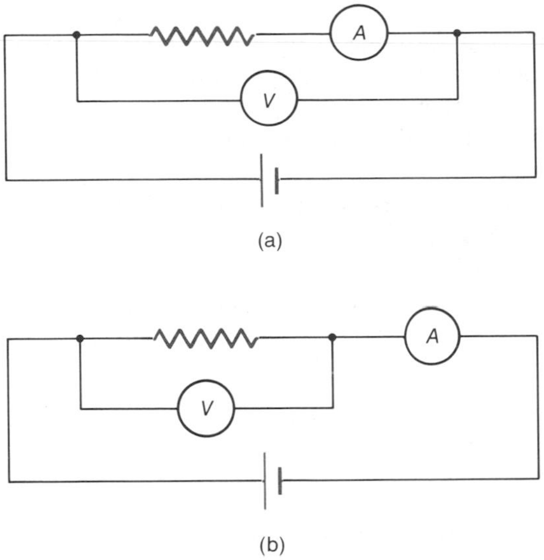

FIGURE 18.11 Circuits to measure resistance using an ammeter and a voltmeter. (a) The ammeter measures the current in the unknown resistor, but the voltmeter measures the voltage drop across the unknown resistor plus the ammeter. (b) The voltmeter measures the voltage drop across the unknown resistor, but the ammeter measures the current in the unknown resistor plus the voltmeter.

arrange the circuit so that *either* the ammeter *or* the voltmeter measures exactly what we require—namely, either the current in the resistor (Fig. 18.11a) or the potential difference across the resistor (Fig. 18.11b). However, the voltmeter of Fig. 18.11(a) measures the potential difference across the resistor and the ammeter combined, and the ammeter of Fig. 18.11(b) measures the current in the resistor and voltmeter combined. In some circumstances the circuits of Fig. 18.11 may give good estimates for the value of the unknown resistor, depending on the relative resistances of the unknown resistor, the ammeter, and the voltmeter.

Special circuits have been devised to carry out resistance measurements that avoid the pitfalls mentioned above. The circuit shown in Fig. 18.12 is called the **Wheatstone bridge** after Sir Charles Wheatstone (1802–1875), who popularized its use for resistance measurement in the growing electrical industry of the late nineteenth century. The basic circuit is shown in Fig. 18.12(a). Four resistors, R_1, R_2, R_3, and R_4, are wired together in the form of a parallelogram. A battery is connected between the diagonally opposite corners A and B, and a galvanometer is connected between the other two corners, C and D.

The bridge is balanced if there is no measurable current in the galvanometer, and balance occurs only for a definite relationship between the resistances in the branches of the bridge.

For the galvanometer current to be zero, Kirchhoff's first law requires that the current in R_1 equal the current in R_2. For the same reason the current in R_3 must equal the current in R_4. We call these currents I_1 and I_2, respectively, as shown in Fig. 18.12(a). If there is no current in the galvanometer, the potential difference between C and D must be zero, and the potential differences across R_1 and R_3 must be equal. From Eq. (18.2a) this condition is:

$$R_1 I_1 = R_3 I_2$$

Similarly, the potential differences across R_2 and R_4 must be equal, giving:

$$R_2 I_1 = R_4 I_2$$

Dividing these two equations gives the required condition for balance.

■ Balance Condition for the Wheatstone Bridge

$$\frac{R_1}{R_2} = \frac{R_3}{R_4} \qquad \textbf{(18.10)}$$

(Refer to Fig. 18.12a for notation.)

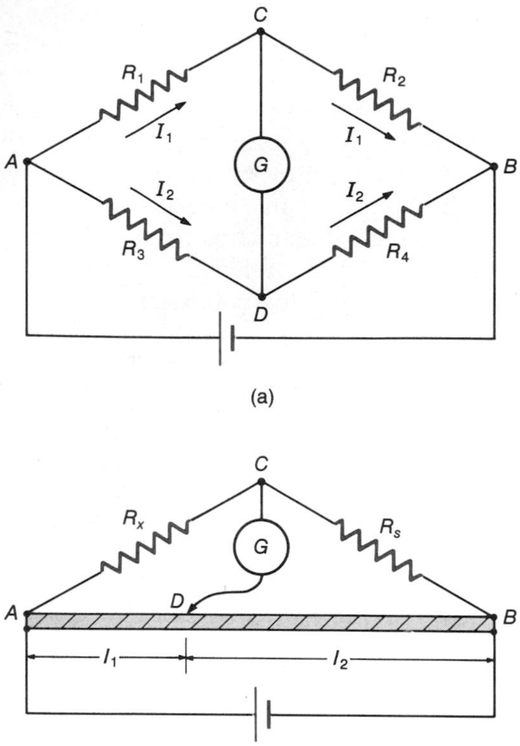

FIGURE 18.12 The Wheatstone bridge for resistance measurement: (a) the basic circuit; (b) a typical slide wire arrangement.

The equation gives a method for measuring a resistance in terms of other known resistances, without relying on ammeter or voltmeter readings. All that we require of the galvanometer is sufficient sensitivity to indicate that there is negligibly small current in the branch *CD*.

We show the usual practical arrangement for a Wheatstone bridge in Fig. 18.12(b). Two branches of the bridge contain the unknown resistance R_x and a standard resistor R_s, whose resistance is accurately known. The other two branches (*AD* and *DB*) are portions of a length of uniform resistance wire. The point *D* is a sliding contact, which can be moved back and forth until balance is achieved. If the wire is uniform in both cross section and resistivity, Eq. (18.3) gives:

$$\frac{\text{Resistance of branch } AD}{\text{Resistance of branch } DB} = \frac{\rho l_1/A}{\rho l_2/A} = \frac{l_1}{l_2}$$

Substitution into Eq. (18.10) gives as the balance condition for the bridge

$$R_x = \frac{R_s l_1}{l_2} \tag{18.11}$$

We can therefore use the circuit arrangement of Fig. 18.12(b) to measure an unknown resistance in terms of two length measurements if a standard resistor is available. Depending on the accuracy required, a standard resistor may be purchased commercially or calibrated by a standards laboratory.

EXAMPLE 18.17

An unknown resistor is compared with a 10.0-Ω standard resistor by using a slide wire Wheatstone bridge. The slide wire is 1 m long, and the balance occurs when the sliding contact is 63.4 cm from the end of the wire to which the unknown is connected. Calculate the value of the unknown resistance.

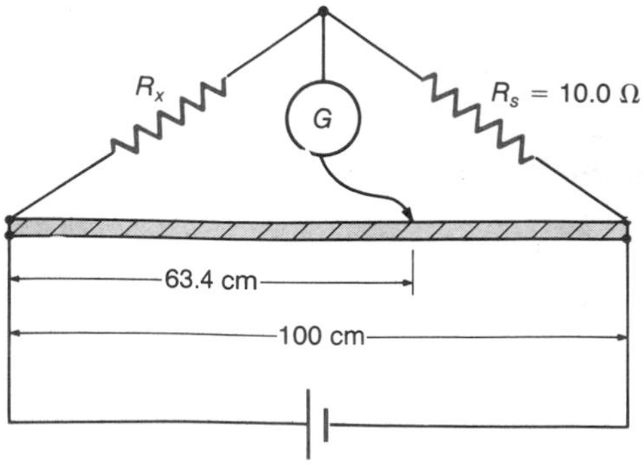

SOLUTION Comparison with the diagram of Fig. 18.12(b) shows:

$$l_1 = 63.4 \text{ cm}$$

$$l_2 = 100 \text{ cm} - 63.4 \text{ cm} = 36.6 \text{ cm}$$

We can now calculate the unknown resistance directly from Eq. (18.11):

This fanciful print from the nineteenth century shows Sir Charles Wheatstone in the role of patron of electrical communication. The top right shows the mythical god Mercury spreading telegraph cables around the earth—a job that was done across the Atlantic by the *Great Eastern* (top left) with a good deal more sweat and tears. The bottom right corner shows an early telegraph instrument for which Wheatstone held the first patent. (Culver Pictures, Inc.)

$$R_x = R_s\frac{l_1}{l_2}$$

$$= 10.0\ \Omega \times \frac{63.4\text{ cm}}{36.6\text{ cm}}$$

$$= 17.3\ \Omega$$ ■

A technique such as the Wheatstone bridge method for measuring resistance is called a **null method** because it relies not on the calibration of an instrument but only on the instrument's capability to record a negligibly small effect. The galvanometer is not used to measure current in the Wheatstone bridge—it simply allows resistance combinations to be selected that yield a negligibly small current in it.

The **potentiometer** provides another example of a null instrument, which we can use to measure voltage difference, and the basic potentiometer circuit is shown in Fig. 18.13(a). A battery of unspecified open-circuit voltage and internal resistance is connected across the ends of a piece of uniform resistance wire. Another battery of open-circuit voltage V and internal resistance R is connected to one end of the resistance wire and through a galvanometer to a sliding contact. The sliding contact is adjusted until the galvanometer shows negligibly small current. When this oc-

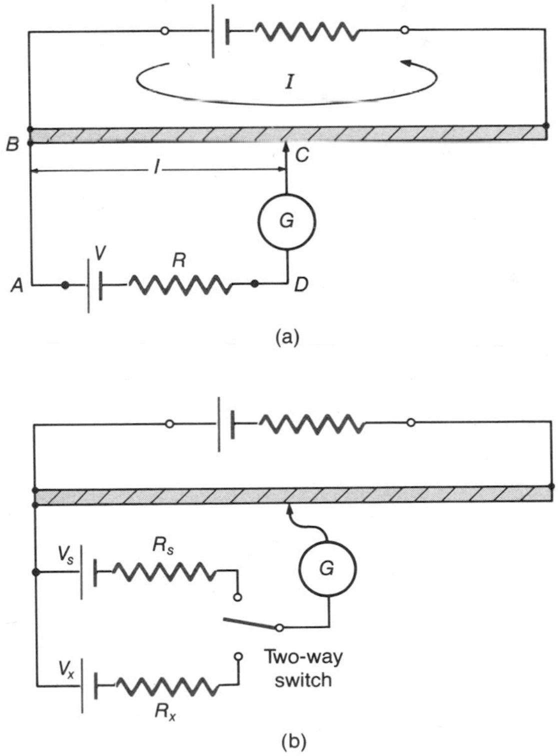

FIGURE 18.13 (a) The basic potentiometer circuit; (b) Using the potentiometer to measure the open-circuit voltage of a battery.

curs, there is some current I in the upper loop of the potentiometer circuit, and no current in the branch containing the galvanometer. Applying Kirchhoff's second law to the loop $ABCD$ gives:

$$V = IR(l)$$

where l is the distance from the end of the resistance wire to the sliding contact, and $R(l)$ is the resistance of this length of the wire. Note that the internal resistance of the battery in the loop $ABCD$ is not included, since the current in it is zero.

An important practical application of the potentiometer enables us to measure the emf of a battery if a standard battery of known emf is available. The circuit is shown in Fig. 18.13(b), with a two-way switch that selects either the standard battery (open-circuit voltage V_s and internal resistance R_s) or the unknown battery (open-circuit voltage V_x and internal resistance R_x). Let the balance distance of the sliding contact from the left end of the resistance wire be l_s with the standard battery in circuit, and l_x with the unknown battery. Then we have:

$$V_s = IR(l_s)$$

and

$$V_x = IR(l_x)$$

Dividing these equations we have:

$$\begin{aligned} V_x &= V_s \frac{R(l_x)}{R(l_s)} \\ &= V_s \frac{l_x}{l_s} \end{aligned} \qquad \textbf{(18.12)}$$

The last step follows since the ratio of the resistances of two different lengths of the same wire is equal to the ratio of the two lengths. Equation (18.12) permits us to measure the emf of a battery in terms of two length measurements if a standard battery is available. The internal resistances of the standard and unknown batteries do not enter the calculations, since there is no current in them when the potentiometer is balanced.

KEY CONCEPTS

Average **electric current** is charge transferred divided by elapsed time. See Eq. (18.1).

The **resistance** of a circuit element is defined as the potential difference across it divided by the current in it. See Eq. (18.2).

An **ohmic** circuit element is one whose resistance is independent of the current (or potential difference across it). Such element is called a **resistor.**

Resistivity is a property of substances that allows resistance to be calculated from geometrical dimensions. See Eq. (18.3).

The resistivities of most substances change with changing temperature. For metals the **temperature coefficient of resistivity** is a useful measure of this change. See Eq. (18.4).

From simple formulas the equivalent resistance of combinations of resistors in *series* and in *parallel* can be calculated. See Eqs. (18.5) and (18.6).

An **ideal battery** maintains a constant voltage between its terminals independently of the current flow.

The **energy** supplied by an ideal battery is the product of the charge supplied and the battery voltage. See Eq. (18.7).

The **power** supplied by an ideal battery is the product of the current and the battery voltage. See Eq. (18.8).

Any one of several equivalent formulas can be used to calculate the **Joule heating** in a resistor. See Eq. (18.9).

Kirchhoff's laws permit calculation of branch currents and junction voltages in complicated electric circuits. The first law expresses charge conservation, and the second law expresses energy conservation. See the display on p. 511.

A **galvanometer** is an instrument that measures electric current. **Shunt resistors** permit a galvanometer to be used for measuring the current over various ranges. **Series resistors** permit voltage measurements over various ranges.

Instruments that measure an electrical quantity by the absence of current in a certain arm of a circuit are called **null instruments.** The **Wheatstone bridge** is a null instrument to measure resistance. The **potentiometer** is a null instrument to measure potential difference.

QUESTIONS FOR THOUGHT

1. "The electric field inside a conductor is always zero." Use the phenomenon of electrical conduction in a wire to show that this statement is not correct. What important change would you make to correct it?

2. Can you devise an experimental arrangement to measure the resistivity of a cubic block of metal several centimeters on edge?

3. You have only a voltmeter and an ammeter available to measure the resistance of a spool of thick copper wire. Which of the two circuits of Fig. 18.11 would you prefer to use and why?

4. What are the advantages and disadvantages associated with series and parallel connections of lights on a Christmas tree?

5. Two wires of the same length and cross-sectional area but of different resistivity are joined in series and connected to a battery. In which of the two wires is Joule heating greater? Does your answer change if the two wires are joined in parallel?

6. You are given an accurate voltmeter of very large internal resistance and standard resistor of known resistance. How would you use these items to determine the open-circuit emf and the internal resistance of a battery?

7. You use a Wheatstone bridge to measure an unknown resistor and leave the bridge switched on after achieving a balance. Some time later you notice that the bridge has become unbalanced. Suggest a possible reason for this.

8. A student wiring up an electric circuit in the laboratory inadvertently interchanges an ammeter and a voltmeter in the circuit. Which instrument is more likely to be destroyed?

9. Is there any circumstance in which the measured potential difference between the terminals of a battery exceeds the open-circuit emf of the battery?

PROBLEMS

(Use the values of resistivities and temperature coefficients of resistance in Table 18.1.)

A. Single-Substitution Problems

1. A battery is charged for 6 h with an average current of 12 A. Calculate the quantity of charge stored in the battery. [18.1]

2. Calculate the current in a wire if 1 billion electrons per second pass a given point. [18.1]

3. A charge of 400 C flows into one end of a wire and out the other during a time interval of 1 min. Calculate the average current in the wire. [18.1]

4. An average current of 3.6 A exists in a wire for 6 min. Calculate the quantity of charge transferred. [18.1]

5. Calculate the value of the resistance that carries a 45-mA current when connected across the terminals of a 12-V battery. [18.2]

6. Calculate the current in a 30-Ω resistor when a 12-V potential difference is maintained between its ends. [18.2]

7. A certain resistor carries a current of 40 mA with a 6-V potential difference between its ends. Calculate the value of its resistance. [18.2]

8. How large a voltage is required to cause a current of 300 mA in a 12-Ω resistor? [18.2]

9. Calculate the resistance of 20 m of copper wire 0.5 mm in diameter. [18.3]

10. A piece of wire 67.3 m long and 2 mm in diameter has a resistance of 1.2 Ω at 20°C. Identify the material of the wire with the help of Table 18.1. [18.3]

11. A spool of nichrome wire has a rectangular cross section 2.5 mm × 0.2 mm. What length of wire is required to form a heating element of resistance 15 Ω? [18.3]

12. A tungsten lamp filament has a resistance of 24 Ω at 25°C. Calculate the resistance of the filament at 2025°C. [18.3]

13. What rise in temperature is required to cause a 6% increase in the resistance of a copper wire? [18.3]

14. Calculate the rise in temperature required to cause a 0.1% increase in the resistance of a constantan wire. [18.3]

15. Three resistors have resistances of 2 Ω, 3 Ω, and 4 Ω.
   a. Calculate the equivalent resistance of all three in series.
   b. Repeat the calculation for all three in parallel. [18.4]

16. What resistance is required in parallel with a 15-Ω resistor to produce an equivalent resistance of 10 Ω? [18.4]

17. Design a circuit to have an equivalent resistance of 35 Ω using only 10-Ω resistors. [18.4]

18. Calculate the maximum current permitted in an 800-Ω resistor that has a power rating of 0.5 W. [18.5]

19. A 12-V ideal battery is connected across a 4-Ω resistor. Calculate the power dissipated in the resistor. [18.5]

20. Calculate the maximum voltage that can be applied to a 2000-Ω resistor that has a power rating of 5 W. [18.5]

B. Standard-Level Problems

21. A piece of aluminum tubing 4.6 m long has inside and outside diameters of 3 mm and 4 mm, respectively. Calculate the current in the tube if the potential difference between the ends is 60 mV. [18.3]

22. A piece of copper tubing 2.5 m long has an outside diameter of 4 mm. A potential difference of 40 mV between the ends causes a current of 8 A. Calculate the inside diameter of the tube. [18.3]

23. A square 5 cm × 5 cm is cut from a square sheet of constantan 0.25 mm thick. Calculate the resistance between opposite edges of the square. Also calculate the resistance between opposite edges of an 8 cm × 8 cm square cut from the same sheet. [18.3]

24. A polyethylene insulator is made from a sheet 0.1 mm thick. What insulation resistance is provided between the opposite faces of a disk 4 cm in radius cut from this sheet?

25. It is desired to provide 5000 GΩ of insulation resistance between two adjacent flat parallel metal plates each 10 cm × 10 cm. What thickness of polyethylene is required?

26. The coil of a platinum resistance thermometer has a resistance of 8.53 Ω at 25°C. Calculate the resistance

of the thermometer coil if it is inserted into a furnace at 1200°C.

27. A platinum resistance thermometer that has a resistance of 10.18 Ω at a temperature of 20°C is inserted into a furnace, and the resistance changes to 47.81 Ω. Calculate the temperature of the furnace.

28. A length of constantan wire in a piece of scientific equipment has a resistance of 80 Ω at 20°C. Calculate the temperature range within which the instrument can be operated, if the resistance change is not to exceed 0.005%.

29. A tungsten wire and a nichrome wire have the same resistance at 25°C. Calculate the ratio of their resistances at 125°C.

30. Calculate the equivalent resistance of the arrangement of resistors shown in the diagram. [18.4]

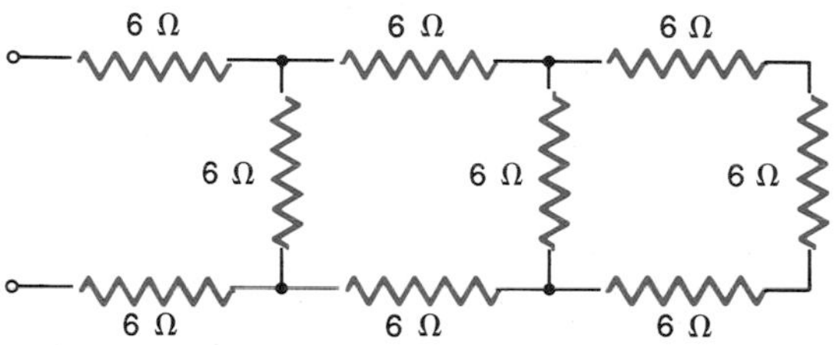

31. Calculate the equivalent resistance of the arrangement of resistors shown in the diagram. [18.4]

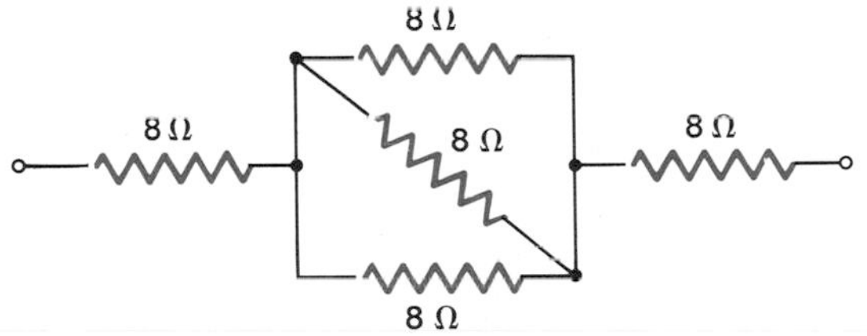

32. Calculate the equivalent resistance of the arrangement of resistors shown in the diagram.

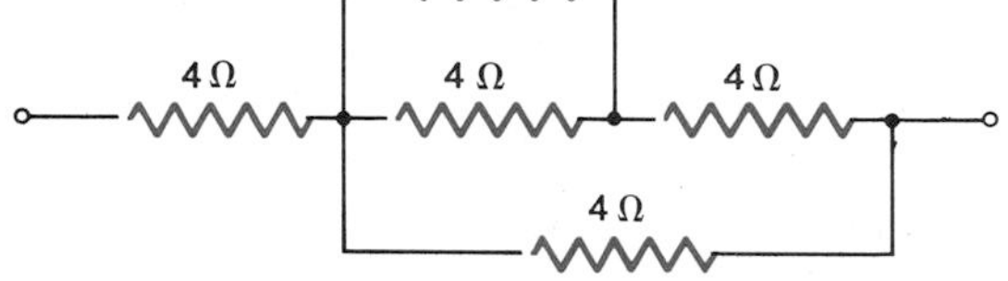

33. A battery with an open-circuit voltage of 12 V and an internal resistance of 0.25 Ω is connected to the arrangement of resistors shown in the diagram. Calculate the current in the battery.

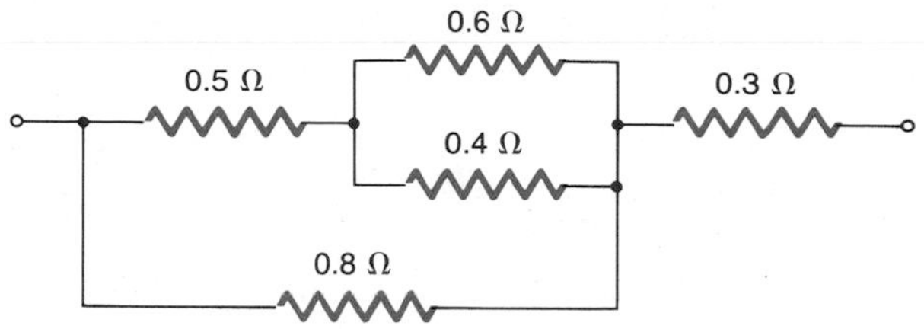

34. A battery with an open-circuit voltage of 6 V is connected to the arrangement of resistors shown in the diagram. The battery current is 8.5 A. Calculate the internal resistance of the battery.

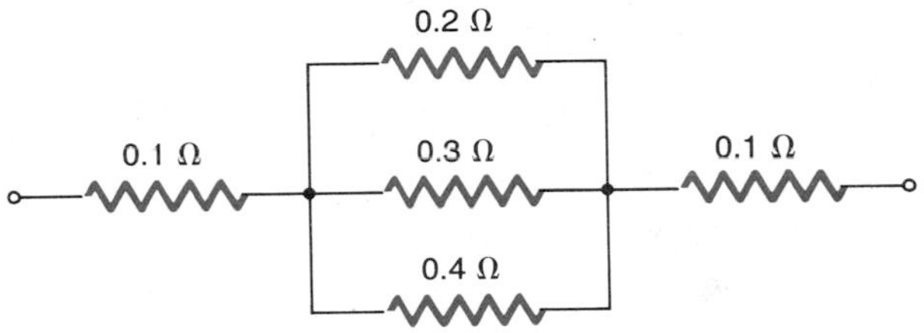

35. A nichrome heating element dissipates 8.2 kW at a temperature of 400°C when the potential difference between its ends is 80 V. How much power would it dissipate at a temperature of 20°C if the potential difference between its ends at the lower temperature is 120 V? [18.5]

36. A 45-W automobile headlight operates from a 12-V battery with the temperature of its tungsten filament about 2000°C.

a. Calculate the resistance of the filament at operating temperature.

b. Use the temperature coefficient of resistivity from Table 18.1 to estimate the resistance of the filament at 25°C. [18.5]

37. The maximum safe current for 10-gauge (2.59 mm diameter) insulated copper wire is 30 A.

a. Calculate the difference in voltage between the ends of a 20-m length of the wire when it carries the maximum rated current.

b. Calculate the power dissipated in the same length of wire for the same current.

38. A copper wire 2.4 m long and 1.2 mm in diameter is joined end to end with a nichrome wire of the same length and diameter. An ideal 12-V battery is connected across the ends of the composite resistance wire.

a. Calculate the power dissipated in each portion of the wire.

b. If the wires were joined side by side instead of end to end, what would be the power dissipation in each one when connected across the 12-V battery?

39. An electric water heater operates from a 12-V automobile battery and boils 0.75 ℓ of water in 18 min from an initial temperature of 25°C. If 30% of the energy supplied by the battery is lost to the container and the surroundings, what is the resistance of the heating coil?

40. A small water heater whose heating element has a resistance of 0.6 Ω operates from a 12-V automobile battery. How long does it take to boil 0.4 ℓ of water from an initial temperature of 20°C if 35% of the heat energy is lost to the water container and surroundings?

41. A battery with an open-circuit voltage of 12 V is connected to a 20-Ω resistor. The power dissipated in the resistor is 5 W. Calculate:

a. the internal resistance of the battery
b. the power dissipated in the battery

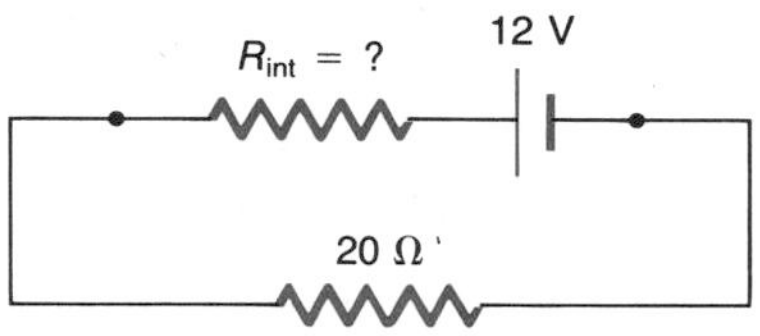

42. A battery with an open-circuit voltage of 10 V and an internal resistance of 0.5 Ω is connected to a 6-Ω resistor. Calculate:

a. the power produced by the battery
b. the power dissipated in the 6-Ω resistor
c. the power dissipated within the battery

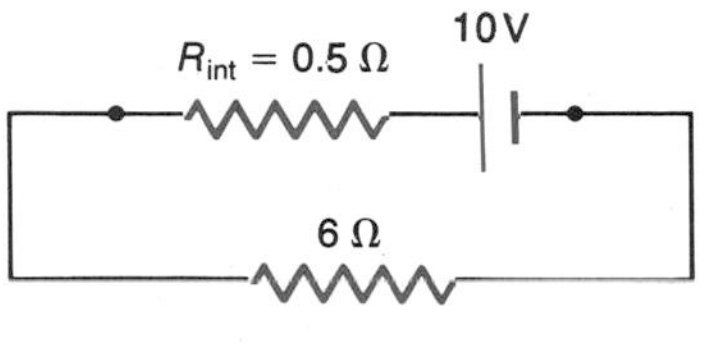

43. A 12-V ideal battery is connected with three 3-Ω resistors as shown in the diagram. Calculate the power produced by the battery.

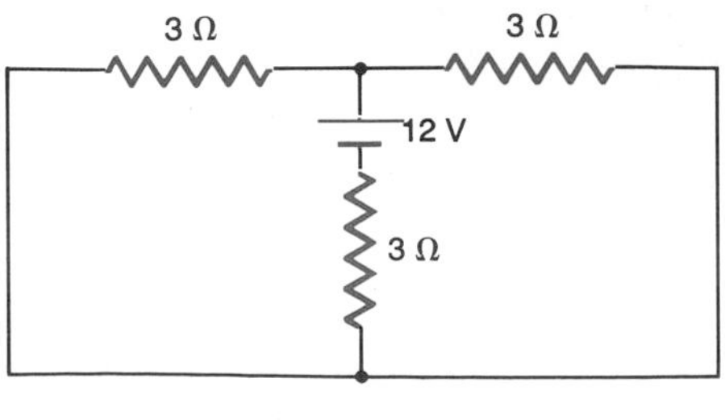

44. Calculate the current in the 3-Ω resistor shown in the accompanying diagram. [18.6]

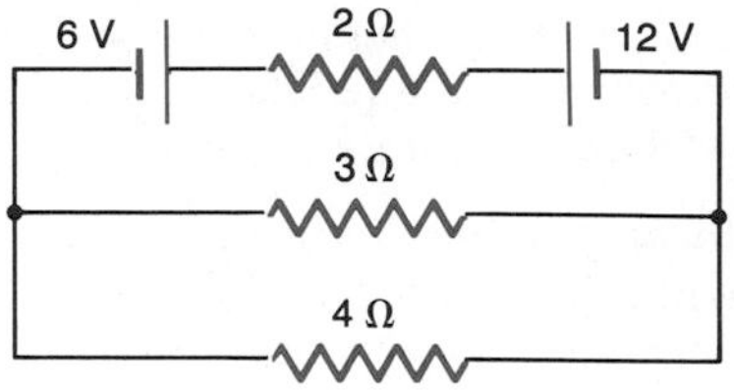

45. Calculate the battery emf ℰ if there is a current of 2 A in the 3-Ω resistor in the direction indicated on the diagram. [18.6]

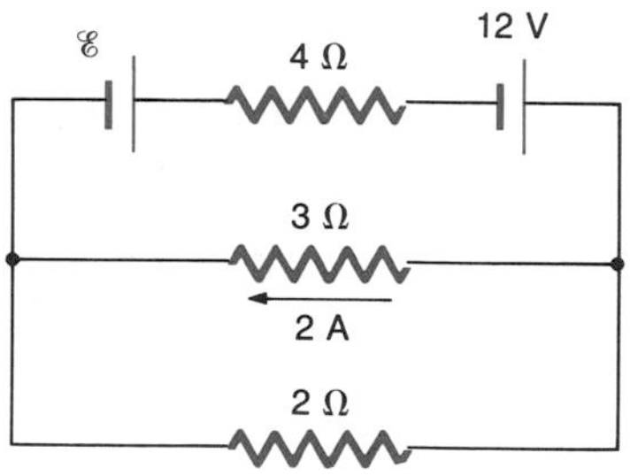

46. Calculate the current in the 3-Ω resistor shown in the accompanying diagram. [18.6]

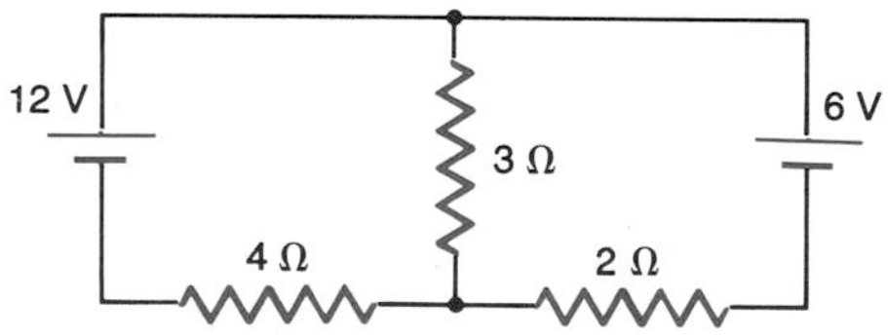

47. Calculate the battery emf ℰ if the current in the 4-Ω resistor is 2 A in the direction indicated.

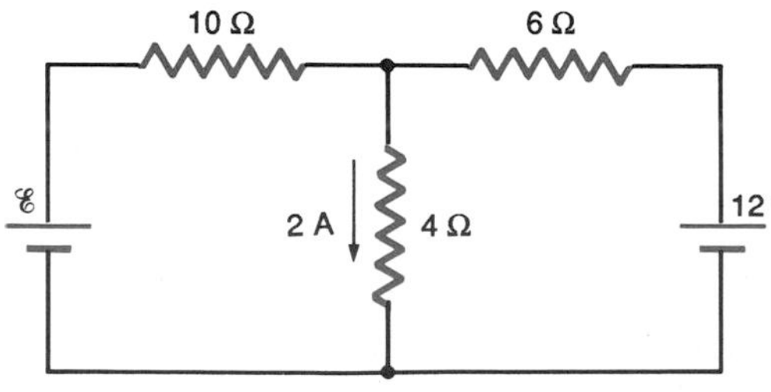

48. Two ideal batteries are connected in circuit with a number of resistors, as shown in the diagram at the top of page 525.

a. Calculate the current in each battery.
b. Calculate the power dissipation in the 4-Ω resistor.

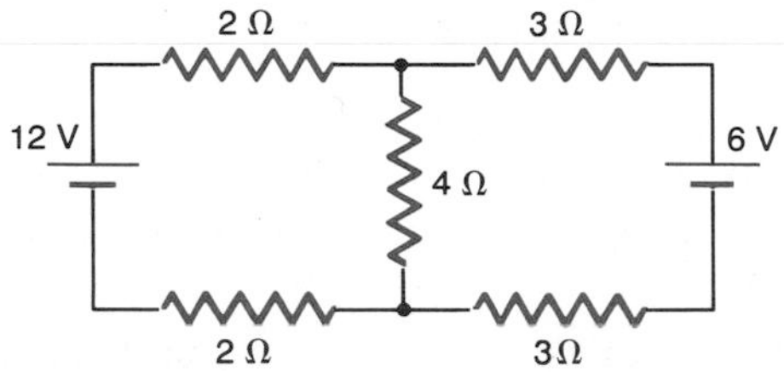

49. Calculate the value for the resistance R that results in a 1-A current in the 3-Ω resistor in the direction shown on the diagram.

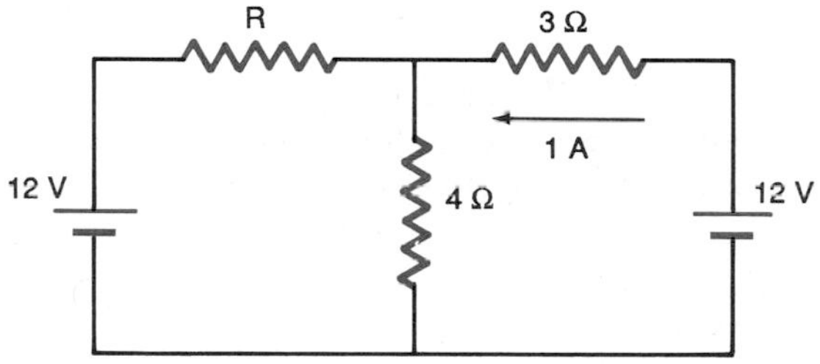

50. Calculate the battery emf's $\mathscr{E}_1$ and $\mathscr{E}_2$ and the resistance R, if the currents in the 8-Ω and 4-Ω resistors are 2 A and 1 A, respectively, in the directions indicated, and the current in the resistor R is 2 A in the direction indicated.

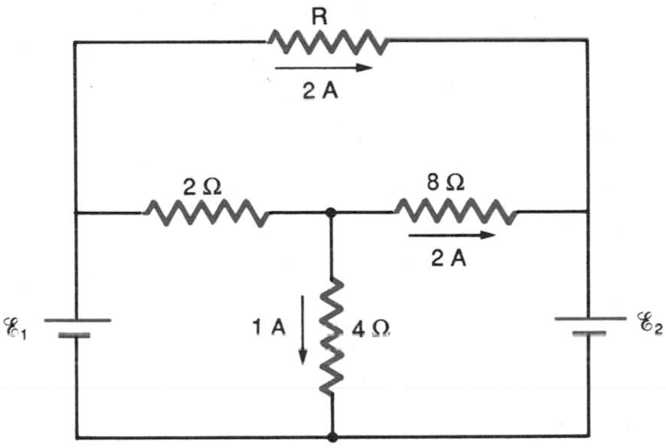

51. a. Calculate the reading of the ammeter in the circuit shown when the switch is open. (Consider the ammeter resistance to be negligible.)

b. When the switch is closed, the ammeter reading decreases by 50%. Calculate the resistance R_x.

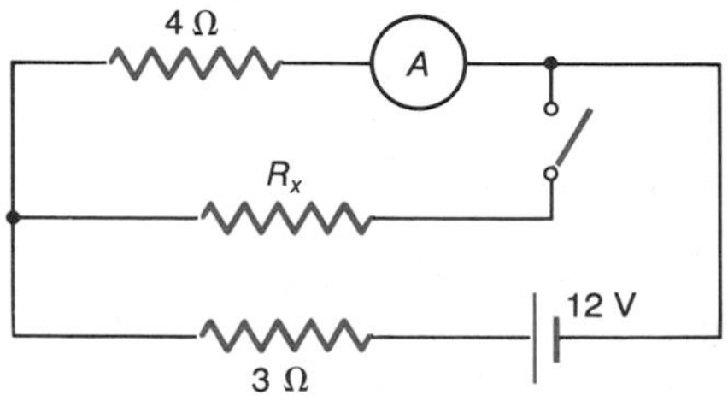

52. For the circuit shown in the diagram calculate:

a. the potential difference across the switch if it is open

b. the current in the switch if it is closed

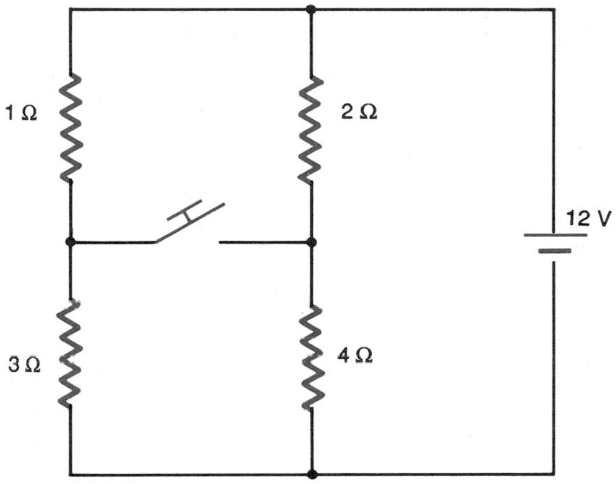

53. A galvanometer gives full-scale deflection with a current of 10 μA and has a resistance of 5000 Ω. Design a series resistor suitable for using the instrument as a voltmeter with a range of 0–100 V. [18.7]

54. Design a shunt resistor suitable for using the galvanometer of problem 53 as an ammeter with a range of 0–50 mA. [18.7]

55. A galvanometer gives full-scale deflection with a current of 20 μA and has a resistance of 4500 Ω. Design series resistors R_1, R_2, and R_3 so that the instrument can be used as a multirange voltmeter, as shown in the diagram.

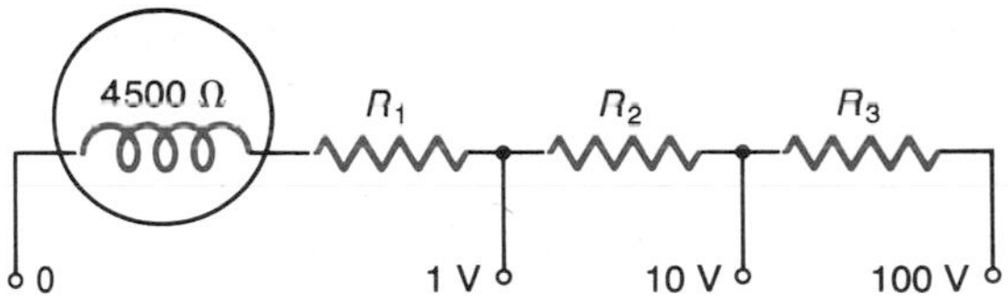

56. A Wheatstone bridge utilizes a 10.00-Ω standard resistor and a uniform resistance wire 100.0 cm long.

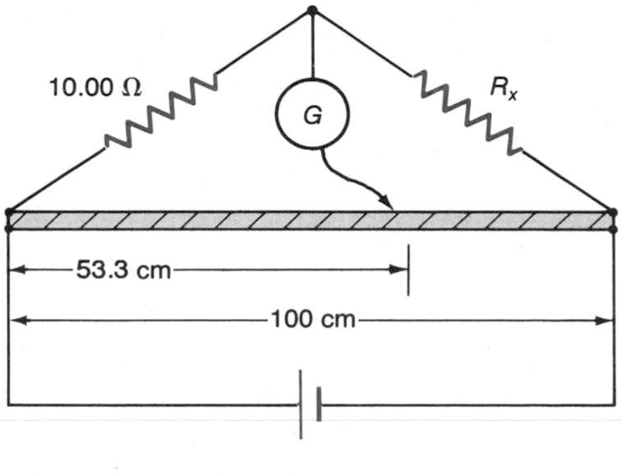

The bridge is balanced when the sliding contact is 53.3 cm from one end. Calculate the value of the unknown resistance.

57. An unknown wire resistor in an oil bath is placed in one arm of a 100-cm slide wire Wheatstone bridge, and a temperature-controlled 10.00-Ω standard resistor is placed in the other arm. If the oil bath is at 15°C, the sliding contact is 46.2 cm from one end; at 110°C, the contact is 55.7 cm from that end. Calculate the temperature coefficient of resistivity of the wire.

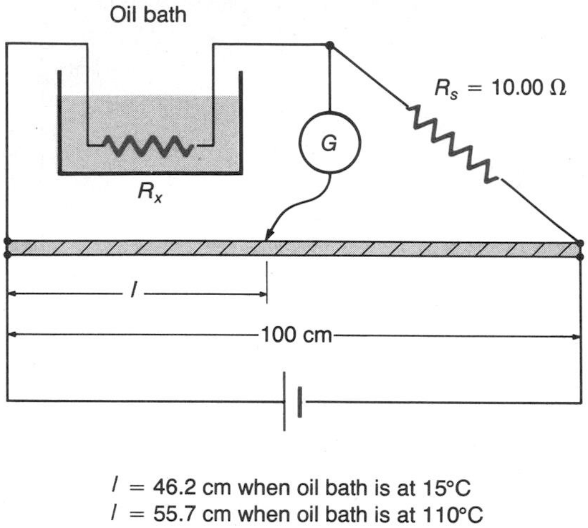

58. The open-circuit voltage of the standard battery in a slide wire potentiometer is 1.0178 V, and it produces a null when the sliding contact is 36.4 cm from one end of the 100-cm resistance wire. When the unknown battery is in circuit, the null occurs with the sliding contact 53.7 cm from the end of the wire. Calculate the open-circuit voltage of the unknown battery.

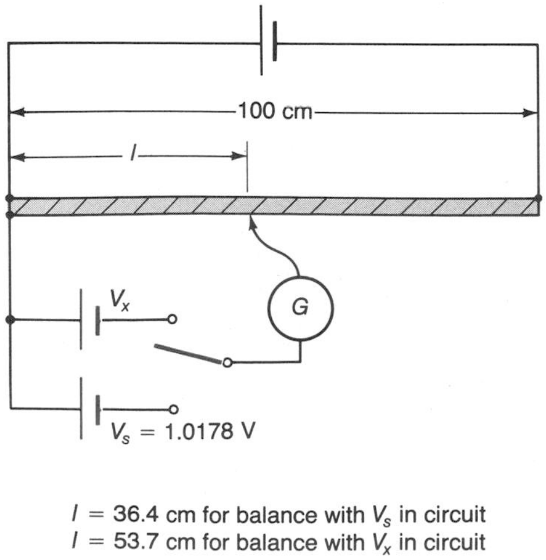

C. Advanced-Level Problems

59. A 50-Ω resistor is to be made from nichrome wire of circular cross section. The volume of the wire resistor is to be 3 cm^3. Calculate the length and the diameter of the wire that will fulfill the required conditions.

60. Two identical batteries have an open-circuit voltage of 12 V and an internal resistance of 0.5 Ω.

a. Should they be connected in series or in parallel to give the largest current in a 1.4-Ω resistor?
b. How should the batteries be connected to give the largest current in a 0.2-Ω resistor?

61. A battery that has an open-circuit voltage of 12 V and an internal resistance of 0.15 Ω is used to charge a 6-V battery that has an internal resistance of 0.25 Ω. The power dissipation within the 6-V battery is not to exceed 20 W.

a. Calculate the value of the series resistor that should be placed in the circuit.
b. How long would it take to increase the charge of the 6-V battery by 30 000 C?

62. A box contains a number of 10-kΩ resistors, each rated at 0.5 W. Design series-parallel combinations of these resistors to satisfy the following requirements:

a. 5 kΩ with a power rating of at least 2 W
b. 10 kΩ with a power rating of at least 2 W
c. 15 kΩ with a power rating of at least 2 W

63. A galvanometer gives a full-scale deflection with a current of 50 μA and has a resistance of 1500 Ω. Design the shunt resistors R_1, R_2, and R_3 so that the instrument can be used as a multirange ammeter, as shown in the diagram.

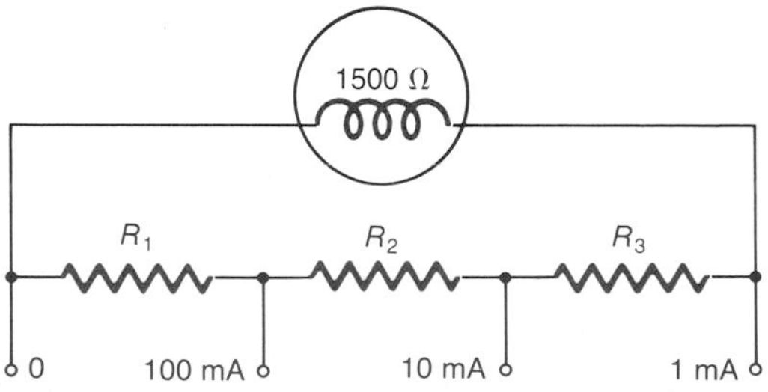

64. A laboratory magnet is energized by a copper coil that draws 80 A from a 200-V battery bank of negligible internal resistance. The coil is cooled by water that enters at 15°C and that must leave at no higher temperature than 75°C. What minimum flow rate must be designed for the cooling water if it is to carry away all the dissipated energy?

65. Two galvanometers each give a full-scale reading with a current of 1 mA. One is converted to a 0–10-V voltmeter by the addition of a series resistor, and the other to a 0–5-mA ammeter by the addition of a shunt resistor. The instruments are then wired into a circuit with an unknown resistor and a battery of unknown voltage, as shown in the diagram. The voltmeter indicates 8.41 V and the ammeter indicates 3.37 mA. Calculate the resistance of the unknown resistor.

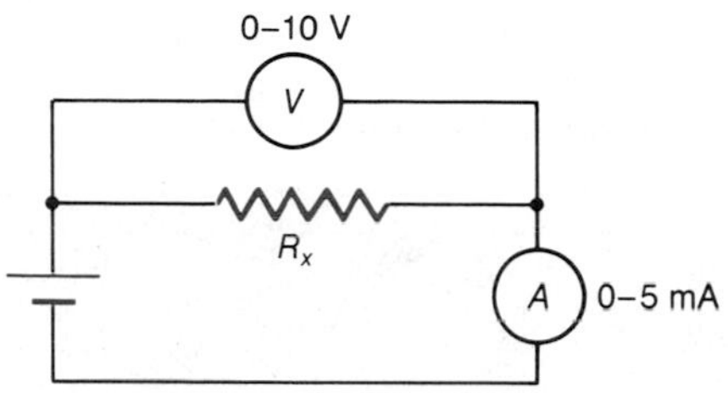

66. Calculate the maximum battery voltage for the circuit shown in the accompanying diagram if the power dissipation is not to exceed 1 W in any resistor.

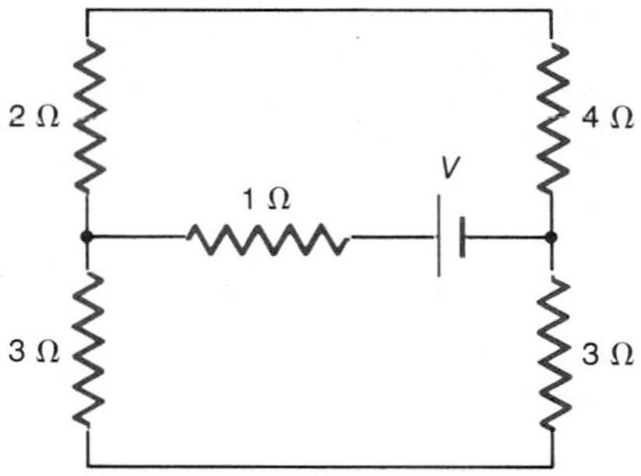

19

ELECTRO-MAGNETISM I

MAGNETIC FORCES AND FIELDS

The existence of magnetic force has been known since early times. Indeed, magnets take their name from the ancient city of Magnesia in the eastern Mediterranean area, which was famous for the production of magnetized ore or lodestone. Not only were forces of attraction or repulsion found between pieces of lodestone, but a piece of lodestone suspended by a thread always pointed in the same direction. In succeeding centuries, this effect was put to use in the mariner's magnetic compass. Even today the compass needle is one of the best known examples of magnetic force. If pressed for other examples, many people might mention the ability of a magnet to pick up small iron objects.

To begin the study of magnetism with magnets certainly seems reasonable, but in fact it leads to difficulties in understanding the phenomenon that is truly the most widespread application of magnetic force: the operation of the electric motor. For this reason, we introduce magnetic force as the force between electric currents, or, alternatively, as the force between moving charges. We begin our study by defining a magnetic field in terms of the force that it causes on currents or moving charges. Later we investigate how other currents or moving charges generate a magnetic field.

It might appear that we have now introduced a third basic type of force, in addition to the gravitational force between masses and the electric force between charges. However, an electric current is a moving electric charge, and we might suspect that electric and magnetic forces are closely related. In fact, they do turn out to be different aspects of the same phenomenon—a force that we simply name electromagnetic. For our present purposes, though, we preserve the distinction and treat magnetic force as a separate entity.

19.1 Magnetic Force

The recognition of **magnetic force** as the force between electric currents dawned slowly. The force between magnets is most obvious, exhibiting striking similarities of form with electrostatic force. If we take two bar magnets whose ends are marked north and south and suspend them by threads attached to their centers, they display forces of attraction and repulsion. A few simple trials will convince us that north and south ends attract each other, while north and north or south and south ends repel each other. We can thus refer to north and south magnetic poles in the ends of the magnets and set up an equation describing the force between them. The equation is an inverse square law of the same form as Newton's law of gravitation and Coulomb's law of electrostatic force.

SPECIAL TOPIC 21 Ampère

André-Marie Ampère was born near Lyon in France in 1775. He was a child prodigy in mathematics, but tragedy touched his life in his teenage years when his father was guillotined by revolutionaries. After the revolution, he became professor of physics and chemistry at Bourg, and later professor of mathematics at Paris. Ampère took as a basic starting point Oersted's discovery that an electric current deflected a compass needle. From this he discovered the basic law of magnetic force between arbitrarily directed current elements. He initiated the concept of magnetic lines of force (later generalized by Faraday), and was the first to use advanced mathematics in the description of electromagnetic phenomena. Ampère died at Marseille in 1836.

(Smithsonian Institution)

At first, electricity and magnetism were thought to have no connection, but the Danish physics professor Hans Christian Oersted (1771–1851) believed that an electric current would cause magnetic force, and in 1820 he arranged a public lecture and demonstration to prove his point. He placed a small bar magnet close to a straight conducting wire and parallel to it. When he connected a battery across the ends of the wire, causing an electric current, the magnet moved. Although the movement of the magnet was feeble and the audience was not deeply impressed, the subject of electromagnetism had been born. From this point it was a small step to discover that two electric currents or moving charges exert a force on each other.

We begin by recalling the steps in the definition of electric field strength. First, we consider an electric field in some region of space that is set up by some unspecified charges. Then we introduce a charge q into the region, and define the electric field $\mathbf{E}$ from the relation $\mathbf{F} = q\mathbf{E}$, where $\mathbf{F}$ is the force that the charge experiences in the field. This defines $\mathbf{E}$ at every point in the region.

We follow a very similar procedure in defining magnetic field, and begin with some unspecified currents (or moving charges) that set up a magnetic field in some region of space. We now introduce into the magnetic field a point charge q traveling with a velocity **v**. Experiment shows that we can calculate the magnetic force on the moving charge in terms of a vector magnetic field strength **B**. The magnetic force has the following characteristics:

1. Its magnitude is proportional to the product of the charge, the magnetic field, and the component of velocity that is perpendicular to the magnetic field. Referring to Fig. 19.1(a), we write this as:

$$F = qv_{\perp}B \tag{19.1}$$

   Note especially that this means that the magnetic force is zero if the velocity of the charge is parallel to the magnetic field, since in that case $v_{\perp}$ would be zero.

2. The direction of the magnetic force is given by the condition that the three vectors ($\mathbf{F}$, $q\mathbf{v}_{\perp}$, $\mathbf{B}$) form a right-handed set. This

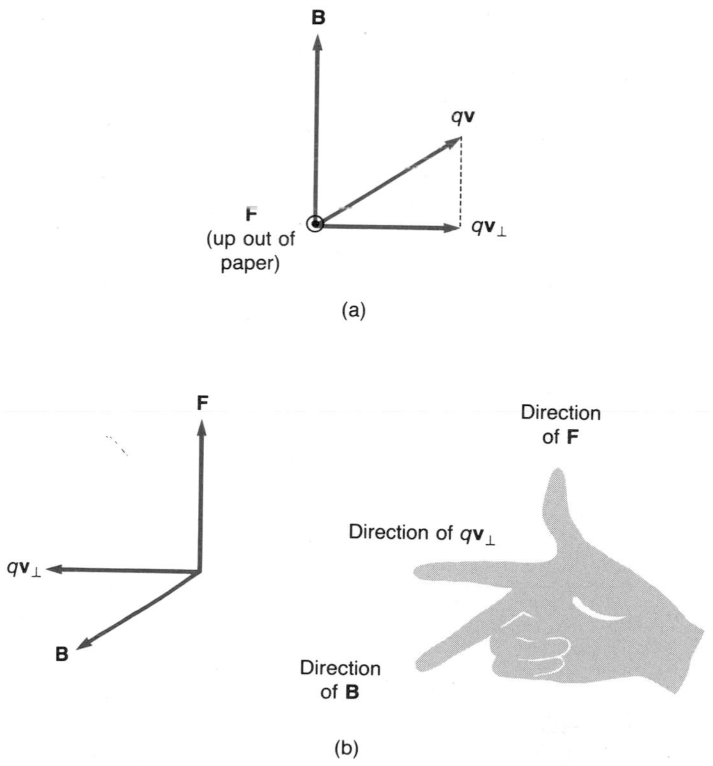

FIGURE 19.1 Illustration of the relation between magnetic force and magnetic field for a moving charge: (a) $q\mathbf{v}_{\perp}$ is the component of $q\mathbf{v}$ that is perpendicular to $\mathbf{B}$; (b) the right hand rule for the vector set ($\mathbf{F}$, $q\mathbf{v}_{\perp}$, $\mathbf{B}$).

means that all three vectors must be perpendicular to each other and have the relative directions indicated in Fig. 19.1(b). The right hand is held with the thumb, first finger, and second finger mutually perpendicular; then the first vector of the set, **F**, points in the direction of the thumb, the second vector, $q\mathbf{v}_{\perp}$, points in the direction of the first finger, and the third vector, **B**, points in the direction of the second finger. Returning to Fig. 19.1(a), this means that the magnetic force vector points up out of the plane of the paper for the relative directions given for **B** and $q\mathbf{v}$.

A shorthand notation allows us to write Eq. (19.1) in a way that includes both the magnitude and direction of the magnetic force vector.

Definition of Magnetic Field

Magnetic force = vector cross product of $q\mathbf{v}$ and **B**

$$\mathbf{F} = q\mathbf{v} \times \mathbf{B} \qquad \textbf{(19.2)}$$

SI unit of magnetic field: N · s/C · m
New name: tesla
Abbreviation: T

The vector cross product is written as a sequence of two vectors separated by a multiplication sign. The rules for evaluating the cross product are:

1. Its magnitude is the product of the magnitude of one of the vectors and the perpendicular component of the other, or alternately the product of the magnitudes of the two vectors and the sine of the angle between them. This magnitude is:

$$|q\mathbf{v} \times \mathbf{B}| = qv_{\perp}B = qvB_{\perp} = qvB \sin \phi$$

Figure 19.2 shows that these quantities are all the same since:

$$qv_{\perp} = qv \sin \phi$$

and

$$B_{\perp} = B \sin \phi$$

2. The direction of the vector cross product is given by the right hand–set rule, explained in Fig. 19.1(b).

The unit of magnetic field strength honors Nikola Tesla (1857–1943), an American engineer who was responsible for many inventions in applied electromagnetism.

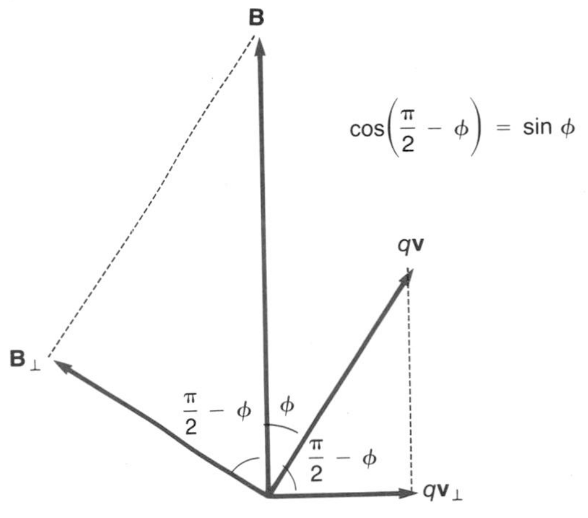

FIGURE 19.2 Illustration of $q\mathbf{v}_{\perp}$ (the component of $q\mathbf{v}$ that is perpendicular to **B**) and $\mathbf{B}_{\perp}$ (the component of **B** that is perpendicular to $q\mathbf{v}$).

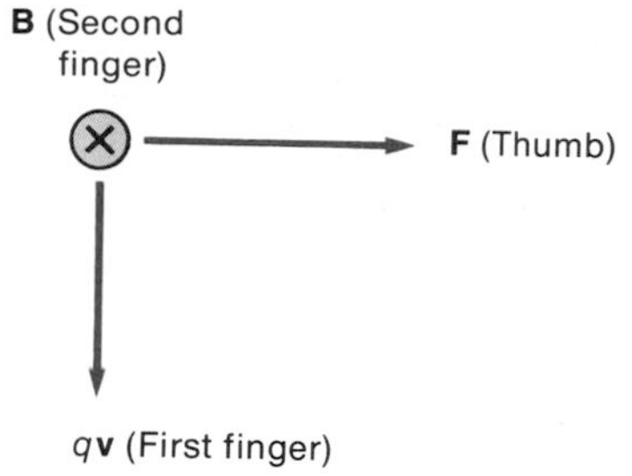

EXAMPLE 19.1

A proton in a particle accelerator is moving horizontally due south with a velocity of 10^7 m/s in a vertical magnetic field of 0.8 T that is directed downward. Calculate the force acting on the proton.

SOLUTION The proton charge is $q = +1.6 \times 10^{-19}$ C. We can calculate the magnitude of the force directly from Eq. (19.1):

$$\begin{aligned} F &= qv_{\perp}B \\ &= 1.6 \times 10^{-19}\ \text{C} \times 10^7\ \text{m/s} \times 0.8\ \text{T} \\ &= 1.28 \times 10^{-12}\ \text{N} \end{aligned}$$

The diagram illustrates the application of the right hand rule to determine the direction of F. The vector set (**F**, q**v**, **B**) is right-handed. The vector q**v** points south, and the vector **B** is vertically downward. The vector **F** is therefore horizontal and pointing to the east. ■

It is important to realize that if q is positive, q**v** points in the direction of **v**; if q is negative, it points in the opposite direction. Let us illustrate by an example.

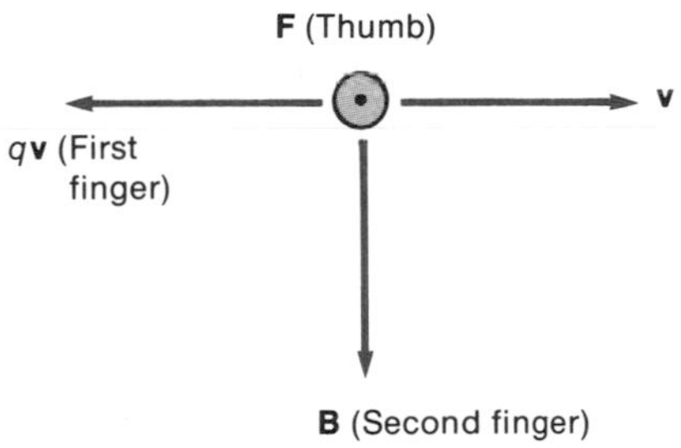

EXAMPLE 19.2

An electron traveling horizontally toward the east with a velocity of 2×10^7 m/s experiences a vertically upward force of 4×10^{-12} N. Calculate the magnitude and direction of the magnetic field required to cause this force.

SOLUTION The charge on the electron is -1.6×10^{-19} C. The negative sign does not alter our calculation of the magnitude of the magnetic field from Eq. (19.1):

$$\begin{aligned} F &= qv_{\perp}B \\ 4 \times 10^{-12}\ \text{N} &= 1.6 \times 10^{-19}\ \text{C} \times 2 \times 10^7\ \text{m/s} \times B \\ \therefore B &= 1.25\ \text{T} \end{aligned}$$

However, we must carefully observe the sign when determining the vector directions. The direction of **F** is vertically upward and that of **v** is to the east; since q is negative, the direction of q**v** is to the west. Using the right hand rule shows that **B** is horizontal and directed toward the south. ■

Consider now a straight piece of conducting wire of length l that is part of a circuit carrying a current I in a constant magnetic field **B**, as shown in Fig. 19.3(a). At any given instant suppose

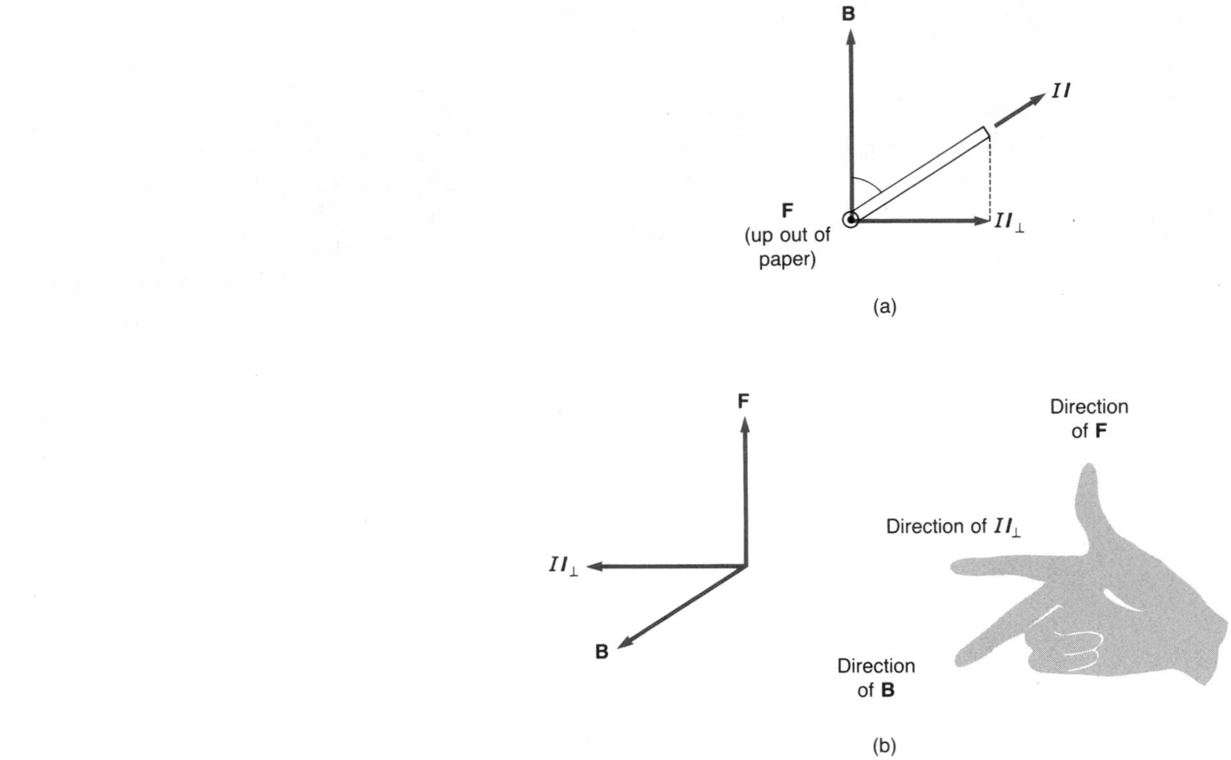

FIGURE 19.3 Illustration of the relation between magnetic force and magnetic field for a current-carrying wire: (a) $I\mathbf{l}_\perp$ is the component of $I\mathbf{l}$ that is perpendicular to **B**; (b) the right hand rule for the vector set (**F**, $I\mathbf{l}_\perp$, **B**).

that the wire contains N conducting electrons, each of charge q and having a drift velocity **v**. The wire as a whole is electrically neutral, but the conducting electrons are moving and will experience a magnetic force. The charge that neutralizes the conducting electrons is stationary and experiences no magnetic force. The total magnetic force on the wire due to the conducting electrons is given by Eq. (19.2):

$$\mathbf{F} = Nq\mathbf{v} \times \mathbf{B}$$

Now let us suppose that each electron drifts the length l of the wire in time t, and that $\mathbf{l}$ is a vector whose magnitude is the length of the wire and whose direction is that of the conventional current I. The electron drift velocity (opposite to the conventional current) is given by:

$$\mathbf{v} = \frac{-\mathbf{l}}{t}$$

Since the electron charge is negative, the conventional current is given by:

$$I = \frac{-Nq}{t}$$

This allows us to write:

$$nq\mathbf{v} = \frac{Nq\mathbf{l}}{t} = I\mathbf{l}$$

Finally, we can write an alternate definition for the magnetic field in terms of the force on a current-carrying wire.

■ Definition of Magnetic Field (Second Form)

Magnetic force = vector cross product of $I\mathbf{l}$ and $\mathbf{B}$

$$\mathbf{F} = I\mathbf{l} \times \mathbf{B} \qquad \textbf{(19.3)}$$

The rules for calculating the vector cross product are the same as those given after Eq. (19.2).

1. The magnitude of **F** is given by:

$$|I\mathbf{l} \times \mathbf{B}| = Il_\perp B - IlB_\perp = IlB \sin \phi$$

where $l_\perp$ is the component of **l** perpendicular to **B**, $B_\perp$ is the component of **B** perpendicular to **l**, and ϕ is the angle between **l** and **B**.

2. The direction of **F** is found from the right hand rule of Fig. 19.3(b); the vectors ($\mathbf{F}$, $I\mathbf{l}_\perp$, $\mathbf{B}$) form a right-handed set.

EXAMPLE 19.3

A wire of length 30 cm experiences a force of 6 N when it carries a current in a magnetic field of 0.8 T that is perpendicular to its length. Calculate the magnitude and relative direction of the current in the wire.

SOLUTION Let the magnetic field be directed upward out of the paper, and let the force direction be from bottom to top on the diagram. Since $I\mathbf{l}$ and **B** are perpendicular to each other, we can write Eq. (19.3) in terms of magnitudes only:

$$F = IlB$$

$$6\ \text{N} = I \times 0.3\ \text{m} \times 0.8\ \text{T}$$

$$\therefore I = 25\ \text{A}$$

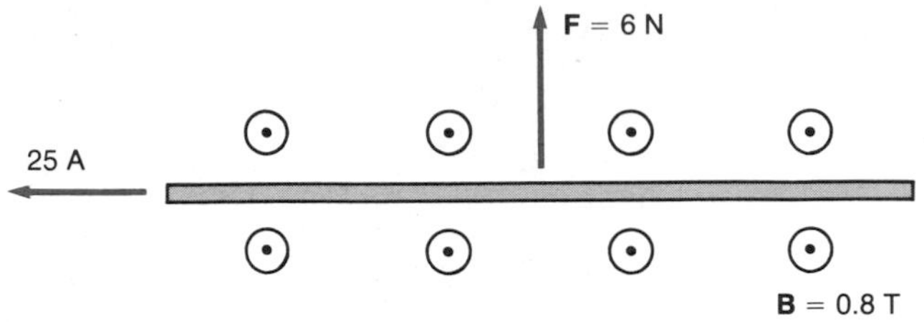

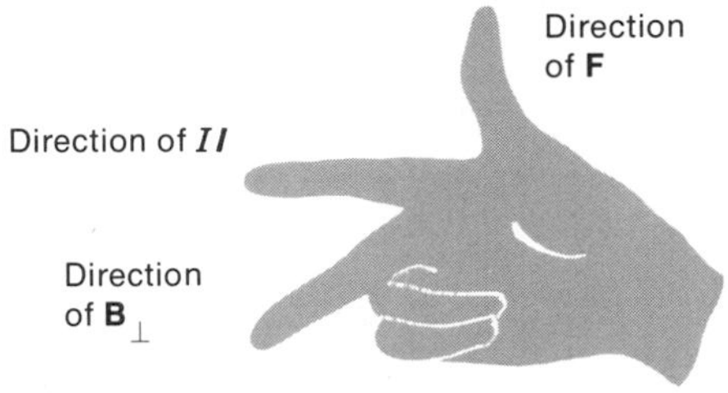

To find the direction of the vector $I\mathbf{l}$, we need to relate the vector set ($\mathbf{F}$, $I\mathbf{l}$, $\mathbf{B}$) by the right hand rule, as shown in the diagram. This shows $I\mathbf{l}$ from right to left on the diagram. ■

19.2 The Magnetic Field of Currents

We have already introduced the idea of a region of space in which there is a magnetic field due to various unspecified currents, and we now turn to the problem of calculating the magnetic field of some simple specified currents.

We begin with a long straight wire carrying a constant current, and can gain insight into the nature of its magnetic field with a simple experiment. We set up a vertical straight wire carrying a current, and place a piece of cardboard in a horizontal position with the wire passing through it, as shown in Fig. 19.4. If we place a handful of iron filings on the cardboard and tap them lightly, the filings quite clearly exhibit a pattern of circles concentric with the wire. Since iron is known to be strongly susceptible to magnetic influences, we take this behavior as evidence that the magnetic field lines are circles concentric with the current-carrying wire. We use the term field line rather than line of force for the magnetic field since, as we saw in the previous section, the force on a moving charge or current-carrying wire does not point in the direction of the magnetic field.

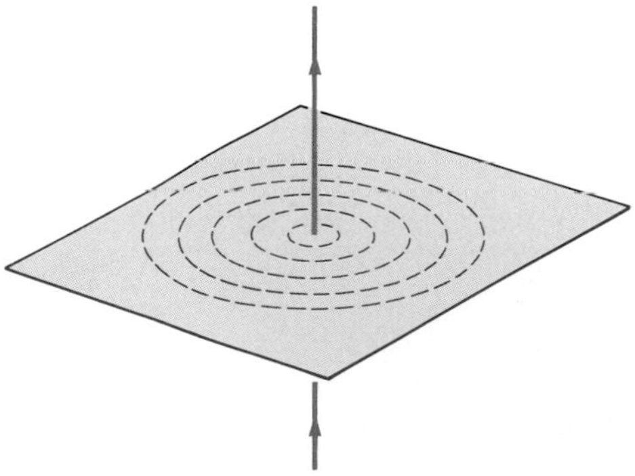

FIGURE 19.4 Experimental demonstration that magnetic force lines form concentric circles around a long straight wire that carries a current.

Another deep-seated difference between the electrostatic and magnetic field lies in the way they relate to the charges and currents that cause them. The electrostatic field originates on positive and terminates on negative charge, but the magnetic field runs in continuous closed loops that encircle the current that causes them. The direction of the field lines is given by the right hand

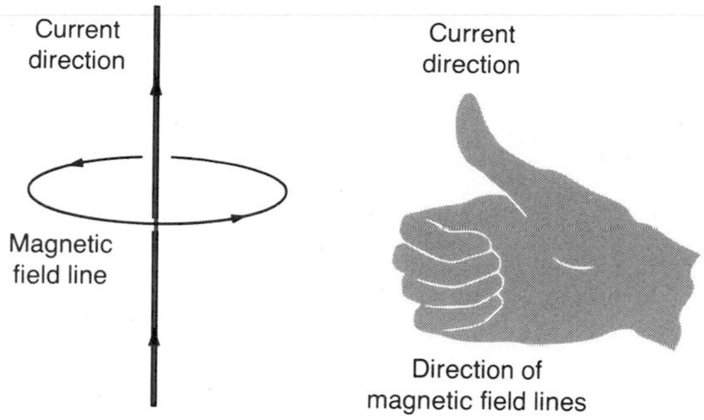

FIGURE 19.5 Illustration of the right hand rule for determining the relation between current direction in a long straight wire and the direction of the circular lines of magnetic field.

rule of Fig. 19.5; with the thumb of the right hand representing current in the wire, the fingers curl in the direction of circulation of the magnetic field lines.

The exact expression for the field of a long straight wire was deduced from experimental investigations by the French physicists Biot and Savart. They found that the strength of the magnetic field is directly proportional to the current in the wire and inversely proportional to the distance from it. Their formula (often referred to as the Biot–Savart law) requires a dimensioned constant to be correct in SI units.

Formula for the Magnetic Field Due to a Current in a Long Straight Wire (Biot–Savart Law)

$$\text{Magnetic field} = \frac{\mu_0}{2\pi} \times \frac{\text{current in wire}}{\text{radial distance from wire}}$$

$$B = \frac{\mu_0}{2\pi}\left(\frac{I}{r}\right) \qquad \textbf{(19.4)}$$

Assigned value of constant μ_0 : $4\pi \times 10^{-7}$ T · m/A
The magnetic field lines circle the wire in accordance with the right hand rule of Fig. 19.5.

The constant μ_0 is called the permittivity of free space, and the proportionality constant of Eq. (19.4) is written as $\mu_0/2\pi$ for reasons similar to those for the assignment of $(4\pi\epsilon_0)^{-1}$ as the proportionality constant in the formula for the electric field due a point charge [Eq. (16.3)].

Biot subsequently deduced a formula for the magnetic field of a small current-carrying element of wire—sometimes also referred to as the Biot–Savart law. An infinite number of small current elements in line with each other constitute a long straight wire, and the methods of integral calculus can then be used to obtain Eq. (19.4). The advantage of the formulation in terms of small current elements is that any circuit can be represented by an infinitude of suitably placed current elements, and the magnetic field obtained by summation of their effects. However, we shall avoid the complexity of this approach since we shall require only the magnetic fields of the long straight wire and the solenoid (see later in this section).

EXAMPLE 19.4

Calculate the magnetic field strength at a point 2 cm distant from a long straight wire that carries a current of 10 A.

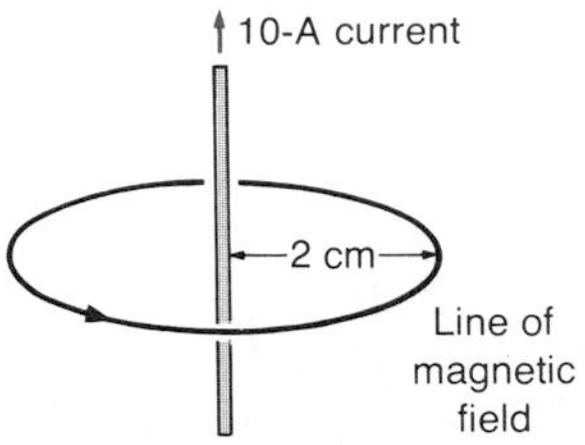

SOLUTION The problem is a straightforward application of Eq. (19.4):

$$\begin{aligned} B &= \frac{\mu_0}{2\pi}\left(\frac{I}{r}\right) \\ &= 2 \times 10^{-7}\ \mathrm{T \cdot m/A} \times \left(\frac{10\ \mathrm{A}}{0.02\ \mathrm{m}}\right) \\ &= 10^{-4}\ \mathrm{T} \end{aligned}$$

The direction of the magnetic field follows from the right hand rule of Fig. 19.4. Our example shows that the tesla is a large unit of magnetic field. ■

We now consider a device that provides a large region of uniform magnetic field. A solenoid is a long current-carrying wire that is closely wound in a helical fashion, as shown in Fig. 19.6. To construct a solenoid, we could begin with a long cardboard cylinder and wind insulated wire around its length. If the diameter of the solenoid is small compared with its length, and if the turns are closely spaced, the interior magnetic field is approximately uniform. At each end the magnetic field lines fringe out, returning by long sweeping paths to the other end to form closed loops. Once again we see the fundamental property of magnetic lines in forming closed loops that encircle the current causing them—in this case the loops of current of the solenoid.

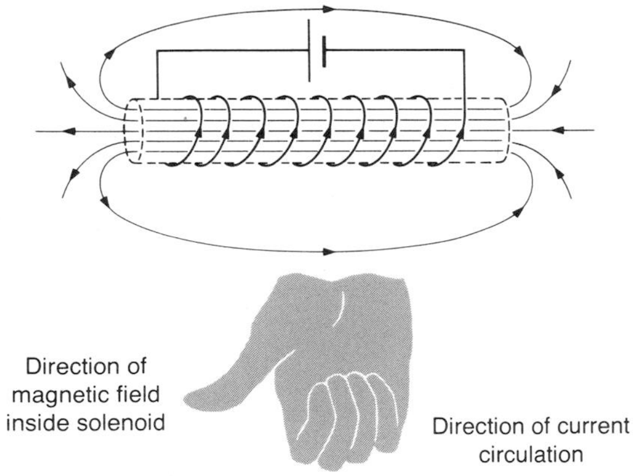

FIGURE 19.6 The magnetic field of a solenoid. The relative directions of the field and the current circulation are given by the right hand rule.

The direction of the magnetic field lines within the solenoid is also given by a right hand rule. If the fingers represent the direction of current circulation in the helix, the thumb points in the direction of the magnetic field within the solenoid. This right hand rule is illustrated in Fig. 19.6.

Calculating the magnetic field strength requires mathematics beyond the scope of this book, but we can express the result with a simple formula.

■ Formula for Magnetic Field Strength Within a Long, Closely Wound Solenoid

$$\text{Magnetic field strength} = \mu_0 \times \text{\# of turns per unit length} \times \text{current}$$

$$B = \mu_0 \frac{n}{l} I \qquad \textbf{(19.5)}$$

n = total number of turns
l = length of the solenoid
I = current

The units of the right-hand side of Eq. (19.5) are the same as those on the right-hand side of Eq. (19.4). Although we have not proved Eq. (19.5), we are at least sure that the result is in the correct units. The magnetic field strength in the region outside the solenoid is not zero, since the lines of force must return to form closed loops. However, the exterior magnetic field strength is negligibly small compared with the interior field strength, except in small regions adjacent to the ends of the solenoid.

The word *solenoid* comes from the Greek word for a channel. The name was conferred by Ampère, who realized that the arrangement channeled the magnetic field through the center, as shown in Fig. 19.6.

EXAMPLE 19.5

Calculate the magnetic field strength within a solenoid 50 cm long composed of 1600 turns of wire carrying a current of 10 A. (Assume the diameter of the solenoid is small compared with its length.)

SOLUTION We require direct use of Eq. (19.5):

$$B = \mu_0 \frac{n}{l} I$$

$$= 4\pi \times 10^{-7}\ \text{N/A}^2 \times \frac{1600}{0.5\ \text{m}} \times 10\text{A}$$

$$= 4.02 \times 10^{-2}\ \text{T}$$

■

19.3 The Current Balance and the Standard Ampere

We are finally in a position to introduce the assignment of the standard ampere, a task that is accomplished using an instrument called a current balance. The operation of this instrument is explained in terms of the concepts and conclusions of the two previous sections.

Before describing a simple laboratory current balance, we calculate the magnetic force between two long straight wires, each of length l, separated by a relatively short distance r, and carrying currents I_1 and I_2 in the same direction [see Fig. 19.7(a)]. The problem is solved by a two-step process:

1. Calculate the magnetic field of wire 1 at the position of wire 2 using Eq. (19.4) and the right hand rule of Fig. 19.4. The result is:

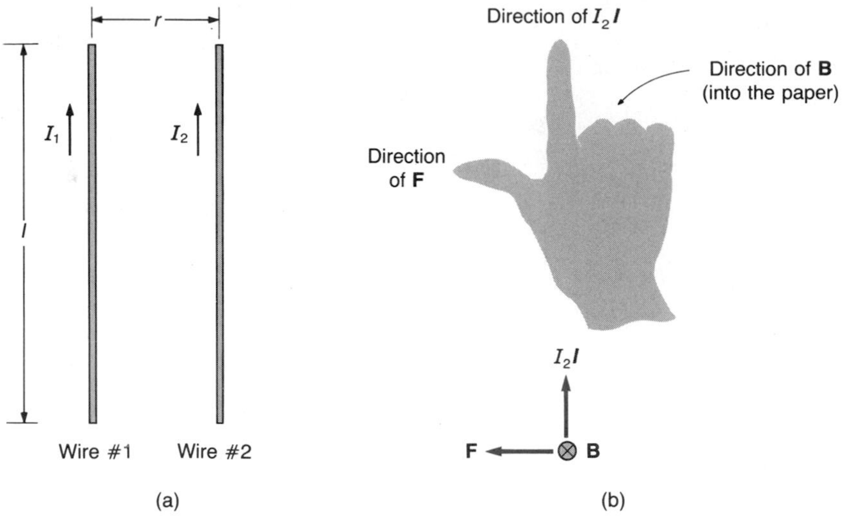

FIGURE 19.7 Diagram to illustrate the calculation of the force between parallel current-carrying wires.

$$B = \frac{\mu_0}{2\pi}\left(\frac{I_1}{r}\right)$$

directed downward into the plane of the paper.

2. Now use Eq. (19.3) and the right hand rule of Fig. 19.3(b) to calculate the force on wire 2 due to the field of wire 1. This gives:

$$\begin{aligned} F &= IlB \\ &= (I_2 l)\left(\frac{\mu_0}{2\pi}\left(\frac{I_1}{r}\right)\right) \\ &= \frac{\mu_0}{2\pi}\frac{I_1 I_2}{r}\,l \end{aligned} \tag{19.6}$$

and the force is a force of attraction toward wire 1, as shown in Fig. 19.7(b). We could have begun by calculating the magnetic field of wire 2 at the position of wire 1, and followed up by finding the force on wire 1 due to the field of wire 2. This alternate calculation would yield a force of attraction of the same magnitude toward wire 2. The magnetic force between two parallel wires carrying currents in the same direction is one of mutual attraction. Using the same calculations would show us that the force between parallel wires carrying currents in opposite directions is one of mutual repulsion.

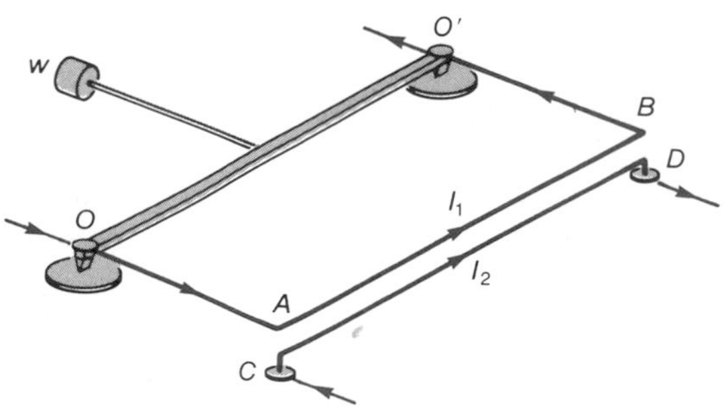

FIGURE 19.8 Diagram of a current balance.

The design features of a simple laboratory current balance are shown in Fig. 19.8. The long wire CD is fixed to the base of the instrument and connected to a battery by wires that approach it at right angles. A pivoted suspension unit balances at the points *OO′*, and conduction arms carry current to the long wire *AB*, which is close to the wire *CD* and parallel to it. The pivoted unit is balanced by a weight *w* on the opposite side. With currents in the same direction, the wires *AB* and *CD* attract each other, necessitating an increase in the weight *w* to maintain the balance of the pivoted unit. If the currents are in opposite directions, the wires repel each other, and the weight *w* must be decreased to maintain the balance.

EXAMPLE 19.6

A current balance of the type shown in Fig. 19.8 has the weight adjusted for balance with no current in the wires. The two parallel wires are each 25 cm long and separated by 2 cm. The parallel wires are 16 cm from the pivot point, and the balance weight is 8 cm on the other side of the pivot point. Calculate the additional balance weight required if there is a current of 10 A in each wire in the same direction.

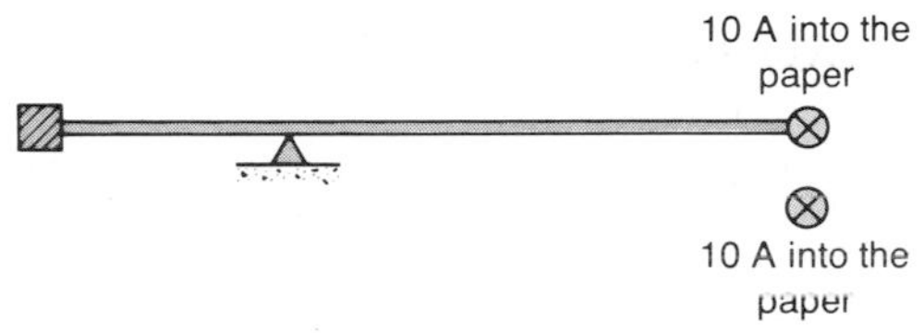

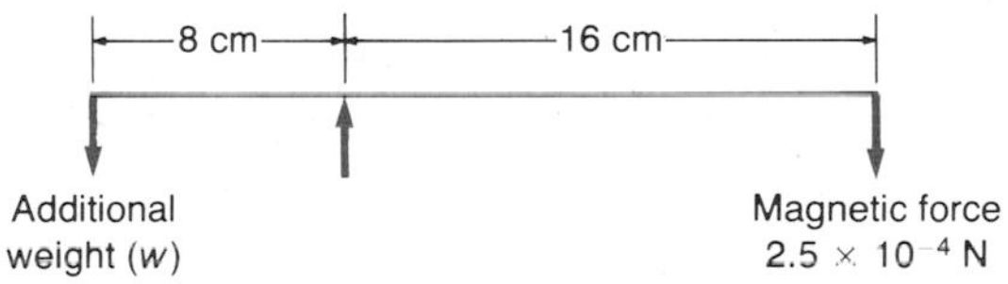

SOLUTION The magnetic force of attraction between the two wires is given by Eq. (19.6):

$$F = \frac{\mu_0}{2\pi}\frac{I_1 I_2}{r}\, l$$

$$= 2 \times 10^{-7}\ \text{T}\cdot\text{m/A} \times \frac{10\ \text{A} \times 10\ \text{A}}{0.02\ \text{m}} \times 0.25\ \text{m}$$

$$= 2.5 \times 10^{-4}\ \text{N}$$

Consulting the force diagram for the pivoted unit, we see that the additional weight can be found with the help of Eq. (5.3), which specifies zero net torque:

$$\Sigma\tau = 0$$

$$2.5 \times 10^{-4}\ \text{N} \times 0.16\ \text{m} - w \times 0.08\ \text{m} = 0$$

$$\therefore w = 5 \times 10^{-4}\ \text{N}$$

To calculate the mass that provides this weight, we use Eq. (4.2):

$$m = \frac{w}{g}$$

$$= \frac{5 \times 10^{-4}\ \text{N}}{9.81\ \text{m/s}}$$

$$= 51\ \mu\text{g}$$

Because the magnetic force between the wires is very small, a careful experimental technique is required to measure it. ■

In the first chapter we explained the assignments of the fundamental SI units of length, time, and mass. Only now can we assign the fourth fundamental unit of the SI system, the ampere—the electric current unit that we have used extensively. To proceed with its assignment, we imagine two long straight parallel wires carrying equal currents and separated by a distance of 1 m. The currents are adjusted until the force between the wires is exactly 2×10^{-7} N/m; with this adjustment the current in the wires is exactly 1 A. The National Bureau of Standards uses a delicate current balance to assign the standard ampere. For reasons of convenience their standard current balance uses two coils of wire to carry the currents rather than two long straight wires, but the physical basis for assigning the unit of current involves two current-carrying wires.

Let us now retrace our steps to understand how the fundamental unit of current is assigned and how the derived unit of charge is defined.

1. The law of magnetic force between parallel currents, Eq. (19.6), is used to assign the ampere. We accomplish this by setting $\mu_0 = 4\pi \times 10^{-7}$ N/A^2, as explained above. This strange value for μ_0 was selected so that the scientific unit of current is the same as the commercial and industrial unit that had been in use for many years.

2. The coulomb is defined as the quantity of charge transported by a current of 1 A during a period of 1 s.

3. For Coulomb's law of force between electric charges, Eq. (16.1), the units of all quantities are now determined, which means that the value of ϵ_0 is determined by experiment.

One may argue that the concept of electric charge is more immediate than that of electric current. Would it not be better to assign an exact value to ϵ_0 and use the coulomb as a fundamental unit? We could then define the ampere as the current due to a passage of 1 C of charge in 1 s and determine the value of μ_0 by experiment. Indeed, the electric and magnetic units were defined in just this manner for some years. However, the high precision that can be attained with the current balance is the major reason for the procedure currently in use.

19.4 Magnets and Magnetic Poles

Some materials are able to produce magnetic fields in the apparent absence of electric currents. Permanent magnets, most of which are made from iron or its alloys, show this property. To

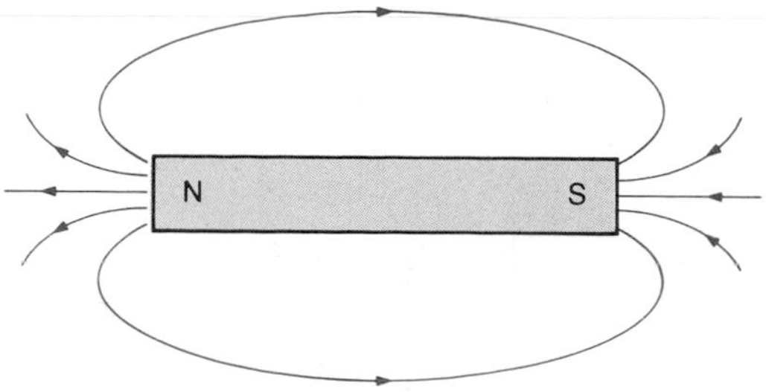

FIGURE 19.9 The external magnetic field of a permanent bar magnet.

make a permanent magnet, we place a bar of suitable alloy in the interior of a solenoid. Switching on the current in the solenoid produces a magnetic field, as shown in Fig. 19.6. If we now reduce the current to zero, the alloy is permanently magnetized. The external magnetic field of a permanent bar magnet, which is illustrated in Fig. 19.9, is very similar to the external field of the solenoid in which it was made. Lines of magnetic field leave one end of the bar (called by tradition the north pole), sweep around, and enter the other end (called the south pole). Since lines of magnetic field strength form closed loops, we presume that the internal magnetic fields are also similar, which is in fact the case. The chief difference between the fields of the solenoid and the bar magnet is the helically circulating current in the solenoid and the apparent absence of current in the bar magnet. Since the theory of permanent magnets ascribes their magnetic fields to currents in the atoms of the constituent material, the absence of currents is apparent rather than real.

The question immediately arises as to why permanent magnets are made from iron alloys rather than, say, aluminum alloys. The answer is that large groups of adjacent atoms in iron can be locked together so that the magnetic fields of the circulating currents in the atoms all reinforce each other. The large groups of atoms are called *domains,* and a material that can produce a magnetic field by reorganizing its domain structure is called *ferromagnetic.* This category mainly includes iron, cobalt, and nickel, as well as some of their alloys. The reorganization of domains to produce a net magnetic field in some direction is called *magnetization.* In its unmagnetized state, a ferromagnetic material contains domains whose typical linear dimensions are in the range of 10 μm to 100 μm. The domains are randomly oriented so as to produce zero net magnetic field in any direction, as illustrated in Fig. 19.10(a). If we now magnetize the material in some direction (for example, by inserting it into a solenoid), the domains that produce a magnetic field in that particular direction grow at the expense of their neighbors, as shown in Fig. 19.10(b). When this occurs, the net effect of the magnetized domains is a magnetic field in the preferred direction. The main property distinguishing ferromagnetic from nonferromagnetic materials is the ability to form oriented domains.

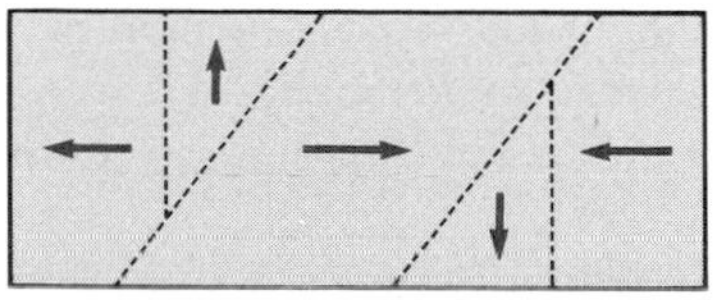

(a)

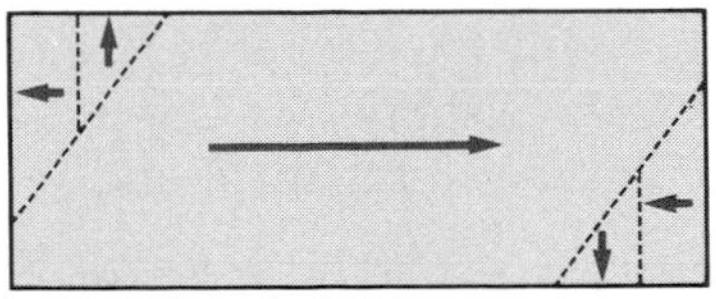

(b)

FIGURE 19.10 Diagram illustrating the growth of domains in a ferromagnetic material. (a) When the magnetization is zero, the magnetic fields due to the domains cancel. (b) When the magnetization is not zero, some domains grow at the expense of their neighbors.

When we magnetize a bar of iron by placing it inside a solenoid, the degree of magnetization produced (and consequently the magnetic field inside the solenoid) depends on the current in the coils of the solenoid. For low values of the solenoid current, the magnetic field strength inside the solenoid is some 500 to 2000 times greater than it would be in the absence of the iron. Figure 19.11 illustrates how the magnetic field increases as the current in an iron-core solenoid increases. For large values of the solenoid current, the magnetic field in the iron becomes approximately constant at a value of about 1.5 T. This phenomenon is called magnetic saturation. For comparison, we also show the magnetic

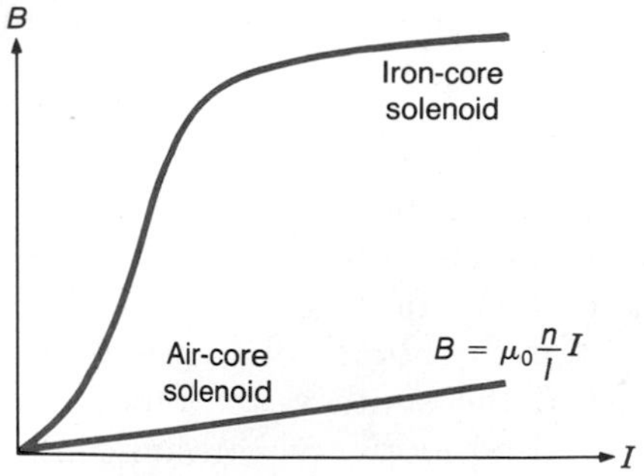

FIGURE 19.11 Inserting an iron core in a solenoid increases the magnetic field strength by a large and variable factor. The line describing the magnetic field in an air-core solenoid is drawn to greatly exaggerated scale—the magnetic field in an air-core solenoid is about 1000 times smaller than in the iron-core solenoid.

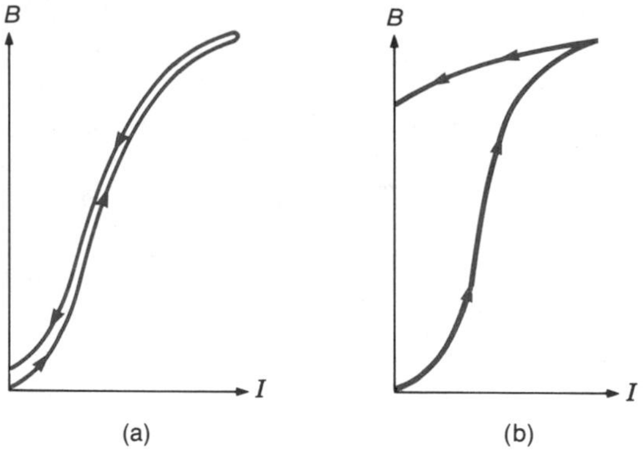

FIGURE 19.12 The difference between soft iron and a permanent magnet alloy lies in the ability to retain magnetization. (a) When the solenoid current is first increased and then reduced to zero, the magnetization of a soft iron core returns almost to zero. (b) The magnetization of a permanent magnet alloy remains high even when the solenoid current is reduced to zero.

field as a function of current for an air-core solenoid. The straight line represents Eq. (19.5).

If the solenoid current is reduced to zero, some ferromagnetic materials (such as soft iron) lose their magnetization. On the other hand, certain suitable alloys retain a large fraction of their magnetization when the solenoid current is turned off. The distinction is illustrated in Fig. 19.12. Alloys that retain substantial magnetization are suitable for being made into permanent magnets. The alloy can be taken out of the solenoid and used in applications that require a permanent magnetic field. On the other hand, soft iron is suitable for use in electromagnets. It is used in applications that require a very large magnetic field when the solenoid current is turned on and a very small field when it is switched off.

Many applications require a strong, uniform magnetic field in an accessible region outside the iron of the solenoid or the permanent magnet. How do we produce a strong, uniform magnetic field in this way? One method is to twist the bar magnet (or the iron-core solenoid) into a horseshoe shape, as shown in Fig. 19.13. With this geometrical configuration the magnetic field lines

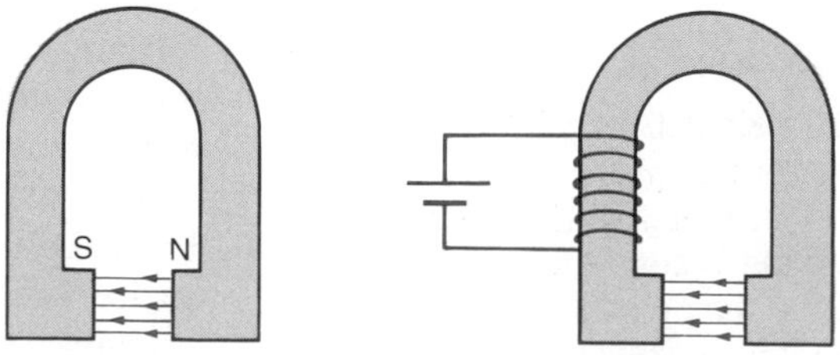

FIGURE 19.13 The use of a permanent magnet or an iron-core solenoid to produce a strong, uniform magnetic field.

follow the iron path through the horseshoe and cross the relatively small gap to give a uniform constant magnetic field in the gap. The magnetic field strength at any other place is relatively very small compared with the field in the gap. Strong uniform fields of this type are used in moving-coil galvanometers, electric motors, and electric generators.

19.5 Magnetic Moment of a Coil

We have seen how to produce a strong magnetic field by using either a permanent magnet or a solenoid, and we have indicated applications of such a field to galvanometers, motors, and generators. However, the strong magnetic field is only one of two basic components common to all three devices. The other component is a coil of conducting wire that is usually, but not always, in the shape of a rectangle and that is free to spin about an axis in its plane. Figure 19.14(a) illustrates the general situation. The coil is rectangular with side dimensions l and b and carries a current I; it is situated in a uniform magnetic field B. The angle between the magnetic field direction and the normal to the plane of the coil is θ. Figure 19.14(b) illustrates the geometry viewed along the axis of the coil.

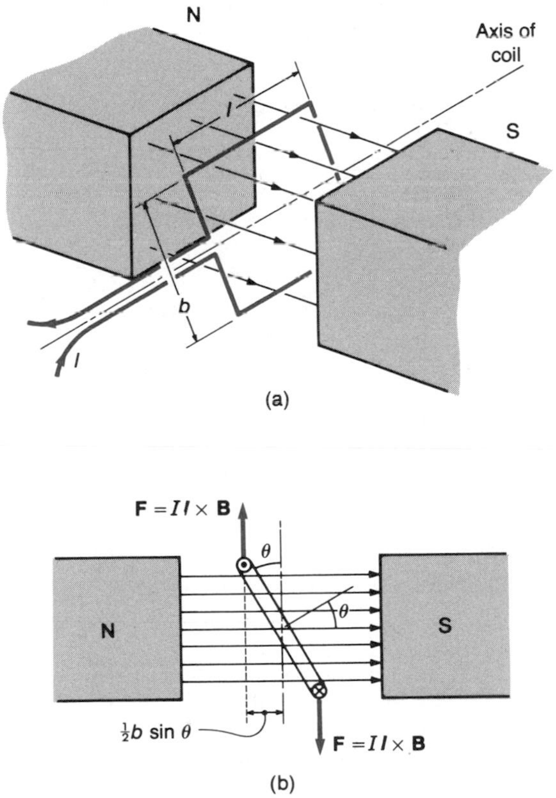

FIGURE 19.14 A current-carrying coil in a uniform magnetic field: (a) general view; (b) viewing along the axis of the coil.

Each side of the rectangular coil experiences a force due to the magnetic field, which can be calculated by using Eq. (19.3):

$$\mathbf{F} - I\mathbf{l} \times \mathbf{B}$$

The directions of the forces on each side of the coil, shown in Fig. 19.14(b), are calculated from the vector sequence rule explained in Fig. 19.3. The forces on the sides of the coil produce a torque about its axis. Since the lever arm of each force is ½b sin θ, we can calculate the torque due to the two forces from Eq. (5.2):

$$\begin{aligned}\tau &= Fl_{\perp}\\ &= 2 \times IlB \times \tfrac{1}{2}b \sin\theta\\ &= IlbB \sin\theta \qquad \textbf{(19.7)}\end{aligned}$$

Each end of the rectangular coil also experiences a force due to the magnetic field. Application of the vector sequence rule of Fig. 19.3 shows us that these forces are parallel to the axis of rotation of the coil. Since they therefore do not contribute to the torque about the rotation axis, we will ignore them in the following discussion.

We can put the formula for torque, Eq. (19.7), in a more memorable and useful form by adopting a two-step approach. Our first step is to define the **magnetic moment** of a current-carrying coil, which we do with reference to Fig. 19.14.

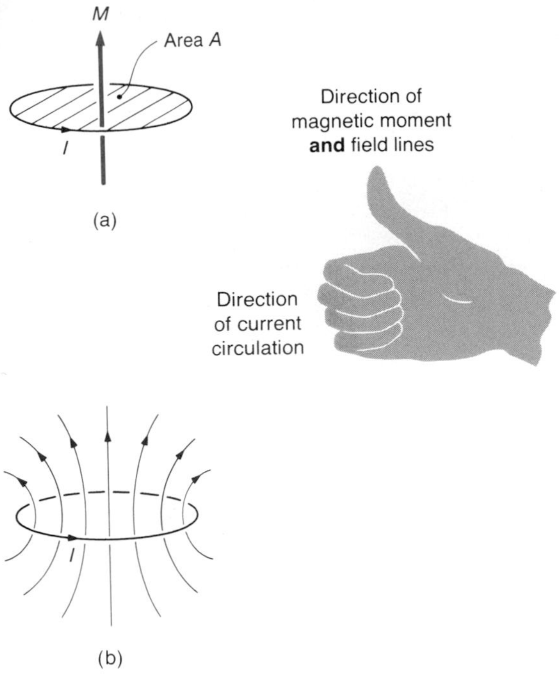

FIGURE 19.15 Illustration of the definition of magnetic moment of a current-carrying coil: (a) the relative directions of the magnetic moment vector and the circulating current: (b) the relative direction of the magnetic field lines and the circulating current.

■ Definition of the Magnetic Moment of a Current-Carrying Coil

$$\text{Magnetic moment of coil} = \text{\# of turns in coil} \times \text{current in coil} \times \text{area of each turn}$$

$$M = nIA \qquad \textbf{(19.8)}$$

SI unit of magnetic moment: ampere-(meter)^2
Abbreviation: $A \cdot m^2$

The magnetic moment is a vector quantity that is perpendicular to the plane of the coil. Its direction relative to the direction of current circulation in the coil is given by the right hand rule, illustrated in Fig. 19.15(a). The magnetic field lines due to the current-carrying coil follow the general pattern of Fig. 19.15(b), threading through the coil in the same direction as the magnetic moment vector.

Let us now relate our definition of magnetic moment to the formula for the torque on a rectangular coil. Using Eq. (19.8), we find for the magnetic moment of a rectangular coil of one turn:

$$M = Ilb$$

If we make this substitution in Eq. (19.7), we can write the torque on the coil in the following way:

■ Formula for Torque on a Current-Carrying Coil in a Uniform Magnetic Field

$$\text{Torque} = \text{magnetic moment of coil} \times \text{magnetic field strength} \times \text{sine of the angle between the directions of the magnetic moment and magnetic field vectors}$$

$$\tau = MB \sin\theta \qquad \textbf{(19.9)}$$

The torque twists the magnetic moment into alignment with the magnetic field.

The torque has its greatest value when the magnetic moment of the coil and the magnetic field are perpendicular to each other, and it is zero when these two vectors are parallel.

We have proved Eq. (19.7) from the basic magnetic force law and shown that it is a special case of Eq. (19.9). Although this does not establish the validity of Eq. (19.9), we state without proof that it is correct for a coil of any shape and any number of turns—we are not restricted to the single turn of rectangular shape that we considered in deriving Eq. (19.7).

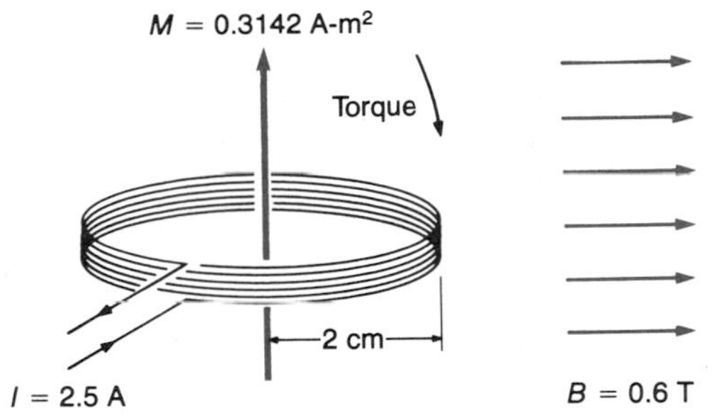

EXAMPLE 19.7

A circular coil of wire of radius 2 cm contains 100 turns and lies with its plane parallel to a uniform magnetic field of 0.6 T. Calculate the torque on the coil if it carries a current of 2.5 A.

SOLUTION To calculate the torque, we first find the magnetic moment of the coil from Eq. (19.8):

$$\begin{aligned} M &= nIA \\ &= 100 \times 2.5\text{ A} \times \pi \times (0.02\text{ m})^2 \\ &= 0.3142\text{ A} \cdot \text{m}^2 \end{aligned}$$

The direction of the magnetic moment vector relative to the current circulation is shown in the diagram, and the angle between the magnetic moment vector and the magnetic field is 90°. We can now use Eq. (19.9) to calculate the torque:

$$\begin{aligned} \tau &= MB \sin\theta \\ &= 0.3142\text{ A} \cdot \text{m}^2 \times 0.6\text{ T} \times \sin 90^\circ \\ &= 0.188\text{ N} \cdot \text{m} \end{aligned}$$

The torque tends to align the magnetic moment with the magnetic field. ■

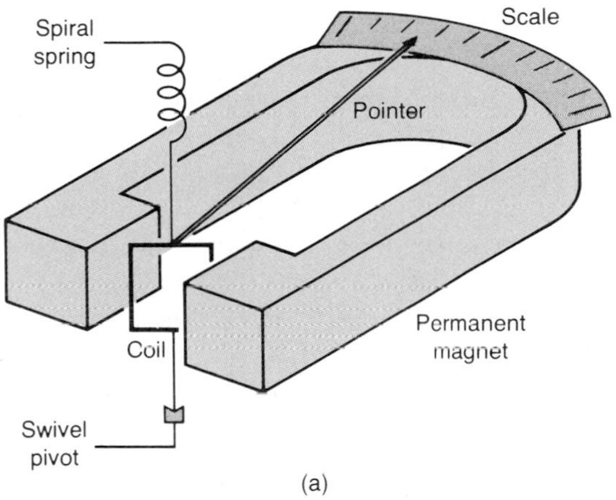

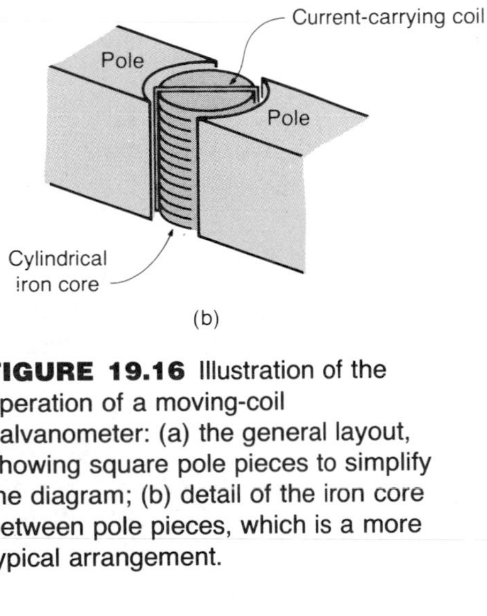

FIGURE 19.16 Illustration of the operation of a moving-coil galvanometer: (a) the general layout, showing square pole pieces to simplify the diagram; (b) detail of the iron core between pole pieces, which is a more typical arrangement.

In a previous chapter we described the circuit connections of ammeters and voltmeters based on a galvanometer instrument. The moving-coil galvanometer is the most commonly used, and its operation is illustrated in Fig. 19.16(a). A closely wound coil is placed in a uniform magnetic field between the poles of a permanent magnet. The coil is mounted on a swivel pivot that leaves it free to turn about a vertical axis. The actual position of the coil is controlled by a sensitive spiral spring. The current to be measured is passed through the coil, creating a magnetic moment. As we have just seen, this produces a torque on the coil that tends to align the magnetic moment and the magnetic field. The torque is counteracted by the spiral spring, and the coil rotates until the torque due to magnetic force exactly balances the opposing torque due to the twisting of the spring. A pointer is rigidly attached to the coil and moves along a scale as the current causes the coil to move to a new equilibrium position. Since the position of the pointer is related to the magnitude of the current in the coil, the instrument can be calibrated to read the value of the current in suitable units. Since galvanometers of this type can be made very sensitive to small currents, they are suitable for use as detectors in the Wheatstone bridge and the potentiometer circuits that we have previously discussed. Because they are also fairly rugged, portable, and inexpensive, they are frequently used in multirange ammeters and voltmeters.

We have omitted one of the features of a moving-coil galvanometer in Fig. 19.16(a) in order to keep the diagram clear. In an actual instrument the coil surrounds a soft iron cylindrical core, as shown in Fig. 19.16(b). The ends of the pole pieces are not square, but shaped to leave small gaps between themselves and the iron core. This design has two desirable effects—it increases the magnetic field strength in the gap, and gives a uniform scale for the current values.

Understanding the operation of motors and generators requires an additional basic law of electromagnetism, a topic that we will discuss in the following chapter.

19.6 Orbit of a Charged Particle in a Magnetic Field

We consider only the case of a charged particle whose velocity is perpendicular to a uniform magnetic field. We see from Eq. (19.2) that the particle experiences a force that is perpendicular to its velocity. This means that the magnitude of the velocity does not change, but only its direction. Since the charge and the magnetic field are constant, the magnetic force remains constant in magnitude and perpendicular to the particle velocity. Recall from Chapter 6 that centripetal force is a force of constant magnitude perpendicular to the velocity of a particle moving in a circular path. The path of the charged particle is therefore a circle in a plane perpendicular to the magnetic field.

Let us consider the case of a particle of charge q and mass m moving in a circular orbit of radius r, as shown in Fig. 19.17. The magnetic field **B** is perpendicular to the plane of the orbit, and the particle velocity is **v**. Since **B** is perpendicular to **v**, Eq. (19.2) gives:

$$\text{Magnetic force} = q\mathbf{v} \times \mathbf{B}$$

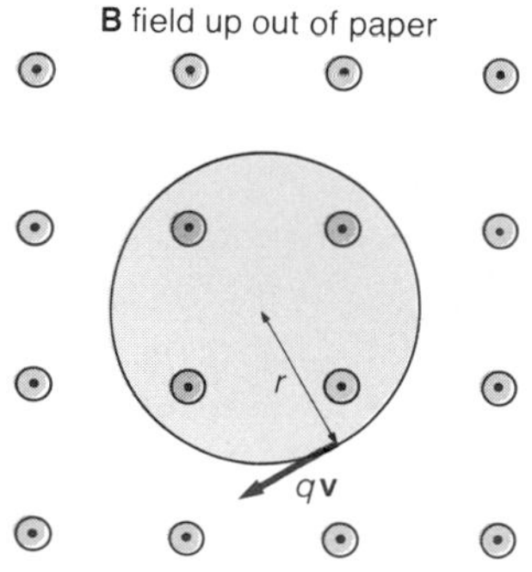

FIGURE 19.17 A charged particle in a circular orbit in a magnetic field.

With the particle moving clockwise around the circle, the right hand rule [see Fig. 19.1(b)] shows that the magnetic force is directed in toward the center of the circle. We can therefore write

$$\text{Magnitude of magnetic force} = \text{particle mass} \times \begin{array}{l}\text{magnitude of}\\ \text{centripetal}\\ \text{acceleration}\end{array}$$

$$qvB = m\frac{v^2}{r}$$

Rearranging yields the following formula for the orbital radius:

■ Formula for Orbital Radius of a Charged Particle Whose Velocity Is Perpendicular to a Magnetic Field

$$r = \frac{mv}{qB} \tag{19.10}$$

EXAMPLE 19.8

Calculate the orbital radius of a proton whose velocity is 10^7 m/s perpendicular to a constant magnetic field of 0.8 T.

SOLUTION Given that the proton mass is 1.67×10^{-27} kg, we can calculate the orbital radius by direct substitution in Eq. (19.10):

$$\begin{aligned} r &= \frac{mv}{qB} \\ &= \frac{1.67 \times 10^{-27}\ \text{kg} \times 10^7\ \text{m/s}}{1.6 \times 10^{-19}\ \text{C} \times 0.8\ \text{T}} \\ &= 0.13\ \text{m} \end{aligned}$$

The relatively small orbital radius results from the low proton velocity. In an actual accelerator, the proton velocity may be close to the velocity of light. In this circumstance we must employ relativistic dynamics, and the orbital radius can be very large. ■

The study of the orbits of charged particles in electric and magnetic fields played a large role in the discovery of the properties of electrons. Decisive evidence regarding the nature of these particles came from experiments on the electrical discharge in a low-pressure gas. Consider the apparatus of Fig. 19.18. A glass tube containing two metal electrodes is evacuated until the pressure falls to about 10^{-2} torr. A voltage of 10 kV or more applied between the two electrodes causes the whole interior of the tube to glow with a faint green light. This green glow is fluorescence of the glass, produced by rays that emanate from the cathode. For this reason the rays were called *cathode rays*—a name that persists for the cathode ray oscilloscope. Figure 19.18 illustrates a classic demonstration of the fact that the rays travel in straight lines from the cathode. An object placed within the tube throws a shadow on the wall opposite the cathode, thereby showing that the rays travel in straight lines from the cathode. By using such an apparatus, it is easily shown that the cathode rays are deflected from their straight-line paths by electric and magnetic fields. In both cases the direction of the deflection is consistent with the assumption that the cathode rays are a stream of negatively charged particles.

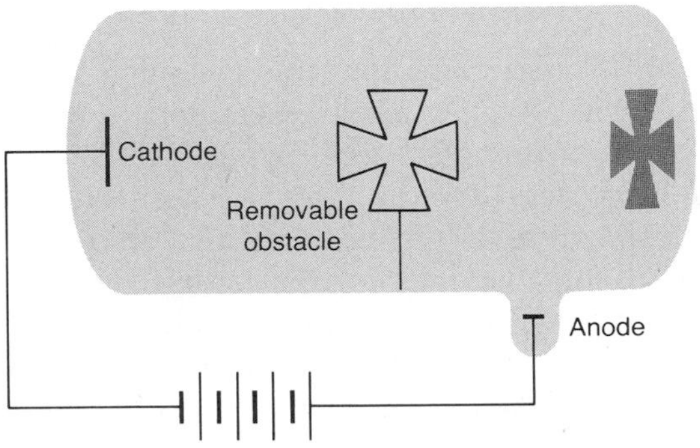

FIGURE 19.18 An electrical discharge tube for demonstrating that cathode rays travel in straight lines.

An English physicist, Sir Joseph Thomson (1856–1940), carefully measured the deflection of cathode rays in electric and magnetic fields. From his results, he determined that the particles had a constant ratio of charge to mass, whose modern value is $e/m = -1.759 \times 10^{11}$ C/kg. In modern terminology we usually refer to cathode rays as electrons, reserving the symbol e to denote the electron charge.

Thomson could not determine either the mass or the charge separately by using his cathode ray deflection experiments. A determination of the electron charge alone was made by an American, Robert Millikan (1868–1953), who won a Nobel Prize for his work on the electron and subsequently played a pioneering role in cosmic ray research. His apparatus for measuring the electron charge employs an atomizer that sprays small oil drops into the region between two plates, which are kept at a constant potential difference by a battery. In passing through the air, the oil drops pick up electrostatic charges and experience a force due to the electric field between the plates as well as a force due to their own weight. By studying the motion of the drops under the combined action of the electric and gravitational forces, Millikan was able to determine the charge on an individual drop. He found that the charge is never smaller than a certain value and is always an integral multiple of the smallest value. If excess electrons are responsible for the charge on the drops, this minimum charge must be the charge on one electron. In combination with Thomson's value for e/m, we can thus calculate the electron mass. Rather than quote the values originally obtained, we simply give the accepted modern values for these quantities:

■ Fundamental Electron Properties

$$\text{Electron charge} = -1.602 \times 10^{-19}\ \text{C}$$

$$\text{Electron mass} = 9.11 \times 10^{-31}\ \text{kg}$$

At first glance our chain of reasoning appears somewhat tenuous. Thomson's value of e/m refers to particles emitted from a metal electrode in a low-pressure discharge, while Millikan's value of e applies to small droplets of oil from an atomizer. However, we can combine these two results to calculate a value for m because of our belief that all material objects are composed of the same basic microscopic particles. The ultimate test of this belief is the mass of confirmatory evidence that comes from many more experiments of different types. We have simply described the two experiments that have historical priority.

The cathode ray tube is an important practical device that makes use of the trajectories of electrons in a vacuum under the

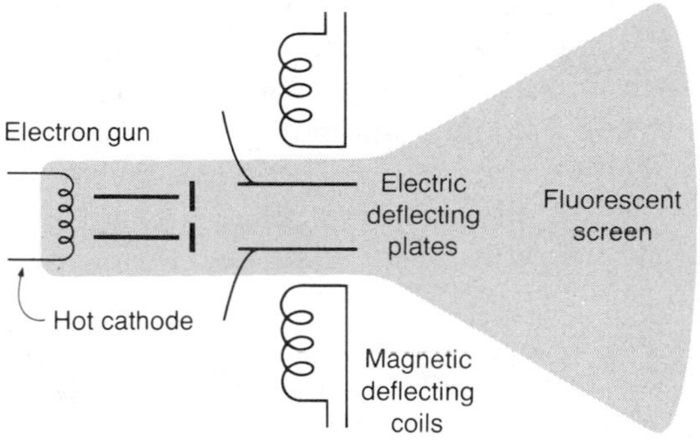

FIGURE 19.19 A typical cathode ray tube showing possible electric deflection plates and magnetic deflection coils. In a practical tube, the deflection is usually electric.

influence of electric and magnetic fields. It is the display unit in an oscilloscope and the picture tube in a TV set. A typical unit is illustrated in Fig. 19.19. A large, evacuated glass container with a relatively thin neck flares out to form a large screen on the end opposite the neck. Electrons are emitted from a hot cathode and accelerated to a high velocity by an electron gun. The acceleration is accomplished by a large positive potential difference between the final part of the gun and the hot cathode. In addition, the design focuses the electrons into a thin beam that travels down the tube and causes a small visible spot on the screen. The spot is made visible by a fluorescent coating on the inside glass surface that emits light when struck by the electron beam. Deflecting the beam from its straight-line path down the tube causes the spot to travel to various parts of the screen, thereby producing patterns and pictures. The deflection may be achieved either electrically or magnetically. In electrical deflection, the electron beam passes between a pair of parallel metal plates. If a potential difference exists between the plates, the beam (of negatively charged electrons) experiences an electric force away from the negative plate and toward the positive. For magnetic deflection, a pair of current-carrying coils produce a magnetic field that is transverse to the path of the beam down the neck of the tube. The beam deflects in a transverse direction at right angles to the magnetic field. Any desired pattern on the screen is produced by suitable signal voltages to electric deflecting plates or by signal currents to magnetic deflecting coils.

19.7 Terrestrial Magnetism

Throughout this chapter we have taken the point of view that electric currents cause magnetic forces and magnetic fields. As a natural result of the study of the earth's magnetic field, the old-fashioned approach to magnetism ascribed the forces and fields to magnetic poles.

A magnetic compass needle is a small permanent magnet mounted so as to be free to rotate about a vertical axis. The practical utility of the compass derives from the fact that one end of the needle always points more or less toward the north. The north-pointing end of the permanent magnet is called a north pole, and the other end a south pole. We have already identified the north and south poles of a bar magnet in Fig. 19.9 with reference to the direction of the magnetic field lines of the magnet. A comparison of the magnetic field lines of Figs. 19.9 and 19.15(b) shows that the magnetic fields of a permanent bar magnet and a current-carrying coil are very similar. Without trying to define it precisely, we see that a permanent bar magnet has a magnetic

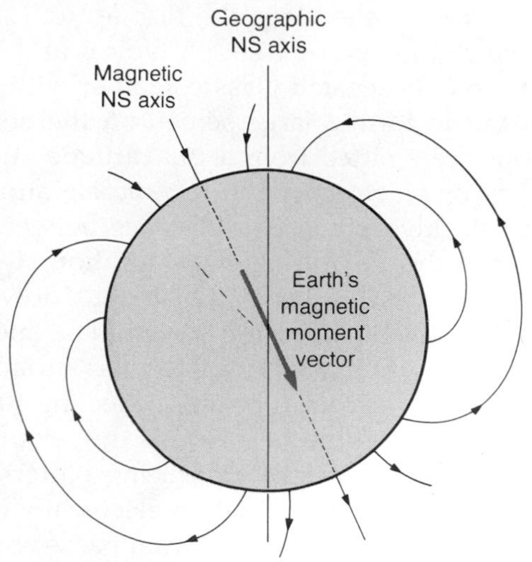

FIGURE 19.20 The earth's magnetic field.

moment, whose vector direction is from the south pole toward the north pole. But a magnetic field exerts a torque on a magnetic moment that tends to align it with the field. Consequently, the magnetic field of the earth must point from south to north if the north pole of a compass needle points north. The general form of the magnetic field of the earth is shown in Fig. 19.20. The details of the earth's magnetism are very complicated, but two major features emerge at a glance.

In the first place, the earth's magnetic field is similar to the fields illustrated in Fig. 19.9 (a permanent bar magnet) and Fig. 19.15(b) (a current-carrying coil). A comparison with Fig. 19.20 shows that the south magnetic pole is buried deep in the Northern Hemisphere and the north magnetic pole deep in the Southern Hemisphere if we regard the earth as a giant bar magnet. However, from another point of view, the earth's magnetic field is caused by giant current loops circulating from east to west deep within the earth's interior. This viewpoint is preferred because every attempt to isolate a magnetic pole has failed. If we take a permanent iron alloy bar magnet and break it in two, we do not obtain a solitary north pole on one piece and a south pole on the other. Instead, we obtain two bar magnets with each piece having its own north and south poles. If we continue the process of breaking up the bar magnet so that we have single atoms of iron, the result is the same. Each atom of iron exhibits a magnetic moment that we could regard as caused by a north-south pole pair or a circulating current loop. No experiment justifies the belief that an iron atom contains north and south magnetic poles, but there is every reason for believing that some of the atomic electrons set up current-carrying loops within the atom.

The other very notable feature of the earth's magnetism is that the axis of the earth's rotation and the direction of the magnetic moment vector are not the same. The places at which the axis of rotation of the earth comes through the surface are called the geographic North and South Poles; the places at which the magnetic moment vector comes through the surface are called the magnetic North and South Poles. (We use this nomenclature, even though the magnetic moment vector points in the opposite direction, as shown in Fig. 19.20.) At present the magnetic North Pole is near Bathurst Island, halfway between the north geographic pole and the Arctic Circle above central Canada; the magnetic South Pole is on the edge of Antarctica, south of Hobart in Tasmania. The positions of the magnetic poles have changed by hundreds of miles within recent history, and the magnetic fields of certain rocks indicate much bigger changes in earlier times. Indeed, scientists believe that the earth's magnetic moment vector has pointed in the opposite direction and has altered direction many times.

Because of the widely differing positions of the magnetic and geographic poles, a compass needle does not in general point to

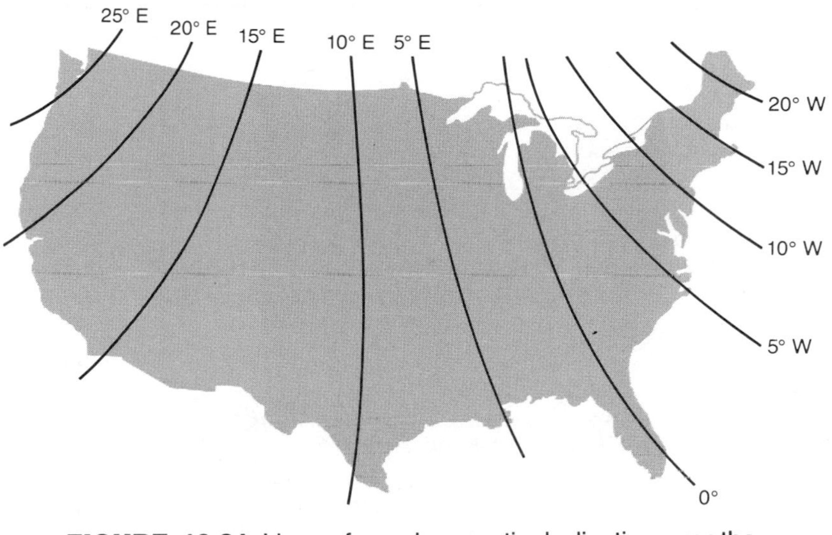

FIGURE 19.21 Lines of equal magnetic declination over the continental United States.

the geographic north. The angle between the geographic north and the direction of the compass needle is called the *magnetic declination.* Since the positions of the magnetic poles change, the magnetic declination varies over time. The approximate values of the magnetic declination over the continental United States at present are shown in Fig. 19.21. The magnetic compass points to true geographic north along a line that runs along the eastern shore of Lake Michigan and down the eastern coast of Florida. In New England, the compass points 10°–20° west of true geographic north, and in the Pacific Northwest it points 20°–25° east of true geographic north. The values of magnetic declination near London give a good indication of how it can change. In the late sixteenth century William Gilbert (Queen Elizabeth's physician) measured the declination at 10° E. It slowly changed to 25° W by the early nineteenth century and has since gone back to the present value of about 6° W.

Figure 19.20 also shows that the earth's magnetic field has both horizontal and vertical components at most locations on the earth's surface. A permanent magnet needle that is free to rotate about a horizontal axis in a vertical plane parallel to the magnetic field direction measures the *magnetic dip angle.* Figure 19.22 shows the orientation of a dip needle at Washington, D.C. The earth's magnetic field strength there is 57 μT, pointing 71° below the horizontal and toward the north. The magnitude of the earth's field is between about 40 μT and 80 μT over most of the earth's surface. It is vertical at the magnetic North and South Poles, horizontal at the magnetic equator, and has both horizontal and vertical components at other locations.

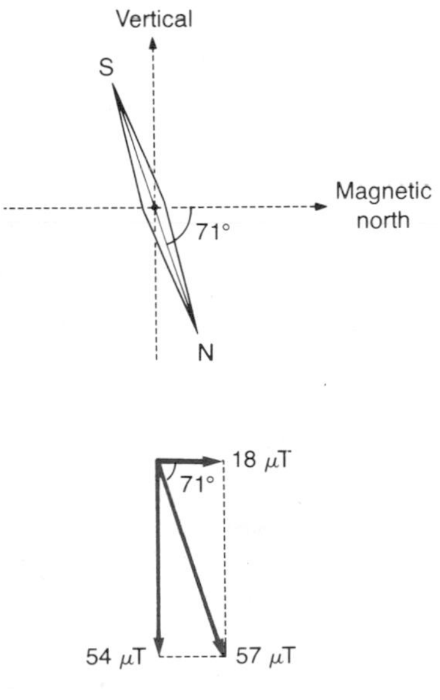

FIGURE 19.22 The earth's magnetic field at Washington, D.C.

KEY CONCEPTS

Magnetic force is the force that exists between moving charges or electric currents.

Magnetic field strength is defined in terms of the magnetic force on a moving charge or on a current-carrying wire; see Eq. (19.2) and Eq. (19.3). The magnetic force is perpendicular to both the magnetic field and the velocity of the moving charge or current direction; see Fig. 19.1(b) and Fig. 19.3(b).

Magnetic field lines form closed loops around the current that causes them; see Fig. 19.4 and Fig. 19.6.

Calculation of the **magnetic field strength** due to an electric current assumes a simple form for the cases of a long straight wire and a solenoid; see Eq. (19.4) and Eq. (19.5).

A **current balance** is an instrument that relates the magnetic force between two adjacent currents to the force between them. See Eq. (19.6) for the case of parallel straight wires.

A current-carrying coil experiences a torque in a magnetic field. The torque is most easily calculated in terms of the **magnetic moment** of the coil; see Eq. (19.8) and Eq. (19.9).

A charged particle whose velocity is perpendicular to a uniform magnetic field describes a circular orbit; see Eq. (19.10).

The **magnetic field** of the **earth** is similar to the field of a giant **bar magnet** or of giant **current loops** located deep in the earth. The magnetic moment vector is along the line of the magnetic north-south axis, and points from north to south; see Fig. (19.20).

QUESTIONS FOR THOUGHT

1. In his historic experiment, Oersted demonstrated the deflection of a small compass needle by the magnetic field due to a current in a long straight wire. Carefully explain how to set up this equipment so that the compass needle does not deflect when the current is switched on.

2. A loosely wound helical copper coil carries a current. Do you think that the passage of the current would compress the coil or expand it? Explain your answer.

3. A beam of electrons within a cathode ray tube forms a single bright spot in the center of the screen. Discuss how you would calculate the magnitude and direction of the magnetic field inside the tube due to the electron beam. Does the magnetic field tend to enlarge or contract the spot on the screen?

4. An electron passes through a certain region of space in a straight line. Can we say with certainty that the magnetic field in the region is negligible?

5. The magnetic field in a cube-shaped room is uniform, vertical, and directed upward. Can you arrange to project an electron to describe the following paths?

a. a horizontal circle
b. a vertical circle
c. a straight line

Explain your answer in each case. (Assume the influence of gravity on the electron is negligible.)

6. Why do electronic technicians often twist wires together that carry equal currents in opposite directions?

7. The earth's magnetic field may be caused by a current loop within the earth or by one in the upper atmosphere. Could careful measurements of the magnitude and direction of the earth's field at points all over the earth's surface help to determine which hypothesis is correct?

8. The magnetization of a horseshoe magnet is preserved much longer by placing an iron bar between the poles when the magnet is not in use. Explain this phenomenon.

9. You are given two iron bars of identical appearance, one of which is a magnet and the other is not. How could you tell which bar is magnetized without suspending either of them as a compass needle?

PROBLEMS

A. Single-Substitution Problems

1. Calculate the magnitude and direction of the velocity of an electron that experiences a force of 5×10^{-12} N to the west in a magnetic field of 0.5 T that is vertically upward. [19.1]

2. Calculate the magnitude and direction of the magnetic force on a proton that has a velocity of 5×10^6 m/s vertically downward in a horizontal magnetic field of 0.3 T that points north. [19.1]

3. A straight wire carrying a current of 2.6 A experiences a force of 5.7 N due to a magnetic field of 0.62 T that is perpendicular to the wire. Calculate the length of the wire. [19.1]

4. A straight wire 150 cm long carries a current of 40 A in a magnetic field of 0.8 T that is at right angles to the wire. Calculate the magnitude of the force on the wire. [19.1]

5. A straight vertical wire 80 cm long is placed in a uniform magnetic field of 1.2 T that points north. Calculate the magnitude and direction of the force on the wire if it carries a downward current of 35 A. [19.1]

6. Calculate the current in a long straight wire if the magnetic field due to it is 60 μT at a point distant 40 cm from the wire. [19.2]

7. A long straight vertical wire carries a downward current of 12 A. Calculate the magnitude and direction of the magnetic field due to the current at a point 20 cm to the east of the wire. [19.2]

8. A long straight wire carries a current of 6 A. At what distance from the wire is the magnetic field due to this current 20 μT? [19.2]

9. A solenoid 75 cm long has a winding of 800 turns and carries a current of 3.5 A. Calculate the magnetic field strength inside the solenoid. [19.2]

10. The winding of a solenoid 120 cm long carries a current of 12 A. How many turns are required to produce a magnetic field of 0.2 T inside the solenoid? [19.2]

11. A solenoid 80 cm long has a winding of 800 turns. What current is required to produce a magnetic field of 26 mT inside the solenoid? [19.2]

12. Two parallel wires, each 2 m long, are 2 cm apart and carry 35-A currents in opposite directions. Calculate the magnitude and direction of the force between the wires. [19.3]

13. Two parallel wires, each 2 m long, are 1 cm apart and carry equal currents. How large should the currents be to cause a force of 100 N between the wires? [19.3]

14. Calculate the magnetic moment of a circular loop of 30 turns and radius 2.5 cm that carries a current of 3.2 A. [19.5]

15. A square coil of 160 turns has a side length of 3 cm and carries a current of 18 A. Calculate:

a. the magnetic moment of the coil
b. the maximum torque on the coil in a uniform magnetic field of 0.4 T [19.5]

16. A square coil of 200 turns of wire has a side length of 16 cm. What current in the coil produces a maximum torque of 3 N · m in a uniform magnetic field of 0.5 T? [19.5]

17. A coil of magnetic moment 3.6 A · m^2 lies in a uniform magnetic field of 0.4 T whose direction makes an angle of 30° with the plane of the coil. Calculate the magnitude of the torque on the coil. [19.5]

18. A positively charged carbon ion (mass 2.00×10^{-26} kg and charge 1.6×10^{-19} C) moves in a circular path of radius 40 cm in a magnetic field of 0.5 T that is perpendicular to the plane of the circular path. Calculate the speed of the carbon ion. [19.6]

19. A proton (mass 1.67×10^{-27} kg and charge $+1.6 \times 10^{-19}$ C) moves in a circular path of radius 1.2 m with a speed of 1.85×10^7 m/s. Calculate the magnetic field strength normal to the plane of the circular path. [19.6]

B. Standard-Level Problems

20. A proton is accelerated from rest through a potential difference of 600 kV and subsequently enters a magnetic field of 0.15 T that is perpendicular to the direction of its path. Calculate the magnitude of the centripetal acceleration of the proton in the magnetic field. [19.1]

21. An electron that has been accelerated from rest by a potential difference of 12 V moves due north into a uniform vertical magnetic field. Calculate the magnitude and direction of the magnetic field if the force on the electron is 8×10^{-14} N toward the east. [19.1]

22. An electron is accelerated through a potential difference of 1.5 kV by an electron gun. It then passes between a pair of parallel conducting plates 3 cm apart at a potential difference of 2 kV, as shown in the diagram. Calculate the magnitude and direction of a magnetic field that will cause the electron to continue moving in a straight line between the plates.

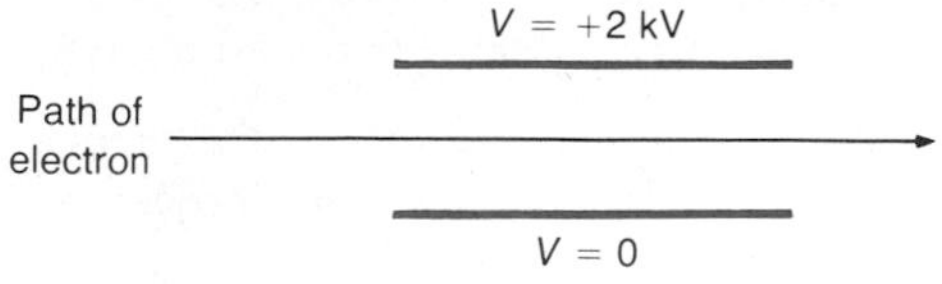

23. A rectangular coil 40 cm × 10 cm carries a current of 12 A and lies with its 10-cm sides parallel to a 0.5-T magnetic field. Calculate the magnitude and direction of the force on each of the two longer sides of the coil.

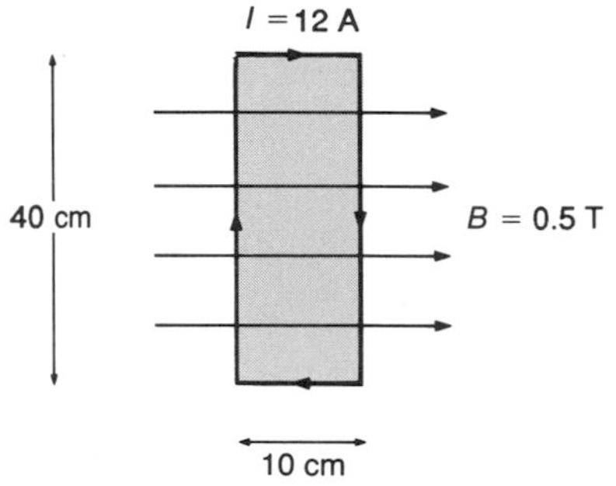

24. A flat square coil of 15-cm side carries an 8-A current and lies with its plane parallel to a uniform magnetic field of 0.5 T. The direction of the magnetic field makes an angle of 45° with each side of the square. Calculate the magnitude and direction of the magnetic force on each side of the coil.

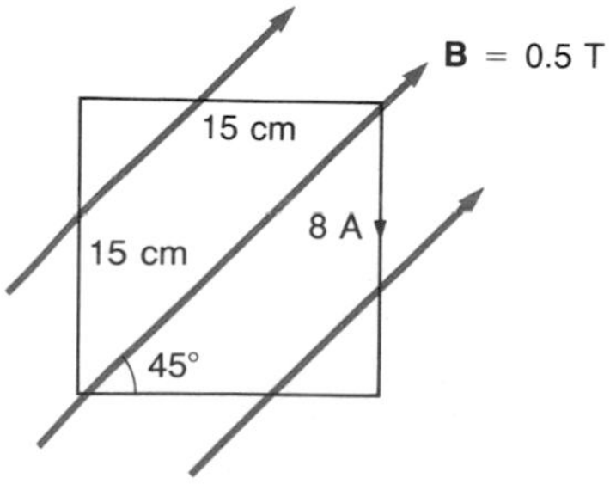

25. Two long straight parallel wires, each carrying a 16-A current, are separated by 24 cm. Calculate the magnitude of the magnetic field midway between the wires

a. if the currents are in opposite directions
b. if the currents are in the same direction [19.2]

26. Two long straight parallel wires are 16 cm apart and carry currents of 20 A and 35 A in the same direction.

a. Calculate the magnitude of the magnetic field at a point midway between the wires.
b. Calculate the position of a point on the line joining the two wires at which the magnetic field is zero. [19.2]

27. Two long straight parallel wires are 12 cm apart and carry equal currents in opposite directions. Calculate the values of the currents if the magnetic field midway between the wires is 25 μT.

28. A long horizontal wire carries a 30-A current from east to west. Another long horizontal wire that carries a 20-A current from south to north is 20 cm above the first wire at the place where they cross. Calculate the magnitude and direction of the magnetic field midway between the wires at the crossover.

29. Two long straight parallel wires that are 40 cm apart carry currents of 4 A and 8 A in opposite directions. Calculate the magnitude of the magnetic field at the point *P* on the diagram that is 32 cm from the 8-A wire and 24 cm from the 4-A wire.

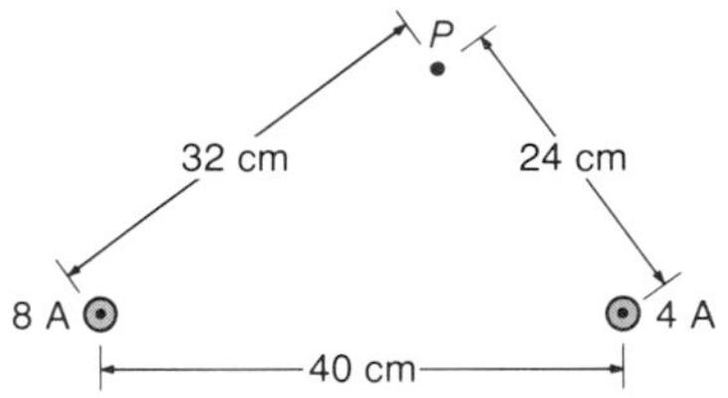

30. The electron gun in a TV tube shoots a narrow beam of electrons at the screen, forming a small bright spot in the center. Electrons arrive at the screen at the rate of 3×10^{15} per second. Calculate the magnetic field at a point 1 cm from the beam. Is the magnetic field clockwise or counterclockwise about the beam as viewed from the front of the screen?

31. A length of 130 m of wire is used to wind a solenoid uniformly on a cylinder 40 cm long and 2 cm in radius. Calculate the magnetic field inside the solenoid due to a 12-A current in the wire.

32. A solenoid is wound on a cylinder 80 cm long and 1.5 cm in radius. What length of wire is required if a current of 5 A is to produce a magnetic field of 3.5 mT inside the solenoid?

33. A solenoid is wound on a cylinder 75 cm long and 3 cm in radius using 800 turns of copper wire that is 0.8 mm in diameter. Calculate the interior magnetic field if the wire is connected to a 12-V battery.

34. A length of 120 m of copper wire 1 mm in diameter is used to wind a uniform solenoid on a cylinder 85 cm long and 2.5 cm in radius. What battery emf is required to produce a magnetic field of 40 mT when connected to the wire?

35. A current balance has parallel wires each 40 cm long and separated by 2.1 cm. The wires are 21 cm from the pivot suspension, and carry 24-A currents in the same direction. What additional weight, at a lever arm distance of 9 cm, is necessary for balance? [19.3]

36. A current balance requires an additional weight of 0.003 N to balance it, due to equal currents in the same direction in the parallel wires. The wires are each 60 cm long and are separated by 2.5 cm; the lever arms for the wires and the balance weight are 18 cm and 11 cm, respectively. Calculate the currents in the wires. [19.3]

37. Three straight parallel wires, each 125 cm long, are 3 cm apart from each other, and carry 18-A currents all in the same direction. Calculate the magnitude and direction of the force on each wire.

38. A flat square coil of wire of 12-cm side length carries a 25-A current. Calculate the force on each side of the coil due to the current in the opposite side.

39. Four straight parallel wires each 1.25 m long are arranged as shown in the diagram with the nearest wires 3.5 cm apart. The wires carry 16-A currents in the directions shown. Calculate the magnitude and direction of the force on each wire.

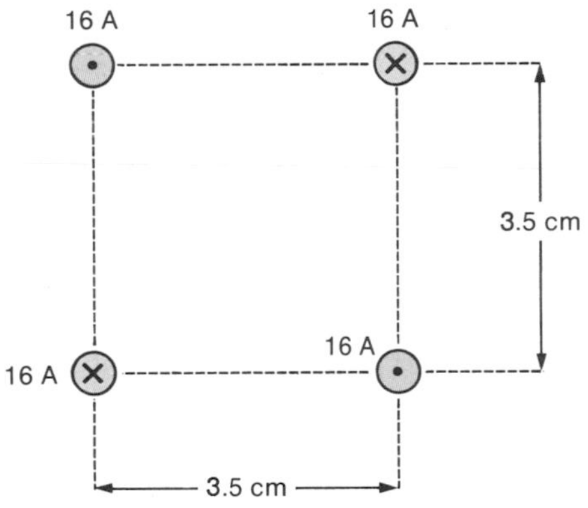

40. A circular coil of 160 turns of wire has a 4-cm radius.

a. Calculate the current that causes a magnetic moment of 4.5 A · m^2.

b. What is the maximum torque on the coil when carrying this current in a uniform magnetic field of 0.25 T? [19.5]

41. A rectangular coil 6 cm × 4 cm of 100 turns of wire carries a current of 8 A. The coil experiences a torque of 8 N · m in a magnetic field whose direction is at 30° with the plane of the coil. Calculate the magnitude of the magnetic field. [19.5]

42. A circular coil of 200 turns of wire has a radius of 1.5 cm and carries a current of 80 mA. The coil is positioned inside a long solenoid with its plane parallel to the axis of the solenoid. The solenoid has 30 turns per centimeter and carries a current of 12 A. Calculate the torque on the coil.

43. A circular coil of 100 turns of wire has a radius of 6 cm and carries a current of 2 A. The coil is positioned in a uniform magnetic field of 0.8 T whose direction makes an angle of 60° with the plane of the coil. Calculate:

a. the magnetic moment of the coil

b. the torque exerted on the coil

44. A proton is accelerated from rest through a potential difference of 650 kV and subsequently enters a region of space where there is a constant magnetic field of 0.3 T perpendicular to the direction of its velocity. Calculate the radius of the circular path in which the proton moves. [19.6]

45. An α-particle (mass 6.64×10^{-27} kg and charge 3.2×10^{-19} C) moves perpendicularly to a constant magnetic field of 0.85 T in a circular path of radius 36 cm. Calculate the energy of the α-particle. [19.6]

46. An electron is accelerated from rest and subsequently describes a circular path of radius 15 cm in a constant magnetic field of 0.5 mT that is perpendicular to the plane of the circular path. Calculate the magnitude of the potential difference that accelerated the electron.

47. A horizontal wire 2 m long points toward the magnetic north and carries a current of 150 A in that direction. The earth's magnetic field is 60 μT with a 55° angle of dip. Calculate the magnitude and direction of the force on the wire due to the earth's field. [19.7]

48. A flat square coil of 80 turns of wire and side length 50 cm has its plane horizontal. The coil carries a current of 3 A, and experiences a torque of 1.75×10^{-3} N · m at a place where the earth's magnetic field is 58 μT. Calculate the angle of dip. [19.7]

C. Advanced-Level Problems

49. A horizontal wire 12 m long points 30° east of the magnetic north and carries a current of 100 A in that direction. The earth's magnetic field is 60 μT and the angle of dip is 60°. Calculate the force on the wire due to the earth's field.

50. A square coil of 400 turns of wire has a side length of 40 cm and is suspended from one edge by a smooth hinge. In a vertical magnetic field of 0.5 T, the plane of the coil makes an equilibrium angle of 30° with the vertical when a current of 2.5 mA flows. Calculate the mass of the coil.

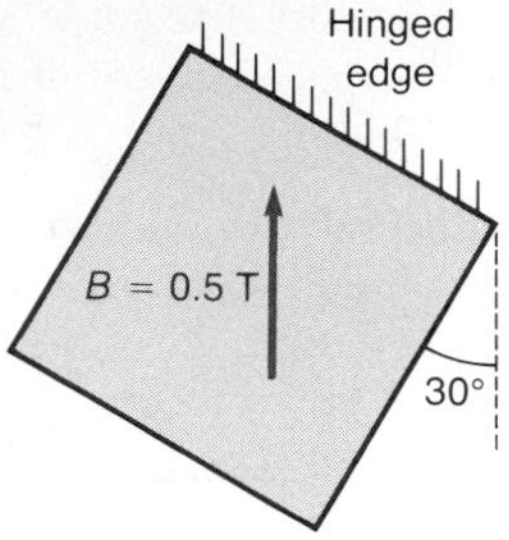

51. At a certain point on the magnetic equator, the earth's magnetic field is 60 μT directed horizontally. A copper rod 1 m long and 5 mm in diameter is held horizontally in the magnetic east-west direction.

a. Calculate the current required in the copper rod for the magnetic force to support its own weight.
b. To check the feasibility of the arrangement, calculate the power dissipation in the copper for this current.

52. Electrons in a vacuum tube are accelerated by a potential difference of 5 kV. The electrons then move alongside a long straight wire carrying a current of 3 A that is parallel to their original direction of travel and 18 cm away from it.

a. Calculate the force on each electron in the beam.
b. Calculate the deviation of each electron from a straight path over a 25-cm distance of travel. (Assume that deviation is sufficiently small that the magnetic field remains constant over the path considered.)

20

ELECTROMAGNETISM II

ELECTROMAGNETIC INDUCTION

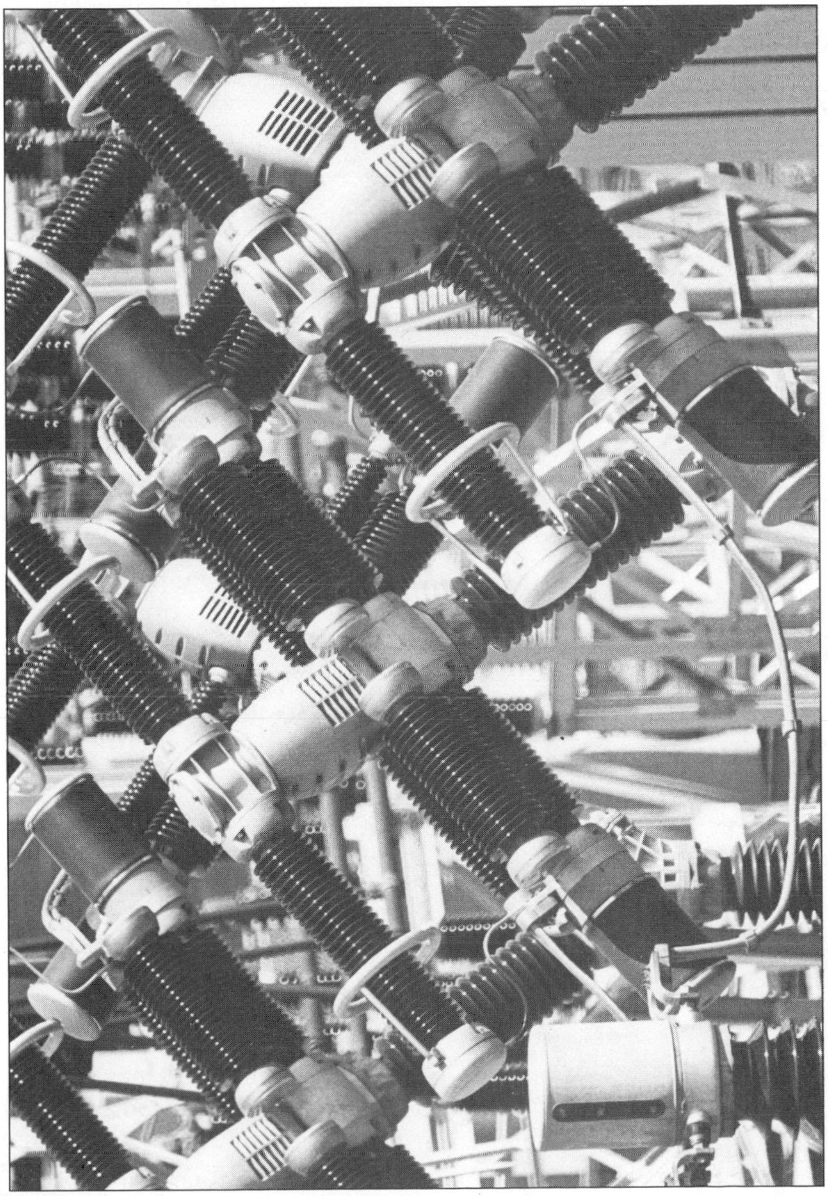

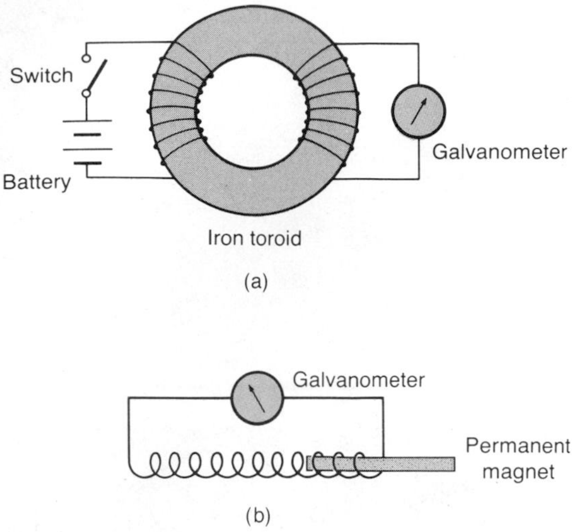

FIGURE 20.1 Illustration of experiment performed by Faraday in his discovery of the law of electromagnetic induction. (a) The apparatus remains stationary, but the magnetic field changes. (b) There is relative motion between the coil and the magnetic field.

We have already seen that Oersted discovered electromagnetism by demonstrating that an electric current causes a magnetic field. Faraday meditated on this remarkable fact and decided that it would be a fitting symmetry of nature if a magnetic field could cause an electric current. He investigated this possibility, but his first attempts were unsuccessful. Figure 20.1 illustrates Faraday's experiment. In Fig. 20.1(a) an iron toroid (a piece of iron in the shape of a doughnut) is magnetized by a current passed through a coil that is wrapped around the toroid, and the magnetic field passes through the turns of a second coil that is connected to a galvanometer. Faraday hoped that the galvanometer would deflect whenever a magnetic field was threading the second coil—that is, whenever the switch was closed and the battery was supplying a current. In this he was disappointed. However, he did find that closing the switch deflected the galvanometer one way and opening it deflected the galvanometer in the opposite direction. If the current remained steady, the galvanometer did not deflect. Faraday had made the first step in the discovery of the great law of electromagnetic induction that now bears his name. The electric current was not caused by the magnetic field, but by a change in the magnetic field.

In the experiment just described, the apparatus remains stationary and the magnetic field changes. Faraday used apparatus similar to that of Fig. 20.1(b) to show that the relative motion between a coil and a magnetic field causes a similar effect. As long as the magnet remains stationary, the galvanometer needle does not move; any movement of the magnet deflects the galvanometer needle, in one direction if the magnet is thrust into the coil and in the opposite direction if the magnet is withdrawn.

Michael Faraday is unique in the world of great physicists in having no formal education in mathematics. He expressed his law of electromagnetic induction with intuitive geometrical constructs rather than mathematical equations. We will explain Faraday's law for situations that involve only simple geometry. Before explaining the law itself, however, we will require the important concept of the flux of a vector.

20.1 Magnetic Flux

The first dictionary meaning for *flux* is "a flowing." If we followed this meaning strictly, we could not apply the term to magnetic fields. However, although a magnetic field does not flow in the usual sense, we use the idea of a flowing fluid to develop our concept. Let us begin with an especially simple case. Consider a uniform magnetic field **B** and an area A whose plane is normal to the direction of **B**, as shown in Fig. 20.2. We define the flux of **B** through the area in question as follows:

$$\text{Flux of magnetic field} = \text{magnetic field strength} \times \text{area}$$

SPECIAL TOPIC 22 FARADAY

Michael Faraday was born near London in 1791, one of ten children of a blacksmith. Until he was apprenticed to a bookbinder, his formal education was limited to reading and writing. His employer encouraged his interest in scientific books, and even allowed him time off to attend scientific lectures. At the age of twenty-two, he accepted a job as a bottle washer in Sir Humphrey Davy's laboratory at the Royal Institution. Twelve years later, as a result of conscientious hard work and genius, he became director of the laboratory; some few years later he became professor of chemistry. Davy was bitter and resentful of the young protégé who had completely outshone him.

His work puts him among the very great figures of science. He discovered the laws of electrolysis, and laid the foundation for electrochemistry; his discovery of electromagnetic induction led to the invention of both the transformer and the electric generator. This latter was especially important since the only previous source of electrical power was the chemical battery—a device whose power production capacity is relatively small.

Faraday was also a distinguished lecturer (impressing even Charles Dickens) and established the famous Christmas lectures that were especially for the benefit of young people. He remained a modest retiring man, and upon his death in 1867 was buried in a quiet ceremony for family only.

(Smithsonian Institution)

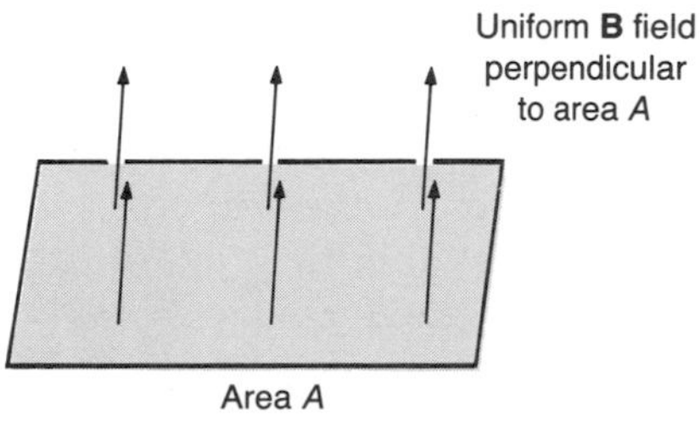

FIGURE 20.2 Diagram to illustrate the flux of a magnetic field through an area normal to the field direction.

The flux of a vector is a scalar quantity that measures the amount of flow of magnetic field strength through an area.

Before generalizing our definition of **magnetic flux,** let us reflect on a few points that concern the flux of any vector:

1. Vector flux has a meaning only with respect to a designated area. There can be no vector flux without reference to an area.
2. For a vector to have flux, it must have a value at all points of the designated area. Since the magnetic field strength has a value at all points of the area in our example, we can calculate its flux. It is not difficult to think of a vector that does not have

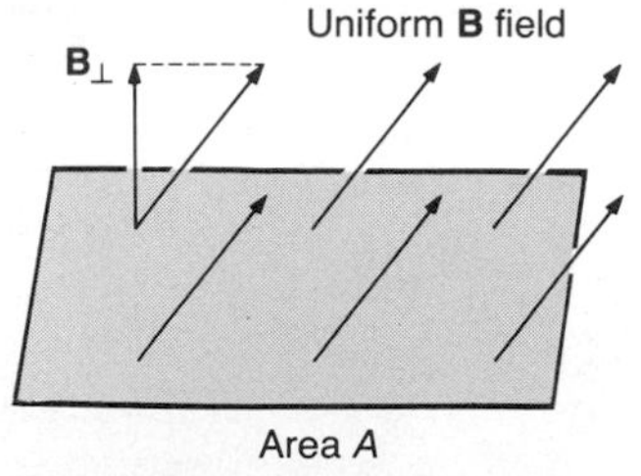

FIGURE 20.3 Diagram to illustrate the calculation of the flux of a uniform magnetic field when the field direction is not normal to the area.

flux. A stone thrown through the area has a velocity vector associated with its motion, but this vector has a value only where the stone is, not at every point of the area. Therefore, we cannot calculate its flux.

3. In our example, the magnetic field strength vector is normal to the designated area. How do we modify our definition if this is not so? Remember that flux means flow through an area. The component of field strength that is normal to an area produces flow *through* that area, whereas the component parallel to the area produces only flow *past* it. We therefore modify our definition of *flux* by specifying the vector component that is normal to the designated area.

Figure 20.3 shows a uniform magnetic field passing through a planar area. Since the direction of the magnetic field is not normal to the area, we take the normal component of the field and then multiply by the area.

■ Definition of Magnetic Flux Due to a Uniform Magnetic Field over a Planar Area

$$\text{Magnetic flux} = \begin{matrix}\text{component of}\\ \text{magnetic field strength}\\ \text{perpendicular to the area}\end{matrix} \times \text{area}$$

$$\Phi = B_{\perp}A \qquad \textbf{(20.1)}$$

SI unit of magnetic flux: tesla-(meter)^2
New name: weber
Abbreviation: Wb

The symbol for flux is the capital Greek letter phi (Φ). The name for the unit honors Wilhelm Weber (1804–1891), a German physicist who was a friend and colleague of the great mathematician Carl Gauss. Weber is known for contributions to many areas of physics but especially to electromagnetism.

EXAMPLE 20.1

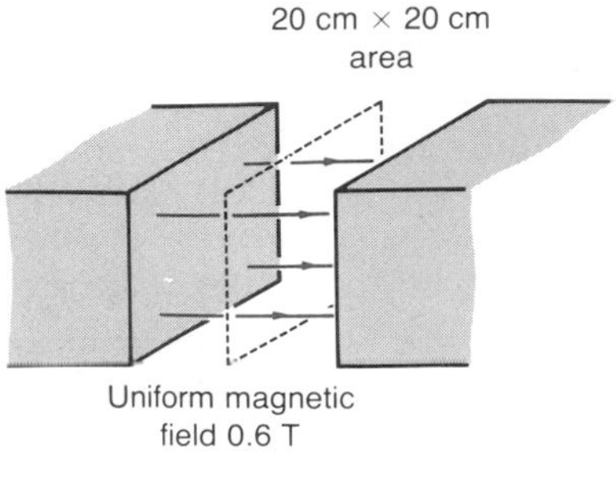

The pole pieces of a magnet are 20 cm × 20 cm in cross section. The magnetic field between them is uniform and of strength 0.6 T. Calculate the flux through an area 20 cm × 20 cm normal to the field direction.

SOLUTION Since the magnetic field strength is normal to the area chosen, the solution is a straightforward application of Eq. (20.1):

$$\Phi = B_{\perp}A$$
$$= 0.6\ \text{T} \times (0.2\ \text{m} \times 0.2\ \text{m})$$
$$= 2.4 \times 10^{-2}\ \text{Wb}$$

Let us consider another example in which the magnetic field is not normal to the designated area.

EXAMPLE 20.2

The earth's magnetic field at Washington, D.C., is 57 μT with an angle of dip 71°. Calculate the flux through a horizontal plane surface 20 m × 10 m.

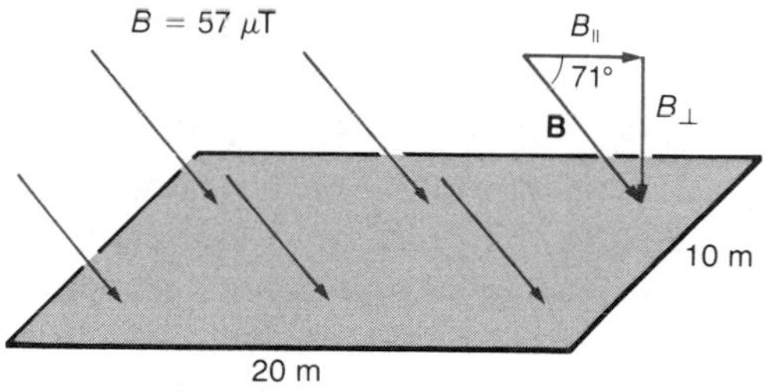

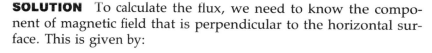

SOLUTION To calculate the flux, we need to know the component of magnetic field that is perpendicular to the horizontal surface. This is given by:

$$B_{\perp} = 57 \times 10^{-6}\ \text{T} \times \sin 71°$$
$$= 54.0 \times 10^{-6}\ \text{T}$$

To calculate the flux we use Eq. (20.1):

$$\Phi = B_{\perp}A$$
$$= 54.0 \times 10^{-6}\ \text{T} \times 20\ \text{m} \times 10\ \text{m}$$
$$= 1.08 \times 10^{-2}\ \text{Wb}$$

20.2 Faraday's Law

We have already seen that Faraday noted that a changing magnetic field can cause an electric current. As a result of his long and thorough investigation, Faraday was able to formulate a general law that we call the law of electromagnetic induction, or Faraday's law. The law depends on the basic experimental fact that *a change in the magnetic flux through a circuit produces an electromotive force around the circuit.* We have already used the term electromotive force (abbreviated "emf") to denote the open-circuit voltage of a battery. The direct effect of a battery emf is the production of current in a circuit to which the battery is connected. A changing magnetic flux through a circuit has the same effect—the emf caused by the changing magnetic flux produces a current of the same magnitude as that caused by a battery of equal voltage; the only difference is that no chemical action is involved.

Let us now proceed with a precise statement of **Faraday's law of electromagnetic induction.**

Faraday's Law of Electromagnetic Induction

$$\text{Average emf induced in a circuit} = -\frac{\text{magnetic flux change through the circuit}}{\text{elapsed time for flux change}}$$

$$\langle \mathscr{E} \rangle = -\frac{\Phi - \Phi_0}{t - t_0} \qquad \textbf{(20.2)}$$

SI unit of emf: weber per second

The weber per second is the same as the volt.

The negative sign in Eq. (20.2) is necessary to give the correct polarity to the induced emf—a matter that we shall take up shortly.

We should note that the definition of flux [Eq. (20.1)] contains both the area and the magnetic field component that is normal to it. A change of flux (and consequently an induced emf) could result from a change of either of these two quantities or both of them. In what follows we have examples of two situations: changing area with constant normal magnetic field, or constant area with changing normal magnetic field.

If the flux changes at a uniform rate, the average symbol can be dropped and the induced emf is constant.

We can justify the identification of the weber per second as a volt in the following way:

1 weber per second = 1 tesla-(meter)^2 per second
[since Eq. (20.1) gives 1 Wb = 1 T · m^2]
= 1 newton · meter per coulomb
[since Eq. (19.2) gives 1 T = 1 N · s/C · m]
= 1 joule per coulomb
[since Eq. (7.1) gives 1 J = 1 N · m]
= 1 volt
[since Eq. (16.6) gives 1 V = 1 J/C]

EXAMPLE 20.3

A circular loop of copper wire of radius 50 cm lies in a horizontal plane. The resistance of the wire is 0.2 Ω. A vertical magnetic field of initial value 0.1 T falls uniformly to zero during a time interval of 0.15 s. Calculate the current in the wire as the magnetic field collapses.

SOLUTION The magnetic field is perpendicular to the plane of the coil. We calculate the initial value of the magnetic flux by using Eq. (20.1):

$$\begin{aligned}\Phi_0 &= B_\perp A \\ &= 0.1\ \text{T} \times \pi \times (0.5\ \text{m})^2 \\ &= 7.85 \times 10^{-2}\ \text{Wb}\end{aligned}$$

The final value of the flux is $\Phi = 0$. If the flux changes at a uniform rate, we can drop the average symbol in Eq. (20.2):

$$\begin{aligned}\mathscr{E} &= -\frac{\Phi - \Phi_0}{t - t_0} \\ &= -\frac{0\ \text{Wb} - 7.85 \times 10^{-2}\ \text{Wb}}{0.15\ \text{s}} \\ &= 0.524\ \text{V}\end{aligned}$$

Since the induced emf is equivalent to a battery emf, we can calculate the current in the coil by using Eq. (18.2b):

$$\begin{aligned}I &= \frac{V}{R} \\ &= \frac{0.524\ \text{V}}{0.2\ \Omega} \\ &= 2.62\ \text{A}\end{aligned}$$

If both a battery emf and an induced emf act in a circuit, then the emf that drives the current is the sum or difference of the two.

■

It remains to clarify the question of the polarity of the induced emf caused by the changing magnetic flux. We use the word *polarity* to mean either clockwise or counterclockwise around a given circuit; the word *direction* might imply that emf is a vector, which it most certainly is not. The polarity of the emf indicates whether it tends to drive conventional current clockwise or counterclockwise in the circuit. Faraday himself answered this question with a series of rules connecting relative directions of motion. The negative sign of Eq. (20.2) gives the polarity of the emf when properly interpreted. However, we can use a simpler rule published by the Russian scientist Emil Lenz (1804–1865) a few years after Faraday's discovery. **Lenz's law** completely substitutes for the negative sign of Eq. (20.2).

■ Lenz's Law of Electromagnetic Induction

> The polarity of the induced emf is such as to oppose the change that caused it.

Let us apply Lenz's law to a specific example.

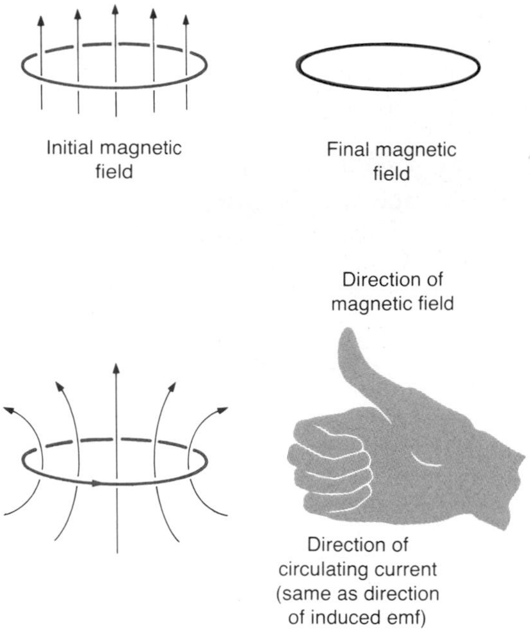

EXAMPLE 20.4

Determine the direction of current in the circular loop of Example 20.3 if the initial magnetic field is vertically upward.

SOLUTION The magnetic field, which is initially vertically upward, is zero in the final situation. The induced emf acts to oppose the change; that is, it causes a current in the circular coil that attempts to maintain the upward magnetic field. By the right hand rule, the direction of the circulating current that causes an upward magnetic field is counterclockwise when viewed from above. Since the induced emf causes this current, the polarity of the induced emf is counterclockwise from above. ■

Up to this point, our examples of electromagnetic induction refer to a fixed loop and a changing magnetic field, but Faraday's law also applies to the situation in which the circuit moves relative to the magnetic field. Consider the device shown in Fig. 20.4, which has a conducting rod forced to slide along a U-shaped conductor. Let l be the distance between the parallel arms of the U-shaped conductor, and let $\mathbf{v}$ be the velocity of the sliding rod. In addition, let a uniform magnetic field $\mathbf{B}$ be directed down into the plane of the paper. As the rod slides to the right, the magnetic flux increases through the circuit formed by the rod and the base and arms of the U-shaped conductor. Since the rod has a velocity $\mathbf{v}$, it slides a distance $v \times 1$ s during a 1-s interval. Therefore the extra area added to the loop during the 1-s interval is given by $l \times v \times 1$ s. Since the magnetic field $\mathbf{B}$ is perpendicular to the plane of the loop, the extra magnetic flux through the loop in 1 s is $B \times l \times v \times 1$ s. Using Eq. (20.2), the induced emf is this flux change divided by 1 s, which gives us a simple formula for the emf induced in the moving rod.

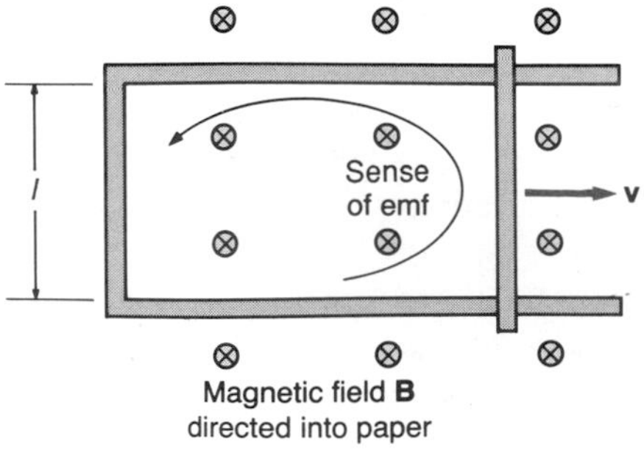

FIGURE 20.4 Diagram to illustrate the calculation of the emf induced in a straight conductor moving perpendicularly to a magnetic field.

■ Formula for emf Induced in a Rod Moving in a Magnetic Field

$$\text{Induced emf} = \begin{array}{c}\text{magnetic field}\\ \text{component}\\ \text{perpendicular}\\ \text{to rod}\end{array} \times \begin{array}{c}\text{length}\\ \text{of rod}\end{array} \times \begin{array}{c}\text{velocity}\\ \text{component}\\ \text{perpendicular to}\\ \text{both these}\\ \text{quantities}\end{array}$$

$$\mathscr{E} = B_{\perp} l v_{\perp} \qquad \textbf{(20.3)}$$

In most of our examples all three quantities are mutually perpendicular. In other cases, the application of Eq. (20.3) requires the components indicated.

We can easily find the polarity of the induced emf by using Lenz's law. The motion of the rod increases the flux through the circuit loop, and the polarity of the induced emf causes a current in the loop that attempts to retain the original value of the flux. As a result, the induced current causes a magnetic field directed up out of the paper in Fig. 20.4. By the right hand rule, the current must be counterclockwise in the loop, as is the polarity of the induced emf.

We can also show that the polarity of the induced emf in the moving rod does not depend on the position of the U-shaped conductor. For example, we could place the open end to the left of the sliding conductor instead of to the right, and Lenz's law would give the same result for the polarity of the emf in the conductor. In this situation the sliding rod is the seat of the emf in a very real sense. In fact, the existence of the induced emf does not depend on the presence of the U-shaped conductor, but only on the motion of the rod; the conductor only provides a circuit for the induced current flow. Let us clarify the situation by means of an example.

EXAMPLE 20.5

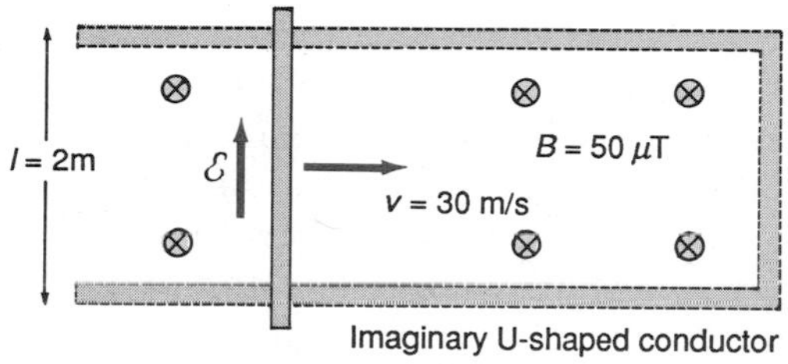

A horizontal conducting copper rod 2 m long points in a north-south direction and has a velocity of 30 m/s due east. The vertical component of the earth's magnetic field is 50 μT downward.

a. Calculate the magnitude and polarity of the emf induced in the rod.

b. Does the induced emf cause current in the rod?

SOLUTION

a. We can calculate the magnitude of the emf directly from Eq. (20.3) since the length of the rod, the direction of the magnetic field, and the velocity of the rod are in mutually perpendicular directions:

$$\begin{aligned}\mathcal{E} &= B_{\perp} l v_{\perp} \\ &= 50 \times 10^{-6}\ \text{T} \times 2\ \text{m} \times 30\ \text{m/s} \\ &= 3\ \text{mV}\end{aligned}$$

To find the polarity of the induced emf, we introduce an imaginary U-shaped conductor as shown on the diagram. Since the flux through the conducting loop decreases as the rod moves, by Lenz's law the emf would cause a current that produces a downward magnetic field within the loop. From the right hand rule this current would be clockwise in the loop; the polarity of the induced emf in the rod is therefore from south to north. This conclusion is independent of the existence of the U-shaped conductor.

b. If the straight conductor is initially at rest, no emf is acting. As the conductor starts moving to the east, an induced emf causes a current toward the north that transfers some electrons from the north to the south end of the rod. The rod is then positively charged at the north end and negatively charged at the south end. By the time the rod reaches a steady velocity of 30 m/s, the induced emf is 3 mV toward the north, which is exactly balanced by an electrostatic potential difference of 3 mV toward the south. As we have just explained, the electrostatic potential difference is caused by a charge transfer brought about by the small current in the initial stage of the motion. As the rod continues to move at a constant velocity, there is no further current. ■

20.3 Direct Current Generators and Motors

Most dc motors and generators have conducting rods arranged on a cylindrical surface that rotates in a magnetic field. However, it is easier to clarify the physical principles by considering rectilinear motion of a conducting rod on a pair of parallel conducting rails. We have already calculated the induced emf when the sliding rod moves with a given velocity [see Eq. (20.3) and Fig. 20.4], but many other questions remain to be answered, such as what force is needed to push the rod and how much electrical power is generated. To answer these questions and to uncover other details of the linear generator, let us consider an example.

EXAMPLE 20.6

A linear generator consists of a conducting rod sliding on a pair of parallel conducting rails 1.2 m apart with a magnetic field of 1.4 T perpendicular to the plane of the rails. The induced emf is 60 V and the generator delivers power to a 10-Ω resistor.

a. Calculate the velocity of the sliding rod.

b. What force is required to push the rod? (Ignore mechanical friction.)

c. How much mechanical power is needed to push the rod?

d. Calculate the amount of electrical power dissipated as heat in the resistor.

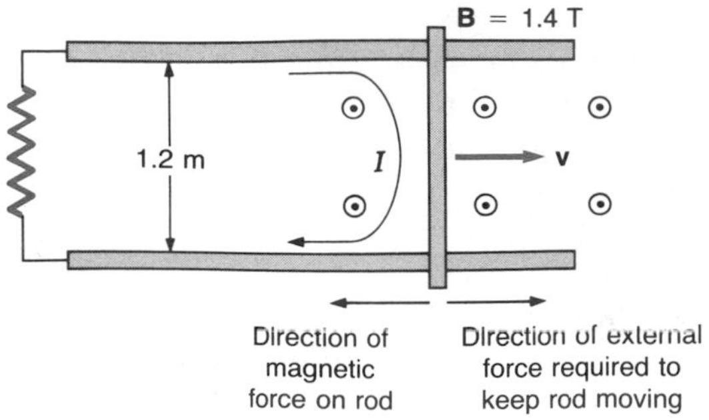

SOLUTION

a. Since the magnetic field is perpendicular to the plane of the loop, we can calculate the velocity of the sliding rod directly from Eq. (20.3):

$$\mathscr{E} = B_\perp l v_\perp$$

$$\therefore v_\perp = \frac{\mathscr{E}}{B_\perp l}$$

$$= \frac{60 \text{ V}}{1.4 \text{ T} \times 1.2 \text{ m}}$$

$$= 35.7 \text{ m/s}$$

b. A conductor that carries a current in a magnetic field experiences a mechanical force given by Eq. (19.3), but to calculate this force we need to know the current in the conductor. Equation (18.2) gives:

$$I = \frac{V}{R}$$

$$= \frac{60 \text{ V}}{10 \ \Omega}$$

$$= 6 \text{ A}$$

Equation (19.3) now gives:

$$\mathbf{F} = I\mathbf{l} \times \mathbf{B}$$

$$= 6 \text{ A} \times 1.2 \text{ m} \times 1.4 \text{ T}$$

$$= 10.1 \text{ N}$$

Using the right hand rule, we find that the magnetic force is to the left on the diagram with velocity and magnetic field directions assumed as shown. Since the rod moves at constant velocity, the net force on it is zero. Consequently, an external force of 10.1 N to the right is required to keep the rod sliding at 35.7 m/s.

c. We can calculate the mechanical power required to move the rod from Eq. (7.6):

$$P = Fv_\parallel$$

$$= 10.1 \text{ N} \times 35.7 \text{ m/s}$$

$$= 360 \text{ W}$$

d. The rate of Joule heating in the resistor is given by Eq. (18.11c):

$$P = \frac{V^2}{R}$$

$$= \frac{(60 \text{ V})^2}{10 \ \Omega}$$

$$= 360 \text{ W}$$

■

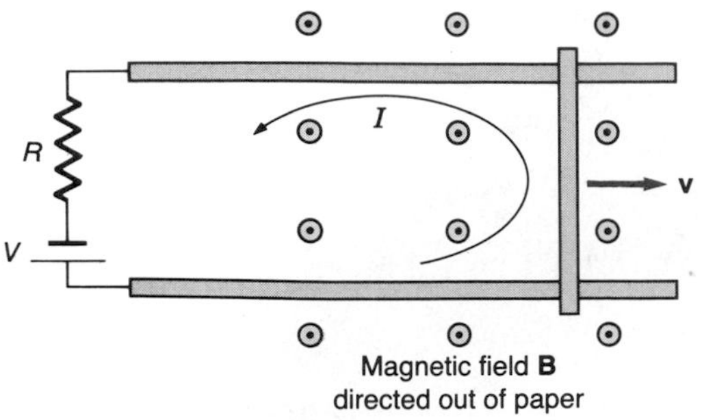

FIGURE 20.5 The linear dc motor.

The linear generator illustrates the interplay of many physical laws: Faraday's law governs the induced emf; Lenz's law gives its polarity; the force law for a current-carrying conductor in a magnetic field gives the force required to push the sliding rod; and Newtonian dynamics gives the external power required to keep the rod moving at constant velocity. Ignoring losses due to mechanical friction, we see finally that the rate of Joule heating in the resistor is exactly equal to the power required to drive the generator. Energy is conserved in the transition from mechanical to electrical power, just as we would expect.

The linear dc motor follows easily from our understanding of the linear generator and is illustrated in Fig. 20.5. A conducting rod is free to slide along a pair of parallel conducting rails, and a magnetic field **B** is perpendicular to the plane of the rails. A battery connected to one end of the rails supplies a voltage V that causes a current in the circuit loop in the direction shown. As we did for the linear generator, we suppose that the resistance of the circuit loop is some constant value R. Because of the magnetic field, the current in the sliding rod produces a force that, by the right hand rule, is to the right on the diagram. The rod moves to the right and attains a steady velocity if the motor delivers a constant mechanical power output.

However, the rod that is moving at a steady velocity in a constant magnetic field produces an induced emf by Faraday's law. The battery emf that causes the motion is ultimately responsible for this situation. It follows from Lenz's law that the induced emf opposes the driving battery. Because *every electric motor is also a generator* that generates an emf to oppose the driving battery, the induced emf in a motor is referred to as a counter emf. Consequently, we calculate the current by using Eq. (18.2):

$$I = \frac{V - \mathscr{E}}{R}$$

where V is the battery emf and $\mathscr{E}$ the counter emf. Multiplying both sides of the equation by I and rearranging give:

$$VI = I^2R + \mathscr{E}I \tag{20.4}$$

We can readily interpret this equation in terms of power supplied and used. The left-hand side is simply the power output of the battery [see Eq. (18.8)]. The first term on the right-hand side is the rate of Joule heating in the resistance of the motor [see Eq. (18.9b)]. Since we ignore energy losses due to mechanical friction, it follows that the second term on the right-hand side represents the useful power output of the motor.

Although we have derived the power equation with a linear motor in mind, nothing in the calculation requires this. The same conclusions hold true for a motor of any construction. A practical motor has many other sources of small power loss, such as mechanical friction in the bearings. However, we have considered only the Joule heat loss to make calculations that illustrate important physical ideas.

EXAMPLE 20.7

A dc motor with an internal resistance of 5 Ω is connected to a 100-V battery. Calculate the power output and the energy dissipated if the motor draws a 5-A current. Repeat the calculation for currents of 10 A and 15 A. Also calculate the counter emf in each case.

SOLUTION Each part of the problem requires the use of Eq. (20.4):

$$\text{Power output from battery} = \begin{matrix}\text{rate of Joule}\\ \text{heating}\end{matrix} + \begin{matrix}\text{mechanical}\\ \text{power output}\end{matrix}$$

$$VI = I^2R + \mathscr{E}I$$

For the 5-A current drain:

$$100\ \text{V} \times 5\ \text{A} = (5\ \text{A})^2 \times 5\ \Omega + \mathscr{E} \times 5\ \text{A}$$

$$500\ \text{W} = 125\ \text{W} + \mathscr{E} \times 5\ \text{A}$$

$$\therefore \mathscr{E} = 75\ \text{V}$$

The power output of the motor is 375 W, and the power dissipated is 125 W. Similar calculations for the other two currents give the following results:

Current	*Power output of battery*	*Power dissipated*	*Counter emf*	*Mechanical power output*
5 A	500 W	125 W	75 V	375 W
10 A	1000 W	500 W	50 V	500 W
15 A	1500 W	1125 W	25 V	375 W

■

Solving Example 20.7 shows features common to all dc electric motors. For a small current there is a large counter emf and a correspondingly low power dissipation. For a larger current, there is generally an increased mechanical power output up to a certain limit accompanied by increased dissipation. If the current increases beyond this limit, the power dissipation increases, but the mechanical power output actually drops. The optimum conditions for operating a motor depend on its design; a motor that is forced beyond its maximum power output overheats, and its power output is reduced.

The linear motor and generator display the physical principles in a straightforward way, but in a practical device the linear arrangement is replaced by a rotary one. The general features of a rotary dc motor or generator are illustrated in Fig. 20.6. A coil

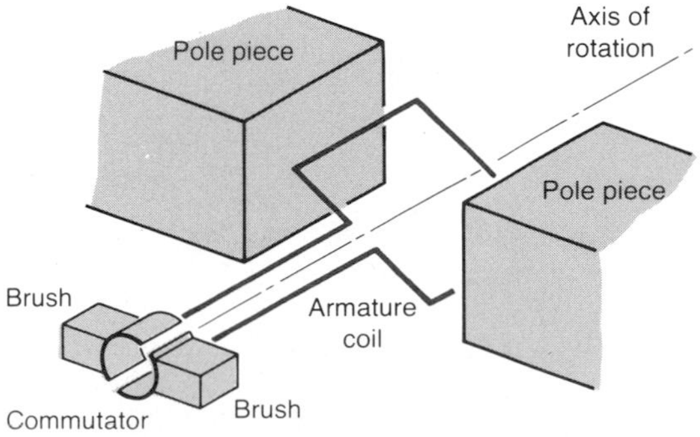

FIGURE 20.6 The rotary dc motor or generator.

rotates about an axis that is perpendicular to the uniform magnetic field between two pole pieces. The pole pieces are the extremities of an arrangement similar to that shown in Fig. 19.13 for producing a uniform magnetic field. The coil is called the *armature;* its ends are connected to a rotating switch called a *commutator,* which is connected to the external circuit by carbon brushes. If the machine is functioning as a generator, one side of the armature coil moves upward in the magnetic field while the other moves down. The commutator ensures that the upward moving side of the coil is always attached to the same carbon brush, and similarly for the downward moving side. In this way the brushes always carry current to the external circuit in the same direction. The armature coil plays the same role in the rotary machine that the sliding rod does in the linear machine. If a battery is connected to the carbon brushes, the rotary machine becomes a motor, and the energy conservation concept of Eq. (20.4) applies exactly as it does for the linear motor.

In actual construction, the armature of a dc machine is wound on a soft iron cylindrical core. The ends of the pole pieces are not flat, but curved to accommodate the cylindrical armature. To simplify the diagram, we omit these details from Fig. 20.6.

20.4 Self Inductance

When we close the switch in a dc circuit, current begins, causing a magnetic flux through the circuit. The flux, which starts at zero, attains its final value when the current reaches its final value. By Faraday's law, the changing magnetic flux causes an induced emf in the circuit, which, by Lenz's law, opposes the battery that is causing the current. Similarly, if we open the switch in a dc circuit that is already carrying current, an induced emf acts in a direction that attempts to maintain the current. For any dc circuit the **self inductance** measures the degree of opposition to the current change. Our definition of self inductance is as follows:

Definition of Self Inductance

$$\begin{array}{c}\text{Average induced emf}\\ \text{in a circuit due to}\\ \text{change in the current}\end{array} = -\text{self inductance} \times \frac{\text{current change}}{\text{elapsed time}}$$

$$\langle \mathscr{E} \rangle = -L\,\frac{I - I_0}{t - t_0} \qquad \textbf{(20.5)}$$

SI unit of self inductance: volt-second per ampere
New name: henry
Abbreviation: H

The negative sign reflects Lenz's law, giving the sense of the emf as opposing the current change. The unit of self inductance is named in honor of Joseph Henry (1797–1878), an American physicist who discovered electromagnetic induction while working in Albany, N.Y. Henry's discovery was independent of Faraday's, but his results were not published and received little recognition.

EXAMPLE 20.8

A circuit whose self inductance is 8 mH carries a steady current of 12 A. The current is reduced to zero in 1.2 ms by opening a switch. Calculate the induced emf in the circuit.

SOLUTION The example is simply an application of Eq. (20.5):

$$\begin{aligned}\langle \mathscr{E} \rangle &= -L \frac{I - I_0}{t - t_0} \\ &= -8 \times 10^{-3}\ \text{H} \times \frac{0\ \text{A} - 12\ \text{A}}{1.2 \times 10^{-3}\ \text{s}} \\ &= +80\ \text{V}\end{aligned}$$

In general, calculating the inductance of a circuit is difficult, but one instance of self inductance is easily calculated. Consider a long thin solenoid (Fig. 20.7) of length l and cross-sectional area A. Let the solenoid contain n turns, each carrying a current I. The magnetic field inside the solenoid is given by Eq. (19.5):

$$B = \mu_0 \frac{n}{l} I$$

Since this field is parallel to the axis, we can calculate the flux inside the solenoid by using Eq. (20.1):

$$\begin{aligned}\Phi &= B_\perp A \\ &= \mu_0 \frac{n}{l} IA\end{aligned}$$

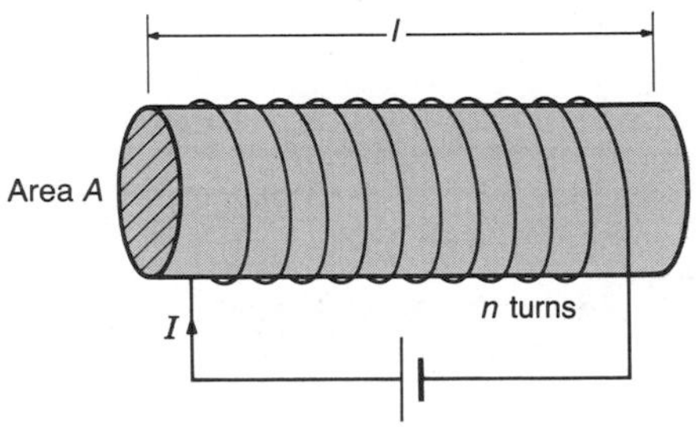

FIGURE 20.7 Diagram for calculation of the self inductance of a solenoid.

If the initial current is I_0, the initial value of the flux is given by:

$$\Phi_0 = \mu_0 \frac{n}{l} I_0 A$$

The use of Faraday's law, Eq. (20.2), shows us that the induced emf *in each turn* of the solenoid is given by:

$$\langle \mathscr{E} \rangle_{1\ \text{turn}} = -\frac{\Phi - \Phi_0}{t - t_0}$$

Substituting the results for Φ and Φ_0 gives us:

$$\langle \mathscr{E} \rangle_{1\text{ turn}} = -\mu_0 \frac{n}{l} A \frac{I - I_0}{t - t_0}$$

Since there are n turns in series, the total induced emf is n times the emf in one turn:

$$\langle \mathscr{E} \rangle_{n\text{ turns}} = -\mu_0 \frac{n^2}{l} A \frac{I - I_0}{t - t_0}$$

Comparison with Eq. (20.5) gives us the formula for the self inductance of a solenoid:

■ Formula for the Self Inductance of a Long Thin Solenoid of *n* Turns Wound on a Length *l* of Cross-Sectional Area *A*

$$L = \frac{\mu_0 n^2 A}{l} \qquad \textbf{(20.6)}$$

Since n is a pure number, Eq. (20.6) shows that μ_0 has the unit of henry per meter. This designation is more commonly used than the previous one of tesla · meter/ampere.

EXAMPLE 20.9

The circuit of Example 20.8 is a uniformly wound solenoid on a cardboard cylinder 50 cm long and 4 cm in diameter. How many turns are required to provide the 8-mH inductance?

SOLUTION We can calculate the number of turns from Eq. (20.6):

$$L = \frac{\mu_0 n^2 A}{l}$$

$$\therefore n = \left\{\frac{Ll}{\mu_0 A}\right\}^{1/2}$$

$$= \left\{\frac{8 \times 10^{-3}\text{ H} \times 0.5\text{ m}}{4\pi \times 10^{-7}\text{ H/m} \times \pi \times (0.02\text{ m})^2}\right\}^{1/2}$$

$$= 1590$$

■

As we have just seen, the self inductance of a circuit arises from the magnetic flux through the circuit itself that is caused by a current in it. However, current in a circuit also causes magnetic

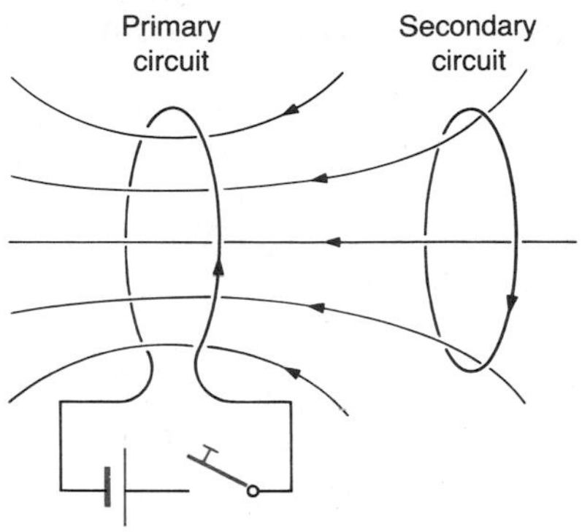

FIGURE 20.8 Closing the switch in the primary circuit causes a magnetic field, which establishes a magnetic flux through the secondary circuit. Before the switch was closed, there was no flux through the secondary. This change in the flux causes an induced emf in the secondary, which sets up a current to oppose the change.

flux through any other nearby circuit. If a switch is closed in one circuit (called the primary circuit), the resulting current establishes a magnetic field that, in general, causes a magnetic flux through the second circuit (called the secondary circuit). We illustrate the situation in Fig. 20.8. The change of flux through the secondary circuit causes an induced emf whose value and polarity can be determined through Faraday's and Lenz's laws.

EXAMPLE 20.10

A solenoid of 200 turns that has a resistance of 3 Ω is wound on a cardboard cylinder 0.5 m long with a cross-sectional area of 6×10^{-4} m^2. A second solenoid of 1000 turns is wound closely on the first but insulated from it. A 12-V battery and a switch are connected to the first solenoid, and there is a steady current after the switch is closed.

a. Calculate the magnetic flux in the solenoid when the steady current is established.

b. Calculate the magnitude and polarity of the average induced emf in the second solenoid if the magnetic flux falls to zero in 0.8 ms after the switch is opened.

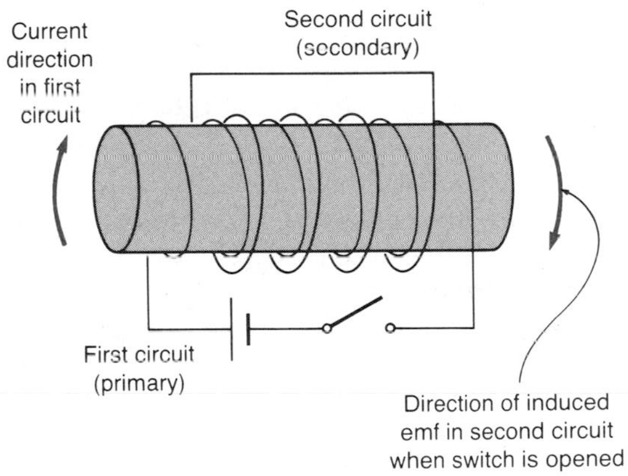

SOLUTION We solve the problem by taking the steps indicated by the order of the questions.

a. If the current has reached a steady value, there is no change in the magnetic field and consequently no induced emf. The steady current is determined entirely by the resistance of the solenoid, and Eq. (18.2) gives:

$$I = \frac{V}{R} = \frac{12\ \text{V}}{3\ \Omega} = 4\ \text{A}$$

Knowing the current, we can calculate the magnetic field by using Eq. (19.5):

$$B = \mu_0 \frac{n}{l} I = 4\pi \times 10^{-7}\ \text{H/m} \times \frac{200}{0.5\ \text{m}} \times 4\ \text{A} = 2.01 \times 10^{-3}\ \text{T}$$

Finally, we calculate the magnetic flux in the solenoid by using Eq. (20.1):

$$\begin{aligned}\Phi &= B_{\perp}A \\ &= 2.01 \times 10^{-3}\ \mathrm{T} \times 6 \times 10^{-4}\ \mathrm{m}^2 \\ &= 1.21 \times 10^{-6}\ \mathrm{Wb}\end{aligned}$$

b. When the current in the first solenoid falls to zero, the magnetic flux also falls to zero. Faraday's law, Eq. (20.2), gives the value of the average induced emf in each turn of the second solenoid:

$$\begin{aligned}\langle \mathscr{E} \rangle_{1\ \mathrm{turn}} &= -\frac{\Phi - \Phi_0}{t - t_0} \\ &= -\frac{1.21 \times 10^{-6}\ \mathrm{Wb} - 0\ \mathrm{Wb}}{8 \times 10^{-4}\ \mathrm{s}} \\ &= -1.5 \times 10^{-3}\ \mathrm{V}\end{aligned}$$

To find the total average induced emf in the second solenoid, we multiply by the number of turns:

$$\begin{aligned}\langle \mathscr{E} \rangle &= 1000 \times 1.5 \times 10^{-3}\ \mathrm{V} \\ &= 1.5\ \mathrm{V}\end{aligned}$$

The polarity of the induced emf attempts to maintain the decreasing magnetic flux, as shown in the diagram. ■

In the previous chapter we discussed the effect of an iron core in a solenoid, which is to increase the magnetic field strength by many hundreds (or even thousands) of times for a given current in the windings. The increased magnetic field strength means increased magnetic flux, which in turn leads to increased values of induced emf's. Let us illustrate with an example.

EXAMPLE 20.11

The pair of solenoids of Example 20.10 are given an iron core that increases the magnetic flux by a factor of 1500. Calculate the average emf induced in the second solenoid when the iron core is in place.

SOLUTION Faraday's law, Eq. (20.2), shows that the average induced emf is directly proportional to the flux change. If the flux change is 1500 times greater, the induced emf is also 1500 times greater. We therefore have:

$$\begin{aligned}\langle \mathscr{E} \rangle &= 1500 \times 1.5\ \mathrm{V} \\ &= 2250\ \mathrm{V}\end{aligned}$$

■

It may seem strange that a 12-V battery in one circuit can cause an average induced emf of 2250 V in an adjacent circuit, but this is exactly the situation in an automobile ignition system, where a 12-V battery causes many thousands of volts at the spark plug. The pair of solenoids (iron core) is the ignition coil; the switch that connects the first solenoid to the battery is a pair of contact points that opens and closes rapidly as the engine rotates; and the large induced emf in the second solenoid is distributed to the spark plugs through heavily insulated cables.

20.5 The Alternating Current Generator

As the name implies, an *alternating current* (abbreviated "ac") periodically reverses direction in its circuit. We have already seen that a battery or a dc generator causes a current in a circuit continually in one direction. One of the simplest devices that produces an alternating current operates directly on Faraday's law of electromagnetic induction. It is called an ac generator, or more simply an **alternator.**

Consider a coil rotating about its axis in a uniform magnetic field between a pair of pole pieces. The axis of rotation of the coil is perpendicular to the direction of the magnetic field, and a rotating ring is connected to each end of the coil, as shown in Fig. 20.9. Sliding contacts take the current from the rotating rings to the external circuit. The alternator uses the two ring contacts (one for each side of the coil) rather than the commutator of the dc generator. Except for this very important difference, the two machines are the same.

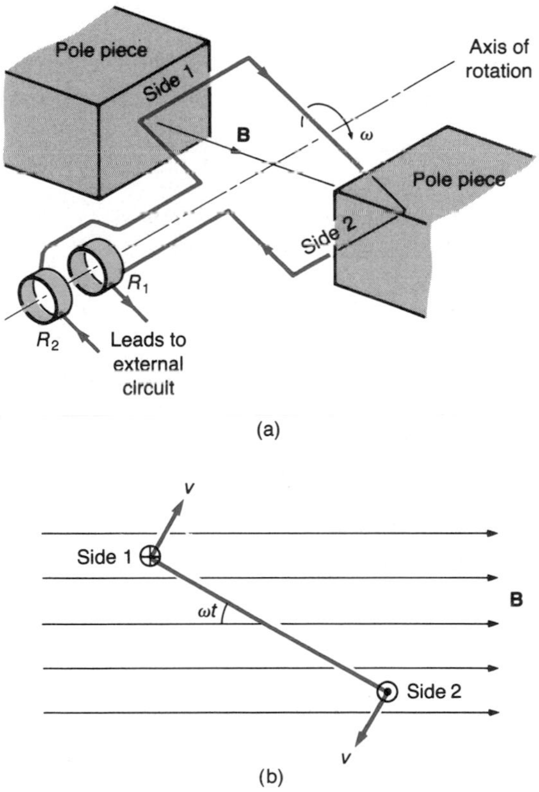

FIGURE 20.9 Illustration of the operation of an alternator: (a) general view of the rotating coil; (b) the rotating coil viewed from its end.

Consider the situation at the instant shown in Fig. 20.9(a). Side 1 of the coil is moving up through the magnetic field, and side 2 downward. The total magnetic flux through the coil is increasing. Faraday's law predicts an induced emf, and Lenz's law tells us that its sense will cause a current that tries to stop the flux increase. The current direction is shown in the figure—namely, out from ring R_1 to the external circuit and in via ring R_2. One half revolution later, side 1 is in the position that side 2 had been in, and vice versa. The current direction marked on the sides of the coil does not change, but current to the external circuit is now out from ring R_2 and in via ring R_1. The current reverses periodically as the coil rotates, producing an alternating current in an external circuit that is connected to the rings.

A typical alternator supplies an alternating voltage that causes the alternating current in the external circuit. To find out more about the alternating voltage, consider the view from the end, shown in Fig. 20.9(b). The coil is rotating with constant angular velocity ω. If we let $t = 0$ when the coil is horizontal, then the angle between the plane of the coil and the horizontal is ωt at some later time t. Each of the two sides moves with velocity v

through a magnetic field. The velocity component that is perpendicular to both the coil sides and the magnetic field is:

$$v_\perp = v \cos \omega t$$

If each side is of length l, we can find the induced emf in each side by using Eq. (20.3):

$$\mathscr{E} = B_\perp l v_\perp$$
$$= Blv \cos \omega t$$

Since the emf in the two sides add, the total emf of the coil is twice this amount. For a coil consisting of n turns, the emf of each turn adds to the total, and the induced emf for the coil is:

$$\mathscr{E} = 2nBlv \cos \omega t$$

Let r be the radial distance from the axis of rotation to the side of the coil. Then Eq. (9.11b) gives:

$$v = r\omega$$

The induced emf becomes:

$$\mathscr{E} = 2nBlr\omega \cos \omega t$$
$$= nBA\omega \cos \omega t$$

where A is the area of the flat coil. Since the magnetic field, the area of the coil, its angular velocity, and the number of turns are all constant for an alternator rotating at a constant angular velocity, we can write the alternator voltage in the following way:

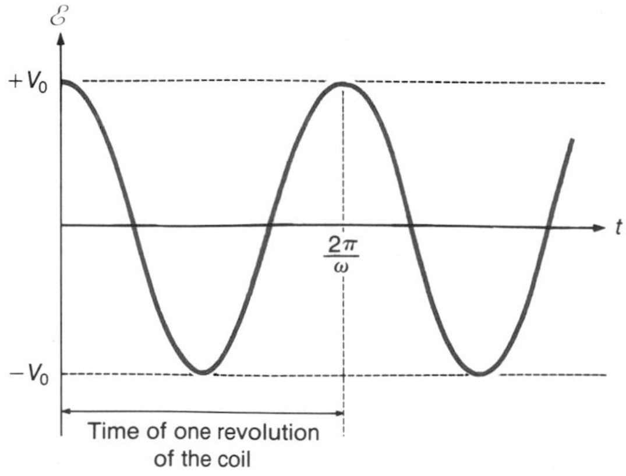

FIGURE 20.10 The induced emf (or output voltage) of an alternator as a function of time.

Expression for the emf of an Alternator

$$\mathscr{E} = V_0 \cos \omega t$$

For a flat coil of n turns

$$V_0 = nBA\omega \qquad \textbf{(20.7)}$$

B = magnetic field strength
A = coil area
ω = angular velocity of the coil

The quantity V_0 is important because it is the peak voltage of the alternator. The alternating emf rises to V_0 and falls to $-V_0$ once during each revolution of the coil, as illustrated in Fig. 20.10, where the alternator emf appears as a function of time. We see that the alternator voltage varies with time in exactly the same way as does the displacement of a particle undergoing simple harmonic motion (see Fig. 11.7). In fact, many of the same concepts apply to both cases. The angular velocity of the rotating coil ω is also called the angular frequency of the alternating emf. The coil rotates through 2π rad before the wave shape of the emf begins to repeat itself. The period of the emf is therefore 2π rad divided by the angular velocity of the coil (angular frequency of the emf), and the frequency of the emf is simply the reciprocal of the period.

■ Relations Between Period, Angular Frequency, and Frequency of Alternating emf

$$\text{Period of alternating emf} = \frac{2\pi}{\text{angular frequency}}$$

$$T = \frac{2\pi}{\omega} \qquad \textbf{(a)}$$

$$\text{Frequency of alternating emf} = \frac{1}{\text{period}} \qquad \textbf{(20.8)}$$

$$\nu = \frac{1}{T} = \frac{\omega}{2\pi} \qquad \textbf{(b)}$$

Let us apply these concepts by working an example.

EXAMPLE 20.12

A flat rectangular coil of area 0.4 m^2 contains 120 turns and rotates at 1500 rev/min in a uniform magnetic field of 0.1 T. Calculate the peak voltage and the frequency of the alternating emf.

SOLUTION Before using Eq. (20.7), we need to calculate the angular velocity of the coil:

$$\omega = 1500 \text{ rev/min} \times \frac{2\pi}{60} \text{ (rad/s)/(rev/min)}$$

$$= 50\pi \text{ rad/s}$$

(When problems involve both frequency and angular frequency, it is often convenient to leave factors of π in the arithmetic.) We can now find the peak voltage directly from Eq. (20.7):

$$V_0 = nBA\omega$$

$$= 120 \times 0.1 \text{ T} \times 0.4 \text{ m}^2 \times 50\pi \text{ rad/s}$$

$$= 754 \text{ V}$$

To calculate the frequency we use Eq. (20.8b):

$$\nu = \frac{\omega}{2\pi} = \frac{50\pi \text{ rad/s}}{2\pi} = 25 \text{ Hz}$$

■

In many household and commercial applications, alternating voltage is applied to the circuit rather than direct. We take up the study of ac circuits in the following chapter.

20.6 Electromagnetic Waves

The discoveries of Oersted and Faraday set the stage for the ultimate discovery in electromagnetism by the English mathematician and physicist James Clerk Maxwell (1831–1879). Maxwell reasoned that a piece was missing from the jigsaw puzzle of electromagnetism. Oersted had shown that an electric current causes a magnetic field. Since magnetic current does not exist, it cannot cause an electric field. Faraday showed that a changing magnetic field causes an electric field, so surely a changing electric field should cause a magnetic field. Maxwell developed this concept in detail and presented his famous set of equations that unify the whole theory of electric and magnetic fields.

Maxwell's equations comprise a complete theory of classical electromagnetic fields. They do not assign priority to either a changing magnetic or electric field—the two are intimately bound together, and one cannot occur without the other. Maxwell showed that his equations predict the existence of **electromagnetic waves.** They also provide a specific formula for the **velocity of electromagnetic waves** in a vacuum that contains the constants in the equations for electric and magnetic forces. Maxwell did not use SI units, but the expression for his formula that we use is equivalent to the one that he discovered.

Formula for the Velocity of Electromagnetic Waves in Free Space

$$\text{Wave velocity} = \left(\begin{matrix}\text{permittivity of}\\ \text{free space}\end{matrix} \times \begin{matrix}\text{permeability of}\\ \text{free space}\end{matrix}\right)^{-1/2}$$

$$c = (\epsilon_0\mu_0)^{-1/2} \qquad \textbf{(20.9)}$$

$$= 2.998 \times 10^8 \text{ m/s}$$

SPECIAL TOPIC 23 MAXWELL

James Clerk Maxwell was born in Edinburgh in 1831, and showed great talent in mathematics from an early age. He graduated from Cambridge and became professor of mathematics at Aberdeen at the age of twenty-five. Together with Boltzmann, he discovered the velocity distribution function for the molecules of an ideal gas. In 1871, he was appointed professor of experimental physics at Cambridge. There he set up the famous Cavendish Laboratory, and worked out the unified theory of electromagnetism, which gathers together the work of Coulomb, Ampère, Oersted, Faraday, and others. Many regard this as the most brilliant feat in physics between Newton and Einstein. Maxwell died at Cambridge before the age of fifty.

(Smithsonian Institution)

The velocity of light had been measured accurately prior to Maxwell's work, and the measured values were in excellent agreement with the predicted velocity for electromagnetic waves. The predicted velocity of electromagnetic waves was the same as the measured velocity of light! This remarkable discovery could surely have only one interpretation—**light is an electromagnetic wave**—and all subsequent research has supported this hypothesis. We will reserve our discussion on the details of light waves for a subsequent chapter.

Maxwell's theory was very complete, predicting many details of the electromagnetic wave. First, it is a wave of electric and magnetic field strength that does not require an identifiable physical medium for propagation, distinguishing it from other types of wave motion that we have considered. Waves on a guitar string require the presence of the string, water waves require water, and sound waves require air, but electromagnetic waves can propagate in empty space.

Second, the electric and magnetic field vectors are perpendicular to each other, and each is perpendicular to the direction of

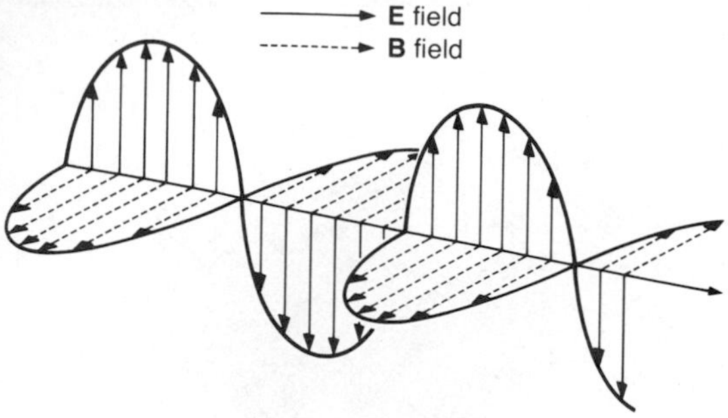

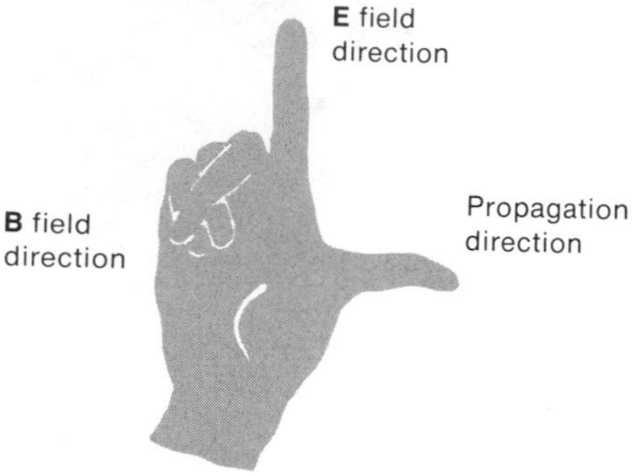

FIGURE 20.11 Relationship between the **E** field, the **B** field, and the direction of propagation of an electromagnetic wave.

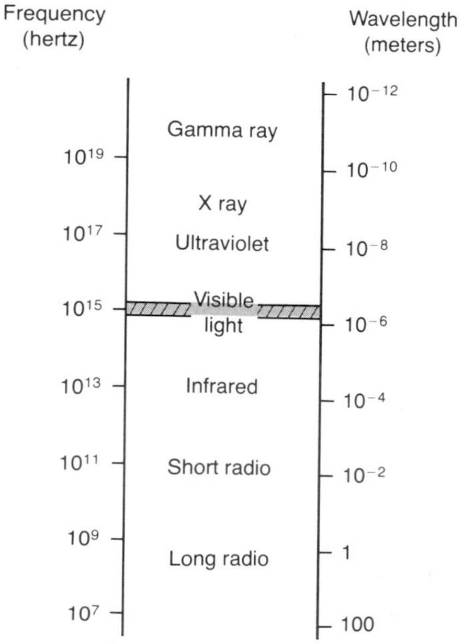

FIGURE 20.12 The known electromagnetic spectrum.

wave propagation as illustrated in Fig. 20.11. The right hand rule relates the three directions. If we let **k** be a vector in the direction of propagation, then (**k, E, B**) form a right hand set as illustrated in Fig. 19.1.

Finally, the electric and magnetic waves are exactly in phase with each other, and the relationship of their magnitudes is constant at each point.

■ Relation Between Electric and Magnetic Field Strengths for an Electromagnetic Wave in Free Space

Electric field strength = velocity of light × magnetic field strength

$$E = cB \qquad \textbf{(20.10)}$$

Maxwell announced his theoretical results in 1865. A quarter of a century later the existence of electromagnetic waves (of the type that we now call radio waves) was demonstrated experimentally by Hertz. The various types of electromagnetic waves and their typical wavelengths and frequencies are shown in Fig. 20.12. They range from very high-frequency gamma rays, through X rays, ultraviolet, visible light, infrared, to very low frequency radio waves.

Recall that wave velocity, wavelength, and frequency are related by Eq. (12.2). Because of the importance of electromagnetic waves in physics, we restate the equation specifically for them.

■ Relation Between Wave Velocity, Wavelength, and Frequency for Electromagnetic Waves in Empty Space

Wave velocity = frequency × wavelength

$$c = \nu\lambda \qquad \textbf{(20.11)}$$

$$\approx 3 \times 10^8 \text{ m/s}$$

The easily remembered value of 3×10^8 m/s for c has an error of less than 0.1%.

EXAMPLE 20.13

A radio station broadcasts at 950 kHz. The maximum electric field strength at a point some miles away is 40 μV/m. Calculate the wavelength of the electromagnetic radiation and the maximum magnetic field strength at the point in question.

SOLUTION We use Eq. (20.11) to calculate the wavelength:

$$c = \nu\lambda$$

$$\therefore \lambda = \frac{c}{\nu}$$

$$= \frac{3 \times 10^8 \text{ m/s}}{950 \times 10^3 \text{ Hz}}$$

$$= 315.8 \text{ m}$$

The maximum electric and magnetic fields are related to each other by Eq. (20.10):

$$E = cB$$

$$\therefore B = \frac{E}{c}$$

$$= \frac{40 \times 10^{-6} \text{ V/m}}{3 \times 10^8 \text{ m/s}}$$

$$= 1.33 \times 10^{-13} \text{ T}$$

■

KEY CONCEPTS

Magnetic flux is the amount of flow of lines of magnetic field strength through a designated area. See Eq. (20.1) for the flux of a uniform magnetic field through a flat area.

Faraday's law of electromagnetic induction equates the induced emf around a circuit to the flux change through the circuit divided by the elapsed time. See Eq. (20.2).

Lenz's law states that the direction of an induced emf is in opposition to the change that caused it.

A simple formula can be used to calculate the emf induced in a straight rod that moves with constant velocity in a uniform magnetic field. See Eq. (20.3).

The **self inductance** of a coil is a measure of the emf induced in it by a given rate of current change. See Eq. (20.5).

An **alternator** consists of a coil of wire rotating in a magnetic field. The peak value of the alternating voltage is equal to the product of the number of turns in the coil, the area of the coil, its angular velocity, and the magnetic field strength. See Eq. (20.7). The angular frequency of the induced emf is equal to the angular velocity of the rotation of the coil.

An isolated magnetic wave or an isolated electric wave is impossible. The two must exist conjointly in an **electromagnetic wave.** The electric field, the magnetic field, and the direction of propagation are mutually perpendicular. The relationship of the three vector directions is given by the right hand rule. See Fig. 20.11.

The **velocity of an electromagnetic wave** in empty space is equal to the square root of the reciprocal of the product of the permittivity and permeability. See Eq. (20.9).

The ratio of the electric field strength to the magnetic field strength for an electromagnetic wave in free space is equal to the velocity of light. See Eq. (20.10).

QUESTIONS FOR THOUGHT

1. What factors influence the emf induced in a straight wire that moves in a constant magnetic field?

2. A circular coil of wire resting with its plane on a horizontal surface carries a current in a clockwise direction when viewed from above. Another coil, with its plane also horizontal, falls vertically down on the first one. What is the sense of the induced emf in the second coil?

3. You are given a steel bar, a length of copper wire, and a sensitive galvanometer. Can you determine whether the steel bar is magnetized and, if so, the direction of the magnetic lines of force?

4. A pendulum bob consists of a flat copper disk that is in the vertical plane of oscillation of the pendulum. If the pole pieces of a strong magnet are placed so that the disk swings through them at the low point of its path, the oscillatory motion of the pendulum stops very quickly. Explain this phenomenon qualitatively.

5. If you were asked to construct a resistor from a piece of resistance wire that is several meters long, how would you arrange the wire so that the resistor had the lowest possible inductance?

6. The alternator that we described had a coil of wire rotating between the pole pieces of a magnet. Could you arrange a stationary coil of wire so that a rotating bar magnet would induce an ac voltage between the ends of the wire?

7. When the starter motor of an automobile is cranked with the headlights on, the lights go very dim at the instant when the starter engages, brighten a little as the starter turns, and achieve full brightness when the engine starts. Explain these changes.

8. How would you classify electromagnetic radiation whose wavelength is about the size of the following objects:

a. the length of a football field
b. the diameter of a basketball
c. the diameter of a toothpick
d. the thickness of a sheet of paper
e. the diameter of an atom

PROBLEMS

A. Single-Substitution Problems

1. Calculate the magnetic flux through a rectangular area 30 cm × 60 cm due to a magnetic field of 0.25 T that is perpendicular to the plane of the area. [20.1]

2. The earth's magnetic field in a certain location is uniform at a value of 60 μT. How large an area (with its plane perpendicular to the field direction) is required for a magnetic flux of 1 Wb? [20.1]

3. The magnetic flux through a loop of wire changes from 0.4 Wb to 0.1 Wb during a time interval of 3 ms. Calculate the average emf induced in the loop. [20.2]

4. The magnetic flux through a coil of wire changes from 0.35 Wb to 0.25 Wb and causes an average induced emf of 25 V. Calculate the time interval for the flux increase. [20.2]

5. A flux change that takes place over a time interval of 12 μs causes an average induced emf of 24 V in a certain coil. Calculate the flux change. [20.2]

6. Two rectangular coils of wire lie in the same plane. The outer coil contains a battery and a switch, and the current is counterclockwise when the switch is closed.

a. Is the induced current in the inner coil clockwise or counterclockwise when the switch is closed?
b. What is the sense of the induced current if the switch is opened? [20.2]

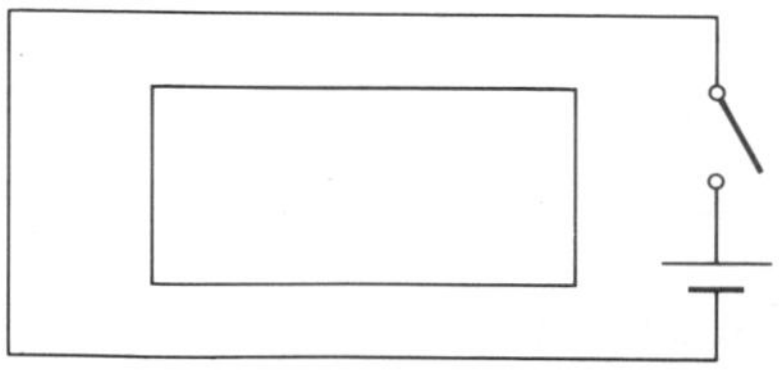

7. A straight wire 20 cm long moves through a uniform magnetic field of 0.06 T with a velocity of 20 m/s. The length of the wire, the direction of the magnetic field, and the velocity of the wire are mutually perpendicular. Calculate the emf induced in the wire. [20.2]

8. A straight wire 45 cm long moves through a uniform magnetic field of 0.35 T. The length of the wire, the direction of the magnetic field, and the velocity of the wire are mutually perpendicular. Calculate the velocity that would produce an induced emf of 8 V. [20.2]

9. A dc motor operating from a 120-V source is producing a counter emf of 100 V. The resistance of the armature coil is 1.4 Ω. Calculate the power output of the motor. [20.3]

10. A dc motor operating from a 200-V source draws a current of 6 A. The resistance of the armature coil is 1.8 Ω. Calculate the counter emf of the motor. [20.3]

11. Calculate the average induced emf in a 250-mH inductor when the current drops from 18 A to 5 A in 3 ms. [20.4]

12. Calculate the self inductance of a solenoid of 800 turns wound on a cylindrical core 60 cm long and 1 cm in radius. [20.4]

13. The current in a circuit changes from zero to 8 A in 5 ms. What value of self inductance is required to cause an average induced emf of 1000 V in the circuit? [20.4]

14. The self inductance of a coil is 125 mH. The current in the coil changes from 2 A to 3.5 A in 15 μs. Calculate the induced emf in the coil. [20.4]

15. Calculate the angular frequency and the period of the 60-Hz alternating emf household supply. [20.5]

16. Calculate the frequency and the angular frequency of an alternating emf that has a period of 1 ms. [20.5]

17. a. The AM radio band has a frequency range of about 530–1600 kHz. Calculate the corresponding range of wavelengths.

b. The FM radio band has a frequency range of about 88–108 MHz. Calculate the corresponding range of wavelengths. [20.6]

18. Calculate the maximum electric field for an electromagnetic wave in free space if the maximum magnetic field is 50 μT. [20.6]

19. A radio wave is traveling horizontally. At a given place and time, the electric field is vertically upward and the magnetic field points northeast. What is the direction of propagation of the radio wave? [20.6]

B. Standard-Level Problems

20. At a certain location the earth's magnetic field is 60 μT and the magnetic flux through a horizontal square area of side 15 cm is 1.25 μWb. Calculate the angle of dip. [20.1]

21. The magnetic flux through a horizontal circular area of radius 20 cm is 7 μWb at a location where the angle of dip is 63°. Calculate the value of the earth's magnetic field. [20.1]

22. A solenoid 60 cm long and 1 cm in radius is uniformly wound with 800 turns of copper wire 1 mm in diameter. The solenoid is connected to a 12-V battery. Calculate the magnetic flux through the interior of the solenoid. The resistivity of copper is 1.72×10^{-8} Ω · m.

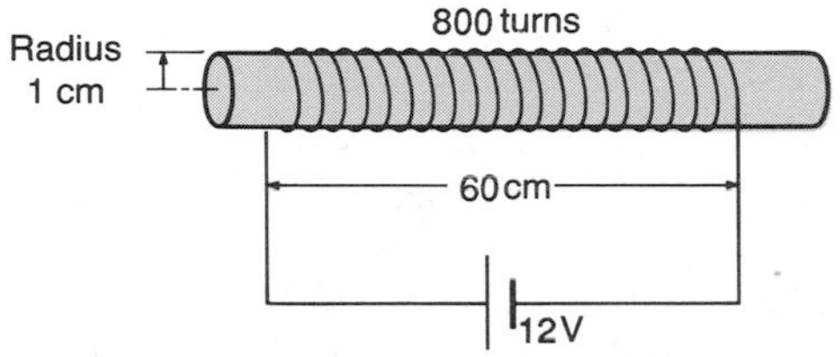

23. A solenoid 60 cm long has a cross-sectional area of 60 cm^2 and carries a current of 8 A in its windings. How many turns are required to produce a magnetic flux of 8 μWb within the solenoid?

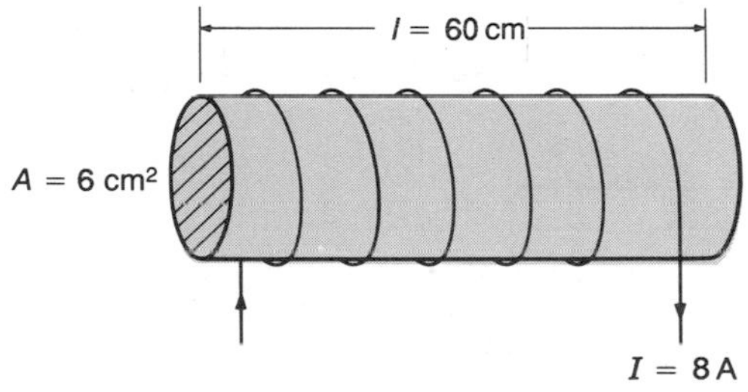

24. A circular loop of tungsten wire has a radius of 28 cm and the radius of the wire is 0.8 mm. A uniform magnetic field perpendicular to the plane of the wire falls to zero in 2.4 ms and causes an average induced current of 1.5 A in the wire. Calculate the original value of the magnetic field. (The resistivity of tungsten is 5.6×10^{-8} Ω · m.) [20.2]

25. A uniform magnetic field of 0.8 mT is perpendicular to the plane of a circular loop of copper wire. The radius of the loop is 50 cm, and the radius of the copper wire is 0.6 mm. Calculate the average current induced in the wire when the magnetic field drops to zero in 3 ms. (The resistivity of copper is 1.72×10^{-8} Ω · m.) [20.2]

26. A square loop of side length 24 cm is made of 80 turns of copper wire 0.4 mm in radius. When the loop is flipped completely around a diameter in 18 ms the average induced current is 0.12 A. Calculate the value of the component of magnetic field perpendicular to the loop. (The resistivity of copper is 1.72×10^{-8} Ω · m.) [20.2]

27. A student performing a measurement on the earth's magnetic field takes a circular coil of 400 turns of wire of radius 18 cm and places the coil with its plane horizontal. When the coil is flipped completely over (about a horizontal diameter) in 0.42 s, the average induced emf is 12 mV.

a. Does this experiment provide the student with the value of the horizontal component, the vertical component, or the magnitude of the earth's magnetic field?
b. What is the magnetic field value found from the experiment?

28. The earth's magnetic field at Washington, D.C., is 57 μT with an angle of dip of 71°. A rectangular conducting coil of wire 2 m × 1 m is hinged along one of the long edges. The axis of the hinge is horizontal and points in the magnetic north-south direction. The coil is held with its plane horizontal and then falls so that its plane is vertical in 0.4 s. Calculate the average emf induced in the coil as it falls.

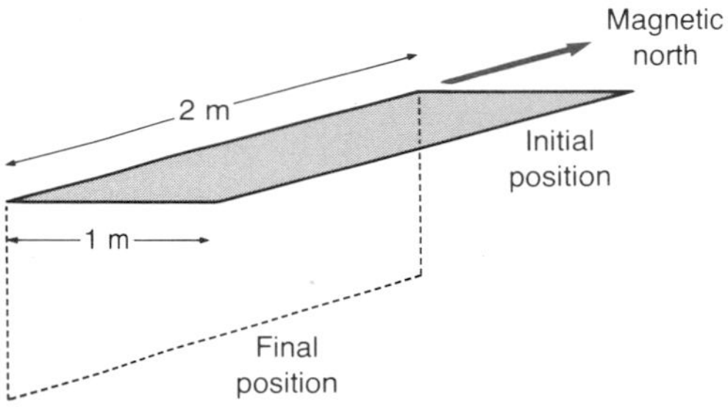

29. At what speed would a metal rod 1 m long have to be moved horizontally (and perpendicularly to its own length) through the earth's magnetic field to generate 1 V between its ends? Take the earth's field at 60 μT and the angle of dip at 60°.

30. An aircraft with a 40-m wingspan is flying horizontally at 800 km/h. The earth's magnetic field is 55 μT and the angle of dip is 62°. Calculate the induced emf between the wing tips.

31. A rectangular coil 40 cm × 30 cm is moved to the right with a steady velocity of 12 m/s (see diagram). The coil has a resistance of 3 Ω. The left end of the coil is in a magnetic field of 0.4 T perpendicular to its plane, and the magnetic field is zero at the right end of the coil.

a. Calculate the magnitude and direction of the induced current in the coil.
b. What force is required to keep the coil moving at its steady velocity?

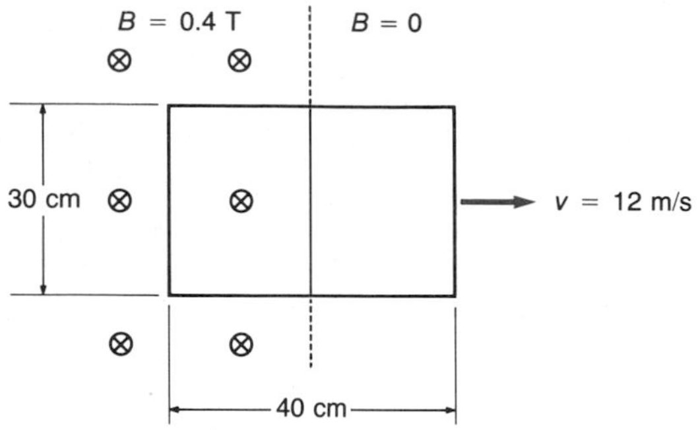

32. A linear generator consists of a pair of wires of negligible resistance 35 cm apart and located in a uniform magnetic field of 0.6 T that is perpendicular to their plane. A conducting rod of resistance of 0.36 Ω is moved along the rails whose ends are connected to a 2.8-Ω resistor.

a. At what speed must the sliding rod move to deliver 20 W of electrical power to the 2.8-Ω resistor?
b. What is the efficiency of the generator operating under these conditions? [20.3]

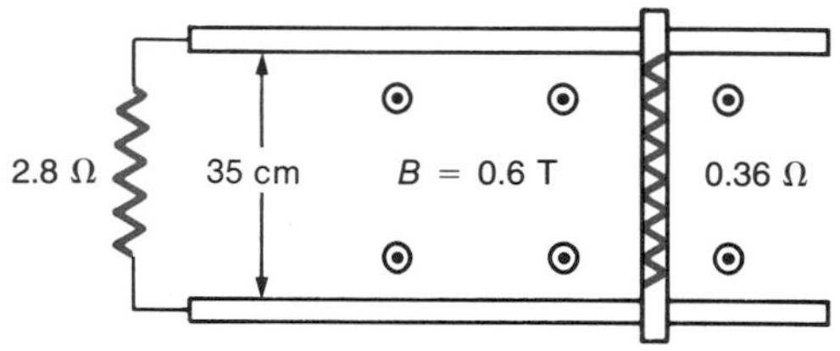

33. A pair of parallel wires of negligible resistance are 20 cm apart and located in a uniform magnetic field of 0.6 T that is perpendicular to their plane. A resistance of 2.5 Ω is connected between the ends of the wires, and a conducting rod of negligible resistance is moved along the loop at 15 m/s.

a. Calculate the magnitude and polarity of the induced current.
b. Calculate the force required to keep the conducting rod moving at its steady velocity. [20.3]

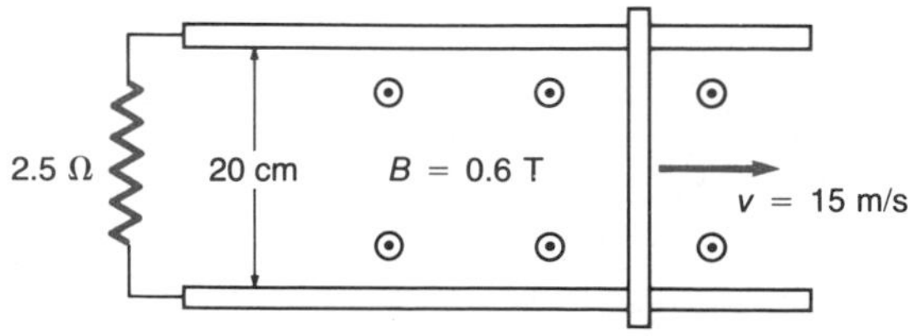

34. A linear motor consists of a pair of parallel wires of negligible resistance 40 cm apart located in a magnetic field of 0.8 T perpendicular to their plane, and having a sliding rod of negligible resistance. The motor is

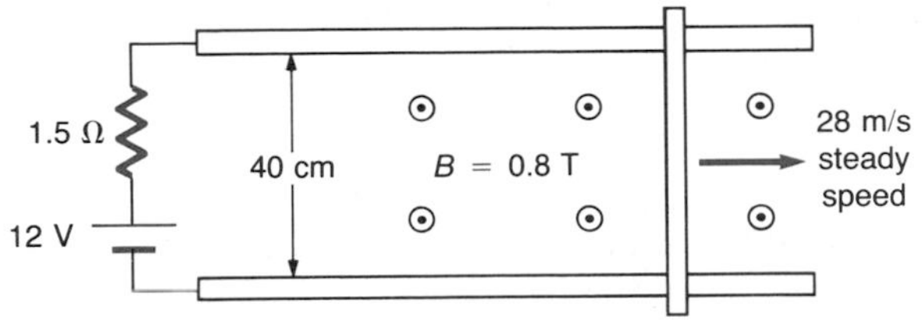

driven by a 12-V battery through a series resistance of 1.5 Ω and the rod ultimately reaches a steady speed of 28 m/s. Calculate:

a. the current before the rod begins to move
b. the counter emf of the motor at steady speed
c. the current at steady speed
d. the power output at steady speed
e. the efficiency of the motor

35. A linear motor consists of a pair of parallel wires of negligible resistance 35 cm apart located in a magnetic field of 0.75 T perpendicular to their plane, and having a rod of negligible resistance. The motor is driven by a 12-V battery through a series resistance of 1.2 Ω, and exerts a force of 1.6 N after it reaches steady speed. Calculate:

a. the current at steady speed
b. the counter emf at steady speed
c. the steady speed of the rod
d. the power output at steady speed
e. the efficiency of the motor

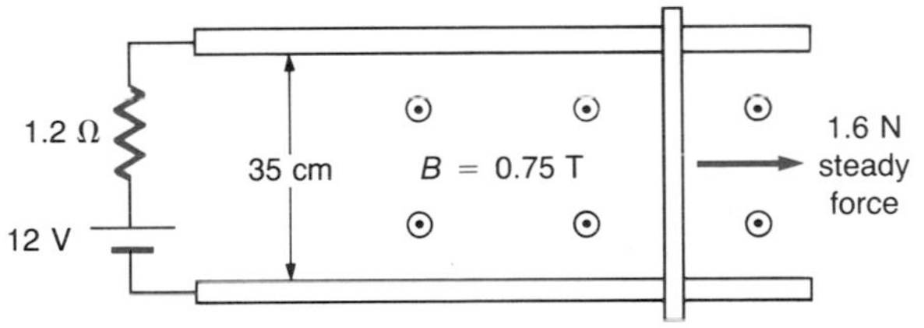

36. An electric motor operates from a 120-V dc source. The armature resistance of the motor is 3 Ω, and the counter emf when the motor is operating at steady speed is 100 V.

a. Calculate the starting current for the motor.
b. Calculate the power drain from the source and the mechanical power output at the steady speed.

37. The starter motor in an automobile has an armature resistance of 0.14 Ω and operates from a 12-V battery. When it is cranking the engine at a steady speed, the current drain from the battery is 50 A.

a. Calculate the initial current drawn by the starter motor.
b. Calculate the counter emf and the mechanical power delivered by the motor at a steady speed.
c. Calculate the power drain on the battery at a steady speed. (Assume that the internal resistance of the battery is negligible, even though this may not be a good assumption.)

38. An electric motor whose armature resistance is 0.6 Ω operates from a 240-V dc supply. The motor is required to lift a mass of 300 kg at a steady speed of 2.5 m/s. Under these conditions calculate:

a. the motor current (take the smaller of the two values)
b. the value of the counter emf
c. the efficiency

39. An air-core solenoid of 1200 turns is wound on a cylinder 65 cm long and 1.5 cm in radius. Calculate the average induced emf if the current in the winding increases from 2 A to 6 A in 3 ms. [20.4]

40. An iron-core solenoid of 800 turns has a cross-sectional area of 5 cm^2. The magnetic field inside the solenoid is 0.6 T when the current is 4 A. The current and the magnetic field fall to zero in a time interval of 10 ms.

a. Calculate the average induced emf.
b. Calculate the self inductance of the solenoid. [20.4]

41. The magnetic field is 0.6 T inside an iron-core solenoid of 8 cm^2 cross-sectional area. A secondary coil of 750 turns is closely wound on the solenoid. Calculate the emf induced in the secondary when the primary current and the magnetic field fall to zero in 3 ms.

42. An air-core solenoid has a primary winding of 1500 turns per meter of length and a cross-sectional area of 5.4 cm^2. How many turns of a secondary winding are required for an average secondary emf of 120 V when the primary current falls from 12 A to 3 A in 1.2 ms?

43. A flat rectangular coil 15 cm × 30 cm rotates at 1400 rev/min in a uniform magnetic field of 0.75 T that is perpendicular to the axis of the coil. How many turns should the coil have to generate a peak voltage of 120 V? [20.5]

44. A flat rectangular coil of 100 turns and area 25 cm × 40 cm rotates at 1800 rev/min. A uniform magnetic field of 0.8 T is perpendicular to the axis of the coil. Calculate the peak voltage and the frequency of the induced emf. [20.5]

45. A flat rectangular coil of 60 turns of area 10 cm × 24 cm rotates in a uniform magnetic field of 0.7 T.

a. At what speed should it be operated to generate a 60-Hz induced emf?
b. What is the peak voltage of the induced emf?

46. An alternator coil consists of 120 turns of area 80 cm × 40 cm in a uniform magnetic field of 0.08 T.

a. At what speed should it be operated to produce a peak emf of 200 V?
b. What is the frequency of the induced emf?

47. A flat rectangular coil 10 cm × 24 cm has 250 turns. An alternating emf of peak value 100 V is produced when the coil rotates at 1400 rev/min in a uniform magnetic field. Calculate the value of the magnetic field strength.

48. A rectangular coil 30 cm × 20 cm having 5000 turns is used to generate an alternating emf by rotation in the earth's magnetic field.

a. How should the axis of the generator be oriented to produce the maximum voltage?
b. At what speed must it be rotated to generate a peak voltage of 1 V if the earth's magnetic field strength is 60 μT?

C. Advanced-Level Problems

49. An engineer wishes to determine the feasibility of operating the lights of railway carriages by using the emf induced in the axles of the train by the earth's magnetic field. For the scheme to work he needs a 40-V output when the train is moving at 25 m/s.

a. How many axles, each 1.6 m long, would need to be put in series to produce this voltage? (Assume that the vertical component of the earth's magnetic field is 50 μT.)
b. List practical disadvantages of this scheme.

50. A solenoid consists of 250 turns of copper wire 2.4 mm in diameter wound on a cardboard cylinder 60 cm long and 3 cm in diameter. What is the potential difference between the ends of the wire if it is carrying a current of 0.1 A that is increasing at a rate of 12 A/s? What is the potential difference if the current is decreasing at the same rate?

51. An instrument to measure wind velocity consists of small metal cups attached to the ends of light rods, each 25 cm long. Assume that the peripheral velocity of the cups is 75% of the wind velocity. A 100-turn rectangular coil of 20 cm × 10 cm is attached to the vertical spindle that supports the cups. What is the peak value of the alternating emf generated by a 120 km/h wind at a place where the earth's magnetic field is 60 μT and the angle of dip is 55°?

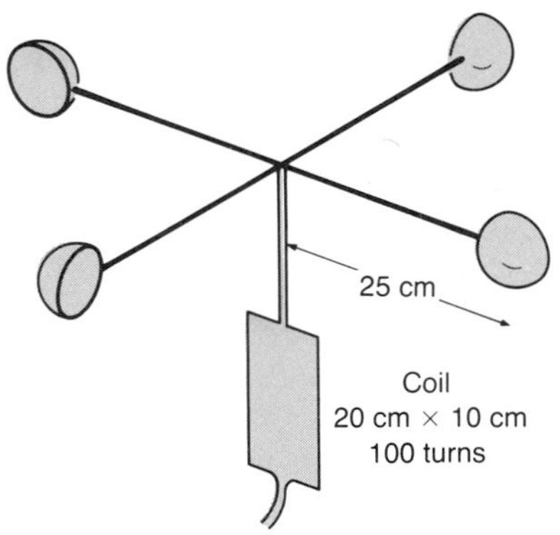

21

ALTERNATING CURRENT CIRCUITS

Most household devices operate on alternating current rather than direct current for reasons related to the technique for transmitting large amounts of electrical power over long distances. Less power is lost when a very high voltage is used for transmission, and is then changed to low household voltage by a transformer. This relatively simple, inexpensive, and efficient device makes possible the widespread distribution and use of electrical power.

Resistors are the chief elements in simple dc circuits, and we will start the study of ac circuits also with resistors. However, inductors and capacitors are also important elements of ac circuits, and we will discuss them as well. The method for analyzing ac circuits is similar in many respects to that used for dc circuits.

21.1 The Resistance Circuit and rms Values

The most familiar source of alternating voltage is the household outlet. When we connect a device that is essentially a resistor to the outlet, we establish a simple ac circuit. An electric toaster is an appliance that behaves as a resistor, and we will begin by analyzing the behavior of such a circuit.

Consider the circuit of Fig. 21.1(a), in which a source of ac voltage is connected to a resistor of resistance R. Let the voltage have peak value V_0 and angular frequency ω. The value of the voltage at any instant is given by Eq. (20.7):

$$v = V_0 \cos \omega t$$

We use the lowercase v for instantaneous voltage and the uppercase with subscript V_0 for peak voltage. To calculate the instantaneous current (lowercase i), we divide the instantaneous voltage by the resistance [compare Eq. (18.2)]:

$$i = \frac{v}{R} = \frac{V_0 \cos \omega t}{R} = I_0 \cos \omega t$$

We use the uppercase with subscript I_0 to denote the peak current, which is equal to V_0/R. The current is an alternating current with the same angular frequency as the alternating voltage. Figure 21.1(b) shows both voltage and current as functions of time. Both have a maximum positive value at the same time, fall to zero at the same time, and achieve a maximum negative value at the same time in every cycle. A current and a voltage that have this relationship are said to be **in phase** with each other. In a simple resistor circuit, the ac current and voltage are in phase.

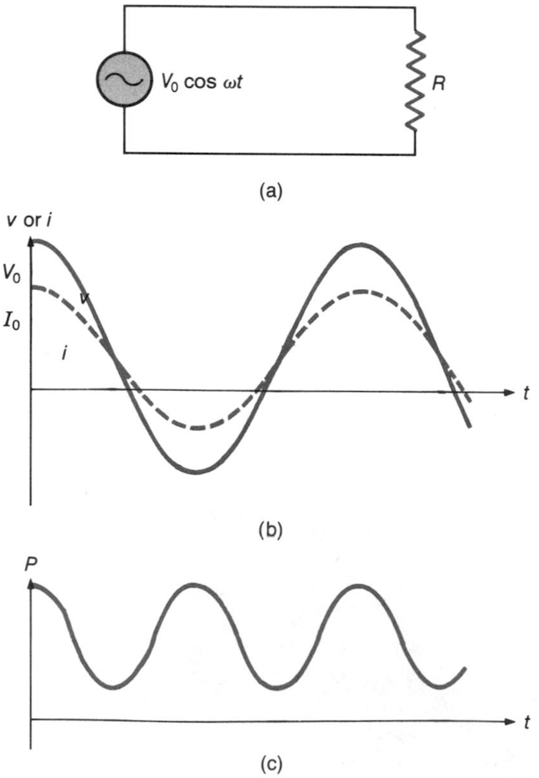

FIGURE 21.1 (a) A source of alternating voltage connected to a resistor; (b) the current and voltage waveforms; (c) the power dissipation in the resistor.

Following the values of voltage and current as they continually change with time is much too cumbersome. Just as in many previous situations that we have encountered in physics, we get a better understanding by dealing with average values. Figure 21.1(b) shows us that the average values for the voltage across the resistor and the current in it are both zero over a long period of time. The current is just as often in one direction as in the other, and the voltage acts just as frequently one way as the other. We are quite correct in saying that the average voltage and the average current in our toaster are zero, but the toaster generates heat to brown the bread, and a pair of zero averages does not explain this result very well. We have come up with the wrong sort of average! To get on the right track we need to consider the Joule heating of the resistor by the alternating current.

Recall from Eq. (18.8) that the power output of a voltage supply at any instant is the product of the current and the voltage. In the case of the electric toaster, the power output goes entirely into Joule heat in the resistor. The instantaneous power dissipation in the resistor is given by:

$$P = iv = I_0 V_0 \cos^2 \omega t$$

Figure 21.1(c) shows the instantaneous power as a function of time. The instantaneous power is never negative, varying from a low of zero (when i and v are simultaneously zero) to a high of $I_0 V_0$ (when i and v have simultaneous peak values). Moreover, the instantaneous power curve is symmetrical between these two extremes, and so the average power lies midway between the extremes of zero and $I_0 V_0$:

$$\langle P \rangle = \tfrac{1}{2} I_0 V_0$$

Recall that the power dissipation in a resistor carrying a current is given by Eq. (18.9a):

$$P = IV$$

We can make the average ac power dissipation formula identical with the dc formula by defining effective current and effective voltage. To accomplish this, we put:

$$I = \frac{I_0}{\sqrt{2}}$$

and

$$V = \frac{V_0}{\sqrt{2}}$$

The average ac power dissipation formula then becomes:

$$\langle P \rangle = IV$$

which is the same as the dc formula.

Formulas for a Resistor in an ac Circuit

The relation between current and voltage is the same as for the dc case:

$$I = \frac{V}{R} \qquad \textbf{(a)}$$

The average Joule heat dissipation in a resistor is the same as for the dc case:

$$\begin{aligned} \langle P \rangle &= IV & \textbf{(21.1)} \\ &= I^2R \qquad \textbf{(b)} \\ &= \frac{V^2}{R} \end{aligned}$$

Alternating currents and voltages are quoted in rms values.

The factor $1/\sqrt{2}$ is not as arbitrary as it might seem. It is the factor that converts a **peak** *value* to a **root-mean-square** (abbreviated "rms") *value* for the special case of sinusoidal functions. According to algebraic convention, the last-mentioned function is performed first. Thus, to calculate the function:

$$\text{root-mean-square } (I_0 \cos \omega t)$$

we first square, then take the mean value, and finally extract the square root. Performing the indicated sequence of operations gives:

$$i = I_0 \cos \omega t$$

Square: $i^2 = I_0^2 \cos^2 \omega t$

Mean: $\langle i^2 \rangle = I_0^2 \times \frac{1}{2}$*

Root: $\sqrt{\langle i^2 \rangle} = I_0 \times \frac{1}{\sqrt{2}}$

*To show that the average value of $\cos^2 \omega t$ is $\frac{1}{2}$, we begin with the trigonometric identity:

$$\cos 2\omega t = 2 \cos^2 \omega t - 1$$

Transposing we have:

$$\cos^2 \omega t = \tfrac{1}{2}(1 + \cos 2\omega t)$$

Since the average value of $\cos 2\omega t$ over many cycles is zero, we have:

$$\langle \cos^2 \omega t \rangle = \tfrac{1}{2}$$

The significant average in ac quantities is the rms—not the mean. Unless we specifically make note to the contrary, *all ac currents and voltages are quoted as rms.* When we refer to the household supply as being 120 V, 60 Hz, we mean 120 V (rms) with a frequency of 60 cycles per second. The uppercase I and V (without subscripts) signify rms values.

The relationship between current and voltage is the same whether we use peak, instantaneous, or rms values. In each case Eq. (18.2) generalizes to $I_0 = V_0/R$, $i = v/R$, or $I = V/R$. All of the equations that describe the function of a resistor in a dc circuit apply to an ac circuit without modifications if we use rms values for the current and voltage.

EXAMPLE 21.1

An electric toaster designed for use with a 120-V ac supply is marked 1250 W. Calculate the current and the resistance of the toaster element.

SOLUTION The current follows immediately from the first of the three alternate forms of Eq. (21.1b):

$$\begin{aligned} I &= \frac{\langle P \rangle}{V} \\ &= \frac{1250\ \text{W}}{120\ \text{V}} \\ &= 10.42\ \text{A} \end{aligned}$$

We can use Eq. (21.1a) to calculate the resistance:

$$\begin{aligned} R &= \frac{V}{I} \\ &= \frac{120\ \text{V}}{10.42\ \text{A}} \\ &= 11.5\ \Omega \end{aligned}$$

The wattage marking on the toaster refers to the situation when the element is hot. Our calculation therefore gives the resistance of the hot element; the cold element has a considerably smaller resistance. It follows that the toaster draws a larger current while it is warming up than when it is hot. ■

21.2 Inductors and Capacitors as ac Circuit Elements

Consider a solenoid consisting of a number of turns of thick conducting wire connected to a source of ac voltage. Let us assume that the solenoid possesses a self inductance but that the resistance of the wire is negligibly small. This device is termed more simply an inductor, and its self inductance determines the value of the alternating current that flows in it.

The circuit is illustrated in Fig. 21.2(a), with the sinusoidal voltage shown as a function of time in Fig. 21.2(b). The current in the inductor has the same frequency as the applied voltage but is *out of phase* with it; that is, the peaks and the zeros of the current and voltage functions do not occur at the same time. The reason is that the changing magnetic flux within the solenoid causes an induced emf that by Lenz's law continually attempts to maintain the status quo. As a result, the sinusoidal current waveform is delayed one-quarter of a cycle behind the voltage waveform. We say that the current in the inductor lags the voltage by one-quarter of a cycle.

For a given voltage, the current is determined by the inductance of the inductor and the angular frequency of the voltage. We can use a dimensional argument to determine the voltage-to-current ratio. To do this we rewrite the inductance unit as follows:

$$\begin{aligned} 1 \text{ henry} &= 1 \text{ volt-second per ampere} \quad \text{from Eq. (20.5)} \\ &= 1 \text{ ohm-second} \quad \text{from Eq. (18.2)} \end{aligned}$$

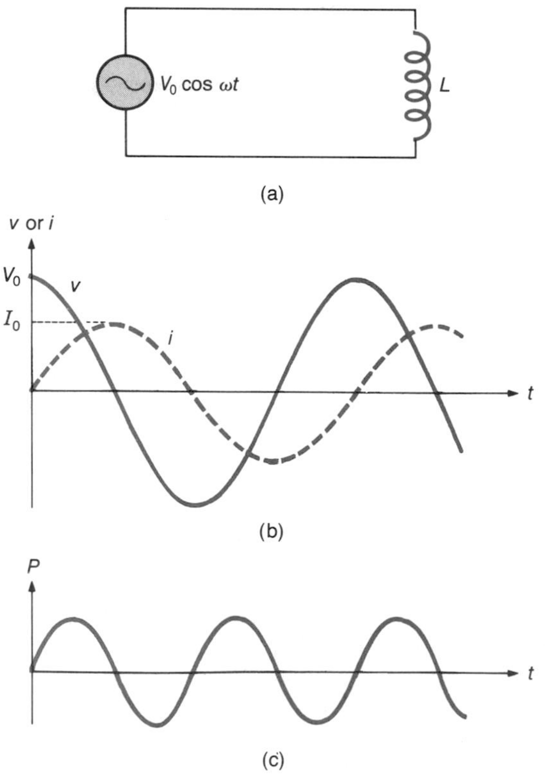

FIGURE 21.2 (a) A source of alternating voltage connected to an inductor; (b) the current and voltage waveforms; (c) the power supplied to the inductor.

Since the angular frequency has the unit of inverse second, the product of inductance L and angular frequency ω has the unit of ohm. We call this quantity the **inductive reactance:**

$$X_L = L\omega$$

The reactance plays a similar role in calculating the current to the role of resistance. Instead of writing $I = V/R$ as in Eq. (21.1a), we write

$$\begin{aligned} I &= \frac{V}{X_L} \\ &= \frac{V}{L\omega} \end{aligned}$$

However, the reactance and resistance are not identical, since the average power dissipation in an inductor is zero. Figure 21.2(c) shows the instantaneous power as a function of time. When the current and voltage are either both positive or both negative, the power is positive, which means that the ac voltage source is supplying energy to the inductor. When the current and the voltage have different signs, the power is negative, which means that the inductor is returning energy to the ac voltage source. No power is dissipated by the inductor, and the average power drain on the source over a long time interval is zero. We can summarize the ac circuit properties of inductors as follows:

■ Formulas for an Inductor in an ac Circuit

The reactance of an inductor to an ac voltage is given by

$$X_L = L\omega \qquad \textbf{(a)}$$

The relation between current and voltage is **(21.2)**

$$I = \frac{V}{X_L} = \frac{V}{L\omega} \qquad \textbf{(b)}$$

The current waveform lags the voltage waveform by one-quarter of a cycle.
The average power consumption is zero.

EXAMPLE 21.2

A 30-mH inductor of negligible resistance is connected to a 120-V, 60-Hz supply. Calculate the current in the inductor.

SOLUTION To calculate the reactance of the inductor, we need to know the angular frequency. From Eq. (20.8b):

$$\nu = \frac{\omega}{2\pi}$$

$$60 \text{ Hz} = \frac{\omega}{2\pi}$$

$$\therefore \omega = 377 \text{ rad/s}$$

The reactance is given by Eq. (21.2a):

$$\begin{aligned} X_L &= L\omega \\ &= 30 \times 10^{-3} \text{ H} \times 377 \text{ rad/s} \\ &= 11.3\ \Omega \end{aligned}$$

Finally, we calculate the rms current from Eq. (21.2b):

$$\begin{aligned} I &= \frac{V}{X_L} \\ &= \frac{120 \text{ V}}{11.3\ \Omega} \\ &= 10.6 \text{ A} \end{aligned}$$

■

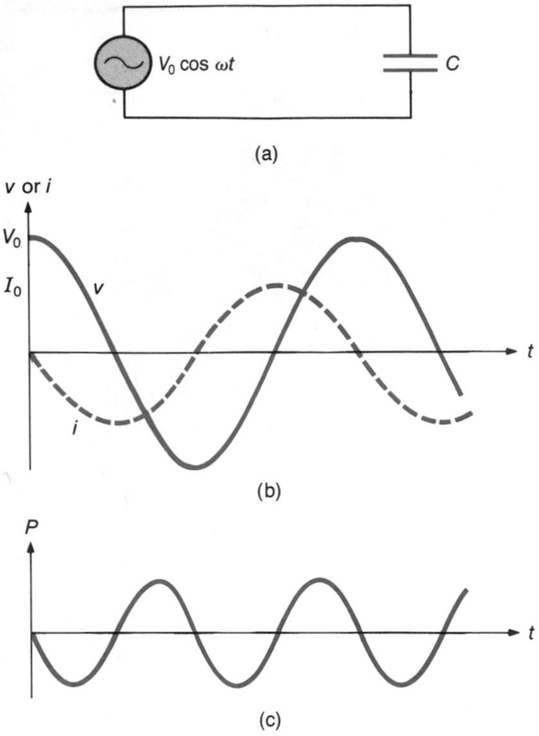

FIGURE 21.3 (a) A source of alternating voltage connected to a capacitor; (b) the current and voltage waveforms; (c) the power supplied to the capacitor.

The capacitor is another important ac circuit element, and our discussion closely follows that for the inductor. Figure 21.3(a) shows a capacitor connected to an ac voltage source. At first sight it seems that current is impossible, since there is no electrical connection across the capacitor plates. However, when one side of the voltage source is positive, the capacitor plate to which it is connected is positively charged, and the other plate is negatively charged. When the sign of the voltage source reverses, the sign of the charge on the capacitor plates must reverse, which entails transport of charge around the circuit. This charge transport occurs just as frequently as the change of sign of the source voltage. We therefore have an alternating current in the circuit whose frequency equals the frequency of the voltage source. A larger capacitor carries a larger charge and hence causes a larger current. Likewise, a larger angular frequency causes a more rapid transfer of charge and also causes a larger current. The sinusoidal current waveform is one-quarter of a cycle in advance of the voltage waveform; that is, the current in a capacitor leads the voltage by one-quarter of a cycle.

To give a dimensional argument for finding the reactance of a capacitor, we begin by rewriting the unit of capacitance:

$$\begin{aligned} 1 \text{ farad} &= 1 \text{ coulomb per volt} \quad \text{from Eq. (17.3)} \\ &= 1 \text{ ampere-second per volt} \quad \text{from Eq. (18.1)} \\ &= 1 \text{ second per ohm} \quad \text{from Eq. (18.2)} \end{aligned}$$

It follows that the product of the capacitance C and the angular frequency ω has the unit of inverse ohm. We call the negative reciprocal of this quantity the capacitive reactance:

$$X_C = \frac{-1}{C\omega}$$

The negative sign is necessary to take care of the fact that the phase relation between current and voltage for capacitors and inductors is different. This point will be clarified in the following section; for the moment we are concerned with magnitudes. The current follows from a relation similar to that for an inductor:

$$\begin{aligned} I &= \frac{V}{X_C} \\ &= -VC\omega \end{aligned}$$

No power is dissipated by the capacitor, and the average power consumption is zero. When the current and the voltage have the same sign, the instantaneous power is positive, and the source is supplying energy to the capacitor. When the signs differ, the instantaneous power is negative, and the capacitor is supplying energy to the source.

■ Formulas for a Capacitor in an ac Circuit

The reactance of a capacitor to an ac voltage is given by

$$X_C = \frac{-1}{C\omega} \quad \textbf{(a)}$$

The relation between current and voltage is: **(21.3)**

$$I = \frac{V}{X_C} = -VC\omega \quad \textbf{(b)}$$

The current waveform leads the voltage waveform by one-quarter of a cycle.
The average power consumption is zero.

EXAMPLE 21.3

A 120-μF capacitor is connected to a 120-V, 60-Hz supply. Calculate the magnitude of the current.

SOLUTION Just as in Example 21.2, the angular frequency of the supply is 377 rad/s. The capacitor reactance is given by Eq. (21.3a):

$$\begin{aligned} X_C &= \frac{-1}{C\omega} \\ &= \frac{-1}{120 \times 10^{-6}\ \text{F} \times 377\ \text{rad/s}} \\ &= -22.1\ \Omega \end{aligned}$$

Since we need only the magnitude of the current, we use Eq. (21.3a) modified for magnitudes:

$$\begin{aligned} I &= \frac{V}{|X_C|} \\ &= \frac{120\ \text{V}}{22.1\ \Omega} \\ &= 5.43\ \text{A} \end{aligned}$$

■

21.3 Impedance of an ac Circuit

Up to this point we have considered circuits containing either a resistor, an inductor, or a capacitor. The situation is more complicated if we have a circuit built up from series and/or parallel combinations of any or all of these basic components. When an ac voltage is applied to such a circuit, we need the answers to two questions:

1. How is the magnitude of the current related to the magnitude of the voltage?
2. What is the relative phase of the voltage and current? (We shall further explain the meaning of this question below.)

A detailed pursuit of these questions would lead us into the study of steady state ac circuit theory. We propose here to supply some of the general answers, and apply them only to the case of circuit elements in series.

Let us begin with the first question. We have seen that the relation between current and voltage is $I = V/R$ for a resistor, $I = V/X_L$ for an inductor, and $I = V/X_C$ for a capacitor. For a circuit consisting of a combination of these elements, the analogous relation is $I = V/Z$, where Z is the impedance of the circuit. We formalize this in a relation similar to our definition of resistance [Eq. (18.2)] as follows:

Definition of Circuit Impedance

$$\text{Impedance} = \frac{\text{voltage}}{\text{current}}$$

$$Z = \frac{V}{I} \tag{21.4}$$

Since impedance is volts divided by amperes, its SI unit is the ohm.

Impedance is a more generalized concept than resistance or reactance. Resistors have resistance, inductors or capacitors have reactance, and the more general circuit (composed of all three) has impedance. The impedance must be some sort of combination of the individual resistances and reactances.

Before proceeding with the second question, let us be sure of what it means. For a resistor alone, the current and voltage are in phase; this means that the peaks of the sinusoidal current occur at the same time as the peaks of the sinusoidal voltage. For an inductor alone, the current lags the voltage by one-quarter of a cycle; this means that the peaks of the sinusoidal current occur one-quarter of a cycle later than the peaks of the sinusoidal voltage. Since one full cycle of a sinusoidal function is 360°, one-quarter of a cycle is 90°. The inductor current lags the inductor voltage by 90°. Similarly, for a capacitor alone, the current leads the voltage by 90°. We now answer our question by stating, without proof, a basic theorem of circuit theory:

Basic Theorem of Circuit Theory

Any combination of resistors, inductors, and capacitors has an equivalent resistance R_{eq} and an equivalent reactance X_{eq} such that the total impedance Z is given by:

$$Z = \sqrt{R_{eq}^2 + X_{eq}^2} \qquad \textbf{(a)}$$

and the relative phase angle ϕ is given by: **(21.5)**

$$\phi = \arc\tan\left(\frac{X_{eq}}{R_{eq}}\right) \qquad \textbf{(b)}$$

The angle ϕ calculated from Eq. (21.5b) is chosen between $+90°$ and $-90°$. A positive angle is a current lag, and a negative angle is a current lead (relative to the voltage). This shows that the relative phase lies somewhere in between the extreme values of $+90°$ corresponding to the case of a pure inductor, and $-90°$ corresponding to the case of a pure capacitor. If ϕ has a positive value, the current lags, and we say that the circuit is basically inductive; if ϕ is negative, the current leads, and we say that the circuit is basically capacitive.

EXAMPLE 21.4

An ac circuit whose equivalent resistance is 6 Ω and whose equivalent reactance is -12 Ω is connected to a 120-V ac supply. Calculate the current in the circuit and its relative phase angle.

SOLUTION We begin by calculating the impedance from Eq. (21.5a):

$$\begin{aligned} Z &= \sqrt{R_{eq}^2 + X_{eq}^2} \\ &= \sqrt{(6\ \Omega)^2 + (-12\ \Omega)^2} \\ &= 13.42\ \Omega \end{aligned}$$

The current is given by Eq. (21.4):

$$\begin{aligned} I &= \frac{V}{Z} \\ &= \frac{120\ \text{V}}{13.42\ \Omega} \\ &= 8.94\ \text{A} \end{aligned}$$

To find the relative phase, we use Eq. (21.5b):

$$\begin{aligned}\phi &= \arctan\left(\frac{X_{eq}}{R_{eq}}\right)\\ &= \arctan\left(\frac{-12\ \Omega}{6\ \Omega}\right)\\ &= -63.4°\end{aligned}$$

Since the phase angle is negative, the current leads the voltage. ■

A calculation of the type we have just made certainly shows how to find the current in a circuit if the equivalent resistance and reactance are known, but it does not show us how to calculate these quantities from the arrangement of the individual elements in the circuit. This is often a difficult problem; the equivalent resistance and reactance of even fairly simple circuits can both be very complicated functions of the individual resistances and reactances of the circuit elements. However, we can give an easy answer for the case of a resistor, an inductor, and a capacitor in series as shown in Fig. 21.4. This is known as a series RLC circuit, and we state the equivalent resistance and reactance without proof.

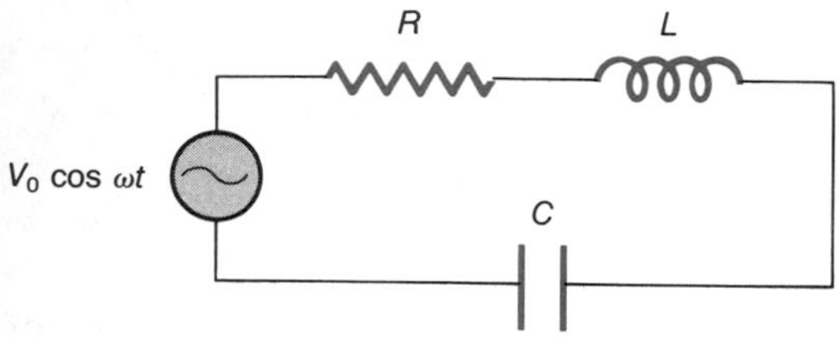

FIGURE 21.4 A series RLC circuit—that is, a resistor, an inductor, and a capacitor connected in series.

■ Equivalent Resistance and Reactance for a Series RLC Circuit

Equivalent resistance = resistance of resistor

$$R_{eq} = R$$

Equivalent reactance = sum of the inductive and capacitive reactances

$$X_{eq} = X_L + X_C \qquad \textbf{(21.6)}$$

It is certainly not unreasonable that the equivalent quantities for a series circuit be given by these simple results. We emphasize that the series circuit is the only one that permits such a simple expression for the equivalent resistances and reactances. Let us consider an example of a resistor in series with one reactive component.

EXAMPLE 21.5

A 30-μF capacitor and an 8-Ω resistor are connected in series to a 50-V, 1000-Hz ac supply. Calculate the magnitude and phase angle of the current.

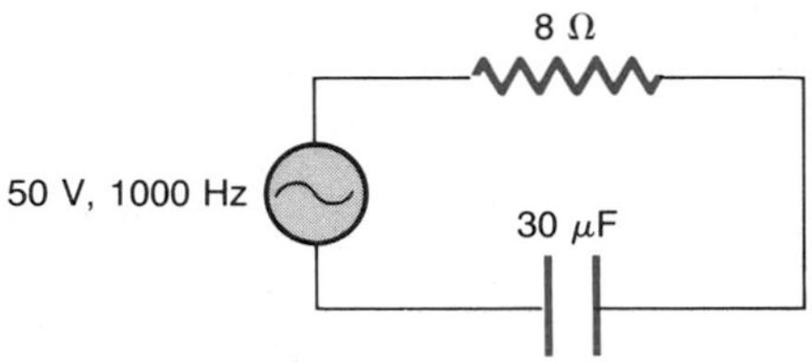

SOLUTION We begin by finding the reactance of the capacitor, which requires that we first find the angular frequency from Eq. (20.8b):

$$\begin{aligned}\omega &= 2\pi\nu \\ &= 2\pi \times 1000 \text{ Hz} \\ &= 6283 \text{ rad/s}\end{aligned}$$

The capacitive reactance is given by Eq. (21.3a):

$$\begin{aligned}X_C &= \frac{-1}{C\omega} \\ &= \frac{-1}{30 \times 10^{-6} \text{ F} \times 6283 \text{ rad/s}} \\ &= -5.305 \ \Omega\end{aligned}$$

The equivalent reactance of the series circuit is given by Eq. (21.6):

$$\begin{aligned}X_{eq} &= X_L + X_C \\ &= -5.305 \ \Omega\end{aligned}$$

and the equivalent resistance is simply the 8-Ω resistance. The circuit impedance follows from Eq. (21.5a):

$$\begin{aligned}Z &= \sqrt{R_{eq}^2 + X_{eq}^2} \\ &= \sqrt{(8 \ \Omega)^2 + (-5.305 \ \Omega)^2} \\ &= 9.600 \ \Omega\end{aligned}$$

The current is then given by Eq. (21.4):

$$\begin{aligned}I &= \frac{V}{Z} \\ &= \frac{50 \text{ V}}{9.600 \ \Omega} \\ &= 5.21 \text{ A}\end{aligned}$$

The relative phase is given by Eq. (21.5b):

$$\begin{aligned}\phi &= \text{arc tan}\left(\frac{X_{eq}}{R_{eq}}\right) \\ &= \text{arc tan}\left(\frac{-5.305 \ \Omega}{8 \ \Omega}\right) \\ &= -33.5^\circ\end{aligned}$$

Since the relative phase angle is negative, the current leads the voltage and (not surprisingly) the circuit is basically capacitive. ■

EXAMPLE 21.6

A 1.5-mH inductor is added in series with the two circuit elements of Example 21.5. Calculate the magnitude and phase of the current with the same ac voltage source.

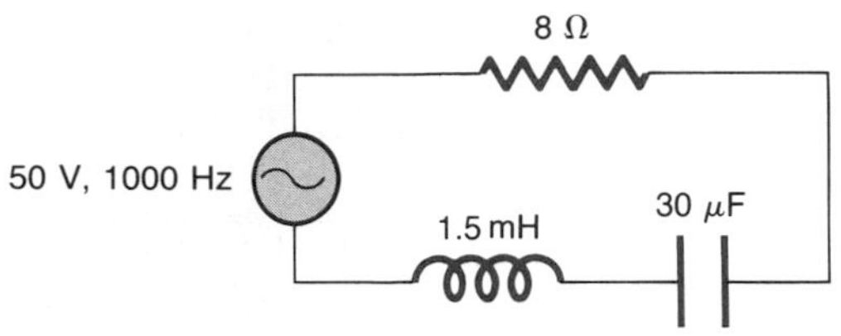

SOLUTION Since the ac voltage is the same as in the previous example, the angular frequency is again 6283 rad/s. We calculate the inductive reactance from Eq. (21.2a):

$$\begin{aligned} X_L &= L\omega \\ &= 1.5 \times 10^{-3}\ \text{H} \times 6283\ \text{rad/s} \\ &= 9.424\ \Omega \end{aligned}$$

The capacitor is the same as that of Example 21.5, so that its reactance is again given by:

$$X_C = -5.305\ \Omega$$

The equivalent reactance of the series circuit is given by Eq. (21.6):

$$\begin{aligned} X_{eq} &= X_L + X_C \\ &= 9.424\ \Omega - 5.305\ \Omega \\ &= 4.119\ \Omega \end{aligned}$$

The circuit impedance follows from Eq. (21.5a):

$$\begin{aligned} Z &= \sqrt{R_{eq}^2 + X_{eq}^2} \\ &= \sqrt{(8\ \Omega)^2 + (4.119\ \Omega)^2} \\ &= 8.998\ \Omega \end{aligned}$$

The current is given by Eq. (21.4):

$$\begin{aligned} I &= \frac{V}{Z} \\ &= \frac{50\ \text{V}}{8.998\ \Omega} \\ &= 5.56\ \text{A} \end{aligned}$$

and the relative phase angle by Eq. (21.5b):

$$\phi = \arc\tan\left(\frac{X_{eq}}{R_{eq}}\right)$$

$$= \arc\tan\left(\frac{4.119\ \Omega}{8\ \Omega}\right)$$

$$= 27.2°$$

The positive relative phase angle means that the current lags the voltage. Although this circuit contains both inductance and capacitance, it is basically an inductive circuit at the frequency considered. ■

The resistors, inductors, and capacitors that we have considered are ideal circuit elements; any real circuit element will surely behave at least a little differently. For example, we might make an inductor by winding insulated copper wire around an iron core. Our work of the previous chapter convinces us that it should function as an inductor, but since it is made from copper wire, it will also have resistance. Intuitively, we feel that it should be a good approximation to represent this real inductor as an ideal inductor in series with an ideal resistor. (The approximation turns out to be well justified provided that the physical dimension of the inductor is very much smaller than the wavelength of an electromagnetic wave of the frequency of the supply.)

EXAMPLE 21.7

An inductance coil carries a current of 40 mA when connected to a 50-V, 1000-Hz supply; when connected to a 12-V battery the current is 300 mA. Calculate:

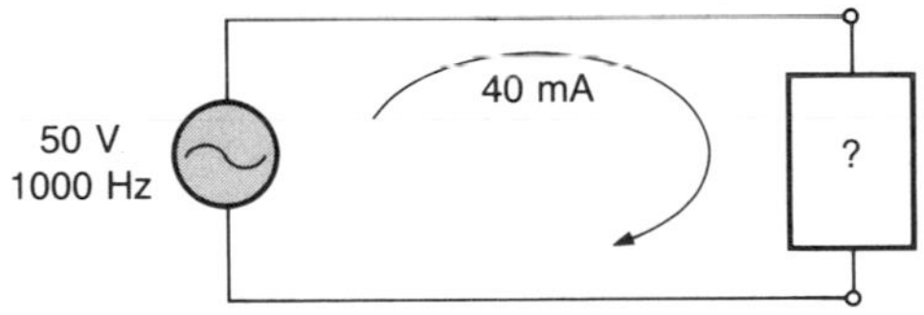

a. The values for the ideal resistor and ideal inductor whose series combination would represent this circuit element

b. The relative phase between voltage and current at this frequency

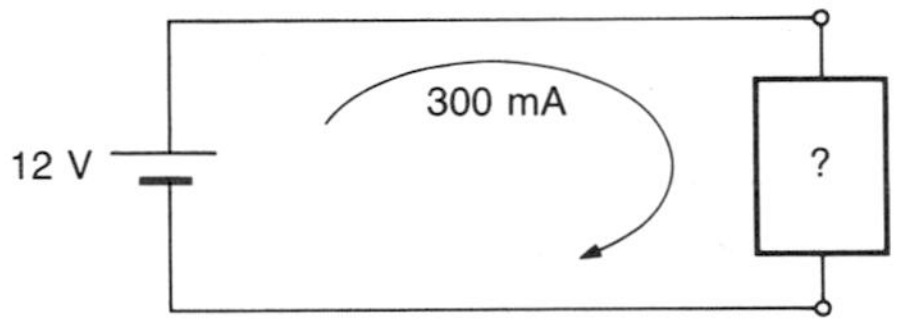

SOLUTION

a. We find the series resistance from the battery connection data. Equation (18.2) gives:

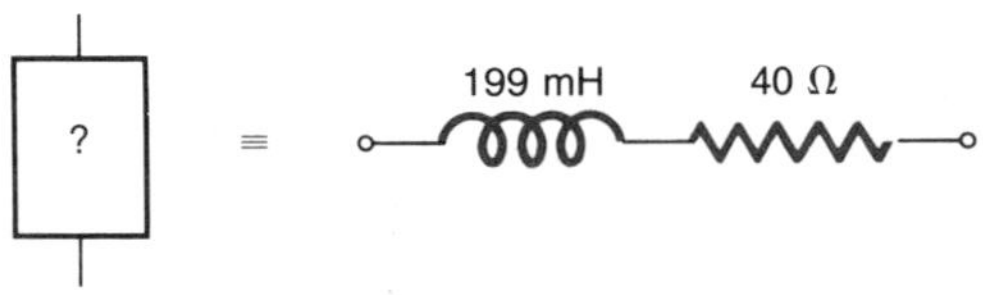

$$R = \frac{V}{I}$$

$$= \frac{12\ \text{V}}{300 \times 10^{-3}\ \text{A}}$$

$$= 40\ \Omega$$

The total impedance is found from the ac supply data using Eq. (21.4):

$$Z = \frac{V}{I}$$
$$= \frac{50 \text{ V}}{40 \times 10^{-3} \text{ A}}$$
$$= 1250\ \Omega$$

We can now use Eq. (21.5a) to find the equivalent inductive reactance:

$$Z = \sqrt{R_{eq}^2 + X_{eq}^2}$$
$$\therefore X_{eq} = \sqrt{Z^2 - R_{eq}^2}$$
$$= \sqrt{(1250\ \Omega)^2 - (40\ \Omega)^2}$$
$$= 1249\ \Omega$$

Since the reactance is due only to an inductor, the equivalent reactance is just the reactance of the inductor, and Eq. (21.2a) gives:

$$L = \frac{X_L}{\omega}$$
$$= \frac{1249\ \Omega}{6283 \text{ rad/s}}$$
$$= 0.199 \text{ H}$$

(The angular frequency corresponding to 1000 Hz is 6283 rad/s as in the two previous examples.) We therefore represent the inductance coil by an ideal inductor of 199 mH in series with a 40-Ω resistor.

b. The relative phase angle is given by Eq. (21.5b):

$$\phi = \text{arc tan}\left(\frac{X_{eq}}{R_{eq}}\right)$$
$$= \text{arc tan}\left(\frac{1249\ \Omega}{40\ \Omega}\right)$$
$$= 88.2^\circ$$

Since this angle is positive, the current lags the voltage by 88.2°. This is very close to a lag angle of 90° for the lag angle of an ideal inductor, and the reactive impedance (1249 Ω) is very close to the total impedance (1250 Ω). For many purposes it may be adequate to represent this coil by a 199-mH ideal inductor so long as we are considering a 1000-Hz supply frequency. At lower frequencies the inductive reactance would decrease and the role of the resistor would be correspondingly greater. It would then be necessary to represent the coil by the ideal inductor in series with the ideal resistor. ■

21.4 Series Resonance

In Example 21.5 we calculated the current in a series combination of a resistor and a capacitor connected to an ac source. In Example 21.6 we calculated the current from the same ac source when an inductor was added to the circuit in series; the current was larger with the additional circuit element in series. The fact that an additional circuit element in series could increase the current (that is, lower the impedance) could be puzzling until we recollect [from Eq. (21.6)] that the capacitor and inductor in a series circuit act in opposition to each other in forming the equivalent reactance of the circuit. Consider now a series RLC circuit connected to an ac supply of fixed potential but variable frequency. We wish to know how the impedance (and consequently the current) changes as we vary the frequency of the ac voltage supply. As the frequency changes, there is no change in R, but $X_L = L\omega$ and $X_C = -1/C\omega$ both change in the manner shown in Fig. 21.5. The equivalent reactance of the circuit can be positive, negative, or zero depending on the values of X_L and X_C. However, Eq. (21.5a) shows that the lowest value of impedance occurs when $X_{eq} = X_L + X_C = 0$. This phenomenon is termed resonance, and the frequency at which it occurs is the resonant frequency of the RLC circuit. The resonant angular frequency, ω_0, is readily determined from the condition:

$$X_L + X_C = L\omega_0 - \frac{1}{C\omega_0} = 0$$

which yields the following result:

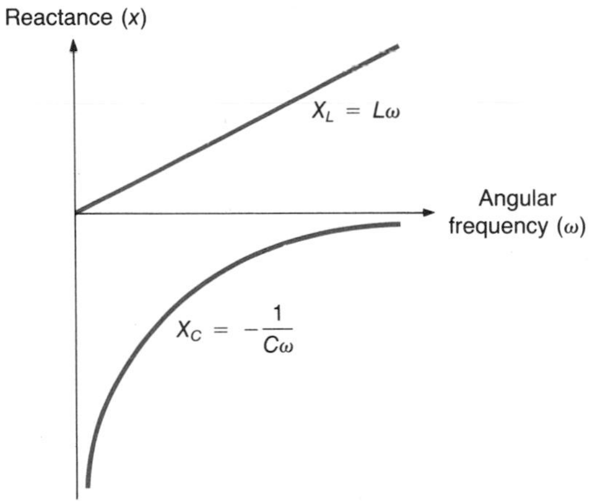

FIGURE 21.5 The graphs of capacitive reactance and inductive reactance as functions of angular frequency.

■ Formula for Resonant Angular Frequency of Series RLC Circuit

$$\omega_0 = \frac{1}{\sqrt{LC}} \qquad \textbf{(21.7)}$$

Since the equivalent reactance is zero at resonance, the impedance of the circuit has a minimum value at resonance, and this value is equal to the circuit resistance. For lower frequencies the impedance is larger, and the circuit is basically capacitive; for higher frequencies the impedance is also larger but basically inductive. This means that the current in a series RLC circuit that is supplied by an ac source of constant voltage but variable frequency has a maximum value at resonance; at frequencies higher or lower than the resonant frequency, the current is smaller than it is at resonance.

EXAMPLE 21.8

A series RLC circuit consists of a 0.1-Ω resistor, a 1.5-mH inductor, and a 30-μF capacitor. The circuit is connected to a 50-V ac supply of variable frequency. Calculate:

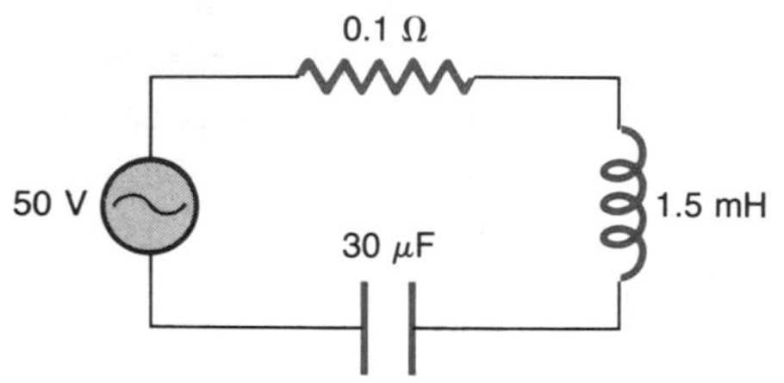

a. the resonant frequency

b. the current at resonance

SOLUTION

a. The resonant angular frequency is given by Eq. (21.7):

$$\omega_0 = \frac{1}{\sqrt{LC}}$$

$$= \frac{1}{(1.5 \times 10^{-3}\ \text{H} \times 30 \times 10^{-6}\ \text{F})^{1/2}}$$

$$= 4714\ \text{rad/s}$$

The resonant frequency is found from Eq. (20.8b):

$$\nu_0 = \frac{\omega_0}{2\pi}$$

$$= \frac{4714\ \text{rad/s}}{2\pi}$$

$$= 750\ \text{Hz}$$

b. At resonance, the equivalent reactance of the circuit is zero, and the impedance is simply the circuit resistance. We calculate the current from Eq. (21.4):

$$I = \frac{V}{Z}$$

$$= \frac{50\ \text{V}}{0.1\ \Omega}$$

$$= 500\ \text{A}$$

For frequencies close to resonance, the equivalent reactance is not zero and it must be included in the calculation of the current. A

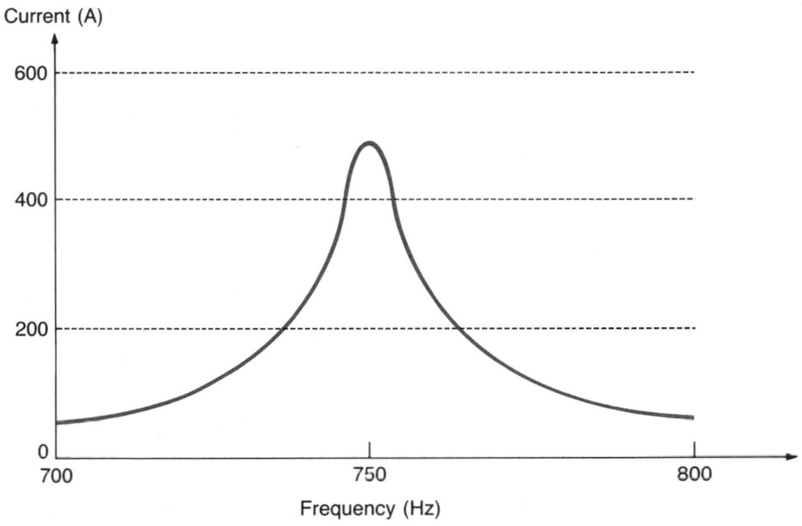

FIGURE 21.6 Computed curve of current vs. frequency for frequencies close to the resonant frequency of the series RLC circuit of Example 21.8.

large number of calculations would be quite laborious, but computed results for frequencies close to resonance are shown in Fig. 21.6. For the particular values of circuit elements chosen for this example, the current rises to a sharp peak at resonance, and falls off quickly to low values on either side of it. ■

EXAMPLE 21.9

Calculate the voltage across the inductor at resonance for the series RLC circuit of Example 21.8.

SOLUTION The current in the circuit at resonance is 500 A, as shown in the previous example. To calculate the voltage across the inductor, we need to know its reactance at the resonant frequency; this is given by Eq. (21.2a):

$$\begin{aligned} X_L &= L\omega_0 \\ &= 1.5 \times 10^{-3}\ \text{H} \times 4714\ \text{rad/s} \\ &= 7.071\ \Omega \end{aligned}$$

(The resonant angular frequency is unchanged from the previous example.) The voltage across the inductor is now given by Eq. (21.2b):

$$\begin{aligned} V_L &= IX_L \\ &= 500\ \text{A} \times 7.071\ \Omega \\ &= 3540\ \text{V} \end{aligned}$$

It is surprising at first sight that a voltage of more than 70 times the supply voltage appears across the inductor at resonance. However, we can make some sense of this by calculating the voltage across the capacitor using Eq. (21.3b):

$$\begin{aligned} V_C &= IX_C \\ &= 500\ \text{A} \times (-7.071\ \Omega) \\ &= -3540\ \text{V} \end{aligned}$$

(The reactance of the capacitor is equal in magnitude but opposite in sign to that of the inductor at resonance.) The sum of the inductive and capacitive voltages is zero; the resulting voltage across the whole circuit is just the voltage across the resistor. The resistor voltage is 50 V, which is equal to the supply voltage (as it must be). The cancellation of the inductor and capacitor voltages can also be looked at from the point of view of phases. Since the elements are in series, the current in each is the same. But inductor current lags inductor voltage by 90°, and capacitor current leads capacitor voltage by 90°. Since the currents are the same, the inductor voltage must lead the capacitor voltage by 180°. This means that the inductor and capacitor voltages are exactly out of phase and cancel each other as implied by the algebraic treatment above. ■

21.5 Power in ac Circuits

An inductor and a capacitor both consume zero power on the average, because they draw power from the ac source and return it periodically. However, a resistor draws energy from the source at an average rate equal to the rate of generation of Joule heat. This means that we have only to calculate the average power dissipation in the resistors to find the average rate of power being supplied by an ac voltage source to the circuit. In fact, if we know the total voltage, the total current, and the relative phase, we do not even have to identify the resistors.

Consider the schematic circuit diagram of Fig. 21.7. An ac generator with voltage output V supplies an unspecified circuit. We assume that the current I and the relative phase angle ϕ are known. By the basic circuit theorem, Eq. (21.5), we can regard the circuit as an equivalent resistance in series with an equivalent reactance. The power dissipation occurs only on the resistance, and is given by Eq. (21.1b):

$$\langle P \rangle = I^2 R_{eq}$$

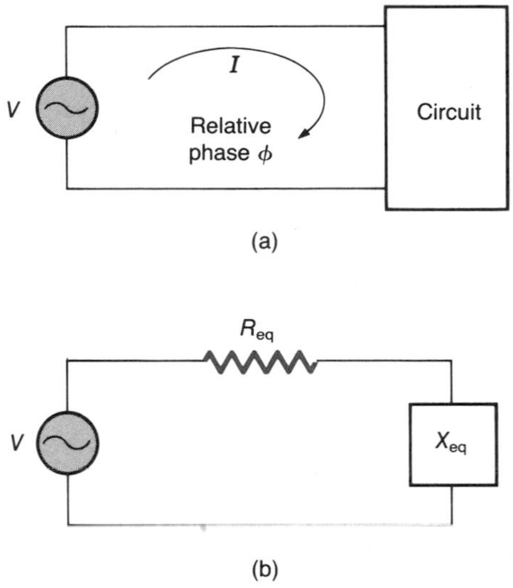

FIGURE 21.7 Establishing an equivalent circuit for power calculations: (a) schematic diagram; (b) equivalent series circuit—the equivalent reactance could be either inductive or capacitive.

But $I = V/Z$ from Eq. (21.4) and

$$\begin{aligned}\langle P\rangle &= \frac{IVR_{eq}}{Z}\\ &= IV\cos\phi^*\end{aligned}$$

This result is quite general for any ac circuit.

■ Formula for the Average Power Supplied to an ac Circuit

Average power = voltage × current × power factor

$$\langle P\rangle = VI\cos\phi \qquad \textbf{(21.8)}$$

The cosine of the relative phase between current and voltage is called the power factor. Its inclusion in Eq. (21.8) is the only difference between it and Eq. (18.8) for a dc circuit. Recall also that voltage and current in Eq. (21.8) are rms values.

EXAMPLE 21.10

An ac supply of 120 V, 60 Hz is connected to a circuit that draws a current of 3 A with a 60° phase lag. Calculate the average power supplied, and construct a possible equivalent circuit.

SOLUTION We can calculate the average power using Eq. (21.8):

$$\begin{aligned}\langle P\rangle &= VI\cos\phi\\ &= 120\text{ V} \times 3\text{ A} \times \cos 60^\circ\\ &= 180\text{ W}\end{aligned}$$

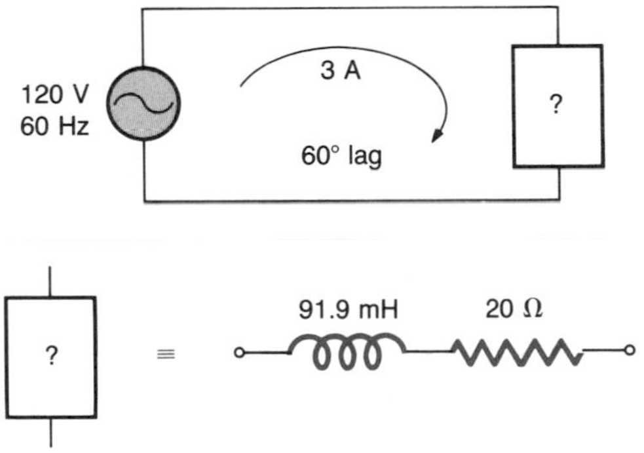

---

*To show that $R_{eq}/Z = \cos\phi$, we begin with the trigonometric identity:

$$\begin{aligned}\cos^2\phi &= \frac{1}{1+\tan^2\phi}\\ &= \frac{1}{1+(X_{eq}/R_{eq})^2} \quad \text{using Eq. (21.5b)}\\ &= \frac{R_{eq}^2}{R_{eq}^2+X_{eq}^2}\\ &= \frac{R_{eq}^2}{Z^2} \quad \text{using Eq. (21.5a)}\end{aligned}$$

and the result follows on taking the square root.

One possible equivalent circuit is the equivalent series resistance and reactance. Since all of the power is dissipated in the equivalent resistor, we must have:

$$\langle P\rangle = I^2R_{eq}$$

$$R_{eq} = \frac{\langle P\rangle}{I^2} = \frac{180\ \text{W}}{(3\ \text{A})^2} = 20\ \Omega$$

To find the equivalent reactance, we use Eq. (21.5b):

$$X_{eq} = R_{eq}\tan\phi = 20\ \Omega \times \tan 60^\circ = 34.62\ \Omega$$

The positive value of the reactance indicates an inductor whose value is given by Eq. (21.2a):

$$L = \frac{X_L}{\omega} = \frac{34.62\ \Omega}{377\ \text{rad/s}} = 0.0919\ \text{H}$$

(The angular frequency corresponding to a frequency of 60 Hz is 377 rad/s, as in Example 21.2.) A possible equivalent circuit at 60 Hz is therefore a 91.9-mH inductor in series with a 20-Ω resistor. ■

If we begin with known values of circuit components, then the current must be calculated before we find the average power. Let us illustrate with an example.

EXAMPLE 21.11

A 40-μF capacitor and a 50-Ω resistor are connected in series across a 120-V, 60-Hz ac supply. Calculate the power supplied to the circuit.

SOLUTION Let us begin by calculating the capacitor reactance; Eq. (21.3a) gives:

$$
\begin{aligned}
X_C &= \frac{-1}{C\omega} \\
&= \frac{-1}{40 \times 10^{-6}\ \text{F} \times 377\ \text{rad/s}} \\
&= -66.3\ \Omega
\end{aligned}
$$

The impedance of the circuit is given by Eq. (21.5a):

$$
\begin{aligned}
Z &= \sqrt{R_{eq}^2 + X_{eq}^2} \\
&= \sqrt{(50\ \Omega)^2 + (-66.3\ \Omega)^2} \\
&= 83.04\ \Omega
\end{aligned}
$$

Equation (21.4) gives the current:

$$
\begin{aligned}
I &= \frac{V}{Z} \\
&= \frac{120\ \text{V}}{83.04\ \Omega} \\
&= 1.445\ \text{A}
\end{aligned}
$$

The final item that we need in order to calculate the power is the phase angle given by Eq. (21.5b):

$$
\begin{aligned}
\phi &= \arctan\left(\frac{X_{eq}}{R_{eq}}\right) \\
&= \arctan\left(\frac{-66.3\ \Omega}{50\ \Omega}\right) \\
&= -52.98^\circ
\end{aligned}
$$

With all required quantities calculated, we can now find the average power from Eq. (21.8):

$$
\begin{aligned}
\langle P\rangle &= VI\cos\phi \\
&= 120\ \text{V} \times 1.445\ \text{A} \times \cos(-52.98^\circ) \\
&= 104\ \text{W}
\end{aligned}
$$

■

21.6 The Transformer and ac Power Transmission

The transformer is the key element in the large-scale distribution of electrical power. An ironic incident occurred while Faraday was discovering the law of electromagnetic induction. A learned soci-

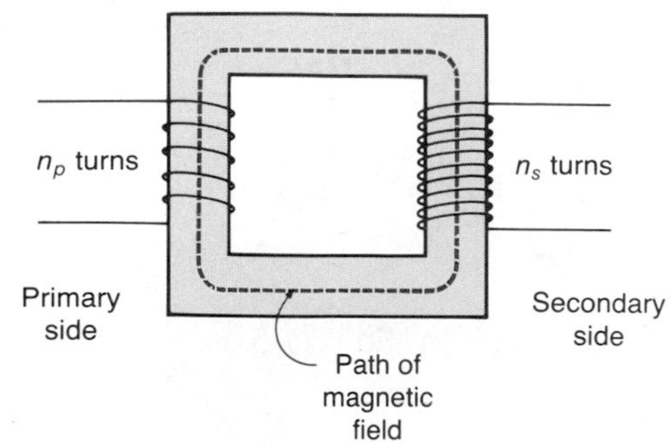

FIGURE 21.8 Diagram of a transformer.

ety awarded a prize medal for an essay that "proved" the practical impossibility of distributing electrical power over long distances. An application of Faraday's law quickly contradicted this "proof."

The **transformer** is a relatively simple, inexpensive, and highly efficient device that changes the voltage of an ac power supply without changing its frequency. The essential parts of a transformer are illustrated in Fig. 21.8. A transformer is a mutual inductor with a primary and a secondary coil that are usually wound on a common continuous iron core. If an ac voltage is connected to the primary coil, the current causes a magnetic flux in the iron core that alternates at the same frequency as the supply voltage. If the magnetic field is wholly within the iron core (a fairly good assumption), then the flux is the same through each turn of both windings. The alternating magnetic flux causes an induced emf in each turn of the primary and secondary coils, which is the same for every turn. The total induced voltage for each coil is therefore proportional to the number of turns. Let V_p and V_s be the primary and secondary voltages, and let n_p and n_s be the numbers of turns on the two coils. The transformer voltage equation then follows.

Equation Relating Primary and Secondary Voltages of a Transformer

$$\text{Voltage ratio} = \text{turns ratio}$$

$$\frac{V_p}{V_s} = \frac{n_p}{n_s} \tag{21.9}$$

When n_s is larger than n_p, the transformer steps the voltage up; if n_s is smaller than n_p, it steps the voltage down.

So long as the secondary coil of a transformer is not connected to a circuit, there is no current in it. Under these conditions the primary coil is approximately a pure inductor, and the current in it lags 90° behind the voltage. However, the situation is quite different if the secondary coil is connected to a circuit that draws a substantial quantity of power, which is the case in most domestic and commercial usages. Consider the circuit of Fig. 21.9, in which a resistor connected across the transformer secondary coil causes a large power dissipation. In this case the current and voltage in both the primary and secondary circuits are approximately in phase according to a detailed transformer theory. With unit power factor in both circuits, Eq. (21.8) gives:

$$\text{Power supplied to resistor in secondary} = V_sI_s$$

$$\text{Power drawn from ac source in primary} = V_pI_p$$

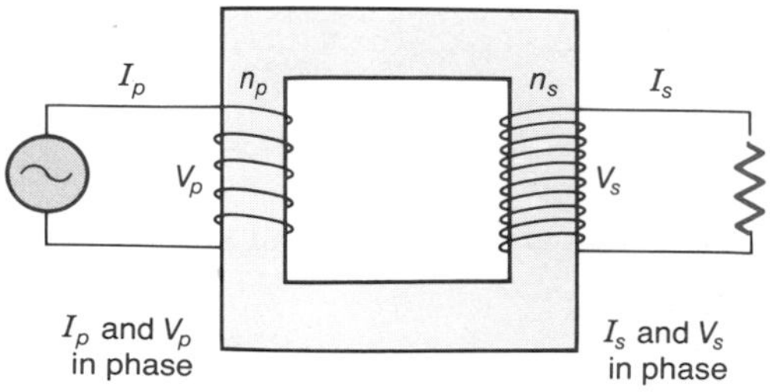

FIGURE 21.9 A transformer connected to a resistor that dissipates a substantial quantity of power.

A well-designed transformer is a very efficient device that wastes only a few percent of the power that it handles. We will ignore this small loss and set the power output of the ac source equal to the power dissipation in the resistor in the secondary circuit. This condition gives:

$$V_p I_p = V_s I_s$$

Using Eq. (21.9) this becomes:

$$n_p I_p = n_s I_s$$

We then have a transformer current equation.

One of the major applications of transformers is in electrical power transmission. A full treatment of electrical power transmission lines is far beyond the scope of this book, but we can do a simple calculation that at least shows the important role of the transformer.

■ Equation Relating Large In-Phase Primary and Secondary Currents in a Transformer

Current ratio = inverse of turns ratio

$$\frac{I_p}{I_s} = \frac{n_s}{n_p} \tag{21.10}$$

EXAMPLE 21.12

A typical electrical power line cable has a resistance of about 1 Ω per 5 mi of cable. An alternator is required to deliver 120 V and 300 A at unit power factor at a point 5 mi from the alternator.

a. Calculate the power loss in the transmission lines and the required output voltage of the alternator.

b. A step-up transformer is used at the alternator end, and a step-down transformer at the load end, so that the power transmission takes place at 60 kV. Calculate the power loss in the transmission lines.

SOLUTION Let us treat the two parts of the problem separately.

a. To calculate the electrical power delivered at the load we use Eq. (21.8):

$$\begin{aligned}\langle P\rangle &= VI\cos\phi \\ &= 120\text{ V} \times 300\text{ A} \times 1 \\ &= 36\text{ kW}\end{aligned}$$

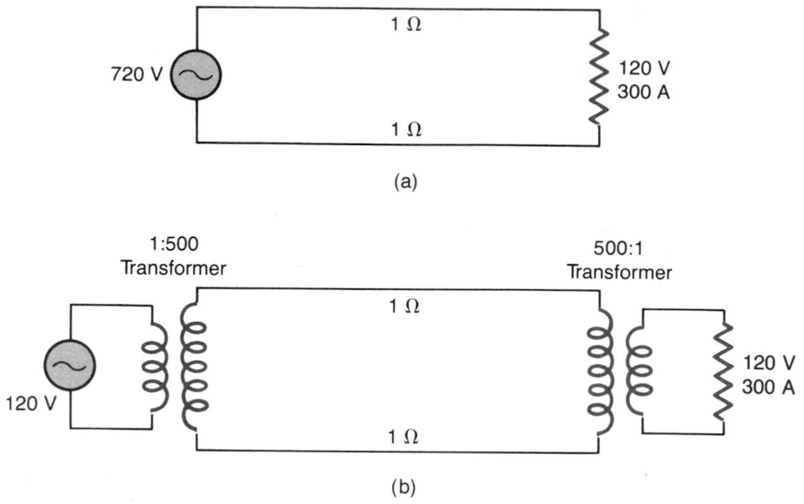

The transmission line conductors each have a resistance of 1 Ω and carry a current of 300 A. The voltage drop along each conductor is given by Eq. (21.1a):

$$\begin{aligned} V &= IR \\ &= 300\text{ A} \times 1\ \Omega \\ &= 300\text{ V} \end{aligned}$$

Since there are two conductors, the total voltage drop between the alternator and the load is 600 V. The output voltage of the alternator must therefore be 720 V to deliver 120 V at the load.

The power loss in each conductor of the line is given by Eq. (21.8):

$$\begin{aligned} \langle P \rangle &= VI \cos \phi \\ &= 300\text{ V} \times 300\text{ A} \times 1 \\ &= 90\text{ kW} \end{aligned}$$

Since there are two lines, the power loss is 180 kW, which is 5 times the amount of power delivered to the load.

b. In the second instance, we use a transformer adjacent to the load that steps down from 60 kV to 120 V. We can calculate the turns ratio from Eq. (21.9):

$$\frac{V_p}{V_s} = \frac{n_p}{n_s}$$

$$\frac{6 \times 10^4\text{ V}}{120\text{ V}} = \frac{n_p}{n_s} = 500$$

To find the current in the transmission line, we use Eq. (21.10):

$$\frac{I_p}{I_s} = \frac{n_s}{n_p}$$

$$\frac{I_p}{300\text{ A}} = \frac{1}{500}$$

$$\therefore I_p = 0.6\text{ A}$$

The voltage drop along each wire of the transmission line is given by Eq. (21.1a):

$$\begin{aligned} V &= IR \\ &= 0.6\text{ A} + 1\ \Omega \\ &= 0.6\text{ V} \end{aligned}$$

This is negligible compared with 60 kV. To produce the 60-kV line voltage requires a 1:500 step-up transformer at the alternator end. The power loss in each conductor of the line is given by Eq. (21.8):

$$\begin{aligned} \langle P \rangle &= VI \cos \phi \\ &= 0.6\text{ V} \times 0.6\text{ A} \times 1 \\ &= 0.36\text{ W} \end{aligned}$$

The two lines together give a 0.72-W power loss, which is also negligible compared with the power of 36 kW delivered to the load. ■

Our treatment of the transmission line wires as resistors in a circuit is a gross oversimplification. Since the wires are adjacent conductors, there is capacitance between them. The current also causes a magnetic flux between the wires that results in inductance in the circuit. A treatment that includes these effects would give a power loss that is several times greater than the one that we calculated, but our calculation at least shows the vital role of the transformer. When a very modest amount of power is delivered over a comparatively short distance without using transformers, an unacceptable amount of power is lost in the transmission lines. When transformers are used, however, the power loss is very small. In actual long-distance power transmission, three phase systems are used with voltages up to several hundred kilovolts.

Transformers are by no means restricted to use in power transmission circuits. They are used wherever a change in ac voltage is necessary. For example, most domestic electrical supply is 120-V alternating current, which would be a dangerously high voltage for operating a miniature train set. The first item in a typical control unit is a transformer that reduces the voltage to about 10 V. A special circuit frequently converts the power to direct current, but the voltage remains at the safe level provided by the transformer.

KEY CONCEPTS

The ac current in a **resistor** is **in phase** with the ac voltage across it; see Fig. 21.1. The relationship between current, voltage, and resistance is the same for the ac case as for dc; see Eq. (21.1a).

The average power dissipation in a resistor is conveniently calculated using **root-mean-square** values of ac current and voltage. The formula is the same as that for the dc case; see Eq. (21.1b).

The alternating current in an **inductor lags** the voltage drop across it by **one-quarter of a cycle;** see Fig. 21.2. The current in the inductor is related to the voltage across it by the inductive reactance; see Eq. (21.2). There is no power dissipation in an inductor.

The alternating current in a **capacitor leads** the voltage drop across it by **one-quarter of a cycle;** see Fig. 21.3. The current in the capacitor is related to the voltage across it by the capacitive reactance; see Eq. (21.3). There is no power dissipation in a capacitor.

The **impedance** of any ac circuit is the ratio of the voltage to the current; see Eq. (21.4).

Any ac circuit is equivalent to a **resistance** and a **reactance in series.** The values of the equivalent resistance and reactance give both the impedance and the relative phase angle; see Eq. (21.5).

For a **series RLC circuit,** the **equivalent resistance** is the value of the resistor, and the **equivalent reactance** is the sum of the reactances of the inductor and the capacitor; see Eq. (21.6).

A series RLC circuit has a **resonant frequency** whose value is the reciprocal square root of the product of the inductance and capacitance; see Eq. (21.7). The series resonant circuit is purely resistive at the resonant frequency.

The **power dissipated** in any ac circuit is the product of the **voltage,** the **current,** and the **power factor.** The power factor is the cosine of the phase angle between current and voltage; see Eq. (21.8).

A **transformer** changes the voltage of an ac power supply, and the **voltage ratio** is equal to the turns ratio; see Eq. (21.9). For large currents at unit power factor, the **current ratio** is the inverse of the turns ratio; see Eq. (21.10).

QUESTIONS FOR THOUGHT

1. Develop an argument based on Faraday's law to show that the sinusoidal current waveform in an inductor lags the sinusoidal voltage waveform.

2. The sinusoidal current waveform in a capacitor leads the sinusoidal voltage waveform. Develop an argument to justify this.

3. When an inductor or a capacitor is connected to any voltage source, the power dissipation is zero. Does this imply that no energy is drawn from the source?

4. A circuit of unknown composition has a positive power factor at a given frequency. Can you predict how the power factor changes with a small change in frequency? Can you say anything about how the power factor changes for a large change in frequency?

5. In what way does the resonant frequency of a series RLC circuit change due to the following:

   a. increased resistance
   b. increased inductance
   c. increased capacitance

6. What is the power factor of a series RLC circuit at resonance?

7. Is it possible for a circuit to have a lagging current at low frequencies, and a leading current at high frequencies? Explain your answer.

PROBLEMS

A. Single-Substitution Problems

1. Calculate the peak voltage that occurs between terminals of the 120-V domestic supply. [21.1]

2. Calculate the value of rms current in a resistor if the peak current is 8 A. [21.1]

3. Calculate the rms current in a 35-Ω resistor that is connected to the 120-V domestic supply. [21.1]

4. What size resistor draws a current of 1 A from the 120-V domestic supply? [21.1]

5. Calculate the reactance of a 1.5-H inductor at (a) 60 Hz, (b) 1 kHz, and (c) 1 MHz. [21.2]

6. Calculate the reactance of a 10-μF capacitor at (a) 60 Hz, (b) 1 kHz, and (c) 1 MHz. [21.2]

7. What size inductor draws a current of 1 A from the 120-V, 60-Hz domestic supply? [21.2]

8. What size capacitor draws a current of 1 A from the 120-V, 60-Hz domestic supply? [21.2]

9. A circuit draws 3.5 A from a 120-V, 60-Hz ac supply. Calculate the impedance of the circuit at the supply frequency. [21.3]

10. Calculate the current drawn from a 20-V, 2-kHz ac supply by a circuit whose impedance is 4.2 Ω at the supply frequency. [21.3]

11. A circuit has equivalent series resistance and reactance of 5 Ω and 12 Ω, respectively, at 60 Hz. Calculate the current and the relative phase angle when it is connected to a 120-V, 60-Hz ac supply. [21.3]

12. A circuit has equivalent series resistance and reactance of 9 Ω and −12 Ω, respectively, at 1 kHz. Calculate the current and the relative phase angle when it is connected to a 40-V, 1-kHz ac supply. [21.3]

13. Calculate the resonant frequency for a 1-mH inductor and a 1-μF capacitor connected in series. [21.4]

14. What size capacitor is required to give a 1-kHz resonant frequency when connected in series with a 3-mH inductor? [21.4]

15. What size inductor is required to give a 5.4-MHz resonant frequency when connected in series with a 0.2-nF capacitor? [21.4]

16. A hot plate has a resistance of 26 Ω when hot and it is producing 1.8 kW. To what supply voltage is the plate connected? [21.5]

17. A 1.2-kW electric heater is connected to a 120-V, 60-Hz ac supply. What current does it draw? [21.5]

18. A 120-V household circuit is protected by a 40-A fuse. How many coffee percolators each requiring 1.1 kW can be operated simultaneously on this circuit? [21.5]

19. An electric bell requires a 10-V, 60-Hz supply. What turns ratio is required on a transformer to operate from a 120-V, 60-Hz ac supply? If the transformer is connected the wrong way around in the circuit, what voltage is supplied to the bell? [21.6]

20. The power supply to a small housing development comes on a 2200-V, 60-Hz transmission line. What turns ratio is required in a transformer to produce 120 V, 60 Hz? [21.6]

B. Standard-Level Problems

21. A 10-Ω resistor carries an rms current of 10 A. Calculate the peak voltage across the resistor. [21.1]

22. Calculate the peak current when a 16-Ω resistor is connected to a 120-V rms ac supply. [21.1]

23. A 35-mH inductor carries a 10-mA current at 250 Hz. What is the peak voltage across the inductor? [21.2]

24. An inductor connected to a 120-V rms, 120-Hz ac supply carries a peak current of 75 mA. Calculate the value of the inductance. [21.2]

25. A 2-μF capacitor is connected to a 180-V peak, 220-Hz ac supply. Calculate the rms current in the capacitor.

26. A capacitor connected to a 120-V, 60-Hz ac supply carries a 100-mA peak current. Calculate the value of the capacitance.

27. A 5-μF capacitor is connected in series with a 30-Ω resistor, and the circuit draws 1.5 A from a 1-kHz ac supply. Calculate the supply voltage and the relative phase angle. [21.3]

28. A 1-mH inductor is connected in series with a 5-Ω resistor. Calculate the current and the relative phase angle when the circuit is connected to a 50-V, 1.5-kHz ac supply. [21.3]

29. A 10-μF capacitor is connected in series with a 200-Ω resistor, and the circuit draws 2.5 A from a 60-Hz ac supply. Calculate the supply voltage and the relative phase angle.

30. A 40-mH inductor is connected in series with a 30-Ω resistor, and the circuit draws 7.5 mA from a 240-Hz ac supply. Calculate the supply voltage and the relative phase angle.

31. A 10-μF capacitor is connected in series with a 15-Ω resistor. Calculate the impedance at (a) 60 Hz, (b) 1 kHz, and (c) 1 MHz.

32. A coil has an inductance of 1.5 H and a resistance of 7500 Ω. Calculate the impedance at (a) 60 Hz, (b) 1 kHz, and (c) 1 MHz.

33. The impedance of a certain coil is 138 Ω at 1 kHz. The inductance of the coil is 20 mH; calculate its resistance.

34. A circuit draws 2.6 A at a leading phase angle of 30° from a 120-V, 60-Hz ac supply. Calculate the equivalent series resistance and reactance at the supply frequency.

35. An RLC circuit consists of a 10-Ω resistor, a 100-mH inductor, and a 40-μF capacitor in series. Calculate the current and the relative phase angle when the circuit is connected to a 120-V, 60-Hz ac supply.

36. An RLC circuit consists of a 15-Ω resistor, an 80-mH inductor, and a 12-μF capacitor in series. The circuit draws 120 mA from a 150-Hz ac supply. Calculate the supply voltage and the relative phase angle.

37. A coil has an inductance of 30 mH and a resistance of 8 Ω.

a. What capacitance is required in series to cause resonance at 1 kHz?
b. What is the current when the circuit is connected to a 50-V, 1-kHz ac supply?
c. What is the peak voltage across the capacitor? [21.4]

38. A series RLC circuit consists of a 2-Ω resistor, a 12-mH inductor, and a 10-μF capacitor.

a. What is the resonant frequency of the circuit?
b. What is the current when the circuit is connected to a 20-V ac supply of resonant frequency?
c. What is the peak voltage across the inductor at resonance? [21.4]

39. A series RLC circuit consists of a 20-Ω resistor, a 100-mH inductor, and a 10-μF capacitor. The circuit is connected to a 25-V ac supply of variable frequency. Calculate:

a. the resonant frequency
b. the current at resonance
c. the current at a frequency of one-tenth of the resonant frequency

40. A series RLC circuit consists of a 10-Ω resistor, a 50-mH inductor, and a capacitor. The circuit is connected to a 50-V ac supply of variable frequency, and the resonant frequency is found to be 20 kHz. Calculate:

a. the value of the capacitance
b. the current at resonance
c. the peak voltage across the capacitor at resonance

41. A certain circuit draws a current of 8 A from a 120-V, 60-Hz ac supply, the current lagging the voltage by 20°. Calculate the average power drain on the supply, and determine the equivalent series resistance and reactance of the circuit. [21.5]

42. A circuit draws 25 A, 2.1 kW from a 120-V, 60-Hz ac supply. Calculate the power factor of the circuit. Determine as much as the data permit of the equivalent series resistance and reactance of the circuit. [21.5]

43. A circuit consisting of a capacitor and a resistor in series draws 50 mA, 150 mW from a 50-V, 3-kHz ac supply. Calculate the values of the resistance and capacitance.

44. An iron-core coil whose resistance is not negligible draws 3 A, 30 W from a 120-V, 60-Hz ac supply. Calculate the resistance and inductance of the coil.

45. An iron-core coil draws a current of 2.8 A from a 120-V, 60-Hz ac supply. When connected across a 12-V battery, it draws a current of 0.6 A. Calculate:

a. the inductance of the coil
b. the power factor of the coil at 60 Hz
c. the average power drain on the ac supply

46. A series RLC circuit consists of a 40-Ω resistor, a 50-mH inductor, and a 75-μF capacitor. Calculate the power factor of the circuit at 100 Hz.

47. A transformer for an electric car set operates from a 120-V, 60-Hz ac supply, and is required to deliver 6 V, 80 W at unit power factor. Calculate the required turns ratio, and the primary current. [21.6]

48. A transformer whose primary winding can safely carry 100 A steps down from 2200 V in the primary to 120 V in the secondary. Calculate the required turns ratio, and the maximum secondary current at unit power factor. [21.6]

C. Advanced-Level Problems

49. A solenoid consists of 2000 turns of copper wire 1 mm in diameter wound on a cardboard cylinder 50 cm long and 2 cm in radius. Calculate the power factor of the solenoid at 3 kHz.

50. A small industrial plant requires 100 kW at 240 V and unit power factor. The power is supplied by an alternator that feeds a 1:8 step-up transformer connected to a two-wire transmission line that has a resistance of 0.6 Ω in each wire. The transmission line is connected to a 20:1 step-down transformer that supplies the factory. Calculate:

a. the power loss in the transmission line
b. the output voltage of the alternator

(Neglect power loss in the transformers as well as effects due to inductance or capacitance in the transmission lines.)

22

RAY OPTICS

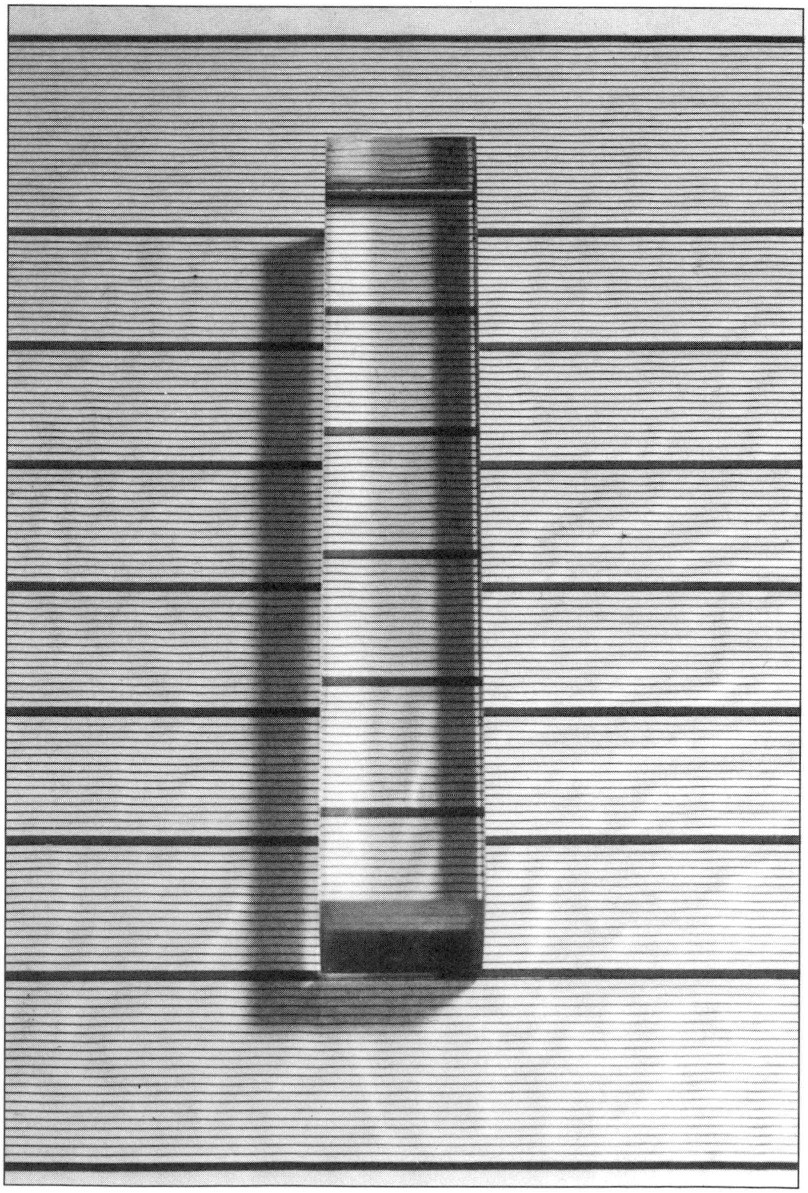

Light has intrigued men since the beginning of history, and there have been abundant speculations about its nature and properties. Ancient thinkers were aware of the finite velocity of sound, noticing that the noise of the blow arrived perceptibly later than the sight of an ax hitting a tree. From similar observational data, they also concluded that the velocity of light was infinite. For instance, the sun rising in the morning seemed to illuminate the whole visible landscape at the same instant.

Apparently, Galileo was the first to propose an experiment to measure the velocity of light. He suggested that two observers should occupy adjacent mountain tops, and use lanterns and shutters in the following way. The first observer, who had a timepiece, was to open the shutter of his lantern at a given instant. The second observer, some kilometers away, was to open his shutter when he saw the light from the first lantern. The original observer was then to note the time at which he saw the light from the second lantern. This should have been later by the time required for light to travel to the distant mountain and back again. However, the speed of light is so great that the experiment yielded only the reaction times of the observers.

We begin by discussing the first successful attempts to measure the velocity of light, since the nature of light was first grasped as a result of knowledge of its velocity.

22.1 The Nature of Light

The Danish astronomer Olaf Romer (1644–1710) made the first quantitative measurement of the velocity of light while working in Paris during the reign of Louis XIV. Galileo had discovered the satellites of Jupiter, and their periods of revolution were accurately known. Romer carefully observed the period of the satellite Io and noticed discrepancies in his data that depended on whether the earth was receding from Jupiter or approaching it. The period of Io is about 42.5 h, and this is the time that should elapse between successive eclipses of the satellite by Jupiter. During this time the earth moves a distance of about 4.54×10^9 m in its orbit. If the earth was in a position such as that marked *A* on Fig. 22.1, then the distance between the earth and Jupiter would change by a negligible amount during successive eclipses of Io. However, if the earth is at position *B*, the light from Io must travel the whole extra distance of 4.54×10^9 m to record the second eclipse. The time for light to travel this distance is about 15 s, so that the measured period of Io would be about 15 s longer with the earth at position *B* than it was at position *A*. With the earth at position *C* the measured period is the same as at position *A*, while at position *D* it is about 15 s shorter since the light now travels a shorter distance between successive eclipses. Romer calculated that the total accumulated time delay between *A* and *C*

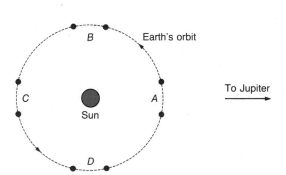

FIGURE 22.1 Diagram illustrating the principle behind Romer's measurement of the velocity of light.

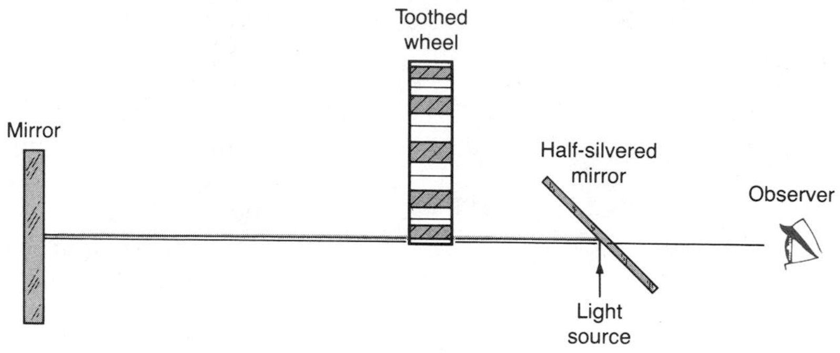

FIGURE 22.2 Diagram illustrating the principle behind Fizeau's measurement of the velocity of light.

was about 22 min (the delay is zero at point *A*, slowly rises to a maximum value at point *B*, and then falls to zero at point *C*). Reasoning that the delay time was in fact the time required for light to travel a distance approximately equal to the diameter of the earth's orbital path around the sun, Romer calculated the velocity of light by dividing the diameter of the earth's orbit by 22 minutes. The value obtained was about 20% low, but it did at least indicate the immensely large magnitude of the velocity of light.

The first measurement of the speed of light here on earth was made about 150 years later by the French physicist Armand Fizeau (1819–1896). A general diagram of his apparatus is shown in Fig. 22.2. A beam of light is directed onto a half-silvered mirror that reflects some of the light and lets some pass through. The reflected light passes through a gap in the rim of a toothed wheel and travels several kilometers to a plane mirror, which reflects the light back through the gap in the toothed wheel to the half-silvered mirror. The portion of the light that comes through the half-silvered mirror is sighted by an observer. Our diagram is greatly simplified—Fizeau certainly used lenses to focus the light in various places. If the toothed wheel begins to rotate, the observer will see nothing if a tooth obscures the light returning from the mirror. If the wheel rotates faster, the returning light may arrive at just the right moment to pass through the next gap around the rim, and the observer once again views the light reflected from the mirror. When this occurs the time taken for the light to travel from the wheel to the mirror and back is exactly equal to the time for a gap to move one space around the circumference of the wheel. Fizeau calculated this time interval by measuring the speed of the wheel when the observer was able to see the reflected light. From this calculation, he found a velocity of 3.12×10^8 m/s for the light beam, compared with the accepted modern value for the velocity of light in free space of about 2.998×10^8 m/s. Since the velocity of light in air differs from the value in free space by about 1/30 of 1%, Fizeau's result was too high by about 4%.

In addition to measuring its velocity, a great deal of experimental work was done to determine the properties and nature of light. Much of Newton's stature derived from his work in optics. He proposed that light was a stream of tiny particles traveling with immense velocity. This concept, known as the corpuscular theory of light, attracted many adherents because of Newton's prestige and reputation. However, an alternate proposal was put forward by one of Newton's contemporaries—the Dutch physicist Christian Huygens (1629–1695), who conceived of light as a wave and was able to explain many of its properties on this basis just as well as they could be explained by the corpuscular theory. Not until the early nineteenth century did the English physician Thomas Young (1773–1829) give clear experimental evidence for the wave nature of light. The final episode in the classical theory of light was written several decades later by Maxwell, who, as we have already seen, identified the physical nature of the light wave as electromagnetic.

We recall from Fig. 20.12 that **visible light** occupies only a tiny portion in the vast electromagnetic spectrum. Of course, it is a very important part of the spectrum for human beings, since we detect visible radiation with our eyes. Moreover, we distinguish the various visible wavelengths and frequencies by their different colors. The wavelengths of radiation that are visible to the human eye range from about 380 nm (violet light) up to about 760 nm (red light). Table 22.1 lists the various colors of the visible spectrum together with the approximate wavelength range that they occupy. The data in the table are subject to personal judgment. For example, there is nothing hard and fast about the wavelength of 550 nm separating green from yellow. Most people would judge the color to be greenish yellow, and some would say it is more one than the other.

TABLE 22.1 Approximate wavelength ranges for light of various colors

Color	*Approximate wavelength range (nm)*
Violet	380–440
Blue	440–470
Green	470–550
Yellow	550–570
Orange	570–620
Red	620–760

The preferred SI unit for the wavelength of visible light is the nanometer. Another frequently used unit is the angstrom (abbreviated "Å"), which is named in honor of Anders Ångström (1814–1874), a Swedish physicist who was famous for his detailed study of the light emitted by the aurora borealis. The angstrom is exactly 10^{-10} m, and it is especially convenient for describing molecular structure, since the atoms that constitute molecules are only a few angstroms apart.

Light is able to propagate through certain materials. It cannot pass through any metal, but it can pass through some dielectrics. The materials capable of propagating light are aptly classified as transparent dielectrics, and they include air, water, glass, polyethylene, and many other substances.

When light travels in a transparent medium such as air or glass, the velocity is always lower than the free space figure quoted in Eq. (20.9). The ratio of the free space velocity to the velocity in the medium is called the **refractive index** of the medium.

■ Definition of Refractive Index of a Medium

$$\text{Refractive index} = \frac{\text{velocity of light in free space}}{\text{velocity of light in a medium}}$$

$$n = \frac{c}{v} \tag{22.1}$$

TABLE 22.2 Refractive indexes of various materials

Material	*Refractive index*
Air at STP	1.0003
Benzene	1.50
Carbon disulfide	1.63
Crown glass	1.52
Diamond	2.42
Ethanol	1.36
Flint glass	1.66
Fused quartz	1.46
Ice	1.31
Methanol	1.33
Plexiglas	1.51
Polyethylene	1.52
Polystyrene	1.59
Sodium chloride	1.53
Water	1.33

Note: The values quoted are averages over the range of wavelengths of visible light.

The refractive index depends on the medium in question and on the wavelength of the light. However, the variation of the refractive index of light through the visible region is no more than a few percent for most materials. Average values of the refractive index are shown in Table 22.2 for a selection of transparent dielectrics.

EXAMPLE 22.1

Green light of wavelength 500 nm in free space enters a slab of crown glass. Calculate the wavelength in the crown glass.

SOLUTION The frequency of the light does not change as it crosses from one medium to another. We can calculate this frequency by using Eq. (20.11) for travel in free space:

$$\begin{aligned} \nu &= \frac{c}{\lambda} \\ &= \frac{3 \times 10^8 \text{ m/s}}{500 \times 10^{-9} \text{ m}} \\ &= 6 \times 10^{14} \text{ Hz} \end{aligned}$$

The velocity of the light in crown glass can be found from Eq. (22.1):

$$\begin{aligned} v &= \frac{c}{n} \\ &= \frac{3 \times 10^8 \text{ m/s}}{1.52} \\ &= 1.97 \times 10^8 \text{ m/s} \end{aligned}$$

To find the wavelength in the crown glass, we use Eq. (20.11):

$$\lambda_{\text{glass}} = \frac{v}{\nu} = \frac{1.97 \times 10^8 \text{ m/s}}{6 \times 10^{14} \text{ Hz}} = 328 \text{ nm}$$

■

22.2 Reflection and Refraction—Snell's Law

We started this chapter on ray optics with a section explaining the wave nature of light. However, a little thought will convince us that many optics problems seem to suggest a ray approach. For example, everyday experience seems to tell us that light cannot travel around corners as sound waves do. In order to relate rays and waves to each other, consider the disturbance caused by a small stone dropped into a still pond of water. The waves move out in concentric circles that are centered on the point where the stone fell, as illustrated in Fig. 22.3(a), where the equispaced concentric circles represent the positions of wave crests at a given time instant. These circles, which we call *wave fronts,* spread out from the center as the crests advance and form circles of larger radius. (We need not use crests to define wave fronts—troughs or any other constant phase positions will do just as well. However, since it is easy to visualize the wave fronts as crests, we will continue to choose crests as the constant phase positions.)

We can describe the situation in another way by drawing a series of straight lines radiating from the center. A few of these lines, which are called *rays,* are illustrated in Fig. 22.3(a). The rays are normal to the wave fronts at all points of intersection.

This beautiful photograph shows ripples with their associated wave crests spreading from a central point of disturbance in water. The same thing is illustrated in an idealized way by the sketch of Fig. 22.3(a). (Fundamental Photographs, New York)

Figure 22.3(b) illustrates an optical analog to the stone dropped into a still pond. A small source of light, such as a candle flame, sends out rays of light in all directions. We can easily visualize the path of the rays by placing a screen with a large hole between the candle and a wall. The bright patch on the otherwise dark wall determines the path of the rays that pass through the hole in the screen. The wave fronts are spherical surfaces that spread outward from the central point occupied by the candle flame.

A *plane wave,* illustrated in Fig. 22.3(c), is a special type of wave whose wave fronts are equispaced parallel planes and whose rays are parallel straight lines. The circular and spherical wave fronts of Figs. 22.3(a) and 22.3(b) become approximately plane if we are a great distance from the central point where the disturbance originates.

The behavior of a plane wave that falls on the plane boundary interface between two different media is basic to the study of ray optics. Let us study this problem by using both ray and wave front diagrams. The ray diagram is shown in Fig. 22.4. A ray of

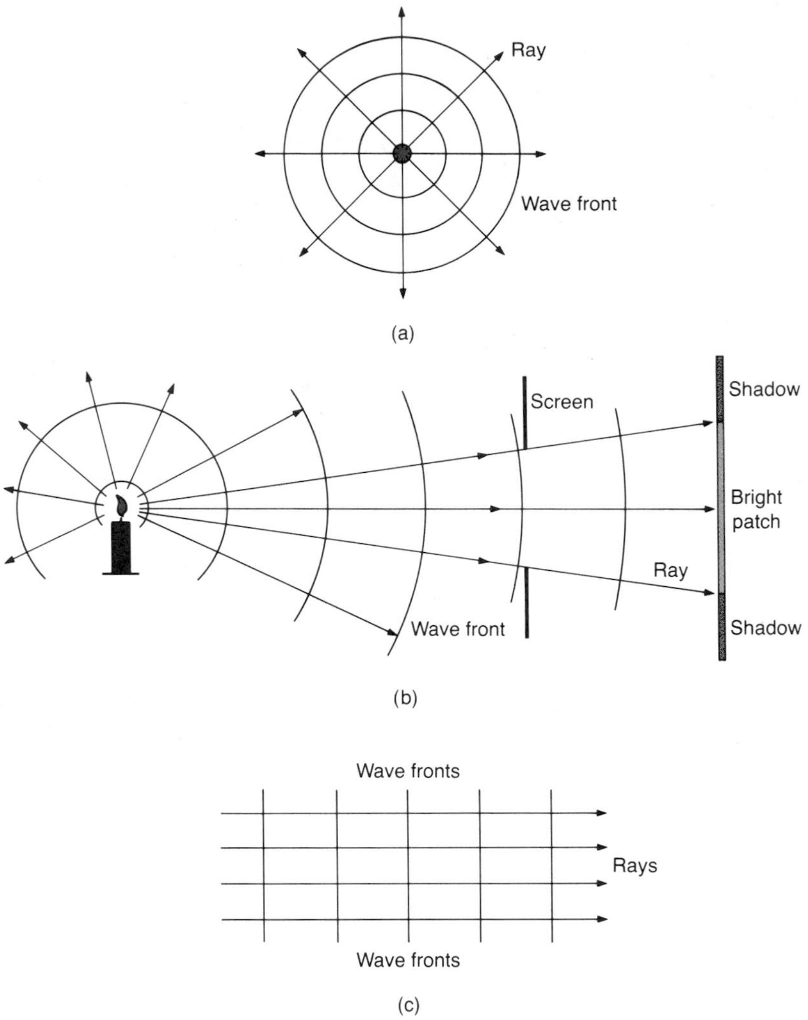

FIGURE 22.3 Illustration of the concept of wave fronts and rays for the following systems: (a) a disturbance on a still water surface spreading from a central point; (b) a candle flame; (c) a plane wave.

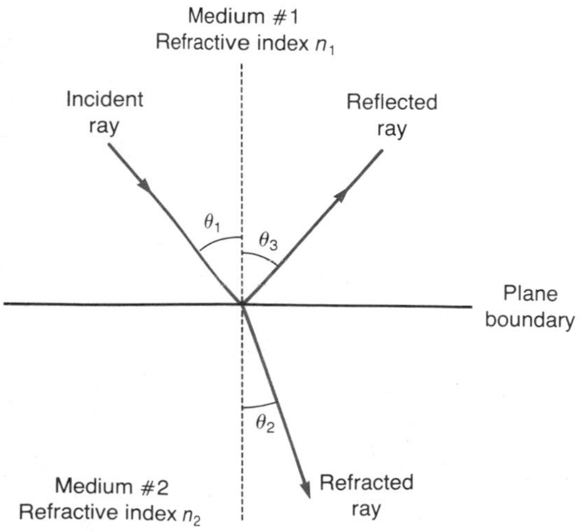

FIGURE 22.4 Reflection and refraction of a light ray that is incident on the plane boundary separating two transparent media.

light that we call the **incident ray** falls on the boundary between the two media, where it is broken into two parts. One is the **reflected ray,** which travels in the same medium as the incident ray. The other is the **refracted ray,** which travels in the second medium. To specify the paths of the three rays, we need to know three angles that fix their direction, which are called the angles of incidence, reflection, and refraction. *All angles are measured between the ray path and the normal to the plane boundary* between the two media.

Let the incident ray traveling in medium #1 (index of refraction n_1) have an angle of incidence θ_1 on the plane boundary. Let the angle of reflection (also in medium #1) be θ_3. Let the angle of refraction of the refracted ray, which travels in medium #2, be

θ_2. Using this notation, we will calculate the paths of the reflected and refracted rays with the help of the **law of reflection** and **Snell's law of refraction.**

Law of Reflection

Angle of incidence = angle of reflection

$$\theta_1 = \theta_3 \qquad \textbf{(22.2)}$$

(Notation is described above.)

Now let us consider reflection and refraction from a wave front point of view. From our definition of the index of refraction [Eq. (22.1)], the wave velocity in medium #1 is given by $v_1 = c/n_1$, and in medium #2 by $v_2 = c/n_2$. The wave frequency ν does not change as the wave passes from one medium to the other. It follows that the wavelengths in the two media are given by:

Snell's Law of Refraction

Refractive index of medium #1 × sine of angle of incidence = refractive index of medium #2 × sine of angle of refraction

$$n_1 \sin \theta_1 = n_2 \sin \theta_2 \qquad \textbf{(22.3)}$$

$$\lambda_1 = \frac{v_1}{\nu} = \frac{c}{n_1 \nu}$$

and

$$\lambda_2 = \frac{v_2}{\nu} = \frac{c}{n_2 \nu}$$

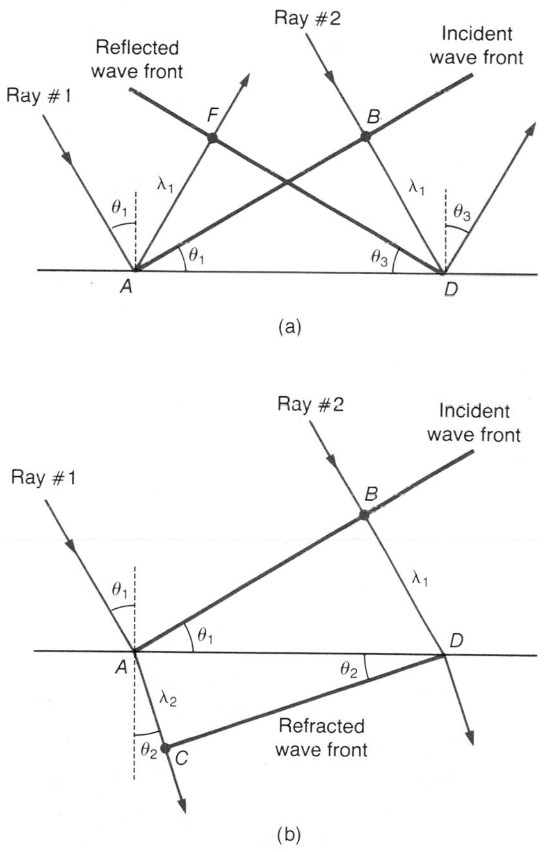

FIGURE 22.5 A wave interpretation of the laws of reflection and refraction: (a) construction of the reflected wave front; (b) construction of the refracted wave front.

We now select two rays in the incident plane wave so that when ray #1 strikes the interface, ray #2 still has to travel a distance λ_1 to reach it. The wave crest marked A is at the interface as shown in Fig. 22.5(a) and Fig. 22.5(b), but the wave crest B on the same incident wave front is a distance λ_1 from the interface. We want to construct the positions of the reflected and refracted wave fronts.

Figure 22.5(a) shows the construction for finding the reflected wave front. Points A and B are crests on the incident wave front. The next crest on ray #2 is at the point D on the interface, since we have chosen the two rays so that the length BD is equal to λ_1. The next crest on the reflected ray #1 is at the point F and the length AF is also equal to λ_1. The reflected wave front subsequent to AB is FD. Since rays and wave fronts are normal to each other, the triangles ABD and AFD are both right triangles. From the geometry of the diagram, it follows that the angle between the incident wave front and the interface is equal to the angle between the reflected wave front and the interface. Furthermore, these angles are also equal to the angles that the incident and reflected rays make with the normal to the interface, which are labeled θ_1 and θ_3 on the diagram, respectively. Since our discussion shows that $\theta_1 = \theta_3$, we see that the ray picture and the wave front picture are equivalent.

Figure 22.5(b) shows the construction of the refracted wave front that is subsequent to the incident wave front AB. The refracted wave front is CD; the length BD in medium #1 is equal to λ_1, but the length AC in medium #2 is equal to λ_2. Since rays and wave fronts are normal to each other, triangles ABD and ACD are both right triangles. Let θ_2 be the angle between the refracted wave front and the interface. Then we have:

$$\frac{\sin\theta_1}{\sin\theta_2} = \frac{\lambda_1/AD}{\lambda_2/AD} = \frac{c/n_1\nu}{c/n_2\nu} = \frac{n_2}{n_1}$$

But the angle between a wave front and the interface is equal to the angle between a ray and normal to the interface. This result for wave fronts is therefore identical with Snell's law for rays, which we stated in Eq. (22.3).

The law of refraction was discovered experimentally by the Dutch astronomer Willebrord Snell (1591–1626). About 50 years later Newton proposed his corpuscular theory of light, and Huygens expounded a wave theory. As we have just seen, Snell's law of refraction follows from a wave theory of light and the definition of the index of refraction. Why did Huygens fail to carry the day with his wave theory of light, when he could derive Snell's law by using only the definition of the refractive index? The answer is that a measurement of the velocity of light in any medium was almost two centuries in the future. The modern definition of refractive index, which was used in Eq. (22.1), was not then available. In fact, Newton had a derivation of Snell's law based on the dynamics of corpuscles that required the corpuscles to travel faster in the medium of higher refractive index, whereas the wave theory requires the propagation velocity to be slower there. If experimental techniques had been sufficiently good to measure the velocity of light, Huygens's theory would have prevailed in his lifetime.

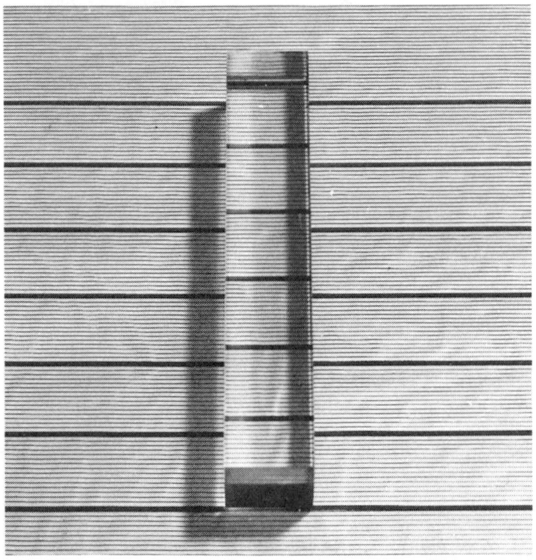

The photo above shows the apparent displacement of ruled parallel lines due to refraction in a glass block placed over them. Below, a beam of light is refracted by water in a transparent tank. (Fundamental Photographs, New York)

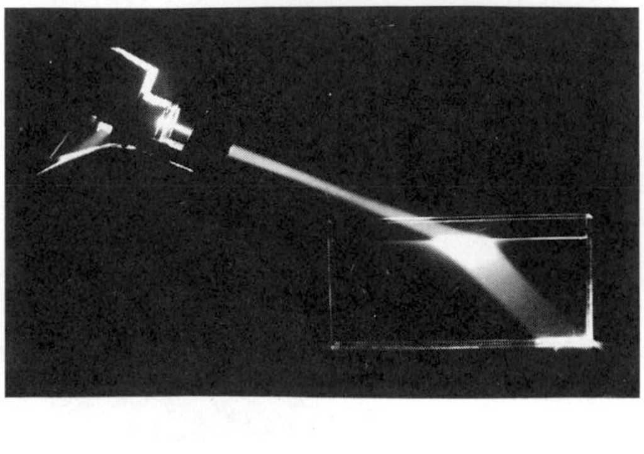

The material having the higher refractive index is referred to as being optically denser. Figure 22.4 shows the incident ray in the optically less dense medium and the refracted ray in the optically denser medium. In this case, the refracted ray is bent toward the normal as it enters the optically denser medium. The opposite holds true if the ray travels from the optically denser to the less dense medium; in this case the refracted ray is bent away from the normal.

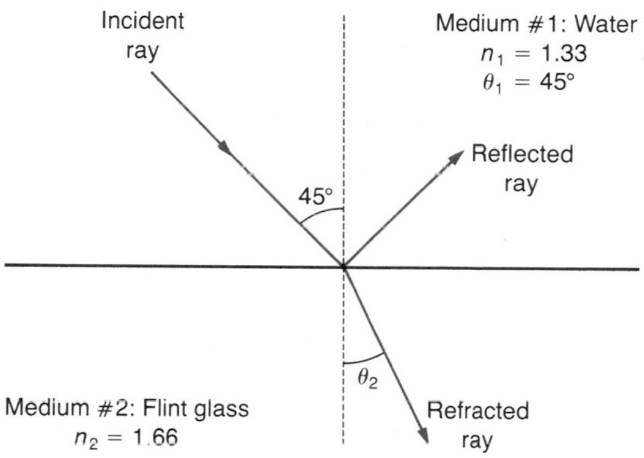

EXAMPLE 22.2

A ray of light traveling through water strikes a flat slab of flint glass at an angle of incidence of 45°. Calculate the path of the ray in the flint glass.

SOLUTION The path of the refracted ray is determined by the angle of refraction. Taking the water as medium #1 and the flint glass as medium #2, we find $n_1 = 1.33$ and $n_2 = 1.66$ from Table 22.2. The angle of refraction is given by Snell's law, Eq. (22.3):

$$n_1 \sin \theta_1 = n_2 \sin \theta_2$$

$$1.33 \times \sin 45° = 1.66 \times \sin \theta_2$$

$$\therefore \theta_2 = 34.5° \qquad \blacksquare$$

Let us also consider an example in which the refracted ray is bent away from the normal.

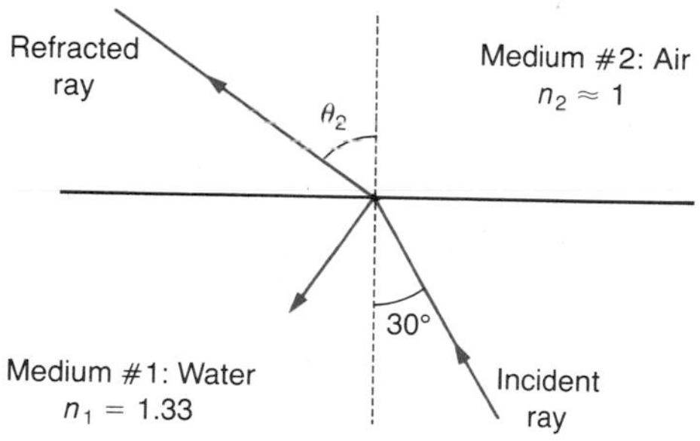

EXAMPLE 22.3

An underwater diver aims the beam of a flashlight to make an angle of incidence of 30° with the surface. At what angle does the beam emerge into the air?

SOLUTION Take the water as medium #1 and the air as medium #2. The refractive indexes from Table 22.2 are $n_1 = 1.33$ and $n_2 \approx 1$. Snell's law, Eq. (22.3), gives us:

$$n_1 \sin \theta_1 = n_2 \sin \theta_2$$

$$1.33 \times \sin 30° = 1 \times \sin \theta_2$$

$$\therefore \theta_2 = 41.7° \qquad \blacksquare$$

22.3 Total Internal Reflection

So far we have not considered the relative fractions of incident light that are reflected and refracted from a plane boundary. By using Maxwell's electromagnetic theory, we can calculate the relative amounts reflected and refracted in terms of the indexes of refraction of the two media. An especially interesting situation arises if light is incident on an interface from the optically denser medium. When the light is perpendicularly incident on the interface (angle of incidence 0°), a very small percentage is reflected and the rest is transmitted into the second medium. As the angle of incidence increases, the percentage of reflected light also increases. At first this increase is slow, but it eventually becomes very rapid. At a certain critical angle of incidence, all of the incident light is reflected and the percentage of light refracted falls to zero. This phenomenon, called **total internal reflection,** is illustrated in Fig. 22.6. The critical angle of incidence corresponds to the refracted light grazing the interface between the two media (angle of refraction 90°). Total internal reflection occurs whenever the angle of incidence is equal to or greater than the critical angle.

We can calculate the value of the critical angle by using Snell's law. For the critical ray, the angle of incidence is θ_c (the critical angle), and the angle of refraction is 90°. Substituting in Snell's law, Eq. (22.3), we have:

$$n_1 \sin \theta_c = n_2 \times 1$$

By rearranging we obtain an equation for calculating the critical angle:

■ Equation for the Critical Angle of Incidence

$$\sin \theta_c = \frac{n_2}{n_1} \qquad \textbf{(22.4)}$$

The light is traveling from medium #1 to medium #2 and $n_1 > n_2$.

Light traveling from the optically less dense into the optically denser medium cannot be totally reflected.

EXAMPLE 22.4

Calculate the critical angle of incidence for light traveling from flint glass to air.

SOLUTION From Table 22.2 we find the indexes of refraction; $n_1 = 1.66$ for flint glass and $n_2 \approx 1$ for air. We can use Eq. (22.4) to calculate the critical angle of incidence:

$$\sin\theta_c = \frac{n_2}{n_1}$$

$$= \frac{1}{1.66}$$

$$\therefore \theta_c = 37.0°$$ ■

It is important to realize that the critical angle is not a property of a single substance, but a property of two substances sharing an interface. Another example clarifies this point.

EXAMPLE 22.5

Calculate the critical angle of incidence for light traveling from flint glass into water.

SOLUTION From Table 22.1 we find $n_1 = 1.66$ for flint glass and $n_2 = 1.33$ for water. Equation (22.4) gives the critical angle of incidence:

$$\sin\theta_c = \frac{n_2}{n_1}$$

$$= \frac{1.33}{1.66}$$

$$\therefore \theta_c = 53.2°$$ ■

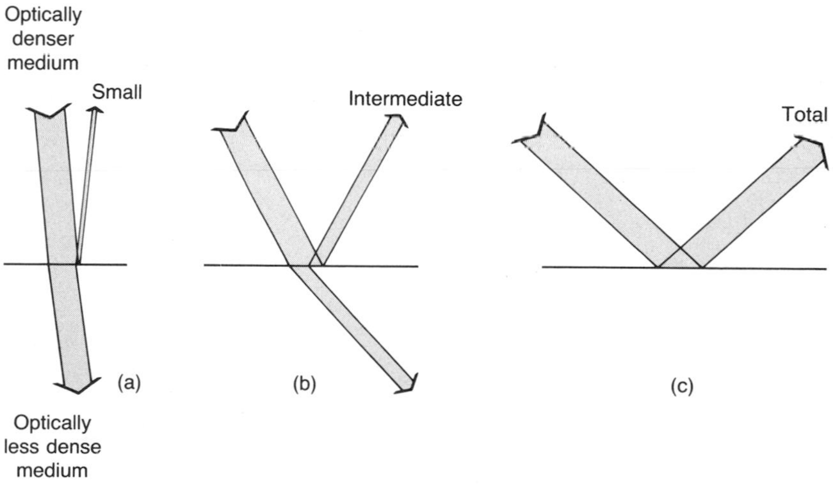

FIGURE 22.6 Diagram to illustrate relative amounts of reflected and refracted light from an optically denser into an optically less dense medium: (a) small angle of incidence—small fraction reflected; (b) moderate angle of incidence—moderate fraction reflected; (c) critical angle of incidence—total reflection.

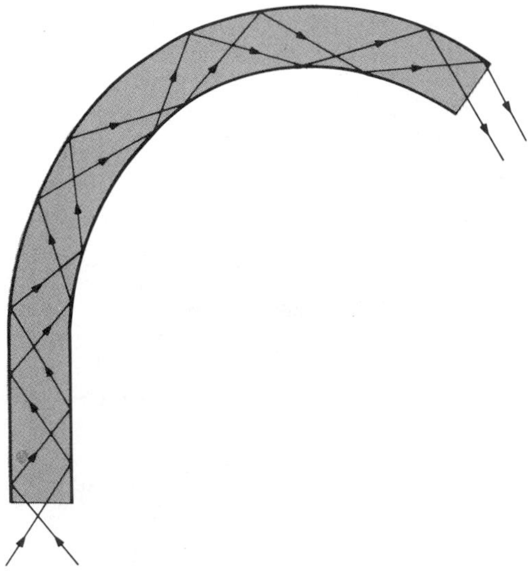

FIGURE 22.7 Total internal reflection of light within a light pipe.

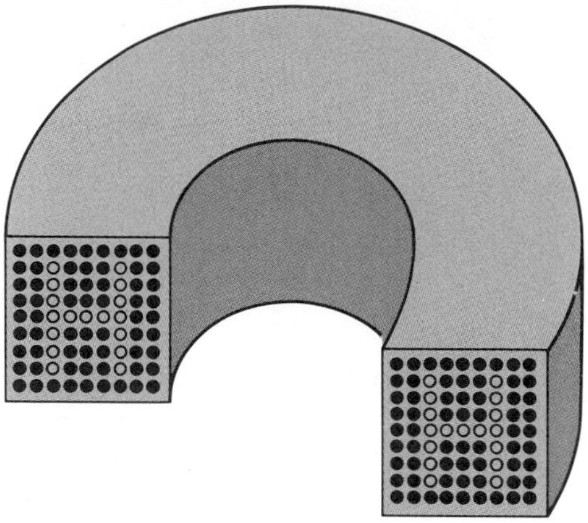

FIGURE 22.8 Transmission of an image using a bundle of light pipes.

Total internal reflection is responsible for the sparkling quality of cut diamonds. The refractive index of diamond is 2.42, and the critical angle for a diamond–air interface is 24.4°. In other words, a beam of light within a diamond must hit a face at a nearly perpendicular incidence in order to emerge. Light entering a cut diamond has relatively few faces from which it can exit. If the diamond is moved slightly, light enters at a different angle and the exit faces change abruptly.

Modern technology has also made use of total internal reflection to carry light around corners. Light is introduced into one end of a thin glass fiber that does not have any sharp bends. As the light progresses, it strikes the walls of the fiber at angles of incidence that always exceed the critical angle for the glass–air interface. As shown in Fig. 22.7, this situation causes the light to follow the path of the fiber just as an electric current follows the wire in a circuit. The light pipe is used in objects ranging from ornamental displays to telephone cables.

If light pipes are put together in an ordered array, pictures can be transmitted from one end to the other. Figure 22.8 shows the operating principle for this technique. The individual light pipes are arranged in an array whose order persists throughout the length of the cable. An image input at one end of the cable (which is achieved by having light incident on some pipes and not on others) produces the same image output at the far end of the cable. Physicians have been able to observe the walls of a patient's stomach through a bundle of light pipes inserted through the throat.

22.4 Ray Diagrams for Mirrors

If light traveling in a transparent medium strikes a polished metal surface, there is no refracted ray into the metal—the incident

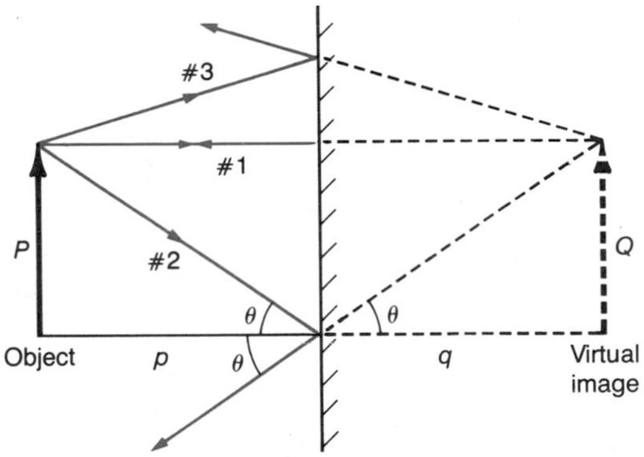

FIGURE 22.9 Ray diagram to locate the image of an object formed in a plane mirror.

light is almost totally reflected, regardless of the angle of incidence. A polished metal surface, optically speaking, is called a *mirror*. Although we instinctively think of mirrors as being made of glass, the glass only serves to protect the thin silvered film of paint behind it. It is the shiny metal surface that is the real mirror.

The simplest type of mirror is a plane reflecting surface. Our everyday experience leads us to believe that the image of any object is behind the mirror by the same distance that the object is in front of it. This is in fact correct, and the truth of this assertion is demonstrated by the ray diagram in Fig. 22.9. We begin by placing an object labeled P at a distance p in front of the mirror. The top of the object is marked by an arrowhead, and we can find the position of the image of the arrowhead by tracing the paths of light rays that come from it. Consider, for example, the ray labeled #1, which is normally incident on the mirror and thus reflected straight back along its original path. The ray labeled #2, which strikes the mirror opposite the base of the object, is reflected at an angle θ equal to the angle of incidence. The reflected rays #1 and #2 do not intersect, since they diverge from the front of the mirror, but the continuations of these rays backwards into the space behind the mirror do intersect. To a person standing in front of the mirror, the reflected rays seem to originate from this point of intersection. We can easily verify that any other ray, such as that labeled #3, is also reflected along a path that appears to come from the same point behind the mirror. We draw an arrowhead at the common intersection point of the continuations of the reflected rays and call it the **virtual image** of the arrowhead in front of the mirror.

Each point on the object has a virtual image point behind the mirror. If we drew them all, we would find the line Q, which is the virtual image of the object P. From the geometry of the diagram we see that the size of the image Q is equal to the size of the object P and also that the image distance q is equal to the object distance p. We can describe the image fully by saying that it is virtual, erect, the same size as the object, and the same distance behind the mirror.

Thus, our definition of a **virtual** image is an image formed by reflected rays that do not intersect. In a sense the image is not really there. If we place a lighted candle in front of a plane mirror, it is useless to try to locate an image of the candle on a screen placed behind the mirror. The type of image that does appear on a screen is called **real** to distinguish it from the virtual image.

The technique that we have used on this plane mirror problem is useful for any type of mirror system, and we shall later extend it to lens systems. To locate the image, we need only draw two rays from a point on the object and find their point of intersection. We select rays whose reflection properties are most obvious, but any rays will do.

Another important class of mirrors is the spherical mirror. A concave spherical mirror reflects from the inner side of a portion of a spherical surface; a convex spherical mirror reflects from the

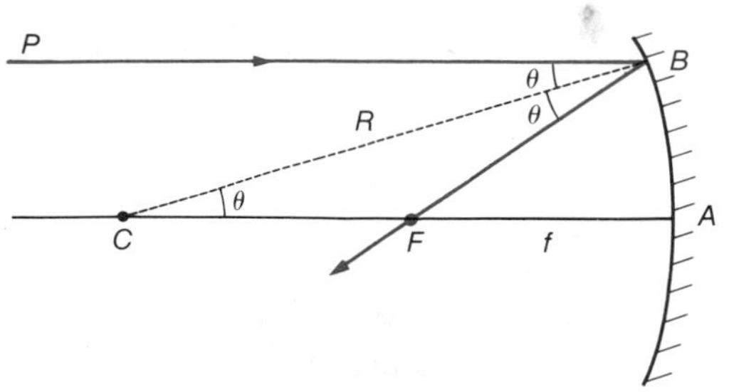

FIGURE 22.10 Reflection of a ray of light parallel to the axis through the focus of a concave mirror.

outer side. Before constructing ray diagrams to locate images, we will investigate the behavior of certain special rays reflected from a spherical mirror. Considering the concave mirror of Fig. 22.10. Let A be the center of the mirror and C the center of curvature of the spherical surface; the line CA is the optic axis of the mirror. Let a light ray coming from point P and traveling parallel to the axis strike the mirror at point B. The ray is reflected as shown in the diagram, and its path cuts the axis at point F. Since C is the center of curvature, the line CB is normal to the surface of the mirror at the point of reflection of the ray. By the law of reflection, the angle PBC is equal to the angle CBF. But since the lines PB and CA are parallel, the angle PBC also equals the angle BCA. (These angles are all marked θ on the diagram.) The triangle CBF is therefore an isosceles triangle, with the length CF equal to the length BF. But the length CB is equal to the radius of curvature of the mirror R. If the light ray is close to the axis CA, then the angle θ is small and the two equal sides of the resultant isosceles triangle are each approximately equal to $\frac{1}{2}R$. Thus, the distance AF is also approximately equal to $\frac{1}{2}R$. We call the point F the **focal point** (or the *focus*) of the mirror, and the distance $AF = f \approx \frac{1}{2}R$ the **focal length.** Since the position of the focus does not depend on the position of the light ray PB (provided that θ is small), any light parallel to the axis is reflected through the focus. Moreover, due to the law of reflection, the same diagram results if the ray travels in the opposite direction. This means that a light ray through the focus is reflected parallel to the axis. Our analysis shows two rays that have relatively simple properties of reflection from the surface of the concave mirror. Since only minor modifications to the statements below are necessary for a convex mirror, we will make no distinction at this point.

■ Data for Ray Optics of a Spherical Mirror

$$\text{Focal length} = \tfrac{1}{2} \times \text{radius of curvature}$$

$$f = \tfrac{1}{2}R \qquad \textbf{(22.5)}$$

A ray parallel to the axis is reflected through the focus.
A ray through the focus is reflected parallel to the axis.

Let us apply this information to find the position of images in a concave mirror with the use of ray diagrams.

EXAMPLE 22.6

A concave mirror has a radius of curvature of 40 cm. Use a ray diagram to determine the image position of an object that is 30 cm from the mirror. Repeat the problem if the object is 10 cm from the mirror.

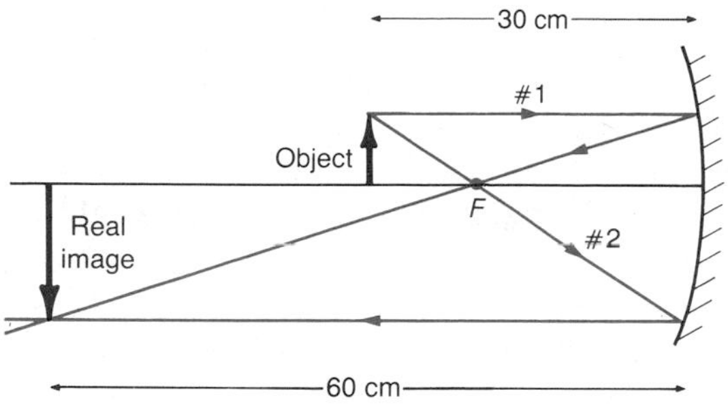

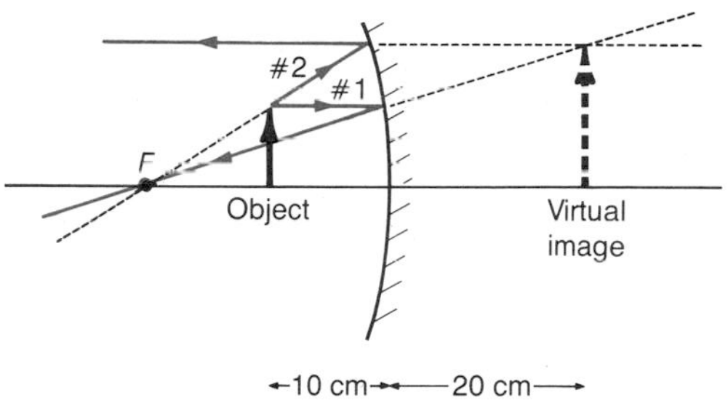

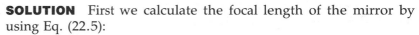

SOLUTION First we calculate the focal length of the mirror by using Eq. (22.5):

$$\begin{aligned} f &= \tfrac{1}{2}R \\ &= \tfrac{1}{2} \times 40 \text{ cm} \\ &= 20 \text{ cm} \end{aligned}$$

The focus is in front of the mirror at the point marked F on the diagram. In the first part of the diagram ray #1, which is parallel to the axis, is reflected through the focus, and ray #2, which goes through the focus, is reflected parallel to the axis. Since the two reflected rays intersect in front of the mirror, the image is real. With suitable precautions to eliminate stray light, the image would show on a screen placed at the image position. Measurements of the diagram show that the image is about 60 cm in front of the mirror and about twice as large as the object. The second part of the problem requires us to draw continuations of the ray trajectories. Ray #1 is reflected through the focus, but ray #2, whose continued trajectory goes through the focus, is reflected parallel to the axis. The two reflected rays do not intersect. We treat the problem just as we did for the plane mirror, drawing continuations of the ray trajectories. The continuations intersect, and the image is virtual and erect. Measurements on the diagram show that the image is about 20 cm behind the mirror and about twice as large as the object. ■

The convex spherical mirror is similar to the concave mirror. All the data in the display for Eq. (22.5) hold good with only slight changes of meaning. The focal length of a convex mirror is one-half of the radius of curvature, but the focus is behind the mirror. For this reason none of the light rays actually go through the focus, although the continuations of their trajectories do. Let us use an example to illustrate the ray-tracing method for a convex mirror.

EXAMPLE 22.7

A convex mirror has a radius of curvature of 40 cm. Use a ray diagram to determine the position of the image of an object that is 20 cm from the mirror.

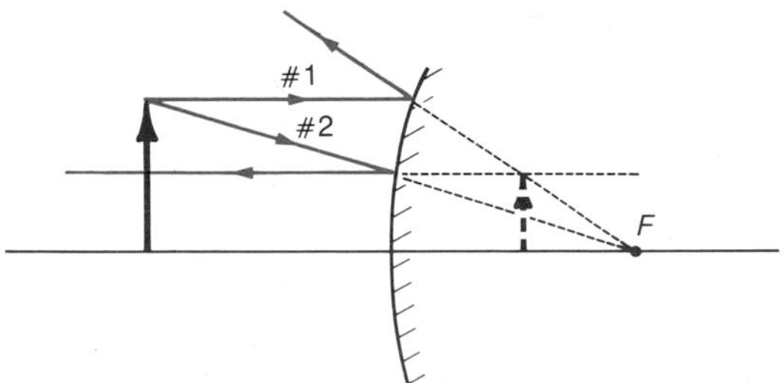

SOLUTION The focal length of the mirror is again 20 cm, as in Example 22.6, but this time the focus is behind the mirror at the point marked F. Ray #1, which is parallel to the axis, is reflected along a path whose continuation goes through the focus. Ray #2, whose continuation goes through the focus, is reflected parallel to the mirror. The reflected rays do not intersect, but their continuations intersect behind the mirror to give an erect virtual image. Measurements show that the image is about 10 cm behind the mirror and about one-half the size of the object. ■

22.5 The Mirror Equation

We can also derive algebraic equations that give both the position and magnification of the image of an object. Consider the concave mirror of focal length f shown in Fig. 22.11. An object of height P is placed at a distance p from the mirror. According to our usual ray diagram (see Example 22.6), it produces an inverted real image of height Q at a distance q from the mirror. The geometry of the diagram is easy to analyze if we can treat the surface of the mirror as being approximately flat. Given the approximation, the horizontally and vertically shaded areas are pairs of similar triangles. Since the corresponding sides of similar triangles have the same ratio, we can write:

$$\frac{P}{Q} = \frac{p - f}{f} \text{ (horizontally shaded triangles)}$$

and

$$\frac{P}{Q} = \frac{f}{q - f} \text{ (vertically shaded triangles)}$$

Equating the right-hand sides and rearranging gives:

$$pq = qf + pf$$

Dividing each side by pqf gives:

$$\frac{1}{f} = \frac{1}{p} + \frac{1}{q} \tag{22.6}$$

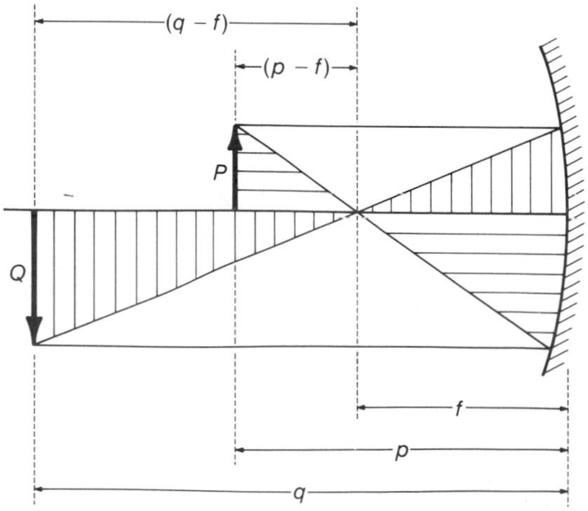

FIGURE 22.11 Diagram illustrating the notation for the mirror equation applied to a concave mirror.

This equation enables us to calculate the image distance q from the object distance p and the focal length f. However, we can also use the geometry of Fig. 22.11 to find how the image size relates to the object size. The image height divided by the object height is termed the **magnification.** Let us adopt the convention that distances measured above the optic axis are positive and those below it are negative. Then:

$$M = \frac{-Q}{P} = \frac{-(q - f)}{f} = -\frac{q}{f} + 1$$

Substituting for $1/f$ from Eq. (22.6) gives:

$$M = -q\left(\frac{1}{p} + \frac{1}{q}\right) + 1 = -\frac{q}{p} \tag{22.7}$$

Let us apply Eqs. (22.6) and (22.7) to solve mirror problems algebraically.

EXAMPLE 22.8

Solve Example 22.6 by using an algebraic method.

SOLUTION Example 22.6 specifies a concave mirror with $f = 20$ cm. In the first part, the object is 30 cm from the mirror; that is, $p = 30$ cm. Substituting in Eq. (22.6) gives:

$$\frac{1}{f} = \frac{1}{p} + \frac{1}{q}$$

$$\frac{1}{20\text{ cm}} = \frac{1}{30\text{ cm}} + \frac{1}{q}$$

$$\therefore q = 60\text{ cm}$$

(Note that there is no need to convert units if every term in the equation is expressed in the same units.) From Eq. (22.7) we find the magnification to be:

$$M = -\frac{q}{p} = -\frac{60\text{ cm}}{30\text{ cm}} = -2$$

The magnification is negative so that the image is inverted. These are the same results obtained graphically by the ray method.

In the second part, the object is 10 cm from the mirror; that is, $p = 10$ cm. Substituting in Eq. (22.6) gives:

$$\frac{1}{f} = \frac{1}{p} + \frac{1}{q}$$

$$\frac{1}{20\text{ cm}} = \frac{1}{10\text{ cm}} + \frac{1}{q}$$

$$\therefore q = -20\text{ cm}$$

Our ray diagram solution to the same problem showed a virtual image 20 cm behind the mirror. The calculation gives the magnitude of the image distance correctly, but the sign is negative. We suspect that the negative sign reflects the fact that the image is virtual. Calculating the magnification from Eq. (22.7) gives

$$M = -\frac{q}{p} = -\frac{-20\text{ cm}}{10\text{ cm}} = 2$$

The magnification is positive so the image is erect; the results again agree with the ray diagram. ■

Our ray diagram for a convex mirror placed the focal point behind the mirror. Although the reflected rays do not intersect at the focus, their backward continuations do. Thus, we can say that

Using solar energy to produce a high temperature requires focusing of the sun's radiant energy. This experimental solar furnace at the White Sands Missile Range in New Mexico collects the solar energy using the large array of movable mirrors on the structure at the right. These mirrors reflect the energy to the concave array on the left that focuses back through a hole in the primary reflecting array into the test room on the upper right of the picture. (U.S. Army Photography)

the focus is virtual. To be consistent, we should assign a negative focal length to a convex mirror when we use the mirror equation.

EXAMPLE 22.9

Solve Example 22.7 using an algebraic method.

SOLUTION Example 22.7 specifies a convex mirror with a focal length of 20 cm. Since the focus is behind the mirror, we must put $f = -20$ cm. The object is 20 cm from the mirror, so $p = 20$ cm. Equation (22.6) gives:

$$\frac{1}{f} = \frac{1}{p} + \frac{1}{q}$$

$$\frac{1}{-20 \text{ cm}} = \frac{1}{20 \text{ cm}} + \frac{1}{q}$$

$$\therefore q = -10 \text{ cm}$$

The negative sign indicates a virtual image 10 cm behind the mirror. This agrees with our ray diagram calculations and confirms our assignment of a negative focal length. The magnification is given by Eq. (22.7):

$$M = -\frac{q}{p} = -\frac{-10 \text{ cm}}{20 \text{ cm}} = \frac{1}{2}$$

Once again the positive value indicates an erect image. ■

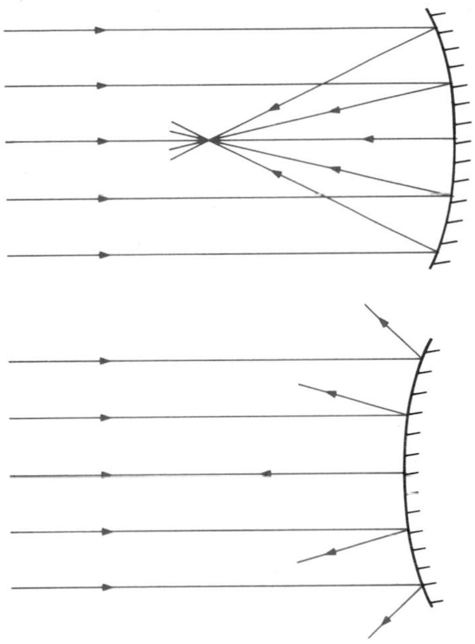

FIGURE 22.12 A concave mirror converges the incident light, and a convex mirror diverges the incident light.

Let us collect the formulas and conventions together for easy reference. To make a smooth carry-over from mirrors to lenses, we have dropped the words *concave* and *convex* and substituted *converging* and *diverging*, respectively. Figure 22.12 shows that a concave mirror converges light rays parallel to the axis onto the focus of the mirror, while a convex mirror diverges the rays as if they came from the focus. Let us apply these results to an example without using a ray diagram.

■ Relation Between Focal Length, Object Distance, and Image Distance for a Spherical Mirror

$$\frac{\text{Reciprocal focal length}}{} = \text{reciprocal object distance} + \text{reciprocal image distance}$$

$$\frac{1}{f} = \frac{1}{p} + \frac{1}{q} \qquad \textbf{(a)}$$

Focal length is positive for a converging mirror and negative for a diverging mirror.
Object distance is positive. **(22.8)**
Image distance is positive for a real image and negative for a virtual image.

$$\text{Magnification} = -\frac{\text{image distance}}{\text{object distance}}$$

$$M = -\frac{q}{p} \qquad \textbf{(b)}$$

Positive magnification signifies an erect image and negative magnification an inverted image.

EXAMPLE 22.10

Calculate the image position and the magnification for an object 2 m in front of a diverging mirror of focal length 20 cm.

SOLUTION The mirror is diverging and therefore has a negative focal length. We set down the data:

$$f = -20 \text{ cm}$$
$$p = 200 \text{ cm}$$

(The distances must be in the same units.) We now use Eq. (22.8a):

$$\frac{1}{f} = \frac{1}{p} + \frac{1}{q}$$
$$\frac{1}{-20 \text{ cm}} = \frac{1}{200 \text{ cm}} + \frac{1}{q}$$
$$\therefore q = -18.2 \text{ cm}$$

The negative sign indicates that the image is virtual and 18.2 cm behind the mirror. The magnification is given by Eq. (22.8b):

$$M = -\frac{q}{p} = -\frac{-18.2 \text{ cm}}{200 \text{ cm}} = 0.091$$

The image is less than 10% of the size of the object and it is erect. A diverging (convex) mirror is often used to monitor the aisles in supermarkets, since it covers a wide field of view and produces a virtual, erect, and diminished image of all the objects before it. ■

22.6 Ray Diagrams for Thin Lenses

We are all familiar with the lens in a camera and the lenses in eyeglasses. In both instances a piece (or pieces) of glass having two spherical surfaces is used to affect the paths of the light rays that pass through it. Light passing through a lens undergoes refraction toward the normal as it enters the first spherical surface and then refraction away from the normal as it exists the second. The details are much more complicated than for reflection from a spherical mirror, and we shall merely state the results while relying heavily on our experience with mirrors.

As stated earlier, light parallel to the axis of a converging mirror is converged to the focal point, while light parallel to the axis is a diverging mirror is diverged as if it were coming from the

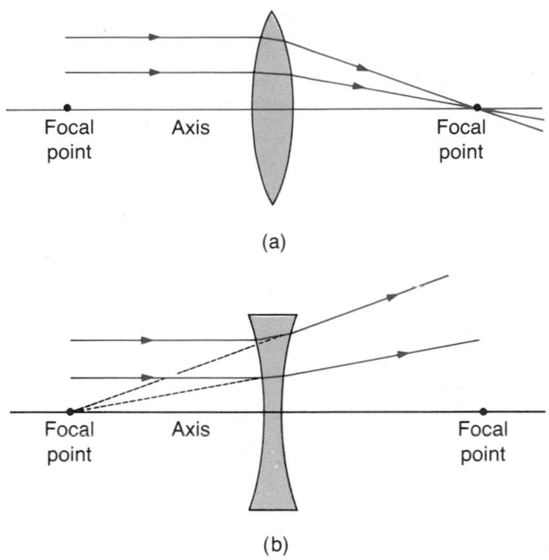

FIGURE 22.13 (a) Light rays converge to the focal point of a converging lens; (b) light rays diverge as if coming from the focal point of a diverging lens.

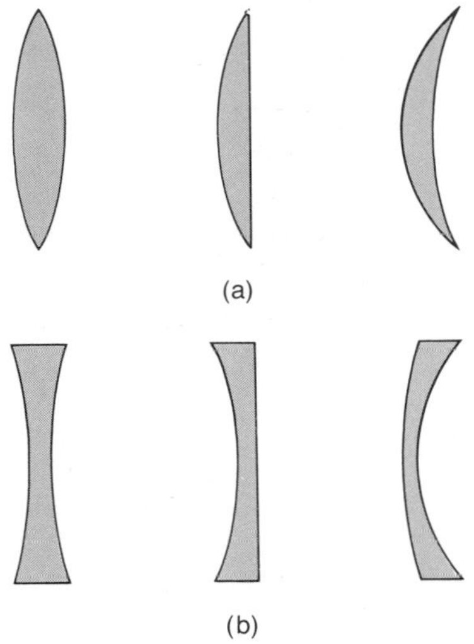

FIGURE 22.14 Various types of lenses: (a) converging lenses; (b) diverging lenses.

focal point. Both of these statements also hold true for converging and diverging lenses. Figure 22.13(a) shows a converging lens with two convex surfaces. Light that is parallel to the axis of the lens passes through the lens and converges at the focal point. In Fig. 22.13(b) light parallel to the axis of a lens with two concave surfaces passes through the lens and diverges as if coming from the focal point.

> Light rays parallel to the axis of a converging lens are refracted through the focal point on the *opposite* side of the lens.
>
> Light rays parallel to the axis of a diverging lens are refracted so that their backward continuations pass through the focal point on the *same* side of the lens.

However, there are differences between a lens and a mirror. A lens can be used in either direction, since the light passes through it. Consequently, it has two focal points on its axis, one on each side of the lens. If the lens is thin, the two focal points are each the same distance from the lens. (By a thin lens we mean one whose surface radii of curvature greatly exceed the lens thickness.) We can now describe more exactly the effect of a thin lens on light rays parallel to its axis.

The lenses illustrated in Fig. 22.13 are not the only possible types of converging or diverging lenses. Each lens surface can be concave, convex, or flat; thus, six lens types are possible, as illustrated in Fig. 22.14. Three types are converging lenses, and three

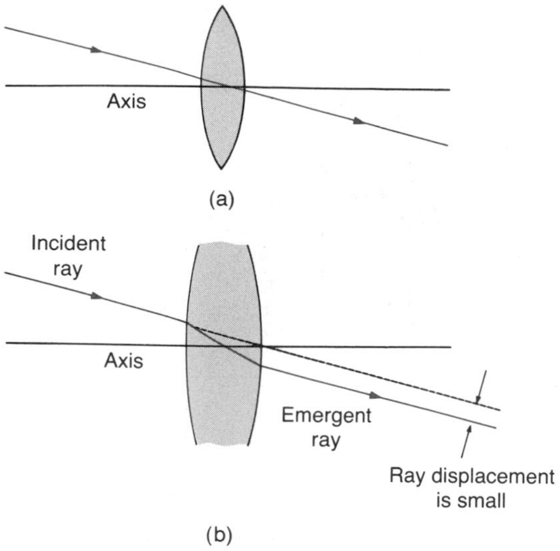

FIGURE 22.15 The path of a light ray that goes through the center of a lens: (a) the ray drawn undeviated and undisplaced; (b) a detailed treatment shows a small displacement, which is ignored for a thin lens.

types are diverging. A simple test distinguishes the two types. A lens is converging if the glass is thinner around the circumference than at the center and diverging if the situation is reversed.

To construct a lens image from a ray diagram, we need to draw two typical rays whose intersection forms the image of the point from which they came. A convenient starting point is to select a ray parallel to the axis and refracted as shown in Fig. 22.13. For the second ray we use one that passes through the center point of the lens, which is undeviated and undisplaced, as shown in Fig. 22.15(a). Strictly speaking, this simple condition for the second ray is not quite realized. Figure 22.15(b) shows the detailed path of the ray through the thick central part of the lens. The incident and the emergent rays are parallel, which means zero deviation; but the emergent ray is slightly displaced from the continued path of the incident ray. If the lens is thin, this displacement is very small, and we are justified in ignoring it.

> A light ray through the center of a thin lens continues undeviated and undisplaced.

We illustrate ray tracing for a thin lens with some examples.

EXAMPLE 22.11

A converging lens has a focal length of 20 cm. Use a ray diagram to determine the position and magnification of the image of an object that is 40 cm from the lens. Repeat the problem if the object is 10 cm from the lens.

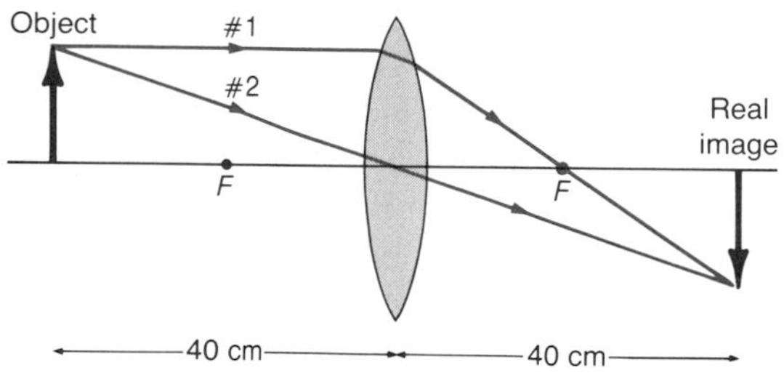

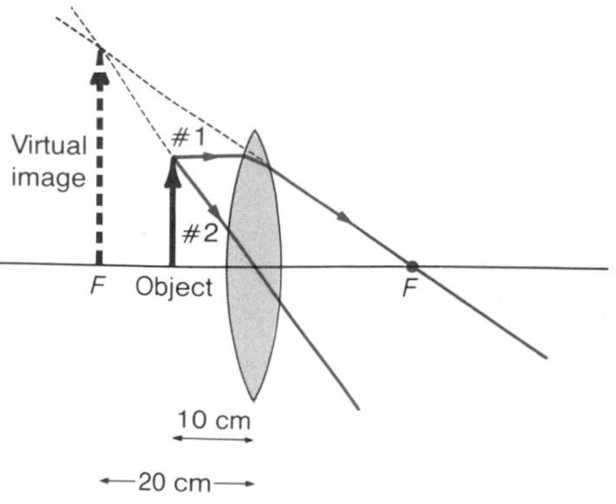

SOLUTION The first diagram shows the object at 40 cm from the lens. Ray #1, which is parallel to the axis, is refracted through the focus on the opposite side of the lens. Ray #2, which goes through the center of the lens, is undeviated, and the rays intersect to form an inverted real image. Once again, a real image is an image that would appear on a screen placed at the image point. Measurements on the diagram show that the image is about 40 cm from the lens and the same size as the object. The second diagram shows the object at 10 cm from the lens. Ray #1 is refracted through the focus on the opposite side of the lens, and ray #2 is undeviated. These rays do not intersect, but their continuations do, creating an image that is virtual and erect. Measurements on the diagram show that it is about 20 cm from the lens and twice the size of the object. ■

As we stated earlier, a real image is on the same side of a mirror as the object, since reflected rays can only intersect in front of the mirror. The reasoning for a lens is similar although the effect is reversed. A real image is on the opposite side of the lens from the object, since refracted rays can only intersect on that side.

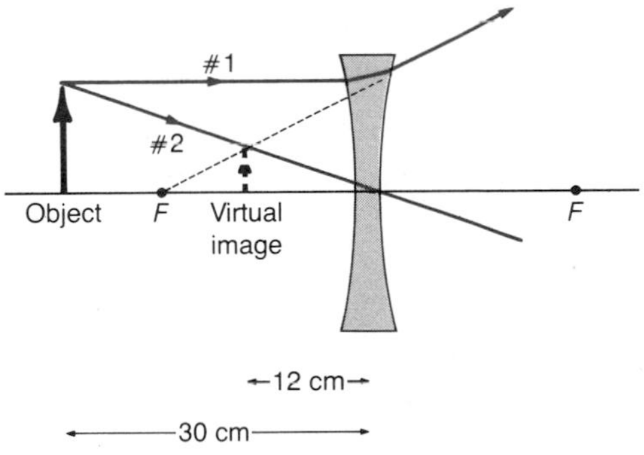

EXAMPLE 22.12

A diverging lens has a focal length of 20 cm. Use a ray diagram to determine the position and magnification of the image of an object that is 30 cm from the lens.

SOLUTION Ray #1, parallel to the axis, is refracted as if it came through the focus on the same side. Ray #2, which goes through the center of the lens, is not deviated. These rays do not intersect, but the continuation of ray #1 intersects ray #2, creating an image that is virtual and erect. Measurements on the diagram show that the image is about 12 cm from the lens and diminished to about 40% of the size of the object. ■

22.7 The Lens Equation

Algebraic equations for locating the image and magnification of a thin lens are similar to those for a mirror. Consider the converging lens of focal length f shown in Fig. 22.16. An object of height P is placed a distance p from the lens. Provided that p is greater than f, our ray diagram technique (see Example 22.11) shows that the image is real and inverted. Let it be a distance q from the lens and of height Q. The horizontally and vertically shaded areas are pairs of similar triangles. We can therefore write:

$$\frac{P}{Q} = \frac{p}{q} \quad \text{(horizontally shaded triangles)}$$

$$\frac{P}{Q} = \frac{f}{q - f} \quad \text{(vertically shaded triangles)}$$

Equating the right-hand sides and cross-multiplying gives:

$$pq = fq + fp$$

Dividing each side by pqf gives:

$$\frac{1}{f} = \frac{1}{p} + \frac{1}{q}$$

The magnification is given directly from the proportionality of the sides of the horizontally shaded triangles:

$$M = \frac{-Q}{P} = -\frac{q}{p}$$

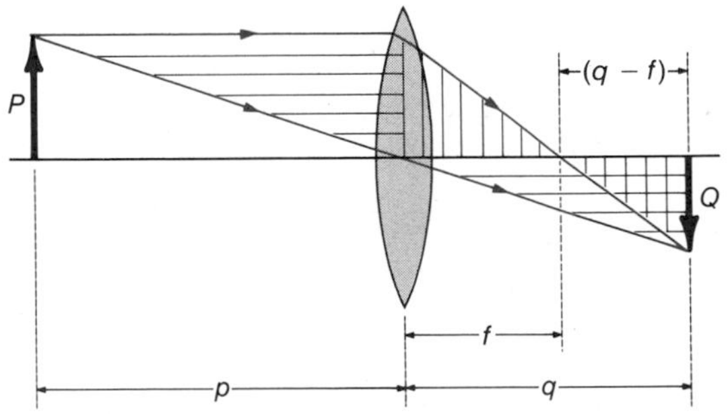

FIGURE 22.16 Diagram illustrating the notation for the lens equation applied to a converging lens.

These equations are the same as those for a converging mirror. Let us assume similar sign conventions and then check the results against our ray diagrams.

Relation Between Focal Length, Object Distance, and Image Distance for a Thin Lens

$$\frac{\text{Reciprocal}}{\text{focal length}} = \frac{\text{reciprocal}}{\text{object distance}} + \frac{\text{reciprocal}}{\text{image distance}}$$

$$\frac{1}{f} = \frac{1}{p} + \frac{1}{q} \qquad \textbf{(a)}$$

Focal length is positive for a converging lens and negative for a diverging lens.
Object distance is positive. **(22.9)**
Image distance is positive for a real image and negative for a virtual image.

$$\text{Magnification} = -\frac{\text{image distance}}{\text{object distance}}$$

$$M = -\frac{q}{p} \qquad \textbf{(b)}$$

Positive magnification signifies an erect image and negative magnification an inverted image.

EXAMPLE 22.13

Solve Example 22.11 by using the algebraic method.

SOLUTION Since the lens is converging, the sign convention of Eq. (22.9) gives us $f = 20$ cm. In the first part of the problem $p = 40$ cm, and we have from Eq. (22.9a):

$$\frac{1}{f} = \frac{1}{p} + \frac{1}{q}$$

$$\frac{1}{20\text{ cm}} = \frac{1}{40\text{ cm}} + \frac{1}{q}$$

$$\therefore q = 40\text{ cm}$$

The image is real and 40 cm behind the lens. For the magnification we use Eq. (22.9b):

$$M = -\frac{q}{p} = -\frac{40\text{ cm}}{40\text{ cm}} = -1$$

and the negative value indicates an inverted image. For the second part of the problem $p = 10$ cm, and Eq. (22.9a) gives:

$$\frac{1}{f} = \frac{1}{p} + \frac{1}{q}$$

$$\frac{1}{20\ \text{cm}} = \frac{1}{10\ \text{cm}} + \frac{1}{q}$$

$$\therefore q = -20\ \text{cm}$$

The image is virtual because of the negative sign and 20 cm in front of the lens. Equation (22.9b) gives the magnification:

$$M = -\frac{q}{p} = -\frac{-20\ \text{cm}}{10\ \text{cm}} = 2$$

The positive value indicates an erect image. All the answers agree with those found by using the ray diagram. ■

EXAMPLE 22.14

Solve Example 22.12 by using the algebraic method.

SOLUTION Since the lens is diverging, our assumed sign convention gives $f = -20$ cm. The object distance is $p = 30$ cm. Equation (22.9a) gives

$$\frac{1}{f} = \frac{1}{p} + \frac{1}{q}$$

$$\frac{1}{-20\ \text{cm}} = \frac{1}{30\ \text{cm}} + \frac{1}{q}$$

$$\therefore q = -12\ \text{cm}$$

The negative sign indicates that the image is virtual, and it is 12 cm in front of the lens. Equation (22.9b) gives us the magnification:

$$M = -\frac{q}{p} = -\frac{-12\ \text{cm}}{30\ \text{cm}} = 0.4$$

Again our calculations agree with the ray diagram results. ■

When two or more lenses are used in conjunction, we can use either a ray diagram or the algebraic approach. First, the image of the object in the first lens is located, while the second lens is simply ignored. The image then becomes the object for the second lens, which determines the final image. Since the ray diagram method can become a little messy, let us use the algebraic approach on some typical examples.

EXAMPLE 22.15

A converging lens of focal length 12 cm is placed 52 cm from another converging lens of focal length 8 cm. Calculate the image position and magnification of an object that is 16 cm in front of the first lens.

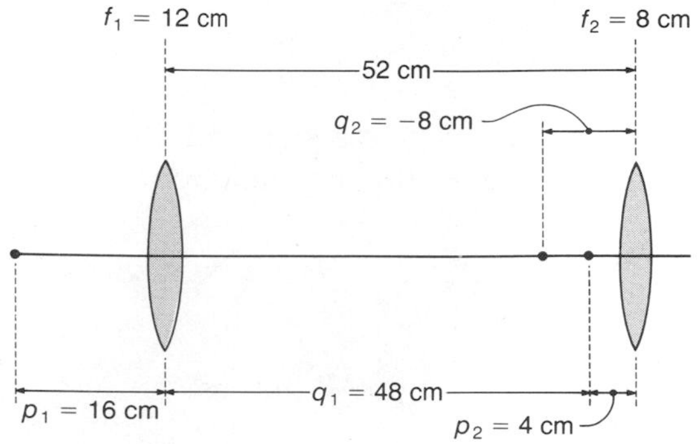

SOLUTION We begin by locating the image of the object in the first lens. Our data are:

$$f_1 = 12 \text{ cm} \qquad p_1 = 16 \text{ cm}$$

Equation (22.9a) gives:

$$\frac{1}{f_1} = \frac{1}{p_1} + \frac{1}{q_1}$$

$$\frac{1}{12 \text{ cm}} = \frac{1}{16 \text{ cm}} + \frac{1}{q_1}$$

$$\therefore q_1 = 48 \text{ cm}$$

The image is real and 48 cm to the right of lens #1. The magnification follows from Eq. (22.9b):

$$M_1 = -\frac{q_1}{p_1} = -\frac{48 \text{ cm}}{16 \text{ cm}} = -3$$

Since the distance between the two lenses is 52 cm, the first image is 4 cm to the left of lens #2. Our data for lens #2 are:

$$f_2 = 8 \text{ cm} \qquad p_2 = 4 \text{ cm}$$

Equation (22.9a) again gives:

$$\frac{1}{f_2} = \frac{1}{p_2} + \frac{1}{q_2}$$

$$\frac{1}{8 \text{ cm}} = \frac{1}{4 \text{ cm}} + \frac{1}{q_2}$$

$$\therefore q_2 = -8 \text{ cm}$$

The final image is virtual and 8 cm to the left of lens #2. The magnification due to lens #2 is given by Eq. (22.9b):

$$M_2 = -\frac{q_2}{p_2} = -\left(\frac{-8 \text{ cm}}{4 \text{ cm}}\right) = 2$$

The total magnification of the two lenses is the product of their individual magnifications:

$$M = M_1 M_2 = -3 \times 2 = -6$$

The negative sign for the total magnification indicates that the final image is inverted. ■

EXAMPLE 22.16

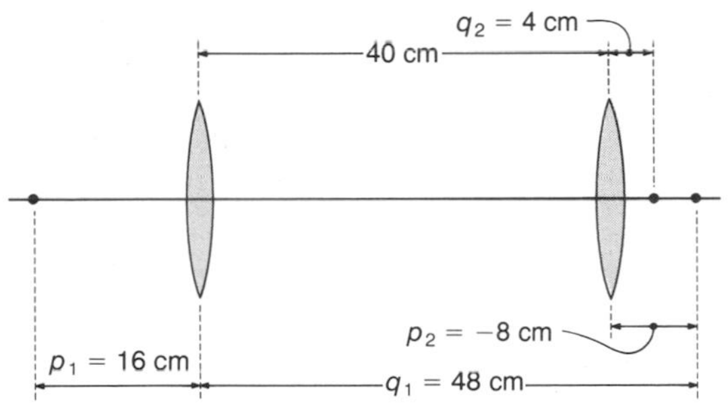

The two converging lenses of the previous example are placed 40 cm apart. Calculate the image position and magnification of an object that is 16 cm in front of the first lens.

SOLUTION The formation of the first image is the same as for Example 22.15—that is, 48 cm to the right of lens #1. However, this time the lenses are only 40 cm apart, so the second object is 8 cm *to the right* of lens #2—the object is on the wrong side of the lens, so to speak. We can handle this by treating it as a virtual object and giving its distance a negative sign. Our data for lens #2 now become:

$$f_2 = 8 \text{ cm} \qquad p_2 = -8 \text{ cm}$$

To locate the final image we use Eq. (22.9a):

$$\frac{1}{f_2} = \frac{1}{p_2} + \frac{1}{q_2}$$

$$\frac{1}{8 \text{ cm}} = \frac{1}{-8 \text{ cm}} + \frac{1}{q_2}$$

$$\therefore q_2 = 4 \text{ cm}$$

The final image is real and 4 cm to the right of lens #2. The magnification of lens #1 is $M_1 = -3$, just as in Example 22.15. The magnification of lens #2 is given by Eq. (22.9b):

$$M_2 = -\frac{q_2}{p_2} = -\left(\frac{4 \text{ cm}}{-8 \text{ cm}}\right) = 0.5$$

The total magnification is therefore:

$$M = M_1 M_2 = -3 \times 0.5 = -1.5$$

The final image is inverted. ■

22.8 The Lensmaker's Equation and Lens Power

We began our discussion of spherical mirrors with the remarkably simple fact that the focal length is equal to one-half of the radius of curvature of the mirror. However, we have yet to calculate the focal length of a lens from the radii of curvature of the surfaces. Because the rays are refracted by the glass of which the lens is made, the refractive index of the glass must enter our formula, as must the radii of curvature of the surfaces. We will omit the details of the calculation and simply quote the celebrated **lensmaker's formula,** which was used by generations of craftsmen who made lenses by hand.

Lensmaker's Equation to Calculate the Focal Length of a Thin Lens in Air

$$\frac{1}{f} = (n - 1)\left(\frac{1}{R_1} + \frac{1}{R_2}\right) \qquad \textbf{(22.10)}$$

f = focal length
n = refractive index of the lens material
R_1, R_2 = radii of curvature of the lens surface
Consider the radius of curvature of a convex surface to be positive and the radius of curvature of a concave surface to be negative.

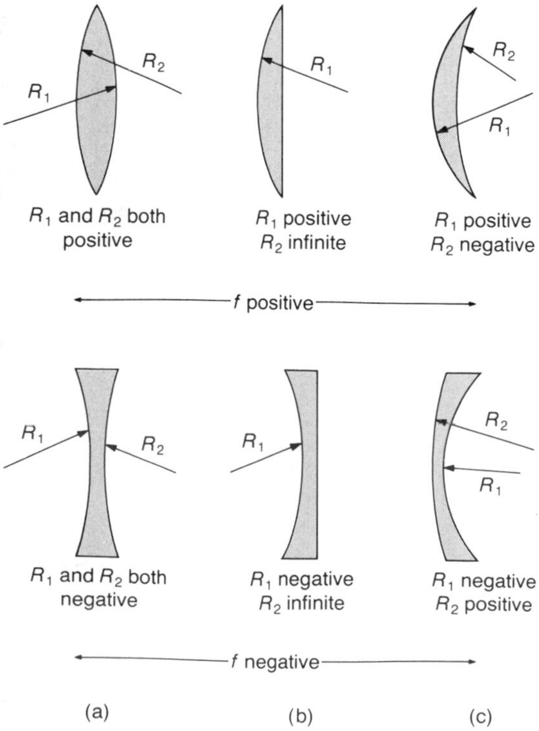

The only difficulty with this formula lies in the assignment of signs to the surface radii of curvature. Let us illustrate our convention with an example.

EXAMPLE 22.17

The six lens types of Fig. 22.14 on page 641 are grouped into converging and diverging lenses. Show that the lensmaker's equation correctly predicts the sign of the focal length for all six types.

SOLUTION Since $(n - 1)$ is always positive, the focal length calculated from Eq. (22.10) has the sign of $[(1/R_1) + (1/R_2)]$. It is easiest to take the lenses in pairs.

Case (a) Two convex or two concave faces mean that R_1 and R_2 are either both positive or both negative, respectively. In the first instance the focal length is positive and the lens is converging, while in the second instance it is negative and the lens is diverging.

Case (b) If one face is flat, the radius of curvature (say R_2) is infinite, and $1/R_2 = 0$. The focal length then has the same sign as the curvature of the remaining face. The lens is converging if this face is convex and diverging if it is concave.

Case (c) When one face is convex and the other concave, then R_1 and R_2 have opposite signs. The quantity $[(1/R_1) + (1/R_2)]$ has the same sign as the *smaller* of R_1 and R_2. The focal length is positive if the convex surface has the smaller radius of curvature, and negative if the concave surface has the smaller radius of curvature. ■

Since the reciprocal focal length of a lens is an important quantity, it has a special name. We call it the **lens power** and assign the unit **diopter** when the focal length is expressed in meters.

Definition of the Power of a Lens

$$\text{Lens power} = \frac{1}{\text{focal length}} \qquad \textbf{(22.11)}$$

SI unit of lens power: $(\text{meter})^{-1}$
New name: diopter

Opticians usually prescribe simple corrective lenses by quoting the required power in diopters.

EXAMPLE 22.18

A corrective lens for a patient is required to have a power of -2 diopters. The lens is made from crown glass, and the convex front surface has a radius of curvature of 35 cm. Calculate the radius of curvature of the rear surface.

SOLUTION We need only use the lensmaker's formula and the definition of the diopter. However, we must be careful to set all lengths in meters, since the diopter is a reciprocal meter. Equation (22.10) gives:

$$\frac{1}{f} = (n - 1)\left(\frac{1}{R_1} + \frac{1}{R_2}\right)$$

$$-2 \text{ diopters} = (1.52 - 1)\left(\frac{1}{0.35 \text{ m}} + \frac{1}{R_2}\right)$$

$$\therefore R_2 = -0.149 \text{ m}$$

The negative sign tells us that the rear surface of the lens is concave, and its radius of curvature is about 15 cm. ■

KEY CONCEPTS

Visible light occupies a small region of the electromagnetic spectrum, and its free space wavelengths are in the hundreds of nanometers; see Table 22.1.

When light travels in a material medium, its velocity is lower and its wavelength longer than the corresponding free space values. The **refractive index** of a medium is the ratio of the free space velocity of light to the velocity of light in the medium; see Eq. (22.1).

When light is incident on an interface between two media, it is partly reflected and partly refracted. The directions of the **reflected rays** and the **refracted rays** are given by the **law of reflection** and **Snell's law of refraction;** see Eq. (22.2) and Eq. (22.3).

The **focal point** of a mirror or a lens is the point at which reflected or refracted rays converge or the point from which continuations of the rays diverge; see Fig. 22.12 and Fig. 22.13.

The image of an object in a mirror or a lens can be found by **ray tracing.** An image point can be located from any two rays.

For convenient ray paths, see Eq. (22.5) and the diagrams on pp. 635 and 641.

The same formulas determine the position and magnification of images in mirrors or lenses. The **reciprocal focal length** equals the **sum of the reciprocal object distance** and **the reciprocal image distance.** The magnification is the negative of the value of the image distance divided by the object distance. A **converging** lens or mirror has a **positive focal length,** and a **diverging** lens or mirror has a **negative focal length.** A **real image** is an image formed on the side of the lens or the mirror to which the refracted or reflected light rays actually travel; a **virtual image** is one formed on the side to which the light rays do not travel. Positive image distances are associated with real images, and negative with virtual images. See Eq. (22.8) and Eq. (22.9).

Positive magnification is associated with an erect image, and negative magnification with an inverted image.

The focal length of a mirror is one-half the radius of curvature of the surface. See Eq. (22.5).

The focal length of a lens is given by the **lensmaker's formula** in terms of the refractive index of the lens material and the radii of curvature of its surfaces; see Eq. (22.10).

The **lens power** in **diopters** is the reciprocal of the focal length in meters; see Eq. (22.11).

QUESTIONS FOR THOUGHT

1. "The apparent depth of an object below the surface of water in a pool depends on the observer's angle of view." Explain this effect with the aid of ray diagrams.

2. Two rectangular plane mirrors are set vertically with two vertical edges in contact, and an object is placed between the mirrors. Explain how the angle between the mirrors determines the number of images that are formed.

3. If you hold a page of printed material in front of a plane mirror, the image is reversed right to left. Why is it not also reversed top to bottom?

4. What type of mirror would you choose to give a magnified erect image for shaving? Approximately what focal length would you suggest for the mirror?

5. Describe a simple experiment for measuring the focal length of a converging lens. Does the same experiment work for a diverging lens? If not, can you suggest a modification to make the experiment work?

6. Is the focal length of a converging glass lens longer or shorter if it is immersed in water? Does the same conclusion hold true for a diverging glass lens immersed in water?

7. An air bubble in a glass slab acts somewhat like a lens for rays that pass through it. Does it behave as a converging or a diverging lens?

8. Can you suggest any arrangements of mirrors or lenses in which a plane mirror forms a real image?

9. Can a double convex lens be used to diverge an incident light beam? Explain your answer.

10. A solar furnace uses a lens or a mirror to focus the sun's rays in a small region of space. What design features would you emphasize to produce (a) high power; (b) high temperature?

PROBLEMS

A. Single-Substitution Problems

(Use Table 22.2 for refractive indexes.)

1. Calculate the frequency of violet light of wavelength 420 nm. Repeat the calculation for red light of wavelength 680 nm. [22.1]

2. Calculate the speed of light in a very dense flint glass of refractive index 1.74. [22.1]

3. The speed of light in a certain transparent plastic material is 1.80×10^8 m/s. Calculate the refractive index of the material. [22.1]

4. A ray of light traveling in air falls on a crown glass surface with a 35° angle of incidence. Calculate the angle of refraction. [22.1]

5. A ray of light traveling in air falls on a crown glass surface and is refracted at an angle of 35°. Calculate the angle of incidence on the glass. [22.2]

6. Calculate the critical angle for a diamond–air interface. [22.3]

7. Calculate the critical angle for a diamond–water interface. [22.3]

8. What must the index of refraction of a material be if its critical angle with an air interface is 40°? [22.3]

9. Calculate the focal length of a spherical convex mirror with a radius of curvature of 75 cm. [22.4]

10. A concave mirror has a focal length of 12 cm. Calculate the position of the image of an object that is 30 cm in front of the mirror. [22.5]

11. A concave mirror has a focal length of 16 cm. Calculate the position of the image of an object that is 6 cm in front of the mirror. [22.5]

12. An object is 35 cm in front of a convex mirror, and the virtual image is 12 cm behind the mirror. Calculate the focal length of the mirror. [22.5]

13. A convex mirror has a focal length of -12 cm. Calculate the position of the image of an object that is 35 cm in front of the mirror. [22.5]

14. An object is placed 15 cm from a converging lens, and its real image is 8 cm from the lens. Calculate the focal length of the lens. [22.7]

15. An object is placed 9 cm from a converging lens of focal length 16 cm. Calculate the position of the image. [22.7]

16. A converging lens of focal length 16 cm produces a real image 24 cm from the lens. What is the position of the object that produces this image? [22.7]

17. A diverging lens of focal length -25 cm produces a virtual image 10 cm from the lens. What is the position of the object that produces this image? [22.7]

18. An object is placed 20 cm from a diverging lens of focal length -8 cm. Calculate the position of the image. [22.7]

19. The virtual image produced by a diverging lens of an object that is 16 cm from the lens is located 10 cm from the lens. Calculate the focal length of the lens. [22.7]

20. Calculate the focal length of a flint glass lens that has two concave surfaces, each with a 20-cm radius of curvature. [22.8]

21. A magnifying lens made from crown glass has a power of $+8$ diopters. The two surfaces are convex and have equal radii of curvature. Calculate the radius of curvature of each surface. [22.8]

B. Standard-Level Problems

22. Red light has a wavelength of 510 nm in water. Calculate the refractive index of glass in which its wavelength is 438 nm. [22.1]

23. Calculate the wavelength in water of light whose wavelength in diamond is 190 nm. [22.1]

24. What distance does light travel in crown glass in the time that it takes to travel 20 cm in water?

25. A ray of light is normally incident on a slab of flint glass with plane parallel faces. What thickness of glass is required to delay the beam by 0.1 ns compared with the situation in which the glass was not there?

26. A beam of light traveling in air is incident at 50° on a layer of oil floating on water. The refractive index of the oil is 1.58. Calculate the angle of refraction of the beam in each liquid. [22.2]

27. A flashlight beam shines through the air at an angle of 40° with the horizontal. At what angle with the horizontal does it reach a skindiver working below the water? [22.1]

28. A ray of light is incident at 20° from air to a crown glass slab 2 cm thick with parallel sides. The ray that emerges into the air on the other side of the glass is

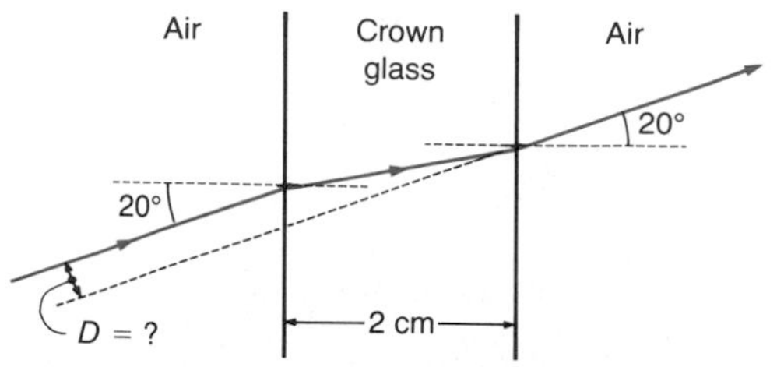

parallel to the incident ray. Calculate the distance between the paths of the two parallel rays in air.

29. A ray of light traveling in air is incident at 40° on a plane glass surface, and the angle of refraction is 25.1°. Calculate the critical angle of incidence for a light ray traveling in this glass to a plane interface with water. [22.3]

30. A sheet of flint glass of uniform thickness covers a flat water surface. A ray of light traveling in the water is incident at 25° on the undersurface of the glass.

a. Calculate the angle of refraction of the light ray in the air.
b. Calculate the range of values of the angle of incidence in the water for which there is no refracted ray in the air. [22.3]

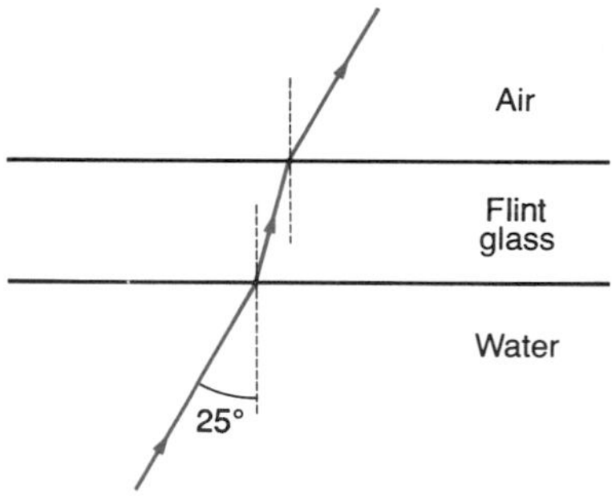

31. A glass light pipe in air has total internal reflection of light provided that the angle of incidence lies between 40° and 90°. If the pipe is placed in water, what range of angles of incidence is required for total internal reflection?

32. A concave mirror has a focal length of 20 cm. Use carefully drawn ray diagrams to locate the position and magnification of the image of an object whose distance in front of the mirror is (a) 10 cm, (b) 20 cm, and (c) 40 cm. [22.4]

33. A convex mirror has a 50-cm radius of curvature. An object 5 cm high is placed 40 cm in front of the mirror.

a. Calculate the position, size, and nature of the image.
b. Repeat the calculations if the object is 15 cm in front of the mirror. [22.5]

34. An object 3 cm high is placed 25 cm in front of a concave mirror with a 40-cm radius of curvature.

a. Calculate the position, size, and nature of the image.
b. Repeat the calculations if the object is 10 cm in front of the mirror. [22.5]

35. A convex mirror has a focal length of −15 cm. Use carefully drawn ray diagrams to locate the position and magnification of the image of an object whose distance in front of the mirror is (a) 5 cm, (b) 15 cm, and (c) 25 cm.

36. A concave mirror has a radius of curvature of 40 cm. Where should an object be placed to satisfy the following conditions?

a. The image is real and twice the size of the object.
b. The image is virtual and twice the size of the object.

37. A convex mirror has a radius of curvature of 50 cm. Where should an object be placed so that the image is virtual and one-third the size of the object?

38. A converging lens has a focal length of 20 cm. Use carefully drawn ray diagrams to locate the position and magnification of an object placed at the following distances from the lens: (a) 40 cm, (b) 18 cm, and (c) 10 cm. [22.6]

39. A diverging lens has a focal length of −25 cm. Use ray diagrams to locate the position and magnification of an object placed at the following distances from the lens: (a) 50 cm, (b) 25 cm, and (c) 10 cm. [22.6]

40. A converging lens has a power of +3.6 diopters. Calculate the position, size, and nature of the image for a 3-cm high object placed at the following distances from the lens: (a) 50 cm, (b) 25 cm, and (c) 5 cm. [22.7]

41. A diverging lens has a power of −5 diopters. Calculate the position, size, and nature of the image of a 2-cm high object placed at the following distances from the lens: (a) 40 cm, (b) 25 cm, and (c) 10 cm. [22.7]

42. A converging lens has a focal length of 30 cm. Where should an object be placed to satisfy the following conditions?

a. The image is real and 3 times the size of the object.
b. The image is virtual and 3 times the size of the object.

43. A diverging lens has a focal length of −27 cm. Where should an object be placed so that the image is virtual and one-fourth the size of the object?

44. Two converging lenses of focal lengths 20 cm and 30 cm are placed 80 cm apart. Calculate the position and magnification for the final image of an object that is 30 cm from the first lens.

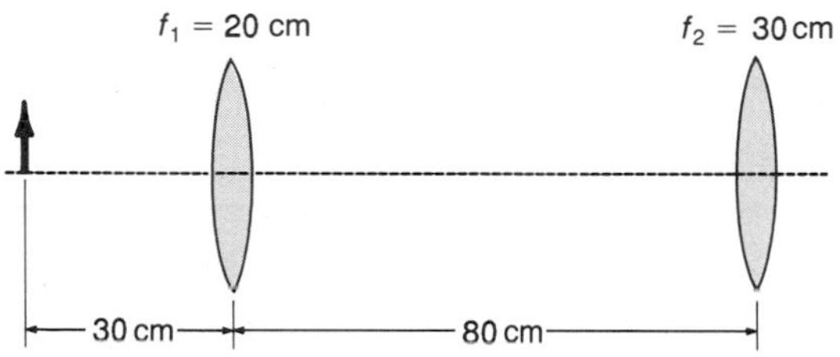

45. A diverging lens of focal length −24 cm is placed 16 cm from a converging lens of focal length 10 cm. Calculate the position and magnification for the final image of an object that is 16 cm in front of the first lens.

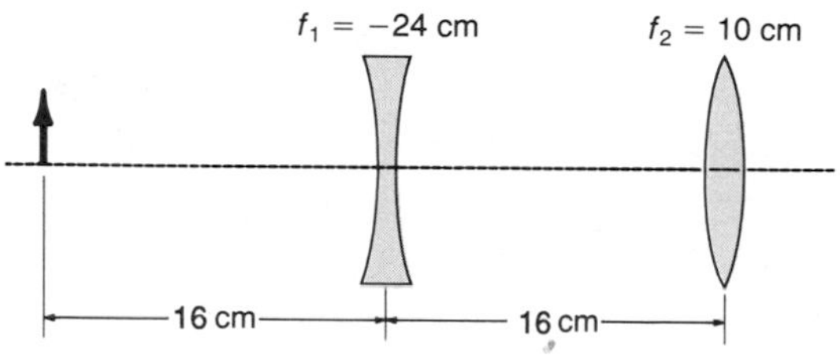

46. A converging lens of focal length 12 cm is placed 10 cm from a diverging lens of focal length −12 cm. Calculate the position and magnification for the final image of an object that is 20 cm in front of the first lens.

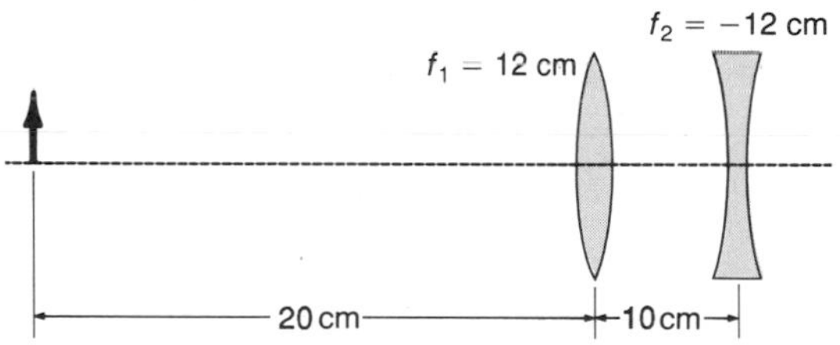

C. Advanced-Level Problems

47. A wheel having 200 teeth around its circumference is set up 300 m away from a plane mirror in a Fizeau-type experiment to measure the speed of light (see Fig. 22.2). At how many revolutions per minute must the wheel turn if light that travels out through one gap in the rim of the wheel returns through the next gap?

48. A transparent slide is placed 2.5 m from a viewing screen. What focal length should the projector lens have if the image is 12 times larger than the slide. The lens is made from crown glass and has one flat surface. Calculate the radius of curvature of the other surface.

49. A converging lens of focal length 8 cm, a diverging lens of focal length −4 cm, and a converging lens of focal length 7 cm are placed in that order, each adjacent pair 2 cm apart. Light rays parallel to the axis fall on the first lens. How far behind the third lens does the parallel light come to a focus?

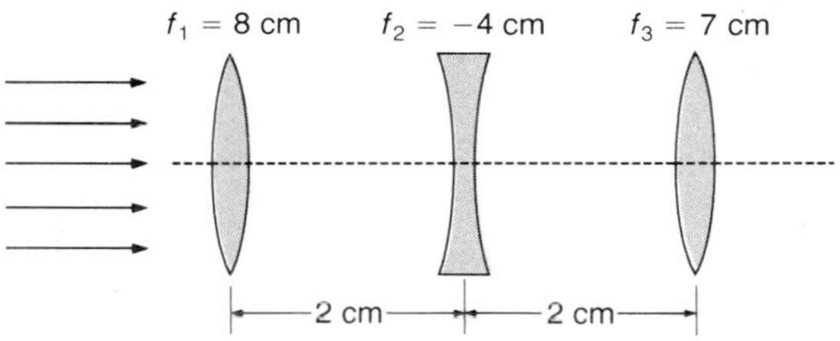

50. A lens of power +5 diopters is placed 50 cm from a concave mirror whose radius of curvature is 20 cm. Light rays parallel to the axis fall on the lens. Calculate the positions of all the points on the axis at which the light is focused by the lens-mirror system.

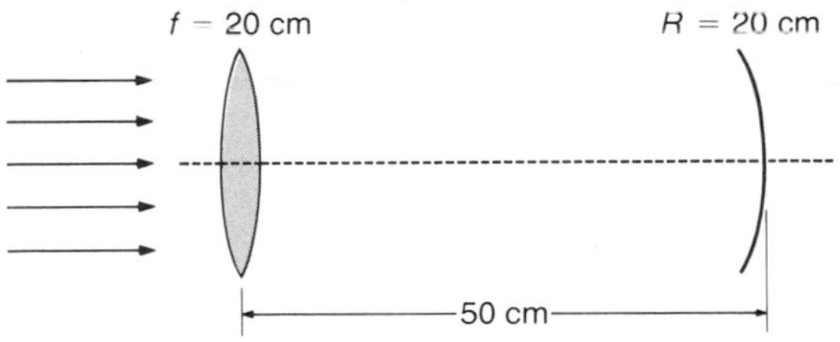

23

OPTICAL INSTRUMENTS

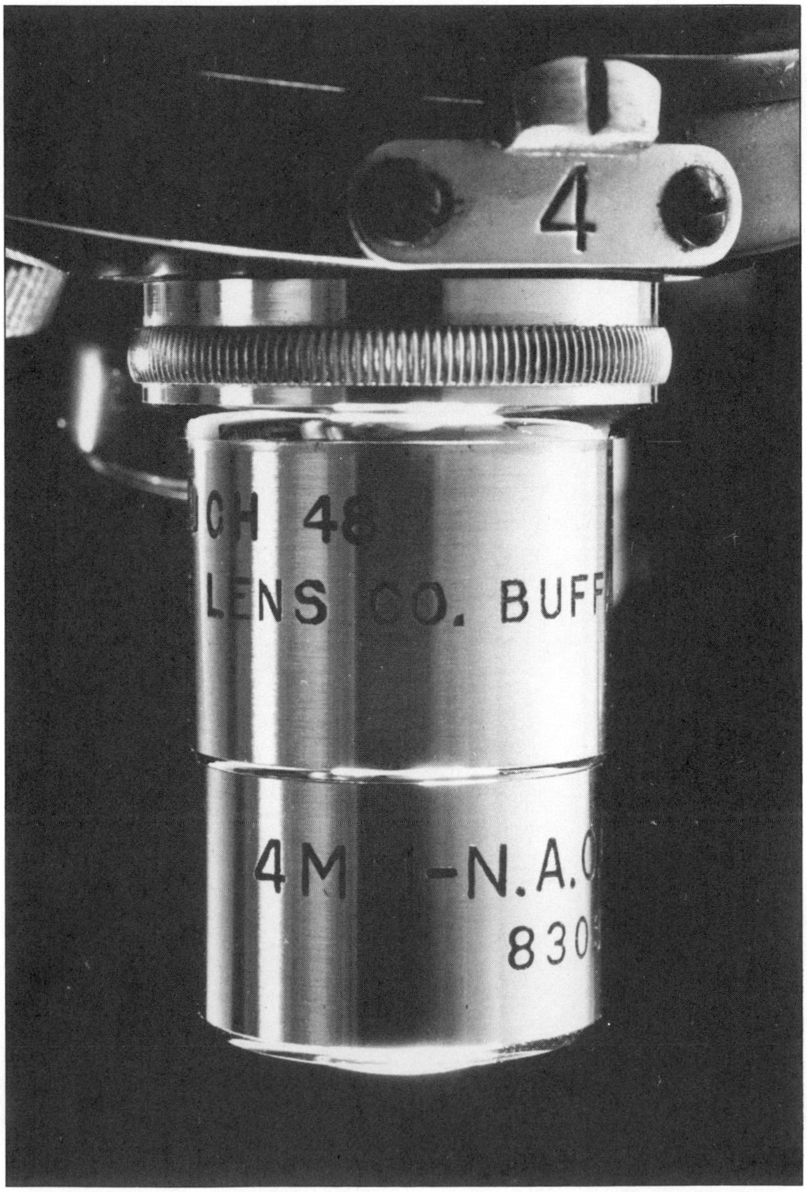

Our previous discussion of the images formed by lenses and mirrors is somewhat abstract. Apart from identifying a real image as an image that would appear on a screen, we have not yet indicated practical uses for optical images. The ultimate use of an optical instrument, which is a device that uses systems of lenses and mirrors, is to form an image to be seen by the eye. A possible alternative to using your eye is to use a camera to produce a photographic record, but the design of a camera is like that of the eye in almost every detail. The same considerations therefore apply to the final viewing of an image in an optical instrument whether it is done by eye or by camera. Let us concentrate on the human eye, and begin our study of optical instruments by discussing the ray optics of vision.

23.1 The Human Eye

The first step in the act of seeing is the formation of a real image on a screen in the eye that converts the light energy of the image to electrical impulses that travel along a nerve fiber to the brain. After the brain processes the data, we see the object of our vision. Our purpose in this section is to investigate the formation of the real image in the eye.

The human eye is roughly spherical and about 2.3 cm in diameter. Figure 23.1 shows the main parts of the eye in cross section. Light enters the front of the eye through the transparent cornea, passes through a circular opening in the iris, and is fo-

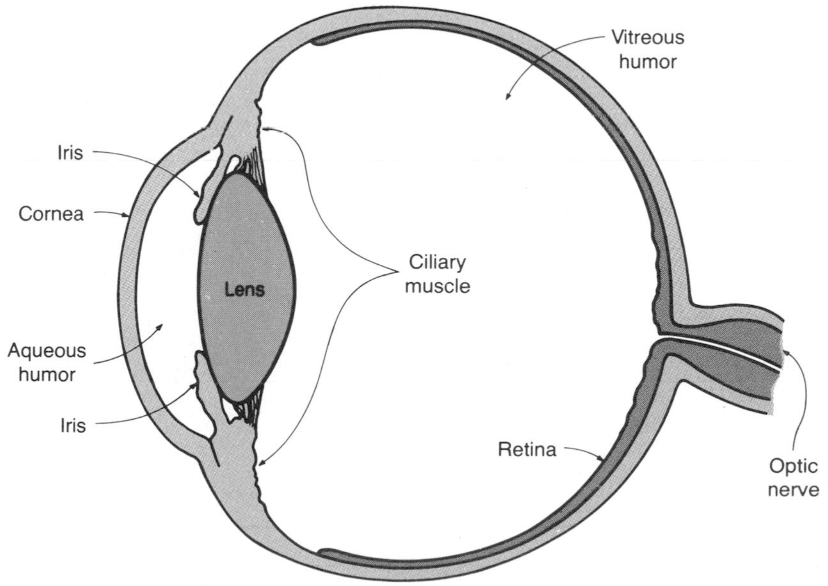

FIGURE 23.1 Cross-sectional diagram of the human eye.

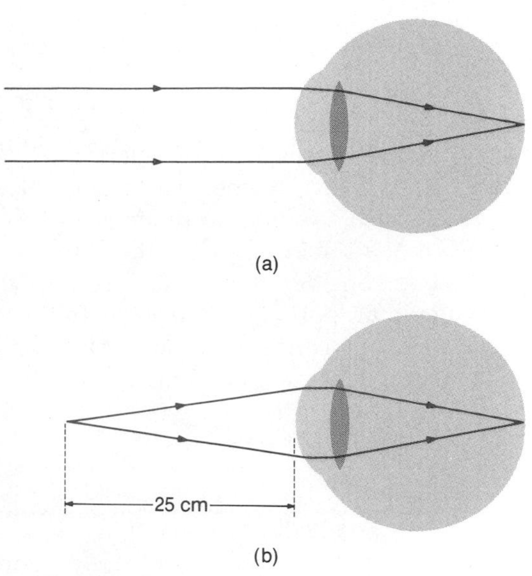

FIGURE 23.2 Diagrams of the ray optics of a normal eye (not to scale). (a) With the ciliary muscles relaxed, an object at a very large distance is focused on the retina. (b) With the lens adjusted to maximum curvature by the ciliary muscles, an object about 25 cm away is focused on the retina.

cused by the lens on the retina in the back of the eyeball. When we say that the light is focused on the retina, we mean that the image of the object from which the light rays come falls exactly on the retina. The region between the cornea and the lens is filled with a transparent fluid called the aqueous humor; the interior of the eyeball between the lens and the retina is filled with a jellylike fluid called the vitreous humor. The iris adjusts to decrease the circular opening when the light is strong and to increase it when the light is dim. The ciliary muscles change the curvature of the lens in order to focus objects at different distances in front of the eye.

The preceding discussion shows that the eye is not a particularly simple optical system. The spherical cornea, which has an almost constant thickness, would have no refracting effect if it had air on both sides. However, since it has air on one side and aqueous humor on the other, there is converging refraction for incident light. For a normal eye the power of the cornea with the aqueous humor behind it is about +40 diopters. The variable lens has an additional converging effect in producing the final focused image on the retina. With the ciliary muscles relaxed, the power of the lens is about +20 diopters; an object at a great distance is focused on the retina as illustrated in Fig. 23.2(a). When the ciliary muscles adjust the lens for maximum curvature, the power of the lens increases to about +24 diopters; an object 25 cm from the eye is focused on the retina as shown in Fig. 23.2(b). The total power of the normal eye, including both the cornea and the lens, is about +60 diopters with the ciliary muscles relaxed, and an object that is in focus under this condition is at the *far point* of the normal eye. With the lens at maximum curvature, the total power is about +64 diopters, and an object in focus under this condition is at the *near point* of the normal eye. As shown in Fig. 23.2, the near point is about 25 cm away and the far point is at a very large distance.

EXAMPLE 23.1

Show that the change in power from +60 diopters to +64 diopters for the normal eye does in fact make the near point 25 cm from the eye.

SOLUTION We regard the relaxed eye as a single lens of power +60 diopters. The lens equation, Eq. (22.9a), gives:

$$\frac{1}{f} = \frac{1}{p} + \frac{1}{q}$$

$$60 \text{ diopters} = \frac{1}{p_{\text{far}}} + \frac{1}{q}$$

With the lens at maximum curvature, the power is +64 diopters, and the lens equation gives:

$$64 \text{ diopters} = \frac{1}{p_{\text{near}}} + \frac{1}{q}$$

(Note that the distance between the lens and the retina does not change, so the image distance is the same in both cases.) Subtracting the first equation from the second we have:

$$4 \text{ diopters} = \frac{1}{p_{\text{near}}} - \frac{1}{p_{\text{far}}}$$

But $1/p_{\text{far}} \approx 0$, since p_{far} is a very large distance. It follows that:

$$4 \text{ diopters} \approx \frac{1}{p_{\text{near}}}$$

$$\therefore p_{\text{near}} \approx 0.25 \text{ m}$$

(Since a diopter is a reciprocal meter, we must use meters for the image and object distances if the power is in diopters.)

Note that the eye forms an inverted image of an object. The image of our whole visual experience is upside-down on the retina! However, the brain interprets the picture right side up. ■

23.2 Corrective Eyeglasses for Visual Defects

Most defects of vision are due to either too much or too little focusing power in the cornea and the lens. All that is required to remedy these situations are simple corrective lenses. Distortions of the cornea or the lens cause more complex irregularities that require correction by lenses with nonspherical surfaces, but we will not discuss such lenses here.

One of the most frequent defects of vision is when the relaxed eye does not focus properly on distant objects. The image may be in front of the retina if the eyeball is too long or if the power of the cornea-lens system is too great. On the other hand, an eyeball that is too short or a cornea-lens system that has too little power places the image of a distant object behind the retina of the relaxed eye. The first of these two conditions is called **myopia,** which is illustrated in Fig. 23.3(a). The relaxed eye has too much focusing power and needs to be corrected by a **diverging lens,** as shown in Fig. 23.3(b). An example will show us how to calculate the required lens power.

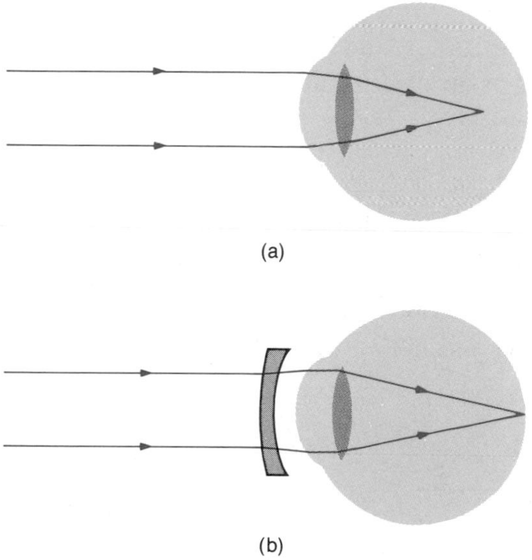

FIGURE 23.3 Illustration of myopia. (a) With the ciliary muscles relaxed, the image of a distant object is formed in front of the retina. (b) A suitable diverging lens results in the proper focus of a distant object.

EXAMPLE 23.2

A myopic person has a far point of 2 m and a near point of 18 cm.

a. What power lens should be prescribed to correct the myopia?

b. Where is the near point when the person wears the corrective lenses?

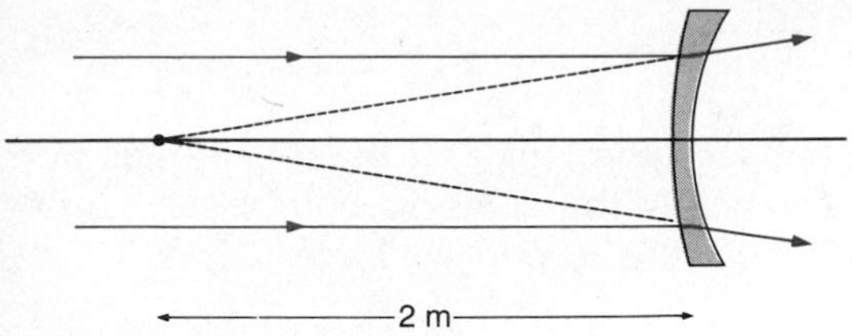

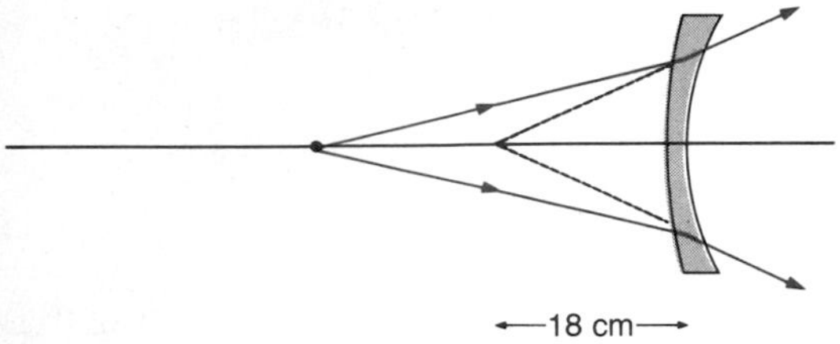

SOLUTION

a. If the far point of the myopic eye is 2 m, the corrective lens must make an image of a distant object 2 m in front of the eye. We therefore have the following data for our problem:

$$p = \infty \quad \text{(the object is at a large distance)}$$

$$q = -2 \text{ m} \quad \text{(the image is virtual)}$$

$$f = ?$$

The lens equation, Eq. (22.9a), gives:

$$\frac{1}{f} = \frac{1}{p} + \frac{1}{q}$$

$$= \frac{1}{\infty} - \frac{1}{2 \text{ m}}$$

$$\therefore f = -2 \text{ m}$$

The required lens power from Eq. (22.11) is:

$$\text{Power} = \frac{1}{f}$$

$$= \frac{1}{-2 \text{ m}}$$

$$= -0.5 \text{ diopter}$$

b. The near point of the unaided eye is 18 cm. We therefore need to calculate the distance of an object that has an image 18 cm in front of the lens. Our data are:

$$p = ?$$

$$q = -0.18 \text{ m} \quad \text{(the image is virtual)}$$

$$f = -2 \text{ m}$$

The lens equation, Eq. (21.9a), gives:

$$\frac{1}{f} = \frac{1}{p} + \frac{1}{q}$$

$$\frac{1}{-2 \text{ m}} = \frac{1}{p} + \frac{1}{-0.18 \text{ m}}$$

$$\therefore p = 19.8 \text{ cm}$$

Without spectacles the range of distances for which objects are sharply focused is from 18 cm to 2 m. With spectacles, objects are sharply focused that are 19.8 cm to a very large distance away.

■

VISIBLE-REGION SPECTRA FOR SELECTED ELEMENTS

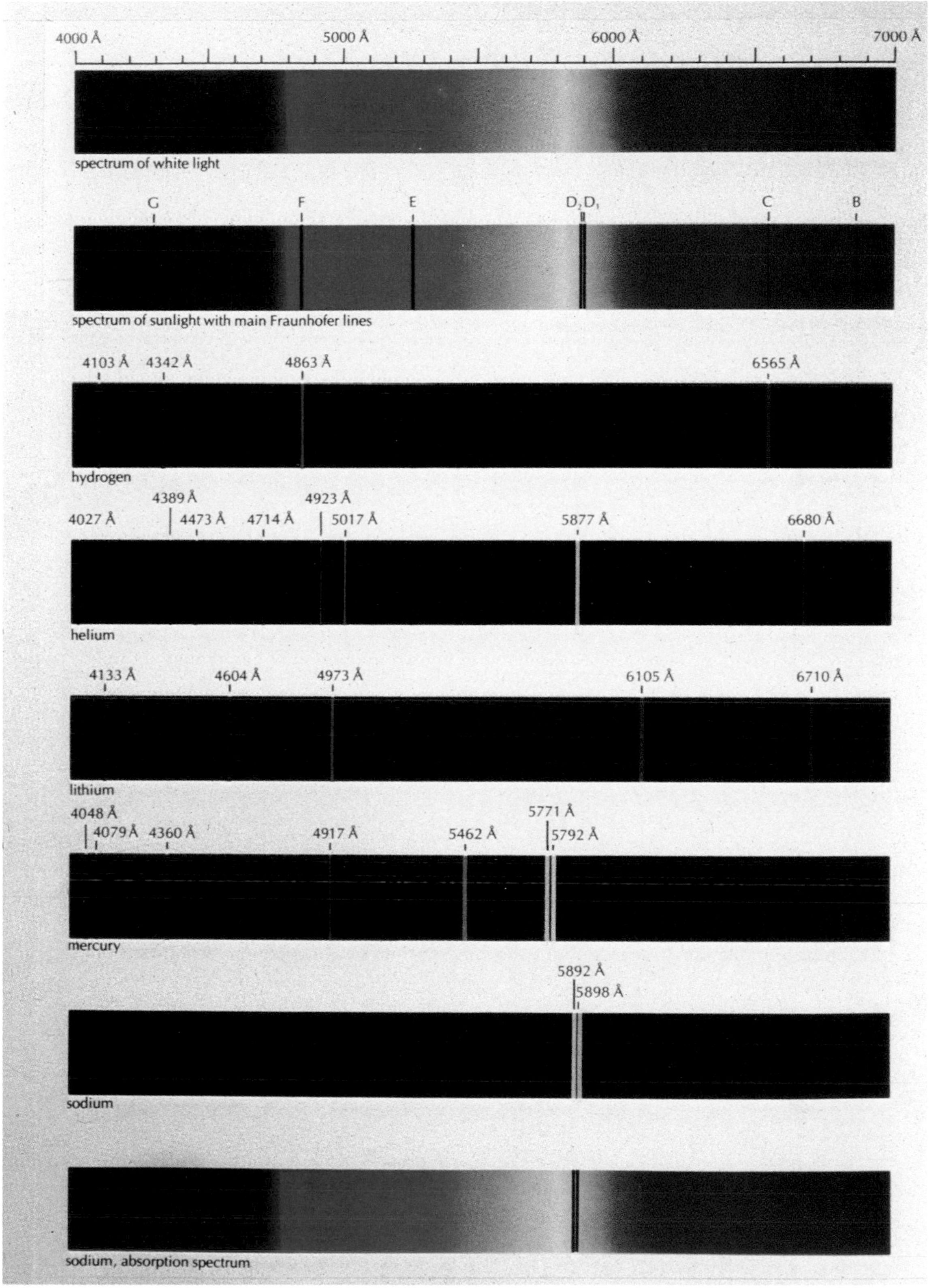

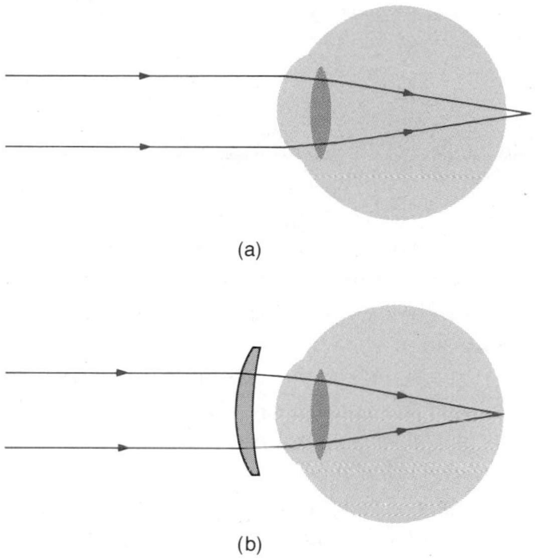

FIGURE 23.4 Illustration of hypermetropia. (a) With the ciliary muscles relaxed, the image of a distant object is formed behind the retina. (b) A suitable converging lens results in the proper focus of a distant object.

The other simple visual defect places the image of a distant object behind the retina of the relaxed eye. This condition, called **hypermetropia,** is illustrated in Fig. 23.4(a). In this case the relaxed eye has too little focusing power and must use some of its accommodation (that is, the ability to alter the focal length of the lens) to focus on distant objects. As a result there is insufficient accommodation to focus an object that is 25 cm away. A **converging lens** is required to correct this condition. An example will illustrate how to calculate the required lens power.

EXAMPLE 23.3

A hypermetropic patient has a near point of 60 cm. Calculate the lens power required to remedy this defect.

SOLUTION Since the unaided eye has a near point of 60 cm, an object at the normal reading distance of 25 cm must form a virtual image in the lens at a distance of 60 cm. Our data for the problem are:

$$p = 25 \text{ cm}$$
$$q = -60 \text{ cm} \quad \text{(the image is virtual)}$$
$$f = ?$$

Equation (22.9a) gives:

$$\frac{1}{f} = \frac{1}{p} + \frac{1}{q}$$
$$= \frac{1}{25 \text{ cm}} + \frac{1}{-60 \text{ cm}}$$
$$\therefore f = 42.8 \text{ cm}$$

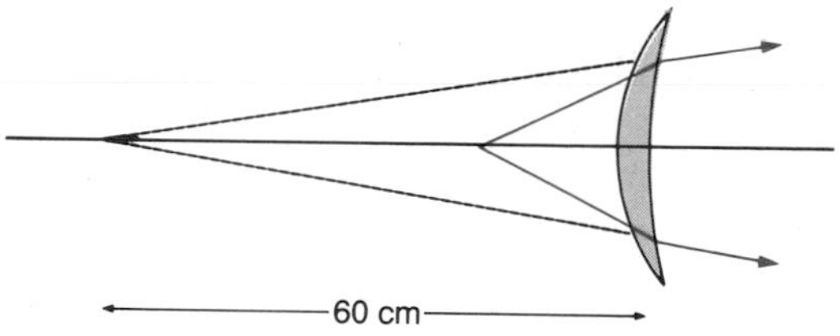

The required lens power is given by Eq. (22.11):

$$\text{Power} = \frac{1}{f}$$
$$= \frac{1}{0.428 \text{ m}}$$
$$= +2.33 \text{ diopters}$$ ■

The ability of the eye to accommodate declines with advancing years. Most middle-aged persons who have normal distant vision are unable to focus on objects as close as 25 cm, a condition known as *presbyopia.* Correction requires a converging lens to move the near point of the eye inward, just as for hypermetropia. However, since no correction is required for distant vision, bifocal spectacles are prescribed, which have converging lenses in the lower portions and either no lenses or glass having no focusing power in the upper portions.

23.3 The Magnifying Glass

So far we have considered the ability of the eye to focus an object's image on the retina. Now we will consider the apparent size of an object of vision. We often use a magnifying glass to examine the fine details of an object, such as a postage stamp or a coin, because the glass enlarges the image on the retina. We will begin by discussing the apparent size of objects seen without the help of a magnifying glass and then describe the effect of the glass.

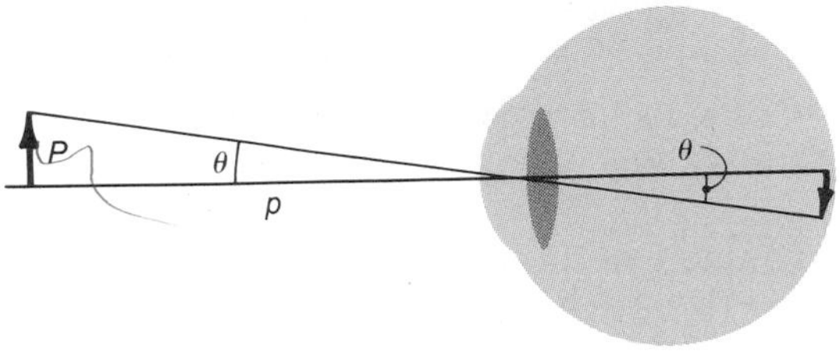

FIGURE 23.5 Illustration of the angular size of an object.

Consider an object of height P that is a distance p from the eye, as shown in Fig. 23.5. The size of the image on the retina is determined by the angle θ; the image is large for a large value of this angle and smaller for a small value. In most circumstances the angle is very much smaller than 1 rad. We can therefore use the approximation of Eq. (1.6) and put the angle θ approximately equal to the object size divided by the object distance. This angle is called the **angular size** of the object.

Definition of Angular Size

$$\text{Angular size} \approx \frac{\text{object size}}{\text{object distance}}$$

$$\theta \approx \frac{P}{p} \qquad \textbf{(23.1)}$$

The object size in Eq. (23.1) is considered to be the dimension of the object in a direction at right angles to the line of vision.

EXAMPLE 23.4

A penny is about 19 mm in diameter, and the lettering of the top inscription is about 1 mm high. Calculate the angular size of the penny and one of the letters of the top inscription if the penny is held at the near point of normal vision (25 cm).

SOLUTION We need only apply Eq. (23.1) twice. For the coin:

$$\theta \approx \frac{P}{p} = \frac{19 \text{ mm}}{250 \text{ mm}} = 0.076 \text{ rad}$$

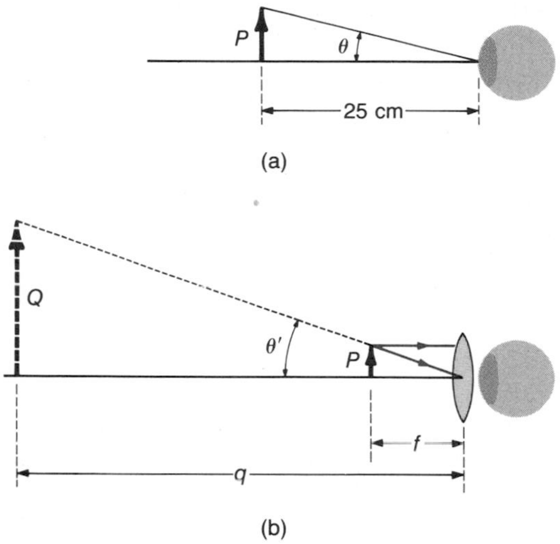

FIGURE 23.6 Illustration of the operation of the magnifying glass: (a) viewing an object that is 25 cm distant with the unaided eye; (b) viewing the virtual image of the same object through a converging lens; the object is at the focal point of the lens, and the virtual image is very distant.

For the small lettering:

$$\theta \approx \frac{P}{p} = \frac{1\ \text{mm}}{250\ \text{mm}} = 0.004\ \text{rad}$$

■

Suppose that we wish to have a larger image of some small object, such as the lettering on a coin. Without an optical aid we can only bring the object closer, thereby reducing the value of p and consequently increasing the value of θ. However, the eye cannot focus on an object that is too close, and the image on the retina is blurred. The simplest optical device that permits us to bring the object closer without blurring the image is a converging lens placed close to the eye. This is a simple magnifying glass, and its operation is illustrated in Fig. 23.6. In Fig. 23.6(a) an object of height P that is 25 cm distant from the unaided eye has an angular size given by:

$$\theta = \frac{P}{25\ \text{cm}}$$

(P must be expressed in centimeters.) In Fig. 23.6(b) the same object is viewed through a converging lens. We set the object almost exactly at the focal point of the lens (but just slightly closer to the lens in order to make sure that the image is virtual). The virtual, erect image will be very large and at a very great distance in front of the lens. The eye forms (on the retina) a real image of this virtual image whose angular size it perceives to be:

$$\theta' = \frac{Q}{q} = \frac{P}{f}$$

This result follows from the similar triangles in Fig. 23.6(b).

The angular magnification of an optical instrument is equal to the angular size of an object when viewed through the instrument divided by the angular size when viewed by the unaided eye. Our definition is somewhat ambiguous because the angular size when viewed by the unaided eye depends on where the object is, and the angular size when viewed through the instrument depends on where the image is. For the unaided eye case, we take the object as 25 cm distant since this gives the largest angular size; with the instrument in use, we take the virtual image at a very great distance, since this corresponds to the comfortable, relaxed-eye usage of an optical instrument. Surprisingly, the choice of the virtual image position (between 25 cm and a very large distance) makes little difference in the result; we will check this later by working an example.

Definition of Angular Magnification of an Optical Instrument

$$\text{Angular Magnification} = \frac{\text{angular size with the instrument}}{\text{angular size with unaided eye}}$$

$$\mathcal{M} = \frac{\theta'}{\theta} \qquad \textbf{(23.2)}$$

We use the script $\mathcal{M}$ for angular magnification in order to distinguish it from the image magnification of the previous chapter.

For the magnifying glass, use of the values calculated above gives us the following result:

Formula for the Angular Magnification of a Converging Lens

$$\mathcal{M} = \frac{25\ \text{cm}}{f} \qquad \textbf{(23.3)}$$

EXAMPLE 23.5

Calculate the apparent angular size of the lettering on the coin of Example 23.4 when viewed through a magnifying glass of focal length 5 cm.

SOLUTION The angular magnification of the glass is given by Eq. (23.3):

$$\mathcal{M} = \frac{25\ \text{cm}}{f} = \frac{25\ \text{cm}}{5\ \text{cm}} = 5$$

From Example 23.4 the angular size with the unaided eye is 0.004 rad, so the apparent angular size with the magnifying glass is given by Eq. (23.2):

$$\theta' = \mathcal{M}\theta = 5 \times 0.004\ \text{rad} = 0.020\ \text{rad}$$ ■

In our discussion of the magnifying glass, we have placed the virtual image at a very great distance since this corresponds to viewing with the relaxed eye. However, the normal eye can focus on a virtual image that is anywhere from 25 cm up to a very great distance in front of it. We now use an example to show that thc position of the virtual image makes only a very small change in the angular magnification.

EXAMPLE 23.6

Calculate the angular modification and the position of the object for a magnifying glass with a 2.5-cm focal length when the distance from the lens to the image is (a) 25 cm, (b) 100 cm, and (c) very large.

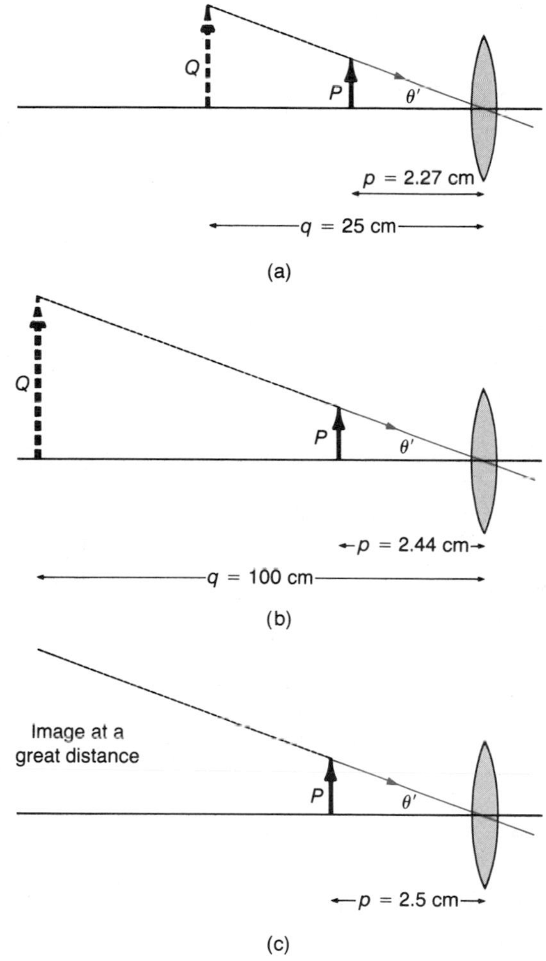

SOLUTION For all cases $\theta = P/25$ cm, and it remains only to find θ' for each image distance.

a. For a 25-cm image distance, $q = -25$ cm, and we use Eq. (22.9a):

$$\frac{1}{f} = \frac{1}{p} + \frac{1}{q}$$

$$\frac{1}{2.5\ \text{cm}} = \frac{1}{p} + \frac{1}{-25\ \text{cm}}$$

$$\therefore p = 2.27\ \text{cm}$$

$$\theta' = \frac{Q}{25\ \text{cm}} = \frac{P}{2.27\ \text{cm}}$$

$$\therefore \mathcal{M} = \frac{\theta'}{\theta} = \frac{P/2.27\ \text{cm}}{P/25\ \text{cm}}$$

$$= 11.0$$

b. For a 100-cm image distance, $q = -100$ cm, and again Eq. (22.9a) gives:

$$\frac{1}{f} = \frac{1}{p} + \frac{1}{q}$$

$$\frac{1}{2.5\ \text{cm}} = \frac{1}{p} + \frac{1}{-100\ \text{cm}}$$

$$\therefore p = 2.44\ \text{cm}$$

$$\theta' = \frac{Q}{100\ \text{cm}} = \frac{P}{2.44\ \text{cm}}$$

$$\therefore \mathcal{M} = \frac{\theta'}{\theta} = \frac{P/2.44\ \text{cm}}{P/25\ \text{cm}}$$

$$= 10.2$$

c. The very far image corresponds to our standard situation of Eq. (23.3); the object distance is given by $p = 2.5$ cm, and the angular magnification is:

$$\mathcal{M} = \frac{25\ \text{cm}}{f} = \frac{25\ \text{cm}}{2.5\ \text{cm}}$$

$$= 10.0$$

■

The example shows that the angular magnification changes very little indeed as the position of the image moves from 25 cm to a very great distance. The object position varies from just inside the focal point to right at the focal point. Persons using a magnifying lens frequently place it so that the image is at a considerable distance in front of the eye. This results in a slight loss of angular magnification, but the distant object can be viewed with a relaxed eye.

23.4 The Compound Microscope

The magnifying power of a single converging lens increases as we shorten the focal length of the lens, but for various practical reasons the focal length cannot be reduced below several centimeters. We see from Eq. (23.3) that this leads to an angular magnification of about 10.

Much larger magnification can be obtained by using two converging lenses in an instrument called the **compound microscope.** The formation of the image by the two lenses is shown in Fig. 23.7. The object P is placed just outside the focal point of the first lens, which is called the **objective.** An inverted real image Q is formed at the focal point of the second lens, which is called the **eyepiece.** The eyepiece functions as a magnifying glass and forms an inverted virtual image R of the real image Q. Finally, the eye itself forms a real image S of the virtual image R. The final image on the retina is erect and therefore seen by the eye as inverted. To put the matter in its simplest terms:

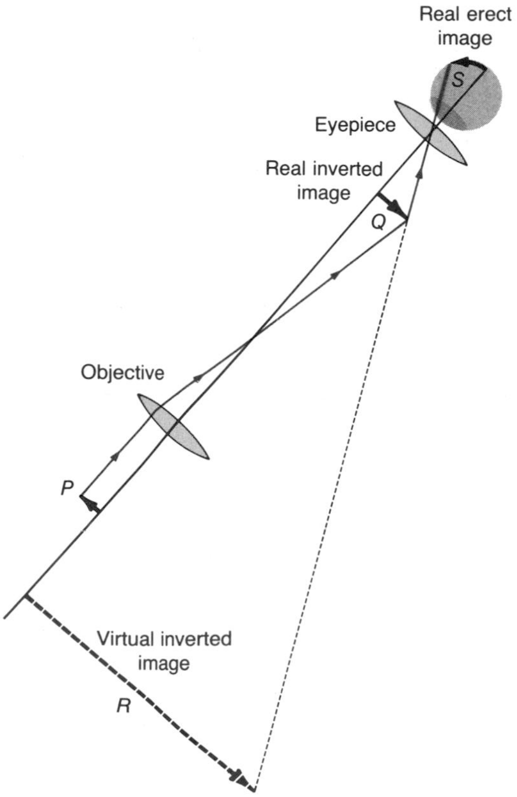

FIGURE 23.7 Formation of an image by a compound microscope.

1. The objective produces a real magnified image.
2. The eyepiece is used as a magnifying glass on the already magnified real image.

It follows from this that the angular magnification of the instrument is the product of the magnification produced by the objective and the angular magnification of the eyepiece; that is,

$$\mathcal{M} = M_o \mathcal{M}_e \tag{23.4}$$

We can easily find the magnification M_o using the methods of Chapter 22. Let the objective of focal length f_o and the eyepiece of focal length f_e be a distance L apart, as shown in Fig. 23.8. Since we require the real image to be at the focal point of the eyepiece, the image distance must be given by:

$$q = L - f_e$$

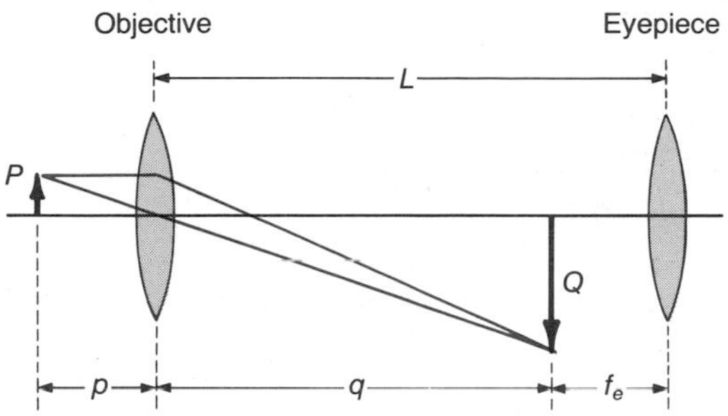

FIGURE 23.8 Diagram to illustrate the calculation of the magnification in the formation of the real image by the objective.

Then using Eq. (22.9a) we have:

$$\frac{1}{f_o} = \frac{1}{p} + \frac{1}{L - f_e}$$

Multiplying throughout by $(L - f_e)$ and transposing gives:

$$-\frac{L - f_e}{p} = -\left\{\frac{L - (f_o + f_e)}{f_o}\right\}$$

But Eq. (22.9b) shows that:

$$M = -\frac{q}{p} = -\frac{L - f_e}{p}$$

is just the magnification of the objective, and we therefore have:

$$M_o = -\left\{\frac{L - (f_o + f_e)}{f_o}\right\} \tag{23.5}$$

Substituting Eqs. (23.3) and (23.5) into Eq. (23.4) gives us the angular magnification of the microscope.

■ Formula for the Angular Magnification of the Compound Microscope

$$\mathcal{M} = M_o \mathcal{M}_e$$

$$= \left\{\frac{L - (f_o + f_e)}{f_o}\right\} \times \frac{25 \text{ cm}}{f_e} \tag{23.6}$$

L = distance between the objective and the eyepiece
f_o = objective focal length
f_e = eyepiece focal length
(All distances must be expressed in centimeters.)

We have quoted the angular magnification as a positive quantity in Eq. (23.6), even though the derivation shows it to be negative. The reason is that the angular magnification of optical instruments is almost always quoted as a positive number with the understanding that it refers to the magnitude of the angular magnification. A negative sign simply means that the final image is inverted.

EXAMPLE 23.7

The objective and the eyepiece of a microscope each have a focal length of 2.5 cm, and the distance between the two lenses is 30 cm. Calculate the angular magnification.

SOLUTION The problem is solved by application of Eq. (23.6):

$$\mathcal{M} = \left\{\frac{L - (f_o + f_e)}{f_o}\right\} \times \frac{25\text{ cm}}{f_e}$$

$$= \left\{\frac{30\text{ cm} - (2.5\text{ cm} + 2.5\text{ cm})}{2.5\text{ cm}}\right\} \times \frac{25\text{ cm}}{2.5\text{ cm}}$$

$$= 100$$

A modern high-quality microscope is usually equipped with turrets carrying various lenses of different focal lengths. By turning the turrets, the user can select different combinations of objective and eyepiece. Each lens mount on both turrets is generally marked with its magnification, which corresponds to each of the two terms of Eq. (23.6). For example, the objective of Example 23.7 would be marked $10\times$ and the eyepiece $10\times$. The angular magnification of the instrument is the product of these numbers—that is, 100.

In a high-quality instrument, both the objective and the eyepiece usually consist of several lenses in close proximity, which permits a higher magnification than is possible with a single lens. It also allows for the correction of various lens defects that we will discuss later.

23.5 The Telescope

A **telescope** is an optical device for viewing distant objects. There are many different types, but let us begin with the simplest refracting type, which is usually called an astronomical telescope. This instrument consists of a converging objective lens with a long focal length and an eyepiece with a relatively short focal length. Calculating the angular magnification differs from the microscope case, since we cannot move the object.

The ray diagram for a simple astronomical telescope is shown in Fig. 23.9. A very distant object P forms an inverted real image Q at the focal point of the objective. The light ray from the distant object that passes through the center of the lens makes an angle θ with the axis, which is the same as the angular size of the object when viewed by the unaided eye. From the geometry of the diagram we have:

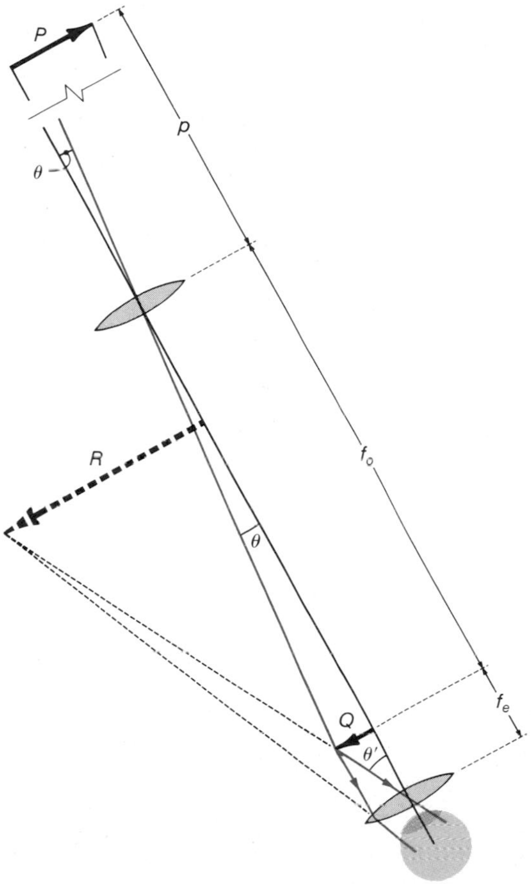

FIGURE 23.9 The formation of an image by an astronomical telescope.

$$\theta \approx \frac{P}{p} = \frac{Q}{f_o}$$

where f_o is the focal length of the objective. The eyepiece plays the role of a simple magnifying glass that is used to view the real image Q. If the image Q is just inside the focal point of the eyepiece, its image R is virtual and inverted. The angular size of the image R is labeled θ' on the diagram, and we have:

$$\theta' \approx \frac{Q}{f_e}$$

We can now use Eq. (23.2) to write a simple expression for the angular magnification of the instrument:

$$\mathcal{M} = \frac{\theta'}{\theta} = \frac{Q/f_e}{Q/f_o} = \frac{f_o}{f_e}$$

Since we see from Fig. 23.9 that the final image is inverted, our result should be negative. However, just as for the microscope, the angular magnification of telescopes is almost always quoted as a positive number.

■ Formula for the Angular Magnification of an Astronomical Telescope

$$\text{Angular magnification} \approx \frac{\text{objective focal length}}{\text{eyepiece focal length}}$$

$$\mathcal{M} = \frac{f_o}{f_e} \qquad \textbf{(23.7)}$$

EXAMPLE 23.8

An astronomical telescope has an angular magnification of 25; the eyepiece has a focal length of 5 cm. Calculate the focal length of the objective and the distance between the two lenses.

SOLUTION We can find the required focal length for the objective from Eq. (23.7):

$$\mathcal{M} = \frac{f_o}{f_e}$$

$$25 = \frac{f_o}{5\text{ cm}}$$

$$\therefore f_o = 125\text{ cm}$$

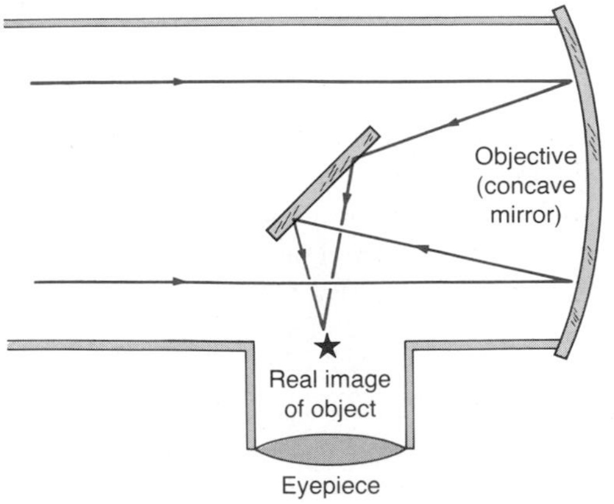

FIGURE 23.10 A reflecting astronomical telescope.

Consulting Fig. 23.9, we see that the distance between the two lenses is equal to the sum of the focal lengths. This means that the telescope is 130 cm long. ■

A faint star must be viewed through a telescope that collects a large amount of the star's emitted light; this requires an objective lens of a very large diameter. As a result, large astronomical telescopes are more frequently of the reflecting type shown in Fig. 23.10. The objective of the telescope is a converging mirror that produces a real image at its focal point. The real image can be viewed through a suitably placed eyepiece. Because the real image is located in front of the concave mirror, a special viewing arrangement is necessary. A small plane mirror inclined at 45° to the main axis reflects the converging rays to the side of the telescope tube, and a transversely placed eyepiece forms the final image. Not only is a large mirror easier to construct than a large lens, but it can be freed of optical defects much more easily. The

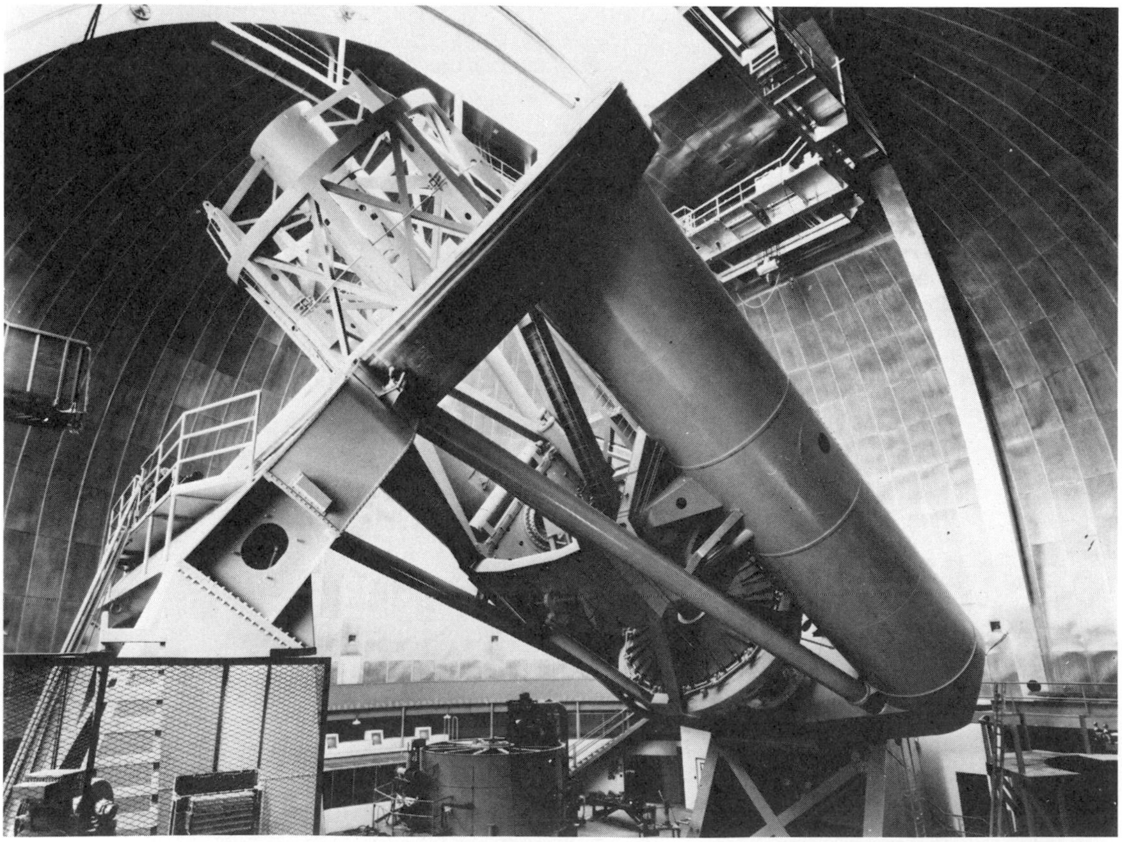

The 200-in. reflecting telescope on Mount Palomar weighs over 500 tons, with the mirror alone weighing nearly 15 tons. The large light-collecting capability of this instrument makes it invaluable for studying faint stars. (Courtesy of the Archives, California Institute of Technology)

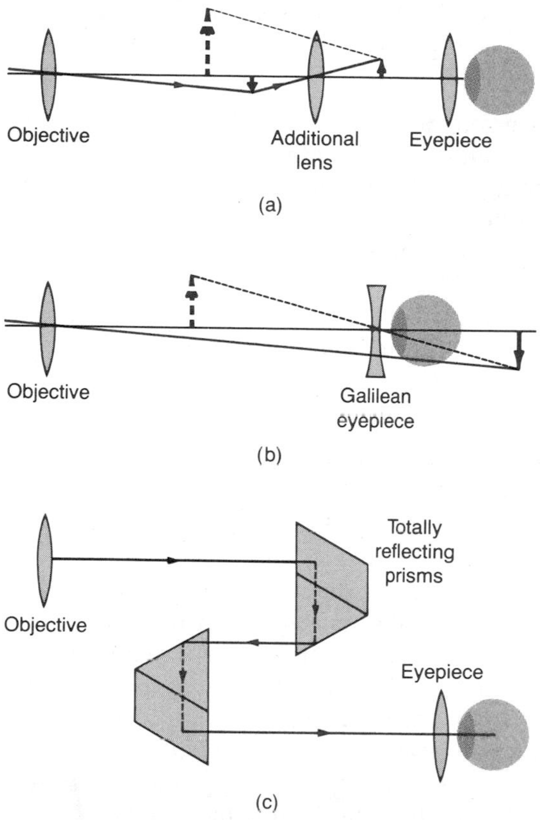

FIGURE 23.11 Types of terrestrial telescope: (a) a simple astronomical telescope using an additional converging lens; (b) an opera glass with a Galilean eyepiece; (c) one side of the optical system of prismatic binoculars.

magnification of a reflecting telescope is still given by Eq. (23.7), where f_o is the focal length of the objective mirror.

Two features make the refracting astronomical telescope generally unsuitable for viewing terrestrial objects. First, the observer sees the object upside down, and second, the instrument is inconveniently bulky because of its great length. Several methods for overcoming these drawbacks are illustrated in Fig. 23.11. Perhaps the most obvious way to have the final image erect (that is, inverted on the retina) is to use an additional converging lens between the objective and the eyepiece, as shown in Fig. 23.11(a). The inverted real image formed by the objective is outside the focal point of the additional lens and therefore produces an erect real image in front of the eyepiece. The eyepiece itself produces an erect virtual image, which causes an inverted image on the retina. Such a telescope does indeed see the world right side up, but it is even longer than the astronomical telescope because of the additional lens. An instrument known as a spyglass was made in this way. The casing frequently consisted of concentric cylindrical tubes that allowed the length to be reduced to a more manageable size when the instrument was not in use.

Another approach to the problem was pioneered by Galileo himself in the early years of the seventeenth century. The system is illustrated in Fig. 23.11(b). A converging objective forms a real image of a distant object, and a diverging eyepiece is placed *in front of* the real image at a distance slightly greater than its focal length. A diverging lens used in this manner is called a Galilean eyepiece. It produces a magnified erect virtual image as shown in the diagram. The system has two advantages—it produces an erect image, and it is shorter than the comparable astronomical telescope (since the eyepiece is inside the focal point of the objective). This arrangement of lenses is frequently used in inexpensive opera glasses of relatively low magnifying power.

EXAMPLE 23.9

An opera glass has an objective with a focal length of 20 cm and a Galilean eyepiece with a focal length of -4 cm.

a. Calculate the magnifying power and the length of the instrument.

b. Check this calculation with a ray diagram. (Assume that the eyepiece is 4.4 cm from the focal point of the objective.)

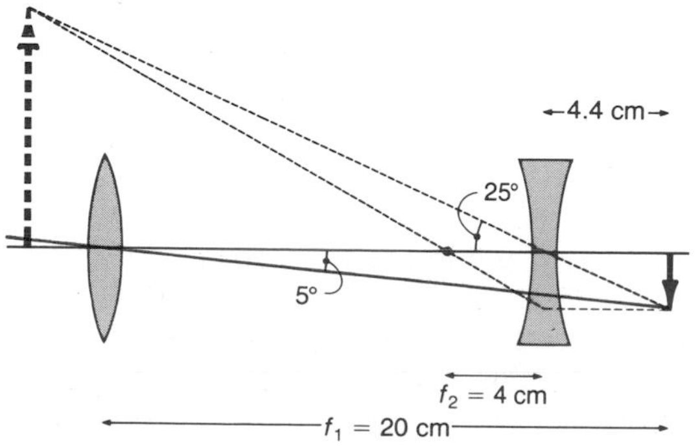

SOLUTION

a. Let us assume that Eq. (23.7) gives the angular magnification correctly. Then

$$\mathcal{M} = \frac{f_o}{f_e} = \frac{20\text{ cm}}{4\text{ cm}} = 5$$

Since the eyepiece is inside the focal point of the objective, the length of the instrument is approximately equal to the difference of the two focal lengths—that is, 16 cm.

b. To draw the ray diagram, we take a convenient angular size for the distant object as seen by the unaided eyes. Let us select 5° for this quantity. The image in the objective would be at the focal point, but this image is behind the eyepiece and forms a virtual object for it. The rays through the center and the focal point of the eyepiece form the virtual image shown. The eye of the observer adjacent to the eyepiece sees an apparent angular size of about 25° as measured from the diagram. This confirms the assumption of our calculation. ■

A third type of terrestrial telescope is illustrated in Fig. 23.11(c). The eyepiece and the objective are relatively close to each other, but the distance traveled by the light rays is much longer because of total internal reflection in the two prisms, which are situated as shown in the diagram. Besides increasing the light path distance between the objective and its focal point, the two prisms also turn the image right side up. When a lens produces an inverted image, it also interchanges right and left. One prism in the binoculars interchanges top and bottom, and the other interchanges right and left; the net result is an erect image with correct disposition of the right and left sides. The diagram shows only the path of a light ray that comes along the axis of the objective, because a ray diagram showing image formation would be very complicated. Prismatic binoculars are compact and rugged, and are used in most situations requiring a terrestrial telescope.

Prismatic binoculars are usually marked with a pair of numbers that give the angular magnification and the objective diameter in millimeters. For example, a marking of 10 × 35 means that the angular magnification is 10 and the diameter of the objective is 35 mm. The objective diameter is given because it indicates the amount of light that comes through the lens to form the image. For viewing in dim light, binoculars marked 7 × 50 would usually be preferred. Even though we sacrifice some magnification, the larger-diameter objective lens creates a brighter image.

23.6 The Prism Spectroscope

Up to this point we have treated the index of refraction as a constant number for any ray of visible light, but recall that the data of Table 22.2 are averages over the range of visible light. Table 23.1, which gives the refractive indexes of crown glass, flint glass, and diamond for light rays of various wavelengths, shows that

TABLE 23.1 Refractive indexes for several transparent solids for light of various wavelengths

Wavelength and color of light	*Crown glass*	*Flint glass*	*Diamond*
660 nm (red)	1.520	1.662	2.410
610 nm (orange)	1.522	1.665	2.415
580 nm (yellow)	1.523	1.667	2.417
550 nm (green)	1.526	1.674	2.426
470 nm (blue)	1.531	1.684	2.444
410 nm (violet)	1.538	1.698	2.458

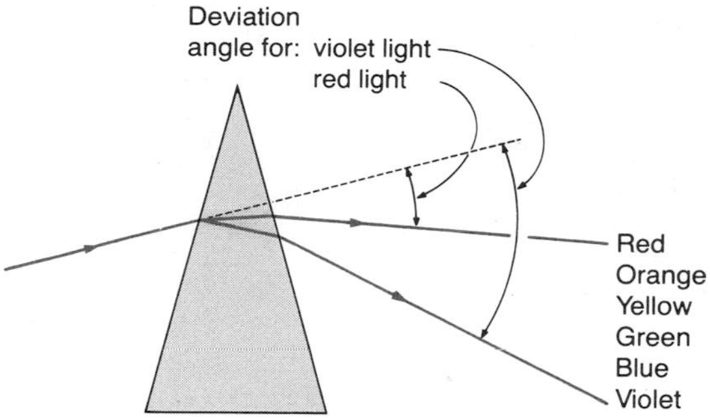

FIGURE 23.12 Dispersion of white light into its component colors by a glass prism.

there are small but noticeable differences in the refractive index as we move from one end of the visible spectrum to the other. We will now investigate some of the important consequences of these differences.

Light that comes from the sun on a clear day around noon is called white light. If a ray of white light shines on the surface of a glass prism, the various color components are refracted differently. This effect, called **dispersion,** is illustrated in Fig. 23.12. The light emerging from the prism is split into different colors traveling in different directions. The total angle through which the light ray is bent, called the **deviation angle,** is smallest for red light and greatest for violet light. Let us investigate dispersion by working a specific example.

EXAMPLE 23.10

A beam of white light strikes the face of a flint glass prism at normal incidence. The prism angle is 30°. Calculate the deviation angles for the red and violet components of the light.

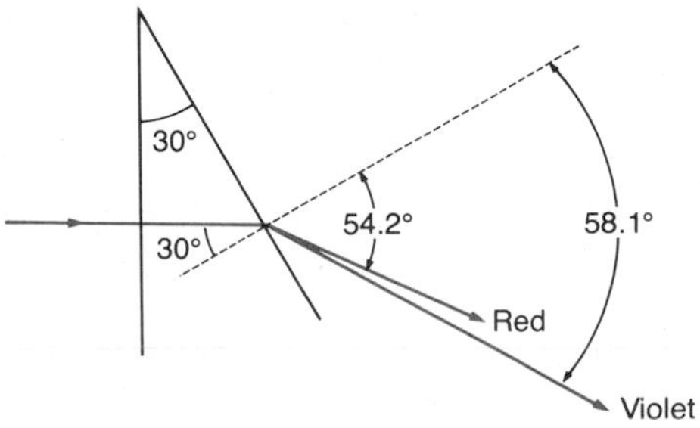

SOLUTION Since the light strikes the face of the prism at normal incidence, no refraction occurs for any component as it enters the prism. When the light exits from the opposite face, the angle of incidence is 30°. We can calculate the angle of refraction for each component separately using Snell's law. For the red and violet components the indexes of refraction for the flint glass are 1.662 and 1.698 from Table 23.1. For the red component, Eq. (22.3) gives:

$$n_1 \sin \theta_1 = n_2 \sin \theta_2$$

$$1.662 \times \sin 30° = 1 \times \sin \theta_2$$

$$\therefore \theta_2 = 56.2°$$

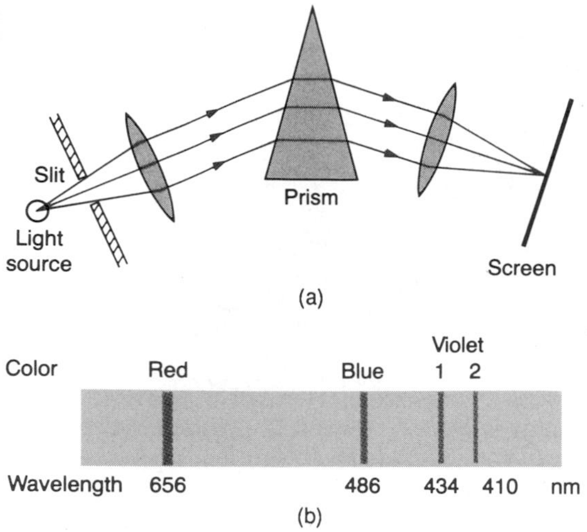

FIGURE 23.13 The operation of a prism spectroscope. (a) Light of a single wavelength shines through a slit, is converged by a lens to a parallel beam, is dispersed by a prism, and is focused by a lens to form an image of the slit on a screen. (b) Light from a hydrogen discharge lamp shows four visible images of the slit on the screen. The colors and wavelengths are marked.

For the violet component we use the same equation:

$$n_1 \sin \theta_1 = n_2 \sin \theta_2$$

$$1.698 \times \sin 30^\circ = 1 \times \sin \theta_2$$

$$\therefore \theta_2 = 58.1^\circ$$

From the geometry of the diagram we see that the angle of deviation of each component is 30° less than the angle of refraction. We conclude that:

$$\text{Angle of deviation for red} = 26.2^\circ$$

$$\text{Angle of deviation for violet} = 28.1^\circ$$

■

The example deals with the normal incidence of white light on the prism face, and the resulting geometry is fairly simple, with deviation occurring only at the second prism face. If light falls at other-than-normal incidence, there is deviation at both faces of the prism. The resulting geometry is not difficult but does lead to complicated diagrams.

In subsequent chapters we shall discuss the spectrum of the light emitted by an atom. The word *spectrum* refers to the range of values of some physical quantity. In this case it is the range of colors (with corresponding wavelengths) of the emitted light. We have already seen how the prism breaks white light up into its various color components, which include all the colors from red through violet to which the eye is sensitive. However, not all light is white. The emissions from mercury vapor or sodium vapor lamps are visibly different from white light. In fact, the vapor of any pure substance emits a spectrum that is its distinctive fingerprint. Figure 23.13(a) illustrates the components of a prism spectroscope. Light of a single wavelength (called *monochromatic light*) passes through a slit; a converging lens produces parallel rays, which are dispersed by a prism and then focused onto a screen. The image on the screen is an image of the slit in the color of the light. Suppose we now remove the monochromatic light source and replace it with a discharge tube. The light now contains various wavelength components, and an image of the slit appears on the screen for every component color in the light source. Moreover, since the prism deviates the different colors by different amounts, the differently colored images appear at different places on the screen. The spectrum from an atomic hydrogen discharge, shown in Fig. 23.13(b), contains four lines in the visible region—one red, one blue, and two violet. Most atomic spectra are more complex than hydrogen, with some containing dozens, even hundreds, of lines. We defer our explanation of the structure of atomic spectra until the chapter on quantum physics.

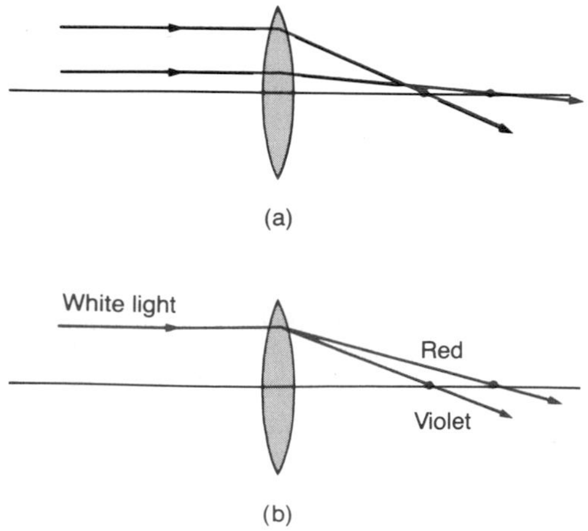

FIGURE 23.14 Two important lens aberrations: (a) spherical aberration; (b) chromatic aberration.

23.7 Lens Aberrations

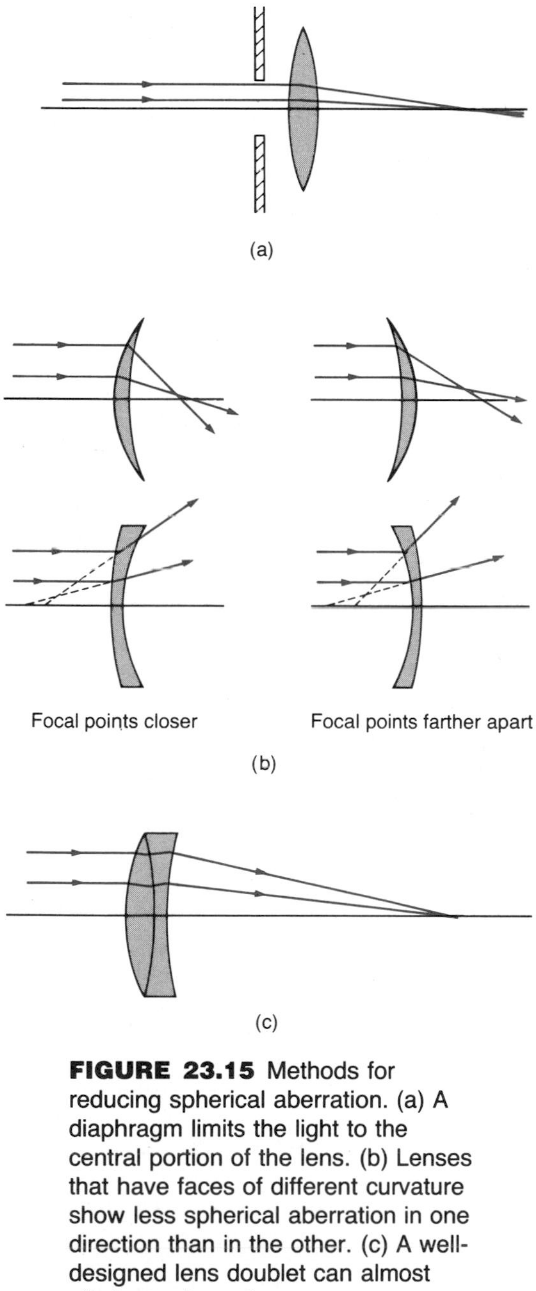

FIGURE 23.15 Methods for reducing spherical aberration. (a) A diaphragm limits the light to the central portion of the lens. (b) Lenses that have faces of different curvature show less spherical aberration in one direction than in the other. (c) A well-designed lens doublet can almost eliminate aberrations.

In discussing the telescope, we mentioned that a lens may have any of a number of defects that may impair its performance. Confining our attention to light rays parallel to the lens axis, we can readily understand two of the most important defects.

Figure 23.14(a) shows two light rays parallel to the axis incident on a converging lens; one ray is close to the axis, and the other is farther out. In a perfect lens both rays pass through the focal point on the other side of the lens, but in a real lens with spherical surfaces the rays near the axis focus at a different point from rays near the outside of the lens. This is called **spherical aberration** since the lack of a single focal point is due to the spherical nature of the lens surfaces. Spherical aberration is one of the factors that cause a lens to form a blurred image.

A second cause of a blurred image is illustrated in Fig. 23.14(b). When white light passes through a lens at any point off the axis, the light is deviated. However, the lens acts like a prism and also disperses the light, causing the focal point for violet light to be closer to the lens than the focal point for red light (the focal points for other color components are in between these two). This effect is called **chromatic aberration.** Any object that emits light components of different colors forms a blurred image since the different color components have different focal points.

These and other aberrations are undesirable in any optical system. Figure 23.15 illustrates various methods for reducing aberrations. Perhaps the most obvious approach is to use a diaphragm that only permits light to come through the central part of the lens, as shown in Fig. 23.15(a). Most cameras are fitted with an adjustable diaphragm for controlling the size of the portion of the lens that is uncovered. The image on the film is more sharply focused when the diameter of the aperture is small, but of course, less light enters the smaller aperture.

If a lens has different curvatures on its two faces, the spherical aberration for parallel light is less in one direction than the other, as shown in Fig. 23.15(b). If the lens is converging, the focal points are closer when the more highly curved surface faces the incoming parallel light. If the lens is diverging, the focal points are closer when the less highly curved surface faces the incoming light. This is one of several reasons for constructing spectacle lenses as shown in the diagrams on the left in Fig. 23.15(b).

Adjacent lenses of different glasses are used in expensive optical instruments and cameras to correct for both spherical and chromatic aberrations. Figure 23.15(c) illustrates such a lens doublet in which a converging crown glass lens is combined with a diverging flint glass lens to produce a single converging unit. Careful design results in the cancellation of the spherical and chromatic aberrations that each lens individually would have. The lens systems of expensive instruments frequently have up to six or more elements to correct aberrations.

KEY CONCEPTS

The normal human eye forms an inverted real image on the retina for objects at least 25 cm distant from it.

If the relaxed eye forms the image of a distant object in front of the retina, the subject has **myopia** and needs **diverging corrective lenses.** If the image is behind the retina under the same conditions, the subject has **hypermetropia** and needs **converging corrective lenses.**

The **angular magnification** of an optical instrument is the angular size of an object viewed through the instrument divided by the angular size seen with the unaided eye; see Eq. (23.1) and Eq. (23.2).

A simple **magnifying glass** is a **converging lens** held directly in front of the viewer's eye; it forms an erect virtual image of an object placed just inside its focal point. The angular magnification is related to the focal length of the lens; see Eq. (23.3).

The **compound microscope** forms enlarged images of close objects. A **converging objective** lens forms a real image of an object placed just outside its focal point, while the **converging eyepiece** lens produces a final virtual image for the viewer. The angular magnification depends on the focal lengths of the two lenses and on their separation; see Eq. (23.6).

The **telescope** is used to view distant objects. A **converging objective** lens or mirror forms a real image of a distant object near its focal point. A **converging eyepiece** behind the real image (or a **diverging eyepiece** in front of it) forms a virtual image for the viewer. The angular magnification is equal to the objective focal length divided by the eyepiece focal length in every case; see Eq. (23.7).

The refractive index of transparent materials is slightly dependent on the wavelength of the light. This results in the phenomenon of **dispersion.** A leading example of dispersion is the different amounts by which different wavelengths of light passing through a prism deviate.

All lenses produce distorted images to some degree. **Spherical aberration** results from the spherical nature of the lens surfaces, and **chromatic aberration** from dispersion in the material of the lens.

QUESTIONS FOR THOUGHT

1. When you swim laps in a pool, it is impossible to focus the lines on the bottom of the pool. However, if you wear goggles this problem disappears. Explain this effect.

2. A student who has not taken physics claims that microscopes and telescopes are really the same instruments, since they are both used to produce enlarged images. Discuss this claim and clarify any errors in it.

3. What features in a pair of binoculars would produce (a) the greatest magnification? (b) a clear image in dim light?

4. Two converging lenses have the same focal length, but one has a larger diameter. Explain why you would expect one to form a sharper image than the other.

5. When a diamond sparkles, you observe different colors from some of the facets. Explain this effect.

6. The lane markings on the bottom of a swimming pool frequently have short transverse portions at each end. If you stand at the edge of the pool, you see a narrow red line at one edge of the transverse marker and a narrow blue line at the other edge. Explain this effect.

PROBLEMS

A. Single-Substitution Problems

1. What power lens should be prescribed for a myopic patient whose far point is 85 cm? [23.2]

2. What power lens should be used in reading glasses for a patient whose near point is 85 cm? [23.2]

3. Reading glasses have a focal length of 30 cm. What is the near point of the patient for whom they are prescribed? [23.2]

4. A myopic person wears glasses of focal length −60 cm. What is the person's far point? [23.2]

5. A soccer ball is 22 cm in diameter. At what distance is its angular size 1°? [23.3]

6. Calculate the angular size of a building 65 m high when viewed from a distance of 750 m. [23.3]

7. What focal length is required for a magnifying glass to provide an angular magnification of 6? [23.3]

8. A specially designed magnifying glass has several components, which produce a focal length of 2.25 cm for the whole unit. Calculate the magnifying power. [23.3]

9. A microscope has objective and eyepiece focal lengths of 10 mm and 20 mm, respectively. Calculate the angular magnification if the length of the instrument is 16 cm. [23.4]

10. A microscope that uses objective and eyepiece lenses of focal lengths 12 mm and 16 mm, respectively, has an angular magnification of 160. What is the distance between the lenses on the instrument? [23.4]

11. A student makes a very inexpensive microscope using two low-power lenses, each of focal length 5 cm, that are placed 20 cm apart. What angular magnification do you expect from this instrument? [23.4]

12. The objective of a telescope has a focal length of 120 cm. Calculate the focal length of an eyepiece that will produce an angular magnification of 40. [23.5]

13. The focal lengths of the objective and the eyepiece in a telescope are 120 cm and 4 cm, respectively. Calculate the angular magnification of the instrument. [23.5]

14. A telescope has an eyepiece with a focal length of 3.2 cm. What should be the focal length of an objective to produce an angular magnification of 40? [23.5]

B. Standard-Level Problems

15. A young person with exceptional accommodation sees distant objects sharply with the eye relaxed and total focusing power +60 diopters. The near point is 15 cm. What is the total focusing power of the eye when accommodated to the near point? [23.1]

16. A young person with exceptional accommodation is able to vary the total eye-focusing power from +60 to +66 diopters. When the eye is relaxed with a power of +60 diopters, distant objects are sharply focused on the retina. What is the shortest distance that an object can be from the eye to form a sharp image on the retina? [23.1]

17. A myopic patient has a near point of 18 cm and a far point of 56 cm.

a. What power lens should be prescribed so that the relaxed eye focuses sharply on distant objects?
b. What is the distance to the nearest object that is in good focus when wearing these spectacles? [23.2]

18. A hypermetropic person wears spectacles with a power of +2.25 diopters, which bring the near point to 25 cm. What would the near point be without spectacles?

19. A person suffering from myopia has a near point of 15 cm but has normal accommodating power (that is, the ciliary muscles can change the focusing power by +4 diopters).

a. At what distance is the far point for this person?
b. What power spectacles should be prescribed to correct the myopia?

20. An elderly person with presbyopia sees distant objects clearly without glasses but has a near point of 55 cm. What power reading glasses should be prescribed to bring the near point to 25 cm?

21. Bifocal lenses of power +2.5 diopters and −0.8 diopter are prescribed for an elderly patient. What are the near and far points of the patient's unaided vision?

22. Calculate the angular magnification of a magnifying glass whose power is +24 diopters. [23.3]

23. A magnifying glass of special design has an angular magnification of 25. What is the power of the lens in diopters? [23.3]

24. A high-quality eyepiece has a power of +50 diopters.

a. Calculate the angular magnification of the eyepiece when the virtual image is 25 cm from the eyepiece.
b. Calculate the angular magnification of the eyepiece when the object is at the focal point of the eyepiece.

25. A compound microscope has objective and eyepiece lenses of focal lengths 6 mm and 15 mm, respectively; the lenses are 18 cm apart.

a. Where must the object be placed so that the final virtual image is 25 cm in front of the eyepiece?
b. What is the angular magnification in these circumstances?

c. What is the angular magnification calculated on the usual assumption that the final virtual image is very distant? [23.4]

26. A low-power compound microscope has an objective lens of focal length 8 mm and an eyepiece of focal length 5 cm; the lenses are separated by 8.6 cm. The object is located 10 mm in front of the objective. Calculate:

a. the position of the final image
b. the angular magnification under these circumstances
c. the angular magnification if the position of the object is adjusted to make the final virtual image very distant [23.4]

27. A microscope has an eyepiece of focal length 4.2 mm, which is 18 cm from the objective. What focal length objective is required to give an angular magnification of 1250?

28. A microscope has an eyepiece marked 20× and an objective with a focal length of 3.5 mm. Calculate the angular magnification if the real image formed by the objective is 15 cm from it.

29. An eyepiece of power +40 diopters is available to make an astronomical telescope.

a. What power is required for the objective to give a total angular magnification of 40?
b. How far apart are the two lenses if the eyepiece forms a virtual image at a great distance when the telescope is focused on an object that is 20 m away?
c. How far is it necessary to move the eyepiece [from its position in part (b)] to focus on a very distant object? [23.5]

30. A student constructs a telescope using two lenses, of powers +1.25 diopters and +25 diopters.

a. What is the angular magnification of the telescope?
b. How far apart are the two lenses if the eyepiece forms a virtual image that is 25 cm away when the telescope is focused on a distant object? [23.5]

31. The objective lens of a telescope has a focal length of 120 cm, and the eyepiece has a focal length of 2.5 cm.

a. Calculate the angular magnification of the instrument.
b. An extra lens with a 10-cm focal length is used between the objective and the eyepiece to convert the instrument to a spyglass (that is, the extra lens ensures that the final image is seen erect). How much longer than the telescope must the spyglass be if the angular magnification remains the same?

32. a. Calculate the angular size of the moon when viewed from the earth with the unaided eye. (Use the data on the inside back cover.)
b. An eyepiece with a 3-cm focal length is available for use in a telescope to view the moon. The angular size of the moon when viewed through the telescope is to be the same as an orange 10 cm in diameter when viewed at a distance of 60 cm by the unaided eye. What focal length should be selected for the objective?

33. A pair of opera glasses (Galilean telescopes) have the objective and the eyepiece 12 cm apart. The eyepiece has a focal length of −5 cm. Calculate the angular magnification of a distant object.

34. A Galilean telescope has an objective lens of focal length 60 cm and an eyepiece of focal length −8 cm.

a. How far apart are the two lenses if the final virtual image of a distant object is also at a great distance?
b. The telescope is pointed at an object 10 m away. How far must the eyepiece be moved to refocus the final virtual image at a great distance?

35. A pair of binoculars is marked 7 × 50, and the eyepieces are known to have a focal length of 3.2 cm.

a. What is the angular magnification of the instrument?
b. What are the diameters and focal lengths of the objectives?

36. A pair of binoculars is marked 12 × 85, and the objective lenses are known to have a focal length of 30 cm.

a. Calculate the focal length of the eyepieces.
b. Calculate the angular size of a ship 200 m long and 5 km distant when viewed through the binoculars.

37. An amateur astronomer grinds a concave mirror for a reflecting telescope that has a radius of curvature of 480 cm. What focal lengths should be chosen for the eyepieces to obtain angular magnifications of 50, 100, and 200?

38. White light falls normally on one face of a flint glass prism having a 30° angle. Use Table 23.1 to calculate the angle between the emerging red and violet rays. [23.6]

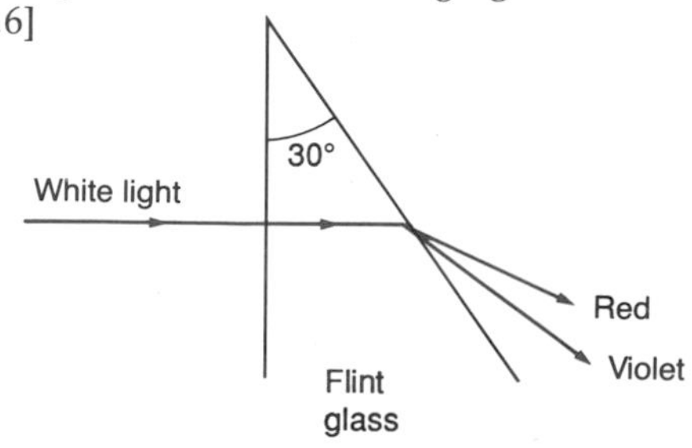

39. A ray of blue light falls on a crown glass prism with a 45° angle. The angle of incidence on the face of the prism is adjusted so that the refracted ray in the glass is parallel to the base of the prism. Use Table 23.1 to calculate the total deviation of the ray. [23.6]

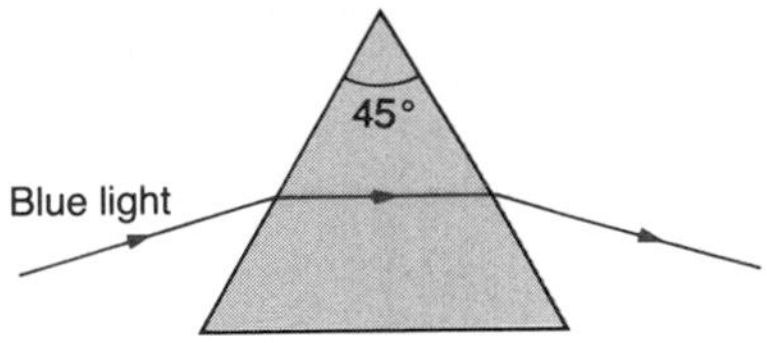

40. A diverging lens of crown glass has one flat surface and one concave surface whose radius of curvature is 38 cm. Use Table 23.1 to calculate the distance between the focal points for red and blue light. [23.7]

41. Use Table 23.1 to calculate the distance between the focal points for blue and red light for a converging flint glass lens. Each surface is convex and has a radius of curvature of 60 cm. [23.7]

C. Advanced-Level Problems

42. An elderly patient has a near point of 80 cm and a far point of 240 cm.

a. Design a bifocal lens system in which one component brings the near point to 25 cm and the other component shifts the far point to a great distance.

b. What is the region of poor vision when wearing the spectacles?

43. The objective lens in a compound microscope has a focal length of 4 mm and is 17 cm from the eyepiece. The final virtual image is to be at a great distance. What should the focal length of the eyepiece be to produce an angular magnification of 500?

44. An amateur astronomer inherits a high-quality eyepiece of focal length 4 mm. He wishes to build an astronomical telescope with 200× magnification and orders a crown glass objective to be made. The lens has two convex surfaces, each having the same radius of curvature. What should the radius of curvature be for the objective? (Use the refractive index for green light in the calculation.)

45. A ray of red light falls on a flint glass prism of angle 45° with an angle of incidence of 60°. Using Table 23.1, calculate the total deviation of the light ray.

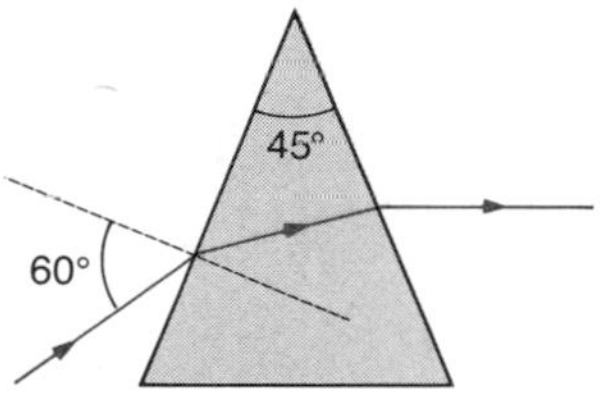

46. A doublet lens consists of plano-convex crown glass and plano-concave flint glass components. The radius of curvature of the convex surface of the crown glass lens is 30 cm. Calculate the radius of curvature for the concave surface of the flint glass lens that is required for the doublet to have the same focal length for both red and blue light. (Use Table 23.1.)

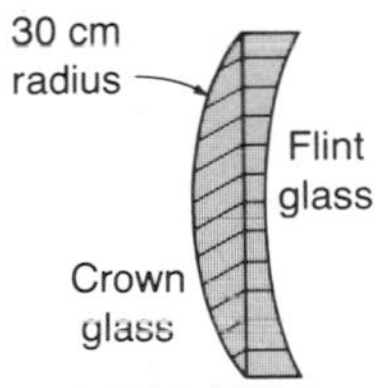

24

WAVE OPTICS

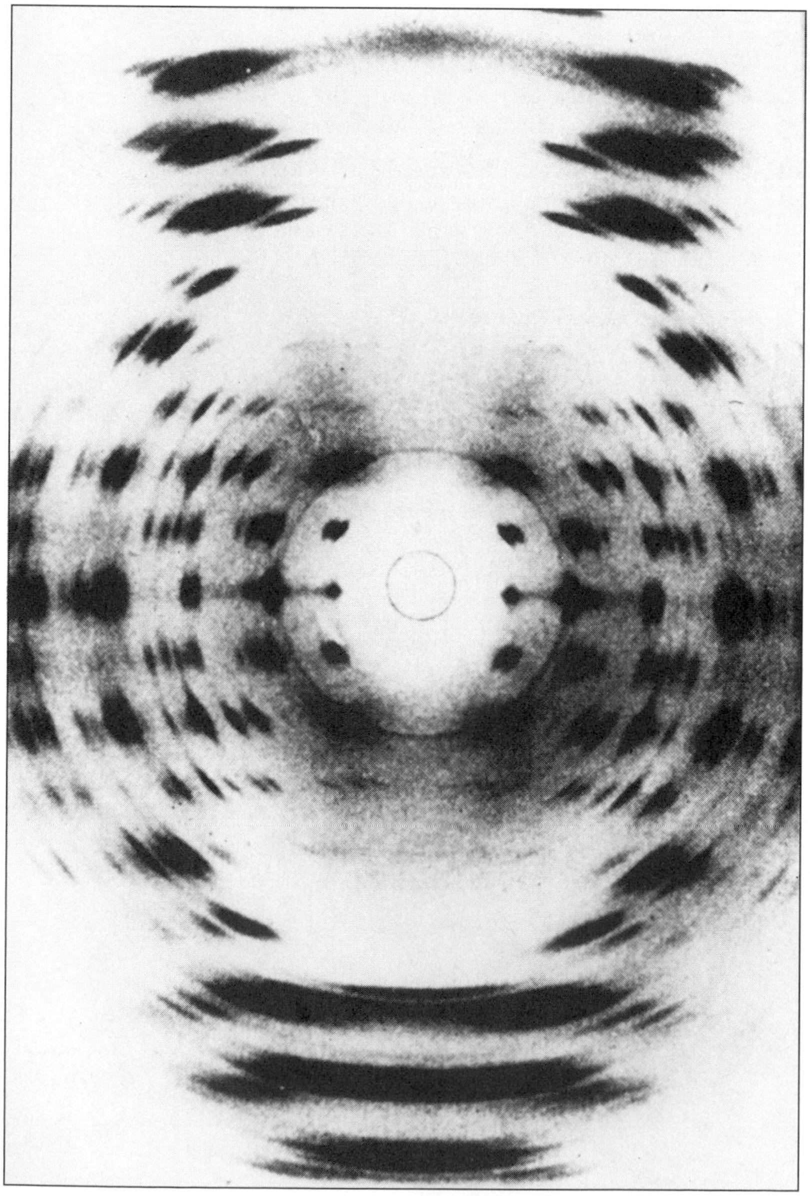

We have already seen that ray optics can explain many of the properties of light, but many interesting and beautiful effects exist that can only be explained by waves, primarily because of the ability of a wave to bend around a corner. However, light certainly does not seem to do so in everyday experience; the effects in question are somewhat out of the ordinary.

We will begin our discussion with a geometrical construction that enables us to visualize the bending of waves in general and to understand the conditions that are required for the phenomenon to be apparent. With the help of this knowledge, we will then propose a relatively simple experiment to show that light does indeed bend around corners—an experiment that was first performed in the early years of the nineteenth century.

24.1 Huygens' Principle

The first major proponent of the wave theory of light was the Dutch scientist Christian Huygens (1629–1695). Newton, a contemporary of Huygens, proposed a corpuscular theory of light, and his great reputation undoubtedly gained adherents to his theory. In addition, the crucial experiments that were to justify Huygens' wave theory had not been performed at that time.

It is said that Huygens gained many of his insights into wave motion by watching ripples on the water in canals. Watching water waves is certainly a good way to visualize wave properties, and we have already used a wave picture to help in the understanding of Snell's law. However, Huygens went a good deal further than our previous example implied by constructing a subsequent wave front when the original one was known. From this study he proposed the following principle:

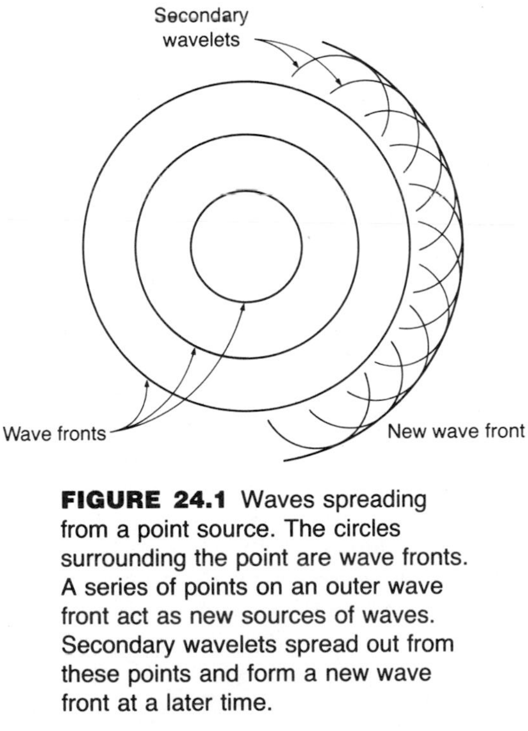

FIGURE 24.1 Waves spreading from a point source. The circles surrounding the point are wave fronts. A series of points on an outer wave front act as new sources of waves. Secondary wavelets spread out from these points and form a new wave front at a later time.

Huygens' Principle

> Every point on a wave front can be regarded as a new point source for waves generated in the direction of the wave propagation.

To see how this principle operates, reconsider our previous example of circular wave fronts spreading from a point disturbance on a calm water surface. Figure 24.1 shows a series of equispaced wave fronts surrounding a point disturbance. Now suppose that one of the wave fronts is known and we wish to construct a subsequent wave front. Huygens' principle instructs us to take every point on the known wave front and turn it into the beginning point for a new disturbance. We do this by selecting a series of points on thewave front and using each one as the center of a new set of concentric wave fronts, which we call secondary wavelets. Figure 24.1 shows a set of wave fronts that represent the first crest of each secondary wavelet. The line that is tangential to all of the secondary wavelet crests is a new wave front of the original disturbance, which in Fig. 24.1 is simply the next concentric wave crest centered on the original disturbance.

You may object that this is a long-winded way of determining the quite obvious position of the new wave front. The objection stands for relatively simple problems, but in many problems the position of the new wave front is not obvious. We will use Huygens' principle to explain the results of the experiment that demonstrated the wave nature of light.

24.2 Interference and Young's Experiment

The first compelling evidence for the wave theory of light did not come from velocity measurements in various media, but rather from an experiment that demonstrated the ability of light to travel around corners.

In order to gain insight into how waves bend around corners, let us consider some specific cases. Figure 24.2(a) shows a plane wave incident on a barrier containing a small hole, which is the only point on the wave front that can send secondary wavelets into the region beyond the barrier. The wave fronts beyond the barrier are therefore circles centered on the hole, and the rays are straight lines traveling outward from the hole. Water waves can penetrate a gap in a barrier and spread throughout the region on the other side. In this manner sound waves traveling through a small opening can be heard by a listener who is not in line with the source of sound and the opening.

However, light does not appear to behave in this way. In fact, Fig. 22.3 implies that light coming through an opening casts a sharp shadow of the opening rather than bending around the edge of the opening, as Fig. 24.2(a) implies that it should. We will let this apparent paradox stand for the moment and consider another example of wave motion.

Figure 24.2(b) shows a plane wave incident on a barrier that contains two small openings. Each opening acts as a point source for secondary wavelets that spread into the region beyond the barrier. Recall that two waves interfere constructively if a crest of one occurs at the same place as the crest of the other, and destructively if the crest of one coincides with a trough of the other. These situations are illustrated in Fig. 12.8. Since the wave fronts in Fig. 24.2(b) represent the positions of crests, the intersection of two wave fronts marks a point where two crests coincide. One of the crests is due to the secondary wavelet from one opening, and the other crest from the other opening. Troughs lie between two consecutive crests, and the points where a crest from one opening

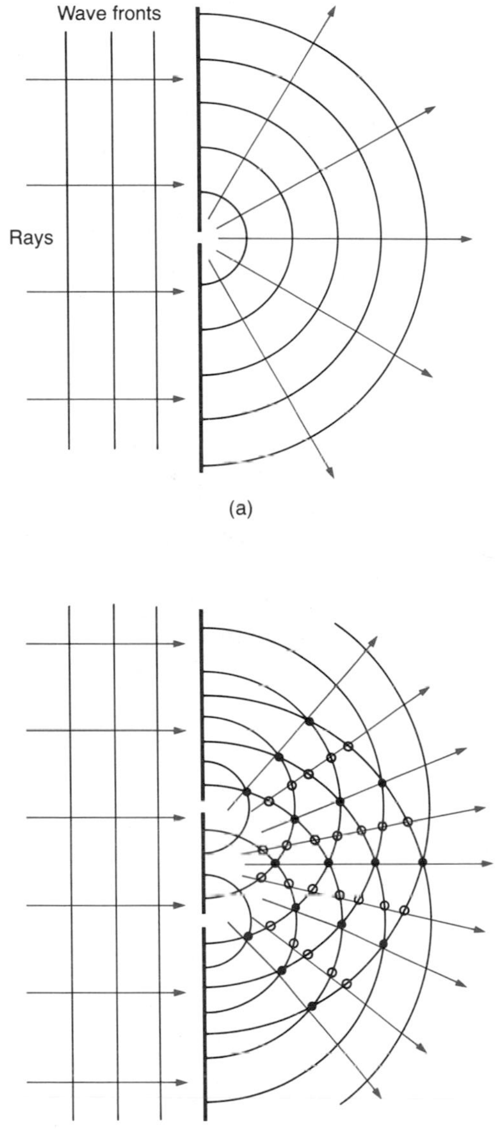

FIGURE 24.2 Wave phenomena when a wave falls on a barrier containing small openings. (a) A plane wave spreads through a single small opening into the region beyond a barrier. (b) A plane wave that falls on a barrier containing two small openings produces a pattern containing both constructive (solid dots) and destructive (open dots) interference.

is halfway between two crests from the other are points of destructive interference.

A noteworthy feature of the diagram is how the points of constructive and destructive interference lie on straight lines radiating outward from a point midway between the two small openings. However, Fig. 24.2(b) shows the wave fronts frozen at some fixed instant of time. What happens to the points of constructive and destructive interference with the passage of time? They move continuously along the trajectories indicated by the straight lines.

If light is a wave motion, then we can interpret Fig. 24.2(b) as follows: A beam of light comprising parallel rays (that is, a plane wave) falls on a barrier containing two narrow slits, or a **double slit.** The disturbance propagated in the region beyond the barrier has rays of maximum brightness (constructive interference) alternating with rays of darkness (destructive interference). A suitably placed screen should reveal a series of light and dark bands. This is exactly the experiment performed by the English physician Thomas Young (1773–1829), whose name we have already seen in connection with the tensile elastic modulus.

The actual experiment that Young performed did not show the spectacular results that we can obtain with modern equipment. Before describing a modern experimental setup, let us examine the assumptions implied in Fig. 24.2(b).

1. The wave that falls on the barrier has a fixed wavelength. The simple nature of our interference pattern would be destroyed by various secondary wavelets of different wavelengths. For this experiment with light, the incident beam should be **monochromatic.**

2. The secondary wavelets that originate from the two small openings are in phase at their point of origin in the openings. Thus, the incident light beam must reach the barrier as a wave front of constant phase. Such a beam is called **coherent.**

3. The openings are small compared with the wavelength of the incident light. (Remember that the wave fronts on the diagram represent consecutive crests, and the distance between them is one wavelength.)

4. The distance between the two openings is not too large compared with the wavelength of the incident light.

Not all of these conditions must be fulfilled to see an interference pattern. Young himself used a beam of sunlight that was certainly not monochromatic. However, our modern version of Young's experiment is easier to interpret if we observe the four conditions. Perhaps the easiest way to produce a monochromatic light beam that presents a wave front of constant phase at the two openings is to use a lower-power laboratory laser. The two small openings can be provided by scratching a pair of closely spaced parallel lines in the emulsion of a piece of exposed and

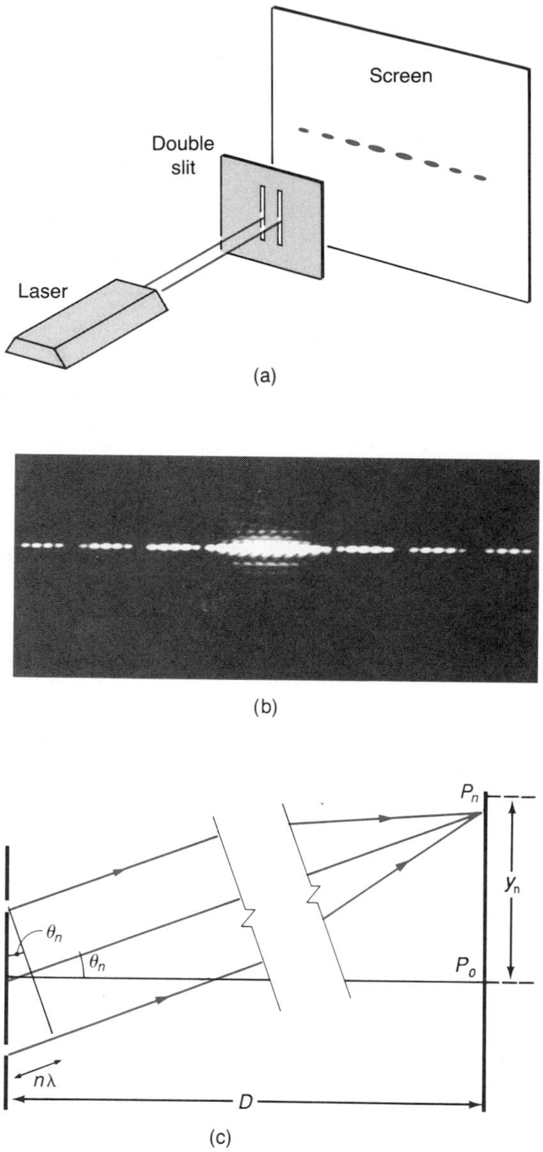

FIGURE 24.3 Young's experiment showing the interference of light: (a) a modern-day apparatus setup; (b) photograph of an interference pattern; (c) diagram to analyze the geometry of the experiment. (Photo courtesy of Mark McKenna © 1989)

developed photographic film. The interference pattern appears on any convenient viewing surface, such as the wall of a room. The apparatus setup is illustrated in Fig. 24.3(a). Figure 24.3(b) shows an actual photograph of the interference pattern obtained from two slits using a helium–neon laser. The intensity of the bright fringes falls off as we go away from the center because of diffraction, which we will discuss in a subsequent section.

Young derived a formula for the spacing of the bright fringes. Figure 24.3(c) shows the geometrical diagram that is required. The slits S_1 and S_2 are separated by a distance d, and the distance from the slits to the screen is D. Rays from the two slits produce constructive interference on the screen when a crest from one slit arrives at the same time as a crest from the other. This occurs if the distances traveled differ by an integral number of wavelengths. The simplest case of constructive interference is at the point P_0, which is opposite the midpoint between the two slits. The crests arrive simultaneously at P_0 to produce a bright fringe because the distances S_1P_0 and S_2P_0 are equal. The ray paths differ by zero wavelengths, and the fringe is called the *zero-order bright fringe*. Now consider a point P_n on the screen for which the ray path S_2P_n exceeds the path S_1P_n by n wavelengths. Crests arrive simultaneously at P_n to produce a bright fringe that is called the nth-order bright fringe. Between P_0 and P_n lie $(n - 1)$ bright fringes of 1st to $(n - 1)$th order. The pattern of Fig. 24.3(b) shows many bright fringes on each side of the central one.

Suppose now that the distance from the slits to the screen is very large compared with the slit separation. This means that the rays S_1P_n and S_2P_n are approximately parallel, as shown in the left part of Fig. 24.3(c). Let θ_n be the angular displacement of the nth-order bright fringe from the zero-order bright fringe, and let y_n be the displacement on the screen (that is, y_n is the distance on the screen between the nth-order and zero-order bright fringes). From the geometry of the diagram we can write a formula for the angular displacement of a bright fringe.

■ Young's Double-Slit Formula for the Angular Displacement of the *n*th-Order Bright Fringe

$$\sin \theta_n = \frac{n\lambda}{d} \approx \frac{y_n}{D} \quad \text{if } \theta_n \ll 1 \tag{24.1}$$

λ = wavelength of the light
d = slit separation
n = 0, 1, 2, . . .

Let us apply Young's formula to an example.

EXAMPLE 24.1

In an interference pattern from two slits, the seventh-order bright fringe is 32.1 mm from the zero-order bright fringe. The double slit is 5 m from the screen, and a traveling microscope shows that the two slits are 0.691 mm apart. Calculate the wavelength of the light.

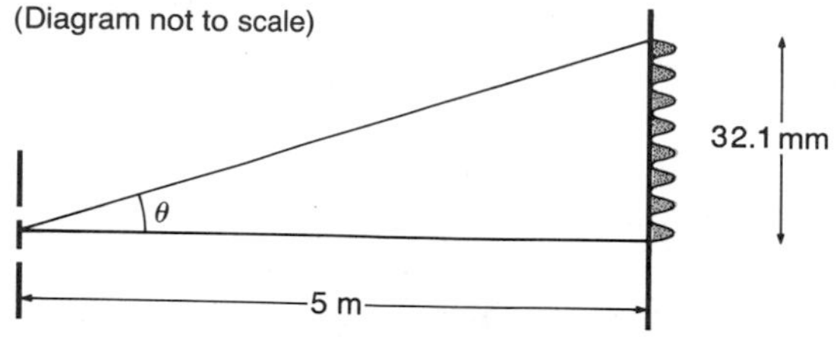

SOLUTION The angular displacement of the seventh-order bright fringe is a very small fraction of a radian. We can therefore use the approximate form of Young's formula Eq. (24.1):

$$\sin \theta_n = \frac{n\lambda}{d} \simeq \frac{y_n}{D}$$

$$\therefore \lambda \simeq \frac{y_n d}{nD}$$

$$= \frac{32.1 \times 10^{-3}\ \text{m} \times 6.91 \times 10^{-4}\ \text{m}}{7 \times 5\ \text{m}}$$

$$= 634\ \text{nm}$$

The smallest length that we must measure to find this result is the slit separation, which is more than 1000 times greater than the wavelength of the light. The interference pattern gives us an important experimental method for measuring the wavelength of light. ■

Our example concerns the double-slit interference pattern of light, but it is important to realize that any wave produces the same effect. Water waves falling on two openings produce an alternating smooth-and-rough pattern in the region beyond the openings. A sound wave coming from two apertures produces quiet and noisy areas. In fact, our belief in the wave nature of light rests on this very point; light produces the same sort of interference patterns that all other waves exhibit.

24.3 The Diffraction Grating

Some of the most beautiful interference-diffraction effects arise from a set of many narrow slits that are side by side. A slit arrangement of this type is called a **diffraction grating.** We now investigate the pattern of a diffraction grating that is illuminated by a monochromatic, coherent beam of light.

Consider the grating of Fig. 24.4 with the small slits all separated by a distance d from their neighbors. The pattern due to illumination by a plane wave, monochromatic, coherent light beam is displayed on a screen that is a distance D from the slits, where $D >> d$. The pattern exhibits an intense bright fringe whenever there is constructive interference of the rays from every slit of the grating. Since the screen is very distant, the rays that interfere constructively to produce the bright fringes are approximately parallel as they leave the slits. For the nth-order bright fringe, each ray path differs from the neighboring ray path by $n\lambda$,

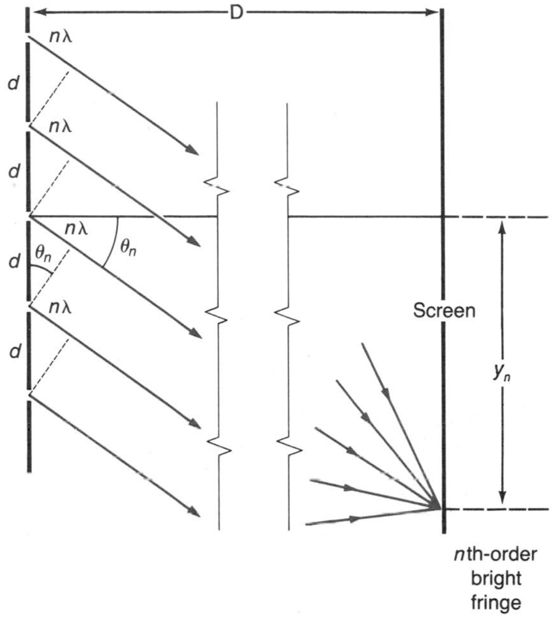

FIGURE 24.4 Diagram to calculate the angular displacement of the *n*th-order bright fringe of a diffraction grating.

where λ is the wavelength of the light. From the geometry of the diagram we can write a formula to locate the bright fringes.

■ Diffraction Grating Formula for the Angular Displacement of the *n*th-Order Bright Fringe

$$\sin \theta_n = \frac{n\lambda}{d} \qquad (24.2)$$

$$\approx \frac{y_n}{D} \qquad \text{if } \theta_n \ll 1$$

λ = wavelength of the light
d = slit separation
$n = 0, 1, 2, \ldots$

Note that the bright fringe formula for a diffraction grating is identical with the bright fringe formula for Young's experiment on two slits, Eq. (24.1). At first sight, we appear to have gained nothing by using a grating, but the gain lies in the much greater sharpness of the grating fringes. For the double-slit pattern the light intensity falls gradually to zero at the midpoint between two bright fringes. For a diffraction grating composed of many slits, the bright fringes are very sharp and are separated from each other by regions of such faint illumination that they are dark for all practical purposes. The reason is that the rays from many slits show a much higher degree of cancellation than the rays from two slits as soon as the bright fringe criterion is not satisfied. We illustrate the diffraction grating interference pattern in Fig. 24.5.

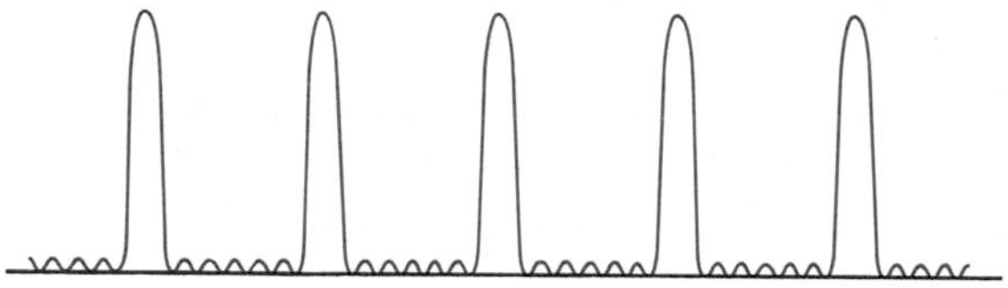

FIGURE 24.5 The interference pattern for a diffraction grating. Sharp maxima are separated by relatively dark regions.

EXAMPLE 24.2

A diffraction grating has 3200 slits per centimeter. Calculate the angular deviation of the third-order bright fringe when the grating is illuminated by a helium–neon laser. (The wavelength for the laser light is 632.8 nm.)

SOLUTION Use of the diffraction formula requires that we know the slit separation. Since there are 3200 slits per centimeter, the slit separation is given by:

$$d = \frac{1}{3200}\ \text{cm}$$

$$= \frac{1}{3200} \times 10^{-2}\ \text{m}$$

$$= 3.125 \times 10^{-6}\ \text{m}$$

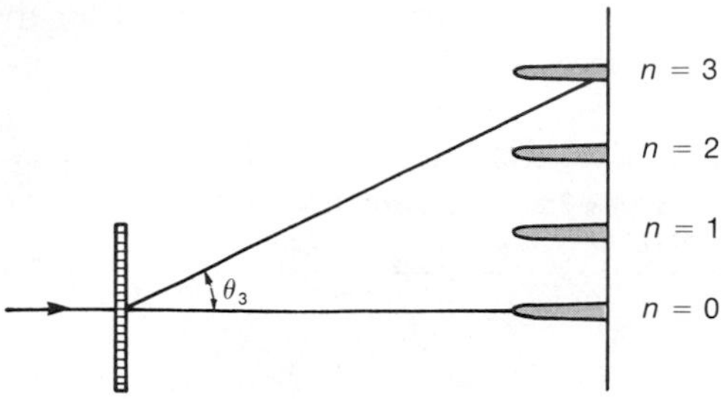

The diffraction formula [Eq. (24.2)] then gives:

$$\sin \theta_n = \frac{n\lambda}{d}$$

$$\sin \theta_3 = \frac{3 \times 632.8 \times 10^{-9} \text{ m}}{3.125 \times 10^{-6} \text{ m}}$$

$$= 0.6075$$

$$\therefore \theta_3 = 37.4°$$

■

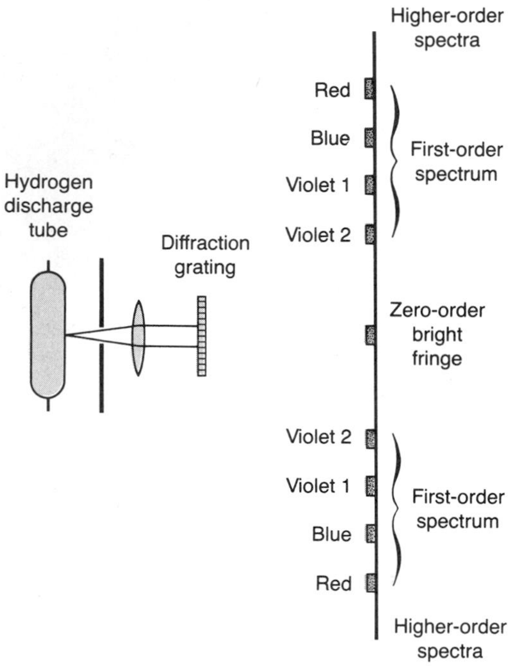

FIGURE 24.6 A diffraction grating used as a spectroscope to identify the components in the light from a hydrogen discharge tube.

High-quality diffraction gratings are made by machines that make closely equispaced lines on a glass plate that are actually scratches in the glass, which effectively block out the light. The unscratched portions between the lines act as slits in the grating. Another technique is to photograph a large array of equispaced black lines on a white background. The picture is diminished greatly in size by a process akin to photographic enlargement (but opposite to it), and the photographic negatives themselves then serve as gratings. A good diffraction grating may contain as many as 10 000 slits per centimeter.

Up to now we have considered only the interference of monochromatic light, but one of the most important practical applications of the diffraction grating is the grating spectroscope. Consider a beam of light that is not monochromatic falling on a diffraction grating. In Fig. 24.6 light from a hydrogen discharge tube is focused into a parallel beam to illuminate a grating. We know from Fig. 23.13(b) that this light contains four component colors—a red line at 656.3 nm, a blue line at 486.1 nm, and two violet lines at 434.0 and 410.1 nm. Since the zero-order bright fringe of the grating is not deviated for any wavelength, it contains all the colors of the incident beam. However, the angular deviation of the first- and higher-order bright fringes is given by the diffraction grating equation, Eq. (24.2):

$$\sin \theta_n = \frac{n\lambda}{d}$$

We see at once that components of longer wavelength are deviated to a greater extent in any order, since $\sin \theta_n$ is larger for larger values of λ. The diffraction grating therefore produces a whole series of hydrogen spectra on the screen, with one spectrum on each side of the center for each order bright fringe. Red light has a longer wavelength than violet light and is deviated to a greater extent in every order, whereas in the prism spectroscope the violet light undergoes the greatest deviation.

EXAMPLE 24.3

A coherent beam of light from a hydrogen discharge tube falls normally on a diffraction grating of 8000 lines per centimeter. Cal-

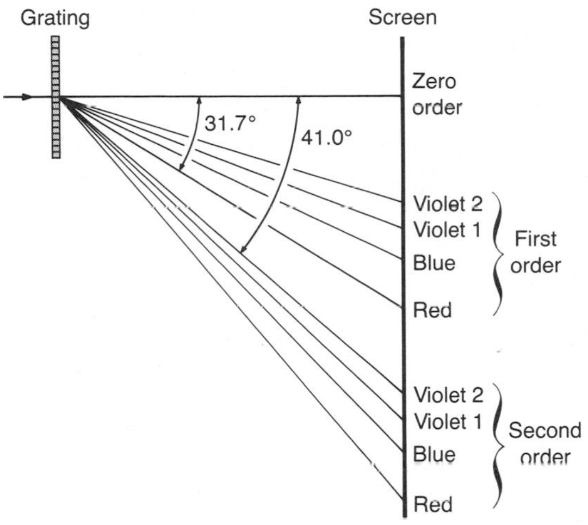

culate the angular deviation of each line in the first-order spectrum. Do any lines of the second-order spectrum overlap the first-order spectrum? (For the hydrogen discharge tube, $\lambda_{red} = 656.3$ nm, $\lambda_{blue} = 486.1$ nm, $\lambda_{violet\ 1} = 434.0$ nm, and $\lambda_{violet\ 2} = 410.1$ nm.)

SOLUTION Since the grating has 8000 lines per centimeter, the grating spacing is given by:

$$d = \frac{1}{8000}\ \text{cm}$$
$$= 1.25 \times 10^{-6}\ \text{m}$$

We can calculate the deviation of each component in turn by using the diffraction grating equation, Eq. (24.2):

$$\sin\theta_n = \frac{n\lambda}{d}$$

Violet 2:
$$\sin\theta_1 = \frac{1 \times 410.1 \times 10^{-9}\ \text{m}}{1.25 \times 10^{-6}\ \text{m}}$$
$$\therefore \theta_1 = 19.2°$$

Violet 1:
$$\sin\theta_1 = \frac{1 \times 434.0 \times 10^{-9}\ \text{m}}{1.25 \times 10^{-6}\ \text{m}}$$
$$\therefore \theta_1 = 20.3°$$

Blue:
$$\sin\theta_1 = \frac{1 \times 486.1 \times 10^{-9}\ \text{m}}{1.25 \times 10^{-6}\ \text{m}}$$
$$\therefore \theta_1 = 22.9°$$

Red:
$$\sin\theta_1 = \frac{1 \times 656.3 \times 10^{-9}\ \text{m}}{1.25 \times 10^{-6}\ \text{m}}$$
$$\therefore \theta_1 = 31.7°$$

To determine whether the second-order spectrum overlaps the first-order, we need only calculate the position of the least deviated line in the second-order spectrum:

Violet 2:
$$\sin\theta_2 = \frac{2 \times 410.1 \times 10^{-9}\ \text{m}}{1.25 \times 10^{-6}\ \text{m}}$$
$$\therefore \theta_2 = 41.0°$$

Since all other second-order lines are deviated even further, it follows that there is no overlap between the first- and second-order spectra. In this example the bright fringes have relatively large angular displacements, so we cannot assume approximate equality for the sine and the tangent. ■

24.4 Interference by Thin Films

The interference patterns that we have studied so far have arisen from scientifically contrived situations such as that involving the diffraction grating. However, many well-known, everyday experiences are due to interference. Some examples are the colored bands seen in the reflection of light from a thin film of oil on water and the colors observed on reflection of light from a soap bubble. These phenomena share a common feature—the interference of light rays reflected from opposite surfaces of a thin transparent film.

Consider a ray of light incident on a thin transparent film as shown in Fig. 24.7. At the point A, the ray is partly reflected and partly refracted, the refracted ray is partly reflected at point B, and this reflected ray is again partly refracted at point C. Depending on the phase relationship, the interference seen by the observer may be either constructive or destructive. Since white light is composed of different wavelengths corresponding to the different colors, the interference may be constructive for some colors and destructive for others. In this way, reflection of white light from a thin transparent film may give rise to colored bands.

Let us begin the detailed discussion with a simple device that shows interference bands on reflection of monochromatic light. In Fig. 24.8, we show a pair of glass plates that are in contact along one pair of edges and held apart along the opposite edges. (This can easily be achieved by taking a pair of microscope slides and holding one pair of edges slightly apart by inserting a thin thread.) We now have a transparent film (or wedge) of air of variable thickness. The path difference for the interfering rays will depend on the part of the wedge they traverse, and one therefore

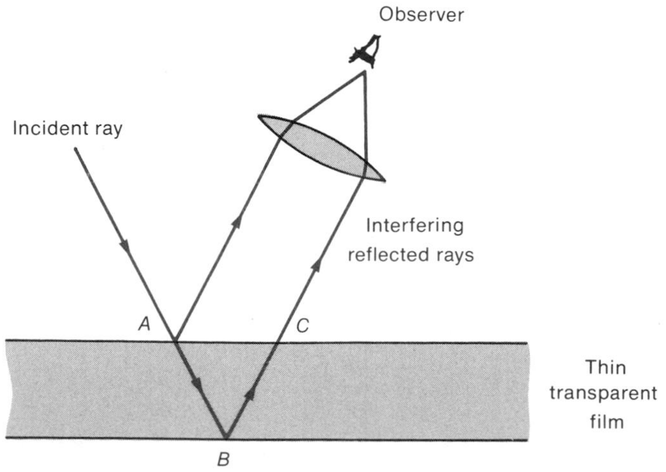

FIGURE 24.7 Interference of light by reflection from a thin transparent film.

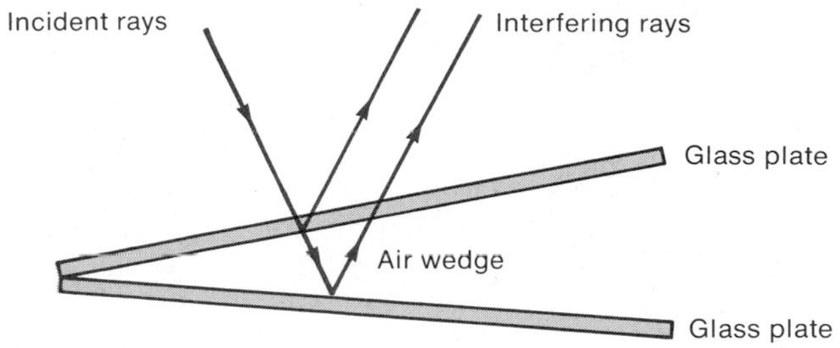

FIGURE 24.8 Interference of light by reflection from the opposite surfaces of a thin air wedge between glass plates.

expects alternate constructive and destructive interference; this is precisely what is observed. At the end where the glass plates are in contact, the thickness of the air wedge is zero, and we might reasonably expect constructive interference since the two interfering waves appear to have the same path length. However, experiment shows that a *dark* fringe occurs, rather than a bright one, where the air-wedge thickness is zero. This can only mean that one of the two reflected waves has suffered a phase change of π upon reflection. This would move all the crests by exactly the right amount to interfere with the troughs of the other wave and, hence, cause destructive interference. More detailed experiments show that the abrupt phase change occurs when the light wave initially traveling in a less optically dense medium is reflected at an interface with a more optically dense medium. For the air wedge in Fig. 24.8, the phase change occurs at the lower surface and not at the upper one. For a transparent film that is optically more dense than the bounding media (for example, an oil film on water), the phase change occurs at the upper surface and not at the lower one.

In view of this discussion we can write a simple formula for *destructive* interference of light that falls at near normal incidence on a transparent film of thickness d. In this case the wave reflected from the lower surface travels an extra distance $2d$ in the film. If this is an integral number of wavelengths, then the phase change of π that occurs in one of the two reflections results in destructive interference. Algebraically, the condition for destructive interference is:

$$2d = n\lambda_m \tag{24.3}$$

where

$$n = 1, 2, 3, \ldots$$
$$d = \text{film thickness}$$
$$\lambda_m = \text{wavelength of the light in the film}$$

Let us consider an example to utilize this result.

EXAMPLE 24.4

Two microscope slides each 7.5 cm long are in contact along one pair of edges while the other edges are held apart by a piece of paper 0.012 mm thick. Calculate the spacing of interference fringes under illumination by light of 632 nm wavelength at near normal incidence.

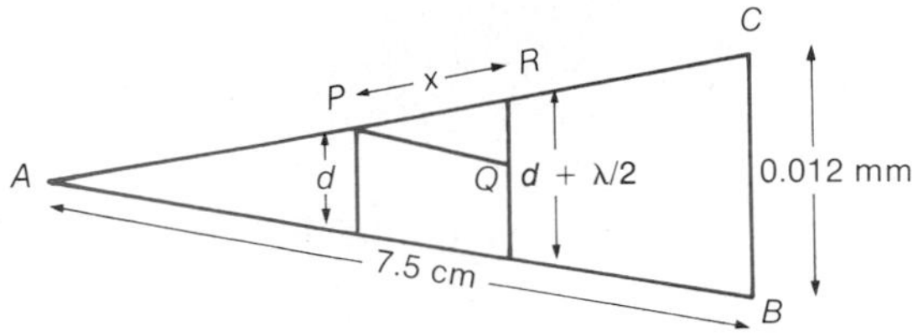

SOLUTION Let a film thickness d shown on the diagram correspond to a dark fringe. Let the next dark fringe occur at a distance x along the film. From Eq. (24.3) the film thickness at this position must be $(d + \lambda/2)$. Draw the line PQ parallel to AB; then QR must be equal to $\lambda/2 = 316 \times 10^{-9}$ m. The triangles ABC and PQR are similar so that:

$$\frac{AB}{BC} = \frac{PQ}{QR}$$

$$\frac{0.075 \text{ m}}{0.012 \times 10^{-3} \text{ m}} = \frac{x}{316 \times 10^{-9} \text{ m}}$$

$$\therefore x = 1.97 \text{ mm}$$

This is the spacing of the successive dark fringes. ■

A well-known example of interference fringes in an air wedge is called "Newton's rings." The phenomenon, first studied by Newton, uses a curved wedge of air between a convex lens and a flat surface, as shown in Fig. 24.9. The resulting interference pattern consists of concentric circular interference fringes with a dark fringe in the center; this occurs because of the phase difference π on reflection from the lower surface.

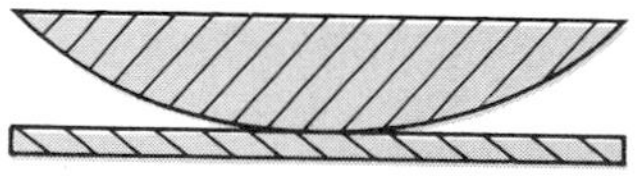

FIGURE 24.9 A curved air film between a convex glass surface and a flat plane to produce Newton's rings.

So far we have been studying the effect of wedge-shaped films in producing fringes by reflecting monochromatic light. Let us now use an example to consider how a film of uniform thickness could produce colors by reflection of white light.

EXAMPLE 24.5

A soap bubble 550 nm thick and of refractive index 1.33 is illuminated at near normal incidence by white light. Calculate the wavelengths of the light for which destructive interference occurs.

SOLUTION The condition for destructive interference is given by Eq. (24.3):

$$2d = n\lambda_m$$

$$2 \times 550 \times 10^{-9} \text{ m} = n\lambda_m \qquad (n = 1, 2, 3, 4, \ldots)$$

$$\therefore \lambda_m = \frac{1.1 \times 10^{-6} \text{ m}}{n} \qquad (n = 1, 2, 3, 4, \ldots)$$

The first four values of λ_m are therefore:

$$\frac{1.1 \times 10^{-6}\ \text{m}}{1} = 1100\ \text{nm}$$

$$\frac{1.1 \times 10^{-6}\ \text{m}}{2} = 550\ \text{nm}$$

$$\frac{1.1 \times 10^{-6}\ \text{m}}{3} = 367\ \text{nm}$$

$$\frac{1.1 \times 10^{-6}\ \text{m}}{4} = 275\ \text{nm}$$

These are wavelengths in the soap solution. To find the corresponding wavelengths in air, we recollect that the speed of light in a transparent medium is lower by a factor equal to the reciprocal refractive index [Eq. (22.1)]. Since the wave frequency is unaltered, the wavelength in a transparent medium is shorter by the same factor; that is, $\lambda_m = \lambda/n$. Solving for λ, the first four wavelengths in air that suffer destructive interference are:

$$1100\ \text{nm} \times 1.33 = 1460\ \text{nm (infrared)}$$

$$550\ \text{nm} \times 1.33 = 731\ \text{nm (red)}$$

$$367\ \text{nm} \times 1.33 = 488\ \text{nm (blue-green)}$$

$$275\ \text{nm} \times 1.33 = 366\ \text{nm (ultraviolet)}$$

In the visible region, the light from both ends of the spectrum is reflected with destructive interference. We would, therefore, find constructive interference for wavelengths midway between these two in the vicinity of 600 nm. We expect the soap bubble to have an orange hue when viewed in white light. ■

The colorful bands seen in the reflection of white light from oil films on water surfaces are also due to interference. The reason for the irregular shapes and colors often seen in such bands is twofold. In the first place, the thickness of the oil film may vary, and secondly, the angle of incidence of the reflected light may be different for different parts of the film. Both of these reasons cause changes in the wavelengths, for which destructive interference occurs, and consequently in the color of the reflected light.

24.5 Diffraction by a Single Slit

Up to this point our treatment of interference has centered on the interference that occurs with wave trains from different slits in a grating or from different surfaces of a thin film. We now turn to investigate the phenomenon that occurs when a monochromatic wave traverses a *single slit*. Does this wave also bend around the corners? Figure 24.2(a) shows a plane wave falling on a narrow slit with the secondary wavelet spreading throughout the whole of the region beyond the slit, but this result is obtained only if the slit is so narrow that only one secondary wavelet originates from it. This is an unreal situation—the slit must always have a finite width to let some light pass. In the real case we must consider a number of secondary wavelets originating from points that are very close to each other. The resulting phenomenon is called **diffraction.**

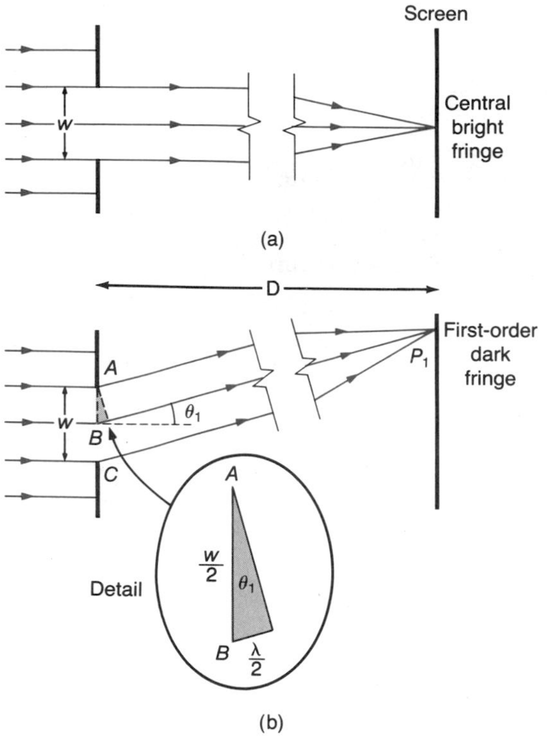

FIGURE 24.10 Diffraction by a single slit. (a) The secondary wavelets interfere constructively, producing a bright fringe at the point directly opposite the slit. (b) A dark fringe occurs whenever the secondary wavelets interfere destructively in pairs.

Let us consider a wave-and-ray diagram like the one that was so useful in analyzing Young's experiment. Figure 24.10(a) shows a parallel beam of light that is normally incident on a barrier containing a single slit of width w. The screen is assumed to be at a great distance from the barrier compared with the slit width. It is essential to recognize the importance of the slit width in differentiating this analysis from that of Young's experiment. In that case (illustrated in Fig. 24.3), we simply classed the slits as very small openings and considered only one secondary wavelet from each slit. Now we must recognize each point on the wave front across the width of the slit as a source of secondary wavelets into the region beyond the barrier. Consider the rays that travel from the secondary wavelet source points to the point on the screen directly opposite the center of the slit, as shown in Fig. 24.10(a). Since we assume that the screen is at a very great distance from the slit, these rays all have the same length. The results in constructive interference at the point in question, and this part of the diffraction pattern is termed the *central bright fringe.*

Now consider the rays that travel from the secondary wavelet centers at some angle to the direction of the incident rays. Figure 24.10(b) shows three points A, B, and C at which secondary wavelets originate. Points A and C are at the edges of the slit, and point B is midway between them. The rays from the three points interfere at a point P_1 on the screen, which is chosen so that the ray path BP_1 exceeds the path AP_1 by one-half wavelength. Thus, the secondary wavelets from points A and B interfere destructively at point P_1. Moreover, for every point between A and B at which a secondary wavelet originates, there is a point between B and C at which a wavelet originates that interferes destructively with the former secondary wavelet. It follows that there is complete destructive interference at the point P_1, which we call the *first-order dark fringe.*

To find the angular deviation of the point P_1 from the original ray direction, consider the detailed inset of Fig. 24.10(b) from which we have:

$$\sin \theta_1 = \frac{\lambda/2}{w/2} = \frac{\lambda}{w}$$

The angular position of the first-order dark fringe corresponds to one-half wavelength difference between the rays from A and B;

in other words, there is one full wavelength difference between the rays from A and C. We can generalize our formula to locate the angular displacement of dark fringes of higher orders by stipulating an integral number of full wavelengths difference between the rays from A and C. The generalized formula to locate the dark fringes of the diffraction pattern then follows.

■ Formula for the Angular Displacement of the *n*th-Order Dark Fringe of a Single Slit

$$\sin \theta_n = \frac{n\lambda}{w} \approx \frac{y_n}{D} \quad \text{if } \theta_n \ll 1 \tag{24.4}$$

λ = wavelength of the light
w = slit width
D = slit to screen distance
y_n = distance from bright central maximum to nth-order dark fringe
n = 1, 2, 3, . . .

At first sight this result is similar to the double-slit formula, Eq. (24.1). The most important difference is that the diffraction formula determines the positions of the dark fringes, while the double-slit formula gives the bright fringe positions. Note also that the center is bright in both cases. For this reason we exclude n = 0 in Eq. (24.4). Finally, in Eq. (24.1) the denominator d is the slit separation; in Eq. (24.4) the denominator w is the slit width.

A full treatment of the diffraction problem requires difficult mathematics, but we can easily summarize the main conclusions. The first two points refer to the central bright fringe and the dark fringes that we have just discussed. The third point relates to the relative brightness of the bright fringes.

1. The diffraction pattern of a single slit exhibits a series of bright and dark fringes. The central bright fringe is directly opposite the slit.

2. A dark fringe occurs whenever the ray path from one edge of the slit is an integral number of wavelengths longer or shorter than the ray path from the opposite edge of the slit.

3. The central fringe is much brighter than the others. The first pair of bright side fringes are only about 4% of the brightness of the central fringe, and the outer bright fringes are even fainter.

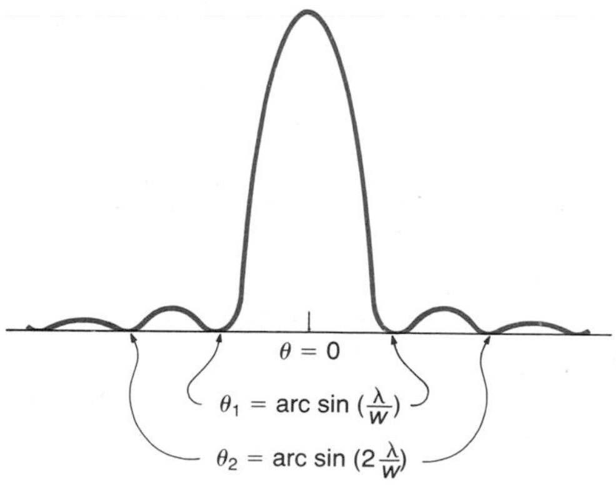

FIGURE 24.11 The diffraction pattern intensity for monochromatic light of wavelength λ passing through a slit of width w.

The diffraction pattern is shown graphically in Fig. 24.11. The light intensity on the screen is plotted as a function of the angular

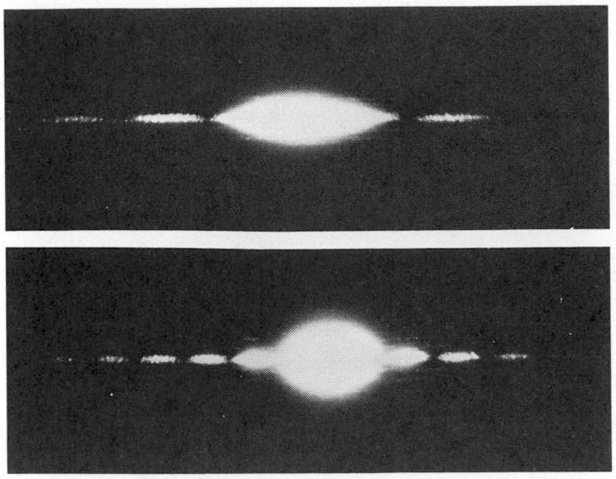

FIGURE 24.12 Diffraction patterns of single slits of two different widths illuminated by a helium–neon laser. (Courtesy of Gary Burgess)

displacement from the point opposite the center of the slit. The very intense central bright fringe is called the **central diffraction maximum.**

To observe the diffraction of light, we need a very narrow slit and a fairly intense light beam. Figure 24.12 shows two diffraction patterns obtained by illuminating slits of different widths with the beam of a helium–neon laser. Both pictures show a very strong central diffraction maximum with **secondary maxima** visible on each side. The position of the dark fringes given by Eq. (24.4) shows that the first-order dark fringe has a greater angular displacement for a smaller slit width. The pattern with the broader central maximum in Fig. 24.12 is therefore generated by the narrower slit.

EXAMPLE 24.6

A diffraction pattern is obtained from a single slit by using a helium–neon laser. The screen is 5 m from the slit, and the central maximum is 4.4 cm wide. Calculate the slit width. (The wavelength of the helium–neon laser is 632.8 nm.)

SOLUTION If the central maximum is 4.4 cm wide, the distance from the center to the first dark fringe is one-half of this, or 2.2 cm. Since the angular displacement is very small compared with 1 rad, we can use the approximation in Eq. (24.4):

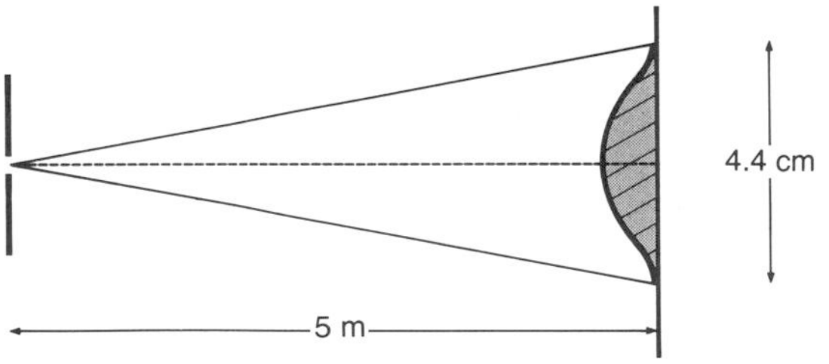

$$\sin\theta_n = \frac{n\lambda}{w} \approx \frac{y_n}{D}$$

$$\therefore\ w \simeq \frac{n\lambda D}{y_n}$$

$$= \frac{1 \times 632.8 \times 10^{-9}\ \text{m} \times 5\ \text{m}}{2.2 \times 10^{-2}\ \text{m}}$$

$$= 0.144\ \text{mm}$$

■

Note that interference and diffraction cannot really be separated. Diffraction occurs because of the interference of secondary wavelets that originate at different points in a slit opening. In our analysis of Young's experiment, we considered only one secondary wavelet from each slit, but even a narrow slit must originate a number of secondary wavelets. It follows that the interference pattern of the double slit should have diffraction effects. Indeed, there are such effects! The brightness of the interference fringes decreases toward the edges because of diffraction, as illustrated in Fig. 24.13. The double-slit formula of Eq. (24.1) predicts a bright fringe for each value of θ_n given by the equation. Moreover, the fringes should all be equally bright, as shown in Fig.

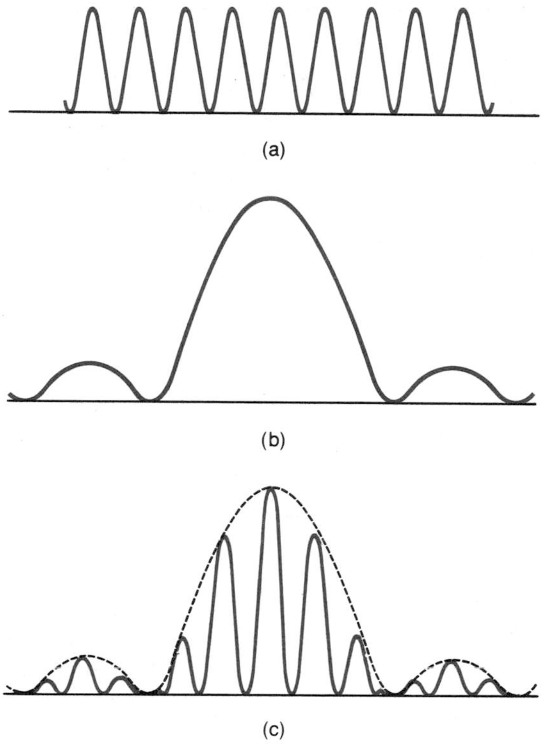

FIGURE 24.13 The structure of the interference-diffraction pattern from a double slit: (a) the interference pattern of the two slits with diffraction not considered; (b) the diffraction pattern of one slit on its own; (c) the composite interference-diffraction pattern.

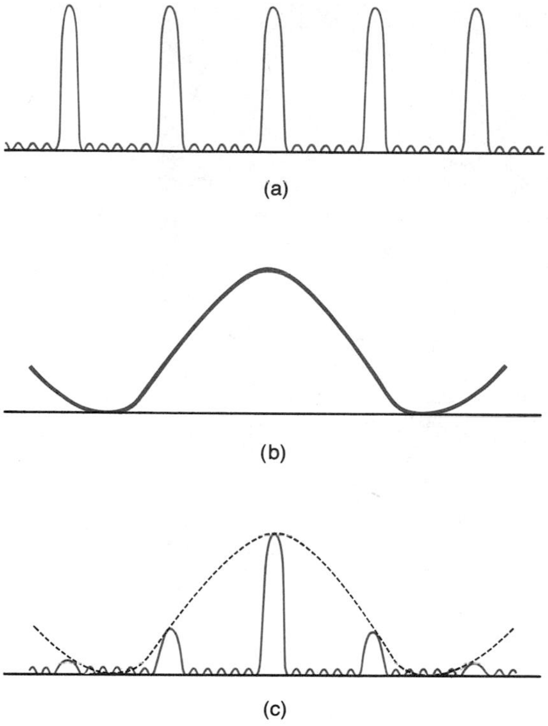

FIGURE 24.14 The structure of the interference-diffraction pattern from a diffraction grating: (a) the interference pattern of the grating with diffraction not considered (the same diagram as Fig. 24.5); (b) the diffraction pattern of one slit on its own (the same diagram as Fig. 24.11); (c) the composite interference-diffraction pattern.

24.13(a). Figure 24.13(b) shows the diffraction pattern that would be obtained from one slit alone. The pattern that is actually observed is a combination of both effects. The interference pattern correctly locates the position of each bright fringe, and the diffraction pattern from one slit shows the intensity of each bright fringe. The construction of the composite interference-diffraction pattern is shown in Fig. 24.13(c).

The combined interference-diffraction pattern is easily seen in Fig. 24.13(b), although the central diffraction pattern is so bright that the interference fringes in it are lost. However, the edges of the pattern show exactly the effect illustrated in Fig. 24.13(c).

Diffraction modifies the interference pattern of a grating in just the same way that it does for two slits—the **interference pattern** determines the position of each bright fringe, and the **diffraction pattern** determines its intensity. The construction of the composite pattern is shown in Fig. 24.14. Comparison of Figs. 24.13(c) and 24.14(c) shows that a grating produces much sharper bright fringes than a double slit. Figure 24.15 shows the interference-diffraction patterns of various gratings illuminated by a helium–neon laser.

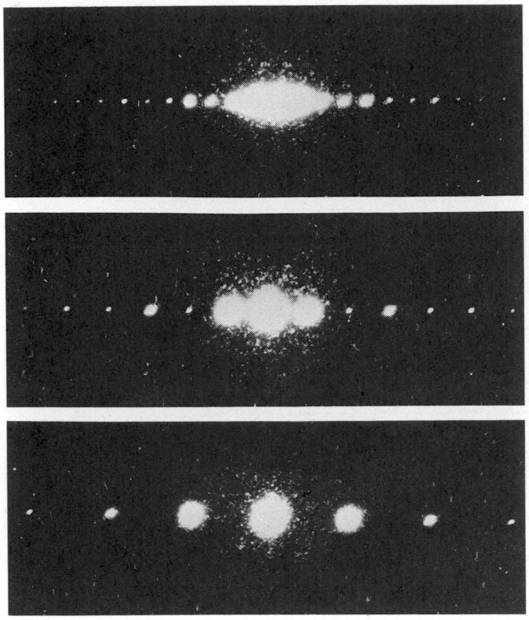

FIGURE 24.15 Interference-diffraction patterns of gratings illuminated by a helium–neon laser. (Courtesy of Gary Burgess)

EXAMPLE 24.7

The interference-diffraction pattern of a grating illuminated by a helium–neon laser shows bright fringes that are 4.4 cm apart on a screen that is 5 m from the diffraction grating. The fifth-order bright interference fringe is missing from the pattern. Calculate the slit separation and the slit width for the grating. (The wavelength of the helium–neon laser is 632.8 nm.)

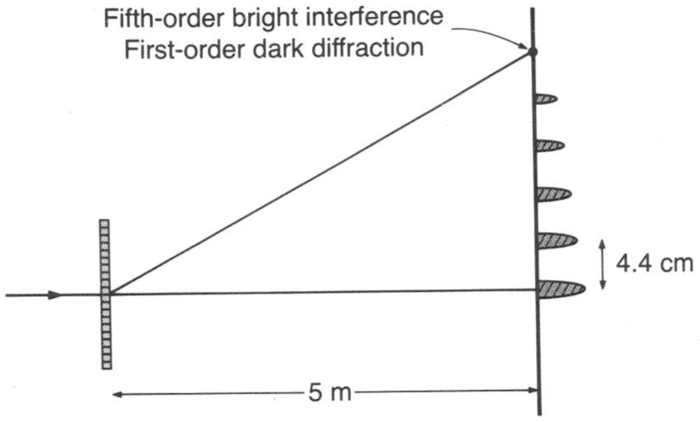

SOLUTION We can calculate the slit separation by using the approximate form of the diffraction grating formula, Eq. (24.2):

$$\sin\theta_n = \frac{n\lambda}{d} \approx \frac{y_n}{D}$$

$$\therefore d \simeq \frac{n\lambda D}{y_n}$$

$$= \frac{1 \times 632.8 \times 10^{-9}\ \text{m} \times 5\ \text{m}}{4.4 \times 10^{-2}\ \text{m}}$$

$$= 7.19 \times 10^{-5}\ \text{m}$$

The fringes decrease in brightness up to the fourth from the center. The fifth fringe is missing, and the sixth fringe and some subsequent fringes appear. We therefore believe that the first dark fringe of the diffraction pattern occurs at about the same position as the fifth bright fringe of the interference pattern. We can use Eq. (24.4) to calculate the slit width:

$$\sin\theta_n = \frac{n\lambda}{w} \approx \frac{y_n}{D}$$

$$\therefore w \simeq \frac{n\lambda D}{y_n}$$

$$= \frac{1 \times 632.8 \times 10^{-9}\ \text{m} \times 5\ \text{m}}{5 \times 4.4 \times 10^{-2}\ \text{m}}$$

$$= 1.44 \times 10^{-5}\ \text{m}$$

■

We can use Eq. (24.4) to show why sound waves travel around corners in everyday experience, while light waves do not appear to do so. Most of the energy of the wave is concentrated in the central diffraction maximum between the first-order dark fringes. The angular width of the central diffraction maximum, which is equal to the angular distance between the first dark fringes on each side of the center, is very small if the wavelength is very small compared with the width of the slit. Therefore, a beam of light traveling through a slit several centimeters wide produces a central diffraction maximum of negligibly small angular width, and we simply see a bright region on the screen that is the same width as the slit. On the other hand, the central diffraction maximum becomes broad when the magnitude of the width of the slit is similar to that of the wavelength. If the slit width is smaller than the wavelength, Eq. (24.4) cannot be satisfied for any integer n, and there are no dark fringes. The central diffraction maximum covers the whole screen. The diffraction of light through a slit that is narrower than one wavelength is not commonplace. However, a sound wave traveling through a slit several centimeters wide is diffracted to all parts of the region beyond the slit. We can hear the sound of such a wave whether or not we are opposite the opening.

24.6 The Resolving Power of Optical Instruments

Our ray optics treatment of optical instruments took no account of the actual wave nature of light. As we have just seen, every wave undergoes diffraction, which sets a limit on the ability of an optical instrument to distinguish objects that are close together. For example, suppose we train a telescope on two stars. Our common sense tells us that we see two stars if they are sufficiently far apart but that we may see only one image if the stars are too close together. With the help of diffraction theory we can describe this effect exactly.

Consider a situation in which two distant point sources of light, such as two stars, illuminate a narrow slit. Each beam of light produces its own diffraction pattern on the screen, as illustrated in Fig. 24.16(a). When the two stars have a relatively large angular separation and the slit is reasonably wide, the two diffraction patterns are narrow and well separated. Under these circumstances we can easily identify two patterns as being distinct. However, if the angular separation of the two stars is sufficiently small, their diffraction patterns overlap, and identifying the two patterns as being distinct becomes a matter of judgment.

A condition for identifying diffraction patterns as being distinct was given by the English physicist Lord Rayleigh (1842–1919), who is famous for his work in many areas of wave motion. The **Rayleigh criterion** is that the center of one diffraction pattern

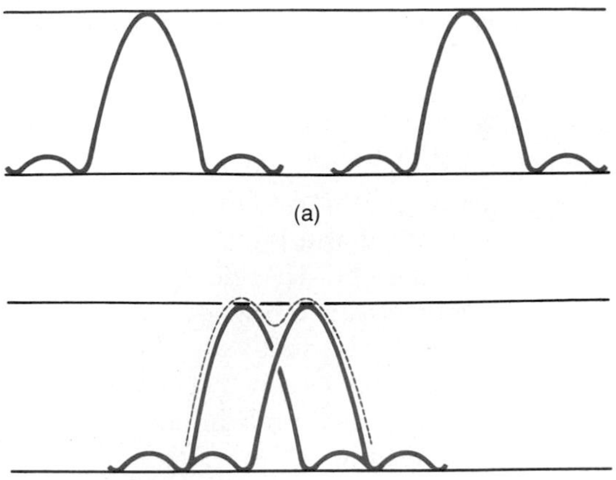

(a)

(b)

FIGURE 24.16 The distinguishability of point light sources depends on their diffraction patterns: (a) a pair of clearly distinguishable patterns; (b) according to Rayleigh's criterion, the total pattern is just recognizable as due to two sources if the center of one falls over the first dark fringe of the other.

must come no closer than the first dark fringe of the other pattern; this means that the angular separation must exceed the half width of the central diffraction maximum. We illustrate this condition in Fig. 24.16(b). The angular separation of the two patterns is equal to the angular separation between the center of one pattern and its first dark fringe. This angular separation is given by Eq. (24.4) for the first dark fringe:

$$\sin \theta_1 = \frac{\lambda}{w}$$

Since we are dealing with very small angles, Eq. (1.6) holds, which specifies that the angle in radians is approximately equal to its sine. This gives:

$$\theta_1 \approx \frac{\lambda}{w} \qquad \textbf{(24.5)}$$

However, the angular separation of the two diffraction patterns is equal to the angular separation of the two objects. We define the **resolving power** of a slit to be the angular separation of two objects whose diffraction patterns are seen as distinct. With this understanding Eq. (24.5) gives us the resolving power in radians of a slit of width w.

Because in most optical instruments light comes through a circular aperture rather than a slit, a numerical factor must be inserted into the mathematical expression for the resolving power. Detailed calculations put the numerical factor at 1.22. We must also replace the slit width by the circular aperture diameter.

■ Rayleigh's Formula for the Resolving Power of a Circular Aperture of Diameter *d* for Light of Wavelength λ

$$\theta_1 \approx \frac{1.22\lambda}{d} \qquad \textbf{(24.6)}$$

The resolving power predicted by this formula may not be attained, but the wave theory of light guarantees that it can never be exceeded.

EXAMPLE 24.8

How far apart are two objects on the moon that can be recognized as distinct by the unaided eye? Take the diameter of the eye pupil as 4 mm and the wavelength of the light as 500 nm.

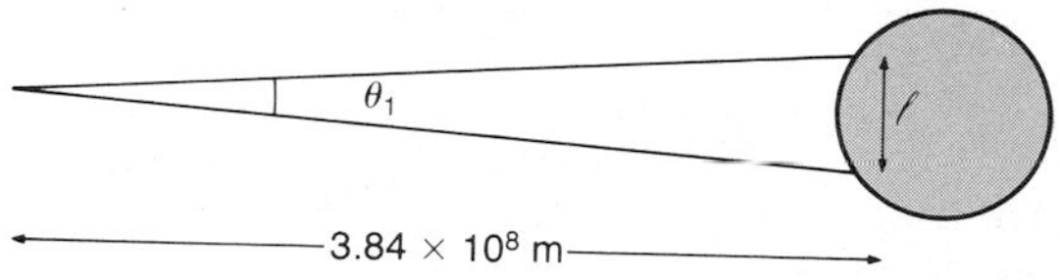

SOLUTION Light beams from two point objects on the moon enter the circular aperture in front of the eye and form diffraction patterns on the retina. The angular resolution of the eye is given by Rayleigh's formula, Eq. (24.6):

$$\theta_1 \approx \frac{1.22\lambda}{d}$$

$$= \frac{1.22 \times 500 \times 10^{-9}\ \text{m}}{4 \times 10^{-3}\ \text{m}}$$

$$= 1.52 \times 10^{-4}\ \text{rad}$$

This is equal to the angular separation of the two objects. The distance from the earth to the moon is about 3.84×10^8 m. If l is the actual separation of the two objects on the moon's surface, we have:

$$1.52 \times 10^{-4}\ \text{rad} = \frac{l}{3.84 \times 10^8\ \text{m}}$$

$$\therefore l = 58.6\ \text{km}$$

■

Diffraction sets a limit! No matter how sharp two features of the lunar landscape may be, we could not possibly recognize them as distinct (by using the unaided eye) if they were closer together than about 60 km.

It is true that the eye is a more complex optical device than just a small circular opening, but the role of the lens is to focus the image on the nearby retina. In our analysis of diffraction we considered the screen to be at a great distance from the small opening. A lens that focuses the diffracted light on a nearby screen leaves the diffraction pattern unaltered. We should, though, have a correction for the fact that the diffracted light inside the eye travels in a medium whose refractive index is about 1.3. However, we will neglect this correction since the whole calculation is only an estimate.

We can now see that there is more to an optical instrument than just its magnifying power. The diameters of the apertures through which the light passes play a vital role in determining the resolving power. Each type of instrument must be considered separately, and we will confine ourselves to the telescope. The circular apertures involved are the objective lens (or mirror), the eyepiece, and the pupil of the eye. In the usual telescope design, the limit is set by the diameter of the objective. A telescope that has a large objective lens (or more usually mirror) has a greater resolving power than one with a smaller objective. A large-objective telescope also collects more light through its larger opening so that faint objects can be detected.

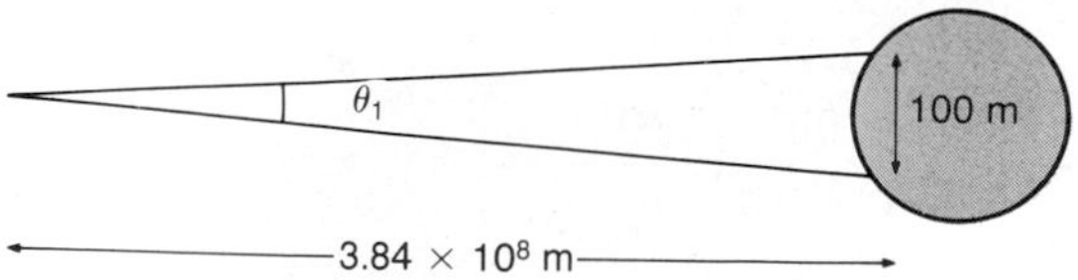

EXAMPLE 24.9

What size objective must a telescope have to distinguish features that are 100 m apart on the moon's surface? Assume that the wavelength of the light is 500 nm.

SOLUTION The required resolving power is the separation of the two objects divided by the distance from the telescope to the moon:

$$\theta_1 = \frac{100 \text{ m}}{3.84 \times 10^8 \text{ m}}$$
$$= 2.60 \times 10^{-7} \text{ rad}$$

We can calculate the required objective diameter by using Rayleigh's formula, Eq. (24.6):

$$\theta_1 = \frac{1.22\lambda}{d}$$
$$\therefore d = \frac{1.22\lambda}{\theta_1}$$
$$= \frac{1.22 \times 500 \times 10^{-9} \text{ m}}{2.60 \times 10^{-7} \text{ rad}}$$
$$= 2.34 \text{ m}$$

■

We stress again that the actual resolving power may not be nearly as good as implied by Rayleigh's criterion. Visual defects, lens and mirror aberrations, and variations in the atmospheric refractive index all cut down the resolving power. In actual practice the observed angle of maximum resolution may be up to 10 times as great as the angle calculated from Rayleigh's criterion.

24.7 Polarized Light

Recall that a wave may be either longitudinal or transverse. In a longitudinal wave, the direction of the disturbance is back and forth along the direction of the wave propagation; in a transverse wave, the disturbance is perpendicular to the propagation direction. However, an infinite number of directions are all perpendicular to a given direction. This fact permits *polarization* of a transverse wave. Let us consider a simple example. A string of a violin resting horizontally on a table is pulled upward and then released. The resulting standing wave is vertically plane polarized, since the displacement of the string is entirely in a vertical plane up or down from the rest position. If we pull the string to one side, the standing wave is horizontally plane polarized. We see that polarization of a transverse wave means the assignment of a particular direction to the transverse physical displacement.

Let us now turn to the more complex case of polarized light. Light is an electromagnetic wave. Figure 20.11 showed that both the electric and magnetic vectors are transverse to the direction of propagation. We must assign priority to one of them, in spite of the fact that electric and magnetic fields are inseparable components of the electromagnetic wave. We will arbitrarily choose the electric field vector in assigning the polarization. If it vibrates only in one plane, we say that the wave is *plane polarized,* and the **plane of polarization** is the plane containing all the electric field vectors of the wave. Figure 20.11 showed a vertically plane polarized wave, while Fig. 24.17(a) shows a horizontally plane polarized wave. However, neither of these situations holds in many circumstances. When the atoms of a typical light source, such as the sun, emit radiation, they do so quite independently of each other. There is no restriction on the electric field vector other than it be perpendicular to the propagation direction. Figure 24.17(b) shows an unpolarized wave, which has many random orientations of the electric field vector.

As we have implied, light is usually unpolarized, and we must use some special technique to produce a polarized beam. Two methods in common use are reflection from a plane interface and transmission through a Polaroid.

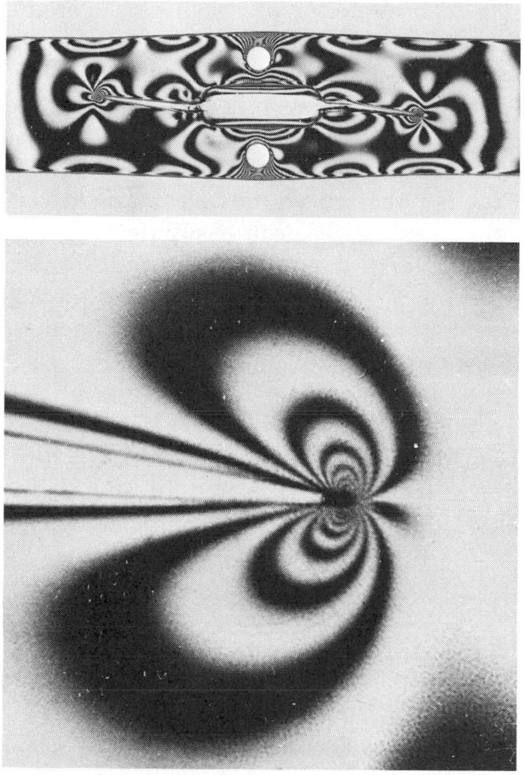

Photoelastic stress analysis uses the properties of polarized light to determine patterns of stress concentration. The upper photograph shows a strip of epoxy with a long hole in the middle and two small holes at each side. When pulled apart by forces applied at the small holes, cracks formed at each end of the long hole. Closely spaced fringes show regions of high stress; the lower photograph shows a detail of the highly stressed region at the tip of the crack. (Courtesy Lawrence Livermore Laboratory)

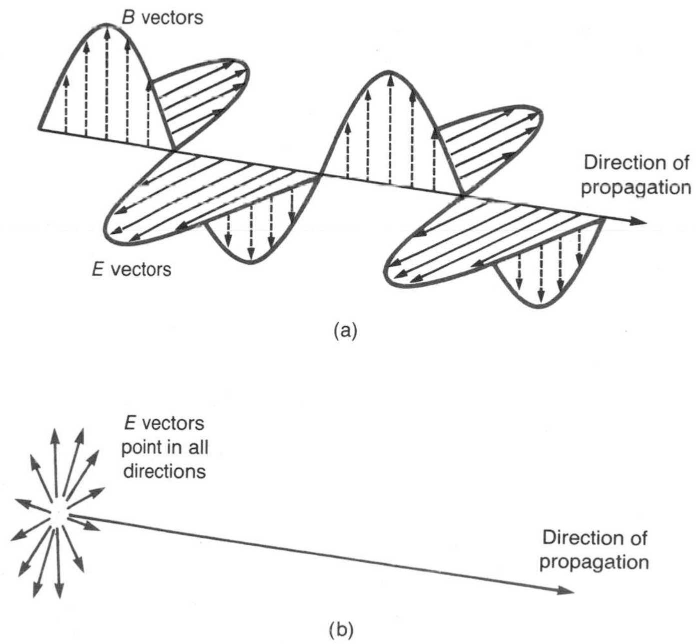

FIGURE 24.17 (a) A horizontally plane polarized light wave. (b) An unpolarized light wave.

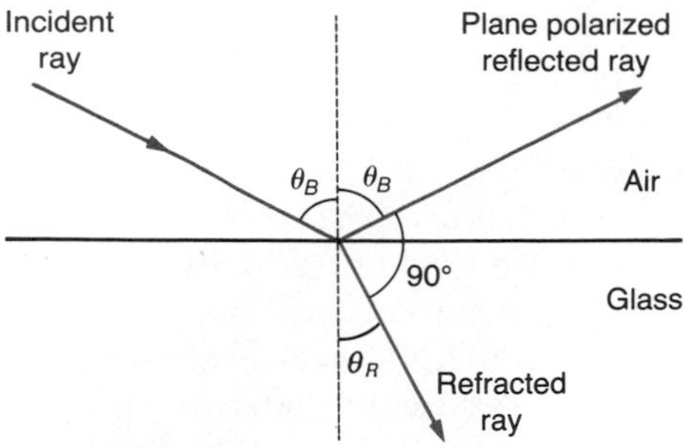

FIGURE 24.18 Plane polarization by reflection at the Brewster angle.

A Scottish physicist, David Brewster (1781–1868), discovered a way to produce plane polarized light by reflecting a light beam from the surface of a medium such as glass. When the reflected and refracted rays are perpendicular, the reflected ray is plane polarized, as illustrated in Fig. 24.18. The angle of incidence that produces a plane polarized reflected ray is called the **Brewster angle,** and the reflected ray is polarized with its electric field vector parallel to the glass–air interface. We can easily derive a formula for the Brewster angle in terms of the refractive index of the glass. Referring to the diagram, θ_B (the Brewster angle) is the angle of incidence (and reflection) and θ_R is the angle of refraction. Snell's law, Eq. (22.3), allows us to write:

$$\sin \theta_B = n \sin \theta_R$$

Because the reflected and refracted rays are at right angles, we have $\theta_B + \theta_R = 90°$, so $\sin \theta_R = \cos \theta_B$. Substituting in Snell's law we find a formula for the Brewster angle.

■ Formula to Calculate the Brewster Angle for Plane Polarization on Reflection from a Surface Whose Refractive Index Is *n*

$$\tan \theta_B = n \qquad \textbf{(24.7)}$$

EXAMPLE 24.10

Calculate the Brewster angle for producing a plane polarized beam of light by reflection in air from a crown glass surface.

SOLUTION The index of refraction of crown glass is 1.52 (from Table 22.2). We simply use Eq. (24.7):

$$\tan \theta_B = n$$

$$= 1.52$$

$$\therefore \theta_B = 57.7°$$

■

If light is incident at some angle other than the Brewster angle, it is not plane polarized. For angles of incidence close to the Brewster angle, most of the reflected light is plane polarized with its electric vector parallel to the interface; a relatively small component is polarized in the perpendicular direction. In such a case we say that the reflected beam is partially plane polarized. Strangely, Brewster remained an adherent of the corpuscular theory of light throughout his long lifetime, even though the phenomenon of polarization is so neatly explained by a wave theory.

The second method for producing polarized light is more complicated but very important from a technical viewpoint. When

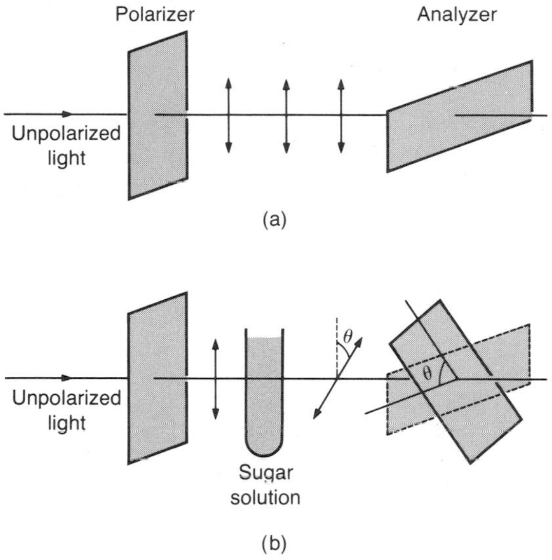

FIGURE 24.19 The use of two Polaroids to analyze polarized light. (a) Crossed Polaroids produce extinction when their polarizing directions are perpendicular. (b) When an optically active substance is inserted between the two Polaroids, the analyzer must be rotated to produce extinction.

light falls on a typical transparent medium, it is refracted in accordance with Snell's law, Eq. (22.3). However, certain special crystalline substances exist that produce two refracted beams that are plane polarized with different planes of polarization. This phenomenon is called *double refraction*. Moreover, one of the two beams is strongly absorbed in some doubly refracting crystals. A Polaroid makes use of this effect by embedding small crystals of iodoquinine sulfate in a plastic plate. The crystals are oriented so that the polarized beam that is not strongly absorbed passes straight through.

We can set up two Polaroid plates, one behind the other, along a beam of light to analyze the polarization of the beam. In Fig. 24.19(a) an unpolarized light beam falls on a Polaroid whose polarizing direction is vertical. This Polaroid, called the *polarizer*, transmits a beam of vertically plane polarized light. A second Polaroid whose polarizing direction is horizontal transmits no light, since the plane of polarization of the incident light is at right angles to its polarizing direction. The second Polaroid is called the *analyzer*, since it determines the plane of polarization of the light. The technical importance of the polarizer-analyzer combination arises from the property of *optical activity* that some substances possess, which is the ability to rotate the plane of polarization of transmitted light. A sugar solution is such an optically active substance. Suppose that a test tube containing sugar solution is inserted into the plane polarized beam emerging from a polarizer, as in Fig. 24.19(b). The sugar rotates the plane of polarization so that the light emerging from it is still plane polarized but not vertically. If the polarizing direction of the analyzer remains horizontal, the light will not be extinguished, since the wave that is incident on the analyzer has an electric field component parallel to its polarizing direction. The analyzer must be turned until its polarizing direction is perpendicular to the new plane of polarization for extinction to occur. The angle that the analyzer is turned from the horizontal position is equal to the angle of rotation of the polarization plane produced by the sugar. But the angle of rotation of the polarization plane is proportional to the path length of the light ray in the sugar and on the concentration of the sugar solution. For container tubes of constant thickness, the rotation required to produce extinction is a direct measure of the concentration of the sugar solution.

We can now understand the operation of Polaroid sunglasses. In many everyday situations, the light that reaches our eyes is largely reflected from horizontal surfaces, such as a road or water surface. Figure 24.20 shows a typical situation in which light is reflected from a horizontal surface. Even if the light is not incident exactly at the Brewster angle, the reflected light is partially polarized. Since the largest electric vector component is parallel to the horizontal reflecting surface, the polarizing direction of the Polaroid lens is vertical in order to extinguish this component. The extinction of the major component of the reflected light reduces the glare experienced by the wearer.

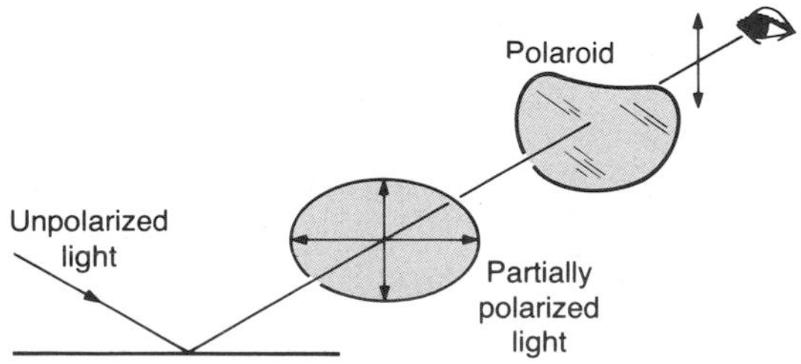

FIGURE 24.20 Polaroid sunglasses extinguish the major component of partially polarized reflected light.

KEY CONCEPTS

Interference and **diffraction** characterize waves that penetrate various types of openings.

A plane, monochromatic, coherent light wave falling on a **double slit** produces a series of **bright interference fringes** on a screen. The fringe separation depends on the wavelength of the light and the distance between the slits; see Eq. (24.1).

A plane, monochromatic, coherent light wave falling on a **diffraction grating** produces a series of sharply delineated bright interference fringes. The fringe separation follows the same formula as for the double slit; see Eq. (24.2).

A plane, coherent light wave falling on a diffraction grating exhibits the spectral content of the wave in various orders; see Fig. 24.6.

The interference of monochromatic light reflected from opposite surfaces of a thin transparent wedge results in bright and dark fringes. White light reflected from opposite surfaces of a thin transparent film shows colored bands. For light at near normal incidence, destructive interference occurs when twice the film thickness equals an integral number of wavelengths; see Eq. (24.3).

A plane, monochromatic, coherent light wave falling on a **single slit** produces a bright **central diffraction maximum** and a series of **secondary maxima** that are much less intense. The position of the dark bands between the maxima depends on the wavelength of the light and the slit width; see Eq. (24.4).

Every **interference pattern** from multiple slits also shows the **diffraction pattern** caused by a single slit; see Fig. 24.13 and Fig. 24.14.

The **Rayleigh criterion** considers two distant light sources to be distinct if their angular separation exceeds the half width of the central diffraction maximum; see Fig. 24.16 and Eq. (24.6).

The **plane of polarization** of a light wave is, by convention, the plane of vibration of the electric vector; see Fig. 24.17(a).

Two principal methods of producing polarized light are reflection from a plane interface and transmission through a Polaroid.

Light reflected from the interface with a transparent medium is plane polarized (for example, the electric vector is parallel to the plane of the interface) when the reflected and refracted rays are perpendicular to each other. The angle of incidence (and reflection) that meets this condition is the **Brewster angle;** see Eq. (24.7).

QUESTIONS FOR THOUGHT

1. Young performed his original double-slit experiment with white light. What did his interference fringes look like?

2. What would you expect if you performed Young's double-slit experiment with a widely spaced pair of slits and a screen very close to the slits?

3. Estimate the slit width required so that no color component of visible light would have a dark diffraction fringe on a screen beyond it.

4. Our discussion of single-slit diffraction implied that the height of the slit was much greater than its width. How would you expect the diffraction pattern to appear if the slit height was only several times its width?

5. What grating spacing would you choose for a diffraction grating to examine the infrared spectrum of a molecule?

6. You shine a laser beam on a diffraction grating and find closely spaced interference maxima with many missing orders. What can you say about the breadth and the spacing of the slits in the grating?

7. Distinguish carefully between the effects of slit width and slit separation in the spectrum formed by a diffraction grating.

8. What modifications would you expect in a spectrum shown by a diffraction grating if the instrument was completely immersed in water?

9. Which of the following phenomena could be used to devise an experiment to measure the wavelength of a monochromatic light beam, and which could not? (a) diffraction; (b) interference; (c) reflection; (d) refraction.

10. Would you use red or blue light to illuminate a microscope slide if you wished to see the greatest possible amount of detail?

11. A radio telescope analyzes the intensity of radio frequency waves and the directions in the sky from which they come. The installations often focus the radio waves by using a large converging mirror. Why is the mirror large, and how does the radio astronomer relate the diameter of the mirror to her observational requirements?

12. How would you test a pair of sunglasses in a store to see if they were polarizing?

PROBLEMS

A. Single-Substitution Problems

1. Monochromatic green light (wavelength 550 nm) illuminates two narrow slits at normal incidence. The slits are 4.16 μm apart. Calculate the angular deviation of the third-order bright interference fringe. [24.2]

2. Two narrow slits 5.34 μm apart are illuminated by monochromatic light at normal incidence. The fourth-order bright interference fringe is 37° from the path of the incident beam. Calculate the wavelength of the light. [24.2]

3. Monochromatic light of wavelength 589 nm falls normally on a pair of narrow slits. The third-order bright interference fringe is 47° from the path of the direct beam. Calculate the slit separation. [24.2]

4. A diffraction grating with 6000 lines per centimeter is illuminated by monochromatic light at normal incidence. The angular deviation of the third-order maximum is 51°. Calculate the wavelength of the light. [24.3]

5. Blue light of 455 nm wavelength falls at normal incidence on a diffraction grating, and the fourth-order maximum is deviated 61°. Calculate the number of lines per centimeter on the grating. [24.3]

6. A diffraction grating having 5000 lines per centimeter is illuminated at normal incidence by a beam of red light (wavelength 650 nm). Calculate the angular deviation of the second-order bright fringe. [24.3]

7. A diffraction grating has 3500 lines per centimeter and is illuminated by light of 589 nm wavelength at normal incidence. A bright interference fringe is observed deviated 55.5°. Calculate the order of the fringe. [24.3]

8. Calculate the wavelengths (in air) that undergo destructive interference when white light at near normal incidence is reflected by a polymer film 375 nm thick and with an index of refraction of 1.47. [24.4]

9. A thin glass film of refractive index 1.52 causes destructive interference by reflection at near normal incidence for light whose wavelength in air is 620 nm. What is the minimum thickness of the film? [24.4]

10. A single slit 2.2 μm wide is illuminated by light of wavelength 550 nm. Calculate the angular position of the first dark fringe on either side of the central maximum. [24.5]

11. Red light of wavelength 650 nm illuminates a single narrow slit. The first dark fringes on either side of the central maximum are separated by 60°. Calculate the width of the slit. [24.5]

12. Calculate the minimum slit width for which no dark diffraction fringe can be observed when the slit is illuminated by a helium–neon laser (wavelength 632.8 nm). [24.5]

13. Monochromatic light illuminates a single slit that is 2.5 μm wide, and produces first-order dark diffraction fringes at 12.6° each side of the central maximum. Calculate the wavelength of the light. [24.5]

14. Calculate the angular resolving power for green light (wavelength 550 nm) of field glasses that have 50-mm diameter objective lenses. [24.5]

15. What diameter objective lens is required for a telescope to achieve a resolving power of 1 s of arc for green light of wavelength 550 nm? [24.5]

16. What angle of reflection is required from a still water surface to produce plane polarized light? [24.6]

17. The Brewster angle for light incident on a clear plastic material is 58.3°. Calculate the refractive index of the plastic. [24.6]

B. Standard-Level Problems

18. Two narrow slits that are 0.035 mm apart are illuminated at normal incidence by monochromatic light. The twelfth-order bright fringe of the interference pattern is 85 cm from the central maximum on a screen 4 m from the slits. What are the wavelength and color of the light? [24.2]

19. Two adjacent narrow slits are illuminated by the beam from a helium–neon laser (wavelength 632.8 nm). The positions of the bright interference fringes are marked on the screen. The slit is then illuminated by monochromatic light of unknown wavelength, and the sixth-order bright fringe is in exactly the same position as the fourth-order bright fringe from the laser. Calculate the unknown wavelength. [24.2]

20. Two narrow slits that are 0.032 mm apart are illuminated at normal incidence by light of the mercury green line (wavelength 546 nm). How far from the slits should a screen be placed so that the 15 bright fringes in the center of the pattern are spread across a region 25 cm wide?

21. A pair of narrow slits 20 μm apart are illuminated at normal incidence by light from the cadmium blue line (wavelength 480 nm). What is the separation of the bright interference fringes at the center of the pattern on a screen that is 150 cm from the slits? What is the separation of the bright interference fringes of twentieth and twenty-first order?

22. A certain diffraction grating has 35 000 lines ruled over a width of 6 cm. Calculate the wavelength of the light for which the angle between the two second-order bright fringes (one on either side of the center) is 60°. What color is this light?

23. A beam from a mercury discharge lamp illuminates a diffraction grating at normal incidence. The second-order green line (wavelength 546 nm) is deviated 61.3°. What is the deviation of the first-order blue line (wavelength 436 nm)? [24.3]

24. Monochromatic blue light of wavelength 450 nm is normally incident on a diffraction grating ruled with 8500 lines per centimeter. What is the highest-order interference pattern that can be produced with this grating? [24.3]

25. A monochromatic beam of blue light (wavelength 450 nm) illuminates a diffraction grating. If the angular deviation of the second-order maximum is 38.2°, calculate the number of lines per centimeter in the grating.

26. Light of wavelength 589 nm falls at normal incidence on a diffraction grating of 3200 slits per centimeter. Calculate the separation of the first- and third-order bright interference fringes on a screen that is 2.5 m from the grating.

27. White light falls at normal incidence on a grating that has 6200 lines per centimeter. Use the wavelengths of Table 22.1 to calculate the angular width of the green region in the second-order spectrum.

28. Two microscope slides each 10 cm long are used to form an air wedge, the separation being achieved at one end by a fine wire 7.5 μm in diameter. Calculate the separation of adjacent dark bands under illumination by green light of 550 nm wavelength at near normal incidence. [24.4]

29. The air wedge of problem 28 is illuminated by monochromatic light at near normal incidence, and a band separation of 4.4 mm is observed. Calculate the wavelength of the light. [24.4]

30. An air wedge is formed between glass microscope slides 12 cm long in contact along one edge and separated by a fine thread 8.2 μm thick along the other edge. Calculate the number of dark bands observed if the wedge is illuminated at near normal incidence by light of wavelength 589 nm (the orange line of sodium).

31. Repeat problem 30 if the wedge space between the glass slides is filled with oil of refractive index 1.62.

32. A thin wedge of water (refractive index 1.33) is formed between flat glass plates and illuminated at near normal incidence by monochromatic light of wavelength 560 nm. The separation of adjacent dark bands is 3.5 mm. Calculate the angle of the wedge.

33. The surface of the lens on a good camera is frequently coated by a thin film of magnesium fluoride (refractive index 1.25) to reduce reflection. The designer selects 500 nm as the wavelength of light for which there should be destructive interference in reflection at normal incidence. Calculate the minimum thickness coating that should be deposited.

34. A soap bubble (index of refraction 1.33) appears green when viewed by white light at near normal incidence. Calculate an approximate minimum thickness for the bubble.

35. A soap bubble 180 nm thick has an index of refraction of 1.33. Find those colors that suffer destructive interference on reflection of white light at near normal incidence.

36. A beam of light from the mercury green line (wavelength 546 nm) illuminates a slit at normal incidence. The two second-order dark fringes are 15.6 cm apart on a screen that is 150 cm from the slit. Calculate the slit width. [24.5]

37. Monochromatic light illuminates a slit 40 μm in width. The central diffraction maximum is 4.7 cm wide on a screen that is 1.72 m from the slit. Calculate the wavelength of the light. What color is the light? [24.5]

38. Light from a helium–neon laser (wavelength 632.8 nm) falls at normal incidence on a narrow slit. The central bright fringe is 115 cm wide on a screen that is 250 cm from the slit. Calculate the slit width.

39. Light from the orange line of sodium (wavelength 589 nm) illuminates a slit that is 12 μm in width. How far away is a screen on which the distance between the first- and second-order dark fringes is 10 cm?

40. Blue light of wavelength 450 nm falls at normal incidence on a pair of slits that are separated by 0.01 mm. Calculate the slit width if the eighth-order bright interference fringe has the same angular deviation as the first dark minimum of the diffraction pattern.

41. The bright star Sirius is one member of a binary pair. The distance to Sirius is 7.8×10^{16} m, and Sirius is separated from its companion by 3×10^{12} m. Calculate the minimum diameter for the objective lens of a small astronomical telescope that will resolve the pair. Take the wavelength of the light at 550 nm. [24.6]

42. The headlights of a small automobile are 95 cm apart. Calculate the maximum distance from which an observer can resolve the two lights on a dark night. Take the wavelength of the light at 550 nm and the diameter of the iris of the eye at 5 mm. [24.6]

43. The reflecting telescope on Mt. Palomar has a diameter of 5.08 m.

a. Calculate its angular resolving power.

b. What is the closest distance at which two objects on the moon could be observed as distinct? (Consider the wavelength of light to be 550 nm and the earth–moon distance to be 3.84×10^8 m.)

44. What is the angle of refraction of a light ray in flint glass ($n = 1.66$) if the reflected ray is plane polarized? [24.7]

45. The critical angle for total internal reflection in a medium that has an interface with air is 43°. What is the Brewster angle for light reflected in air by a plane face of the medium? [24.7]

C. Advanced-Level Problems

46. Light of wavelength 550 nm passes through a slit of width 0.08 mm. The diffraction pattern is seen on a screen that is 2 m from the slit.

a. What is the width of the central bright fringe?

b. What is the width of the central bright fringe when the experiment is carried out under water?

47. Light from a helium–neon laser (wavelength 632.8 nm) is normally incident on a double slit. The two slits, each 1.8 μm wide, are separated by 0.041 mm. How many bright interference fringes are contained within the central maximum of the diffraction pattern?

48. A diffraction grating is used to produce continuous spectra from an incident beam of white light.

a. Show that the third-order spectrum always overlaps the second-order spectrum no matter what the grating spacing is.

b. What are the wavelengths and colors of the light in the two spectra at the edges of the overlap region?

(Take the range of wavelengths of various colors from Table 22.1.)

49. A diffraction grating that has 1800 lines per centimeter is illuminated by monochromatic light at normal incidence. A strong interference maximum is located at a deviation of 28.6° from the line of the original beam.

a. What are the possible values for the wavelength of the incident light? (Assume the light is in the visible region.)

b. Another strong maximum is located at 21.0°. In what way does this restrict the possible values of the wavelength?

50. Light from a hydrogen discharge lamp is examined with a diffraction grating. The second-order red line is deviated 63.5°, and the blue line is missing in the third-order spectrum. Calculate the number of lines per centimeter and the minimum width of each slit. The known wavelengths of the lines are 656 nm for the red and 486 nm for the blue.

25

SPECIAL RELATIVITY

The second part of the nineteenth century was a period of great complacency for physics. The successes of Newtonian mechanics and Maxwell's electromagnetic theory led many scientists to believe that all of the important discoveries had been made, and only loose ends remained to be tied.

However, the closely knit structure began to come apart around the turn of the century. New experiments showed that electromagnetic waves possessed a particle character, and further experiments showed that particles possessed a wave character. To compound the difficulties that these experiments presented, it became obvious that something was seriously wrong with Newtonian mechanics for objects moving at velocities close to that of light. Finally, the search for the luminiferous ether (which was supposed to be the medium in which light propagates) turned out so badly that the very existence of the ether became suspect.

All of these developments were interrelated, but let us take them one at a time. In this chapter we will follow the developments that led to Einstein's formulation of the theory of relativity.

25.1 The Luminiferous Ether and the Michelson–Morley Experiment

Maxwell proposed the electromagnetic theory of light in the middle of the nineteenth century. Every wave motion known at the time required a medium for wave propagation, on whose properties the wave velocity depended. Why should electromagnetic waves be different? It was only natural to propose a medium for their propagation and to search for its properties. The unknown medium was named the luminiferous ether from Latin words that mean a rarefied carrier of light.

The ether needed to be a very unusual substance. It had to be very rigid to vibrate at the high frequencies of light waves and yet very thin to permit passage of the planets through space without causing any drag. Moreover, it had to permeate transparent substances such as water and glass, since light propagates through them. Fizeau demonstrated that light travels faster through flowing water in the direction of water flow than in the opposite direction. This important experiment was successfully explained by assuming that the water drags the ether along with a certain fraction of its own velocity. Various experiments were suggested to measure the velocity of the earth through the ether, but they required observational accuracy that did not seem possible. The crucial experiment that broke the apparent impasse was designed and carried out by the first American Nobel Prize winner, Albert Michelson (1852–1931). Michelson used an optical interferometer that he had designed to measure very small changes in length. Let us first briefly describe Michelson's instrument.

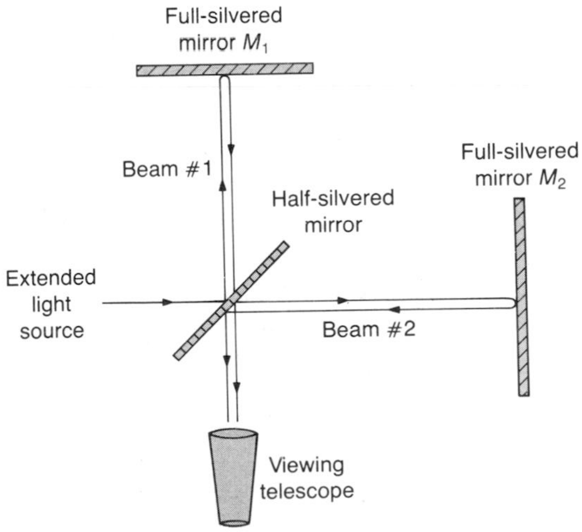

FIGURE 25.1 Schematic diagram of the Michelson interferometer.

The essential parts of the Michelson interferometer are shown in Fig. 25.1. A broad beam of light falls on a half-silvered mirror, which reflects part of the light to the mirror M_1 and transmits the remainder to the mirror M_2. The diagram shows one beam from the extended source, which is split into beam #1 and beam #2 by the half-silvered mirror. Upon reflection from M_1 and M_2, the light returns to the half-silvered mirror, and components of the reflected beams arrive together at the viewing telescope. If the light from the two beams is in phase, constructive interference will produce an intensity maximum; if the two beams are out of phase, destructive interference will produce an intensity zero. However, the light that is incident on the half-silvered mirror is not a single ray as Fig. 25.1 shows. Rather, a number of rays from the extended source follow paths that are only approximately parallel to each other. The result is a series of equispaced bright and dark fringes.

Now if we move the mirror M_2 either backward or forward along a line parallel to the beam #2, the phase relationship between the rays reflected from the mirror's changes. If we move the mirror back one-quarter of a wavelength, we increase the distance traveled by beam #2 by one-half of a wavelength. If this beam originally interfered constructively with the beam from mirror #1, it will now interfere with it destructively. As a result, each bright fringe will shift to the position originally occupied by the adjacent dark fringe. In fact, we can say that the whole fringe pattern will move across the field of view. Further movement of either mirror will shift the fringe pattern in a way dependent on the magnitude and direction of the movement.

Michelson used his interferometer to relate the length of the standard meter rod in Paris to the wavelengths of very pure lines in the spectrum of cadmium. He found that $\lambda = 643.84722$ nm for the wavelength of the red line in the cadmium spectrum, a figure that differs from the modern value by about one part in a billion. Michelson's Nobel Prize was awarded primarily for his work in this area.

However, Michelson's most famous experiment, which was done in collaboration with Edward Morley (1838–1923), attempted to measure the velocity of the earth through the ether with the aid of the interferometer. To this end they mounted the interferometer on a large stone (to eliminate mechanical strain in the instrument) and floated the stone in a bowl of mercury so that the whole apparatus could be rotated easily. Let us follow the line of reasoning that led Michelson and Morley to expect that they could measure the earth's velocity through the ether. Consider the interferometer shown in Fig. 25.1. The movement of the earth through the ether should produce an ether wind, just as you feel a wind when riding in a convertible on a calm day. Let us suppose that the ether wind is blowing parallel to beam #1 in the interferometer and therefore transversely to beam #2. Because of the differing relative ether wind directions, the velocity of light should differ on the two paths of the beam; in fact, it should be different in the two directions associated with beam #1. However, regardless of the ether wind direction, a pattern of interference fringes is observed in the telescope. Since rotation of the whole apparatus through 90° reverses the beam that is parallel to the ether wind and the one that is transverse to it, Michelson and Morley calculated that the rotation should produce a fringe shift. A knowledge of the earth's orbital velocity around the sun permits estimation of the minimum value of the ether wind velocity and thereby of the expected magnitude of the fringe shift. From their calculations they expected to find a shift of 0.04 of a fringe. Instead their observations showed no detectable shift at all—certainly less than 0.01 of a fringe.

The **Michelson–Morley experiment** shows that the velocity of light does not depend on the motion of the observer. This means that there is no identifiable medium of propagation—in other words, *there is no ether.* This result was so startling and so much at variance with accepted ideas that it was not generally believed for some time. The publication of Einstein's first paper on relativity in 1905 cast a new light on the interpretation of the Michelson–Morley experiment. In fact, Einstein's theory made their experiment so important that it has been repeated on many occasions with instruments of greatly improved accuracy. But night and day, winter and summer, the result is always the same—there is no detectable motion of the earth through the ether.

25.2 Einstein's Theory—Time Dilation and Length Contraction

The new kinematics and the dynamics that followed it began with the work of several scientists around the turn of the century. The theory became known as **special relativity** to distinguish it from the earlier Galilean relativity that we studied in a previous chapter.

The Dutch physicist and Nobel Prize winner H. A. Lorentz (1853–1928), the Irish physicist G. F. Fitzgerald (1851–1901), the German physicist (and former teacher of Einstein) Hermann Minkowski (1864–1909), and the French mathematician J. H. Poincaré (1854–1912) all contributed to the early development of the subject. However, the analysis from basic principles was provided by Albert Einstein (1879–1955), who will probably be judged the greatest scientific genius of this century. Einstein's basic principles for the development of special relativity appear deceptively simple, and it is a measure of Einstein's profound intellect that he could develop his whole theory from the following two principles alone:

1. A law of physics that is valid in one inertial frame of reference is also valid in any other inertial frame.

SPECIAL TOPIC 24 EINSTEIN

Albert Einstein was born in Ulm, Germany, in 1879. He did not show much promise as a student, and after graduation he became a clerk at the Swiss Patent Office in Berne.

In this position, without any academic supervision, he began his theoretical work. In 1905, five of his papers were published, including three on topics of major importance. One concerned the application of Planck quanta to explain the photoelectric effect, and another worked out the detailed theory of Brownian motion. However, in retrospect, the third was really the epoch-making development. He set forth his basic postulates for the theory of special relativity, and brought coherence to all of the previous work in this area. His work not only tied up loose ends, however, but also gave rise to the famous mass–energy relationship.

Despite these feats, it was some years before he obtained a premier professorship. Due to the efforts of Planck he became professor of physics in Berlin. In 1915, he published another famous paper on general relativity that still remains the basis of modern cosmological theory. In 1921, he was awarded the Nobel Prize in physics for his work on the photoelectric effect—an irony of history, since his work in special and general relativity far outshines that for which the prize was awarded. Indeed many believe that his work on relativity marked him as the greatest physicist since Newton.

Einstein was lecturing in California when Hitler came to power. Since he was Jewish, he could not return to Germany. He accepted a post at the Institute for Advanced Studies in Princeton, New Jersey, where he spent the remainder of his life hunting for a theory that would embrace both gravitation and electromagnetism. He died at Princeton in 1955.

2. The speed of light in empty space is always the same in every inertial reference frame, and is independent of the relative motion of the light source and the observer.

Let us recall our previous concept of an inertial frame from Newtonian dynamics. A coordinate system in which Newton's laws are true is termed inertial, and any coordinate system moving at uniform velocity with respect to the first is also inertial. If there is no ether pervading all space, we have no reason to prefer one inertial frame over another—they all provide equivalent frameworks for stating the laws of physics.

Einstein's second principle makes the outcome of the Michelson–Morley experiment quite logical. Since the velocity of light is always the same along each arm of the interferometer, there is no reason to expect any shift of the fringes when the instrument is rotated.

In addition, we must give up the old ideas of absolute space and absolute time. The basic observational data are the time and space coordinates of an event relative to some reference frame. Different observers in different reference frames may come to different conclusions about the values of the time and space coordinates for the same event. The basic task in relativistic kinematics is to give a general relation between event coordinates as measured by different observers. However, let us address the easier task of showing how two observers determine the time interval between two events or the length of some object.

Consider two inertial frames S and S' that have relative velocity v, as shown in Fig. 25.2(a). On the diagram we have placed a velocity vector **v** alongside the rocket ship that is the frame S'. This does not mean that we regard S as being at rest and S' as moving with velocity **v**, but only that a relative velocity exists between them. We could just as well put the velocity vector **v** on S and point it in the opposite direction. In short, no frame can be chosen that is at rest with regard to all possible inertial frames. To do so would assign priority to one particular inertial frame, and Einstein's first basic principle forbids this. We further suppose that observers in both frames have identical clocks and identical distance-measuring equipment.

Now suppose that the observer in S has two parallel mirrors A and B that are a distance d apart on the y-axis, as shown in Fig. 25.2(b). At $t = 0$ a light signal leaves mirror A, travels to mirror B, and, after reflection, returns to A. The emission of the light signal from mirror A and its return to the same place are two definite events. Since the light signal travels a total distance $2d$

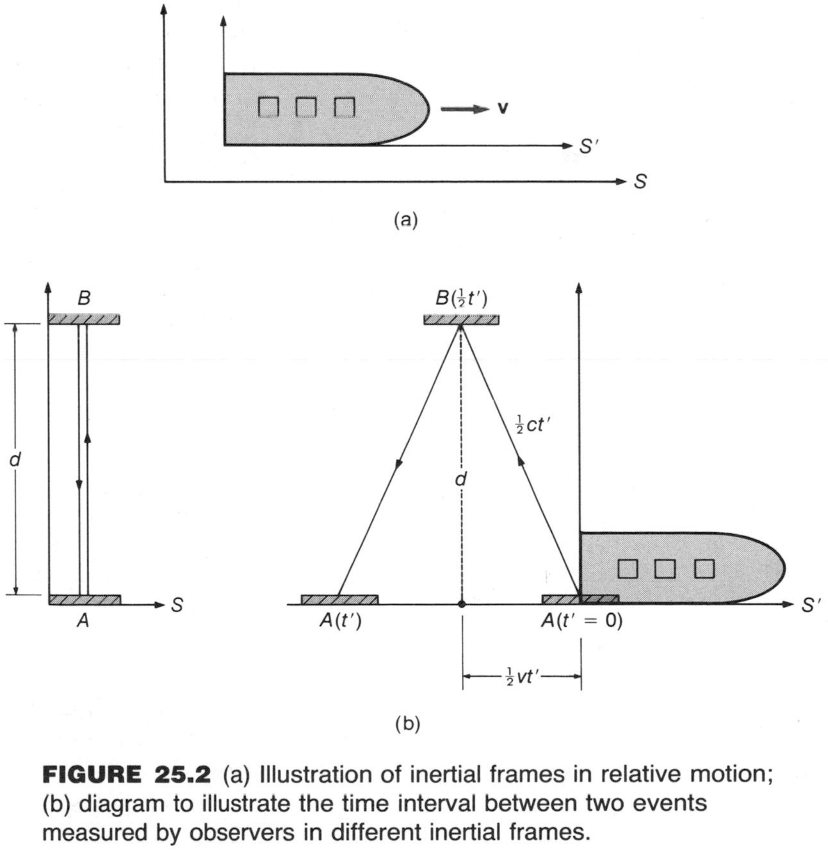

FIGURE 25.2 (a) Illustration of inertial frames in relative motion; (b) diagram to illustrate the time interval between two events measured by observers in different inertial frames.

with velocity c, the observer in S obtains the time interval between these two events from:

$$t_0 = \frac{2d}{c}$$

We use the symbol t_0 to denote the **proper time interval** between two events. This proper time interval is the time between two events that occur at the same place in a given frame. The events in this case are the emission of the light signal from A and its return to the same place in the frame S.

The observer in S' does not see these two events occurring at the same place. Suppose that the origins of coordinates of S and S' coincide when the light signal leaves A. At this time the observer in S' measures $t' = 0$ on his clock, and when the light signal returns to A, he measures a time t' on his clock. Since the reflection at mirror B is the halfway point for the constant-velocity trip of the light beam, the observer in S' measures $\frac{1}{2}t'$ for the time coordinate of this reflection. However, the observer S' moves with constant velocity $\mathbf{v}$ (along the x-axis) relative to the mirrors A and B. Upon reflection of the light beam the mirror B is distant $\frac{1}{2}vt'$ from the origin of S'. The separation of the two mirrors along the y direction remains unchanged at d. By Einstein's second basic principle, the observer in S' also measures c for the velocity of light. The distance traveled by the light signal in this frame as it goes from mirror A to mirror B is therefore $\frac{1}{2}ct'$. Using the Pythagorean theorem on the diagram of Fig. 25.2(b), we have the following relation:

$$\left(\tfrac{1}{2}ct'\right)^2 = d^2 + \left(\tfrac{1}{2}vt'\right)^2$$

Substituting $d = \frac{1}{2}ct_0$ from the measurement of the observer in S, we have:

$$\left(\tfrac{1}{2}ct'\right)^2 = \left(\tfrac{1}{2}ct_0\right)^2 + \left(\tfrac{1}{2}vt'\right)^2$$

Rearranging terms and taking the square root yield:

$$t' = \frac{t_0}{(1 - v^2/c^2)^{1/2}} \qquad \textbf{(25.1)}$$

Because of later developments, it is now convenient to write Eq. (25.1) in a different form. As a first step, we define the **relativistic gamma factor.** The formula for calculating the gamma factor is as follows:

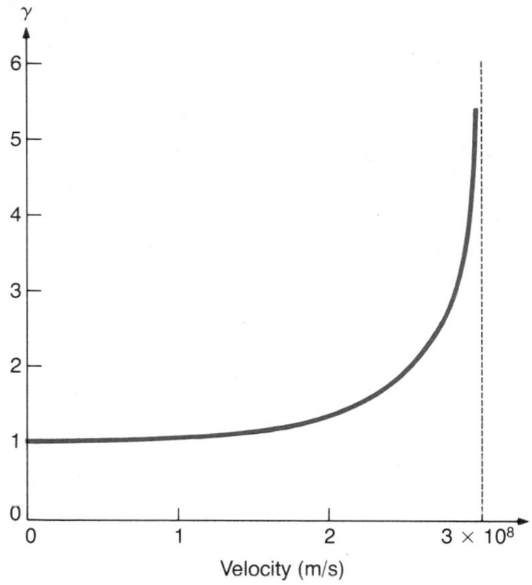

FIGURE 25.3 The relativistic gamma factor as a function of velocity.

■ Formula for Calculating the Relativistic Gamma Factor

$$\gamma = \left(1 - \frac{v^2}{c^2}\right)^{-1/2} \qquad (25.2)$$

v = relative velocity of the inertial frames
c = velocity of light

The variation of γ with a change in v is shown in Fig. 25.3. A good approximation is $\gamma \approx 1$ for values of v between 0 and $0.1c$. For values of v between $0.1c$ and $0.9c$, the γ factor increases in value to $\gamma \approx 2$. For higher values of v, the value of γ increases very rapidly, approaching infinity as v approaches c.

We can now rewrite Eq. (25.1) in a simpler form. Using the γ factor from Eq. (25.2), we have the following formula:

■ Relativistic Time Dilation Formula

Time interval measured by relatively moving observer = relativistic gamma factor × proper time interval

$$t' = \gamma t_0 \qquad (25.3)$$

We call the result a **time dilation** formula, since a relatively moving observer always measures a time interval between two events that is longer than the proper time interval between them by an extent that depends directly on the gamma factor. Figure 25.3 shows that the gamma factor is close to unity for values of the relative velocity that are considerably less than the velocity of light. Consequently, time dilation is not observed until the relative velocity of observers is a substantial fraction of the velocity of light. For low relative velocities we simply find the Galilean result, in which the time interval between two events is the same for all observers.

EXAMPLE 25.1

An electron travels with a velocity of $0.99c$ along a drift tube that is 10 m long. Calculate the transit time as measured by an observer who is stationary with respect to the drift tube, and also by an observer who moves with the electron.

SOLUTION Let the drift tube be stationary in the frame S', and let the frame S be attached to the moving electron. Now consider two events. The first event is when the electron is at one end of

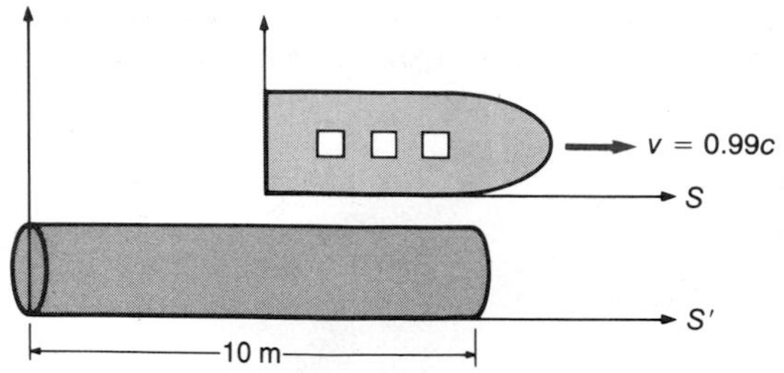

the tube, and the second event is when it reaches the other end. These two events are at the same place in the frame S—namely, at the coordinate origin of S. It follows that the time interval between them measured by the observer in S is the proper time interval t_0. The time interval measured by the observer in S' is dilated by the gamma factor for the relative motion of the frames. To calculate the gamma factor, we use Eq. (25.2):

$$\gamma = \left(1 - \frac{v^2}{c^2}\right)^{-1/2}$$
$$= (1 - 0.99^2)^{-1/2}$$
$$= 7.09$$

But we can readily calculate the dilated time interval t' measured by the observer in S' who records the electron as traveling 10 m with a velocity of $0.99c$. The dilated time is given by:

$$t' = \frac{10 \text{ m}}{0.99 \times 3 \times 10^8 \text{ m/s}}$$
$$= 3.37 \times 10^{-8} \text{ s}$$

We can now use Eq. (25.3) to calculate the proper time interval measured by the observer in S:

$$t_0 = \frac{t'}{\gamma}$$
$$= \frac{3.37 \times 10^{-8} \text{ s}}{7.09}$$
$$= 4.75 \times 10^{-9} \text{ s}$$

■

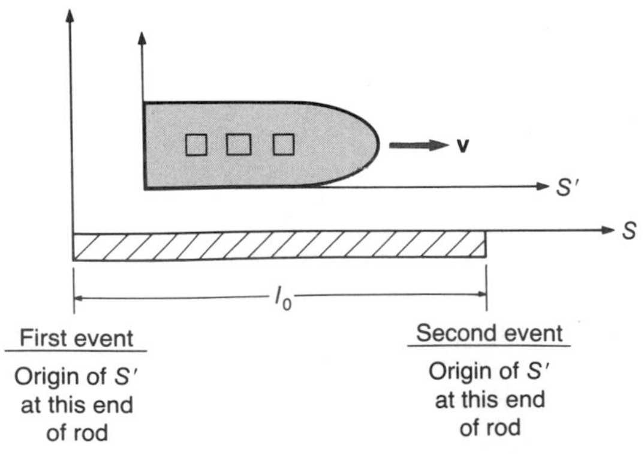

FIGURE 25.4 Diagram to illustrate the measurement of the length of a rod by observers in differential inertial frames.

Let us now investigate how different observers measure the length of a rod. Consider a rod at rest in the frame S and placed parallel to the x-axis with one end at the origin of coordinates, as shown in Fig. 25.4. Let l_0 be the length of the rod measured by the observer in S. We call this the **proper length** of the rod—that is, its length in a frame in which it is at rest. The observer in S', who moves parallel to the rod with velocity v, cannot measure its length by using measuring rods. To carry out the length measurement, he times two events. The first event is the coincidence of the coordinate origins of the frames S and S' at the near end of the rod, and the second is the coincidence of the coordinate origin of S' with the far end of the rod. Both events occur at the same place in S'—that is, at the origin of coordinates. The time interval between the two events as measured by S' is the proper time interval t_0. Since the relative velocity of the rod is $\mathbf{v}$, the observer S' infers that the length of the rod is:

$$l' = t_0 v$$

The observer in S measures the dilated time γt_0 between the events in question. Moreover, he sees that the origin of S' travels a distance l_0 at velocity $\mathbf{v}$ during this time interval. The observer in S therefore concludes that:

$$l_0 = \gamma t_0 v$$

Combining these two results gives us the relativistic length contraction formula.

■ Relativistic Length Contraction Formula

$$\text{Length measured by relatively moving observer} = \frac{\text{proper length}}{\text{gamma factor}}$$

$$l' = \frac{l_0}{\gamma} \qquad (25.4)$$

The contraction is measured by an observer moving parallel to the rod. There is no measured length contraction if the observer moves perpendicular to the rod. For low relative velocities, the gamma factor is close to unity, and we once again obtain the Galilean result that the length of the rod is the same for all observers.

The preceding discussion warns us to be careful with the concepts of proper time, proper length, dilated time, and contracted length. The proper time and length do not necessarily belong to the observer in S any more than the dilated time and contracted length belong to the observer in S'. The proper time interval between two events is the time interval measured by an observer who sees them occurring at the same place, and the proper length is the length of a rod measured by an observer at rest with respect to the rod. Our assignment of letters S and S' to the frames carries no implications of priority. In fact, we can interchange the letters without changing any of our reasoning on time dilation and length contraction.

In a way, length contraction and time dilation represent opposite sides of the same coin. To see how this is so, let us recount some of the details of a dilemma that arose in connection with experimental observations on muons, which are unstable particles that are created high in the atmosphere by cosmic rays and that decay rapidly into electrons and neutrinos. There is a range of muon decay times, but the average value is 2.2 μs. If the muon travels at close to the speed of light, the average distance that it travels between the times of creation and decay will be given approximately by:

$$(3 \times 10^8 \text{ m/s}) \times (2.2 \times 10^{-6} \text{ s}) = 660 \text{ m}$$

However, muons that are created many kilometers high in the atmosphere reach sea level before decaying, traveling a distance that is many times greater than the calculated average. Let us use an example to show how the dilemma was resolved with relativistic kinematics.

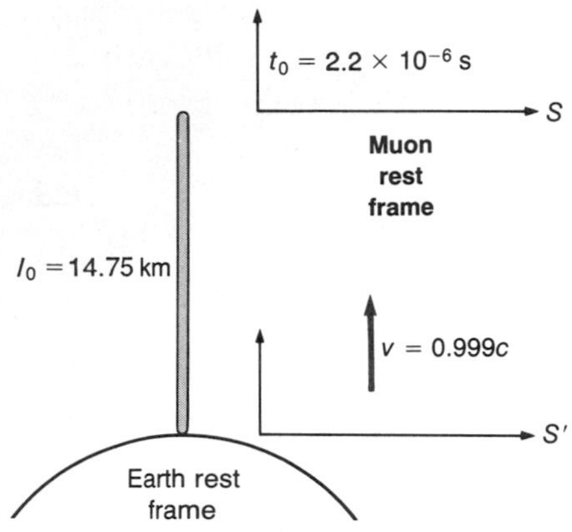

EXAMPLE 25.2

Muons with an average lifetime of 2.2 μs are created in the atmosphere and travel vertically downward at 99.9% of the speed of light. The average muon decays on reaching sea level. Calculate:

a. The average lifetime of the muon measured by an earthbound observer

b. The probable height at which the muon was created

c. The average distance traveled by the muon measured by an observer in an inertial frame attached to it

SOLUTION For a correct relativistic treatment, we must carefully distinguish between measurements made in the two inertial frames. The relative velocity of the two frames is $0.999c$, and we can calculate the gamma factor from Eq. (25.2):

$$\begin{aligned}\gamma &= \left(1 - \frac{v^2}{c^2}\right)^{-1/2} \\ &= (1 - 0.999^2)^{-1/2} \\ &= 22.37\end{aligned}$$

a. Let us refer to the inertial frame of the muon as S. The creation and decay of the muon occur at the same point in this frame. The time between two events measured in S is therefore the proper time interval $t_0 = 2.2 \times 10^{-6}$ s. The time between the two events measured by the earthbound observer in the frame S' is given by Eq. (25.3):

$$\begin{aligned}t' &= \gamma t_0 \\ &= 22.37 \times 2.2 \times 10^{-6}\ \text{s} \\ &= 4.921 \times 10^{-5}\ \text{s}\end{aligned}$$

b. The earthbound observer can calculate the probable height at which the muon was created by multiplying its speed by his measurement of its lifetime:

$$\begin{aligned}\text{Height (measured by observer in } S') &= \text{muon speed} \times \text{muon lifetime (measured by observer in } S') \\ &= 0.999 \times 3 \times 10^{8}\ \text{m/s} \times 4.921 \times 10^{-5}\ \text{s} \\ &= 14.75\ \text{km}\end{aligned}$$

c. The place in the atmosphere where the average muon was created and the sea level location where it decays are the opposite ends of an imaginary rod 14.75 km long that is at rest in the earth frame S'. The observer in S' measures the proper length of the rod as $l_0 = 14.75$ km, but its length measured in S (the frame of the muon) is given by Eq. (25.4):

$$\begin{aligned} l' &= \frac{l_0}{\gamma} \\ &= \frac{14.75 \times 10^3 \text{ m}}{22.37} \\ &= 659.4 \text{ m} \end{aligned}$$

Note the contrasting but consistent conclusions of the observers in the two reference frames. Although the two observers measure quite different values for time intervals and distances, they both measure the same value for the relative speed of the earth and the muon.

Observer in frame S:

$$\begin{aligned} \text{Time interval} &= \text{proper time interval} = 2.2 \times 10^{-6} \text{ s} \\ \text{Length} &= \text{contracted length} = 659.4 \text{ m} \\ \text{Relative speed of the two frames} &= \frac{659.4 \text{ m}}{2.2 \times 10^{-6} \text{ s}} \\ &= 2.997 \times 10^8 \text{ m/s} \end{aligned}$$

Observer in frame S':

$$\begin{aligned} \text{Time interval} &= \text{dilated time} = 4.921 \times 10^{-5} \text{ s} \\ \text{Length} &= \text{proper length} = 14.75 \text{ km} \\ \text{Relative speed of the two frames} &= \frac{14.75 \times 10^3 \text{ m}}{4.921 \times 10^{-5} \text{ s}} \\ &= 2.997 \times 10^8 \text{ m/s} \end{aligned}$$

■

Let us conclude our discussion of relativistic kinematics with a simplified version of the relative velocity formula. Consider the situation of Fig. 25.5; an observer S observes an object moving parallel to the x-axis with velocity v_x. A second observer S' moves parallel to the x-axis with velocity v. Our question is to decide on the velocity of the object as measured by S'.

We have already answered this question in Chapter 3 from the point of view of Galileo and Newton. The result is:

$$v'_x = v_x - v \tag{25.5}$$

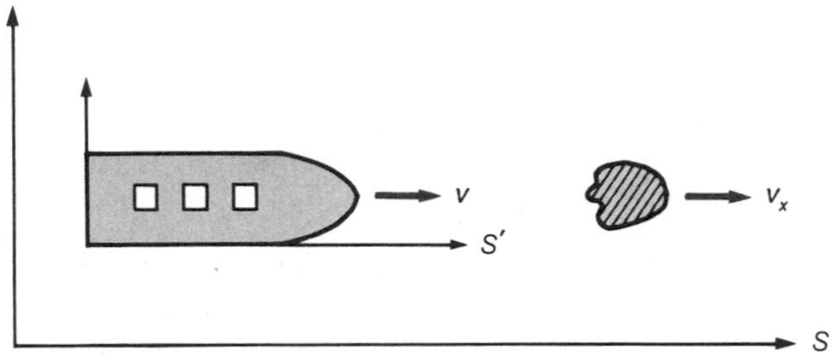

FIGURE 25.5 Two observers S and S' measure the velocity of an object that is moving parallel to the direction of their own relative motion.

since the relative velocity of the observers simply subtracts from (or adds to) the velocity in question. This is the result of Galilean relativity, and it is adequate when the magnitudes of all velocities are very small compared with the speed of light.

However, there is immediate conflict between Eq. (25.5) and Einstein's relativity postulates. If the object whose speed is being measured is a light signal, then the Galilean equation shows the two observers measuring velocities that differ by their relative velocity. The detailed resolution of this problem is beyond the scope of this book, but we can quote the Einstein formula that corresponds to the Galilean Eq. (25.5):

■ Relativistic Velocity Transformation Equation

$$v'_x = \frac{v_x - v}{1 - v_x v/c^2} \tag{25.6}$$

v is the velocity of S' relative to S
v_x is velocity of object measured by S
v_x' is velocity of object measured by S'

We see at once that the equation is in agreement with the Einstein postulate. Suppose the object in question is a light signal; then observer S observes $v_x = c$. From Eq. (25.6) we have:

$$v_x' = \frac{c - v}{1 - cv/c^2} = c$$

so that observer S' also observes $v_x' = c$ for any value of the relative velocity of S and S'.

A more general treatment in which the velocity of the object is not parallel to the relative velocity of the two observers gives a more complex algebraic solution. However, Eq. (25.6) remains correct for the x components of the velocities, as also does the invariance of the speed of a light signal.

Another interesting consequence of the velocity transformation law is well introduced with an example.

EXAMPLE 25.3

Two neutrons approach each other along a straight line with velocities $0.9c$ and $-0.9c$ as measured by a laboratory observer. What velocity does an observer attached to one neutron assign to the other neutron?

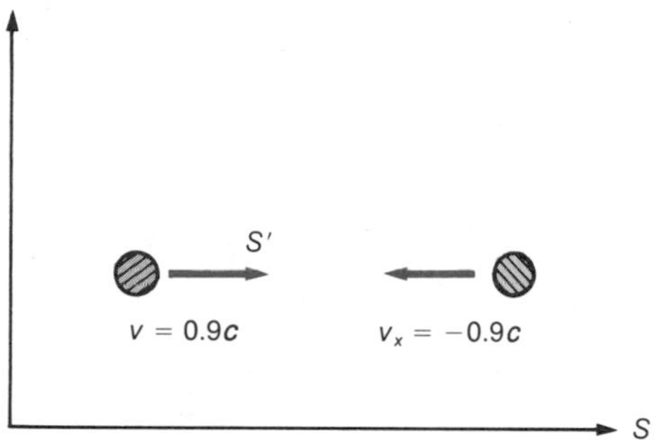

SOLUTION The situation is shown in the diagram. The observer S' moves with the first neutron so that the relative velocity of the two observers is $v = 0.9c$. The second neutron has a velocity $v_x = -0.9c$ as measured by S. Its velocity as measured by S' is given by Eq. (25.6):

$$
\begin{aligned}
v_x' &= \frac{v_x - v}{1 - v_x v/c^2} \\
&= \frac{-0.9c - 0.9c}{1 - \dfrac{(-0.9c)(0.9c)}{c^2}} \\
&= -\frac{1.8c}{1.81} \\
&= -0.994c
\end{aligned}
$$

It is easy to check that this is truly a relative velocity; that is, if we had put the observer S' on the second neutron, then he would measure $v_x' = +0.994c$ for the velocity of the first neutron.

Note that the Galilean Eq. (25.5) would give $v_x' = -1.8c$, but the Einstein Eq. (25.6) reduces the measured velocity of the material particle (the neutron) to a magnitude below the speed of light. We shall return to this point in the following section. ■

25.3 Relativistic Dynamics

Newtonian dynamics appeared to be faultless for about two centuries. However, the death of the ether theory also spelled doom to the conceptual framework presented by Newton. When the velocity of some object is a substantial fraction of the velocity of light, we need a completely new approach.

In our study of Newtonian dynamics, we defined the force on an object as the product of its inertial mass, which is an inherent constant property of the object, and the acceleration. When the force acting on an object can be calculated explicitly (for example, from the law of gravitation or from Coulomb's law), we have a complete physical theory of the motion. Concepts such as momentum and energy followed the analysis of force and acceleration.

However, force and acceleration play little role in relativistic dynamics, while the conservation laws for energy and momentum are of fundamental importance. In fact, the energy conservation law is expanded to include a form of energy not known to classical physics. In his study of relativistic dynamics, Einstein arrived at the remarkable conclusion that energy must possess inertia. However, since mass is our classical measure of inertia, mass and energy must be equivalent for us to associate inertia with energy. Einstein enshrined this result in his famous formula for **mass–energy equivalence.**

■ Einstein's Formula for the Equivalence of Mass and Energy

$$\text{Energy} = \begin{matrix}\text{relativistic}\\ \text{gamma factor}\end{matrix} \times \text{mass} \times \begin{matrix}\text{(velocity}\\ \text{of light)}^2\end{matrix}$$

$$E = \gamma mc^2 \qquad \textbf{(25.7)}$$

Note that the type of energy calculated from Eq. (25.7) is not specified; it is the total energy of the mass. To understand this fully, we will consider a special case of the energy equation for a mass at rest. When $v = 0$, we have $\gamma = 1$ from Eq. (25.2), and the energy thus calculated from Eq. (25.7) is called the **rest energy.** We write a separate formula for this important special case. This is an entirely different situation from that of classical physics. Classically, we set the energy of a mass at rest equal to zero, which is quite justified for the range of problems considered there. However, a mass at rest does possess energy that must be taken into account in certain applications of the conservation law. Let us work some examples to get a feeling for the magnitudes involved.

■ Formula for the Rest Energy of a Mass

$$\text{Rest energy} = \text{mass} \times (\text{velocity of light})^2$$

$$E_0 = mc^2 \qquad \textbf{(25.8)}$$

EXAMPLE 25.4

Calculate the rest energy of an electron in electron volts.

SOLUTION From Eq. (25.8) we have:

$$
\begin{aligned}
E_0 &= mc^2 \\
&= 9.11 \times 10^{-31}\ \text{kg} \times (3 \times 10^8\ \text{m/s})^2 \\
&= 8.20 \times 10^{-14}\ \text{J} \\
&= 8.20 \times 10^{-14}\ \text{J} \div 1.6 \times 10^{-19}\ \text{J/eV} \\
&= 0.511\ \text{MeV}
\end{aligned}
$$

The rest energy, which seems to be very large indeed, is equal to the kinetic energy that an electron would gain after acceleration by a potential difference of 0.511 MV. As a further check on magnitudes, let us try another example. ■

EXAMPLE 25.5

Calculate the rest energy of 1 kg of coal, and compare it with the energy of combustion.

SOLUTION From Eq. (25.8) we have:

$$
\begin{aligned}
E_0 &= mc^2 \\
&= 1\ \text{kg} \times (3 \times 10^8\ \text{m/s})^2 \\
&= 9 \times 10^{16}\ \text{J}
\end{aligned}
$$

From Table 15.1 we find that the energy of combustion of the coal is about 3×10^7 J. The rest energy is about *3 billion times greater.* ■

How can classical physics ignore such immense quantities of energy and still give correct answers? The reason is that we can select any convenient zero of energy. *If the rest energy remains constant,* it is a convenient zero and does not enter our calculations. We shall see that when we use the relativistic approach, we must consider the rest energy in cases where it does not remain constant.

Let us now obtain a correct concept of kinetic energy from a relativistic viewpoint. In Newtonian dynamics we define kinetic energy as the energy due to the motion of a mass. In relativistic dynamics we use the same approach, setting the kinetic energy equal to the total energy less the rest energy. With the help of Eqs. (25.7) and (25.8) we arrive at the following result:

■ Relativistic Definition of Kinetic Energy

$$
\begin{aligned}
\text{Kinetic energy} &= \text{total energy} - \text{rest energy} \\
\text{KE} &= \gamma mc^2 - mc^2 \\
&= (\gamma - 1)mc^2
\end{aligned}
\tag{25.9}
$$

SPECIAL TOPIC 25 GENERAL RELATIVITY AND BLACK HOLES

The special theory of relativity considers observers in inertial reference frames. The inertial frames have constant relative velocities, and no frame has priority over another. There is no room in this theory for either accelerated reference frames or gravitational fields. In 1916 Albert Einstein proposed his general theory of relativity, which could perhaps be called more descriptively a theory of gravitation.

In special relativity the relationship between space and time can be formalized mathematically by introducing a four-dimensional space referred to as the space-time manifold, or as Minkowski space. There are three spatial dimensions and one temporal dimension in Minkowski space, and it is subject to Euclidean geometry. Einstein assumed that real space-time corresponds not to Euclidean but to Reimannian geometry. Riemannian space is curved in contradistinction to Euclidean space which is flat. Einstein further assumed that the curvature of space-time is a measure of the gravitational field, and he set up an equation to calculate this curvature in terms of the distribution of mass. Furthermore, the path of motion of a mass in a gravitational field is assumed to be a geodesic in curved space. A geodesic in curved space is the analog of the straight line in Euclidean space; namely, it is the shortest distance between two points in the space. For example, any great circle is a geodesic of the two-dimensional curved surface of a sphere; geodesics in higher dimensional curved space are much less easy to imagine.

Calculation of the motion of mutually interacting masses is then a most difficult mathematical problem since the masses cause the curvature of the space-time, which in turn causes the path of motion. However, for the case of a small mass interacting with a much larger mass, the gravitational field of the small mass is neglected, and it describes a geodesic in space-time that is curved by the gravitational field of the larger mass. In order to calculate planetary orbits, one would calculate the curvature of space-time caused by the gravitational field of the sun and then find the appropriate geodesics in this curved space. Whereas Newtonian theory predicts elliptic orbits for this problem, Einstein's theory predicts elliptic orbits whose axis slowly rotates; this phenomenon is called precession of

We have not attempted to elaborate on all the details of Einstein's dynamics; instead, we have merely written relativistically correct formulas for the total energy, the rest energy, and the kinetic energy of a massive particle.

Many experimental checks of Einstein's new dynamics were made after the publication of his theory. W. Bertozzi devised a particularly elegant experimental verification of Eq. (25.9). He used the apparatus of Fig. 25.6(a) to measure the kinetic energy of electrons as a function of their velocity. Short bursts of electrons acquire large kinetic energy from a high-voltage Van de Graaf accelerator. The velocity of the electrons is measured electronically as they pass through a drift tube. Figure 25.6(b) shows the experimental points together with curves representing the Newtonian and relativistic kinetic energy formulas. Two conclusions emerge at once. First, the Newtonian formula for kinetic energy is grossly inaccurate for high-velocity electrons. Second, even when the kinetic energy is very great, the velocity of elec-

SPECIAL TOPIC 25 (continued)

the perihelion. The largest effect would be for the planet Mercury, and the predicted value of its precession is about 42″ of arc per century. This very small value is in good agreement with that part of the procession that is not otherwise accounted for.

As we suspect from this example, the predictions of Newton's and Einstein's theories of gravitation are almost the same for weak gravitational fields. The gravitational field at the surface of a uniform spherical mass M of radius r is weak if the dimensionless number $2GM/rc^2$ is very much smaller than unity. (G is the universal gravitation constant, and c is the speed of light in empty space.) Substitution of the appropriate values shows that this dimensionless number is about 1.4×10^{-9} at the surface of the earth, and about 4×10^{-6} at the surface of the sun. The gravitational fields in the solar system are exceedingly weak, and this was clear from the values of the precession of the perihelion of Mercury.

However, not all cosmic phenomena are necessarily associated with weak gravitational fields. It was pointed out in the late eighteenth century by the French astronomer and mathematician, Pierre Laplace, that some of the most massive objects in the universe might be invisible. Einstein's theory predicts many properties of these "black holes." Among other conclusions, the theory shows that a distribution of matter for which the dimensionless number $2GM/rc^2$ is unity or larger permits nothing to escape from it. Absolutely nothing—not even light—emerges from a black hole. If a star collapses to a radius for which the dimensionless number reaches unity, then it will devour any material thing (including light) that comes its way, and will permit no escape. The observation of a black hole must therefore depend on its effect on other objects. It is possible that one member of some binary star systems may be a black hole, and it would be expected to strip matter from its visible companion. Just prior to disappearance, this material should emit characteristic X rays; there is some evidence for this in the binary system Cygnus X-1. If one accepts general relativity, there seems to be no doubt that black holes must be possible, and evidence for their existence will probably continue to accumulate.

trons does not exceed the velocity of light. This is a far-reaching and general result that has been amply confirmed by experiment. Increasing the electron energy or measuring the velocity of other high-energy particles never yields a result in excess of the velocity of light, which is the ultimate velocity for material particles.

Let us now direct check one of the data of Fig. 25.6(b) by using the relativistic energy formulas.

EXAMPLE 25.6

The kinetic energy of the most energetic electron data point in Fig. 25.6 is 4.5 MeV. Calculate the relativistic gamma factor and the velocity of this electron.

SOLUTION Equation (25.9) gives:

$$KE = (\gamma - 1)mc^2$$

We need only to keep the energies on both sides of the equation in the same units, since γ is a pure number. From Example 25.4 we have $mc^2 = 0.511$ MeV for an electron. Substituting this value in Eq. (25.9), we have:

$$\text{KE} = (\gamma - 1)mc^2$$

$$\therefore \gamma = \frac{\text{KE}}{mc^2} + 1$$

$$= \frac{4.5 \text{ meV}}{0.511 \text{ MeV}} + 1$$

$$= 9.81$$

To calculate the electron velocity, we use the formula for the gamma factor, Eq. (25.2):

$$\gamma = \left(1 - \frac{v^2}{c^2}\right)^{-1/2} \qquad 9.81 = \left(1 - \frac{v^2}{c^2}\right)^{-1/2}$$

Squaring both sides we have:

$$96.16 = \frac{1}{1 - v^2/c^2}$$

Taking reciprocals of both sides, we get:

$$\frac{1}{96.16} = 1 - \frac{v^2}{c^2}$$

$$\therefore \frac{v}{c} = 0.989$$

$$\therefore v = 2.97 \times 10^8 \text{ m/s}$$

The electron velocity is close to the velocity of light, as the data point on Fig. 25.6 implies. Additional calculations show that Einstein's formula also fits well with the other data points. ■

We will complete our discussion by showing how the Newtonian and relativistic formulas give essentially the same results for the kinetic energy at lower velocities.

EXAMPLE 25.7

Calculate the kinetic energy of an electron moving at one-tenth the velocity of light. Use both the Newtonian and the relativistic formulas.

SOLUTION Since $c = 3 \times 10^8$ m/s, the electron velocity is 3×10^7 m/s. The Newtonian formula, Eq. (7.2), gives:

$$\begin{aligned}\text{KE} &= \tfrac{1}{2}mv^2 \\ &= \tfrac{1}{2} \times 9.11 \times 10^{-31}\ \text{kg} \times (3 \times 10^7\ \text{m/s})^2 \\ &= 4.10 \times 10^{-16}\ \text{J} \\ &= 4.10 \times 10^{-16}\ \text{J} \div 1.6 \times 10^{-19}\ \text{J/eV} \\ &= 2.562\ \text{keV}\end{aligned}$$

Using the electron rest energy calculated in Example 25.4 and the relativistic formula for the kinetic energy [Eq. (25.9)], we have:

$$\begin{aligned}\text{KE} &= (\gamma - 1)mc^2 \\ &= \left\{\frac{1}{[1 - (0.1)^2]^{1/2}} - 1\right\} \times 0.511\ \text{MeV} \\ &= 2.574\ \text{keV}\end{aligned}$$

■

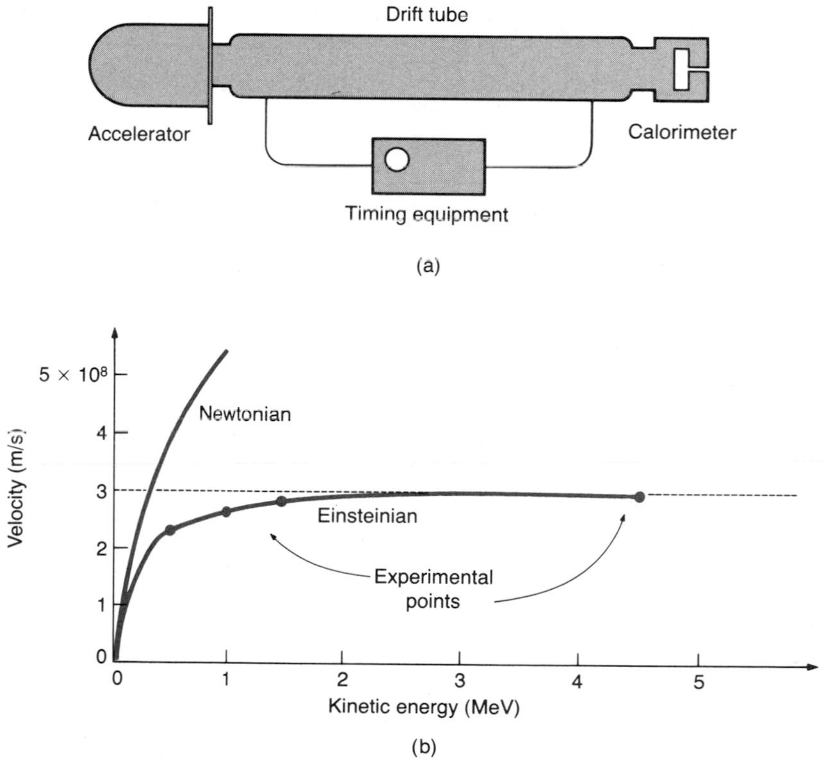

FIGURE 25.6 The measurement of electron kinetic energy as a function of velocity: (a) the experimental apparatus; (b) results of the experiment (after W. Bartozzi, *Am. J. Phys.* 32 (1964): 551). The two curves show Newtonian and Einsteinian kinetic energy formulas. (By permission of American Association of Physics Teachers)

The Newtonian and relativistic formulas for kinetic energy differ by less than 1% when the velocity is one-tenth that of light. This is a convenient changeover point. For lower velocities we use Newtonian dynamics, and for higher velocities relativistic dynamics.

Although we will reserve serious consideration of relativistic energy conservation for our discussion of nuclear physics, examining a simple instance is instructive. We have already seen how Joule and other scientists of the mid-nineteenth century gave a new dimension to the energy conservation law by discovering that heat is a form of energy. A heat energy term must be included where appropriate in the expression of energy conservation. Einstein's work made the notion of energy still broader by including mass energy, as the relativistic formulas imply. This means that we must also include a mass energy term in the conservation law. Perhaps you wonder why this escaped the attention of previous experimental investigators! We have already mentioned that we can set the energy of a mass at rest as the energy zero, provided that it remains constant. An example illustrates the extent of the energy change in an everyday situation.

EXAMPLE 25.8

A kilogram of ice at 0°C absorbs 33.5×10^4 J (the latent heat of fusion) to become water at 0°C. By how much does the mass of the water differ from the mass of the ice?

SOLUTION Let us write an energy equation that includes the energy of the mass at rest:

Rest energy of ice + latent heat energy = rest energy of water

Using Eq. (25.8) for the rest energy and substituting L for the latent heat, we write the conservation of energy as follows:

$$M_{\text{ice}}c^2 + L = M_{\text{water}}c^2$$

$$\therefore M_{\text{water}} - M_{\text{ice}} = \frac{L}{c^2}$$

$$= \frac{33.5 \times 10^4 \text{ J}}{(3 \times 10^8 \text{ m/s})^2}$$

$$= 3.72 \times 10^{-12} \text{ kg}$$

The water is more massive by almost 4 parts in 10^{12}! Needless to say, such a small effect is not detected. We usually say that 1 kg of ice melts into 1 kg of water when the latent heat is added, but strictly speaking this is not correct—the water has a larger supply of energy and is therefore more massive. ■

We conclude our discussion of relativistic dynamics by setting up a relativistically correct expression for calculating the momentum of a massive particle. Although we had to revise our approach to energy substantially, we need only modify our definition of momentum slightly by multiplying the Newtonian formula by the relativistic gamma factor.

Relativistic Definition of Linear Momentum

Linear momentum = gamma factor × mass × velocity

$$\mathbf{p} = \gamma m \mathbf{v} \qquad \textbf{(25.10)}$$

Recall that γ approaches unity for velocities that are low compared with the velocity of light. Thus, our formula approaches the Newtonian formula for the special case of low velocities.

It is convenient to have a direct relation between the energy and momentum of a particle to avoid unnecessary and tedious calculation of the velocity. We begin by writing the difference between the square of the energy and the square of the rest energy.

$$\begin{aligned} E^2 - E_0^2 &= \gamma^2 m^2 c^4 - m^2 c^4 \quad \text{from Eqs. (25.7) and (25.8)} \\ &= \left(\frac{1}{1 - v^2/c^2} - 1\right) m^2 c^4 \quad \text{from Eq. (25.2)} \\ &= \left(\frac{v^2/c^2}{1 - v^2/c^2}\right) m^2 c^4 \\ &= \gamma^2 m^2 v^2 c^2 \quad \text{from Eq. (25.2)} \\ &= p^2 c^2 \quad \text{from Eq. (25.10)} \end{aligned}$$

Let us write this important result in display.

Relativistic Equation Connecting Momentum, Energy, and Rest Energy

$$p^2 c^2 = E^2 - E_0^2 \qquad \textbf{(25.11)}$$

A simple example illustrates the utility of our momentum-energy formula.

EXAMPLE 25.9

Calculate the momentum of a 1-MeV electron.

SOLUTION When we speak of a 1-MeV electron, we refer to the kinetic energy of the particle. In Example 25.4 we calculated the rest energy of an electron to be 0.511 MeV. We can therefore write the total energy as:

$$\begin{aligned} E &= E_0 + \text{KE} \\ &= 0.511\text{ MeV} + 1.0\text{ MeV} \\ &= 1.511\text{ MeV} \end{aligned}$$

To calculate the momentum, we simply use Eq. (25.11):

$$\begin{aligned} p^2c^2 &= E^2 - E_0^2 \\ &= (1.511\text{ MeV})^2 - (0.511\text{ MeV})^2 \\ &= (2.022\text{ MeV})^2 \\ \therefore pc &= 1.422\text{ MeV} \\ \therefore p &= 1.422\text{ MeV}/c \end{aligned}$$

Note the natural way in which MeV/c arises as a unit of momentum. In relativistic calculations it is usually convenient to retain such designations rather than make the cumbersome conversion to SI units. ■

KEY CONCEPTS

The **Michelson–Morley experiment** indicates that there is no medium for electromagnetic wave propagation.

Einstein based **special relativity** on two postulates:

1. The laws of physics have the same essential content in all inertial frames.
2. The measured speed of light in free space is independent of the relative motion of the source and the observer.

The **relativistic gamma factor** plays a key role in the kinematics and dynamics of special relativity; see Eq. (25.2).

For an observer in motion relative to two events occurring in the same place in another frame, the **time interval** between the events is **dilated** by the gamma factor; see Eq. (25.3).

For an observer in motion along the direction of a long rod, its **length** is **contracted** by the gamma factor; see Eq. (25.4).

The **relativistic velocity transformation law** permits comparison of an object's velocity as measured by different observers. See Eq. (25.6).

The **total energy** of a moving point mass is given by **Einstein's** famous **energy formula;** see Eq. (25.7).

The **rest energy** of an object is the energy that it has by virtue of its mass alone. It is equal to the mass multiplied by the square of the velocity of light; see Eq. (25.8).

The **kinetic energy** of a moving object is equal to the total energy less the rest energy; see Eq. (25.9). For a slowly moving object, the relativistic and classical formulas for kinetic energy give the same result; see Example 25.6.

The speed of light is an *ultimate and unattainable limit* for a massive particle.

In applying the energy conservation law, **both** the **rest energy** and **kinetic energy** of each massive particle must be considered.

The **momentum** of a point mass is equal to the classical value multiplied by the gamma factor; see Eq. (25.10).

An important relation connects **momentum, total energy,** and **rest energy;** see Eq. (25.11).

QUESTIONS FOR THOUGHT

1. Do time dilation and length contraction depend on the relative positions of two observers or only on their relative velocities? Formulate an argument to justify your answer.

2. In the imaginary experiment used to calculate time dilation, the light clock was aligned perpendicularly to the direction of the relative velocity of the two observers. Does one obtain the same result if the light clock is aligned parallel to the observers' relative velocity?

3. What are the extreme limits for the velocity and the energy of an atom in an ideal gas? Does doubling the gas temperature make any difference to either of your answers?

4. As the sun radiates electromagnetic waves, its total energy content (and consequently its total mass) decreases. Does the sun also lose momentum in this way?

5. All experiments point to the conclusion that the velocity of light is an unattainable velocity for a massive body. Which of the kinematic and dynamic equations of relativity directly support this result?

6. Is it possible that an astronaut could reach a star system 1000 light years away during a period of 10 years of his own lifetime? Estimate the total energy of an astronaut and a small spaceship if this feat is to be achieved. How does your answer compare with the energy of combustion contained in a full tank of gasoline in an average automobile?

PROBLEMS

(Use data from the inside back cover where necessary.)

A. Single-Substitution Problems

1. Calculate the relativistic gamma factor for the following values of relative velocity: (a) $0.90c$, (b) $0.99c$. [25.2]

2. At what relative velocity does the relativistic gamma factor achieve the following values: (a) 5, (b) 10, (c) 20? [25.2]

3. At what speed must an observer travel to measure the length of an object as one-third its proper length? [25.2]

4. At what speed must an observer travel to measure time intervals at twice the proper time interval? [25.2]

5. Calculate the rest energy of an alpha particle in megaelectron volts. (Consider the alpha particle mass to be 6.68×10^{-27} kg.) [25.3]

6. How large a mass possesses a rest energy of 1×10^9 J? [25.3]

7. Calculate the total energy of an electron that is accelerated in a 3-MV accelerator. [25.3]

8. At what speed must a particle travel so that its total energy is twice the rest energy? [25.3]

9. Calculate the total energy of an electron with a momentum of 3 MeV/c. [25.3]

10. Calculate the momentum of an electron whose total energy is 4.511 MeV. [25.3]

11. Assume that electrical energy costs 5¢ per kilowatt-hour. What would be the value of a 1-kg mass if its rest energy could be completely utilized? [25.3]

B. Standard-Level Problems

12. Calculate the γ value for an object moving with a speed equal to 1000 times the speed of the earth in its orbit around the sun. (Use the astronomical data on the inside back cover.) [25.2]

13. What is the length of a meter stick that is measured by an observer moving past it longitudinally with a velocity two-thirds that of light? [25.2]

14. Two events that occur at the same place in a certain inertial frame are separated by a 10-s time interval. What is the relative velocity of an observer who measures a time interval of 12 s between the two events?

15. A certain observer moving parallel to the length of a meter rule (that is, a rod whose proper length is 1 m) measures its length to be 95 cm. What percentage change in the observer's speed would be required for the length measurement to be 94 cm?

16. A beam of positive pions (average proper lifetime 26 ns) is moving with a speed of $0.94c$ relative to a certain observer. What is the average lifetime of the pions as measured by the observer? What distance of travel does the observer assign to a pion that survives for exactly the average lifetime?

17. The nearest star, Alpha Centauri, is about 4 light-years away from the earth. A spaceship makes the trip with a velocity of $0.95c$. What is the trip time as measured by the spaceship's clocks?

18. A free neutron has a half-life of about 10.8 min. Can a neutron that leaves the sun with velocity $0.98c$ reach the planet Jupiter (7.8×10^{11} m away) before its half-life elapses?

19. A moon-bound spaceship traveling at 8×10^4 m/s passes an incoming ship moving with the same speed in the opposite direction. What would their captains measure for the relative approach velocity if equipment of the utmost precision were available?

20. An observer in a laboratory measures the velocity of two protons as $0.8c$ and $0.9c$ in the same straight line. What is their relative velocity as measured by an observer moving with one of the protons?

21. Calculate the accelerating voltage required for a proton to have a total energy three times its rest energy. (The proton rest energy is about 938.3 MeV.) [25.3]

22. Calculate the velocity of a proton whose kinetic energy is twice its rest energy. [25.3]

23. An electron is accelerated from rest through a potential difference of 255 kV. What is the velocity of the electron?

24. Calculate the rest energy of a singly ionized particle whose total energy is four times the rest energy after acceleration through a total potential difference of 1.48 GV.

25. An accelerator is constructed to accelerate an electron from rest to a velocity of $0.98c$. What potential difference is required to achieve this?

26. A proton has a velocity of $0.6c$. What is the percentage error incurred in calculating its kinetic energy from the Newtonian formula?

27. A proton is accelerated from a velocity of $0.90c$ to $0.94c$, and then subsequently to $0.98c$. By what factor does the energy required to provide the second acceleration exceed that required to provide the first acceleration?

28. A neutron with a kinetic energy of 240 MeV has its kinetic energy doubled. Calculate its percentage increase in velocity. (The neutron rest energy is 939.6 MeV.)

29. **a.** Calculate the kinetic energy of an electron whose momentum is 2 MeV/c.
b. Calculate the electron velocity as a percentage of the velocity of light.

30. **a.** Calculate the momentum of an electron whose kinetic energy is 3 MeV.
b. What answer does Newtonian dynamics give to this question?
(Express both results in MeV/c.)

C. Advanced-Level Problems

31. Approximately what distance would a commercial jet liner fly before its clocks were in 1-s error compared with earth clocks? Would you expect the jet liner clocks to be fast or slow?

32. The energy received from the sun amounts to about 1350 W/m^2 at the distance of the earth's orbit. Calculate the sun's approximate rate of loss of mass.

33. A laboratory observer sees two neutrons, each with kinetic energy of 350 MeV, approaching each other along the same straight line. Calculate the kinetic energy of one neutron as seen by an observer on the other.

34. The diameter of our galaxy is about 10^5 light-years, which means that light takes about 10^5 years to cross the galaxy. How long does it take a proton of 3×10^{16} eV energy to cross the galaxy as measured by

a. an observer at rest in the galaxy?
b. an observer moving with the proton?
c. What is the diameter of the galaxy as measured by an observer moving with the proton?

26

QUANTUM PHYSICS

WAVES AS PARTICLES

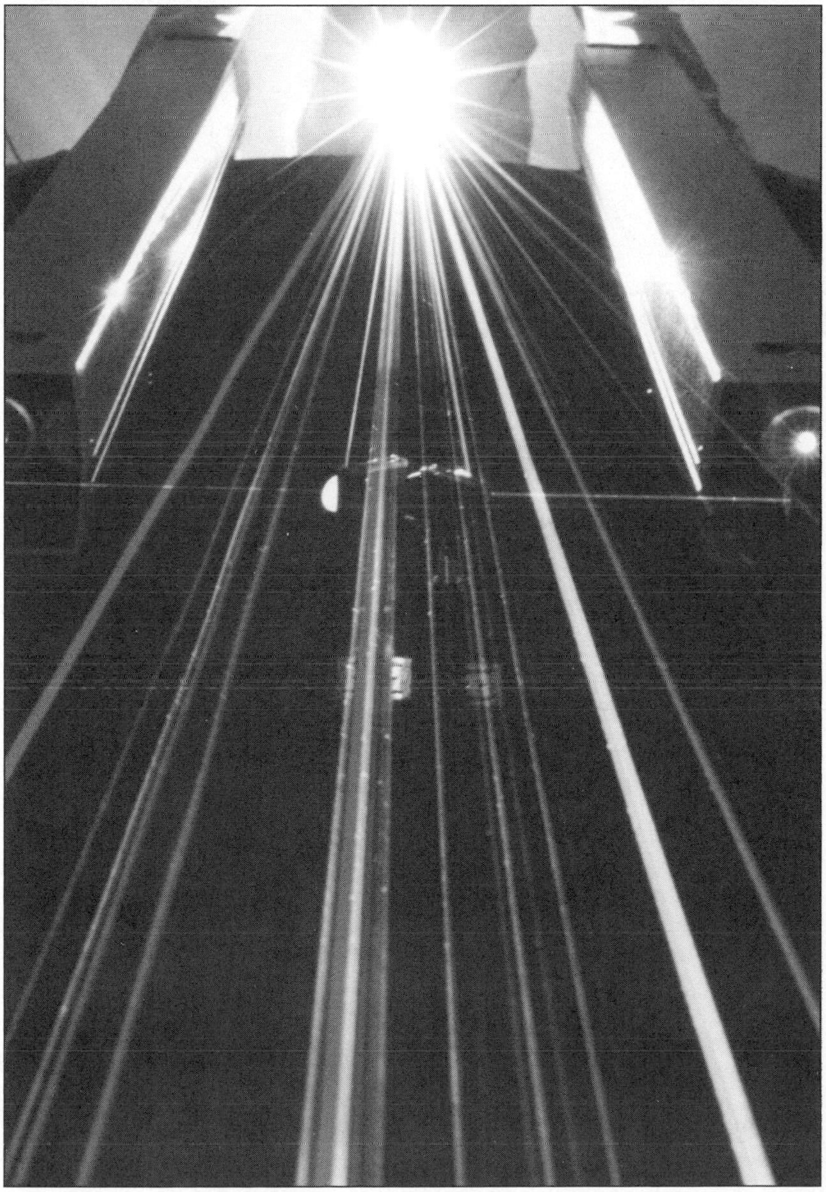

The theory of relativity was only a part of the revolution in physics that occurred in the early years of the twentieth century. The other part of the story is usually called quantum physics. As we have already mentioned, the new developments were all interrelated. We will see this in the role that relativistic concepts play in the following discussion.

Quantum physics had its beginning in the study of thermal radiation—that is, the electromagnetic radiation emitted from a hot object. Stefan's law states that the total radiated energy is proportional to the fourth power of the absolute temperature of a blackbody radiator. This is a most intriguing relationship, and it is not surprising that physicists began to research the details of blackbody radiation in an attempt to find a deeper explanation of this remarkable law. The work involved many false starts and unproductive ideas, but it finally gave birth to quantum physics.

We will briefly discuss blackbody radiation and then describe the photoelectric effect and the scattering of X rays by electrons. These phenomena reinforce and confirm the quantum nature of electromagnetic radiation.

26.1 Blackbody Radiation and Planck's Hypothesis

We have already studied the process of radiation as a mechanism for the transfer of heat energy. In this connection, let us recall Eq. (14.13) (Stefan's law):

$$Q = e\sigma T^4 At$$

e = emissivity
σ = Stefan's constant
T = absolute temperature
A = area of the radiating body
t = time interval

A perfectly reflecting surface emits no thermal radiation and has an emissivity of zero. The highest possible emissivity value is unity for a blackbody radiator, which absorbs all the thermal radiation that falls on it without any reflection or transmission.

Examples of surfaces that approximate perfect reflectors are easy to find, and include any clean, highly polished metal surface. However, the concept of a blackbody is a little more elusive. The name arose in English-speaking countries because a mat black surface absorbs all the radiation that falls on it. Being a perfect absorber, it is also a perfect radiator, having unit emissivity.

However, another excellent practical instance of a blackbody is a large cavity with a small hole in its wall. Any radiation that enters the hole is absorbed in the interior of the cavity, and the

SPECIAL TOPIC 26 PLANCK

Max Planck was born in Kiel in 1858, and received his doctorate from the University of Munich in 1879. In 1898, he was named professor of physics at the University of Berlin, and held that post until retirement in 1926.

His fame as a physicist rests on his introduction of the quantum energy for electromagnetic radiation. The idea was so revolutionary that even Planck himself was inclined to regard it as a mathematical trick to explain the facts of thermal radiation. However, first Einstein and then Bohr applied his work to explain the photoelectric effect and the spectrum of hydrogen. With the great importance of his work no longer in doubt, Planck received the Nobel Prize for physics in 1918.

Both of his sons were killed in World War II, but he lived through it, and died in Gottingen in 1947.

(Courtesy Burndy Library, AIP Niels Bohr Laboratory)

radiation emitted from the hole is blackbody radiation. Most European physicists therefore refer to blackbody radiation as cavity radiation—a term that is certainly more descriptive of the concept. Note that the material of the cavity walls has no effect on the nature of the radiation. If the interior of the cavity is all at the same temperature, the radiation emitted from the small hole depends only on this temperature.

In the light of these remarks, it comes as no surprise that the radiation from the door of a blast furnace is closely related to blackbody radiation at the temperature of the furnace interior. Somewhat more surprising is how many well-known objects emit radiation that closely approximates blackbody. For example, the radiation from the sun is close to blackbody radiation at the surface temperature of the sun. A red-hot block of metal and the embers of a fire both emit radiation that is approximately blackbody at the temperature of the emitter. In all of these examples

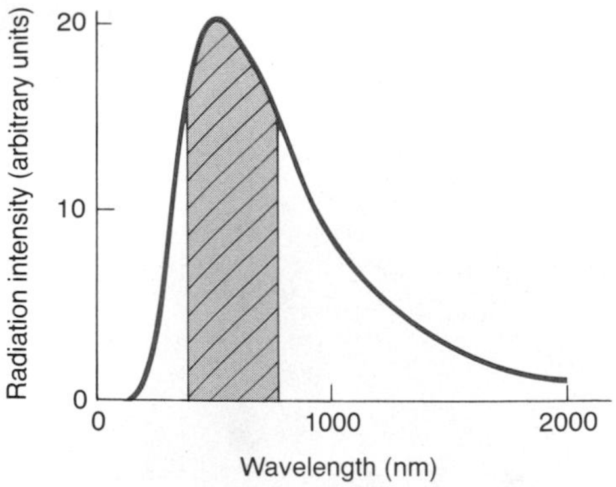

FIGURE 26.1 The blackbody radiation spectrum of the sun. The shaded area is visible light; the shorter wavelengths are ultraviolet, and the longer ones infrared.

the blackbody emitter glows brightly, but since light is an electromagnetic wave, we naturally conclude that blackbody thermal radiation is electromagnetic. This is true whether the radiation is in the visible part of the spectrum or not.

But what is the spectrum of blackbody radiation? Is it a series of lines such as the atomic spectra that we observe in low-pressure discharges, or does it include a continuous range of frequencies? Let us answer these questions by referring to the spectrum of radiation from the sun, which, as we have already mentioned, is approximately blackbody. Figure 26.1 shows the spectrum of the sun's radiation, which is the blackbody radiation curve for an object whose surface temperature is 5700 K. Varying fractions of electromagnetic radiation of every wavelength are contained in the radiant energy. The curve displays the intensity of radiation in arbitrary units; consequently, the amount of radiant energy between any two selected wavelengths is proportional to the area under the curve between those two wavelengths. For example, the shaded area under the curve is proportional to the energy radiated by the sun in the visible light region (380–760 nm). By inspection we see that the largest fraction of the sun's radiant energy is in the infrared and the smallest fraction in the ultraviolet, and careful measurements show that the energy fractions are about 25% ultraviolet, 30% visible, and 45% infrared.

However, the sun is not the only blackbody radiator. The glowing embers in a fire also emit radiation that is approximately blackbody radiation, but the sun and the glowing embers differ noticeably in two features. The embers have a temperature that is a good deal lower than 5700 K, and they glow with a red color rather than the sun's bright gold color. These differences reflect a basic law of thermal radiation. As the temperature falls, the radiation becomes strongly red and infrared; as it rises, the radiation occurs predominantly in the blue, violet, and ultraviolet. Astronomers have known for centuries that stars differ in color—some are red, some white, and others blue—and we now know that the color reflects the temperature of the star. Our sun is a medium-temperature star; the red stars are much cooler, and the blue ones are very much hotter.

Figure 26.2 expresses these facts more quantitatively, showing the spectra for blackbody radiators of equal area at temperatures of 3000 K, 3500 K, and 4000 K. The most obvious feature of the graph is the much smaller area under the lowest-temperature curve. Since the area under the curve is proportional to the total radiant energy output, this reflects Stefan's law. In fact, the area under each curve is proportional to the fourth power of the associated temperature. The next feature of interest is that the peak of the curve shifts toward longer wavelengths as the temperature falls. At 4000 K almost nothing is in the ultraviolet, an appreciable fraction is in the visible region, and the largest fraction is in the infrared; the peak of the curve is at a wavelength of about 725 nm, which is at the extreme red end of the visible spectrum.

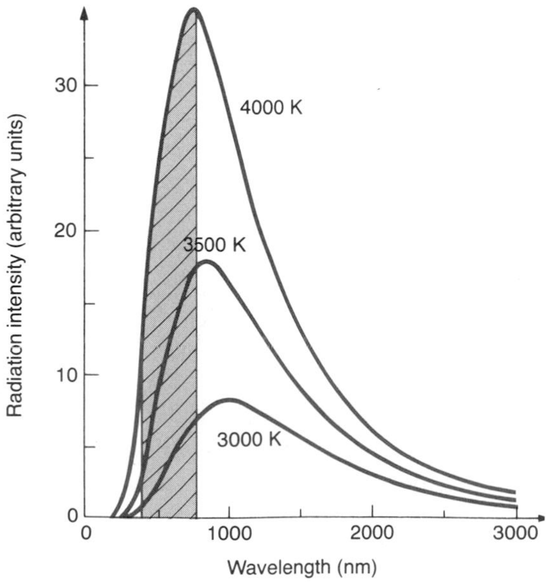

FIGURE 26.2 The blackbody radiation spectra for three different temperatures. The region of visible light wavelengths is shaded.

When the temperature is 3000 K an even larger fraction of the radiation is in the infrared, and the curve peaks at about 960 nm in the infrared.

A simple relation for determining the wavelength at the peak of the radiation curve was discovered by the German physicist Wilhelm Wien (1864–1928). The law expresses how the peak shifts with changing temperature and is therefore called **Wien's displacement law.**

■ Wien's Displacement Law for Blackbody Radiation

$$\text{Wavelength at the peak} \times \text{absolute temperature} = \text{Wien's constant}$$

$$\lambda_{peak}\, T = 2.898 \times 10^{-3}\ \text{m} \cdot \text{K} \qquad \textbf{(26.1)}$$

The value of Wien's constant is determined by laboratory measurements on blackbodies at various temperatures. Obviously the surface temperature of the sun, which is about 5700 K, was not measured by an astronaut with a thermometer! Let us show how Wien's displacement law is used to calculate the sun's surface temperature.

EXAMPLE 26.1

Measurement reveals that the sun's radiation spectrum is blackbody in nature with a peak at a wavelength of 510 nm. Calculate the sun's surface temperature.

SOLUTION We simply apply Wien's displacement law, Eq. (26.1), to do this calculation:

$$\lambda_{peak}\, T = 2.898 \times 10^{-3}\ \text{m} \cdot \text{K}$$

$$510 \times 10^{-9}\ \text{m} \times T = 2.898 \times 10^{-3}\ \text{m} \cdot \text{K}$$

$$\therefore T = 5680\ \text{K}$$

■

As interesting as Wien's law is, it does not lead to a deeper understanding of thermal radiation. The first attempt in this direction was a notable failure. Lord Rayleigh combined Maxwell's theory of electromagnetism with statistical considerations to produce a famous formula that was incorrect! Why should an incorrect formula cause a stir? Maxwell's theory had successfully explained all the known phenomena of ray and wave optics, so its failure to explain thermal radiation was hard to believe. The situation was even more confusing because Rayleigh's formula was correct for very long wavelengths in the far infrared but hopelessly wrong in the visible and ultraviolet regions.

Against this background a German physicist, Max Planck (1858–1947), came forward with a most startling suggestion. Observing that Rayleigh's calculation permitted radiation of a given frequency and wavelength to have any arbitrary energy, Planck made a simple modification in the calculations that yielded exactly the right formula for the radiation curve. Planck pointed out that the material of the cavity walls emitted the blackbody radiation from a cavity. He referred to the entities that emit the radiation as resonators and found that he could correct Rayleigh's calculation by a modification affecting the energy transfer from the resonators to the blackbody radiation. The modification called for the energy transfer to occur by way of discrete jumps; slow continuous increase in the energy content of the radiation was ruled out. The discrete energy jumps were called quantum jumps, and postulating them was a very bold step—there was nothing anywhere in classical physics to give the slightest clue that energy could not change continuously. Probably from a reluctance based, at least in part, on the lack of anything comparable in classical physics, Planck associated the energy quanta only with the resonators in the cavity walls. The identification of energy quanta with the radiation itself is due to Einstein in his explanation of the photoelectric effect, which we consider in the following section. The size of the energy quantum (postulated by Planck for the resonators, and extended by Einstein to radiation) is proportional to the frequency; we express this as the Planck–Einstein law.

■ Planck–Einstein Energy Quantization Law

Quantum energy = Planck's constant × frequency

$$E = h\nu \qquad \textbf{(26.2a)}$$

The value of Planck's constant is:

$$h = 6.626 \times 10^{-34}\ \text{J} \cdot \text{s}$$
$$= 4.136 \times 10^{-15}\ \text{eV} \cdot \text{s}$$

We give the value of Planck's constant in SI units and also in electron-volt units since the latter are often very useful for calculations in atomic physics.

Since radiation is specified by wavelength at least as often as it is by frequency, it is convenient to have the Planck–Einstein relation expressed in a form containing the wavelength. We achieve this with the help of Eq. (20.12) to give:

■ Planck–Einstein Energy Quantization Law (Alternate Form)

$$E = \frac{hc}{\lambda} \qquad \textbf{(26.2b)}$$

The value of the product hc is:

$$hc = 1.988 \times 10^{-25}\ \text{J} \cdot \text{m}$$
$$= 1.241 \times 10^{-6}\ \text{eV} \cdot \text{m}$$

Planck determined the value of the constant in the quantum energy relationship by fitting the theoretical radiation curves to the experimental ones.

About 25 years after Planck's work, the quantum energy of radiation was named the photon. Let us use an example to calculate some representative photon energies.

EXAMPLE 26.2

Calculate the photon energies for the following types of electromagnetic radiation:

a. a 600-kHz radio wave

b. a 500-nm green light wave

c. a 0.1-nm X ray

Give the quantum energies in electron volts. For comparison, calculate the mean thermal kinetic energy of a molecule of an ideal gas at room temperature.

SOLUTION

a. Since the radio wave is specified by frequency, Eq. (26.2a) is appropriate:

$$E = h\nu$$
$$= 4.136 \times 10^{-15}\ \text{eV} \cdot \text{s} \times 600 \times 10^{3}\ \text{Hz}$$
$$= 2.48 \times 10^{-9}\ \text{eV}$$

b. The light wave is specified by wavelength, so that Eq. (26.2b) is more suitable:

$$E = \frac{hc}{\lambda}$$
$$= \frac{1.241 \times 10^{-6}\ \text{eV} \cdot \text{m}}{550 \times 10^{-9}\ \text{m}}$$
$$= 2.26\ \text{eV}$$

c. Equation (26.2b) is also convenient for the X ray:

$$\begin{aligned} E &= \frac{hc}{\lambda} \\ &= \frac{1.241 \times 10^{-6}\ \text{eV} \cdot \text{m}}{0.1 \times 10^{-9}\ \text{m}} \\ &= 12.4\ \text{keV} \end{aligned}$$

Finally, to give some basis for comparison, we will calculate the mean thermal kinetic energy of an ideal gas molecule at room temperature. Take $T \approx 300$ K and use Eq. (13.7):

$$\begin{aligned} \left\langle \tfrac{1}{2}mv^2 \right\rangle &= \tfrac{3}{2}kT \\ &= \tfrac{3}{2} \times 1.38 \times 10^{-23}\ \text{J/K} \times 300\ \text{K} \\ &= 6.21 \times 10^{-21}\ \text{J} \\ &= 6.21 \times 10^{-21}\ \text{J} \div 1.60 \times 10^{-19}\ \text{J/eV} \\ &= 3.88 \times 10^{-2}\ \text{eV} \end{aligned}$$

■

Since all the energies calculated are very small fractions of a joule, the electron volt is easier to work with in quantum calculations. The gas molecule has an energy of several hundredths of an electron volt, and the photon energy for the radio wave is more than a million times smaller. The photon of visible light is more energetic than the gas molecule by about 50 times, and the X ray by about a quarter of a million times. The significance of energy quantization depends on the size of the energy quantum. For almost all purposes quantization of the radio wave is insignificant, and quantization of the X ray is highly significant. Visible light is in a middle region where quantization may or may not be important.

Ironically, Planck and other leading physicists resisted the literal interpretation of the quantum law. Maxwell's electromagnetic theory had firmly established that light was an electromagnetic wave, and the corpuscular theory of light was stone-dead. Thus, most physicists could not believe that the energy of a light wave consisted of corpuscular bundles. As we will see in the following section, Albert Einstein took the first step in firmly establishing the corpuscular viewpoint; subsequent research has more than amply confirmed the reality of the radiation quantum. As with many scientific advances, each success creates further problems. Light reveals the character of an electromagnetic wave, and it just as certainly exhibits energy quanta. Reconciling these apparently conflicting viewpoints has been a major problem for modern theoretical physics.

SPECIAL TOPIC 27 PHOTOSYNTHESIS

Aristotle knew that sunlight was needed for the greening of plants, but it was not until the advent of quantum mechanics that scientists could begin the detailed explanation of why this was so.

Photosynthesis is the process by which green plants use light energy from the sun, carbon dioxide from the air, and water from the environment to produce sugar molecules and oxygen. This process is the basis of every food chain on our planet, and the sheer magnitude of the conversion is staggering. It is estimated that some 200 billion tons of carbon are taken from the air each year (in the form of carbon dioxide) and combined with hydrogen from water to form organic compounds.

Many organic chemical reactions take place under the influence of thermal agitation at temperatures not too far removed from 300 K, and this is the case with many biological reactions. However, photosynthesis is essentially quantum in the same way that the photoelectric effect is quantum. In the photoelectric case, the energy of the light quantum goes into overcoming the work function and supplying kinetic energy to the ejected electron—an amount of energy that is simply not available to an individual electron from room temperature thermal agitation. In the case of photosynthesis, individual electrons receive energy from the light quanta (to an amount not available from thermal agitation) and use this energy to proceed with the production of organic chemicals.

The substance in green plants that absorbs the light quanta is called chlorophyll, and absorption is in the red and blue-violet parts of the spectrum; the green light is not absorbed, and hence the plant appears green. The initiation of the complicated organic reactions requires chlorophyll, but it alone is not sufficient for them. Enzymes also are required, and their part in the process is controlled by thermal agitations. This results in the highly different rates of photosynthesis that are observed in the same plants exposed to the same light but kept at different temperatures; the plant at the higher temperature has a greater rate of photosynthesis. If the photosynthetic reaction were driven only by absorbed light, this would not be so.

In addition to light and temperature, other factors such as the carbon dioxide content of the air, water supply, and mineral trace elements all play a modifying or sometimes an essential role in photosynthesis.

26.2 The Photoelectric Effect

The quantum nature of light had its origin in the theory of thermal radiation and was strongly reinforced by the discovery of the **photoelectric effect.** The history of the photoelectric effect begins with the experiments by which Hertz confirmed Maxwell's prediction of the existence of electromagnetic waves. To generate the electromagnetic waves he used a spark gap between a pair of metal spheres. A large voltage between the spheres causes a spark to jump the gap, which gives off electromagnetic radiation. The situation is similar to the operation of a faulty household switch that causes a harsh sound on a radio. A spark in the switch emits radiation that is detected by the radio. Although he

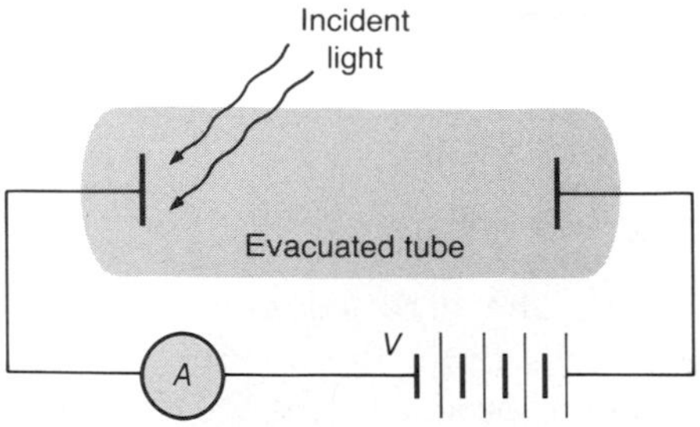

FIGURE 26.3 Apparatus to investigate the photoelectric effect.

had no radio receiver, Hertz used a primitive form of detector to perform an experiment that is essentially the same as the detection of a spark by a radio.

Hertz also noticed that a spark gap sparked more readily under an applied voltage when it was exposed to the radiation of another spark. This discovery triggered numerous investigations. It was found that a negatively charged metal sphere but not a positively charged sphere could be discharged by illuminating it with ultraviolet light. Apparently, the light possesses the ability to remove negative charge from a metal, but not positive charge. We can easily interpret these experiments from the modern viewpoint. The light simply ejects some electrons from the solid object and thereby removes the excess negative charge. Since the positively charged metal sphere already has a large electron deficit, it is not discharged by the light. This interpretation is confirmed by studying the motion of the emitted particles in electric and magnetic fields.

An apparatus to study the effect of changes in the nature of light and electrode surfaces is shown in Fig. 26.3. A glass tube contains two electrodes of the same material, one of which is irradiated by light. The electrodes are connected to a battery and a sensitive current detector measures the current flow between

During the past decade many devices have been made commercially available that depend on an explicit knowledge of quantum physics for their design; this laser disc player is a case in point. The first disc players operated by the principles of classical mechanics, and later versions required electromagnetic theory for their understanding. (Courtesy of Pioneer Video, Inc.)

them. The current flow is a direct measure of the rate of emission of electrons from the irradiated electrode. The electrons that are ejected by the light have a certain amount of kinetic energy, and the maximum kinetic energy of ejection can be determined by reversing the battery and measuring the potential difference across the tube that is just sufficient to stop the flow of electrons.

Changing the frequency and intensity of the light, the voltage across the tube, and the nature of the electrode surfaces leads to the following conclusions:

1. For a given electrode material, no photoemission exists at all below a certain frequency of the incident light. The frequency at which emission begins is called the **threshold frequency,** and it is different for different electrode materials.
2. The rate of electron emission is directly proportional to the intensity of the incident light.
3. Increasing the intensity of the incident light does not increase the kinetic energy of the photoelectrons. Increasing the frequency of the light does increase the photoelectron kinetic energy even for very low intensity levels.
4. There is no measurable time delay between irradiating the electrode and the emission of photoelectrons, even when the light is of very low intensity.
5. The photoelectric current is profoundly affected by the nature of the electrodes and chemical contamination of their surfaces.

The second and fifth conclusions are not difficult to reconcile with Maxwell's electromagnetic theory of light, but the other three conflict with any reasonable interpretation of the classical theory. Since the rate of energy supply to the electrode surface is proportional to the intensity of the light, we would expect to find a time delay in photoelectron emission for a very low intensity light beam. The delay would allow the light to deliver adequate energy to the electrode surface to cause the emission. Moreover, classical theory could not explain the existence of a threshold frequency, since light energy was light energy regardless of the frequency of the beam. Finally, an intense beam supplies energy at a great rate and would surely increase the maximum kinetic energy of the photoelectrons.

Einstein resolved the dilemma by determining that Planck's quantization hypothesis applied not only to the *emission* of radiation by a material object but also to its *transmission* and its *absorption* by another material object—in this case, the photoelectrode. The electrode can absorb energy from the incident light only in integral multiples of the energy quantum. All the facts of photoelectric emission are readily explained from the following assumptions:

1. The photoemission of an electron from an electrode occurs when an electron absorbs a photon (the energy quantum) of the incident light.
2. The photon energy is calculated from Planck's quantum relationship, Eq. (26.2a).
3. A minimum energy is required to release an electron from the surface of an electrode. The minimum energy is characteristic of the electrode material and the nature of its surface; it is termed the **work function.**

We can now write an energy equation for photoelectric emission. The photon energy is entirely expended on the electron that absorbs it. Part of the energy releases the electron from the electrode surface, and the remainder supplies kinetic energy to the ejected electron. Because the work function is the minimum energy required to eject the electron, the photon energy is equal to the sum of the work function and the maximum electron kinetic energy.

TABLE 26.1 Representative values of the average work function for clean metal surfaces determined from photoelectric data

Metal	*Work function (eV)*
Barium	2.52
Cesium	1.91
Chromium	4.37
Gold	4.82
Platinum	6.30
Potassium	2.26
Silver	4.73
Sodium	2.29
Tungsten	4.58

■ Photoelectric Energy Conservation Equation

$$\text{Photon energy} = \begin{array}{c}\text{work}\\ \text{function}\end{array} + \begin{array}{c}\text{maximum photoelectron}\\ \text{kinetic energy}\end{array}$$

$$h\nu = \phi + \tfrac{1}{2}mv_{max}^2 \qquad \textbf{(26.3)}$$

The usual symbol for the work function is the Greek letter phi (ϕ), and some typical average values for clean metal surfaces are listed in Table 26.1. For the case of a metal electrode we can draw an energy diagram that gives a pictorial representation of Eq. (26.3), as shown in Fig. 26.4. Distance in the vertical direction represents energy. The left-hand side of the diagram shows electrons in the metal electrode occupying all energies up to a certain value, called the Fermi energy level in honor of the Italian physicist Enrico Fermi (1901–1954). The right-hand side of the diagram refers to the vacuum outside the metal electrode. The lowest energy for an electron in the vacuum exceeds the Fermi energy level by the work function ϕ. A photon of energy $h\nu$ that removes an electron from the Fermi energy level provides an excess energy of ($h\nu - \phi$), which appears as the kinetic energy of the ejected electron, as illustrated in Fig. 26.4. We have labeled this kinetic energy $½mv_{max}^2$, since it corresponds to the maximum kinetic energy that the photon can give to the emitted electron. If the photon emits an electron that is below the energy of the Fermi level, then the electron requires more energy than the work function to lift it to the vacuum energy level. This implies a smaller surplus kinetic energy for the emitted electron.

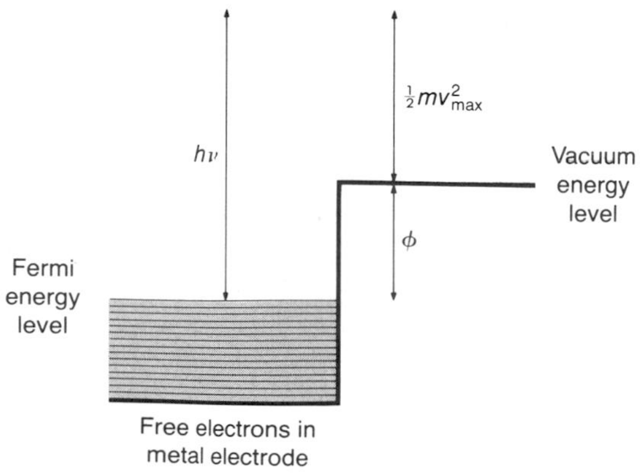

FIGURE 26.4 Diagram to illustrate the energy equation of photoelectric emission.

EXAMPLE 26.3

Ultraviolet light of wavelength 150 nm falls on a chromium electrode. Calculate the maximum kinetic energy and the corresponding velocity of the photoelectrons.

SOLUTION In using the energy equation for the photoelectric effect, it is convenient to express all energies in electron volts. To calculate the photon energy of the ultraviolet light, we use Eq. (26.2b):

$$E = \frac{hc}{\lambda} = \frac{1.241 \times 10^{-6}\ \text{eV} \cdot \text{m}}{150 \times 10^{-9}\ \text{m}} = 8.27\ \text{eV}$$

Table 26.1 shows that the work function of chromium is 4.37 eV. The photoelectric energy equation, Eq. (26.3), gives:

$$h\nu = \phi + \tfrac{1}{2}mv_{\text{max}}^2$$

$$8.27\ \text{eV} = 4.37\ \text{eV} + \tfrac{1}{2}mv_{\text{max}}^2$$

$$\therefore \tfrac{1}{2}mv_{\text{max}}^2 = 3.90\ \text{eV}$$

Although the electron volt is the natural unit for calculations in atomic physics, we must convert to SI units to calculate v_{max}:

$$\tfrac{1}{2}mv_{\text{max}}^2 = 3.90\ \text{eV} \times 1.6 \times 10^{-19}\ \text{J/eV} = 6.24 \times 10^{-19}\ \text{J}$$

Substituting for the electron mass, we have:

$$\tfrac{1}{2} \times 9.11 \times 10^{-31}\ \text{kg} \times v_{\text{max}}^2 = 6.24 \times 10^{-19}\ \text{J}$$

$$\therefore v_{\text{max}} = 1.17 \times 10^{6}\ \text{m/s}$$

Since the kinetic energy is substantially less than the rest energy, the Newtonian expression for the kinetic energy can be correctly used here. ■

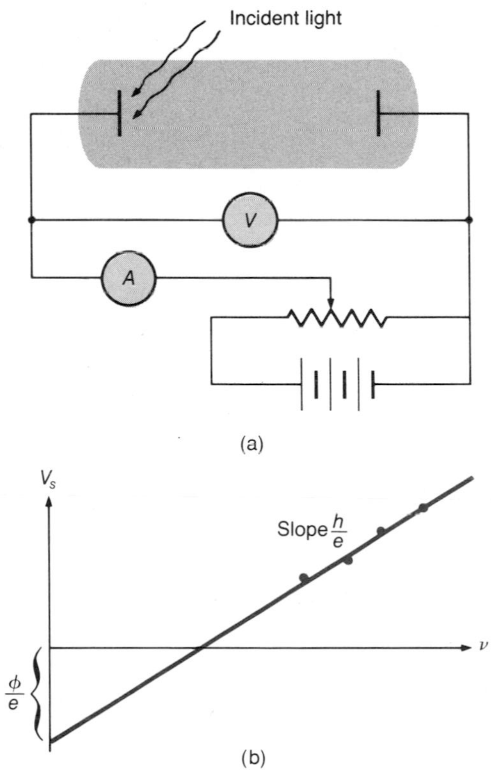

FIGURE 26.5 Measurement of Planck's constant and the work function resulting from using the photoelectric effect: (a) the experimental setup; (b) values of the stopping potential for incident light of various frequencies.

The energy equation for the photoelectric effect is the basis for an important experiment that confirms Einstein's hypothesis regarding the quantization of the absorbed light. It also serves to measure the work function of a metal electrode. Consider the apparatus of Fig. 26.5(a). Light of frequency ν illuminates the photoelectrode. A variable voltage source is used to make the other electrode negative with respect to the photoelectrode, and the photocurrent is measured by an ammeter. For low values of the variable voltage, the ejected photoelectrons have sufficient kinetic energy to overcome the retarding potential and cause a current in the phototube. For a sufficiently large value of the retarding potential, the photocurrent falls to zero. The voltage that is just sufficient to cause zero photocurrent is called the stopping potential. The stopping potential energy eV_S is equal to the maximum kinetic energy $\frac{1}{2}mv_{max}^2$. Making this substitution in Eq. (26.3) and rearranging, we have:

$$V_S = \frac{h}{e}\nu - \frac{\phi}{e}$$

Figure 26.5(b) plots the values of the stopping potential for various frequencies of the incident light. The result is a straight line of slope h/e and intercept $-\phi/e$. Knowing the electron charge, we can then calculate both Planck's constant and the work function of the metal surface.

The value of Planck's constant derived from this experiment agrees with the value required to fit the curve of a blackbody radiator. Einstein's interpretation of the photoelectric effect not only avoids the difficulties associated with the classical viewpoint but also confirms the quantum nature of light. Interestingly, Einstein received the Nobel Prize for his work on the photoelectric effect and not for his more famous contributions to the theory of relativity.

26.3 The Compton Effect

Although further confirmation of the quantum nature of light might not seem necessary, the American physicist Arthur H. Compton (1892–1962), working at Washington University, provided it in a startling way in 1922. Having observed a small change in the wavelength of X rays that were scattered by a piece of carbon, Compton explained the phenomenon by using a simple dynamical model.

Recall that in a previous chapter we analyzed the dynamics of the collision of two freight cars by using the conservation laws for energy and momentum. Compton used exactly the same technique to analyze the collision between a photon and an electron. There could hardly be a better example of the wide-ranging applicability of the basic laws.

In order to apply the energy and momentum conservation laws, Compton required relativistically correct expressions for the energy and momentum of both the photon and the electron. For the electron, Eqs. (25.7) and (25.10) give the required expressions. For the photon, the Planck–Einstein relation, Eq. (26.2), gives the correct expression for the energy, but we still have no relation for the photon momentum. Note that Eq. (25.10):

$$\mathbf{p} = \gamma m\mathbf{v}$$

does not help us in this regard. When the velocity is equal to the velocity of light, the relativistic gamma factor approaches infinity. Presumably we should set the photon mass at zero, but we still cannot calculate the momentum. However, the relativistic energy–momentum relation, Eq. (25.11):

$$p^2c^2 = E^2 - E_0^2$$

removes these difficulties. If we set $E_0 = 0$ for the photon (this is the same as setting the photon mass at zero), we are left with the conclusion that the **photon momentum** is equal to its energy divided by the velocity of light.

Formula for the Photon Momentum

$$\text{Photon momentum} = \frac{\text{photon energy}}{\text{photon velocity}}$$

$$p = \frac{E}{c} = \frac{h\nu}{c} = \frac{h}{\lambda} \qquad \textbf{(26.4)}$$

Every experimental result confirms this formula for photon momentum. Although it had been known for some time that electromagnetic radiation carried momentum equal to the energy of the radiation divided by the velocity of light, the quantum theory and relativity theory combine nicely to give the momentum of a photon as a particle participating in dynamical collisions.

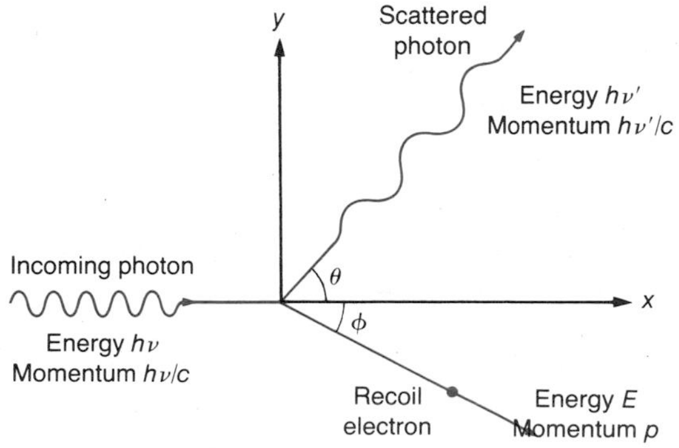

FIGURE 26.6 Diagram illustrating the conservation of momentum and energy in a photon–electron collision.

We can now apply the conservation laws to the photon–electron collision. Figure 26.6 shows a photon of energy $h\nu$ and momentum $h\nu/c$ traveling along the positive direction of the x-axis and colliding with a stationary electron of mass m whose energy is mc^2. After the collision, the photon has a reduced energy $h\nu'$ and a momentum $h\nu'/c$ in a direction that makes an angle θ with the x-axis. The electron has a momentum p that makes an angle ϕ on the opposite side of the x-axis. Since momentum is a vector quantity, its components in both the x and y directions are conserved. Conservation of the x component of momentum gives:

$$\frac{h\nu}{c} = \frac{h\nu'}{c}\cos\theta + p\cos\phi$$

Conservation of the y component of momentum gives:

$$0 = \frac{h\nu'}{c} \sin \theta - p \sin \phi$$

If the electron has an energy E after the collision, energy conservation (a scalar conservation law) gives:

$$h\nu + mc^2 = h\nu' + E$$

These three equations relate the physical quantities in a rather complicated way. However, let us suppose that we know the energy $h\nu$ of the incoming photon and that we have a detector set up to locate the scattered photon at some definite angle θ. In addition, we are not concerned with the electron but are primarily interested in the energy change of the scattered photon; that is, we wish to solve the set of equations for ν'. Since this calculation results in messy algebra, we will simply describe the steps of the solution. Isolating the terms $p \cos \phi$ and $p \sin \phi$ in the first and second equations, squaring, and adding, we obtain:

$$p^2c^2 = (h\nu)^2 + (h\nu')^2 - 2h\nu h\nu' \cos \theta$$

Isolating E in the third equation and squaring, we have:

$$E^2 = (h\nu - h\nu')^2 + 2(h\nu - h\nu')mc^2 + (mc^2)^2$$

Subtracting these two equations and making use of Eq. (25.9), we have after some rearrangement:

$$\frac{1}{\nu'} - \frac{1}{\nu} = \frac{h}{mc^2}(1 - \cos \theta)$$

Multiplying through by c and using Eq. (20.12), we can express the result as a wavelength shift:

$$\lambda' - \lambda = \frac{h}{mc}(1 - \cos \theta)$$

Following our usual procedure, we will set the formula for the **Compton effect** in words as well as in symbols.

■ Compton Formula for the Wavelength Shift of a Photon Scattered by an Electron

$$\text{Wavelength shift} = \text{Compton wavelength} \times \left(1 - \text{cosine of photon-scattering angle}\right)$$

$$\lambda' - \lambda = \frac{h}{mc}(1 - \cos \theta) \qquad \textbf{(26.5)}$$

The quantity h/mc contains only fundamental constants and has the dimension [L]. In Eq. (26.5) we refer to it as the **Compton wavelength,** since its size gives a rough measure of the wavelength change to expect in the photon. We can easily calculate the value of the Compton wavelength:

$$\text{Compton wavelength} = \frac{h}{mc}$$

$$= \frac{6.63 \times 10^{-34}\ \text{J} \cdot \text{s}}{9.11 \times 10^{-31}\ \text{kg} \times 3 \times 10^{8}\ \text{m/s}}$$

$$= 0.00243\ \text{nm}$$

Since visible light has a wavelength of several hundred nanometers, its Compton scattering produces a relatively minute effect, but the Compton scattering of X rays can have a substantial effect.

EXAMPLE 26.4

X rays of wavelength 0.0710 nm are scattered by free electrons. Calculate the energy of an X ray that is scattered through 120° and the kinetic energy of the recoil electron.

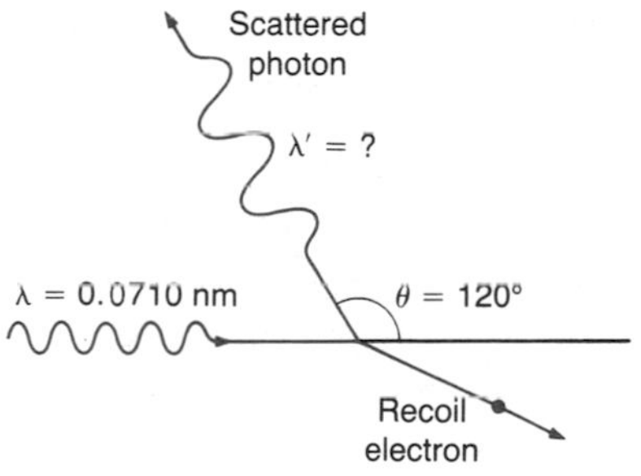

SOLUTION We first calculate the wavelength of the scattered X ray using the Compton formula, Eq. (26.5), and our value for the Compton wavelength:

$$\lambda' - \lambda = \frac{h}{mc}(1 - \cos\theta)$$

$$\lambda' - 0.0710\ \text{nm} = 0.00243\ \text{nm} \times (1 - \cos 120^\circ)$$

$$\therefore \lambda' = 0.07464\ \text{nm}$$

The energies of the X rays are given by Eq. (26.2b); for the scattered X ray we have:

$$E' = \frac{hc}{\lambda'}$$

$$= \frac{1.241 \times 10^{-6}\ \text{eV} \cdot \text{m}}{74.64 \times 10^{-12}\ \text{m}}$$

$$= 16.63\ \text{keV}$$

For the incident X ray, the result is:

$$E = \frac{hc}{\lambda}$$

$$= \frac{1.241 \times 10^{-6}\ \text{eV} \cdot \text{m}}{71.0 \times 10^{-12}\ \text{m}}$$

$$= 17.47\ \text{keV}$$

The difference between the two X-ray energies is the kinetic energy of the recoil electron:

$$\text{Recoil electron energy} = 17.47\ \text{keV} - 16.63\ \text{keV}$$
$$= 0.84\ \text{keV}$$

■

Compton used X rays in his original investigation. He observed not only the wavelength change in the scattered X ray but also the kinetic energy and the angle of recoil of the electron. All these quantities were in excellent agreement with the theory.

Let us pause and review the evidence for the quantum nature of electromagnetic waves. Planck showed that we can explain the blackbody radiation spectrum by assuming that the radiant energy is emitted in quanta. Einstein fully established the energy quantum theory by using it to explain the energy relationship in the photoelectric effect. Going much further, Compton associated momentum with the radiation quanta and thereby provided a basis for incorporating the quanta into dynamics.

All this raises a perplexing question. Surely the point particle model and the wave model are poles apart. Must we not use either one or the other to represent some physical reality such as light? The answer is that the point particle and the wave models are oversimplifications. A more complete theory requires the elaborate mathematical apparatus of wave mechanics. However, in two extreme cases the simpler models provide excellent approximations. For electromagnetic waves of very long wavelength, such as long radio waves, the wavelike character is easily established. Such waves show all the phenomena of interference and diffraction that we regard as incontrovertible evidence of a wave nature. We believe in the energy quantization of long radio waves only because other electromagnetic waves of much shorter wavelength exhibit energy quantization. The situation is quite the opposite for high-energy γ rays, which appear as quantized particles in Compton scattering and similar situations. We believe that they have a wave nature because the longer-wavelength electromagnetic waves possess it.

However, the clinching argument for dual nature is for the waves in the middle of the electromagnetic spectrum. Infrared, visible light, ultraviolet, and X rays all plainly exhibit both wave and particle natures. In all these cases, we can arrange experiments showing interference and diffraction, which prove the wave nature, as well as experiments showing energy quantization, which prove the particle nature. Since electromagnetic waves appear to be of the same basic nature throughout the entire spectrum, we infer that both the wave and particle nature exist for all wavelengths of the spectrum.

KEY CONCEPTS

The wavelength of the most intense radiation from a blackbody decreases as the temperature increases; see **Wien's displacement law,** Eq. (26.1).

In classical physics, an electromagnetic wave could have any energy imaginable. In quantum physics, the *energy of an electromagnetic wave* is emitted or absorbed only in *discrete quanta*.

The **photon** (or **electromagnetic energy quantum**) is equal to Planck's constant multiplied by the frequency; see Eq. (26.2).

The ejection of electrons from matter by photons is called the **photoelectric effect.** The velocity of the ejected electrons is calculated from an energy equation; see Eq. (26.3).

The **photon momentum** is equal to its energy divided by the velocity of light; see Eq. (26.4).

The **Compton effect** analyzes the collision of a photon with a free electron in terms of the conservation of energy and momentum; see Eq. (26.5).

QUESTIONS FOR THOUGHT

1. Many hot bodies that emit radiant energy are not blackbodies. List some examples.

2. Cavities between the glowing coals in a fire seem to be brighter than the coals themselves. Explain why this is so.

3. A cavity with a small hole in the wall is maintained at constant temperature. Under these conditions we cannot see any details of the interior by looking in through the hole. Explain why this is so.

4. Some stars are red, and some have a distinctly blue color. Which ones have the higher surface temperature? Explain your answer.

5. Do you require two different monochromatic light sources to measure the work function of a photocathode, or will one suffice?

6. Consider a Compton-type collision between a photon and a proton. How would you modify the theory?

7. In discussing the collision between photons and electrons in the Compton effect, we used relativistic dynamics, but in the photoelectric effect we used only classical dynamics. Justify these choices.

PROBLEMS

A. Single-Substitution Problems

1. A block of iron is heated to 900°C. Calculate the wavelength at which the radiated energy is a maximum. [26.1]

2. What is the surface temperature of a star if its blackbody radiation has an energy maximum at a wavelength of 85 nm? [26.1]

3. Calculate the wavelength at which the thermally radiated energy from the human body is a maximum. (Take the body temperature at 37°C.) [26.1]

4. The temperature in the interior of the sun is believed to be about 15 000 000 K. What is the wavelength for the energy maximum of blackbody radiation at this temperature? [26.1]

5. The background electromagnetic radiation in the universe is believed to be blackbody, corresponding to a temperature of about 2.8 K. Calculate the wavelength for the energy maximum of this radiation. [26.1]

6. Calculate the photon energy of light from a helium–neon laser whose wavelength is 632 nm. [26.1]

7. Calculate the frequency of photons for each of the following quantum energies: (a) 10^{-6} eV, (b) 3.0 eV, (c) 10^{6} eV. To what region of the electromagnetic spectrum do these photons belong? [26.1]

8. Calculate the photoelectric threshold frequency for chromium. (Refer to Table 26.1.) [26.2]

9. The photoelectric threshold frequency for a certain metal surface is 6.1×10^{14} Hz. Calculate the work function of the metal. [26.2]

10. Can green light of wavelength 520 nm cause photoemission from a cesium surface? [26.2]

11. Calculate the photon momentum for each of the photon energies in problem 7. [26.3]

12. Calculate the wavelength of a photon that has momentum of 10^{-26} kg m/s. [26.3]

13. Calculate the momentum of a photon of green light of wavelength 540 nm. [26.3]

B. Standard-Level Problems

14. The temperature of a blackbody radiator is decreased so as to halve the radiant power output. By what factor does the wavelength corresponding to the peak of the radiation curve change? [26.1]

15. A blackbody radiator is radiating 55 kW/m^2. Calculate the wavelength corresponding to the peak of the radiation curve. [26.1]

16. Calculate the rate of emission of photons by a 60-W sodium vapor lamp. (Assume that all photons have the same wavelength of 589 nm.)

17. Calculate the rate of emission of photons by a 10-kW, 115-MHz FM radio transmitter.

18. In a very dark room the human eye is able to detect a flash of visible light of about 50 photons directed through the iris. Calculate the energy of the light flash in joules. (Assume $\lambda = 500$ nm.)

19. The wavelength of light from a helium–neon laser is 632 nm, and the beam power is 3 mW. How many photons per second pass through a fixed area across the path of the beam?

20. Calculate the wavelength of a photon whose energy is equal to the average translational kinetic energy of an ideal gas molecule at 27°C. To what region of the electromagnetic spectrum does this photon belong?

21. A ruby laser with a wavelength of 694 nm operates for 0.25 μs with a beam power of 1.45 MW. How many photons are emitted during this time interval?

22. Visible light covers the wavelength range between 380 nm and 760 nm. Calculate the corresponding photon energy range in electron volts. List the metals from Table 26.1 that eject photoelectrons when illuminated by visible light. [26.2]

23. Calculate the maximum velocity of electrons photoelectrically emitted from a silver surface by ultraviolet light of photon energy 6 eV. [26.2]

24. The maximum velocity of photoelectrons emitted from a gold surface is 1.75×10^{6} m/s. Calculate the photon energy of the radiation.

25. Monochromatic light has a photon energy equal to the threshold for emission from platinum. Calculate the maximum velocity of electrons that would be ejected from a barium surface by light of this wavelength.

26. The photoelectric threshold wavelength for a certain surface is 284 nm. Calculate the wavelength of light required to produce photoelectrons whose maximum velocity is 6×10^{5} m/s.

27. The photoelectric threshold wavelength for a gold electrode is 260 nm. Calculate the maximum velocity of the photoelectrons when the electrode is illuminated by ultraviolet light of wavelength 100 nm.

28. Ultraviolet light of wavelength 262 nm falling on an electrode emits photoelectrons with a maximum velocity of 3.6×10^{5} m/s. Calculate:

a. the work function in electron volts
b. the wavelength of the photoelectric threshold

29. The work function of a potassium surface is 2.26 eV. Calculate:

a. the stopping potential for incident light of wavelength 380 nm
b. the maximum velocity of the photoelectrons

30. The experimental arrangement of Fig. 26.5 is used for different wavelengths of incident light. The stopping potentials from a certain surface are:

$$V_S = 0.38 \text{ V} \quad \text{when} \quad \lambda = 750 \text{ nm}$$

$$V_S = 1.59 \text{ V} \quad \text{when} \quad \lambda = 435 \text{ nm}$$

Calculate the value of Planck's constant and the work function of the surface by using these measurements. (Assume that $e = -1.6 \times 10^{-19}$ C.)

31. Ultraviolet light of wavelength 200 nm falls on a metallic photocathode, and the stopping potential is 1.85 V. Use the data of Table 26.1 to give a tentative identification of the metal of the photocathode.

32. A certain ruby laser (wavelength 694 nm) emits a light pulse of 10^{22} photons during a time interval of 1.2 μs. Calculate:

a. the average power output during the pulse
b. the momentum of the light pulse [26.3]

33. A ruby laser (wavelength 694 nm) operates in a pulsed mode with an average power output of 4 GW during the pulse. Calculate the required pulse time so that the momentum of the pulse should be the same as a beam of 10^{15} electrons, each of 2 MeV kinetic energy. [26.3]

34. A 75-keV X ray has its direction of motion reversed when Compton-scattered by an electron. Calculate the recoil kinetic energy of the electron.

35. A photon is Compton-scattered by an electron in a direction perpendicular to its original direction of motion, and the wavelength is doubled in the scattering process. Calculate the energy of the original photon.

36. In a certain Compton-scattering process, the recoil electron has a kinetic energy of 1.5 MeV. The incident photon has a wavelength of 0.000620 nm. Calculate:

a. the wavelength of the scattered photon
b. the angle between the directions of the incident and scattered photons

37. The incident and scattered photons in a Compton-scattering experiment have wavelengths of 0.050 nm and 0.052 nm, respectively. Calculate:

a. the kinetic energy of the scattered electron
b. the angle between the directions of the incident and scattered photons

38. An X ray of wavelength 0.00078 nm is scattered by a free electron. The scattered X ray is observed at an angle of 30° with the direction of the incident X ray. Calculate:

a. the wavelength of the scattered X ray
b. the kinetic energy of the recoil electron

27

QUANTUM PHYSICS

PARTICLES AS WAVES

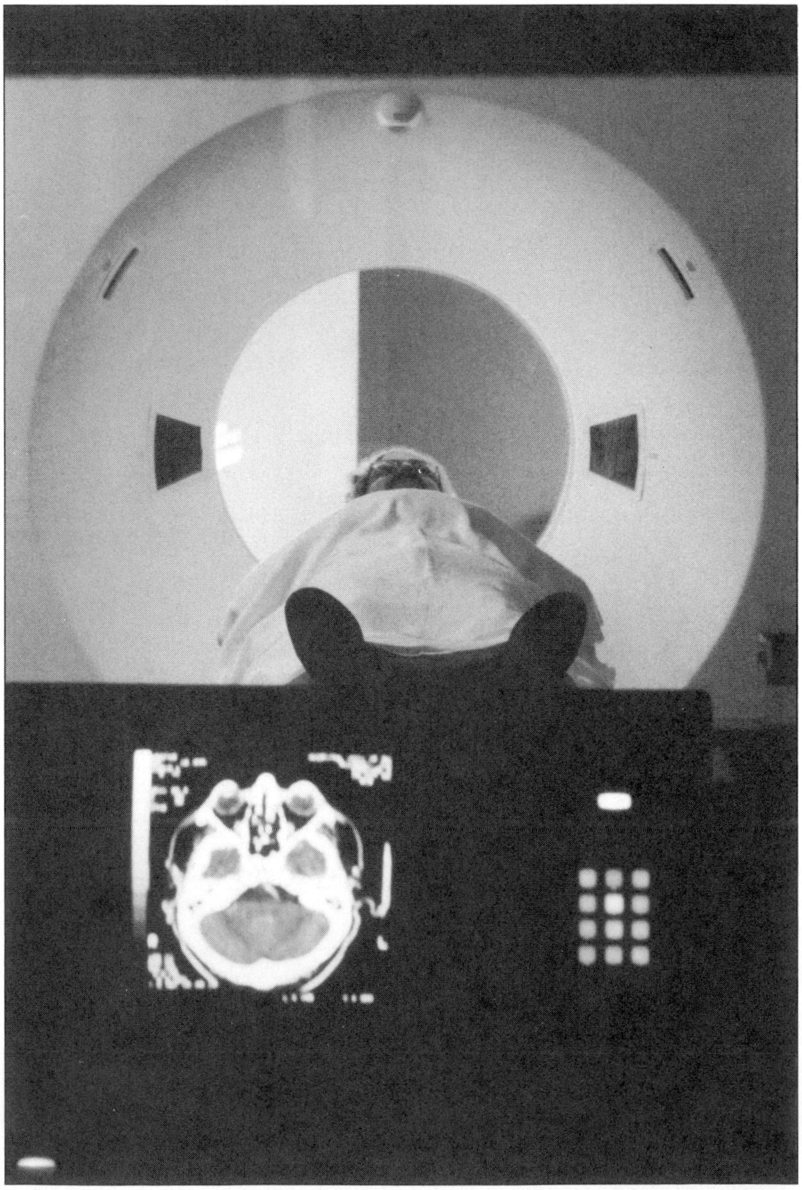

In our first discussion of quantum physics, we saw that electromagnetic waves possess a corpuscular property. This finding spawned the belief that perhaps the particles of classical physics possessed a wave nature—this despite the fact that the theory had little basis except for strong hints from the theory of relativity and a belief in the beautiful symmetry of nature.

However, the development of this idea unlocked one of the major puzzles in physics. For a long time the spectra of gas discharges had been known to show lines rather than the continuous spectrum associated with blackbody radiation. Moreover, different gases had different line spectra. Each gas produced its own characteristic set of lines. Quantum physics led to an atomic model that could be used to explain characteristic spectra in detail. The achievement of the quantum physicists ranks alongside that of Einstein, and their detailed model of the workings of the atom represents a major advance in science.

27.1 The Wave Nature of Electrons

In the previous chapter we traced the development of the quantum character of electromagnetic waves. Now we will turn to the consequences of the discovery that the particles of classical physics also possess a wave nature. The first person to propose this idea was the French scientist Louis de Broglie (1892–), who was only a graduate student at the time.

De Broglie's insight arose from his study of relativity. He noted that the formula for the photon momentum, Eq. (26.4), can also be written in terms of wavelength. Since Eq. (20.12) gives $c = \nu\lambda$, we have:

$$p = \frac{h\nu}{c} = \frac{h}{\lambda}$$

If this relation is true for massive particles as well as for photons, the view of matter and light would be much more unified. In certain circumstances each could behave as a wave, and in other instances each could behave as a particle. De Broglie's basic starting point was the assumption that the momentum–wavelength relation is true for both photons and massive particles.

De Broglie Wave Equation

$$\text{Particle momentum} = \frac{\text{Planck's constant}}{\text{particle wavelength}}$$

$$p = \frac{h}{\lambda} \tag{27.1}$$

At first sight, to claim that a particle such as an electron has a wavelength seems somewhat absurd. The classical concept of an electron is a point particle of definite mass and charge, but de Broglie argued that the wavelength of the wave associated with an electron might be so small that it had not been previously noticed. If we wish to prove that an electron has a wave nature, we must perform an experiment in which electrons behave as waves. In other words, we must demonstrate interference and diffraction for beams of electrons. At this point, recall that interference and diffraction of light become noticeable when light travels through slits whose width and separation are comparable with the wavelength of the light. Therefore, let us first work an example to determine the magnitude of the expected wavelength for some representative objects.

EXAMPLE 27.1

Calculate the de Broglie wavelengths for the following objects:

a. An electron whose kinetic energy is 600 eV

b. A golf ball of mass 45 g traveling at 40 m/s

SOLUTION

a. To apply Eq. (27.1), we need to calculate the momentum of an electron whose kinetic energy is 600 eV. First convert the energy to joules:

$$\begin{aligned}\text{Kinetic energy} &= 600\ \text{eV} \times 1.6 \times 10^{-19}\ \text{J/eV} \\ &= 9.6 \times 10^{-17}\ \text{J}\end{aligned}$$

We now use the Newtonian formula, Eq. (7.2), to calculate the electron velocity:

$$\begin{aligned}\text{KE} &= \tfrac{1}{2}mv^2 \\ \therefore v &= \left(\frac{2 \times \text{KE}}{m}\right)^{1/2} \\ &= \left(\frac{2 \times 9.6 \times 10^{-17}\ \text{J}}{9.11 \times 10^{-31}\ \text{kg}}\right)^{1/2} \\ &= 1.452 \times 10^{7}\ \text{m/s}\end{aligned}$$

(Since the kinetic energy is much smaller than the rest energy, we do not need relativistic dynamics.) The momentum is given by Eq. (8.3):

$$\begin{aligned}p &= mv \\ &= 9.11 \times 10^{-31}\ \text{kg} \times 1.452 \times 10^{7}\ \text{m/s} \\ &= 1.323 \times 10^{-23}\ \text{kg} \cdot \text{m/s}\end{aligned}$$

Finally, we calculate the wavelength from the de Broglie formula Eq. (27.1):

$$\lambda = \frac{h}{p}$$

$$= \frac{6.626 \times 10^{-34}\ \mathrm{J \cdot s}}{1.323 \times 10^{-23}\ \mathrm{kg \cdot m/s}}$$

$$= 0.0501\ \mathrm{nm}$$

b. For the golf ball we find:

$$p = mv$$

$$= 45 \times 10^{-3}\ \mathrm{kg} \times 40\ \mathrm{m/s}$$

$$= 1.8\ \mathrm{kg \cdot m/s}$$

The de Broglie formula gives:

$$\lambda = \frac{h}{p}$$

$$= \frac{6.626 \times 10^{-34}\ \mathrm{J \cdot s}}{1.8\ \mathrm{kg \cdot m/s}}$$

$$= 3.68 \times 10^{-34}\ \mathrm{m}$$ ■

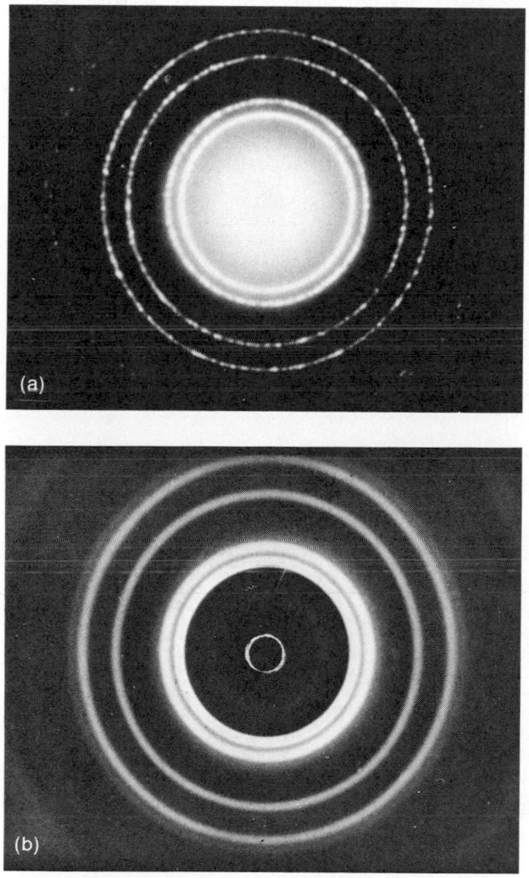

FIGURE 27.1 Diffraction patterns produced by transmission through a thin aluminum foil: (a) 600-eV electrons; (b) 0.071-nm X rays. (From PSSC Physics film "Matter Waves," Education Development Center, Inc., Newton, MA)

We recall that diffraction of light waves is significant when the slit width is about the same as or smaller than the wavelength. Thus, when a beam of light passes through a broad window, diffraction is not noticeable. The situation is even more extreme when a golf ball passes between two tree trunks—the wavelength is truly insignificant compared with the separation of the trees.

We must now consider whether we could observe diffraction of electrons whose wavelength is a small fraction of a nanometer. For a grating to show observable diffraction, the slit separation should be comparable to the wavelength, but we cannot rule a series of lines that are only a small fraction of a nanometer apart! That is less than the separation of the atoms in solid materials. Yet this comment supplies the answer. The clean, flat surface of a single crystal of a solid already has lines ruled on it—the lines of atoms that are arranged in an orderly array. If electrons truly have a wave nature, we should see interference and diffraction effects when a beam of monochromatic electrons (electrons having the same energy) illuminates a clean crystal surface.

Experimental confirmation of the de Broglie equation was supplied by two American physicists, Clinton Davisson and Lester Germer, who were studying the reflection of electrons from a clean nickel crystal surface at Bell Telephone Laboratories. They found that the reflected beam of electrons consisted of "bright and dark fringes," just as if the nickel surface were a diffraction grating. Of course, the diffraction pattern of the electrons was not detected visually, as it is in the case of light waves. Rather, specially designed detectors picked up the variation in intensity in the electron diffraction pattern. The experiment provided direct observational proof of the wave nature of electrons.

Independent confirmation came at the same time from the work of George P. Thomson at the University of Aberdeen. He passed electrons through a thin gold foil and obtained diffraction patterns, again indicating the wave nature of electrons. Subsequently other particles, such as protons and neutrons, were also found to diffract under suitable conditions. The de Broglie relation is believed to express a basic property of all objects that possess momentum.

In Example 27.1 we found that an electron with 600 eV of kinetic energy has an associated wavelength of 0.0501 nm. An electromagnetic wave with this wavelength is in the X-ray region of the electromagnetic spectrum. How does the diffraction of an electron compare with that of an X ray when both have the same wavelength? We answer this question in Fig. 27.1 on page 757, which shows two diffraction patterns produced by waves of similar wavelength transmitted through thin polycrystalline aluminum foil. Figure 27.1(a) shows the diffraction pattern produced by electrons having 600 eV of kinetic energy, whose matter wavelength is 0.050 nm; Fig. 27.1(b) shows the diffraction pattern produced by X rays with a wavelength of 0.071 nm. The two wavelengths, which are of the same order of magnitude, have strikingly similar diffraction patterns. In each instance the location of the diffraction fringes experimentally verifies the correctness of the de Broglie formula.

27.2 Energy Quantization

Hard on the heels of de Broglie's postulate concerning the wave nature of material objects and its experimental verification came a whole new theory to deal with the statics and dynamics of these waves. The theory, called quantum mechanics, is due chiefly to the German physicists Erwin Schrödinger (1887–1961) and Werner Heisenberg (1901–   ). Although each of these men approached the subject from a different point of view, their theories are equivalent mathematical descriptions of the same reality. In both cases the mathematics involved is beyond the scope of this book. Both approaches involve **energy quantization** of a confined particle and the **Heisenberg uncertainty principle.** We can easily analyze an example that will help us to understand energy quantization.

In a previous chapter we studied some general ideas of wave motion. Many of the applications concerned waves on stretched strings, and it is helpful to keep this example in mind when studying matter waves. The type of wave that exists on a stretched guitar string is called a standing wave, or a stationary wave. Because the string is held firmly at each end, the wave can only travel back and forth between the two ends. We identified the possible modes of vibration of the guitar string by noting that a node existed at each end of the string. The different modes have different numbers of nodes at other locations on the string itself.

Let us now examine the stationary de Broglie matter waves that have the same form as the fundamental mode and the harmonics of a guitar string. We show such wave patterns in Fig. 27.2 (these are the same patterns as those of Fig. 12.4 for the guitar string). Since a node exists at each end, the matter waves are confined to the region between the nodes. We therefore assume that the waves represent a classical particle, such as an electron, a neutron, or an atom confined within a one-dimensional box of side length l. Such a box cannot be a very good model of reality because all real boxes are three-dimensional, but it does lead us to very important conclusions about matter waves.

The first column of data in Fig. 27.2 gives the wavelengths of the standing waves, which we determine from inspection of the geometrical properties. Every wave has a node at both ends. The stationary wave of longest wavelength has no other node, and other stationary waves have one or more nodes at intermediate positions. In the case of the guitar string, we referred to the various possible stationary waves as modes of the string. In quantum dynamics we speak of stationary wave modes as **quantum states** of the particle. The longest-wavelength quantum state is the **ground state,** and the other states are **excited states.**

The second data column lists the momentum of the various stationary waves as calculated from the de Broglie equation. You may object that a stationary wave is traveling neither to the right nor to the left. How, then, can it have momentum? The answer is that we can regard a stationary wave as composed of two equal waves, one traveling to the right and the other to the left. For this reason we place a "$\pm$" sign before the value for the momentum.

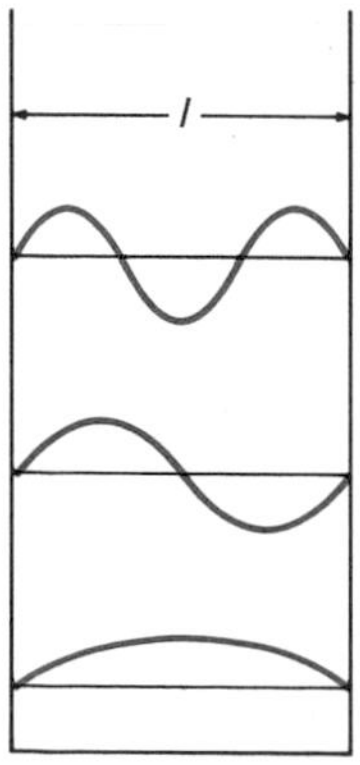

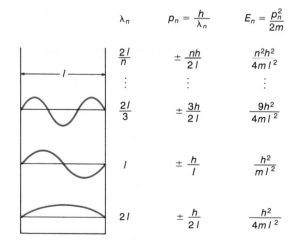

λ_n	$p_n = \dfrac{h}{\lambda_n}$	$E_n = \dfrac{p_n^2}{2m}$
$\dfrac{2l}{n}$	$\pm\dfrac{nh}{2l}$	$\dfrac{n^2h^2}{4ml^2}$
$\vdots$	$\vdots$	$\vdots$
$\dfrac{2l}{3}$	$\pm\dfrac{3h}{2l}$	$\dfrac{9h^2}{4ml^2}$
l	$\pm\dfrac{h}{l}$	$\dfrac{h^2}{ml^2}$
$2l$	$\pm\dfrac{h}{2l}$	$\dfrac{h^2}{4ml^2}$

FIGURE 27.2 The matter waves of a particle confined within a fixed one-dimensional region of length l.

Recall that the kinetic energy of a point mass can be calculated if its momentum is known. Equations (7.2) and (8.3) give for the kinetic energy and momentum, respectively:

$$\text{KE} = \tfrac{1}{2}mv^2$$

$$p = mv$$

It follows that we can write the kinetic energy in terms of the momentum:

$$\text{KE} = \frac{p^2}{2m}$$

Applying this result, we calculate the kinetic energies for each stationary matter wave, obtaining the results given in the third column of Fig. 27.2.

We can summarize the dynamical properties of the stationary matter waves as follows:

■ Quantum Formulas for the Momentum and Kinetic Energy of a Particle of Mass *m* Confined Within a One-Dimensional Region of Length *l*

$$p_n = \pm\frac{nh}{2l} \quad \text{(a)}$$

$$KE_n = \frac{n^2h^2}{8ml^2} \quad \text{(b)} \qquad (27.2)$$

where $n = 1, 2, 3, \ldots$

The number n, which may have any positive integral value, labels the possible quantum states of the particle and is called the **quantum number** for the problem under consideration. The ground state, given by $n = 1$, has the lowest energy possible. The excited states are given by $n = 2, 3, 4, \ldots$, and they have progressively higher energies.

EXAMPLE 27.2

An electron is confined within a one-dimensional region 0.2 nm wide. Calculate the energy of the ground state in electron volts.

SOLUTION The possible stationary matter waves for the system are shown in Fig. 27.2. The quantum energies are given by Eq. (27.2b):

$$KE_n = \frac{n^2h^2}{8ml^2}$$

For the ground state, $n = 1$, which gives:

$$\begin{aligned} KE_1 &= \frac{(6.63 \times 10^{-34}\ \text{J} \cdot \text{s})^2}{8 \times 9.11 \times 10^{-31}\ \text{kg} \times (0.2 \times 10^{-9}\ \text{m})^2} \\ &= 1.51 \times 10^{-18}\ \text{J} \\ &= 1.51 \times 10^{-18}\ \text{J} \div 1.6 \times 10^{-19}\ \text{J/eV} \\ &= 9.40\ \text{eV} \end{aligned}$$

We have chosen 0.2 nm for the size of the box in this example because it is the approximate diameter of a typical atom. There is no implication that a one-dimensional square potential box is a good model to represent the rather complex three-dimensional potential that holds an electron in an atom, but it might give some feeling of the order of magnitude associated with the energy quantization. Finally, we note that 9.40 eV is a very substantial amount of energy compared with the mean thermal energy per particle in an ideal gas at room temperature. (In Example 26.2, we calculated that to be 3.88×10^{-2} eV.) ■

27.3 The Heisenberg Uncertainty Principle

In 1926 the German physicist Werner Heisenberg proposed a principle that has come to be regarded as basic to the theory of quantum mechanics. It is called the "uncertainty principle," and it limits the extent to which we can possess accurate knowledge about certain pairs of dynamical variables. Many pairs of variables go together to make up uncertainty relations, but we shall choose a component of linear momentum and its associated position variable as a good example of the concept.

To set up an approximate form for the Heisenberg uncertainty principle for these two variables, we refer back to the situation illustrated in Fig. 27.2. Recall that the magnitude of the momentum is determined exactly for each standing wave, but the direction could be to the right or left. For example, the ground state momentum could be either $+h/2l$ or $-h/2l$, depending on the direction. The momentum is uncertain by the amount of the difference between these two values—that is, by h/l. Since we think of the matter wave as representing a particle, its position is also uncertain by the amount of the spread of the wave that represents the particle. In this case the position uncertainty is the width of the box l. If we take the product of the two uncertainties for the ground state, we obtain the following:

$$\begin{matrix}\text{Uncertainty} \\ \text{in momentum}\end{matrix} \times \begin{matrix}\text{uncertainty} \\ \text{in position}\end{matrix} = \frac{h}{l} \times l = h$$

Repeating the calculation for excited states gives a progressively larger momentum uncertainty and an unchanging position uncertainty. We are thus led to the conclusion that the *product of the uncertainties in momentum and position must always exceed Planck's constant* for a particle contained within a definite region in one dimension.

Generalizing this proposition requires only two modifications:

1. Both momentum and position are vectors. When dealing with a real three-dimensional situation, we take the uncertainties of the components of each vector in the same direction.
2. Our sample calculation is restricted to the simplest interpretation of what we mean by uncertainty. A more elaborate statistical interpretation gives the lower limit of the uncertainty product as $h/4\pi$ rather than h.

With these two changes, we can now write the required modification for the product of the uncertainties.

■ Heisenberg Uncertainty Principle

The product of the uncertainty in a momentum component and the corresponding uncertainty in a position component cannot be less than $h/4\pi$.

$$\Delta p_x \times \Delta x \geqslant \frac{h}{4\pi} \qquad (27.3)$$

We have arbitrarily chosen the x direction for components, but the same result also holds for y and z components. The symbol Δ (capital Greek letter delta) immediately preceding the algebraic symbols represents the uncertainty in the quantity that the symbol represents. Thus, Δp_x means the uncertainty in the x component of momentum, and Δx the uncertainty in the x component of position.

The momentum–position uncertainty relations are of general validity in quantum mechanics and are in no way restricted to the stationary wave example that we used to introduce them. Heisenberg himself described a simple imaginary experiment to illustrate his principle. He imagined a single electron moving in a perfect vacuum and sensitive detection equipment that could track the path of the electron by bouncing single photons from it. In order to minimize the effect of the photon collisions with the electron, low-frequency photons would have to be used whose energy and momentum [see Eqs. (26.2) and (26.4)] were small. In this way the electron would suffer only negligible deviation from its path when the detecting photons collided with it. However, a new problem would then present itself. A photon of low frequency has a long wavelength [see Eq. (20.12)] and therefore experiences a large diffraction in its encounter with the small electron. Because of the large diffraction effect, we could not use low-frequency photons to give an accurate indication of the electron position. However, if photons of higher frequency were used, their larger energy and momentum would disturb the electron from its path and make it impossible to determine its momentum accurately. Thus, Heisenberg showed that the uncertainty relation, Eq. (27.3), limits our ability to measure the electron's momentum and position simultaneously.

EXAMPLE 27.3

An electron moving along the x-axis has a measured velocity of (6000 ± 1) km/s. Calculate the minimum uncertainty in a simultaneous measurement of its position.

SOLUTION The electron velocity is uncertain by 2×10^3 m/s. The momentum uncertainty is therefore equal to the electron mass multiplied by the velocity uncertainty:

$$\begin{aligned}\Delta p_x &= m\,\Delta v_x \\ &= 9.11 \times 10^{-31}\ \text{kg} \times 2 \times 10^3\ \text{m/s} \\ &= 1.82 \times 10^{-27}\ \text{kg} \cdot \text{m/s}\end{aligned}$$

The Heisenberg principle, Eq. (27.3), now gives us the position uncertainty:

$$\Delta p_x \times \Delta x \geqslant \frac{h}{4\pi}$$

$$1.82 \times 10^{-27}\ \text{kg} \cdot \text{m/s} \times \Delta x \geqslant \frac{6.63 \times 10^{-34}\ \text{J} \cdot \text{s}}{4\pi}$$

$$\therefore\ \Delta x \geqslant 29.0\ \text{nm}$$

The relatively precise velocity measurement prevents us from locating the electron position without an error of at least 29 nm. This is about 100 interatomic distances in a solid or liquid substance. ■

Let us briefly review how quantum dynamics differs from classical dynamics. Classically, both the momentum and position of a point particle can be determined to whatever degree of accuracy that the measuring apparatus permits. From the quantum viewpoint, the product of the momentum and position uncertainties must be at least as great as $h/4\pi$. Which viewpoint is correct? Strictly speaking, the quantum viewpoint is the correct one, but in everyday life situations the two are not in opposition. Let us consider an example to illustrate this point.

EXAMPLE 27.4

A golf ball of mass 45 g has a measured velocity of (40.000 ± 0.001) m/s. Calculate the uncertainty in a simultaneous measurement of its position.

SOLUTION The velocity measurement represents enormously high precision. We can calculate the momentum uncertainty in the same way as in the previous example:

$$\begin{aligned}\Delta p_x &= m\,\Delta v_x \\ &= 45 \times 10^{-3}\ \text{kg} \times 0.002\ \text{m/s} \\ &= 9 \times 10^{-5}\ \text{kg} \cdot \text{m/s}\end{aligned}$$

The Heisenberg principle, Eq. (27.3), gives:

$$\Delta p_x \times \Delta x \geqslant \frac{h}{4\pi}$$

$$9 \times 10^{-5} \text{ kg} \cdot \text{m/s} \times \Delta x \geqslant \frac{6.63 \times 10^{-34} \text{ J} \cdot \text{s}}{4\pi}$$

$$\therefore \Delta x \geqslant 5.86 \times 10^{-31} \text{ m}$$

There is no conceivable way of measuring position to such accuracy. For all practical purposes, the precision of measurement of the momentum and position of the golf ball are limited only by the measuring instruments. The situation is quite the opposite with the electron in Example 27.3, where the precision is limited by the wave nature of matter, and no measuring instrument can ever improve it. ■

27.4 The One-Electron Atom

There is no way in nature that an electron can be cooped up in one dimension between two rigid reflecting walls. One of the simplest real examples of the confinement of an electron within a small region of space is the hydrogen atom. A central proton holds the relatively light electron within a region of space whose dimension is of order of 0.1 nm. However, there are two extremely important differences between the hydrogen atom and the one-dimensional example of the previous section. First, the force of containment is the Coulomb force of attraction between the proton and the electron rather than rigid walls. Second, the stationary matter wave does not have nodes at the end points, as in Fig. 27.2. In fact, its actual shape cannot be described in simple mathematical terms or even sketched, since the wave shape exists in three dimensions rather than in one.

The three-dimensional nature of the containment raises another issue. In our previous one-dimensional example, the quantum number n relates to the number of nodes in the one-dimensional wave pattern. In a similar way, the three-dimensional wave pattern requires three quantum numbers to characterize its complex nodal arrangement. Without delving into the mathematics, we can quote the results and emphasize the physical relevance of the quantum numbers. The three quantum numbers determine the energy, the angular momentum, and the angular momentum component relative to some assigned direction.

The total energy of the electron in a hydrogen atom includes the electron's kinetic energy and the mutual electrostatic potential energy of the electron and the proton. Since the proton is a positive charge and the electron is negative, the electrostatic potential energy is negative. The kinetic energy of an electron is necessarily

SPECIAL TOPIC 28 BOHR

Niels Bohr was born in Copenhagen in 1885. He received his doctorate from the University of Copenhagen in 1911, and studied for a few years in England. Returning to Copenhagen as professor of physics, he proposed the first quantum model of the hydrogen atom in order to explain Balmer's formula for the spectrum of hydrogen. The model proved inadequate, but it did set scientists on the correct path to the new quantum mechanics. For this work Bohr won the Nobel Prize in physics in 1922.

During the following decades, Bohr directed the Institute for Atomic Studies, which was supported by Carlsberg Brewery of Copenhagen. It became a world famous center for the study of atomic physics. He was associated with the nuclear bomb project during World War II, but worked hard for the development of peaceful uses for nuclear energy. He died in Copenhagen in 1962.

(Courtesy AIP Niels Bohr Laboratory)

positive. For any quantum state in which the electron is bound to the vicinity of the proton, its total energy turns out to be negative. The appropriate quantum dynamical formula gives the total electron energy in terms of one of the quantum numbers.

Quantum Formula for the Energy of an Electron Bound in a Hydrogen Atom

$$E_n = -\frac{1}{n^2}\frac{me^4}{8\epsilon_0^2 h^2} \qquad \textbf{(27.4)}$$

n = *principal quantum number* with possible values 1, 2, 3, . . .
e = electron charge
m = electron mass
ϵ_0 = permittivity of free space
h = Planck's constant

Let us work an example to get a precise idea of the electron energies in the hydrogen atom.

EXAMPLE 27.5

Calculate the ground state energy and the energies of the first two excited states for the electron in a hydrogen atom. Express your answers in electron volts.

SOLUTION To calculate the ground state energy, we use Eq. (27.4) with the principal quantum number set at unity:

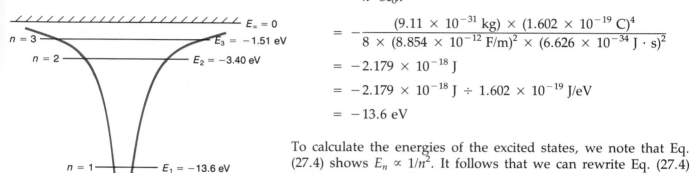

$$E_1 = -\frac{1}{n^2}\frac{me^4}{8\epsilon_0^2 h^2}$$

$$= -\frac{(9.11 \times 10^{-31}\ \text{kg}) \times (1.602 \times 10^{-19}\ \text{C})^4}{8 \times (8.854 \times 10^{-12}\ \text{F/m})^2 \times (6.626 \times 10^{-34}\ \text{J} \cdot \text{s})^2}$$

$$= -2.179 \times 10^{-18}\ \text{J}$$

$$= -2.179 \times 10^{-18}\ \text{J} \div 1.602 \times 10^{-19}\ \text{J/eV}$$

$$= -13.6\ \text{eV}$$

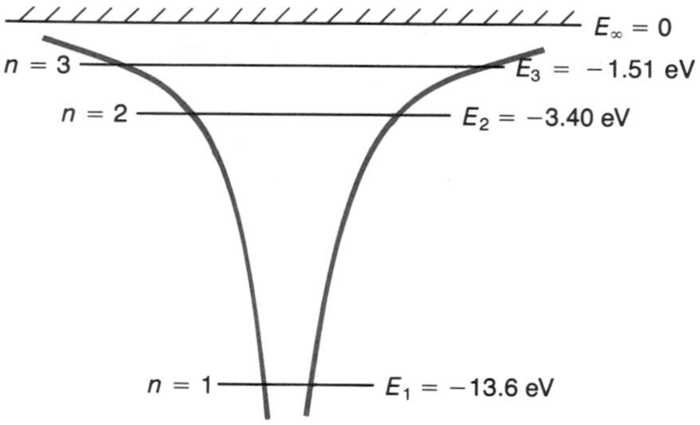

To calculate the energies of the excited states, we note that Eq. (27.4) shows $E_n \propto 1/n^2$. It follows that we can rewrite Eq. (27.4) for hydrogen as follows:

$$E_n = -\frac{2.179 \times 10^{-18}\ \text{J}}{n^2}$$

$$= -\frac{13.6\ \text{eV}}{n^2}$$

By doing this we avoid unnecessary, tedious calculations. The energies of the first and second excited states are as follows:

$$E_2 = -\frac{13.6\ \text{eV}}{2^2} = -3.40\ \text{eV}$$

$$E_3 = -\frac{13.6\ \text{eV}}{3^2} = -1.51\ \text{eV}$$

When the principal quantum number is very large, the energy is zero; therefore, all of the bound states have energies between 0 and -13.6 eV. ■

The quantum numbers for angular momentum and the component of angular momentum enable these quantities to be calculated.

Quantum Formulas for the Angular Momentum and the Component of Angular Momentum of an Electron in a Hydrogen Atom

$$\text{Angular momentum} = \sqrt{l(l+1)}\,\frac{h}{2\pi} \qquad \textbf{(a)}$$

l = *angular momentum quantum number* with possible values 0, 1, 2, . . . , but restricted so that $l < n$. **(27.5)**

$$\begin{matrix}\text{Component of angular momentum} \\ \text{in an assigned direction}\end{matrix} = \frac{m_l h}{2\pi} \qquad \textbf{(b)}$$

m_l = *angular momentum component quantum number* with possible values 0, ±1, ±2, . . . , but restricted so that $|m_l| \leq l$.

A vector diagram offers a useful aid to visualizing the relationship between the quantized total angular momentum and its quantized components. Figure 27.3 shows such a diagram drawn for the case in which the angular momentum quantum number is given by $l = 2$. Using Eq. (27.5a), we see that the magnitude of the angular momentum vector is $\sqrt{6}h/2\pi$; taking all dimensions in units of $h/2\pi$, the diagram is a sphere of radius $\sqrt{6}$. If we knew only the magnitude of the angular momentum, then it would be represented by a vector from the origin to any point on this spherical surface. However, the choices are limited since the component of angular momentum is also quantized. We select the vertical as the direction of the quantized components. The possible component quantum numbers are given by Eq. (27.5b) as $m_l = 0, \pm 1, \pm 2$; these values are marked on the vertical axis. The five horizontal circles shown in Fig. 27.3 give all of the points on the sphere for which the component quantum conditions are obeyed. This means that the total angular momentum is represented by a vector drawn from the origin to any point on one of these five circles. For example, let us suppose that the angular momentum component quantum number is $m_l = +1$. In this case, the tip of the angular momentum vector can lie anywhere on the second circle down; or to put the same thing in another way, the angular momentum vector can lie anywhere on a cone whose axis is vertical and whose base is the circle in question.

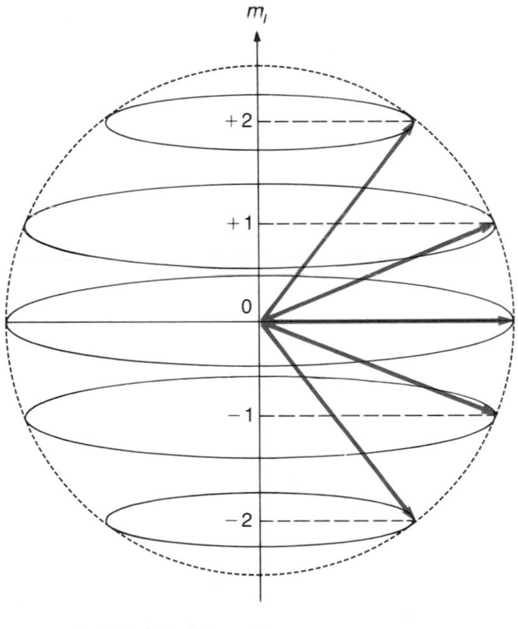

FIGURE 27.3 Vector diagram to illustrate the possible values of the quantum number m_l for the case $l = 2$.

In the hydrogen atom the two angular momentum quantum numbers play no role in determining the energy. This means in general that there may be many different standing wave patterns, all having the same energy. Such an energy level is termed *degenerate*. The *degree of degeneracy* is equal to the number of different quantum number sets belonging to the same energy.

Before we investigate the degeneracies of the hydrogen energy levels, we need to recognize one more set of quantum numbers that are intrinsic to the electron. Experiments on electrons in atoms and on free electrons moving in electric and magnetic fields show that they possess an **intrinsic angular momentum,** which is termed **spin.** The quantum formulas for spin angular momentum are similar to those of Eq. (27.5).

■ Quantum Formulas for Electron Spin and the Component of Spin Angular Momentum

$$\text{Spin angular momentum} = \sqrt{s(s+1)}\frac{h}{2\pi} \qquad \textbf{(a)}$$

s = *spin quantum number,* which is restricted to $s = \frac{1}{2}$. **(27.6)**

$$\begin{array}{c}\text{Component of spin angular momentum}\\ \text{in an assigned direction}\end{array} = \frac{m_s h}{2\pi} \qquad \textbf{(b)}$$

m_s = *spin component quantum number* with possible values $+\frac{1}{2}$ and $-\frac{1}{2}$.

Figure 27.4 shows the vector diagram for the spin angular momentum of an electron. Since $s = \frac{1}{2}$ always, Eq. (27.6a) gives the spin angular momentum as $\frac{1}{2}\sqrt{3}h/2\pi$; taking all dimensions in units of $h/2\pi$, the diagram represents a sphere of radius $\sqrt{3}/2$. The possible values of $m_s(\pm\frac{1}{2})$ are marked on the vertical axis, and the spin angular momentum vector lies anywhere on one of the two cones that are described similarly to the previous case.

Since the quantum number s can have only the value $\frac{1}{2}$ for an electron, it is usually not mentioned as one of the electron quantum numbers. Thus, of the remaining four, n, l, and m_l describe the standing matter wave in the Coulomb field, and m_s is an intrinsic property of the electron.

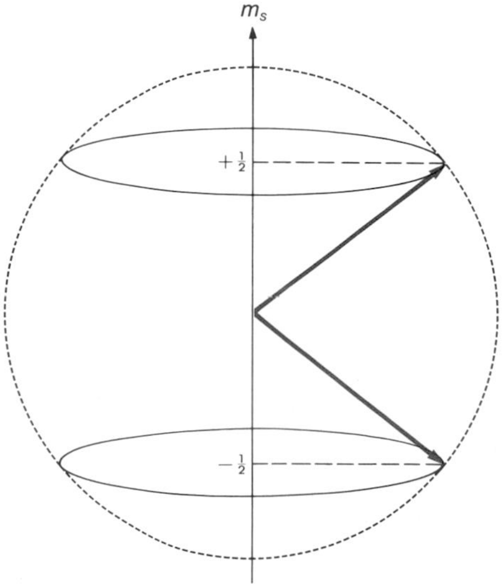

FIGURE 27.4 Vector diagram to illustrate the possible values of the quantum number m_s for an electron.

EXAMPLE 27.6

Determine the degeneracies of the ground state and the first two excited states of the hydrogen atom.

SOLUTION We take the principal quantum numbers $n = 1, 2, 3$, and use the rules of Eqs. (27.5) and (27.6) to find all the possible values for the quantum numbers l, m_l, and m_s. The degeneracy of a level is the number of different possible sets of l, m_l, and m_s. We can tabulate the results as follows:

n	l	m_l	m_s	Degeneracy
1	0	0	$\pm\frac{1}{2}$	2
2	0	0	$\pm\frac{1}{2}$	8
—	1	0, ±1	$\pm\frac{1}{2}$	
3	0	0	$\pm\frac{1}{2}$	18
—	1	0, ±1	$\pm\frac{1}{2}$	
—	2	0, ±1, ±2	$\pm\frac{1}{2}$	

The ground state of hydrogen is 2-fold degenerate. The first two excited states are 8-fold and 18-fold degenerate. The degeneracies of higher excited states can be determined by extending the simple counting procedure. ■

Because of the importance of the angular momentum quantum number, a special notation has come into use to designate its value. The notation arose from the appearance of lines in the optical spectra of alkali atoms, which were called *sharp, principal, diffuse,* and *fundamental.* These types were associated with the quantum numbers $l = 0, 1, 2, 3$. Accordingly, the following letters are used to signify angular momentum quantum numbers:

Value of angular momentum quantum number l	*Letter designation*
0	*s*
1	*p*
2	*d*
3	*f*
4	*g*
5	*h*

After *f* the letter series continues alphabetically except for the omission of the letter *j*.

A hydrogen atom in its lowest energy state is in a 1*s* state. The principal quantum number is quoted explicitly, and the angular momentum quantum number is signified by the letter. If the atom acquires sufficient energy to raise the electron to the first excited state, it is either in a 2*s* or a 2*p* state. The quantum numbers for components of angular momentum and spin are not mentioned. In a subsequent section, we will examine the role that these quantum numbers play in determining the electron structure of many-electron atoms.

27.5 Characteristic Optical Spectrum of Hydrogen

Our introduction of the wave mechanical model of the hydrogen atom as a consequence of the de Broglie relation is a logical way to proceed, but it is quite out of step with how the atomic model actually developed. That development ran parallel to the long and interesting history of the investigation of the optical spectrum of hydrogen. In 1885 the Swiss physicist Johann Balmer (1825–1898) found a remarkably simple formula for the wavelengths of the visible spectral lines of atomic hydrogen. Balmer's formula is:

$$\frac{1}{\lambda} = \text{constant} \times \left(\frac{1}{2^2} - \frac{1}{n^2}\right)$$

where $n = 3, 4, 5, \ldots$. The first few values of n give the wavelengths of the red, blue, and violet lines of the spectrum; higher values give wavelengths for lines in the ultraviolet. Surely, this fascinatingly simple formula must have some underlying explanation. The conviction was strengthened in 1908 when the German physicist F. Paschen (1865–1947) discovered a series of infrared spectral lines from atomic hydrogen, which he was able to calculate from the formula:

$$\frac{1}{\lambda} = \text{constant} \times \left(\frac{1}{3^2} - \frac{1}{n^2}\right)$$

where $n = 4, 5, 6, \ldots$. The constant in Paschen's formula had the same value as the constant in Balmer's formula.

In 1913 the Danish physicist Niels Bohr (1885–1962) proposed the first quantum theory of the hydrogen atom. In Bohr's model the electron moves in circular orbits around the central proton in much the same way as the planets of the solar system orbit the sun but with one important difference. In the atom, only certain preferred orbits are allowed. Bohr postulated that the only allowed orbits have an angular momentum equal to an integer multiplied by $h/2\pi$. Using classical dynamics, Bohr obtained the quantum energy formula, Eq. (27.4), for the energy of a point particle electron moving in a circular orbit. Although we now regard Eq. (27.4) as giving the quantum energy of a stationary electron wave, the result is usually called the Bohr formula in honor of its originator.

Bohr used his quantum energy formula to explain the regularities in the spectral lines. His explanation referred to changes in electron orbits, but we restate it using the stationary electron wave model. When an atom is in its ground state, the electron standing wave pattern corresponds to the lowest possible energy. If an atom that is initially in the ground state acquires more energy from an outside source, it can move to an excited state in which the electron has a principal quantum number belonging to a stationary wave pattern of higher energy. The electron can get rid of this surplus energy if it can make a transition to an unoccupied stationary wave pattern of lower energy.

When an electron makes a transition from a higher to a lower energy quantum state, one photon of electromagnetic radiation is emitted. Since the total energy is conserved, the photon energy must be equal to the difference between the two quantum states in question.

■ Bohr Equation Relating Photon Energy to Atomic Energy Levels

Photon energy = difference between atomic energy levels

$$h\nu_{ni \to nf} = E_{ni} - E_{nf} \tag{27.7}$$

The notation implies that an electron, initially in a quantum energy state, n_i, makes a transition to an unoccupied quantum energy state, n_f.

Although we have described the basis for the Bohr formula with the hydrogen atom in mind, the same result holds true for the transition of an electron stationary wave pattern from a higher to a lower energy mode in any atom.

The various quantum energy levels are characteristic of the atom in which they occur. Since the energy of the emitted photon is the difference in energy between two characteristic quantum levels, the photon is characteristic of the atom from which it is emitted. Thus, the emission spectrum is a true fingerprint of the atom concerned.

The type of emitted radiation depends on how the atoms are excited. We will first consider the characteristic optical spectrum that gives visible radiation. One of the easiest ways to excite the optical spectrum of an element is to pass an electrical discharge through a tube containing the element as a low-pressure gas. The orange-colored lamps that are sometimes seen in street illumination are sodium vapor lamps. The outermost electron in the configuration is excited to a higher level by the electrical discharge. The various emitted spectral lines each correspond to possible transitions from an excited quantum state to a lower unoccupied quantum state.

Calculating the wavelengths of the spectral lines for any atom except hydrogen is very difficult. However, the relatively simple Bohr quantum energy formula, Eq. (27.4), allows us to calculate the frequencies and wavelengths of the lines emitted from a hydrogen discharge lamp. We begin with the Bohr equation for photon energy, Eq. (27.7):

$$h\nu_{ni \to nf} = E_{ni} - E_{nf}$$

We now use the result for the quantum energy levels of atomic hydrogen that was derived in Example 27.5:

$$E_n = -\frac{13.6 \text{ eV}}{n^2}$$

This gives:

$$h\nu_{n_i \to n_f} = -13.6 \text{ eV} \times \left(\frac{1}{n_i^2} - \frac{1}{n_f^2}\right)$$

Divide both sides of the equation by Planck's constant $h = 4.136 \times 10^{-15}$ eV · s to give:

$$\nu_{n_i \to n_f} = -3.29 \times 10^{15} \text{ Hz} \times \left(\frac{1}{n_i^2} - \frac{1}{n_f^2}\right)$$

Finally, we divide by the velocity of light $c = 3 \times 10^8$ m/s to obtain:

$$\frac{\nu_{n_i \to n_f}}{c} = \frac{1}{\lambda_{n_i \to n_f}}$$

$$= -1.097 \times 10^7 \text{ m}^{-1} \times \left(\frac{1}{n_i^2} - \frac{1}{n_f^2}\right)$$

This celebrated result is called Rydberg's formula in honor of the Swedish physicist who proposed it as a modification of the Balmer formula. However, its first justification in terms of an atomic model was given by Bohr, as we have just seen. We can remove the negative sign from the front of the equation by interchanging the n_i and the n_f in the parentheses.

Rydberg's Formula for the Wavelength of the Spectral Lines of Hydrogen

$$\frac{1}{\lambda_{n_i \to n_f}} = R\left(\frac{1}{n_f^2} - \frac{1}{n_i^2}\right) \qquad \textbf{(27.8)}$$

where $R = 1.097 \times 10^7 \text{ m}^{-1}$ is the Rydberg constant.

We see that there are as many lines in the hydrogen spectrum as there are combinations of integers n_i and n_f. This apparent chaos is systematized by assigning a series of spectral lines to common values of the final quantum state. Each series is specified by a fixed value of n_f and named for its discoverer in the hydrogen spectrum.

1. $n_f = 1$ in Eq. (27.8) specifies the Lyman series, named after the American physicist Theodore Lyman (1874–1954), who discovered the series at Harvard three years after Bohr predicted its existence. All members of the Lyman series are in the ultraviolet.

2. $n_f = 2$ in Eq. (27.8) specifies the Balmer series. This series includes the only visible lines in the hydrogen spectrum, and it was the first of the hydrogen series to be discovered. As we have seen, Balmer found the correct formula for the series wavelengths.

3. $n_f = 3$ specifies the Paschen series in the infrared. Its discovery also predated Bohr's work.

4. $n_f = 4$ and $n_f = 5$ specify the Brackett and Pfund series, respectively, both of which are in the far infrared. These two series were discovered in the decade following Bohr's work.

Figure 27.5 shows the first four hydrogen series on a quantum energy level diagram. The horizontal lines on the diagram represent energy levels, and the vertical distance between two horizontal lines is proportional to their energy difference. The lowest line represents the ground state, which is 13.6 eV below the energy zero; the second line represents the first excited state, which is 3.40 eV below the energy zero; and so on for the higher states. The vertical lines joining states represent electron jumps (with corresponding photon emission) as indicated.

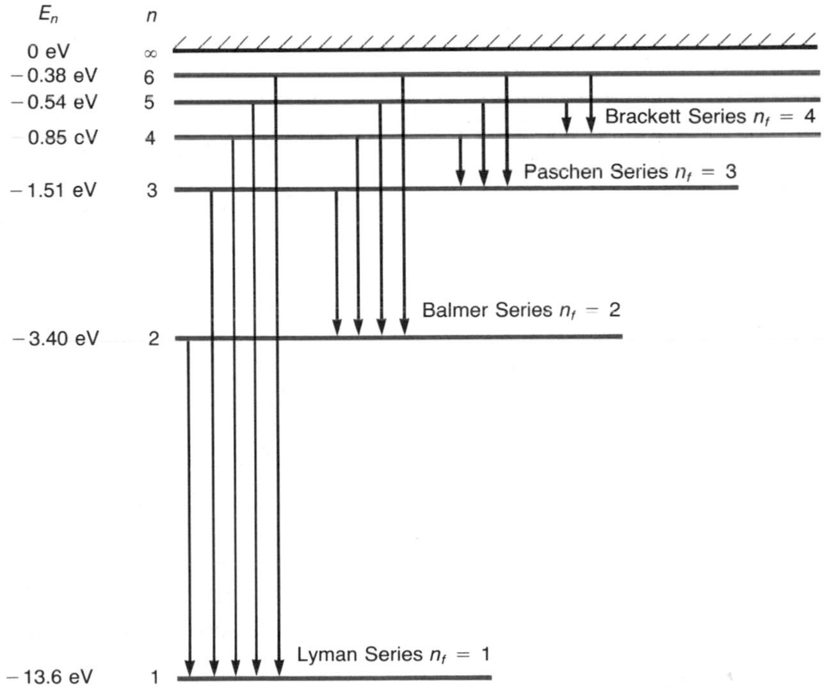

FIGURE 27.5 Energy level diagram (not to scale) illustrating the transitions that give rise to the spectral series in hydrogen. There are an infinite number of quantum energy levels (and corresponding spectral lines) between $n = 6$ at -0.38 eV and $n = \infty$ at 0 eV.

EXAMPLE 27.7

Calculate the wavelengths of the first four lines in the Balmer series. (These are the visible lines in the hydrogen spectrum that we have discussed in the chapters on optics.)

SOLUTION We have simply to use Rydberg's formula, Eq. (27.8), with $n_f = 2$. The Balmer series results from using values of $n_i > 2$. Calculating the wavelengths in order for $n_i = 3, 4, 5, 6$, we have:

$$\frac{1}{\lambda_{3\to2}} = 1.097 \times 10^7 \text{ m}^{-1} \times \left(\frac{1}{2^2} - \frac{1}{3^2}\right)$$

$$\therefore \lambda_{3\to2} = 656.3 \text{ nm (red)}$$

$$\frac{1}{\lambda_{4\to2}} = 1.097 \times 10^7 \text{ m}^{-1} \times \left(\frac{1}{2^2} - \frac{1}{4^2}\right)$$

$$\therefore \lambda_{4\to2} = 486.2 \text{ nm (blue)}$$

$$\frac{1}{\lambda_{5\to2}} = 1.097 \times 10^7 \text{ m}^{-1} \times \left(\frac{1}{2^2} - \frac{1}{5^2}\right)$$

$$\therefore \lambda_{5\to2} = 434.1 \text{ nm (violet 1)}$$

$$\frac{1}{\lambda_{6\to2}} = 1.097 \times 10^7 \text{ m}^{-1} \times \left(\frac{1}{2^2} - \frac{1}{6^2}\right)$$

$$\therefore \lambda_{6\to2} = 410.2 \text{ nm (violet 2)}$$

■

Finally, we should mention some of the reasons why we prefer the electron stationary wave model to the Bohr orbit model. First, Bohr was not able to give any convincing reason for the existence of a certain set of circular orbits and the absence of all others. He simply made this postulate and obtained very successful and correct formulas. On the other hand, the stationary wave model identifies the quantum states as vibrational modes of the atom, which belong to the atom in the same way that the fundamental mode and its harmonics belong to a guitar string. The fact that only a certain select group of quantum states appears presents no difficulty. Second, spectroscopic evidence shows that the electron in hydrogen has zero angular momentum in its ground state. However, since the angular momentum of a point particle cannot be zero in any orbit, the orbiting particle model cannot be correct. Other arguments also support the standing wave model as opposed to the orbiting particle model.

27.6 Many-Electron Atoms

As we have seen, the simplest atom is hydrogen, in which the standing matter wave of one electron surrounds one proton. An obvious way to classify other atoms is to arrange them in order of the number of electrons they contain. Since the central nucleus of a neutral atom contains the same number of protons as there are surrounding electrons, our arrangement is also in order of the number of protons in the nucleus.

Calculating the standing wave patterns and the corresponding quantum energies for atoms containing more than one electron is an extremely complex matter. We begin by imagining that each atom has available to it the same types of energy levels that the hydrogen atom has. By this we mean that we can assign values of the quantum numbers n, l, m_l, and m_s to each electron in the atom. However, we no longer have the simple situation in which the electron energy depends only on the principal quantum number n, as it does in hydrogen. The energy of an electron depends markedly on the angular momentum quantum number l and the quantum numbers of the other electrons in the atom. Perhaps you feel that this does not matter too much! After all, an atom in its ground state surely has all the electrons in the lowest energy level. But this is not true! The atom has all the electrons in the *lowest allowed energy configuration*, which is not the same thing as having all the electrons in the lowest energy level.

The difference between the two situations is made clear with the help of the **Pauli exclusion principle.** The principle is named for the Austrian physicist Wolfgang Pauli (1900–1958), who is famous for his contributions to many branches of modern physics. The principle limits the electron occupancy of quantum energy levels as follows:

■ Pauli Exclusion Principle

> No two electrons in the same atom can have the same set of quantum numbers.

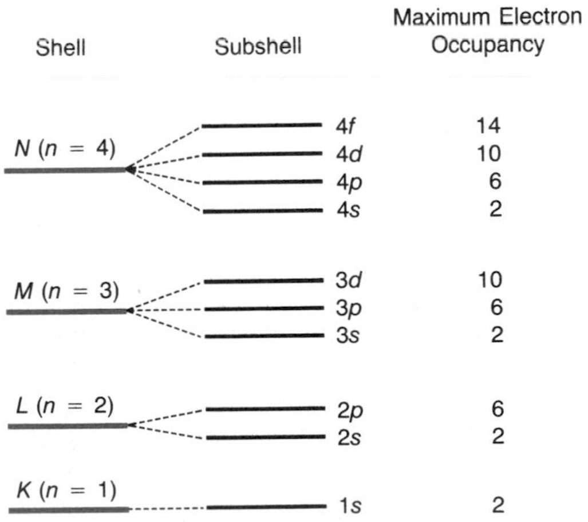

FIGURE 27.6 The shell and subshell scheme for assigning quantum numbers to electrons in an atom.

We will construct the electron configuration of many-electron atoms by first setting up the scheme of Fig. 27.6. The principal quantum number n is the chief determinant of the energy, and the values of the principal quantum number assign an electron to a certain **shell** in accordance with the following scheme:

Value of principal quantum number n	*Letter designation of shell*
1	*K*
2	*L*
3	*M*
4	*N*
5	*O*
6	*P*
7	*Q*

Within a given shell, the energy varies for different values of the angular momentum quantum number l, which determine **subshells** within the shells. For example, the six levels labeled 3p are the $l = 1$ subshell within the M shell ($n = 3$).

To write the electron configuration of an atom, we simply begin to fill the subshells, starting with the lowest energy and being careful to obey the Pauli exclusion principle. The electron configuration for hydrogen in its ground state is 1s. The quantum number set is:

$$n = 1, \quad l = 0, \quad m_l = 0, \quad \text{and} \quad m_s = \pm\tfrac{1}{2}$$

There are two possibilities for m_s, since there is only one electron in the level.

Helium has two electrons per atom, and we can fit the additional electron in the 1s state without violating the Pauli principle. The quantum numbers for the two electrons are:

$$n = 1, \quad l = 0, \quad m_l = 0, \quad \text{and} \quad m_s = +\tfrac{1}{2}$$
$$n = 1, \quad l = 0, \quad m_l = 0, \quad \text{and} \quad m_s = -\tfrac{1}{2}$$

The electron configuration for helium in its ground state is $1s^2$. The superscript indicates that two electrons have the quantum numbers $n = 1$ and $l = 0$.

Lithium has three electrons per atom. Since we cannot construct three different quantum number sets all having $n = 1$, we must move on to the next higher principal quantum number to write the quantum numbers for all three lithium electrons. They are:

$$n = 1, \quad l = 0, \quad m_l = 0, \quad \text{and} \quad m_s = +\tfrac{1}{2}$$
$$n = 1, \quad l = 0, \quad m_l = 0, \quad \text{and} \quad m_s = -\tfrac{1}{2}$$
$$n = 2, \quad l = 0, \quad m_l = 0, \quad \text{and} \quad m_s = \pm\tfrac{1}{2}$$

Thus, there are two possibilities for m_s for the third electron. The electron configuration for lithium in its ground state is $1s^2 2s$.

We can write the electron configuration of successive elements by assigning in turn the lowest available values for n and l that permit us to write different quantum number sets. By this we mean that no two complete sets can be the same. The numbers of electrons that have given values for the n and l quantum numbers are shown in Table 27.1 for the first 24 elements. We see from the table that our scheme breaks down after being successful for the first 18 elements. For the 19th element (potassium), we have an electron in the 4s subshell before any have been added to the 3d subshell. For this reason we must use data tables to write the electron configuration of elements beyond argon.

TABLE 27.1 Shell structure showing the electron configuration for the first 24 elements of the periodic table

Element	*Symbol*	*Atomic number*	*K shell* 1*s*	*L shell* 2*s*2*p*	*M shell* 3*s*3*p*3*d*	*N shell*[a] 4*s*
Hydrogen	H	1	1			
Helium	He	2	2			
Lithium	Li	3	2	1		
Beryllium	Be	4	2	2		
Boron	B	5	2	2 1		
Carbon	C	6	2	2 2		
Nitrogen	N	7	2	2 3		
Oxygen	O	8	2	2 4		
Fluorine	F	9	2	2 5		
Neon	Ne	10	2	2 6		
Sodium	Na	11	2	2 6	1	
Magnesium	Mg	12	2	2 6	2	
Aluminum	Al	13	2	2 6	2 1	
Silicon	Si	14	2	2 6	2 2	
Phosphorus	P	15	2	2 6	2 3	
Sulfur	S	16	2	2 6	2 4	
Chlorine	Cl	17	2	2 6	2 5	
Argon	Ar	18	2	2 6	2 6	
Potassium	K	19	2	2 6	2 6	1
Calcium	Ca	20	2	2 6	2 6	2
Scandium	Sc	21	2	2 6	2 6 1	2
Titanium	Ti	22	2	2 6	2 6 2	2
Vanadium	V	23	2	2 6	2 6 3	2
Chromium	Cr	24	2	2 6	2 6 5	1

[a]The *N* shell also has 4*p*, 4*d*, and 4*f* subshells, but they do not contain electrons for the atoms listed.

EXAMPLE 27.8

Use the data in Table 27.1 to write the electron configuration for titanium. Which shells and subshells are full and which are partly full?

SOLUTION Table 27.1 shows that the electron configuration of titanium is $1s^2 2s^2 2p^6 3s^2 3p^6 3d^2 4s^2$. The K and L shells are full, and the M and N shells are partly full. The 1s, 2s, 2p, 3s, 3p, and 4s subshells are full, while the 3d subshell is partly full. The values $n = 3$ and $l = 2$ give $m_l = 0, \pm 1, \pm 2$ and $m_s = \pm \frac{1}{2}$. There are five values of m_l, and two values of m_s for each value of m_l. This gives ten possible sets of different quantum numbers in the 3d subshell, and only two of the sets are occupied. ■

The electron configuration is highly significant in determining the chemical properties of atoms. The systematic display of the elements is called the **periodic table.** Figure 27.7(a–c) shows a

Group

Period	I	II	III	IV	V	VI	VII	VIII
1	1 H $1s$							2 He $1s^2$
2	3 Li $2s$	4 Be $2s^2$	5 B $2s^2\,2p$	6 C $2s^2\,2p^2$	7 N $2s^2\,2p^3$	8 O $2s^2\,2p^4$	9 F $2s^2\,2p^5$	10 Ne $2s^2\,2p^6$
3	11 Na $3s$	12 Mg $3s^2$	13 Al $3s^2\,3p$	14 Si $3s^2\,3p^2$	15 P $3s^2\,3p^3$	16 S $3s^2\,3p^4$	17 Cl $3s^2\,3p^5$	18 Ar $3s^2\,3p^6$
4	19 K $4s$	20 Ca $4s^2$	31 Ga $4s^2\,4p$	32 Ge $4s^2\,4p^2$	33 As $4s^2\,4p^3$	34 Se $4s^2\,4p^4$	35 Br $4s^2\,4p^5$	36 Kr $4s^2\,4p^6$
5	37 Rb $5s$	38 Sr $5s^2$	49 In $5s^2\,5p$	50 Sn $5s^2\,5p^2$	51 Sb $5s^2\,5p^3$	52 Te $5s^2\,5p^4$	53 I $5s^2\,5p^5$	54 Xe $5s^2\,5p^6$
6	55 Cs $6s$	56 Ba $6s^2$	81 Tl $6s^2\,6p$	82 Pb $6s^2\,6p^2$	83 Bi $6s^2\,6p^3$	84 Po $6s^2\,6p^4$	85 At $6s^2\,6p^5$	86 Rn $6s^2\,6p^6$
7	87 Fr $7s$	88 Ra $7s^2$						

Configurations show outer electrons only.

(a)

21 Sc $3d\,4s^2$	22 Ti $3d^2\,4s^2$	23 V $3d^3\,4s^2$	24 Cr $3d^5\,4s$	25 Mn $3d^5\,4s^2$	26 Fe $3d^6\,4s^2$	27 Co $3d^7\,4s^2$	28 Ni $3d^8\,4s^2$	29 Cu $3d^{10}\,4s$	30 Zn $3d^{10}\,4s^2$
39 Y $4d\,5s^2$	40 Zr $4d^2\,5s^2$	41 Nb $4d^4\,5s$	42 Mo $4d^5\,5s$	43 Tc $4d^6\,5s$	44 Ru $4d^7\,5s$	45 Rh $4d^8\,5s$	46 Pd $4d^{10}$	47 Ag $4d^{10}\,5s$	48 Cd $4d^{10}\,5s^2$
57 La $5d\,6s^2$	72 Hf $5d^2\,6s^2$	73 Ta $5d^3\,6s^2$	74 W $5d^4\,6s^2$	75 Re $5d^5\,6s^2$	76 Os $5d^6\,6s^2$	77 Ir $5d^7\,6s^2$	78 Pt $5d^9\,6s$	79 Au $5d^{10}\,6s$	80 Hg $5d^{10}\,6s^2$
89 Ac $6d\,7s^2$	104 Ku $6d^2\,7s^2$	105 Ha $6d^3\,7s^2$							

Configurations show outer electrons only.

(b)

58 Ce $4f\,5d\,6s^2$	59 Pr $4f^3\,6s^2$	60 Nd $4f^4\,6s^2$	61 Pm $4f^5\,6s^2$	62 Sm $4f^6\,6s^2$	63 Eu $4f^7\,6s^2$	64 Gd $4f^7\,5d\,6s^2$
90 Th $6d^2\,7s^2$	91 Pa $f\,6d^2\,7s^2$	92 U $5f^3\,6d\,7s^2$	93 Np $5f^4\,6d\,7s^2$	94 Pu $5f^6\,7s^2$	95 Am $5f^7\,7s^2$	96 Cm $5f^7\,6d\,7s^2$

65 Tb $4f^9\,6s^2$	66 Dy $4f^{10}\,6s^2$	67 Ho $4f^{11}\,6s^2$	68 Er $4f^{12}\,6s^2$	69 Tm $4f^{13}\,6s^2$	70 Yb $4f^{14}\,6s^2$	71 Lu $4f^{14}\,5d\,6s^2$
97 Bk $5f^8\,6d\,7s^2$	98 Cf $5f^{10}\,7s^2$	99 Es $5f^{11}\,7s^2$	100 Fm $5f^{12}\,7s^2$	101 Md $5f^{13}\,7s^2$	102 No $5f^{14}\,7s^2$	103 Lw $5f^{14}\,6d\,7s^2$

(c)

Configurations show outer electrons only.

FIGURE 27.7(a) The eight principal groups of the periodic table are determined by similar electron configurations in the *s* and *p* subshells.

FIGURE 27.7(b) The transition elements occur in the fourth and subsequent periods after the filling of the outermost *s* subshell. They show the gradual filling of the *d* subshell that is one shell lower than the just completed *s* subshell.

FIGURE 27.7(c) The lanthanide and the actinide series occur in the sixth and seventh periods immediately after the first transition element of that period. They show the gradual filling of the *f* subshell that is one shell lower than the *d* shell being filled by transition elements.

simplified version of the periodic table. The vertical columns of elements have very similar chemical properties. All the elements of the first column (the alkalis) have one electron in an *s* subshell; all the final column elements (the inert gases) have completely filled *p* subshells. The horizontal rows of the table are called periods. The subshells of the first three periods (up to the 18th element, argon) are filled in the order shown in Fig. 27.6. In the fourth, fifth, and sixth periods we have ten *transition elements* that fit between the second and third columns of the main table. Within each period, this corresponds to the filling of a *d* subshell with ten electrons. After lanthanum (element 57) and actinium (element 89), 14 elements correspond to the filling of an *f* subshell with 14 electrons.

27.7 X-Ray Spectra

An electrical discharge through a low-pressure gas makes relatively small energies available for excitation. The outermost atomic electron is excited, and the resulting spectrum is mainly in the visible and infrared regions. A mode of excitation that can eject electrons from inner levels results in X-ray emission. Let us first briefly describe the usual method of exciting inner shell electrons and then interpret the situation in terms of quantum energy levels.

X rays were discovered late in the nineteenth century by the German physicist Wilhelm Roentgen (1845–1923). Roentgen was studying low-pressure discharges in a gas and noticed that the anode of his discharge tube was the source of energetic and penetrating radiation. (The penetrating properties of X rays are of great practical importance and have been utilized by the medical profession almost since the year of their discovery.) X rays are produced when high-velocity electrons strike a solid target. A typical X-ray tube is illustrated in Fig. 27.8. A beam of electrons

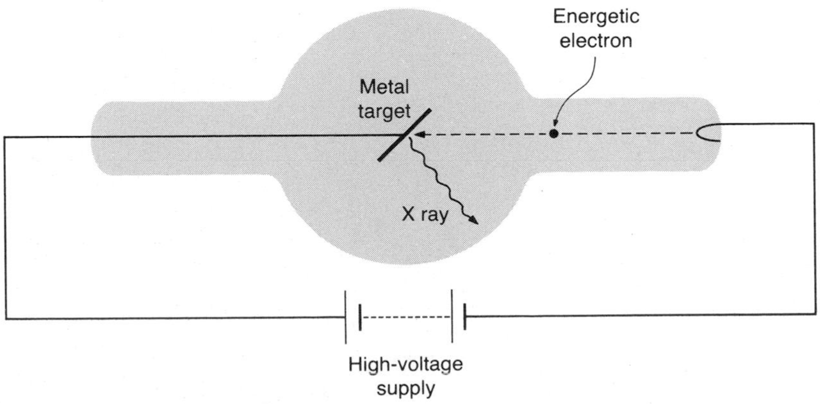

FIGURE 27.8 An X-ray tube.

emitted from a hot cathode in a vacuum is accelerated by a high voltage and strikes a metal target anode. The impact energy of the beam of electrons ejects some electrons from the inner shells of the atoms in the metal target, enabling electrons from the outer shells to make transitions to the available vacancies. The energy lost by the electrons that make the transitions is given up as X rays, whose emission accompanies the transitions.

The notation for naming X rays is shown in Fig. 27.9. Each series of X-ray lines is named for the shell in which the excitation process causes the vacancy. The *K* series results from an electron in an upper shell making a downward transition to a *K* shell vacancy, the *L* series from a downward transition into an *L* shell vacancy, and so on. A Greek letter subscript (α, β, γ, and so on) indicates the position of the line in the series. The K_α line is the first line in the *K* series, which results from an *L* shell electron making a transition to a vacancy in the *K* shell; the K_β line results from an *M* shell electron making a transition to a *K* shell vacancy; and so on.

Extensive research on X rays by the English physicist Henry Moseley (1887–1915) showed that the wavelengths of K_α and L_α lines could be calculated from empirical formulas similar to Rydberg's formula except for the addition of the factor $(Z - 1)^2$ in the K_α wavelength formula and the factor $(Z - 7.4)^2$ in the L_α wavelength formula. **Moseley's formulas** are as follows:

■ Moseley's Formulas for the Wavelengths of K_α and L_α X-Ray Lines

$$\frac{1}{\lambda_{K\alpha}} = R(Z - 1)^2\left(\frac{1}{1^2} - \frac{1}{2^2}\right) \quad \text{(a)}$$

$$\frac{1}{\lambda_{L\alpha}} = R(Z - 7.4)^2\left(\frac{1}{2^2} - \frac{1}{3^2}\right) \quad \text{(b)} \qquad \textbf{(27.9)}$$

R = Rydberg constant
Z = atomic number

Before Moseley's work the position of some elements in the periodic table was uncertain. Some felt that the atomic weight determined the order of the atomic elements, but Moseley correctly identified the atomic number Z in his formulas as a measure of the charge on the atomic nucleus. Thus, measurement of the wavelengths of K_α X-ray lines gives a straightforward method for ordering the elements correctly.

EXAMPLE 27.9

Calculate the wavelengths of the K_α and L_α X-ray lines of silver. The atomic number of silver is 47.

SOLUTION To calculate the X-ray wavelengths, we need only to apply Moseley's formulas. From Eq. (27.9a):

$$\frac{1}{\lambda_{K\alpha}} = R(Z - 1)^2\left(\frac{1}{1^2} - \frac{1}{2^2}\right)$$

$$= 1.097 \times 10^7 \text{ m}^{-1} \times (47 - 1)^2 \times (1 - \tfrac{1}{4})$$

$$\therefore \lambda_{K\alpha} = 0.0574 \text{ nm}$$

From Eq. (27.9b) we have:

$$\frac{1}{\lambda_{L\alpha}} = R(Z - 7.4)^2\left(\frac{1}{2^2} - \frac{1}{3^2}\right)$$

$$= 1.097 \times 10^7 \text{ m}^{-1} \times (47 - 7.4)^2 \times (\tfrac{1}{4} - \tfrac{1}{9})$$

$$\therefore \lambda_{L\alpha} = 0.418 \text{ nm}$$ ■

When high-energy electrons strike a target, the emitted X rays are not confined to the characteristic lines. A typical measurement of X-ray intensity as a function of wavelength is shown in Fig. 27.10(a). The total X-ray emission is composed of two separate parts. The wavelengths of the characteristic X-ray lines that we have just discussed depend on the material of the target. The continuous part of the X-ray radiation, known as **bremsstrahlung** (German for "braking radiation"), does not depend on the target material, but only on the magnitude of the high voltage applied to the tube. As the name implies, bremsstrahlung has its origin in the braking effect of the target on the electrons in the incident

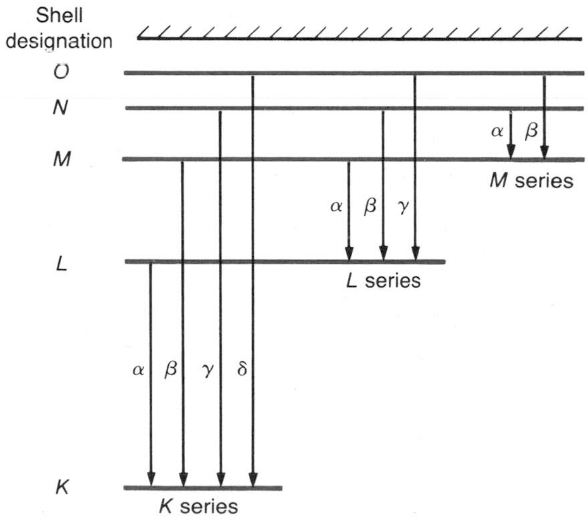

FIGURE 27.9 Energy level diagram illustrating the notation for naming X rays.

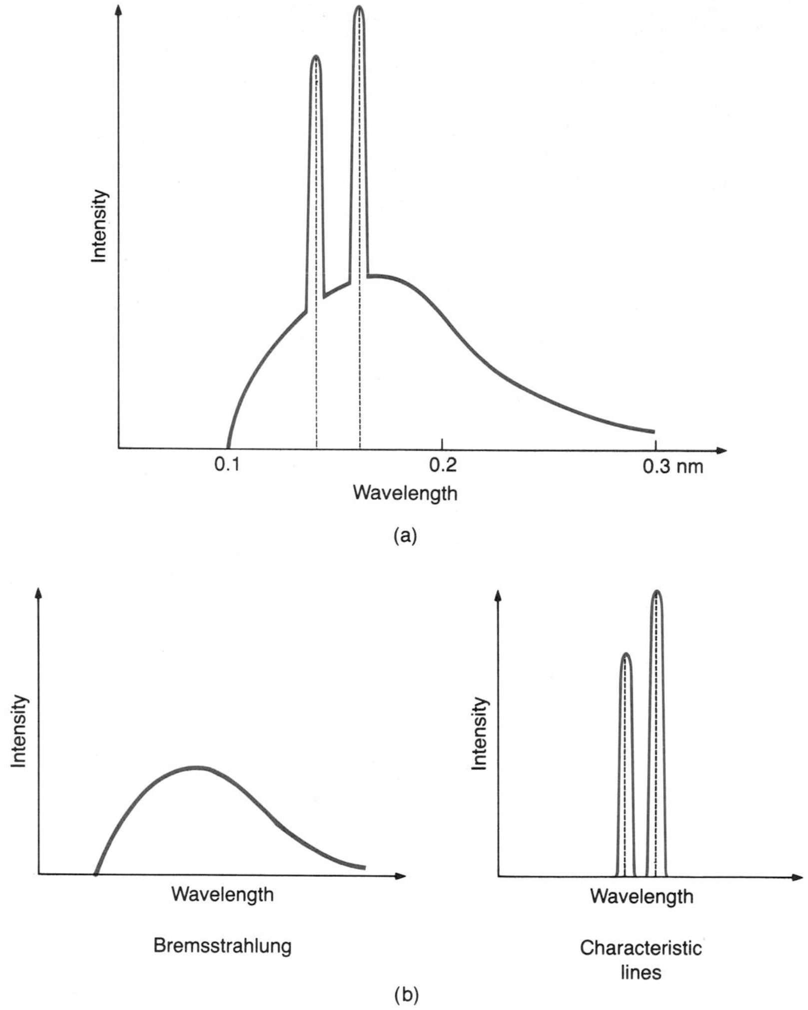

FIGURE 27.10 (a) Measured intensity of X-ray emission from a copper target as a function of wavelength; (b) the total spectrum is composed of characteristic X-ray lines and bremsstrahlung.

beam. The target slows these electrons, and some of the energy lost is given off directly as X rays. The most energetic X-ray photon that is possible has a quantum energy equal to the kinetic energy of an electron in the beam. The kinetic energy in turn arises from the transport of an electron charge through the high-voltage potential difference. We therefore have:

$$\text{Maximum photon energy} = h\nu_{\text{max}} = \frac{hc}{\lambda_{\text{min}}}$$

$$\text{Kinetic energy of electron} = \tfrac{1}{2}mv^2 = eV$$

Combining these results, we can write a formula for the wavelength of the most energetic photon.

■ Formula for Minimum Bremsstrahlung Wavelength

$$\lambda_{min} = \frac{hc}{eV} \qquad (27.10)$$

h = Planck's constant
c = velocity of light
e = electron charge
V = voltage across the tube

EXAMPLE 27.10

Calculate the high voltage across the X-ray tube that produced the emission spectrum of Fig. 27.10(a).

SOLUTION The minimum bremsstrahlung wavelength in Fig. 27.10(a) occurs at 0.1 nm. We need only use Eq. (27.10), setting λ_{min} at 0.1 nm $= 10^{-10}$ m, and using the value $hc = 1.988 \times 10^{-25}$ J · m from Eq. (26.2b):

$$\lambda_{min} = \frac{hc}{eV}$$

$$\therefore V = \frac{hc}{e\lambda_{min}}$$

$$= \frac{1.988 \times 10^{-25}\ \text{J} \cdot \text{m}}{1.6 \times 10^{-19}\ \text{C} \times 10^{-10}\ \text{m}}$$

$$= 12.4\ \text{kV}$$

■

We can now explain the reason for the strange shape of the X-ray emission spectrum of Fig. 27.10(a). Some electrons in the incident beam give up their energy for excitation of the atoms of the target, and the excited target atoms then emit characteristic X-ray lines. Most of the other incident electrons, which are stopped or slowed down when they hit the target, give only part of their kinetic energy to X rays corresponding to the hump in the bremsstrahlung spectrum. The balance of their energy is converted to heat in the anode target. Occasionally a few electrons give all their kinetic energy to X rays of the minimum wavelength. At the other end of the spectrum, some electrons give most of their energy to heating the target, while a small fraction of their energy goes to X rays.

A specially constructed tube is not the only way to produce X rays, which can occur as an unintentional by-product in any component that produces high-velocity electrons. For instance, the picture in a TV tube is formed on the screen by the impact of energetic electrons. Since the voltage (and consequently the electron kinetic energy) is very high in color TV, unwanted X rays may be emitted from the material of the screen. For this reason sufficient absorbing material should always be placed between the screen and viewers.

27.8 The Laser

The word *laser* is an acronym for "light amplification by stimulated emission of radiation." The output of a laser is a monochromatic, coherent light beam. We recall that a monochromatic beam is one composed of a single wavelength, and that a coherent beam is one whose component photons are in phase with each other.

To understand laser action, we should first examine the distinction between spontaneous and stimulated emission of radiation. If we pass an electrical discharge through a gas, radiation from the heated discharge path is usually quite visible. This occurs because electrons in some of the atoms become excited into high energy states and emit photons when they fall back to lower energy states. This is spontaneous emission, illustrated on the energy level diagram of Fig. 27.11(a). Stimulated emission also involves a transition between energy levels, but it is triggered by the presence of another photon whose energy is equal to that of the emitted photon. When this occurs, the stimulating and the emitted photons are in phase. It follows that we can produce an intense coherent beam of light if we can build a device that emits stimulated rather than spontaneous photons.

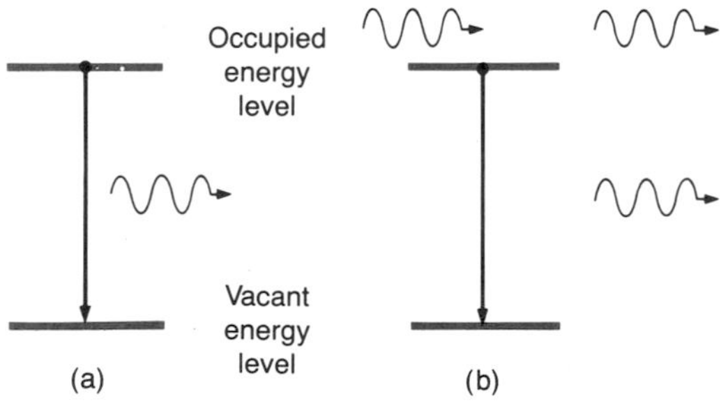

FIGURE 27.11 The distinction between spontaneous and stimulated emission. (a) An excited atom spontaneously emits a photon as the electron makes a transition from the higher to the lower energy state. (b) The same transition is stimulated by the presence of another photon whose energy is equal to the energy of the emitted photon; the two photons are in phase.

However, not all pairs of atomic energy levels are suitable for laser action. The upper energy level must be relatively stable against spontaneous emission, so that the majority of the transitions occur under stimulation. Moreover, we must produce a situation called *population inversion* before appreciable stimulated emission can occur. To understand why, we must realize that electron transitions between the energy levels can be a two-way street. We have just described stimulated emission in which an electron in the upper level makes a transition to the vacant lower level, emitting a photon in the process. But what happens if the lower level contains the electron and the upper one is vacant? In that case a photon of the appropriate energy can stimulate the electron into the upper level, transferring to the electron all of its energy. The photon itself disappears from the scene, and we refer to this type of stimulated process as *absorption*.

In any given situation involving photons and energy levels the outcome is the result of the two competing processes of stimulated emission and absorption. If we heat a gas by an electrical discharge, the relative electron populations of the two levels de-

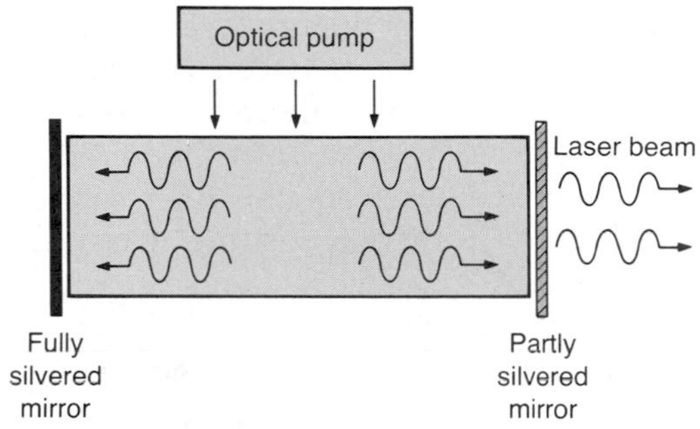

FIGURE 27.12 A schematic diagram of a laser. An optical pumping device produces a population inversion in the energy levels of the atoms responsible for laser action. The stimulated photons travel back and forth between parallel mirrors. A fraction of these photons escape through a partly silvered mirror at one end of the tube and form the laser beam.

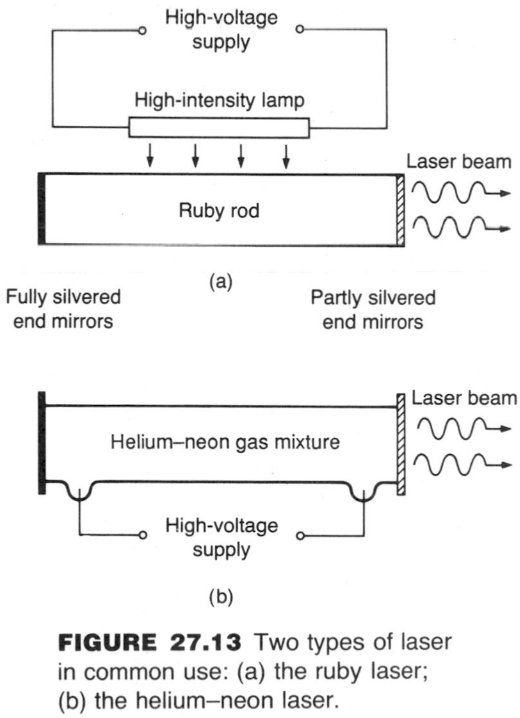

FIGURE 27.13 Two types of laser in common use: (a) the ruby laser; (b) the helium–neon laser.

pend on thermal agitation, and considerably fewer electrons are excited to the upper energy level than to the lower one. This means that there are many more candidates for absorption (in the lower energy level) than for stimulated emission (in the upper energy level). In this situation, absorption is the dominant process, and we cannot have laser action. The preponderance of the electron population in the lower level is characteristic of all systems in thermal equilibrium at some temperature. What we need is some type of pump to concentrate the electrons in the upper level by disturbing the thermal equilibrium. This concentration in the upper level is the population inversion necessary for laser action.

Our final requirement is some way of organizing the photons so that they continually pass the excited atoms in the direction required for the eventual laser beam. The usual method for accomplishing this is to have the laser action occur between a pair of parallel mirrors, as shown in Fig. 27.12. We refer to the region between the mirrors as the laser cavity, and the atoms that undergo stimulated emission are located here. Photons directed perpendicular to the mirrors travel back and forth, stimulating more of their kind until an intense stream of photons is in the cavity. The mirror at one end is only partly silvered, which permits some of the photons to escape from the cavity and form the usable laser beam. Photons that are emitted in any other direction simply escape from the sides of the cavity without undergoing the intensification that results for photons moving perpendicular to the mirrors.

Our schematic diagram shows the optical pump as an external device feeding energy into the laser cavity. Some designs use this type of arrangement, while in others the optical pumping occurs entirely within the cavity. Figure 27.13 illustrates two types of lasers that are in common use. The ruby laser uses a ruby rod with its ends cut flat perpendicularly to the axis of the rod and then polished. One end is fully silvered and the other partly silvered so that the rod itself forms the laser cavity. Ruby is essentially a transparent crystal of aluminum oxide (Al_2O_3) containing about 0.05% of chromium impurity. The chromium ions, which are responsible for the red color, provide the two energy levels that are involved in the stimulated emission. The optical pump is frequently a high-intensity lamp surrounding the ruby rod, and the power required to pump electrons into the upper energy level comes from the high-voltage supply to the lamp. The ruby laser is capable of high-power operation, and the wavelength of the beam is 694.3 nm.

The cavity for the helium–neon gas laser is a glass tube whose end plates are fully and partly silvered mirrors. The tube has two electrodes through which a high-voltage supply initiates an electrical discharge in the low-pressure helium–neon mixture within it. The energy levels involved in the laser action are those of the neon atoms, and the pumping to the upper energy level is caused by the relatively complex collision process between the excited helium and neon atoms. The optical pump is therefore provided by interatomic collisions within the cavity. The power required to produce the population inversion comes from the high-voltage supply, which causes the discharge in the gas mixture. Laser action in the helium–neon tube occurs for a visible wavelength of 632.8 nm as well as for several wavelengths in the infrared region of the spectrum. This type is very frequently used when a rugged, inexpensive, portable instrument is required that produces a relatively low-power laser beam.

Laser technology is a rapidly developing field. Many other types of lasers of various frequencies and power capabilities are suitable for special purposes. However, they all operate on the basic principles outlined here.

KEY CONCEPTS

From a classical viewpoint, a point mass particle has no spatial dimensions. From a quantum viewpoint, the particle properties are governed by an associated **matter wave.**

The **wavelength of the matter wave** is equal to Planck's constant divided by the particle momentum; see Eq. (27.1).

The **Heisenberg uncertainty principle** limits the product of the uncertainties in the particle momentum component and the corresponding position component to be at least as great as Planck's constant divided by 4π; see Eq. (27.3).

Representation of a particle by a matter wave requires **energy quantization** for the confined particle.

One quantum number describes the matter wave of a particle confined within a one-dimensional square well. The quantum number also appears in a formula for the quantum energies of the particle; see Eq. (27.2b).

Three quantum numbers describe the matter wave of a particle confined within a three-dimensional Coulomb field. The **principal quantum number** determines the energy. The **angular momentum** and **angular momentum component quantum numbers** determine those two quantities; see Eqs. (27.4) and (27.5).

In addition, the electron possesses **intrinsic angular momentum quantum numbers,** called spin quantum numbers. The **spin quantum number** of an electron is fixed at $\frac{1}{2}$, but the **spin component quantum number** can be $+\frac{1}{2}$ or $-\frac{1}{2}$; see Eq. (27.6).

When an atomic electron makes a transition from a higher to a lower energy level, it emits a photon whose energy is equal to the energy difference between the levels; see Eq. (27.7).

Rydberg's formula determines the wavelengths of the lines in the spectrum of atomic hydrogen; see Eq. (27.8).

The **Pauli exclusion principle** prohibits the occupancy of an electron energy state by two electrons having identical quantum number sets.

The **periodic table** of chemical elements arises from filling successive energy shells and subshells in accordance with the Pauli principle. An energy shell is determined by the principal quantum number, and a subshell by the angular momentum quantum number; see Fig. 27.6.

Moseley's formula determines the wavelengths of the K and L series X rays of the chemical elements; see Eq. (27.9).

The shortest **bremsstrahlung** wavelength is determined by equating the photon energy to the kinetic energy of the electron that causes it; see Eq. (27.10).

QUESTIONS FOR THOUGHT

1. The ground state energy for an electron in a hydrogen atom is negative. Is this energy potential or kinetic or a combination of both? What is the arbitrarily selected zero of energy?

2. Why do you think that the Balmer series in hydrogen was the first to be discovered? Surely the Lyman series with transitions into the ground state should have been found first.

3. Some of the spectral series in atomic hydrogen have their shortest wavelength line in the ultraviolet, and others have it in the infrared. Can all of these series have lines in the visible region of the spectrum, or does the shortest wavelength line of the series cause a restriction? Do any other factors exclude the possibility of visible lines from a given spectral series?

4. How is a photon of a given wavelength different from an electron whose matter wave has the same wavelength?

5. What type of experiment would you devise to investigate the possible wavelike nature of a stream of particles?

6. How would you ascertain the possible particlelike properties of a stream of radiation known to have wavelike properties?

7. The atoms of a crystalline substance vibrate about their mean positions in a random manner due to thermal agitation. It therefore seems reasonable to suppose that the atomic motion would cease at the absolute zero of temperature. Comment on this supposition in the light of the uncertainty principle.

PROBLEMS

A. Single-Substitution Problems

1. Calculate the wavelength of the matter wave associated with an electron moving at 3×10^6 m/s. [27.1]

2. The wavelength of the matter wave associated with a neutron is 2.2 nm. Calculate the velocity of the neutron. [27.1]

3. Calculate the fractional change in the matter wavelength of a particle when the kinetic energy is tripled. (Assume that nonrelativistic dynamics is sufficient.) [27.1]

4. Calculate the ground state energy for a proton (mass 1.67×10^{-27} kg) that is confined within a one-dimensional region 1.8×10^{-15} m wide. [27.2]

5. Within what width in one dimension should an electron (mass 9.11×10^{-31} kg) be confined so that its ground state energy will be 4 eV? [27.2]

6. Calculate the uncertainty in the velocity of an electron if the uncertainty in its position is 3×10^{-10} m. [27.3]

7. The uncertainty in a proton's velocity is 40 km/s. Calculate the uncertainty in its position. [27.3]

8. Calculate the energy of an electron in the quantum states $n = 4$ and $n = 5$ in the hydrogen atom. [27.4]

9. A certain atom in an excited state of energy -3.7 eV makes a transition to the ground state of energy -5.6 eV. Calculate the wavelength of the emitted radiation. [27.5]

10. Calculate the wavelengths of the first four lines of the Lyman series in atomic hydrogen. To what region of the spectrum do these lines belong? [27.5]

11. Calculate the wavelengths of the first four lines of the Paschen series in atomic hydrogen. To what region of the spectrum do these lines belong? [27.5]

12. List the quantum numbers for each electron in a neon atom ($Z = 10$) in its ground state. [27.6]

13. Write the electronic configuration for a chlorine atom ($Z = 17$) in its ground state. [27.6]

14. Calculate the wavelengths of the K_α and L_α characteristic X-ray lines in cobalt ($Z = 27$). [27.7]

15. Calculate the atomic number of the element whose K_α line has a wavelength of 0.2756 nm. [27.7]

16. What accelerating voltage should be used in a TV set to guarantee that no X ray of wavelength shorter than 0.1 nm be generated? [27.7]

17. What is the wavelength of the most energetic X ray emitted from the screen of a TV set if the accelerating potential is 17.5 kV? [27.7]

B. Standard-Level Problems

18. A proton starts from rest and is accelerated through a potential difference until the wavelength of its associated matter wave is 0.01 nm. Calculate the value of the potential difference. [27.1]

19. Calculate the wavelength of the matter wave associated with an electron that has fallen through a potential difference of 50 V starting from rest. [27.1]

20. A typical atom has a diameter of about 0.3 nm. Calculate the kinetic energy of an electron that has an associated matter wavelength of this magnitude.

21. A typical nucleus has a diameter of about 3×10^{-15} m. Calculate the kinetic energy of a proton that has an associated matter wavelength of this magnitude.

22. In helium gas at a certain temperature, the de Broglie wavelength for an atom that has the mean kinetic energy is 6×10^{-11} m. Calculate the gas temperature. (The mass of the helium atom is 6.68×10^{-27} kg.)

23. Calculate the de Broglie wavelength of a hydrogen molecule that has the mean kinetic energy corresponding to a gas temperature of 50°C. (The mass of the hydrogen molecule is 3.34×10^{-27} kg.)

24. When a radioactive nucleus emits an α particle (mass $= 6.68 \times 10^{-27}$ kg), the kinetic energy of the α particle is about 6 MeV. Calculate the associated matter wavelength for the α particle.

25. Calculate the kinetic energy of an α particle (mass $= 6.68 \times 10^{-27}$ kg) whose associated matter wavelength is 3×10^{-15} m (approximately equal to a nuclear diameter).

26. Calculate the ground state energies in electron volts for the following two systems:

a. A mass of 10 g confined within a one-dimensional region 1 cm wide
b. A neutron (mass 1.67×10^{-27} kg) confined within a one-dimensional region 3×10^{-15} m wide [27.2]

27. An oxygen atom (mass $= 2.66 \times 10^{-26}$ kg) is confined in a one-dimensional region 1 cm wide. Calculate its ground state energy as a fraction of its mean kinetic energy ($\frac{1}{2}kT$; it is one-dimensional) at room temperature of 300 K.

28. Calculate the quantum number for a nitrogen atom (mass $= 2.33 \times 10^{-26}$ kg) that has an energy of 0.1 eV when confined to a one-dimensional region 1 cm wide.

29. Two hydrogen atoms in the ground state traveling with equal speed in opposite directions collide, and both come to rest; all of the kinetic energy loss goes into producing the excited state $n = 2$ in one of the atoms. Calculate the temperature, if the kinetic energy of each atom was $\frac{3}{2}kT$. [27.4]

30. How many different photons can be emitted by a hydrogen atom in undergoing single or successive transitions from the state $n = 5$ to the ground state $n = 1$? How many of these photons are in the infrared? The visible? The ultraviolet? [27.5]

31. Find the quantum numbers for the initial and final states that give rise to an ultraviolet line of wavelength 95.06 nm in the spectrum of atomic hydrogen. [27.5]

32. Calculate the ratios of the wavelengths of the red and blue lines in the Balmer series in hydrogen as whole integers.

33. Which lines in the Lyman and Balmer series of atomic hydrogen possess sufficient energy to eject an electron from a potassium photoelectrode? (The work function of potassium is 2.26 eV.)

34. The wavelength of the light from a helium–neon laser is 632.8 nm. Calculate the momentum of one photon of this light and also the energy difference in electron volts between the two states that provided it.

35. a. To how many electrons does the Pauli exclusion principle limit the occupancy of the $n = 4$ atomic shell?
b. How many of these electrons are in the $n = 4$, $l = 3$ subshell? [27.6]

36. How many electrons can occupy a g subshell?

37. Which element has the electron configuration $1s^2 2s^2 2p^6 3s^2 3p^6 3d^6 4s^2$? [27.6]

38. Which of the inert gases completes a p subshell in the O shell?

39. The wavelength of the K_α X ray from molybdenum (atomic number 42) is 0.0711 nm. Use only Moseley's formulas to predict the wavelengths of the K_α and L_α X rays from manganese (atomic number 25). [27.7]

40. Calculate the wavelengths and photon energies of the K_α and L_α characteristic X rays from silver (atomic number 47).

41. An unknown substance emits L_α X rays with a wavelength of 0.0964 nm. Use Moseley's formulas to determine the atomic number of the unknown substance.

28

RADIOACTIVITY

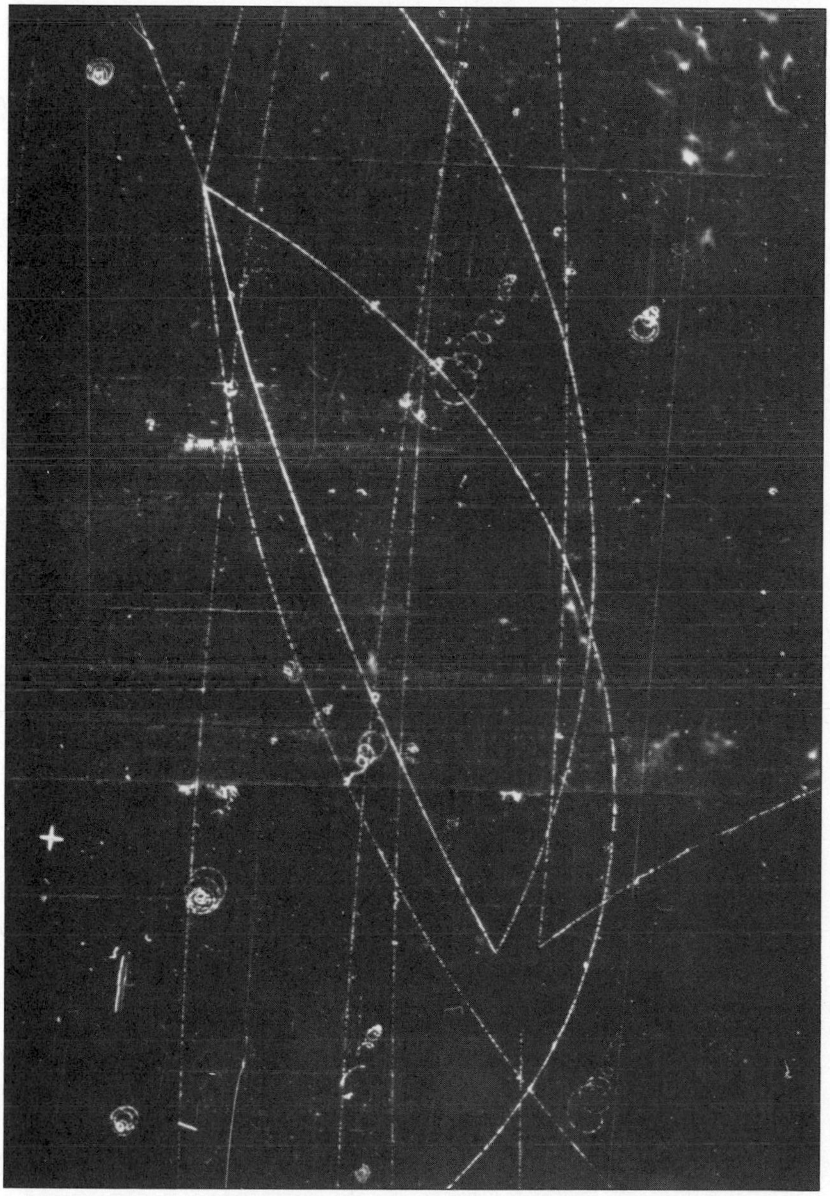

We have seen that many discoveries in physics were made in the years around 1900, but none captured the public imagination more than the discovery of invisible rays. In a previous chapter we mentioned X rays, and their ability to penetrate matter is well known. However, other forms of invisible radiation with varying powers of penetration were also discovered during the same period. These were named α, β, and γ rays, and were subsequently found to be associated with quantum energy jumps within the atomic nucleus.

In this chapter we will briefly review the discovery of radioactivity and the changes that radioactive nuclei undergo.

28.1 Natural Radioactivity

Most of the naturally occurring nuclei are stable, but some of them undergo spontaneous disintegration—a phenomenon known as **radioactivity,** which was discovered by the French scientist Henri Becquerel (1852–1909) at the end of the nineteenth century. During a research project Becquerel developed a photographic plate that had been stored in a drawer in his workbench and found the image of a key superimposed on the picture. He recognized the key as one that had been lying on top of the box that contained the photographic plate. Reflecting on this strange occurrence, he remembered that some uranium ore lay in boxes on the bench top and speculated whether the uranium ore was emitting penetrating radiation. Many scientists worked on this strange phenomenon, and the materials capable of emitting the mysterious radiation were termed **radioactive.**

The next major advance was brought about by the husband–wife team of Pierre and Marie Curie, who isolated two radioactive substances. Beginning with several tons of uranium pitchblende ore, they painstakingly extracted a few milligrams of radium and polonium. In 1903 Becquerel and the Curies shared the Nobel Prize in physics for their pioneering work in radioactivity. Soon after, they and other workers isolated many more radioactive substances.

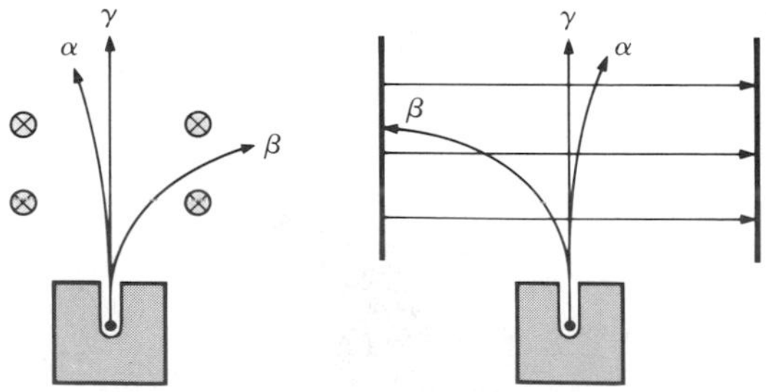

(a) (b)

FIGURE 28.1 The deflection of α, β, and γ rays by electric and magnetic fields: (a) a magnetic field pointing into the paper; (b) an electric field pointing from left to right.

The nature of the rays coming from radioactive substances was determined by the British physicist Lord Rutherford (1871–1937) in an extensive series of experiments. He and his collaborators found three quite distinct types of radiation, which they named α, β, and γ rays. Some unstable nuclei emit α rays, some emit β rays, and γ rays usually accompany both types of emission.

To illustrate the different properties of the rays in electric and magnetic fields, let us consider the simplified experiment of Fig. 28.1. A mixture of radioactive nuclei emit all three types of radia-

SPECIAL TOPIC 29 MARIE CURIE

Marie Sklodowska was born in Warsaw in 1867, at a time when Poland was under foreign occupation. Sklodowska worked hard to save money to attend the Sorbonne in Paris, where she was admitted in 1891. She graduated at the top of her class.

Sklodowska married the French chemist Pierre Curie in 1895, and for the next decade they worked tirelessly with radioactive materials. They isolated and named polonium and radium, and in 1903 they shared the Nobel Prize in physics with Becquerel. Pierre was killed in a traffic accident in 1906, but Marie continued her work and received the Nobel Prize in chemistry in 1911. She is the only person ever to win two Nobel Prizes in science. She died in France in 1934 of leukemia caused by overexposure to radiation.

(Courtesy AIP Niels Bohr Laboratory)

tion from a long hole drilled in a lead block. As the rays emerge from the hole, we test their behavior in both magnetic and electric fields. In both experiments the respective deflections indicate that the α rays carry a positive charge, and the β rays a negative charge; the γ rays are undeflected by either field. Other experiments have shown that the α particles are the nuclei of helium atoms and that the β particles are fast-moving electrons. The γ rays are electromagnetic radiation similar to X rays but usually of somewhat shorter wavelength.

The nature of the atomic nucleus was not well understood when these discoveries were made. It is easier for us to interpret the α, β, and γ radiations if we begin with the modern view of the composition of the nucleus. In discussing atomic physics we have already mentioned the *proton,* a positively charged nuclear particle that is almost 2000 times as massive as the electron. The nucleus of a neutral atom contains the same number of protons as there are atomic electrons surrounding it. In addition, every nucleus except hydrogen contains a number of *neutrons.* The neutron is uncharged but has a mass approximately equal to the pro-

SPECIAL TOPIC 30 RUTHERFORD

Ernest Rutherford was born in New Zealand in 1871 to a Scots farming family. He won scholarships to the local university and then to Cambridge in 1895, where he worked at the Cavendish laboratory. Apart from short overseas trips, Rutherford did not leave England again.

He began by identifying (and naming) the three types of radiation from the newly discovered radioactive substances, and by discovering the proton. However, his greatest work was in showing that most of the mass of an atom was in a tiny, dense nucleus. This is the nuclear atomic model that is universally accepted today; it replaced a model that considered the charge and the mass to be more or less uniformly distributed. For this work he received a Nobel Prize in 1908.

Rutherford's further work included the development of particle counters and the first nuclear reaction. In 1931, he was made Baron Rutherford of Nelson (after his birthplace in New Zealand); he died in 1937, and was buried next to Newton and Kelvin in Westminster Abbey.

ton mass. We use the term *nucleon* to denote either a proton or a neutron. The number of massive particles in a nucleus (that is, the number of protons plus the number of neutrons) is called the *nucleon number.* The information regarding nuclear composition is usually displayed by prefixing the nucleon number and the *charge number* (the number of protons) to the chemical symbol:

Nucleon number / Charge number — Chemical symbol $\quad {}^{A}_{Z}\mathrm{X}$

For the following reasons, the nucleon number A is also called the mass number:

1. It is a good approximation to the mass of an atom with this nucleus measured in atomic mass units (see the following chapter for an explanation of the atomic mass unit).
2. It is a good approximation to the mass in grams of one mole of a substance having atoms with this nucleus.

The charge number Z is also called the atomic number, since it is the same as the numerical position of the element in the periodic table. For this reason, it is not really necessary to use both the charge number and the chemical symbol, since they both contain the same information. However, we retain the charge number as a help in showing charge conservation in nuclear processes.

Nuclei that have the same charge number but different nucleon numbers are called *isotopes.* For example, helium has two stable isotopes, ${}^{3}_{2}\mathrm{He}$ and ${}^{4}_{2}\mathrm{He}$. Each contains two protons, but the first contains one neutron and the second two neutrons. Our previous identification of the α particle as a helium nucleus must be qualified further. It is the helium nucleus of nucleon number 4, namely $^{4}_{2}He$. Isotopes do not differ significantly in chemical interactions, which are basically electrical in nature and depend much more strongly on the charge than on the mass.

In nuclear physics, the term *nuclide* is usually preferred for an atomic nucleus characterized by its nucleon number and charge number, so we will reserve the term *isotope* for a member of a set of two or more nuclides of the same chemical species.

We can now write equations that describe the change in nucleon and charge numbers resulting from radioactive decay. The emission of an α particle decreases the nucleon number by 4 and the charge number by 2. On the other hand, β-particle emission does not change the nucleon number, since the number of massive nuclear particles does not change, but it does increase the charge number by 1, since the nucleus loses an electron. Since a γ ray is electromagnetic radiation, its emission does not change either the nucleon or the charge number. We can write both α and β decays in the form of decay equations, as shown in the following example.

EXAMPLE 28.1

It is known that the nuclide $^{214}_{83}Bi$ undergoes α decay followed by β decay. Write an equation to show this, and identify the product nuclide of the decay.

SOLUTION When a radioactive decay occurs, both the total nucleon number and the total charge number remain unchanged. We therefore write the α decay as follows:

$$^{214}_{83}\mathrm{Bi} \rightarrow {}^{210}_{81}\mathrm{X} + {}^{4}_{2}\mathrm{He}$$

The total nucleon number remains the same, since:

$$214 = 210 + 4$$

The total charge number also remains the same, since:

$$83 = 81 + 2$$

Turning to the periodic table, we identify element 81 as thallium. The complete decay equation is:

$$^{214}_{83}\mathrm{Bi} \rightarrow {}^{210}_{81}\mathrm{Tl} + {}^{4}_{2}\mathrm{He}$$

The thallium now decays by β emission. Proceeding in the same way, we write the equation as follows:

$$^{210}_{81}\mathrm{Tl} \rightarrow {}^{210}_{82}\mathrm{Pb} + {}^{0}_{-1}\mathrm{e}$$

The electron is designated as ${}_{-1}^{0}e$ since its nucleon number is zero, and its charge number is -1. Once again the total nucleon and charge numbers do not change, and element 82 is identified as lead. ■

We can view our example as the application of two conservation laws to the decay process—the conservation of nucleon number and the conservation of charge number. However, this is very far from being the whole story, since many other conservation laws must be obeyed in these processes. A belief in the universal validity of energy conservation led Fermi to conclude that another particle should be on the right-hand side of the β-decay equation, which Fermi called the *neutrino.* We will not consider the neutrino for the time being, since we are considering only nucleon and charge conservation.

28.2 Detection of Radioactivity

Of course, we cannot see α, β, or γ radiation, and methods of detection depend on the interaction that the radiation has with matter. Most of the interaction results in **ionization**—that is, the stripping away of electrons from neutral atoms or molecules. When a charged particle passes through a material medium, it leaves a trail of ionization behind it, since its electric field acts to separate electrons from the atoms and molecules of the medium. Most detectors use the ionization to produce an audio or a visual display.

An α particle is relatively massive and carries a double charge but is fairly easily absorbed by matter. A sheet of paper, a very thin metal foil, or 10 cm of air is sufficient to absorb α particles emitted by radioactive species. The β particle is much less massive and more penetrating. Roughly speaking, it will travel hundreds of times farther through air than an α particle of comparable energy, and the trail of ionization that it leaves behind is correspondingly less intense. The γ ray is a high-frequency photon that is more penetrating still. Roughly speaking, it will penetrate matter hundreds of times more deeply than a β particle of comparable energy. The usual absorption processes for a γ-ray photon are Compton scattering and the photoelectric effect, both of which require electrons. Materials such as lead that have a high density of electrons are the most effective for absorbing γ rays.

Since the different types of radiation leave very different trails of ionization, the design of detectors may depend on the nature of the radiation to be detected. The following list indicates only the basic principle on which each detector functions.

Photographic Emulsion

A photographic plate served as the detector in the accidental discovery of radioactivity. The type of ionization process that results when photosensitive material is exposed to radiation blackens the emulsion when the material is developed. This method of detection is still in widespread use. The badges worn by workers in radiation areas are plates of photographic film which are periodically developed to measure the amount of the wearer's exposure to radiation during the period. Photographic plates are also used in various research projects to record the passage of radiation and ionizing particles.

Scintillation Counter

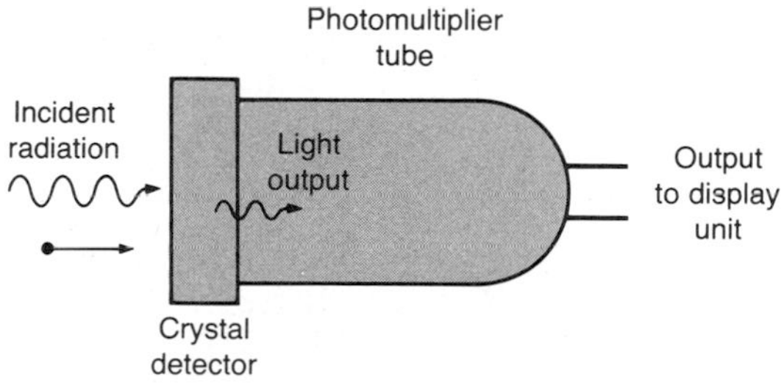

FIGURE 28.2 Schematic diagram of a modern scintillation counter.

In his early experiments, Rutherford detected α rays with a screen coated with a layer of small zinc sulfide crystals. When struck by an α particle, the zinc sulfide emits a flash of light, which Rutherford observed through a magnifying glass. The same principle is used in a TV screen! When fast electrons strike the layer of fluorescent material, it emits visible light. The modern-day version of the scintillation counter uses a single crystal of some suitable substance to absorb the incident particles of radiation and emit visible light. Sodium iodide is frequently chosen as the crystal. A typical arrangement is shown in Fig. 28.2. The emitted light passes to a photomultiplier tube, which is an electronic device that converts light into a pulse of electric current. A sensitive tube can convert one photon into a substantial current pulse. The current pulse then activates the display unit, which could be an electronic counter that records the total number of events or a small speaker that gives an audible signal whenever an event is recorded.

The scintillation counter is a rugged instrument that efficiently detects both γ rays and charged particles. It is also an energy-sensitive detector, since the magnitude of the output is usually proportional to the energy of the incident radiation.

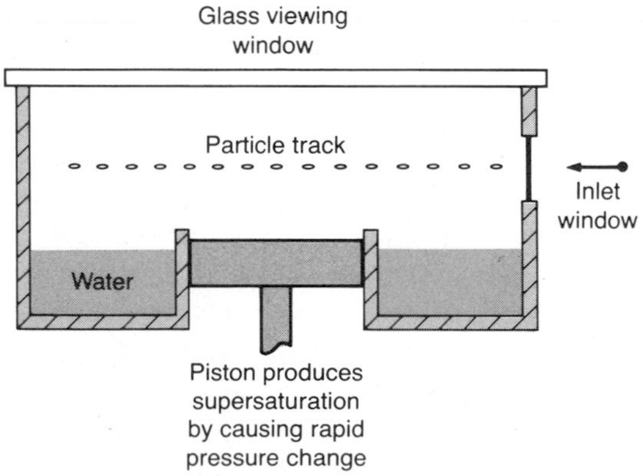

FIGURE 28.3 Schematic diagram of a cloud chamber detector.

Cloud and Bubble Chambers

Certain systems can be set up in a thermodynamically unstable state. By this we mean that a phase change should occur under the existing conditions of temperature and pressure but for some reason does not. The passage of a charged particle can then be displayed if the phase change occurs along the track of ionization. The first such device was the cloud chamber, illustrated in Fig. 28.3. Air in an enclosure is kept saturated with water vapor, and the sudden movement of a piston causes a pressure change that results in supersaturation. By this we mean that the water vapor should condense to form droplets but does not. The passage of a charged particle causes an ionization track in the air, and the water begins to condense into droplets on the ions. The resulting vapor trail, which is visible through the glass window, can be photographed to form a permanent record of the track of the particle.

Bubble chambers operate in a similar fashion. Liquid hydrogen is superheated to the point where it should vaporize but does not. The ionization track of a particle causes the vaporization to begin, and the trail is visible as tiny bubbles of gas. A magnetic field is often used to cause curvature in the track of the charged particle. Some typical track photographs are shown in the photograph on page 789 that opens this chapter.

Cloud and bubble chambers not only record the existence of the ionizing particle but also display its path. However, because they are bulky and delicate, their use is confined to laboratories.

Ionization Chamber

When ionization occurs in air, the air becomes electrically conductive. Roentgen investigated this effect in his early experiments on X rays by using an apparatus similar to that shown in Fig. 28.4. The radiation enters the window and causes ionization within the chamber. The electrons drift toward the positive electrode, and the positive ions toward the metal container. This causes a current in the circuit that registers on a sensitive ammeter. The ionization chamber gives a measure of the strength of the radiation entering its air space, but it does not record the passage of each particle individually.

Geiger-Müller Counter

The Geiger-Müller counter evolved from the ionization chamber. Instead of air at atmospheric pressure, the Geiger tube (shown in Fig. 28.5) contains argon at a pressure of about 0.1 atm. Also, the potential difference between the central electrode and the walls of the tube is about 1 kV instead of the relatively low voltage used in the ionization chamber. These changes amplify the ionization process. Ionizing radiation enters the tube and ionizes the low-pressure argon along its path. Because of the high voltage between the electrodes and the low gas pressure, the argon ions achieve sufficient energy to cause additional ionization as they rush toward the electrodes. This type of gas discharge is called an *avalanche*. The current pulse registered by the avalanche discharge, which is much larger than that due only to the ions formed directly by the radiation, causes a voltage pulse across the series resistor, which activates the display unit and records the

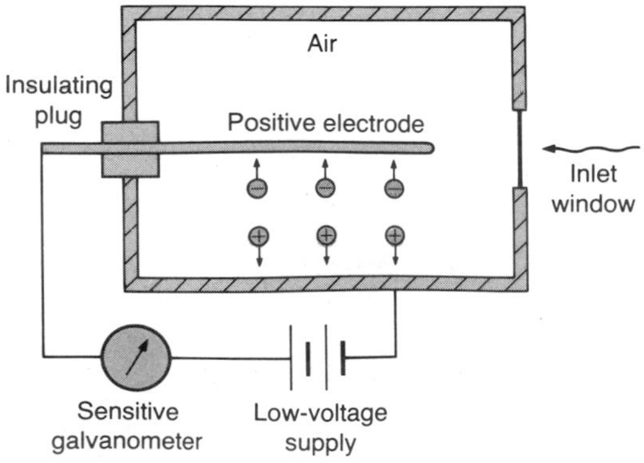

FIGURE 28.4 Schematic diagram of an ionization chamber.

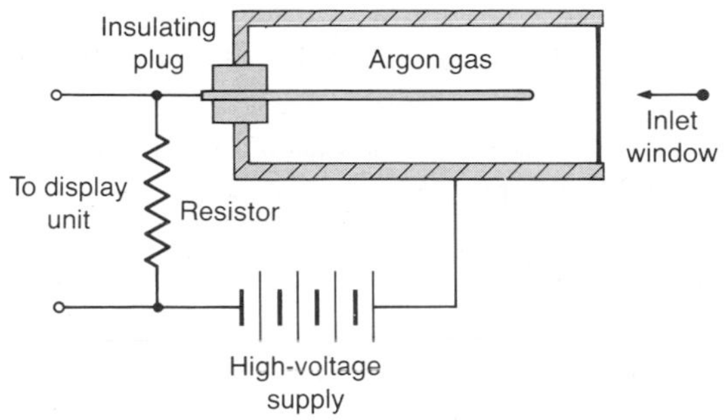

FIGURE 28.5 Schematic diagram of a Geiger-Müller counter.

passage of the ionizing particle. The end window of a Geiger-Müller counter would have to be exceedingly thin to admit α particles, but β particles are readily admitted and counted. Although γ rays can enter both the end window and the side walls, they do not cause much ionization in the gas. Their detection is usually indirect. They eject photoelectrons from the inside walls of the tube, which cause an avalanche that is counted in the usual way.

The Geiger-Müller counter is rugged and portable. It is especially suitable for detecting β radiation; however, it cannot count the β rays if they come at too fast a rate, because it takes about a millisecond for an avalanche to form, register its current pulse, and clear the ions from the tube. If a second ionizing particle arrives during the avalanche caused by a previous particle, it simply adds to the avalanche and does not cause a clearly identifiable discharge of its own. Consequently, the two are not counted separately. This effect limits the Geiger-Müller counter to about 1000 counts per second.

Solid-State Counter

This detector is similar in principle to a Geiger-Müller counter except that the ionization takes place in a solid-state device rather than in a gas. The time for formation, buildup, and disappearance of the avalanche is about $\frac{1}{100}$ of the corresponding time for the avalanche in the low-pressure gas; thus, a solid-state detector can register a count rate that is about 100 times greater than the maximum rate for a Geiger-Müller counter.

28.3 The Disintegration Constant and the Half-Life

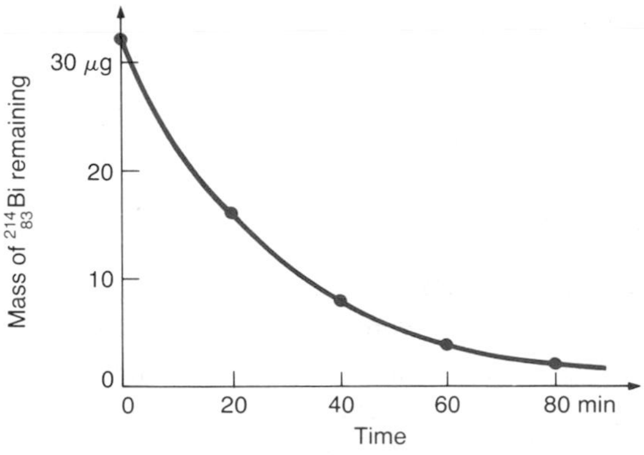

FIGURE 28.6 The decay of $^{214}_{83}$Bi. The original quantity of the isotope is 32 mg and the half-life is 20 min.

Each radioactive nuclide decays at a characteristic rate; some decay extraordinarily rapidly, and others very slowly indeed. The decay rate is not affected by changes in external physical conditions, such as temperature or pressure. Each individual nuclide decays at its own rate, regardless of its environment. The usual measure of the rate of decay is the **half-life** of the nuclide. The half-life is the period of time during which one-half of a given number of radioactive nuclei decay. We can see the suitability of this concept by looking at the decay of some particular species. Figure 28.6 shows experimental data on the decay of 32 μg of $^{214}_{83}$Bi. The quantity of the radioactive species remaining at a given time is shown by the points on the diagram. The half-life of $^{214}_{83}$Bi is 20 min, and each point represents exactly one-half the quantity that existed 20 min previously. The smooth line through these points is called an *exponential decay function*.

To set up an algebraic expression for exponential decay, we consider a quantity N_0 of the radioactive substance present at the initial time. At a later time t, the quantity of radioactive substance is reduced to $N(t)$. We express the decay algebraically as follows:

Formula for Radioactive Decay

$$N(t) = N_0 e^{-\lambda t} \qquad \textbf{(28.1)}$$

λ = decay constant

The notation $e^{-\lambda t}$ means that we first calculate $-\lambda t$ and then consult either tables or a calculator to find the exponential function of this input variable. The quantities $N(t)$ and N_0 can be expressed in any convenient unit provided that the same units are used on both sides of the equation. The unit of the decay constant λ is inverse time, and again we use any convenient unit provided that it is the same as the unit used in elapsed time t.

EXAMPLE 28.2

The decay constant of $^{214}_{83}Bi$ is 0.0346/min. Calculate the amount that remains 30 min after preparation of a fresh sample of 32 μg of the radioactive substance.

SOLUTION We have only to apply Eq. (28.1):

$$\begin{aligned} N(t) &= N_0 e^{-\lambda t} \\ &= 32\ \mu g \times e^{-0.0346/\text{min} \times 30\ \text{min}} \\ &= 32\ \mu g \times e^{-1.038} \\ &= 32\ \mu g \times 0.354 \\ &= 11.3\ \mu g \end{aligned}$$

We have set the problem out in great detail to show the calculation of the exponential function. The units associated with the decay constant are the same as those used in the elapsed time, and the value of $e^{-1.038}$ is found either from tables or from an electronic calculator. ■

We have described the rate of radioactive decay in two ways: by the half-life and the decay constant. It follows that there must be a relation between these two quantities, and we can find it by recalling that one-half of the original quantity remains after the elapse of one half-life. With this in mind, we make the following substitutions in Eq. (28.1):

$$\frac{N(t)}{N_0} = \frac{1}{2}$$

and

$$t = T_{1/2}$$

to give:

$$\tfrac{1}{2} = e^{-\lambda T_{1/2}}$$

Taking the natural logarithm of both sides gives:

$$\ln(\tfrac{1}{2}) = -\ln 2 = -\lambda T_{1/2}$$

(See Appendix A.4 for a discussion of the natural logarithm.) From a calculator (or from tables) we find the desired result:

■ Relation Between the Radioactive Half-Life and the Decay Constant

$$\begin{aligned} \lambda T_{1/2} &= \ln 2 \\ &= 0.693 \end{aligned} \qquad \textbf{(28.2)}$$

EXAMPLE 28.3

The half-life of the radionuclide $^{90}_{38}Sr$ is 28.9 yr. What percentage of an original quantity of this radioactive substance remains after 10 yr?

SOLUTION We first calculate the decay constant from Eq. (28.2):

$$\begin{aligned} \lambda T_{1/2} &= 0.693 \\ \lambda \times 28.9 \text{ years} &= 0.693 \\ \therefore \lambda &= 2.398 \times 10^{-2}/\text{year} \end{aligned}$$

(Note that all times remain in years; we can do this safely since time is the only physical quantity occurring in the equation.) We now use the radioactive decay law, Eq. (28.1):

$$\begin{aligned} N(t) &= N_0 e^{-\lambda t} \\ &= N_0 \times e^{-(2.398 \times 10^{-2}/\text{year}) \times (10 \text{ years})} \\ &= N_0 \times 0.787 \end{aligned}$$

It follows that 78.7% of the original quantity remains after 10 years. ■

It is convenient to have a standard unit for the rate of radioactive decay. The SI unit is the becquerel (abbreviated "Bq"), named in honor of the discoverer of radioactive emissions. The becquerel is equal to one disintegration per second. However, this is a very small unit for practical use. The common practical unit is the curie (abbreviated "Ci"), named in honor of Madame Curie. A curie of radioactive material is the quantity that gives 3.7×10^{10} disintegrations per second. The curie was originally defined as the number of disintegrations per second in 1 g of freshly prepared $^{226}_{88}Ra$. Although this basis for the unit has been dropped, it gives a useful practical measure of the curie.

The rate of decay of a given sample of radioactive material is frequently called the **strength** of the sample. We can use the decay constant (measured in inverse seconds) to calculate the strength of a radioactive sample only if the half-life of the sample is much greater than 1 s. When this condition is met, the strength of the sample is equal to the product of the decay constant and the number of radioactive nuclides present in the sample. If the half-life is much greater than 1 s, we can regard the number of radioactive nuclides as approximately constant over a 1-s time interval. For a very short half-life this is not true, and the formula does not apply.

$$\text{Strength of radioactive source (in becquerel)} = \text{decay constant (per second)} \times \text{number of radioactive nuclides}$$

$$A = \lambda N \qquad \textbf{(28.3)}$$

The half-life must be very much greater than 1 s to apply this formula.

EXAMPLE 28.4

Calculate the strength in becquerels and curies of a freshly prepared sample of 30 μg of $^{214}_{83}Bi$. The half-life is 20 min.

SOLUTION Since the half-life greatly exceeds 1 s, we can use Eq. (28.3) to calculate the strength of the source. However, we need the decay constant in inverse second units and the number of radioactive nuclides in the sample. Let us treat these questions in order. To calculate the decay constant, we use Eq. (28.2) with $T_{1/2} = 1200$ s:

$$\lambda T_{1/2} = 0.693$$

$$\therefore \lambda = \frac{0.693}{1200 \text{ s}}$$

$$= 5.77 \times 10^{-4}/\text{s}$$

One mole of any substance is the relative atomic mass in grams, which contains Avogadro's number of atoms. It follows that 214 g of $^{214}_{83}Bi$ contain 6.02×10^{23} atoms. We therefore calculate the number of radioactive nuclides in 30 μg as follows:

$$N = \frac{30 \times 10^{-6}\ \text{g}}{214\ \text{g}} \times 6.02 \times 10^{23}$$
$$= 8.44 \times 10^{16}$$

Finally, we use Eq. (28.3) to calculate the strength of the radioactive sample:

$$A = \lambda N$$
$$= 5.77 \times 10^{-4}/\text{s} \times 8.44 \times 10^{16}$$
$$= 4.87 \times 10^{13}\ \text{Bq}$$

Since 1 Ci $= 3.7 \times 10^{10}$ Bq, the strength of the sample in curies is given by:

$$A = 4.87 \times 10^{13}\ \text{Bq} \div 3.7 \times 10^{10}\ \text{Bq/Ci}$$
$$= 1320\ \text{Ci}$$

This is an exceedingly intense radioactive source in spite of its small mass. ■

The strength of a radioactive source measured in disintegrations per unit time is not an accurate measure of its effect on living things because different types of radiation produce different effects. Thus, we need some measure of the rate at which energy is deposited in irradiated tissue. The subject of radiation dosage is discussed in a later section.

28.4 The Natural Radioactive Series

Most of the naturally occurring radioactive elements are contained in minerals that are about the age of the earth. Since this time interval is several billion years, some radioactive nuclides must have half-lives of comparable magnitude or longer.

Table 28.1 shows the steps by which a radioactive thorium nuclide decays into a stable isotope of lead. The nuclide $^{232}_{90}Th$ has a half-life that is longer than the estimated age of the earth, which is usually put at about 5×10^9 years. This means that more than half the quantity of $^{232}_{90}Th$ present at the formation of the earth is still with us. A nuclide that is formed by radioactive decay is called a daughter product. The table shows that the half-lives of the daughter products of $^{232}_{90}Th$ range from a small fraction of a

TABLE 28.1 The thorium radioactive series

Radioactive isotope	*Type of decay*	*Half-life*
$^{232}_{90}Th$	α	1.41×10^{10} years
$^{228}_{88}Ra$	β	85.75 years
$^{228}_{89}Ac$	β	6.13 h
$^{228}_{90}Th$	α	1.91 years
$^{224}_{88}Ra$	α	3.64 days
$^{220}_{86}Rn$	α	55.6 s
$^{216}_{84}Po$	α or β	0.15 s
$^{212}_{82}Pb$	β	10.6 h
$^{216}_{85}At$	α	0.3 ms
$^{212}_{83}Bi$	α or β	60.6 min
$^{212}_{84}Po$	α	0.3 μs
$^{208}_{81}Tl$	β	3.1 min
$^{208}_{82}Pb$	Stable	—

second up to several years. The present existence of the daughter products is a result of the ongoing decay of the parent member of the series. Even though the daughter products have relatively short half-lives, they are continually replenished by the decay of nuclides that are above them in the series.

The decay processes of the thorium series are illustrated in Fig. 28.7. Each point on the grid represents a nuclide whose charge number and nucleon number are marked along the axis. The α decays are represented by the solid lines; note that the nucleon number diminishes by 4 and the charge number by 2. In a β decay the nucleon number remains unchanged, and the charge number increases by 1. The β decays are shown as dashed lines. The emission of γ radiation takes place simultaneously with most of the decay processes illustrated. However, because γ decay does not cause a change in either the nucleon or the charge number, it does not appear on the diagram.

Some of the nuclides in the series are unstable to both α and β decay. For such nuclides, a certain fraction decays in one mode, and the balance in the other mode. In each case the daughter product of the α decay undergoes β decay, and the daughter product of the β decay undergoes α decay. The second daughter is therefore the same, no matter which decay path is followed. For example, there are two decay modes for $^{212}_{83}Bi$ as follows:

$$^{212}_{83}Bi \rightarrow {}^{208}_{81}Tl + {}^{4}_{2}He$$

$$^{208}_{81}Tl \rightarrow {}^{208}_{82}Pb + {}^{0}_{-1}e$$

$$^{212}_{83}Bi \rightarrow {}^{212}_{84}Po + {}^{0}_{-1}e$$

$$^{212}_{84}Po \rightarrow {}^{208}_{82}Pb + {}^{4}_{2}He$$

In each case the second daughter of the double decay is the stable nuclide $^{208}_{82}Pb$.

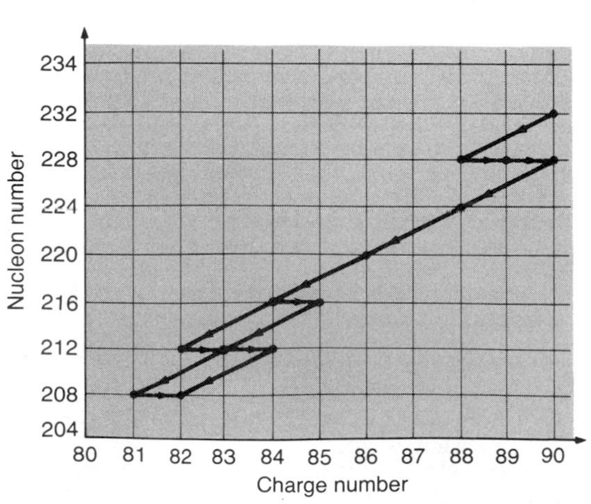

FIGURE 28.7 Graphical display of the steps in the decay of radioactive thorium to a stable isotope of lead.

The nucleon number on the vertical axis of Fig. 28.7 ranges from 208 to 232, but not every intermediate number appears. The nucleon numbers are in fact separated from each other by intervals of 4. In the thorium decay series illustrated, each nucleon number is a multiple of 4. We can easily see the reason for this when we reflect that only α decay changes the nucleon number, and it causes a reduction of exactly 4.

Two other radioactive series occur naturally. The uranium series has $^{238}_{92}U$ as the parent nuclide, and its half-life is 4.5×10^9 years. Each member of the series has a nucleon number that leaves a remainder of 2 when divided by 4. The actinium series is named after one of the daughter products. The parent nuclide is $^{235}_{92}U$, whose half-life is 7.1×10^8 years. Each member of the actinium series has a nucleon number that leaves a remainder of 3 when divided by 4. For both of these series the parent isotope has a half-life that is comparable with the age of the earth.

Notice that no radioactive series occurs naturally in which the nucleon number leaves a remainder of 1 when divided by 4. We suspect that no sufficiently long-lived nuclide meets this condition, and none has been discovered.

28.5 Induced Nuclear Transmutations

The discovery of the spontaneous disintegration of nuclei in natural radioactive decay led physicists to think about the possibility of disintegration of normally stable nuclei by bombarding them with other nuclei.

Rutherford first demonstrated the validity of this idea. He used α particles from a radium nuclide to bombard nitrogen nuclei. As a result of a long series of careful experiments, he was able to show that a fast proton was expelled in the nuclear disruption. We can write the Rutherford reaction as follows:

$$^{14}_{7}N + ^{4}_{2}He \rightarrow ^{17}_{8}O + ^{1}_{1}H$$

On the left-hand side, $^{4}_{2}He$ is the bombarding particle and $^{14}_{7}N$ is the target nucleus. The products of the reaction on the right-hand side are the emitted proton and the product nucleus $^{17}_{8}O$. Note the conservation of nucleon number and charge number in the reaction. The total nucleon number is 18 (14 + 4 on the left-hand side and 17 + 1 on the right), and the total charge number is 9 (7 + 2 on the left-hand side and 8 + 1 on the right).

Transmutation of other nuclei by α-particle bombardment with emission of a proton can be represented by similar equations. Examples are:

$$^{10}_{5}B + ^{4}_{2}He \rightarrow ^{13}_{6}C + ^{1}_{1}H$$

$$^{27}_{13}Al + ^{4}_{2}He \rightarrow ^{30}_{14}Si + ^{1}_{1}H$$

$$^{32}_{16}S + ^{4}_{2}He \rightarrow ^{35}_{17}Cl + ^{1}_{1}H$$

In each case, the total nucleon number and charge number are conserved. The reactions listed are called alpha-proton reactions, abbreviated "(α,p)." The first particle is the bombarding particle, and the second the emitted particle. In all (α,p) reactions, the product nucleus has a charge number increased by 1 unit and a nucleon number increased by 3 units over the target nucleus.

Not all nuclear reactions that are caused by a bombarding α particle produce an emitted proton. In 1932 James Chadwick, an English physicist, discovered the neutron by bombarding a beryllium nucleus with α particles. Chadwick's reaction is:

$$^{9}_{4}Be + ^{4}_{2}He \rightarrow ^{12}_{6}C + ^{1}_{0}n$$

This is an example of an alpha-neutron (α,n) reaction. Other neutron-producing reactions are:

$$^{7}_{3}\text{Li} + ^{4}_{2}\text{He} \rightarrow ^{10}_{5}\text{B} + ^{1}_{0}\text{n}$$

$$^{27}_{13}\text{Al} + ^{4}_{2}\text{He} \rightarrow ^{30}_{15}\text{P} + ^{1}_{0}\text{n}$$

Once again, nucleon number and charge number are conserved.

The neutron was not known before Chadwick's experiments. The determination that the emitted particle had no charge and a mass approximately equal to the proton mass was a major breakthrough in nuclear physics. Since the neutron is uncharged, it is not deflected by electric or magnetic fields. Having no electrical interaction with either electrons or nuclei, it causes a negligible amount of ionization when it passes through matter. For this reason it is very difficult to detect directly, and indirect methods are usually employed. If a neutron bombards a boron nucleus, the following reaction occurs:

$$^{10}_{5}\text{B} + ^{1}_{0}\text{n} \rightarrow ^{7}_{3}\text{Li} + ^{4}_{2}\text{He}$$

One of the products is an α particle that can be detected by conventional means. Suppose that the tube of a Geiger-Müller counter contains boron compounds, either coating the inside wall or in the low-pressure gas. Incident neutrons react with the boron to produce an α particle within the chamber, which is readily detected by the ionization that it causes. The ionization therefore provides indirect evidence of the passage of the neutron. The reaction is a neutron-alpha (n,α) reaction, and it obeys our usual conservation laws on nucleon and charge number.

Various bombarding particles can produce many types of nuclear reactions. Protons ($^{1}_{1}$H), deutrons ($^{2}_{1}$H), and γ rays have also been commonly used for bombardment, in addition to the α particles and neutrons already discussed. Some typical reactions are:

$$^{27}_{13}\text{Al} + ^{1}_{1}\text{H} \rightarrow ^{24}_{12}\text{Mg} + ^{4}_{2}\text{He}$$

$$^{27}_{13}\text{Al} + ^{2}_{1}\text{H} \rightarrow ^{25}_{12}\text{Mg} + ^{4}_{2}\text{He}$$

$$^{27}_{13}\text{Al} + \gamma \rightarrow ^{26}_{13}\text{Al} + ^{1}_{0}\text{n}$$

The three examples given are (p,α), (d,α), and (γ,n) reactions, respectively.

We can construct a diagram to help visualize the possibilities for product nuclei that result from low-energy nuclear reactions involving one bombarding particle and one emitted particle. Figure 28.8 plots the charge number Z vertically and the neutron number $(A - Z)$ horizontally. We choose these axes because they are the ones commonly displayed in charts of the nuclides. The central shaded square represents the target nucleus, and the surrounding squares show the relative positions of product nuclei that arise from the type of reaction displayed in them.

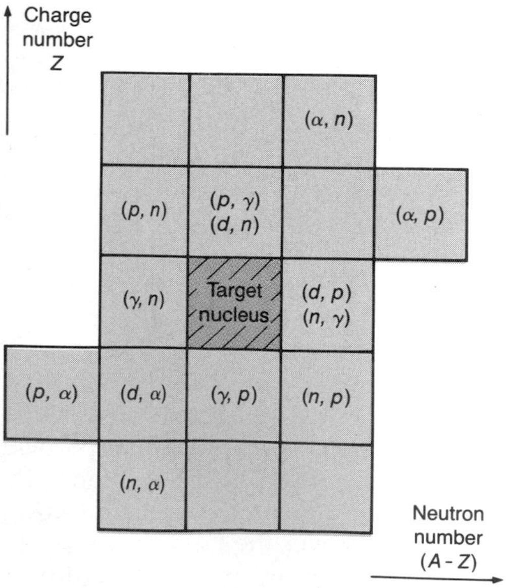

FIGURE 28.8 Diagram representing the shift in charge number (or proton number) Z and neutron number $(A - Z)$ for low-energy nuclear reactions that involve one bombarding particle and one product particle.

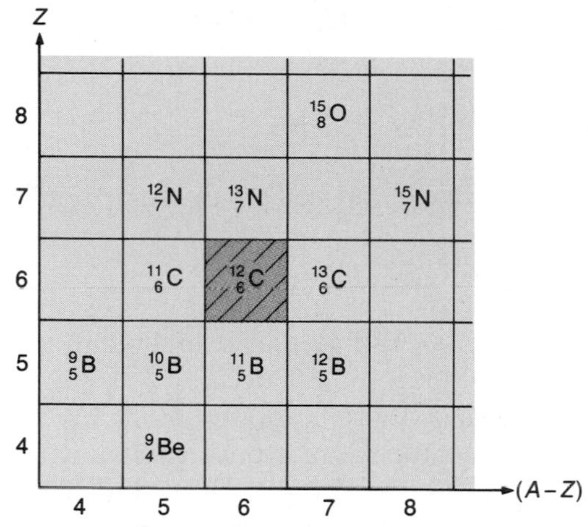

EXAMPLE 28.5

What product nuclei might be formed from low-energy nuclear reactions with a $^{12}_{6}C$ target?

SOLUTION We begin by drawing a chart like Fig. 28.8 with $^{12}_{6}C$ in the central position. We then fill in the identities of all the nuclides that result from the reactions shown in Fig. 28.8. For example, the (α,n) reaction produces $^{15}_{8}O$, the (p,n) reaction produces $^{12}_{7}N$, and so on. ■

We do not wish to imply that every possible low-energy reaction actually occurs, although in fact most of them do. Moreover, we do not claim that Fig. 28.8 exhausts the possibilities. There can be other bombarding particles than the ones shown, and there can also be reactions at higher energy that involve several emitted particles. From a given target nuclide, almost any adjacent nuclide can be formed by using a suitable incident particle.

28.6 Artificial Radioactivity

Artificially produced radioactive substances were discovered in 1934 by the French physicists Curie and Joliot, who were investigating the nuclear reactions that occur when light elements are bombarded by α particles. In addition to the neutrons and protons that were expected as products, they observed *positively charged electrons*, which are called *positrons*. A positron is a fundamental particle having the same mass as an electron and a charge of the same magnitude but of opposite sign. The emission of positrons continued for some time after the α-particle bombardment ceased. Curie and Joliot discovered that various product nuclei from the reactions were unstable and disintegrated with the emission of positrons. Let us return to one of our previous nuclear reactions:

$$^{27}_{13}\mathrm{Al} + ^{4}_{2}\mathrm{He} \rightarrow ^{30}_{15}\mathrm{P} + ^{1}_{0}\mathrm{n}$$

The $^{30}_{15}P$ is unstable and decays with a half-life of 3.50 min:

$$^{30}_{15}\mathrm{P} \rightarrow ^{30}_{14}\mathrm{Si} + ^{0}_{1}\mathrm{e}$$

This type of decay is called a β^+ decay. From now on we will refer to our previous type of β decay as β^- decay in order to distinguish the two types. Both β^- and β^+ decays produce an additional particle of the neutrino family. We have omitted this particle from the decay equations since it plays no role in the conservation of nucleon number or charge number.

After the initial discovery of artificial radioactivity, many other product nuclides were found to be radioactive, some of which

Prior to the rise of Roman power, the Etruscans dominated the north Italian peninsula. The discovery of the statues in the early part of the twentieth century was considered to be of great archeological importance since few Etruscan relics had survived. But the enthusiasm was short lived—radio carbon dating proved that the statues were fabricated in the twentieth century! (The Metropolitan Museum of Art, Purchased by subscription, 1896)

undergo β^- decay. An example of this type begins with the nuclear reaction:

$$^{27}_{13}\text{Al} + ^{1}_{0}\text{n} \rightarrow ^{24}_{11}\text{Na} + ^{4}_{2}\text{He}$$

The product nucleus $^{24}_{11}\text{Na}$ is unstable and has a half-life of 15.0 h:

$$^{24}_{11}\text{Na} \rightarrow ^{24}_{12}\text{Mg} + ^{0}_{-1}\text{e}$$

This decay is of exactly the same type as that exhibited by the naturally occurring nuclides that undergo β^- decay.

In addition to β^- and β^+ decays, there is another mode involving electrons. A nucleus that has too large a positive charge for stability can reduce the charge number by β^+ decay, but it can also reduce the charge number by absorbing an electron. Electrons are available for this purpose adjacent to the nucleus in the inner atomic shells. The decay of a nuclide that seizes an atomic electron is called **electron capture.** The nuclide $^{7}_{4}\text{Be}$ decays exclusively by electron capture. The decay reaction is:

$$^{7}_{4}\text{Be} + ^{0}_{-1}\text{e} \rightarrow ^{7}_{3}\text{Li}$$

with a half-life of 53.6 days. Some nuclides that have too large a positive charge sometimes decay by β^+ emission, and sometimes by electron capture. The nuclide $^{124}_{53}\text{I}$ is a case in point. Its two decay modes are:

$$^{124}_{53}\text{I} \rightarrow ^{124}_{52}\text{Te} + ^{0}_{1}\text{e}$$

$$^{124}_{53}\text{I} + ^{0}_{-1}\text{e} \rightarrow ^{124}_{52}\text{Te}$$

with a half-life of 4.5 days.

You may wonder how electron capture is detected, since the decaying nuclide emits nothing. Once again, we rely on secondary evidence. When the nuclide absorbs an atomic electron, it leaves a vacancy in an electron shell—usually the *K* shell, but sometimes the *L* shell. The atom then emits a characteristic X ray that provides the evidence of the electron capture.

Uranium has the highest charge number ($Z = 92$) of any naturally occurring nuclide. In a short but famous paper in 1934, Fermi suggested a possible method for producing elements with a higher charge number than uranium. Such elements became known as *transuranium elements,* because they lie beyond uranium in the periodic table. Six years of research passed before a transuranium element was positively identified. The new element was named *neptunium* ($Z = 93$). The initial step in producing it is the (n,γ) reaction:

$$^{238}_{92}\text{U} + ^{1}_{0}\text{n} \rightarrow ^{239}_{92}\text{U} + \gamma$$

The product nuclide $^{239}_{92}\text{U}$ undergoes β^- decay with a half-life of 23 min:

$$^{239}_{92}U \rightarrow ^{239}_{93}Np + ^{0}_{-1}e$$

However, the nuclide $^{239}_{93}Np$ is also radioactive. It undergoes β^- decay with a half-life of 2.3 days to produce the next transuranium element, *plutonium* ($Z = 94$). The decay is written:

$$^{239}_{93}Np \rightarrow ^{239}_{94}Pu + ^{0}_{-1}e$$

At this point no new elements are produced by the decay process. The nuclide $^{239}_{94}Pu$ is radioactive with a half-life of 24 400 years, but it emits an α particle to produce an isotope of uranium:

$$^{239}_{94}Pu \rightarrow ^{235}_{92}U + ^{4}_{2}He$$

The product nuclide $^{235}_{92}U$ is the parent nuclide of the naturally occurring radioactive actinium series.

Many additional transuranium elements have been created in the laboratory by various nuclear reactions. In addition to the two already mentioned, the list includes americum, curium, berkelium, californium, einsteinium, fermium, mendelevium, nobelium, lawrencium, kurchatorium, and hahnium whose atomic numbers range from 95 through 105. Beyond these are several more (as yet unnamed), and the list may well extend even further in the future.

The number of known radioactive nuclides is around 1500. Approximately 150 are α emitters, and the remainder decay by β^- or β^+ emission or electron capture. As mentioned previously, γ emission is a concomitant process.

The technique of radiocarbon dating uses the properties of an interesting radioactive nuclide, $^{14}_{6}C$. Protons bombarding the upper atmosphere from space produce free neutrons as a result of high-energy collision processes. The free neutron is radioactive and undergoes β^- decay with a half-life of about 11 min. However, some of the neutrons produce $^{14}_{6}C$ by bombarding the $^{14}_{7}N$ nuclide that occurs in atmospheric nitrogen:

$$^{14}_{7}N + ^{1}_{0}n \rightarrow ^{14}_{6}C + ^{1}_{1}H$$

The nuclide $^{14}_{6}C$ is radioactive, undergoing β^- decay with a half-life of 5730 years.

In living organisms most of the carbon is $^{12}_{6}C$, but a certain amount of $^{14}_{6}C$ is always present. This occurs because the $^{14}_{6}C$ formed in the upper atmosphere combines with atmospheric oxygen to form carbon dioxide, which is washed down by the rain and then enters the food cycle on the earth's surface. As each organism feeds and discharges its wastes, it acquires a small, relatively constant fraction of $^{14}_{6}C$. Moreover, the fraction is believed to have remained constant for each body part of an organism over a long period of time. Thus, bones of a person living today are thought to contain the same small fraction of $^{14}_{6}C$ as did the bones of a person living thousands of years ago.

However, when an organism dies, it is usually removed from the food chain. Consequently, the carbon is not replenished, and the $^{14}_{6}C$ nuclides decay with their 5730-year half-life. The age of ancient finds can be estimated by comparing their count rates with those of living counterparts.

EXAMPLE 28.6

A human bone from an ancient grave site gives 10.2 counts per minute per gram of carbon. The corresponding count from a similar living bone is 16 counts per minute per gram of carbon. Estimate the era in which the dead person lived.

SOLUTION For a given radionuclide, the count rate is proportional to the number of radionuclides present. We can therefore write:

$$\frac{N(t)}{N_0} = \frac{10.2}{16}$$

We can also find the decay constant of $^{14}_{6}C$ from Eq. (28.2):

$$\lambda T_{1/2} = 0.693$$

$$\therefore \lambda = \frac{0.693}{5730 \text{ yr}}$$

$$= 1.209 \times 10^{-4} \text{ per year}$$

(We will keep all time units in years.) The elapsed time now follows from Eq. (28.1):

$$\frac{N(t)}{N_0} = e^{-\lambda t}$$

$$\frac{10.2}{16} = e^{-(1.209 \times 10^{-4}/\text{year}) \times t}$$

$$\therefore t = 3720 \text{ years}$$

■

28.7 Nuclear Stability

We have examined the radioactive decay of various nuclides in some detail, but we have not discussed why some nuclides decay and some are stable. At its deepest level this is a complex and only partly resolved question. However, various approaches to the problem greatly increase our understanding. The most significant insight comes from the use of the energy conservation principle in its relativistic form, which we will examine in the next chapter. For the moment, we will confine ourselves to certain empirical criteria that describe general conditions of **nuclear stability.**

Figure 28.9 plots the stable nuclides. Each small black square represents a stable nuclide of given neutron number ($A - Z$) and

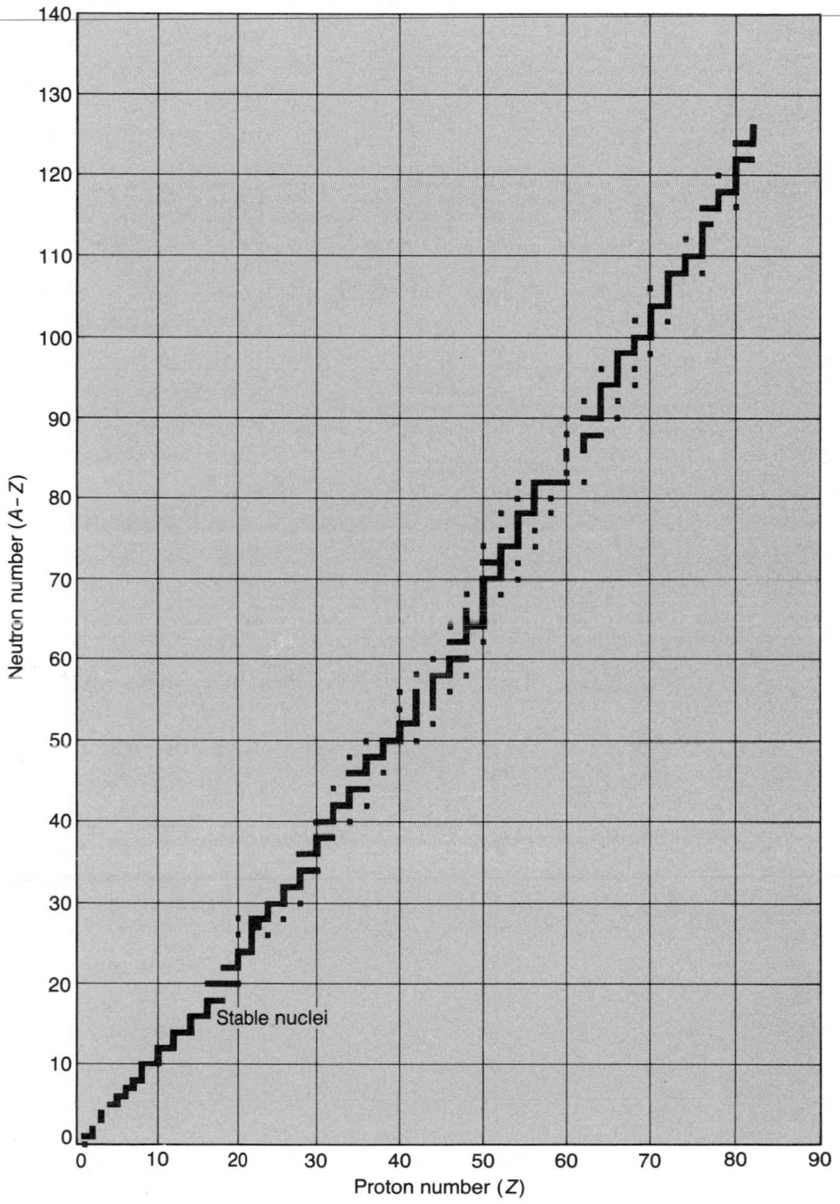

FIGURE 28.9 The neutron number ($A - Z$) plotted against the proton number (Z) for the stable nuclides. (From Kaplan, *Nuclear Physics*, © 1963, 2/e, Addison-Wesley Publishing Company, Inc., Chapter 9, Fig. 9.9, page 216, "Neutron Proton Plot of the Stable Nucleides." Reprinted with permission.)

proton number (Z). We can see some of the nuclear stability criteria by looking at the general features of Fig. 28.9, and others by examining its details. First, the size of stable nuclides has an upper limit. The stable nuclide of bismuth, $^{209}_{83}\text{Bi}$, has the largest neutron number ($A - Z = 126$) and the largest proton number ($Z = 83$) of any stable nuclide. Our chart therefore shows an upper limit for nuclear stability.

Second, the numbers of neutrons and protons are approximately equal for light nuclides. As we move up the chart, the neutron number slowly becomes larger than the proton number. For the heaviest stable nuclides, the neutron number exceeds the proton number by about 50%. We can specify the relative neutron and proton numbers for the light nuclides. For proton numbers up to $Z = 15$ (phosphorus), no stable nuclide has a neutron number that exceeds the proton number by more than 2. For only two nuclides (^1_1H and ^3_2He) does the proton number exceed the neutron number. Consider, for example, the element neon, whose proton number is $Z = 10$. Since this is a lower proton number than that for phosphorus, we know that the neutron number cannot exceed $(A - Z) = 12$. Neither can the neutron number be lower than 10. The possible stable nuclides, then, are $^{20}_{10}\text{Ne}$, $^{21}_{10}\text{Ne}$, and $^{22}_{10}\text{Ne}$. A detailed chart of the nuclides shows that all three are stable.

The third outstanding feature of stable nuclides is not obvious from Fig. 28.9. This is the overwhelming tendency for at least one of the neutron or proton numbers to be even. Table 28.2 shows the data for stable nuclides. There are 279 stable nuclides, of which about 60% have both even proton and neutron numbers. These are called even-even nuclides. Of the remaining 40%, all but 4 have either the proton or the neutron number even. The four odd-odd stable nuclides are ^2_1H, ^6_3Li, $^{10}_5\text{B}$, and $^{14}_7\text{N}$.

We summarize these criteria in display:

TABLE 28.2 Distribution of neutron and proton numbers of the stable nuclides

Proton number	*Neutron number*	*Number of stable nuclides*
Even	Even	168
Even	Odd	56
Odd	Even	51
Odd	Odd	4

■ Nuclear Stability Criteria

1. The upper limit to nuclear size: There are no stable nuclides whose proton number exceeds 83 or whose neutron number exceeds 126.
2. Approximate equality of neutron and proton numbers: For elements up to and including phosphorus ($Z = 15$), the neutron number does not exceed the proton number by more than 2. Only for the nuclides ^1_1H and ^3_2He does the proton number exceed the neutron number.
3. Preference for even numbers of protons or neutrons: About 60% of stable nuclides are even-even; of the remainder, only ^2_1H, ^6_3Li, $^{10}_5\text{B}$, and $^{14}_7\text{N}$ are odd-odd.

EXAMPLE 28.7

There is only one stable nuclide of aluminum ($Z = 13$). Use the stability rules to identify it.

SOLUTION Let us apply the stability criteria in order:

1. Since $Z < 83$ for aluminum, a stable nuclide is possible.
2. Since $Z < 15$, the neutron number must be 13, 14, or 15. This gives the nuclides ${}^{26}_{13}Al$, ${}^{27}_{13}Al$, and ${}^{28}_{13}Al$.
3. The three possibilities give odd-odd, odd-even, and odd-odd nuclides. Since aluminum is not one of the stable odd-odd nuclides, ${}^{26}_{13}Al$ and ${}^{28}_{13}Al$ are unstable. It follows that ${}^{27}_{13}Al$ is the stable nuclide. ■

EXAMPLE 28.8

Identify the two stable nuclides from the following:

$$ {}^{62}_{29}Cu,\ {}^{63}_{29}Cu,\ {}^{64}_{29}Cu,\ {}^{65}_{29}Cu,\ {}^{66}_{29}Cu $$

In what way should the unstable nuclides decay?

SOLUTION Let us again apply the stability criteria:

1. Since $Z < 83$ for copper, a stable nuclide is possible.
2. Since $Z > 15$, the second stability condition does not apply.
3. The neutron numbers for the nuclides listed are 33, 34, 35, 36, and 37. Three of these are odd-odd and not stable. The remaining two are the stable copper nuclides ${}^{63}_{29}Cu$ and ${}^{65}_{29}Cu$.

None of the relatively light nuclides decay by α emission. We therefore have to choose between β^- or β^+ emission and electron capture. Because of their position relative to stable nuclides, there are too many protons in some nuclides and too many neutrons in others. The nuclide ${}^{62}_{29}Cu$ has too few neutrons for stability, since both stable nuclides of copper have more neutrons. We therefore expect β^+ decay:

$$ {}^{62}_{29}Cu \rightarrow {}^{62}_{28}Ni + {}^{0}_{1}e $$

Another possibility for this nuclide is electron capture, which also results in ${}^{62}_{28}Ni$ as the product nuclide:

$$ {}^{62}_{29}Cu + {}^{0}_{-1}e \rightarrow {}^{62}_{28}Ni $$

A detailed nuclide chart confirms our hypothesis. Both decay modes to the stable ${}^{62}_{28}Ni$ are possible.

The nuclide $^{66}_{29}\text{Cu}$ has too many neutrons. It should experience β^- decay:

$$^{66}_{29}\text{Cu} \rightarrow {}^{66}_{30}\text{Zn} + {}^{0}_{-1}\text{e}$$

The product nuclide $^{66}_{30}\text{Zn}$ is also stable.

Finally, we note that $^{64}_{29}\text{Cu}$ lies between two stable nuclides. We cannot say that it has either too many protons or too many neutrons. Perhaps it decays by all three modes! The nuclide chart confirms our supposition, and the possible decays are:

$$^{64}_{29}\text{Cu} \rightarrow {}^{64}_{28}\text{Ni} + {}^{0}_{1}\text{e} \quad (\beta^+ \text{ decay})$$

$$^{64}_{29}\text{Cu} + {}^{0}_{-1}\text{e} \rightarrow {}^{64}_{28}\text{Ni} \quad (\text{electron capture})$$

$$^{64}_{29}\text{Cu} \rightarrow {}^{64}_{30}\text{Zn} + {}^{0}_{-1}\text{e} \quad (\beta^- \text{ decay})$$

All the product nuclides are stable. ■

We do not wish to imply that all problems of nuclear stability and decay can be handled by applying the simple rules given. The examples have been carefully chosen so that they work out easily. However, the stability rules do give us some insight, even though they are far from complete. As we have already mentioned, relativistic energy conservation gives us a far more complete picture, and we will investigate its application in the following chapter.

28.8 Biological Effects of Radiation

The effects of radiation on biological material depend on the type and energy of the radiation, the quantity and the rate of the absorbed dose, and the type and function of the exposed tissue. The whole situation is clearly very complex, and we must begin with a simplified approach to the physics.

There are two objects to consider: the source of the high-energy radiation and the living organism that absorbs the radiation. We already have a precise definition of the strength of a radioactive source. The SI unit of source activity (see Section 28.3) is the becquerel, which is equal to one disintegration per second. We also have the curie as a practical unit (1 Ci $= 3.7 \times 10^{10}$ Bq).

However, these precise measures may be somewhat misleading for the situation that we are considering. Consider the nuclide $^{60}_{27}\text{Co}$, which is widely used in clinical applications. It decays by β^- emission (maximum electron energy 314 keV), which is followed immediately by two γ rays (energies 1.173 MeV and 1.332 MeV). Thus, three particles (one electron and two photons) are emitted for each disintegration of the source. The number of particles emitted per second is therefore 3 times the disintegration rate, and even that may not be the whole story. If the cobalt source is separated from the patient by more than a meter of air, the electron is absorbed and only the γ rays get through.

The *absorbed dose* is a primary measure of the effect of the radiation on the organism. The SI unit of absorbed dose is the *gray* (abbreviated "Gy"), which corresponds to an energy deposition in the absorbing tissue of 1 J/kg. Because study of radiation biology began before the general acceptance of SI units, an alternate metric unit is frequently used, the *rad,* which is the energy deposition of 0.01 J/kg. Thus, it follows that 1 Gy = 100 rad. Since most of the literature refers to the absorbed dose in rads, we will follow that practice.

Human beings have no immediate sensation of exposure to high-energy radiation. We might expect that the deposition of energy into living tissue would cause a perceptible rise in temperature, but we can use a simple example to show that this is not the case.

EXAMPLE 28.9

A dose of 1000 rad is almost certainly fatal to a human being. Estimate the temperature rise that would accompany this dose if all of the deposited energy appears as heat.

SOLUTION From the definition of the rad, we know that 1000 rad are equal to 10 J/kg. We will assume that the average specific heat of the human body is about the same as that for water, 4180 J/kg · C°. The temperature rise is then given by Eq. (14.2):

$$Q = mC'(T - T_0)$$

$$10\text{ J} = 1\text{ kg} \times 4180\text{ J/kg} \cdot \text{C}^\circ \times (T - T_0)$$

$$\therefore T - T_0 = 0.00239\text{ C}^\circ$$

■

A fatal dose of radiation causes an undetectable temperature change. You may wonder why the radiation is so damaging when the total energy transfer is relatively so small. The answer is that the energy is not deposited uniformly throughout the irradiated tissue. Instead, the high-energy radiation imparts relatively much energy to individual atoms at random points in the living tissue. In some instances this destroys important biological molecules, which then cease to perform their specific functions.

TABLE 28.3 Relative biological effectiveness (RBE) of various radiation types

Type of radiation	*RBE*
200-keV X ray	1.0[a]
Low-energy γ rays (<1 MeV)	0.7
High-energy γ rays (>5 MeV)	0.6
β particles	1
α particles	10–20
Fast neutrons and protons	2–10

[a]By definition.

However, the damaging effect of radiation involves more than just the energy deposited per unit mass of tissue. It also depends on the types of radiation (photons, neutrons, α particles, electrons, and so on) and on the energy. For example, α particles are more damaging than protons of the same energy, since the α particles travel a shorter distance in stopping. The ionization that they cause is more intense and localized, and it is more difficult for the body to repair intense local damage than damage of the same total energy content that is more spread out. Since the damaging effect varies for the different radiation types, a standard for comparison is necessary. The *relative biological effectiveness* (RBE) of radiation is defined by comparing its effect with an equal absorbed dose of 200-keV X rays. Table 28.3 lists the approximate RBE values for different radiation types. For a given absorbed dose, γ rays are somewhat less effective in causing biological damage than 200-keV X rays, while α particles are very much more effective.

These comments lead us to the definition of the equivalent biological unit of dose, which is known as the *rem* (rad-equivalent-man). We define the rem from the following relation:

$$\begin{matrix}\text{Equivalent biological} \\ \text{dose in rem}\end{matrix} = \text{RBE} \times \begin{matrix}\text{absorbed} \\ \text{dose in rad}\end{matrix}$$

One rem of any type of radiation produces the same biological damage as 1 rad of 200-keV X rays.

TABLE 28.4 Average equivalent biological dose of radiation received by a U.S. resident

Source of radiation	*Average equivalent biological dose* (mrem/yr)
Cosmic rays	44
Radioactive minerals	40
Internal radioactive nuclides	18
Medical diagnostic tests	72
Fallout	
Bomb tests (1970)	4
Nuclear plants	0.003
Occupational (average over population)	1
Total	179

EXAMPLE 28.10

A patient undergoing therapy with $^{60}_{27}Co$ receives 800 rad of γ rays whose RBE is 0.7. Calculate the equivalent biological dose in rem.

SOLUTION We have only to apply the definition of the rem:

$$\begin{aligned}\begin{matrix}\text{Equivalent} \\ \text{biological dose}\end{matrix} &= \text{RBE} \times \text{absorbed dose in rad} \\ &= 0.7 \times 800 \text{ rad} \\ &= 560 \text{ rem}\end{aligned}$$

■

The human race has always been subject to low-level background radiation from natural sources, such as cosmic rays, radiation from radioactive minerals in the earth's crust, and radiation from internal radionuclides such as $^{14}_{6}C$ and $^{40}_{19}K$. The natural background radiation varies sharply from place to place around the world. High altitude, close proximity to radioactive minerals, or abnormally high radioactivity in drinking water all enhance the background level. In some instances the radiation level is very

TABLE 28.5 Effects of a whole-body radiation by γ rays

γ-ray dose (rad)	*Effect*
0–25	None observable
25–100	Slight blood change and slight damage to lymph nodes, spleen, and bone marrow
100–300	More extensive damage, vomiting, and fatigue; recovery within a few weeks
300–600	Severe damage, hemorrhaging, and infection with a high chance of death
600–1000	Death almost a certainty

TABLE 28.6 Maximum permissible doses (MPDs) for whole-body exposure set in 1971 by the National Council on Radiation Protection

Exposed population	*MPD*
General population	
Any individual	0.5 rem/yr
Per capita average	0.17 rem/yr
Radiation workers	
Annual	5 rem/yr
3-mo period	3 rem/3 mo
Pregnant worker	0.5 rem/9 mo

much larger than normal, and the local population suffers from an abnormally high cancer rate.

In addition to the natural background, there are various man-made sources of radiation. These include medical diagnostic tests, radioactive fallout from nuclear testing and nuclear power stations, and various occupational exposures. Table 28.4 lists the average annual dose received by a person in the United States from all sources, both natural and artificial. The total is about 179 mrem/yr. Slightly more than half of this arises from natural sources over which we have little or no control; almost all of the remainder arises from X rays used by the medical and dental professions.

Most of us show no ill effect that can be discerned due to this exposure. Presumably it is too little to cause any damage that the body cannot heal. However, we do know that massive doses of radiation are crippling or even fatal, so the question arises as to how much is too much. Our evidence in this respect is limited to survivors of the nuclear attack on Japan and persons accidentally exposed to massive doses of radiation. Table 28.5 shows the general effects that are known to follow from whole-body radiation by γ rays over a short time interval. A short-term dose of 25 rad of γ rays whose RBE is about 1 is an equivalent biological dose of 25 rem. This is over 200 times greater than the average annual dose listed in Table 28.4.

How much exposure to radiation, then, should be considered safe? The question is a vexing one, since we have insufficient data. Over the years the National Council on Radiation Protection has revised its estimates of the maximum permissible dose (MPD) downward. Forty years ago the MPD was 0.1 rem per day. In 1971 the council reduced this greatly to the values shown in Table 28.6. More recently the National Academy of Sciences has urged a further substantial reduction in the MPD. Clearly, the technological use of radiation conveys benefits as well as associated risks. A reasonable balance, both from the individual and societal point of view, is not easy to find.

KEY CONCEPTS

Radioactive nuclides decay in the following possible modes:

1. α decay, in which a $^4_2$He nuclide is emitted
2. β^- decay, in which an electron is emitted
3. β^+ decay, in which a positron is emitted
4. γ decay, in which a photon is emitted
5. electron capture

Detectors for the emitted radiation usually function by monitoring the **ionization** that the radiation causes in a medium that it traverses.

Radioactive species obey an exponential decay law. The rate of decay is conveniently measured by either the **half-life** or the **decay constant;** see Eqs. (28.1) and (28.2).

The **strength** of a radioactive source (count rate) is equal to the decay constant multiplied by the number of radioactive nuclides; see Eq. (28.3).

Transmutations of nuclear species occur when a target nucleus is bombarded by an incident particle. **Nucleon number** and **charge number** are conserved in the process.

Whether a nuclide will demonstrate **nuclear stability** or radioactive decay can be determined in many instances from three criteria; see the display on p. 810.

QUESTIONS FOR THOUGHT

1. You are told that one nucleus of given nucleon and charge numbers decays into another whose nucleon and charge numbers are also given. Is this information sufficient to predict the number of α particles in the decay scheme? Is it sufficient to predict the number of β decays?
2. What experimental evidence can you cite to support the theory that radioactivity is a nuclear rather than an atomic phenomenon?
3. Give reasons why the phenomenon of radioactivity cannot be explained on the basis of Newton's laws of motion.
4. A nucleus absorbs an electron but emits nothing. How can we detect the existence of electron capture?
5. Information concerning nuclear reactors often refers to radioactive water. In what ways do you think water might be radioactive?
6. How would you distinguish between the cloud chamber tracks of an electron and a proton?
7. What advantages does the hydrogen bubble chamber have over the cloud chamber?
8. Radioactive waste material from a reactor contains a mixture of components whose half-lives greatly vary in length. About what range of half-lives presents the least problem for safe disposal, and what range causes the greatest difficulty?
9. List the factors that could cause error in the radiocarbon dating method.
10. The earth is believed to be several billion years old. In naturally occurring minerals, we still find radioactive species whose half-lives are only a tiny fraction of the age of the earth. How can this happen?

PROBLEMS

A. Single-Substitution Problems

1. Write the proton number, neutron number, and nucleon number for each of the following nuclides:

   $^{3}_{2}He$, $^{36}_{17}Cl$, $^{132}_{54}Xe$, $^{225}_{89}Ac$ [28.1]

2. Beginning with the nuclide $^{232}_{90}Th$, nuclear disintegrations occur in the following order: α decay, β^- decay, β^- decay, and α decay. Write equations for the decays and identify the daughter nuclides in each case. [28.1]
3. The nuclide $^{212}_{83}Bi$ may disintegrate by either α or β^- decay. Write the decay equations and identify the daughter product in each case. [28.1]
4. Write equations that express the α decay of the following nuclides:

   $^{210}_{84}Po$, $^{211}_{84}Po$, $^{212}_{84}Po$ [28.1]

5. Write equations that express the β^- decay of the following nuclides:

   $^{211}_{82}Pb$, $^{212}_{82}Pb$, $^{214}_{82}Pb$ [28.1]

6. The amount of $^{44}_{19}K$ in a given sample is found to be $\frac{1}{32}$ of its original value after 1 h 50 min. What is the half-life of $^{44}_{19}K$? [28.3]
7. The half-life of $^{20}_{11}Na$ is 0.3 s. What fraction of a given amount of this nuclide remains after 0.5 min? [28.3]
8. The half-life of $^{75}_{32}Ge$ is 82 min. Calculate its decay constant. [28.3]
9. The decay constant of $^{51}_{24}Cr$ is 2.493×10^{-2}/day. Calculate its half-life. [28.3]
10. Calculate the number of disintegrations per minute from a radioactive sample of strength 7.5 μCi. [28.3]

11. What is the strength in curies of a radioactive sample that undergoes 1000 decays per hour? [28.3]

12. A radioactive sample undergoes 5.41×10^5 disintegrations per second. What is the strength of the sample in curies? [28.3]

13. The following equations represent low-energy nuclear reactions:

$$^{58}_{28}Ni + ^{1}_{1}H \rightarrow X + ^{1}_{0}n$$

$$^{23}_{11}Na + ^{2}_{1}H \rightarrow X + ^{1}_{1}H$$

$$^{27}_{13}Al + ^{1}_{0}n \rightarrow X + \gamma$$

$$^{31}_{15}P + \gamma \rightarrow X + ^{1}_{0}n$$

In each case identify the product nucleus X. [28.5]

14. The following equations represent low-energy nuclear reactions:

$$^{35}_{17}Cl + ^{1}_{0}n \rightarrow ^{35}_{16}S + x$$

$$^{55}_{25}Mn + ^{1}_{1}H \rightarrow ^{55}_{26}Fe + x$$

$$^{25}_{12}Mg + \gamma \rightarrow ^{24}_{11}Na + x$$

$$^{7}_{3}Li + ^{1}_{1}H \rightarrow ^{4}_{2}He + x$$

Identify the product particle x in each case. [28.5].

15. Write equations to describe the decay of the following nuclides by positron emission:

$^{44}_{21}Sc$, $^{49}_{24}Cr$, $^{51}_{25}Mn$, $^{62}_{29}Cu$ [28.6]

16. Write equations to describe the decay of the following nuclides by electron capture:

$^{26}_{13}Al$, $^{36}_{17}Cl$, $^{45}_{22}Ti$, $^{55}_{26}Fe$ [28.6]

17. Should the following nuclides be stable or unstable:

$^{18}_{9}F$, $^{26}_{14}Si$, $^{80}_{35}Br$, $^{234}_{90}Th$ [28.7]

18. Use the general conditions for nuclear stability to determine which of the following nuclides are stable:

$^{15}_{6}C$, $^{16}_{8}O$, $^{20}_{10}Ne$, $^{120}_{60}Nd$ [28.7]

B. Standard-Level Problems

19. How many neutrons are there in 1 mol of chlorine? (Chlorine contains 75% of the isotope $^{35}_{17}Cl$ and 25% of $^{37}_{17}Cl$.) [28.1]

20. How many grams of phosphorus contain 10^{25} neutrons? (There is one stable isotope of phosphorus, $^{31}_{15}P$.) [28.1]

21. The half-life of $^{223}_{88}Ra$ is 11.4 days. What percentage of a freshly prepared sample of this radioactive substance remains after 7 days? [28.3]

22. How long does it take for the strength of a sample of $^{224}_{88}Ra$ (half-life 3.64 days) to decline to 5% of its original value? [28.3]

23. The half-life of $^{226}_{88}Ra$ is 1620 years. Show that the strength of a freshly prepared sample of 1 g of this substance is approximately 1 Ci.

24. A certain soil sample contains 75 μg of $^{90}_{38}Sr$ (half-life 28.9 years). How many years must elapse before the amount of $^{90}_{38}Sr$ falls to 1 μg?

25. A certain radioactive sample gives 378 counts per minute. After 24 h the count rate has fallen to 58 per minute. Calculate the half-life and the decay constant of the radioactive substance.

26. A solution contains the radioactive nuclide $^{24}_{11}Na$, which undergoes β^- decay with a half-life of 15.0 h. The count rate at a certain time is 710 per minute. What is the daughter nuclide, and what is the count rate 48 h later?

27. The radionuclide $^{131}_{53}I$ has a half-life of 8.06 days. A quantity of this radioactive material having an activity of 1.6 μCi is introduced into a patient's thyroid gland. Approximately how many counts per second should be registered 14 days later if all the radioactive material is retained in the gland?

28. Tests on living wood show 15.3 counts per minute per gram of carbon. An excavated wooden tool handle shows 12.6 counts per minute per gram of carbon. Estimate the age of the tool handle. (The half-life of $^{14}_{6}C$ is 5730 years.)

29. Calculate the decay rate per minute per gram of carbon that would be expected of an excavated bone that is 10 000 years old. (Use the data of problem 28, with the decay rate of living bone taken to be the same as for wood.)

30. Calculate the fraction of $^{14}_{6}C$ in the carbon of living wood that is implied by the data of problem 28.

31. How many grams of $^{14}_{6}C$ would be required for an activity of 1 Ci? (Use the data of problem 28.)

32. Write equations to express the β^+ decay of the following nuclides:

$^{8}_{5}B$, $^{15}_{8}O$, $^{18}_{10}Ne$, $^{52}_{26}Fe$, $^{106}_{47}Ag$

Do you expect the daughter nuclides to be stable or unstable? [28.7]

33. Write equations to express each of the following nuclear decays:

a. ${}^{32}_{14}Si$ emits an electron
b. ${}^{22}_{11}Na$ emits a positron
c. ${}^{26}_{13}Al$ captures an electron

Do you expect the daughter nuclides to be stable or unstable? [28.7]

34. Write equations to represent electron capture by the following nuclides:

$${}^{7}_{4}Be,\quad {}^{22}_{11}Na,\quad {}^{26}_{13}Al,\quad {}^{36}_{17}Cl$$

Do you expect the daughter nuclides to be stable or unstable?

35. A certain source of ${}^{60}_{27}Co$ produces γ rays whose RBE is 0.7. The absorbed dose in tissue under treatment is 75 rad/min. What exposure should be provided for an equivalent biological dose of 280 rem? [28.8]

36. The crew of an airliner flying at their operational height for 20 h each week receive an average of 0.65 millirems per hour during this time. They fly for 40 weeks of the year. Express their annual biologically equivalent dose as a fraction of the maximum permissible dose for the general population and for radiation workers. [28.8]

37. A radiation technologist receives a dose of 0.12 rem each time a radium source is loaded. How many times is he permitted to perform this task in (a) 3 months; (b) 1 year? Repeat the calculations for the case in which the worker is a pregnant woman.

29

CONSERVATION LAWS IN NUCLEAR AND PARTICLE PROCESSES

In the previous chapter we examined nuclear decay and transmutation with particular emphasis on charge and nucleon number conservation. The deeper we delve into nuclear and particle physics, the more important conservation laws become. The law of charge conservation is valid without restriction, as far as it is known. The law of nucleon number conservation gives way to a more general conservation law that includes the former as a special case. However, our main objective in this chapter is to apply the energy conservation law to nuclear problems.

The extension of the energy conservation principle to nuclear reactions was a major triumph for the theory of relativity. In the previous century the work of Joule had broadened the concept of energy to include heat energy, thereby giving a new dimension to energy conservation. Einstein further extended the scope of the law with his famous equation for rest energy, which we introduced in a previous chapter. In this chapter, we study its application to various nuclear processes.

Finally, we close the chapter with some remarks on conservation laws in general, and on the role they play in our understanding of particle physics.

29.1 The Mass Spectrograph

We have already seen (in Example 25.8) that the melting of ice into water produces a mass difference of about one part in a trillion. There is no known way to check this experimentally, and the relativistic energy formulas would remain idle speculation if we could not produce larger fractional mass changes that can be experimentally verified. The processes of nuclear decay and transmutation beautifully confirm the relativistic mass–energy formulas. The mass differences that arise in nuclear processes, which are typically of the order of several parts in 10 000, are not very large, but they can be detected by precision measuring equipment. We begin with a brief survey of the principles of operation of such an apparatus, the **mass spectrograph.**

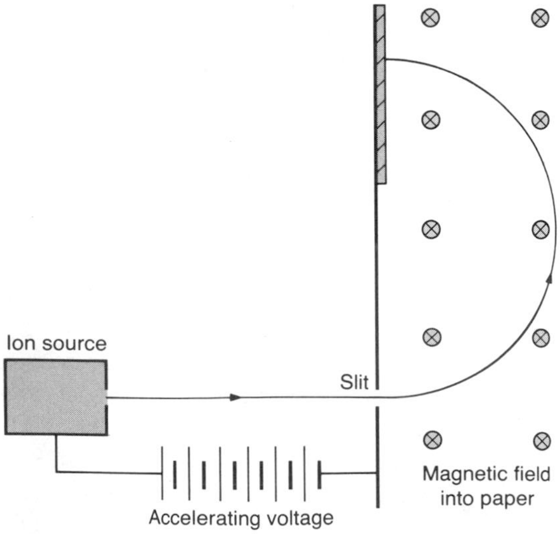

FIGURE 29.1 Schematic diagram of a mass spectrograph. (The whole equipment works in vacuum.)

Figure 29.1 shows the essential features of a mass spectrograph. An ion source produces positive ions of the species whose mass m is to be determined. The ions (each carrying a charge q) emerge from the source at low velocity and are accelerated toward the slit by the large accelerating voltage. The kinetic energy of the ions on reaching the slit is equal to the work done on them by the accelerating voltage:

$$\tfrac{1}{2}mv^2 = qV$$

On passing through the slit, the ion enters a magnetic field directed into the plane of the paper and continues in a circular path at constant speed, striking a photographic plate (or other suitable detector) and leaving a mark when the plate is developed. The radius of the path of the ion in the magnetic field is given by Eq. (19.10):

$$r = \frac{mv}{qB}$$

Combining the two equations, we obtain a formula for the mass of the ion.

■ Formula for Ionic Mass in Terms of Quantities Measured on the Mass Spectrograph

$$m = \frac{B^2qr^2}{2V} \qquad (29.1)$$

B = magnetic field
q = ion charge
r = measured path radius
V = accelerating voltage

EXAMPLE 29.1

A certain mass spectrograph uses an accelerating voltage of 1 kV and a magnetic field of 0.2 T. Calculate the mass of singly charged gold ions that move in a path of 32.04-cm radius.

SOLUTION We need only substitute the data of the problem into Eq. (29.1):

$$m = \frac{B^2qr^2}{2V}$$

$$= \frac{(0.2\ \text{T})^2 \times 1.602 \times 10^{-19}\ \text{C} \times (0.3204\ \text{m})^2}{2 \times 10^3\ \text{V}}$$

$$= 3.29 \times 10^{-25}\ \text{kg}$$ ■

As a practical example of the use of a mass spectrograph, this calculation is somewhat misleading. First, we require much higher accuracy than is implied by the number of significant digits quoted. Most tables of atomic masses quote between six and nine digits; hopefully, only the last digit is uncertain. Obtaining such accuracy from the instrument described in the example would be difficult, even if it was most carefully designed. Commonly two mass spectrographs are used in tandem to produce the high precision required.

Second, the kilogram is a very inconvenient unit for nuclear masses, since relating a measurement made on a mass spectro-

TABLE 29.1 Masses of selected atoms in atomic mass units

Atom	*Atomic mass (u)*	*Atom*	*Atomic mass (u)*
${}_{-1}^{0}e$	0.000 548	${}_{7}^{15}N$	15.000 108
${}_{-0}^{1}n$	1.008 665	${}_{8}^{16}O$	15.994 915
${}_{1}^{1}H$	1.007 825	${}_{8}^{17}O$	16.999 133
${}_{1}^{2}H$	2.014 102	${}_{8}^{18}O$	17.999 160
${}_{1}^{3}H$	3.016 050	${}_{9}^{18}F$	18.000 936
${}_{2}^{3}He$	3.016 030	${}_{9}^{19}F$	18.998 405
${}_{2}^{4}He$	4.002 603	${}_{10}^{22}Ne$	21.991 385
${}_{3}^{6}Li$	6.015 125	${}_{11}^{22}Na$	21.994 437
${}_{3}^{7}Li$	7.016 004	${}_{13}^{27}Al$	26.981 539
${}_{3}^{8}Li$	8.022 487	${}_{13}^{28}Al$	27.981 905
${}_{4}^{7}Be$	7.016 929	${}_{88}^{226}Ra$	226.025 360
${}_{4}^{8}Be$	8.005 308	${}_{88}^{228}Ra$	228.031 139
${}_{4}^{9}Be$	9.012 186	${}_{90}^{228}Th$	228.028 750
${}_{4}^{10}Be$	10.013 534	${}_{92}^{232}U$	232.037 168
${}_{5}^{10}B$	10.012 939	${}_{92}^{234}U$	234.040 904
${}_{5}^{11}B$	11.009 305	${}_{92}^{235}U$	235.043 915
${}_{6}^{12}C$	12.000 000	${}_{92}^{238}U$	238.048 608
${}_{6}^{13}C$	13.003 354	${}_{93}^{237}Np$	237.048 056
${}_{6}^{14}C$	14.003 242	${}_{94}^{238}Pu$	238.049 511
${}_{7}^{14}N$	14.003 074		

NOTE: Atomic electrons are included in the values quoted.

graph to the mass of the standard kilogram is very difficult. Instead, it is much easier to relate two different lines on the photographic plate of the spectrograph. This is exactly the approach taken in assigning a new unit suitable for nuclear masses. The new unit is known as the **atomic mass unit** (abbreviated "u"). For the mass of ${}_{6}^{12}C$ (including the atomic electrons) we assign exactly 12.000 000 u. The masses of other atoms are then determined by measuring the position of the line they cause on the photographic plate relative to that of ${}_{6}^{12}C$.

Table 29.1 lists the masses of selected atoms; all values quoted include the masses of the atomic electrons. In some cases the number of significant digits may not be justified, but our applications do not rely heavily on the values of the last couple of digits. With the help of accurate nuclear masses and the theory of relativity, we can now study energy conservation in nuclear processes.

EXAMPLE 29.2

It is convenient to have a value for the rest energy that corresponds to the atomic mass unit. Given that 1 u $= 1.6605 \times 10^{-27}$ kg, calculate the corresponding rest energy in mega-electron volts.

SOLUTION We use Eq. (25.8):

$$\begin{aligned} E_0 &= mc^2 \\ &= 1.6605 \times 10^{-27}\ \text{kg} \times (2.998 \times 10^8\ \text{m/s})^2 \\ &= 1.4925 \times 10^{-10}\ \text{J} \\ &= 1.4925 \times 10^{-10}\ \text{J} \div 1.6022 \times 10^{-19}\ \text{J/eV} \\ &= 931.5\ \text{MeV} \end{aligned}$$

■

The result is so useful that we will write it in another way.

■ **Useful Mass–Energy Conversion Relation**

$$1\ \text{u} \times c^2 = 931.5\ \text{MeV} \qquad \textbf{(29.2)}$$

29.2 Energy Conservation in Nuclear Reactions

In our study of nuclear reactions, we applied only two conservation principles—that is, nucleon number and charge number conservation. Various other conservation laws apply to the reaction, but we will focus on energy conservation. To begin with, let us demonstrate that mass is not conserved in a nuclear reaction. For example, consider Chadwick's reaction:

$$^{9}_{4}\text{Be} + ^{4}_{2}\text{He} \rightarrow ^{12}_{6}\text{C} + ^{1}_{0}\text{n}$$

For the combined mass of the bombarding particle and the target nucleus, we have:

$$\begin{array}{rl} 9.012\ 186\ \text{u} & (^{9}_{4}\text{Be}) \\ \underline{+4.002\ 603\ \text{u}} & (^{4}_{2}\text{He}) \\ 13.014\ 789\ \text{u} & \end{array}$$

A similar calculation for the emitted particle and the product nucleus gives:

$$\begin{array}{rl} 12.000\ 000\ \text{u} & (^{12}_{6}\text{C}) \\ \underline{+1.008\ 665\ \text{u}} & (^{1}_{0}\text{n}) \\ 13.008\ 665\ \text{u} & \end{array}$$

The interior of this accelerator at the University of California in Berkeley shows the long tube along which charged particles are accelerated to the experimental chamber. (Courtesy Lawrence Berkeley Laboratory)

The combined mass of the bombarding particle and the target nucleus exceeds that of the products by 0.006 124 u, which is a much larger amount than we could attribute to error in the last couple of digits.

Our experience with problems involving relativistic energy conservation leads us to believe that we should write an energy balance equation for the *total energy.* The kinetic energy plus the rest energy of the particles on one side of the reaction must equal the corresponding sum for the particles on the other side. Using the values we have just calculated, we have:

$$\begin{array}{c}\text{Kinetic}\\ \text{energy}\\ \text{of } {}^{9}_{4}\text{Be}\\ \text{and } {}^{4}_{2}\text{He}\end{array} + 13.014\ 789 \text{ u} \times c^2 = \begin{array}{c}\text{kinetic}\\ \text{energy}\\ \text{of } {}^{12}_{6}\text{C}\\ \text{and } {}^{1}_{0}\text{n}\end{array} + 13.008\ 665 \text{ u} \times c^2$$

Transposing and using the conversion factor of Eq. (29.2), we have:

$$\begin{array}{c}\text{Kinetic}\\ \text{energy}\\ \text{of } {}^{12}_{6}\text{C}\\ \text{and } {}^{1}_{0}\text{n}\end{array} - \begin{array}{c}\text{kinetic}\\ \text{energy}\\ \text{of } {}^{9}_{4}\text{Be}\\ \text{and } {}^{4}_{2}\text{He}\end{array} = 0.006\ 124 \text{ u} \times c^2 \times \left(\frac{931.5 \text{ MeV}}{1 \text{ u} \times c^2}\right)$$

$$= 5.70 \text{ MeV}$$

The kinetic energy of the product particles exceeds the kinetic energy of the initial particles by 5.70 MeV. We call such a reaction **exoergic,** since it produces kinetic energy from the rest energy surplus of the initial particles. If the rest energy of the product particles exceeds that of the initial particles, the reaction is **endoergic.** In this case the surplus rest energy of the product particles results in a loss of kinetic energy.

Remember that there is no change of total energy. When we speak of energy release in a nuclear reaction, we mean the amount of rest energy that is converted into other energy forms, which are usually the kinetic energy of particles and the energy of electromagnetic quanta. Given this understanding, we refer to the energy release of a nuclear reaction as the ***Q* value** of the reaction. To derive an expression for the Q value, we write the conservation law for total energy:

$$\text{Total initial energy} = \text{total final energy}$$

$$m_I c^2 + E_I = m_P c^2 + E_P$$

The subscript I refers to the initial situation (before the reaction), and the subscript P refers to the products of the reaction. The terms $m_I c^2$ and $m_P c^2$ are the rest energies of all the initial and product massive particles. The terms E_I and E_P refer to all other energy forms before and after the reaction, respectively. These energies include, for example, the kinetic energy of massive particles and the energy of photons. Transposing our energy conservation equation, we can write it as follows:

$$(m_I - m_P)c^2 = E_P - E_I$$

From this equation we can define the Q value of the reaction.

■ Definition of the Q Value of a Nuclear Reaction

$$\begin{aligned} Q &= (m_I - m_P)c^2 \qquad \text{(a)} \\ &= E_P - E_I \qquad \text{(b)} \end{aligned} \qquad (29.3)$$

Subscript I refers to initial particles.
Subscript P refers to product particles.

Positive Q: exoergic reaction
Negative Q: endoergic reaction

A positive Q value means energy release ($E_P > E_I$), and the reaction is exoergic. For a negative Q value, the opposite is true, and the reaction is endoergic.

EXAMPLE 29.3

A 4.78-MeV α particle bombards a stationary nitrogen nuclide in the Rutherford reaction:

$${}^{14}_{7}\text{N} + {}^{4}_{2}\text{He} \rightarrow {}^{17}_{8}\text{O} + {}^{1}_{1}\text{H}$$

Calculate the kinetic energy of the product particles.

SOLUTION We begin by calculating the Q value of the reaction, inserting the values for the nuclide masses from Table 29.1 into Eq. (29.3a):

$$\begin{aligned} Q &= (m_I - m_P)c^2 \\ &= [m({}^{14}_{7}\text{N}) + m({}^{4}_{2}\text{He}) - m({}^{17}_{8}\text{O}) - m({}^{1}_{1}\text{H})]c^2 \\ &= (14.003\ 074\ \text{u} + 4.002\ 603\ \text{u} \\ &\quad - 16.999\ 133\ \text{u} - 1.007\ 825\ \text{u})c^2 \\ &= -0.001\ 281\ \text{u} \times c^2 \\ &= -0.001\ 281\ \text{u} \times c^2 \times \left(\frac{931.5\ \text{MeV}}{1\ \text{u} \times c^2}\right) \\ &= -1.19\ \text{MeV} \end{aligned}$$

The reaction is endoergic, and we can calculate the kinetic energy of the products from Eq. (29.3b):

$$Q = E_P - E_I$$
$$-1.19 \text{ MeV} = E_P - 4.78 \text{ MeV}$$
$$\therefore E_P = 3.59 \text{ MeV}$$

■

An endoergic reaction cannot occur if the initial particles have negligible kinetic energy, since the kinetic energy of the product particles would then be negative. In the case of the Rutherford reaction, the bombarding α particles require substantial kinetic energy before the reaction can begin. The effect is further complicated by electrostatic repulsion between the positive nitrogen nucleus of the target and the positive helium nucleus of the bombarding particle.

If the bombarding particle is a neutron, the electrostatic repulsion does not exist. A very slow neutron can initiate a nuclear reaction if the Q value is positive.

EXAMPLE 29.4

Determine whether the radioactive nuclide ${}^{14}_{6}\text{C}$ can be produced by bombarding ${}^{14}_{7}\text{N}$ with slow neutrons.

SOLUTION The nuclear reaction to be considered is:

$${}^{14}_{7}\text{N} + {}^{1}_{0}\text{n} \rightarrow {}^{14}_{6}\text{C} + {}^{1}_{1}\text{H}$$

Using the masses of Table 29.1, we find the Q value from Eq. (29.3a):

$$\begin{aligned}
Q &= (m_I - m_P)c^2 \\
&= [m({}^{14}_{7}\text{N}) + m({}^{1}_{0}\text{n}) - m({}^{14}_{6}\text{C}) - m({}^{1}_{1}\text{H})]c^2 \\
&= (14.003\ 074 \text{ u} + 1.008\ 665 \text{ u} \\
&\quad - 14.003\ 242 \text{ u} - 1.007\ 825 \text{ u})c^2 \\
&= 0.000\ 672 \text{ u} \times c^2 \\
&= 0.000\ 672 \text{ u} \times c^2 \times \left(\frac{931.5 \text{ MeV}}{1 \text{ u} \times c^2}\right) \\
&= 0.626 \text{ MeV}
\end{aligned}$$

The reaction is exoergic, so it can be initiated by a very slow neutron. ■

In all of our examples we have used the atomic masses listed in Table 29.1. However, these values include the masses of the atomic electrons, while our reactions concern only the nucleus. To obtain nuclear masses, we would have to subtract the appropriate number of electron masses from each value in the table. There are two reasons for not making this correction. First, the number of electron masses to be subtracted from each side of our nuclear reactions is the same, since the proton number is also the electron number for a neutral atom. Consequently, the correction cancels out of the equations. Second, the energy of the atomic electrons is also involved, and the overall accuracy is better when the atomic masses are used rather than the nuclear masses.

29.3 Energy Conservation in Radioactive Decay

The decay of a radioactive nuclide obeys the energy conservation law in much the same way that the nuclear reactions do. However, in this case the only initial particle is the radioactive nuclide—no bombarding particle is required to cause the disintegration. In every instance the product particles are the daughter nuclide and the emitted α or β particle. Calculation of the Q value determines whether a nuclide is stable against decay and provides us with the kinetic energy of the decay products. We will illustrate by determining the stability of some specific nuclides.

EXAMPLE 29.5

Determine whether the nuclides $^{238}_{94}\text{Pu}$ and $^{16}_{8}\text{O}$ are stable against α decay.

SOLUTION Let us take the two cases in turn. If the nuclide $^{238}_{94}\text{Pu}$ undergoes α decay, the decay equation is:

$$^{238}_{94}\text{Pu} \rightarrow {}^{234}_{92}\text{U} + {}^{4}_{2}\text{He}$$

From Eq. (29.3a) we have for the Q value:

$$\begin{aligned} Q &= (m_I - m_P)c^2 \\ &= [m(^{238}_{94}\text{Pu}) - m(^{234}_{92}\text{U}) - m(^{4}_{2}\text{He})]c^2 \\ &= (238.049\ 511\ \text{u} - 234.040\ 904\ \text{u} - 4.002\ 603\ \text{u})c^2 \\ &= 0.006\ 004\ \text{u} \times c^2 \\ &= 0.006\ 004\ \text{u} \times c^2 \times \left(\frac{931.5\ \text{MeV}}{1\ \text{u} \times c^2}\right) \\ &= 5.59\ \text{MeV} \end{aligned}$$

Since the Q value is positive, the initial nuclide is unstable against α decay.

If the nuclide $^{16}_{8}O$ undergoes α decay, the decay equation is:

$$^{16}_{8}O \rightarrow {}^{12}_{6}C + {}^{4}_{2}He$$

The corresponding Q value is:

$$\begin{aligned}
Q &= (m_I - m_P)c^2 \\
&= [m(^{16}_{8}O) - m(^{12}_{6}C) - m(^{4}_{2}He)]c^2 \\
&= (15.994\ 915\ u - 12.000\ 000\ u - 4.002\ 603\ u)c^2 \\
&= -0.007\ 688\ u \times c^2 \\
&= -0.007\ 688\ u \times c^2 \times \left(\frac{931.5\ MeV}{1\ u \times c^2}\right) \\
&= -7.16\ MeV
\end{aligned}$$

The initial nuclide, $^{16}_{8}O$, is stable against α decay. ■

The positive Q value for α decay is the total kinetic energy of the product particles. When $^{238}_{94}Pu$ decays, 5.59 MeV of kinetic energy is distributed to the α particle and the recoil nucleus $^{234}_{92}U$. We can calculate the actual distribution of the energy with the help of the linear momentum conservation law.

EXAMPLE 29.6

Calculate the kinetic energy of the α particle in the decay of the nuclide $^{238}_{94}Pu$.

SOLUTION The decay equation is:

$$^{238}_{94}Pu \rightarrow {}^{234}_{92}U + {}^{4}_{2}He$$

Let v_R be the velocity of the recoil nucleus and v_α the velocity of the α particle. We assume that both velocities are sufficiently low to permit use of the Newtonian formulas for momentum and kinetic energy. Linear momentum conservation gives:

$$0 = m(^{234}_{92}U)v_R + m(^{4}_{2}He)v_\alpha$$

The plutonium nucleus is initially at rest so that the plutonium and helium nuclei must move in opposite directions. To simplify the appearance of the algebra, we put:

$$m_u = m(^{234}_{92}U) \qquad \text{and} \qquad m_\alpha = m(^{4}_{2}He)$$

Then the momentum conservation equation yields:

$$v_R = -\frac{m_\alpha v_\alpha}{m_u}$$

We have already noted that the Q value of the decay is the total kinetic energy of the products, that is:

$$\tfrac{1}{2}m_u v_R^2 + \tfrac{1}{2}m_\alpha v_\alpha^2 = Q$$

Substituting the value for v_R gives:

$$\frac{1}{2}\left(\frac{m_\alpha}{m_u}\right) m_\alpha v_\alpha^2 + \tfrac{1}{2}m_\alpha v_\alpha^2 = Q$$

Solving this equation for $\tfrac{1}{2}m_\alpha v_\alpha^2$ we have:

$$E_\alpha = \tfrac{1}{2}m_\alpha v_\alpha^2 = \frac{Q}{1 + m_\alpha/m_u}$$

Since $m_\alpha \ll m_u$, the α particle carries off most of the kinetic energy from the decay. Using the Q value from Example 29.5, and setting the ratio of the atomic masses as approximately equal to the ratio of the mass numbers, we have:

$$E_\alpha = \frac{5.59 \text{ MeV}}{1 + 4/234} = 5.50 \text{ MeV}$$

Since the kinetic energy of the α particle is very much smaller than its rest energy, the use of Newtonian formulas is justified.

■

Careful experimental work shows that the α particle emitted by $^{238}_{94}Pu$ has the kinetic energy that we have calculated from the energy and momentum conservation laws. For the universality of these laws to predict correctly the kinetic energy of an α particle in nuclear decay is a considerable triumph. The method is exactly the same as that used in Chapter 8 to deal with the problem of colliding freight cars.

We can use the same method to investigate stability against β^- or β^+ decay or electron capture. However, in these instances we must consider the electron mass involved. Let us use an example to illustrate.

EXAMPLE 29.7

Investigate the stability of the nuclide $^{10}_{4}Be$ against β^- decay.

SOLUTION We begin by writing the decay equation:

$$^{10}_{4}\text{Be} \rightarrow {}^{10}_{5}\text{B} + {}^{0}_{-1}\text{e}$$

In calculating the Q value for the reaction, we must remember that the values of Table 29.1 include four electrons in $^{10}_{4}Be$ and five electrons in $^{10}_{5}B$. Therefore, we must not add the mass of the β^- electron, or else it will be counted twice. The Q value for the reaction is:

$$\begin{aligned} Q &= (m_I - m_P)c^2 \\ &= [m(^{10}_{4}Be) - m(^{10}_{5}B)]c^2 \\ &= (10.013\ 534\ u - 10.012\ 939\ u)c^2 \\ &= 0.000\ 595\ u \times c^2 \\ &= 0.000\ 595\ u \times c^2 \times \left(\frac{931.5\ MeV}{1\ u \times c^2}\right) \\ &= 0.554\ MeV \end{aligned}$$

The positive Q value shows that the nuclide $^{10}_{4}Be$ is unstable against β^- decay. ■

For electron capture and β^+ decay, we must keep track of the electron masses. The positive Q value indicates instability against decay and gives the amount of energy available to the decay products. For any of the β-decay processes (i.e., β^- or β^+ decay or electron capture), a particle of the neutrino family is included in the decay products, and no simple theory has been determined that assigns the relative energies of the products.

29.4 Nuclear Binding Energy

We picture the nucleus as being composed of a certain number of protons and neutrons. In discussing radioactive decay, we have spoken a great deal about nuclear decay processes involving the emission of helium nuclei and electrons. But what about the possible decay of a nucleus into all of its component protons and neutrons? Instinctively we feel that this cannot happen—certainly the stable nuclides will not decay in this way, and even the radioactive nuclides decay in a manner that is much less catastrophic than breaking up into individual nucleons.

Nevertheless, the concept is important. The Q value for decay into component nucleons is negative for all stable and radioactive nuclides, but its value provides a measure of the extent to which the nuclide resists such a decay. We call the magnitude of the Q value the **nuclear binding energy.** This figure is the energy that must be supplied to the nuclide to break it into its component nucleons, and we can calculate the binding energy in terms of the proton and neutron masses and the mass of the nuclide. To do this, we subtract the rest energy of the nucleus from the rest energy of the constituent nucleons. Since our atomic mass tables include the electron masses, it is convenient to write the binding energy formula in terms of atomic rather than nuclear masses.

Definition of Nuclear Binding Energy

$$\text{Binding energy} = \begin{array}{c}\text{energy required to separate}\\ \text{a nucleus into individual nucleons}\end{array}$$

$$E_B = [Zm(^1_1\text{H}) + (A - Z)m(^1_0\text{n}) - m(^A_Z X)]c^2$$

$m(^1_1\text{H})$ = hydrogen atomic mass **(29.4)**
$m(^1_0\text{n})$ = neutron mass
Z = proton number
A = nucleon number
$m(^A_Z X)$ = atomic mass of $^A_Z X$

Let us look at some examples of binding energy.

EXAMPLE 29.8

Calculate the binding energies of $^{16}_{8}\text{O}$ and $^{238}_{92}\text{U}$.

SOLUTION We need only apply Eq. (29.4) for each nuclide in turn. For $^{16}_{8}\text{O}$ we have:

$$\begin{aligned}
E_B &= [Zm(^1_1\text{H}) + (A - Z)m(^1_0\text{n}) - m(^A_Z X)]c^2 \\
&= [8 \times 1.007\ 825\ \text{u} + (16 - 8) \times 1.008\ 665\ \text{u} - 15.994\ 915\ \text{u}]c^2 \\
&= 0.137\ 005\ \text{u} \times c^2 \\
&= 0.137\ 005\ \text{u} \times c^2 \times \left(\frac{931.5\ \text{MeV}}{1\ \text{u} \times c^2}\right) \\
&= 127.6\ \text{MeV}
\end{aligned}$$

For $^{238}_{92}\text{U}$ we have:

$$\begin{aligned}
E_B &= [92 \times 1.007\ 825\ \text{u} + (238 - 92) \\
&\quad \times 1.008\ 665\ \text{u} - 238.048\ 608\ \text{u}]c^2 \\
&= 1.936\ 382\ \text{u} \times c^2 \\
&= 1.936\ 382\ \text{u} \times c^2 \times \left(\frac{931.5\ \text{MeV}}{1\ \text{u} \times c^2}\right) \\
&= 1804\ \text{MeV}
\end{aligned}$$

The binding energies of the two nuclides are vastly different. However, if we divide by the number of nucleons, we find for $^{16}_{8}\text{O}$:

$$\frac{E_B}{A} = \frac{127.6\ \text{MeV}}{16} = 7.975\ \text{MeV/nucleon}$$

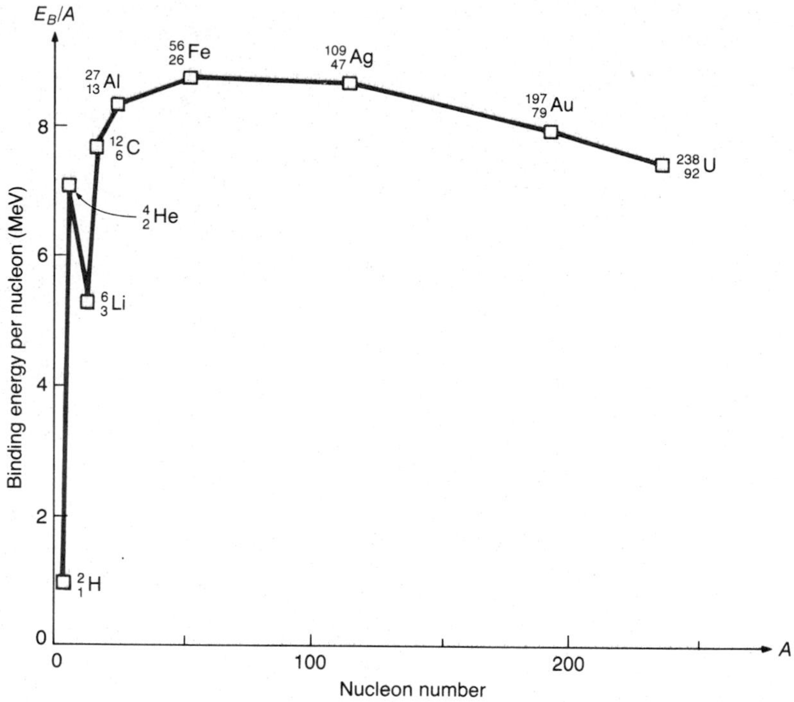

FIGURE 29.2 The binding energy per nucleon for some stable nuclides.

and for $^{238}_{92}$U:

$$\frac{E_B}{A} = \frac{1804 \text{ MeV}}{238} = 7.579 \text{ MeV/nucleon}$$

The binding energy per nucleon is strikingly similar, even though the two elements are at opposite ends of the periodic table. ■

Further calculations confirm that the binding energy per nucleon is approximately constant for all nuclei. Figure 29.2 shows the binding energy values for some stable nuclides. Apart from a few values at low nucleon numbers, all the binding energies lie between 7 and 9 MeV/nucleon. If we had extended our calculations to radioactive nuclides, the results would have been similar. Out of more than 1000 stable and radioactive nuclides, only about 20 have binding energies outside the range of 7 to 9 MeV/nucleon. All of these low-binding-energy nuclides have nucleon numbers of 12 or less.

The smallest nuclide, $^{2}_{1}$H, is given a special name—the deuteron. It is of great interest in the search for clues to the nature of the nuclear force, and it also very directly confirms our interpretation of the binding energy. The binding energy of the deuteron is about 2.22 MeV. The deuteron can be broken into its constituent neutron and proton by γ-ray bombardment, and the threshold γ-ray energy for the disintegration to occur is 2.22 MeV. This phenomenon is a kind of nuclear photoelectric effect in which the binding energy plays the role of the work function.

29.5 Fission and Nuclear Reactors

In the previous section we saw that the binding energy per nucleon is approximately constant for most nuclides, as displayed in Fig. 29.2. However, notice that the departure from a constant value is not random but systematic. The binding energy per nucleon is greatest for nucleon numbers between about 50 and 100 and decreases for nucleon numbers outside this range. This slow variation of binding energy per nucleon is of immense practical importance, providing the basis for energy-producing nuclear fission and fusion processes.

The discovery of nuclear fission was an accidental by-product of the search for transuranium elements. We have already seen that bombarding a nuclide with a neutron increases the atomic number if the product nuclide undergoes β^- decay. However, this simple approach ran into unusual difficulties with certain isotopes of uranium. In an epoch-opening paper in 1939, Hahn and

A nuclear power station uses nuclear fission to provide the power to raise steam. The steam drives a turbine which in turn drives an alternator to provide electrical power. (Courtesy U.S. Department of Energy)

Strassman showed that the bombardment of uranium by slow neutrons produced nuclides near the middle of the periodic table, which they identified as isotopes of xenon ($Z = 54$), barium ($Z = 56$), lanthanum ($Z = 57$), krypton ($Z = 36$), and strontium ($Z = 38$). Clearly, the uranium nucleus was breaking into two large fragments in a type of nuclear reaction that was previously unknown. The reaction, called nuclear **fission,** has been the subject of extensive research. Some of the main findings about the fission reaction are as follows:

1. Dozens of heavy nuclides undergo fission under a variety of circumstances. The most famous example is the fission of ${}^{235}_{92}\mathrm{U}$ by slow neutrons.
2. The typical products of fission are:
   a. Two large fragments
   b. Several neutrons
   c. An assortment of γ rays
3. The large fragments are relatively neutron rich and decay to a stable nuclide by a series of β^- emissions.
4. Occasionally a fission process produces a light nuclide in addition to the heavy fragments.
5. The Q value for a typical fission reaction is about 200 MeV.

The heavy fragments from a fission reaction are not always the same. Two typical reactions for the fission of ${}^{235}_{92}\mathrm{U}$ by slow neutrons are as follows:

$$
{}^{235}_{92}\mathrm{U} + {}^{1}_{0}\mathrm{n} \rightarrow {}^{141}_{56}\mathrm{Ba} + {}^{92}_{36}\mathrm{Kr} + 3 \times {}^{1}_{0}\mathrm{n}
$$

$$
{}^{235}_{92}\mathrm{U} + {}^{1}_{0}\mathrm{n} \rightarrow {}^{140}_{54}\mathrm{Xe} + {}^{94}_{38}\mathrm{Sr} + 2 \times {}^{1}_{0}\mathrm{n}
$$

In each case the heavy fragments are highly radioactive and decay in a series of β^- emission processes. For example, ${}^{140}_{54}\mathrm{Xe}$ undergoes four β^- decays to the stable nuclide ${}^{140}_{58}\mathrm{Ce}$.

The energy release in fission is truly enormous, as illustrated in the next example.

EXAMPLE 29.9

How much coal would have to be burned to produce the same amount of energy that is liberated by the complete fission of 1 kg of ${}^{235}_{92}\mathrm{U}$?

SOLUTION About 200 MeV of energy are liberated by the fission of each nucleus of ${}^{235}_{92}\mathrm{U}$. Since 1 mol of ${}^{235}_{92}\mathrm{U}$ has a mass of 235 g and contains 6.02×10^{23} atoms, we have

$$\text{Number of } {}^{235}_{92}\text{U nuclei in 1 kg} = 6.02 \times 10^{23} \times \frac{1000 \text{ g}}{235 \text{ g}}$$
$$= 2.562 \times 10^{24}$$
$$\text{Fission energy released} = 2.562 \times 10^{24} \times 200 \times 10^{6} \text{ eV}$$
$$= 5.124 \times 10^{32} \text{ eV}$$
$$= 5.124 \times 10^{32} \text{ eV} \times 1.6 \times 10^{-19} \text{ J/eV}$$
$$= 8.20 \times 10^{13} \text{ J}$$

According to Table 15.1, the energy of combustion of coal is about 3×10^{7} J/kg. Our result is as follows:

$$\text{Quantity of coal required} = \frac{8.20 \times 10^{13} \text{ J}}{3 \times 10^{7} \text{ J/kg}}$$
$$= 2.73 \times 10^{6} \text{ kg}$$ ■

The use of nuclear power for both civilian and military purposes is well known. A plant that produces energy from the fission process is called a *nuclear reactor,* and it generates large amounts of energy by establishing a *chain reaction.* This occurs because the fission process produces several neutrons, each of which can initiate a new fission process. Figure 29.3 shows the beginning of a chain reaction. Fissionable nuclei, such as ${}^{235}_{92}$U, are represented by open circles, and the figure shows each fission producing two heavy fragments (represented by closed circles) and two neutrons. Since each neutron can produce another fission, the reaction grows at a fast rate.

Our diagram of a chain reaction shows every neutron initiating another fission process, but this is usually far from the truth. The several neutrons produced in a fission decay have energies of the order of 1 MeV, and such fast neutrons are not particularly effective in causing fission in ${}^{235}_{92}$U. Moreover, they tend to escape from the reactor substance. To counteract this, the reactor contains material, called a *moderator,* whose purpose is to slow the neutrons. Graphite, water, and beryllium are frequently chosen for this purpose.

In addition, the average number of neutrons from one fission process that go on to cause another fission process must be controlled. A reactor is in a *critical* configuration when, on the average, exactly one neutron from each fission process causes another fission. In this situation the fission reaction proceeds smoothly at a constant rate. If less than one product neutron causes fission, the reactor is in a subcritical state and the reaction process dies away. When more than one product neutron causes fission, the reaction builds up in an uncontrolled way. Fortunately, a material such as cadmium can absorb neutrons and remove them from the chain reaction. Rods of such absorbing material (called control

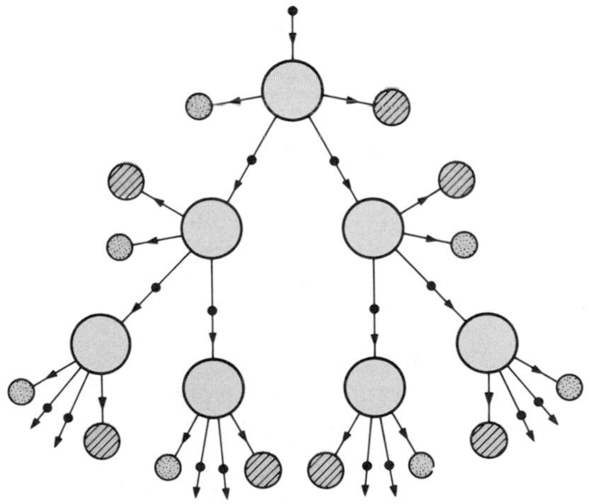

FIGURE 29.3 Illustration of a chain reaction. The open circles represent the fissionable nuclei, and the shaded circles represent fission fragments. The black dots are neutrons.

rods) are adjusted at various positions within the reactor. When the control rods are inserted well into the nuclear fuel material, most of the neutrons are absorbed and the reactor is subcritical. As the rods are withdrawn, more neutrons become available to cause fission, and at a certain stage the reactor becomes critical.

The world's first nuclear reactor was built under the stands in the University of Chicago football stadium in 1942. The construction was supervised by Enrico Fermi, who had emigrated from Italy to the United States a few years previously. The reactor was a giant pile of graphite bricks weighing 1400 tons in all. Alternate rows of graphite bricks contained lumps of uranium or uranium oxide, the graphite acting as a moderator to slow down the fast neutrons. Strips of cadmium between the rows of graphite bricks absorbed the neutrons and effectively prevented a chain reaction from taking place. Withdrawal of the cadmium strips caused a sharp increase in the neutron density within the pile, which went critical quicker than had been expected. Because of the construction of Fermi's first model, a nuclear reactor was called a pile for some years. Initially Fermi's reactor operated at only 0.5 W, but it was subsequently dismantled and rebuilt with improved safety shields. In its revised form, the pile produced tens of kilowatts of nuclear-fueled power.

The energy produced in a nuclear reactor is dissipated as heat in the reactor material. In the case of Fermi's nuclear pile, the carbon bricks become hot from the energy dissipation. However, if water is used as a moderator, the heat energy from the reactor can be extracted by arranging for a continuous flow of water through the critical area. The heat energy from the reactor can be used to raise steam, drive a steam turbine, and produce electrical power.

Figure 29.4 shows the component parts of a water-cooled nuclear reactor. The fuel rods contain fissionable material and are arranged within the reactor pressure vessel. The control rods can be moved into the fissionable material to shut the reactor down or withdrawn to produce the critical condition. Water within the pressure vessel serves both as a moderator and coolant, transferring heat energy to the heat exchanger so that steam can be produced. The steam drives a turbine and returns through a condenser to the heat exchanger. Pumps complete the circulation of both the reactor cooling water and the water used as the working substance for the turbine.

Nuclear power plants have undesirable environmental effects, as in fact do all types of power generating plants. The thermal pollution that results from the waste heat energy transferred to the condenser water is common to all operations that use a heat engine to drive an electrical generator. However, the nuclear plant poses some hazards of its own. One is the risk of the chain reaction getting out of hand, causing a meltdown of the reactor. High-grade engineering can reduce this danger as much as is desired. However, the second type of hazard—the spent fuel rods,

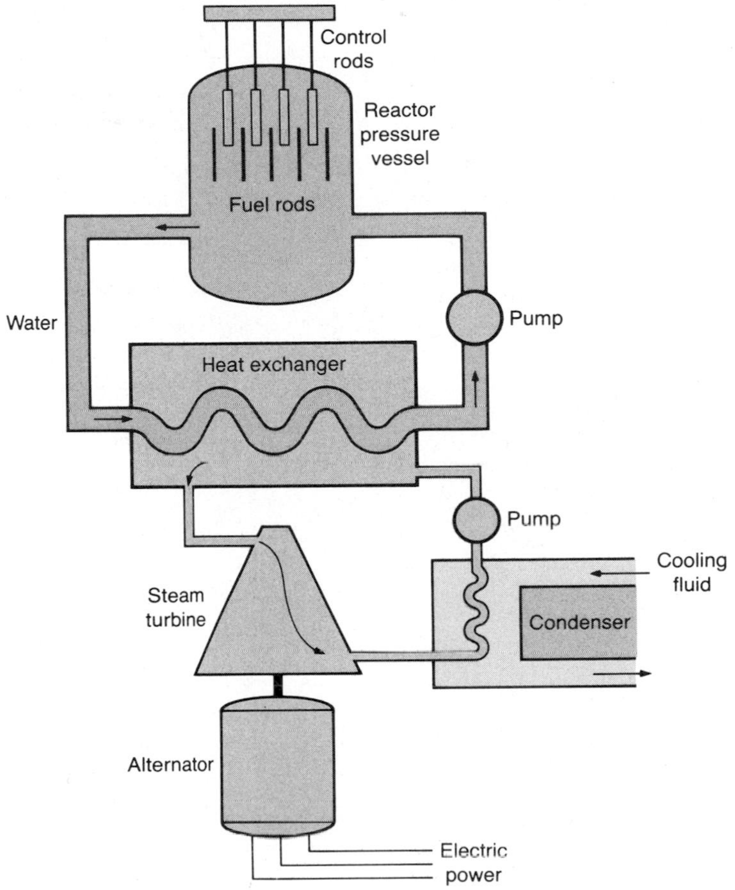

FIGURE 29.4 Components of a nuclear generator plant.

which are highly radioactive and which remain dangerous for many years—cannot be reduced by good design. Disposing of this nuclear waste safely remains one of the most vexing problems in the use of nuclear reactors.

29.6 Nuclear Fusion

When a heavy nucleus undergoes fission, the fragments possess more binding energy per nucleon than the original nucleus. Figure 29.2 clarifies this point; the original nucleus at the right-hand end of the diagram has a binding energy of about 7.5 MeV/nucleon. The heavy fragments are in the center of the diagram and have binding energies approximately in the range 8.2 to 8.7 MeV/nucleon. Glancing to the left of the same diagram, we see another possible means of producing energy. Some of the light nuclei, which have low binding energies per nucleon, could be fused together, thereby releasing energy. We call this process nuclear **fu-**

sion. To investigate fusion, we will work an example involving a deuteron (the nuclide ${}^2_1$H) and a triton (the nuclide ${}^3_1$H).

EXAMPLE 29.10

Calculate the Q value of the following reaction:

$$ {}^2_1\text{H} + {}^3_1\text{H} \rightarrow {}^4_2\text{He} + {}^1_0\text{n} $$

SOLUTION We have only to apply Eq. (29.3) and use the atomic masses from Table 29.1:

$$
\begin{aligned}
Q &= (M_I - M_P)c^2 \\
&= (2.014\ 102\ \text{u} + 3.016\ 050\ \text{u} - 4.002\ 603\ \text{u} - 1.008\ 665\ \text{u})c^2 \\
&= 0.018\ 884\ \text{u} \times c^2 \\
&= 0.018\ 884\ \text{u} \times c^2 \times \left(\frac{931.5\ \text{MeV}}{1\ \text{u} \times c^2}\right) \\
&= 17.6\ \text{MeV}
\end{aligned}
$$

■

At first sight this process seems to be much less efficient than fission, since a typical fission reaction produces about 200 MeV. However, the fusion process studied involves only 5 nucleons, while fission involves about 235. The energy yield in this particular fusion is therefore about 4 MeV/nucleon, while in fission it is less than 1 MeV/nucleon.

This wonderfully efficient source of energy production suffers from a serious drawback—it does not happen spontaneously in small systems! We can cause nuclear fission by arranging for a slow neutron to react with certain heavy nuclides, but the deuteron and the triton in the fusion reaction of Example 29.10 repel each other electrostatically. If they approach at slow velocity, they deflect and never meet each other. Then let us have them approach at high velocity! Although we could obtain the high velocities required by using an accelerator, nuclear scattering occurs more frequently than fusion. We would end up spending more energy accelerating the particles than we would obtain from the occasional fusion.

Another possibility for creating high-energy collisions is to produce a hot gas of the particles, so that the frequent random thermal collisions can provide many opportunities for the fusion reaction. Unfortunately, the gas temperature required is enormous, as we can see by the following example.

EXAMPLE 29.11

Experimentation shows that kinetic energies of about 10 keV are required for the fusion reaction to begin. Estimate the gas temperature that produces such an energy by thermal motion.

SOLUTION In a previous chapter we related the mean kinetic energy of a gas molecule to the absolute temperature [Eq. (13.7)]:

$$\text{Mean kinetic energy} = \tfrac{3}{2}kT$$

We have only to substitute in this equation, taking care to use SI units:

$$10^4 \text{ eV} \times 1.6 \times 10^{-19} \text{ J/eV} = \tfrac{3}{2} \times 1.38 \times 10^{-23} \text{ J/K} \times T$$

$$\therefore T = 7.7 \times 10^7 \text{ K}$$

■

But this is almost 100 million kelvin! At these high temperatures, most atoms are completely denuded of electrons, leaving only charged particles—the positively charged nuclei and the negatively charged electrons. Matter consisting of positively and negatively charged particles is termed plasma. The only known natural occurrence of such a high-temperature plasma is in the interior of stars. Containment of such a hot plasma on earth raises serious difficulties. The walls of a solid vessel would both cool the plasma (by absorbing energy from its particles) and contaminate it (by vaporization of surface atoms from the walls). These problems do not occur in a star because the huge gravitational force acts as a kind of a container that holds the plasma at the enormous pressures and temperatures involved. Containment in a huge gravitational field is out of the question on earth, but the charged particles of the plasma make possible another type of container.

Since charged particles experience a force in a magnetic field that is perpendicular to their velocities, a suitably designed magnetic field might act as a container. This effect does occur, but some outstanding problems need to be solved before fusion can be a practical reality. We need to contain a plasma of sufficiently high density and temperature for a long enough time interval to permit fusion processes to occur. In addition, plasma instabilities sometimes cause the magnetic containment system to fail unexpectedly.

We can express the problem of producing a controlled fusion reaction in simple terms. We need to make a small-sized star and extract the energy from the fusion. In the hydrogen bomb, the correct conditions for fusion are obtained by exploding a fission bomb. Although this technique produces a single output of energy of obviously sufficient magnitude, it is of little use for generating electrical power. For that purpose we need either a continuous energy output or at least small, controllable bursts of energy. The rewards are high for the production of a successful fusion reactor. Abundant fuel in the world's oceans could replace the scarce and costly uranium required for the fission reactor. In addition, the waste product of the fusion reaction is the stable gas helium. However, this does not mean that fusion has no radioactive wastes. Neutrons and γ rays from the fusion reaction would cause radioactivity in the plant material. However, it is highly probable that good engineering would create a much less difficult situation than the one associated with plutonium production in a fission reactor.

Let us conclude this section with a few comments about the fusion process in stars. The sun is a medium-sized star that radiates energy at the rate of about 4×10^{26} W. Geological and astronomical evidence indicates that the current power output has been proceeding for at least several billion years. If the sun were made of pure carbon and oxygen, the energy from their combustion could maintain the sun's power output for only a few thousand years. Although the interior of a star cannot be directly observed, the present physical model, which attributes solar energy output to fusion, gives a picture of the physics of the sun that is consistent with what is observed. Near the center of the sun, the temperature is some tens of millions of degrees kelvin in a plasma consisting mostly of protons and free electrons. The electromagnetic radiation in this inferno is chiefly in the X-ray and γ-ray region of the spectrum. The radiation takes some millions of years to reach the sun's surface. By this time multiple Compton scatterings have reduced the frequency so that the escaping radiation is mostly ultraviolet, visible, and infrared, and the energy lost is continually made up by the fusion processes deep in the sun's core. Of course, not even fusion gives an eternal supply of energy, and the sun will one day become a lifeless star that has exhausted its energy supply.

One more item remains in our discussion of stellar fusion. The primordial matter of the universe is hydrogen and not its more massive isotopes. However, our example of fusion used a deuteron and a triton rather than two protons. We need a reaction (or a set of reactions) that begins with protons. One such sequence is the **proton–proton chain.** One form of the chain is as follows:

$$^{1}_{1}\mathrm{H} + {}^{1}_{1}\mathrm{H} \rightarrow {}^{2}_{1}\mathrm{H} + {}^{0}_{1}\mathrm{e}$$

$$^{1}_{1}\mathrm{H} + {}^{2}_{1}\mathrm{H} \rightarrow {}^{3}_{2}\mathrm{He} + \gamma$$

$$^{3}_{2}\mathrm{He} + {}^{3}_{2}\mathrm{He} \rightarrow {}^{4}_{2}\mathrm{He} + 2 \times {}^{1}_{1}\mathrm{H}$$

As a result of this series of reactions, six protons are transformed into one α particle, two protons, two positrons, a γ ray, and an energy release of about 26 MeV.

Another possible set of reactions is the **carbon–nitrogen cycle,** which was proposed to account for energy production in the sun and in other stars of the main sequence. To begin this cycle of reactions, we need protons and the nuclide $^{12}_{6}\mathrm{C}$. The latter is not used up but is continually regenerated in the cycle. The reaction cycle is:

$$^{12}_{6}C + ^{1}_{1}H \rightarrow ^{13}_{7}N + \gamma$$
$$\downarrow$$
$$^{13}_{6}C + ^{0}_{1}e$$
$$^{13}_{6}C + ^{1}_{1}H \rightarrow ^{14}_{7}N + \gamma$$
$$^{14}_{7}N + ^{1}_{1}H \rightarrow ^{15}_{8}O + \gamma$$
$$\downarrow$$
$$^{15}_{7}N + ^{0}_{1}e$$
$$^{15}_{7}N + ^{1}_{1}H \rightarrow ^{12}_{6}C + ^{4}_{2}He$$

The net result is that four protons are transformed into an α particle, two positrons, and three γ rays.

Both the proton–proton chain and the carbon–nitrogen cycle probably occur in the sun, with the former being the more important energy source. The carbon–nitrogen cycle is believed to have a more important role in stars that have higher central temperatures than the sun.

In the proton–proton chain and the carbon–nitrogen cycle a neutrino is produced in each process involving an electron or a positron. If our model is correct, the sun should give a great output of neutrinos in addition to the electromagnetic radiation. However, recent research has shown the neutrino output to be much lower than anticipated, thus posing a relatively serious dilemma for astrophysicists. Either the measurement of the sun's output of neutrinos is in error, or the currently accepted model for the physics of the interior of the sun is faulty.

29.7 Elementary Particles

The question of which material things are truly elemental is one that people have been trying to answer for a very long time. The ancient Greeks named earth, water, air, and fire as the four basic elements from which all things were composed. With the discovery of atoms and the periodic table in the nineteenth century, the list of elementary particles grew to 92—the atoms from hydrogen to uranium. In the twentieth century, photons were discovered, and atoms were found to be composed of a nucleus (consisting of protons and neutrons) with surrounding electrons. For a couple of years in the early 1930s the number of elementary particles was back to four: the photon, the electron, the proton, and the neutron. Then, beginning with the neutrino (postulated by Pauli in 1934 to preserve energy conservation in β decay) and the muon (discovered in 1937 in photographic emulsions exposed to cosmic radiation) came the discovery of a staggering array of subatomic

TABLE 29.2 Fundamental interaction forces

Name	*Relative strength*	*Field particle*
Gravitational	10^{-40}	Graviton ?
Weak	10^{-13}	$W^{\pm}$ and Z particles
Electromagnetic	10^{-2}	Photon
Strong	1	Pion

particles. In the following discussion, we do not attempt to give a step-by-step history of how the current model was reached, but rather to simply summarize some of its essential features.

The basic criterion for separating particles into families or categories is the type of force by which they interact with each other. Physicists recognize four basic types of forces, which are listed in Table 29.2 along with their approximate relative strengths and the corresponding field particles. By a field particle, we mean the particle whose exchange is regarded as the source of the force in question. For example, quantum electrodynamics shows that we can regard the electromagnetic force as due to the exchange of virtual photons; the photons are called "virtual" because one of the versions of the uncertainty principle makes it impossible to conduct an experiment to observe them directly. Nevertheless, we regard the photon as the field particle of the electromagnetic force. In a similar way, the exchange of other virtual field particles mediates their corresponding fields. The graviton (mediator of gravitational force) has not been detected experimentally, but the mediating particles for the weak and strong forces have been found.

The families of particles shown in Table 29.3 are divided according to the types of forces they experience. To begin with, we see that there is no division on the basis of gravitational force since all particles experience it. Our first family contains only one particle, the photon, which participates in the electromagnetic interaction; its symbol is γ. The next family is the leptons, which interact via the weak interaction as well as by the electromagnetic (if they are charged) and the gravitational. The leptons are the electron, muon, and tauon (with symbols e^-, μ^-, and τ^-), together with their neutrinos (symbolized by ν_e, ν_μ, and ν_τ). The final category is the hadrons, which participate in the strong interaction as well as in all the other types of interaction. The hadrons are divided into two subfamilies. The first is the mesons, which comprise three pions (π^+, π^0, π^-) and two kaons (κ^+, κ^0). The second subfamily is the baryons, comprising the nucleons (p, n), one lambda (Λ^0), three sigmas (Σ^+, Σ^0, Σ^-), two xis (Ξ^0, Ξ^-), and one omega (Ω^-). There are literally hundreds of hadrons; in Table 29.3 we have listed only those that are stable or have a mean lifetime longer than about 10^{-19} s. The mean lifetimes are also listed in the table.

TABLE 29.3 Subatomic particles

Category	*Particle*	*Symbol*	*Antiparticle*	*Charge number (Q)*	*Baryon number (B)*	*Strangeness number (S)*	*Rest energy (MeV)*	*Mean lifetime (s)*
PHOTON	Photon	γ	Self	0	0	0	0	Stable
LEPTON	Electron	e^- or β^-	e^+ or β^+	−1	0	0	0.511	Stable
	e-Neutrino	ν_e	$\bar{\nu}_e$	0	0	0	0 ?	Stable
	Muon	μ^-	μ^+	−1	0	0	105.7	2.2×10^{-6}
	μ-Neutrino	ν_μ	$\bar{\nu}_\mu$	0	0	0	0 ?	Stable
	Tauon	τ^-	τ^+	−1	0	0	1748	5×10^{-13}
	τ-Neutrino	ν_τ	$\bar{\nu}_\tau$	0	0	0	0 ?	Stable
HADRON								
MESON	Pion	π^+	π^-	+1	0	0	139.6	2.6×10^{-8}
		π^0	Self	0	0	0	135.0	8×10^{-17}
		π^-	π^+	−1	0	0	139.6	2.6×10^{-8}
	Kaon	K^+	K^-	+1	0	+1	493.7	1.2×10^{-8}
		K^0	$\bar{K}^0$	0	0	+1	497.7	8.8×10^{-11}
BARYON	Nucleon	p	$\bar{p}$	+1	+1	0	938.3	Stable
		n	$\bar{n}$	0	+1	0	939.6	925
	Lambda	Λ^0	$\bar{\Lambda}^0$	0	+1	−1	1116	2.5×10^{-10}
	Sigma	Σ^+	$\bar{\Sigma}^+$	+1	+1	−1	1189	8×10^{-11}
		Σ^0	$\bar{\Sigma}^0$	0	+1	−1	1192	6×10^{-20}
		Σ^-	$\bar{\Sigma}^-$	−1	+1	−1	1197	1.5×10^{-10}
	Xi	Ξ^0	$\bar{\Xi}^0$	0	+1	−2	1315	2.9×10^{-10}
		Ξ^-	$\bar{\Xi}^-$	−1	+1	−2	1321	1.7×10^{-10}
	Omega	Ω^-	$\bar{\Omega}^-$	−1	+1	−3	1673	8.2×10^{-11}

NOTE: The charge, baryon, and strangeness numbers all refer to the particles; the corresponding numbers for the antiparticles are of opposite sign.

Every particle in the table is either uncharged or carries a positive or negative charge whose magnitude equals the magnitude of the electron charge. This is indicated by the charge number, Q, which is -1 for the electron charge and $+1$ for charge of opposite sign. Every particle also has an antiparticle (occasionally itself) whose mass is equal to the particle mass, but whose charge is of opposite sign to the particle charge. Particle rest energies are listed in the table as well. The remaining columns in the table are explained in the following section.

29.8 Conservation Laws

In their decays and interactions the particles of Table 29.3 obey various conservation laws, some well known in classical physics and others newly discovered in particle physics. Indeed, conservation laws are a major means of bringing order to the apparent chaos; they provide a systematic theory of why some decays and interactions occur while others have never been found. The lifetime of an unstable particle is an indicator of the type of interaction that is causing the decay; the weak interaction goes along with particles that have relatively long lifetimes—on the order of 10^{-10} s or longer. The electromagnetic interaction is much stronger, and goes along with particles that have shorter lifetimes—usually in the range of 10^{-19} s to 10^{-16} s. The strong interaction occurs with particles of even shorter lifetimes—on the order of 10^{-23} s; particles of such short lifetime are not listed in Table 29.3.

To begin with, we cite the conservation laws on energy, linear momentum, angular momentum, and charge; these laws are found to be strictly obeyed in all types of decays and reactions. In addition, we have seen conservation of nucleon number in radioactive decay and low-energy nuclear reactions. This is not a fundamental conservation law, however, and it must be replaced in high-energy processes. We replace it with a new conservation law, the **conservation of baryon number,** which includes conservation of nucleon number as a special case. We assign the baryon number $+1$ to particles in the baryon subcategory of hadrons, and -1 to their corresponding antiparticles. Both the meson subcategory of hadrons and the leptons have baryon number zero.

Consider how baryon number conservation applies to the reaction between a proton and a neutron. One possible reaction between these particles is the low-energy reaction:

$$\mathrm{p} + \mathrm{n} \rightarrow \mathrm{d} + \gamma$$

We have already investigated this reaction, and it obeys all of the conservation laws. However, it is not the only possibility; for example, we might try the reaction:

$$p + n \rightarrow p + p + \bar{p}$$

This, of course, obeys charge number conservation, but baryon number conservation is violated:

$$1 + 1 \neq 1 + 1 - 1$$

This reaction has never been observed.

On the other hand, it is not difficult to construct a reaction that does conserve baryon number:

$$p + n \rightarrow p + p + \bar{p} + n$$

Once again charge number is conserved, but this time baryon number is also conserved:

$$1 + 1 = 1 + 1 - 1 + 1$$

This reaction occurs if the reacting particles have sufficient kinetic energy to provide the rest energy required for the creation of the antiproton and the proton.

Another example of baryon number conservation is provided by proton–antiproton annihilation:

$$p + \bar{p} \rightarrow \pi^{+} + \pi^{0} + \pi^{-}$$

Charge number is conserved, and a baryon number conservation check gives:

$$1 - 1 = 0 + 0 + 0$$

Note that there is no conservation law on meson or photon numbers; thus, these particles can be created or annihilated in arbitrary numbers.

There is, however, a number conservation law on leptons, and it comes in three parts, one for each branch of the lepton family. We assign lepton numbers +1 to the particles and −1 to the antiparticles, and require conservation of lepton number within each family separately. Let us return once more to beta decay, which we have written up to this point as:

$$n \rightarrow p + \beta^{-}$$

Charge and baryon numbers are conserved here, but a check on lepton number shows the violation:

$$0 \neq 0 + 1$$

We need to add an antiparticle from the electron branch of the lepton family, and it must be $\bar{\nu}_e$ since adding e^{+} would violate charge number conservation. Finally β^{-} decay becomes:

$$n \rightarrow p + \beta^- + \bar{\nu}_e$$

To see how the various lepton number conservation laws hold separately, consider the decay of a muon into an electron of the same charge plus neutrinos:

$$\mu^+ \rightarrow e^+ + \nu_e + \bar{\nu}_\mu$$

Conservation of lepton number for the electron branch of the family gives:

$$0 = -1 + 1 + 0$$

and for the muon branch of the family:

$$-1 = 0 + 0 - 1$$

We see that the requirement for neutrinos to satisfy conservation laws goes well beyond the energy conservation consideration that originally led to their discovery.

EXAMPLE 29.12

Check the following decays for conservation of charge, baryon, and lepton numbers.

a. $n \rightarrow p + \mu^- + \bar{\nu}_\mu$

b. $n \rightarrow p + \pi^-$

Do any other conservation laws forbid decays that are otherwise permitted?

SOLUTION Let us check the conservation laws in order.

a.	$n \rightarrow p + \mu^- + \bar{\nu}_\mu$	
Charge:	$0 = 1 - 1 + 0$	(Yes)
Baryon:	$1 = 1 + 0 + 0$	(Yes)
Lepton (muon family):	$0 = 0 + 1 - 1$	(Yes)

The reaction is allowed by each law tested.

b.	$n \rightarrow p + \pi^-$	
Charge:	$0 = 1 - 1$	(Yes)
Baryon:	$1 = 1 + 0$	(Yes)
Lepton:	(Not applicable)	

The reaction is allowed by each law tested.

The conservation laws not tested include linear momentum, angular momentum, and energy. We can always satisfy linear momentum in a decay by having the fragments travel in different directions with velocities that produce a zero total final momentum. Given the spin angular momentum quantum numbers of the particles involved, it can also be shown that both of these decay modes conserve angular momentum. However, satisfaction of energy conservation in a decay demands that the rest energy of the decaying particle exceed the sum of the rest energies of the product particles. We see from Table 29.3 that this is not true for either of the decay processes considered; they are therefore forbidden, and β^- decay remains the only decay mode for the free neutron. ■

Our final number conservation law is really qualified conservation. When the K, Λ, and Σ particles were discovered in the 1950s, it was found that they were produced only in pairs. The following reaction was frequently observed:

$$\pi^- + p \rightarrow K^0 + \Lambda^0$$

On the other hand, a reaction such as:

$$\pi^- + p \rightarrow K^0 + n$$

was never observed in spite of the fact that it breaks no previous conservation law. This made the new particles strange, and the anomaly was removed by assigning a strangeness quantum number *(S)*, as indicated in Table 29.3; the antiparticles received an *S* value of opposite sign to that assigned to the particles. Checking the observed reaction for conservation of strangeness number gives:

$$\pi^- + p \rightarrow K^0 + \Lambda^0$$
$$0 + 0 = 1 - 1 \qquad \text{(Yes)}$$

while for the unobserved reaction we have:

$$\pi^- + p \rightarrow K^0 + n$$
$$0 + 0 \neq 1 + 0 \qquad \text{(No)}$$

But conservation of strangeness number is required only in processes that proceed via the strong interaction (such as the creation of the strange particles discussed above). Decay of strange particles can proceed by the weak interaction (as indicated by a relatively long lifetime) in such a way that strangeness number changes by ± 1, but not by a larger amount. For this reason, we refer to a qualified conservation of the strangeness number.

Consider, for example, the decay of the negative Ξ particle, which proceeds in two steps, as follows:

$$\Xi^- \rightarrow \Lambda^0 + \pi^-$$
$$\searrow$$
$$p + \pi^-$$

Checking the strangeness numbers for the first decay gives:

$$-2 \rightarrow -1 + 0 \qquad \text{(Yes; } S \text{ changes by } +1)$$

and for the second decay we have:

$$-1 \rightarrow 0 + 0 \qquad \text{(Yes; } S \text{ changes by } +1)$$

The decay $\Xi^- \rightarrow p + \pi^- + \pi^-$ is not observed. Although it obeys the previous conservation laws, it would require a strangeness change of $+2$, which is not allowed.

EXAMPLE 29.13

Check the following decays for conservation of charge, baryon, and strangeness numbers.

a. $\Xi^0 \rightarrow \Sigma^+ + \pi^-$

b. $\Omega^- \rightarrow \Sigma^0 + \pi^-$

Do other conservation laws forbid decays that are otherwise permitted?

SOLUTION We check the conservation laws in order.

a.	$\Xi^0 \rightarrow \Sigma^+ + \pi^-$	
Charge:	$0 = 1 - 1$	(Yes)
Baryon:	$1 = 1 + 0$	(Yes)
Strangeness:	$-2 \rightarrow -1 + 0$	(Yes; change of $+1$)

The decay is allowed by the weak interaction, as far as we have tested.

b.	$\Omega^- \rightarrow \Sigma^0 + \pi^-$	
Charge:	$-1 = 0 - 1$	(Yes)
Baryon:	$1 = 1 + 0$	(Yes)
Strangeness:	$-3 \rightarrow -1 + 0$	(No; change of $+2$)

The decay is not allowed by the weak interaction since the strangeness number changes by 2.

Linear momentum and angular momentum can be conserved for the decay of part (a), but examination of the data in Table 29.3 shows that the rest energy of the decaying particle is smaller than the sum of the rest energies of the two products. The decay of part (a) is therefore forbidden by energy conservation. ■

29.9 Quarks

Returning to our categories of particles, the photon is considered to be elementary and so too are the leptons. All attempts to find an internal structure in leptons have failed, however, and experiments to determine their size have shown them to be smaller than about 10^{-18} m.

On the other hand, there are hundreds of hadrons, and experiments do indicate that they have internal structure. In 1964 two American physicists, Murray Gell-Mann and George Zweig, proposed that hadrons are made up of more elemental particles called **quarks.** Originally, they proposed three quarks, called **up, down,** and **strange;** the number has since grown to six, with the three new quarks called **charm, bottom,** and **top.** The major innovation of quark theory is the assignment of charge numbers that are fractions of the electron charge. All quarks carry charges whose magnitude is either ⅓ or ⅔ the magnitude of the electron charge. Table 29.4 lists the six quarks, with their names and symbols; these names are called **flavors.** Also shown in the table are the charge, baryon, and strangeness quantum numbers.* Antiquarks (designated by a bar above the quark symbol) have charge, baryon, and strangeness numbers of opposite sign to the corresponding quark.

All of the nonstrange hadrons (that is, the pions, nucleons, and their antiparticles) are composed of u and d quarks and their antiparticles. In fact, these quarks—together with the electron, the electron-neutrino, and their antiparticles—are regarded as first-generation fundamental particles. All pions are a quark–antiquark pair; for example, π^- is a $d\bar{u}$ pair having charge number $-\frac{1}{3} - \frac{2}{3} = -1$ and baryon number $\frac{1}{3} - \frac{1}{3} = 0$. Its antiparticle is π^+, which is $u\bar{d}$, with charge number $\frac{2}{3} + \frac{1}{3} = 1$ and baryon number $\frac{1}{3} - \frac{1}{3} = 0$. Nucleons, on the other hand, are composed of three quarks; for example, n is udd, with charge number $\frac{2}{3} - \frac{1}{3} - \frac{1}{3} = 0$ and baryon number $\frac{1}{3} + \frac{1}{3} + \frac{1}{3} = 1$. Its antiparticle is $\bar{n}$, which is $\bar{u}\bar{d}\bar{d}$, with charge number $-\frac{2}{3} + \frac{1}{3} + \frac{1}{3} = 0$ and baryon number $-\frac{1}{3} - \frac{1}{3} - \frac{1}{3} = -1$. By making different combinations of u and d quarks as quark–antiquark pairs or as three quarks, we can easily construct all of the remaining nonstrange hadrons.

---

*There are also charm, bottom, and top quantum numbers not listed in Table 29.4 and not needed in this discussion.

TABLE 29.4 Some properties of quarks

Flavor	*Symbol*	*Antiquark*	*Charge number*	*Baryon number*	*Strangeness number*
Up	u	$\bar{u}$	$+\frac{2}{3}$	$\frac{1}{3}$	0
Down	d	$\bar{d}$	$-\frac{1}{3}$	$\frac{1}{3}$	0
Strange	s	$\bar{s}$	$-\frac{1}{3}$	$\frac{1}{3}$	-1
Charm	c	$\bar{c}$	$+\frac{2}{3}$	$\frac{1}{3}$	0
Bottom	b	$\bar{b}$	$-\frac{1}{3}$	$\frac{1}{3}$	0
Top	t	$\bar{t}$	$+\frac{2}{3}$	$\frac{1}{3}$	0

NOTE: Charge, baryon, and strangeness quantum numbers have the opposite sign for the antiquarks.

The s and c quarks, together with the next pair of leptons—the muon and its neutrino—are regarded as the second generation of fundamental particles. As shown in Table 29.4, the s quark has strangeness $S = -1$, and its antiquark $\bar{s}$ has strangeness $S = 1$. Without even having to introduce the c quark, this gives sufficient flexibility to form all of the remaining strange hadrons listed in Table 29.3. Replacing one of the nonstrange quarks by an s quark leads to combinations such as dds, which is the Σ^-. Its charge number is $-\frac{1}{3} - \frac{1}{3} - \frac{1}{3} = -1$, its baryon number is $\frac{1}{3} + \frac{1}{3} + \frac{1}{3} = 1$, and its strangeness number is $0 + 0 - 1 = -1$. If we replace two of the nonstrange quarks by an s quark, we come up with combinations such as uss, which is the Ξ^0. Its charge number is $\frac{2}{3} - \frac{1}{3} - \frac{1}{3} = 0$, its baryon number is $\frac{1}{3} + \frac{1}{3} + \frac{1}{3} = 1$, and its strangeness number is $0 - 1 - 1 = -2$. Finally, we could replace all three nonstrange quarks by s quarks to obtain sss (which has been called the king of the strange particles) and is Ω^-. Its charge number is $-\frac{1}{3} - \frac{1}{3} - \frac{1}{3} = -1$, its baryon number is $\frac{1}{3} + \frac{1}{3} + \frac{1}{3} = 1$, and its strangeness number is $-1 -1 -1 = -3$. All of these details are in agreement with the list of particle properties of Table 29.3.

Quarks do not seem to be capable of existing as independent particles. Moreover, they also seem to be capable of only certain types of combination; that is, all of the mesons are quark–antiquark systems and all of the baryons are three-quark systems. Other combinations would have nonintegral electronic charge, and have never been found with independent existence.

In addition to the quality that we have called flavor, quarks also have another quality that is called **color.** Every quark possesses a color; the usual assignment of colors is red, green, and blue (antiquark colors are correspondingly antired, antigreen, and antiblue). Baryons are made up of three quarks of different colors, and for this reason they are termed white (or colorless). Mesons are similarly colorless if composed of a certain color quark and an antiquark of its anticolor. The force that binds quarks together is called the **color force,** and its field particle is called the **gluon.** This means that the interaction between quarks is regarded as being due to the exchange of gluons, which mediate the color force in a way that is somewhat similar to the mediation of the electromagnetic force by the exchange of photons. There is an important difference, however, since gluons can exchange color between quarks, but photons cannot exchange charge between charged particles. However, the similarities are sufficient that the theory of the color force is called quantum chromodynamics, an analogy to the very successful quantum electrodynamics theory of electromagnetic interaction. The color force between quarks is now regarded as being the truly basic strong interaction. The strong interaction between hadrons that is mediated by pions (as mentioned in Table 29.2) is now regarded as a much weaker derivative of the immensely strong force mediated by the gluons.

Throughout most of the text the presentation has been fairly dogmatic because the basic principles are well understood in relation to many physical problems. Newtonian dynamics, electromagnetic theory, and thermodynamics are cases in point. If we stay within the limits of a well-traveled domain, our methods and conclusions are not open to serious doubt. However, physics is a science of discovery—it deals continually with the discovery of new principles. One way in which discoveries occur is by the attempt to extend the frontiers of the well-traveled domain. The whole of elementary particle physics is a major frontier of science and will probably remain so for some time to come.

KEY CONCEPTS

The **mass spectrograph** gives precise measurements of nuclear masses by measuring the path curvature of ions in a magnetic field.

The **atomic mass unit** is chosen so that the mass of $^{12}_{6}C$ is exactly 12.000 000 u. The rest energy of 1 u is approximately 931.5 MeV.

The ***Q* value** of a nuclear reaction has two equivalent definitions:

1. The net release of the energy in the reaction
2. The rest energy of the initial particles less the rest energy of the product particles

See Eq. (29.3).

The energy release in a radioactive decay is equal to the Q value of the decay. The energy is distributed among the decay fragments so as to conserve linear momentum.

Nuclear binding energy is the energy required to separate a nucleus into individual nucleons; see Eq. (29.4).

The binding energy per nucleon is greatest for elements in the middle of the periodic table and less for elements at each end. The **fission** of a heavy nucleus into lighter fragments and the **fusion** of two light nuclei into a heavier one both produce energy.

The basic types of interaction are **gravitational, weak, electromagnetic,** and **strong.** Elementary particles are divided into categories: the **photon, leptons,** and **hadrons.** Hadrons are subdivided into **mesons** and **baryons.** See Tables 29.2 and 29.3.

Reaction and decay processes for fundamental particles obey **conservation laws.** The conservation laws for strong interactions are for **energy, linear momentum, angular momentum, charge, baryon number, lepton number,** and **strangeness number.** For weak interactions, the strangeness number may change by unity.

Hadrons are composed of **quarks** and **antiquarks;** quarks are characterized by **flavor** and **color.** See Table 29.4.

QUESTIONS FOR THOUGHT

1. A free neutron decays into a proton and an electron in β^- decay. Explain why a free proton cannot decay into a neutron and a positron by β^+ decay. However, both processes can occur inside a nucleus. Why is a proton in a nucleus different from a free proton as far as β^+ decay is concerned?

2. How can both fission and fusion produce energy when one process separates nuclei and the other joins them together?

3. Distinguish clearly between nuclear mass number and nuclear mass. Are either of the two the same as the chemist's "atomic weight"?

4. When an electron and positron annihilate, either two or three photons are created. Why cannot the annihilation produce only one photon?

5. In Chapter 28 we noted that the decay rate of radioactive nuclides does not depend on the temperature. Is this strictly true, or would you qualify the statement?

6. The Einstein mass–energy relation assigns immense amounts of energy to relatively small masses of any substance. Does any basic law prevent one from inventing a device to turn most of this energy to useful purposes?

7. How do you decide whether a particle is stable against decay into a number of other given particles? Can you express your decision as one simple criterion?

8. Which of the conservation laws of particle physics is the replacement for Newton's laws of motion?

PROBLEMS

(Use Tables 29.1 and 29.3 in solving the problems.)

A. Single-Substitution Problems

1. A singly charged ion of ${}^{12}_{6}C$ moves in a path of radius 26.3 cm, under the influence of a magnetic field of 0.4 T, in a mass spectrograph. Calculate the accelerating voltage. [29.1]

2. Calculate the mass of a singly charged ion that moves in a path of radius 21.3 cm in a mass spectrograph. The accelerating voltage is 2.5 kV and the deflecting magnetic field is 0.3 T. [29.1]

3. A singly charged chromium ion of mass 8.36×10^{-26} kg is accelerated in a mass spectrometer by an accelerating voltage of 1.8 kV. Calculate the value of the magnetic field required to cause a path of 30 cm radius. [29.1]

4. Calculate the Q value of the reaction

$${}^{11}_{5}B + {}^{1}_{1}H \rightarrow {}^{8}_{4}Be + {}^{4}_{2}He \qquad [29.2]$$

5. Calculate the Q value of the reaction

$${}^{16}_{8}O + {}^{2}_{1}H \rightarrow {}^{14}_{7}N + {}^{4}_{2}He \qquad [29.2]$$

6. Calculate the binding energy of the deuteron (${}^{2}_{1}H$). [29.4]

7. Calculate the binding energy per nucleon of ${}^{3}_{2}He$. [29.4]

8. Calculate the binding energy per nucleon of ${}^{4}_{2}He$. [29.4]

B. Standard-Level Problems

9. Calculate the Q values for the following reactions:

a. ${}^{10}_{5}B + {}^{4}_{2}He \rightarrow {}^{13}_{6}C + {}^{1}_{1}H$

b. ${}^{7}_{3}Li + {}^{4}_{2}He \rightarrow {}^{10}_{5}B + {}^{1}_{0}n$

c. ${}^{7}_{3}Li + {}^{1}_{1}H \rightarrow {}^{6}_{3}Li + {}^{2}_{1}H$

Which of the reactions are exoergic and which are endoergic? [29.2]

10. Calculate the energy of the γ ray emitted in the following reactions:

a. ${}^{1}_{1}H + {}^{1}_{0}n \rightarrow {}^{2}_{1}H + \gamma$

b. ${}^{2}_{1}H + {}^{1}_{0}n \rightarrow {}^{3}_{1}H + \gamma$

c. ${}^{27}_{13}Al + {}^{1}_{0}n \rightarrow {}^{28}_{13}Al + \gamma$

In every case assume that the kinetic energies of the neutron and the product nucleus are negligible. [29.2]

11. Calculate the Q values for each of the following reactions:

a. ${}^{18}_{8}O + {}^{1}_{1}H \rightarrow {}^{18}_{9}F + {}^{1}_{0}n$

b. ${}^{16}_{8}O + {}^{2}_{1}H \rightarrow {}^{14}_{7}N + {}^{4}_{2}He$

c. $^{14}_{7}\text{N} + ^{2}_{1}\text{H} \rightarrow ^{15}_{7}\text{N} + ^{1}_{1}\text{H}$

Which of the reactions are exoergic and which are endoergic?

12. Calculate the minimum energy required for a photon to break $^{7}_{3}\text{Li}$ into an α particle and a triton.

13. A slow neutron bombarding $^{6}_{3}\text{Li}$ causes the following reaction:

$$^{6}_{3}\text{Li} + ^{1}_{0}\text{n} \rightarrow ^{4}_{2}\text{He} + ^{3}_{1}\text{H}$$

Calculate the kinetic energy of each of the product nuclides.

14. Slow neutrons are detected because of the α particles that are emitted in the reaction

$$^{1}_{0}\text{n} + ^{10}_{5}\text{B} \rightarrow ^{7}_{3}\text{Li} + ^{4}_{2}\text{He}$$

Calculate the kinetic energy of each of the product nuclides.

15. It is conceivable that $^{4}_{2}\text{He}$ could decay into $^{3}_{2}\text{He}$ by emitting a neutron. Show that such a decay is energetically impossible, and calculate its Q value. [29.3]

16. The nuclides $^{8}_{4}\text{Be}$, $^{12}_{6}\text{C}$, and $^{16}_{8}\text{O}$ could conceivably decay into two, three, and four α particles, respectively. Which of the nuclides are stable, and which are unstable against these decays? [29.3]

17. Is the nuclide $^{18}_{8}\text{O}$ stable against α decay? Is it stable against β^- decay?

18. Is the nuclide $^{22}_{11}\text{Na}$ stable against β^+ decay? Is it stable against electron capture?

19. Calculate the maximum energy of the electrons emitted in the β^- decay of $^{14}_{6}\text{C}$. (Ignore the recoil kinetic energy of the daughter nuclide.)

20. The nuclide $^{3}_{1}\text{H}$ is unstable against β^- decay. Identify the daughter nuclide and calculate the maximum kinetic energy of the β^- particle. (Ignore the recoil kinetic energy of the daughter nuclide.)

21. The nuclide $^{226}_{88}\text{Ra}$ emits an α particle of kinetic energy 4.78 MeV. Calculate:

a. the recoil velocity of the daughter nuclide
b. the kinetic energy of the daughter nuclide
c. the mass of the daughter nuclide

22. The nuclide $^{237}_{93}\text{Np}$ emits an α particle of kinetic energy 4.87 MeV. Calculate the Q value of the decay, and hence the atomic mass of $^{233}_{91}\text{Pa}$.

23. The average Q value for fission of $^{235}_{92}\text{U}$ is 200 MeV. Calculate the mass of $^{235}_{92}\text{U}$ required to produce 1000 MW of fission power for 30 days. (The mass of the $^{235}_{92}\text{U}$ atom is about 3.9×10^{-25} kg.) [29.5]

24. A slow neutron produces fission in the nuclide $^{235}_{92}\text{U}$ with an energy release of 200 MeV. Calculate the total mass of all the products.

25. A nuclear-powered warship requires 85 000 hp to drive it at 30 knots. Assuming an overall 10% efficiency for the conversion of mass to propulsive energy, what is the mass reduction of the fuel in driving the ship 20 000 nautical miles?

26. The Q value for fusion of a deuteron ($^{2}_{1}\text{H}$) and a triton ($^{3}_{1}\text{H}$) is 17.6 MeV. Calculate the mass of material that produces 1000 MW of fusion power for 30 days. (The mass of the deuteron is 3.34×10^{-27} kg, and the mass of the triton is 5.01×10^{-27} kg.) [29.6]

27. Calculate the energy released in the proton–proton chain. [29.6]

28. Consider a volume of seawater that is a cube 1 km on edge. (This is a tiny fraction of the water in the world's oceans.) Now suppose that the protons in the hydrogen could all undergo a fusion reaction that produces electric power with 20% overall efficiency. For how long a period of time could the protons supply 1 kW continuously to every one of 200 million people? (Assume that 25 MeV of energy are released for each $^{4}_{2}\text{He}$ nuclide produced.)

29. The sun radiates energy at the rate of about 4×10^{23} kW. Calculate the sun's rate of loss of mass in kilograms per second. What fraction of its mass does the sun lose in a century?

30. Check the following decays for possible violations of conservation laws:

a. $\text{p} \rightarrow \text{n} + \beta^+ + \nu_e$
b. $\text{p} \rightarrow \text{n} + \mu^+ + \nu_\mu$
c. $\overline{\text{n}} \rightarrow \overline{\text{p}} + \beta^+ + \nu_e$ [29.8]

31. Check the following reactions for possible violations of conservation laws:

a. $\text{p} + \overline{\text{p}} \rightarrow \text{n} + \overline{\text{n}} + \gamma$
b. $\text{n} + \overline{\text{p}} \rightarrow \gamma + \gamma$
c. $\text{n} + \overline{\text{n}} \rightarrow \gamma + \gamma$ [29.8]

32. Check the following reactions for possible violations of the conservation laws for the strong interaction:

a. $\text{p} + \text{p} \rightarrow \Lambda^0 + \text{K}^+ + \text{n} + \pi^+$
b. $\text{p} + \text{p} \rightarrow \Lambda^0 + \overline{\Lambda}^0$
c. $\text{p} + \pi^+ \rightarrow \Sigma^0 + \text{K}^0 + \pi^+$

33. Check the following decays for possible violations of the conservation laws for the weak interaction:

a. $\Lambda^0 \rightarrow \text{p} + \text{K}^-$
b. $\Sigma^0 \rightarrow \text{n} + \text{K}^0$
c. $\Omega^- \rightarrow \overline{\text{p}} + \overline{\text{K}}^0$

A

ARITHMETIC AND ALGEBRA

A.1 SCIENTIFIC NOTATION

In physics problems, we frequently deal with numbers that are either very large or very small. For example, the mean distance from the earth to the sun is about 149 600 000 000 m, and the distance between the sodium and chlorine ions in common salt is about 0.000 000 000 235 m. We have expressed both numbers in standard decimal notation, and they are very unwieldy. We can express both numbers more conveniently in *scientific notation*. To do this, we write a number between 1 and 10, and multiply by a power of 10. Using this scheme, the earth–sun distance is 1.496×10^{11} m. The quantity 10^{11} means 10 multiplied by itself 11 times, and we refer to the number 11 as the exponent. Negative exponents indicate division by the appropriate power of 10. Thus, the interionic distance in common salt is 2.35×10^{-10} m, where $10^{-10} = 1/10^{10}$. Scientific notation presents both numbers in a very convenient form. To transfer back to standard decimal notation, we shift the decimal point by a number of places equal to the power of 10, to the right if the power is positive and to the left if it is negative.

The common arithmetic operations are addition, subtraction, multiplication, division, and taking powers or roots. When carrying out these operations on numbers that are written in scientific notation, we must follow certain simple rules. The examples that follow are typical physics problems, but we will concentrate only on the arithmetic procedures involving scientific notation.

EXAMPLE A1.1

Calculate the difference in frequency between two radio waves whose individual frequencies are $\nu_1 = 8.59 \times 10^7$ Hz and $\nu_2 = 6.23 \times 10^6$ Hz.

SOLUTION The frequency difference is given by:

$$\nu_1 - \nu_2 = 8.59 \times 10^7\ \text{Hz} - 6.23 \times 10^6\ \text{Hz}$$

If we enter this problem on a calculator that features scientific notation, the calculator carries out the subtraction correctly with the numbers just as we have written them. But if we wish to do the problem without a calculator, we must prepare the data by writing each number with the same power of 10. Usually it is convenient to select the smaller power, but either way is correct. Our frequency difference is:

$$\begin{aligned}\nu_1 - \nu_2 &= 8.59 \times 10^7\ \text{Hz} - 6.23 \times 10^6\ \text{Hz}\\ &= 85.9 \times 10^6\ \text{Hz} - 6.23 \times 10^6\ \text{Hz}\\ &= 79.67 \times 10^6\ \text{Hz}\\ &= 7.97 \times 10^7\ \text{Hz}\end{aligned}$$

In the final step, we have rounded off the significant digits and adjusted the power of 10 to place the initial portion of the number between 1 and 10. To add numbers in scientific notation, we follow the same method—the power of 10 must be the same in each of the numbers to be added. ■

There is no need to adjust the powers of 10 when multiplying or dividing numbers in scientific notation; we simply add the powers of 10 in multiplication and subtract them in division. Some examples illustrate the point.

EXAMPLE A1.2

In Eq. (26.2), we see that the quantum energy of a photon is equal to Planck's constant (6.63×10^{-34} J · s) multiplied by the photon frequency. Calculate the quantum energy of a photon whose frequency is 8.12×10^{14} Hz.

SOLUTION The quantum energy is given by Eq. (26.2)·

$$\begin{aligned} E &= h\nu \\ &= 6.63 \times 10^{-34}\ \text{J}\cdot\text{s} \times 8.12 \times 10^{14}\ \text{Hz} \\ &= 53.8 \times 10^{-20}\ \text{J} \\ &= 5.38 \times 10^{-19}\ \text{J} \end{aligned}$$

In the first step we multiply 6.63 and 8.12 to get 53.8 and calculate the power of 10 by adding exponents:

$$10^{-34} \times 10^{14} = 10^{(-34+14)} = 10^{-20}$$

In the final step, we alter the power of 10 to make the initial portion of the number fall between 1 and 10. ■

EXAMPLE A1.3

The radius of the path of a charged particle moving in a magnetic field is given by Eq. (19.10). Calculate the path radius for an electron (charge 1.602×10^{-19} C and mass 9.11×10^{-31} kg) moving with a velocity of 4.2×10^{7} m/s in a magnetic field of 4.8×10^{-3} T.

SOLUTION We calculate the path radius from Eq. (19.10):

$$\begin{aligned} r &= \frac{mv}{qB} \\ &= \frac{9.11 \times 10^{-31}\ \text{kg} \times 4.2 \times 10^{7}\ \text{m/s}}{1.602 \times 10^{-19}\ \text{C} \times 4.8 \times 10^{-3}\ \text{T}} \\ &= 4.98 \times 10^{-2}\ \text{m} \end{aligned}$$

We obtain this result by first multiplying the numbers before the powers of 10:

$$\frac{9.11 \times 4.2}{1.6 \times 4.8} = 4.98$$

The second step is to calculate the exponent:

$$10^{(-31+7)-(-19-3)} = 10^{-2}$$ ■

To take the square root of a number in scientific notation, we take the square root of the first part of the number and divide the power of 10 by 2. Let us illustrate this with an example.

EXAMPLE A1.4

Use Eq. (7.2) to calculate the velocity of an electron whose kinetic energy is 1.64×10^{-19} J. (The electron mass is 9.11×10^{-31} kg.)

SOLUTION The kinetic energy is given by Eq. (7.2):

$$\begin{aligned} \text{KE} &= \tfrac{1}{2}mv^2 \\ 1.64 \times 10^{-19}\ \text{J} &= \tfrac{1}{2} \times 9.11 \times 10^{-31}\ \text{kg} \times v^2 \\ \therefore v^2 &= \frac{2 \times 1.64 \times 10^{-19}\ \text{J}}{9.11 \times 10^{-31}\ \text{kg}} \\ &= 3.60 \times 10^{11}\ \text{m}^2/\text{s}^2 \end{aligned}$$

Using the procedure indicated, we have:

$$\begin{aligned} v &= \sqrt{3.60} \times 10^{11/2}\ \text{m/s} \\ &= \sqrt{3.60} \times \sqrt{10} \times 10^{5}\ \text{m/s} \\ &= 1.897 \times 3.162 \times 10^{5}\ \text{m/s} \\ &= 6.00 \times 10^{5}\ \text{m/s} \end{aligned}$$

We can avoid this awkward and painful process by selecting an even power of 10 in the value for v^2. By this method we have:

$$\begin{aligned} v^2 &= 3.60 \times 10^{11}\ \text{m}^2/\text{s}^2 \\ &= 36.0 \times 10^{10}\ \text{m}^2/\text{s}^2 \\ \therefore v &= 6.0 \times 10^{5}\ \text{m/s} \end{aligned}$$

The numbers in this example are contrived, but it always makes life easier if we make the exponent even before taking a square root. ■

A.2 LINEAR ALGEBRAIC EQUATIONS

A linear equation contains only the first power of the unknown quantities. Perhaps one of the most common equations in physics is a single equation containing only one unknown quantity. We write such an equation:

$$ax + b = 0$$

where x is the unknown quantity, and both a and b are known constants. To solve for x, we first subtract b from each side of the equation to give:

$$ax = -b$$

Now we divide both sides by a for the result:

$$x = -\frac{b}{a}$$

The solution of a single linear equation is a simple matter, and we do not need to memorize a formula. Let us work an example.

EXAMPLE A2.1

The equations of uniformly accelerated motion provide us with formulas that relate various kinematic quantities. Calculate the acceleration of an object that accelerates uniformly from an initial velocity of 4 m/s to a final velocity of 18 m/s in an elapsed time of 2 s.

SOLUTION We solve the problem by using Eq. (2.8):

$$v = v_0 + at$$

$$18 \text{ m/s} = 4 \text{ m/s} + a \times 2 \text{ s}$$

The equation is linear in the single unknown quantity a. We subtract 4 m/s from each side to isolate the term containing the unknown:

$$14 \text{ m/s} = a \times 2 \text{ s}$$

Dividing each side by 2 s gives us our result:

$$\therefore a = 7 \text{ m/s}^2$$

■

If the problem to be solved contains two unknown quantities, we need two equations to obtain the solution. We write such a pair of equations:

$$ax + by = c$$

$$dx + fy = g$$

where x and y are the unknown quantities, and a, b, c, d, f, and g are known constants. We could proceed to solve this pair of equations in terms of the known constants, but an actual example illustrates the method more simply.

EXAMPLE A2.2

The use of Kirchhoff's laws to solve electric circuit problems often leads to simultaneous linear equations. In one such problem we obtain the equations:

$$3\,\Omega \times I_1 + 5\,\Omega \times I_2 = 6 \text{ V}$$

$$6\,\Omega \times I_1 - 2\,\Omega \times I_2 = -6 \text{ V}$$

Solve for the unknown currents I_1 and I_2.

SOLUTION We reduce the two equations containing two unknowns to a single equation with one unknown quantity by using a technique called Gaussian elimination. To do this, we multiply either equation by a suitable number so that addition or subtraction will eliminate one of the variables. For example, we can multiply the first equation by 2 and then subtract the second equation from it:

$$6\,\Omega \times I_1 + 10\,\Omega \times I_2 = 12 \text{ V}$$

$$\underline{6\,\Omega \times I_1 - 2\,\Omega \times I_2 = -6 \text{ V}}$$

$$12\,\Omega \times I_2 = 18 \text{ V}$$

$$\therefore I_2 = 1.5 \text{ A}$$

Substituting this value of I_2 in the second equation gives:

$$6\,\Omega \times I_1 - 2\,\Omega \times 1.5 \text{ A} = -6 \text{ V}$$

$$6\,\Omega \times I_1 - 3 \text{ V} = -6 \text{ V}$$

$$6\,\Omega \times I_1 = -3 \text{ V}$$

$$\therefore I_1 = -0.5 \text{ A}$$

We could just as easily eliminate I_2 from the original pair of equations. To do this, we multiply the second equation by 2.5 and add it to the first, obtaining exactly the same answers. ■

If three or more variables are present in simultaneous linear equations, we proceed in exactly the same manner by first eliminating one variable from the system of equations. However, most elementary physics problems do not lead to sets of equations containing more than two variables.

A.3 QUADRATIC EQUATIONS

A quadratic equation contains both first and second powers of an unknown quantity. The standard form of quadratic equation is:

$$ax^2 + bx + c = 0$$

where x is the unknown quantity, and a, b, and c are known constants. As we see below, there are two possible solutions, which are called the roots of the quadratic equation. To find a formula for the roots, we use a method known as completing the square.

We begin by dividing the equation by a to give:

$$x^2 + \frac{b}{a}x + \frac{c}{a} = 0$$

Now add and subtract $b^2/4a^2$ as follows:

$$x^2 + \frac{b}{a}x + \frac{b^2}{4a^2} + \frac{c}{a} - \frac{b^2}{4a^2} = 0$$

The first three terms in this equation are a perfect square, $(x + b/2a)^2$. Taking the remaining terms to the right-hand side, we have:

$$\left(x + \frac{b}{2a}\right)^2 = \frac{b^2 - 4ac}{4a^2}$$

Now take the square root of each side:

$$x + \frac{b}{2a} = \frac{\pm\sqrt{b^2 - 4ac}}{2a}$$

Subtracting b/a from each side yields our result:

$$x = \frac{-b \pm \sqrt{b^2 - 4ac}}{2a}$$

Many people choose to memorize this formula for the roots of a quadratic equation in order to avoid the tedious derivation of it. An example of a quadratic equation arises in the kinematics of uniform accelerated motion.

EXAMPLE A3.1

An object accelerates uniformly at 6 m/s^2 from an initial velocity of 12 m/s. Its displacement is 96 m. Calculate the elapsed time.

SOLUTION We use Eq. (2.10) to perform the required calculation:

$$x = v_0t + \tfrac{1}{2}at^2$$

$$96 \text{ m} = 12 \text{ m/s} \times t + \tfrac{1}{2} \times 6 \text{ m/s}^2 \times t^2$$

The unknown quantity is t, and we write the equation in standard quadratic form as follows:

$$3 \text{ m/s}^2 \times t^2 + 12 \text{ m/s} \times t - 96 \text{ m} = 0$$

The quadratic equation formula gives:

$$t = \frac{-b \pm \sqrt{b^2 - 4ac}}{2a}$$

$$= \frac{-12 \text{ m/s} \pm \sqrt{144 \text{ m}^2/\text{s}^2 + 4 \times 3 \text{ m/s}^2 \times 96 \text{ m}}}{6 \text{ m/s}^2}$$

$$= \frac{-12 \text{ m/s} \pm \sqrt{1296 \text{ m}^2/\text{s}^2}}{6 \text{ m/s}^2}$$

$$= \frac{-12 \text{ m/s} \pm 36 \text{ m/s}}{6 \text{ m/s}^2}$$

$$= 4 \text{ s} \quad \text{or} \quad -8 \text{ s}$$

In Chapter 2 we discussed the physical significance of two answers to the problem. Our purpose here is simply to follow the steps involved in solving the equation. ■

A.4 LOGARITHMS AND EXPONENTIALS

The logarithmic and exponential functions are inverse functions. Thus, if $y = 10^x$, then $x = \log_{10} y$, where the notation $\log_{10}$ means logarithm to the base 10.

Logarithms to the base 10 occur in the definition of the sound intensity level given by Eq. (12.11) and repeated for convenience:

$$\beta = 10 \log_{10}(I/I_0)$$

The sound intensity is I, the threshold intensity is I_0 (10^{-12} W/m^2), and the sound intensity level in decibels is β. Let us illustrate the details of this formula with an example.

EXAMPLE A4.1

a. Calculate the intensity level of a sound whose intensity is 0.0056 W/m^2.

b. Calculate the intensity of a sound whose intensity level is 56 dB.

SOLUTION

a. We simply use the definition of the decibel:

$$\beta = 10 \log_{10} (I/I_0)$$

$$= 10 \log_{10} \left(\frac{0.0056 \text{ W/m}^2}{10^{-12} \text{ W/m}^2}\right)$$

$$= 10 \log_{10} (5.6 \times 10^9)$$

$$= 10 \times 9.748$$

$$= 97.5$$

b. For this part of the problem, we must use the definition in an inverse way:

$$\beta = 10 \log_{10} (I/I_0)$$

$$56 = 10 \log_{10} \left(\frac{I}{10^{-12}\ \text{W/m}^2}\right)$$

$$5.6 = \log_{10} \left(\frac{I}{10^{-12}\ \text{W/m}^2}\right)$$

For our definition of the logarithmic and exponential functions as inverses, this means that:

$$10^{5.6} = \frac{I}{10^{-12}\ \text{W/m}^2}$$

$$3.981 \times 10^5 = \frac{I}{10^{-12}\ \text{W/m}^2}$$

$$\therefore I = 3.98 \times 10^{-7}\ \text{W/m}^2$$

(*Note:* On many calculators $\log_{10}$ is obtained by pressing the button marked *log*, and 10^x is obtained by pressing *inv* followed by *log*.) ■

Logarithms and exponentials to the natural base are related by the $y = e^x$ and $x = \ln y$ where $e = 2.71828...$ is the natural base. This base is called *natural* because the slope of the tangent to the graph of $y = e^x$ at any point is e^x evaluated at that point.

Logarithms to the natural base occur in our description of radioactive decay. Repeating Eq. (28.1):

$$N(t) = N_0 e^{-\lambda t}$$

where $N(t)$ is the quantity of radioactive substance present at time t, N_0 is the initial quantity, and λ is the decay constant. An example illustrates the use of this formula.

EXAMPLE A4.2

The decay constant of $^{30}_{15}\text{P}$ is 0.272/min.

a. What fraction of a freshly prepared sample of this substance remains after 10 min?

b. What time must elapse for the amount of $^{30}_{15}\text{P}$ to reduce to 1% of its initial value?

SOLUTION

a. We proceed straight from the radioactive decay formula:

$$N(t) = N_0 e^{-\lambda t}$$

$$= N_0 e^{-(0.272/\text{min}) \times 10\ \text{min}}$$

$$= N_0 e^{-2.72}$$

$$= N_0 \times 0.06587$$

$$\therefore \frac{N(t)}{N_0} = 0.06587$$

b. This problem requires us to use the inverse property of the logarithmic function:

$$N(t) = N_0 e^{-\lambda t}$$

$$0.01 = e^{-(0.272/\text{min}) \times t}$$

From our definition of the exponential and logarithmic functions as mutual inverses, this means that:

$$\ln(0.01) = -(0.272/\text{min}) \times t$$

$$-4.605 = -(0.272/\text{min}) \times t$$

$$\therefore t = 16.9\ \text{min}$$ ■

A.5 A USEFUL TRIGONOMETRIC APPROXIMATION

Many physics problems involve trigonometry with small angles. By a small angle we mean one whose measure in radians is very much less than unity. (Remember that 1 rad $\simeq$ 57.29°; thus, angles smaller than about 10° qualify as small for many purposes.) A very useful approximation is available for small angles, and we illustrate as follows. Referring to the diagram, let *OPQ* be the sector of a circle, and let the angles *OAP* and *OQB* be right angles. From the definition of angular measure in radians [Eq. (6.1)] and the definitions of sine and tangent [Eq. (3.1)], we have:

$$\theta = \frac{\text{arc } PQ}{\text{radius } OQ}$$

$$\sin\theta = \frac{\text{length } AP}{\text{radius } OP}$$

$$= \frac{\text{length } AP}{\text{radius } OQ}$$

$$\tan\theta = \frac{\text{length } BQ}{\text{radius } OQ}$$

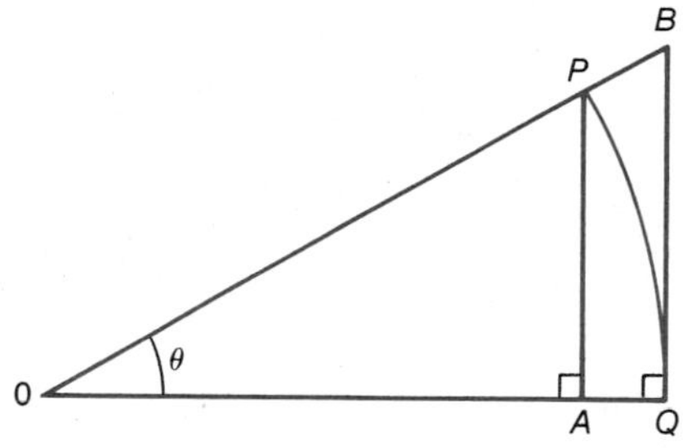

The circular arc PQ, the length AP, and the length BQ are all approximately equal for a small angle θ. To the extent to which this intuitive geometric approximation is valid, we have:

$$\theta \simeq \sin\theta \simeq \tan\theta \quad \text{if} \quad \theta \ll 1 \text{ rad}$$

Let us check this approximation for an angle of 10°. Using a calculator (or tables), and keeping four significant digits, we find:

$$10° = 0.1745 \text{ rad}$$

$$\sin 10° = 0.1736$$

$$\tan 10° = 0.1763$$

All of these values agree to within about 1.5%; for smaller angles, the agreement becomes progressively better.

B

LIST OF SYMBOLS WITH USUAL MEANINGS

Latin letter	*Meaning as an algebraic symbol*	*Meaning as a unit abbreviation or prefix*
a	acceleration	—
A	area amplitude of SHM activity of radioactive sample nuclear mass number	ampere
B	bulk modulus magnetic field strength	—
Bq	—	becquerel
c	velocity of light	prefix "centi-" (10^{-2})
C	specific heat capacitance	celsius coulomb
Ci	—	curie
d	distance between two points slit separation	—
e	emissivity electron charge	—
E	energy electric field strength	—
$\mathcal{E}$	emf	
f	friction force focal length	—
F	force	farad
g	acceleration of gravity	gram
G	gravitation constant	prefix "giga-" (10^9)
h	height Planck's constant	—
H	—	henry
Hz	—	hertz

Latin letter	*Meaning as an algebraic symbol*	*Meaning as a unit abbreviation or prefix*
I	moment of inertia electric current sound intensity	—
J	—	joule
k	force constant in SHM Boltzmann's constant	prefix "kilo-" (10^3)
K	dielectric constant	kelvin
l	lever arm length angular momentum quantum number	—
L	angular momentum latent heat inductance	—
m	mass angular momentum component quantum number	meter prefix "milli-" (10^{-3})
M	mass magnification mutual inductance magnetic moment	prefix "mega-" (10^6)
n	an integer index of refraction principal quantum number # of moles	prefix "nano-" (10^{-9})
N	# of molecules	newton
p	momentum pressure object distance	—

Latin letter	*Meaning as an algebraic symbol*	*Meaning as a unit abbreviation or prefix*
P	power object height	—
Pa	—	pascal
q	electric charge image distance	—
Q	quantity of fluid flow quantity of heat image height	—
r	radius	—
R	displacement (in two dimensions) reaction force universal gas constant resistance radius Rydberg constant	—
s	spin quantum number	second
S	arc length shear modulus	—
t	time	—
T	tension force period (of rotation or vibration) temperature	tesla
u	—	atomic mass unit
U	energy	—
v	velocity	—
V	volume potential difference	volt
w	weight slit width	—
W	work	watt
Wb	—	weber
x	displacement	—
X	reactance	—
y	displacement	—
Y	Young's modulus	—
Z	nuclear charge number impedance	—

Greek letter	*Meaning as an algebraic symbol*	*Meaning as a unit abbreviation or prefix*
α (alpha)	angular acceleration thermal expansion coefficient (linear) temperature coefficient of resistivity	—
β (beta)	sound intensity level thermal expansion coefficient (volumetric)	—
γ (gamma)	ratio of principal specific heats relativistic gamma factor	—
ϵ (epsilon)	electric permittivity	—
η (eta)	efficiency	—
θ (theta)	angle	—
κ (kappa)	thermal conductivity	—
λ (lambda)	wavelength radioactive decay constant	—
μ (mu)	friction coefficient magnetic permeability	prefix "micro-" (10^{-6})
ν (nu)	frequency	—
π (pi)	3.14159 . . .	—
ρ (rho)	density resistivity	—
σ (sigma)	Stefan constant	—
τ (tau)	torque	—
φ (phi)	phase angle work function	—
Φ (phi)	magnetic flux	—
ω (omega)	angular velocity	—
Ω (omega)	—	ohm

C

TRIGONOMETRIC TABLES

Angle degrees	Radians	Sine	Cosine	Tangent
0	0	0	1	0
1	0.0175	0.0175	0.9998	0.0175
2	0.0349	0.0349	0.9994	0.0349
3	0.0524	0.0523	0.9986	0.0524
4	0.0698	0.0698	0.9976	0.0699
5	0.0873	0.0872	0.9962	0.0875
6	0.1047	0.1045	0.9945	0.1051
7	0.1222	0.1219	0.9925	0.1228
8	0.1396	0.1392	0.9903	0.1405
9	0.1571	0.1564	0.9877	0.1584
10	0.1745	0.1736	0.9848	0.1763
11	0.1920	0.1908	0.9816	0.1944
12	0.2094	0.2079	0.9781	0.2126
13	0.2269	0.2250	0.9744	0.2309
14	0.2443	0.2419	0.9703	0.2493
15	0.2618	0.2588	0.9659	0.2679
16	0.2793	0.2756	0.9613	0.2867
17	0.2967	0.2924	0.9563	0.3057
18	0.3142	0.3090	0.9511	0.3249
19	0.3316	0.3256	0.9455	0.3443
20	0.3491	0.3420	0.9397	0.3640
21	0.3665	0.3584	0.9336	0.3839
22	0.3840	0.3746	0.9272	0.4040
23	0.4014	0.3907	0.9205	0.4245
24	0.4189	0.4067	0.9135	0.4452
25	0.4363	0.4226	0.9063	0.4663
26	0.4538	0.4384	0.8988	0.4877
27	0.4712	0.4540	0.8910	0.5095
28	0.4887	0.4695	0.8829	0.5317
29	0.5061	0.4848	0.8746	0.5543
30	0.5236	0.5000	0.8661	0.5774
31	0.5411	0.5150	0.8572	0.6009
32	0.5585	0.5299	0.8480	0.6249

Angle degrees	Radians	Sine	Cosine	Tangent
33	0.5760	0.5446	0.8387	0.6494
34	0.5934	0.5592	0.8290	0.6745
35	0.6109	0.5736	0.8192	0.7002
36	0.6283	0.5878	0.8090	0.7265
37	0.6458	0.6018	0.7986	0.7536
38	0.6632	0.6157	0.7880	0.7813
39	0.6807	0.6293	0.7771	0.8098
40	0.6981	0.6428	0.7660	0.8391
41	0.7156	0.6561	0.7547	0.8693
42	0.7330	0.6691	0.7431	0.9004
43	0.7505	0.6820	0.7314	0.9325
44	0.7679	0.6947	0.7193	0.9657
45	0.7854	0.7071	0.7071	1.0000
46	0.8029	0.7193	0.6947	1.0355
47	0.8203	0.7314	0.6820	1.0724
48	0.8378	0.7431	0.6691	1.1106
49	0.8552	0.7547	0.6561	1.1504
50	0.8727	0.7660	0.6428	1.1918
51	0.8901	0.7771	0.6293	1.2349
52	0.9076	0.7880	0.6157	1.2799
53	0.9250	0.7986	0.6018	1.3270
54	0.9425	0.8090	0.5878	1.3764
55	0.9599	0.8192	0.5736	1.4281
56	0.9774	0.8290	0.5592	1.4826
57	0.9948	0.8387	0.5446	1.5399
58	1.0123	0.8480	0.5299	1.6003
59	1.0297	0.8572	0.5150	1.6643
60	1.0472	0.8661	0.5000	1.7321
61	1.0647	0.8746	0.4848	1.8040
62	1.0821	0.8829	0.4695	1.8807
63	1.0996	0.8910	0.4540	1.9626
64	1.1170	0.8988	0.4384	2.0503
65	1.1345	0.9063	0.4226	2.1445

Angle degrees	*Radians*	*Sine*	*Cosine*	*Tangent*
66	1.1519	0.9135	0.4067	2.2460
67	1.1694	0.9205	0.3907	2.3559
68	1.1868	0.9272	0.3746	2.4751
69	1.2043	0.9336	0.3584	2.6051
70	1.2217	0.9397	0.3420	2.7475
71	1.2392	0.9455	0.3256	2.9042
72	1.2566	0.9511	0.3090	3.0777
73	1.2741	0.9563	0.2924	3.2709
74	1.2915	0.9613	0.2756	3.4874
75	1.3090	0.9659	0.2588	3.7321
76	1.3265	0.9703	0.2419	4.0108
77	1.3439	0.9744	0.2250	4.3315
78	1.3614	0.9781	0.2079	4.7046

Angle degrees	*Radians*	*Sine*	*Cosine*	*Tangent*
79	1.3788	0.9816	0.1908	5.1446
80	1.3963	0.9848	0.1736	5.6713
81	1.4137	0.9877	0.1564	6.3138
82	1.4312	0.9903	0.1392	7.1154
83	1.4486	0.9925	0.1219	8.1443
84	1.4661	0.9945	0.1045	9.5144
85	1.4835	0.9962	0.0872	11.430
86	1.5010	0.9976	0.0698	14.301
87	1.5184	0.9986	0.0523	19.081
88	1.5359	0.9994	0.0349	28.636
89	1.5533	0.9998	0.0175	57.290
90	1.5708	1.0000	0	∞

D

REFERENCES

This reference list is by no means complete; the entries have been chosen on two criteria:

1. They do not contain mathematics that would be difficult for a student using this book.
2. They are in readily available journals.

HISTORY AND BIOGRAPHIES

Adler, C. G., and Coulter, B. L. 1978. "Galileo and the Tower of Pisa Experiment." *Am. J. Phys.* 46:199.

Andrade, E. N. da C. 1954. "Robert Hooke." *Sci. American* 191(6): 94.

Bloch, F. 1976. "Heisenberg and the Early Days of Quantum Mechanics." *Physics Today* 29(12):23.

Bork, A. 1987. "Newton and Comets." *Am. J. Phys.* 55:1089.

Brehme, R. W. 1976. "New Look at the Ptolemaic System." *Am. J. Phys.* 44:506.

Cohen, I. B. 1949. "Galileo." *Sci. American* 181(2):40.

Cohen, I. B. 1955. "Newton." *Sci. American* 193(3):73.

Cohen, I. B. 1981. "Newton's Discovery of Gravity." *Sci. American* 244(3):166.

de Santillana, G. 1965. "Alessandro Volta." *Sci. American* 212(1):82.

Drake, S. 1973. "Galileo's Discovery of the Law of Free Fall." *Sci. American* 228(5):84.

Drake, S. 1980. "Newton's Apple and Galileo's Dialogue." *Sci. American* 243(2):150.

Fadner, W. L. 1988. "Did Einstein Really Discover '$E=mc^2$'?" *Am. J. Phys.* 56:114.

Foley, V., and Soedel, W. 1986. "Leonardo's Contributions to Theoretical Mechanics." *Sci. American* 255(3):108.

Franklin, A. 1976. "Principle of Inertia in the Middle Ages." *Am. J. Phys.* 44:529.

Heilbron, J. L. 1976. "Franklin's Physics." *Physics Today* 29(7):32.

Heilbron, J. L. 1985. "Bohr's First Theories of the Atom." *Physics Today* 38(10):25.

Kondo, H. 1953. "Michael Faraday." *Sci. American* 189(4):90.

Miller, F., Jr. 1966. "Kepler's Third Law and the Mass of the Moon." *Am. J. Phys.* 34:53.

Morrison, P., and Morrison, E. 1957. "Heinrich Hertz." *Sci. American* 197(6):98.

Motz, L. 1975. "The Conservation Principles and Kepler's Laws of Planetary Motion." *Am. J. Phys.* 43:575.

Newman, J. R. 1955. "James Clerk Maxwell." *Sci. American* 192(6):58.

Pohl, R. W. 1960. "Discovery of Interference by Thomas Young." *Am. J. Phys.* 28:530.

Postl, A. 1972. "Kepler's Anniversary." *Am. J. Phys.* 40:660.

Raman, V. V. 1973. "Copernicus and His Prescient Revolution." *Am. J. Phys.* 41:1341.

Rawlins, D. 1987. "Ancient Heliocentrists, Ptolemy, and the Equant." *Am. J. Phys.* 55:235.

Schagrin, M. L. 1974. "Early Observations and Calculation of Light Pressure." *Am. J. Phys.* 42:927.

Sharlin, H. 1961. "From Faraday to the Dynamo." *Sci. American* 204(5):107.

Swenson, L. S. 1987. "Michelson and Measurement." *Physics Today* 40(5):24.

Uritam, R. A. 1974. "Medieval Science, the Copernican Revolution, and Physics Teaching." *Am. J. Phys.* 42:809.

Weisskopf, V. F. 1977. "The Frontiers and Limits of Science." *American Scientist* 65:405.

Wilson, D. G. 1986. "A Short History of Human Powered Vehicles." *American Scientist* 74:350.

Wilson, M. 1954. "Joseph Henry." *Sci. American* 191(1):52.

Wilson, M. 1960. "Count Rumford." *Sci. American* 203(4):158.

Wilson, S. S. 1981. "Sadi Carnot." *Sci. American* 245(2):134.

UNITS AND STANDARDS

Astin, A. V. 1968. "Standards of Measurement." *Sci. American* 218(6):50.

Beams, J. W. 1971. "Finding a Better Value for G." *Physics Today* 24(5):34.

Brouwer, D. 1951. "The Accurate Measurement of Time." *Physics Today* 4(8):6.

Carrigan, R. A., Jr. 1978. "Decimal Time." *American Scientist* 66:305.

Giacomo, P. 1984. "The New Definition of the Meter." *Am. J. Phys.* 52:607.

Gray, W. T., and Finch, D. I. 1971. "How Accurately Can Temperature Be Measured?" *Physics Today* 24(9):32.

Hellwig, H., Evenson, K., and Wineland, D. 1978. "Time, Frequency, and Physical Measurement." *Physics Today* 31(5):23.

Macurdy, L. B. 1951. "Standards of Mass." *Physics Today* 6(1):7.

Norman, E. B. 1986. "Are Fundamental Constants Really Constant?" *Am. J. Phys.* 54:317.

Wilson, R. E. 1953. "Standards of Temperature." *Physics Today* 6(1):10.

MECHANICS

Angrist, S. W. 1968. "Perpetual Motion Machines." *Sci. American* 218(1):114.

Brehme, R. W. 1985. "On Force and the Inertial Frame." *Am. J. Phys.* 53:953.

Burger, W. 1984. "The Yo-Yo: A Toy Flywheel." *American Scientist* 72:137.

Hildebrand, M. 1987. "The Mechanics of Horse Legs." *American Scientist* 75:594.

Lin, H. 1982. "Fundamentals of Zoological Scaling." *Am. J. Phys.* 50:72.

Mak, S. 1987. "Extreme Value Problems in Mechanics Without Calculus." *Am. J. Phys.* 55:929.

Mark, R. 1987. "Reinterpreting Ancient Roman Structures." *American Scientist* 75:142.

O'Dell, C. R. 1987. "The Physics of Aerobatic Flight." *Physics Today* 40(11):24.

Palmer, F. 1951. "Friction." *Sci. American* 184(2):54.

Post, R. F., and Post, S. F. 1973. "Flywheels." *Sci. American* 229(6):17.

Thomsen, J. S. 1984. "Maxima and Minima without Calculus." *Am. J. Phys.* 52:881.

TERRESTRIAL GRAVITY

Celnikier, L. M. 1983. "Weighing the Earth with a Sextant." *Am. J. Phys.* 51:1018.

Gamow, G. 1961. "Gravity." *Sci. American* 204(3):94.

Heiskanen, W. A. 1955. "The Earth's Gravity." *Sci. American* 193(3):164.

Iona, M. 1978. "Why Is g Larger at the Poles?" *Am. J. Phys.* 46:790.

Klostergaard, H. 1976. "Determination of Gravitational Acceleration Using a Uniform Circular Motion." *Am. J. Phys.* 44:68.

Press, W. H. 1980. "Man's Size in Terms of Fundamental Constants." *Am. J. Phys.* 48:597.

Wesson, P. S. 1980. "Does Gravity Change with Time?" *Physics Today* 33(7):32.

Wild, J. F. 1973. "Simple Non-Coriolis Treatments for Explaining Terrestrial East-West Deflections." *Am. J. Phys.* 41:1057.

ASTRONOMY AND ASTROPHYSICS

Alfven, H. 1986. "The Plasma Universe." *Physics Today* 39(2):2.

Bethe, H. A., and Brown, G. 1985. "How a Supernova Explodes." *Sci. American* 252(5):60.

Bok, B. J. 1981. "The Milky Way Galaxy." *Sci. American* 244(3):92.

Boss, A. P. 1985. "Collapse and Formation of Stars." *Sci. American* 252(1):40.

Dicus, A., Letaw, J. R., Teplitz, D. C., and Teplitz, V. L. 1983. "The Future of the Universe." *Sci. American* 248(3):90.

Flannery, B. P. 1977. "Stellar Evolution in Double Stars." *American Scientist* 65:737.

Geller, M. J. 1978. "Large Scale Structure in the Universe." *American Scientist* 66:176.

Harrison, E. R. 1974. "Why the Sky Is Dark at Night." *Physics Today* 27(2):30.

Helfand, D. J. 1978. "Recent Observations of Pulsars." *American Scientist* 66:332.

Hodge, P. W. 1981. "The Andromeda Galaxy." *Sci. American* 244(1):92.

Hodge, P. 1984. "The Cosmic Distance Scale." *American Scientist* 72:474.

Hutchings, J. B. 1985. "Observational Evidence for Black Holes." *American Scientist* 73:52.

Larson, R. B. 1977. "The Origin of Galaxies." *American Scientist* 65:188.

Maran, S. P., and Boggess, A. 1980. "Ultra Violet Astronomy Enters the Eighties." *Physics Today* 33(9):40.

Peters, C. P. 1974. "Black Holes: New Horizons in Gravitational Theory." *American Scientist* 62:575.

Pollack, J. B. 1978. "The Rings of Saturn." *American Scientist* 66:30.

Ruffini, R., and Wheeler, J. A. 1971. "Introducing the Black Hole." *Physics Today* 24Z(1):30.

Shu, F. H. 1973. "Spiral Structure, Dust Clouds, and Star Formation." *American Scientist* 61:524.

Silk, J. 1987. "The Formation of Galaxies." *Physics Today* 40(4):28.

Smarr, L. L., and Press, W. H. 1978. "Our Elastic Space-Time: Black Holes and Gravitational Waves." *American Scientist* 66:72.

Stecker, F. W. 1978. "Gamma Ray Astronomy and the Origin of Cosmic Rays." *American Scientist* 66:570.

van Horn, H. 1979. "The Physics of White Dwarfs." *Physics Today* 32(2):23.

Wahr, J. 1985. "The Earth's Rotation Rate." *American Scientist* 73:41.

Wheeler, J. C., and Nomoto, K. I. 1985. "How Stars Explode." *American Scientist* 73:240.

Zelik, M. 1978. "The Birth of Massive Stars." *Sci. American* 238(4):110.

PHYSICS IN SPORTS

Abbott, A. V., Brooks, A. N., and Wilson, D. G. 1986. "Human Powered Watercraft." *Sci. American* 255(6):132.

Alexander, R. McN. 1984. "Walking and Running." *American Scientist* 72:348.

Brody, H. 1979 and 1981. "Physics of the Tennis Racket." *Am. J. Phys.* 47:482 and 49:816.

Brody, H. 1986. "The Sweet Spot of a Baseball Bat." *Am. J. Phys.* 54:640.

Curry, S. M. 1976. "How Children Swing." *Am. J. Phys.* 44:924.

Drela, M., and Langford, J. S. 1985. "Human Powered Flight." *Sci. American* 253(5):144.

Fox, G. T. 1973. "On the Physics of Drag Racing." *Am. J. Phys.* 41:311.

Frohlich, C. 1980. "The Physics of Somersaulting and Twisting." *Sci. American* 242(3):154.

Frohlich, C. 1985. "Effect of Wind and Altitude on Record Performance in Foot Races, Pole Vault, and Long Jump." *Am. J. Phys.* 53:726.

Goldenbaum, G. C. 1988. "Equilibrium Sailing Velocities." *Am. J. Phys.* 56:209.

Gross, A. C., Kyle, C. R., and Malewicki, D. J. 1983. "The Aerodynamics of Human Powered Land Vehicles." *Sci. American* 249(6):142.

Hunter, L. 1984. "The Art and Physics of Soaring." *Physics Today* 37(4):34.

Jones, D. E. H. 1970. "The Stability of the Bicycle." *Physics Today* 23(4):34.

King, A. 1975. "Project Boomerang." *Am. J. Phys.* 43:770.

Laws, K. 1985. "The Physics of Dance." *Physics Today* 38(2):24.

Lin, H. 1978. "Newtonian Mechanics and the Human Body: Some Estimates of Performance." *Am. J. Phys.* 46:15.

McFarland, E. 1986. "How Olympic Records Depend on Location." *Am. J. Phys.* 54:513.

McMahon, T., and Greene, P. 1978. "Fast Running Tracks." *Sci. American* 239(6):148.

Vos, H. 1985. "Straight Boomerang of Balsa Wood and Its Physics." *Am. J. Phys.* 53:524.

Walker, J. D. 1975. "Karate Strikes." *Am. J. Phys.* 43:845.

ENERGY

Bartlett, A. A. 1978. "Forgotten Fundamentals of the Energy Crisis." *Am. J. Phys.* 46:876.

Cohen, B. L. 1984. "Cost per Million BTU of Solar Heat, Insulation, and Conventional Fuels." *Am. J. Phys.* 52:614.

Duguay, M. A. 1977. "Solar Electricity: The Hybrid System Approach." *American Scientist* 65:422.

Dyson, F. 1971. "Energy in the Universe." *Sci. American* 225(3):50.

Glaser, P. E. 1977. "Solar Power from Satellites." *Physics Today* 30(2):30.

Gray, C. L., and von Hippel, F. 1981. "The Fuel Economy of Light Vehicles." *Sci. American* 244(5):48.

Hubbert, M. K. 1971. "The Energy Resources of the Earth." *Sci. American* 225(3):60.

Kreith, F., and Meyer, R. T. 1983. "Large Scale Use of Solar Energy with Central Receivers." *American Scientist* 71:598.

Kulcinski, G., et al. 1979. "Energy for the Long Run: Fission or Fusion?" *American Scientist* 67:78.

Meinel, A. B., and Meinel, M. P. 1972. "Physics Looks at Solar Energy." *Physics Today* 25(2):44.

Pelka, D. G., Park, R. T., and Singh, R. 1978. "Energy from the Wind." *Am. J. Phys.* 46:495.

Penney, T. R., and Bharathan, D. 1987. "Power from the Sea." *Sci. American* 256(1):86.

Pollard, W. G. 1976. "The Long Range Prospects for Solar Energy." *American Scientist* 64:424.

Ross, M. 1980. "Efficient Use of Energy Revisited." *Physics Today* 33(2):24.

Rubin, M. H. 1978. "Figures of Merit for Energy Conversion Processes." *Am. J. Phys.* 46:637.

Sorensen, B. 1981. "Turning to the Wind." *American Scientist* 69:500.

Summers, C. M. 1971. "The Conversion of Energy." *Sci. American* 225(3):148.

Walsh, W. J. 1980. "Advanced Batteries for Electric Vehicles." *Physics Today* 33(6):34.

Wieder, S., and Jaoudi, E. 1977. "Solar Energy—Its Measurement." *Am. J. Phys.* 45:981.

ACOUSTICS AND MUSIC

Benade, A. H. 1960. "The Physics of Wood Winds." *Sci. American* 203(4):144.

Blackman, E. D. 1965. "The Physics of the Piano." *Sci. American* 213(6):88.

Fletcher, H. 1946. "The Pitch, Loudness, and Quality of Musical Tones." *Am. J. Phys.* 14:215.

Fletcher, N. H., and Thwaites, S. 1983. "The Physics of Organ Pipes." *Sci. American* 248(1):94.

Henry, G. 1954. "Ultrasonics." *Sci. American* 190(5):54.

Hutchins, C. M. 1962. "The Physics of Violins." *Sci. American* 207(5):78.

Hutchins, C. M. 1981. "The Acoustics of Violin Plates." *Sci. American* 245(4):170.

Krumhansl, C. L. 1985. "Perceiving Tonal Structure in Music." *American Scientist* 73:372.

Rossing, T. D. 1982. "The Physics of Kettledrums." *Sci. American* 247(5):172.

Schroeder, M. R. 1980. "Toward Better Acoustics for Concert Halls." *Physics Today* 33(10):24.

Shankland, R. S. 1973. "Acoustics of Greek Theaters." *Physics Today* 26(10):30.

Sundberg, J. 1977. "The Acoustics of the Singing Voice." *Sci. American* 236(3):82.

Tan, B. T. G. 1982. "Fundamental Resonant Frequency of a Loudspeaker." *Am. J. Phys.* 50:348.

Wightman, F. L., and Green, D. M. 1974. "The Perception of Pitch." *American Scientist* 62:208.

Zych, D. A., and Earle, T. 1980. "Room Reverberation Time." *Am. J. Phys.* 48:32.

HEAT AND THERMODYNAMICS

Bartlett, A. A. 1976. "Introductory Experiment to Determine the Thermodynamic Efficiency of a Household Refrigerator." *Am. J. Phys.* 44:555.

Benzinger, T. 1961. "The Human Thermostat." *Sci. American* 204(1):134.

Bryant, L. 1969. "Rudolph Diesel and His Rational Engine." *Sci. American* 221(2):108.

Budzinski, W. V., and Neeson, J. F. 1984. "Static Method for Determining the R-Value of Thermal Insulation." *Am. J. Phys.* 52:1093.

Cohen, E. G. D. 1977. "Toward Absolute Zero." *American Scientist* 65:752.

Eastman, G. 1986. "The Heat Pipe." *Sci. American* 218(5):38.

Kelley, J. 1956. "Heat, Cold and Clothing." *Sci. American* 194(2):109.

Klein, M. J. 1974. "Carnot's Contributions to Thermodynamics." *Physics Today* 27(8):23.

Margaria, R. 1972. "The Sources of Muscular Energy." *Sci. American* 226(3):84.

Pierce, J. 1975. "The Fuel Consumption of Automobiles." *Sci. American* 232(1):92.

Rees, W. G., and Viney, C. 1988. "On Cooling Tea and Coffee." *Am. J. Phys.* 56:434.

Velards, M. G., and Normand, C. 1980. "Convection." *Sci. American* 229(2):80.

Walker, G. 1973. "The Stirling Engine." *Sci. American* 229(2):80.

Wilson, D. 1978. "Alternative Automobile Engines." *Sci. American* 239(1):39.

ELECTRICITY AND MAGNETISM

Barthold, L., and Pfeiffer, H. 1964. "High Voltage Transmission." *Sci. American* 210(5):38.

Becker, J. 1970. "Permanent Magnets." *Sci. American* 223(6):92.

Churchill, E. J., and Noble, J. D. 1971. "A Demonstration of Lenz' Law." *Am. J. Phys.* 39:285.

Cohen, D. 1975. "Magnetic Fields of the Human Body." *Physics Today* 28(4):35.

Heirtzler, J. 1962. "The Longest Electromagnetic Waves." *Sci. American* 206(3):128.

Moore, A. 1972. "Electrostatics." *Sci. American* 226(3):46.

Parker, E. N. 1971. "Universal Magnetic Fields." *American Scientist* 59:578.

Rubin, L. G., and Wolff, P. A. 1984. "High Magnetic Fields for Physics." *Physics Today* 37(8):24.

Scher, A. 1961. "The Electrocardiogram." *Sci. American* 205(5):132.

OPTICS

Arell, A., and Kolari, S. 1978. "Experiments on a Model Eye." *Am. J. Phys.* 46:613.

Bahuguna, R. D., Western, A. B., and Lee, S. 1988. "Young's Double-Slit Experiment Using Speckle Photography." *Am. J. Phys.* 56:531.

Bloembergen, N. 1975. "Lasers: A Renaissance in Optics Research." *American Scientist* 63:16.

Carleton, N. P., and Hoffman, W. F. 1978. "The Multiple Mirror Telescope." *Physics Today* 31(9):30.

Chynoweth, A. G. 1976. "The Fiber Lightguide." *Physics Today* 29(5):28.

Cook, J. 1973. "Communication by Optical Fibers." *Sci. American* 229(5):28.

Ewing, J. J. 1978. "Rare Gas-Halide Lasers." *Physics Today* 31(5):32.

Fraser, A., and Mack, W. 1976. "Mirages." *Sci. American* 234(1):102.

Land, E. H. 1959. "Experiments in Color Vision." *Sci. American* 200(5):84.

Levinstein, H. 1977. "Infrared Detectors." *Physics Today* 30(11):23.

Mathur, S. S., and Bahuguna, R. D. 1977. "Reading with the Relaxed Eye." *Am. J. Phys.* 45:1097.

Melchior, H. 1977. "Detectors for Lightwave Communication." *Physics Today* 30(11):322.

Nussenzveig, H. M. 1977. "The Theory of the Rainbow." *Sci. American* 236(4):116.

Price, W. 1976. "The Photographic Lens." *Sci. American* 235(2):72.

Shankland, R. S. 1974. "Michelson and His Interferometer." *Physics Today* 27(4):36.

Smith, F. D. 1968. "How Images Are Formed." *Sci. American* 219(3):96.

Thomas, D. E. 1980. "Mirror Images." *Sci. American* 243(6):206.

RELATIVITY

Bronowski, J. 1963. "The Clock Paradox." *Sci. American* 208(2):134.

Einstein, A. 1982. "How I Created the Theory of Relativity." *Physics Today* 35(8):45.

Martins, R. de A. 1978. "Length Paradox in Relativity." *Am. J. Phys.* 46:667.

Sastry, G. P. 1987. "Is Length Contraction Really Paradoxical?" *Am. J. Phys.* 55:943.

Shankland, R. S. 1964. "The Michelson–Morley Experiment." *Sci. American* 211(5):107.

Stauton, L. P., and van Dam, H. 1980. "Graphical Introduction to the Special Theory of Relativity." *Am. J. Phys.* 48:807.

QUANTUM PHYSICS

Boys, D. W., Cox, M. E., and Mykolajenko, W. 1978. "Photoelectric Effect Revisited (or an Inexpensive Device to Determine h/e)." *Am. J. Phys.* 46:133.

Darrow, K. K. 1952. "The Quantum Theory." *Sci. American* 186(3):47.

Gamow, G. 1958. "The Principle of Uncertainty." *Sci. American* 198(1):51.

Gamow, G. 1959. "The Exclusion Principle." *Sci. American* 201(1):74.

Gehrenbeck, R. K. 1978. "Electron Diffraction: Fifty Years Ago." *Physics Today* 31(1):30.

Ivey, H. F. 1957. "Electroluminescence." *Sci. American* 197(2):40.

Kidd, R., Ardini, J., and Antom, A. 1985. "Compton Effect as a Double Doppler Shift." *Am. J. Phys.* 53:641.

Matthias, B. T. 1957. "Superconductivity." *Sci. American* 197(5):92.

Schroedinger, E. 1953. "What Is Matter?" *Sci. American* 189(3):52.

Schwartz, B. B., and Foner, S. 1977. "Large Scale Applications of Superconductivity." *Physics Today* 30(7):34.

NUCLEAR PHYSICS AND REACTORS

Agnew, H. M. 1981. "Gas Cooled Nuclear Power Reactors." *Sci. American* 224(6):55.

Brownell, G. L., and Shalek, R. J. 1970. "Nuclear Physics in Medicine." *Physics Today* 23(8):32.

Coppi, B., and Rems, J. 1972. "The Tokomak Approach in Fusion Research." *Sci. American* 227(1):65.

Cowan, G. 1976. "A Natural Fission Reactor." *Sci. American* 235(1):36.

Emmett, J., Nuckolls, J., and Wood, L. 1974. "Fusion Power by Laser Implosion." *Sci. American* 230(6):24.

Fleming, S. 1980. "Detecting Art Forgeries." *Physics Today* 33(4):34.

McIntyre, H. C. 1975. "Natural-Uranium Heavy-Water Reactors." *Sci. American* 233(4):17.

Pehl, R. H. 1977. "Germanium Gamma-Ray Detectors." *Physics Today* 30(11):50.

Ralph, E. K., and Michael, H. N. 1974. "Twenty-Five Years of Radiocarbon Dating." *American Scientist* 62:553.

Renfrew, C. 1971. "Carbon-14 and the Prehistory of Europe." *Sci. American* 225(4):63.

Saeborg, G. T., and Bloom, J. L. 1970. "Fast Breeder Reactors." *Sci. American* 223(5):13.

Stickley, C. M. 1978. "Laser Fusion." *Physics Today* 31(5):50.

Weinberg, A. M. 1981. "The Future of Nuclear Energy." *Physics Today* 34(3):48.

Yonas, G. 1978. "Fusion Power with Particle Beams." *Sci. American* 239(5):50.

ELEMENTARY PARTICLES

Brown, L. M. 1978. "The Idea of the Neutrino." *Physics Today* 31(9):23.

Fritzsch, H. 1983. *Quarks.* New York: Basic Books.

Gaillard, M. K. 1982. "Toward a Unified Picture of Elementary Particle Interactions." *American Scientist* 70:506.

Georgi, H. 1981. "A Unified Theory of Elementary Particles and Forces." *Sci. American* 244(4):48.

Golomb, S. W. 1982. "Rubik's Cube and Quarks." *American Scientist* 70:257.

Greenberg, O. W. 1988. "The Quest for Elementary Particles of Matter." *American Scientist* 76:361.

Harari, H. 1983. "The Structure of Quarks and Leptons." *Sci. American* 248(4):56.

Isgur, N., and Karl, G. 1983. "Hadron Spectroscopy and Quarks." *Physics Today* 36(11):36.

Kernan, A. 1986. "The Discovery of the Intermediate Vector Bosons." *American Scientist* 74:21.

Lederman, L. 1986. "Unification, Grand Unification, and the Unity of Physics." *Am. J. Phys.* 54:594.

LoSecco, J. M., Reines, F., and Sinclair, D. 1985. "The Search for Proton Decay." *Sci. American* 252(6):54.

McLerran, L., and Svetitsky, B. 1987. "Making a Quark Plasma." *American Scientist* 75:490.

Rebbi, C. 1983. "The Lattice Theory of Quark Confinement." *Sci. American* 248(2):54.

Sulak, L. R. 1982. "Waiting for the Proton to Decay." *American Scientist* 70:616.

Weinberg, S. 1975. "Light as a Fundamental Particle." *Physics Today* 28(6):32.

Weinberg, S. 1977. "The Forces of Nature." *American Scientist* 65:171.

Weinberg, S. 1981. "The Decay of the Proton." *Sci. American* 244(6):64.

Weisskopf, V. F. 1983. "The Origin of the Universe." *American Scientist* 71:473.

E

ANSWERS TO ODD-NUMBERED PROBLEMS

CHAPTER 1

1. 5630 km **3.** 8850 m **5.** 41.3 mm × 92.1 mm **7.** 4460 m^2 **9.** 56.8 ℓ **11.** 1057 **13.** 6.70 cm^3 **15.** 9.76 ℓ **17.** 1.43 gal **19.** 1.93 in. **21.** 337 kg **23.** 2.35 **25.** copper **27.** 5.39×10^9 kg **29.** 1.11×10^6 **31.** Exact results are: **(a)** 5520 kg/m^3 **(b)** 28.9 h

CHAPTER 2

1. 9.90 s **3.** 13.8 m **5. (a)** 8380 m **(b)** no **7.** 37.6 yd **9. (a)** 6300 m **(b)** yes **11.** 4.41 m/s^2 **13.** 2.23 m/s^2 for the automobile **15.** 7.07 s **17.** 4.68 m/s **19.** 2.21 s **21.** 78.5 m **23.** 125 km/h **25.** 266 m **27. (a)** 2.60 s **(b)** 9.86 m/s **29. (a)** 8.70 m/s **(b)** 0.184 m/s^2 **31.** 39.5 s **33.** 789 m **35. (a)** 0.914 m/s^2 **(b)** 1 min 44 s **37. (a)** 19.8 m/s **(b)** 28.0 m/s **(c)** 23.9 m/s **39.** 1.19 s and 4.27 s; ±15.1 m/s **41.** 24.8 m **43.** 17.5 m **45. (a)** 0.0209 m/s^2 **(b)** 12.1 m/s **(c)** 9 min 42 s **(d)** 3528 m **47.** 1425 m **49. (a)** 11.0 s **(b)** 310 m **51.** 0.216 m/s^2

CHAPTER 3

1. (a) 1150 m **(b)** 55.2° N of E **3.** 9.57 mi at 38.8° E of N **5.** 6.08 m at −80.5° with the positive direction of the *x*-axis **7.** *x* component is 14.3 m, and *y* component is 0.657 m **9.** 3.91 m/s at 153° with the positive direction of the *x*-axis **11.** 2.90 m/s^2 at 27.2° S of E **13.** 8.86 m/s **15.** 28.0 m/s **17.** 813 m at 7.06° N of E **19.** 8.16 yd **21.** Average velocity is 3.35 m/s along the line from P_0 to P; average acceleration is 0.418 m/s^2 in the direction of her velocity at *P*. **23.** 9.81 m/s^2 vertically downward **25.** 41.8 m **27.** 20.1 m/s **29.** 9 **31. (a)** 5.44 s **(b)** 1630 m **33.** No; the ball is 6.70 m above ground level when it reaches the fielder. **35. (a)** 3.68° S of W **(b)** 584 mi/h **(c)** 1 h 43 min **(d)** 3 h 40 min **37. (a)** 50 s **(b)** 72.1 m at an angle of 56.3° with the bank **39. (a)** 76.0° **(b)** 19.2 m **41.** 86.0 m

CHAPTER 4

1. 3.04 m/s^2 **3.** 280 N **5.** 5.40 N **7.** 60 N **9.** 34.3 N **11.** 13 700 N **13.** 170 N **15.** 0.394 **17.** 39.2 N downward **19.** 9210 N **21.** 7210 N **23. (a)** 346 N **(b)** 886 N **25. (a)** 60.6 N **(b)** 3.40 m/s^2 **27. (a)** 37.2 s **(b)** 75 000 N **29.** 1.92 s **31.** 687 N; 500 N **33. (a)** 84.2 m/s **(b)** 116 ton **35.** 29.9 N in the upper cord; 22.1 N in the lower cord **37. (a)** 2.58 m/s^2 **(b)** 86.7 N and 36.1 N **39.** 2.00 N **41. (a)** 26.1 N **(b)** 0.443 **43.** 27.6 N **45. (a)** 400 N **(b)** 160 N **47. (a)** 6.58 kg **(b)** 0.211 **49. (a)** 39.9 N **(b)** 6.25 m/s^2 **51. (a)** 131 m **(b)** 57.7 m **53. (a)** 0.675 **(b)** 0.558 **(c)** No; it cancels from the equations. **55. (a)** 15.6 m/s **(b)** 3060 N **57. (a)** 721 N **(b)** 481 N **59.** 10.0 kg **61. (a)** 68.7 N **(b)** 65.8 N **(c)** 2.44 s **63. (a)** 0.414 **(b)** 116 N

CHAPTER 5

1. (a) 127 N **(b)** 281 N **3.** 298 N **5.** no **7.** 1.75 N·m **9.** 2.20 kg **11.** 13.9 cm from the center of the sphere **13.** 6.82 cm from the center of the larger sphere **15.** 448 N and 34.0° **17. (a)** 16.2 m **(b)** 1990 N **19.** 16.8 N with the vertical wall and 20.5 N with the sloping wall **21.** 156 cm from the other end **23.** 87.3 cm from the other end **25.** 5.22 m **27.** 1.03 m **29.** 5270 N upward at the support nearest the diver; 4290 N downward at the other support **31. (a)** 5.58 ft from the lighter child at the heavier end of the beam **(b)** 4.74 ft from the heavier child at the heavier end of the beam **33.** 7 cm from the center of the disk **35.** 5.45 cm from the 28-cm wire and 10.8 cm from the 24-cm wire. **37. (a)** 100 lb **(b)** 100 lb **39. (a)** 574 N **(b)** 574 N at 20° with the vertical **41.** 0.396 **43.** 1250 N and 1050 N **45.** Muscle force is 232 N; horizontal and vertical joint forces are 223 N and 31.9 N, respectively. **47.** 4.24 m **49.** 7.74 cm from the short side on a line between the midpoint and the intersection of the other two sides

CHAPTER 6

1. 5.42 m/s **3.** 3.91 m/s^2 **5.** 29.0 m/s **7.** 0.267 N **9.** 6.67×10^{-9} N **11. (a)** 3.58×10^{22} N **(b)** 2.00×10^{20} N **13.** 238.4 min **15.** 6.45×10^{23} kg **17.** 5.93×10^{-3} m/s^2 **19.** 499 m **21.** 3.26 rev/min **23.** 0.604 **25.** 32.4 rev/min **27. (a)** 106 N **(b)** 1070 N **29.** 25.9 m/s^2 **31.** 3.46×10^8 m from the earth's center **33.** 172 min **35.** 5.07 earth radii **37.** 7.25×10^{22} kg **39.** 728 m **41.** 24.6° **43. (a)** 7.26° **(b)** 71.2 m/s **(c)** 51.9 m/s **45.** 4.97 N

CHAPTER 7

1. 18.0 kJ **3.** 212 kJ **5.** 4.00 kJ **7.** 589 kJ **9.** 1.10 kJ **11.** 6.20 m/s **13.** 31.4 W **15.** 4.30 MJ **17.** 110 kg **19.** 175 W **21.** 9.20 m/s **23.** 48.9 m **25.** 123 J **27. (a)** 1.97 J **(b)** 14.8 m **29. (a)** 93.7 J **(b)** 11.2 J **(c)** 30.5 m/s **31.** 10.5 m/s; 277 J **33.** 0.300 **35.** 19.8 m/s **37. (a)** 28.9 m/s **(b)** 42.5 m **39.** 82.3 hp **41.** 1310 hp **43.** 100 W **45.** 52.6 hp **47.** 45.0 hp **49.** 117 kW **51.** 3.22 hp **53. (a)** 2.67 **(b)** 1.60 m **55. (a)** 12.6° **(b)** 2.04; 44.6% **57.** three pulleys in the upper and two pulleys in the lower block **59.** 20 **61.** 82.3 hp

CHAPTER 8

1. 880 kg m/s **3.** 500 μg m/s **5.** 5210 N **7.** 4.33 m/s **9.** 11 100 kg **11.** 460 m/s **13.** 45.0 N **15.** 19.2 N **17.** 0.686 **19.** 1.00 m/s toward the lighter diver **21.** 2.86 m/s at 19.1° S of W **23.** 16.2 m/s at 22.6° N of E **25.** 500 g with velocity at 127° to the velocity of the 400-g fragment **27. (a)** 7.92 m/s **(b)** 969 kJ **29. (a)** 0.750 kg m/s **(b)** 0.397 J **31.** 1.20×10^5 m/s for the helium nucleus, and 3.20×10^5 m/s for the neutron **33.** 12.9 metric tons **35.** 16.2 m **37.** 283 m/s **39. (a)** 29.1 cm **(b)** 3590 J **41.** 750 W **43.** 0.572 **45. (a)** 3.01 m/s at 56.2° with the original direction of motion **(b)** 5.39%

CHAPTER 9

1. 90.9 rad/s **3.** 3180 rev/min **5.** 1.68 s **7.** 0.360 kg·m^2 **9.** 1.74×10^{-3} kg·m^2 **11.** 30.0 rad/s^2 **13.** 9.13 rad/s **15.** 182 kg·m^2 **17. (a)** 11.5 rad/s^2 **(b)** 367 rev **19. (a)** 4.91 rad/s^2 **(b)** 6.40 s **21. (a)** 63.0 rad/s **(b)** 78.9 rev **23. (a)** 72.0 rad/s **(b)** 91.7 rev **25.** 2.50% **27. (a)** 0.0576 kg·m^2 **(b)** 0.115 N·m **29. (a)** 6.70 s **(b)** 16.8 m **31.** 11.5 rad/s^2; 92.3 N **33.** 1.00 rad/s^2; 5.83 kg·m^2 **35.** 142 J **37.** 2910 J **39.** 9.80 hp **41.** 15.3 h **43. (a)** 5.22 h **(b)** 8.94 m **45. (a)** 0.918 N·m **(b)** 0.206 hp **47. (a)** 457 rev/min **(b)** 6020 J **49.** 1.81 kg **51. (a)** 5.66% **(b)** accurate to about 5% **53. (a)** 11.0 m/s **(b)** 1530 J

CHAPTER 10

1. 99.5 m **3.** 4710 Pa **5.** 104 N **7.** 17.0 cm **9.** 1.21 m **11.** 47.4% methanol and 52.6% water **13. (a)** 88.3 N **(b)** 118 N **15.** 9.13 cm **17.** 9.08 m/s **19.** 48.5 N **21. (a)** 2940 Pa **(b)** 0.191 N **23. (a)** 80.0 kPa **(b)** 58.7 kPa; 26.7 kPa **25.** 0.826 **27.** 102 cm **29.** 89.5% **31.** 17 **33.** 2.70 and 0.774 **35. (a)** 3210 N **(b)** 568 N **37.** 23.9 N **39.** 18.0 cm **41.** 25.6 kPa **43.** 1.67 N **45. (a)** 13.3 m **(b)** 0.0715 gal/min **47. (a)** 11 800 ℓ/min **(b)** 626 kPa **49.** 151 kPa **51.** 99.1 m/s **53.** The level falls by 15.3 cm.

CHAPTER 11

1. 1.73×10^7 N/m^2 **3.** 0.664 mm **5.** 248 metric tons **7.** 0.380% **9.** 2.80×10^5 N/m **11.** 2.27 ms **13.** 21.1 g **15.** 2530 N/m **17.** 22.4 cm **19.** 9.816 m/s^2 **21.** 188 N; 0.174 μm **23.** 1.23×10^5 N **25. (a)** 2.91 mm **(b)** 0.0455 μm **27.** 2.31×10^7 N/m^2 for the steel; 7.98×10^6 N/m^2 for the aluminum **29.** 3110 atm **31.** 4.67% **33.** 114 atm **35.** 2.44×10^{-4} rad **37.** 7.33×10^9 N/m^2 **39. (a)** 1570 N/m **(b)** 75.0 cm **41.** 0.331 s **43. (a)** 14.0 kg **(b)** 3 cm above the center point **45. (a)** 0.969 Hz **(b)** 0.875 Hz

47. 203 g **49.** At 70.7% of the amplitude **51. (a)** 0.634 s **(b)** 50.5 cm **(c)** ±4.77 m/s **53. (a)** 6.72×10^4 N/m **(b)** 6040 m/s^2 **55.** 30.7 times per minute **57.** 1.00024 s **59. (a)** a decrease of 1.36×10^{-2} cm^3 **(b)** a decrease of 4.17×10^{-2} cm^3 **61. (a)** 11.4 cm **(b)** 4.35 Hz

CHAPTER 12

1. 77.3 cm **3.** 167 GHz **5.** 358 N **7.** 1480 m/s **9.** 50.0 cm **11.** 572 m/s **13.** 66.4 cm **15.** 437 Hz or 443 Hz **17.** 27.2 m/s **19.** 275 Hz **21.** 6.31×10^{-5} W/m^2 **23.** 63.0 dB **25.** 283 Hz, 566 Hz, and 849 Hz **27. (a)** 81 Hz **(b)** 1.85 m **29. (a)** 67.2 Hz **(b)** 53.3 Hz **31. (a)** 1.33 m **(b)** 192 Hz and 320 Hz **33. (a)** closed **(b)** 261 Hz **(c)** 32.6 cm **35.** 2.78 beats per second **37. (a)** 5.25 Hz **(b)** 22.8 Hz **39.** 149 N and 161 N **41.** 40.8 m/s receding **43.** 3.17571 m **45. (a)** 16.0 s **(b)** 35.3 s **47.** 7.78 dB **49.** 92.8 dB **51.** 0.316% **53.** 10.6 m/s approaching **55.** 4.16 Hz

CHAPTER 13

1. 7.48×10^5 Pa **3.** 27.7 ℓ **5.** 4.79 ℓ **7.** 52.6 cm **9.** 25.2°C **11.** 28.0 g **13.** 438 m/s **15.** 214 kPa **17.** 13.2 g **19.** 19.2 **21.** 3.22×10^9 **23.** 70.9 mPa **25.** 7.10 g **27.** 113 cm^3 **29.** 8.45×10^{-4} in. **31.** 0.591 cm **33.** 0.461 in. **35.** 4.63 cm^3 **37.** 4.65 mℓ **39. (a)** 25.9°C **(b)** 50.7°C **41.** 31.8 C° heating **43.** 2.59×10^8 Pa = 2560 atm **45.** 11 200 K **47.** 500 m/s **49.** 1.58×10^6 m/s **51.** 84.2 kPa **53. (a)** 1.90×10^6 N **(b)** 0.367 mm **55. (a)** 2200 kg **(b)** 1270 kg

CHAPTER 14

1. +3500 J **3.** 2.48 kcal **5.** 314 kJ **7.** 21.3°C **9.** 136 kJ **11.** 100 kJ **13.** 1040 m/s **15.** 12.1 W **17.** 1.26 kW **19.** 64.8 K **21.** 153 kJ **23.** 0.141 C° **25.** 1.02 C° **27.** 29.4% **29.** 80.0°C **31.** 0.571 cal/g·C° **33.** 6.34°C **35.** 62.6 g **37. (a)** 240 s **(b)** 1500 s **(c)** 230 s **39. (a)** 159 h **(b)** 1020 h **41.** 2830 J; 0 J; 2830 J **43.** 0.0600 m^3 **45. (a)** 152°C **(b)** 63.2 kJ **47.** 320 m/s and 298 m/s **49.** 9.47×10^4 Btu **51.** 716 kW **53.** 346 min **55.** 215 kcal/h **57. (a)** 5.45 kW **(b)** 966°C **(c)** No; 9.38 W **59. (a)** 4.11 Btu **(b)** 22.1 Btu **61.** 136 kcal/h **63. (a)** 132 g of ice and 418 g of water at 0°C **(b)** 600 g of water at 35.7°C **65.** 277 K

CHAPTER 15

1. 33.3% **3.** 3.00 **5.** 7.29 kJ **7.** 55.0% **9.** −12.2°C **11.** −491 J/K **13.** 23.1% **15.** 11.3 J **17.** 368 kJ **19. (a)** 280 K; 0.333 **(b)** 384 K; 0.456 **(c)** 840 K; 1.00 **21.** 138°C **23.** 8.62; 1.92 tons of refrigeration **25.** 7 min 50 s **27. (a)** $T_A = 240.5$ K; $T_B = 420.9$ K; $T_C = 841.7$ K; $T_D = 481$ K **(b)** 9.45% **(c)** 71.4% **29.** 1.82 J/kg increase **31.** 0.535 kg; 345 J/kg increase **33.** 14.9% **35.** 0.230 **37. (a)** 37.7% **(b)** 532 kg/s **(c)** 50.6%

CHAPTER 16

1. 25.0 N repulsion **3.** 19.0 cm **5.** 2.16×10^6 N/C toward the point charge **7.** +21.3 nC **9.** 0.800 J **11.** −17.8 μC **13.** 144 kV **15.** −107 nC **17.** −33.3 μC **19.** 6.58 μC **21. (a)** 68.9 N **(b)** 72.0 N (both forces are inward along the diagonal) **23.** 3150 **25.** 0.384 mg **27. (a)** 9.60×10^6 N/C **(b)** 2.51×10^6 N/C (both forces are toward the right) **29. (a)** 0 N/C **(b)** 3.22×10^6 N/C toward the center of the square **31.** −5.00 μC **33.** 0.461 J **35.** −208 kV between the charges of opposite sign; −512 kV between the charges of the same sign **37. (a)** 197 N at an angle of 24.0° with the 40-cm side; **(b)** 11.6 MV/m **39.** 0.304 μC

CHAPTER 17

1. 3.54 nC **3.** 2.70 MV/m **5.** 96.0 V **7.** 4.80 μC **9.** 1.70 nF **11.** 5.55 μF **13.** 4.24 **15.** 196 cm^2 **17. (a)** 33.3 nF **(b)** 0.300 μF **19.** 2.85 V **21.** 0.850 μC **23.** 63.6 cm **25.** 29.5 μJ **27.** 556 μF **29.** 19.6 m **31.** 8.46 cm **33. (a)** 12.0 μC **(b)** 0.720 mJ **(c)** 0.313 mJ **(d)** 52.2 V **35.** 0.500 μF **37. (a)** 6.71 μF **(b)** 51.4 μC; 12.9 V **39. (a)** 2.91×10^6 m/s **(b)** 2.70×10^6 m/s

41. (a)

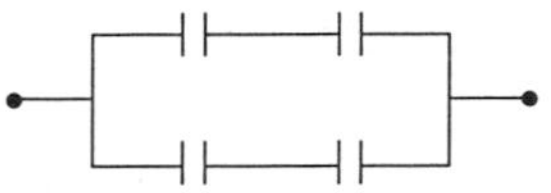

(b)

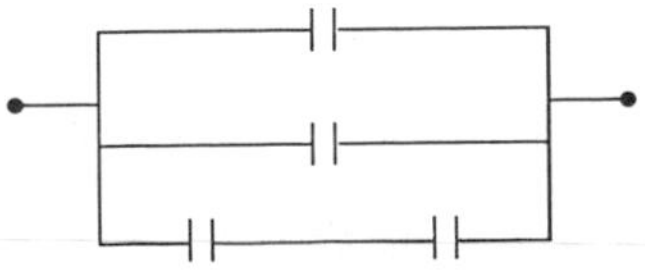

CHAPTER 18

1. 2.59×10^5 C **3.** 6.67 A **5.** 267 Ω **7.** 150 Ω **9.** 1.75 Ω **11.** 6.82 m **13.** 15.8 C° **15. (a)** 9.00 Ω **(b)** 0.923 Ω **17.** three resistors in series in series with two resistors in parallel **19.** 36.0 W **21.** 2.54 A **23.** 1.96×10^{-3} Ω; 1.96×10^{-3} Ω **25.** 0.500 mm **27.** 988°C **29.** 1.39 **31.** 18.7 Ω **33.** 12.8 A **35.** 21.3 kW **37. (a)** 1.96 V **(b)** 58.8 W **39.** 0.462 Ω **41. (a)** 4.00 Ω **(b)** 1.00 W **43.** 32.0 W **45.** 14.0 V **47.** 21.3 V **49.** 2.40 Ω **51. (a)** 1.71 A **(b)** 1.71 Ω **53.** 9.995 MΩ **55.** $R_1 = 45.5$ kΩ, $R_2 = 450$ kΩ, $R_3 = 4.50$ MΩ **57.** 4.90×10^{-3} (C°)$^{-1}$ **59.** length, 11.7 m; diameter, 0.572 mm **61. (a)** 0.271 Ω **(b)** 55.9 min **63.** $R_1 = 0.789$ Ω, $R_2 = 7.11$ Ω, $R_3 = 71.1$ Ω **65.** 3320 Ω

CHAPTER 19

1. 6.25×10^7 m/s to the north **3.** 3.54 m **5.** 33.6 N to the east **7.** 12.0 μT toward the south **9.** 4.69 mT **11.** 20.7 A **13.** 1580 A **15. (a)** 2.59 A·m^2 **(b)** 1.04 N·m **17.** 1.25 N·m **19.** 0.161 T **21.** 0.244 T vertically downward **23.** 2.40 N perpendicular to the plane of the coil and oppositely directed **25. (a)** 53.3 μT **(b)** 0.00 T **27.** 3.75 A **29.** 6.01 μT **31.** 39.0 mT **33.** 3.12 mT **35.** 5.12×10^{-3} N **37.** 4.68×10^{-3} N toward the centroid of the triangle **39.** 1.29×10^{-3} N outward from the center of the square **41.** 4.81 T **43. (a)** 2.26 A·m^2 **(b)** 0.905 N·m **45.** 7.22×10^{-13} J = 4.51 MeV **47.** 0.0147 N toward the west **49.** 0.0649 N **51. (a)** 28 700 A **(b)** 725 kW

CHAPTER 20

1. 45.0 mWb **3.** 100 V **5.** 0.288 mWb **7.** 0.240 V **9.** 1.43 kW **11.** 1.08 kV **13.** 0.625 H **15.** 377 rad/s; 16.7 ms **17. (a)** 566 m to 187.5 m **(b)** 3.41 m to 2.78 m **19.** northwest **21.** 62.5 μT **23.** 796 turns **25.** 4.38 A **27. (a)** the vertical component **(b)** 61.9 μT **29.** 19.2 km/s **31. (a)** 0.480 A clockwise **(b)** 0.0576 N **33. (a)** 0.72 A clockwise **(b)** 0.0864 N **35. (a)** 6.09 A **(b)** 4.68 V **(c)** 17.8 m/s **(d)** 28.6 W **(e)** 39.0% **37. (a)** 85.7 A **(b)** 5.00 V; 250 W **(c)** 600 W **39.** 2.62 V **41.** 120 V **43.** 24 **45. (a)** 3600 rev/min **(b)** 380 V **47.** 0.114 T **49. (a)** 20,000 axles **(b)** very long train; low power when traveling slowly **51.** 6.88 mV

CHAPTER 21

1. 170 V **3.** 3.43 A **5. (a)** 565 Ω **(b)** 9.42 kΩ **(c)** 9.42 MΩ **7.** 0.318 H **9.** 34.3 Ω **11.** 9.23 A; 67.4° **13.** 5.03 kHz **15.** 4.34 μH **17.** 10.0 A **19.** 12:1; 1.44 kV **21.** 141 V **23.** 778 mV **25.** 352 mA **27.** 65.6 V; 46.7° leading **29.** 830 V; 53.0° leading **31. (a)** 266 Ω **(b)** 21.9 Ω **(c)** 15.0 Ω **33.** 57.0 Ω **35.** 3.96 A; 70.7° leading **37. (a)** 0.844 μF **(b)** 6.25 A **(c)** 1.67 kV **39. (a)** 159 Hz **(b)** 1.25 A **(c)** 25.2 mA **41.** 902 W; 14.1 Ω resistor and 13.6 mH inductor **43.** 60.0 Ω and 53.1 nF **45. (a)** 100 mH **(b)** 0.467 **(c)** 157 W **47.** 20:1; 667 mA **49.** 0.0231

CHAPTER 22

1. 7.14×10^{14} Hz; 4.41×10^{14} Hz **3.** 1.67 **5.** 60.7° **7.** 33.3° **9.** −37.5 cm **11.** 9.60 cm behind the mirror **13.** 8.94 cm behind the mirror **15.** 20.6 cm in front of the lens **17.** 16.7 cm from the lens **19.** −26.7 cm **21.** 13.0 cm **23.** 346 nm **25.** 4.55 cm **27.** 54.8° **29.** 61.4° **31.** 58.7° to 90°
33. (a) 15.4 cm behind the mirror, 1.92 cm high, virtual
(b) 9.37 cm behind the mirror, 3.12 cm high, virtual
35. (a) 3.75 cm behind the mirror, 0.75
(b) 7.50 cm behind the mirror, 0.50
(c) 9.37 cm behind the mirror, 0.375
37. 50.0 cm from the mirror
39. (a) 16.7 cm in front of the lens, 0.333
(b) 12.5 cm in front of the lens, 0.500
(c) 7.14 cm in front of the lens, 0.714
41. (a) 13.3 cm in front of the lens, 0.667 cm high, virtual
(b) 11.1 cm in front of the lens, 0.889 cm high, virtual
(c) 6.67 cm in front of the lens, 1.33 cm high, virtual
43. 81.0 cm from the lens **45.** 16.4 cm behind the converging lens; −0.385 **47.** 150 000 rev/min **49.** 14.0 cm

CHAPTER 23

1. −1.18 diopters **3.** 150 cm **5.** 12.6 m **7.** 4.17 cm **9.** 162× **11.** 10× **13.** 30× **15.** +66.7 diopters **17. (a)** −1.78 diopters **(b)** 26.5 cm **19. (a)** 37.5 cm **(b)** −2.67 diopters **21.** 66.7 cm and 125 cm **23.** +100 diopters **25. (a)** 6.23 mm from the objective **(b)** 465× **(c)** 442× **27.** 8.37 mm **29. (a)** +1.00 diopters

(b) 108 cm **(c)** 5.26 cm closer to the objective **31. (a)** 48× **(b)** 40.0 cm **33.** 3.4× **35. (a)** 7× **(b)** 50 mm and 22.4 cm **37.** 4.80 cm, 2.40 cm, and 1.20 cm **39.** 26.8° **41.** 14.6 mm **43.** 23.7 mm **45.** 38.0°

CHAPTER 24

1. 23.4° **3.** 2.42 μm **5.** 4810 **7.** 4th **9.** 0.204 μm **11.** 1.30 μm **13.** 545 nm **15.** 13.8 cm **17.** 1.62 **19.** 422 nm **21.** 3.60 cm; 5.47 cm **23.** 20.5° **25.** 6870 **27.** 7.35° **29.** 660 nm **31.** 46 **33.** 0.100 μm **35.** 479 nm (blue-green) **37.** 546 nm (yellow-green) **39.** 2.02 m **41.** 1.74 cm **43. (a)** 7.57×10^{-6} degree **(b)** 50.7 m **45.** 55.7° **47.** 43 **49. (a)** 665 nm, 532 nm, or 443 nm **(b)** restricts it to 665 nm

CHAPTER 25

1. (a) 2.29 **(b)** 7.09 **3.** 0.943*c* **5.** 3750 MeV **7.** 3.51 MeV **9.** 3.04 MeV **11.** 1250 M$ **13.** 74.5 cm **15.** 9.29% speed increase **17.** 1.31 yr **19.** 159 999.989 m/s **21.** 1880 MV **23.** 2.23×10^8 m/s **25.** 2.05 MV **27.** 3.29 **29. (a)** 1.55 MeV **(b)** 96.9% **31.** 10^{15} m; slow **33.** 1660 MeV

CHAPTER 26

1. 2.47 μm **3.** 9.35 μm **5.** 1.04 mm **7. (a)** 242 MHz **(b)** 7.25×10^{14} Hz **(c)** 2.42×10^{20} Hz **9.** 2.52 eV **11. (a)** 5.34×10^{-34} J·s/m **(b)** 1.60×10^{-27} J·s/m **(c)** 5.34×10^{-22} J·s/m **13.** 1.23×10^{-27} J·s/m **15.** 2.92 μm **17.** 1.312×10^{29} photons per second **19.** 9.54×10^{15} photons per second **21.** 1.27×10^{18} **23.** 6.68×10^5 m/s **25.** 1.15×10^6 m/s **27.** 1.64×10^6 m/s **29. (a)** 1.01 V **(b)** 5.94×10^5 m/s **31.** chromium **33.** 98.3 ns **35.** 511 keV **37. (a)** 935 eV **(b)** 79.8°

CHAPTER 27

1. 0.242 nm **3.** change by a factor 0.577 **5.** 0.307 nm **7.** 7.88×10^{-13} m **9.** 653 nm **11.** 1.88 μm, 1.28 μm, 1.10 μm, 1.01 μm; infrared **13.** $1s^2 2s^2 2p^6 3s^2 3p^5$ **15.** 22 **17.** 7.09×10^{-11} m **19.** 0.173 nm **21.** 91.1 MeV **23.** 9.91×10^{-11} m **25.** 22.8 MeV **27.** 9.96×10^{-18} **29.** 39 400 K **31.** 1 and 5 **33.** all of the Lyman lines and all except the first of the Balmer lines **35. (a)** 32 **(b)** 14 **37.** Fe **39.** 0.207 nm and 2.08 nm **41.** 90

CHAPTER 28

1. proton numbers: 2, 17, 54, and 89; neutron numbers: 1, 19, 78, and 136; nucleon numbers: 3, 36, 132, and 225
3. ${}^{212}_{83}\text{Bi} \rightarrow {}^{208}_{81}\text{Tl} + {}^{4}_{2}\text{He}$
${}^{212}_{83}\text{Bi} \rightarrow {}^{212}_{84}\text{Po} + {}^{0}_{-1}\text{e}$
5. ${}^{211}_{82}\text{Pb} \rightarrow {}^{211}_{83}\text{Bi} + {}^{0}_{-1}\text{e}$
${}^{212}_{82}\text{Pb} \rightarrow {}^{212}_{83}\text{Bi} + {}^{0}_{-1}\text{e}$
${}^{214}_{82}\text{Pb} \rightarrow {}^{214}_{83}\text{Bi} + {}^{0}_{-1}\text{e}$
7. 2^{-100} **9.** 27.8 days **11.** 7.51×10^{-12} Ci
13. ${}^{58}_{29}\text{Cu}$; ${}^{24}_{11}\text{Na}$; ${}^{28}_{13}\text{Al}$; ${}^{30}_{15}\text{P}$
15. ${}^{44}_{21}\text{Sc} \rightarrow {}^{44}_{20}\text{Ca} + {}^{0}_{1}\text{e}$
${}^{49}_{24}\text{Cr} \rightarrow {}^{49}_{23}\text{V} + {}^{0}_{1}\text{e}$
${}^{51}_{25}\text{Mn} \rightarrow {}^{51}_{24}\text{Cr} + {}^{0}_{1}\text{e}$
${}^{62}_{29}\text{Cu} \rightarrow {}^{62}_{28}\text{Ni} + {}^{0}_{1}\text{e}$
17. unstable (odd-odd nucleus); unstable ($Z > A/2$); unstable (odd-odd nucleus); unstable ($Z > 83$) **19.** 1.114×10^{25} **21.** 65.3% **25.** 8.87 h and 0.0781/h **27.** 17 800 s^{-1} **29.** 4.60/min **31.** 0.224 g
33. a. ${}^{32}_{14}\text{Si} \rightarrow {}^{32}_{15}\text{P} + {}^{0}_{-1}\text{e}$; unstable (odd-odd)
b. ${}^{21}_{11}\text{Na} \rightarrow {}^{21}_{10}\text{Ne} + {}^{0}_{1}\text{e}$; stable
c. ${}^{26}_{13}\text{Al} + {}^{0}_{-1}\text{e} \rightarrow {}^{26}_{12}\text{Mg}$; stable
35. 5.33 min **37. (a)** 25 times **(b)** 41 times; and 4 times in 9 months for the pregnant woman.

CHAPTER 29

1. 44.4 kV **3.** 0.144 T **5.** 3.11 MeV exoergic **7.** 2.57 MeV **9. (a)** 4.06 MeV (exoergic) **(b)** −2.79 MeV (endoergic) **(c)** −5.03 MeV (endoergic) **11. (a)** −2.44 MeV (endoergic) **(b)** 3.11 MeV (exoergic) **(c)** 8.61 MeV (exoergic) **13.** 2.05 MeV for helium, and 2.73 MeV for hydrogen **15.** Q = −20.6 MeV (endoergic) **17.** for α decay, Q = −6.23 MeV (endoergic) (stable); for β^- decay, Q = −1.65 MeV (endoergic) (stable) **19.** 0.156 MeV **21. (a)** 2.74×10^5 m/s **(b)** 0.0861 MeV **(c)** 222.01753 u **23.** 31.6 kg **25.** 16.9 g **27.** 25.7 MeV **29.** 44.4×10^9 kg/s; 1 part in 1.42×10^{11} **31. (a)** no violations **(b)** charge conservation is violated **(c)** no violations **33. (a)** violates energy conservation **(b)** violates energy conservation, and changes strangeness by +2 **(c)** violates angular momentum conservation, and changes strangeness by +2.

INDEX

PHYSICAL CONSTANTS

Absolute zero of temperature	$-273.15°C$
Atomic mass unit	1.6605×10^{-27} kg
	931.5 MeV/c^2
Avogadro's number	6.022×10^{23} molecules/mol
Boltzmann constant	1.381×10^{-23} J/K
	8.620×10^{-5} eV/K
Electric permittivity constant	8.854×10^{-12} F/m
Electron charge	-1.602×10^{-19} C
Electron mass	9.109×10^{-31} kg
	0.511 MeV/c^2
Gas constant	8.314 J/K · mol
Gravitation constant	6.672×10^{-11} N · m^2/kg^2
Magnetic permeability constant	$4\pi \times 10^{-7}$ H/m
Neutron mass	1.6750×10^{-27} kg
	939.6 MeV/c^2
Planck's constant	6.626×10^{-34} J · s
	4.136×10^{-15} eV · s
Proton mass	1.6726×10^{-27} kg
	938.3 MeV/c^2
Rydberg constant	1.097×10^{7} m^{-1}
Speed of light (vacuum)	2.998×10^{8} m/s
Stefan's constant	5.670×10^{-8} W/m^2 · K^4

SOLAR SYSTEM DATA

Body	*Mass (kg)*	*Radius (m)*	*Mean distance from sun (m)*
Sun	1.99×10^{30}	6.96×10^{8}	——
Mercury	3.28×10^{23}	2.42×10^{6}	5.79×10^{10}
Venus	4.87×10^{24}	6.10×10^{6}	1.08×10^{11}
Earth	5.98×10^{24}	6.37×10^{6}	1.49×10^{11}
Moon	7.35×10^{22}	1.74×10^{6}	——
Mars	6.46×10^{23}	3.37×10^{6}	2.28×10^{11}
Jupiter	1.90×10^{27}	6.99×10^{7}	7.78×10^{11}
Saturn	5.69×10^{26}	5.85×10^{7}	1.43×10^{12}

Mean earth–moon distance = 3.84×10^{8} m